HYDRAULIC ENGINEERING '93

VOLUME 2

Proceedings of the 1993 Conference

Sponsored by the
Hydraulics Division of the
American Society of Civil Engineers

in cooperation with the
Environmental Engineering Division
Irrigation and Drainage Division
Water Resources Planning and Management Division
Waterway Port Coastal and Ocean Division
of ASCE

Organization Committee
S.T. Su, Conference/Symposium General Chairman
Hsieh Wen Chen, Conference Technical Program Chairman
Chin Y. Kuo, Symposium Technical Program Chairman

Hydraulics Division Program Committee
S.T. Su, Chairman
George V. Cotroneo, Vice Chairman
William Espey, Secretary
Marshall Jennings, Past Chairman

Hydraulics Division Executive Committee
Steven R. Abt, Chairman
Edward R. Holly, Vice Chairman
David S. Biedenharn, Secretary
Catalino B. Cecilio, MGD Representative
Arlen D. Feldman
Linda S. Weiss, New Correspondent

San Francisco, California
July 25-30, 1993

Edited by Hsieh Wen Shen, S.T. Su, and Feng Wen

Published by the
American Society of Civil Engineers
345 East 47th Street
New York, New York 10017-2398

ABSTRACT

This proceedings, *Hydraulic Engineering,* contains papers presented at the 1993 National Conference on Hydraulic Engineering held in San Francisco, California, July 25-30, 1993. The basic theme of these papers in the reduction of man-made and natural disasters through hydraulic engineering. It covers such topics s: 1) Disaster and hazard reduction; 2) wetland and tidal hydraulics; 3) mechanics of debris flows; 4) sediment transport; 5) bridge scour; 6) three- dimensional flow modeling; 7) computational hydraulics; 8) California water issues; and 9) probabilistic approaches to hydraulics. Engineers who are involved with these hydraulic engineering issues will find this proceedings an excellent source of information.

The Society is not responsible for any statements made or opinions expressed in its publications.

Photocopies. Authorization to photocopy material for internal or personal use under circumstances not falling within the fair use provisions of the Copyright Act is granted by ASCE to libraries and other users registered with the Copyright Clearance Center (CCC) Transactional Reporting Service, provided that the base fee of $2.00 per article plus $.25 per page copied is paid directly to CCC, 27 Congress Street, Salem, MA 01970. The identification for ASCE Books is 0-87262-920-1/93 $2.00 + $.25. Requests for special permission or bulk copying should be addressed to Permissions & Copyright Dept., ASCE.

Copyright ©1993 by the American Society of Civil Engineers,
All Rights Reserved.
ISBN 0-87262-920-1
ISSN 1070-1559
Manufactured in the United States of America.

CONTENTS

SESSION CW-1
WATER MANAGEMENT IN CALIFORNIA
Moderator: WILLIAM GIANELLI, Consultant

California Water Issues Bay-Delta Hearings Impacts on State Water Project
EDWARD F. HUNTLEY, and EDWARD D. WINKLER, Division of Planning, California Department of Water Resources 1

Issues in Balancing Environmental Costs and Benefits in Water Resources Planning
M. MCRAE, J. HASHEM and E. HAITES, Barakat & Chamberlin, Inc. 8

Establishing a Drought Reserve Policy for the Monterey Peninsula Water Management District
J.R. COFER and DARBY W. FUERST, Monterey Peninsula Water Management District 15

SESSION GW-1
GROUND WATER I
Moderator: Y.C. KIM, California State University, Los Angles, and R.G. QUIMPO, University of Pittsburgh

Well Response in a Leaky Aquifer and Computational Interpretation of Pumping Tests
ZBIGNIEW J. KABALA, Dept. of Soil and Environmental Science, University of California, Riverside 21

Concentration History During Pumping from a Leaky Aquifer with Stratified Initial Concentration
DANIEL J. GOODE, PAUL A. HSIEH, ALLEN M. SHAPIRO, WARREN W. WOOD and THOMAS F. KRAEMER, U.S. Geological Survey, Menlo Park 29

Parameter Estimates for a Groundwater Model
M.R. CONDON, ROBERT G. TRAVER, and W.B. FERGUSSON, Dept. of Civil Engineering, Villanova University 36

Simulation of Buoyant, Miscible Liquid Plumes in Heterogeneous Aquifers
SALWA RASHAD, JOHN HOOPES, DAVID OLSON and TSWN-SYAU TSAY, Civil and Envir. Engineering Dept., University of Wisconsin 42

An Overview of Methods of Delineate Capture Zones of Pumping Wells
BENJAMIN LEVY, KIM HENRY, and CHRISTIAN HILLER, ENSR
Consulting and Engineering 48

SESSION HR-1
HAZARD REDUCTION I
Moderator: M.T. TSENG, U.S. Army Corps of Engineers, and S.S. FAN, FERC

Regulatory Constraints on Vegetation Control in Channels—A Case Study of the Santa Ynez River, Santa Barbara County, California
RUDOLF OHLEMUTZ, Santa Barbara County Flood Control and Water Conservation District 54
Regulatory Obstacles to Emergency Response—A Case Study of the Paint Fire, June 1990, Santa Barbara, California
RUDOLF OHLEMUTZ, Santa Barbara County Flood Control and Water Conservation District 59
A Semi-Discrete Element Approach to the River Ice Breakup and Jamming Process
QIZHONG GUO and CHARLES C.S. SONG, Dept. of Civil and Envir. Engineering, Rutgers University 65
Adaptive Control of Unavoidable Hazardous Releases
NIKOLAOS D. KATOPODES and M. PIASECKI, Dept. of Civil and Envir. Engineering, University of Michigan 70

SESSION HS-1
DAM DESIGN, SAFETY, AND OPERATION
Moderator: J. RUFT, Colorado State University, and W. MIH, Washington State University

Mechanism of Air Entrainment in Spillway Aerators
AMIR REZA ZARRATI, Dept. of Civil Engineering, AmirKabir University of Technology 75
Simulation of Rapid Reservoir Drawdown for Flood Control, Cowlitz Falls Project
ROBERT H.A. JANSSEN and FREDERICK A. LOCHER, Bechtel Corporation 81
The Influence of Slope and Surface Roughness on Trapezoidal Free Overfall Characteristics
R.J. KELLER and DARYL F. WALTERS, Dept. of Civil Engineering, Monash University 87
Nonlinear Hydrodynamic Pressures on Dam Faces with Arbitrary Reservoir Shapes
BANG-FUH CHEN, Dept. of Marine Environment, National Sun Yat-Sen University 93
Performance of Prototype Aerating Weirs Downstream From TVA Hydropower Dams
GARY E. HAUSER, WILLIAM D. PROCTOR and TONY A. RIZK, TVA Engineering Laboratory 99

SESSION ST-1
WATERSHEDS AND RESERVOIRS
Moderator: H. CHANG, San Diego State University, and J.B. BRADLEY, WEST Consultants

Watershed Erosion
RAÚL PACHECO-CEBALLOS, Ingenieros Consultores Ltda. 104
Sediment Yields from a Watershed in Taiwan
WEN C. WANG, CHANG-TAI TSAI and H.W. SHEN, Multech Engineering Consultants, Inc. 111
Hydraulic Desiltation for Noncohesive Sediment
HSIEH WEN SHEN, JIHN-SUNG LAI and DIHUA ZHAO, Dept. of Civil Engineering, University of California, Berkeley 119
Optimization Modeling for Sedimentation in Alluvial Rivers Considering Uncertainties
CARLOS CARRIAGE and LARRY W. MAYS, Dept. of Civil Engineering, Arizona State University 125

SESSION WL-1
PHYSICAL PROCESSES IN COASTAL WETLANDS
Moderator: P. GOODWIN, Philip Williams & Associates, Ltd., and C EVERTS, Moffat & Nochol Engineers

Flow Patterns in Constructed Wetlands
ROBERT H. KADLEC, Dept. of Chemical Engineering, University of Michigan 131
Fundamental Principles of Tidal Wetland Restoration
R.B. KRONE, Dept. of Civil and Environmental Engineering, University of California, Davis 137
Integrated Description of Wetland Hydrology and Ecology by Mathematical Models
KARSTEN HAVNØ, and JESPER DØRGE, Danish Hydraulic Institute 143
Structural and Ice Effects on Salt Water Marsh Hydrology
THOMAS P. BALLESTERO, JOSEPH P. MARRONE and DEBORAH M. TROTTIER, Dept. of Civil Engineering, University of New Hampshire 150

SESSION PD-1
RESEARCH NEEDS IN HYDRAULIC ENGINEERING
(PANEL DISCUSSION)
Moderator: JEFFERY P. HOLLAND, U.S. Army Corps of Engineers

Research Needs in Hydraulic Engineering
ASCE HYDRAULICS DIVISION RESEARCH COMMITTEE 156

SESSION CW-2
BAY DELTA DECISION
Moderator: B.J. MILLER, Consultant

Is There a Future for Desalting in Meeting California's Water Needs?
    E.O. KARTINEN JR., Boyle Engineering Corporation 162
Coping with Agricultural Shortages
    J.S. JENKS, Consulting Engineer 168
Economic Costs to the State Water Project of Environmental Protection and Mitigation Measures
    RANDALL BROWN and RAYMOND HOAGLAND, California Department of Water Resources 174

SESSION GW-2
GROUND WATER II
Moderator: H. NOURI, California Institute of Technology, and Z.J. KABALA, University of California, Riverside

Designing Self-Cleaning Wet Wells for Wastewater Pumping
    ROBERT L. SANKS, GARR M. JONES and CHARLES E. SWEENEY, Dept. of Civil and Agricultural Engineering, Montana State University .. 180
Real Time Forecast of Landfill Leachate Flow
    THOMAS P. BALLESTERO and M.A. HOLANDA DE CASTRO, Water Resources Research Center, University of New Hampshire 186
Nonlinear Flow in Embankments
    NAZEER AHMED, Dept. of Civil and Environmental Engineering, University of Nevada-Las Vegas 192
Modeling Ground Water Mounding in a Heterogeneous Unconfined Aquifer
    TSWN-SYAU TSAY, JOHN HOOPES, DAVID OLSON, SALWA RASHAD, and M.Y. JIANG, Civil and Environmental Engineering Dept., University of Wisconsin 198

SESSION HR-2
HAZARD REDUCTION II
Moderator: C.B. CECILIO, PG&E, and R. KRONE, University of California, Davis

Reducing the Potential Downstream Impacts of a Dam Failure
    PETER W. SOLTYS, Water Resources Group, Woolpert Consultants ... 204
Assessment Procedures for Lahars, Mudflows, Debris Flow and Debris Torrents
    ROBERT MACARTHUR, DOUGLAS L. HAMILTON and WILLIAM E. BRANCH, Resource Consultants and Engineers, Inc. 210
Disaster Reduction in Dam and Reservoir Design
    K.P. SINGH, Office of Surface Water Resources and Systems Analysis, Illinois State Water Survey 216

The Coyote Dam Outlet Works Replacement Project: Hazard Reduction
D.E. HOOK, Santa Clara Valley Water District 222

SESSION HS-2
OUTLET AND INTAKE STRUCTURE
Moderator: F. LOCHER, Bechtel, and E. CARTER, Harza

Velocity Reduction at a Submerged Pipe Outlet
CHARLES E. RICE and KEM C. KADAVY, USDA, ARS 228
Hydraulic Performance of Culvert End Sections Designed for Collision Safety
JEFFERY A. BARTLEY, and BRUCE M. MCENROE, Dept. of Civil Engineering, University of Kansas 234
Compacted Spillway
ABDELKAWI KHALIFA, Civil Engineering Dept., United Arab Emirates University 240
Equations to Predict Critical Submergence at Horizontal Hydraulic Intakes
JOHN E. HITE, JR. and WALTER C. MIH, Locks and Conduits Branch, Hydraulic Structures Div., US Army Engr. Waterways Experiment Station 247
Hydraulic Analysis of the McCook Outlet Manifold
RICHARD L. STOCKSTILL, US Army Engr. Waterways Experiment Station 253

SESSION ST-2
RIVER SEDIMENT CASE STUDIES
Moderator: D.L. HAMILTON, Resource Consultants and Engineers, and G.D. GLYSSON, U.S. Geological Survey

Total Sediment Loads of Tropical Rivers
L. POSADA G. and C.F. NORDIN JR., Dept. of Civil Engineering, Universidad Nacional de Columbia 258
Sedimentation and Flood Protection for the Lower Yellow River
TA WEI SOONG, Office of Hydraulics and River Mechanics, Illinois State Water Survey 263
Sacramento River Environmental Requirements
W.C. GAINES, U.S. Army Engineers District, Sacramento ....
Dune Profiles Before and After Storm Events in Coastal Massachusetts
LISA J. WOLF, JAMES D. BOWEN, and KENNETH A. HICKEY, ENSR Consulting and Engineering 269

SESSION WL-2
SIMULATIONS OF WATER QUALITY IN WETLANDS
Moderator: J. OBEYSEKERA, South Florida Water Management District, and S. TU, PG&E

Refined Modelling of Water Quality Constituents in a Semi-Enclosed Coastal Wetland Basin
ROGER A. FALCONER and LIU SUIQING, Dept. of Civil Engineering, University of Bradford 275
Development of the Lake Okeechobee Watershed Phosphorus Transport Model
RICHARDSON A. WAGNER and LARRY A. ROESNER, Camp Dresser & McKee Inc. 281

Orange County Florida Landfill Dilute Leachate Wetland Treatment
and Restoration System
  LARRY N. SCHWARTZ, J.G. LADNER and S.J. KEELY, Camp
  Dresser & McKee Inc. 287
Environmental Control of Wetland Plant Communities
  MICHAEL DUEVER, Ecosystem Research Unit, National
  Audubon Society 293
Surface Flows over Intertidal Marshes
  FLORA C. WANG, Dept. of Oceanography and Coastal Sciences,
  Louisiana State University 299

SESSION GW-3
GROUND WATER II
Moderator: W. WALDROP, TVA, and D. PETERSON, CH2M HILL

Development of a Two-Layered Groundwater Model for the
Taipei Basin
  N.S. HSU, M.J. HORNG, C.M. WU, and WILLIAM YEH, Civil
  Engineering Dept., National Taiwan University 305
Development of San Gorgonio Pass Groundwater Flow Model
  ALADDIN SHAIKH and R. B. BELL,
  Boyle Engineering Corporation 311
Modelling Saltwater Intrusion Control Measures in the West
Coast Basin
  DONALD SCHROEDER and BRUCE JACOBS, Camp Dresser &
  McKee, Inc. 317

SESSION HR-3
HAZARD ASSESSMENT
Moderator: M.A. FODA, University of California, Berkeley, and F. TSAI, FEMA

A Model for Low-Drag Landslides
  DAGANG ZHANG and M.A. FODA, Civil Engineering Dept.,
  University of California, Berkeley 322
Uncertainties of Tailwater in an Inundation Study: Dam Upgrade
Optimization for Hazard Reduction
  S. SAMUEL LIN and RICHARD O. DAMERON, Commonwealth of
  Virginia, Dept. of Conservation and Recreation, Division of Soil and
  Water Conservation 328
Debris Flow Velocity Estimation Methods for Natural
Hazard Assessment
  DOUGLAS HAMILTON, ZHANG SHUCHENG and
  ROBERT C. MACARTHUR, Resource Consultants and Engineers 334
Federal Levee Effects on Flood Heights Near St. Louis
  GARY R. DYHOUSE, Hydrologic Engineering Section, Corps of
  Engineers, St. Louis District 340

Case History: 100 Year Flood in San Francisco
JAMES WALSH, Bureau of Engineering, City and County of
San Francisco 347

SESSION HS-3
ANALYSIS AND MODIFICATIONS TO AVERT DAM FAILURES
Moderator: H. COPP, Washington State University, and J.GEORGE, Water Experiment Station

Analysis of Cabinet Gorge Dam
JOHN Z. GIBSON, Washington Water Power Company 353
Dam Safety Modification for Clear Creek Dam
J. TROJANOWSKI, Concrete Dams Branch, Bureau of Reclamation ... 359
New Austrian Dam Side-Channel Spillway
MICHAEL A. STEVENS and SAL A. TODARO, Consultant 365
Bartlett Dam Fuseplug Auxiliary Spillway
BRUCE C. MULLER, JR., Concrete Dams Branch, Bureau
of Reclamation 372

SESSION ST-3
ASCE MANUAL 54 EXPANSION
Moderator: R.C. MACARTHUR, Resource Consultants and Engineers, Inc. and P.C. KLINGEMAN, Oregon State University

Current Legal Issues in Sedimentation
JAMES E. SLOSSON, Gerard Shuirman Slosson & Associates 378
Updating Chapter II—Sediment Transportation Mechanics
HASAN NOURI, ROBERT C. MACARTHUR and VITO A. VANONI,
Rivertech Inc. 383
Computational Modeling of Sedimentation Processes
COMMITTEE ON COMPUTER MODELING 388
Engineering Geomorphology
S.A. SCHUMM and M.D. HARVEY, Resource Consultants &
Engineers, Inc. 394
Scour Analysis at Highway Structures
E.V. RICHARDSON and J. R. RICHARDSON, Resource Consultants &
Engineers, Inc. 400

SESSION WL-3
WETLANDS—MANAGEMENT
Moderator: R.A. FALCONER, Bradford University, and L. ROSENER, Camp Dresser & McKee, Inc.

Variability of Hydraulic Response of Constructed Wetlands
M. T. WATERS, D.H. PILGRIM, T.J. SCHULZ and I.D. PILGRIM,
Centre for Wastewater Treatment, The University of New South Wales .. 406
Multi-Disciplinary Strategies for Flood-Plain Restoration at the River Rhine
R. ROETTCHER, E. RITTERBACH and G. ROUVÉ, Inst. fur
Wasserban und Wasserwitschaft 412

Development Strategies for Botswnan's Okavango Delta
  T. SCUDDER, Dept. of Anthropology, California Institute
  of Technology 418
Wetland Management in Britain: a Comparative Approach
  J.S. PETHICK, Institute of Estuarine & Coastal Studies, University
  of Hull 424
The Florida Everglades Nutrient Removal Project
  S. NEWMAN, J. ROY and J. OBEYSEKERA, Dept. of Research, South
  Florida Water Management District 430
Establishing Environmental Standards for Water Projects Based on
Limited Data
  DAVE VOGEL, Vogel Environmental Services .... *

SESSION CW-4
COPING WITH PROBLEMS
Moderator: ROBERT POTTER, California Deptartment of Water Resources

Mathematical Modeling of the Sacramento-San Joaquin Delta
  MOHAMMAD RAYEJ and FRANCIS CHUNG, California Dept. of
  Water Resources 436
Application of Four Point Model to the Sacramento-San Joaquin Delta
  PARVIZ NADER, California Dept. of Water Resources 442
Accounting for Antecedent Conditions in Seawater Intrusion
Modeling—Applications for the San Francisco Bay-Delta
  RICHARD A. DENTON, Contra Costa Water District 448
The DWR Delta Assembly Project
  ARTHUR HINOJOSA and RALPH FINCH, Division of Planning,
  California Dept. of Water Resources 454

SESSION GW-4
GROUNDWATER IV
Moderator: W. YEH, University of California, Los Angeles, and
  T. BALLESTERO, University of New Hampshire

Removal of Tidal Fluctuations from Pumping Test Data
  B.S. LEVY, ENSR Consulting and Engineering 460
A Unified Optimization-Simulation Aquifer Management Model
  B.T. REELY and A.K. TYAGI, Envirotech Services Inc. 466
An Optimal Parameter Estimation Model for Groundwater
Resource Management
  Y.S. TSAO, K.L. HWANG, and S.C. LIN, National I-Lan Institute of
  Agriculture and Technology 471
Field-Scale Research at the TVA Columbus Groundwater Research
Test Site
  W.R. WALDROP and J.M. BOGGS, TVA Engineering Laboratory 477

---

*Manuscript not available at time of printing

Simulation of Subsurface Drainage of Highway Pavements
B.M. MCENROE and S. ZOU, Dept. of Civil Engineering, University
of Kansas 483

SESSION BS-1
NATIONAL BRIDGE SCOUR EVALUATION PROGRAM I
Moderator: D. HALVERSON, Minnesota DOT, and A. WADDOUPS, Federal
Highway Administration

Quality-Control & Quality-Assurance Plan for Bridge Channel-Stability Assessments in Massachusetts
GENE W. PARKER and HARLOW PINSON, USGS Water Resource
Division, Massachusetts-Rhode Island District 489
Screening of Bridges in New Jersey Bridge for Scour
SALIM M. BAIG, New Jersey DOT, ANELLO F. MONACO and
JITENDRA C. PATEL, New Jersey Dept. of Transportation 495
Quality Control in Evaluating Scour at Bridges
EDWARD J. KENT, Whitman & Howard Inc. 501
New Jersey Bridge Scour Evaluation Program
PAUL WOJCIK, TAMS Consultants 507
An Efficient Method for Assessing Channel Instability Near Bridges
BRET A. ROBINSON and R.E THOMPSON, JR., U.S. Geological
Survey, Water Resource Division 513
Bridge Scour Analysis in New Jersey: Which Scour Factors Matter Most?
THOMAS W. ANELLA and GEORGE OLIGER, Parsons
Brinckerhoff, Inc. 519
In-Depth Scour Evaluations for Bridges in Pennsylvania
EREZ SELA and GEORGE R. OLIGER, Parsons Brinckerhoff, Inc. ... 525

SESSION HR-4
HURRICANE ANDREW
Moderator: A. MILLEDGE, South Florida Water Management District,
and J. OBEYSEKERA, South Florida Management District

Hurricane Andrew
ALLEN MILLEDGE, South Florida Water Management District .... *
Hurricane Andrew in South Florida: Preparing a Water Management System for Disaster
VINIO FLORIS and CATHLEEN ANCLADE, South Florida Water
Management District 531
Hurricane Andrew in South Florida: Steps to Recovery and Lessons Learned
VINIO FLORIS and CATHLEEN ANCLADE, South Florida Water
Management District 537

---

*Manuscript not available at time of printing

Protecting the Boca Raton Outfall Before and After Hurricane Andrew
JONATHAN A. FRENCH, W.A. JOHNSON, J.A. MILLS and
G.S. MARSH, Camp Dresser & McKee International Inc. 543

SESSION HS-4
ASPECTS OF RECLAMATION'S DAM SAFETY PROGRAM—PART I
Moderator: S. HIGINBOTHAM, Bureau of Reclamation, and D SCHREIBER,
Grant, Schreiber & Associates

Reclamation's Review of Operation and Maintenance Examination Program
V. A. HOFFMAN and D.E. KRAUSE, U.S. Bureau of Reclamation ... 549
The Dam Safety Process
C.J. VEESAERT, U.S. Bureau of Reclamation 556

SESSION WL-4
WETLANDS—TIDAL CHARACTERISTICS OF WETLANDS
Moderator: R.E. NECE, University of Washington, and K. NG, Bechtel

The Tidal Inlet Characteristics of a Small Californian Estuary
PETER GOODWIN, J. NIELSON and C. KELLY CUFFE, Philip
Williams & Associates, Ltd. 562
Wetland Restoration In the Dutch Dune Area
B. H. TANGENA and B. KORF, N.V. PWN Water Supply Company of
North-Holland 568
A Hydrodynamic Model for a Tidal Wetland
JEFF A. LEWANDOWSKI, R.J. SOBEY and PETER GOODWIN,
Black & Veatch 574
Mechanics of Riprap Movement in Tidal Flow
C. GALVIN, Coastal Engineer 580
Annual Oscillation of Mean Monthly Water Levels at U.S. Ports
TASK COMMITTEE ON EFFECTS OF ANNUAL TIDES, TIDAL
HYDRAULICS COMMITTEE, C. GALVIN, Coastal Engineer 584

SESSION BS-2
NATIONAL BRIDGE SCOUR EVALUATION PROGRAM II
Moderator: C. HARRIS, CALTRANS, and C. DUNN,
Federal Highway Administration

Comparison of Two Methods of Screening Bridges for Scour
DAVID S. HUNTER, EBASCO Services, Inc., M.K. HIXSON and
SALIM M. BAIG, New Jersey DOT 586
Case Studies of Bridge Scour in Western New York
KENNETH R. AVERY and M.A. HIXSON, Bergmann Associates 592
Scour Analysis For Bridges Over Missouri and Mississippi Rivers
S.L. MCCASKIE, C.C. CHANG and R.G. CHANTOME,
Sverdrup Corp. 598
Risk Analysis of Bridge Failure
GEORGE W. ANNANDALE, HDR Engineering, Inc. 604
Stream Stability and Scour Training in Support of the NBIS
P.F. LAGASSE, J.D. SCHALL and E.V. RICHARDSON, Resource
Consultants & Engineers, Inc. 611

Implementation of the NBIS Scour Evaluation Program District 2, Florida
P.F. LAGASSE, E.V. RICHARDSON and NIZAR JETHA, Resource Consultants & Engineers, Inc. 617

SESSION CW-5
MODELING THE EFFECTS OF PROJECT OPERATIONS ON BAY DELTA
Moderator: JACK CASSIDY, Hydraulics/Hydrology, Betchel Corporation

Modeling the Operation of a Water Quality Reservoir and Its Effect on the Sacramento-San Joaquin Delta
RICHARD A. DENTON, G. GARTRELL and A. W. NELSON, Contra Costa Water District 623

Quantification of Uncertainties in Water Quality Model with Application to the Sacramento-San Joaquin Delta
GREGORY GARTRELL, Contra Costa Water District 629

San Francisco Bay and Delta Oil Spill Fate Studies, Part I: Hydrodynamic Simulation
PARMESHWAR L. SHRESTHA, CAMILLA M. SAVIZ, GERALD T. ORLOB, IAN P.KING, RODNEY J. SOBEY and R. GLENN FORD, Dept. of Civil Engineering, Virginia Tech. 635

San Francisco Bay and Delta Oil Spill Fate Studies, Part II: Oil Spill Simulation
R. GLENN FORD, RODNEY J. SOBEY, PARMESHWAR L. SHRESTHA, CAMILLA M. SAVIZ, GERALD T. ORLOB and IAN P. KING, Ecological Consulting, Inc. 641

SESSION EH-1
SAN FRANCISCO BAY AREA
Moderator: R. CHENG, U.S. Geological Survey, and A. FINDIKAKIS, Bechtel

Modeling the Fate and Transport of Toxic Heavy Metals in South San Francisco Bay
PARMESHWAR L. SHRESTHA and G. T. ORLOB, Dept. of Civil And Environmental Engineering, University of California, Davis 647

New Methodology for Optimization of Freshwater Inflows to Estuaries
YIXING BAO and LARRY W. MAYS, Department of Civil Engineering, College of Engineering and Applied Sciences, Arizona State University 653

Relationships Between Flow and Benthic Communities
DAVID A. COBB, Bechtel Co. 659

Tidal Propagation in a Distorted Model
JIANLU XU and R.J. SOBEY, Dept. of Civil Engineering, University of California, Berkeley 665

SESSION HS-5
ASPECTS OF RECLAMATION'S DAM SAFETY PROGRAM—PART II
Moderator: S. HIGINBOTHAM, Bureau of Reclamation, and G. PICKERING, Waterways Experiment Station

Bureau of Reclamation Downstream Hazard Classification
DOUGLAS J. TRIESTE, U.S. Bureau of Reclamation 671

Using Threat to Life Studies to Guide Dam Safety Decisions
WAYNE J. GRAHAM, Surface Water Branch, Bureau of Reclamation 678

Economic Aspects of the Bureau of Reclamation Safety of
Dams Program
   ROBERT W. WALKER, Surface Water Branch, Bureau
   of Reclamation 684
Reclamation's Design Process of Early Warning Systems for Dam
Safety
   DAVID B. FISHER, Surface Water Branch, Bureau of Reclamation 690
Emergency Action Plan
   PATRICIA HAGAN-CHAGNON, Surface Water Branch, Bureau
   of Reclamation 696

SESSION ST-4
GRAVEL-BED RIVERS
Moderator: J.R. ADAMS, Illinois State Water Survey, retired, and
D.T. WILLIAMS, WEST Consultants, Inc.

U.S. Geological Survey Bed Load Sampling Policy
   G. DOUGLAS GLYSSON, U.S. Geological Survey 701
Incipient Motion in Gravel-Bed Rivers
   PETER C. KLINGEMAN and HABIB MATIN, Dept. of Civil
   Engineering, Oregon State University 707
Sediment Budgets in Gravel Bed Streams
   JEFFREY B. BRADLEY and DAVID T. WILLIAMS, WEST
   Consultants, Inc. 713

SESSION WL-5
WETLANDS
Moderator: J. PETHICK, University of Hull, and F. WEN, University of
California, Berkeley

Extreme Events and Coastal Wetlands
   JEFFERY HALTINER, Philip Williams & Associates, Ltd. 719
A Model of Mixing in a Stratified Tidal Flow
   STEPHEN MONISMITH, DEREK FONG and MARK STACEY,
   Environmental Fluid Mechanics Laboratory, Stanford University 725
Yolo Bypass Wetlands—Impact Investigation
   MICHAEL K. DEERING, U.S. Army Corps of Engineers,
   Sacramento District 731
Weaver Bottoms Backwater Rehabilitation
   JON S. HENDRICKSON and DENNIS D. ANDERSON, U.S. Army
   Corps of Engineers, St. Paul District 737

SESSION BS-3
UNCERTAINTIES IN QUANTIFYING STREAM STABILITY AND SCOUR I
Moderator: J.S. JONES, Federal Highway Administration, and M. ZELLER,
Simons, Lee & Assoc.

Flow and Scour Near an Abutment
   HSIEH WEN SHEN, CHRISTIAN T. CHAN, JIHN-SUNG LAI, and
   DIHUA ZHAO, Dept. of Civil Engineering, University of
   California, Berkeley 743
The Fallacy of Local Abutment Scour Equations
   J.R. RICHARDSON and E.V. RICHARDSON, Resource Consultants &
   Engineers, Inc. 749

Scour Prediction Model at Bridge Abutments
G.K. YOUNG, GKY and Assoc., M. PALAVICCINI, Catholic
University of America, and R.T. KILGORE, GKY and Assoc. 755
Bridge Abutment Scour in a Floodplain
TERRY W. STURM, Georgia Institute of Technology, and
NAZER JANJUA, Pakistan Council of Research in Water Resources ... 761
Bridge Abutment Scour in Compound Channels
B.W. MELVILLE, Dept. of Civil Engineering, Auckland University,
R. ETTEMA, University of Iowa 767
Plans for a Sensitivity Analysis of Bridge-Scour Computations
DAVID D. DUNN, and P.N. SMITH, U.S. Geological Survey 773

SESSION CA-1
COMPUTER APPLICATION I
Moderator: E.R. HOLLEY, University of Texas, Austin, and F.T. WATTS, Idaho State University

Computational Hydraulics—the Systems Approach
GERALD J. BARIL and GLEN DROGIN, Civil & Highway
Engineering, Gannett Fleming Engineers 779
Computer Aided Design and Cost Estimation of Gabion Lined Channel
DAVID T. WILLIAMS and GRAY R. OSENDORF, WEST
Consultants, Inc. 785
Water Surface Profile Computations—How Many Sections Do I Need?
DAVID B. THOMPSON and T.D. ROGERS, Dept. of Civil
Engineering, Texas Tech University 791
Simulation of River Bed Evolution Below Tsenwen Reservoir in Taiwan
CHANG TAI TSAI and BOR-CHYI TSAI, Dept. of Hydraulics and
Ocean Engineering, National Cheng-Kung University 797
On the Applicable Ranges of Kinematic and Diffusion Models in
Open Channels
GYE-WOON CHOI, G.H. KIM and S.J. AHN, Water Resources
Research Institute, Korea Water Resources Corporation 803

SESSION CW-6
WATER QUALITY MODELING I
Moderator: GERALD T. ORLOB, University of California, Davis

Predicting Water Quality at Municipal Water Intakes—Part 1:
Application to the Contra Costa Canal Intake
RICHARD A. DENTON, Contra Costa Water District 809
Predicting Water Quality at Municipal Water Intakes—Part 2:
Application to the Southern Sacramento-San Joaquin Delta
GREGORY GARTRELL, Contra Costa Water District 815
Simulating THM Precursors Transport with DWRDSM
PAUL H. HUTTON and CHRISTOPHER ENRIGHT, Division of
Planning, Dept. of Water Resources 821
Particle Tracking Model for the Sacramento-San Joaquin Delta
GILBERT V. BOGLE, TARA A. SMITH and FRANCIS I. CHUNG,
Water Engineering and Modeling 827

Levee Breach Inundation Study, Sacramento County, California
DAVE PETERSON, Municipal Services Department, CH2M Hill 833

SESSION EH-2
DENSITY CURRENTS
Moderator: V. ALAVIAN, TVA Engineering Laboratory, and S. MONISMITH, Stanford University

Density Currents in Pollutant Transport and Mixing
GERHARD H. JIRKA and P.J. AKAR, DeFrees Hydraulics Laboratory,
School of Civil and Environmental Engineering, Cornell University 838
Criticality of Density Intrusions
STEVEN J. WRIGHT and DIANA RAEZ-RIVADENERIA, Department
of Civil and Environmental Engineering, University of Michigan 845
**Hydraulic Performance of a Flexible Curtain Used for Selective
Withdrawal—A Physical Model and Prototype Comparison**
TRACY VERMEYEN and PERRY JOHNSON, U.S. Department of the
Interior, Bureau of Reclamation 2371
**A Flexible Curtain Structure for Control of Vertical Reservoir Mixing
Generated by Plunging Inflows**
PERRY JOHNSON and TRACY VERMEYEN, U.S. Department of the
Interior, Bureau of Reclamation 2377

SESSION HS-6
PLUNGE POOLS—DESIGN AND ANALYSIS CONCEPTS
Moderator: R. WITTLER, Bureau of Reclamation, and M. SKINNER, Colorado State University

Plunge-Pool Aeration Due to Inclined Jets
ASSEM AFIFY and GILBERTO E. URROZ, Hydraulics and Sediment
Research Institute 851
Design of Pre-Excavated Scour Hole Below Flip-Bucket Spillways
NOSRATOLLAH AMANIAN and GILBERTO E. URROZ, Utah Water
Research Laboratory, College of Engineering, Utah State University 856
Design of Riprap Stilling Basin for Overhanging Pipe
M. SHAFAI-BAJESTAN and M.L. ALBERTSON,
Shahid-Chamran University 861

SESSION ST-5
CHANNEL PROCESSES
Moderator: R.R. COPELAND, U.S. Army Corps of Engineers

Bed Sediments Size Changes, Atchafalaya Rivers
C.W. SOILEAU, R.W. RENTCHLER, F.L. OGDEN, and
C.F. NORDIN, New Orleans District, U.S. Army Corps of Engineers .. 869
A Study of Lateral Bed Slopes Developed at Bend
DIANA YU MA and H.W. SHEN, Dept. of Civil Engineering,
University of California, Berkeley 875
Bank Stability Analyses verses Field Observations
DANIEL E. MARCH, STEVEN R. ABT, and COLIN R. THORNE,
Dept. of Civil Engineering, Colorado State University 881

Effect of Grain Size on Sediment Transport Calculations
D. GESSLER, C.C. WASTSON and N. RAPHELT, Dept. of Civil
Engineering, Colorado State University 887
Nonuniform Transient Sediment Transport Modeling
KEH-CHIA YEH, CHIAN-MIN WU, JINN-CHUANG YANG and
SHIAN-JANG LI, Dept. of Civil Engineering, National Chiao
Tung University 893

SESSION BS-4
UNCERTAINTIES IN QUANTIFYING STREAM STABILITY AND SCOUR II
Moderator: D. POTTER, Arkansas DOT, and R.E. TRENT, Federal
Highway Administration

Computation of Flow Past a Cylinder Mounted on a Flat Plate
CESAR MENDOZA-CABRALES, Dept. of Civil Engineering,
Columbia University 899
The Separated Flow Around a Circular Bridge Pier
F. WEN, D. SEYMOUR and H.W. SHEN, Dept. of Civil Engineering,
University of California, Berkeley 905
Top Width of Pier Scour Holes in Free and Pressure Flow
E.V. RICHARDSON, Resource Consultants & Engineers, Inc. and
L. ABED, Ministry of Public Works and Water Resources, Egypt 911
Preliminary Studies of Pressure Flow Scour
J. STERLING JONES, Federal Highway Administration,
DAVID BERTOLDI and ED UMBRELL, GKY and Associates 916
The Influence of Exposed Footings on Pier Scour Depths
LISA M. FOTHERBY, Colorado State University, Engineering Research
Center and J. STERLING JONES, Federal Highway Administration 922
Economics of Floods, Scour and Bridge Failures
JENNIFER RHODES, University of Maryland and ROY TRENT,
Federal Highway Administration 928
Scour Around Bridge Piers in Oklahoma Streams in 1986
A.K. TYAGI, School of Civil and Environmental Engineering, Oklahoma
State University 934

SESSION EH-3
STRATIFIED FLOW I
Moderator: A. LAW, Bechtel, and P. ROBERTS, Georgia Institute of Technology

A Model for Vertical Transport at a Sheared Density Interface
GREG D. SULLIVAN and E. JOHN LIST, Contra Costa
Water District 939
Operational and Structural Optimization of Hydraulic Structures for
Light Liquid Removal
A.J. KUCK, G. STROMBERG and G. ROUVÉ, Institute for Hydraulic
Engineering and Water Resources Management, Aachen University
of Technology 945
Mixing Characteristics of a Transitional Stratified Two-Layer Flow
C. LIU and F. WEN, Dept. of Civil Engineering, University of
California, Berkeley 951

Observations of Artificial Destratification
S. G. SCHLADOW, Centre for Water Research, University of
Western Australia 957

SESSION HS-7
EROSION PROTECTION OF HYDRAULIC STRUCTURES
Moderator: D.H. PILGRIM, University of New South Wales, B. FLETCHER,
Waterways Experiment Station

Emergency Protection, San Luis Rey River Aqueducts
ERGUN BAKALL, JEFF MONCRIEF, JON WALTERS, and
HOWARD CHANG, San Diego County Water Authority 962
Design and Performance of Emergency Spillway Channel Erosion Protection Grenada Lake Dam, Grenada, Mississippi
PAUL BARNES and JOHN E. HITE, JR., Design Branch, Engineering,
US Army Engineer District 968
The Study of Riprap as Scour Protection for Bridge Abutments
A. TAMIM ATAYEE, J.E. PAGÁN-ORTIZ, S. JONES and
R.T. KILGORE, GKY & Associates, Inc. 973
Erosion Protection at Hydraulic Structure—A Report from the Task Committee
RODNEY J. WITTLER, FRED WATTS, JOHN HITE JR. and
GILLBERTO URROZ, U.S. Bureau of Reclamation 979

SESSION ST-6
SEDIMENT TRANSPORT MECHANISM I
Moderator: C.T. YANG, Bureau of Reclamation

Video Analysis of Gravel Saltation
Y. NIÑO, MARCELO H. GARCÍA and LUIS AYALA, Department of
Civil Engineering, University of Illinois, Urbana-Champaign 983
Scour Development at Isolated River-Bed Obstacles
P.C. KLINGEMAN and C-C HUANG, Dept. of Civil Engineering,
Oregon State University 989
N Values for Shallow Flow in Rough Channels
F. J. WATTS, Department of Civil Engineering, University of Idaho ... 995
Problems with Numerical Modeling of Gravel-Bed River
R.R. COPELAND and W.A. THOMAS, US Army Engineer Waterways
Experiment Station 1001

SESSION EC-1
ECOLOGICAL HYDRAULICS I
Moderator: R.T. MILHOUS, National Ecology Research Center, and
A.J. ODGAARD, University of Iowa

An Experimental Study of Bivalve Siphonal Jets in a Turbulent Boundary Layer Flow
CATHERINE A. O'RIORDAN, S.G. MONISMITH and J.R. KOSEFF,
Environmental Fluid Mechanics Laboratory, Department of Civil
Engineering, Stanford University 1007
An Assessment of Fish Entrainment and Impingement Potential for an Offshore Cooling Water Intake in a Tropical Bay
ANDREW M. DASINGER and RONALD SUTTON, ENSR Consulting
and Engineering 1013

Using a Numerical Model to Evaluate Striped Bass Management—
Scenarios in the Sacramento-San Joaquin Delta, CA
 H.F.N. WONG, J.F. ARTHUR, M.D. BALL and L.J. HESS, US
 Bureau of Reclamation 1019
Numerical Models of Phytoplankton Dynamics for Shallow Estuaries
 L.L. VIDERGAR, J.R. KOSEFF, and S.G. MONISMITH,
 Environmental Fluid Mechanics Laboratory, Stanford University 1025

SESSION BS-5
UNCERTAINTIES IN QUANTIFYING STREAM STABILITY AND
SCOUR III
Moderator: N. BORMANN, Gonzaga University,
and STANLEY DAVIS, Consultant

Pier Scour Equations Used in China
 GAO DONG GUANG, LILIAN POSADA G. and CARL F. NORDIN,
 Dept. of Civil Engineering, Colorado State University 1031
Local Scour at Skewed Bridge Piers
 E.A. MOSTAFA, A.A. YASSAN, R. ETTEMA and B.W. MELVILLE,
 Dept. of Irrigation & Hydraulics, University of Alexandria 1037
Estimating Pier Scour with Artificial Neural Networks
 ROY TRENT, N. GAGARIN, and JENNIFER RHODES, Federal
 Highway Administration 1043
An Artificial Neural Network for Computing Sediment Transport
 ROY TRENT, ALBERT MOLINAS and N. GAGARIN, Federal
 Highway Administration 1049
Supply of Large Woody Debris in a Stream Channel
 TIM DIEHL and B.A. BRYAN, U.S. Geological Survey 1055
The Control and Monitoring of Local Scour at Bridge Piers
 COLIN PAICE, and RICHARD HEY, Environmental Sciences,
 University of East Anglia 1061
Comparison of Theoretical and Historical Scour
 M.A. HIXSON, K.R. AVERY, New York State Thruway Authority ... 1067

SESSION CA-2
COMPUTER APPLICATION II
Moderator: W.T. THOMAS, U.S. Army Corps of Engineers, and M.H. HSU,
National Taiwan University

Calibration of Manning's Roughness for a River Reach
 CASSIE C. KLUMPP and D.C. BAIRD, U.S. Bureau of Reclamation .. 1073
Hydrodynamic Modeling for Channel Barrier Design
 P. L. SHRESTHA, J.J. DEVRIES and R. B. KRONE, Dept. of Civil
 and Environmental Engineering, University of California, Davis 1079
A 2-D Numerical Model for High Velocity Channels
 R.C. BERGER and RICHARD L. STOCKSTILL, U.S. Army Engr.
 Waterways Experiment Station 1085
Numerical Solution of Transient Closed-Conduit Flow Equations by the
Method of Lines Along Characteristics
 NOSRAT MAGHSOUDI, HydroTel Intl. 1091

Visualization and Analysis of Multi-Dimensional Velocity Measurements in Carquinez Strait, California
D. AGOSTINI, J. EVANS, F. WEN and P. SMITH, Civil Engineering Department, University of California, Berkeley 1096

SESSION EH-4
STRATIFIED FLOW II
Moderator: G. JIRKA, Cornell University, and L. LEE, Bechtel

* **Fluid Mechanics Aspects of Ocean Outfalls**
PHILIP J.W. ROBERTS, School of Civil Engineering, Georgia Institute of Technology 1102
A 2-D Lake Model with Artificial Destratification
S. G. SCHLADOW, Centre for Water Research, University of Western Australia 1108
Stratification Models Sensitivity to Solar Radiation Data
MONICA F.A. PORTO, University de São Paulo 1113
Selective Withdrawal in a Rotating Stratified Fluid
STEPHEN G. MONISMITH, N.R. MCDONALD and JÖRG IMBERGER, Environmental Fluid Mechanics Laboratory, Stanford University 1119

SESSION HS-8
OUTLET AND INTAKE STRUCTURES
Moderator: L. HARRISON, PG&E, and F.J. WATTS, University of Idaho

Multi Objective Detention Outlet Control Structure
DARRELL KIM BEATLEY and JAMES N. WIGFIELD, Espey, Huston & Associates, Inc. 1125
Velocity Downstream of a Submerged Pipe Outlet
KERRY M. ROBINSON, CHARLES E. RICE and KEM C. KADAVY, Agricultural Research Service, USDA, ARS 1131
Pressure Relief Under Hydraulic Jump Stilling Basins
C.D. SMITH and Z. GUI, Department of Civil Engineering, University of Saskatchewan 1137
General Implicit Representation of Hydraulic Structures in Numerical Flow Models
LEWIS L. DELONG and J.M. FULFORD, U.S. Geological Survey 1143
LaPlace Valve Stroking to Control Water Hammer
DONALD SERPAS, Dept. of Hydraulic Engineering, University of São Paulo 1149

SESSION EC-2
ECOLOGICAL HYDRAULICS II
Moderator: G.M. KONDOFF, University of California, Berkeley, and T. WAKEMAN, US Army Corps of Engineers

Low-Flow Habitat in Flood Control Channels
J. CRAIG FISCHENICH, STEVEN R. ABT and CHESTER C. WATSON, US Army Engineer Waterways Experiment Station 1155
Management of Rice Fields for Wetlands, Water, and Rice Production
ELIZABETH S. ANDREWS and PHILIP B. WILLIAMS, Philip Williams & Associates, Ltd. 1161

Modeling the Impacts of Plankton Entrainment in a Tropical Bay
J. D. BOWEN, D.P. GALYA and M.T. VILLARS, ENSR Consulting
and Engineering 1167
The Flushing Flow Problem on the Trinity River, CA
G.M. KONDOLF and P.R. WILCOCK, Department of Landscape
Architecture, University of California, Berkeley 1172

SESSION ST-7
SEDIMENT TRANSPORT MECHANISM II
Moderator: M.E. FALTAS, Michael Baker Jr. Inc.

Drag Characteristics of Coarse Sediment in Clay Suspension
HYOSEOP WOO, and HYOUNGSUP KIM, Korea Institute of
Construction Technology 1178
Fractal Dimension of Aggregated Sediments
H. S. KIM and P. D. SCARLATOS, Dept. of Ocean Engineering,
Florida Atlantic University 1184
Mass Transport in Mud Layer Induced By Wave Action
ISMAEL PIEDRA CUEVA, Institute of Fluid Mechanics and Env.
Engineering, Uruguay 1189
Sediment Disposal and Transport in Central SF Bay
THOMAS H. WAKEMAN, A.E. MATHIESEN and
G.W. CHATFIELD, US Army Corps of Engineers 1194
Shape Effect on Bedload Transport in Pipes
C. NALLURI and A. A. GHANI, Department of Civil Engineering,
University of Newcastle upon Tyne 1200

FIELD TRIP TO SAUSALITO US ARMY CORPS OF ENGINEERS' BAY MODEL
Organized by S.T. SU, Harza Kaldveer

SESSION BS-6
COASTAL AND TIDAL BRIDGE SCOUR
Moderator: J. MORRIS, Federal Highway Administration, and B. EDGE, Edge and Associates

Scour at Highway Structures in Tidal Waters
E. V. RICHARDSON, J. R. RICHARDSON, Resource Consultants &
Engineers, Inc., and B. EDGE, Edge & Associates 1206
Model Technology for Estimating Storm-Inducing Currents
MARY A. CIALONE and H. LEE BUTLER, Coastal Engineering
Research Center, US Army Engineer Waterway Experiment Station 1212
Indian River Inlet: Is There A Solution?
H. LEE BUTLER and JEFF LILLYCROP, Coastal Engineering Research
Center, US Army Engineers Waterway Experiment Station 1218
Bridge Hydraulic Design in Tidal Situations
C. R. NEILL and E. K. YAREMKO, Northwest Hydraulic
Consultants, Ltd. 1224
McCormick Bridge Scour Evaluation—A Case Study of A Tidal Bridge
MICHAEL P. MARONEY, Browwell & Carrier, Inc. 1230

A Tidal Inlet Bridge Scour Assessment Model
M.S. VINCENT and MARK A. ROSS, Center for Modeling Hydrologic and Aquatic Systems, Dept. of Civil Engineering, University of South Florida 1236

SESSION CA-3
COMPUTER APPLICATION III
Moderator: D. FREAD, National Weather Service, and J.Y. LU, National Chung-Hsing University

A Mathematical Model of Flow in Mildly Sinuous, Deep Channels
AMARTYA KUMAR BHATTACHARYA, and SRIJIB K. KAR, Department of Civil Engineering, Indian Institute of Technology 1242
Two-Dimensional Hydrodynamic Modeling with a Computer Graphics System
DANIEL H. HOGGAN and D.E. TWISS, Hydraulic Design Section, U.S. Army Engineer District, Sacramento 1248
Numerical Modeling of Unsteady Compound Channel Flow
R.S.M.M. RASHID and M. HANIF CHAUDHRY, Dept. of Civil and Environmental Engineering, Washington State University 1254
"Halloween" Wave Transformation at Virginia Coast
JEROME P.-Y. MAA, Virginia Institute of Marine Science, School of Marine Science, College of William and Mary 1260
Use of Floodplain PCB Concentrations to Calibrate a River Hydraulics Model
JAMES M. HASSETT and LEONARD T. WRIGHT, Division of Environmental and Resource Engineering, SUNY-CESF 1266

SESSION DF-1
OVERVIEW, RESEARCH NEEDS, AND PROCESSES
Moderator: C.L. CHENG, U.S. Geological Survey, and P.A. CARLING, Institute of Freshwater Ecology

* Assessment and Prediction of Debris-Flow Hazards
GERALD F. WIECZOREK, U.S. Geological Survey, Geologic Division 1272
Research Needs for Debris Flow Disaster Prevention
TIMOTHY R.H. DAVIES, Dept. of Natural Resources Engr., Lincoln University 1284
Debris Flows in Grand Canyon National Park, Arizona: Magnitude, Frequency and Effects on the Colorado River
T.S. MELIS and ROBERT H. WEBB, U.S. Geological Survey 1290
Debris Flows and Mass Wasting in Volcanic Torrents
HIROSHI SUWA, Disaster Prevention Research Institute, Kyoto University 1296
Interpreting Debris-Flow Hazard from Study of Fan Morphology
KELIN X. WHIPPLE, Dept. of Geological Sciences, University of Washington 1302

SESSION EH-5
FISH SCREEN
Moderator: S. TU, PG&E, and WHITMAN, National Marine Fishery

Hydraulic Modelling of Fish Screens
A. JACOB ODGAARD, Y. WANG, and R.A. ELDER, Institute of
Hydraulic Research, University of Iowa 1308
Fish Screen Developments Columbia River Dams
D.E. WEITKAMP and REX A. ELDER, Parametrix, Inc. 1314
Hydraulic Aspects of a Low-Velocity, Inclined Fish Screen
FREDERICK A. LOCHER, VICTOR C. BIRD and
A. JACOB ODGAARD, Bechtel Corp. 1320
Survival of Atlantic Salmon Smolts Bypassed Through Ice-Log Sluices Determined by the HI-Z Turb'N Tag
PAUL G. HEISEY, DILIP MATHUR, G. A. NARDACCI and
MILTON ANDERSON, RMC Environmental Services, Inc. 1326
Ask Young Clupeids if Kaplan Turbines Are Revolving Doors or Blenders
DILIP MATHUR and PAUL HEISEY, RMC Environmental
Services, Inc. 1332

SESSION HS-9
CHANNEL WORKS AND DRAINAGE
Moderator: R.H. FRENCH, Desert Research Institute, and K.C. YIH, National Chiao Tung University

Regulation of Flow Downstream of Weirs
BOUALEM HADJERIOUA, TONY RIZK, EMMETT M. LAURSEN
and GARY HAUSER, Dept. of Civil Eng. & Eng. Mech., University
of Arizona 1338
Flow Resistance Properties of Flexible Linings
GEORGE K. COTTON, CRSS Civil Engineers, Inc. 1344
Calculon Weir Structure—Effective Use of Gabions
JOHN G. HARMAN, Progressive Engineering Consultants, Inc. 1350
Storm Drainage Channel Rehabilitation
MARK S. HOLSTAD and KENNETH W. WYLIE, Greiner, Inc. 1355
Determination of Hydraulic Roughness for Concrete-Lined, Supercritical Channels
SCOTT E. STONESTREET, M.E. MULVIHILL and
RONALD R. COPELAND, U.S. Army Corps of Engineers 1361

SESSION BS-7
COUNTERMEASURES FOR STREAM INSTABILITY AND SCOUR I
Moderator: L. HARRISON, Resource Consultants & Engineers, Inc., and
J.R. RICHARDSON, Resource Consultants & Engineers, Inc.

Cable-Tied Concrete Block Erosion Protection
J.A. MCCORQUODALE, A MOAWAD and A.C. MCCORQUODALE,
Dept. of Civil Environmental Engineering, University of Windsor 1367

A Method of Managing Floating Debris
  SELDEN SAUNDERS and M. LEONARD OPPENHEIMER, Saunders
  Product Development 1373
The Influence of Protective Material on Local Scour Dimensions
  LISA M. FOTHERBY, Colorado State University 1379
Tetrapods as a Scour Countermeasure
  DAVID BERTOLDI and ROGER KILGORE, GKY & Associates, Inc.   1385
Variations Encountered in Design Analysis of Local Scour at
Drop Structures
  NOEL E. BORMANN and MICHAEL ZELLER, Dept. of Civil
  Engineering, Gonzaga University 1391
Threats to Bridge Stability from Scour Related Failures of
Drop Structures
  NOEL E. BORMANN, Gonzaga University, and MICHAEL ZELLER,
  Simons, Li and Associates 1397

SESSION DF-2
MECHANICS, RHEOLOGY, AND RHEOMETRY
Moderator: C.C. MEI, M.I.T., and J.S. O'BRIEN, FLO Engineering, Inc.

Mechanics of Debris Flows
  STUART B. SAVAGE, Dept. of Civil Engineering & Applied
  Mechanics, McGill University 1402
Kinetic-Theory Approach to the Nevado del Ruis 1985 Debris Flow
  JUAN A. GARCIA and STUART B. SAVAGE, Dept. of Civil
  Engineering & Applied Mechanics, McGill University 1408
Continuum-Mechanics-Based Rheological Formulation for Debris Flow
  CHENG-LUNG CHEN and C.H. LIN, U.S. Geological Survey, Water
  Resources Division 1414
Rheometry of Natural Sediment Slurries
  JON J. MAJOR, U.S. Geological Survey 1415

SESSION EH-6
TRANSPORT AND DISPERSION I
Moderator: R. KELLER, Monash University, and P. RYAN, Bechtel

Salt Transport in a Tidal Canal, West Neck Creek, Virginia
  JERAD D. BALES and S.C. SKROBIALOWSKI, U.S.
  Geological Survey 1422
Hydrodynamic and Toxic Contaminant Dispersion in the Lower St.
Marys River
  E.M. YUEN, and J.A. MCCORQUODALE, Civil Engineering
  Department, Lawrence Technological University 1428
Modeling Low Flow Transport of Nonconservative Pollutants
in Streams
  IL WON SEO and DAE YOUNG YU, Department of Civil Engineering,
  Seoul National University 1434

Wind Induced Circulation in Shallow Lakes
IAN P. KING, PARMESHWAR L. SHRESTHA and
GERALD T. ORLOB, Department of Civil Engineering, University of
California, Davis 1440
Prediction of Filtrate Turbidity by Parameter Estimation
SADATAKA SHIBA, Department of Chemical Engineering,
Osaka University 1446

SESSION HM-1
HYDRAULIC MEASUREMENT I
Moderator: A.C. MILLER, Pennsylvania State University, and P.D. SCARLATOS,
Florida Atlantic University

Repeatability and Oblique Flow Response Characteristics of Current Meters
JANICE M. FULFORD, K.G. THIBODEAUX and W.R. KAEHRLE,
U.S. Geological Survey 1452
Bed Shear Stress in Unsteady Open-Channel Flows
IEHISA NEZU, H. NAKAGAWA, Y. ISHIDA and A. KADOTA,
Department of Civil Engineering, Kyoto University 1458
Limitations of Reduced Scale Testing of Parshall Flumes
STEVEN J. WRIGHT, Department of Civil and Environmental
Engineering, University of Michigan 1464
Urban Storm Water Instrumentation: a Field Observation
R. BRAD JENNINGS, USEPA Water Management Div. 1470
Water-Level, Velocity and Dye Measurements in the Chicago Tunnels
K.A. OBERG and A.R. SCHMIDT, U.S. Geological Survey, Water
Resources Division 1476

SESSION HS-10
HYDRAULIC STRUCTURES I
Moderator: B.K. LEE, Harza Engineering Company

A Simple and Reliable Method for Analysis of Water Supply Distribution
TSUN-HOU KUAN, Public Works City of Los Angeles 1482
Friction Loss Equations for Hydropower Facilities
ED A. TOMS, E. JUNE BUSSE and C.A. THOMPSON, ATC
Engineering Consultants, Inc. (ECI) 1488
Development of Enhanced Tools for the Integrated Analysis of Reservoir and Power System Operations
BARBARA MILLER, P. OSTROWSKI, JR. and VAHID ALAVIAN,
TVA Engineering Laboratory 1494
Hydraulic Modeling of High Unit Discharge Energy Dissipators
DAVID R. MOORE, DONALD H. BURN, PHILIP D. WANG and
DENNIS E. LEMKE, Department of Civil Engineering, University
of Manitoba 1500

Model Study of Center Hill Fuse Plug Spillway
BOBBY P. FLETCHER and PAUL A. GILBERT, Hydraulics
Laboratory, US Army Engineers Waterways Experiment Station 1505

SESSION ST-8
SEDIMENT TRANSPORT MECHANISM III
Moderator: N.G. BHOWMIK, Illinois State Water Survey, and R. MACARTHUR, Resource Consultants & Engineers, Inc.

An Application of Nonhomogeneous Poisson Process in Sediment Infiltration into Gravel Bed
FU-CHUN WU, H.W. SHEN, Dept. of Civil Engineering, University of California, Berkeley 1511

Variation of Froude Number with Discharge for Large-Gradient Streams
KENNETH L. WAHL, U.S. Geological Survey, Water Resources Division 1517

Prediction of Gravel Transport Using Parker's Algorithm
DAVID R. DAWDY and WEN C. WANG, Consultant 1523

Sacramento River Environmental Requirements
W. CRAIG GAINES, US Army Corps of Engineers, Sacramento District 2383

SESSION BS-8
COUNTERMEASURES FOR STREAM INSTABILITY AND SCOUR II
Moderator: W. LINDSEY, CALTRANS, and J.R. RICHARDSON, Resources Consultants & Engineers, Inc.

Emergent Techniques in Scour Monitoring Devices
J. R. RICHARDSON, Resource Consultants & Engineers, Inc., and
J. Price, ETI, Inc. 1529

Riprap Design at Bridges-Factor of Safety Approach
GEORGE K. COTTON, Cess Civil Engineers, Inc. 1534

Riprap Coverage Around Bridge Piers
JAMES F. RUFF and JAMES R. NICKELSON, Colorado State University 1540

Predicting Critical Scour Stage At Bridges
JOHN N. PAINE, DARRELL KIM BEATLEY, Espey, Huston & Assoc. Inc., and JAMES N. WIGFIELD, Virginia DOT 1546

Riprap Incipent Motion and Shield's Parameter
ROGER T. KILGORE and G.K. YOUNG, GKY and Associates 1552

Scour Retrofit Case Studies for Arizona
GEORGE K. COTTON and JIRI VITEK, CRSS Civil Engineers, Inc. .. 1558

SESSION CA-4
EDUCATION IN COMPUTATIONAL HYDRAULICS
Moderator: N.D. KATOPODES, Dept. of Civil and Environmental Engineering, University of Michigan

Applicability of Two Simplified Flood Routing Methods: Level-Pool and Muskingum-Cunge
D.L. FREAD and K.S. HSU, U.S. Department of Commerce, National Oceanic and Atmospheric Administration, National Weather Service 1564

Selection of Δx and Δt Computational Steps for Four-Point Implicit
Nonlinear Dynamic Routing Models
D.L. FREAD and J.M. LEWIS, National Weather Service 1569
Advanced HEC-2 Modeling on Forester Creek
L.T.M. VOMERO, WEST Consultants, Inc. 1574
Modeling Critical Depth in Open Channels
ROBERT G. TRAVER, Dept. of Civil Engineering,
Villanova University 1580
A Numerical Model for Learning Concepts of Streamflow Simulation
L.L. DELONG, U.S. Geological Survey 1586

SESSION DF-3
ANALYSIS, SIMULATION AND EXPERIMENTS
Moderator: S.B. SAVAGE, McGill University, and T. TAKAHASHI, Disaster
Prevention Research Institute

A Comparison Between Two Kinematic Wave Solutions for Movement
of Debris Flows
M. ARATTANO and WILLIAM Z. SAVAGE, U.S. Geological Survey,
Geologic Division 1592
Roll Waves in Mud Flow
C.O. NG and CHIANG C. MEI, Dept. of Civil and Environmental
Engineering, Massachusetts Institute of Technology 1598
Friction in Debris Flows: Inferences from Large-Scale
Flume Experiments
RICHARD M. IVERSON and R.G. LAHUSEN, U.S. Geological
Survey, David A. Johnston Cascades Volcano Observatory 1604
Analysis of Pulsing Phenomenon in Viscous Debris Flow
WAN ZHAOHUI and HUA JINGSHEN, Dept. of Sedimentation
Engineering, Institute of Water Conservancy and Hydroelectric
Power Research 1610
A Study on Debris Flow Surges
ZHAOYIN WANG, Institute of Water Conservancy and Hydroelectric
Power Research, China 1616

SESSION EH-7
TRANSPORT AND DISPERSION II
Moderator: G. ORLOB, University of California, Davis, and K. NG Bechtel

Comparisons Between Experimental and Numerical Studies on Laminar
Flow with Tracer Transport
CHRISTIAN FORKEL, HELMUT DANIELS, JENS BIRKHÖLZER and
GERHARD ROUVÉ, Institute for Hydraulic Engineering and Water
Resources Management, Aachen University of Technology 1622
Comparison of Advective Transport Algorithms with an Application in
Suisun Bay
JON BURAU, STEPHEN MONISMITH and JEFFREY KOSEFF,
Environmental Fluid Mechanics Laboratory, Stanford University 1628
An Optimization Approach to River Adjustments
ROBERT MILLAR and MICHAEL QUICK, Department of Civil
Engineering, University of British Columbia 1635

Gully Intrusion on Reclaimed Disposal Sites
SCOTT A. HOGAN, C.J. PANLEY, S.R. ABT and T.L. JOHNSON,
Engineering Research Center, Colorado State University 1641
Sediment Deposition in Jennings Randolph Reservoir, Maryland and
West Virginia
MARGARET M. BURNS and ROBERT MACARTHUR, US Army
Corps of Engineers 1647

SESSION HM-2
HYDRAULIC MEASUREMENT II
Moderator: Y.K. TUNG, Wyoming Water Resources Center,
and G. CONTRONOC, Acres International

Laboratory Study of the Characteristics of Shallow Open Channel Flow
Using Fiber-Optic Laser Doppler Velocimetry
JAU-YAU LU and LI-CHUAN CHEN, Department of Civil Engineering,
National Chung-Hsing University 1653
Distorted Physical Models for Mixing Studies
JIANLU XU and H.W. SHEN, Department of Civil Engineering,
University of California, Berkeley 1659
Investigation of Saltating Particle Motions Using Flow
Visualization Technique
HONG-YUAN LEE and IN-SONG HSU, Dept. of Civil Engineering and
Hydraulic Research Laboratory, National Taiwan University 1665
Acoustic Doppler Current Profiler for Inlet Current Measurements
YEN-HSI CHU, M. METCALF, J.B. SMITH, T. PUCKETTE and
G.K. NERSESIAN, Waterways Experiment, U.S. Army Corps
of Engineers 1671
Turbulent Velocity Fluctuations in Natural Rivers
NANI G. BHOWMIK and RENJIE XIA, Hydrology Division, Illinois
State Water Survey 1677

SESSION HS-11
HYDRAULIC STRUCTURES I
Moderator: M.H. CHAUDHRY, Washington State University, and H.Y. LEE,
National Taiwan University

Determination of Pulsating Pressures for Baldhill Dam, N.D. Spillway
BOBBY P. FLETCHER and GREG EGGERS, Hydraulics Laboratory,
Waterways Experiment Station 1683
Hydraulic Model Studies of Y-Branch
B. K. LEE, H. W. COLEMAN, J. H. KIM and H. I. KWON, Harza
Engineering Company 1689
Model Study of Rio Hondo Flood Control Channel Los
Angeles, California
JOHN E. HITE, JR., SCOTT E. STONESTREET, and M.E.
MULVIHILL, Locks and Conduits Branch, Hydraulic Structures Div.,
US Army Engr. Waterways Experiment Station 1695
Confirmation of Seismic Safety at Stafford Dam
CHRIS D. DEGABRIELE and GLEN A. ROYCROFT, North Marin
Water District 1701

An Improved Method for Measuring System Performance of Hydraulic Infrastructure Systems
SUE-JEN WU and RU-LIN HSU, Dept. of Hydraulic Engineering,
Sinotech Engineering Consultants, Inc. 1708

SESSION BS-9
COMPUTER AND PHYSICAL MODELLING PROCESS ON BRIDGE SCOUR
Moderator: D. FROELICH, University of Kentucky, and R.T. KILGORE, GKY & Associates

Bridge-Scour Analysis Using the Water-Surface Profile (WSPRO) Model
DAVID S. MUELLER, U.S. Geological Survey, Water
Resources Division 1714
Innovations and Practical Procedures for Hydraulic Model Applications in Bridge Scour Evaluations
JEFFREY S. GLENN, Whitman & Howard Inc. 1720
Bri-Stars Model for Alluvial River Simulation
A. MOLINAS, Hydrau-Tech. Inc. 1726
Practical Comparison of One-Dimensional and Two-Dimensional Hydraulic Analyses for Bridge Scour
M.A. PORTS, T.G. TURNER, Parsons Brinckerhoff Inc., and
D.C. FROEHLICH, University of Kentucky 1732
Model Study of Local Scour Downstream Bridge Piers
LAILA ABED and M.M. GASSER, Hydraulics & Sediment
Research Institute 1738
Selecting Sediment Transport Equation for Scour Simulation at Bridge Crossing
HOWARD H. CHANG, CARROLL HARRIS, B. LINDSAY,
STEVE S. NAKAO and RAY KIA, San Diego State University 1744
Model Study of the Confluence and Wake Vortex Effect on Local Scour Downstream Bridge Pier
LAILA ABED, Water Research Center, Egypt .... *
Effect of Fenders on Local Pier Scour
LAILA ABED and E.V. RICHARDSON, Ministry of Public Works and
Water Resources, Egypt 1750

SESSION DF-4
ROUTING AND MAPPING FROM INITIATION TO TERMINATION
Moderator: R.D. JARRETT, U.S. Geological Survey, and Z. WAN, Institute of Water Conservancy and Hydroelectric Power Research, China

Debris Flow Initiation and Termination in a Gully
TAMOTSU TAKAHASHI, Disaster Prevention Research Institute,
Kyoto University 1756
Hydraulic Modeling and Mapping of Mud and Debris Flows
JIM S. O'BRIEN, FLO Engineering, Inc. 1762

---

*Manuscript not available at time of printing

An Empirical Model for the Volume-Change Behavior of Debris Flows
SUSAN H. CANNON, U.S. Geological Survey, Geologic Division 1768
Mapping Debris-Flow Hazard in Honolulu Using a DEM
STEPHEN D. ELLEN and R.K. MARK, U.S. Geological Survey,
Geologic Division 1774
Prediction of Occurrence and Runoff Analysis of Debris Flow
MUNEO HIRANO and T. MORIYAMA, Department of Civil
Engineering, Hydraulics and Soil Mechanics, Kyushu University 1780
Sediment Deposition from Debris Flow on a Gentle and Wide Slope
H. HASHIMOTO and M. HIRANO, Dept. of Civil Engineering,
Hydraulics and Soil Mechanics, Kyushu University 1786

SESSION EH-8
ENVIRONMENTAL MANAGEMENT
Moderator: S. HUI, Bechtel, and J. LINDLEY

Integration of Environmental Management with Reservoir and Power System Operations
VAHID ALAVIAN, TVA Engineering Laboratory 1792
Decision Support for Predicting the Hydraulics and Water Quality Characteristics of Natural Rivers
CHRISTIAN JOKIEL, P. RULAND and G. ROUVÉ, Institute for
Hydraulic Engineering and Water Resources Management, Aachen
University of Technology 1799
Quantity and Quality of Dry Weather Flow in the Las Vegas Valley, Nevada
STEVE A. MIZELL and RICHARD H. FRENCH, Desert Research
Institute, Water Resources Center 1806
Protected Streamflow and Water Uses
KRISHAN P. SINGH, Office of Surface Water Resources & Systems
Analysis, Illinois State Water Survey 1812

SESSION HS-12
HYDRAULIC STRUCTURES III
Moderator: C.H. LING, U.S. Geological Survey, and C. MENDOZA-CABRALES, Columbia University

Application of Environmental Regulations on Design of Hydraulic Structures for Open Cooling Water System
JAGDISH K. VIRMANI, MAHMOOD NAGHASH and
ADNAN M. ALSAFFAR, Bechtel Corporation 1818
The Concept of 'Local Euler Number' as an Aid for Sizing Pitot Tubes
JOSÉ ROBERTO BONILHA, Bscola Politécnica da USP, UNICAMP
Universida de Campinas 1824
Methods for Prediction of Maximum Scour at Coastal Structures
JIMMY E. FOWLER, Coastal Engineering Research Center, U.S. Army
Engineer Waterways Experiment Station 1830
Design of a Curved Baffle Energy Dissipation Structure
R. J. BERGQUIST and CHARLES C. HUTTON, ECI 1836

Simulation of Rapid Reservoir Drawdown for Flood Control, Cowlitz
Falls Project
    R.H.A. JANSSEN, F.A. LOCHER, Bechtel Corp. 1842

SESSION BS-10
FIELD MEASUREMENTS OF BRIDGE SCOUR PROCESSES I
Moderator: M.N. LANDERS, U.S. Geological Survey, and
C. MENDOZA-CABRALES, Columbia University

Pier Scour on the South Saskatchewan River
    F. AHMED, M.A. SABUR and D.D. ANDRES, Dept. of Civil
    Engineering, University of Alberta 1848
Scour At a Bridge over the Weldon River, Iowa
    E. E. FICHER, U.S. Geological Survey 1854
Measurement of Bridge Scour At the SR-32 Crossing of the Sacramento
River at Hamilton City, California, 1987–92
    J.C. BLODGETT and CARROLL HARRIS, U.S. Geological Survey ... 1860
Estimating Bridge Scour in New York from Historical U.S. Geological
Survey Streamflow Measurements
    G.K. BUTCH, U.S. Geological Survey 1866
Relation of Local Scour to Hydraulic Properties at Selected Bridges in
New York
    G.K. BUTCH, U.S. Geological Survey 1872
Overmining Causes Undermining (It's a Mad Mad River)
    CATHERINE CROSSETT, CALTRANS 1876
Bridge Scour Prediction Methods Applicable to Streams
in Pennsylvania
    DENNIS JOHNSON and ARTHUR C. MILLER, Department of Civil
    Engineering, Pennsylvania State University 1882
Scour Inspection Using Ground Penetrating Radar
    W.A. HORNE, Clough, Harbour & Associates 1888

SESSION DF-5
DEBRIS FLOW
Moderator: T.R.H. DAVIS, Lincoln University, and G.F. WIECZOREK, U.S.
Geological Survey

Debris Flow and Hyperconcentrated Flows—A UK perspective
    PAUL A. CARLING, Institute of Freshwater Ecology,
    Windermere Laboratory 1894
Differentiation of Debris-Flow and Flash-Flood Deposits: Implications
for Paleoflood Investigations
    CHRISTOPHER F. WAYTHOMAS, and ROBERT D. JARRETT, U.S.
    Geological Survey, Water Resources Division 1900
Time-Dependent Landslide Probability Mapping
    RUSSELL H. CAMPBELL, and RICHARD L. BERNKNOPF, U.S.
    Geological Survey, Geologic Division 1902
Operation of a Real-Time Warning System for Debris Flows in the San
Francisco Bay Area, California
    RAYMOND C. WILSON, ROBERT K. MARK, and
    GARY BARBATO, U.S. Geological Survey, Geologic Division 1908

Structural and Non-Structural Debris-Flow Countermeasures
 TAKAHISA MIZUYAMA, Laboratory of Erosion Control, Department
 of Forestry, Kyoto University 1914

SESSION PH-1
APPLICATION IN HYDROLOGY
Moderator: F.C. WANG, Louisiana State University, and G. TABIOS, University of California, Berkeley

Evaluation of Hydraulic Structure Reliability Considering Uncertainties in Hydrologic Models
 Y.K. TUNG, BING ZHAO and J.C. YANG, Wyoming Water Resources
 Center, University of Wyoming 1920
Analysis Uncertainty of IUH of NASH
 J.C. YANG, SHYH-YANN TARNG and Y.K. TUNG, Dept. of Civil
 Engineering, National Chiao-Tung University 1927
Hydrologic Design: Extending Traditional Methods
 DELBERT D. FRANZ, Linsley, Kraeger Associates, Ltd. 1933
Importance of Hydraulic-Model Uncertainty in Flood-Stage Estimation
 SATVINDER SINGH, CHARLES S. MELCHING, New Jersey Dept. of
 Envir. Protection and Energy 1939
Rainfall Depth-Duration-Frequency Analysis For Major Cities In Taiwan
 BAOLIN WU, MING-HSI HSU, V. YIH, S. KING and C.M. WU,
 Contra Costa County Flood Control District 1945

SESSION HS-13
HYDRAULIC STRUCTURES V
Moderator: M.E. JENNINGS, U.S. Geological Survey, and W. ESPEY, Espey Houston

Large-Scale Embankment Overtopping Protection Tests
 KATHLEEN H. FRIZELL and JAMES F. RUFF, U.S. Bureau
 of Reclamation 1951
Progressive Failure of an Overtopped Embankment
 GHASSAN ALQASER and JAMES F. RUFF, Engineering Research
 Center, Colorado State University 1957
A New Type of Concrete Trapezoid Dam
 YUI TSANG CHAN, Civil Engineering Dept., China Light & Power
 Co. Ltd. 1963
Assessing the True Value of Flood Control Reservoirs: The Experience of Folsom Dam in the February, 1986 Flood
 PHILIP B. WILLIAMS, Philip B. Williams & Associates, Ltd. 1969
Chatuge Hydroproject Aerating Infuser Physical Model Study
 TONY A. RIZK and GARY E. HAUSER, Tennessee
 Valley Authority 1975

SESSION 3D-1
3D MODELLING I
Moderator: S.Y. WANG, University of Mississippi

Recent Developments in Three-Dimensional Numerical Estuarine Models
RALPH T. CHENG, PETER E. SMITH and VINCENZO CASULLI, U.S. Geological Survey 1982

TRIM-3D: A Three-Dimensional Model for Accurate Simulation of Shallow Water Flow
VINCENZO CASULLI, ENRICO BERTOLAZZI and RALPH T. CHENG, Dipartimento di Matematica, Universita' di Trento 1988

Parameterization of Turbulence for Three-Dimensional Circulation Modeling
Y. PETER SHENG, Dept. of Coastal & Oceanographic Engr., University of Florida 1994

Three-Dimensional Free Surface Modeling (Is there a Best Approach?)
BILLY H. JOHNSON, H. LEE BUTLER and CHARLIE BERGER, U.S. Army Engineers Waterways Experiment Station 1997

Galveston Bay 3-D Model Study Channel Deepening Lessons Learned in Management of a Large Modeling Study
J.H. SCHMIDT, R.T. MCADORY, W.D. MARTIN and R.C. BERGER, U.S. Army Engr. Waterways Experiment Station 2003

SESSION EH-9
ENVIRONMENTAL HYDRAULICS I
Moderator: J. ODGAARD, University of Iowa, and J.S. WANG, Bechtel

Oxygenation Experiments in the Wave Breaking Zone
E.I. DANIL, and C. MOUTZOURIS, Department of Civil Engineering, Laboratory of Harbour Works, National and Technical University of Athens 2008

Integrated Planning Analysis for an Aggregate Mine
EDWARD E. WALLACE and ROBERT C. MACARTHUR, Meridian Consulting Engineers Inc. 2014

Bank Stabilization with Environmental Features on the Middle Rio Grande
DREW C. BAIRD, J.P. WILBER and R.W. SLATER, U.S. Bureau of Reclamation 2020

Two Options for Disposal of Desalination Reject Water
LOUIS J. ARMSTRONG, P.R. MINEART and R.H. CROSS III, Woodward-Clyde Consultants 2026

Ballast Water Treatment Effluent Dispersion Studies
PETER A. MANGARELLA, JOSEPH M. COLONELL, Water Resources Group, Woodward-Clyde Consultants 2032

SESSION BS-11
FIELD MEASUREMENTS OF BRIDGE SCOUR PROCESSES II
Moderator: PETE HAENI, U.S. Geological Survey, and ROLLINS HOTCHKISS, University of Nebraska

Evaluation of an Existing Scour Hole at the Castleton Bridge, a Tidal Crossing
S.A. KHONDKER, EBASCO and M.A. HIXSON, EBASCO Infrastructure 2038
Development of Bridge-Scour Instrumentation for Inspection and Maintenance Personnel
DAVID S. MUELLER and MARK N. LANDERS, U.S. Geological Survey, Water Resource Division 2045
Using Graphysical Data to Assess Scour Development
G. PLACZEK, F.P. HAENI, U.S. Geological Survey, and R. TRENT, Federal Highway Administration 2051
Local Scour Measurements at Bridge Piers in Alberta
D. WILLIAMSON, Bridge Engineering Branch, Alberta Transportation and Utilities 2057
Instrumentation for Detailed Bridge-Scour Measurements
M.N. LANDERS, D.S. MUELLER, U.S. Geological Survey, and R. TRENT, Federal Highway Administration 2063
Local Scour at Bridge Piers in Alberta-Case History
A. HUMPHRIES, Bridge Engineering Branch, Alberta Transportation and Utilities 2069
Reference Surfaces for Bridge Scour Depths
M. LANDERS and DAVID S. MUELLER, U.S. Geological Survey 2075

SESSION PH-2
APPLICATION IN HYDRAULICS
Moderator: W.C. WANG, Multech Engineering Consultants, Inc.

Probability of Bridge Failure due to Scouring
WOLFGANG KRON, and ERICH PLATE, Institute for Hydrology and Water Resources Planning, University of Karlsruhe 2081
Uncertainty of Bridge Scour Estimates
PEGGY A. JOHNSON and BILAL M. AYYUB, Dept. of Civil Engineering, University of Maryland 2087
Reliability Analysis of Levee System of a River
K. MIZUMURA, Dept. of Civil Engineering, KANAZAWA Institute of Technology 2092
Uncertainty Analysis of the FEMA Method for Alluvial Fans
BING ZHAO and LARRY W. MAYS, Dept. of Civil Engineering, Arizona State University 2098
Risk and Uncertainty in Flood Damage Reduction Project Design
MING T. TSENG, EARL EIKER and D. W. DAVIS, Hydraulics and Hydrology Branch, Engineering Division, US Army Corps of Engineers 2104

SESSION 3D-2
3D MODELLING II
Moderator: A.F. BLUMBERG, HydroQual, Inc.

Three-Dimensional Variable Resolution Hydrodynamic and Transport Modeling of the Chesapeake Bay System
JOHN M. HAMRICK, Virginia Institute of Marine Sciences 2110
A Finite-Difference Model for 3-D Flow in Bays and Estuaries
PETER E. SMITH and BRUCE E. LAROCK, Water Resources Division, U.S. Geological Survey 2116
Horizontal Gradients in Sigma Transformed Bathymetries with Steep Bottom Slopes
G.S. STELLING, and JAN A.T.M. VAN KESTER, Delft Hydraulics .. 2123
2-D Vertical and 3-D Modelling of Mud Transport in Tidal Flows
B.A. DEVANTIER and L.C. VAN RIJN, Dept. of Civil Engineering and Mechanics, Southern Illinois University at Carbondale 2135
Three-Dimensional Numerical Modeling for Transport Studies
W.H. MCANNALLY, R.C. BERGER, and A.M. TEETER, U.S. Army Corps of Engineers 2141

SESSION EH-10
ENVIRONMENTAL HYDRAULICS II
Moderator: J. HOOPES, University of Wisconsin, and F. CHUNG, California Dept. of Water Resources

Mixing Character and Meandering Mechanism of a Plane Jet Bounded in a Shallow Water Layer
DAOYI CHEN and G.H. JIRKA, DeFrees Hydraulics Laboratory, School of Civil and Environmental Engineering, Cornell University 2147
Selective Withdrawal for Reducing Turbid Water in a Reservoir
YUKIHIDE TASHIRO, MASAHIKO YATSUGI, J. TANO, and N. MATSUO, Civil Engineering Department, Kyushu Electric Power Co., Inc. 2153
Shifts in Solute Transport Direction Induced by Transient Flow
F.X. MARKERT, RAFAEL G. QUIMPO, IT Corp. 2159
A New In-Stream Aerator
J. GULLIVER, Dept. of Civil and Mineral Engineering, University of Minnesota 2165
Modeled Hydraulic and Salt Transport Patterns in the Sacramento-San Joaquin Delta
R.T. BROWN, P.WISHEROPP, D. SMITH, and R. RACHIELE, Jones & Stokes Associates 2171

SESSION CA-6
COMPUTER APPLICATION IV
Moderator: J.Y. LU, National Chung-Hsing University and H.W. Coleman, Harza

A Diffusion Hydrodynamic Model of a Shallow Estuary: Verification
T.V. HROMADKA II, G.L. GUYMON, M.H. KHAN, and M. COLLINS, Boylr Engineering Corp. 2178
Structure of Coastal Upwelling on a Sloping Bottom
YAN ZANG, ROBERT L. STREET and JEFFREY R. KOSEFF, Environmental Fluid Mech. Lab., Department of Civil Eng., Stanford University 2184

Field Examination of a Distribution System Water Quality Model
S.G. ELMAALOUF and Y.C. KIM, Public Works Dept., City of
Los Angeles 2190
The Courant Number and Unsteady Flow Computation
CHINTU LAI, U.S. Geological Survey 2196
River Engineering Flows and Some Related Structures
RAWYA M. KANSOH, Civil Engineering Department,
Alexandria University 2202

SESSION HS-14
HYDRAULIC STRUCTURES IV
Moderator: L.W. MAYS, Arizona State University, and A. ALSAFFAR, Bechtel

The EUR Water Station in Rome—Italy
F. CIACCHELLA, G. MARTINO and P. MASSARINI, Water
Production Plants Sector, ACEA 2207
A Culvert Analysis Program for Indirect Measurement of Discharge
JANICE M. FULFORD, U.S. Geological Survey 2213
Sediment and Water Quality Control Devices in Small Watersheds
HASAN NOURI, Rivertech Inc. 2219
Flow Diversion in a Steep, Coarse-Bed Stream
EDWARD F. SING, U.S. Army Engineer District, Sacramento 2389
Vortex Spillway: Test Study on a Hydraulic Model
G. MARTINO, Azienda Comunale Energia ed Ambiente, ACEA 2225

SESSION BS-12
COMPARISONS OF MEASURED AND COMPUTED BRIDGE SCOUR
Moderator: C. NORDIN, Colorado State University, and P. WOJCIK, TAMS

Evaluation of Historical Scour at Selected Stream Crossings in Indiana
DAVID S. MUELLER and ROBERT L. MILLER, U.S.
Geological Survey 2231
Bridge Scour Evaluations in Washington State
J.P. JOHNSON, C.R. NEILL and R.P. HOVDE, Northwest
Hydraulic Consultants 2237
Relation of Channel Stability to Scour at Highway Bridges Over Waterways in Maryland
E. J. DOHENY, U.S. Geological Survey 2243
Bridge Scour and Change in Contracted Section, Razor Creek, Montana
S.R. HOLNBECK, C. PARRETT and T.N. TILLINGER, U.S.
Geological Survey 2249
Shale Scour at BNRR Yellowstone River Bridge, MT
GARY LEWIS, HDR Engineering, Inc. 2255
Analysis of Local Scour at Bridge Piers
HAN-BIN LIANG and JORGE ROMERO-LOZANO, Philip Williams &
Associates, Ltd. 2261

SESSION PH-3
SYSTEM OPTIMIZATION
Moderator: C.S. MELCHING, U.S. Geological Survey, and RICHARD DENTON, Contra Costa Water District

Variability in Solutions of Constrained Optimization Problems: Ocean Outfall Design Case
MAILI WANG and KEVIN LANSEY, Dept. of Civil Engineering and Engineering Mechanics, University of Arizona 2267

Optimizing Water Transfers in Urban Water Supply Planning
JAY R. LUND, and MORRIS ISRAEL, Dept. of Civil and Environmental Engineering, University of California, Davis 2273

Optimizing a Reservoir Operation Policy from the Properties of Reliability and Resiliency
K.S. TICKLE and I.C. GOULTER, University of Central Queensland .. 2279

Hydraulic and Water Quality Reliability and Resiliency for Water Distribution Systems Under Random Demands
MAO FANG and JAMES G. UBER, Dept. of Civil & Envir. Engineering, University of Cincinnati 2285

The Seattle Forecast Model: A Tool for Water Resources Management in the Seattle Area
LAURA MARIÑO, Hydrocomp Inc. 2291

SESSION 3D-3
3D MODELLING III
Moderator: P.Y.P. SHENG, University of Florida at Gainesville

Modeling Vertical Stratification in the Pamlico River Estuary Using Modern Hydrodynamics
ALAN F. BLUMBERG, C. KIRK ZIEGLER and BRADLEY NISBET, HydroQual, Inc. 2297

Three-Dimensional Modeling of Tides and Wind-Waves
SHIAO-KUNG LIU, System Research Inst. 2301

Three-Dimensional Modeling of the Coastal Region Offshore from Sydney, Australia
IAN P. KING, WILLIAM L. PEIRSON and BRUCE A. CATHERS, Dept. of Civil and Environmental Engineering, University of California, Davis 2307

Simulation of 3D Free Surface Flows in Hydraulic Structure with Submerged Flow Passages
SAM S.Y. WANG, JOSEPH LETTER JR., and PHIL COMBS, Center for Computational Hydroscience and Engineering, University of Mississippi 2313

Galveston Bay 3-D Model Study Channel Deepening Circulation and Salinity Results
R.C. BERGER, W.D. MARTIN, R.T MCADORY, and J.H. SCHMIDT, U.S. Army Engr. Waterways Experiment Station 2318

Modeling the Tides of Massachusetts and Cape Cod Bays
H.L. JENTER, R.P. SIGNELL, ALAN BLUMBERG, U.S.
Geological Survey 2323

SESSION EH-11
ENVIRONMENTAL HYDRAULICS III
Moderator: J.S. WANG, Bechtel Corp., and G. SULLIVAN,
Contra Costa Water District

Storm Water Regulations—Aircraft Deicer/Anti-icers Operations
W.H. ESPEY and G.I. LEGARRETA, Espey, Houston & Associates ... 2333
GIS and SWMM Applications in Developing the Lake Houston Watershed Management Program
DUKE G. ALTMAN, R.G. MONTGOMERY, T.L. KING, and
C.W. PATTERSON, Espey, Houston & Associates Inc. 2339
Probability and Impact of an Observed Rare Sequence of Floods
LEO BEARD, W.H. ESPEY, PHIL COMBS, and
BEN M. LITTLEPAGE, Espey, Houston & Associates Inc. 2345
Wastewater Treatment for Better Environment
RAWYA MONIR KANSOH, Civil Engineering Dept.,
Alexandria University 2350
Numerical Modelling of Salinity Transport in a Shallow Well-Mixed Tidal Reach
G.S. REDDY, and S.N. GHOSH, Civil Engineering Dept., Indian
Institute of Technology 2358
New York Bight Three-Dimensional Water Quality Model
R.W. HALL and M.S. DORTCH, USAE Waterways
Experiment Station 2365

Subject Index 2395
Author Index 2401

Mass transport in mud layer induced by wave action.

Ismael Piedra Cueva[1]

Abstract

This paper analyzes the mass transport velocity in a two layers system produced by wave action. In a previous paper, Piedra Cueva (1993), a theory for linear water waves propagating in a two layered viscoelastic fluid system was described. The mass transport in the mud layer and in the water boundary layer is obtained solving the momentum and mass conservation equations in the Lagrangian coordinate system. It was found that the Eulerian coordinate system is not appropiate because it does not guarantee the continuity of the mass transport velocity through the water-mud interface. The results show that the mass transport velocity can be increased one order of magnitude in relation to the velocity given by the rigid bed theory and can be opposite to the wave propagation direction, depending on the physical mud properties. This result is very important when considering near bed sediment transport.

Introduction

If the horizontal (u) and vertical (w) wave induced velocities components at some point oscillate with a phase difference other than $\pi/2$, so that the time average of their product is non-zero, there will be a net transfer of x-momentum across a surface element with the normal in z-direction during one cycle of oscillation, and thus it will produce a non-zero average force on the fluid. The consequent steady motion of the fluid may be weak, but, inasmuch as it leads to extensive migration of fluid elements in an apparent purely oscillatory system, its effect is important when considering near bed sediment transport. Shibayama, Takikawa & Horikawa (1986), Tsuruya, Nakano & Takahama (1987), and Sakakiyama & Bijker (1989) have studied the phenomenon theoretically. While all the theories above have been based on the application of the Eulerian coordinate system, an alternative is to employ the Lagrangian coordinate system which is more attractive, for the boundary conditions can be conveniently stated. This line of work is developed further in this paper, which focuses on the mass transport velocity inside of the water boundary layer and the mud layer. This velocity is worked out employing Lagrangian coordinates, since it was found that employing the Eulerian system it is not possible to obtain continuity of the Stokes drift through the water-mud interface.

---

[1] Prof. Associate, Institute of Fluid Mechanics and Env. Engineering, Montevideo, Uruguay, CC 30.

First order solution

For the purpose of calculating the second order mass transport velocity inside of the mud layer and the water boundary layer, the movement inside of the two layers should be obtained first. In a previous paper, Piedra Cueva (1993), a theory for linear water waves propagating in a two layered viscoelastic fluid system was described.

Mass transport velocity in the Eulerian coordinate system

The mass transport velocity (mean particle velocity) in the Eulerian coordinate system is given by Eq. 1:

$$\overline{u}_L = \overline{u}^{(2)} + \overline{\left(\int u^{(1)} dt\right) \frac{\partial u^{(1)}}{\partial x}} + \overline{\left(\int w^{(1)} dt\right) \frac{\partial u^{(1)}}{\partial z}} \tag{1}$$

where subscripts L means Lagrangian velocity, the superscripts (1) and (2) mean first and second order Eulerian velocities respectively, and the over bar means time average. The summation of the third and fourth term on the right-hand side, which is indicated by u_s, is known as Stokes drift. The first term on the right hand side of Eq. 1, is the mean Eulerian second order velocity. The basic theory to describe the governing equation of the mean Eulerian velocity was presented by Batchelor (1967) and Mei (1983). The application of the Eulerian coordinate system to calculate second order Lagrangian velocities, is not appropiate due to the fact that it is not possible to guarantee the continuity of the Stokes drift through the water-mud interface since the velocity gradients (in Eq. 1) are not continuous.

Mass transport velocity in Lagrangian coordinate.

Let a and c be the initial horizontal and vertical coordinates of a fluid particle. In the a-c plane, the coordinate origin is located at the interface in such a way that the position of the free surface is simply c=h1 and the position of the interface is c=0 at all times and is no longer unknown. The mass conservation equation is:

$$\frac{\partial(x,z)}{\partial(a,c)} = x_a z_c - x_c z_a = 1 \tag{2}$$

and the x and z momentum equations are:

$$\ddot{x} = -\frac{1}{\rho}\frac{\partial(P,z)}{\partial(a,c)} + \nu\nabla^2 \dot{x} \quad ; \quad \ddot{z} = -\frac{1}{\rho}\frac{\partial(x,P)}{\partial(a,c)} + \nu\nabla^2 \dot{z} \tag{3}$$

Eqs. 2 and 3 can be linearized introducing the perturbation series (Mei, 1983):

$$x = a+x1+x2+\ldots; \quad z = c+z1+z2+\ldots; \quad P = (p-\rho\ g.c)+p1+p2+\ldots$$

Substituting these expressions in Eqs. 2 and 3, and collecting first order terms, Eq. 4 is obtained, where the Laplacian is written in Lagrangian coordinate.

$$\begin{aligned}\ddot{x}1 &= -\frac{1}{\rho}P1_a - gz1a + \nu\nabla^2 \dot{x}1 \\ \ddot{z}1 - gx1_a &= -\frac{1}{\rho}P1_c + \nu\nabla^2 \dot{z}1 \\ x1_a + z1_c &= 0\end{aligned} \tag{4}$$

The second order equations can be obtained collecting second order terms:

$$\dot{x}1_a \cdot \dot{z}1_c - \dot{x}1_c \cdot \dot{z}1_a = 0 \tag{5}$$

$$\ddot{x}2 + \frac{p2_a}{\rho} + gz2_a - \nu(\ddot{x}2_{aa}+\ddot{x}2_{cc}) = G$$

$$G = -\ddot{x}1 \cdot x1_a - \ddot{z}1 \cdot z1_a + \nu\,[x1_a(\dot{x}1_{ac}+3 \cdot \dot{x}1_{cc}) +2 \cdot \dot{x}1_{aa}\cdot z1_c+ \tag{6}$$
$$+\dot{x}1_a(z1_{ac}-x1_{cc}) -2\dot{x}_{ac}(z1_a+x1_c) -\dot{x}1_c(z1_{aa}-x1_{ac}) +z1_a(\dot{z}1_{aa}+\dot{z}1_{cc})$$

For the second order, only the x-momentum equation is needed. The stress tensor is stated as T=-pI+S, where tensor S is related to the deformation tensor by the relation $2\mu D$. The normal and tangential components of the stress to the first and second order are:

$$(Tn)\,t = \frac{\dot{x}1_c + \dot{z}1_a}{\|N\|^2}$$

$$(Tn)\,n = \frac{p1-2\rho\nu\,\dot{z}1_c}{\|N\|^2} \tag{7}$$

$$(\overline{Tn})\,\bar{t} = \frac{\ddot{x}2_c + \ddot{z}2_a - 3\dot{x}1_a \cdot z1_a - x1_c \cdot \dot{x}1_a + 2\dot{x}1_c x1_a}{\|N\|^2} \tag{8}$$

where N is the norm of the vector.

First order solution.
Formally, the first order problem is identical to the first order linearized Navier Stokes equations in Eulerian form. In this way, it is possible to obtain the Lagrangian coordinates from the first order solution. For the water boundary layer:

$$x1 = X_w(c)\,e^{i\theta} \qquad z1 = Z_w(c)\,e^{i\theta} \qquad \theta = ka - \sigma t \tag{9}$$

$$X_w(c) = i\frac{\hat{U}_\infty}{\sigma}(1+B1\,e^{-qc}) \qquad Z_w(c) = [b+\frac{k\hat{U}_\infty}{\sigma}[c-\frac{B1}{q}(e^{-qc}-1)] \tag{10}$$

and for the mud layer Eq. 11 is obtained, where C_b y D_b are two coefficients.

$$X_b(c) = \frac{i}{\sigma}D_b\,[\,l\cosh k(c+h2) - l\cosh l(c+h2)\,] +$$
$$+ \frac{i}{\sigma}C_b\,[-k\sinh k(c+h2) - l\sinh l(c+h2)\,]$$
$$Z_b(c) = -\frac{D_b}{\sigma}[\,-l\sinh k(c+h2) + k\sinh l(c+h2)\,] - \tag{11}$$
$$- \frac{C_b}{\sigma}[\,-k\cosh k(c+h2) + k\cosh l(c+h2)]$$

Second order solution
Following Mei (1983), taking the time average of Eq. 6 and ignoring the spatial attenuation in the direction of wave propagation,

that is $\overline{z2}_a$, the second order equation (Eqs. 12 and 13) is obtained.

$$-\nu \frac{d^2}{dc^2}\overline{x}2 = -\frac{1}{\rho}\frac{\partial \overline{P2}}{\partial a} + \overline{G} \qquad (12)$$

$$\overline{G} = \Re \left[\frac{\sigma}{2} FF^*\nu \left[k^2 k^* (|X|^2 + |Z|^2) - 4k^*X^* \frac{d^2X}{dc^2} + iX^* \frac{dZ}{dc} \right. \right.$$

$$\left. \cdot (-2k^{*2} + kk^*) + ikZ\frac{dX^*}{dc} \cdot (-2k^* + k) - \left|\frac{dX}{dc}\right|^2 \cdot (2k^* + k) - \qquad (13) \right.$$

$$\left. -k^*Z^* \frac{d^2Z}{dc^2} \right] - \frac{1}{2} FF^*\sigma^2 i k^* \cdot (|X|^2 + |Z|^2) \right]$$

where $F.F^* = \exp(-2.kj.a)$. For the water boundary layer region, Eq. 13 can be simplified by using dimensional analysis. Eqs. 12 and 13 are solved with the following boundary conditions:
a) at the outer edge of the water boundary layer, the vanishing of the shear stress is required. The condition for $\overline{x}2_c$ is obtained from Eq.8, after averaging time.
b) at the interface (c=0), the continuity of the velocity $\overline{x}2|_w = \overline{x}2|_b$ and shear stress is required.
c) at the bottom c=-h2, the vanishing of velocity is required $\overline{x}2|_b = 0$.

Results

The system of Eqs. 12-13 written for both layers with the boundary conditions, was solved using the shooting method. The results presented here were obtained neglecting the second order pressure gradient.
First, the mud response to changes in the kinematic viscosity and mud elasticity G is analyzed. The velocities shown in these figures are dimensionless with the Lagrangian velocity in the outer edge of the water boundary layer (ULr) as given by the rigid bed theory. Figure 1 shows the Lagrangian velocity profiles for two elasticity values and for various values of viscosity. For zero elasticity, the velocities increase when the kinematic viscosity decrease. The maximum value in the mud layer is about one half of ULr. This fact agrees well with the experimental results presented by Sakakiyama and Bijker (1989). The mass transport inside the water boundary layer can change its direction depending of the mud viscosity. When the elasticity modulus is G=10 N/m², the velocity direction inside of the mud layer depends on the mud viscosity. The mud velocity increases continuously when the viscosity decreases. For viscosities lower than $\nu_b = 0.01$ m²/s, the velocity direction is opposite to the water wave propagation. This velocity can reach values five times the rigid bed velocity. Taking the mud elasticity into account can change considerably the magnitude and direction of the mud and water layers velocities.

Conclusion

A theoretical model is brought forward to determine the Lagrangian velocity (mass transport) in the water boundary layer and the mud layer based on a viscoelastic mud model. Applying the Lagrangian coordinate system, it is possible to obtain the Lagrangian

velocities inside both layers straightforward. The Eulerian coordinate system was found not to be appropriate, since it is not possible to guarantee the continuity of the mass transport through the water-mud interface. The mass transport was found to depend strongly on the wave frequency and on the thickness, viscosity and elasticity of the mud. The magnitude of mass transport can be one order of magnitude higher than the one stated by Longuet-Higgins, who considers the rigid bed theory; and the velocity direction can be opposite to the wave propagation. This fact is due mainly to the modelling of the mud as a viscoelastic body.
These results highlight two aspects. The first refers to the sediment transport by wave action. Considering the hight sediment concentrations of the muddy bottoms, the volumes of sediment transported by wave action can be very important when considering siltation in navigation channels. The second refers to the determination of the rheological properties of mud. The importance of mud elasticity as regard as mass transport was shown. Thus, it seems of primary importance to determine the proper rheological mud models and -if possible- their parameters by in situ tests.

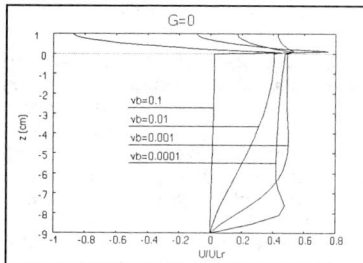
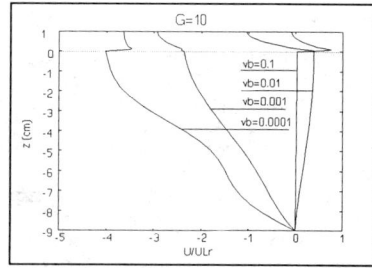

Fig. 5. Velocity profiles for various viscosities and elasticities G.

References

Nagai,T, Yamamoto,T & Figueroa,L (1984). A laboratory experimentation on the interactions between water waves and soft clay beds. Coastal Engineering in Japan, vol 27, pp 279-291.
Piedra Cueva, I. (1993). On the response of a muddy bottom to surface water waves. J. of Hydraulic Research, to appear.
Sakakiyama,T & Bijker,E (1989). Mass transport velocity in mud layer due to progressive waves. J.of Port, Coastal and Ocean Engineering, vol.115, No 5, pp 614-632.
Shibayama,T, Takikawa,H & Horikawa,K (1986). Mud mass transport due to waves. Coastal Engineering in Japan, vol 29, pp 151-161.
Swan,C & Sleath,J.F.A (1990). A second approximation to the time-mean Lagrangian drift beneath a series of progressive gravity waves. Ocean Engineering, vol.17, No 1/2, pp 65-79.
Tsuruya,H, Nakano,S & Takahama,J (1987). Interactions between surface waves and a multy-layered mud bed. Report of the Port and Harbour Research Institute (Japón), vol 26, No 5.

SEDIMENT DISPOSAL AND TRANSPORT IN CENTRAL SF BAY

T.H. Wakeman, A.E. Mathiesen and G.W. Chatfield[1]

Abstract

The relocation of dredged material from navigation channels in the Francisco Bay is crucial to the movement of deep-draft shipping and shallow-draft recreational vessels. The margins of the Bay are shallow, as are portions of its northern and southern reaches, and must be continually excavated to provide adequate depths for vessel transit. This requirement for construction and maintenance of local channels began with the passage of the River and Harbor Act of 1868. Since that time, the majority of dredged material has been relocated to several dispersive, aquatic disposal sites within the Bay. The primary site is near Alcatraz Island in Central San Francisco Bay. In 1982, this site was discovered to be mounding. Continued bathymetric surveys through the eighties verified that the site was not behaving as a fully dispersive site and that a stable mound had formed. Currently, the San Francisco Bay region has limited disposal options other than this aquatic disposal site. Understanding the site's sediment consolidation and transport characteristics is essential to its management and longevity. Presently, surveys, calculations and modeling are being used to further investigate and quantify the sustained sediment yield from the site. The purpose of this paper is to describe the engineering studies conducted, their findings, and management consequences.

Introduction

The construction and maintenance of navigation channels in the San Francisco Bay began with the passage of the River and Harbor Act of 1868. With respect to this Bay, the purpose of the legislation was to provide a deepwater port in an otherwise naturally shallow embayment to handle the inflow of supplies and equipment for the miners in the Sierra Nevada foothills. Subsequently, as the population in the region grew, so did the need for new channels,

---

[1] Civil, Hydraulic and Civil Engineer, respectively, San Francisco District, US Army Corps of Engineers, 211 Main Street, San Francisco, CA 94105-1905

SEDIMENT DISPOSAL AND TRANSPORT

ports and marinas. The primary source of sediments (80 to 90%) to the Bay system is erosion from the 163,000 km^2 Central Valley, producing approximately 8M m^3 in 1960 (Krone, 1979). During the period between the 1950s to the 1970s, there was an annual requirement for approximately 7.5M m^3 of maintenance dredging. More recently this volume is closer to 4.5M m^3. In addition to the on-going maintenance requirements for the Bay's channels and harbors, there is also a continuing demand for channel improvements to support ever changing commercial and military needs. Presently, future new work for Congressionally authorized civil works projects (including Oakland and Richmond harbors and John F. Baldwin ship channel, Phase III) and new work for US Navy facilities construction is estimated to be approximately 14.5M m^3.

Disposal History

Prior to 1972, material dredged from San Francisco Bay was disposed at eleven known aquatic disposal sites within the Bay as well as several undocumented sites. In 1972, the US Army Corps of Engineers (USACE), San Francisco District, responded to comments from the Committee on Tidal Hydraulics (USACE, 1965) and the State of California by limiting dredged material disposal activities to five sites. The District further reduced the number of aquatic disposal sites in 1978, to only three dispersive sites in the Bay for authorized discharges. This action was taken to decrease the amount of dredged material potentially being redredged throughout the Bay and to provide for better regulatory control.

One of the three disposal sites was a hole near Alcatraz Island in Central San Francisco Bay, ranging from 25 to 35 m in depth. This site has been used for over fifty years and is the primary disposal site for the region -- receiving maintenance and new work materials from projects located in both Central and South Bay. Between 1984 and 1988, based on dredging records during that period, the disposal site received approximately 3M m^3 annually, although a concentrated slug of material (4.6M m^3) was discharged from a single project in late 1985/early 1986. The approach of disposing of the majority of dredged material at Alcatraz was justified as being economically and environmentally sound for several years. However, in November of 1982, mounding at the Alcatraz disposal site was detected. After directing on-going disposal away from the mound, it was noted that natural currents failed, over two winter seasons, to reduce the mound. Accumulation continued after dredging was used to lower the peak of the mound in 1984 and 1985. As investigations progressed, it became apparent that the site could not accept dredged material in an unlimited fashion.

Sediment Mounding and Transport

Historically, dredged material discharged at the Alcatraz site quickly left the area either during initial dispersion or subsequent erosion. Both processes are functions of the strong tidal currents at the disposal site. Immediately after discharge, material descends at a rate determined by bulk density and initial momentum (Trawle and Johnson, 1986). During the descent considerable dilution can occur with hydraulically dredged sediments to the point where individual particle settling rates govern descent, and lateral currents carry material from the site before it impacts the bottom. Low density material that does deposit in the disposal area is easily eroded during the next period of high tidal currents (Trawle and Johnson, 1986). Denser, more consolidated materials discharged at the disposal site, excavated by mechanical means (ie., clamshelled) descends more rapidly to the bottom and deposits at a density near to the in situ density at the dredging site (SAIC, 1987a/b). Disposal monitoring studies at Alcatraz have demonstrated that pumping by a hopper dredge reduced density and internal shear strength of in situ sediments sufficiently that it accumulated at a rate about one-half that of clamshelled material (SAIC, 1987a/b).

Material, not dispersed shortly after discharge, accumulates in the disposal area until eroded by tidal currents or consolidates. In a study by Teeter (1987), it was determined that the site has a capacity for eroding material of 1.35 g/cc at the rate of 375,000 m^3 per month. When the density to be eroded was 1.40 and 1.45 g/cc, erosion rates dropped to 84,700 m^3 and 760 m^3 per month, respectively. The ultimate fate of this eroded material must be estimated from circumstantial evidence because quantitative data are lacking. Useful information is available from previously conducted field work looking at Central Bay water quality and geomorphic conditions. First, all suspended sediment plumes tracked during field investigations (SAIC, 1987a/b) at Alcatraz moved in an east-west direction. The suspended material did not disperse significantly in a north-south direction. Second, geomorphic evidence that is useful includes an investigation of erosion and accretion patterns gleamed from historic surveys (Beeman, 1992). Currently, two new efforts are underway to collect field data on suspended particulates in Central Bay (LTMS, 1992). One of these efforts is supporting development of a numerical model to predict long term sediment transport from the Alcatraz site (LTMS, 1992).

Site Management

Dredged material retained at the site and environs, based on bathymetric surveys and logs of disposal quantities using two templates from March 1983, has been essentially constant since 1987 (Figure 1). Prior to 1987, the site was severely overload and filled rapidly (Figure 2). From 1987 through 1991, management of the disposal site utilized monthly surveys to calculate retention. Since retention seemed to be stable, further analyses were not performed. In 1991, peak elevation was added as a second management parameter, and data analyses demonstrated that the site was not static (Figure 3). Between 1985 and 1992, peak elevation increased approximately 2.5 m. It was determined that loading rate influnced mound elevation, particulary from barge disposal of clamshelled materials.

In order to maintain disposal capacity at the Alcatraz site, the

District revisited data and performed calculations which compared dispersion rates of material at Alcatraz versus type of dredging equipment and type of sediment. These data indicated that clamshell dredging has contributed most to the consolidation of material at Alcatraz -- denser material being disposed and not dispersing as readily as pumped material. Short term disposal of large volumes of material at Alcatraz regardless of other factors has also contributed to site accretion. The management approach seems to work best for maintaining the Alcatraz site is small constant disposal amounts, all of which is dispersed (Figure 4). Previously guidelines allowed monthly disposal of up to 0.7M m^3, a large volume which on two occassions in 1992 was directly credited for increasing peak elevation. Therefore the District revised its loading rate allowances at the site in 1993. In Public Notice 93-3, Febuary 1, 1993, the District proposed monthly guideline for disposal of dredged material at Alcatraz of: (a) 305,000 m^3 total volume; and (b) 114,500 m^3 maximum volume of clamshelled material. Yearly disposal of dredged material at Alcatraz was proposed as 3M m^3 total volume. Under the public notice, these guidelines were effective March 1993 for a period of one year after which they will be revisited and adjusted based on the data gathered during this period.

References

Beeman and Associates. 1992. Sediment Budget Study for San Francisco Bay. Prepared under Contract DACW07-89-D-0029 for USACE, San Francisco District. Portland, OR. 25 pp w/appendices.

Krone, Raymond B. 1979. Sedimentation in the San Francisco Bay System. In Conomos, ed., San Francisco Bay: The Urbanized Estuary. Amer. Assoc. Advan. Sci., San Francisco, CA, p. 85-95.

LTMS (Long-Term Management Strategy). 1992. Modeling Needs Assessment. Prepared for LTMS by USACE, San Francisco as part of LTMS Technical Rept. Series. San Francisco, CA. 14 pp. w/appendices.

SAIC (Science Applications International Corporation). 1987a. Alcatraz Disposal Site Survey, Phase 1: San Raphael Clamshell/Scow Operations. Prepared for USACE, San Francisco Dist., Newport, RI.

SAIC. 1987b. Alcatraz Disposal Site Survey, Phase 2: Richmond Channel Hopper Dredge Operations. Prepared for USACE, SF Dist., Newport, RI.

Teeter, Allen M. 1987. Alcatraz Disposal Site Investigation Report #3, San Francisco Bay - Alcatraz Disposal Site Erodability, HL-86-1. US Army Engineer Waterways Experiment Station, Vicksburg, MS. 17pp.

Trawle, M.J. and Johnson, B.H. 1986. Alcatraz Disposal Site Investigation Report #1, HL-86-1. US Army Engineer Waterways Experiment Station, Vicksburg, MS. 29 pp. w/appendices.

USACE. 1965. San Francisco Bay, California, Disposal of Dredge Spoil. Report of Committee on Tidal Hydraulics, Vicksburg, MS. 53 pp.

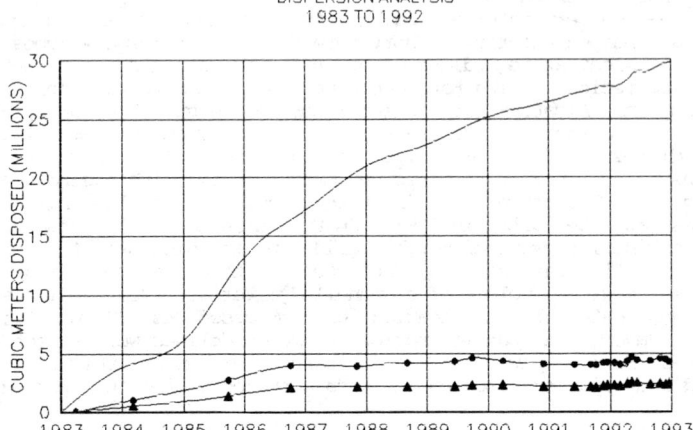

FIGURE 1

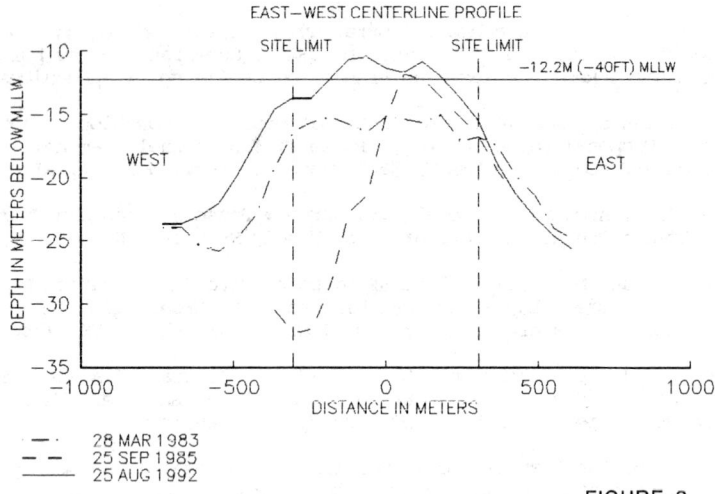

FIGURE 2

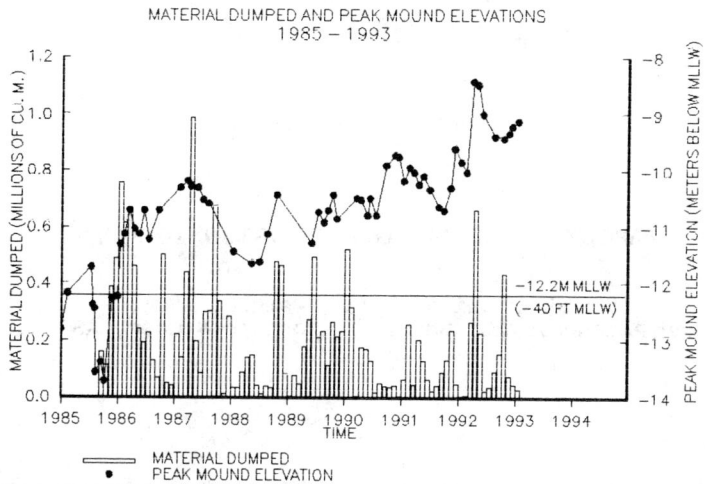

FIGURE 3

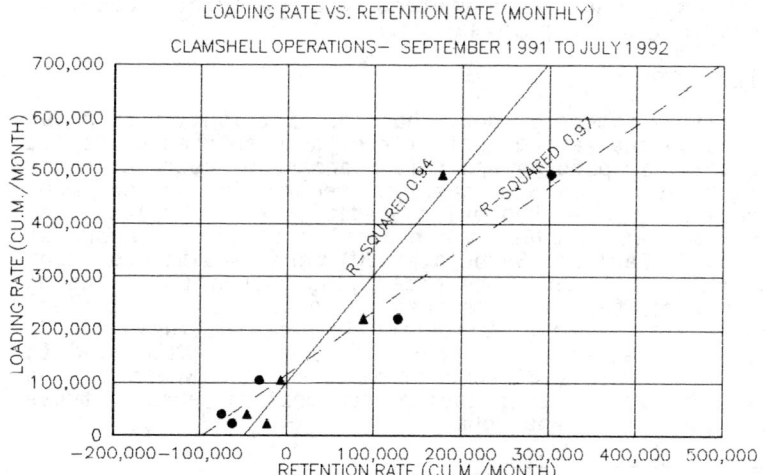

FIGURE 4

SHAPE EFFECT ON BED-LOAD TRANSPORT IN PIPES

Chandramouli Nalluri[1], Member, ASCE,
and Aminuddin Ab. Ghani[2], Associate Member, ASCE

Abstract

The nature of part-full flows in pipe sewers produces different channel shapes resulting in non-uniform shear stress distribution on the wetted perimeter. New experimental data were utilised to check the applicability of several available methods based on different shape parameters to predict bed-load concentration in pipe sewers.

Introduction

It is commonly known that the distribution of shear stress on the wetted perimeter of an open channel is not uniform. In pipe sewers this phenomenon occurs at part-full flows. In the case of sewers for no sediment deposition, the hydraulic radius, R (Mayerle et.al 1991, May et.al 1989) and hydraulic depth, D_h (Arora et al, 1984; Paul and Sakhuja, 1990) were used to represent the shape effects. The underlying assumption in using R is that the shear stress on pipe wall is uniform contrary to the real situation in part-full sewers. In using D_h instead of R, Arora et al (1984) attempted to take account the effects of free surface which is the main factor in introducing non-uniform shear stress distribution in open channel.

---

[1] Senior Lecturer, Department of Civil Engineering, University of Newcastle upon Tyne, NE1 7RU, U.K.

[2] Research Student, Department of Civil Engineering, University of Newcastle upon Tyne, NE1 7RU, U.K.

In this paper new experimental data covering different pipe sizes at part-full flows with no sediment deposition were used to check the validity of using R and D_h to represent the shape effects on transport of sediment as bed-load.

Experimental Equipment and Procedure

Experiments were carried out in smooth (154, 305 and 450mm dia.) as well as artificially roughened (305mm dia.) pipe channels. Artificial roughnesses were made of sand of diameters (=d_{50}) 0.5 and 1.0mm (tests in progress). Graded sand and gravel of diameters ranging from 0.5 to 8.3mm with average density of 2550 kg/m³ were used. All experiments were carried out under part-full uniform flows covering proportional flow depths (y/D) ranging from 0.15 to 0.80. The range of flow Reynolds numbers (=4VR/ν) was 1.3×10^4 to 4.6×10^5. The sediment concentration varied from 1 to 1450 ppm by volume.

Analysis of Data

Multiple-regression analyses were used to fit the present experimental (smooth pipes) data to a function given by Mayerle et al (1991):

$$\frac{V}{\sqrt{gd_{50}(S_s-1)}} \equiv f(C_v, D_{gr}, \frac{R}{d_{50}}, \lambda_s) \qquad (1)$$

where C_v is the volumetric sediment concentration, S_s is the relative density of sediments, λ_s is the Darcy-Weisbach's friction factor with sediment, and D_{gr} (=d_{50} $[g(S_s-1)/\nu^2]^{1/3}$) is the non-dimensional grain diameter.

The data collected from 305mm dia. pipe channel yielded the best-fit relationship:

$$\frac{V}{\sqrt{gd_{50}(S_s-1)}} \equiv 4.16 C_v^{0.23} D_{gr}^{0.03} (\frac{R}{d_{50}})^{0.61} \lambda_s^{0.03} \qquad (2)$$

with adjusted determination (adj. r^2) coefficient = 0.98.

The friction factor, λ_s, with sediment transport could be evaluated from a function proposed by Nalluri and Kithsiri (1992):

$$\lambda_s \equiv f(\lambda_c, C_v, D_{gr}) \qquad (3)$$

λ_c being clear-water Darcy-Weisbach's friction factor of the channel which can be obtained by Barr's equation (Featherstone and Nalluri, 1988). The resulting best-

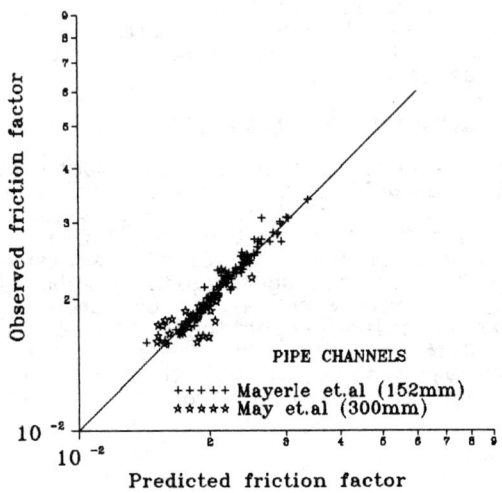

Figure 1. Verification of Eqn. 4 with other data

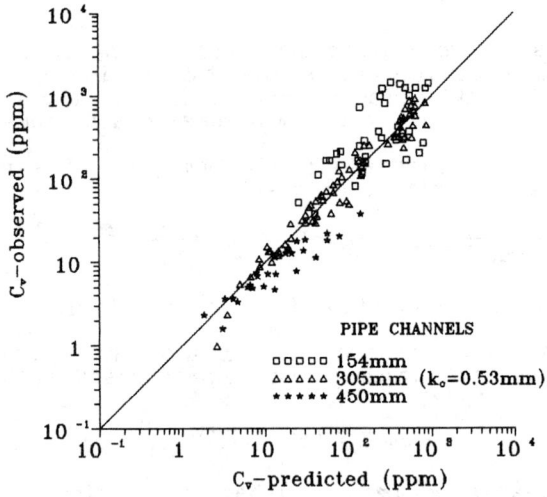

Figure 2. Verification of Eqn. 2 (R-model) with authors' independent data

fit equation was based on all of authors' data (adj. r^2 = 0.95). Equation 4 was tested with the independent

$$\lambda_s \equiv 1.26 \lambda_c^{1.01} C_v^{0.02} D_{gr}^{0.01} \approx 1.26 \lambda_c \quad (4)$$

data of Mayerle et al (1991) and May et al (1989) - see Figure 1 - resulting in a very good agreement.

Further analyses of data were carried out to fit a function used by Paul and Sakhuja (1990), intended for various channel shapes :

$$C_v \equiv f\left(\frac{q}{\nu}, \frac{y}{d_{50}}, \lambda_s, \frac{D_h}{y}\right) \quad (5)$$

where q is the unit discharge (=Q/T, Q and T being flow discharge and water surface width respectively), ν is the kinematic viscosity of water, and S_c is the slope parameter (=S/(S_s-1) in which S is the channel slope). D_h is defined as the ratio of flow area, A, to water surface width, T.

The best-fit equation (adj. r^2 = 0.97) was obtained from the 305mm dia. pipe channel data :

$$C_v \equiv 0.016 \left(\frac{q}{\nu}\right)^{0.03} \left(\frac{y}{d_{50}}\right)^{-0.36} S_c^{1.96} \lambda_s^{-2.18} \left(\frac{D_h}{y}\right)^{-1.23} \quad (6)$$

Verification of Transport Models

Validation of Eqns. 2 and 6 were carried out using independent data collected by the authors from smooth and rough boundary pipe channels of 154mm, 305mm and 450mm dia. These data were used to compute predicted parameter, C_v, and plotted against its observed values.

Figures 2 and 3 show that Eqns. 2 and 6 predict reasonably well the authors' data collected from smooth 154mm and 450mm dia. channels as well as the data from rough (k_o =0.53mm, k_o being the equivalent sand roughness of channel boundary) 305mm dia. channel.

In order to verify their validity for channel shapes other than circular, both models (Eqns. 2 and 6) were also used to the available data from smooth rectangular channels (Mayerle et al, 1991). Figures 4 and 5 show the predicted parameter, C_v, in comparison with its measured values. Despite the poor correlation it may be concluded that the model based on hydraulic radius is in better agreement.

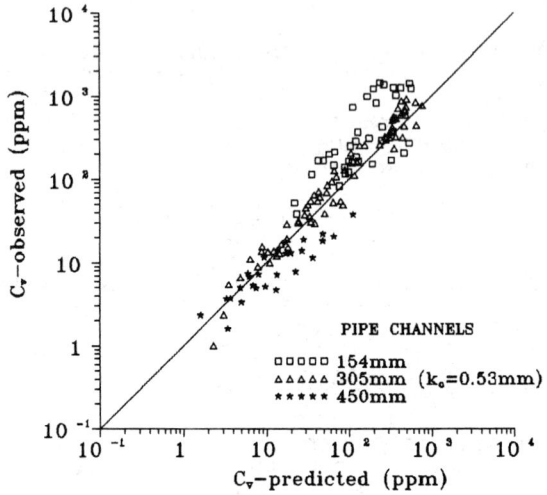

Figure 3. Verification of Eqn. 6 (D_h-model) with authors' independent data

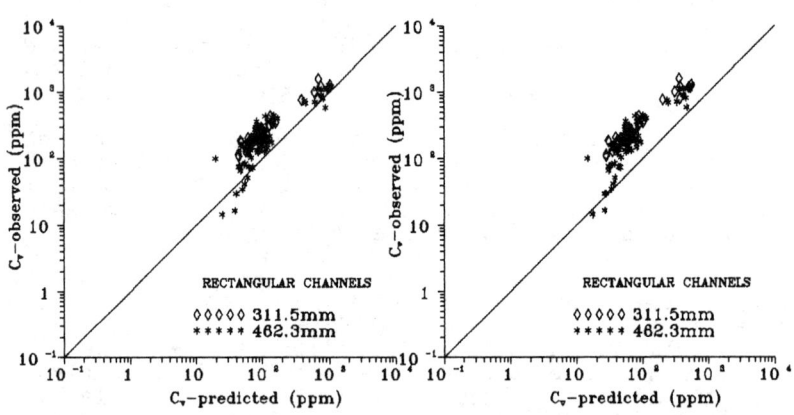

Figure 4. Predicted concentrations using Eqn. 2 (R-model)

Figure 5. Predicted concentrations using Eqn. 6 (D_h-model)

Conclusion

The above analyses, within the range of current studies by the authors, suggest that either R or D_h may be used to represent shape effects for bed-load transport in pipe sewers. Preliminary investigations of both models indicate the possibilty of extending their uses over various channel shapes. It is suggested that this possibility to be examined once such data are available; however, model based on hydraulic radius seems to be more promising.

Appendix: References

Arora, A.K., Ranga Raju, K.G., and Garde, R.J. (1984) "Criteria for deposition of sediment transported in rigid boundary channels". Proc. First Int. Conf. on Channels and Channel Control Structures, University of Southampton, England.

Featherstone, R.E. and Nalluri, C. (1988), Civil Engineering Hydraulics, Blackwell Scientific Publication, London.

May, R.W.P., Brown, P.M., Hare, G.R., and Jones, K.D. (1989) "Self-cleansing conditions for sewers carrying sediment". Hydraulics Research Ltd., Wallingford, England, Report SR 221.

Mayerle, R., Nalluri, C., and Novak, P. (1991) "Sediment transport in rigid bed conveyances". Jnl. of Hydraulic Research, Vol. 29, No. 4, pp. 475-495.

Nalluri, C. and Kithsiri, M. M. A. U. (1992) "Extended data on sediment transport in rigid bed rectangular channels". Jnl. of Hydraulic Research, Vol.30, No. 6, pp. 851-856.

Paul, T.C., and Sakhuja, V.S. (1990) "Why sediments deposit in lined channels". Jnl. of Irrigation and Drainage engineering, ASCE, Vol. 116, No. 5, pp. 589-602.

SCOUR AT HIGHWAY STRUCTURES IN TIDAL WATERS

E.V. Richardson[1], J.R. Richardson[2] and B.E. Edge[3]

Abstract

Scour analysis of bridges over tidal waterways uses the same scour equations as for the rivers. However, determining the hydraulic variables to use in the equations is complex. A major difference, as explained in this paper, is the effect on velocity and discharge of an increase in cross sectional area.

Introduction

Scour analysis of bridges crossing tidal waterways is very complex. Tidal waterways is a generic term for inlets, estuaries, tidally affected streams, bays and channels between islands or islands and the mainland. The analysis must consider the magnitude of the design storm surge (50-,100- and 500-year storm tide), the characteristics (geometry) of the tidal waterway, hydraulic variables resulting from the design storm surge, and the effect of any constriction of the flow due to the tidal waterway or bridge. In addition, the analysis must consider the long-term effects of the normal tidal cycles on long-term aggradation or degradation, contraction scour, local scour and stream instability.

A storm tide or storm surge in coastal waters may be caused by astronomical tides, wind action, rapid barometric pressure changes or seismic action or a combination of these events. In addition, the change in

---

[1] Senior Associate, [2] Senior Engineer, Resource Consultants & Engineers, PO Box 270460, Ft. Collins, CO 80527
[3] President, Edge and Assoc., 79 Anson St., Charleston, SC 29401

elevation resulting from the storm surge may be increased by resonance in harbors and inlets, whereby the tidal range in an estuary, bay or inlet is larger than on the adjacent coast. Astronomical tides are rhythmic diurnal or semi-diurnal variations in sea level that result from gravitational attraction of the moon and sun and other astronomical bodies acting on the rotating earth. Principal tidal terms are defined in another paper in these proceeding (Neill and Yaremko, 1993), in HEC 18 (Richardson et al., 1993) and in Shore Protection Manual (Corps of Engineers 1977).

The total scour at a bridge crossing can be evaluated using the scour equations recommended for inland rivers and the hydraulic variables for the tidal waterway. However, it should be emphasized that the scour equations and subsequent results need to be carefully evaluated considering the unsteady flow conditions, the density of sea water, inland runoff, the mixing or non mixing of sea and fresh water in the tidal waterway, vertical and lateral velocity distribution and magnitude of the astronomical tide and storm surge at the crossing. Tidal effects will be largest at a crossing located near the ocean and smallest to no effect inland where there is still tidal effect but no reversal of the flow through the bridge. In the latter the tide acts as a variable downstream control.

There is one major difference between riverine scour at highway structures and scour resulting from tidal forces. In determining scour depths for riverine conditions a design discharge is used (discharge associated with a 50-, 100- and 500- return period). For tidal conditions a design storm surge elevation is used and from that the discharge is determined. That is, for the riverine case the discharge is fixed whereas for the tidal case the discharge may not be. In the riverine case as the area of the stream increases the velocity and shear stress on the bed decrease because of the fixed discharge. In the tidal case as the area of the waterway increases the discharge may also increase and the velocity and shear stress on the bed may not decrease appreciably. Thus, long term degradation and contraction scour can continue until sediment inflow equals sediment outflow or the discharge driving force (difference in elevation between the ocean and the bay or estuary or between islands or islands and the mainland) reduces to a value that the discharge no longer increases.

This paper will describe a three level approach to analyzing tidal waterways similar to the approach outlined in HEC-20 (Lagasse et al., 1991), describe methods for obtaining the value of hydraulic variables (Level 2 analysis) and present an example problem that illustrates the increase in discharge as the waterway area increases.

Three Level 1 Approach

Level 1 is a **qualitative evaluation** of the long-term stability of the tidal waterway (vertical and lateral stability) and the potential for waterway response to change; estimation of the magnitude of the tides, storm surges, and flow in the tidal waterway; and qualitative determination of whether the hydraulic analysis depends on tidal or river conditions, or both (Richardson et al., 1993). **Level 2** is the engineering analysis necessary to obtain the velocity, depths, and discharge for tidal waterways to be used in determining long-term aggradation, degradation, contraction scour and local scour. The hydraulic variables obtained from the Level 2 analysis are used in existing riverine scour equations to obtain total scour. Using these riverine scour equations, which are for steady state equilibrium conditions for unsteady, dynamic tidal flow will usually result in estimating deeper scour depths than will actually occur (conservative estimate), but this represents the state of knowledge at this time. **Level 3** analysis is the use of physical and 2-dimensional computer models for complex tidal situations. In Level 2 analyses, unsteady 1-dimensional or quasi 2-dimensional computer models are often used to obtain the hydraulic variables needed for the scour equations.

Level 2 Analysis

The general procedures for Level 2 analysis are to (1) determine the magnitude of the design tides, storm surges and floods; (2) determine the hydraulic variables at the highway crossing; (3) determine the scour components; and (4) evaluate the results.

The magnitude of the design tides, storm surges and floods can be determined from the historical tidal and stream flow data maintained by the Federal Emergency Management Agency (FEMA), National Oceanic and Atmospheric Administration (NOAA), U.S. Army Corps of Engineers (COE), and U.S. Geological Survey (USGS).

Solution of the equations of motion and continuity to obtain the hydraulic variables of discharge, velocity and depth range from simple equations to numerical methods using computers. For many bridge crossings the simple equations give satisfactory answers.

The simple equations and procedure by Neill (1973) and Neill and Yaremko (1993) or a weir equation given in HEC-18 (Richardson et al., 1993) for constricted tidal inlets can be used to evaluate the hydraulic conditions at bridges influenced by tidal flows. A step-wise procedure for using Neill's method is given in HEC-18.

DYNLET1 (COE, 1991) and ACES-INLET (COE, 1991) are two unsteady quasi 2-dimensional mathematical models which can be used to model the hydraulic characteristics at the bridge. ACES-INLET is restricted to analysis of tidal inlets with up to two inlets to a bay; whereas, DYNLET1 can be used for multiple tidal inlets, tidal estuaries, tidally affected streams and bridge crossings of passages between islands.

The weir equation after Bruun (1990) and van de Kreeke (1967) is as follows:

$$V_{max} = C_d (2g \, \Delta h)^{1/2} \tag{1}$$

$$Q_{max} = A' V \tag{2}$$

V_{max} = maximum velocity in the inlet, fps
Q_{max} = maximum discharge in the inlet, cfs
C_d = coefficient of discharge ($C_d < 1.0$)
g = acceleration due to gravity, 32.2 ft/s^2
Δh = maximum difference in water surface elevation between the bay and ocean, ft
A' = net cross-sectional area in the waterway, at mean water surface elevation, ft^2

$$C_d = (1/R)^{1/2} \tag{3}$$

$$R = K_o + K_b + \frac{2g \, n^2 \, L_c}{1.49^2 \, h_c^{4/3}} \tag{4}$$

R = coefficient of resistance
K_o = head loss coefficient on the ocean side
K_b = head loss coefficient on the bay side
n = Manning's roughness coefficient
L_c = length of the waterway, ft
h_c = average depth of flow in the waterway at mean water elevation, ft

Example Problem

The problem is to determine the long-term potential degradation that may occur because construction of jetties has cut off the delivery of bed sediments from littoral drift to the inlet. For this situation, the magnitude of long-term degradation can be approximated by assuming clear-water contraction scour and using the average difference in water surface between the ocean and bay for normal tides to compute hydraulic variables using the weir equation. This will give the magnitude of long term degradation but not the time. The daily tidal movement will remove bed sediments because of the deficiency in sediment inflow into the waterway.

The length of the inlet is 1,500 ft, the width is 400 ft, Manning's n is 0.03, depth at mean water level is 20 ft. and area A' is 8,200 ft^2. The D_{50} of the bed material is 0.30 mm or 0.00098 ft. and the D_m (1.25 D_{50}) is 0.375 mm or 0.00123 ft.

$$R = 0.7 + 1.0 + \frac{2g\,(0.03)^2\,1,500}{1.49^2\,(20)^{4/3}} = 2.42 \tag{5}$$

$$C_d = \left(\frac{1}{2.42}\right)^{1/2} = 0.643 \tag{6}$$

$$V_{max} = (0.643)\,(2g\,\,0.6)^{0.5} = 4.0 \text{ ft/s} \tag{7}$$

$$Q_{max} = V_{max}\,A' = 4.0\,(8,200) = 32,800 \text{ cfs} \tag{8}$$

Potential long-term degradation is determined using a modification of Laursen's clear-water contraction scour equation (Richardson et al., 1993)

$$\frac{y_s}{y_1} = 0.13\left[\frac{Q}{D_m^{\frac{1}{3}} y_1^{\frac{7}{6}} W}\right]^{\frac{6}{7}} - 1 \tag{9}$$

y_s = depth of scour, ft
y_1 = depth of flow in the waterway, ft
Q = discharge in the waterway, cfs

D_m = effective mean diameter of the bed material (1.25 D_{50}), ft
D_{50} = median diameter of bed material. Use a weighted average of the material in the scour zone, ft
W = bottom width of the waterway, ft

$$\frac{y_s}{20} = 0.13 \left[\frac{32,800}{0.00123^{\frac{1}{3}} 20^{\frac{7}{6}} 400} \right]^{\frac{6}{7}} - 1 = 18.4 \text{ ft.} \qquad (10)$$

Adding the scour depth to the original depth and recalculating the velocity and discharges using the weir equation it is easily demonstrated the velocity and discharge will increase and the scour will continue.

References

Bruun, P. (1990). "Tidal Inlets on Alluvial Shores," Chapter 9, Vol 2, Port Engineering, 4th edition, Gulf Publishing, Houston, Texas.

Corps of Engineers (1977). "Shore Protection Manual." V. III, CERC, Waterways Experiment Station, Vicksburg, MS.

Corps of Engineers, (1991). "DYNLET1: Dynamic Implicit Numerical Model of One-Dimensional Tidal Flow Through Inlets," Technical Report CERC-91-10, Waterways Experiment Station, Vicksburg, MS.

Corps of Engineers (1990). "Flood Hydrograph Package User's Manual," HEC-1, Hydrologic Engineering Center, Davis, CA.

Lagasse, P.F., Schall, J.D., Johnson, F., Richardson, E. V., Richardson, J.R. Chang, F. (1991). Stream Stability at Highway Structures." FHWA-IP-90-014, HEC 18.

Neill, C.R. (1973). "Guide to Bridge Hydraulics," (Editor) Roads and Transportation Association of Canada, University of Toronto Press, Toronto, Canada.

Neill, C.R. and Yaremko, E.K. (1993). "Bridge Hydraulic Design in Tidal Waters." Proceedings ASCE Hyd. Div. Spec. Conf., San Francisco, CA.

Richardson, E.V., Harrison, L.J., Richardson, J.R. and Davis, S.R. (1930). "Evaluating Scour at Bridges." FHWA HEC-18, Washington, D.C.

van de Kreeke, J. (1967). "Water-Level Fluctuations and Flow in Tidal Inlets," J. Waterways, ASCE, 93(4).

Model Technology for Estimating Storm-Induced Currents

Mary A. Cialone[1] and H. Lee Butler[2]

Abstract

This paper demonstrates the application of the one-dimensional shallow-water equation model DYNLET1 at Brunswick Harbor, Georgia to simulate storm-induced velocities precisely at a bridge pier. A simple stochastic procedure to estimate frequency-indexed currents due to tropical storm-tide events is also presented. DYNLET1 results are compared with field data for normal tide conditions. The model is then used to compute storm-induced currents and probability exceedance curves are constructed.

Introduction

Knowledge of currents caused by an extreme storm event is important to various aspects of coastal and hydraulic engineering. Accurate estimates of storm-induced currents are critical to the proper design of bridge foundations spanning inlets and inland waterways subject to such extreme conditions. These data are used to estimate time-dependent forces acting on bridge piers causing sediment transport and scour. The objective of this paper is to demonstrate how off-the-shelf model technology can be employed to provide this information.

A PC-based, comprehensive one-dimensional shallow-water equation model DYNLET1 (Amein and Kraus 1991), is applied at Brunswick Harbor, Georgia. DYNLET1 predicts flow conditions in channels with varied geometry and can be used to resolve bays and tributaries if the flow field is not strongly two dimensional. Model simulation and demonstration of a simple stochastic procedure to estimate frequency-indexed currents due to tropical storm-tide events are presented.

---

[1]Member, ASCE. Research Hydraulic Engineer, Coastal Engineering Research Center, US Army Engineer Waterways Experiment Station, Vicksburg, MS 39180

[2] Member, ASCE. Chief, Research Division, Coastal Engineering Research Center, US Army Engineer Waterways Experiment Station, Vicksburg, MS 39180

DYNLET1

DYNLET1 is a one-dimensional model for computing tidal and storm-induced flows in a system of inlets and bays. The model uses a network of *channels* which meet at *junctions* (Figure 1). Cross-sections or *nodes* are defines along the channels at each location where there is a significant change in cross-sectional geometry. DYNLET1's ability to precisely defined a cross-section and the velocity field at user-specified locations on the cross-section are critical to this application because it allows for bridge piers to be resolved in a cross-section (Figure 2) and for the computation of velocities precisely at a pier.

The model employs the shallow-water hydrodynamic equations for one-dimensional depth-averaged flow. They consist of the equations for conservation of mass, momentum, and energy (e.g., French 1985), and may be written as,

$$\frac{\partial Q}{\partial t} + \frac{\partial}{\partial y}\left(\frac{Q^2}{A}\right) = -gAS_f + gB\tau_s - gAS_e - gA\frac{\partial z}{\partial y} \quad (1)$$

$$\frac{\partial Q}{\partial y} + \frac{\partial A}{\partial t} - q = 0 \quad (2)$$

where Q = volume flow rate, t = time, y = horizontal distance (along a channel), A = cross-sectional area, g = acceleration due to gravity, S_f = friction slope, B = width of top of channel cross section, τ_s = surface shear stress due to wind, S_e = transition loss rate with distance, z = water surface elevation, and q = lateral inflow or outflow per unit channel length per unit time. The equations are valid if the assumption of a hydrostatic pressure distribution holds. They are applicable to tidal flow, flows in lakes and reservoirs, river flow, and wave motion where the wavelength is significantly greater than the water depth. Equations 1 and 2 are solved numerically using an enhanced version of the implicit finite-difference technique of Amein (1975).

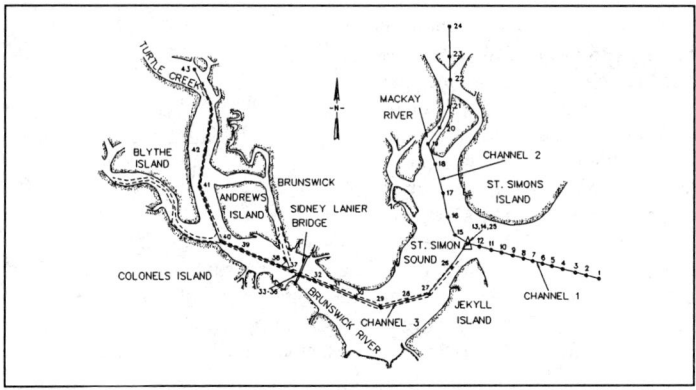

Figure 1. Grid network

The main input requirements for DYNLET1 are a grid network, cross-sectional geometry and friction factors, and external boundary conditions. The key input for a storm simulation is the external boundary condition consisting of a time-series of water surface elevation at the ocean boundary. The procedure for (1) selecting appropriate storm events for simulation and (2) constructing storm boundary conditions is outlined in the following section.

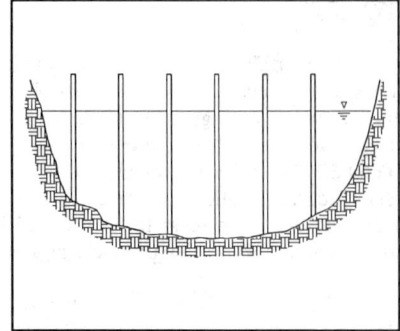

Figure 2. Typical bridge cross-section

Stochastic Procedure

Selection of storms to simulate for the purpose of estimating frequency-indexed velocities is based on knowing only surge-plus-tide stage at specific return periods (Figure 3). A given stage can be achieved by varying storm intensities, size (R=radius of maximum winds), forward speed (f), track, and its combination with the tide. The objective is to develop a set of storm parameters {R,f} which, when combined with tidal possibilities, form a total set of storm-tide events approximating the full spectrum of conditions that may occur. The initial step is to select values for R and f which represent maximum and minimum storm duration (D), determined by dividing estimates of the maximum and minimum R by the minimum and maximum f, respectively (Figure 4).

Central pressure variation is accounted for in the convolution of each surge hydrograph with tide. To introduce tidal phase variation, the peak total water elevation, S_{tot}, usually known from National Oceanographic and Atmospheric Administration (NOAA), Federal Emergency Management Agency (FEMA), or Corps of Engineers (COE) stage-frequency curves can be approximated by the combination of peak storm surge (S_p) and a single constituent mean tide (H_t) at one of four locations in the tidal cycle: mid-tide rising, high tide, mid-tide falling, and low tide. Since surge intensity varies with central pressure deficit, the hypothetical time evolution of surge height can be represented by

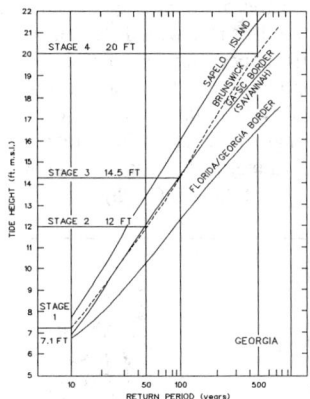

Figure 3. Stage-frequency curve

$$S(t) = S_p(1 - e^{-(D/2t)}) \qquad (3)$$

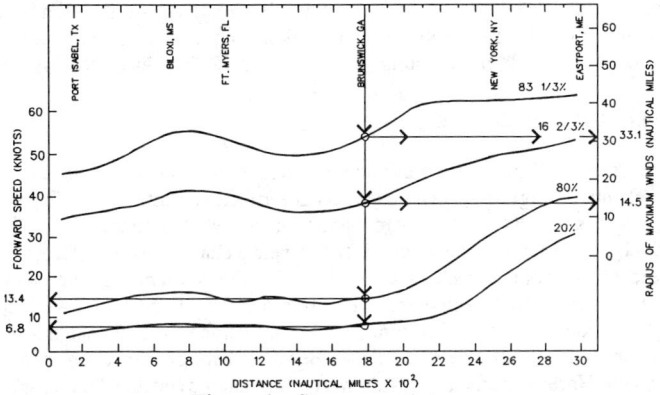

Figure 4. Storm parameters

where S(t) is surge level at time t. Two surge hydrographs for maximum and minimum durations are developed using Equation 3. Surge plus tide can be obtained by adding H_t and S(t) at the 4 specified phases of the tide. The next step is to adjust S_p such that the maximum water elevation for a given combination is equal to S_{tot}. This procedure is equivalent to backing out a surge-only hydrograph which, when combined with a particular position in the tide, gives the known stage specified by NOAA/FEMA/COE. By running DYNLET1 for the 2 surge hydrographs combined with 4 tide positions gives 8 storm-plus-tide events which produce the specified stage at selected return periods. For each storm-tide combination, velocity at specified locations in the model are recorded for later statistical analysis. Exceedance probabilities can then be attached to the velocities through a rank-order process.

To use the procedures discussed above certain tide and storm data are required. The key tide data required are estimates of mean tide ranges which are given in NOAA Tide Tables (1983) and rarely change from year to year. Tropical storm data required in the procedure were developed by NOAA (1975) and Figure 4 displays probability distributions of R and f, identified with a given percentile of occurrence for Gulf of Mexico and Atlantic seaboard coastal locations.

A PC-based Storm SELection (SSEL) program is run to create a set of elevation time-series according to the procedures discussed above. Each time-series is used as an ocean forcing boundary condition in DYNLET1 for the 8 hypothetical storm-plus-tide events. For each known stage and location where currents are to be analyzed, the 8 peak flood and ebb velocities obtained from DYNLET1 are ranked from 1 to 8, with the smallest flood (ebb) velocity ranked as 1. The cumulative probability, P, for a particular velocity is given by M/9, where M is the order number of the event. This range of velocity gives an estimate of the strength of current one may expect for a given stage return period, covering the range of storm durations and possible tide combinations associated with that stage. This approach is limited in that a very small number of possible combinations are used in order to minimize the number of DYNLET1 simulations required. A more conservative estimate could be made by selecting more extreme {R,f} values (NOAA 1975).

Brunswick Harbor Application

Brunswick Harbor is represented in the model by 43 cross-sections or nodes in three channels (Figure 1). The channels are joined at a junction inside St. Simon Sound. Nodes 13, 14, and 25 are "junction nodes" for channels 1, 2, and 3, respectively, and therefore have identical geometric characteristics. Cross-sectional data were taken from NOAA chart 11506 with the reference datum being mean lower low water (mllw). Values of Manning's coefficient of friction were specified at every point on every cross-section. A constant value of 0.025 was used to obtain the best comparison with measured data. Because there are no severe constrictions at Brunswick Harbor, transition losses were eliminated by setting the transition loss coefficient to zero at every cross section. At the ocean boundary, Node 1, measured water surface elevation data obtained near the Sidney Lanier Bridge during a 1992 field study were shifted in phase to approximate conditions occurring at the ocean boundary node. At the end nodes in the Mackay River and Turtle Creek (Nodes 24 and 43), a discharge boundary condition was specified and set equal to zero due to lack of stream discharge or surface elevation data upstream. For the tidal calibration, a time-series of water surface elevation was reproduced near the Sidney Lanier Bridge. The model reproduces velocities at the bridge as well (Figure 5).

For the storm simulations, R and f values selected for Brunswick were 33.1 and 14.5 n. mi and 13.4 and 6.8 knots, respectively (Figure 4), yielding durations of 9.7 and 2.2 hrs. NOAA data for stage frequency at the Florida-Georgia border are given in Figure 3 (Ho 1974). For example computations, stages at return periods of 10, 50, 100, and 500 years, namely, 7.1, 12.0, 14.5, and 20.0 ft, respectively, were selected. The mean range from NOAA Tide Tables for the ocean gage is 6.5 ft giving an amplitude of 3.25 ft. The tide is semi-diurnal. SSEL was run to create the forcing

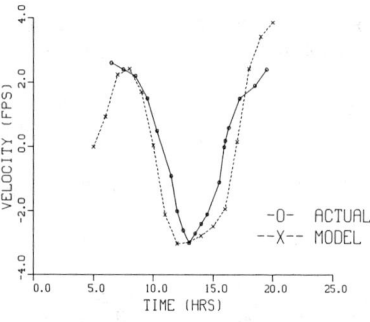

Figure 5. Velocity calibration

boundary conditions for 32 DYNLET1 simulations (4 stages x 8 hydrographs) and velocities at one selected bridge pier were saved for later analysis. As discussed previously, any given stage for a particular return period could be produced by combinations of varying storm intensities, durations, and tidal phase. While only 8 events at a particular stage were run, results indicate a linear-with-time estimate of probability exceedance was appropriate (Figure 6).

Conclusions

Accurate estimates of storm-induced currents are critical to the proper design of bridge foundations spanning inlets and inland waterways subject to such extreme conditions. A PC-based one-dimensional model (DYNLET1) capable of accurately defining bridge piers in a cross-section and predicting velocities precisely at a bridge pier was used in this study.

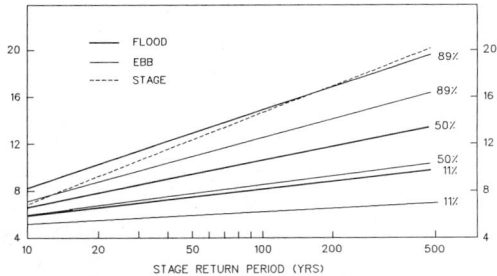

Figure 6. Probability-exceedance

Secondly, a simple stochastic procedure to estimate frequency-indexed currents due to tropical storm-tide events was developed and applied at Brunswick Harbor, Georgia. Selection of storms to simulate for the purpose of estimating frequency-indexed velocities was based on available historical tide and storm parameter data and the surge-plus-tide stage at specific return periods for the project site. Probability-exceedance curves were constructed.

Acknowledgements

We are indebted to Dr. Norman Scheffner for his outstanding effort in developing the statistical approach used in this study and Dr. Michael Amein for his assistance in the Brunswick Harbor application. Ms. Marsha Darnell and Ms. Holley Messing were instrumental in preparing graphics for this paper. The study was sponsored by the U.S. Department of Transportation, Federal Highway Administration. Permission was granted by the Chief of Engineers to publish this information.

References

Amein, M. 1975. "Computation of Flow through Masonboro Inlet, N. C.", J. of the Waterways, Harbors and Coastal Engineering Div., Vol 1, No. WW1, pp 93-108.

Amein, M., and Kraus, N. C. 1991. "DYNLET1: Dynamic Implicit Numerical Model of One-Dimensional Tidal Flow through Inlets," Technical Report CERC 91-10, Coastal Engineering Research Center, USAE Waterways Experiment Station, Vicksburg, MS.

French, R. H. 1985. *Open Channel Flow*, McGraw-Hill Book Co., New York, NY.

Ho, Francis P., Storm Tide Frequency Analysis for the Coast of Georgia, U.S. Dept. of Commerce, NOAA, NWS, TM NWS HYDRO-19, Silver Spring, MD, Sept. 1974.

U.S. Dept. of Commerce, Tide Tables 1983, High and Low Water Predictions, East Coast of North and South America Including Greenland, NOAA, NOS, Washington, DC, 1982.

U.S. Department of Commerce, Some Climatological Characteristics of Hurricanes and Tropical Storms, Gulf and East Coasts of the United States, NOAA, NWS, NOAA TR NWS 15, Washington, DC, May 1975.

Indian River Inlet: Is There a Solution?

H. Lee Butler[1] and W. Jeff Lillycrop[2]

Abstract

   Man-made changes at Indian River Inlet, Delaware, have occurred over the past 6 decades. During this period the tidal prism of the Indian River - Rehoboth Bays system has increased along with scouring of the inlet. This paper gives a brief history of the inlet engineering and discusses the immediate problem of protecting a major bridge spanning the inlet. Results from the application of a one-dimensional shallow-water equation model are used to demonstrate the variation of currents from typical tidal events to extreme storm events. The paper does not attempt to provide an answer to the question posed in the title, but suggests a direction if a solution is to be found.

Introduction

   Indian River Inlet is located approximately half-way between Cape Henlopen at the entrance to Delaware Bay and the Maryland state line (Figure 1). The inlet is the only opening to the Atlantic for the dual-bay system of Indian River Bay and Rehoboth Bay. Both bays are quite shallow with an average depth of approximately 7 feet. The inlet is stabilized by two rubble mound jetties and is spanned by a state highway bridge with two piers in the inlet channel.

   Construction of jetties and bay bulkheads, mining of beach nourishment material from the flood tidal delta, construction of the present bridge and removal

---

[1]Member, ASCE. Chief, Research Division, Coastal Engineering Research Center, US Army Engineer Waterways Experiment Station, Vicksburg, MS 39180

[2] Member, ASCE. Research Hydraulic Engineer, Coastal Engineering Research Center, US Army Engineer Waterways Experiment Station, Vicksburg, MS 39180

of an older bridge are the primary engineering projects that have taken place over the past 6 decades. Besides the expected offset between the adjacent shorelines, the net result has been a marked increase in scour within the inlet throat.

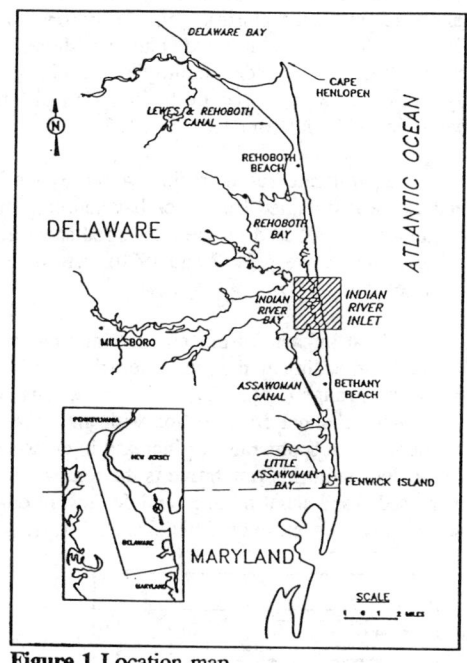

The objective of this paper is to provide the reader with a brief summary of the engineering practice at a "problem" inlet, discuss the range of velocities which occur throughout the year and for extreme events using an off-the-shelf numerical tool, and suggest a direction for the engineering approach to solve the "problem".

Historic Summary

The Federal navigation project at Indian River Inlet was approved in 1937 for the purpose of: 1) improving navigation through the inlet, 2)

Figure 1 Location map

increasing bay salinity and reducing stagnation (to improve the fishing industry), and 3) increasing the tide range (Anders, et al, 1989). Prior to the Federal project the inlet was ephemeral, typically breaking through the beach during periods of heavy rainfall, and migrating alongshore until closed by littoral processes. The current location of the inlet coincides with the historical location of the Indian River valley. Today, modern sediments (approximately 30 to 40 ft deep) cover silts and clays deposited during the Holocene period.

Construction began in 1938 and during that year dredging of the inlet to -14 ft NGVD was completed. Two rubble mound jetties, 1,550 ft long and 500 ft apart, were completed in 1939. Erosion of the adjacent bay shorelines began immediately after inlet stabilization. In 1941, steel sheetpile bulkheads were constructed to protect the bay shoreline from erosion (Figure 2). The

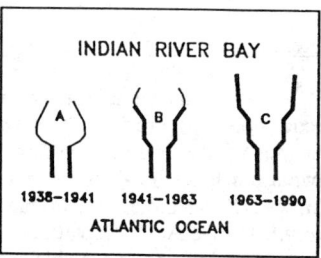

Figure 2 Historical inlet changes

design created a venturi effect which likely increased inlet flow velocity. Completion of the bulkheads simply transferred the bay shoreline erosion problem landward, requiring construction of additional bay shore-line protection in 1963. By 1989, the entire north and south bay shorelines near the inlet were protected by revetments. An older, multi-piered bridge constructed in 1941 was replaced by a modern, 2-pier structure in 1965. Final removal of the old bridge was begun in 1972 and completed in 1976.

To mitigate beach erosion caused by the jetty placement, nourishment material was obtained from back-bay sources. Initiated in 1957, the practice continued and focused on the inlet flood shoal in 1972. A total of 3.6 million cubic yards (between 1957 and 1990) were removed and pumped onto the north ocean beach.

What impact did this engineering practice in combination with nature have on tidal and sedimentation characteristics? It is apparent an increase in tidal prism, tide range (Figure 3), and inlet velocity occurred. A history of a typical inlet throat channel cross-section with time is shown in Figure 4. A definite acceleration in scour rate (or increase in cross-section area) began around 1974. Coincident with this timeframe is the removal of the multi-piered old bridge, increased flood shoal mining, and downward erosion of the inlet through the sand veneer covering more erodible silts and clays.

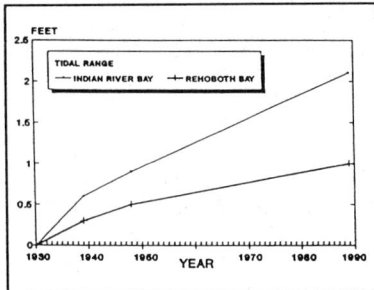

Figure 3 Change of tide range

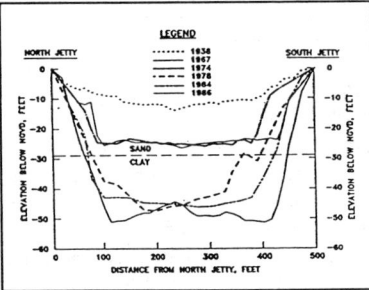

Figure 4 Inlet scour

Present Conditions

Presently the inlet throat averages over 40 ft in depth and contains 3 large scour holes (Figure 5). The seawardmost scour hole has formed near the tip of the north jetty, and has contributed to loss of about 250 ft of jetty. The other two scour holes lie along the midline of the inlet in the vicinity of the present bridge. Scour around the bridge piers recently reached severe proportions. As a result, the Delaware Department of Transportation undertook an emergency repair by placing stone rubble around the bridge piers to prevent further erosion.

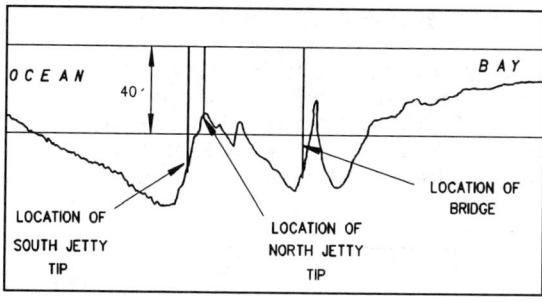

Two shallower areas separate the scour holes. The seawardmost shallow area corresponds to a position in the south jetty which is frequently overtopped by waves and may be partially permeable, allowing sand transport into the inlet throat. The other shallow area separates the scour holes in the vicinity of the present bridge. Subbottom profiles and side scan sonar indicate that foundation materials left from the old bridge are the cause of the shallow depth in this area. The end results are highly variable inlet cross-sections as well as the highly variable inlet bottom profile.

Model Application

A PC-based, comprehensive one-dimensional shallow-water equation model DYNLET1 (Amein and Kraus, 1991) was applied at Indian River Inlet on a grid network shown in Figure 6. The model was calibrated and verified to measured surface elevation and velocity data obtained at several stations throughout the dual-bay system.

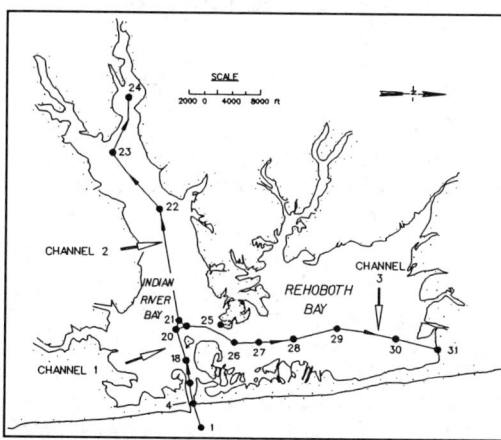

Figure 6 Grid Network

The original inlet design expected maximum currents on the order of 4 ft/sec. It is not uncommon today for peak velocities under spring tide conditions to exceed 8 ft/sec. DYNLET1's ability to account for structures in a cross-section, in this case bridge piers, and predict velocity adjacent to the structure make it an excellent tool for estimating tide and storm-induced currents (Cialone and Butler, 1993).

Applying a measured time series of water elevation for Hurricane Gloria (Jarvinen and Gebert, 1986), an estimate of current velocities at the present bridge cross-section is given in Figure 7. Stations 24 and 7 are located between a bridge pier and the north and south shoreline, respectively. The peak ebb velocity for this storm approached 12 ft/sec in the deepest part of the channel between the bridge piers. Similar velocities would be experienced during most severe extratropical storms and calculations for stronger hurricanes indicated peak velocities would well exceed 13 ft/sec.

Discussion

Past studies indicated an initial moderate scour rate may be a result of the inlet-bulkhead design (Anders, et al, 1989). However, scour rate acceleration in the mid-1970's appears to be related to other factors discussed herein. Attention to storm effects was not a focus of previous studies. Over 10 hurricanes have struck the area along with several yearly extratropical storms in the past 6 decades. A comprehensive tidal prism study (Raney, 1989) using a two-dimensional numerical model concluded the inlet has not yet reached an equilibrium condition.

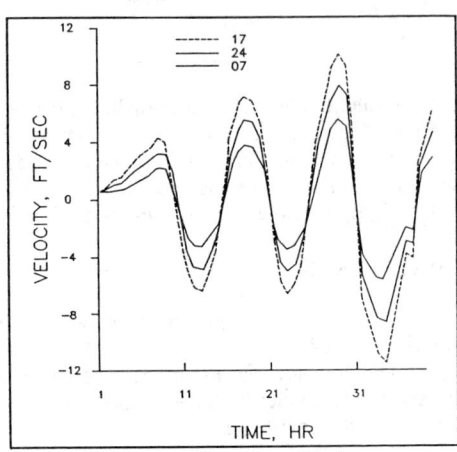

Figure 7 Velocity, Hurricane Gloria

Is there a solution for this complicated, "problem" inlet? If there is, it will probably be in the form of a holistic approach. Fixing only one part of the problem might exacerbate another problem or shift the problem to a different location in the system. Any proposed engineering solution should be examined for its impact to the total system.

Acknowledgements

We appreciate the assistance of Ms. Marsha Darnell in preparing the graphics for this paper. The topics covered in the paper were sponsored by the US Army Engineer District, Philadelphia, and the US Department of Transportation, Federal Highway Administration. Permission was granted by the Chief of Engineers to publish this information.

References

Anders, Fred J., Lillycrop, W. Jeff, and Gebert, Jeff, 1989. "Effects of Natural and Man-made Changes at Indian River Inlet, Delaware," Proc. Beach Preservation Technology 1989, St. Petersburg, FL.

Amein, Michael, and Kraus, Nicholas C., 1991. "DYNLET1: Network Model for Tidal Inlet Dynamics," Proc. ASCE 2nd Int. Conf. on Estuarine and Coastal Modeling, Nov 13-15, Tampa, FL.

Cialone, Mary A., and Butler, H. Lee, 1993. "Model Technology for Estimating Storm-Induced Currents," Proc. ASCE '93 Natl. Conf. on Hyd. Eng., July 25-30, San Francisco, CA.

Jarvinen, Brian, and Gebert, Jeff, 1986. "Comparison of Observed Versus Slosh Model Computed Storm Surge Hydrographs Along the Delaware and New Jersey Shorelines for Hurricane Gloria, September 1985," NOAA Tech. Memorandum NWS NHC 32, National Hurricane Center, Coral Gables, FL.

Raney, D.C., Doughty, J.O., 1990. "Tidal Prism Numerical Model Investigation and Analysis Task of the Indian River Inlet Scour Study, Delaware," Univ. of Alabama, Bureau of Eng. Res., Rept. No. 500-183.

Bridge Hydraulic Design in Tidal Situations

Charles R. Neill and Eugene K. Yaremko[1]

Abstract

The analysis of bridge hydraulics and scour potential in tidal waters, in comparison with inland rivers, requires consideration of various additional factors. This paper condenses a simplistic approach presented in a Canadian publication of 1973, and summarizes three cases from the authors' experience.

Introduction

Some tidal terms are defined in Figure 1. Water-level changes in coastal waters may be due to astronomical tides, storm surges, resonant oscillations, and seismic waves. For bridge crossing design, rigorous analysis of tidal crossings is often unwarranted, but in difficult cases a tidal specialist may be required. Complicated situations can be studied with mathematical or physical models.

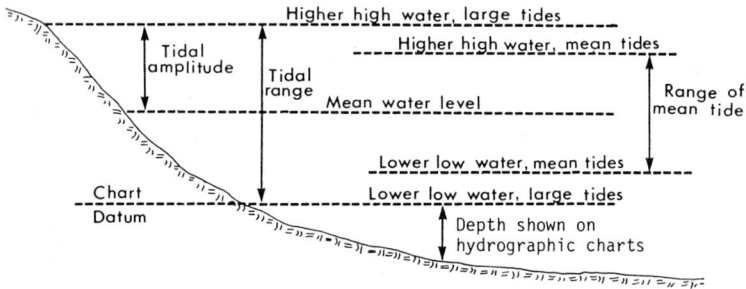

Figure 1. Principal tidal terms (simplified).

---

[1] Principal Engineers, Northwest Hydraulic Consultants, 4823 - 99 St., Edmonton, Alberta T6E 4Y1, Canada.

The "Guide to Bridge Hydraulics" (RTAC 1973), which was mainly concerned with inland rivers, divided tidal crossings into three categories: inlets from the open sea to a closed lagoon or bay; river estuaries; and passages open to the sea in both directions. Hydraulic peculiarities of tidal crossings include: discharges may be altered by constricted bridge openings or flood control works; currents and ice may flow in both directions; ice frozen to piers and piles may cause large vertical forces; structures may be subject to wave action; scour can be enhanced by periodic current reversals; and fine-grained deltaic sediments may be susceptible to seismic settlement.

Tidal Inlets

Crossings of tidal inlets without significant freshwater flow may involve complete closure, full bridging, or partial closure. The choice may depend on navigation requirements, foundation conditions, economics and environmental impacts.

Complete closure by a highly impermeable causeway normally produces an interior water level close to mean sea level. If a stagnant pool is unacceptable, a permeable causeway or the provision of small culverts may be considered.

Complete bridging requires consideration of scour, ice action and waves. Because worst natural scour is determined by frequent tidal conditions rather than rare river floods, local observations at appropriate times should enable reasonably reliable prediction, if additional allowance is made for local scour due to piers etc.

Partial closure introduces the following hydraulic complications: (1) velocities and scour potential are increased; (2) the phase difference between exterior and interior levels is increased; and (3) the interior tidal range and wave heights may be reduced. If the bed of the constriction is more or less inerodible, discharges and interior water levels over a tidal cycle may be computed by tidal routing. However, if the bed is easily erodible and the constriction is not very severe, increased scour will tend to compensate for the reduced width. As a first approximation it can then be assumed that the maximum natural tidal discharge will not be reduced. This maximum discharge is preferably determined by field gagings, but can be estimated approximately from the formula:

$$Q_{max} = \frac{\pi \Omega}{T}$$

where Ω is the tidal prism or interior volume between low and high tides, and T is the tidal period or interval between successive lows or highs. The derivation of the formula is indicated in Figure 2.

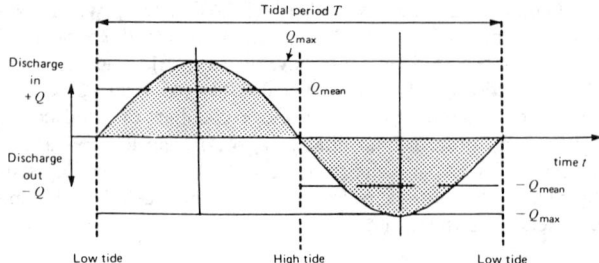

Figure 2. Discharge curve for the mouth of a tidal inlet over one tidal cycle. Assuming a sinusoidal variation, the equation of the discharge curve is $Q = Q_{max} \sin(2\pi t/T)$. The area between the curve and $Q = 0$ over half a tidal cycle represents the tidal prism volume Ω, i.e $\Omega = \int_0^{T/2} Q dt$. Also $Q_{max} = \pi\Omega/T$, $Q_{mean} = 2\Omega/T$, and $Q_{max}/Q_{mean} = \pi/2$. The assumption of sinusoidal variation is correct only for inlets with vertical shorelines, but is acceptable if the relative difference between the inlet surface areas at low and high tide is small.

The maximum tidal discharge occurs approximately midway between low and high tides. The corresponding mean velocity through the constricted opening is given by

$$V_{m_{max}} = \frac{\pi \Omega}{TA}$$

where A is the cross-sectional area below mean sea level. According to CERC, (1966), the maximum velocity at the centre of the inlet is approximately 33% higher. With discharges and velocities determined, scour may be estimated as for river bridges.

Estuaries

Estuaries involve various possible combinations of river and tidal flows and salinity effects on currents and sediment. Under certain conditions fresh and salt water layers may flow simultaneously in opposite directions. Suspended and dissolved river sediments tend to flocculate and precipitate in salt water to form a soft muddy bottom. Unconstricted estuary crossings require consideration of: (1) design river flood in conjunction with high tide plus storm surge, for design high water; (2) design river flood in conjunction with low tide, for velocities and scour; and possibly (3) design river flood in conjunction with maximum tidal outflow. Maximum tidal flows if ungaged can be estimated on the basis of the inland tidal prism as for tidal inlets: often this is negligible under river flood conditions. Constricted estuaries pose complex problems including changes in tidal levels, backwater effects in river floods, changes in currents, and effects on scour, deposition and ice regime.

Open Tidal Passages

Flows in tidal passages result from phase differences between tides at the two ends. Discharges are determined by differences in water level, cross-sections and hydraulic roughness. Some passages can be closed without unacceptable effects, but often an opening must be left for navigation and fish and to avoid changes in tidal regime. Partial closure may be acceptable and economically attractive. The analysis of a partial closure is complex and requires expertise in tidal hydraulics.

Design Cases

Selected aspects of three tidal crossing studies are summarized below. However, these cases do not cover all the hydraulic aspects of tidal crossings as referred to above, since river flood flows have tended to dominate the analyses in cases within the authors' experience.

Case 1 - Rapid Transit Rail Bridge. The studies concerned a crossing of the Fraser River near Vancouver, British Columbia, 26 km from the coast. Scour estimates were based on river flow and water level gage records, local surveys, soil drilling, and for final design on scale model testing. The site tidal range is 95% of the coastal range and the mean water level is only 0.7 m above sea level. At the lower river flows, tidal discharges are relatively high because of a large tidal lake upstream. Key data are as follows:

River discharges in m^3/s, based on upstream gage records: minimum = 650; mean = 3500; mean annual maximum = 10 000; 200-year maximum (design flood) = 20 000.

Peak tidal discharges in m^3/s, based on local stage and velocity records: 6000 for river flow of 1000, decreasing to 1000 for river flow of 15 000.

Range of water levels: mean annual range = 3.7 m, extreme historical range = 4.9 m.

Cross-sectional mean velocities including tidal effects: mean annual flood up to 1.9 m/s; 200-year flood up to 2.7 m/s; maximum upstream velocity in winter flood tide = 1 m/s.

Maximum point velocities: mean annual = 2.7 m/s; 200-year = 3.2 m/s; reported in 1936 during nearby bridge construction = 3.8 m/s (channel constricted by cofferdams).

River bed materials: alluvial sand with thin seams of silt and fine gravel.

Channel cross-sectional dimensions at mean water level: width = 500 m; area = 6300 m^2; mean depth = 12.6 m, max. depth = 20.5 m. The high ratio of max/mean depths is due to plan curvature.

The following causes of scour below normal river bed levels were considered for design: (1) temporary accommodation to extreme floods (natural scour); (2) permanent adjustment to proposed substructures (contraction scour); and

(3) local scour at piers due to vortex systems. All were estimated by semi-empirical methods, supplemented by later scale model studies for alternative foundation geometries and protection arrangements.

Temporary accommodation to the design flood was estimated on the basis of a "regime" cross-sectional area. Allowing for the rise in stage, the net area increase translated to an average scour of 2.5 m. Permanent adjustment to waterway reduction by piers and abutments was estimated to produce an additional 1.5 m of scour, giving an average bed lowering for design conditions of 4 m.

The proposed pier design included ship impact barriers 20 m wide. Potential local scour was estimated as around 30 m below ambient bed. There were supporting reports that during nearby bridge construction in 1936, a cofferdam 11 m wide produced local scour of 17 m. Riprap aprons were therefore designed to surround the pier foundations, the surface elevation of the riprap being set at 5 m below normal bed levels to allow for natural and contraction scour.

Case 2 - Highway Bridge with Ice-jam Potential. The bridge crosses the estuary of the Miramichi River at Newcastle, New Brunswick, about 20 km from the coast, and fully spans the active channel. Concerns were related mainly to environmental effects of approach embankments over a marsh area and to ice-jam flood phenomena. Key data are as follows:

River discharges in m^3/s: mean = 260, median annual maximum = 2000, 100-year 6000.

Max. tidal discharge about 400 m^3/s, not significant at high river floods. Tides extend to 20 km upstream of site.

Water levels: LLW = −0.7 m, extreme high tide +1.9 m, storm surge max. +3.8 m, ice-jam max. +4.5 m in 1970.

Channel cross-sectional properties at mean water level: width = 580 m; area = 3800 m^2; mean depth = 6.5 m; max. depth = 13 m (triangular section); 100-year mean velocity = 1.6 m/s.

Bed materials: silts and soft clays overlying gravel.
Solid ice thickness: 0.5 to 1.0 m.

Pier shafts are 2.4 m wide over pile caps 9 m wide, with top of pile caps flush with normal bed. Potential local scour at the deeper piers is countered by riprap aprons extending 13 m all around the pile caps.

Potential maximum ice-jam levels were estimated by applying recent analysis methods (Watt et al 1989, Beltaos and Wong 1991) to a range of river flows and assumed downstream jamming sites. Primary attention was given to a historical maximum of +4.5 m caused by a winter flood and ice break-up in February 1970. Computations indicated that with certain adverse combinations of assumptions, water levels could reach +6 to +8 m.

Case 3 - Bridge Replacement over Delta Channel. The crossing is over a distributary of the Fraser River Delta in Vancouver, British Columbia, 600 m downstream of its bifurcation from another distributary and 6 km from the coast. The main hydraulic concern was over potential scour. The natural channel has been constricted by bank protection works and marinas. Comparison of bathymetric surveys by navigation and bridge agencies showed substantial bed-level fluctuations, up to 4 m at certain points on the existing alignment. Key data are as follows:

Design flood peak from river inflow 1100 m^3/s, plus tidal outflow 100 m^3/s, for total of 1200 m^3/s.
LLW elevation −2.2 m.
Effective channel width as constricted = 206 m.
Bed material: medium sand.

In terms of channel regime, the channel even as constricted seems very wide for the estimated design flows. Flows are believed to have been reduced as a result of navigation dredging in other distributaries. Regime estimates using the effective channel width indicated potential general scour depths (discounting local pier effects) of around 8 m, corresponding to a minimum scour elevation of −10.2 m under LLW conditions. Historical bathymetric surveys indicated almost identical minimum levels in the vicinity of the existing bridge. The status of the existing pier foundations with respect to local pier protection was not well known, but the deepest surveyed points did not appear to be associated with local pier scour.

References

Beltaos, S. and Wong, J. 1991. "Ice jam configuration: second generation model." Paper presented to ASCE Cold Regions Engineering Conference, February.

CERC (Coastal Engineering Research Centre), U.S. Army, 1966. "Shore protection, planning and design." Technical Report No. 4, 3rd edition.

RTAC (Roads and Transportation Association of Canada - now TAC) 1973. "Guide to bridge hydraulics". Ed. C.R. Neill, publ. University of Toronto Press. (Out of print.)

Watt, W.E. et al, 1989. "Hydrology of floods in Canada, a guide to planning and design." National Research Council Canada, Assoc. Committee on Hydrology, Ottawa.

Acknowledgements

The first part of the paper is based on material in the Guide to Bridge Hydraulics. Case 1 is based on a report of 1984 to B.C. Transit, Vancouver. Case 2 is based on work conducted in 1990 for WMS Associates, Fredericton, N.B. and N.B. Transportation. Case 3 is based on work conducted in 1992 for Reid Crowther and Partners and B.C. Transport and Highways.

McCORMICK BRIDGE SCOUR EVALUATION - A CASE STUDY OF A TIDAL BRIDGE

Michael P. Maroney, P.E. Member ASCE[1]

Abstract

Although the Federal Highway Authority has implemented procedures for bridge scour analysis in their HEC-18 publication, they do not indicate a procedure for scour depth determination for tidal waterways. The first tidal bridge scour analysis performed in Florida was for the McCormick Bridge in Jacksonville, and the experiences in production of the analysis required research capabilities, knowledge of the mechanics and equations involved in the scour process, and good engineering judgement.

Introduction

The State of Florida has more beach front per square mile of land mass than any other State in the United States. Its narrow, peninsular shape allows the majority of Floridians (and visitors to the State) the opportunity to enjoy Ocean and Gulf front recreational activities. However, this shape and its location makes the State especially susceptible to significant tropical storms and hurricanes. The majority of the large cities within Florida are located adjacent to tidal water bodies, but the main centers of commerce are generally located well inland and the residential and tourist-related centers are the facilities located at the beach.

The State also has a large percentage of narrow, tidal waterways proximate, and parallel to the beach front. Some of these narrow waterways are natural, some of them were constructed by the U.S. Army Corps of Engineers. The connection of these waterways forms the Intracoastal Waterway (ICWW) system which forms a barrier between the mainland municipalities and the beach front communities. Since the beach is a very desirable area to live, a large portion of Floridians daily commute from the beach to the corporate centers, and a larger

---

[1]Civil Design-Infrastructure Department Head, Bromwell & Carrier (BCI) Inc., P. O. Box 5467, Lakeland, FL 33807-5467

number of tourists vacation at the beach. This has lead to the construction of many bridges over the Intracoastal Waterway.

These bridges range in age from 10 to over 70 years in age, and therefore the technology used for their design significantly varies as well. It is certainly a matter of concern (especially in the wake of Hurricane Andrew) that these bridges be analyzed for their ability to withstand tropical storms so they are able to convey traffic from the beach to the mainland in times of storm-related emergencies. At best, the loss of a bridge leads to increased traffic congestion. At worst, the bridge provides the only evacuation route from the beach and its loss results in populations stranded on low lying stretches of land and subject to the brunt of the storm.

In response to this need for analysis (subsequent to three non-tidal scour-related bridge failures in the State), and in conjunction with the Federal Highway Program on bridge stability analysis, the State of Florida Department of Transportation (FDOT) contracted for analysis of several bridges within the State, some of which traversed the ICWW. The first tidal bridge to be analyzed was the McCormick Bridge.

Case Study Parameters And Data Collection

The McCormick Bridge is part of U.S. 90 in Jacksonville, Florida. The roadway is a major thoroughfare in the Jacksonville area, and the bridge conveys traffic from the mainland to the beaches communities over the Intracoastal Waterway (ICWW). The waterway is tidal in nature with connections to the Atlantic Ocean 6 miles north and 40 miles south of the bridge; however, the ICWW floodway narrows severely 2 miles south of the bridge from a cross-sectional distance of approximately 5000 feet just south of the bridge to a distance of approximately 350 feet (at Cabbage Creek, the southerly section of the ICWW). Therefore the contributing watershed north of Cabbage Creek primarily flows northward towards the bridge into the ICWW's expanded conveyance width.

The waterway also conveys the runoff from more than 50 square miles of the Southside of Jacksonville, the area which has experienced the fastest growth over the last 10 years. Although the ICWW discharges into the Atlantic Ocean 6 miles north of the bridge location, the tidal effect decreases significantly upstream due to the large conveyance capacity of the ICWW and the St. Johns River, which discharges at the same ocean outfall.

The existing bridge is a two-leaf bascule facility providing 90 feet of horizontal clearance at the main channel. The bridge was originally constructed in 1947, and the waterway cross section depths under the bridge have significantly deepened over the years. Both the bascule sections of the bridge and the main spans are supported by concrete pilings driven into the sand and limestone strata under the waterway. The study involved determination of the ultimate scour depths for the 100-year and 500-year storms, evaluation of member stability at ultimate scour depths, and recommendations to alleviate scour occurring at the existing support substructures.

According to available bridge inspection reports, profile channel bottom depths under the bridge have consistently increased with each successive bridge inspection report throughout the life of the structure. The two previous FDOT bridge inspection reports examined the structure and verified that scouring areas do exist, primarily around the sheet piling encircling the legs of the two bascule piers, and that approximately 25 feet of material has scoured from around the piers in the main channel during the life of the structure. Field inspections of the waterway surface estimated a velocity of 2 knots maximum at the bridge.

Scour is a result of some or all of four factors outlined in the Federal Highway Association publication HEC-18. Scour evaluations for this structure were performed utilizing procedures established in the aforementioned HEC-18 publication as well as FDOT's Guidelines For Scour Evaluation Studies (which has been subsequently revised). Data was acquired on the existing bridge facility in accordance with the procedures of HEC-18. The bridge is approximately 350 feet in length. The area around the bridge is predominantly wetlands with upland fill areas located at the touchdown points of the bridge. Upstream and downstream of the bridge, the width of the waterway is over 5,000 feet (with the inclusion of the shallow, wetland areas. The construction methodology employed at the time the bridge was constructed was to fill the shallow wetland areas on either side of the waterway and span 350 feet of the waterway and fill the roadway sections at the bridge's termini. Therefore, the bridge provides a significant constriction to the waterway.

Analysis

Once the available data was acquired, the scour evaluation proceeded. However, one of the first parameters that had to be determined was an estimate of stream flow velocities for the normal, 100-year and 500-year event. This proved to be a major roadblock due to the lack of equations and procedures for determining tidal velocities in waterways. In addition, this bridge had both tidal and riverine components, and a determination was required for which condition produced the maximum scour depths. HEC-18 indicates that tidal velocity determination is difficult to calculate with the currently available modelling aids, namely WSPRO and HEC-2. Therefore it was logical that the waterway be modeled as a combination of tidal and hydrologic effects, then the hydrologic velocity data would be compared with the tidal velocity estimate to determine a worst case scenario. Maximum tidal velocities at the ocean outfall (i.e. Mayport, Florida) are listed for ebb and flood events in the USGS Tidal Current Tables-

1992, which indicated that the maximum velocity occurred during the ebb flow and resulted in a velocity of 2.9 knots (4.9 feet/second). This value is less at the bridge due to energy losses occurring on the six miles between the bridge and the outfall.

The first attempt at scour calculations resulted in unacceptably high velocity determinations and associated scour depths. Since the bridge was constructed over 40 years ago, the structure had experienced stream flows due to several tropical storms and associated hurricanes. Scour depths greater than 3 times the present scour depth would have occurred, according to the calculations. This was simply not the case. Calculations were revised, additional hydraulic analyses (using HEC-2) were performed, historical profile depth records were reviewed, and new scour depths were determined. Although the contraction coefficient and the associated contraction scour calculations were sizable (especially with no overbank flow available), the highest scour calculations resulted from the pier scour equation. The original HEC-18 procedures indicated that the Chang equation was suitable, but this equation is based upon anticipated flow velocities in excess of those experienced at the bridge, and higher than expected scour depths were calculated.

It is interesting to note that, as this analysis was proceeding, so was the Federal and State criteria for scour analysis. At three to six month intervals during the project study, significant revisions occurred to the evaluation procedure manuals. This resulted in complete revisions to the document at times when the document was reviewed by the FDOT. HEC-18 revised its procedures during this period, and FDOT temporarily abandoned its scour evaluation manual.

Revisions of the calculations involved an analysis of the Flood Insurance Rate Study for the Jacksonville area produced by the Federal Emergency Management Agency (FEMA), which contained flow rates for storm events in the major creek outfall for the area which discharged immediately south (upstream) of the bridge. It was first thought that the tidal surge from a hurricane event would produce the highest stream velocities under the bridge. But, as indicated previously, there is significant dampening of the available energy in the waterway. As a hurricane surge progresses into the ICWW in this area, the large areas of tidal, estuarine floodplain provide temporary attenuation. The surge in the ICWW is further inhibited by discharges from the tributary streams discharging into the ICWW in opposition to the surge flow direction. Therefore, it was determined through comparative runs with the HEC-2 model that the largest anticipated velocities in the waterway will be due to a severe rainfall event with a normal tidal cycle at the Ocean, and these velocities were calculated at 5 to 6 feet per second.

Through this trial-and-error type of analysis, elements such as bridge location and upstream/downstream conditions were shown to be critically important factors in scour prevention. Once reasonable velocities were calculated, the output was used for scour equations outlined in HEC-18 and the scour depths for particular scour factors was determined. The Colorado State pier scour equation was used which takes the velocity into consideration and therefore produces more reasonable scour depths. These depths were subsequently analyzed for adequate pile imbed and the unsupported pile length during the scour event. It was discovered that the scour depths did not completely undermine the supporting pile structures, but the unsupported length of these pile structures resulted in excessive flexure of the members and posed a catastrophic failure condition for the bridge superstructure.

Scour Alleviation Recommendations

As a final measure of the study, three recommendations for scour alleviation were proposed:

1) It was recommended that the values of the scour depths calculated were excessive and perhaps overestimated due to the "state of the art" procedures governing at that time. The potential failure condition in the superstructure should be alleviated by installation of lateral supports, and the FDOT should adopt a "wait and see" attitude whereby they would monitor bottom elevations immediately following large storm events to determine the rate of scour. This recommendation is also due to the recent bridge inspection reports which indicated that the rate of scour at the bridge may be decreasing and the waterway may be approaching an equilibrium condition in scour depth. Additional riprap would also be placed around the pilings in an attempt to reduce scour potential. The FDOT was warned, however, that this may lead to bridge structure problems in the future.

2) It was recommended that the scour potential be eliminated by installation of gabbion type "Reno Mattresses" on the stream bottom. This procedure had been recently attempted under the Acosta Bridge in the St. Johns River in downtown Jacksonville, and subsequent analysis showed that the installation of these mats did eliminate scour depth increase.

3) Finally, a recommendation was made for correction of the problem which was the actual culprit and cause of the excessive scour, namely the velocities resulting from extreme waterway constriction. It was recommended that box culverts be constructed as "relief bridges" on each side of the existing structure in order to increase the available flow area. The report noted that another bridge constructed in the 1980's five miles upstream of the McCormick Bridge spans more that 5,000 feet from bank to bank, indicating that the McCormick Bridge is substantially undersized.

The FDOT approved implementation of recommendation #2 and an investigation into the feasibility of recommendation #1. The construction and subsequent study is currently ongoing.

During the scour analysis of the McCormick Bridge, it was discovered that the prediction of scour depth for bridge structures, especially tidal bridges, requires not only a thorough knowledge of the scour procedures outlined in <u>HEC-18</u>, it also requires an engineer with insight and creativity, and the ability to interpret historical trends and recommend scour prevention measures based upon these historic trends. While it is true that the primary concern is for public safety and that this should drive the analysis process, there are many areas where the engineer can overestimate conditions and the resultant analysis exaggerate the level of reparations necessary for the bridge to withstand scour from significant storm events. In the case of Florida this is especially critical due to the number of bridge structures currently existing and the lack of unlimited funds for analysis and repair. The engineer performing the scour analysis must be aware that moneys expended for over-designed scour protection measures at one structure eliminate implementation of scour protection measures at another.

Appendix - References

Florida Department of Transportation. "Draft Guidelines For Evaluating and Designing For Scour At Bridges", April 30, 1991

U.S. Department of Commerce, National Oceanic and Atmospheric Administration, National Ocean Service. Tidal Current Tables 1992 - Atlantic Coast of North America.

U.S. Department of Transportation, Federal Highway Administration, Office of Research and Development. Publication No. FHWA-IP-90-017, September 30, 1988 (and revised edition February, 1991). Hydraulic Engineering Circular No. 18, "Evaluating Scour at Bridges".

A TIDAL INLET BRIDGE SCOUR ASSESSMENT MODEL

By Mark S. Vincent[1], Student Member, ASCE
and Mark A. Ross[2], Member, ASCE

Abstract
    Undermining of bridge structures constructed in tidal inlets is a common problem. Scour mechanisms in tidal inlets are often complex combinations of local contraction and regional tidal inlet dynamics. A numerical model for quantitatively evaluating and assessing the scour and deposition magnitudes associated with contraction and inlet geomorphological changes has been developed. The model uses a two-dimensional, dynamic numerical hydraulic model coupled to a movable bed sediment scour and deposition model developed specifically for this need. The model provides for subgrid features such as bridge piers and limited inhomogeneity in sediment grain size. A simplistic representation for armoring associated with sediment sorting or manmade structures is also provided. Initial applications to Clearwater Pass and Johns Pass in West Central Florida indicate reasonable comparisons between model predictions and observations of scour/deposition in the inlets, warranting additional testing of this approach.

Introduction
    The Federal Highway Authority estimates that at least 20% of the nation's 483,000 bridges are susceptible to scour. In coastal states like Florida, many of these bridges are located in tidally controlled inlets, estuaries and bays. The complex hydrodynamics of these systems, involving reversing tidal flows, wave contributions, littoral drift and wind setup, render the standard riverine approaches to bridge scour prediction inappropriate and inadequate. An ongoing program is

---

[1] Graduate Research Assistant, Center for Modeling Hydrologic and Aquatic Systems, Department of Civil Engineering and Mechanics, University of South Florida. 4202 Fowler Ave., Tampa, Florida, 33620.

[2] Associate Director, Center for Modeling Hydrologic and Aquatic Systems, Department of Civil Engineering and Mechanics, University of South Florida. 4202 Fowler Ave., Tampa, Florida, 33620.

being conducted at the University of South Florida, Center for Modeling Hydrologic and Aquatic Systems (USF/CMHAS), to develop, refine, and test a numerical model (Scour Model) which can be readily used by bridge hydraulic engineers to evaluate tidal bridge facilities.

Hydrodynamic Model
The foundation of the Scour model consists of a previously developed hydrodynamic model, which has been more recently coupled to wave and sediment dynamic subroutines. The hydrodynamic model is a well documented two dimensional, finite difference, vertically integrated, explicit approach which has been used at USF/CMHAS for many circulation, water quality and inlet stability studies (Ross et al. 1984).

An important feature of the Scour model is the numerical representation of contraction or expansion from features of subgrid dimensions, such as bridge piers, bascules or channels. The model scales the flow rates up or down to compensate for features which influence the flow between adjacent grids. This provision allows for the incorporation of contraction scour processes, which are extremely important in the vicinity of tidal bridges.

Waves
In many exposed tidal systems, the hydrodynamics of short period waves significantly increase the effective bed shear stress and thus the rate of sediment transport. The contribution of waves to the sediment dynamics of tidal systems is incorporated through subroutines employing Airy or Linear wave theory. Discussions of the equations and logic by which the Scour model solves for wave heights, lengths, orbital lengths, and velocities is provided by Vincent et al., (1992).

Friction Factors and Shear Stress
In addition to fluid velocities, sediment transport equations are also a function of the shear stress at the bed, which in turn is a function of the combined friction factor at the bed. The Scour model determines the friction due to currents and the friction due to waves independently, following the approaches of Manning-Strickler (Henderson, 1966) and Swart (1976). From these two terms the combined friction factor can then be found using the method of Jonsson (1966). These equations are presented in equations 1, 2, 3.

The friction factor due to currents:

$$f_c = 0.122\left(\frac{k_s}{d}\right)^{\frac{1}{3}} \qquad (1)$$

The friction factor due to waves:

$$f_w = .00251 \exp\left(5.21\left(\frac{\xi}{k_s}\right)^{-0.19}\right) \qquad (2)$$

The combined friction factor:

$$f_{cw} = \frac{u_c f_c + u_b f_w}{u_c + u_b} \qquad (3)$$

Once the combined friction factor due to waves and currents has been determined, the combined shear stress at the bed is calculated by the quadratic stress law as:

$$\tau_{cw} = \frac{\rho f_{cw} u_{cw}^2}{2} \qquad (4)$$

Sediment Transport

A comprehensive screening and literature review of applicable sediment transport equations was conducted to ensure the selection of a competent and tested method for the Scour model. Consistently ranked as among the most accurate in investigations was the Engelund and Hansen equation (Sleath, 1984). This approach was selected due to its consistent performance in laboratory and field tests, as well as success in tidal inlet modeling endeavors by Ross (1990) and Zarillo and Park (1987). The total sediment load can be expressed by the Engelund and Hansen equation as:

$$q_t = 0.05 \, \overline{u_c}^2 \left(\frac{D_{50}}{g(S_s-1)}\right)^{\frac{1}{2}} \left(\frac{\tau_{cw}}{(\rho_s-\rho_w)gD_{50}}\right)^{\frac{3}{2}} \qquad (5)$$

Sediment Flux Tests

The Scour model employs a series of tests to determine the condition of transport, scour or deposition within each grid. For example, a modified Hjulstrom curve (Herbich et al. 1984) is used by the model to provide the relationship between threshold velocities needed for erosion, transport and deposition, as a function of sediment grain size.

A second test for potential scour is conducted by the equilibrium depth promoted by Bruun (1978). This test, which is also used in the Corps of Engineers HEC 6 model, can be expressed as:

$$d_{EQ} = \left(\frac{q}{10.21 D_{50}^{\frac{1}{3}}} \right)^{\frac{6}{7}} \tag{6}$$

The model also allows for the natural armoring of the inlet bed due to non-homogeneous sediments following the method of Borah (1989) as:

$$T_{AL} = \frac{D_{AR}}{(1-\varepsilon) F_{AR}} \tag{7}$$

At specified time increments a continuity equation relates the volumetric rate of transport to the change in bed depth over time following:

$$\frac{\partial q_{s_x}}{\partial x} + \frac{\partial q_{s_y}}{\partial y} = -(1-\varepsilon) \frac{\partial z}{\partial t} \tag{8}$$

Boundary Conditions
Boundary conditions and input data used by the Scour model include initial bathymetry, tidal driving function at the open water boundaries, wind speed and direction, offshore wave height and period, sediment grain size and specification of subgrid features (e.g., pier dimensions). Model output includes arrays of all pertinent data (i.e. bathymetry or scour depths, velocities etc.) at user specified time increments.

Application
To date the Scour Model has been employed in several tidal inlet engineering investigations along littoral drift shorelines of the West Coast of Florida. Ross (1990) employed the prototype version of the model for a bridge and channel design study for Clearwater Pass. The Johns Pass Bridge maintenance/replacement study was performed using a modified, moveable bed version of the model (Vincent et al., 1992). The Johns Pass investigation provided annual scour/deposition rates for seven structural alternatives. This investigation indicated that contraction scour from the bridge was a primary factor in the historical erosion of up to 6.3 meters of sediment, which had undermined several bridge pier foundations and piles. In addition, the model simulations indicated that past remediation measures such as the installation of additional flow constricting crutchbents, further aggravated scour rates.

Summary and Recommendations
The potential utility of the Scour model as a bridge assessment tool has been explored through the application to Clearwater Pass and Johns Pass. The model reasonably describes the complex hydrodynamic and sediment dynamic processes which are necessary to evaluate bridge scour in tidal environments.

Despite these initial promising results, considerable refinements are needed to enhance the competency of this modeling technology. Unfortunately, many of the model equations were developed by researchers' experiments using small flumes with idealistic conditions of uniform sediment grains, plain beds, and uniform steady flow. Moreover, these equations do not consider the compounding effects of sediment in motion.

An urgent need is seen here for model improvements to incorporate the transport of graded sediments, as well as spatial and temporal variability in sediment mixtures, bed forms and bed armoring. The model currently handles contraction scour and aggradation/degradation, but the incorporation of numerical representation of local scour due to flow separation is needed.

A present limitation affecting the advancement of this technology, is the surprising deficit of accurate hydraulic and bathymetric data of the inlets investigated. The present data is insufficient to fully calibrate and validate the Scour numerical model. Additional data collection, simulations and analysis will needed in this regard. Therefore, only qualitative and comparison results can be obtained at this time.

Although considerable theoretical and modeling improvements are warranted, the USF/CMHAS Scour model shows merits as a tool to be used in management and design studies, as well as a vehicle for understanding basic processes and responses of tidal inlets. Upon the completion of these exercises, one goal of this research will be to transfer this technology to the engineering community via a user friendly interface driven model. The Fortran 77 based program is capable of running on high speed mini computers as well as desk top PC's ubiquitous in most engineering consulting and research facilities.

Acknowledgements
The Authors gratefully acknowledge the Florida Department of Transportation for funding this research (contract no. C-3534) and Dr. Bernard E. Ross for his significant contributions.

Appendix I. References

Borah, D. K., (1989). Scour Depth Prediction Under Armoring Conditions. Journal of the Hydraulic Division, ASCE, 115(10), 1421-1425.
Bruun, P., (1978). Stability of Tidal Inlets, Theory and Engineering. Amsterdam: Elsevier Scientific Publishing Company.
Henderson, F., (1966). Open Channel Flow. New York: The Macmillan Company,
Herbich, J. B., R. E. Schiller, R. K. Watanabe, and W. A. Dunlap. (1984). Seafloor Scour: Design Guidelines for Ocean-Founded Structures. New York: Marcel Dekker, Inc.
Jonsson, I. G., (1966). Wave Boundary Layers and Friction Factors. In Proceedings

of the 10th Conference on Coastal Engineering, Tokyo, Japan. New York: American Society of Civil Engineers, 127-148.

Ross, B. E., (1990). Clearwater Pass Study. Tampa: Hydrosystems Associates.

Ross, B. E., M. Ross and P. Jerkins. (1984). Documentation of the Two-Dimensional Hydraulic Model, Wasteload Allocation Study, Vols. 1 and 2. Tampa: Department of Civil Engineering and Mechanics, University of South Florida.

Sleath, J. F., (1984). Sea Bed Mechanics. New York: J. Wiley & Sons.

Swart, D. H., (1976). Predictive Equations Regarding Coastal Transports. In Proceedings of Fifteenth Coastal Engineering Conference held in Honolulu, Hawaii 11-17, 1976, 1113-1132. New York: ASCE.

Vincent, M., (1992). A Numerical Scour Deposition Model for Tidal Inlets. Masters Thesis, University of South Florida, Tampa, Florida.

Vincent, M., M. Ross, and B. Ross. (1992). Johns Pass Bridge Scour Assessment Model. Tampa: The Center for Modeling Hydrologic and Aquatic Systems, University of South Florida, Tampa, Florida.

Zarillo, G. A. and M. J. Park., (1987). Sediment Transport Prediction in a Tidal Inlet Using a Numerical Model: Application to Stony Brook Harbor, Long Island, New York, USA. Journal of Coastal Research, 3(4), 429-444.

Appendix II. Notation

d	Depth of water
d_{eq}	Equilibrium depth
D_{AR}	Grain size of bed armoring material
D_{50}	The sediment grain diameter such that 50 percent of the bed material is finer
ε	Porosity of the bed sediment
F_{AR}	Fraction of the armoring size component of the sediment
f_c	Current friction factor
f_{cw}	Combined friction factor due to currents and waves
f_w	Wave friction factor
g	Acceleration due to gravity
k_s	Nikuradse roughness number
ρ_s, ρ_w	Densities of sediment and water
q_{sx}, q_{sy}	Sediment transport flux in the x and y directions
q	Water transport per unit width
S_s	The density of the sediment relative to water
τ_{cw}	The shear stress due to currents and waves
t	Time variable
T_{AL}	Thickness of the erodible non-armoring sediment (active layer)
u_b	Orbital velocity at the bed
$\bar{u}_c$	The depth integrated velocity
u_{cw}	Resultant current from currents and waves
ξ	Wave excursion number
z	Vertical coordinate direction

A Mathematical Model of Flow in
Mildly Sinuous, Deep Channels

by

Amartya Kumar Bhattacharya,[1] A.M.ASCE

and

Srijib K. Kar [2]

ABSTRACT

A mathematical study of flow in deep, weakly - meandering channels has been carried out and formulations for the tangential and radial velocities, valid in the entire flow domain, and satisfying all boundary conditions have been developed.

INTRODUCTION

There are a number of mathematical models of shallow meanders in the literature (Odgaard 1986a, 1986b, 1989a, 1989b; Ikeda and Nishimura 1986; Smith and McLean 1984) but to the knowledge of the writers, there exists no model capable of describing the flow in deep meandering channels. Unlike in shallow meanders, in deep meanders there does not exist a central zone of flow where the effect of the side-walls is insignificant. A viable model for deep meanders has to be valid in the entire flow domain and has to satisfy all boundary conditions. That is what the model described in this paper attempts to do.

MATHEMATICAL MODEL

Herein, a body-fitted curvilinear co-ordinate system is used for the meandering stream. The s-axis is along the channel centreline; positive in the streamwise direction. The n-axis is normal to the s-axis so that the n-co-ordinate is the cross-stream co-ordinate. The z-axis is normal to both the s-axis and the n-axis

---

[1] Research Scholar, Department of Civil Engineering, Indian Institute of Technology, Kharagpur-721302, India.

[2] Associate Professor, Department of Civil Engineering, Indian Institute of Technology, Kharagpur-721302, India.

and is directed upwards from the channel bed. The channel bed is assumed to have no lateral slope. The depth is assumed to be comparable to the width. The channel centreline is given by a sine-generated curve and the width of the channel is constant.

The tangential velocity, u, is assumed to be given by the equation
$$u = u_0 + u_1$$
where u_0 = zeroth-order tangential velocity, the zeroth-order state being a straight channel whose width is the same as that of the sinuous channel; and

u_1 = first-order tangential velocity.

u_0 and u_1 are described by the equations

$$u_0 = \bar{u}_0 \left(\frac{n_1+1}{n_1}\right)\left(\frac{z}{d}\right)^{\frac{1}{n_1}} \ldots \quad (2)$$

and
$$u_1 = \bar{u}_0 \left[c\left(\frac{n}{r_{cm}}\right)^3 + \left(\frac{n}{r_{cm}}\right)\right]\left[a_1 \sin\frac{2\pi}{L}s + a_2 \cos\frac{2\pi}{L}s\right]\left(\frac{n_2+1}{n_2}\right)\left(\frac{z}{d}\right)^{\frac{1}{n_2}} \quad (3)$$

where $\bar{u}_0$ = depth-averaged zeroth-order tangential velocity;

r_{cm} is the centreline radious of curvature at the apex of the sine-generated curve;

L = wavelength measured along the centreline;

d = depth;

s and n are the streamwise and cross-stream coordinates respectively;

z is the co-ordinate normal to the channel bed; and

n_1, n_2, c, a_1 and a_2 are co-efficients.

while the harmonic function, $\left[a_1 \sin\frac{2\pi}{L}s + a_2 \cos\frac{2\pi}{L}s\right]$, is of the same form as the harmonic function used by Ikeda and Nishimura (1986) in their eq.(6), the co-efficients used herein are different from the co-efficients used by Ikeda and Nishimura because a different flow situation is treated here. It is to be noted that $\bar{u}_0$ has a variation only in the cross-stream direction. Also, because $\bar{u}_0$ has a value of zero at the sidewalls, u vanishes at these boundaries. Different power law co-efficients were used in the zeroth-order tangential velocity formulation and the first-order tangential velocity formulation in an attempt to simulate experimentally observed irregularities in the vertical distribution of the tangential velocity. Such irregularities have been reported in Kar (1977).

The radial velocity has no zeroth-order component.
The first-order radial velocity is divided into a depth
-averaged component, $\bar{V}$, and a residual component, V_R.
To find the depth-averaged radial velocity, the depth
averaged tangential velocity, $\bar{u}$, which is given by the
equation

$$\bar{u} = \bar{u}_o + \bar{u}_o \left[c \left(\frac{n}{r_{cm}} \right)^3 + \left(\frac{n}{r_{cm}} \right) \right] \left[a_1 \sin \frac{2\pi}{L} s + a_2 \cos \frac{2\pi}{L} s \right] \cdot (4)$$

is used with the depth-averaged continuity equation,

$$\frac{\delta \bar{u}}{\delta s} + \frac{1}{r_c} \frac{\delta}{\delta n} (\bar{v} r) = 0 \quad \cdots \quad \cdots (5)$$

where r = radius of curvature; and

r_c = radius of curvature at the sidewalls.

Two formulations are developed for the depth-averaged
radial velocity. In the first formulation, the mean value
of the depth-averaged zeroth-order tangential velocity
(which is the same as the reach-averaged velocity) is
used and the following equation for $\bar{V}$ is obtained:

$$\bar{V} = \frac{r_c}{r} U \frac{2\pi}{L} \left[\frac{c}{4 r_{cm}^3} \left\{ \left(\frac{b}{2} \right)^4 - n^4 \right\} + \frac{1}{2 r_{cm}} \left\{ \left(\frac{b}{2} \right)^2 - n^2 \right\} \right] \left[a_1 \cos \frac{2\pi}{L} s - a_2 \sin \frac{2\pi}{L} s \right] \cdots (6)$$

where b = width of the channel;

U = reach-averaged tangential velocity; and
the other quantities are as have been described previously. In the second formulation, an equation describing
the lateral variation of the depth-averaged zeroth-order
tangential velocity is used. The equation,

$$\bar{u}_o = \left(\frac{\hat{m}+1}{\hat{m}} \right) U \left[1 - \left(\frac{2n}{b} \right)^{\hat{m}} \right] \quad \cdots \quad (7)$$

where $\hat{m}$ is a coefficient and is even, is introduced in
eq.(4) before the depth-averaged continuity equation is
applied. The equation

$$\bar{V} = \frac{r_c}{r} \left(\frac{\hat{m}+1}{\hat{m}} \right) U \frac{2\pi}{L} \left[\frac{c}{4 r_{cm}^3} \left\{ \left(\frac{b}{2} \right)^4 - n^4 \right\} + \frac{1}{2 r_{cm}} \left\{ \left(\frac{b}{2} \right)^2 - n^2 \right\} - \left(\frac{2}{b} \right)^{\hat{m}} \frac{c}{r_{cm}^3 (\hat{m}+4)} \left\{ \left(\frac{b}{2} \right)^{\hat{m}+4} - n^{\hat{m}+4} \right\} - \left(\frac{2}{b} \right)^{\hat{m}} \frac{1}{r_{cm}(\hat{m}+2)} \left\{ \left(\frac{b}{2} \right)^{\hat{m}+2} - n^{\hat{m}+2} \right\} \right] \left[a_1 \cos \frac{2\pi}{L} s - a_2 \sin \frac{2\pi}{L} s \right] \cdots (8)$$

is obtained. It is noted that in either case, the boundary conditions at the two sidewalls give the same constant of integration.

The residual component of the radial velocity, V_R, is formulated as follows:

$$V_R = \frac{\gamma_c}{\gamma} \frac{\bar{u}^2}{U} \frac{d}{\chi \gamma_{cm}} \left[C_0 + C_1 \left(\frac{z}{d}\right) + C_2 \left(\frac{z}{d}\right)^2 + C_3 \left(\frac{z}{d}\right)^3 \right] \left[\cos \frac{2\pi}{L} s + M \sin \frac{2\pi}{L} s \right] \quad \cdots (9)$$

where χ = von Kármán's co-efficient; and C_0, C_1, C_2, C_3 and M are co-efficients. A cubic parabola has been chosen to describe vertical variation of the radial velocity as a recent study (Bhattacharya and Kar 1992) has revealed that this form of description is appropriate. The co-efficients (C_0, C_1, C_2 and C_3) are subject to constraints (one of which may be imposed by taking boundary condition at the bed; another by equating depth-average of the residual component of the radial velocity to zero) which enable the number of unknowns to be reduced. M is a measure of the phase lag of V_R with respect to the channel planform and clearly $M > 0$.

The formulation for V_R bears some resemblence to the formulation of Ikeda and Nishimura (1986). However, the harmonic function used to describe the streamwise variation is cast rather differently. Also, the magnitude modification factor used by Ikeda and Nishimura (1986) is not needed in the present model. Unlike the formulation of Ikeda and Nishimura (1986), where an artificial function was needed to make the residual component of the radial velocity vanish at the sidewalls, in the present formulation no artificial function is necessary.

CONCLUSION

Eq.(1), (2) and (3) may be used to find the tangential velocity distribution. Eq.(6) or eq.(8) may be used to find the distribution of the depth-averaged radial velocity and eq.(9) may be used to find the residual component of the radial velocity. Though eq.(8) is more complex than eq.(6), both the equations are designed for the full width of the flow. n_1 and $\hat{m}$ can be found from straight-channel-considerations. Further studies are being currently conducted on the other coefficients ($\mathcal{L}$, a_1, a_2, n_2, M and two out of C_0, C_1, C_2 and C_3) but it appears that at least some of them must be specified in terms of ranges of values after extensive experimentation.

APPENDIX I - REFERENCES

Bhattacharya, A.K. and Kar, S.K. (1992). "On the vertical distribution of the velocity and pressure in deep, sinuous channels". Proc., 19th National Conference on Fluid Mechanics and Fluid Power, Bombay, India, pp.D 2-1 —D2-10.

Ikeda, S. and Nishimura, T. (1986). "Flow and bed profile in meandering sand-silt rivers". J.Hydr.Engrg., ASCE, 112(7), 562-579.

Kar, S.K.(1977). "A study of distribution of boundary shear in meander channel with and without flood plain and river flood plain interaction", thesis presented to the Indian Institute of Technology, Kharagpur at Kharagpur, India, in portial fulfillment of the requirements for the degree of Doctor of Philosophy.

Odgaard, A.J. (1986a). "Meander flow model. I : Development". J.Hydr.Engrg., ASCE, 112(12), 1117-1136.

Odgaard, A.J. (1986b), "Meander flow model. II : Applications". J.Hgdr.Engrg. , ASCE, 112(12), 1137-1150,

Odgaard, A.J. (1989a). "River-meander model. I : Development". J. Hydr.Engrg. ASCE, 115(11), 1433-1450.

Odgaard, A.J. (1989b). "River-meander model. II : Applications". J.Hydr. Engrg., ASCE, 115(11), 1451-1464.

Smith, J.D. and McLean, S.R. (1984). "A model for flow in meandering streams". Water Resour. Res. , 20(9), 1301-1315.

APPENDIX II - NOTATION

a_1, a_2 = co-efficients defined through equation (3);
b = channel width;
C_0, C_1, C_2, C_3 = co-efficients defined through equation (9);
d = depth;
L = meander wavelength measured along the centreline;
M = co-efficient defined through equation (9);
$\hat{m}$ = co-efficient defined through equation (7);
n = cross-stream co-ordinate;
n_1, n_2 = co-efficients defined through equations (2) and (3) respectively;
r = radius of curvature;
r_c = radius of curvature at the centreline;
r_{cm} = centreline radius of curvature at the apex of the sine-generated curve;
s = streamwise co-ordinate;
U = reach-averaged tangential velocity;
u = tangential velocity;
u_0 = zeroth-order tangential velocity;
u_1 = first-order tangential velocity;
$\bar{u}$ = depth-averaged tangential velocity;
$\bar{u}_0$ = depth-averaged zeroth-order tangential velocity;

$\bar{v}$ = depth-averaged radial velocity;
v_R = residual component of radial velocity, i.e. point radial velocity minus the depth-averaged radial velocity;
z = co-ordinate normal to the channel bed;
$\measuredangle$ = co-efficient defined through equation (3);
χ = von Kármán's co-efficient.

TWO-DIMENSIONAL HYDRODYNAMIC MODELING
WITH A COMPUTER GRAPHICS SYSTEM

By Daniel H. Hoggan, M.ASCE, and Donald E. Twiss[1]

ABSTRACT

For more than a decade various divisions and district offices of the U.S. Army Corps of Engineers have been active in the development of graphic information system databases and computer-aided design applications. In the Sacramento District, Hydraulic Design Section (HDS), one of the focuses in this activity has been on the use of digital terrain models to solve hydraulic design problems. One recent application of this new technology by the HDS was in the design of a flood bypass channel for an oxbow in the Napa River using a two-dimensional hydrodynamic river model operating on a computer graphics system.

INTRODUCTION

The partnership between two-dimensional hydrodynamic modeling and computer graphics provide the hydraulic modeler with a great advantage over older and less visible hydraulic design and analysis capabilities. Instead of presenting volumes of tabular data, a picture presenting the model and its results can be automatically generated. Graphical presentation of information can be understood at a glance rather than studying and mentally converting numerical data sets into three dimensional images. The hydraulic model and its results can be shared instantly over a computer network. Relationships between topography, project features, geology, real estate, cultural and all other sets of graphical information can be viewed simultaneously.

2-D MODELING SYSTEM

A schematic of the overall modeling system is shown in Figure 1. The major hydraulic component RMA-2V is linked to various computer graphics components as explained below.

Hydraulic Software - RMA-2V(Norton, King and Orlob, 1973), the two-dimensional finite element hydrodynamic component of the Corps' TABS (U.S. Army Corps of Engineers, 1990) multidimensional open channel flow software was used for this hydraulic analysis. RMA-2V solves the Reynolds form of the Navier-Stokes equations for turbulent flows. Friction is calculated with Manning's equation, and eddy viscosity coefficients are used to define turbulence characteristics. A velocity form of the basic equation is used with side boundaries treated as either slip (parallel flow)

---

[1]Professor, Civil and Environmental Engineering Department and Utah Water Research Laboratory, Utah State University, Logan, Utah 84322-8200 (IPA appointee with Corps) and Hydraulic Engineer, Hydraulic Design Section, U.S. Army Engineer District - Sacramento, 1325 J Street, Sacramento, California, 95814-2922, respectively.

HYDRODYNAMIC MODELING

or static (zero flow). The model automatically recognizes dry elements and corrects the mesh accordingly. Boundary conditions may be water-surface elevations, velocities, or discharges, and may occur inside the mesh as well as along the boundaries.

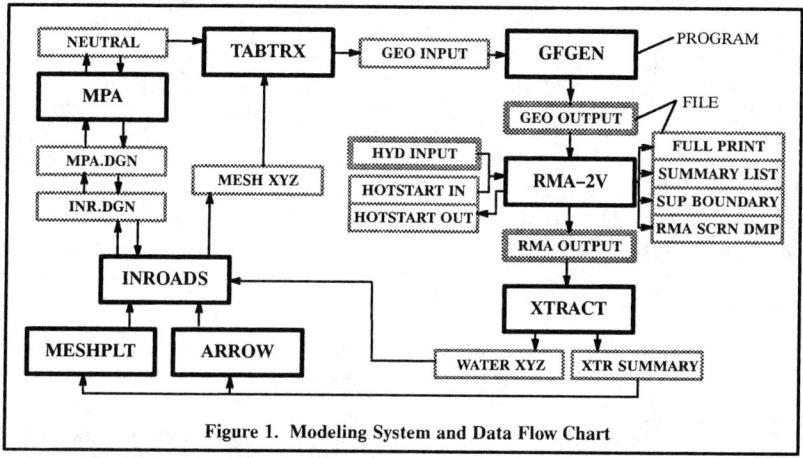

Figure 1. Modeling System and Data Flow Chart

Computer Graphics System – The computer hardware consisted of unix-based workstations with 48 megabytes of random access memory, a color plotter with 400 dots per inch resolution, and a black and white scanner with 800 dots per inch resolution from Intergraph, Inc., Huntsville, Alabama. Four Intergraph software packages were used in the analysis: MICASPLUS ANALYSIS (MPA), used to create the finite element mesh. INROADS used to display and manipulate the terrain data. IRAS32 used to place scanned images in the background such as aerial photographs. MICROSTATION 32 (MS32) the graphics interface for all the other software. The RMA-2V program along with the file manipulation programs GFGEN and XTRACT were ported from the Corps' Cray computer at Waterways Experimental Station in Vicsburg, Mississippi to run on the Intergraph workstation. TABTRX (Copeland and Roney, 1990) another file manipulation program developed by the HDS in Sacramento provides the filter between the structural finite element mesh and the hydraulic finite element mesh.

Finite Element Mesh Generation – The computer graphics system greatly facilitates the construction of the finite element mesh. The elements and nodes of the mesh are laid out with a mouse in real world coordinates directly with all the other graphical information in the computer model. A portion of the mesh constructed for the Napa River is shown below in Figure 2.

MPA is used for constructing the finite element mesh. MPA is a structural modeling system designed to build two- and three-dimensional geometric models for finite element structural analysis. MPA provides a three-dimensional space in which a finite element model is built. It includes an inventory of elements to build the model and a system of boundary and loading conditions to define model behavior. It also provides tools to modify the model and perform predesign operations.

An aerial photograph of the model area can be used as a backdrop for such features as contours, roads, buildings, hazardous and toxic waste sites, geologic features, or any other data previously mapped. Alignment generation utilities in INROADS, line and shape manipulation utilities in MS32, along with plane generation options in MPA are used to generate the finite element mesh defining the hydraulic (channel and banks) model. Tolerance limits on element properties such as aspect ratios (length to width of sides) and internal angles are checked automatically by the

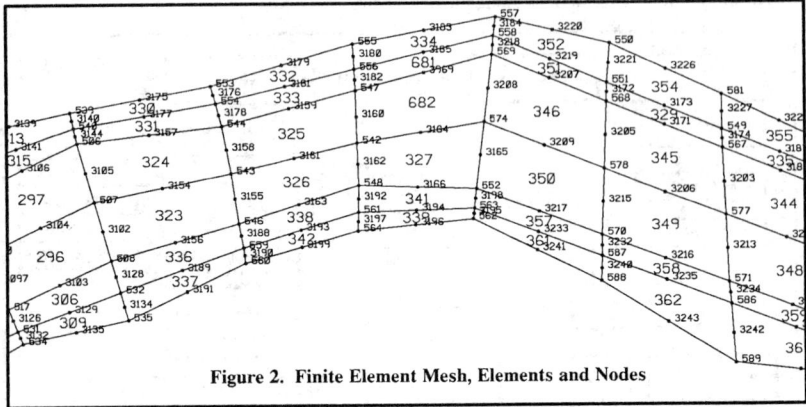

Figure 2. Finite Element Mesh, Elements and Nodes

program to assure that elements are constructed within suitable bounds. See the complete finite element mesh generated for the Napa River in Figure 3 below. MPA provides interfaces with other analysis packages through a neutral file system. Geometric data and the element connector table are extracted from the structural finite element mesh model and converted to an ASCII format by a neutral file translator in MPA.

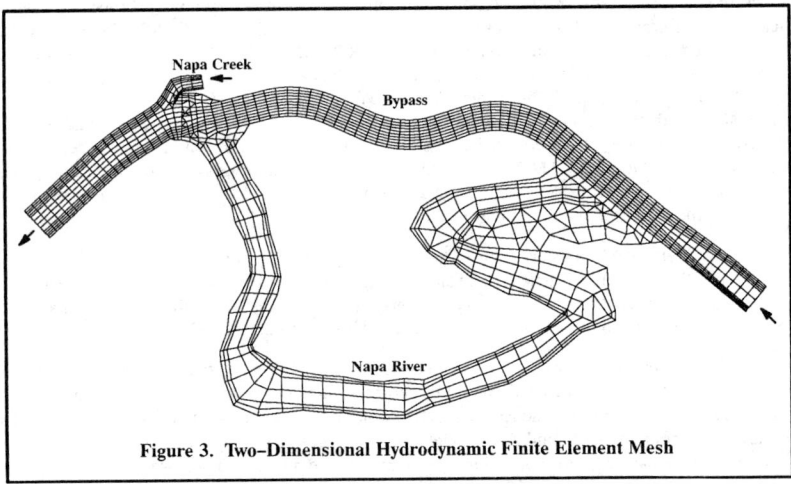

Figure 3. Two-Dimensional Hydrodynamic Finite Element Mesh

INROADS contains utilities to generate, manipulate and display digital terrain models. Digital terrain models may consist of survey data, digitized map data. or design surface data. Design surfaces can be developed by pushing templates along an alignment as in channel design or by numerous other methods such as placing a specific shape into the existing ground surface as in a detention basin or reservoir design. This "surface" like a topographical map, is triangulated forming a topological triangle network, or three-dimensional, triangulated model of the terrain. The points from the MPA finite element mesh are merged into the terrain model. These points then acquire the correct elevations from the terrain model and are saved as an XYZ point file. This XYZ point file contains the correct elevations for the nodes in finite element mesh.

TABTRX developed at the Corps of Engineers, Sacramento District reads the finite element mesh definition from the MPA neutral file and adds the correct ground elevations from the INROADS XYZ point file. TABTRX then generates the geometry input file for GFGEN.

Hydraulic Modeling – Portions of the TABS2 System, a generalized computer program system for open-channel flow and sedimentation is used for the hydraulic modeling.

GFGEN is the finite element geometry file generator for the TABS-2 system. It converts the geometry input file from TABTRX into a mathematical matrix format (in binary code) for direct input to RMA-2V. GFGEN also reorders the sequence in which the elemental equations are assembled for the simultaneous equation solver, this can greatly increase the computational efficiency of RMA-2V.

RMA-2V computes two dimensional water surface elevations and horizontal velocity components for subcritical, free-surface flows. For input a hydraulic file is required along with the geometry output file from GFGEN. The hydraulic file contains continuity check lines, dry/wet element elimination criteria, turbulent exchange coefficients, Manning's "n" values, initial conditions, and boundary conditions. This input can be changed from run to run, thus providing a means of analyzing the effects of using alternative model parameters.

Processing Output – Post-processing of output is performed with the Intergraph civil design software package INROADS. The hydraulic data sets are fully compatible with all other graphical information since it is in real world coordinates and fully defined in three dimensional space (See Figure 4.).

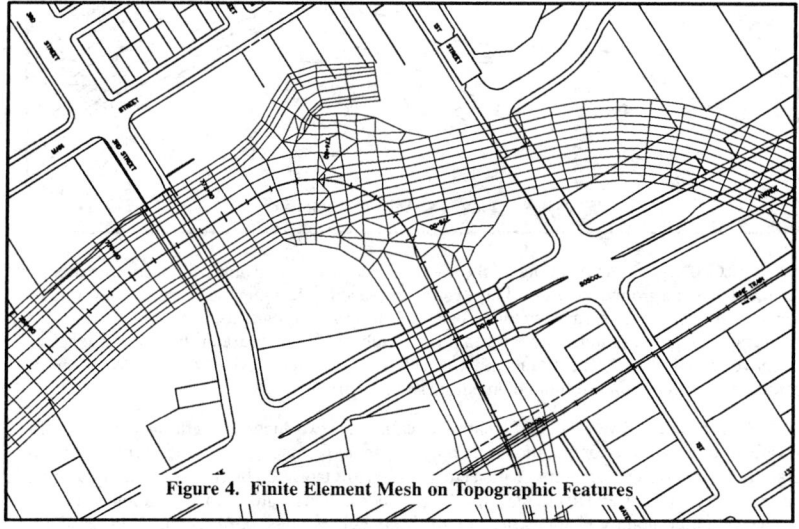

Figure 4. Finite Element Mesh on Topographic Features

XTRACT is a summary extract program that reads the complete solution file from RMA-2V and writes an ASCII summary table of the water surface elevations, velocity vectors, and other information. XTRACT was modified at the HDS to write an ASCII XYZ file of the water surface. This file can be read by INROADS to generate a triangulated network of the water surface.

MESHPLT is a user command developed by HDC to work within MS32 to plot the finite element mesh in the project drawing (see Figure 4). It reads the XTRACT file and draws the finite element mesh along with the element and node numbers and a boundary around the mesh to aid in the display of model ground and water surface contours. MESHPLT can also be used to help debug the mesh if necessary by running XTRACT prior to RMA-2V.

ARROW is another user command developed by HDC to work in MS32 it also reads the XTRACT file and plots the velocity vectors on the project drawing. See Figure 5 for a plot of the velocity vectors.

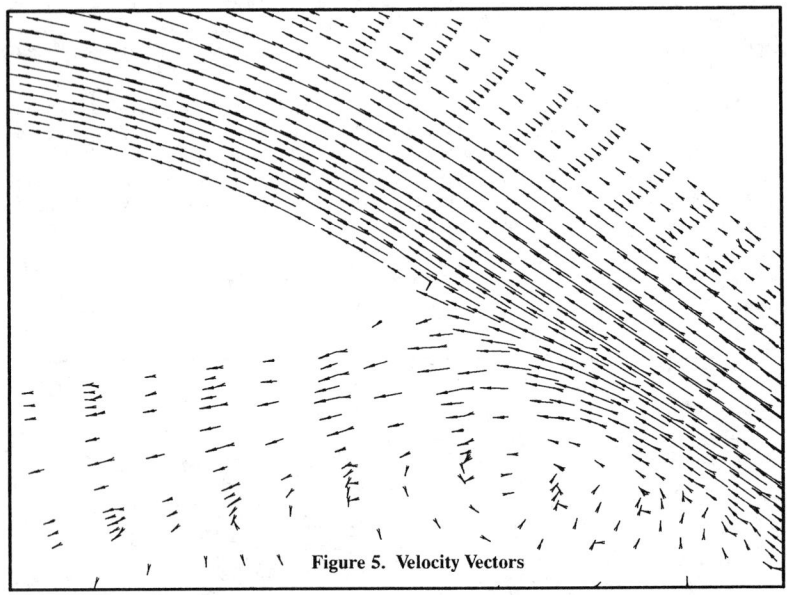

Figure 5. Velocity Vectors

INROADS generates drawings of the water surface profiles and flow cross sections, water surface contours and ground contours. These can be displayed with automatically generated cross sections, profiles, plan views, and they can be over laid on aerial photographs. The combination of the Intergraph computer graphics system and the hydraulic design software of the Corps provides an effective analytical tool. The water surface can also be used to determine volumes. Water surface contours are plotted on the finite element mesh in Figure 6.

Solution of the RMA-2V mathematical equations for two dimensional flow typically requires several iterations of the computations under a given set of parameters and several runs of the program using adjusted parameters. Satisfactory results are indicated by convergence of computed output such as velocity and depth of flow values at nodes. The results of each iteration is displayed on the computer monitor enabling the user to observe the progress of the computations. When a predetermined convergence criterion is met, such as .01 ft for depth, the computations are automatically terminated. In the event that the computations do not converge, it is possible to obtain an interim plot of output, such as velocity vectors, to determining where in the mesh the obstruction to convergence is occurring so that appropriate adjustment to the model can be made.

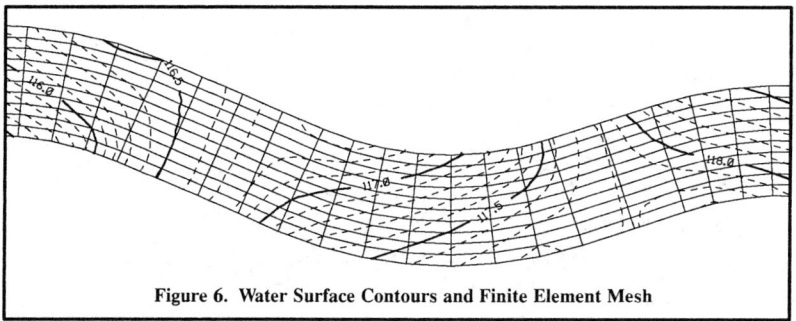

Figure 6. Water Surface Contours and Finite Element Mesh

SUMMARY

Integrated Approach – The great efficiency in utilization of computer graphics and hydraulic analysis cannot be overstressed. This is especially true in water resource project formulation and design. Basic survey information along with aerial photographs can be loaded into the graphical information system. The terrain can then be sculptured to depict the proposed project features. A finite element hydraulic model can then be produced. The hydraulic analysis/design along with that of other disciplines such as geology and soil mechanics can share the same information simultaneously. Alignments created and used to develop the hydraulic model also provide cross sections, profiles, and plan views to be utilized through construction and continuing into operation and maintenance. Volumes for cut and fill are automatically developed along with hydraulic model output. Three dimensional models of the geology and soil types can be displayed along with the hydraulic information required by other disciplines such as environmental or structural design. Each time the same resources and information are shared without duplication a savings of effort and an improvement in quality through project coherence is achieved. This paper presents the views of the authors and not necessarily those of the U.S. Army Corps of Engineers.

REFERENCES

Copeland, Lori and Roney, Michael, 1990, Integration of CADD with Hydraulic and Hydrodynamic Mathematical Models, unpublished paper presented at U.S. International Graphic Users Group, Annual Conference, New Orleans, LA., October.

Norton,W.R., King I.P., and Orlob G.T. 1973, A Finite Element Model for Lower Granite Reservoir, Prepared for U.S. Army Engineer District, Wala Wala Washington, Water Resources Engineers, Walnut Creek California.

US Army Corps of Engineers, 1990, Generalized Computer Program System For Open-Channel Flow and Sedimentation, "TABS System, Volumes 1-2", By Hydraulics Laboratory, Department of the Army, Waterways Experiment Station, February.

NUMERICAL MODELING OF UNSTEADY COMPOUND CHANNEL FLOW

By

R.S.M. Mizanur Rashid[1] and M. Hanif Chaudhry [2], Member ASCE

ABSTRACT

The unsteady open channel flow equations describing conservation of mass and momentum (St. Venant equations) are integrated numerically to determine the depth of flow and the rate of discharge along the length of a compound channel at different times. Preissmann four-point implicit finite difference scheme is used. The suitability of two different approximations for the channel cross section are investigated: (1) The flow velocity over the flood plains is negligible as compared to the velocity in the main channel, i.e., the flood plain acts as storage only and does not contribute to the momentum transfer; area in the continuity equation represents the entire cross-sectional area whereas in the momentum equation it is only the main channel cross-sectional area; the flood plains and the main channel are separated by a vertical line at their interface and the division line is not included in the wetted perimeter. (2) The entire channel section contributes to momentum flux, i.e., without neglecting the flow velocity over the flood plains; the flood plain and the main channel are considered as a single flow section and the momentum coefficient is used to take care of the non uniform velocity distribution. The computed results are compared with the experimental data obtained by the authors. Approximation (1) compares better with the experimental results than approximation (2).

INTRODUCTION

Most of the natural channels have compound sections, consisting of a deep

---

1. Graduate student, Dept. of Civil and Evn. Engnrg., Washington State University., Pullman, WA 99164-2910.

2. Prof. of Civil Engrng. and Director of International Development projects, College of Engineering and Architecture, Washington State Univ., Pullman, WA 99164-2910.

and narrow main channel flanked by flood plain on one or both sides. During a flood, flow overtops the banks of the main channel and flows over the flood plains. The peak and the arrival time of the peak of a flood wave are needed to design structures and to issue warning to protect human life and property.

The channel flows are mostly unsteady and may be mathematically expressed by the St Venant equations. A mathematical analysis of compound channels is a complex problem due to momentum transfer between the flows in the main channel and the flood plain (Sellin 1964, Zheleznyakov 1971). Unsteady flow in rivers have been simulated by numerically integrating the St Venant equation by several investigators, e.g. Amein (1968 and 1970); Mahmood and Yevjevich; Cunge et al. (1980); and Samuels (1985). But the incorporation of the complex momentum transfer phenomena in the mathematical model is yet to be adequately addressed. Myers (1975, 1978); Rajaratnam et al. (1981); Knight et al. (1983, 1984); Wormleaton et al. (1982); Baired et al. (1984); and Holden et al (1989) have expressed the momentum transfer mechanism for steady flow in terms of an apparent shear stress at the flood plain and main channel interface. However, a general relationship to compute apparent shear stress mapping all the ranges of velocities and channel geometry is yet to be achieved even for steady flow. Moreover, the magnitude and distribution of shear stress may be different in unsteady flow than in steady flow. Therefore, if a suitable approximation for the channel cross-section can be developed that allows the modeling of flood plain flow with reasonable accuracy even though the apparent shear stresses are neglected, it will result in saving modeling effort and cost significantly. The effectiveness of two approximations for this purpose is investigated herein.

A numerical model is developed to simulate compound channel flow. The model solves the St. Venant equations by using Preissmann four-point implicit scheme. Two different approximations for the channel cross-sections are investigated: (1) The flow velocity over the flood plain is negligible as compared to the velocity in the main channel, i.e., the flood plain acts as a storage only and does not contribute to the momentum transfer. (2) The entire channel section is assumed to have uniform average flow velocity. In approximation (1), the main channel and the flood plains are separated by a vertical line at their interface. The vertical line is not included in computing the wetted perimeter. However, the area of the entire cross-section is used in the continuity equation whereas only the main channel area is used in the momentum equation. In approximation (2), the entire cross-section is used both in the continuity and the momentum equation and a momentum coefficient is used to allow for non-uniform velocity distribution.

In this paper, the governing equations are presented. The computed results are compared with the experimental data. To conserve space, the experimental procedures are not described here. For details on the laboratory flume and experimental procedure, see Rashid (1991).

GOVERNING EQUATIONS

St. Venant equations describing unsteady one-dimensional free surface flow are
(Cunge et al. 1980):

Continuity Equation

$$B\frac{\partial y}{\partial t} + \frac{\partial Q}{\partial x} = 0 \quad (1)$$

Momentum equation

$$\frac{\partial Q}{\partial t} + \frac{\partial}{\partial x}\left(\beta\frac{Q^2}{A}\right) + gA\frac{\partial y}{\partial x} + gA\left(S_f - S_0\right) = 0 \quad (2)$$

where B = top water surface width; y = depth of flow; S_f = slope of the energy grade line; S_0 = channel bottom slope; and β = momentum coefficient for non-uniform velocity distribution. For computational details on the computation of β, see Cunge et al. (1980). The following simplified form of the continuity equation is adapted from Cunge et al. (1980)

$$\frac{\partial y}{\partial t} + \frac{1}{B_{st}}\frac{\partial Q}{\partial x} = 0 \quad (3)$$

where B_{st} is the combined width of the main channel and the flood plains. The momentum equation is the same as equation (2) except that only the main channel cross-sectional area is used for the flow area instead of the entire cross-sectional area. Preissmann scheme is described in Chaudhry (1993). Details of the initial and boundary conditions, computation of Manning n, variation of Manning n with depth of flow and additional computational and experimental results are given in Rashid (1991).

COMPARISON OF COMPUTED AND EXPERIMENTAL RESULTS

Five tests were simulated numerically. However, to conserve space, the computed results for only one test are compared with the experimental results. Input data for the computer model comprises of channel cross section, channel bottom slope, channel roughness, initial depth of flow and discharge, and upstream and downstream boundary conditions. Slope of the channel bottom was constant at .0021. Figure 1 compares the experimental and computed results. Stations 2, 3, 4, and 5 are located 2.13, 4.87, 6.39 and 10 m downstream of the upstream end of the flume respectively.

The computed rising limb at all the stations has a phase shift while all the falling limbs are in good agreement with the experimental data. The difference between the measured and computed peak depth of flow keeps on increasing in the downstream direction. The computed flow depth is always higher than the measured on the experimental facility. The apparent shear stress at the interface between the main channel and flood plain is not taken into cosideration in this study which confirms Wormleaton and Merrett's conclusion (1990) that computated depths by neglecting these stresses are higher.

The computed results show that the two approximations give almost the same depth of flow at station 2. However, as computations proceed in the downstream direction, computed results by approximation 1 compares better with the experimental results than that of approximation 2. The depth of flow computed by approximation 1 is higher even if the flow velocity over the flood plain is neglected. This shows that the overestimation due to excluding the apparent shear stresses is more than that due to loss caused by neglecting the flow velocity in the flood plains. However, the simulated results with approximation 1 are within 2% of the experimental results which is satisfactory. The maximum difference between the experimental and the computed results by approximation 2 is 8% which clearly shows the affect of neglecting the apparent shear stresses.

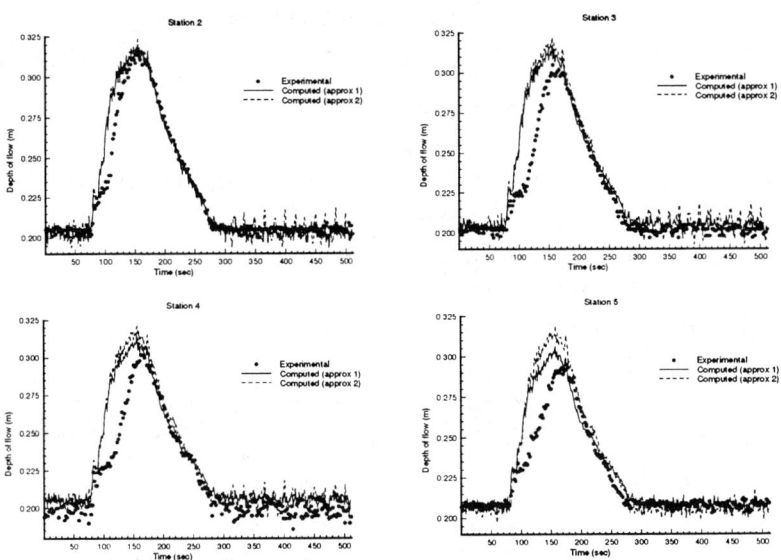

Figure 1. Comparison of computed and measured flow depth.

In this study, the maximum ratio of the depth of flow over the flood plain and the main channel was 0.60 and the width ratio between them was 2. Simulated results will be valid for any depth ratio below 0.60 and the width ratio above 2. In reality, flood plain to main channel depth ratio hardly exceeds this range. The results will be valid even for higher depth ratio if the width ratio increases (Mckee et al. 1985).

SUMMARY AND CONCLUSIONS

The compound channel cross-section is simplified by using two different approximations: (1) The flood plain acts as a storage only and does not contribute to the momentum transfer. (2) The entire channel section contributes to momentum flux, i.e., without neglecting the flow velocity over the flood plains. A mathematical model is developed to verify the suitability of these assumptions by comparing the computed results with the experimental data. Approximation 1 simulates peak depth of flow within 2% of the experimental peak; on the other hand this assumption saves modeling effort and costs by simplifying the channel cross-section. In approximation 2 the maximum deviation between the experimental and computed peak depth of flow is 8%. Therefore, approximation 1 is recommended to determine the maximum depth of flow at least for situations which have similar main channel and flood plain depth and width ratios used in this study.

Appendix I. References

Amein, M., and Fang, C. S. (1970). "Implicit Flood Routing in Natural Channels", *J. Hydr. Div.*, ASCE, 96(5), 2481-2500.

Amein, M. (1968). "An Implicit Method For Numerical Flood Routing" , *Water Resour. Res.*, 4(8), 719-726.

Chaudhry, M. H. (1993). *Open Channel Flow*, Prentice-Hall Inc, Englewood Cliffs, N. J.

Baird, J. I., and Ervine, D. A. (1984). "Resistance to Flow in Channels With Overbank Flood-plain Flow", *Proc. 1st Intl. Conf. Channels and Channel Control Structure.*, Computational Mechanic Center, Southampton, England, and Springer Verlag, Heidelberg, Germany.

Cunge, J. A., Holly, F. M., and Verwey, A. (1980). *Practical Aspects of Computational River Hydraulics*, Pitman Advanced Publishing Program, Boston, MA.

Holden, A. P., and James, C. S. (1989). "Boundary Shear Distribution on Flood Plains", *J. Hydr. Res.*, 27(1), 75-88.

Knight, D. W., and Demetriou, J. D. (1983). "Flood Plain And Main Channel Flow Interaction", *J. Hydr. Div.*, 109(8), 1073-1092.

Knight, D. W., Demetriou, J. D., and Hamed, M. E. (1984). "Stage discharge relationships for compound channels", *Proc. 1st. Intl. Conf. on Hydraulic Design in Water Resources Engineering: Channels and Channel Control Structures.*, University of Southampton, April, 4.21-4.36.

Mahmood, K., and Yevjevich, V. (1975). *Unsteady Flow In Open Channels*, Water Resources Publications, Volume I.

Mckee, P. M., Elsawy, E. M., and Mckeogh, E. J. (1985). "A Study of The Characteristics of Open Channels With Flood-Plains", *21st IAHR Congr.*, Melborne, Australia, August, 361-366.

Myers, R. C., and Elsawy, E. M. (1975). "Boundary Shear in Channel With Flood Plain", *J. Hydr. Div.*, ASCE, 101(7), 933-946.

Myers, R. C. (1978). "Momentum Transfer In A Compound Channel", *J. Hydr. Res.*, ASCE, 16(2), 139-150.

Rajaratnam, N., and Ahmadi, R. (1981). "Hydraulics of Channels With Flood-Plains", *J. Hydr. Res.*, 19(1).

Rashid, R. S. M. (1991). "Numerical Modeling of One-Dimensional Flood Plain Flows", MS thesis, Washington State University, Pullman, Washington.

Samuels, P. G., (1985). "Modeling Open Channel Flow Using Preissmann Scheme", *Proc. 2nd Intl. Conf. on the Hydraulics of Floods and Flood Control.*, British Hydraulics Research Association.

Sellin, H. J. (1964). "A Laboratory Investigation into The Interaction Between The Flow In The Channel of A River and That Over Its Flood Plain", *La Houille Blanche.*, N7, 703-801.

Wormleaton, P. R., Allen. J., and Hadjipanos, P. (1982). "Discharge Assessment in Compound Channel Flow", *J. Hydr. Div.*, ASCE, 108(9), 975-994.

Wormleaton, P. R., and Merrett, D. J. (1990). "An Improved Method of Calculation for Steady Uniform Flow in Prismatic Main Channel/flood plain Sections", *J. Hydr. Res.*, 28(2), 157-174.

Zheleznyakov, G. V. (1971). "Interaction of Channel and Flood Plain Streams", *14th Congress of the Intl. Assoc. for Hydr. Res.*, Paris, France, Vol. 5, 144-148.

"HALLOWEEN" WAVE TRANSFORMATION NEAR VIRGINIA COAST

Jerome P.-Y. Maa[1]

Abstract

The "Halloween" Northeaster of October 29-31, 1991 generated a unique severe sea. The storm waves propagated through two wave stations which are 87 km apart. The wave spectra of the storm peak at these two stations were quite different. We found that the bottom friction is the most important factor that affects wave spectrum transformation between these two stations.

Introduction

In late Oct. 1991, the most powerful northeaster in the last 50 years (Dolan and Davis 1992) caused severe sea that pounded the US east coasts. There are two wave stations, a moored buoy station, 44014, and a Coastal-Marine Automated Network station, CHLV2, located on the continental shelf off the Virginia Coast, see Fig. 1. Although only station 44014 provides directional wave spectra, both recorded the complete history of this storm. The wave spectra at storm peak at these two stations are quite different and may represent a unique feature of wave transformation at the continental shelf. Two grids (see Fig. 1) were used to study the wave transformation process. Grid 1 was used to find the wave height for each wave component at station CHLV2. Grid 2 was used to find the possible deep water wave incident angles.

The most important wave transformation processes are wave refraction, diffraction, and shoaling. In the past two decades, many numerical models that simulate the three major processes have been developed. In this study, we used the RCPWAVE model (Ebersole et al. 1986) to examine the effect of these three major processes. We were also interested in the effect of bottom friction on the wave

---

[1]Asst. Prof., Virginia Institute of Marine Science/School of Marine Science, College of William and Mary, Gloucester Point, VA 23062.

transformation because of the long travel distance, 87 km.

Wave Records

At station 44014, the wave energy reached its maximum on Oct. 31, 03 hr, 1991. At station CHLV2, the maximum occurred 1 to 2 hrs later because of the time required for the wave to travel 87 km between these two stations. The wave spectra of storm peak are given in Fig. 2. At station 44014, the significant wave height, H_s, is 8.05 m. At station CHLV2, the H_s is 3.9 m.

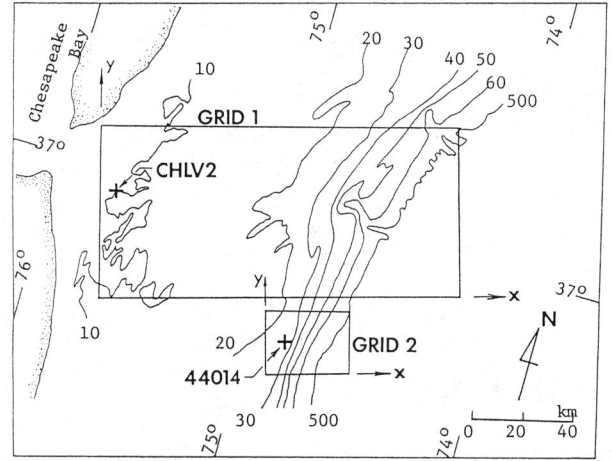

Fig. 1. The Study Area (depth contours in fathom). Wave Stations are Marked by '+'.

The wave directional information from station 44014 shows the waves came from 57 to 85 azimuth degrees. Since the low frequency waves (i.e., frequency, f < 0.06 Hz) start refracting at a water depth deeper than that where the buoy was anchored (48 m), the measured wave directions for the low frequency wave components were already affected by the bottom, and therefore, were local wave direction.

Wave Transformation

Ideally we should use a wave spectrum transformation model (e.g., Briggs et al. 1989) to study the causes of the drastic difference of wave spectra between the two stations. These models are fairly complicated and the process are not fully understood yet. Thus, instead of using a complicated model, we selected a practical approach (using the RCPWAVE model) to study the spectrum transformation.

Because of the computer limitation and the requirement of having a large study area (141.92 km x 71.2 km, see Fig. 1), we had to select a large grid size, 160 m x 200 m. Thus, wave diffraction may not be well simulated for the high frequency wave components.

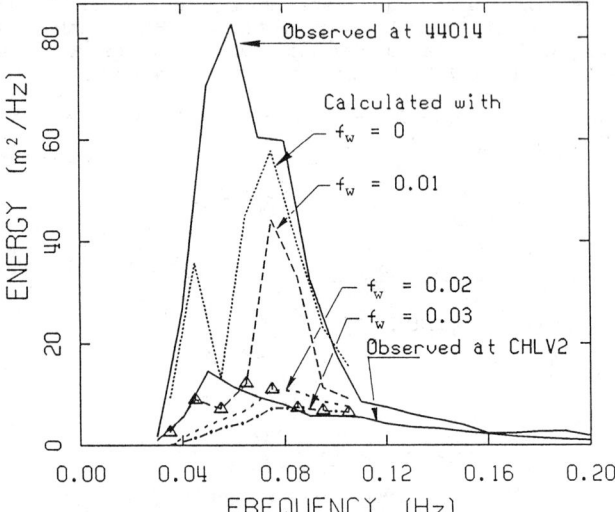

Fig. 2. Comparison of Measured and Calculated Wave Spectra for the Peak Storm Waves.

Bottom Friction

We must consider the effect of bottom friction on wave transformation because of the long wave travel distance. Maa and Kim (1992) adopted a simple procedure to count the wave energy loss for a monochrome wave. Their procedure will be applied here with some modifications. In this study, we used a constant wave friction factor, f_w, when a wave started "touching" the bottom. This selection is based on the following reasons: (1) For this severe sea, ripples can not survive. Only the grain friction (usually small) and movable bed friction (the major factor in this study) contribute to the wave energy loss. Maa and Kim (1992) demonstrated that f_w is reasonably constant and small (on the order of 0.01-0.05) for this condition. (2) It is difficult to estimate f_w for each wave component. The bed responses (e.g., sediment mobility) reflect the results of total wave energy instead of a single component. For a given sea, there exists only one bed form (from a statistical point of view), and different wave components may respond differently for the same bed mobility in terms of energy dissipation. Thus, using a constant f_w is a practical approach at this time.

Approach

We divide the wave spectrum observed at station 44014 (see Fig. 2) into eight frequency bands. For each band, there is a representative wave period, T_i, band width, Δf_i, spectrum density, S_i. The representing wave height, H_i, for that frequency band can be calculated as $H_i = 2(2\Delta f_i S_i)^{1/2}$ (Dean and Dalrymple 1984). Wave components with a frequency greater than 0.11 Hz are excluded because of the small wave energy in that domain.

The first step is to estimate the true wave incident angles. By assuming a deep water wave direction, A_i, using the wave information (T_i and H_i), the bathymerty in Grid 2, and the RCPWAVE model, we calculated the wave angle at station 44014 for each frequency band. Fig. 3 shows the comparsion of the calculated wave directions (based on the assumed wave incident direction) and the observed wave directions. We found the deep water wave direction to be 57 azimuth degrees for those high frequency wave components ($f_i > 0.06$ Hz). For the low frequency components, the A_i's are different. Notice that the observed wave direction for the 0.03 Hz wave component from the spectrum analysis is not accurate because of the wave energy is too small, 1.73 m²/Hz, see Fig. 2, and thus it is not plotted in Fig. 3.

We then studied the wave transformation for each wave band using the T_i, H_i, A_i, and the bathymetry in grid 1, (not include the effects of bottom friction). Figure 4 shows the wave trajectories for a frequency band (T=18.2 s). This figure reveals that the Norfolk Canyon (at water depth from 100 to 500 m) causes low frequency component waves to diverge. As the wave period decreases, the influence of the Norfolk Canyon decreases and becomes negligible when the wave period is less than 14 s. The wave height at station CHLV2 for each component can be found easily from the output file generated by the RCPWAVE model.

Using Eq. 1, we may calculate the wave energy for each frequency band (see Fig. 2) given the wave height of that frequency band. In general, the calculated wave energy is much larger than that recorded at station CHLV2. The only exception is for the 0.055 Hz component because of the wave divergence. For wave frequencies greater than 0.075 Hz, the calculated spectrum is almost equal to that at station 44014. These results indicates that the effects of wave refraction, diffraction, and shoaling are not significant except for the 0.055 Hz component.

When using a constant f_w to check the effect of bottom friction, we do not know what the f_w should be for each component wave. Thus, we found the best fit through trial and error. We calculated the wave heights at station CHLV2 for all the wave components with a given f_w and then calculated the wave spectrum to compare with the measurements. The three selected f_w are 0.01, 0.02, and 0.03. The calculated wave spectra are displayed in Fig. 2. It reveals that using a small f_w to calculate wave energies at station CHLV2 for long period waves matched the observations. For the shorter period waves, we need a relatively large f_w. Using the three f_w for different frequency band, we can have a wave spectrum (marked by △ in Fig. 2) that is close to the observed.

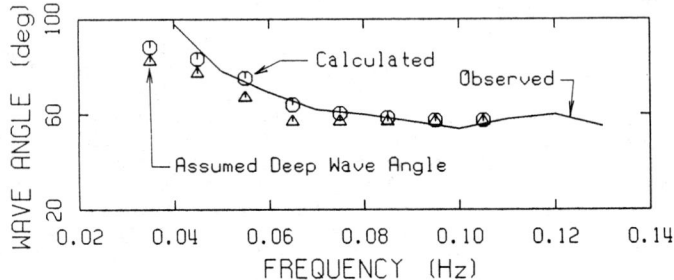

Fig. 3. Comparison of Calculated & Observed Wave Direction at Station 44014 for Oct. 31, 03 hr, 1991, GMT.

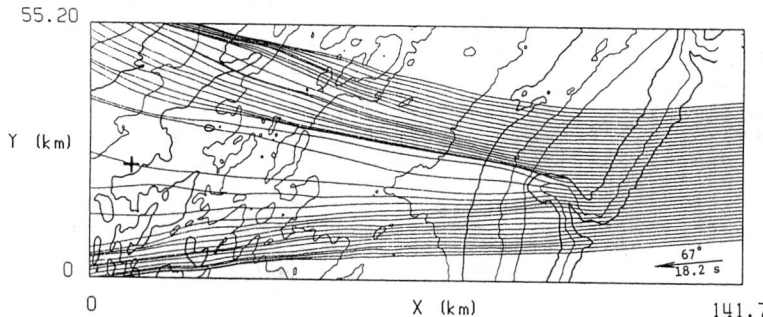

Fig. 4. Calculated Wave Trajectories for a Selected Wave Components (T = 18.2 s) on Oct. 31, 03 hr. GMT.

Discussions and Conclusions

It is impractical to acount for the effect of bottom friction for random waves using a single representative f_w for the entire frequency band, especially when the wave energy is composed from a broad range of frequency. In term of energy loss, each wave component responds differently for a given bottom condition which is controlled by the total wave energy. In this study, we could use a constant f_w for a wave component because of the sea severity. For small or moderate random waves, this assumption may not work because the existence of ripples. In that case, the approach given by Madsen et al. (1988) may be applied.

Because of the extraordinary long period wave, the influence of wind energy input to the continuous growth of wave height is limited and neglected. Lack of the wave-wave interaction process in our calculation may be a possible reason that the calculated wave energy for the 18.2 s wave component is low.

We found that the wave refraction, diffraction, and shoaling do not significantly affect the wave spectrum transformation. The major cause of the difference between an offshore and a nearshore wave station is the wave energy dissipation caused by the bottom friction. Based on this limited study, we concluded that f_w for a long period wave is smaller than that for a short period wave. The variation of f_w, however, is not significant because all ripples were wiped out by the severe sea.

Acknowledgments

I thank Mr. R. Lukens and Dr. A. Kuo for their support of computing resources. This paper is a contribution (No. 1785) of the Virginia Institute of Marine Science/School of Marine Science, College of William and Mary.

References

Briggs, M.J., Grace, P. and Jensen, R.E. (1989) <u>Directional Spectral Wave Transformation in the Nearshore Region, Report 1, Directional Spectral Performance Characteristics</u>. TR-CERC-89-14, CERC, WES, USACE, Vicksburg, MISS 39180.

Dean, R. G., and Dalrymple, R. A. (1984) <u>Water Wave Mechanics for Engineers and Scientists</u>. Prentice Hall, Englewood Cliffs, NJ.

Dolan, R., and Davis, R.E. (1992) "Rating Northeasters," <u>Mariners Weather Log, Winter 1992</u>.

Ebersole, B. A., Cialone, M. A., and Prater, M. D. (1986) <u>Regional Coastal Processes Numerical Modeling System, Report 1, A Linear Wave Propagation Model for Engineering Use</u>. TR-CERC-86-4, CERC, WES, USACE, Vicksburg, MISS 39180.

Maa, J. P.-Y., and Kim, C.-S. (1992) "The Effect of Bottom Friction on Breaking Waves using RCPWAVE Model." <u>J. of Waterway, Port, Coastal and Ocean Engrg.</u>, 118(4), 387-400.

Madsen, O.S., Poon, Y.K., Graber, H.C. (1988) "Spectral Wave Attenuation by Bottom Friction: Theory." <u>Proceedings</u>, 21th ICCE, ASCE, 1, 492-504.

Use of Floodplain PCB Concentrations to Calibrate
A River Hydraulics Model

James M. Hassett[1], M. ASCE and Leonard T. Wright[2]

Abstract

A river floodplain was found to contain measurable quantities of polychlorinated biphenyls (PCBs), presumably deposited during historic flood events. A hydraulic model (HEC-2) was used to examine the relationship between the distribution of PCBs in the floodplain soils and the water surface elevation of the floods. This study demonstrated the utility of using the presence of conservative tracers in floodplain sediments as useful information in the calibration and validation of river hydraulic models.

Introduction

The estimation of flood flows and floodwater inundation are problems of historical and current concern (1). Of the factors influencing the extent of flood inundation, the frictional interactions between the flood flow and the flood plain are considered the most difficult to estimate, since the floodplain surfaces are extremely heterogeneous (9). There have been several approaches used to estimate the flow patterns and energy losses associated with overbank flood flow, including the use of physical models (5,6), the analysis of vegetative effects (7), analysis of historic floods (3) and the development and use of various numerical models (8).

---

[1] Associate Professor, Division of Environmental and Resource Engineering, 312 Bray Hall, SUNY-CESF, Syracuse, NY 13210
[2] Project Engineer, Moffa & Associates, 5710 Commons Drive, P.O. Box 54, Syracuse, NY 13214

We report herein the preliminary results of a study using the distribution of PCBs in a floodplain to estimate the water surface elevation from historical floods along a 22.5 km river reach in the northeastern United States[2].

Study Area

The river reach consists of oxbows, meanders and riparian wetlands. There are no retention structures, although there are 13 bridge crossings. The upstream 10 km of the river flows through an urbanized area consisting mainly of park land or industrial sites. The land use in the downstream-most 18 km is either agricultural pasture land or wildlife management area. The river flows into an impoundment behind a dam at the most downstream end of the study reach.

Flow records from a gaging station located at the upper end of the study reach are available. The river has been gaged at that site since 1936, with stage and flow records available in 15 minute time increments. Table 1 gives selected values for the peak annual flows at that site.

Table 1. Selected Values of Peak Daily Flows - Upstream Gaging Station

Year	Flow (m^3/s)	Year	Flow (m^3/s)
1949	123	1969	58.9
1938	88.1	1977	58.1
1987	81.0	1990	57.0
1984	77.9	1980	56.7

The 1949 flood of 123 m^3/s corresponds roughly to the 1/100 year flood; the 1/10 year flood is approximately 58 m^3/s. A spillway structure located at the downstream end of the study reach provided stage-discharge data for the modeling exercise described below.

PCB Profiles

Soil samples for PCB analysis were obtained from seven flood plain and river cross sections along the study reach. PCB concentrations were determined in both the top and bottom 15 cm of the soil core. The samples

---

[2]Responsibility for PCB contamination and resulting clean-up costs are under discussion; we have therefore been asked to avoid specific references to the site.

were taken at points determined during the physical survey; thus PCB concentrations could be related to elevation. Figure 1 is a schematic of one of the downstream cross sections and shows that PCB concentration in the floodplain generally decreases with elevation increase, a finding true for the other downstream cross sections as well.

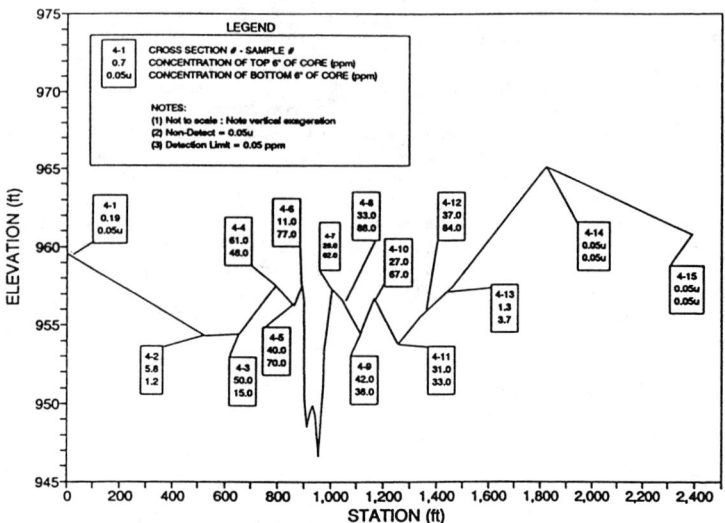

Figure 1. River Cross Section Downstream of Suspected Source of PCBs

The river flow at the time of sampling was approximately 2.1 m^3/s; the water surface elevation at this flow was used as a vertical datum for each cross section to determine the elevation of each soil sampling location. A plot of PCB concentration versus elevation above the water surface approximated the form of the exceedance-flow curve for the flood flows from the upstream gaging station. The point of inflection on both the exceedance-flow curve and the curve of Figure 1 were the most identifiable feature common to both curves and were selected for further analysis. The flow value from the exceedance curve was 57 m^3/s; this corresponded to a water surface elevation of approximately 3.05 m above the water surface elevation surveyed at the flow of 2.1 m^3/s. Table 2 gives the water surface elevations for the downstream cross sections thus estimated.

Table 2. Water Surface Elevations for Surveyed
Cross Sections as Deduced from Elevation of
PCB-Contaminated Soil Samples

Cross Section	Flow (m^3/s)	Elevation (m)
2	57.0	295.09
3	57.0	294.30
4	57.0	293.26
6	57.0	292.07
7	57.0	292.01

The Water Surface Elevation Model

The HEC-2 Water Surface Elevation Model was used to model the water surface elevation of the study reach (4). In practice, Manning's n is often used as an adjustable fitting parameter since it has been found difficult to estimate this parameter for natural streams and channels (13).

The model was first run using estimates of Manning's n for the main channel and left and right overbank determined from standard tables, as appears for example in ref. 2. The results from this model will be identified as the "best estimate" model.

The HEC-2 model was then rerun varying Manning's n so as to obtain the water surface elevations presented in Table 2. No attempt was made to keep the values of Manning's n within the range of the standard tables. The results from this model will be identified as the "calibrated" model. A comparison of these two model runs is presented in Table 3.

Table 3. Comparison of "Best Estimate" (BE) and "Calibrated" (C) Models - Manning's n

Cross Section	Channel n		Left and Right Overbank n	
	BE	C	BE	C
2	0.020	0.040	0.058	0.12
3	0.020	0.040	0.043	0.086
4	0.020	0.018	0.043	0.039
6	0.015	0.065	0.15	0.65
7	0.015	0.065	0.15	0.65

Table 4. Comparison of "Best Estimate" (BE) and "Calibrated" (C) Models - Water Surface Elevation

Cross Section	Target Elevation (m)	BE(m)	C(m)
2	295.09	294.46	295.32
3	294.3	293.48	294.19
4	293.26	293.13	293.18
6	292.07	291.04	292.30
7	292.01	290.91	291.80

Model Validation

Two methods were used to validate the model. First, aerial photographs were obtained for a 1990 flood event, the instantaneous peak flowrate for which was 108 m^3/sec. The photographs were used to determine the extend of floodplain inundation and hence the water surface elevation. Vegetation in the floodplain, however, often obscured the edge of the water making the water surface elevation estimate no better than $\pm$1.0 ft.

The second validation used the depth-discharge data for the upstream gauging station. Both the best estimate and calibrated models were used to predict the rating curve. Results are given in Figure 2, from which it can be seen that the calibrated model predicts the rating curve much better than the best estimate model.

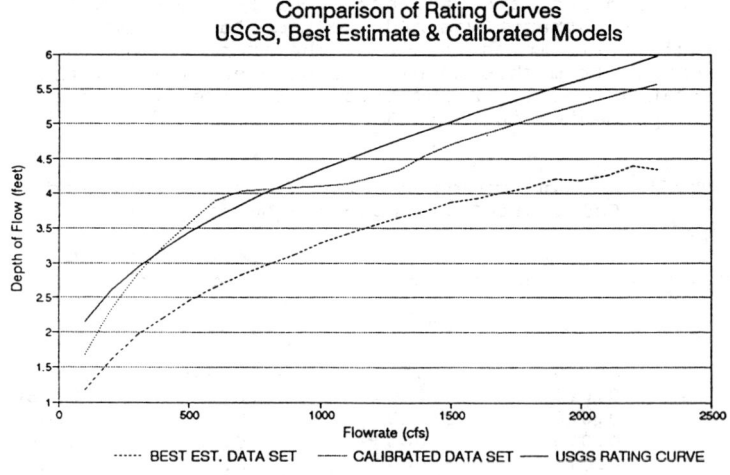

Figure 2. Comparison of HEC-2 generated versus USGS Rating Curve for Upstream Gaging Station

Discussion

The hydraulic model calibrated via the PCB distribution produced values of Manning's n that were relatively high compared to published tables of n values for the various types of land cover associated with the floodplain. The PCB-calibrated n values did however produce a HEC-2 model which better predicts water surface elevations at the upstream cross section. This study shows the utility of using the presence of conservative tracers in floodplain sediments as additional information in calibration and validation of river hydraulics models.

Appendix A - References

1. Baker, V.R., Kochel, R.C. and Patton, P.C., Flood Geomorphology, Wiley, New York, NY, 1988, pp. 97-155.

2. Brater, E.F. and King, H.W., Handbook of Hydraulics, McGraw-Hill, New York, NY, 1982, pp. 7-1 - 7-74.

3. Harper, J.M., "Floodplain Modification Due to a Catastrophic Flood, Potomac River Basin, WV", Eos (AGU), 71(17): 509, 1990 (Abstract).

4. Hoggan, D.H., Computer-Assisted Floodplain Hydrology & Hydraulics, McGraw-Hill, NY, 1989, pp. 283-504.

5. Kiely, G.K., McKeogh, E.J. and Thomas, G., "Flow Patterns in Straight and Meandering Rivers with Flood Plains", Eos (AGU), 71(17); 509, 1990 (Abstract).

6. Knight, D.W., "River Flood Hydraulics and the SERC Flood Channel Facility Experiments", Eos (AGU), 71(17); 509, 1990 (Abstract).

7. Kouwen, N., "A Review of Resistance Erosion and Sedimentation in Vegetated Channels", Eos (AGU), 71(17): 510, 1990 (Abstract).

8. Miller, A.J. and Wolman, M.G., "2D Simulation of Flood Flow Patterns in Mountain Valleys", Eos (AGU) 71(17): 510, 1990 (Abstract).

9. Stevens, G.T., Mueller, A.M. and Strausser, C.N., "A New Approach to the Elusive Manning's n", in Proceedings of Conference - Rivers '83, C.M. Elliot, ed., ASCE, 1983, pp. 586-596.

Assessment and Prediction of Debris-flow Hazards

Gerald F. Wieczorek[1], M.ASCE

Abstract

Study of debris-flow geomorphology and initiation mechanisms has led to better understanding of debris-flow processes. This paper reviews how this understanding is used in current techniques for assessment and prediction of debris-flow hazards.

Introduction

Debris flows are a form of rapid mass movement of granular solids, vegetation, water, and air with flow properties that vary with water content, sediment size, and sorting. Rheologically, they are classified as viscous slurry flows; they contain more sediment than hyperconcentrated streamflow or waterfloods and more water than other types of mass movement (Pierson and Costa, 1987). Debris flows can be initiated on hillslopes or in channels by either intense rainfall, snowmelt, or rapid runoff and cause substantial damage and loss of life worldwide (Costa, 1984).

During 1982, intense rainfall lasting for about 32 hours triggered more than 18,000 landslides, most of which mobilized as debris flows, from hillsides in the San Francisco Bay region, California, that damaged at least 100 homes and killed 14 residents (Ellen et al. 1988). However, other events including volcanic eruptions and earthquakes have triggered catastrophic debris flows. On November 13, 1985, pyroclastic flows and surges from a relatively small volcanic eruption, melted snow and ice on the summit of Nevada del Ruiz in Colombia, producing large volumes of meltwater that initiated debris flows in steep channels. The debris flows killed more than 23,000 inhabitants at the base of the volcano (Pierson et al. 1990).

This paper presents recent developments for assessing debris-flow hazards extending from source areas to channels and fans, and discusses debris-flow triggering mechanisms, rainfall thresholds, and warning systems.

---

[1]U.S. Geological Survey, 922 National Center, Reston, VA 22092

Debris-flow Source Area

Post-storm investigations show that a majority of debris flows initiate as shallow landslides in topographic concavities or hollows. Field observations following the January 1982 storm in the San Francisco Bay area indicated that about two-thirds of debris flows initiated within hollows rather than from planar or convex slopes (Reneau and Dietrich, 1987; Ellen, 1988, p. 122-123). Topographic concavities on steep hillsides where thick colluvial soils accumulate and ground water converges favor slope instability (Hack and Goodlet, 1960; Dietrich and Dunne, 1978; Pierson, 1980; Humphrey, 1982). Toes or flanks of active landslide masses (Morton et al. 1979), steep stream channels with abundant loose alluvium (Bovis and Dagg, 1988), and volcanoes with recently deposited ash also are likely sources of debris flows (Waldron, 1967).

Potential debris-flow source areas can be identified on aerial photographs or from field mapping based on the presence of colluvial-filled swales (Johnson and Sitar, 1986). In heavily forested areas, aerial-photo interpretation alone is inadequate, and extensive field survey is necessary to identify colluvial swales. The presence and thickness of colluvium is best determined by excavation or borings, but it can be estimated using geophysics or other techniques (Dengler et al. 1987; Dengler and Montgomery, 1989). Thickness of colluvium is important because thick accumulation promotes instability and because the distance a debris flow travels beyond the source area depends in part on initial volume (Ellen et al. 1993).

In California, debris flows initiate most commonly on slopes ranging from 26 to 45 degrees in steepness (Campbell, 1975; Ellen, 1988). This range reflects a large natural variation in thickness of colluvium, as well as other geologic, hydrologic, and topographic characteristics. Mapping topographic swales, together with categorizing slope steepness, has been used as an initial step in evaluating debris-flow hazards (Fowler, 1984; Smith, 1988; Montgomery et al. 1991).

In debris-flow source areas, a scar generally remains where shallow translational landslides (soil or debris slides) have mobilized into debris flows. Biogenic transport, colluvial infilling by soil falls and creep, and revegetation gradually disguise scars (Lehre, 1987; Sidle, 1987). Repeated episodes of debris flow and subsequent infilling tend to develop and accentuate hillslope hollows and unchanneled basins (Alger and Ellen, 1987).

Digital elevation models (DEMs) have been used to identify potential debris-flow source areas (Ellen and Mark, 1988; Dietrich and Montgomery, 1992). Using a DEM with 10-m spacing, Ellen et al. (1993) identified potential debris-flow source areas near Honolulu, Hawaii, in part on the basis of slope steepness. Curvature measured in the horizontal plane or in vertical planes perpendicular to hillslope gradient can be used to define hollows. Digital portrayal of hollows is most accurate with DEMs prepared at grid spacings of 15 m or finer. Preliminary results from the

San Francisco Bay area show mapped debris-flow sources to be significantly correlated with hollows mapped from 10- and 15-m DEMs (Cassandra Rogers, Univ. of California at Berkeley, and S.D. Ellen, USGS, oral communs., 1993).

Mark (1992) used slope steepness from a 30-m DEM in conjunction with other variables, including precipitation, geology, and vegetative cover, in a spatial statistical model to prepare a map of debris-flow probability of San Mateo County, California. The map shows the estimated probability of debris-flow occurrence for a 30-hr storm identical to the storm of January 1982 (Fig. 1). Because rainfall intensity and duration characteristics influence the location of debris-flow initiation (Wieczorek, 1987), the probability depicted by Mark (1992) in San Mateo County does not necessarily reflect expected debris-flow distribution in storms having different rainfall characteristics. Campbell and Bernknopf (this volume) assess time-dependent probability for landsliding as a function of duration and rate of rainfall.

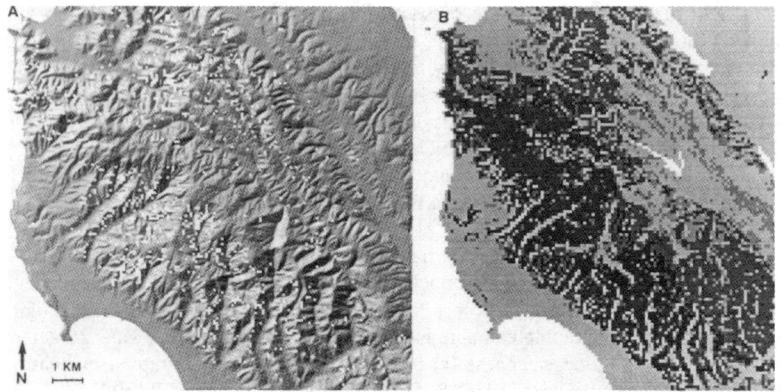

Figure 1 - Maps from part of San Mateo County, California, showing A) Debris-flow source areas (bright squares) from January 1982 storm, and B) Relative degrees of debris-flow probability (darker shades indicate higher probability) modified from Mark (1992).

Debris-flow Channel And Fan

A general conceptual model for debris-flow motion suggests that on steep slopes and channels downslope from source areas, debris flows tend to increase in velocity and volume as they incorporate material from the bottom or sides of channels. In some cases, debris flows have enlarged significantly, increasing their volume many times (McCarter and Kaliser, 1985; Santi and Mathewson, 1988; Jibson, 1989). Once debris flows encounter flatter slopes, such as those on alluvial fans, they tend to decelerate, lose erosive power, and start depositing material.

Using a model for volume-change behavior of debris flows during travel, such as proposed by Cannon and Savage (1988), in conjunction with hydrological routing techniques, Ellen et al. (1993) and Ellen and Mark (this volume) used simulations of debris flows in a DEM to map hazard from debris-flow travel paths near Honolulu, Hawaii. A volume-change model was used to estimate potential debris flow-volume along a path as a function of slope gradient and degree of channel confinement (Cannon, 1989). Beginning with a mean initial volume of 120 m^3, debris-flow paths were routed through the topography until the flows terminated when volume became zero. Relative hazard was assessed by counting the number of flows that crossed each cell (Fig. 2). This one-dimensional model best applies to flows on planar hillsides or in low-order drainages and is less accurate where flows emerge from a confining valley and spread onto alluvial fans. Other one-dimensional models for unsteady debris-flow routing based on the hypothetical breaching of a debris blockage have been developed to provide a downstream hydrograph of the flow (MacArthur et al. 1990; Arattano and Savage, 1992). Such models are useful for forecasting timing of arrival or height of a debris-flow front.

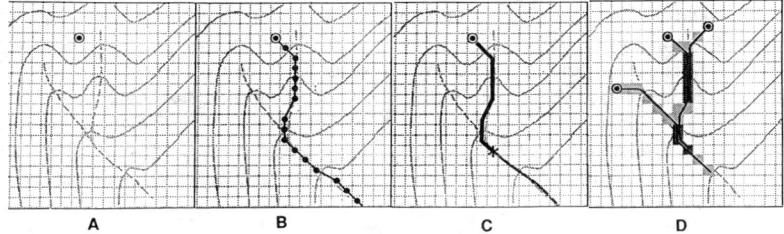

Figure 2 - Schematic maps showing procedure for simulating debris flows and mapping hazard using digital topography (Ellen and Mark, this volume). Gray lattice represents 10-m digital elevation grid; solid gray lines, elevation contours; broken gray lines, drainages. A) Selection of initiation point from grid of initiation frequency, B) Calculation of path downslope from initiation point, C) Debris flow simulated using volume of slope failure and volume-change relation (Cannon, this volume) stops at cell where volume decreases to zero, D) Total strikes from three simulated debris flows. Dark gray cells, two strikes; light gray cells, one strike.

Once a debris flow emerges from a canyon, the problem of determining its path across an alluvial fan becomes complicated by the ability of the flow to spread, to plug its channel, and to alter its direction. The debris-flow hydrograph, rheology, and fan topography are some factors important for debris-flow runout and deposition on fans (Whipple, 1992). Two-dimensional mathematical models for flows on fans have been calibrated based on individual events to determine flow depths, velocities, impact forces and areas of deposition (Mizuyama et al. 1987; Mizuyama and Ishikawa, 1990; O'Brien and Fullerton, 1990; Takahashi, 1991). The transferability of these methods is being tested.

Depending on whether debris-flow source, channel, or fan areas are examined, and on the methodology used, recurrence intervals may vary significantly (Orme, 1990). Based on radiocarbon dating of colluvial deposits in swales of central California, Reneau et al. (1990) found a periodicity of several thousand years for the emptying and infilling of hollows. The frequency of debris flows at the confluence of several hollows may be higher than at individual source areas. Farther from the source, particularly toward the distal ends of alluvial fans, the frequency of debris flows may decrease because only infrequent larger flows reach the ends of fans (Lips and Wieczorek, 1990). Large flows do not necessarily stay confined within channels on an alluvial fan and may take an independent course as well as spread laterally. This somewhat random directional component tends to reduce the frequency with which debris flows reach points on the distal parts of fans. Recurrence intervals of debris flows worldwide range mostly from several tens to several hundreds of years (Costa, 1984). Estimating the probability of debris-flow inundation on alluvial fans remains problematic because of difficulty in determining recurrence and in locating the shifting area of deposition.

The mechanical process of debris flow, which determines the shape of deposition, thickness, grain-size distribution, as well as flow characteristics including velocity, depth, and forces, can be examined using flume experiments. Takahashi (1991) described using a laboratory flume to verify a mathematical model for shape of debris-flow deposition on an alluvial fan. In Oregon, a reinforced concrete flume, 95-m long, 2-m wide, and 1.2-m deep, is being used to study debris-flow rheology and characteristics of debris-flow deposition (Costa, 1992; Iverson and LaHusen, 1992).

Debris-flow Triggering Mechanisms

Debris flows are triggered by rapidly adding moisture to unconsolidated rock and soil debris. The moisture comes most commonly from intense rainfall (Cannon, 1988) and rapid snowmelt on hillslopes (Wieczorek et al. 1989), and less commonly from rapid runoff in steep channels from rapid drainage of lakes (Watters, 1983) or landslide damming of a channel and subsequent breaching of the dam (Bovis and Dagg, 1992). Ground-water contribution from underlying bedrock (Wilson and Dietrich, 1987), concentration of ground water by topographic irregularities, and macropore infiltration can contribute to slope failures that generate debris flows.

Observation and monitoring of debris-flow processes has led to development of conceptual models for debris-flow mobilization from shallow landslides on hillsides (Ellen and Fleming, 1987) and to identification of critical climatic thresholds for triggering debris flows (Campbell, 1975). Other models theorize how debris flows can mobilize in steep channels by a combination of frictional hydraulic drag and pore pressure rise in stream-bed materials (VanDine, 1985; Bovis and Dagg, 1988) or by rapidly moving landslides impacting loose stream-bed materials (Bovis and Dagg, 1992).

Based on observations of debris flows triggered by intense rainfall, Campbell (1975) proposed a general model in which shallow temporary perched ground-water levels parallel to the ground surface develop over less permeable hillside materials during intense rainfall. The transient increase in porewater pressure causes a reduction in effective shear strength of shallow saturated surficial materials which can trigger shallow translational landslides that transform into debris flows.

This simple model associating debris flows with high porewater pressure from infiltration during intense rainfall has generally been verified by measurements of positive pore pressures during storms that triggered debris flows (Sidle, 1984; Reid et al. 1988; Johnson and Sitar, 1990) and experimentally by Harp et al. (1990). This model, however, does not account for all geologic complexities. Bedrock may contribute ground water to overlying soils, and peak ground-water levels may not coincide with peak rainfall intensity (Wilson and Dietrich, 1987). Ground-water or seepage flow on hillsides can be more complicated than flow parallel to the ground surface. Topographic irregularities, such as breaks in slope, and geologic heterogeneities, such as geologic contacts or different thicknesses of soil, may create seepage flow that is not parallel to the ground surface; vectors of seepage directed out of the slope may locally be responsible for slope failures (Iverson and Major, 1986; Reid and Iverson, 1992). Springs observed in many debris-flow source areas (Mathewson et al. 1990) suggest that ground-water-flow irregularities are not uncommon and that simple models do not represent all field conditions.

The presence of macropores or piping systems on hillsides, including root channels and animal burrows, allows rapid infiltration of rainfall into unsaturated soils (Bevin and Germann, 1982; Amen, 1990) and rapid buildup of water tables and positive pore pressures (Pierson, 1983). Macropores have been found to be common in colluvial hollows (Ziemer and Albright, 1987), and their presence at many sites of debris-flow initiation suggests high porewater pressures leading to instability (Pierson, 1983; Harp et al. 1990). Flow through macropores adds a further complication to representing ground-water flow with simple models.

Rainfall Thresholds and Warning Systems

Monitoring of rainfall and correlation with times of debris flows has led to recognition of rainfall thresholds and improved the capability of real-time evaluation and warning of debris-flow hazards. The amount of antecedent rainfall and the rainfall intensity necessary for triggering debris flows from shallow landslides in southern California was determined empirically by comparison of rainfall with documented times of debris flows (Campbell, 1975). Measurement of the intensity and duration of rainfall has been used as the basis for empirical thresholds for the triggering of debris flows worldwide (Caine, 1980), in northern California (Cannon and Ellen, 1985), in North Carolina (Neary and Swift, 1987), in Puerto Rico (Jibson, 1989), and in Hawaii (Wilson et al. 1992).

The U.S. Geological Survey, in cooperation with the U.S. National Weather Service (NWS), has developed a real-time warning system that was used to issue the first public regional landslide warning in the United States during the storms of 12-21 February 1986 in the San Francisco Bay region (Keefer et al. 1987). In addition to rainfall thresholds, the warning was based on monitoring of regional rainfall data from telemetering rain gages and piezometers, and NWS precipitation forecasts. The well-documented times of debris flows and other landslides coincided with times of two separate warnings that were issued during the February 1986 storms. These and subsequently issued warnings have been conveyed to local officials responsible for emergency services.

To provide better calibration for rainfall-landslide relationships, models have been developed that numerically simulate piezometric response to rainfall (Okimura, 1983; Kobashi and Suzuki, 1987; Wilson, 1989; Wilson, this volume). These models are patterned on a physical analog of a tank or barrel that is being filled at a rate equal to the rainfall intensity and being drained at a rate proportional to the level of water retained in the barrel. A piezometric response measured during the storms of 12-20 February 1986 in the San Francisco Bay region is simulated fairly accurately by the model (Wilson, 1989). A somewhat more sophisticated method using a one-dimensional, vertical, transient, unsaturated finite-difference model for predicting pore pressures in debris-flow source areas has been presented by Buchanan et al. (1990). Even more elaborate two-dimensional finite element numerical models have been used to evaluate piezometric levels at sites of individual debris-flow initiation where sufficient hydrologic and geologic detail is available (Reid et al. 1988).

Automated detection and warning systems have been developed that can detect ground vibrations of passing debris flows, infer the relative magnitude of the flow, and transmit alarms to receivers downstream allowing time for evacuation of people and property. Seismic sensors placed next to channels have been used to detect and warn of approaching debris flows (lahars) at Mount St. Helens in Washington, Redoubt Volcano in Alaska, Cotopaxi in Ecuador, and Pinatubo Volcano in the Philippines (LaHusen, 1990, p. 30-31; Hadley and LaHusen, 1991).

Summary

Recently developed methods of assessing debris-flow hazards based on improved understanding of debris-flow processes are described here. These methods, together with technological advancements in monitoring, storage, and transmitting data, have resulted in an emerging capability for real-time prediction and warning of debris-flow hazards.

Appendix. References

Alger, C.S., and Ellen, S.D. (1987). "Zero-order basins shaped by debris flows, Sunol, California, USA," *International Association of Hydrological Sciences*,

Publication No. 165, 111-119.
Amen, B.B. (1990). "The hydrologic role of the unsaturated zone of a forested colluvium-mantled hollow, Redwood National Park, California," *Redwood National Park Technical Report Number 26*, Arcata, Calif.
Arattano, M., and Savage, W.Z. (1992). "Kinematic wave theory for debris flows," U.S. Geological Survey, Open-File Report 92-290.
Bevin, K., and Germann, P. (1982). "Macropores and water flow in soils," *Water Resources Research*, 18(5), 1311-1325.
Bovis, M.J., and Dagg, B.R. (1988). "A model for debris accumulation and mobilization in steep mountain streams," *Hydrological Sciences Journal*, 33(6), 589-604.
Bovis, M.J., and Dagg, B.R. (1992). "Debris flow triggering by impulsive loading: mechanical modeling and case studies," *Canadian Geotechnical Journal*, 29, 345-352.
Buchanan, P., Savigny, K.W., and De Vries, J. (1990). "A method for modeling water tables at debris avalanche headscarps," *Journal of Hydrology*, 113, 61-88.
Caine, N. (1980). "The rainfall intensity-duration control of shallow landslides and debris flows,"*Geografiska Annaler*, 62A, 23-27.
Campbell, R.H. (1975). "Soil slips, debris flows, and rainstorms in the Santa Monica Mountains and vicinity, southern California," U.S. Geological Survey Professional Paper 851.
Cannon, S.H. (1988). "Regional rainfall-threshold conditions for abundant debris-flow activity," *Landslides, floods, and marine effects of the storm of January 3-5, 1982, in the San Francisco Bay region, California*, U.S. Geological Survey Professional Paper 1434, 35-41.
Cannon, S.H. (1989). "An evaluation of the travel-distance potential of debris flows," Utah Geological and Mineral Survey Miscellaneous Publication 89-2.
Cannon, S.H., and Ellen, S.D. (1985). "Rainfall conditions for abundant debris avalanches, San Francisco Bay region, California," *California Geology*, 38(12), 267-272.
Cannon, S.H., and Savage, W.Z. (1988). "A mass change model for the estimation of debris flow runout," *Journal of Geology*, 96, 221-227.
Costa, J.E. (1984). "Physical geomorphology of debris flows," *Developments and Applications of Geomorphology*, Springer-Verlag, Berlin, 268-317.
Costa, J.E. (1992). "Characteristics of a debris fan formed at the U.S. Geological Survey debris-flow flume, H.J. Andrews Experimental Forest, Blue River, OR," *EOS*, 73(43), 227.
Dengler, L., Lehre, A.K., and Wilson, C.J. (1987). "Bedrock geometry of unchannelized valleys," *International Association of Hydrological Sciences*, Publication No. 165, 81-90.
Dengler, L. and Montgomery, D.R. (1989). "Estimating thickness of colluvial fill in unchanneled valleys from surface topography," *Bulletin of the Association of Engineering Geologists*, 26, 333-342.
Dietrich, W.E., and Dunne, T. (1978). "Sediment budget for a small catchment in

mountainous terrain," *Zeitschrift für Geomorphologie*, 29, 191-206.
Dietrich, W.E., and Montgomery, D.R. (1992). "A digital terrain model for predicting debris flow source areas," *EOS*, 73(43), 227.
Ellen, S.D. (1988). "Description and mechanics of soil slip/debris flows in the storm," *Landslides, floods, and marine effects of the storm of January 3-5, 1982, in the San Francisco Bay region, California*, U.S. Geological Survey Professional Paper 1434, 63-112.
Ellen, S.D., and Fleming, R.W. (1987). "Mobilization of debris flows from soil slips, San Francisco Bay region, California," *Debris flows/avalanches: Process, recognition and mitigation*, Geological Society of America, Reviews in Engineering Geology, 7, 31-40.
Ellen, S.D., and Mark, R.K. (1988). "Automated modeling of debris-flow hazard using digital elevation models," *EOS*, 69(16), 347.
Ellen, S.D., Wieczorek, G.F., Brown, W.M. III, and Herd, D.G. (1988). "Introduction," *Landslides, floods, and marine effects of the storm of January 3-5, 1982, in the San Francisco Bay region, California*, U.S. Geological Survey Professional Paper 1434, 1-5.
Ellen, S.D., Mark, R.K., Cannon, S.H., and Knifong, D.L. (1993). "Map of debris-flow hazard in the Honolulu District of Oahu, Hawaii," U.S. Geological Survey, Open-File Report 93-213.
Fowler, W.L. (1984). "Potential debris flow hazards of the Big Bend Drive Drainage Basin, Pacifica, California," thesis presented to Department of Applied Earth Sciences, Stanford University, Stanford, CA, in partial fulfillment of the requirements for Masters degree.
Hack, J.T., and Goodlett, J.C. (1960). "Geomorphology and forest ecology of a mountain region in the central Appalachians," U.S. Geological Survey Professional Paper 347.
Hadley, K.C., and LaHusen, R.G. (1991). "Deployment of an acoustic flow-monitor system and examples of its application at Mount Pinatubo, Philippines," *EOS*, 72(44), 67.
Harp, E.L., Wells, W.G.III, and Sarmiento, J.G. (1990). "Pore pressure response during failure in soils," *Geological Society of America Bulletin*, 102, 428-438.
Humphrey, N.H. (1982). "Pore pressures in debris flow initiation," *Report 45*, State of Washington, Water Research Center, Pullman, Wash.
Iverson, R.M., and Major, J.J. (1986). "Groundwater seepage vectors and the potential for hillslope failure and debris flow mobilization," *Water Resources Research*, 22(11), 1543-1548.
Iverson, R.M., and LaHusen, R.G. (1992). "Momentum transport in debris flows: large-scale experiments," *EOS*, 73(43), 227.
Jibson, R.W. (1989). "Debris flows in southern Puerto Rico," *Landslide processes of the eastern United States and Puerto Rico*, Geological Society of America, Special Paper 236, 29-56.
Johnson, K.A., and Sitar, N. (1986). "Techniques for identification of source areas for debris flows," *Report No. UCB/GT/86-01*, Dept. of Civ. Engrg., Univ. of California, Berkeley, Calif.

Johnson, K.A., and Sitar, N. (1990). "Hydrologic conditions leading to debris-flow initiation," *Canadian Geotechnical Journal*, 27, 789-801.
Keefer, D.K., Wilson, R.C., Mark, R.K., Brabb, E.E., Brown, W.M. III, Ellen, S.D., Harp, E.L., Wieczorek, G.F., Alger, C.S., and Zatkin, R.S. (1987). "Real-time landslide warning during heavy rainfall," *Science*, 238, 921-925.
Kobashi, S., and Suzuki, M. (1987). "The critical rainfall (danger index) for disasters caused by debris flows and slope failures," *International Association of Hydrological Sciences*, Publication No. 165, 201-211.
LaHusen, R.G. 1990. "Debris-flow detection system," The eruption of Redoubt Volcano, Alaska, December 14, 1989-August 31, 1990, U.S. Geological Survey Circular 1061.
Lehre, A.K. (1987). "Rates of soil creep on colluvium-mantled hillslopes in north-central California," *International Association of Hydrological Sciences*, Publication No. 165, 91-100.
Lips, E.W., and Wieczorek, G.F. (1990). "Recurrence of debris flows on an alluvial fan in Central Utah," American Society of Civil Engineers, *Proceedings of International Symposium Hydraulics/Hydrology of Arid Lands*, San Diego, Calif., 555-560.
MacArthur, R.C., Hamilton, D.L., and Mason, R.C. (1990). "Numerical simulation of mudflows from the hypothetical failure of a debris blockage lake below Mount St. Helens, Washington," American Society of Civil Engineers, *Proceedings of International Symposium Hydraulics/Hydrology of Arid Lands*, San Diego, Calif., 416-421.
Mark, R.K. (1992). "Map of debris-flow probability, San Mateo County, California," U.S. Geological Survey Miscellaneous Investigations Series Map I-1257-M.
Mathewson, C.C., Keaton, J.R., and Santi, P.M. (1990). "Role of bedrock ground water in the initiation of debris flows and sustained post-storm stream discharge," *Bulletin of the Association of Engineering Geologists*, 27(1), 73-78.
McCarter, M.K., and Kaliser, B.N. (1985). "Prototype instrumentation and monitoring programs for measuring deformation associated with landslide processes," *Proceedings of Specialty Conference on Delineation of Landslide, Flash Flood, and Debris Flow Hazards in Utah*, Utah State University, 30-49.
Mizuyama, T., Yazawa, A., and Ido, K. (1987). "Computer simulation of debris flow depositional processes," *International Association of Hydrological Sciences*, Publication No. 165, 179-190.
Mizuyama, T., and Ishikawa, Y. (1990). "Prediction of debris flow prone areas and damage," American Society of Civil Engineers, *Proceedings of International Symposium Hydraulics/Hydrology of Arid Lands*, San Diego, Calif., 712-717.
Montgomery, D.R., Wright, R.H., Booth, T. (1991). "Debris flow hazard mitigation for colluvium-filled swales," *Bulletin of the Association of Engineering Geologists*, 28(3), 303-323.
Morton, D.M., Campbell R.H., Barrows A.G. Jr., Kahle J.E., and Yerkes, R.F. (1979). "Wright Mountain mudflow: spring 1969," Landsliding and mudflows

at Wrightwood, San Bernardino County, California, Part II, *California Division of Mines and Geology*, Special Report 136, 7-21.

Neary, D.G., and Swift, L.W. Jr. (1987). "Rainfall thresholds for triggering a debris avalanching event in the southern Appalachian Mountains," *Debris flows/avalanche: process, recognition, and mitigation*, Geological Society of America Reviews in Engineering Geology, 7, 81-92.

O'Brien, J.S., and Fullerton, W.T. (1990). "Two-dimensional modeling of alluvial fan flows," American Society of Civil Engineers, *Proceedings of International Symposium Hydraulics/Hydrology of Arid Lands*, San Diego, Calif., 262-273.

Okimura, T. (1983). "Rapid mass movement and groundwater level movement," *Zeitschrift für Geomorphologie*, 46, 35-54.

Orme, A.R. (1990). "Recurrence of debris production under coniferous forest, Cascade foothills, northwest United States," *Vegetation and Erosion, Processes and Environments*, John Wiley & Sons, 67-84.

Pierson, T.C. (1980). "Piezometric response to rainstorms in forested hillslope drainage depressions," *Journal of Hydrology*, New Zealand, 19(1), 1-10.

Pierson, T.C. (1983). "Soil pipes and slope stability," *Quarterly Journal of Engineering Geology*, London, 16, 1-11.

Pierson, T.C., and Costa, J.E. (1987). "A rheologic classification of subaerial sediment-water flows," *Debris flows/avalanches: Process, recognition and mitigation*, Geological Society of America, Reviews in Engineering Geology, 7, 1-12.

Pierson, T.C., Janda, R.J., Thouret, J. and Borrero, C.A. (1990). "Perturbation and melting of snow and ice by the 13 November 1985 eruption of Nevado del Ruiz, Colombia, and consequent mobilization, flow and deposition of lahars," *Journal of Volcanology and Geothermal Research*, 41, 17-66.

Reid, M.E., Nielsen, H.P., and Dreiss, S.J. (1988). "Hydrologic factors triggering a shallow hillslope failure," *Bulletin of the Association of Engineering Geologists*, 25(3), 349-361.

Reid, M.E., and Iverson, R.M. (1992). "Gravity-driven groundwater flow and slope failure potential 2. Effects of slope morphology, material properties, and hydraulic heterogeneity," *Water Resources Research*, 28(3), 939-950.

Reneau, S.L., and Dietrich, W.E. (1987). "The importance of hollows in debris flow studies; examples from Marin County, California," *Debris flows/avalanches: Process, recognition and mitigation*, Geological Society of America, Reviews in Engineering Geology, 7, 165-180.

Reneau, S.L., Dietrich, W.E., Donahue, D.J., Jull, A.J.T., and Rubin, M. (1990). "Late Quaternary history of colluvial deposition and erosion in hollows, central California Coast Ranges," *Geological Society of America Bulletin*, 102, 969-982.

Santi, P.M. and Mathewson, C.C. (1988). "What happens between the scar and the fan? The behavior of a debris flow in motion," *Twenty-fourth Annual Symposium on Engineering Geology and Soils Engineering*, Washington State Univ., Pullman, Wash., 73-88.

Sidle, R.C. (1984). "Shallow groundwater fluctuations in unstable hillslopes of coastal Alaska," *Zeitschrift für Gletscherkunde und Glazialgeologie*, 20, 79-95.
Sidle, R.C. (1987). "A dynamic model of slope stability in zero-order basins," *International Association of Hydrological Sciences*, Publication No. 165, 101-110.
Smith, T.C. (1988). "A method for mapping relative susceptibility to debris flows, with an example from San Mateo County," *Landslides, floods, and marine effects of the storm of January 3-5, 1982, in the San Francisco Bay Region, California*, U.S. Geological Survey Professional Paper 1434, 185-194.
Takahashi, T. (1991). *Debris Flow*, A.A. Balkema, Rotterdam.
VanDine, D.F. (1985). "Debris flows and debris torrents in the southern Canadian Cordillera," *Canadian Geotechnical Journal*, 22, 44-68.
Waldron, E.H. (1967). "Debris flow and erosion control problems caused by the ash eruptions of Irazu Volcano, Costa Rica," U.S. Geological Survey Bulletin 1241-I.
Watters, R.J. (1983). "A landslide induced waterflood-debris flow," *Bulletin of the International Association of Engineering Geology*, 28, 177-182.
Whipple, K.X. (1992). "Predicting debris-flow runout and deposition on fans: the importance of the flow hydrograph," *International Association of Hydrological Sciences*, Publication No. 209, 337-345.
Wieczorek, G.F. (1987). "Effect of rainfall intensity and duration on debris flows in central Santa Cruz Mountains, California," *Debris flows/avalanches: Process, recognition and mitigation*, Geological Society of America, Reviews in Engineering Geology, 7, 93-104.
Wieczorek, G.F., Lips E.W., and Ellen, S.D. (1989). "Debris flows and hyperconcentrated floods along the Wasatch Front, Utah, 1983 and 1984," *Bulletin of the Association of Engineering Geologists*, 26(2), 191-208.
Wilson, C.J., and Dietrich, W.E. (1987). "The contribution of bedrock groundwater flow to storm runoff and high pore pressure development in hollows," *International Association of Hydrological Sciences*, Publication No. 165, 49-59.
Wilson, R.C. (1989). "Rainstorms, pore pressures, and debris flows: a theoretical framework," *Landslides in a semi-arid environment*, Publication of the Inland Geological Society of Southern California, 2, 101-117.
Wilson, R.C., Torikai, J.D., and Ellen, S.D. (1992). "Development of rainfall warning thresholds for debris flows in the Honolulu District, Oahu," U.S. Geological Survey, Open-File Report 92-521.
Ziemer, R.R., and Albright, J.S. (1987). "Subsurface pipeflow dynamics of north-coastal California swale systems," *International Association of Hydrological Sciences*, Publication No. 165, 71-80.

RESEARCH NEEDS FOR DEBRIS FLOW DISASTER PREVENTION

T.R. Davies[1]

INTRODUCTION

The current International Decade for Natural Disaster Reduction (I.D.N.D.R.) is intended to focus the attention of scientists and others on the clear need to reduce the human suffering, death and material damage caused by natural disasters. Recognition of this need acknowledges that previous attempts at disaster reduction have not been sufficiently successful; consequences of natural disasters seem to be getting worse with time, and it is widely admitted that there is a strong positive correlation between investment in hazard prevention works (e.g., avalanche protection, flood control) and the subsequent damage costs resulting from the hazard. It is thus timely to question the fundamental assumptions of conventional disaster prevention philosophies, to attempt to develop improved strategies for prevention, and to identify the scientific research needed to make these strategies work. This paper attempts such a synthesis in the context of debris flow disasters.

DEBRIS FLOW HAZARDS

A 'typical' debris flow event in a mountain valley results from very intense rainfall or snowmelt, and consists of one or more surge waves carrying mud, rocks, boulders, logs, etc. (Davies et al, 1992). The surges typically can be up to 6 m high, travel at up to 11 m/s, and be twice as dense as water; local channel bed erosion of 15 m depth has been recorded (Haeberli et al, 1992). Of the order of half a million cubic metres of sediment can be involved in an event, most of which can be deposited on the valley fan to depths of several metres (Pierson, 1980). Impact forces of several thousand N/m^2 have been recorded in confined channels. The effect of such an event on human habitation, communications, agriculture and industry can be devastating. Massive concrete structures can be moved, abraded, undermined or buried; houses can be bodily moved or destroyed; factories can be filled with debris, and fields or plantations scoured or buried. Bridges are easily overwhelmed if their clearance is insufficient. A large event can completely relocate a river channel on a fan, and surges entering a river from a side valley can block it and cause flooding upstream (Haeberli et al, 1992).

Clearly structural measures to protect against debris flows are difficult to design and expensive to build, due to the very intensity of the phenomenon, and its ability to undermine or bury a structure, as well as destroying it by impact or abrasion.

ANALYSIS

We have already seen that there are severe limitations to the reliability of structures intended to modify the behaviour of debris flows. Even though the design of water flood control

---

[1] Department of Natural Resources Engineering, Lincoln University, Canterbury, New Zealand.

structures is much easier, the present trend in attempting to mitigate water floods is away from structural works and towards 'soft' countermeasures such as hazard zoning, land use planning and flood forecasting; this is because the latter strategies are much more reliable and effective in mitigating the effects of infrequent events, and have very much smaller environmental, social and cultural impacts (Young and Davies, 1989; Davies and Hall, 1992). It is therefore very unlikely that structural countermeasures will be effective in reliably reducing the scale of debris flow disasters, given that debris flows are much less common and more destructive than water floods.

Short-term experience with structural countermeasures is often good, but experience in 1987 in Switzerland shows that a major event can destroy structures, releasing the sediment they have held in place for many years, and causing the volume of sediment involved to be much greater than it would have been without the event. The history of the Val Varuna catchment illustrates this point (Haeberli et al, 1992); Fig. 1 shows the estimated debris flow volumes historically recorded. Construction of check-dams began in the early 19th century and was completed by the century's end. The mean sediment yield is about 2000 m^3/yr, and it is noticeable that, although short-term yields can vary from this, over the longer term a period of inactivity is followed by a large event and the mean is restored. The longest period of inactivity corresponded to the presence of 28 check-dams in the 2 km of the channel most liable to deep erosion, and the 1987 event following this period, although of comparable rainfall to that of 1834, released almost an order of magnitude more sediment than any previous event when these check-dams were destroyed. This strongly suggests that, in the longer term, natural erosion can only be modified temporarily by structures. Since it is impracticable to build a structure that will control a very large event, when this event occurs it will restore the long-term mean erosion rate. The effect of structural erosion control is to change the progress of erosion from a series of frequent, small events to less frequent, but much larger, events.

The ability of structural measures to increase the predictability of damage due to debris flows is thus not great, and is almost certainly insufficient to be effective in reducing the severity of debris flow disasters; indeed, the opposite might well be true. The possibilities that remain are to be able to predict the occurrence and behaviour of unmodified debris flows, and to modify human behaviour to match the degree of predictability that can be achieved.

PREDICTABILITY OF DEBRIS FLOWS

In principle, any area which has in the geologically recent past been affected by a debris flow can be so affected again at some time in the future; and the time between recurrences can be very short, even for major events, so that any event that has occurred in the past can recur in the near future. Hence, in order to delineate areas susceptible to debris flows, one can simply seek for evidence of previous flows, such as characteristic channel morphologies, sediment deposits (Eisbacher and Clague, 1984), tree age distributions (Strunk, 1992), and tree scars (Hall, 1993). This process might be assisted by attempts to model the deposit extent of a particular debris flow in a particular place (Takahashi, 1991), but given the present state of understanding of the phenomenon and the complexity of local circumstances, the empirical approach would seem to engender more confidence.

Predicting when a debris flow will occur at a given location is impossible in the long term. Occurrence of a debris flow requires

(i) sufficient accumulation of loose material by weathering on slopes and from small mass movements into channels, and
(ii) sufficient intensity of rainfall to trigger the process (Davies et al, 1992). It is thus weather dependent.

In the shorter term it would be possible to monitor sediment accumulation in the gully to determine a potential hazard, and if this were high, to monitor antecedent moisture and local weather for the probability of intense runoff in the catchment; in this way a forecast could be made of the high probability of an event, maybe up to 24 hours ahead. Although limited, such a degree of predictability does allow significant human use of known debris flow susceptible areas, but it is important to note that less than 12 hours' warning renders evacuation liable to occur in darkness and in bad weather, which in itself constitutes an additional hazard.

STRATEGY FOR DISASTER PREVENTION

Davies and Hall (1992) reported the introduction of a forecasting-warning-evacuation strategy for a small holiday community subject to log-jam initiated flash floods similar in effect to debris flows. The significant components of this strategy are

(i) establishment of antecedent catchment conditions that would allow intense rain to trigger the phenomenon - in this case a 15-day antecedent rainfall was established empirically;
(ii) systematic monitoring of developing weather patterns so that evacuation of susceptible areas can occur in favourable conditions well before the situation becomes critical;
(iii) involvement of the local community in the monitoring and warning procedures;
(iv) explanation to the affected people of the investigations that led to these procedures being adopted.

The main problem with such 'soft' countermeasures is the inevitability of false alarms. The community affected needs to understand the hazard they suffer, and the inability of technology to control it, in order to willingly undertake evacuation knowing that the debris flow might not occur; and that this is the smallest price they can pay for reducing the risk of death by debris flow to close to zero.

Quite clearly this strategy does not reduce damage to buildings, roads or crops, and might be seen as inferior to structural countermeasures in this regard. Given the serious doubts about the longer-term reliability of structural debris-flow works referred to earlier, however, this view is highly questionable. When buildings are destroyed by debris flow it would seem sensible not to rebuild them on the same site. The advantages of the strategy in saving lives are obvious - it is inexpensive, reliable, and environmentally clean. It also leads, in time, to a situation where a decreasing number of highly susceptible activities occur in hazard areas. Questions might well be raised as to its political acceptability, the response to which is to ask whether the politicians concerned are prepared to take responsibility for deaths occurring if such evacuation systems are not used where they are feasible.

INFORMATION NEEDS

In order to implement a strategy of soft countermeasures against debris flow disaster, research is needed to increase our ability to:

(i) **Identify catchments in which debris flows have occurred and delineate areas affected.**

The signatures of debris flow activity in catchments and on fans are relatively well known by geomorphologists, but less so by engineers. More attention needs to be paid to dendrochronology as a means of deducing the extent, age and intensity of debris flow activity (Eisbacher and Clague, 1984; Jackson et al, 1987; Strunk, 1992). Until recently there was a tendency for debris flow events to be reported as 'flash floods', and there is still considerable vagueness about the distinction; we suspect that a significant number of damaging surge events in small channels might be initiated by log-jam failures, and this is an area that needs both field and laboratory investigation. The signature of such an event probably differs from that of a debris flow (Higgs, 1974; Hall, 1993).

(ii) **Antecedent and triggering conditions.**

Given the present state of quantitative understanding of how debris flows are initiated, it is probably over-optimistic to expect much theoretical input to establishing which combinations of antecedent catchment (moisture, sediment, vegetation) conditions and rainfall (or snowmelt) intensity could cause a debris flow in a given catchment. However, a warning/evacuation strategy is crucially dependent on reliable identification of such conditions. This must, therefore, be done empirically on a catchment-by-catchment basis, but in the many catchments for which insufficient records exist this will not be possible and less reliable inferences will need to be made. This whole area is a fertile and urgent research field.

(iii) **Monitoring.**

Once critical antecedent and triggering conditions are established, it is necessary to monitor the former and be able to forecast the latter as far in advance as possible. Monitoring catchment moisture status by cumulative rainfall or snowmelt measurement is relatively straightforward. Sediment accumulation is more difficult, necessitating as it does access to remote gullies. Recent advances in PC-based photogrammetry might allow regular aerial photography to be interpreted sufficiently cheaply and precisely to detect bed-level rise in small channels, and if so a large number of sites could be surveyed rapidly. Likewise, vegetation status is probably amenable to interpretation from aerial or even satellite imagery. In each case, however, ground truth will be needed for the initial calibration of remote imagery. Forecasting intense precipitation or snowmelt events will always be somewhat fallible. Recent technical advances in radar measurement of precipitation intensity hold the promise of monitoring the progress of a belt of intense frontal rain towards a catchment, but forecasting the location of convectional rain is much less reliable and more probabilistic.

(iv) **Public Involvement.**

Research to date (Hall, 1993; Morrison, 1993; Pyle, 1992) suggests that a crucial determinant of success is public participation in (or 'ownership of') the decision to implement a 'soft' strategy and in the monitoring-forecasting-warning-evacuation process. If the public can be fully and clearly informed about the hazard that exists, about the inability of science to predict its occurrence and magnitude, and about the very unreliable long-term protection afforded by structural measures; and if it is simply explained that the proposed system is the only way to guarantee saving lives in the long term, then public acceptance of false alarms is likely to be high. Little research has been done on public perceptions of geomorphic hazard and the associated risks, however, and there is a need to apply to the debris flow situation some of the progress recently made in flood risks and hazards (e.g., Pyle and Gough, 1991; Morrison, 1993).

CONCLUSION

A 'soft' disaster reduction strategy is proposed based on identification of hazard location, extent and maximum intensity; monitoring of antecedent catchment moisture, sediment and vegetation conditions; and forecasting of intense precipitation or snowmelt, so that a warning-evacuation procedure can be implemented that will reliably prevent deaths due to debris-flow.

In order to increase the reliability and public acceptability of such a strategy, research is urgently needed to:

(i) better identify catchment areas susceptible to debris flow (or flash flood);
(ii) better understand the geotechnical/hydrological conditions necessary for initiation of a debris flow;
(iii) improve ability to monitor conditions in inaccessible catchments regularly and reliably;
(iv) more reliably forecast the occurrence of intense rain or snowmelt in a given catchment at least 12 (and preferably 24) hours in advance;
(v) better understand public perception of debris flow and flash flood hazards; devise methods of informing the public realistically about the hazards present; involve the public in the decision-making and implementation processes, and in the operation of the 'soft' strategy, so that public response to evacuation decisions remains consistently good.

REFERENCES

Davies, T.R.H., and Hall,R.J. (1992). A realistic strategy for disaster prevention. Interpraevent 1992, Band 3, 381-390.
Davies, T.R.H. et al (1992). Debris flow behaviour - an integrated overview. Proceedings, Erosion, Debris Flows and Environment in Mountain Regions, Chengdu Symposium, IAHS Publication No.209, 217-225.
Eisbacher, G.H., and Clague, J.J. (1984). Destructive mass movements in high mountains; hazard and management. Paper 84-16, Geological Survey of Canada, 230 p.
Haeberli, W.; Rickenmann, D., and Zimmerman, M. (1992). Murgänge 1987; Dokumentation und Analyse, Teil 1. VAW, ETH-Zürich, Switzerland. 264 p.
Hall, R.J. (1993). M.E.(Nat.Res.) Thesis, Lincoln University, Canterbury, New Zealand, in prep.
Higgs, H.C. (1974). Assessing flash flood potential from channel measurements. Proceedings, Flash Floods Symposium, Paris. IAHS Publication No.112, 52-56.
Jackson, L.E. Jr; Kostaschuck, R.A., and MacDonald, G.M. (1987). Identification of debris flow hazard on alluvial fans in the Canadian Rocky Mountains. Reviews in Engineering Geology, Volume VII, Geological Society of America, 115-124.
Morrison, K.D. (1993). Floodplain management - a postmodern approach. Ph.D. thesis, Lincoln University, New Zealand, in prep.
Pierson, T.C. (1980). Erosion and deposition by debris flows at Mt Thomas, New Zealand. Earth Surfaces Processes, Vol.5, 227-247.
Pyle, E. (1992). Sustainable water management. M.Appl.Sc. thesis, Lincoln University, New Zealand, 149p.
Pyle, E., and Gough, J.D. (1991). Environmental Risk Assessment for New Zealand; a Guide for Decision-makers. Information Paper No.29, Centre for Resource Management, Lincoln University, New Zealand, 46 p.
Strunk, H. (1992). Reconstructing debris flow frequency in the Southern Alps back to AD 1500 using dendromorphological analysis. Proceedings, Erosion, Debris Flows and Environment in Mountain Regions, Chengdu Symposium, IAHS Publication No.209, 299-306.
Takahashi, T. (1991). Debris Flow. Balkema, Rotterdam, 165 p.
Young, J.R., and Davies, T.R.H. (1989). Realistic criteria for flood control design. Proceedings, Hydrology and Water Resources Symposium, Christchurch, N.Z., 227-231.

DEBRIS FLOW DISASTER PREVENTION 1289

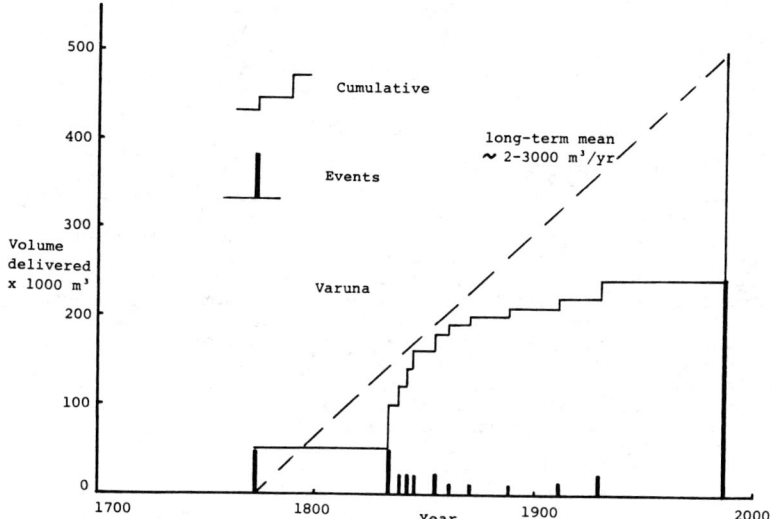

Figure 1. Sediment Yield Events and Cumulative Yield from Debris Flows in Val Varuna, Switzerland, 1772-1987

DEBRIS FLOWS IN GRAND CANYON NATIONAL PARK, ARIZONA: MAGNITUDE, FREQUENCY AND EFFECTS ON THE COLORADO RIVER

By Theodore S. Melis[1] and Robert H. Webb[1], Non-Members, ASCE

Abstract

Debris flows are recurrent sediment-transport processes in 525 tributaries of the Colorado River in Grand Canyon, Arizona. Initiated by slope failures in bedrock and(or) colluvium during intense rainfall, Grand Canyon debris flows are high-magnitude, short-duration floods. Debris flows in these tributaries transport very large boulders into the river where they accumulate on debris fans and form rapids. The frequency of debris flows range from less than 1 per century to 10 or more per century in these tributaries. Before regulation by Glen Canyon Dam in 1963, high-magnitude floods on the Colorado River reworked debris fans by eroding all particles except large boulders. Because flow regulation has substantially decreased the river's competence, debris flows occurring after 1963 have caused increased accumulation of finer-grained sediments on debris fans and in rapids.

Introduction

Debris flows occur in 525 tributaries of the Colorado River in Grand Canyon between Lees Ferry (mile 0) and Diamond Creek (mile 225). Most tributaries range in size from less than 1 to 800 km^2; 98 percent are ephemeral. Initiated by slope failures during intense rainfall, debris flows transport coarse-grained sediments to the Colorado River where they accumulate as debris fans (Howard and Dolan, 1981). Boulders that cannot be moved by the river form large rapids (Webb and others, 1988 and 1989) that attract 20,000 whitewater enthusiasts each year (Stevens, 1990). Before regulation of the river by Glen Canyon Dam in 1963, high-magnitude floods reworked debris fans by eroding particles smaller than boulders, resulting in bouldery debris fans. Debris flows occurring after 1963 deposited finer-grained sediments on debris fans that persist owing to reduced annual peak discharges below Glen Canyon Dam. The purpose of this paper is to describe the process of debris flow in Grand Canyon and the relations between recent debris flows, debris fans, and flow in the Colorado River.

---

[1]Hydrologists, U.S. Geological Survey, Water Resources Division, Desert Laboratory, 1675 W. Anklam Road, Tucson, Arizona 85745

Initiation of Debris Flows

Debris flows in Grand Canyon require the co-occurrence of intense rainfall and slope failures. Debris flows are typically initiated during intense thunderstorms occurring between July and October. Rainfall intensities commonly exceed 25 mm·hr^{-1} during storms that cause debris flows. Total rainfall for historic debris flows ranged from 27 to 355 mm (Webb and others, 1989).

Debris flows are initiated from failures in either bedrock slopes or hillslope colluvium or a combination of both. Five initiation mechanisms have been identified in Grand Canyon: 1) bedrock failures occurring in high-elevation Permian Hermit Shale, shales and sandstones of Permian Supai Group strata and Cambrian Muav Limestone; 2) runoff in channels pouring hundreds of meters over Mississippian Redwall Limestone falls onto unconsolidated colluvium overlying Muav Limestone and Cambrian Bright Angel Shale during intense storms, causing "the firehose effect" (Johnson and Rodine, 1984) which is the most common initiation mechanism in Grand Canyon; 3) runoff generated outside Grand Canyon, especially in drainages having large, high-elevation subcatchments on the North Rim, falls over cliffs of Kaibab Limestone mobilizing unconsolidated materials overlying Hermit Shale and Supai Group strata; 4) failures may occur directly under intense rainfall, or by bank failure adjacent to steep, incised channels in unconsolidated alluvium overlying Muav Limestone and(or) Bright Angel Shale; and 5) failures in cliff-forming strata, such as Kaibab and Redwall Limestones and Permian Coconino Sandstone, may occur in conjunction with other types of failures that provide fine-grained sediments for the debris-flow matrix. Owing to combinations of these initiation mechanisms, variable lithologies and drainage conditions in Grand Canyon, debris flow initiation is complex in these tributaries.

Characteristics of Recent Grand Canyon Debris Flows

Debris flows in Grand Canyon travel 1 to 20 km from initiation points to the Colorado River. They usually contain 15 to 20 percent water by-weight and poorly-sorted sediments including clay (less than 2 percent), sand (approximately 30 percent) and boulders (10 percent) as large as 5 m (Webb and others, 1989). In Grand Canyon, transport of large boulders is facilitated by abundant boulders in source areas, steep side canyon gradients, high peak discharges and relatively short transport distances from source areas to the Colorado River.

Several recent events illustrate typical debris-flow activity in Grand Canyon. On September 24, 1990, debris flows occurred in three tributaries between miles 61 and 65 on the Colorado River. These debris flows, initiated by rainfall intensities ranging from 25 to 50 mm·day^{-1}, occurred following a week of rainfall averaging 24 mm·day^{-1}. The primary initiation mechanism for the debris flows was a combination of bedrock and hillslope failure mobilized by the firehose effect. All debris flows in this reach of river were high-magnitude, short-duration pulses that traveled 1 to 2 km and aggraded debris fans. Instantaneous peak discharges estimated for two of the debris flows from superelevated mudlines preserved on bends, ranged from 300 to 350 m^3·s^{-1} near the Colorado River.

In tributary 62.6-R (our convention for unnamed canyons is "river mile-side") we documented the evolution of debris-flow peak-discharge at five sites. This debris flow initiated after runoff and rockfall sediments struck massive colluvial deposits overlying Muav Limestone and Bright Angel Shale (firehose effect). The peak discharge estimated 0.2 km downstream of this location, designated site A, was 100 $m^3 \cdot s^{-1}$. The flow bulked-up in the steep channel below site A, to 300 $m^3 \cdot s^{-1}$ as it flowed another 0.6 km. Peak discharge was maintained for 0.4 km until reaching the mouth of the 1.5 km^2 drainage. Between its initiation site and the river, the flow's average particle size increased, while its water content decreased and average flow velocity increased erratically. A large streamflow flood followed the debris flow pulse, extensively reworking channel deposits.

Peak debris-flow discharge at tributary 63.3-R was estimated at 300 to 350 $m^3 \cdot s^{-1}$ in a single channel bend 0.1 km upstream of the confluence. Within this bend we also found evidence of an earlier debris flow with a peak discharge of 800 to 1,000 $m^3 \cdot s^{-1}$. We were unable to determine the exact duration of either the 1990 flow or the older flow here, however, twigs preserved in mudcoats yielded an age of 5,400 years B.P. for the older debris flow.

Total sediment volume transported from tributary 62.6-R to the Colorado River was determined by surveying and trenching the new debris fan deposit; we measured its volume to be 1,200 m^3. A maximum duration for the debris flow peak in tributary 62.6-R was estimated at 3 to 6 seconds. The 1990 debris flows created a new rapid at the mouth of tributary 62.5-R and increased the severity of the former riffle at tributary 63.3-R. Debris flow in tributary 62.5-R transported several large boulders into the river, one of which weighs approximately 250 megagrams (Mg). Several other smaller boulders contributed to the formation of a new rapid at that location. At tributary 62.6-R, the 1990 debris-flow deposit buried approximately 30 percent of an existing sand bar on the downstream side of the debris fan. Several large boulders were also deposited on the debris fan at tributary 62.6-R, but caused no changes in the hydraulics of the river. At tributary 63.3-R, large boulders were also deposited on the debris fan and in the river, increasing the severity of an existing riffle.

<u>Frequency of Debris Flows</u>

We analyzed ground and aerial photography to determine historical frequency of debris flows in Grand Canyon. By replicating over 700 ground photographs, and analyzing several others, we obtained frequency information for approximately 130 of 525 tributaries. In addition, aerial photography from 1935, 1965, 1973, 1984, and 1989 was analyzed. The frequency of debris flows is highest between river mile 61.5 and 77.0, where 95 percent of tributaries viewed had at least one debris flow in the last century. The lowest frequency occurred between mile 132 and 160; only half the tributaries in this reach had experienced a debris flow in the last century. Debris-flow frequencies have also been determined for several drainages using dendrochronology and radiometric dating of organic materials. We have obtained a complete historical record and a partial prehistoric record (5,400 years B.P.) of debris-flow activity in drainages between miles 61 and 64. Tree-ring analyses of debris-flow scarred catclaw trees (*Acacia greggii*) indicate that at least three debris flows occurred in tributary 62.6-R during the last 130 years. On average, the frequency of debris flows in any tributary is about one to four per century.

Grand Canyon Debris Fans and Rapids

Before regulation by Glen Canyon Dam, debris fans were periodically eroded, or reworked, by floods of the Colorado River. Reworking of fans created residual surfaces comprised of boulders; gravel and cobble eroded from the debris fan are often deposited downstream as bars or mid-channel islands. These downstream deposits, termed debris bars, are well-sorted and imbricated. Primary debris-flow deposits, reworked debris fans, and debris bars have distinctive particle-size and lithologic distributions. We compared particle-size data from the reworked pre-dam debris fan at tributary 62.6-R with the 1990 debris-flow deposit. Particle-size data for the reworked fan and 1990 deposit were compared to the particle-size distribution of a debris bar (island) located 0.2 km downstream of the debris fan. Our analyses of these particle size data revealed that approximately 15 to 20 percent of the particles contained in the 1990 debris flow deposit could not have been transported by pre-dam floods based on reworked fan textures. Most of the 1990 deposit remains uneroded by dam releases, not only because flood magnitudes have been decreased, but because most of the new deposit is located above the range of normal flow releases. Particle-size data for the debris bar deposit indicate that reworked particles 0.2 to 0.5 m in diameter were formerly transported only short distances downstream from debris fan surfaces.

The relation between reworked debris fans and debris bars downstream is illustrated by particle-size and lithologic data from the reworked debris fan and debris bar associated with tributary 62.6-R. These data show a downstream shift in the particle-size distribution and lithologic composition from debris fan to debris bar. Lower-elevation lithologies that greatly resist erosion during debris flow transport, such as Redwall and Muav Limestone, Cambrian Tapeats Sandstone, and Older Precambrian rocks, generally dominate reworked debris fans in central and western Grand Canyon since they generally contribute larger particles to fan deposits. Other resistant strata, such as Permian Kaibab Limestone and Coconino Sandstone, dominate reworked debris fans when they are at lower elevations relative to the Colorado River. However, percentages of these rocks on debris fans diminish as those formations rise high above river elevation. This occurs quickly as Grand Canyon deepens and widens downstream from Lees Ferry. As a result of changing tributary geology and morphology, debris-flow transported sediments from higher elevation lithologies tend to dominate debris bars rather than debris fans along the river's course in Grand Canyon.

In contrast to Leopold's (1969) conclusions concerning rapid formation, we believe the majority of rapids in Grand Canyon result directly from boulders deposited in the river by debris flows from tributary canyons. Only two rapids in Grand Canyon were created by rockfalls (Webb and others, 1988). Secondary riffles and rapids, which form downstream from the parent debris fan, result from outwash debris bars downstream from large rapids. Most reworked debris fans are comprised mostly of the small percentage of coarsest particles deposited by recurring debris flows. The rate at which boulders exceeding river competence are transported to the river by debris flows controls rates of fan development. Two important factors affecting quasi-equilibrium conditions for any tributary, channel reach, and debris fan are reach-specific channel morphology, and stratigraphic position and lithology of source clasts relative to their erosion during debris flow transport.

Debris flow frequency, induration of source clasts, and transport distance all affect the particle size and lithologic distributions of debris flows reaching the Colorado River and, hence, resulting debris fans. Reach-dependent channel geometries at the river and individual tributary characteristics combine to create debris fans and bars with distinctive geometries and compositions; these fans also affect volumes of stored fine sediments along the river (Webb and other, 1991). Extremely deep, narrow, V-shaped channel reaches, such as the Precambrian gorges of Grand Canyon, favor formation of rapids over debris fans. Such channel geometries lack suitable sites for debris fan formation because of higher flow velocities and excessively steep channel sides. Flows on the Colorado River, therefore, have different erosional effects on debris flow deposits depending on the type of river channel dominating a given reach.

A Debris Fan Sediment Budget

Particle-size data suggest that boulders having b-axis diameters of 0.5 to 1.0 m caused net aggradation on debris fans before regulation by Glen Canyon Dam. These particles provided an immobile foundation for further accumulated debris. Regulated flows now allow a larger percentage of finer-grained sediments contained by debris flows to accumulate on existing fan surfaces. We formulated a conceptual model for pre-dam aggradation and development of debris fans on the Colorado River using a simplified sediment budgeting approach. This approach is based on particle size relationships between debris flows, fans, and bars. Sediments transported in the 1990 debris flows were very poorly sorted with 80 to 90 percent of the sediment less than 0.5 m in diameter. Particle-size distributions measured on three reworked debris fans at tributaries 62.5-R, 62.5-R, and 63.3-R provided an estimate for the competence of floods prior to regulation by Glen Canyon Dam. By re-incorporating the eroded sediments back into these three reworked debris fans, we estimated that 100 to 400 debris flows, ranging from 1,200 to 4,000 m^3 total volume per flow, were required to aggrade each of these three fans to their present sizes. Our approach assumed that debris fans formerly resulted from many relatively small-magnitude debris flows, each contributing a small percent of large boulders that exceeded the pre-dam competence of the Colorado River. Larger magnitude debris flows, similar to the one at 5,400 years B.P., may have occurred more frequently during some earlier stage of the history of these debris fans, but the amount of colluvial source-sediment for such flow magnitudes may have become limiting as the debris fans formed. This would only be true, however, if production rates of source sediments fell behind rates of evacuation owing to debris flows.

Discussion and Conclusions

Debris fans on the Colorado River result from the cumulative effects of magnitude and frequency of debris flows from tributaries and magnitude and frequency of floods on the Colorado River. Using repeat photography and measurement of particle size distributions, we concluded that debris fans in Grand Canyon were very stable prior to regulation in the absence of debris flows from tributary canyons. Before regulation, newly aggraded debris fans would be reworked relatively quickly by floods in the Colorado River. Though the total sediment flux to and away from debris fans may have been great before regulation, total aggradation rates of fans were probably low since only a limited number of large boulders are transported by most debris flows. Since total sediment flux from tributary debris flows is unaffected by regulation of the Colorado River, natural aggradation of debris fans continues, while sediment-transport rates away from debris fans have decreased owing to reduced peak discharges. The effects of flow regulation on reworking of debris fans are further amplified by lower river stages that reduce levels of inundation of debris fans by river flows. Prior to regulation many fans were inundated annually because their low-angle geometries were below the elevation of the average annual discharge (2,478 $m^3 \cdot s^{-1}$). Other debris fans, such as the high-angle fan at Prospect Canyon (179.4-L), were only slightly inundated even by the highest pre-dam flood stages.

Graf (1980) documented increased stability of rapids below Flaming Gorge Dam on the Green River in Utah. Based on our findings, we expect a similar and continuing trend on the Colorado River below Glen Canyon Dam. Although the 1990 debris flows described in this paper were not extremely large or of long duration, these flows were responsible for discharging some very large boulders into the Colorado River. These new boulders increased the severity of one riffle and formed a new rapid while providing a stable foundation for further aggradation of future debris-flow deposits. Following regulation, continued debris flows along the Colorado River in Grand Canyon are now aggrading debris fans with sediments finer than boulders.

Acknowledgments This study was funded by the U.S. Bureau of Reclamation, Glen Canyon Environmental Studies Program. John Parker and William Phillips reviewed earlier versions of this manuscript, we appreciate their helpful suggestions and comments.

References Cited

Graf, W.L., 1980, The Effects of Dam Closure on Downstream Rapids: Water Resources Research, Vol. 16, no. 1, p. 129-136.
Howard, Alan, and Dolan, Robert, 1981, Geomorphology of the Colorado River in Grand Canyon: Journal of Geology, Vol. 89, p. 269-297.
Johnson, A.M., and Rodine, J.R., 1984, Debris flow, in Brunsden, D., and Prior, D.B., editors, Slope instability: Wiley, NY, 620 p.
Leopold, L.B., 1969, The rapids and pools-- Grand Canyon, in The Colorado River Region and John Wesley Powell: U.S. Geological Survey Professional Paper 669, p. 131-145.
Stevens, L., 1990, The Colorado River in Grand Canyon, A Guide: Flagstaff, Arizona, Red Lake Books, 107 p.
Webb, R.H., Pringle, P.T., Reneau, S.L., and Rink, G.R., 1988, The Monument Creek debris flow of 1984: Implications for formation of rapids on the Colorado River in Grand Canyon National Park: Geology, Vol. 16, p. 50-54.
Webb, R.H., Pringle, P.T., and Rink, G.R., 1989, Debris flows in tributaries of the Colorado River in Grand Canyon National Park, Arizona: U.S. Geological Survey Professional Paper 1492, 39 p.
Webb, R.H., Melis, T.S., and Schmidt, J.C., 1991, Historical Analysis of Debris Flows, Recirculation Zones, and Changes in Sand Bars Along the Colorado River in Grand Canyon, Arizona: EOS Supp., Trans. Amer. Geo. Union, Vol. 72, no. 44, p. 219.

Debris Flows and Mass Wasting in Volcanic Torrents

Hiroshi Suwa[1]

Abstract

Observation of debris flow and topographic survey for mass wasting in the volcanic torrents clarified that (1) mass wasting processes such as rock fall from the sidewall and toppling of sidewall of gully are distinctive depending on the topographic, and hydrologic conditions, (2) eruptive activity of volcano remarkably promotes the occurrence of debris flow and (3) debris flow transforms from a fully developed one to a hyperconcentrated one as it runs down to gentler slope.

Introduction

To make the evacuation plans from debris-flow hazard and the effective countermeasures against it, actual processes of debris flow shoud be well known. Field observation of debris flow have been carried out at several sites in the world. Among them the frequency of debris flow is high on volcano, particularly in the midst of and just after the eruptive activity. Processes of debris-flow occurrence would change in time and in space. The relationship between debris-flow occurrence and mass wasting in the volcanic torrents is discussed.

Observation of Debris Flow in the Torrents

The last eruption in 1962 promoted debris flow in the torrents of Yakedake volcano (altitude 2445 m above sea level), Japan, particularly in the Kamikamihori gully on which headwaters the fissure eruption located. The catchment area for the middle reach observation site of debris flow (1580 m a.s.l.) is 0.8 km^2 and the total length of

---

[1]Associate Professor, Disaster Prevention Research Institute, Kyoto University, Gokasho, Uji, Kyoto 611, Japan.

this gully is 2.5 km. Heavy rainfall generates debris flows, but no runoff is found without a high intensity of rainfall (Suwa,1989).

198 years after the former eruption, Unzen volcano began to erupt in 1990. Rainfall runoff began to generate

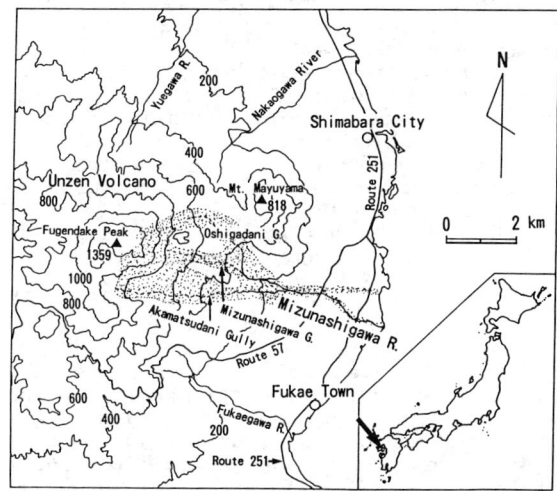

Fig.1. Map of Unzen Volcano

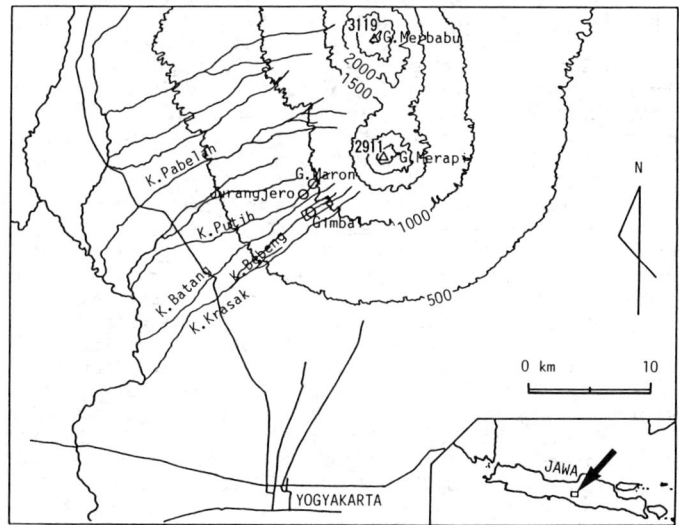

Fig.2. Map of Merapi Volcano

debris flow very easily from the mid May 1991 on the slope of Mizunashigawa river (Fig.1). Pyroclastic flow due to the collapse of lava dome began to occur frequently from the end of May 1991. Observation of debris flow was carried out near the intersection point of the river with the route 57. The catchment area for this point is 12 km^2 and the total length is 7.6 km, and no runoff is found without heavy rainfall.

Merapi volcano generates pyroclastic flows every several years due to the collapse of lava dome. In rainy season, the high intensity of rainfall runoff frequently generates debris flow in many torrents. Observation of debris flow has been carried out at Gimbal in Bebeng river. Catchment area for this observation site is 6 km^2 and total length of Bebeng river is 14 km(cf.Fig.2). Water flow is always found in the lower reach below 1000 m a.s.l. without surface runoff of rainfall.

Mass Wasting in the Gully

The geology of three volcanos is andesite, and central cones of lava dome are surrounded with many layers of tephra, pyroclastic flow and debris-flow deposits.

Frequency and magnitude of debris flow at Kamikamihori valley were very high just after the 1962 eruption, because of low permeability of ash-covered slope. Recently those effects have almost diminished. But rock fall induced firstly by the repetition of freezing and thawing of soil water in early spring and secondly by erosive action of surface runoff in early summer supplies a large amount of debris from sidewalls to the gully bottom. In summer and autumn, certain amount of debris is carried away by debris flows depending on their transporting capacity. Mean volume of debris annually carried away by debris flows is almost balanced with the volume supplied from the sidewalls(Suwa et al.,1988).

Ash-covered slope of Mizunashigawa river is very effective for debris-flow since May 1991. Frequent pyroclastic flows have scattered huge amount (over $10^8 m^3$) of hot debris and ash over the vast area shown as dotted zone higher than 100 m a.s.l. in Fig.1, and lava is steadily supplied to the dome still in March 1993.

The upper reach of Bebeng river is located in the vast area of the 1969 pyroclastic flow deposits. Sidewalls of the gully are very steep due to rather high cohesion of tephra. Various scales of toppling of sidewall are found particularly at the position of undercut slope. Toppled debris sometimes fully dams up the stream. Then a following big flood would easily carry it away. Such processes of mass wasting are very frequent in this gully. Transportation of debris by traction and suspension is always found active at the observation site even without surface runoff of rainfall.

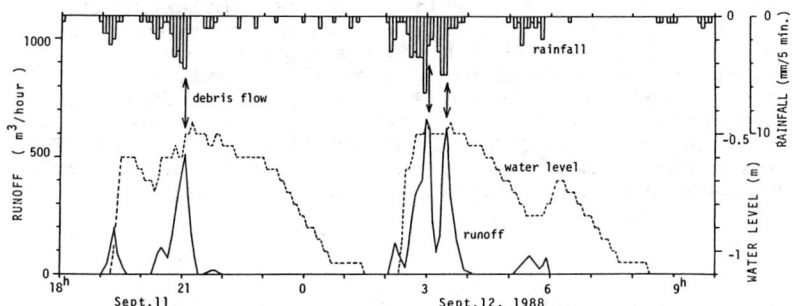

Fig.3. Rainfall, surface runoff and perched water level on the impervious sheet under the bottom at headwaters of Kamikamihori gully

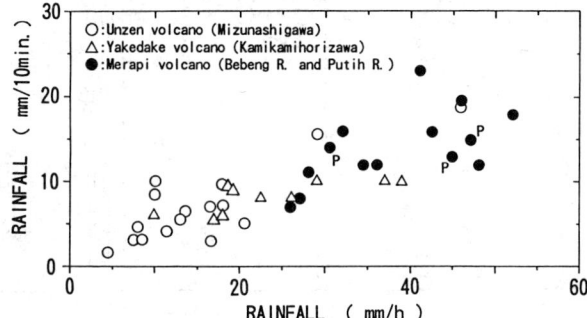

Fig.4. Comparison of rainfall condition for debris flows

Hydrologic Condition of Debris-Flow Occurrence

Hydrologic condition in the gully bottom of Kamikamihori as shown in Fig.3 tells us that a rapid appearance of flood would generate debris flow only if the water content in the surface layer of deposits is high enough with a preceding rainfall. Incorporation of debris on the steep slope would probably convert a muddy water flow to a typical debris flow in which all the debris clasts are fully dispersed at high concentration.

The processes of debris-flow occurrence at Mizunashigawa river is similar to that at Kamikamihori gully. Effectiveness of hot deposits of pyroclastic flow for the occurrence of debris flow is left to be studied.

Relationship between the times of debris flow and

peak intensity of rainfall at Bebeng river showed that in major cases both times coincided well but in minor cases did not. The process in former case is similar to that at Kamikamihori. Time lag of several tens minutes from the peak intensity of rainfall to the debris-flow occurrence indicates that a toppling of sidewall firstly dams up the stream, then break down of the dam generates debris flow.

Rainfall condition for debris flow in three torrents is different each other as shown in Fig.4. Debris flow surely occurs more easily on the slope of Unzen volcano where the eruptive activity is still going on.

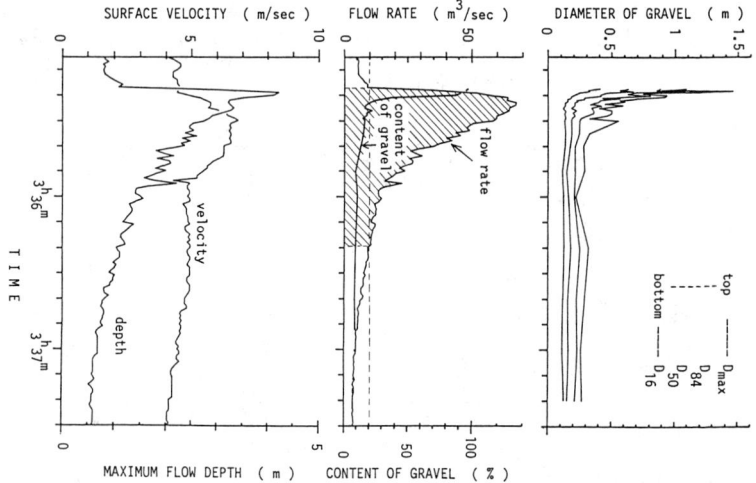

Fig.5. Composition and hydraulic factors of the 5th surge of 12 September 1988 debris flow at Kamikamihori gully

Motion of Debris Flow

A typical hydrograph of boulderly debris flow is shown in Fig.5 which ran down at the observation site of Kamikamihori where the slope angle is about 5 degree. Focusing of large boulders to the massive front and the appearance of peak values in the order of flow depth, flow rate and surface velocity are always found (Suwa,1988). Sloppy type of debris flows are found at the observation sites at Unzen and Merapi volcanos where the slope angle is about 3 degree. At Mizunashigawa river, debris flows of high temperature are frequently found from the hot deposits of pyroclastic flow on the source slope. At Bebeng river the 8 January 1992 debris flow consisted of 10 surges of debris flow and hyperconcentrated flow as shown in Fig.6, which figure shows the lateral movement of stream centre in the width of 75 meters due to the depositive slope for debris flow.

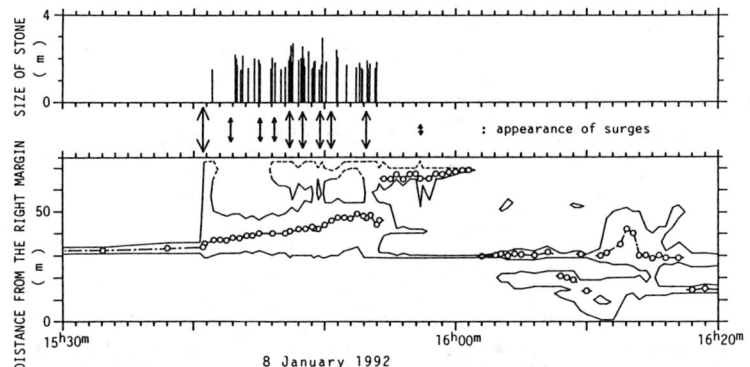

Fig.6. Appearance of boulder larger than 1.5 m and debris-flow surge (top), and changes in flow position, width and the position of maximum velocity as shown with open circles (bottom)

Conclusions

A similar type of boulderly debris flow occurs in the torrents of three volcanos. Debris flow seems to transform from a typical debris flow to a hyperconcentrated flow as it runs down to gentler slope. The rainfall condition for debris flow is highly affected by the stage of volcanic activity. Mass wasting processes are distinctive each other depending on the topographic, hydrologic conditions and past volcanic activities.

References

Suwa, H.(1988) Focusing Mechanism of Large Boulders to a Debris-Flow Front: Trans. Japan. Geomorph. Union, 9(3), 151-178.
Suwa, H. and Okuda, S.(1988) Seasonal Variation of Erosinal Processes in the Kamikamihori Valley of Mt. Yakedake, Northern Japan Alps: CATENA, SUPPLEMENT 13, 61-77.
Suwa, H.(1989) Field Observation of Debris Flow: Proc. Japan-China(Taipei) Joint Seminar on Natural Hazard Mitigation, Kyoto, 343-352.

Interpreting Debris-Flow Hazard from Study of Fan Morphology
Kelin X. Whipple[1]

Abstract

The deposits, stratigraphy, and surface morphology of debris-flow fans are a record of past debris-flow activity, and as such can provide useful information about debris-flow hazards. The morphology of the fan surfaces reflects both characteristic debris-flow inundation patterns and the frequency of channel avulsions; surface morphology is a sensitive indicator of the type of debris-flow hazard. A conceptual model describing the linkages between depositional processes and surface form on debris-flow fans, developed in earlier work, is applied to fans in three field sites to illustrate the use of fan morphologic characteristics in the analysis of debris-flow hazards. This approach should be taken as a first step in any debris-flow hazard mitigation project, particularly where extensive historical records do not exist.

Introduction

Debris flows are important agents of sediment transport and geomorphic change in mountainous regions throughout the world and have excited a great deal of attention, particularly because of their hazard potential. For this reason most research on debris flows has focused on either the process of debris-flow initiation or the fluid-mechanical description of the flow in an effort to predict flow occurrence and run-out distance (Takahashi, 1981; Costa, 1984). Successful prediction and mitigation of the hazard posed by debris flows, however, requires an understanding of the entire suite of processes active on fans, including the interaction of debris flows with fluvial processes and the channel system. This interaction is important because it governs both short-term and long-term patterns of inundation and deposition: without at least a qualitative understanding of the controls on characteristic inundation patterns and the frequency of sudden channel avulsions, responsible predictions of the potential debris-flow hazard can not be made. Fortunately, we have, in the deposits and morphology of the debris-flow fan itself, a record of past debris-flow events. A rapid assessment of the type of debris-flow hazard most prevalent on a fan can be made

---
[1]Department of Geological Sciences, AJ-20, University of Washington, Seattle, Washington 98195, USA

Table 1: Field Conditions

Field Site	Lithology	Climate	Initiation Mechanism
Owens Valley, west	granodiorite	glacial	failure of saturated morainal and supra-glacial debris
Owens Valley, east	meta-sedimentary and meta-volcanic	arid	failure of debris in talus cones or bedrock chutes during brief, intense thunderstorm activity
Xiaoxiang River	sedimentary and meta-sedimentary	humid-monsoonal	massive failure of unstable gully sideslopes during sustained monsoonal rains

from observations of the morphological characteristics of the fan surface. The purpose of this paper is to illustrate how analysis of the deposits and landforms of debris-flow fans can be used to improve hazards recognition and, therefore, the development of mitigation strategies.

Approach: Process and Form

Whipple and Dunne (1992) have developed a conceptual model of debris-flow fan construction which provides a framework within which the linkages between the physical properties of the debris flows (volume, hydrograph shape, sediment concentration, flow recurrence interval) and the morphology of the fan can be evaluated. According to them, the surface morphology of a debris-flow fan reflects: the nature of the fluvial system, the frequency of debris-flow occurrences, typical debris-flow volumes, typical inundation and depositional patterns, and the frequency of sudden lateral shifts of the active channel(s). These last two important factors are set, for a given fluvial system, by the frequency distributions of debris-flow rheologies, volumes, and hydrographs delivered to the fan. Taken together, inundation patterns and channel shifting behavior generally describe the nature of the debris-flow hazard. Morphological characteristics that can be related to different associations of debris-flow and channel behavior (i.e., different types of hazard) include: the number and spacing of abandoned channels, the heights of channel-margin levees, the texture of interfluve surfaces, and the irregularity and convexity of cross-fan topographic profiles (Whipple and Dunne, 1992).

Debris-flow properties are largely determined by source area conditions which set flow initiation mechanisms and debris granulometry. Consequently, great variations in both debris-flow fan morphology and the nature of the debris-flow hazard exist between fans derived from different lithologic assemblages and under different climates. Specifically, inter-fan differences can be ascribed to: (1) differences in debris-flow rheology related to differing debris-flow grain-size distributions (e.g., O'Brien and Julien, 1988; Phillips and Davies, 1991); (2) differences in characteristic debris-flow hydrographs, related to both differing initiation mechanisms and differing grain-size distributions (Dunne and Fairchild, 1984; Fairchild, 1987); (3) differences in debris-flow snout erodibility related to differing concentrations of boulders in the debris flows; and (4) different hydrological conditions associated with the occurrence of debris flows.

Table 2: Debris-Flow Hazards

Field Site	Effective Channelization	Channel Longevity	Type of Hazard
Owens west	strong	short (10's - 100's yr)	channel blockage by bouldery debris-flow snouts and sudden avulsion, narrow corridor of inundation
Owens east	moderate	long (100's - 1000s yr)	overbank flooding by thick, viscous, slow-moving debris flows, extensive near fan-head
Xiaoxiang River Valley	weak	v. short (1 - 10 yr)	massive overbank flooding by large volumes of low-viscosity debris flows, fan-wide inundation

Field Site Descriptions

This paper illustrates the application of the Whipple and Dunne (1992) conceptual model to fans in three field settings: two sites in Owens Valley, California, and one in the Xiaoxiang River valley in China. The three field sites cover a range of geologic and climatic conditions (Table 1) and illustrate a range of debris-flow depositional modes (from well-channelized to unconfined) which result in distinct fan morphological characteristics and are associated with distinct hazards (Table 2).

Fans on the western slope of Owens Valley have surfaces characterized by many well-preserved abandoned channels with little relief on narrow, sharply-defined levees, smooth interfluves, narrow, bouldery debris-flow snouts, and gently convex to flat cross-fan profiles (Fig.s 1a and 2a). Fans on the east side of Owens Valley, in contrast, have surfaces characterized by fewer well-preserved abandoned channels, substantial relief on broad levees, uneven, tiered interfluves (with tiers defined by the margins of wide debris-flow lobes), and more pronounced cross-fan convexity (Fig.s 1b and 2b). The debris-flow fans studied in the Xiaoxiang River Valley lack well-preserved abandoned channels and well-defined levees. They have smooth interfluves textured with on-lapping, thin, areally extensive debris-flow lobes, and only subtle cross-fan convexity (Fig. 1c; see also Fig. 5 of Li et al, 1983).

Geologic Conditions, Fan Morphology, and Debris-flow Hazard

This section aims to (1) explain the morphological differences between fans in the three field areas, and (2) demonstrate the use of morphological interpretations to gauge differences in the debris-flow hazard.

Western slope, Owens Valley, California. These fans were built under glacial climatic conditions, with debris flows originating as failures of morainal or supra-glacial debris or, possibly, as a consequence of glacial outburst floods (Table 1). The debris flows were generally well channelized on the fans owing to vigorous erosion and efficient channel maintenance by sustained meltwater discharges between debris-flows. Debris flows in this area tend to segregate into a dense, bouldery, relatively immobile frontal snout and a more fluid, dilute tail, perhaps as a consequence of the physical properties of the debris itself. Although a correlation with grain-size distribution is unproven, I suggest that the degree of spatial inhomogeneity is set by a combination of: (1) the initiation mechanism(s); (2) the availability and source of additional water inputs; and (3) the mobility of water within

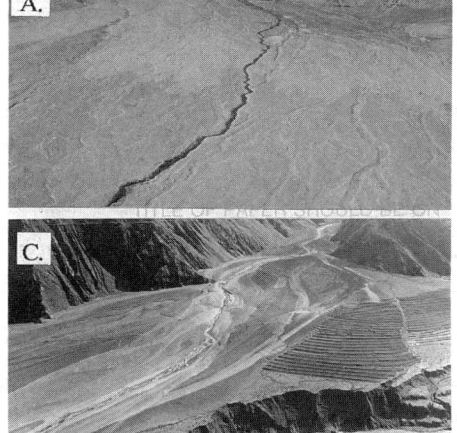

Figure 1. Oblique aerial views of debris-flow fans: A. Owens Valley, west, B. Owens Valley, east. C. Ciaqing Gully, China.

the debris flow, which is a function of grain-size distribution. The idea is that the sand- and granule-rich granitic debris is more susceptible to both draining of the steep frontal bore and dilution of the body by stream water. Certainly dense frontal snouts are common on fans draining granitic source rocks in Owens Valley, and either much less so or absent on the other fans studied.

Although dense frontal snouts were common, they apparently constituted a small proportion of the total sediment delivered to the fan: large-scale overbank deposition of thick, bouldery flows was rare. However, these relatively immobile flows often plugged channels, causing sudden avulsions (Whipple and Dunne, 1992). Frequent channel avulsions have important consequences to both hazards and fan morphology (Fig. 1a). The relatively fluid body and tail of the debris flows constitute the larger share of the sediment flux, have low apparent viscosities, can be effectively channelized, are carried far down-fan, and spread thinly on smooth, low-relief interfluves (see Fig. 1a).

Generally, deposition on this type of fan is limited to a narrow corridor along the active channel: fan-wide deposition rarely, if ever, occurs. However, because of the prevalence of dense, bouldery frontal debris-flow snouts, sudden channel avulsion is a major threat (Table 2). Morphological attributes indicative of fans with this hazard include: (1) many closely-spaced well-preserved abandoned channels with abrupt headward terminations on the fan; (2) channels with narrow, bouldery levees of limited height; and (3) subdued cross-fan convexity (see Fig.s 1a and 2a).

Eastern slope, Owens Valley, California. Debris flows on these fans are generally less-well channelized than on the fans described above. Fan-wide deposition is rare, but debris-flows commonly spread over large portions of the upper fan. This is largely a consequence of marked contrasts in typical debris-flow hydrographs generated in the unglaciated, meta-sedimentary and meta-volcanic source terrain. The debris flows initiate as landslides during intense thunderstorms. The debris is silt- and

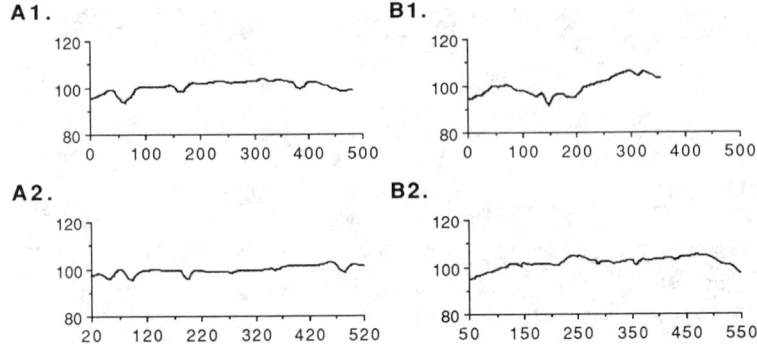

Figure 2. Transverse topographic profiles, scales in meters. A1. Owens west, upper fan (middle ground, Fig. 1a). A2. Owens west, mid-fan (foreground, Fig. 1a). B1. Owens east, upper fan (middle ground, Fig. 1b). B2. Owens east, mid fan (foreground, Fig. 1b).

clay-rich and the low saturated hydraulic conductivity is thought to preclude much interaction with stream waters and to limit important in more granular debris flows. Consequently, the bulk of sediment delivered to the fans comes in the form of relatively viscous debris flows which can not be effectively channelized (Whipple and Dunne, 1992; see Fig. 1b). Broad levees and uneven, tiered interfluves and steeper fan profiles result.

Wide spread debris-flow inundation is a greater hazard on these fans. Sudden channel avulsion is not a frequent occurrence, however, because dense, bouldery frontal debris-flow snouts are uncommon and because the dilute flood-water discharges that commonly follow desert debris flows (Hooke, 1967) effectively scour freshly deposited debris-flow sediments and re-establish channel courses. The morphological consequences of the reduced frequency of channel avulsions are: (1) fewer well-preserved abandoned channels (Fig. 1b); (2) significant heights on broad channel-margin levees (Fig. 2b); and (3) pronounced cross-fan convexity (Fig. 2b).

Jiangjia and Ciaqing Gullies, Xiaoxiang River Valley, China. The geologic and climatic conditions that prevail in the Jiangjia and Ciaqing gullies lead to the formation of frequent, massive, fluid debris flows. Deeply incised gullies have cut steep, unstable hillslopes into the weak bedrock, forest cover has been almost entirely removed, and heavy monsoonal rains cause massive failure and mobilization of debris as often as 20 times a year (Li et al, 1983). The high fines content of the regolith and the great volumes of water available combine to produce huge volumes of mobile, low-strength debris. In general, owing to the vast amounts of sediment deposited annually, channels can not be maintained on these fans (see Li et al, 1983). For example, an extensive, five-meter-deep fan-head trench canyon on upper Ciaqing fan was almost entirely obliterated in just two years (observation communicated by T. Davies, Lincoln College, N.Z., 1992). Channels are not blocked by bouldery, immobile flows; they are simply obliterated by massive deposition. As a result, inundation patterns are largely unconstrained by channels and fan-wide deposition is the norm, rather than the exception. Entire fan surfaces are resurfaced as often as every seven years (Kang Z., Chengdu Acad. Sin., 1992). The debris-flow hazard is severe on these fans. Distinctive morphological characteristics include: (1) no abandoned channels preserved on the fan surface; (2) smooth interfluves, with little relief on the margins of individual debris-flow lobes; (3) a lack of debris-flow levees; and (4) minimal cross-fan convexity (see Fig. 1c).

Conclusions
A rapid assessment of the type of debris-flow hazard most prevalent on a fan can be made from observations of the morphological characteristics of the fan surface. Critical observations include: the number and spacing of abandoned channels, the texture and irregularity of interfluves, the form of channel-margin levees, and the degree of cross-fan convexity. These features can be readily identified and measured on the ground or from remotely sensed imagery with sufficient spatial resolution (e.g., aerial photographs, TMS, SPOT, AVIRIS) and are useful indicators of the patterns of debris-flow inundation and channel shifts characteristic of the fan (Whipple and Dunne, 1992). Analysis of fan morphology is a simple and inexpensive approach which can yield important, and sometimes overlooked, information about debris-flow hazards. This type of assessment should always be made before initiating any expensive computational analyses. Subsequently, a detailed study of debris-flow deposits and fan stratigraphy should be conducted to help guide computational analyses, if any are deemed necessary.

Acknowledgments
This research was supported by NSF Grant EAR-9004843, awarded to the author and T. Dunne. The author is grateful to the Chengdu Institute of Mountain Disasters for their hospitality at the Dongchuan Debris-Flow Observation station, and to Chris Phillips for providing the photo in figure 1c.

References Cited
Costa, J.E. (1984). "Physical geomorphology of debris flows." *Developments and applications of geomorphology*, Costa, J.E., and Fleisher, P.J., eds., Springer-Verlag, Berlin, 269-315.
Dunne, T. and Fairchild, L.H. (1984). "Estimation of flood and sedimentation hazards around Mt. St. Helens (1)." Shin-Sabo, J. Japanese Erosion Control Eng. Soc., 36 (4), 12 - 22.
Fairchild, L.H. (1987). "The importance of lahar initiation processes." Geological Soc. Am., Rev. Eng. Geology, 7, 51 - 61.
Hooke, R. Le B. (1967). "Processes on arid-region alluvial fans." J. Geology, 75, 438-460.
Li, J., Yuan J., Bi C., and Luo D. (1983). "The main features of the mudflow in Jiang-Jia Ravine." Z. Geomorph. N.F., 27, 325-341.
O'Brien, J.S., and Julien, P.Y. (1988). "Laboratory analysis of mudflow properties." J. Hydraul. Eng., ASCE, 114(8), 877 - 887.
Phillips, C.J., and Davies, T.R. (1991). "Determining rheological parameters of debris flow material." Geomorphology, 4, 101 - 110.
Takahashi, T. (1981). "Debris flow". Ann. Rev. Fluid Mech., 13, 57 - 77.
Whipple, K.X., and Dunne, T. (1992). "The influence of debris-flow rheology on fan morphology, Owens Valley, California." Geological Soc. Am. Bull., 104, 887 - 900.

HYDRAULIC MODELLING OF FISH SCREENS

A. Jacob Odgaard[1] M. ASCE, Yalin Wang[2], Marc Serre[3],
and Rex A. Elder[4], F. ASCE

Abstract

Hydraulic model studies coupled with available knowledge of fish behavior can be an important component in the development of fish-screening systems. To reproduce in a model the flow field through and around a fish screen, the model screen must have the same headloss coefficient as the prototype screen. A series of tests is currently being conducted at IIHR to obtain headloss coefficients for various fish screen configurations. The screens include perforated plates, bar screens, and combinations of perforated plates and bar screens at various angles with the flow. The paper gives a summary of the results obtained to date.

Introduction

The main hydraulic factor affecting fish guidance at a screen is the velocity distribution immediately upstream from the screen. This distribution must have certain attributes. The velocity component normal to the screen, V_N, to prevent impingement, must be less than a certain maximum value, which is based on swimming ability of the fish. The velocity component parallel to the screen, V_P, which is often referred to as the sweeping velocity component, serves to guide the fish along the screen face to the bypass entrance. To obtain these attributes, the screen must have certain pressure-loss characteristics, and it will normally have to be placed at an angle with the flow. Examples are seen in Figures 1 and 2.

Studies in hydraulic models can be an important component in developing and optimizing the screen configuration. To reproduce in a model the flow field around a fish screen, the model screen must have the same pressure-loss coefficient as the prototype screen.

Pressure-loss coefficients are available in the literature for many different types of screens. Most coefficients, however, are for relatively simple configurations. Most available data on perforated plates are for plates with sharp-

---

[1] Prof. Civ. and Env. Engrg., and Res. Engr., Inst. of Hydr. Res., Coll. of Engrg., Univ. of Iowa, Iowa City, IA 52242.
[2] Postdoct. Assoc., Inst. of Hydr. Res., Coll. of Engrg., Univ. of Iowa, Iowa City, IA 52242.
[3] Engineer, Engineering Data System Inc., Dubuque, IA 52002.
[4] Cons. Hydr. Engr., 2180 Vistazo East, Tiburon, CA 94920.

edged holes. Commercially available plates have holes that are slightly rounded on one side and therefore have different loss characteristics. Also, fish screens often consist of more than one screen. Information is needed on pressure loss through combined screens.

The effort underway at IIHR is an attempt to create a data bank of pressure-loss coefficients for typical screen configurations and to develop an improved understanding of the flow through screens.

Results

The pressure loss across the screen, Δh, is related to approach velocity v as

$$\Delta h = K \frac{v^2}{2g} \quad (1)$$

in which g = acceleration due to gravity and K = pressure-loss coefficient. Values of K are being obtained by installing the screens in a 25.4-cm-wide, 41.9-cm-high, 12-m-long pressurized conduit and measuring, at various approach velocities and angles, the difference in head between the upstream and downstream gradelines at the screen location. Figure 3 shows coefficients obtained for commercially available perforated plates and bar screens. The data shown are for screens placed perpendicular to the flow. The curve in Figure 3 is a theoretical relationship for K.

Several theoretical relationships for K have been proposed over the years (Law and Liversey, 1978). Most are obtained by applying the momentum and energy equations to a control volume around a single hole (orifice), which yields the formula for an abrupt expansion. The formulas proposed are generally of the form

$$K = A \left(\frac{1 - c\alpha}{c\alpha} \right)^2 \quad (2)$$

in which α = ratio of hole area to total screen area (porosity), c = contraction coefficient, and A = calibration factor. An alternative formula is developed in this study by also accounting for the headloss associated with the contraction of the flow lines:

$$K = \left(\frac{1}{\alpha} - 1 \right) \left(\frac{2}{c\alpha} - 1 \right) \quad (3)$$

With rounded upstream edges of the screen openings, the value of c is essentially one. As seen in Figure 3, this equation simulates well the screen headloss.

Data are currently being taken with screens at angles to the flow of less than 90 degrees. The data are being compared with formulas available in the literature and alternative formulations developed as a part of this study.

Data on combined screens are also being obtained (Figure 4). An example of a combined screen is the diversion screen being field tested at the Wanapum

Development on the Columbia River, WA (Figure 1). This screen consists of a bar screen on the upstream face and perforated plate around 46% located 300 mm downstream from the bar screen. The bars run parallel with the flow and their spacing is 3.2 mm. The open area of the bar screen (porosity) is about 57%. Gage, hole diameter and pitch of the perforated plate are 10, 15.9 mm, and 22.2 mm (staggered), respectively. Samples of this screen configuration have been tested in the headloss channel. The samples were positioned at angles 12° and 38° to the vertical, which are the average angles at which the flow approaches the normal to the screen over the lower half and upper half, respectively, of the prototype screen.

One of the lessons learned from these tests is that the K-value of composite diversion screens, such as the one depicted in Figure 4, is essentially equal to the sum of the K-values of each component of the screen (bar screen and perforated plate and stiffeners and support frame). Another significant observation is that bar screens at an oblique angle to the flow cause the flow lines to curve away from the screen before going through the screen as shown schematically in Figure 5. At an approach flow velocity of 1.03 m/s, the velocity vector at the 38° screen combination is turned upward from the horizontal by 9° on the average just before going through the screen. This feature is the result of the local change in vertical pressure gradient created by the inclined screens. The angle decreases with distance upstream from the screen surface, for the 38° screen combination, from 9° at a distance of 6 mm to 3° at a distance of 130 mm.

The model of the Wanapum fish screen was a 1:16 scale model. It utilized the same bar screen as the prototype. In one test the porosity of the downstream plate was also 46% but the plate was thinner and the hole size smaller. (The ratio of thickness to hole diameter was about the same in model and prototype). The distance between bar screen and perforated plate was 20 mm. Tests in the headloss channel verified that this screen had the same headloss characteristics at model velocities as the prototype screen had at prototype velocities (same K value).

The screen combination proposed for the intake of the Potter Valley Hydroelectric Project, CA (Figure 2) also consists of a bar screen on the upstream face and perforated plates on the downstream face. The porosity and bar spacing are 47% and 3.2 mm respectively. In this case the bars are oriented transversely to the direction of flow, and the porosity of the perforated plates varies from 50% at the upstream end of the screen to 9% at the downstream end. The variation in perforated-plate porosity was designed to ensure uniform throughflow velocity along the screen. The model used for testing this screen was a 1:16 scale model (Odgaard and Wang 1992). It utilized the same bar screen as was proposed for the prototype screen. The same screen could be used because headloss tests in the aforementioned channel showed that the loss coefficient was the same (K = 4.5) at both model and prototype velocities. The screen was tested for screen angles ranging from 7° to 12°, with the horizontal.

Conclusions

Obviously, headloss coefficients play an important role in the design of fish screening systems. The data being obtained in this study supplement the data available in the literature. They help improve our understanding of flow around fish screens and could lead to more innovative designs in the future.

Acknowledgments

The laboratory studies described are supported by the Public Utility District No. 2 of Grant County, Washington, under Contract No. 430-209. The District is represented by Stephen Brown.

Rex A. Elder developed the study program and is providing guidance on data acquisition and analysis. Donald Weitkamp, Parametrix, Inc., Kirkland, WA, is providing fishery biological input to the study.

Finally, credit should be given to the Iowa Institute of Hydraulic Research shop staff who, under supervision of James Goss constructed the test facility and models.

References

Law, E.M., and Liversey, J.L., *"Flow Through Screens,"* Ann. Rev. Fluid Mech., 1978, (10) 247-266.

Odgaard, A.J., and Wang, Y., *"Hydraulic Model Study for Potter Valley Intake Fish Screen,"* IIHR Limited Distribution Report No. 201, Nov. 1992.

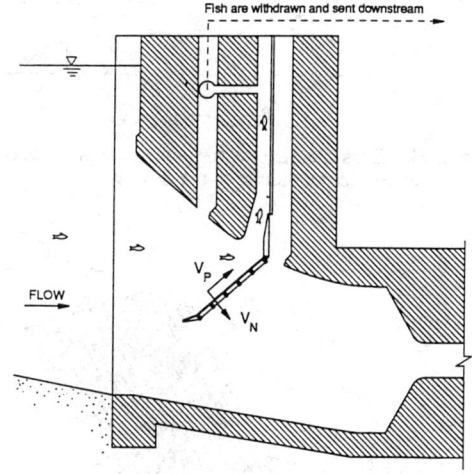

Figure 1. Fish screening in turbine intake.

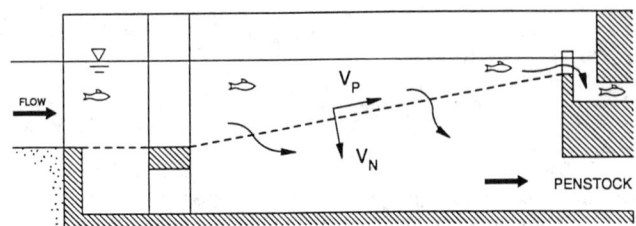

Figure 2. Fish screening upstream of turbine intake.

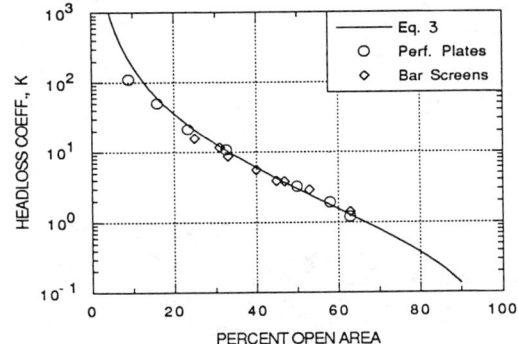

Figure 3. Headloss coefficients for perforated plates and bar screens placed perpendicular to the flow.

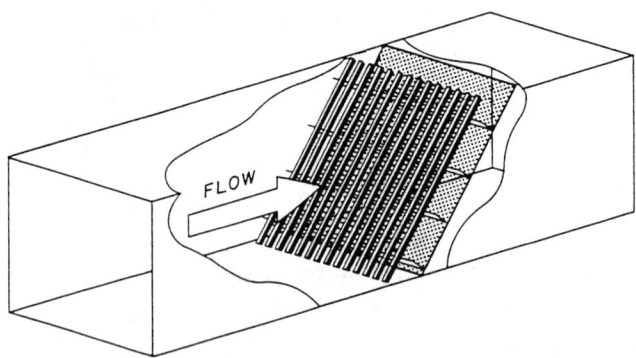

Figure 4. Schematic showing composite fish screen in headloss test channel.

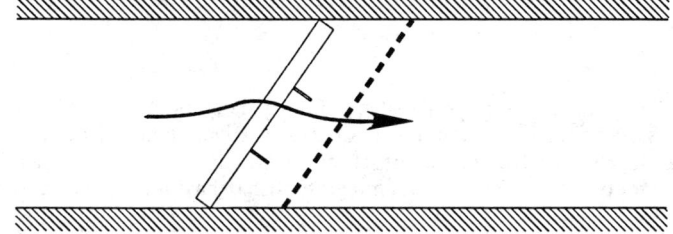

Figure 5. Schematic showing streamline through screens.

FISH SCREEN DEVELOPMENTS
COLUMBIA RIVER DAMS

Don E. Weitkamp[1], and Rex A. Elder, F.ASCE[2]

ABSTRACT

This paper describes developments to date for a fish diversion and bypass system being tested for two Columbia River dams. Because of the concern for the negative impacts of screen systems used to divert young salmon, we have followed an approach that integrates fish behavior and hydraulic design principles.

Model and field testing of a fixed bar screen system indicates this system has hydraulic parameters desirable for effective diversion of the fish into the turbine gatewells. Development of a new orifice has demonstrated an efficient option for getting the young salmon out of the gatewells.

INTRODUCTION

This paper describes the approach we have taken to develop a fish diversion and bypass system for two Columbia River Dams and the results of this development. Our approach has been guided by a desire to follow an interactive program that incorporates fish behavior together with engineering design. Our objective has been to pass fish downstream of each dam with as little stress as possible. Two previous papers have discussed some of these modeling efforts (Elder et al 1987, Odgaard et al 1990).

---

[1] Parametrix, Inc., 5808 Lake Wash. Blvd. NE, Kirkland, WA 98033

[2] Consulting hydro. engr., 2180 Vistazo East, Tiburon, CA 94920

To accomplish this objective, we have relied heavily on hydraulic model testing followed by field testing, at prototype scale, of favorable alternatives. This testing began with a review of information available from the many years of empirical testing by the Corps of Engineers at various Snake and Columbia River Dams. In hydraulic model testing at the Iowa Institute of Hydraulic Research, University of Iowa, we have investigated both physical conditions and behavior response of fish in the larger models. The following describes how we have developed fish diversion and bypass systems for Wanapum and Priest Rapids Dams on the Columbia River in central Washington State. We have applied the approach of integrating fish behavior with engineering design to each component of the diversion/bypass system. Testing included all components of the system beginning with the upstream point at which fish contact the dam (trashracks) and extending to the downstream discharge location.

TRASHRACKS

Existing trashracks at the dams have large horizontal support bars that were not designed for minimum disturbance of flow downstream from the trashrack screen. Hydraulic model test showed large horizontal-axis eddies extending downstream from each horizontal member past the diversion screen.

Our concerns were that the turbulence could lead to a change in the vertical distribution of fish between the trashrack and the diversion screen, and that the turbulence could increase fish disorientation leading to injury or poor guidance. A hydrodynamic trashrack was designed for the upper portion of the intake in the area passing flows intercepted by the diversion screen (Figure 1). The hydrodynamic trashrack has streamlined horizontal members that are angled to parallel the descending flow of the intake opening, rather than parallel to a horizontal plan as in the existing trashracks. Both the streamlining and the angle are important to eliminating the eddies.

DIVERSION SCREEN

Based on biological considerations, we determined the wedge-wire bar screen would be preferable at Wanapum and Priest Rapids rather than traveling screens used at Corps of Engineers Dams. The smooth surface of narrow bars parallel to the flow was less likely to injure fish contacting the screen than the

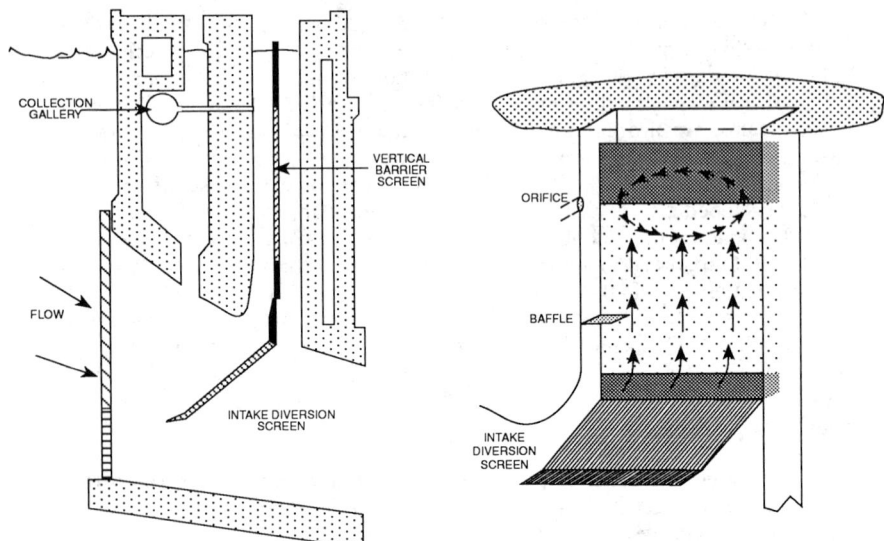

Figure 1. Fish diversion screen and bypass system, Wanapum Dam, Columbia River.

Figure 2. Flow pattern in prototype with gatewell baffle to control flow in upper portion of gatewell.

relatively rough surface of a mesh screen. Engineering considerations also favored the bar screen which is mechanically simple compared to the traveling screens (motors, drive mechanisms, sprockets, etc.).

A major concern with the diversion screen is the direction and speed of approaching flows. We desired to produce flows that approach the screen at an angle toward the upper (downstream) end with a velocity that was not likely to impinge fish. Hydraulic model observations indicated the bar screen had less flow directed toward the bottom of the screen than the traveling screen. Thus the bar screen was more likely to divert more of the fish approaching the screen than was a traveling screen.

Our second concern was to develop favorable velocities passing through the screen. Existing information indicated these should be in the range of 0.6-1.5 m/sec (2-5 ft/sec). The initial bar screen had approach velocities (normal to the screen measured about 10 cm (4 in) away of about 1.0-1.05 m/sec (3.3-

3.5 ft/sec). These approach velocities killed or injured a high percentage of sockeye salmon (about 30%).

We found bar screen spacing in these cases was not adequate to control current speed. To reduce flow through the screen a perforated plate was added to the downstream side. Approach velocities were reduced to about 1.46 m/sec (3.1 ft/sec). which appears adequate to divert young salmon without detectible injury. This is not true for the very small zero-edge migrants (30-60 mm).

The final diversion screen design has wedge-wire bar screen with bars 2.4 mm (0.093 in) wide and bar spacing of 3.2 mm (0.125 in). The downstream perforated plate has a porosity of about 40%. The screen is installed at an angle of 50° from the vertical. The 7.5 m (24 ft) long screen has the lower (upstream) end (1.2 m, 4 ft) angled upward at 70° from the vertical to allow installation of a screen this long.

GATEWELL ENTRANCE

The narrow-bottom opening of the existing gatewell required modification for installation and operation of the diversion screen. The objectives were to provide a smooth, but high velocity flow into the gatewell. The high speed of this flow (0.15-0.38 m/sec, 6-8 ft/sec) was designed to discourage fish from exiting the gatewell through the opening, once they were diverted.

The shape of the upstream wall at the opening was derived from hydraulic model testing. The smooth transition of this curve was necessary to eliminate an eddy just inside the entrance. We speculated diverted fish would tend to hold and become injured or stressed in this eddy, or even escape back upstream. In model testing a few very small trout and fathead minnows did tend to find and hold in this eddy. Entrance velocities as measured in the hydraulic model and verified in the field are between 0.15 and 0.38 m/sec (6-8 ft/sec) at the narrowest portion of the gatewell entrance.

VERTICAL BARRIER SCREEN

Once diverted into the gatewell the fish are separated from about 17 m^3/sec (500 cfs) of the flow by a vertical barrier screen. By agency requirements, the vertical barrier screen was to have openings no larger than 4.8 mm (3/16 in). Velocities normal to the screen were not to exceed 0.3 m/sec (1 ft/sec).

Because fish were not likely to contact this screen we used perforated plate rather than wedge-wire to minimize costs. To maintain constant size of open areas, the upstream side of the vertical barrier screen has 4.8 mm (3/16 in) openings with a uniform 51% porosity. However, outflow conditions through the screen were not uniform. Thus, a second layer of perforated plate was added to the downstream side with different porosities at different elevations to provide balanced flow rates. Prototype velocities through the vertical barrier screen are in the range of 15-45 cm/sec (0.6-1.75 ft/sec). In both hydraulic model testing and prototype testing, we have seen no evidence that fish are contacting the screen and being injured. In the prototype, the descaling rate (sublethal injury) of fish is commonly lower than fish collected from adjacent units not having a screen system.

ORIFICE

Although there are several different alternatives for conveying fish out of the gatewell into a collection system, the resource agencies require a submerged orifice system. The expressed theory is that the fish will seek and find this opening in a sufficiently short time to make migration delays acceptable. Existing orifice systems employ a sharp edge orifice with a larger pipe or opening downstream. The preferred size is a 30.5 cm (12 in) opening. In hydraulic models of this system, we observed a strong vortex and turbulence downstream of the orifice that appeared to be undesirable. Fish entering the orifice in the model tended to tumble out of control in the high velocity turbulence.

A bell mouth orifice located in the upstream corner and having a smooth transition with the gatewell walls was selected from a number of alternatives based on hydraulic model observations. This 30.5 cm (12 in) opening expands to a 40.6 cm (16 in) diameter pipe at a 3.5° sidewall angle. We found only a slight swirl generated by a long radius band downstream, and almost no turbulence in the hydraulic model of this design. Fish passed through this model with no tumbling and apparently in control.

We have observed behavior of fish in the immediate vicinity of the orifice both in the hydraulic models and in the prototype by video camera. In all cases the fish appear to be reluctant to enter the orifice, generally darting away when they sense the orifice's flow net. Fish that enter the orifice appear to be out of control and nearly always enter tail first. Most fish appear to enter the orifice when moving up in the gatewell rather than when moving down.

The orifice is located in the upstream corner of the gatewell at an elevation of 1.7 m (5.5 ft) below minimum reservoir level. The corner location is ideal because this is where the fish appear to be concentrated and also provides a favorable flow-net configuration in the gatewell according to hydraulic model measurements.

GATEWELL BAFFLE

Video tape observations indicate since the direction of flow past the orifice appears to influence the rate at which fish enter the orifice, we have investigated a means for controlling this flow. Hydraulic model tests show that the vertical barrier screen porosity has little influence on the gatewell flows pattern for the Wanapum and Priest Rapids gatewells. The flow pattern in the gatewell establishes a large circular eddy in the vertical plane in the upper portion of the gatewell (Figure 2). This circulation is unstable and can occur in either direction.

To stabilize the direction of the gatewell flow we investigated a baffle (horizontal plate) blocking a portion of the gatewell near the bottom. A plate extending across the gatewell and 0.6 m (2 ft) out from the end wall is adequate to direct the gatewell flow. With the baffle, the flow in the upper region of the gatewell will be upward on the side opposite the baffle and downward on the baffle side. this configuration will be tested in the prototype during 1993.

REFERENCES

Elder, R.A., A.J. Odgaard, D. Weitkamp, and D. Zeigler. 1987. Development of turbine intake downstream migrant diversion screen. Pages 522-531 in Waterpower '87, Proceedings of the International Conference on Hydropower, ASCE Portland, Oregon.

Odgaard, A.J., R.A. Elder, and D.Weitkamp. 1990. Turbine-intake fish-diversion system. Pages 1310-1316 of the Journal of Hydraulic Engineering, ASCE, 116(11).

HYDRAULIC ASPECTS OF A LOW-VELOCITY, INCLINED FISH SCREEN
by
Frederick A. Locher[1], Victor C. Bird[2] and A. Jacob Odgaard[3]

Abstract

The Potter Valley Hydroelectric Project diverts up to 310 cfs from Van Arsdale Reservoir on the Eel River in north-central California and delivers it to Potter Valley for power production and irrigation. A state-of-the-art inclined fish screen has been designed to provide satisfactory screening of Steelhead Trout, and Coho and Chinook Salmon. The development of the hydraulic aspects of the intake, the inclined screen, transition, and fish by-pass system are described. This facility will be a unique application of a low-velocity screen to a hydroelectric intake.

1.0 Introduction

The Potter Valley Hydroelectric Project, owned and operated by Pacific Gas and Electric Company (PG&E), is located near Potter Valley in Mendocino County, California. This 9.2 megawatt project was completed in 1928, and consists of Cape Horn Dam on the Eel River which forms Van Arsdale Reservoir, an intake connecting to a tunnel and penstock, and Potter Valley Powerhouse. As part of the relicensing requirements, PG&E modified the original intake in 1972 and installed a horizontal "Bates" travelling screen. Because of the poor performance in screening fish, mechanical malfunctions, and excessive operating costs, the "Bates" screen was taken out of service in 1976.

To comply with the regulatory requirements for an operating fish screen at the intake, PG&E initiated a major program for the design of a new

---

[1] Chief Hydraulic Engineer, Geotechnical and Hydraulic Engineering Services, Bechtel Corp., P. O. Box 193965, San Francisco, CA 94119
[2] Civil Engineer, Hydro Engineering & Construction, Pacific Gas & Electric Company, One California St., San Francisco, CA 94106
[3] Professor of Civil and Envoronmental Engineering and Research Engineer, Iowa Institute of Hydraulic Research, Iowa City, IA 52242

fish screen facility. Figure 1 depicts an isometric view of the Inclining Horizontal Fish Screen facility which was developed by PG&E in close cooperation with the California Department of Fish and Game (CDF&G), the U. S. Fish and Wildlife Service (USFWS), and the National Marine Fisheries Service (NMFS). In the screening mode, the flow passes through the bar racks, and over a gravel trap into two bays 24 m (80 feet) long and 2.4 m (8 feet) wide containing the fish screens. The fish screens are mounted on a truss approximately 24 m (80 feet) long which can be raised and lowered 1.2 m (4 ft). This changes the screen angle from 7 to 11° to maintain a constant by-pass flow for the fish and debris. At maximum powerhouse flow, 4.4 m^3/s (155 cfs) per bay passes through the screen and into the tunnel. The fish flow of about 0.23 m^3/s (8 cfs) per bay passes through a transition at the end of the screen, over a weir, and into a fish pump sump where a screw pump is used to pump the fish into a secondary fish screen. The fish are separated from the debris and then transported by gravity flow in a pipe to the fish ladder at Cape Horn Dam. In the by-pass mode, the gates to the fish screen bays are closed, and the flow passes through the gravel trap and into the tunnel. The hydraulic design of this facility, including the intake, fish screen, transition to the fish pump sump, and the fish by-pass system are presented.

2.0 Criteria for Design
2.1 Screen The basic criteria for design of the screen were:

1. Velocity. CDF&G criteria require that the maximum discharge through the screen shall not exceed 0.33 cfs per square foot of screen opening, an average velocity normal to the screen of 0.1 m/s (0.33 ft/s.)

2. Recirculation. There shall be no recirculation through the screen.

3. Separation and Eddies. There shall be no regions within the structure prior to the fish screen in which flow separation or eddies occur that could provide dwelling places or habitat for predators.

4. Screen. The screen shall be wedgewire screen with 3 mm (1/8-inch) opening, and the wedgewires shall be oriented perpendicular to the flow.

5. Screen Porosity. CDF&G criteria state that the "screens shall have a minimum open area of 1.5 square feet per cubic foot/second." This criterion and the approach flow velocity criterion implies that the screen porosity should be 50%. A range of 40 to 60% was deemed acceptable.

2.2 Operating Conditions. The intake should function satisfactorily for a range of reservoir levels from the spillway crest to flows up to four feet

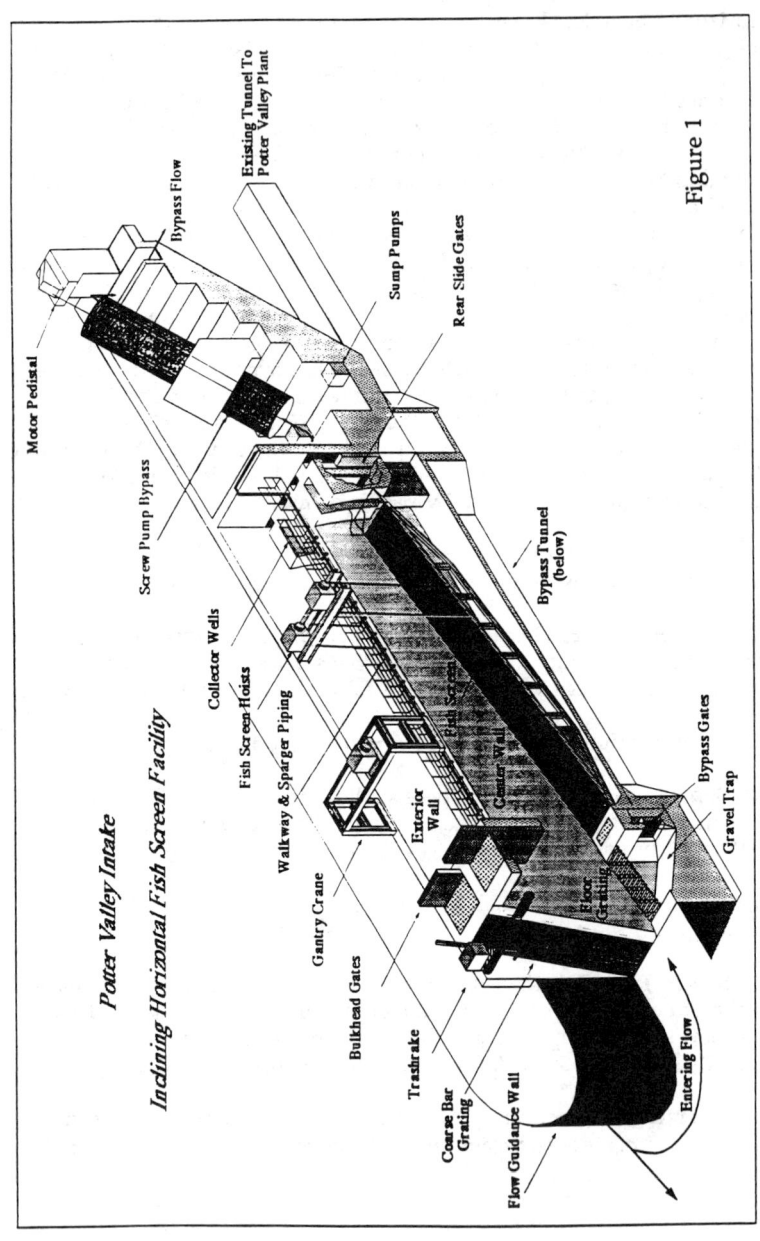

Figure 1

over the dam crest, which corresponds to a reservoir inflow from about 4.4 m³/s (155 cfs) to about 212 m³/s (7500 cfs). Flashboards 1.2 m (4 ft) high are installed in the summer months. During this period, all flow, except for about 10 cfs for the fish ladder, are diverted through the intake. During low flow periods, only one bay will be required to screen the fish. The facility should operate satisfactorily with only one bay in service.

3.0 Intake Hydraulics

3.1 Hydraulic Model. The proposed structure will be aligned with the existing tunnel and intake, which orients the new facility at 135° to Van Arsdale Reservoir. Since the approach flow to the fish screens must be uniform to avoid "hot spots" on the screen, it was concluded that a hydraulic model of the proposed structure would be required to establish the design of the intake and to ensure uniformity of flow through the screens. The hydraulic model was constructed at the Iowa Institute of Hydraulic Research using a 1: 16 scale. A two-dimensional, finite element model of the reservoir (FESWMS) was used to determine the upstream boundary of the physical model. Because this is a run-of river project, it was not clear how far upstream the reservoir/river model should extend. The numerical model in conjunction with field data determined that a section about 183 m (600 ft) upstream from the dam or 107 m (350 ft) upstream from the intake would not be influenced by changes in intake geometry. A two-component, colorburst Laser-Doppler Anemometer was used to obtain the velocity profiles in the model.

3.2 Guide Wall. Because of the re-entrant flow and the requirement that the intake operate satisfactorily with a significant percentage of the flow passing over Cape Horn Dam, a guidewall was developed to reduce separation at the intake. The configuration shown on Figure 1 did separate for all flow conditions, but the flow reattached well before the bar rack. The separation zone was highly three-dimensional and turbulent and was not judged to be a holding area for predators. Similarly, the flow pattern in front of the intake with flow over Cape Horn Dam was also quite turbulent, and was not a holding area for predators.

3.3 Bar Racks. The bar racks used in the model had a depth of bar to spacing ratio of 2:1. In the prototype, the bars will be 9.5 mm (3/8") thick with 38 mm (1.5") clear opening and 152 to 203 mm (6 to 8 inches) deep. The model test results clearly showed that using the bar rack as a flow straightener was necessary to remove angularity from the approach flow. Final measurements at the entrance to the screen showed a maximum deviation of ± 12 % of the mean velocity through the bay. The flow into each bay was nearly equal.

4.0 Screens

A series of tests showed that the head loss coefficient for the prototype screens remained the same at model and prototype velocities. Thus, the Hendrick B9 prototype screen was used in the model tests. Perforated plates were mounted on the downstream side of the Hendrick B9 to vary the head loss along the screen. A variation in head losses in necessary to maintain a uniform flow through the 24 m (80 foot) long screen. The variation in porosity of the perforated plates was from 50 % at the downstream end to 4% at the upstream end.

5.0 Transition to Fish Pump Chamber

Several designs of a transition from the 2.4 m (8 ft) wide fish screen bay to the 610 mm (2 ft) wide slot through which the fish must pass into the screw pump chamber were tested. Gradual transitions significantly reduced the area of the screen and made it difficult to obtain a uniform flow through the screen. The final configuration consisted of a short transition section patterned after contracting sections for wind tunnels. The concept is based on tests with orifices which showed that fish could not sense the acceleration to the orifice until it was too late. The fish were then carried through the orifice, or in this case, the rapid contraction, into fish collection facilities. The transition section did not show any undesirable separation zones when tested in the model.

6.0 Air Backwash System

Several mechanical arrangements of cleaning were considered, including scrapers, brushes and modified rakes. None of these alternatives met the criteria for simplicity in construction, ease of operation and overall reliability. It was concluded that an air backwash system would be the best system. An experimental flume was built to test a full-scale section of the screen and develop an air backwash system. The details are presented in Reference 1.

7.0 Screw Pump Chamber

During development of the design for the Inclined Fish Screen facility, PG&E, CDF&G, USFWS and NMFS agreed that an impeller-type of fish pump was no longer acceptable, and that an alternative fish pump was required. PG&E recommended a modification of an Archimedes screw pump manufactured by Internalift Enclosed Screw Pump, CPC Corp., Sturbridge, MA. This pump has no internal moving parts that could damage fish once the fish are inside because the helical "flights" are welded to the outside cylinder, and the cylinder and flights rotate as a unit to lift the water.

To test the feasibility of using a screw pump, a 762 mm (30") diameter, 3 m (10 ft) long demonstration pump constructed of plexiglas to permit viewing was installed in a test facility. Tests were run with juvenile Steelhead, Chinook Salmon and Coho Salmon with sizes ranging from 1000/lb (est) to 7.5/lb. The fish were introduced at the bottom of the screw pump by a pipe which terminated in a faceplate covering the end of the screw pump. The tests showed that 1) the ends of the flights had to be extended to the end of the outside cylinder, 2) the leading edge of the flights had to be smooth or blunt, 3) there should be a smooth coating applied to the inside of the screw pump, and 4) any gap between the end of the screw and the faceplate covering the bottom of the screw had to be eliminated. After these modifications had been made, the tests showed that the pump had no adverse effects on the fish which were tested. There was no immediate or delayed mortality observed, and it was concluded that the use of a 3 m (10 ft) diameter, 13.7 m (45-ft) long screw pump in the full-scale facility will be an acceptable approach to pumping the fish to the separator facilities and gravity return system.

8.0 Conclusions

1. Close cooperation between PG&E and the regulatory agencies resulted in a state-of-the-art concept for an Inclining Horizontal Fish Facility for the Potter Valley Project.

2. Successful development of the concept required testing of a novel application of an Archimedes screw pump for pumping the fish from the facility to the separator and gravity return pipe.

3. A hydraulic model of Van Arsdale Reservoir and the intake was essential to ensure uniform approach flow conditions and uniformity of flow through the fish screens.

4. An air backwash system was developed through testing of a section of the prototype screen and air sparger system.

Acknowledgements

Mr. Gene Geary, Fisheries Biologist, PG&E, assisted in the evaluation and testing of the screw pumps, air sparger, intake and fish screens. Mr. David Menasian of Steiner Environmental Consulting, Potter Valley, CA did the detailed design of the test facility for the air sparger tests.

Reference

Locher, F. A., Ryan, P. J., Bird, V. C, and Steiner, P, "Debris Removal from a Low-Velocity Inclined Fish Screen", Proc. 13th Annual USCOLD Lecture, Chattanooga, Tennessee, May 17-21, 1993

Survival of Atlantic Salmon Smolts
Bypassed through Ice-log Sluices
Determined by the HI-Z Turb'N Tag

By

Paul G. Heisey, Dilip Mathur, George A. Nardacci[1]
and Milton Anderson[2]

Abstract

The HI-Z Turb'N Tag was used to assess injury and mortality incurred by Atlantic salmon (Salmo salar) smolts bypassed through ice-log sluiceways at two stations. Both sluiceways were similar in height (52 and 60 ft) and spillage (approximately 300 cfs). Total survival 48 h after passage was at least 96% at both facilities. However injury and scale loss were markedly different at the two facilities (42 vs 3%). The facility with a deep, unobstructed plunge area and minimal intercepting structures inflicted least injuries.

Introduction

Downstream fish passage facilities have been recommended and/or installed at many hydropower stations, especially projects that pass valuable migratory species (e.g., salmon and American shad). Young of these species must emigrate from their natal

---

[1] RMC Environmental Services, Inc., Utility Consulting Division, 1921 River Road, P. O. Box 10, Drumore, PA 17518
[2] 11 Poplar Street, New England Power Company, Gloucester, MA 01930

rivers and are subjected to potential injury and mortality at hydro projects. A considerable amount of effort has been expended determining the effect of passage through turbines at these facilities. The most recent review of these studies has been prepared by EPRI (1992). Turbine related mortalities have been reported as high as 82% (Taylor and Kynard 1985), however recent studies (Heisey et al. 1992, 1993) indicate some mortality rates may have been overestimated. Primarily because of reported high turbine-related mortality, fish bypass facilities have been installed at many hydro stations and/or extensive spills have been initated during periods of fish emigration.

Ferguson (1992), Ledgerwood et al. (1991), Poe et al. (1991) have provided reviews of fish bypass facilities which indicate bypassed fish can experience some injury and mortality. The objectives of this paper are to (1) describe the effects of passage through two ice-log sluiceways (sluices) used to bypass Atlantic salmon smolts, and (2) provide insight into design features that may increase or decrease injury.

Study Sites

The two sluices are located on the Connecticut River at New England Power Company's Bellows Falls and Wilder Hydroelectric stations. The Wilder sluice is 52 ft high and normally passes 300 cfs during salmon emigration (Figure 1). Spillage is controlled by lowering a regulating gate which directs water onto a concrete apron approximately 15 ft below the top of the gate. The water then cascades down a sloping concrete channel to the tailrace. Two concrete pillars, associated with a fish ladders, are adjacent to the discharge from the sluice. There was some concern that fish may be striking these pillars.

The height (60 ft) and discharge (275-340 cfs) of the Bellows Falls sluice are similar to Wilder's; however, the water passage route is considerably different (Figure 1). The spillage water gradually falls 30 ft while passing through a 500 ft long concrete channel and then plunges 30 ft into a deep, unobstructed pool in the tailrace. The turbulent conditions at the discharge raised concerns about possible injuries to fish at the Bellows Falls sluice.

Bellows Falls Sluice

Wilder Sluice

Methods

The effects of passage through both sluices were ascertained by the HI-Z Turb'N Tag recovery technique (Turb'N Tag). Details about this technique have been presented in Heisey et al. (1992). Atlantic salmon smolts (5.7 to 14.1 in fork length) were passed at spillages of 200, 300, and 500 cfs at Wilder and approximately 300 cfs at Bellows Falls. A total of 100 test specimens were released at the top of the sluice for each condition except the 300 cfs spillage at Wilder (45 fish). Equivalent numbers of control specimens were released just downstream of the sluice exit. Fish were equipped with one or two Turb'N Tags and a radio tag was also attached to specimens released at Bellows Falls. The Turb'N Tag(s) inflated after both test and control specimens entered the tailrace and buoyed each fish to the surface. A boat crew recaptured the fish, removed the tag(s) and examined each fish. Live specimens were transferred to an onshore tank or floating net pen to assess long-term (48 h) effects of passage. Specimens that died within 1 h after passage were designated short-term mortalities. The short-term and long-term mortality rates of passed fish were estimated using the formula given in Fleiss (1981).

$$P_e = \frac{P_2 - P_1}{1 - P_1}$$

P_e = sluice related mortality
P_1 = fractional mortality of control fish
P_2 = fractional mortality of test fish

$$SURVIVAL = 1 - P_e$$

Results

A total of 245 and 100 salmon smolts were passed through Wilder and Bellows Falls sluices, respectively. Some 99 and 95% of the fish released were recaptured at these respective sites. Twelve of the 245 test fish at Wilder incurred mortality (short and long-term combined) while only 2 control specimens died. The overall survival rate, adjusted for controls, was 96%. An identical survival rate occurred at Bellows Falls.

Survival rates at Wilder varied slightly for the three flow conditions tested. Short-term survival was lowest (93%) at 300 cfs spillage and highest (99%) at 200 cfs. Few additional fish died during the delayed assessment period and the overall survival (after 48 h) was 91% for 300 cfs spillage and 97% for 200 and 500 cfs release.

Although the survival rates were the same at both sluices, injury and scale loss were much higher at Wilder (Table 1). Forty-two percent of all test specimens passed at Wilder had some scale loss and/or injuries. This contrasts to only 3% for the fish passed at Bellows Falls. Control specimens were in excellent condition (<1% with any scale loss or injury) at both sites. Scale loss and injury at Wilder appeared to be inversely related to spillage rate. Some 59% of the fish incurred scale loss and/or injuries at 200 cfs spillage, but this dropped to 26% at 500 cfs spillage.

TABLE 1. Injury and scale loss (%) to Atlantic salmon smolts passed through sluices at Bellows Falls and Wilder Hydroelectric stations.

	Wilder				Bellows Falls
Spillage (cfs)	200	300	500	Combined	275-340
Minor Scale Loss	22%	20%	11%	17%	1%
Major Scale Loss	20%	0%	4%	10%	0%
Lacerations	7%	4%	0%	4%	0%
Bruises	24%	23%	14%	20%	2%
Percent Fish Injured*	59%	36%	26%	42%	3%

* Less than 1% for control specimens

Discussion and Conclusions

The configuration of the two sluices indicated that salmon smolts should incur minimal injuries when bypassed through sluices with minimal obstructions and discharged into deep plunge pools. Sluices which direct water flow and fish onto concrete aprons and other structures may inflict considerable scale loss and injury, however, these injuries may not be lethal. Increased depth of water over aprons and other structures and/or some simple structural modifications may greatly improve the condition of bypassed fish at sluices.

Literature Cited

Burnham, K. P., D. R. Anderson, G. C. White, C. Brownie and K. H. Pollock. 1987. Design and analysis methods for fish survival experiments based on release-recapture. Amer. Fish. Soc. Monogr. 5: 437 pp.

EPRI. 1992. Fish entrainment and turbine mortality review and guidelines. Electric Power Research Institute, Palo Alto, CA. EPRI TR-101231.

Ferguson, J. W. 1992. Analyzing turbine bypass systems at Hydro facilities. Hydro Review 11(3): 46-56.

Fleiss, J. L. 1981. Statistical methods for rates and proportions. John Wiley & Sons, New York, NY. 321 p.

Heisey, P. G., D. Mathur, and T. Rineer. 1992. A reliable tag-recapture technique for estimating turbine passage survival: Application to young-of-the-year American shad (Alosa sapidissima). Can. Jour. Fish. Aquat. Sci. 49: 1826-1836.

Heisey, P. G., D. Mathur, and D. Robinson. 1993. Survival of juvenile American shad in passage through a low-head hydroelectric project. Trans. Amer. Fish. Soc. (Accepted subject to revision).

Ledgerwood, R. D., E. M. Dawley, L. G. Gilbreath, P. J. Bentley, B. P. Sandford, and M. H. Schiewe. 1991. Relative survival of subyearling chinook salmon that have passed through the turbines or bypass sytem of Bonneville Dam second powerhouse, 1990. Report to the U.S. Army Corps of Engineers, Contract E86900104, 90 p. (Available from Northwest Fisheries Science Center, 2725 Montlake Boulevard East, Seattle, WA 98112-2097).

Monten, E. 1985. Fish and turbines: fish injuries during passage through power station turbines. Norstedts Tryckeri, Stockholm, Sweden. 111 p.

Poe, T. P., H. C. Hansel, S. Vigg, D. E. Palmer, and L. A. Prendergast. 1991. Feeding of predaceous fishes on out-migrating juvenile salmonids in John Day reservoir, Columbia River. Trans. Am. Fish. Soc. 120(4): 405-420.

ASK YOUNG CLUPEIDS IF KAPLAN TURBINES ARE REVOLVING DOORS OR BLENDERS

Dilip Mathur and Paul G. Heisey[1]

Abstract

Previous estimates of low survival (18 to 38%) of young clupeids (American shad, blueback herring, and alewife) in passage through hydro turbines have led many to conclude that turbines are fish blenders. In recent studies survival of young clupeids ranged from 96 to 100% (97.5 ± 4.5%) at three different power stations equipped with Kaplan type turbines; at a Francis turbine the estimated survival was 89 ± 9.8%. These results demonstrate that Kaplan turbines rotating at slow speeds are benign exit routes.

Introduction

Fish mortality and injury resulting from turbine passage is a major environmental issue at hydro projects. Throughout North America, there is a strong impetus to protect migratory fish, especially juveniles. Consequently, hydro project owners attempting to license a new project, or relicense existing facilities, must consider the potential for turbine operation to injure or kill migratory fishes.

A variety of factors may affect survival of fish at hydro projects: design of turbine, wicket gate

---

[1] RMC Environmental Services, Inc., Utility Consulting Division, 1921 River Road, P. O. Box 10, Drumore, PA 17518

setting, species and size of fish, rotational speed of turbine blades, turbine blade angle, and number of blades. Consequently, a wide range of survival rates are quoted (20% to 100%). The owners or developers of hydro projects tend to view the passage of fishes through turbines as a benign process while the regulatory agencies responsible for managing the fisheries resource view this entrainment process as extremely detrimental. In the absence of reliable quantitative estimates of turbine-related injury and mortality, divergent and unsubstantiated opinions flourish. A reliable assessment is needed to put these opinions into perspective.

The objective of our paper is to provide results of recent studies on juvenile clupeids conducted at several low head hydro projects equipped with Kaplan type turbines. These studies utilized a novel tag-recapture technique which overcame most of the problems associated with previous mark-recapture studies (Heisey et al. 1992).

Turbine Types and Operation

Juvenile clupeids emigrating from rivers on the east coast of the United States generally encounter two types of turbines: Francis and Kaplan. Figure 1 shows schematics of these turbines. The number of blades (buckets) in a typical Francis turbine generally varies from 14 to 20. Kaplan type turbines usually have 3 to 8 blades. The clearance between blades is greater than that of Francis runners. Kaplan turbines have adjustable blades that are coordinated with wicket gate positions to obtain higher efficiencies. Table 1 provides hydraulic and physical characteristics of the turbine types studied for this presentation. Water flow paths in a typical Kaplan and Francis turbine are shown in Figure 1. In most of the propeller runners water moves through the turbine parallel to the axis of the runner. When a turbine is operating at maximum efficiency, the guide vanes and wicket gates are closely aligned and the flow through the runner is relatively smooth.

Method for Estimating Survival

The Turb'N Tag recapture technique (U.S. Patent No. 4,970,988), used in our studies consists of fitting each fish with a self-inflating balloon tag (Heisey et al. 1992). After a short observation period each tagged test fish is introduced into a turbine while

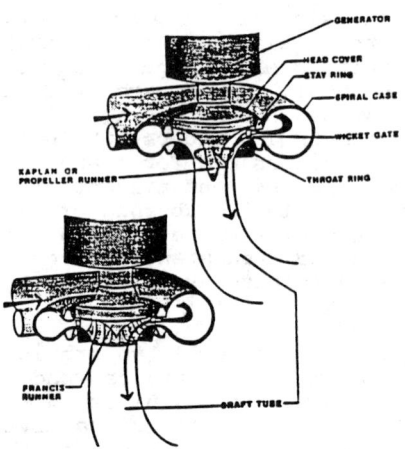

Figure 1. Schematics of typical Francis and Kaplan turbines showing water flow paths. Reproduced from Eicher Associates (1987).

control fish are introduced into the tailwaters through a portable induction apparatus. The tag is chemically activated prior to release of the fish. Upon passage the Turb'N Tag gradually inflates, due to the release of gases within the tag, enabling surface recapture of specimens in the tailrace, usually within 10 minutes.

Immediately after recapture, each fish is placed in an on-board holding tank, the tag removed and each fish is examined for injury, scale loss and classified live or dead.

Results and Discussion

The short-term survival of young American shad (95-140 mm fork length) was 97% at the Safe Harbor Station (Table 2). The estimated survival of juvenile blueback herring (77-105 mm fork length) was 96% at the Crescent Project. At the Hadley Falls Project the estimated survival of young American shad (55-110 mm fork length) was 100%. The overall short-term survival at the three projects was estimated at 97.5 ± 4.5%.

In contrast, the survival of young American shad was slightly lower in passage through another low-head

TABLE 1. Hydraulic and physical characteristics of the low-head hydro projects where turbine-related mortality of juvenile clupeids was estimated. Source: EPRI (1992).

	Holtwood	Safe Harbor	Hadley Falls	Crescent
Turbine Type	Francis	Kaplan	Kaplan	Kaplan
No. Blades/buckets	16	5	5	5
Rated Head (ft)	62	55	50	27
Flow/Unit (cfs)	3500	8300	4200	1500
RPM	95	77-109	128	144
Runner Diameter (in)	164	220-240	N/A	108
Water Passage Diameter (in)	> 164	220-241	N/A	108
Blade tip Speed (ft/s)	31	80-105	N/A	68
Spacing of Wicket Gates (in)	N/A	30	N/A	27

TABLE 2. Comparison of short-term (1 h) and long-term (24 h) survival rates (± 95% confidence limits) of juvenile clupeids in passage through a single vertical shaft Francis and Kaplan turbines. N = number of test fish; equal numbers of control fish were also released to isolate mortality due to turbine passage. Source: EPRI (1992).

	Holtwood (Francis)	Safe Harbor (Kaplan)	Hadley Falls (Kaplan)	Crescent (Kaplan)
N	100	299	200	125
Fish Size (mm)	85-163	95-140	55-110	77-105
1 h Survival	89±9.8%	97±3.8%	100±5.5%	96±8%
24 h Survival	78±16.5%	100±7.7%**	100%*	100±12.4%**

* Due to vandalism and high control mortality 24 h survival considered unreliable and confidence limits not given.
** Overall survival equals survival at 1 h

project equipped with vertical Francis turbine (Table 2). The estimated survival was 89%.

Although sufficient tests at Francis turbines have not been conducted on young clupeids to establish predictive relationships, results to date suggest that survival rates are usually lower at Francis turbines than at Kaplan type turbines (EPRI 1992). Overall, in the present study, the difference in short-term survival rates between the Kaplan and Francis turbines was approximately 8% (89% vs 97%). However, this difference increased to 22% (78% vs 100%) at 24 h after passage through turbines. These differences may be related to the differences in spacing between the wicket gates, number of blades, or other internal structural components of the turbines tested in the present study.

The consistency of high survival rates, accompanied with relatively low variability at the studied sites provides high confidence in suggesting that Kaplan turbines are benign and may offer a safe passage route for emigrating young clupeids.

The high survival observed at Kaplan turbines suggests that direct contact of young clupeids with the rotating turbine blades is a low probabilistic event. This may be due to the slow runner speed and wide clearances between turbine runner blades. Cada (1990) speculated that turbine-related mortality of juvenile fishes might be low, perhaps no more than 5%. The three studies with Kaplan turbines cited in the present paper showed short-term mortality to be less than 5% with little or no additional mortality at 24 h.

Literature Cited

Cada, G. F. 1990. A review of studies relating to the effects of propeller-type turbine passage on fish early life stages. Amer. Jour. Fish. Mgt. 10:418-426.

Eicher Associates, Inc. 1987. Turbine-related fish mortality: review and evaluation of studies. EPRI Research Project 2694-4, Palo Alto, CA.

EPRI. 1992. Fish entrainment and turbine mortality review and guidelines. EPRI TR-101231 Project 2694-01, Palo Alto, CA.

Heisey, P. G., D. Mathur, and T. Rineer. 1992. A reliable tag-recapture technique for estimating turbine passage survival: application to young-of-the-year American shad (*Alosa* *sapidissima*). Can. Jour. Fish. Aquat. Sci. 49:1826-1834.

Regulation of Flow Downstream of Weirs

By Boualem Hadjerioua[1], Tony Rizk[2], Emmett M. Laursen[3] and Gary Hauser[2]

Abstract

Reregulation weirs are one means adopted by the Tennessee Valley Authority (TVA) to provide continuous improved minimum flow and wetted area for aquatic life downstream of hydroprojects during off-generation periods. A series of low-level pipes through the weirs, where some are fitted with regulating float-actuated valves, maintain essentially constant releases over a full range of weir pool elevations. Previous designs of such float-controlled mechanisms were based on physical modeling. This paper describes such a flow-regulating system and presents a quick and reliable analytical technique to determine float movement and pipe discharge rate at different headwater conditions. Data obtained from a full-scale model of the TVA South Holston labyrinth weir pipe and valve assembly were used to develop this analysis. Results of the analytical model were found to compare well with experimental data.

Introduction

Block-loaded operation of peaking hydroelectric plants, where the plants operate for only a few hours each day, results in essentially zero flow downstream for periods of several hours each day. TVA plans to construct reregulating structures downstream of some river when the peaking plant to provide a sustained minimum flow in the

[1]Graduate Student, University of Arizona, Dept. of Civil Eng., and Eng. Mech., Tucson, Arizona.

[2]Tennessee Valley Authority Engrg. Lab., Norris, Tennessee.

[3]Professor Emeritus, University of Arizona, Dept. of Civil Eng. and Eng. Mech., Tucson, Arizona.

turbines are not operating.
TVA is using various types of weirs with controlled and uncontrolled low-level pipes at dams where minimum flow objectives must be met (Hauser, et al.1991). The use of control valves and pipes in a reregulating structure was first introduced by TVA in 1984 at the Clinch River weir below Norris Dam (Harshbarger, 1985).

Experimental Data

A full-scale physical model of a section of the South Holston reregulating weir, shown in Figure 1, was tested at the TVA Engineering Laboratory. For each length of rod L1 tested, float movement and pipe discharge rates were measured at several different headwater elevations. Data from that experimental work was used to verify the analysis described in this paper, which generalizes design for these control systems for similar applications.

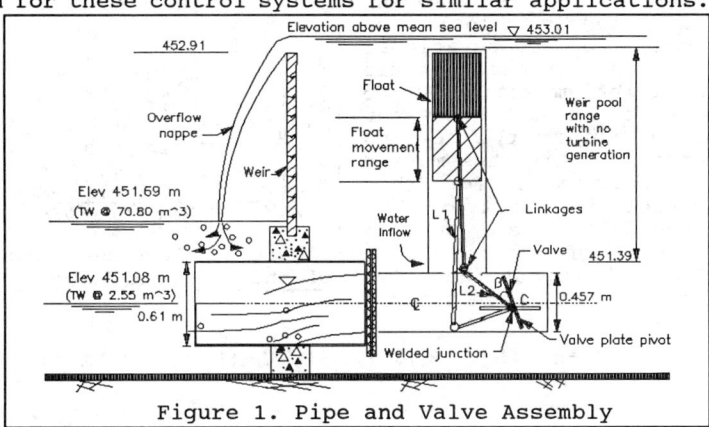

Figure 1. Pipe and Valve Assembly

Mathematical Problem Formulation

The flow regulation problem can be defined as follows: For any given rod length, L1, determine:
1. The submergence ($^*L_{fsub}$) at which the float starts to move; this gives starting time of valve opening.
2. The relation between headwater levels upstream of the weir and discharge through the valved pipe for a range of headwater levels.

The analysis was developed assuming that:

* All symbols are defined in Appendix II.Notation

a. The system is frictionless.
b. An average constant coefficient of pipe discharge can be used for the whole range of heads.

Problem Definition

The objectives of this study is to obtain a mathematical description of the relationships among the headwater, valve opening, and discharge through the regulating pipe. As the pool lowers, the valve opens sufficiently slowly so that, for computational purposes, a quasi-static equilibrium state may be assumed at each headwater level during the opening period. This assumption is reasonable for this system in light of the physical data. Summing moments of the forces in Figure 2, about the valve plate pivot, results in

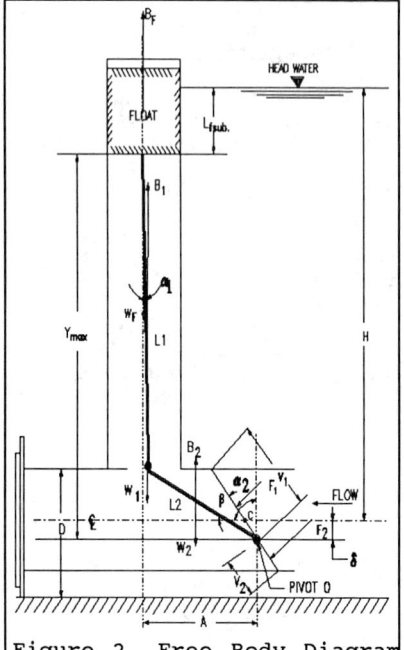

Figure 2. Free Body Diagram of the Pipe and Valve

$$(B_F - W_F) M_F + (B_1 - W_1) M_1 + (B_2 - W_2) M_2 + (F_2) M_{p2} - (F_1) M_{p1} = 0 \quad (1)$$

where

M_F, W_F, and B_F are, respectively, moment arm, weight and buoyancy of the float.
M_1, W_1, and B_1 are, respectively, moment arm, weight and buoyancy of rod L1.
M_2, W_2, and B_2 are, respectively, moment arm, weight and buoyancy of rod L2.
M_{p1} and M_{p2} are, respectively, moment arms of the forces F_1 and F_2.

Computation of Float Movement

The float movement is defined by:

$$Y_{mov} = Y_{max} - Y \qquad (2)$$

where Y is computed from Equation (1) for each weir headwater level, distance from the "horizontal" centerline of the pivot to the bottom of the float.

Computation of Flow Through a Valved Pipe

When power is generated, the water level rises in the reregulating pool above the weir, raising the float in the riser pipe and shutting the valve. The projected area of the valve plate is about equal to the interior pipe area. When the weir pool level lowers, the float drops and opens the valve. As the angle α_2 increases, the valve plate rotates, and additional flow passes through the pipe. The pipe discharge equation is:

$$Q_{flow} = V * A = C_d * A_{flow} \sqrt{2gH} \qquad (3)$$

where

A_{flow} = cross-sectional area of the pipe less projected area of the valve;
V = flow velocity;
C_d = coefficient of discharge
H = vertical distance from headwater level to pipe centerline; and
g = acceleration of gravity.

After algebraic manipulation and considering leakage, Equation (3) yields:

$$Q_F = C_d \sqrt{2g} \frac{\pi}{4} \sqrt{H} [D^2 - [(D-2\epsilon)(V_1 \cos(\alpha_2-\beta) + (V_2 \cos(\alpha_2-\beta))]] + C_{leak}\sqrt{H}$$

(4)

where ϵ is the thickness of the nut holding the valve to the pipe, and C_{leak} is the coefficient of leakage determined experimentally.

Results

When the valve is closed, the only flow through the pipe is due to minor leakage. The moment of the hydrostatic pressure forces were estimated for the upper and lower portions of the valve plate about its pin. The discharge coefficient, determined experimentally, was 0.489 for the opening mode and 0.6 for the fully open valve. As an example, application of the model is illustrated where L1 is 0.762 meter. Figures 3 and 4 show experimental and analytical results for float movement and discharge respectively, through the pipe for the range of headwater elevations. Comparison of similar data for other lengths of L1 also indicate good agreement between analytical values and experimental data.

Conclusion

The mathematical analysis describing the behavior of a float-actuated valve developed in this study performed reasonably well when compared to experimental data obtained from a full-scale physical model of the pipe and valve assembly for the TVA South Holston labyrinth weir. The analytical procedure presented provides a quick and reliable technique to determine movement of a float and pipe discharge rate at different headwater levels, pipe diameters, and connector rod lengths. The mathematical analysis provides a reliable design and eliminates the need for laboratory studies, thus saving time and money. Such low-flow control devices offer a relatively simple mechanism for reliable and automatic flow regulation to meet minimum-flow criteria using weirs downstream of hydro-plants that are operated intermittently for power.

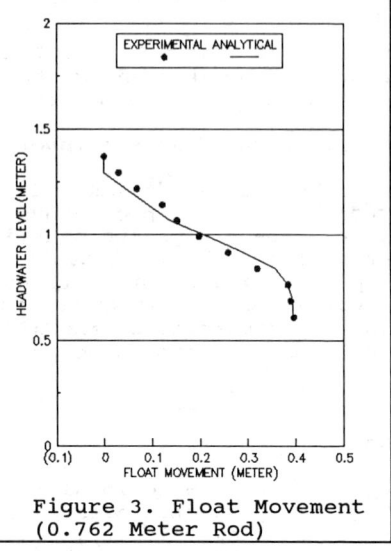

Figure 3. Float Movement (0.762 Meter Rod)

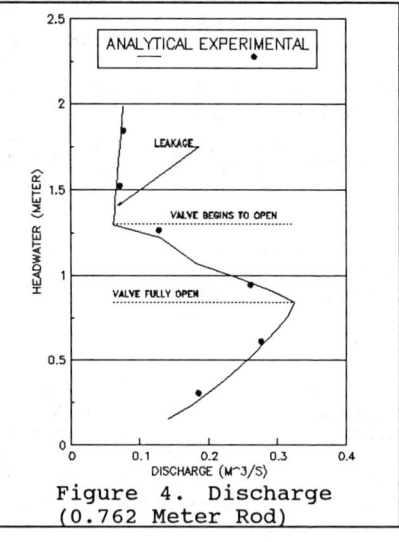

Figure 4. Discharge (0.762 Meter Rod)

REGULATION OF FLOW 1343

APPENDIX I. REFERENCES

1. Harshbarger, E. Dean (1985). "Field Evaluation of the Clinch River, Hydraulic Characteristics," WR28-1-590-116, TVA Engineering Laboratory, Norris, Tennessee.

2. Hauser, G. E., R. M. Shane, J. A. Niznik, and W. G. Brock (1991). "Innovative Reregulation Weirs for Improving Minimum Flow and Dissolved Oxygen in Dam Releases," Proceedings of the ASCE National Conference on Hydraulic Engineering, Nashville, Tennessee.

3. Shane, Richard M. (1985). "Experimental Clinch River Flow Reregulation Weir, Field Evaluation Interim Report," WR28-4-590-118, TVA Engineering Laboratory, Norris, Tennessee.

APPENDIX II. NOTATION

The following symbols are used in this paper:

A_{Flow}	= float cross-sectional area of the flow;
B_F, B_1 and B_2	= buoyancy of the float, rod L1 and L2 respectively;
F_1, F_2 and D	= pressure forces, and pipe diameter;
H	= headwater above pipe centerline;
LF, M_1 and M_2	= moment arm of the float, rod L1, rod L2 respectively;
L_{fsub}	= critical float submergence;
L_{p1} and L_{p2}	= moment arm for the pressure forces F_1 and F_2 respectively;
L1 and L2	= vertical and horizontal rod lengths, respectively;
Q_F	= total discharge including leakage;
V_1 and W_2	= upper and lower length of the valve from the eccentric axis of rotation, respectively;
W_F, W_1 and W_2	= weight of the float, rod L1 and L2 respectively;
Y	= distance from the "horizontal" centerline of the pivot to the float;
Y_{max}	= distance from the valve pivot to the bottom of the float (closed position);
Y_{mov}	= float movement;
α_1	= angle between the "vertical pipe" centerline and rod L1;
α_2	= angle between the pivot and rod L2;
β	= angle between the valve and rod L2 and;
δ	= distance from centerline of the horizontal pipe to axis of rotation.

Flow Resistance Properties of Flexible Linings

by George K. Cotton[1], Member

ABSTRACT

Flow resistance for flexible channel linings is a function of the relative roughness of the lining in accordance with boundary-layer principles. As with vegetation, flexible linings have a tendency to deform under imposed boundary shear stress. The erosion control properties of flexible channel linings depend largely on the creation of a buffer zone above the soil surface. A standardized method for testing flexible linings is discussed.

INTRODUCTION

A variety of flexible channel lining materials are commonly used for erosion protection in drainage channels (Chen and Cotton, 1988). These lining provide a dense uniform cover for channel soils until a permanent cover of vegetation can establish. The design approach for these linings has generally relied on the concept of permissible velocity. This method has a number of limitations, particularly when the large number of combinations of soil and lining types is considered. The erosion control properties of these linings depend on both the resistance imparted to the flow and the boundary shear that results at the soil surface. Erosion will occur beneath the lining when the boundary shear on the soil surface exceeds the threshold for soil particle movement. A flexible lining can reduce the effective boundary shear stress at the soil surface by providing a buffer zone.

FLOW RESISTANCE THEORY

The resistance to flow of a flexible channel lining is based on boundary layer theory that relates the relative roughness to dimensionless flow velocity (Chow, 1959). The form of the resistance equation is

$$\frac{U}{U_*} = a + b \log\left(\frac{R}{k_s}\right) \quad \quad (1)$$

where: U is the mean flow velocity; U_* is the shear velocity, $\sqrt{gRS}$; a, b are coefficients that vary for channel lining types; R is the hydraulic radius; S is the channel slope; g is the acceleration of gravity; and k_s is the roughness height.

---

[1]Senior Water Resources Engineer, CRSS Civil Engineers, Inc. Phoenix, Arizona

Equation 1 has two distinct advantages for flexible linings. First, these linings are typically used in a low range of relative roughness (typically, R/ks < 100) where the resistance coefficient is strongly non-linear. Second, because the lining is flexible the roughness height, ks, will vary depending on the boundary shear.

FLOW RESISTANCE DATA

The first FHWA data set was developed by Mississippi State University (MSU) from 1965 to 1966 (McWhorter et al., 1968). The MSU study provides an extensive data base for the development of resistance equations for flexible channel linings. The primary disadvantage of the MSU data is that many of the linings studied proved to be impractical channel stabilization measures. Four lining types that have stood the test of time are jute net, excelsior mat, fiberglass mat, and straw secured with netting. *Jute net* consists of jute yarn, approximately 6 mm in diameter, woven into a net with openings that are about 10 by 20 mm. *Excelsior mat* consists of curled wood with wood fibers, 80 percent of which are 0.15 m or longer, with a consistent thickness and an even distribution of fiber over the entire mat. *Fiberglass mat* consists of continuous fibers drawn from molten glass, coated and lightly bound together into a mat with asphaltic materials. *Straw with net* consists of a straw mulch spread at the rate of approximately 2.2 tons per hectare and secured with a plastic net.

Table 1. Test results for various linings for the first FHWA study.

Mississippi State Univ. Study				Measured Ranges		
Lining Type	Number of Runs	Soil Type[1]	Channel Shape[2]	Velocity (mps)	Slope (m/m)	Depth (m)
Jute Net	13	4,6,9,11	a	0.18-1.66	.001-.100	.013-.232
Fiberglass Mat (3/8")	4	11,4	a	0.12-1.20	.010-.100	.024-.231
Fiberglass Mat (1/2")	4	11,4	a	0.12-1.79	.025-.125	.030-.262
Excelsior Mat	24	2,4,6,8,10,11	a,b,c	0.17-1.72	.010-.125	.024-.287
Straw with Net	13	2,6,8,10,11	a,b,c	0.14-1.60	.025-.100	.034-.290

(1) Description of Soils

Soil Number	Plasticity Index	D50 (mm)	% Clay
2	NP	11.3	0
4	NP	0.30	0
6	29	0.10	36
8	15	.074	31
9	6	.043	15
10	47	.047	60
11	23	.051	34

(2) Description of Channels

Channel Shape	Description (B:width; Z:side slope)
a	Rectangular, B=0.6m
b	Triangular, Z=3:1
c	Rectangular, B=1.1m
d	Trapezoidal, B=0.3m,:1

The second FHWA data set was developed at the USGS Bay St. Louis laboratory in 1985 (Thibodeaux, 1985). This study re-tested the four lining types still in use from the MSU study and tested two commercial products: Hold-Gro and Enkamat. *Hold-Gro* is typical of a range of products that can be classified as *Woven Paper Net* that consists of paper strips that are interwoven with plastic strands. *Enkamat* is typical of a class of products called *Synthetic Mat* that have a three-dimensional structure of heavy plastic filaments that are molded to have a mat thickness ranging from 6 mm to 20 mm.

Table 1(cont.). Second FHWA study.

Bay St. Louis Study					
Lining Type	Number of Runs	Channel Shape[2]	Velocity (mps)	Slope (m/m)	Depth (m)
Jute Net	4	d	1.13-2.19	.030-.115	.094-.195
Fiberglass Mat (1/2")	4	d	0.24-1.43	.005-.030	.037-.140
Excelsior Mat	7	d	0.27-0.94	.030-.090	.055-.253
Straw with Net & Asphalt	5	d	0.18-1.28	.010-.090	.052-.232
Hold-Gro	4	d	0.52-1.31	.010-.090	.030-.274
Enkamat	3	d	1.25-2.01	.060-.115	.119-.171

REGRESSION ANALYSIS

The author conducted a linear regression analysis on the MSU data sets using the following model:

$$\frac{U}{U_*} = a + b \log(R) \quad \text{...............(2)}$$

Coefficients for this regression model are given in Table 2. Since, the MSU tests did not record roughness element height, a method of inferring roughness element height was developed. As can be seen from Table 2, the value of the coefficient "b", which is the slope of the semi-log curve, varies only slightly for the various lining types. The average value of coefficient "b" is 7.80. Kouwen and Li (1980) reported values of the coefficient "a" from 0.42 to 0.82 for vegetation over a range of stem deflection (from erect to prone) with a corresponding variation in the coefficient "b" from 5.23 to 9.90. Using an interpolated value of a'=0.72 for "a" that corresponds to a "b" of 7.80, the effective roughness height was calculated by the following equation:

$$k_s = 10^{(a'-a)/b} \quad \text{...............(3)}$$

$$\frac{U}{U_*} = a' + b \log\left(\frac{R}{k_s}\right) \quad \text{...............(4)}$$

Table 2. Coefficients for a + b log(R) Regression Model

Lining Type	a	b	r^2	n	a'	k_s(m)
Jute Net	12.1	8.04	0.868	86	0.72	.012
Fiberglass Mat	9.00	8.21	0.782	129	0.72	.030
Excelsior	7.39	7.10	0.803	84	0.72	.035
Straw with Net	7.99	7.83	0.788	99	0.72	.036

Interestingly, as shown by Table 3, the estimated roughness height, ks, is roughly 1.85 times the actual lining thickness.

Table 3. Comparison of Lining Thickness to Roughness Height

Lining Type	Lining Thickness (mm)	Roughness Height (mm)	Ratio Thickness/Height
Jute Net	6.4	12	1.81
Fiberglass Mat	12.8	30	2.33
Excelsior	19.2	35	1.83
Straw with Net	19.2	36	1.87

An additional regression model was also tested that had the following form:

$$\frac{U}{U_*} = a + b \log(R) + c \log(S) \quad\quad (5)$$

Coefficients for this regression model are summarized in Table 4. Using some algebraic manipulation and the same method of estimating roughness element height as with equation 4, the following form of equation 5 was derived:

$$\frac{U}{U_*} = a' + b' \log\left(\frac{R}{c'\tau d'}\right) \quad\quad (6)$$

where τ is the boundary shear stress in Pa.

Table 4. Coefficients for a + b log(R) + c log(S) Regression Model

Lining Type	a	b	c	r^2	n	a'	b'	c'	d'
Jute Net	11.5	6.30	-1.48	0.902	85	0.72	7.78	.00393	0.38
Fiberglass Mat	9.56	6.90	-1.50	0.831	128	0.72	8.40	.00962	0.36
Excelsior	8.31	6.08	-1.27	0.833	83	0.72	7.35	.01070	0.35
Straw with Net	11.1	7.46	-0.44	0.788	98	0.72	7.90	.02470	0.11

Equation 6 provides a relationship between roughness height and boundary shear. The increase in flexible lining roughness height with applied shear stress is similar to the condition where vegetation has become fully prone and is observed to wave vigorously. In fact, a primary cause of lining failure is sufficient displacement of the lining from the channel surface to expose the underlying soil to erosion.

RELATIONSHIPS FOR NEW LINING TYPES

The Bay St. Louis datasets were too small to perform a regression analysis. Instead, an envelope approach that was based on the regression analysis of the larger MSU datasets was used. It was assumed that equation 1 was the correct model and that coefficient "a" would equal 0.72 and roughness height was approximately 1.85 times the actual lining thickness. Coefficients for the enveloped resistance relationships are summarized in Table 5.

Table 5. Envelope Relationships $U/U_* = a + b \log(R/k_s)$

Lining Type	a	b	ks (m)
Hold-Gro	0.72	8.00	0.0012
Enkamat	0.72	8.13	0.0185

CHANNEL STABILITY

In general, the limiting condition for flexible lining materials is the erosion of the underlying soil. Failure of the lining itself due to tearing or washout is uncommon, particularly for manufactured linings. Table 6 summarizes the critical shear stress for each combination of soil and lining type in the MSU study. Figure 1 shows the relationship between critical shear stress for the soil and critical shear stress for the lining.

Table 6. Summary of Critical Shear Stress Conditions, Pa

Soil Type	Jute	Excelsior	Straw+Net	Fiberglass Mat (Sgl)	Fiberglass Mat (Dbl)	Soil
2	-	96.7	101.	-	-	8.6
4	16.3	59.3	-	22.0	-	2.2
6	22.5	53.6	50.7	-	60.3	3.8
8	-	66.5	46.4	-	-	3.6
9	37.3	-	-	-	-	3.1
10	-	87.1	58.9	-	-	10.5
11	25.4	83.3	79.0	36.4	75.6	6.2

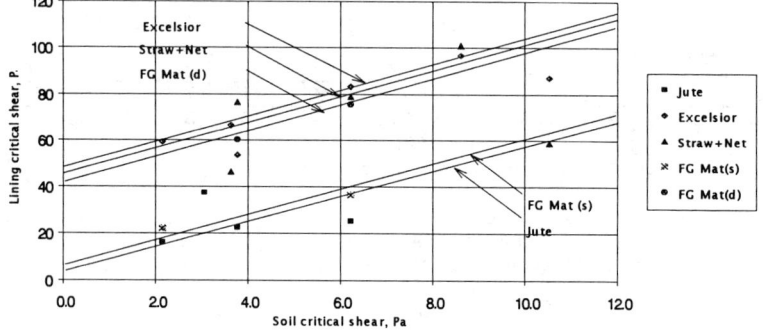

Figure 1. Relationship of lining critical shear to soil critical shear

Figure 1 shows that the shear at the soil surface is substantially reduced compared to the shear stress at the flexible lining. The limited data from the MSU study suggests the following relationship between critical shear at the soil and lining:

$$\tau_{sc} = \theta \, (\tau_{lc} - \tau_l) \quad \quad (7)$$

where τ_{sc} is the critical shear on the soil, τ_{lc} is the critical shear on the lining, and τ_l is the shear parameter for the lining (all shear stresses in psf). For $\theta=0.18$, the lining shear parameter for the four MSU lining types the are as follows:

Lining Type	Jute	Excelsior	Straw+Net	Fiberglass Mat (Sgl)	Fiberglass Mat (Dbl)
τ_{lc}	4	48	46	6	40

TESTING RECOMMENDATIONS

It can be seen from equations 6 and 7 that a test procedure must generate sufficient data to describe flow resistance parameters and the lining and soil interaction. Two facts about flexible lining behavior have been demonstrated that can greatly reduce the number of tests required to develop accurate resistance and shear stress relationships. First, the flow resistance created by the lining is largely independent of the underlying soil type. Flow resistance parameters can therefore be established using a single soil type as a substrate. Measurements of lining hydraulics should include mean flow velocity, flow depth, channel slope, and roughness height. Second, there is substantial buffering of boundary shear by the lining. The MSU data indicates that about 80 to 95 percent of the boundary shear is deflected from the soil surface by the lining after the lining shear, τl, is surpassed. The MSU data also indicates that the lining shear parameter, τl, is related to lining thickness and probably other mechanical properties such as stiffness and density.

Testing for erosion control must cover a reasonable range of permissible soil shear conditions. A useful range is about 1 to 10 Pa, which would be roughly equivalent to non-cohesive sediment gradations with a mean size ranging from 0.1 to 10 mm. It is the opinion of the author that testing should be conducted using non-cohesive sediments. Advantages of non-cohesive sediments are: material gradation and mean size can be controlled; the uniformity and density of the channel bed can be controlled; and most importantly, the motion of sediment grains is easily observed.

Testing should be conducted on uniformly compacted soil that has been completely saturated prior to the runs. Since sediment transport will be disrupted by the channel lining, the motion of sediment grains must be observed. Lining failure occurs when grain motion is observed over most of a test section. An objective method of determining lining failure is to randomly select a number of points within the test section and observe grain motion. A lining failure will be noted when a specified number of the points in the test section have grain motion. The number of points to be measured and the required number to specify a failure conditon should be based on a predetermined confidence limit, such as 90 percent.

REFERENCES

Chen, Y.H. and Cotton, G.K., 1988, "Design of Roadside Channels with Flexible Linings," Hydraulic Engineering Circular No. 15, U.S. Department of Transportation, Publication No. FHWA-IP-87-7.

Chow, V.T., 1959, Open Channel Hydraulics (New York, McGraw-Hill Book Company)

Kowen N. and R.M. Li, 1980, "Biomechanics of vegetated channel linings", J. Hydr. Div., ASCE, 106(6), 1085-1103.

McWhorter, J.C., T.G. Carpenter, and R.N. Clark, 1968, "Erosion Control Criteria for Drainage Channels," Mississippi State University, State College, Mississippi.

Thibodeaux, K.G., 1985, "Performance of Temporary Ditch Linings," Interim Reports 1 to 17, U.S. Geological Survey, Gulf Coast Hydroscience Center.

CALCULON WEIR STRUCTURE - EFFECTIVE USE OF GABIONS

John G. Harman, PE, Assoc. Member

Progressive Engineering Consultants, Inc.
Baltimore, MD

ABSTRACT

Providing onsite stormwater management practices in an industrial area is a compromise between construction cost and site planning. To provide the storm water management facility, we designed a horseshoe-shaped gabion weir wall accompanied with an earthen dam in the site floodplain. The length of the weir was set for safe passage of the 100 year frequency storm from 128Ha, (317 Acre) watershed. Construction in the floodplain was minimized, placing the storm water management facility in the floodplain also allowed more room for parking on the site.

Presented at

1993 National Conference on Hydraulic

Engineering and International Symposium

on Engineering Hydrology

San Francisco, CA July, 1992

Progressive Engineering Consultants, Inc
1506 Joh Ave.
Baltimore, MD   21227

CALCULON WEIR STRUCTURE

The Calculon site is a Light Industrial site located in Rockville, Maryland, approximately .40km, (25 mi.) northwest of Washington, DC. The area of the site is 3.3Ha (8.1 Acre), of which 1 Ha, (2.5 Acres) is floodplain, on the natural stream on the rear of the property. Prior to construction, the site consisted of fallow pasture and light woods. The property lies within an office-industrial park with 14 lots of similar land uses.

In 1979, the engineering design was conducted. Site infrastructure construction followed that fall, with completion in the Spring of 1980. Prevailing methods of providing storm water management best management practices at the time were: storage in underground corrugated metal pipes, storage in large open detention facilities (dry ponds), or storage within stone filled Infiltration trenches. The City of Rockville regulations also provided for a waiver of onsite storage if regional stormwater storage was available and payment of a fee based on the site zoning. The regional stormwater management facility was at capacity and accordingly, the waiver of onsite storm water managmenet was denied.

The runoff calculations were prepared for the industrial site and storage requiements were determined. The City of Rockville regulations required the Retention of the ten year frequency post-development runoff to match the two year pre-development runoff rate. This was more restrictive than the State and County regulations, which only required two years storm retention. Water quality detention was not required at this time.

The analysis of the appropriate Best Management Practice was performed next. The use of u nderground storage in corrugated metal pipe was not highly regarded because of the high cost in relation to the other alternatives. Underground stone filled infiltration trenches were ranked low because of cost and a poor soil infiltration rate. The use of an open detention facility was selected as mot feasible, but that design also required a large amount of storage space.

Siting the pond had to take into account the size of the building, parking requirements, utilities and the large floodplain area. Placement of the pond in an area adjacent to the floodplain placed a large restriction on the building area and area available for parking.

After discussion with the City Engineer, a pond site within the floodplain area was selected. This resolved site planning restrictions but added to the difficulty of the stormwater design by adding the requirement of conveying the 100 Year frequency storm from the entire watershed of 12.8 Ha, (317 Acres). This allowed the natural undisturbed floodplain to serve as the storage area.

The TR-55 hydrology for the watershed was developed next, for the predevelopment condition and the future post-development condition. The watershed contained a mixture of 85 Ha, (210 Acres), of farmland and; 31 Ha (77 Acres), of Industrial park and 12 Ha, (30 Acres), of woodland. The hydrology for the predevelopment condition was relatively straight forward and received quick approval by the City. The hydrology for the post development condition was more difficult because of different views between the designer, City Engineer and County Engineer on the extent the farmland tract would develop and the estimated modifications to the natural stream channel.

PREDEVELOPMENT HYDROLOGY

WATERSHED CN=72   TC=0.92hr

Q_2= 4.8 m3/s   Q10 = 11.8 m3/s   Q100=20.9m 3/s-ess

POST DEVELOPMENT   HYDROLOGY

WATERSHED   CN= 85   TC = 0.60hr

Q_2 =1.7 m3/s   Q10 = 22.8 m3/s   **Q100 = 40.7**m3/s

(Q_2 = 413 cfs)   (Q10 = 805 cfs)   (Q100 = 1437cfs)

Rainfall distribution Type II

70% Type B Soils in Watershed
8% Type C Soils in Watershed
22% Type D Soils in Watershed

Next, design of the weir structure and dam was begun. The use of concrete riser structures with pipe outlets was discarded because of the number required made their use cost prohibitive. The use of a triangular weir rather than a broad crested weir provided advantages because of improved weir coefficient, but the estimated construction expense made it prohibitive. The use of a broad crested weir, constructed of gabions was selected. The rectangular shape was set to maximize the weir length within the available area.

CALCULON WEIR STRUCTURE

To reduce erosion inside the weir reno mats (thin gabions) were utilized.

In order to contain the storm retention pool within the floodplate portion of the lot, a earthen dam and levee were provided. To provide extra storage volume and satisfy the City Engineer, the weir was increased in height by 0.3m,(1 foot) above the onsite requirements. This was accomplished by the addition of a gabion of that thickness. Additional gabions were utilized at the junction of the weir and earthen dam to serve as a retaining wall. To increase the amount of impermeability of the gabion wall, a double thickness of filter fabric was provided between the gabion walls. The weir wall design used the manufactured dimensions of the gabions in order to avoid having to cut them in the field. The 46 cm (18 in.) control pipe was constructed of corrugated metal with a plate headwall welded to it. A rebar trash rack was provided at the inflow end of the pipe. The gabions were galvanized coated.

The constructed design blends well with the natural floodplain. The reno mats have been self-seeded with grass since construction. The gabion weir wall has withstood thirteen years of service with little repair. We estimate the construction cost of $45,000 (1970 prices).

References:

1) Kings Hydrology, Third Edition, King and Brater
2) Urban Hydrology for small watersheds, USDA Soil Conservation Service

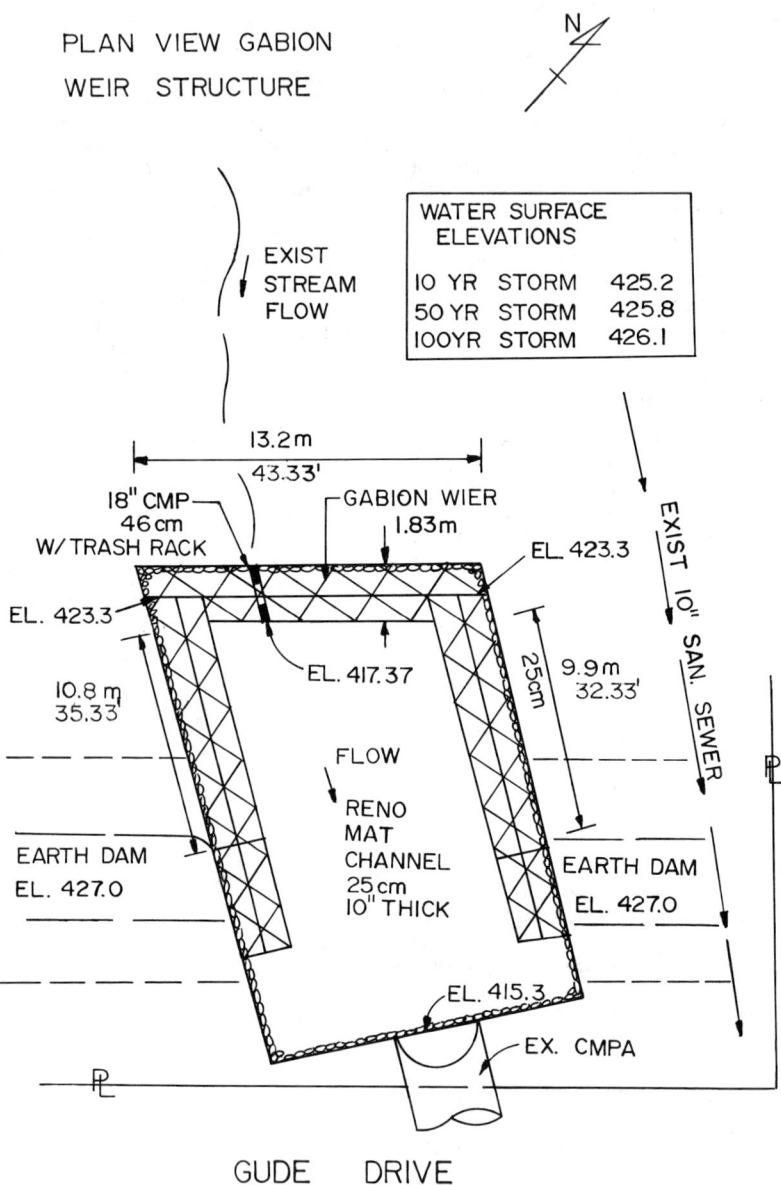

Storm Drainage Channel Rehabilitation

Mark S. Holstad, PE[1], Member, ASCE, and
Kenneth W. Wylie, PE[2], Member, ASCE

Abstract: 30 km (19 miles) of existing concrete lined storm drainage channels were studied due to repeated failures. Alternate failure causes were hypothesized and evaluated. Physical testing was performed to evaluate hypotheses. Major failure considerations were determined to be alkali aggregate reactivity and uplift pressures at expansion joints. A grading system was developed to rate channel segments. Construction is underway to implement new joints and overlays.

Introduction: Approximately 113 km (70 miles) of trapezoidal concrete lined storm drainage channels have been constructed in the Albuquerque metropolitan area. The City of Albuquerque, the Albuquerque Metropolitan Arroyo Flood Control Authority and the New Mexico State Highway & Transportation Department each maintain particular channels. The channels roughly follow the alignment of old arroyos, with typical slopes ranging from 1 to 6 percent. Design velocities exceed 12 m/s (40 ft/s) in some channels. The channels were designed and constructed as numerous projects, starting in the early 1960's and continuing today. Various designs of channel lining and joints were utilized. The first channels constructed by the City of Albuquerque have experienced substantial distress resulting in repeated washouts. Therefore, the City of Albuquerque contracted for a study of these channels. The purposes of the study were to inventory the channels, to evaluate their conditions, to analyze the types and causes of the various failures being experienced, to evaluate alternative approaches to repairing or preventing these failures, and to identify priority reaches of channel for implementation of these alternative repairs.

---

[1] Project Manager, Greiner, Inc., 5971 Jefferson NE, Albuquerque, NM 87109
[2] Owner, K. W. Wylie & Associates, 3200 Carlisle NE, #116, Albuquerque, NM 87110

History of Trapezoidal Concrete Flood Control Channels in Albuquerque: Trapezoidal channels in Albuquerque have undergone an evolution which is almost full cycle (see Reference 3 for much of the information included in this section). The first channels constructed in the 1960's were continuously reinforced using 150 to 200 mm (6 to 8 inches) of concrete with a substantial percentage of rebar. These channels had only construction joints, and some provision for subgrade drainage. In the late sixties and early seventies, designs changed considerably. Concrete thickness was reduced, welded wire fabric was used for reinforcement and extensive use was made of weep holes and joints of all types. It is this era from the late sixties to the early eighties during which the channels that are the subject of this study were built. Later the City returned to the construction of continuously reinforced channels. The current design is characterized primarily by a concrete lining thickness of 150 to 200 mm (6 to 8 inches), continuously reinforced with a minimum steel-to-concrete ratio of 0.5%. Joints are kept to an absolute minimum, and these are of a special design developed in Albuquerque. This current design has functioned well withstanding some major storms.

The first serious failure in the City system occurred in 1975 on the Hahn Arroyo (see picture to the right), built in 1970. Several sections of concrete lining were washed out during a severe storm. Since that time substantial deterioration of these older channels has been noted and there have been numerous additional washouts. The causes were not known, although there have been innumerable theories.

Study Approach: The study approach consisted of the following:
- Field observations to inspect actual conditions
- Failure mechanisms hypothesized
- Physical testing to evaluate hypothesized failure mechanisms after the field observations
- Interviews with a wide range of public officials, contractors and others
- Collection of literature and study of as-builts and previous repairs
- Comparisons of alternative repair methods and materials
- Development of a recommended program

Field Observations: Field observations were conducted on bicycles, with frequent stops for observations, evaluations, photos and commentary using voice dictation equipment. Evaluations were made using a rating system developed for this study, based on ACI 201.3R-86, "Guide for Making a Condition Survey of Concrete Pavements."

Hydraulic Considerations: The force of rushing water is awesome. Under the wrong conditions this force is able to overcome resisting forces. The U.S. Bureau of Reclamation (References 1 and 2) has performed two studies that deal with the uplift force that can develop at offsets in spillway linings where a portion of the flow is deflected and a stagnation pressure is produced beneath the lining. The following figure is taken from these papers and illustrates specific conditions and resultant uplift pressures.

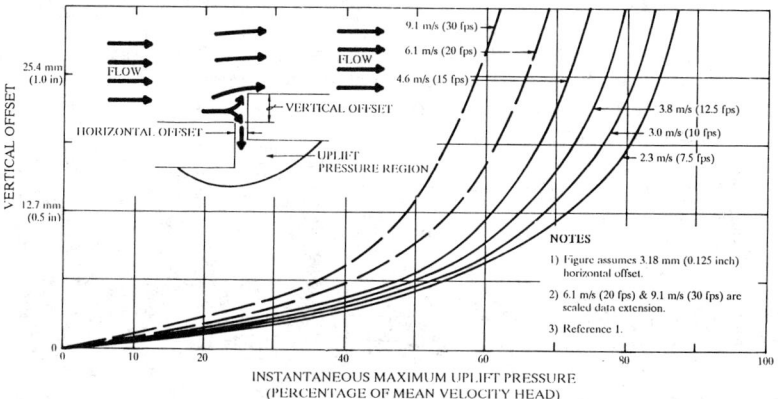

INSTANTANEOUS MAXIMUM UPLIFT PRESSURE

Various conditions are compared in the following table. This table is based on typical channel and flow conditions for two channels (the Hahn Arroyo and the Embudo Arroyo) which have experienced past washouts and are in the greatest need of repair. According to the above results, a washout should not be possible for velocities less than 4.6 m/s (15 fps). This indeed has been the experience of channels in the Albuquerque area.

Uplift Forces versus Resisting Forces for Various Velocities:

Flow Velocity (m/s)/(fps)	Velocity Energy (kpa)/(psi)	Uplift Force (kPa)/(psi)	Resisting Force (kPa)/(psi)
0/0	0/0	0/0	3.6/0.52
4.6/15	10.4/1.51	7.1/1.03	8.07/1.17
7.6/25	29.0/4.21	16.8/2.44	14.3/2.08
9.1/30	41.8/6.06	22.5/3.27	15.5/2.25
10.7/35	56.8/8.24	28.4/4.12	18.5/2.69

Physical Testing Program: From the field observations, condition survey ratings and hypotheses of failure mechanisms, a physical testing program was developed. The results of this testing revealed:
- Original construction of the channels lacked somewhat in quality. This was evidenced by misaligned water stops, joint filler, and dowels at expansion joints, as well as deep thickened sections without an adequate transition to normal channel depth. The primary cause for cracking on either side of the expansion joints was the restraint from these abrupt transitions with further distress from reinforcing mesh located at the channel bottom in these areas.
- Depth of scour was minimal indicating that the existing channel bottoms are in no danger of abrading to any serious depth.
- Petrographic analyses discovered that alkali-aggregate reactivity was significantly present in 9 out of 10 samples.
- Compressive strengths were all above 36,500 kPa (5,300 psi), with the exception of the samples from one channel segment.
- Results of thermal and moisture movement testing revealed no abnormal length change characteristics.

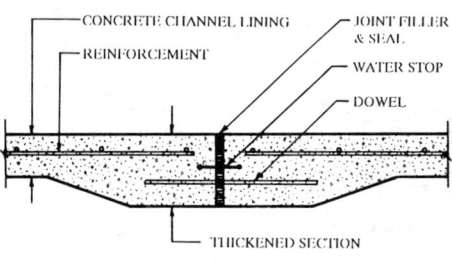

TYPICAL EXPANSION JOINT DETAIL

The photograph above illustrates a typical joint with misaligned water stop. Alkali aggregate reactivity is apparent in the right side of the joint. The detail shows the typical design used in the study channels.

Failure Mechanisms: Failure mechanisms are considered to be those primary factors that can directly result in a washout of channel lining sections caused by floodwaters getting under the lining. From the field observations, interviews, and literature research, two primary failure mechanisms became apparent. They are: 1) full depth spalling and 2) offsets at joints and cracks, which include partial depth spalls on the upstream side of the joint. The full depth spalling allows a direct path for floodwaters to gain access under the lining. Offsets create the condition of deflecting a portion of the flow into the joint or crack thereby creating a stagnation pressure and uplift forces. Both full and partial depth spalls can be caused by any of a multitude of factors acting individually or in various combinations. Alkali aggregate reactivity appears to be one of the main culprits. The expansive forces created by the reaction around reactive aggregate pieces cause an internal fracturing of the concrete and resultant spalling. Reactivity problems are the worst at open, unsealed joints and cracks where water has free access.

Another factor that contributes to spalling is the buildup of incompressible material in open, unsealed joints and cracks. When temperatures rise in the warmer months and concrete expands, compressive forces are created because of the inability of joints and cracks to close resulting in spalling occurs.

Channel Condition Assessment: A numerical scoring system was developed to consolidate the substantial quantity of channel condition data that was developed. This allowed the comparison of the condition of all channel sections. A total of 21 issues were evaluated for each channel section. Two grades were assigned for each issue, one grade for the quality (severity) of that issue and another grade for the extent or frequency of that issue. These grades were primarily based on field observations. A weight was also assigned to each of the issues to reflect its impact on a possible channel failure. The score was then determined for each issue for each channel by multiplying the Severity grade by the Extent/Frequency grade by the Weight. The channel score was determined by summing the 21 issue scores for the channel segment.

Alternative Rehabilitation Methods: In the nearly 30 km (19 miles) of investigated channel, a range of conditions and problems was found. Continuously reinforced channels had virtually no problems. Problems in conventionally reinforced channel segments ranged from lack of sealed joints to sporadic cracking and spalling to severe and frequent cracking, spalling, and buckling. Because of the variety of conditions that exists, a variety of rehabilitation needs and alternative methods for meeting those needs also exists.
• Clean and Reseal Joints: Any joints that are true working joints, (expansion joints, control joints, and cracks actually experiencing movement) should be properly cleaned and sealed. Cost - $7.38/m ($2.25/ft)
• Conventionally Repair Spalled Joints and Cracks: Joints or cracks that contain spalled and de-laminated portions of concrete should be repaired with a full depth patch. Reactivity may be the cause thus warranting the full depth repair. Cost - $215.29/m^2 ($20.00/ft^2)
• Repair Spalled Joints and Cracks Using Slurry Infiltrated Fiber Concrete (SIFCON): As an alternative, SIFCON may be used to produce a tougher joint much more resistant to spalling and crushing. SIFCON is made by filling the repair cavity with steel fibers and a highly fluid cement grout. Typical properties of SIFCON are 103,000 kPa (15,000 psi) compressive strength and 34,000 kPa (5,000 psi) flexural strength. Cost -

$430.57/m^2$ ($40.00/ft^2$)
- Overlay with Continuously Reinforced Concrete (CRC) Lining: In channel segments with excessive problem joints or deteriorating concrete, the channel bottom can be overlaid with an unbonded, CRC overlay. This overlay results in a hydraulic capacity reduction of 18.2%. Cost - $69.97/m^2$ ($6.50/ft^2$)
- Overlay with Slurry Infiltrated Fiber Concrete (SIFCON) Lining: This is an alternative to the CRC overlay consisting of a 38 mm (1½ inch) thick, partially bonded SIFCON overlay. This overlay results in a hydraulic capacity reduction of 3.6%. Cost - $67.28/m^2$ ($6.25/ft^2$)
- Remove and Replace Existing Channel Bottom: In order to create a well-defined invert and alleviate any concern of further deterioration of the existing bottom concrete, the channel bottom can be removed and replaced. Cost - $82.88/m^2$ ($7.70/ft^2$)
- Remove and Replace Entire Existing Channel: This method requires removal of the existing channel and construction of a new CRC channel. Cost - $296.02/m^2$ ($27.50/ft^2$)

Economic Analysis: A Present Worth comparison was made of joint repairs versus liner repairs, either Continuously Reinforced Concrete (CRC) or Slurry Infiltrated Fiber Concrete (SIFCON). This analysis found that joint repairs are less than 20% the cost of a new liner and therefore joint repairs should be used wherever possible. However, some channels are so badly deteriorated that joint repair is not a viable option.

Conclusions: A total of 30 km (19 miles) of channel constructed in Albuquerque primarily in the 1970's were studied. Uplift forces caused by stagnation pressures at damaged or non-maintained expansion joints were found to be the primary cause of washouts. Alkali aggregate reactivity is typical in local concrete of this era and has substantially contributed to deterioration of expansion joints. Joint repairs and SIFCON lining were determined to be the most cost effective means of upgrading existing jointed concrete channels. An $800,000 construction project is underway to rehabilitate the most critical channel sections. SIFCON (slurry infiltrated fiber concrete) is being used in the lining and some joint repairs.

Bibliography:
1) "Analysis of Spillway Failures by Uplift Pressure," Thomas E. Hepler and Perry L. Johnson, USBR, August 1988.
2) Memorandum, July 15, 1976, "Research into Uplift on Steep Chute Lateral Linings," with study by Perry L. Johnson, USBR- Hydraulics Branch of the Division of General Research.
3) "Trapezoidal Concrete Flood Control Channels in Albuquerque: Lessons Learned," Larry A. Blair, Albuquerque Metropolitan Arroyo Flood Control Authority (AMAFCA), July 1984.

Determination of Hydraulic Roughness for Concrete-Lined, Supercritical Channels

Scott E. Stonestreet[1], M ASCE, Michael E. Mulvihill[1,2], M ASCE, and Ronald R. Copeland[3], M ASCE

Abstract

A channel verification program has been initiated by the US Army Engineer District, Los Angeles to determine the hydraulic roughness of flow in a concrete channel. Specifically, this study focuses on the effective roughness of concrete with supercritical flow. The purpose of this paper is to present preliminary results of this channel verification program.

Introduction

Current Corps of Engineers design criteria recommends that flood control channels be designed based on the effective roughness height 'k' for capacity computations. Specifically, the design guidance, EM 1110-2-1601 (US Army, 1991), recommends a k value of 2.13 mm (0.007 ft) for concrete channels. Based on the theoretical uniform-flow equation for rough channels presented in Chow (1959), a relationship between Manning's n-value and the effective roughness height is as follows:

$$n = \frac{R^{1/6}}{19.58 + 18.01 \log(R/k)} \quad (SI) \quad (1)$$

---

[1]Hydraulic Engineer, US Army Corps of Engineers, Los Angeles District, PO Box 2711, Los Angeles, CA 90053

[2]Professor of Civil Engineering, Loyola Marymount University, 7101 W. 80th St., Los Angeles, CA 90045

[3]Research Hydraulic Engineer, Hydraulics Laboratory, US Army Engineer Waterways Experiment Station, 3909 Halls Ferry Road, Vicksburg, MS 39180

where   n = Manning's roughness value,
          R = hydraulic radius (m), and
          k = effective roughness height (m).

In the Los Angeles District, the hydraulic radii of existing concrete channels ranges up to 5.64 m. Using equation (1), a maximum n-value of 0.0164 is computed for a hydraulic radius of 5.64 m. This value is significantly larger than the 0.014 originally used to design most of the existing channels. Thus, the purpose of this investigation is to determine appropriate roughness coefficients based on prototype data.

Rio Hondo Flood Control Channel

A reach of the Rio Hondo flood control channel immediately below the Corps' operated Whittier Narrows Flood Control Basin was selected for the initial detailed prototype study (Figure 1). A series of staff gages were painted along this 823 meter-long reach of concrete, trapezoidal channel. The channel has a constant base width of 27.4 meters with sideslopes of 3 horizontal to 1 vertical. The slope of the invert is approximately 0.002345. This reach of channel was constructed in 1956 and was designed with an n-value of 0.014 to convey 1133 m^3/s with freeboard varying from 0.457 to 0.975 m. Froude numbers for the original design discharge range from 0.7 to 1.6.

The conditions of the Rio Hondo study reach are typical of most of the concrete-lined channels designed and constructed by the Los Angeles District. The finish on the concrete is usually obtained by either metal trowel or wood float. Weathering does not pose a problem given the relatively mild climate lacking the freeze-thaw action of other parts of the country. Additionally, the concrete-lined flood control channels have negligible aquatic growth and are dry most of the time.

Sediment is not a significant factor in the Rio Hondo in terms of bedform roughness. The Whittier Narrows Dam traps all but the finest material which is transported as wash load through the study reach. Field investigations of the

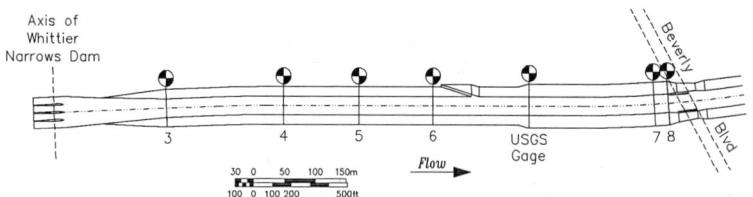

Figure 1. Verification reach of Rio Hondo flood control channel

channel indicate that significant abrasion of the concrete has not occurred during the past 37 years.

Depths are determined by visual observation of the staff gages and are recorded on video tape for later reference. The mean or average depths are used in this study which takes into account the effect of wave activity or flow oscillations typical of supercritical flow. Discharges are determined at a USGS gaging station located in the study reach and are verified with reservoir release data.

Due to the storm activity during the past two winters, several significant reservoir releases have occurred and observations of the resulting profiles made. Discharges for the profiles ranged from 151 to 801 m³/s. The hydraulic radii corresponding to these discharges range from 0.905 to 2.35 m. Calculated Manning's n-values for these discharges range from 0.00985 to 0.01195. The computed Froude numbers for the observed discharges range from 1.41 to 1.95. All of the observations are of supercritical flow throughout the entire reach.

Figure 2 shows a comparison of Manning's n to the hydraulic radius for the Rio Hondo data. A regression curve was developed for the data based on the discharge information from the USGS gage and has the following form:

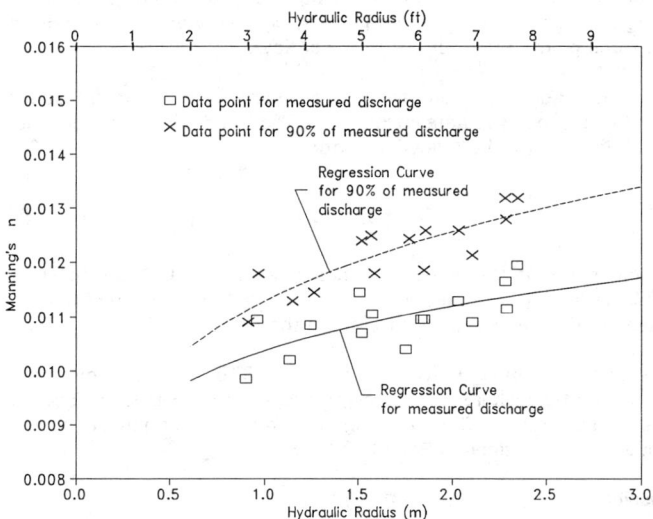

Figure 2. Manning's n versus hydraulic radius for Rio Hondo data

$$n = 0.01037R^{0.1103} \qquad (2)$$

This curve is shown as the lower curve on Figure 2. The Manning's n varies from about 0.0099 to 0.0120 for the range of hydraulic radii observed in the prototype. It is interesting to note that an effective k-value of 0.0610 mm is required to match the prototype data.

The upper curve was developed as a test of the sensitivity of the n-values to the discharge. This analysis based the n-values on discharges which were equal to 90 percent of the USGS gage discharges. For this analysis, the Manning's n varies from about 0.0109 to 0.0132. An effective k-value of about 0.244 mm is required to match the prototype data. In either case, the average k-values are an order of magnitude lower than the 2.13 mm currently recommended in EM 1110-2-1601.

LAD Prototype Data

In 1982, the Los Angeles District compiled all of the available prototype data on roughness coefficients for concrete-lined channels located in the District (US Army, 1985). A total of about 28 data points were collected for 10 different locations. These locations did not include the Rio Hondo verification reach discussed above. Unfortunately, the conditions of the concrete at two of the locations negated the data which left approximately 20 data points. The hydraulic radii for the data ranged from 0.122 to 2.27 m and discharges ranged from 2.66 to 394 m³/s. A review of the data indicates that all but three points were for supercritical flow.

Figure 3 shows a comparison of Manning's n to the hydraulic radius for all of the LAD prototype data including the Rio Hondo data. A regression curve was developed with the following form:

$$n = 0.01127R^{0.03793} \qquad (3)$$

and indicates that the n-values range from about 0.0102 to 0.0121 for hydraulic radii varying from 0.122 to 2.34 m. In this case, an effective k-value of about 0.168 mm is required to match the prototype data.

As shown on Figure 3, k-values of 0.0305 and 0.914 mm are used to develop enveloping curves for the prototype data. Similar to above, the k-values from all of the LAD data are significantly lower than those recommended for capacity calculations in EM 1110-2-1601.

Discussion

The above results for the Rio Hondo represent only 15 data points and are insufficient to draw any hard conclusions regarding the appropriate

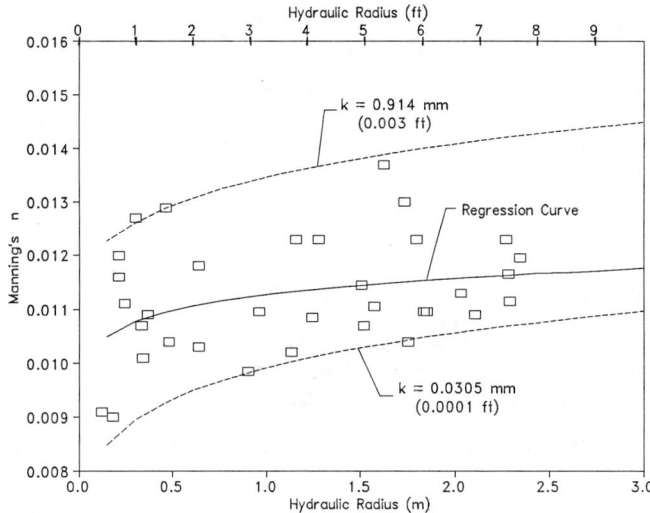

Figure 3. Manning's n versus hydraulic radius for all Los Angeles District data

hydraulic roughness associated with supercritical flow in concrete-lined channels. However, these preliminary results, combined with other prototype data, provide insight to the hydraulic roughness of prototype concrete channels located in the Los Angeles District. Additionally, the hydraulic roughness tends to be less than that recommended for capacity calculations in current Corps of Engineers design guidance.

The Rio Hondo data form a starting point for further data collection and investigation. Additional hydraulic channel verification locations will soon be established at other locations where discharges are well measured and where significant discharges occur on a relatively frequent basis. This verification program will also focus on hydraulic phenomena such as waves and unstable flow as well as hydraulic roughness in supercritical, concrete prototype channels. Channel observers are equipped with wet weather gear and video cameras in order to document the flow during floods.

Appendix I. Conversion Factors, Units of Measurement

To Convert	To	Multiply By
millimetre (mm)	inch (in)	0.0394
metre (m)	foot (ft)	3.28
cubic metre per second (m^3/s)	cubic foot per second (cfs)	35.3

Appendix II. References

Chow, Ven Te. 1959. Open Channel Hydraulics. New York: McGraw-Hill.

US Army, Corps of Engineers, Committee on Channel Stabilization. 1985. Technical Report No. 14, Arizona Canal Diversion Channel, Selection of Roughness Coefficients for Designing the Concrete-Lined Channel.

US Army Corps of Engineers. 1991. Hydraulic Design of Flood Control Channels, EM 1110-2-1601.

Cable-tied Concrete Block Erosion Protection

J. A. McCorquodale[1], M.ASCE, A. Moawad[2] and A. C. McCorquodale[2]

Abstract

An articulated concrete erosion protection system is investigated as a possible revetment for bridge piers to be used in place of the traditional rip-rap revetment. The system consists of concrete blocks in the shape of truncated pyramids interconnected by stainless steel cable. Hydraulic model studies are used to develop guidelines for dimensions of the revetment.

Introduction

There are several articulated concrete erosion protection systems which have been used in place of the traditional rip-rap revetment. This paper describes the laboratory and field experience with one of these products, namely, Cable Concrete™. The product consists of concrete blocks in the shape of truncated pyramids interconnected by stainless steel cable or polypropylene rope as illustrated in Fig. 1. Normally, a geotextile is bonded to the underside of the mat. The mats are available in standard sizes of 4 ft x 8 ft with total surface densities ranging from 20 to 70 lbs/ft^2 with corresponding block heights from 2.5 to 8.5 inches.

The following is a partial list of factors that have been identified as important in the erosion at bridge structures: stream velocity and turbulence, flow depth, bed slope, waves, seepage, subsoil characteristics (bank, and bed materials), sediment transport (load and characteristics), freezing-thawing, ice, water temperature, channel alignment, vegetation, debris, age, size and shape of obstructions. A comprehensive study of scour at bridge piers has been completed at Colorado State University; this study resulted in the CSU formula for bridge pier scour, i.e.

$$y_s/D = 2.0 \, K_1 \, K_2 \, \{y_1/D\}^{0.35} \, F_{r1}^{0.43} \qquad (1)$$

---

[1]Professor and [2]Graduate Students, Department of Civil and Environmental Engg, University of Windsor, Windsor, Canada N9B 3P4

y_s = depth of scour; D = pier diameter; K_1 and K_2 are respectively the pier shape and alignment coefficients; y_1 = approach depth; F_{r1} = Froude number of the approach flow [U.S. Department of Transportation 1991].

This paper emphasizes the results of the model testing of local scour protections at bridges. The stability of the blocks has been established along with suitable revetment dimensions. Friction coefficients have been determined for subcritical and supercritical flow conditions. Approximately 10 years of field experience exists for this product.

Laboratory Testing

Experimental setup:- The experiments were carried in a recirculating flume [600 mm deep, 1200 mm wide and 13.50 m long]. The tailwater was controlled by an adjustable inclined weir. The test section of the flume was contracted by training walls and the floor was gradually raised to allow for the placing of a 150 mm layer of 0.1 mm desilicated sand. The contracted section was 450 mm deep, 750 mm wide and 1.60 m long. A 115 mm circular pipe was placed vertically at the center of the flume. A point gauge mounted on a three-dimensional traversing system, was used to measure the water surface and the scour profiles. The model concrete mats were 120 mm by 460 mm each with a block height of 19 mm and shape as illustrated in Fig. 1. These mats have been tested in scale models with L_R from 4 to 11.

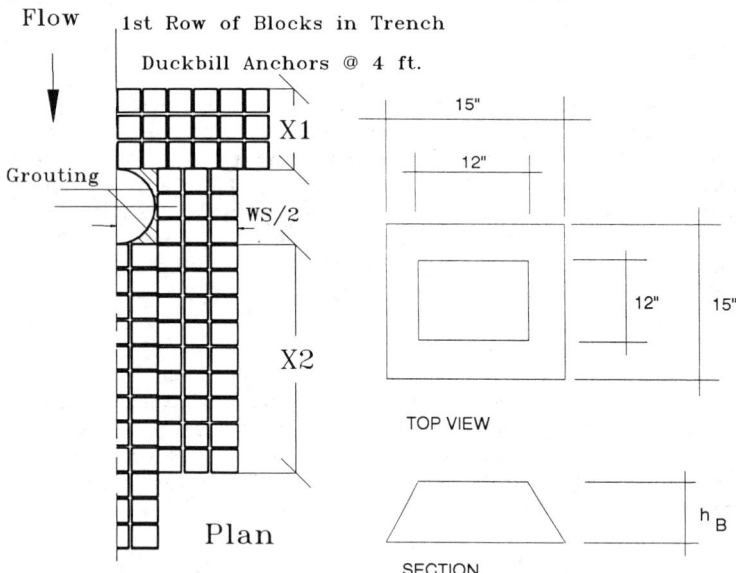

Figure 1. Block Details and Mat Arrangement Around Bridge Pier.

Test Procedure:-The following types of tests have been completed to date: a) high velocity shear tests to check ultimate mat stability;
b) high velocity flow around circular obstructions such as bridge piers;
c) friction tests to estimate the Manning's n.

The test procedure for the bridge pier scour tests was as follows:
a) the pier was positioned centrally into test sections;
b) the mobile bed material was placed and compacted level at the selected depth;
c) the test section was slowly flooded to the desired depth of flow;
d) the pump discharge was adjusted while the depth of flow was maintained;
e) the testing was continued for 2 hours or until equilibrium scour occurred;
f) the scour hole was mapped.

The type of scour is considered to be "clear" water scour since there was not a continuous supply of sediment to the test section. However, the flows and bed shears for the critical tests were such that the bed was in a general meta-stable state.

The erosion protection studies were made for the same hydraulic test conditions as the other scour studies. The mats were placed flush with the original bed in the arrangement shown in Fig. 1. The gaps between the mats and the pier were grouted. The testing procedure and duration were the same as for the other scour tests. Any scour or movement of the mats was recorded at the end of each test.

Some Test Results

General Design Criteria:- A summary of the allowable velocity, shear and suggested friction coefficients are given in Table 1.

Table 1. Block Size Selection Criteria for Moderate Bed Slopes [< 1%] with adequate anchoring and well compacted and drained subgrade.

BLOCK CLASS	SHEAR STRESS N/m²	lbs/ft²	ALLOW. VELOCITY m/s	ft/sec	MANNINGS n Clean	In-filled	H_B Inches
CC20	182	3.8	4.0	13.0	0.017-0.02	0.02-0.03	2.5
CC35	325	4.9	5.2	17.0	0.02-0.025	0.02-0.03	4.5
CC45	390	8.2	6.1	20.0	0.023-.029	0.02-0.03	5.5
CC70	625	13.0	7.6	25.0	0.028-.033	0.02-0.03	8.5

A:Good Placement-all blocks flush with each other and inflow nappe.

Figure 2 shows the suggested block size for protection around circular bridge piers.

Bridge Pier Scour Erosion without protection:- Figure 3(a) shows a typical local scour pattern at a $F_{r1} = 0.3$ and $y_1/D = 1$. The scour depths are compared with the CSU formula in Fig. 4. Because of the good agreement between the upper limit of the observed scour depths and the CSU formula, this equation was used as a basis for the dimensioning of erosion protection.

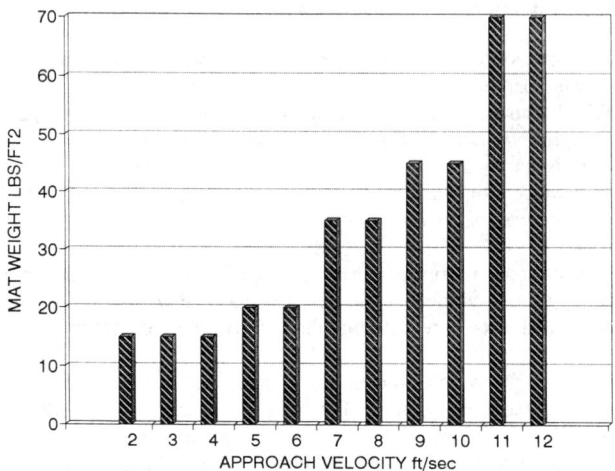

Figure 2. Mat Size Selection Criterion for Circular Bridge Piers.

Erosion With Protection:- Figure 1 shows one of the erosion protection arrangements that was tested. Figure 3(b) shows the erosion pattern after 2 hours of testing under the same approach conditions as in Fig. 3(a). Along the flanks of the protection the depth of scour was about one block height. A similar minor scour appeared at the transition from the revetment to the mobile bed (downstream) and a less amount of scour at the upstream end where the mobile bed leads to the revetment. When the mats were tested without textile there was some entrainment of subgrade from under the mats in the upstream quadrants as indicated in Fig. 3(b). This erosion reached a maximum of about one block height. Geotextile completely eliminated this problem. As indicated in Fig. 3(b) there is a tendency for deposition in the wake zone of the pier. Although this is not a problem it does suggest that the mat block height could be reduced without jeopardizing the safety of the revetment.

Design Implications from the Laboratory Tests

Friction for subcritical flow that apply to the articulated revetment are given in Table 1. The actual Manning's n will depend on whether the crevices between the blocks are: clean, filled with sediment or have vegetation growing in them. The n's for supercritical flow were about 30 % higher than those in Table 1.

Figure 3(a) Local Scour at Bridge Pier Looking Downstream [$F_{r1}= 0.3$; $y_1 = 1$].

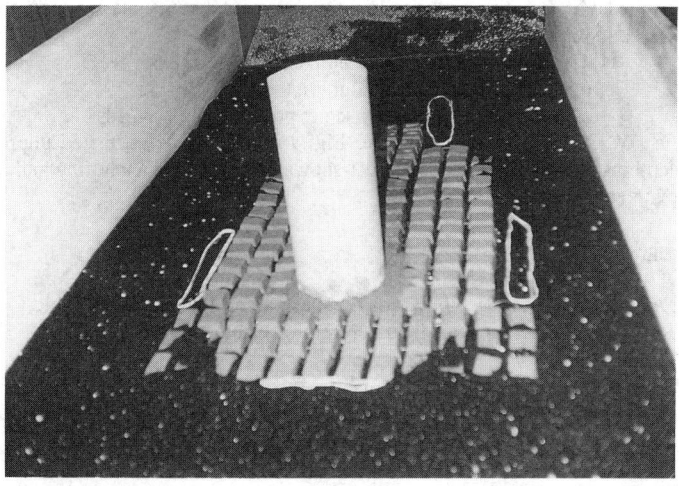

Figure 3(b) View of Concrete Mat Revetment Looking Downstream [$F_{r1}=0.3$; $y_1=1$].

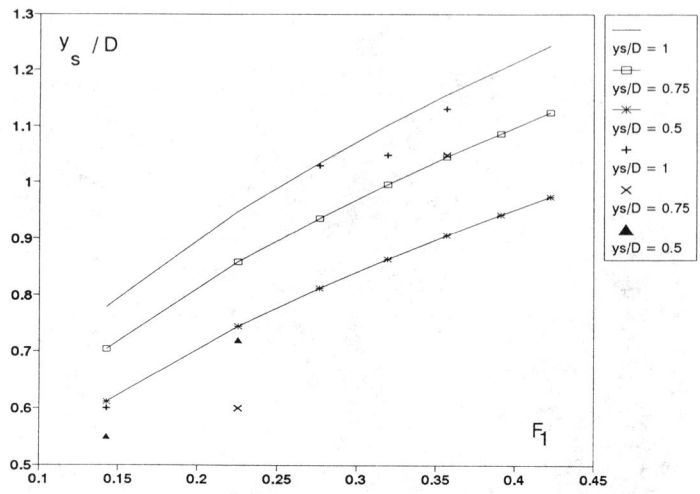

Figure 4. Comparison of Model Scour Depth with CSU Formula.

The criteria for block stability on mild slopes without obstructions are given in Table 1. The corresponding criteria for circular bridge piers is given in Fig. 2. Based on the laboratory testing the following revetment dimensions are suggested for scour protection around circular bridge piers:

 a) Width of scour protection based, $W_s = 2.5d_s + D$
 b) Upstream extent of scour protection, $X_1 = 1.25d_s$
 c) Downstream extent of scour protection, $X_2 = 3d_s$

in which W_s, X_1 and X_2 are defined in Fig. 1 and d_s is the estimated unprotected scour depth using CSU formula. Figure 1 shows the suggested anchoring and leading edge treatment for the revetment.

References

McCorquodale, J. A., 1991, Guide for the Design and Placement of Cable Concrete Mats, Report Prepared for the Manufacturers of Cable Concrete.

U.S. Department of Transportation, 1991, Evaluating Scour at Bridges, Federal Highway Administration, Hydraulic Engg Circular No. 18, Publication No. FHWA-IP-90-017.

A METHOD OF MANAGING FLOATING DEBRIS

Selden Saunders[1] and M. Leonard Oppenheimer[2]

Introduction

When floating debris becomes entangled with a bridge pier, scouring of the streambed is intensified. Small amounts of debris causes a slight increase in contraction of the stream and a resulting scour increase. As the debris builds up, the trapped material is pushed downward along the front face of the pier. This dramatically alters the flow pattern at the foot of the pier. High energy flow impacts the bed, new scour patterns are established, and dangerous erosion of the bed results.

Another dangerous pattern is established when a large log or other debris forms a bridge between adjacent piers. The total amount of debris which can be accumulated is very much higher, and contraction scour can be severe. The bridge span may become inundated and destroyed.

Conventional methods of protecting piers from floating debris are inadequate. For example, the use of an upstream piling or other barrier can actually exacerbate the problem. The upstream barrier accumulates debris which may then be released all at once. The debris raft so formed is turned into the pier by the backwater behind the barrier. Thus the total accumulation can be greater than in the absence of the barrier.

This paper describes a novel debris deflector which is submerged far enough below the water surface so that it is not impacted by floating debris. The deflector is a lunate shaped hydrofoil designed to generate counter-rotating streamwise vortices in its wake. It is positioned so that these vortices migrate to the surface of the water ahead of the pier. The near surface flow induced by the vortices deflects the debris safely around the pier.

---

[1] Saunders Product Development, 208 S. Pulaski St., Balt. MD 21223
[2] Polygon, PO Box 185, Stevenson, MD 21153

Description of the device

The hydrofoil is shown in plan, elevation and left profile views in Figures 1 through 3. A tether or pylon mounting system positions the device a depth d below the surface of the water and a distance z_0 upstream of the pier. The foil is inclined at an angle such that the force on the foil is downward. The reaction on the water causes a local motion upward toward the surface. Vorticity shed from the leading edges of the foil rolls up into vortices as indicated schematically in Figure 2. The debris trajectory is shown in Figure 1. After interacting with a vortex, the debris assumes a new path at an angle delta away from the pier. When the debris reaches the pier, it has been displaced sideways a distance D.

The Physics of the Flow

The motion in a plane normal to the flow downstream of the device is observed to be quasi-two-dimensional. Except near the vortex core the motion can be approximated by a vorticity distribution $\omega(x,y)$ as shown in Figures 3 and 4. The resulting streamlines are $\psi(x,y)$ and the kinetic energy of the vortices is

$$E = \frac{1}{2}\iint \omega \psi \, dxdy$$

These vortices are continuously created by the induced drag or drag-due-to-lift force

$$D_i = E / U$$

where U is the mean stream velocity approaching the device. See Saffmann, 1992, chapter 14. Thus, part of the kinetic energy of the stream is converted into the counter-rotating streamwise vortex pair by the force of the tether.

Each vortex induces an upward flow at the other vortex core. Both vortices migrate toward the surface. In addition, the vortices spread slowly, transverse to the flow and away from the piers.

Suppose one of these vortices to be enclosed in a control volume, a section of which is shown in Figure 3. If no external forces act on the fluid in this volume, then the angular momentum of the vortex is conserved. Dissipative forces within the volume cause the energy to decay and the vorticity to diffuse. The larger the size of the turbulent eddies present in the stream and in the core of the vortex, the faster the diffusion. Observations in the Turner-Fairbanks flume indicate that the vorticity remains highly concentrated for a distance of about 20 times the span b of the hydrofoil when $b/h = 0.6$, h being the depth of the water.

Consider a piece of floating debris which is subject to the flow induced by a vortex. The debris crosses the top surface of the control volume as shown in Figure 3. The cross flow component of the vortex causes a drag force to act upon the debris. The reaction on the water reduces the angular momentum within the control volume.

A drag force will continue to act until this angular momentum is
reduced to zero. Most of the interaction is observed to take place
within two to three mean rotations of the vortex.

The drag force D causes an acceleration of the body away from
the pier

$$D = -\frac{dI_d}{dt}$$

where I_d is the virtual mass of the debris with surface S_d'.
Integrating the above equation gives the trajectory. The new
trajectory can be approximated by a straight line at an angle as
shown. By moving the device far enough upstream from the pier, very
large displacements of the debris can be achieved. As a practical
matter, the device must be kept reasonably close to the pier to
accommodate changes in the stream direction which occur during
flooding.

Relevant Dimensions

The problem is characterized by a bridge pier width w and by
the size of the debris. It is best to consider the debris as an
elongated structure of average diameter D_d and length L. The vortex
produced by the device has a characteristic diameter D_v of order
b, where b is the span of the hydrofoil. If $D_d >> D_v \cong b$, then the
vortex will not impart a net motion to the debris. Hence, take the
value of $b \geqslant 2 D_d$. Further, $b \cong w$ insures that the vortex is
positioned correctly with respect to the pier. This size device
will be effective for debris with diameters less than the width of
the pier. This is the usual case. For larger masses of debris see
the section on oscillation below.

The parameter L enters the problem when the angular momentum
about a vertical axis is considered. It is important to keep the
total angular impulse low, otherwise an elongated piece of debris
such as a log can be rotated into a bridging position between
adjacent piers. This can happen with conventional barriers. In the
present case, this angular impulse is limited by the way in which
the vortices migrate to the surface. By limiting the aspect ratio
of the hydrofoil to moderate values, this problem can be avoided.

Experimental Results

A lunate hydrofoil was constructed with a span b of 28 cm and a
root chord of 30 cm as shown in the figures. This device was tested
in the Turner-Fairbanks Hydraulic Research Lab flume with a water
depth of 45 cm. The pier was of rectangular section with width 15
cm. Preliminary experiments showed significant deflections for
debris as long as 80 cm with mean diameters up to 5 cm.

The hydrofoil was fixed so that it could not oscillate and was
tested at various positions upstream of the pier at an angle of
attack of 35 degrees. By very carefully placing a simulated log
perpendicular to the device, it was possible to get debris trapped

on the face of the pier. In most cases, this debris was dislodged by further impacts. This is in sharp contrast to the case when the hydrofoil was absent. Using the same logs without the deflector, it was quite easy to accumulate large amounts of debris on the face of the pier.

Advantages of an Oscillating Device

If the hydrofoil is tethered so that it oscillates transversely to the stream in a limit cycle, the effectiveness of the device is greatly increased. The flow pattern sweeps back and forth across the pier as shown in Figure 1. This flow tends to destabilize any debris which might have accumulated on the face of the pier. In addition, impacts of additional debris are more likely to have angular momentum about a vertical axis such that the original debris is dislodged.

Other Applications

Streambed scour can also occur at power plant or municipal water supply inlets when floating debris builds up beyond the capabilities of existing trash booms. In these cases, the installation of a device similar to that described will ameliorate the problem. Note that one half of the device will function if it is cut at its plane of symmetry. The resulting hemi-lune is then mounted adjacent the side wall of the main passage upstream of the boom.

Conclusions

The device described offers a promising approach to a new method of managing floating debris. It modifies the surface flow of water from a submerged location. The modified flow reduces the accumulation of debris on a pier or other structure. In the event that some debris does adhere to the pier, an oscillating flow acts to dislodge it.

The device is inexpensive and is easily installed. It protects the pier from catastrophic conditions which occur during flooding. The cost of routinely cleaning accumulated debris from the pier is also reduced. The device itself requires little maintenance. It thus represents a cost-effective solution to many scour problems caused by floating debris.

Acknowledgment

We would like to acknowledge the assistance of Dr. Roy E. Trent, in the many discussions concerning this project.

Reference

Saffman, P.G. 1992. *Vortex Dynamics.* Cambridge Univ. Press.

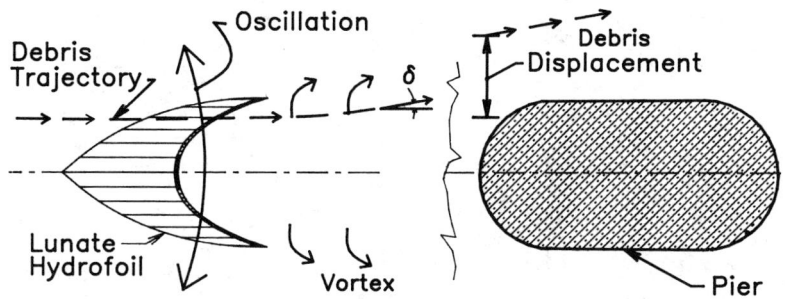

PLAN VIEW OF HYDROFOIL PROTECTING PIER
THE TETHER ALLOWS THE PATTERN TO OSCILLATE AS SHOWN

FIGURE 1

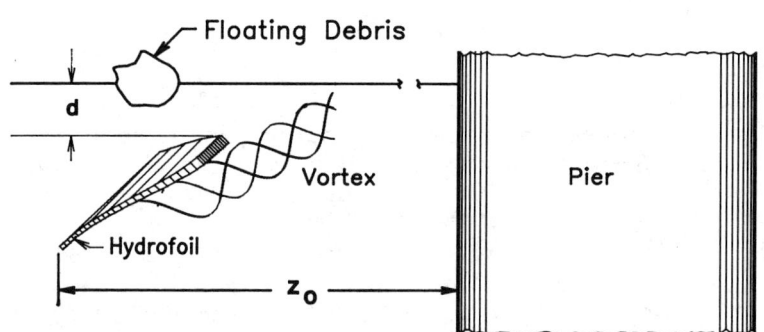

SIDE VIEW OF HYDROFOIL SUBMERGED A DEPTH d
FLOATING DEBRIS DOES NOT IMPACT THE HYDROFOIL

FIGURE 2

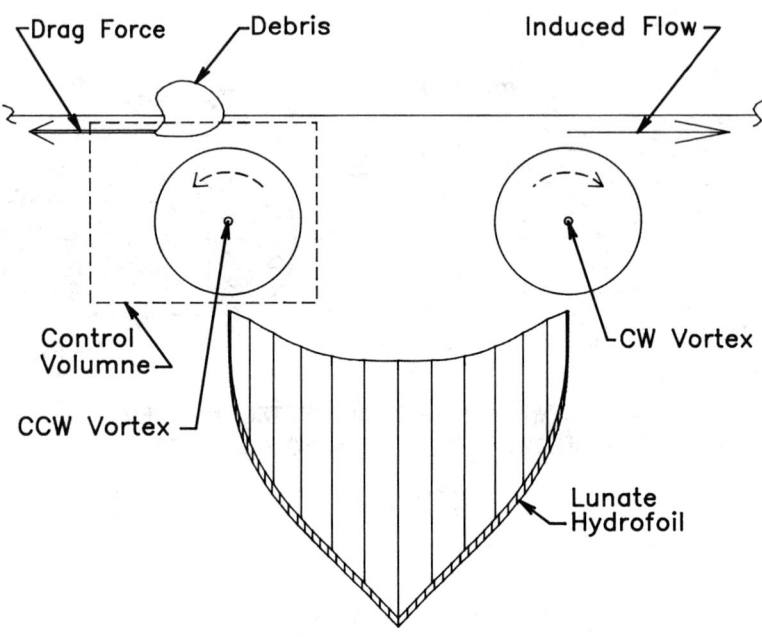

Vortices are shown as sections in a plane downstream of the hydrofoil and normal to flow

FIGURE 3

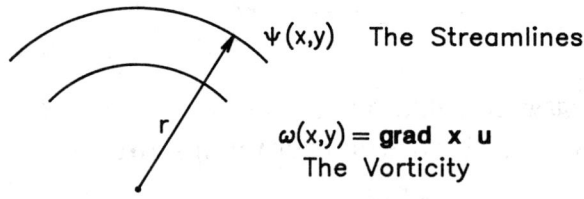

FIGURE 4

The Influence of Protective Material on Local Scour Dimensions

Lisa M. Fotherby[1]

Abstract

Protective materials installed around the base of a bridge pier can reduce or prevent local scour. The performance of four materials: grout mats, grout bags, footings, and riprap, were compared in a laboratory flume study. Performance was evaluated on the basis of scour dimensions around each material. The results of testing indicated that the depth and size of the scour hole varied with the protection material applied and the three materials tested were effective at reducing scour even beyond the limits of the protective layer.

Introduction and Experimental Setup

Local scour is the erosion of stream bed materials from around an obstruction in a channel as a result of secondary flow currents. If the obstruction is a bridge pier, scour undermining the foundation of the bridge pier may, in some instances, provoke a structural failure of the bridge. Riprap is commonly placed around bridge piers in an effort to reduce scour. The use of riprap is effective, but, is not always economically or physically feasible.

Three alternative protection materials for bridge piers: grout mats, grout bags, and footings, are shown in Figure 1. The performance of the three materials were compared to riprap in this study. Grout mats and grout bags are fabricated from geotextiles and filled with grout in situ. A grout mat is a unified layer composed of inter-connected compartments. The 1:20 scale model mats tested in this study represented a prototype with rectangular compartments .30 m by .61 m in areal extent and .12 m in thickness. Grout bags are individual units which range in size. Model grout bags at 1:20 scale represented prototype units with a width of .85 m, a height of .55 m, and lengths that vary from 2.29 m to 4.27 m. Footings tested were symmetrical in shape, having an extension at the front of the pier equal to lateral extensions on the sides (Figure 2).

---

[1]Research Assistant, Colorado State University, Fort Collins, Colorado, 80521

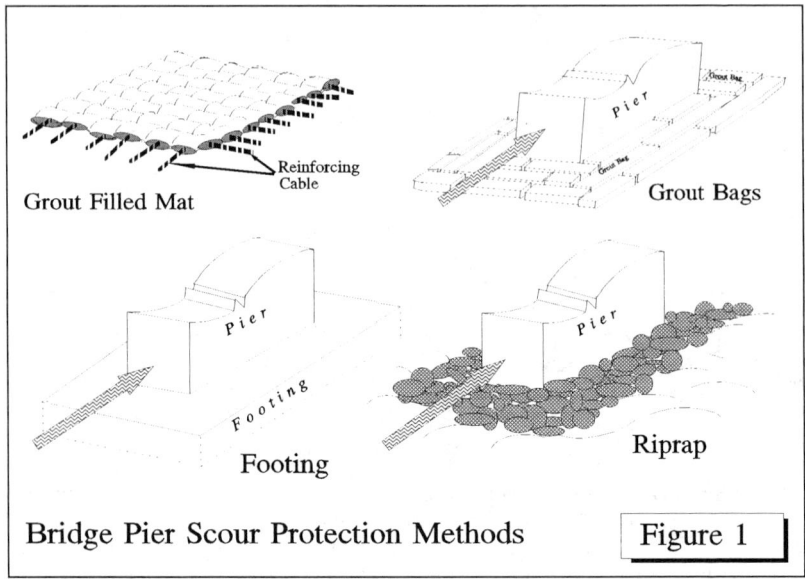

Bridge Pier Scour Protection Methods — Figure 1

The materials were installed around a 1:20 scale rectangular model pier for testing. The pier was 15.2 cm by 30.5 cm and aligned longitudinally with the flow. All materials were installed with the top surface of the material matching the elevation of the channel bed. The materials were tested at areal applications of 2W, 1.5W, 1W, and .5W, where W represents pier width. A 2W designation signified that the test material extended laterally outward from the pier in all directions, a distance of two times the width of the pier (Figure 2).

The model pier was mounted in a sediment recess 1.5 m in length and containing sand, d_{50} = .3mm. The sediment recess was located midway in a 1.8 m by 21.3 m flume. Runs were conducted at flow depths of 15.2 cm and 30.5 cm, and with a constant flow velocity of .3 mps. A state of incipient motion for the sand particles was established at a velocity of .3 mps, creating a maximum scour environment. Flow rates were measured with a venturi meter and flow velocities were checked with an electromagnetic probe.

Results
―――――

Table 1 presents a summary of scour depths (d) and lateral dimensions (l) for scour holes; in an unprotected sand bed, around riprap, and around the three alternative materials tested. The lateral extent of scour is a measurement of the average radius of the front, semi-circular region of the scour hole. The distance is measured horizontally from the pier wall. Lateral extent (l) has been made

TABLE 1 - SUMMARY OF SCOUR HOLE DIMENSIONS

Material	Run No.	Lateral Extent of Protection	Lateral Extent of Scour			Depth of Scour		
			(l) cm	(l/lo)	Ave.	(d) cm	(d/do)	Ave.
Sand	1.1	None	26.4	.90		16.3	.97	
	1.2	None	26.9	.92		16.3	.97	
	1.3	None	29.2	1.00		17.8	1.06	
	1.4	None	30.7	1.05		16.8	1.00	
	1.5	None	30.5	1.04		16.8	1.00	
	1.6	None	31.8	1.09		17.0	1.01	
	Average		29.2 = l_o			16.8 = d_o		
Riprap	1.7	.5W	24.6	.84		6.4	.38	
	1.14	.5W	24.9	.85		5.1	.30	
	1.17	.5W	22.4	.77	.82	6.1	.36	.35
	1.8	1W	23.1	.79		3.3	.20	
	1.13	1W	22.1	.76		2.3	.14	
	1.16	1W	22.4	.77	.77	4.3	.26	.20
	1.9	1.5W	23.9	.82		2.5	.15	
	1.12	1.5W	29.0	.99		2.8	.17	
	1.15	1.5W	24.1	.83	.88	2.8	.17	.16
	1.10	2W	29.2	1.00		2.3	.14	
	1.11	2W	25.7	.88	.94	1.8	.11	.13
Grout Mats	2.1	1W	25.9	.89		2.5	.15	
	2.2	1W	25.1	.86		4.3	.26	
	2.5	1W	21.6	.74		4.1	.24	
	2.8	1W	22.9	.78	.82	3.6	.21	.22
	2.6	1.5W	22.1	.76		2.0	.12	
	2.7	1.5W	21.3	.73	.75	1.5	.09	.11
	2.3	2W	21.3	.73		2.3	.14	
	2.4	2W	20.6	.70	.72	1.8	.11	.13
Grout Bags	2.16	1W	23.6	.81		4.8	.29	
	2.21	1W	20.1	.69		5.6	.33	
	2.22	1W	23.1	.79		5.1	.30	
	2.29	1W	25.4	.87	.79	4.1	.24	.29
	2.23	1.5W	28.7	.98		1.0	.06	
	2.26	1.5W	24.6	.84	.91	2.0	.12	.09
Footings	3.4	.5W	28.4	.97		10.4	.62	
	3.6	.5W	27.7	.95		10.2	.61	
	3.25	.5W	30.7	1.05	.99	11.4	.68	.64
	3.7	1W	32.0	1.10		6.6	.39	
	3.12	1W	33.5	1.15	1.12	5.6	.33	.36
	3.13	1.5W	34.0	1.16		4.3	.26	
	3.28	1.5W	32.0	1.10	1.14	4.6	.27	.26
	3.16	2W	26.2	.90		4.3	.26	
	3.31	2W	37.3	1.28	1.09	4.3	.26	.26

dimensionless by $l_o = 29.2$ cm, the average radius for a scour hole in an unprotected sand bed. Depth of scour has been made dimensionless by dividing depths (d) by the average scour depth for a pier in an unprotected sand bed ($d_o = 16.8$ cm).

Average scour hole depths from Table 1 are plotted in Figure 3. Scour depths for grout mats and grout bags were similar to depths for riprap. Thus, mats, grout bags, and riprap, appear relatively equal in performance for limiting the scour hole depth. Footings limit scour depth but do not provide the same degree of protection.

Figure 4 presents the range of lateral scour dimensions for each material. The graph indicates that the lateral extent of the scour hole is noticeably reduced in the presence of riprap, grout mats, or grout bags. The use of a footing enlarges the lateral extent of the scour hole.

A paired t-test was used to test the hypothesis that mean scour depths associated with grout mats, grout bags, and footings are statistically different from the mean scour depth associated with riprap. A second paired t-test was used to test the hypothesis for mean lateral dimensions of scour. Results of the two t-tests are shown in Table 2. The percentage indicates the probability that the differences in means were significant and not due to random data scatter. A higher percentage demonstrates greater significance. The results of the t-tests are consistent with conclusions drawn from Figures 3 and 4. T-test results indicate that mean scour depths for grout mats and grout bags are not statistically different from that of riprap, while mean scour depths for footings vary significantly from that of riprap. T-test results for lateral extent of scour are inconclusive for comparisons of riprap to grout mats and riprap to grout bags. There is, however, a significant difference between the lateral extent of a riprap scour hole and a footing scour hole.

TABLE 2

RESULTS OF PAIRED T-TESTS FOR AVERAGE SCOUR DIMENSIONS[*]

Compared to Riprap	Scour Depth t Value	% Probability	Lateral Extent of Scour t Value	%Probability
Grout Mats	.490	66%	1.32	83%
Grout Bags	.171	< 60%	2.69	88%
Footings	4.289	98.7%	5.02	99.2%

[*] Assuming a one-sided test and n-1 degrees of freedom.

Values for depth and lateral extent of the scour hole are re-organized in Figure 5. The plot resembles the longitudinal section of a scour hole located upstream from a pier. The face of the pier is located at the y axis. The profile of a scour hole in an unprotected sand bed was provided by Parolla (1990). Protection material is expected to prevent scour at the location of the material but the figure demonstrates that scour depths have been reduced beyond the boundaries of the

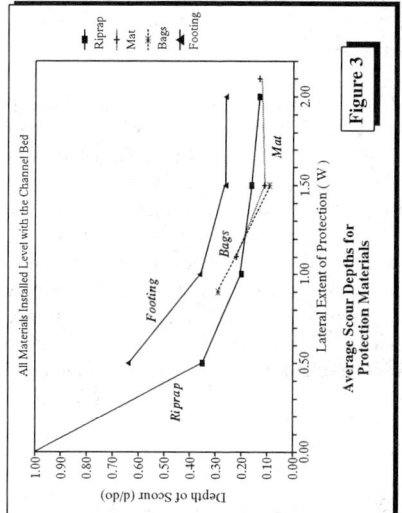

Figure 2

Lateral Extent of a 2W Protection Pad

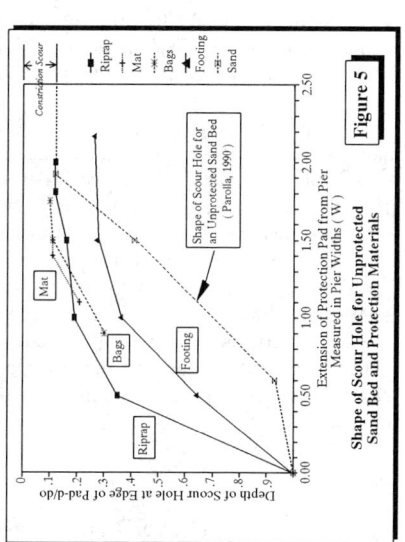

Figure 3

Average Scour Depths for Protection Materials

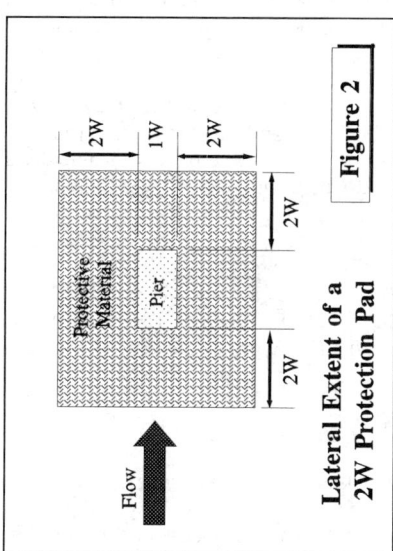

Figure 4

Lateral Extent of Scour Hole Resulting from Different Protection Materials

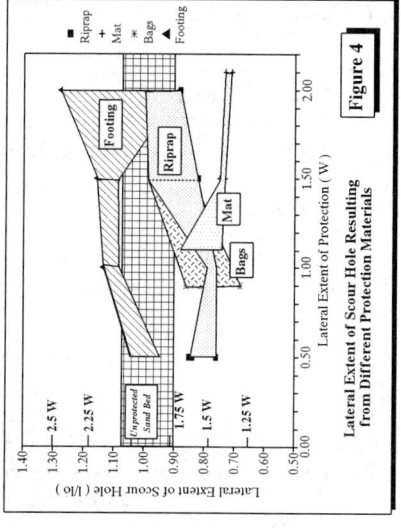

Figure 5

Shape of Scour Hole for Unprotected Sand Bed and Protection Materials

protection material. Scour depth, for example, when grout bags are installed to a distance of 1W around the pier was a maximum of $.27\ d_o$ (refer to Figure 5) at the edge of the material [Note: Maximum scour was usually found at the edge of the protection material]. A footing, the least effective protection material in this study, reduced scour at the 1W location to $.37\ d_o$. If no protection material was applied, the scour depth at the same location was $.70\ d_o$.

The reduction in scour depth beyond the boundaries of the protection material may be a combined result of: a) the geometric shape at the base of the pier, and b) the roughness elements of the material. Schneible (1951) found that one of the most effective scour reduction shapes at the base of a cylinder is a 90 degree angle. Protective material prevents the formation of a scour hole and retains a 90 degree angle between the channel bed and the pier. The square angle converts the downward erosional flow at a pier to a horizontal flow. Roughness elements reduce scour depth through reduction of erosive flow energy and for the same reason may also be significant in reducing the lateral extent of a scour hole. Footings are relatively smooth in comparison to the other protection materials tested. Footings were the least effective at reducing scour depth and had the greatest lateral dimension for a scour hole.

Conclusions

1) Riprap, grout mats, and grout bags reduce scour depth to a similar degree. Footings also reduce scour depth but are less effective. Riprap, grout mats and grout bags reduce the lateral size of the scour hole, and footings increase the lateral size of the scour hole.

2) Protection materials not only prevent scour at the location where they provide coverage but also reduce scour depth at positions beyond the edge of the protective layer. The scouring vortex cannot establish the depth it would have obtained at a position beyond the boundaries of the material with no protective layer present.

Acknowledgements

Laboratory testing was funded by a Federal Highway Administration (FHWA) Graduate Research Fellowship and conducted at the FHWA Turner-Fairbanks Research Center. Appreciation is extended to FHWA project advisor, J. Sterling Jones.

References

Parolla, A., 1990, The Stability of Riprap Used to Protect Bridge Piers, Ph.D. dissertation, Pennsylvania State University, State College, PA, May, 112p.

Schneible, D.E., 1951, An Investigation of the Effect of Bridge-Pier Shape on the Relative Depth of Scour, M.S. thesis, University of Iowa, Iowa City, IA, June, 4p.

Tetrapods as a Scour Countermeasure

by

David Bertoldi[1] and Roger Kilgore[2]

Abstract

Few materials have been studied for their effectiveness in protecting bridge piers and abutments from local scour other than riprap. Recent research conducted at the FHWA research facility in McLean, Virginia, focused on the feasibility of using tetrapods as a scour countermeasure. The geometry of the tetrapod is believed to enhance its ability to resist movement and in turn increase its scour protection characteristics. The interlocking capability of the tetrapod is thought to account for higher stability when compared to riprap. The results of these experiments show that when compared to an equivalent size D_{50} of riprap, tetrapods may exhibit a higher degree of stability.

Introduction

Tetrapods were originally developed to protect ocean shorelines. The geometry of the tetrapod not only provides dissipation of wave energies, but may also provide greater stability over traditional scour countermeasures (i.e. riprap). Various tetrapod and riprap experiment configurations were evaluated for scour protection. Tetrapods and riprap were tested for stability with unobstructed and obstructed (pier) flow. Placement of the armoring in a recessed section versus placement on the fixed bed surface was also tested. This paper will provide an overview of the laboratory experiments comparing tetrapods with riprap and provide some insight into various measures of stability for both. General comments on the economic ramifications involved when incorporating either tetrapods or riprap as a scour countermeasure are also included. A more complete discussion of tetrapods and other armor alternatives may be found in an FHWA lab report (GKY, 1993).

The research for this study was performed in a 1.8-meter wide by 21-meter long flume capable of circulating up to .42 m^3/s. Most of the experiments in this study used a fixed bed surface. This fixed bed surface allowed for precise measurements of velocity and flow depth when testing the stability of tetrapods and riprap. Figure 1 shows the geometry of the tetrapod. For this study the overall height was 1.9 centimeters while the arms were 120 degrees equilateral to each other. The average mass of each tetrapod is 3.3 grams with a specific gravity of 1.85.

---

[1]Research Engineer and [2]Vice President, GKY and Associates, Inc., 5411-E Backlick Road, Springfield, VA 22151.

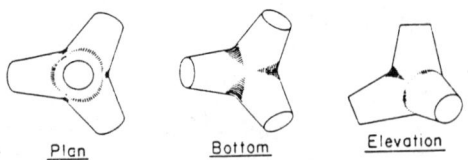

Figure 1. Tetrapod geometry.

Unobstructed Flow Tests

A number of experiments were performed using various configurations to determine the incipient velocity of tetrapods. The test section dimensions were 20.3 cm by 15.2 cm and both recessed and flat surface sections were evaluated on a fixed bed. Figure 2 shows the recessed and surface experiment configurations. The tetrapods correspond to a sphere of the same specific gravity with a diameter of 15 millimeters. In order to compare performance of tetrapods and riprap, a "spherical" D_{50} was calculated based on the known average mass and density of each material. These spherical D_{50}s are used in all computations presented in this paper.

Figure 2. Tetrapod experiment setup.

"Failure" for the tetrapods or riprap was assumed to occur when more than one or more armoring unit "blew out" from its original position. Two types of fixed bed experiments were carried out: armor units on a flat surface, and in a recessed area. The actual D_{50} for the riprap in these experiments is 1.11 cm. This riprap has an average mass of 5.81 grams and a specific gravity of 2.65 for an equivalent sphere diameter of 1.61 centimeters.

Stability of riprap and tetrapods can be evaluated using the sediment number and Shields' parameter. The sediment number is calculated as:

$$N = \frac{V^2}{D_{50}(S_g - 1)g} \quad (1)$$

where:

N	=	sediment number (dimensionless);
V	=	average velocity (m/s);
D_{50}	=	median armor unit size (m);
S_g	=	specific gravity (dimensionless); and
g	=	acceleration due to gravity (9.81 m/s²).

Figure 3 summarizes a comparison of tetrapods and riprap using sediment number versus relative roughness (D_{50}/y). The riprap date of Parola (1991) and Neill (1967) are included to supplement riprap data obtained in these experiments. Inspection of the graph reveals significant scatter in the data and no clear distribution in performance between the tetrapods and riprap.

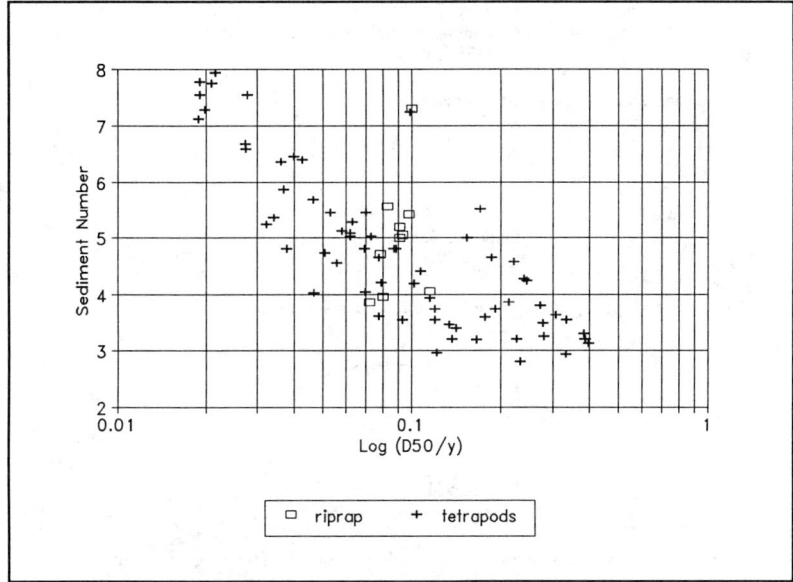

Figure 3. Stability comparison using the sediment number.

An alternative comparative framework is provided by Shields' parameter and the Froude number. The Froude number is defined as:

$$F_r = \frac{V}{\sqrt{gy}} \qquad (2)$$

where:
F_r = Froude number; and
y = depth (m).

The Shields' parameter, SP, is calculated from the following relation:

$$SP = \frac{\tau_c}{D_{50}(S_g-1)\gamma_w} \qquad (3)$$

where:
- SP = Shields' parameter;
- τ_c = Critical shear stress;
- γ_w = Unit weight of water;
- D_{50} = Median armor unit size; and
- S_g = Specific gravity.

The critical shear stress is determined by:

$$\tau_c = \gamma_w y S_e \tag{4}$$

where:
- y = Depth of flow; and
- S_e = Energy slope.

The energy slope, S_e, is calculated using Manning's equation:

$$S_e = \frac{V^2 n^2}{y^{1/3}} \tag{5}$$

where:
- y = Depth; and
- n = Manning's roughness coefficient.

The roughness, n, from equation 4 is calculated from the following equation:

$$n = \frac{y^{1/6}}{3.82\left(2.25 + 5.23\log\left(\frac{y}{D_{50}}\right)\right)} \tag{6}$$

Figure 4 summarizes a comparison of tetrapods and riprap using Shields' parameter versus the Froude number. The data from Neill (1967) and Parola (1991) are included in this comparison as well. Shields' parameter for tetrapods is higher, on average, than the riprap for given values of Froude number. Statistical evaluation of these data is appropriate but was not conducted because of the limited tetrapod data. In addition, none of the tetrapod runs were conducted at Froude numbers greater than 0.8. FHWA is sponsoring additional research to expand this data base encouraged in part by the results shown in figure 4. However, no definitive conclusions are offered on comparative stability as part of this paper.

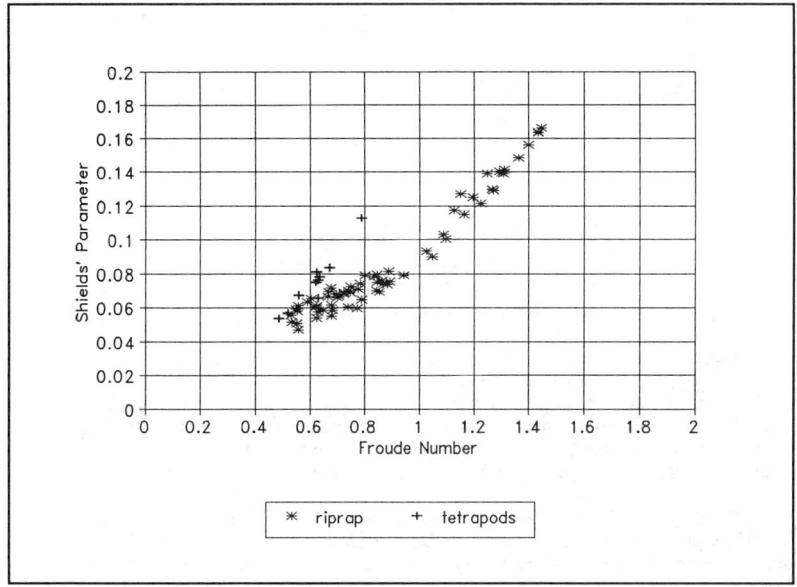

Figure 4. Stability comparison using Shields' parameter.

Obstructed Flow

Another set of experiments explored the stability of tetrapods around a 15 cm by 30.5 cm rectangular pier under fixed bed conditions. These experiments reveal some insight into tetrapod behavior when subjected to downflow. Here, two apron widths were tested: 7.6 cm and 15.2 cm (the width is defined as the distance from the sides and the front of the pier that the mat extends.) The different apron widths are shown in figure 5. In addition, three different tetrapod apron thicknesses or layers were tested: .33 tetrapods/cm^2, .49 tetrapods/cm^2, (1.5 x .33) and .58 tetrapods/cm^2 (1.75 x .33). Failure location for tetrapods varied depending on the apron width. That is, failure typically occurred on the leading edge of the apron as a result of the approach flow and not at the pier as a result of the downflow. As the lateral extent of the apron increased, the local stream lines off of the face of the obstruction created a greater tendency to inhibit failure. This is due to the dissipation of bed shear stress as the apron extended farther out laterally. This can be supported by experiments that fixed the outside edge of the apron to the bed. Here, failure was always initiated at the edges of the obstruction because failure at the perimeter was not permitted. Higher spherical stability numbers for the larger mat widths can be attributed to the higher failure velocities, and therefore, a higher effectiveness can be associated with the lateral extent of the tetrapod apron.

Method of placement of tetrapods in the laboratory or in the field influence their overall stability. That is, when more than one layer is used, placing tetrapods in their most efficient interlocking position provides greater stability as opposed to "dropping" them into an arbitrary position irrelevant to the previous layer of tetrapods.

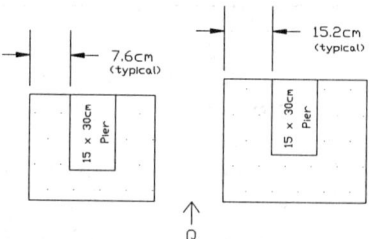

Figure 5. Top view of different apron widths.

Economic considerations of Tetrapods and Riprap

Depending on the size of the application, tetrapods may be fabricated on site to avoid expensive shipping costs. Specially constructed molds can be made (usually from steel) that allow for easy pouring of concrete into the top of the mold. The curing time for this may vary depending on the size of the tetrapod as well as the temperature. The cost of fabrication of tetrapods on site may compare favorably with the transportation costs of riprap in regions where the appropriately sized riprap is not readily available.

Installation of tetrapods as a scour countermeasure requires careful placement around the bridge pier or abutment. Also, a filter fabric is recommended underneath all tetrapod armoring unit applications. If the tetrapods are large, machinery may be needed to lift and lower them into position. Divers can ensure that these tetrapods are properly placed in their most efficient interlocking positions. Unlike tetrapods, riprap is typically "dropped" into position around bridge piers and abutments. This will not always guarantee a stable scour countermeasure protection armoring unit but may be less expensive.

Summary and Conclusions

Given equal specific gravity, physical size and mass, tetrapods may reasonably be expected to be more stable than conventional riprap. Preliminary results reported in this paper suggest that this is the case, but do not quantitatively support such a claim. When used in conjunction with filter fabric, tetrapods may provide greater stability and scour protection than conventional riprap armoring units. Whether this increased stability warrants their use is dependent in the specific application. A trade off in cost versus stability may be the deciding factor when choosing tetrapods over riprap as a scour countermeasure. However, during major flood events, this trade off may be wise considering the cost of repairing structural damage to the bridge as well as the replacement of riprap.

References

GKY and Associates, Inc., "Lab Report on Scour Protection Alternatives at Bridge Piers," FHWA, 1993 (in progress).

Neill, C.R., "Mean Velocity Criterion for Scour of Coarse Uniform Bed Material," Journal of the International Association for Hydraulic Research, 1967.

Parola, Arthur C., "The Stability of Riprap Used to Protect Bridge Piers," Federal Highway Administration Report No. FHWA-RD-91-063, July 1991.

VARIATIONS ENCOUNTERED IN DESIGN ANALYSIS OF LOCAL SCOUR AT DROP STRUCTURES

Noel E. Bormann[1] and Michael E. Zeller[2], Members, ASCE

Abstract

Proper design of drop structures can prevent failure due to scour at upstream structures, like bridges. An important aspect of drop-structure design is the prediction of the local scour that will occur downstream of the drop structure. Predictions of local scour obtained from applying four equations to two sets of hydraulic conditions that have occurred in channelization projects in Arizona are compared. The prediction of each equation is discussed in terms of the conditions under which it was developed, and how the predictive ability may be impacted if those conditions are not met. Other considerations that engineers should be cognizant of, when analyzing local scour, are also introduced: duration of design event, sediment supply to the local scour hole, and armoring of the scour hole.

Introduction

On many large ephemeral streams, flood control is achieved using channelization and diking. Following channelization, these streams often exhibit general scour of the channel bed resulting from increased sediment-transport capacity. Drop structures (also known as grade-control structures) can be incorporated into the channelization scheme to control this general

---

[1]Asst. Prof., Dept. of Civil Engineering, Gonzaga University, East 502 Boone Ave., Spokane, WA 99258.

[2]Principal Eng., Simons, Li and Assoc. Inc., P.O. Box 2712, Tucson, AZ 85702-2712.

scour. Grade-control may also be employed at bridges where other causes of general-scour concerns exist (e.g., from plan-form channel-stability measures). Hydraulic engineers must design foundations that will survive both the general scour and local scour at the structures. Where grade-controls are used to protect bridges or other structures from general scour, the local scour at the grade-control must be properly analyzed to prevent foundation failure of upstream structures.

Practicing engineers are often bewildered by the variety of methods available to predict the extent of local scour at grade-controls. The results calculated from the various equations available in the literature may vary by more than an order of magnitude. Designers are forced to select a scour-prediction equation without clear guidance on the applicability of the many equations. This paper presents a comparison of four local-scour equations in two different situations, and a discussion of the use of these equations in a design analysis of grade-controls.

Flow Conditions and Other Hydraulic Considerations

The flow geometry associated with all local scour phenomena has large influences on scour extent. This geometric dependence has been a major reason for the large number of scour equations, each equation being developed for a different flow geometry, and other specific conditions. For the comparisons in this paper, the two situations analyzed will be as shown in Figures 1 and 2. In both situations, the channel is assumed to be sufficiently wide in relation to depth that the flow can be treated as two-dimensional. The two situations shown embody many other assumptions that the designer must judge: the hydraulics of the channel during the design event, the effect of the time variation of flow magnitude on the hydraulics at the structure; the bed material distribution, and how the scour process affects that distribution; the importance of the variation of the sediment transport rate, in time, for the scour process; and others.

The importance of both the duration and time distribution of flow conditions at the structure is easily understood. The range of sizes in the bed sediment affects the rate of armor formation in the scour hole that can serve to reduce the scour extent. Changes in the hydraulic behavior, as both the rate of flow and the bed level change, during the design event can cause the maximum scour potential to be associated with a flow other than the peak design flow. This is particularly true if the sediment-transport rate changes markedly. It is also important to emphasize that each stream channel must be approached as a unique situation, no two situations will be identical. It may help the design engineer to recognize the concept of a "fluvial system", as presented by S.A. Schumm. The fluvial system must then be considered as a whole.

Figure 1 represents a situation typical of the Rillito Creek near Tucson, Arizona.

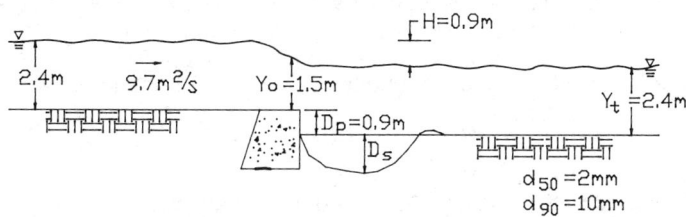

Figure 1. Hydraulic Condition for Case 1.

In Figure 2, a situation occurring on the Santa Cruz River near Tucson, Arizona, is presented. This rather extreme unit discharge presents many problematic issues that designers must face. In both cases, the hydraulics are given for the 100-year peak-flow conditions, assuming the channel bed has reached an equilibrium condition.

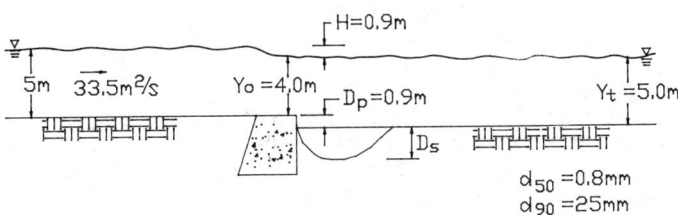

Figure 2. Hydraulic Condition for Case 2.

Local-Scour Equations

Prediction of local scour for the two cases will be made using four equations: Veronese (1937), Mason and Arumugam (1985), Laursen and Flick (1983), and Bormann and Julien (1991), all using SI units. These equations represent a range in complexity and in applicability to these two situations.

Both the Veronese and Mason and Arumugam methods were developed for free jets entering the tail water at a nearly vertical angle, as at a large dam outlet. The Laursen and Flick method is based on dimensional analysis and model studies of drop structures in Arizona. The Bormann and Julien method is based on jet diffusion and particle stability considerations, with large-scale model results used for confirmation and parameter estimation.

The Veronese method, as reported in Mason and Arumugam (1985), is:

$$D_s + Y_T = 0.202 H^{0.225} q^{0.54} d_m^{-0.42} \tag{1}$$

Mason and Arumugam (1985) developed an equation from both laboratory and prototype scour measurements at large structures (like dams), exhibiting free jets entering the tail water at a nearly vertical angle:

$$D_s + Y_T = (6.42 - 3.1 H^{0.1}) \frac{q^{0.6 - H/300} H^{0.05 + H/200} Y_T^{0.15}}{g^{0.3} d_m^{0.1}} \tag{2}$$

The Laursen and Flick (1983) equation for a vertical structure face is:

$$\frac{(D_s + Y_T)}{Y_c} = 8 \left[\frac{V_c}{\omega_o}\right]^{0.75} - \frac{6 + \frac{V_c}{\omega_o}}{\sqrt{1 + \frac{2H}{Y_c}}} \tag{3}$$

The Laursen and Flick (1983) equations are for conditions where tail water was no higher than the crest of the grade-control. For the cases compared here, that condition does not hold true. However, it is argued here that having partial submergence on the structure should reduce the amount of scour observed in comparison to that predicted. The estimate of the fall velocity is made using typical data of fall velocity as a function of sediment size in FHWA (1992).

The equations given in Bormann and Julien (1991) are applicable to both free jets entering the tail water at any angle, and to jets submerged to any degree, with a structure face angle between vertical and 1:3.

$$D_s + D_p = K q^{0.6} \frac{V_o}{g^{0.8} d_s^{0.4}} \sin\beta' \tag{4}$$

where,

$$K = C_d^2 \left[\frac{\gamma \sin\phi}{B \sin(\phi + \alpha)(\gamma_s - \gamma)g} \right]^{0.8} \quad (5)$$

The angle α is estimated to be equal to β'. For free jets, the impingement angle is used to estimate β', and for submerged jets:

$$\beta' = 0.316 \sin\lambda + 0.15\ln\left[\frac{D_p + Y_o}{Y_o}\right] + 0.13\ln\frac{Y_T}{Y_o} - 0.05\ln\left[\frac{V_o}{\sqrt{gY_o}}\right] \quad (6)$$

Comparison of Predictions

The results of the application of each of the scour-prediction equations to the two conditions using both the d_{50} and d_{90} bed sediment sizes are given in Table 1. Due to the armoring process, the use of d_{90} is recommended as the sediment size. The variation in results shown is not unusual for local-scour predictions.

Table 1. Comparison of Scour Predictions, in meters, for Two Conditions.

Case	Veronese (1)	Mason and Arumugam (2)	Laursen and Flick (3)	Bormann and Julien (4,5,6)
1 d_{90}/d_{50}	2.3/6.8	9.4/11.4	60.5/100.8	3.4/7.3
2 d_{90}/d_{50}	1.2/21.3	20.2/30.5	102.2/611.3	6.5/28.3

The important consideration is, of course, which of the predictions given in Table 1 is most nearly correct. At this time, designers of drop structures are forced to use judgement and experience in resolving the variation apparent in Table 1. Experience suggests that local scour for Case 1 should not exceed 4-6m, and should not exceed 6-8m for Case 2. This experience is based upon the survival of structures near Tucson, Arizona (similar to those shown in Figures 1 and 2) for which the design flow rates have been either approached or exceeded twice within a 10-year period of recent history.

CONCLUSIONS

The variation encountered when predicting local scour at drop structures can be disconcerting to the design engineer. The proper use of engineering judgement becomes very important, especially when predictions must be made for a situation that is not similar to those used in the development of the predictive equations. For engineers in this all to common predicament, the following guidance may be helpful.

First, be sure that you understand the overall situation which you must address. Second, make yourself aware of any applicable "real-world" data that is available so unreasonable predictions can be recognized. Third, when (almost inevitably it seems) "real-world" data is not available, use simplified physical analysis (e.g., incipient motion or scour volume) to evaluate a prediction for reasonableness. Finally, if the predicted scour is large, appears to be reasonable, and presents a significant problem, evaluate the effectiveness of design modifications like energy dissipators. Be aware, however, that any additional design features must also be designed to survive general degradation and any general scour.

REFERENCES

FHWA, (1989). *Highways in the River Environment.* U.S Department of Transportation.

Bormann, N.E. and Julien, P.Y. (1991). "Scour Downstream of Grade-Control Structures." Journal of the Hydraulics Division, ASCE, 117(5), 579-594.

Laursen, E.M. and Flick, M.W. (1983). Scour at Sill Structures. FHWA/AZ 83/184. Arizona Department of Transportation, 206 South 17th Avenue, Phoenix, AZ 85721.

Mason, P.J. and Arumugam, K. (1985). "Free Jet Scour Below Dams and Flip Buckets." Journal of the Hydraulics Division, ASCE, 111(2), 220-235.

NOTATION

B = The exponent on relative roughness eqn. (0.5)
C_d = Jet diffusion coefficient (1.8)
D_p = Bed drop at structure (m)
D_s = Local scour depth (m)
d_m = Mean sediment dia. (m)
d_s = Bed sediment sieve dia. (m)
H = Drop of EGL at structure (m)
q = Unit discharge (m^2/s)
V_c = Flow velocity at Y_c (m/s)
V_o = Flow velocity at Y_o (m/s)

Y_c = Critical depth of flow (m)
Y_o = Depth entering tailwater(m)
Y_T = Tail water depth (m)
α = Angle of scour hole face
β' = Angle of impinging jet
γ = Unit weight of water (N/m^3)
γ_s = Unit weight of sed. (N/m^3)
λ = Angle of structure face (rad.)
ϕ = Internal shear angle of d_s
ω_o = Fall velocity of sediment size d_s (m/s)

THREATS TO BRIDGE STABILITY FROM SCOUR RELATED FAILURES OF DROP STRUCTURES

Noel E. Bormann[1] and Michael E. Zeller[2], Members, ASCE

ABSTRACT

Catastrophic failure of bridge foundations may occur when channel bed elevations are lowered unexpectedly. Bridges to be constructed in channels that are known to have bed degradation may be protected from failure by designing foundations that will survive the anticipated bed elevation changes, by protecting the foundation from experiencing the bed elevation change with a drop structure placed downstream of the bridge, or a combination of these approaches. Should protective measures themselves fail, bridges are exposed to new types of threats. These threats are functions of the time rate of scour at the failed drop structure, the scour volume, and the hydraulic characteristics of the flow event in time. Qualitative assessment of these factors and the evaluation of the resulting threats to bridges will help engineers manage bridge construction, maintenance and to prepare for measures that may be necessary to protect bridges from scour threats.

INTRODUCTION

In many cases where channel bed changes are predicted to be unacceptably large, drop structures (also called grade-control structures) are employed to reduce the magnitude of the bed change. Often, requirements

---

[1]Asst. Prof., Dept. of Civil Engineering, Gonzaga University, East 502 Boone Ave., Spokane, WA 99258.

[2]Principal Eng., Simons, Li and Assoc. Inc., P.O. Box 2712, Tucson, AZ 85702-2712.

beyond the need to protect a single structure, such as concerns about tributary channel stability or liability for sediment related damages, will also lead to the use of drop structures. Because of the discontinuity of the bed slope at a drop structure, the hydraulics of the flow there are complicated. The flow boundaries are also changed in time due to the expected lowering of the downstream bed, and the occurrence of local scour. These and other issues associated with the performance of the drop structure must be properly accounted for in the design of the drop structure.

However, if an existing or proposed drop structure is relied on to protect other structures from scour, those other structures will be at risk if the drop structure should fail for any reason.

IDENTIFICATION OF THREATS

Any structure constructed near a stream channel is subject to the behavior of the Fluvial System (Schumm, 1977), where the emphasis should be placed on the concept of "system". Consideration must be given to a reach of channel extending both up and downstream of the structure. This need is addressed in the current FHWA HEC-18 guidelines and in the practice of many consultants (Huber, 1991). However, a large number of designers are not aware of the importance of the channel that is out of sight of their site. It is not unusual to hear a bridge designer, even now, speak of "stabilizing the channel for a couple of hundred feet upstream to improve the approach." Analysis of the susceptibility of a structure to scour damage must include the potential failure of any and all protective measures potentially affecting a structure. A part of this analysis must include the location and physical characteristics of protective drop structures that could impact the safety of other facilities.

Bridge foundations, other structures, or tributary channels that are protected by drop structures face scour related threats with several components: (1) the risk that the drop structure will fail as a result of general channel scour downstream, (2) the risk that the drop structure will fail as a result of local scour at the drop structure, and (3) the risk that the channel will shift and bypass the drop structure, making the drop structure ineffective. These types of failures are not independent. For example, if general scour of the channel bed exceeds the amount the drop structure was designed for, the drop height over the structure will be greater than anticipated and the amount of local scour caused by the flow can be larger than the design value. Failure of the drop structure from any cause threatens previously protected structures if the channel bed. The magnitude of the threat (i.e., the probability of failure) in conjunction with the possible damage related to failure (i.e., the risk) will determine the level of analysis undertaken for a particular situation.

EXTENT OF CHANNEL INVOLVED IN THREAT ASSESSMENT

Recently, a description of the movement of headcuts that incorporates the effects of local scour and sediment type has been completed (Stein, 1990). A further example of the consideration of time rates of scour hole movement in the case of a spillway failure is given with field measured data in Demissie, et al (1988). A clear implication of presently available methods and investigations deals with the relationship between the duration of the flow causing damage and the time required for the scour hole to form, cause the drop structure failure, and then move upstream to the nearest protected structure. The time for the bed discontinuity to move toward a protected structure will be larger for: (a) flood events with a short duration of high flows; (b) channel bed material that is, or becomes, more scour resistant (due to scour induced armoring, rock, or cohesive material); (c) changes in the hydraulics that reduce erosive potential (sediment deposition downstream can increase the water depth reducing scour potential); or (d) if the drop structure that failed is a longer distance from the protected structure.

The large bed discontinuity moving upstream from the Lake Charlestown spillway was observed to move at a rate of approximately 100m per day (Demissie, et al., 1988). In this situation the downstream channel had capacity to store the eroded bed sediment without limiting the scour potential. Such downstream conditions is only typical of a dam. To determine the extent of channel to be evaluated then, the duration of the flow causing damage would also need to be estimated. A resulting choice for evaluation extent might be 1-2 km up and downstream of the protected structure. Note that this distance is distinct from the need to evaluate the lateral stability of the channel. The lateral evaluation might include 10km or more of the channel.

On ephemeral streams that flow only a portion of each year, but that have large flood peaks of short duration, the threat to upstream bridges would be less than that for a stream in a humid region that has floods of long duration. Notwithstanding this guidance, in the 1980's several bridge foundations failed in Arizona on ephemeral streams. The cause of the failures were in-stream gravel mining pits that caused an uncontrolled bed elevation change to travel upstream and undermine the bridge foundations. These failures emphasize the need to examine a adequate length of the channel system to identify potential threats.

QUALITATIVE EVALUATION OF THREATS

Quantitative probabilistic descriptions of scour and the resulting impacts are given by Blazejewski (1991), Johnson (1992), and others. Similar methods are not yet available for a general scour situation at a drop structure.

A probabilistic method for drop structures would need to address the variation in hydraulic behavior as general and local scour occurs and the movement of the scour hole upstream with variable hydraulics and sediment properties. Future research dealing with drop structures and scour progression may extend to incorporate the present deficiencies.

Although the research in this area does not, at this time, provide a quantitative approach to a failure mode evaluation and criticality analysis (FMECA) for drop structures and protected structures a qualitative understanding, however, is useful. To properly develop a qualitative understanding a conceptual framework of the river system is required. The qualitative understanding of the channel requires the consideration of: (a) changes in land use within the watershed resulting in changes of the volume, timing and location of tributary water to the channel; (b) changes in the amount, source, size and location of sediment in the channel; (c) behavior of the drop structure and bridge structure under the influence of the changing hydraulic conditions; and (d) the method used in the design analysis for the drop structure.

Exact relationships for determining the impacts from the many processes involved in these considerations on the ability of drop structure to survive general and local scour in the channel are not available, but useful information can be developed by imaginative engineers. Basic methods that are appropriate for these evaluations are numerical hydraulic and sediment transport models (HEC-2, HEC-6 or others) that can provide estimates of potential scour volumes and scour rates for a given situation. Because hydraulic analysis is required to properly design structures, the threat evaluation could be easily incorporated into the computations needed for design.

By projecting the watershed and channel conditions for the expected life of the structure, hydraulic engineers can make evaluations that will allow the risk to the bridge and other structures to be managed effectively.

CONCLUSIONS

Using a qualitative evaluation that is encompassed by a sound conceptual framework, the threat to bridges or other structures resulting from a scour failure at a protective drop structure can be evaluated. An evaluation of the threat requires an examination of data that is normally developed for bridge design, and may also easily be incorporated into a scour susceptibility study. A further application of a threat evaluation may be to establish the specific criteria the design of protective drop structures should address.

A significant feature of failure of a protective drop structure is that the failure is not instantaneously transmitted as a failure to protected structures. The interval of time between the time the drop structure fails and when the bed change occurs at the threatened structures makes several modifications to the threat possible. Modifications that could be considered include: (a) placing protective drop structures at a distance downstream of the protected structure; (b) limitation of activites within the channel; (c) use of "failure resistant" design features (e.g. buried rip-rap aprons normally not exposed to flow) in design of protective drop structures; (d) scour resistant design for structures with large risks associated with failure; and (e) preparation for repair of drop structures before the bed discontinuity can reach the protected structure.

REFERENCES

Blazejewski, R. (1991). *Prediction of Local Scour in Cohesionless Soil Downstream of Outlet Works.* Translated and edited by Zeidlers. Agricultural University, Poznan, Poland.

Demissie, M. et al. (1988). "Scour Channel Development After Spillway Failure." Journal of Hydraulic Engineering, ASCE, 114:8, 844-860.

Huber, F. (1991) "Update: Bridge Scour," *Civil Engineering,* ASCE, 61:9, 62-63.

Johnson, P.A. and Ayyub, B.M. (1992) "Assessing Time-Variant Bridge Reliability Due to Pier Scour". Journal of Hydraulic Engineering, ASCE, 118:6, 887-903.

Schumm, S.A. (1977). *The Fluvial System.* John Wiley & Sons, Inc., New York, New York.

Stein, O.R. (1990). "Mechanics of Headcut Migration in Rills," Ph.D Dissertation, Colorado State University, Fort Collins, Colo.

Mechanics of Debris Flows

S.B. Savage[1]

Abstract

The paper deals with the mechanics aspects of the modelling of debris flows and will focus on steady, fully developed flows. After a brief review of the types of constitutive equations that have been devised to model these flows, such as the viscoplastic Bingham models, Bagnold dispersive stress models, and generalized viscoplastic models, we shall describe some granular flow kinetic theory models and 'granular dynamics' computer simulations.

1. Introduction

In the present paper we use the term *debris flow* in the somewhat narrow sense (following Japanese researchers) to refer to gravity-driven mass flows comprised of interstitial water and high concentrations of particulate solids including sands, silt, gravel, cobbles and boulders. The rheological behavior of these liquid-solid mixtures is extremely complex and a good understanding of them is still lacking. Although numerous studies have addressed the problem of devising constitutive equations appropriate for debris flows, there is still much to be done before the flow field properties can be predicted with the same level of confidence that we associate with more common single-phase Newtonian fluids. Because of the opacity of these flows and because of the lack of suitable intrusive instrumentation, there is a scarcity of detailed experimental information about flow velocities, concentrations, and stresses, although some data has been obtained for flows next to transparent channel walls. This kind of information is needed for both the development and the verification of new constitutive theories. The attention of this short paper will be focused on the simplest situation of steady, fully developed flows (*uniform flows* in the hydraulics sense), rather than on problems of flow initiation, temporal development, roll waves, etc. There have been a number of reviews of various aspects of debris flows; geomorphological aspects were treated

---

[1] Department of Civil Engineering & Applied Mechanics, McGill University, Montreal H3A 2K6, Canada

by Innes (1983) and Costa (1984); Iverson and Denlinger (1987) gave a concise but perceptive examination of the physical aspects; Chen (1987) provided an extensive review of Japanese work; and Takahashi's (1991) monograph covers a broad range of topics.

2. Review of models based on integration of momentum equations

2.1. Viscoplastic continuum

The viscoplastic continuum or Bingham model was put forth about the same time by Johnson (1965, 1970) and Yano and Daido (1965). In the Bingham model, the shear stress is given

$$\tau = \tau_b + \eta_b \frac{du}{dy} \qquad (2.1)$$

where du/dy is the shear-rate, τ_b is the yield stress and η_b is the Bingham viscosity. In Johnson's (1965) modification, called the Coulomb-viscous model, the yield stress τ_b is replaced by $c + p_n \tan \phi$, where c is the cohesion, p_n is the normal stress and ϕ is the internal friction angle. The theory permits velocity profiles which have rigid plugs and thin shear regions near the boundaries, but it is unable to reproduce some types of observed velocity profiles and it takes no account of the particle interactions and variations in particle concentrations. The Bingham model has been generalized to incorporate a power law dependence on shear-rate.

2.2. Dilatant fluid models

Bagnold's classical work (1954) is one of the major contributions to our present understanding of debris flows. He defined a *Bagnold number*

$$N = \lambda^{1/2} \rho_s \sigma^2 (du/dy)/\mu_f, \qquad (2.2)$$

where ρ_s is the solids mass density, μ_f is the viscosity of the interstitial fluid, σ is the particle diameter, du/dy is the velocity gradient, $\lambda = \sigma/s$ is the *linear concentration* of particles and s is the free space between particles. Different flow regimes can be classified according to the values of N. In the *macroviscous* regime, corresponding to $N < 40$, viscosity is dominant and the shear and normal stresses are linear functions of the velocity gradient, du/dy. Bagnold attributed the normal 'dispersive' stresses to a statistically preferred anisotropy in the spatial particle distributions. In the *grain–inertia* regime, corresponding to $N > 450$, the interstitial fluid plays a minor role and the major effects are due to particle–particle interactions. Bagnold argued that the main mechanism for momentum transfer is the succession of glancing collisions as the grains of one layer overtake those of the adjacent slower layer, giving rise to stresses that depend upon the *square* of the shear rate. By assuming a uniform solids concentration over the depth, Bagnold was able to integrate the momentum equations and obtain the velocity distribution for a free surface fluid-particle flow.

For slower flows, the collisional stresses are insufficient to support the particle overburden. Longer term rubbing particle contacts occur and generate quasistatic, Coulomb-type stresses. Tsubaki and Hashimoto (1983) have extended

Bagnold's ideas to include both collisional and quasi-static stress contributions in the normal stresses. By allowing the solids concentration to vary over the depth, they obtain velocity profiles that can contain an inflection point and give better agreement with some of the observations than Bagnold's predictions.

Takahashi's earlier debris flow models (1981) were similar to Bagnold's constant concentration analysis, but his more recent work (1991) was generalized to take account of the variations of concentration and internal friction angle over the depth, and interstitial fluid turbulence. Takahshi's monograph (1991) also compared macro-viscous experimental results with several analyses intended for this flow regime.

2.3. Generalized viscoplastic models

Chen's generalized model (1988a, 1988b) contains features of the above models and includes cohesion and quasi-static Coulomb type contributions to both the shear and normal stresses. The power law dependence on shear-rate is permitted to vary in order to model behavior ranging from the macro-viscous regime to the grain-inertia regime. With the appropriate choice of coefficients involved in the model, good agreement is obtained with the velocity profiles measured by Takahashi (1981, 1991) and Tsubaki (1983).

3. Models using the particle fluctuation energy equation

During shearing motions, the individual solid particles experience vigorous random fluctuations (Ogawa, et al. 1980) analogous to those in gases at the molecular level. The root-mean-square of these velocity fluctuations has been termed the *pseudo* or *granular temperature*. The analogy has been exploited to develop kinetic theories of granular flows (for example, Lun, et al. 1984) patterned after analyses for dense gases and liquids. In the case of granular flows, one must account for energy dissipation during collisions and sliding contacts between rough, inelastic, granules. Note that the stresses are not directly linked to the particle velocity fluctuations in any of the models discussed in the previous section. However, the kinetic theories yield constitutive expressions for stresses, energy fluxes, etc. that depend on the magnitude of the granular temperatures. The theories also yield evolution equations for the granular temperatures, i.e. the kinetic energies associated with the particle velocity fluctuations and spins. Thus, one solves the fluctuation energy equation in conjunction with the mass and momentum equations to determine the complete flow field. Richman and Marciniec (1990) have applied kinetic theory analyses to study dry granular flows down inclines.

3.1. Quasi-static stresses

The kinetic theories assume instantaneous binary collisions and are appropriate for the grain-inertia regime at moderate concentrations. There are debris flows in which particles come into contact for extended periods and generate sustained normal and Coulomb frictional forces, but it is not obvious how to rigorously incorporate them in the kinetic theories. A rather crude linear patching together of results from analyses for the quasi-static and grain-inertia flow regimes has been used by Savage (1983) and in the more detailed analyses of Johnson, Nott and Jackson (1990), and Anderson and Jackson (1992).

3.2. Interstitial fluid effects

Jenkins and McTigue (1990) have given heuristic arguments to establish the form of the constitutive relations for concentrated suspensions at low shear rates corresponding to the macro-viscous regime. They assumed that the particle phase of the concentrated suspension could be treated as a continuum fluid for which the usual conservation of mass and momentum equations hold. They argue, in a heuristic way, that the velocity fluctuations are governed by a fluctuation energy equation of the same form as that developed in the 'grain-inertia' kinetic theories. By dimensional arguments, they establish the form of the transport coefficients for low particle Reynolds numbers, that are consistent with the experimental observations in Bagnold's macro-viscous regime. However, the dependencies on solids fraction ν and the numerical constants of proportionality are lacking. Further, more detailed computations, analogous to those performed in the grain-inertia kinetic theories, may provide this information.

At the other extreme flow regime, a two-phase flow analysis has been developed by Ding and Gidaspow (1990) to treat bubbling fluidized beds. The solids viscosities and stresses were obtained from the granular flow kinetic theory of Lun, et al. (1984). Particle drag effects were included in the momentum equations for both phases, and an additional term was added to the particle pseudo-thermal energy equation to account for dissipation due to particle fluctuations in the viscous interstitial fluid. The analysis is suggestive of what might be developed to handle interstitial fluid effects for debris flows in the grain-inertia regime. Garcia Aragon (1992) has pursued these ideas and devised a debris flow analysis which includes the interstitial fluid and quasi-static effects in the kinetic theory in simple ways. The approach requires some development, since for some conditions the results are found to be unduly sensitive to the particles' coefficient of restitution e, and small values of e for the wall particles are sometimes required to obtain solutions.

4. 'Granular dynamics' modelling

A potentially useful technique that may be exploited to study debris flows is that of 'granular dynamics' computer simulations (Walton 1992a). They are patterned after the molecular dynamics approaches used to study dense fluids. The normal and tangential forces that arise during collisional interactions between rough, inelastic, particles can be modelled in simple ways by the use of springs and dashpots, by springs with different loading and unloading constants, and with Coulomb-type friction coefficients. The accelerations (linear and angular) of each particle are calculated by accounting for the resultant forces and torques acting on it. Recently, Walton (1992b) and Savage (1993) have considered granular dynamics simulations of fully developed, uniform depth flow of dry granular materials down inclined chutes.

Savage's (1993) interaction model used a standard velocity-independent coefficient of restitution e, a coefficient of friction μ associated with the tangential interparticle forces, and uniform sized spherical particles. No flows were possible for inclinations of the rough bed less than about 24°, which corresponds to the measured angle of repose of spherical particles. For slightly greater bed inclinations, sustained flow occurs with no slip at the bed and the velocity profiles have an inflection point. With increasing bed inclination, the profile becomes more full, and slip develops at the bed. At bed inclinations somewhat above the angle of repose, the solids fraction can be somewhat lower close to the bed, but it increases with increasing depth to a maximum (with a value that is near that for the close random packing density), and then decreases fairly rapidly to zero at the upper interface of the flowing solids. There can be some saltation in the uppermost layers such that the upper interface is not distinct. For shallower, more vigorous flows, the solids fraction is lower and tends to decrease monotonically with distance from the bed. It was found that the streamwise velocities increase with increasing layer depth, consistent with the behavior found by Walton (1992b) in similar simulations done for plane inclined lower boundaries. In general, the simulations are quite consistent with experimental results and, to a lesser extent, with analyses based upon kinetic theory approaches.

4.1. Interstitial fluid effects

For dry materials on rough beds, flow will not occur for bed inclinations less than the angle of repose ϕ. However, the buoyancy developed by the presence of an interstitial fluid permits the particulate matter to flow on inclines considerably less than ϕ. Thus, a simple first attempt to account for the effects of interstitial fluid would be to assume that the constant pressure surfaces in the interstitial fluid are parallel to the bed. The particle body force component perpendicular to the bed is reduced by the interstitial fluid buoyancy force, but that parallel to the bed would be unaffected. This rather crude approximation is successful in reducing the bed inclinations, where flow is possible, to values that are close to those observed in laboratory experiments with spherical glass beads and viscous liquids (Malekzadeh and Savage, 1993).

The next step, appropriate for flows in the macro-viscous regime, would be to attempt to include in the force interaction the effects of viscous fluid being squeezed out when particles come together. This can be done in a simple way by making use of results of a lubrication-type analysis of converging and diverging spheres. Work along these lines is presently underway.

Acknowledgements. This work was supported by an Operating Grant from the Natural Sciences and Engineering Research Council of Canada (NSERC).

REFERENCES

Anderson, K.G. and Jackson, R., 1992, A comparison of the solutions of some proposed equations of motion of granular materials for fully developed flow down inclined planes. *J. Fluid Mech.*, **241**, 145-168.

Bagnold, R.S., 1954, Experiments on a gravity-free dispersion of large solid spheres in a Newtonian fluid under shear. *Proc. Roy. Soc*, Ser. A, **225**, 49-63.

Chen, C.L., 1987, Comprehensive review of debris flow modelling concepts in Japan. *Reviews in Engineering Geology*, **VII**, 13-29.

Chen, C.L., 1988a, Generalized visco-plastic modelling of debris flow. *J. Hydr. Eng., ASCE*, **114**, 237-258.

Chen, C.L., 1988b, General solutions for visco-plastic modelling of debris flow. *J. Hydr. Eng., ASCE*, **114**, 259-282.

Costa, J.E., 1984, Physical geomorphology of debris flows. *Developments and Applications of Geomorphology*, (ed. J.E. Costa and P.J. Fleisher), Springer, Berlin, 268-317.

Ding, J. and Gidaspow, D., 1990, A bubbling fluidization model using kinetic theory of granular flow. *AIChE Journ.* **36**, 523-538.

Garcia Aragon, J.A., 1992, Experimental and analytical investigations of granular-fluid mixtures down inclines. Ph.D. Thesis, McGill Univ., Montreal.

Innes, J.L., 1983, Debris flows. *Prog. Phys. Geogr.*, **7**, 469-501.

Iverson, R.M. and Denlinger, R.P., 1987, The physics of debris flows – a conceptual assessment. *Erosion and Sedimentation in the Pacific Rim.*, IAHS Pub. no 165, 155-165.

Jenkins, J.T. and McTigue, D.F., 1990, Transport processes in concentrated suspensions: The role of particle fluctuations. *Two Phase Flows and Waves* (eds. D.D. Joseph & D.G. Schaeffer), Springer, New York, 70-79.

Johnson, P.C., Nott, P. and Jackson, R., 1990, Frictional-collisional equations of motion for particulate flows and their application to chutes. *J. Fluid Mech.*, **210**, 501-535.

Johnson, A.M., 1965, A model for debris flow. Ph.D. Thesis, Penn. State Univ.

Johnson, A.M., 1970, *Physical Processes in Geology*. W.H. Freeman, San Francisco.

Lun, C.K.K., Savage, S.B., Jeffrey, D.J. and Chepurniy, N., 1984, Kinetic theories for granular flow: inelastic particles in Couette flow and slightly inelastic particles in a general flow field. *J. Fluid Mech.*, **140**, 223-256.

Malekzadeh, M.J. and Savage, S.B., 1993, Experimental study of gravity flow of liquid-solid mixtures down an inclined chute. (in preparation).

Ogawa, S., Umemura, A., and Oshima, N., 1980, On the equations of fully fluidized granular materials. *ZAMP*, **31**, 483-493.

Richman, M.W. and Marciniec, R.P., 1990, Gravity-driven granular flows of smooth, inelastic spheres down bumpy inclines. *J. Appl. Mech.*, **57**, 1036-1043.

Savage, S.B., 1983, Granular flows down rough inclines – Review and extension. *Mechanics of Granular Materials: New Models and Constitutive Relations*, (eds. J.T. Jenkins & M. Satake), Elsevier, 261-282.

Savage, S.B., 1993, Investigation of some particulate flows by granular dynamics simulations. *ASME Applied Mechanics Conf.*, Charlottesville, Va, June 6-9, 12 pp.

Takahashi, T., 1981, Debris flow. *Ann. Rev. Fluid. Mech.*, **13**, 57-77.

Takahashi, T., 1991, *Debris Flow*. A.A. Balkema, Rotterdam.

Tsubaki, T. and Hashimoto, H., 1983, Interparticle stresses and characteristics of debris flows. *J. Hydrosc. & Hydr. Eng.*, **1**, 67-82.

Walton, O.R., 1992a, Numerical simulation of inelastic, frictional particle-particle interactions. *Particulate Two-Phase Flows* (ed. M.C. Rocco), Butterworth-Heinemann, Boston, 29 pp.

Walton, O.R., 1992b, Numerical simulation of inclined chute flows of monodisperse, inelastic, frictional spheres. *Advances in Micromechanics of Granular Materials*, (eds. H. Shen, M. Satake, M. Mehrabadi, C.S. Chang and C.S. Campbell), Elsevier, Amsterdam, 453-461.

Yano, K. and Daido, A., 1965, Fundamental study on mud-flow. *Bull. Disaster Prev. Res. Inst.*, Kyoto Univ., Japan, **14**, 69-83.

KINETIC-THEORY APPROACH TO THE NEVADO DEL RUIZ 1985 DEBRIS FLOW.

Juan A. Garcia [1] and Stuart. B. Savage [2]

ABSTRACT

The Nevado del Ruiz 1985 debris flow was the worst disaster of its kind in this century, killing 22,000 people. By making an idealization of the debris flow material and the flow regime, some insights about the mechanical characteristics of the flow were derived. Equations of motion and boundary conditions for a flowing granular-fluid mixture, based on the kinetic theory for granular flow were extended to allow for drag forces resulting from the interstitial fluid that cushions interparticle collisions and particle-wall collisions. Frictional stresses produced when long term contacts occur and fluid turbulent fluctuations were introduced in the model. The numerical model predicts that the amount of material transported is the same order of magnitude as that estimated by field observation.

INTRODUCTION.

A debris flow can be idealized as a mixture of a viscous interstitial fluid made up of water and the fine matrix and a granular material consisting of coarse particles and boulders. This kind of highly concentrated granular-fluid mixture behaves in a very different way than a dilute suspension. Most of the available literature is directed to the description of dilute suspensions (Batchelor, 1972). Above some value of volume concentration, particle interactions become the main mechanism of momentum transfer. In this kind of granular-fluid mixture the dynamics of the collisions is modified by the presence of the interstitial fluid.

---

[1] INRS-eau, C.P 7500, Ste. Foy, Quebec, Canada, G1V 4C7.

[2] Dept. of Civil Eng. and Applied Mech., McGill University, 817 Sherbrooke W., Montreal, Canada, G1K 7P6.

Two kinds of interparticle interactions are considered in the model that will be developed here; collisional interactions that occur when contacts are of short duration and long term frictional contacts produced during rubbing and sliding. The kinetic theory for granular flow (for example Jenkins and Savage, 1983, Lun et al., 1984), was developed to model the collisions in dry granular flow where the effect of the interstitial fluid, air, is negligible. When the interstitial fluid is highly viscous, a modification of the equations of the kinetic theory for granular flow is needed. The simplest change is to include a viscous dissipation term in the collisional kinetic energy flux equation.

The Nevado del Ruiz debris flow occurred in a narrowly confined valley 65 km long that extends to the canyon mouth where the destroyed town of Armero was located. An assumption of uniform flow is therefore possible. The model will be developed for the flow down an incline making an angle θ with the horizontal. A uniform, steady flow regime will be analyzed.

1. GOVERNING EQUATIONS

From the momentum equation, considering that for large particle Reynolds and Bagnold numbers, the average velocity of the particles is similar to that of the fluid, we obtain the following equation in the normal to the flow direction

$$\frac{\partial \sigma_{yy}}{\partial y} = -\nu (\rho_s - \rho_f) g\cos\theta \qquad (1)$$

where σ_{yy} is the total normal stress, ρ_f, ρ_s are the fluid and solids densities respectively, ν is the particle concentration by volume and g is the acceleration of gravity.

Also from the momentum equation we find the following equation in the streamwise direction

$$\frac{\partial \tau_{yx}}{\partial y} = [(1-\nu)\rho_f + \nu\rho_s)] g\sin\theta \qquad (2)$$

where τ_{yx} is the total shear stress. In these equations the normal and shear stress components include collisional, frictional, viscous and turbulent terms defined following Garcia (1992). The calculations for the collisional stresses and the flux of fluctuation energy follow the approach of Lun et al. (1984). The kinetic energy equation is modified including the cushioning of the collisions by the viscous drag.

To define the drag coefficient we assume that it varies independently with the particle Reynolds number ($Re_f = \rho_f dU_r/\mu_f$) and the concentration, where μ_f is the fluid viscosity and d is particle diameter. The velocity U_r is the most probable relative speed between the particle and the fluid in the neighbourhood.

$$C_D = f(Re_p) \, C_n(\nu) = (\frac{24}{Re_f} + \frac{4}{Re_f^{1/2}} + 0.4)(1 + 2.5\nu + 5.2\nu^2) \qquad (3)$$

2. RESULTING EQUATIONS

The resulting equations are a set of four nonlinear differential equations containing four variables, the velocity u, the concentration v, the granular temperature T and the flux of fluctuation energy (Garcia, 1992). The boundary conditions were formulated by means of an analysis based on the kinetic theory for granular flow following the approach of Richman (1988). His equations were modified to include the turbulent and frictional stresses near the wall and the viscous energy dissipation in the collisional exchange of energy near the wall. This allowed a numerical solution of the equations (Garcia, 1992).

3. APPLICATION TO THE NEVADO DEL RUIZ DEBRIS FLOW.

The most striking characteristic of the 1985 Nevado del Ruiz debris flow was its transport capacity. Pierson et al. (1990) estimated a total deposit of 48,000,000 m³ in two hours flow duration. Granulometric analysis of the Armero deposit show that the fine matrix content is 12% on average, obtained from samples of particles less than 2 mm in diameter (Frye, 1987). To apply the numerical model we assume an interstitial fluid viscosity of 0.5 poises (corresponding to the fine matrix content, (O'Brian et al., 1984)). Field observations show that nearly 40% of the deposit is coarser than 10 cm (Pierson et al. 1990). A particle Reynolds number $Re_1 = 20000$ is obtained with a mean particle diameter of 1 m. Using this value of Re_1 and a slope angle of 10 degrees the numerical model produced the results shown in figures 1 to 4. From figure 4 we conclude that dispersive stresses are dominant over the whole flow depth. Bagnold proposed the formula $Q_s = \beta u \tau_c / \tan\phi$, for bed load transport assuming that dispersive stresses are the main mechanism of momentum transfer. The factor β represents the amount of energy transfer in the layer where bed load is transported, τ_c is the collisional shear stress and ϕ is the internal friction angle of the granular material. In order to apply the results of the numerical model, to obtain an estimate of the sediment carried by the debris flow, we modified Bagnold's bed load transport equation to consider the energy transfer in the whole flow depth. The equation becomes

$$Q_B = \frac{1}{\tan\phi} \int_0^h \tau_c u \, dy \qquad (4)$$

Considering the collisional shear stress profile (figure 4) and the velocity profile (figure 1), we can make a graphical integration. The results show that the flow is able to carry $Q_s = 115,000$ kg/s.m. At the mouth of the Lagunillas canyon, the channel width is 90 m. According to the reported flow duration of 2 hours, the total material transported is 76,000,000 kg. With a mean density of 2500 kg/m^3 the total volume transported is 30,000,000 m^3. This value is the same order of magnitude as that estimated in Armero's deposit.

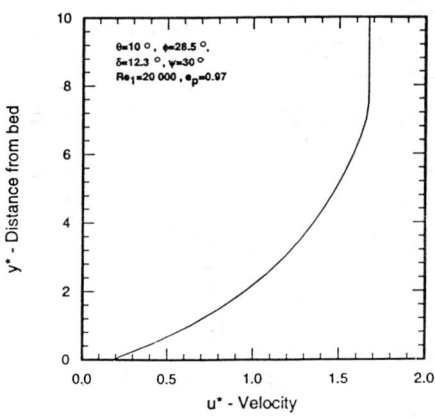

Figure 1

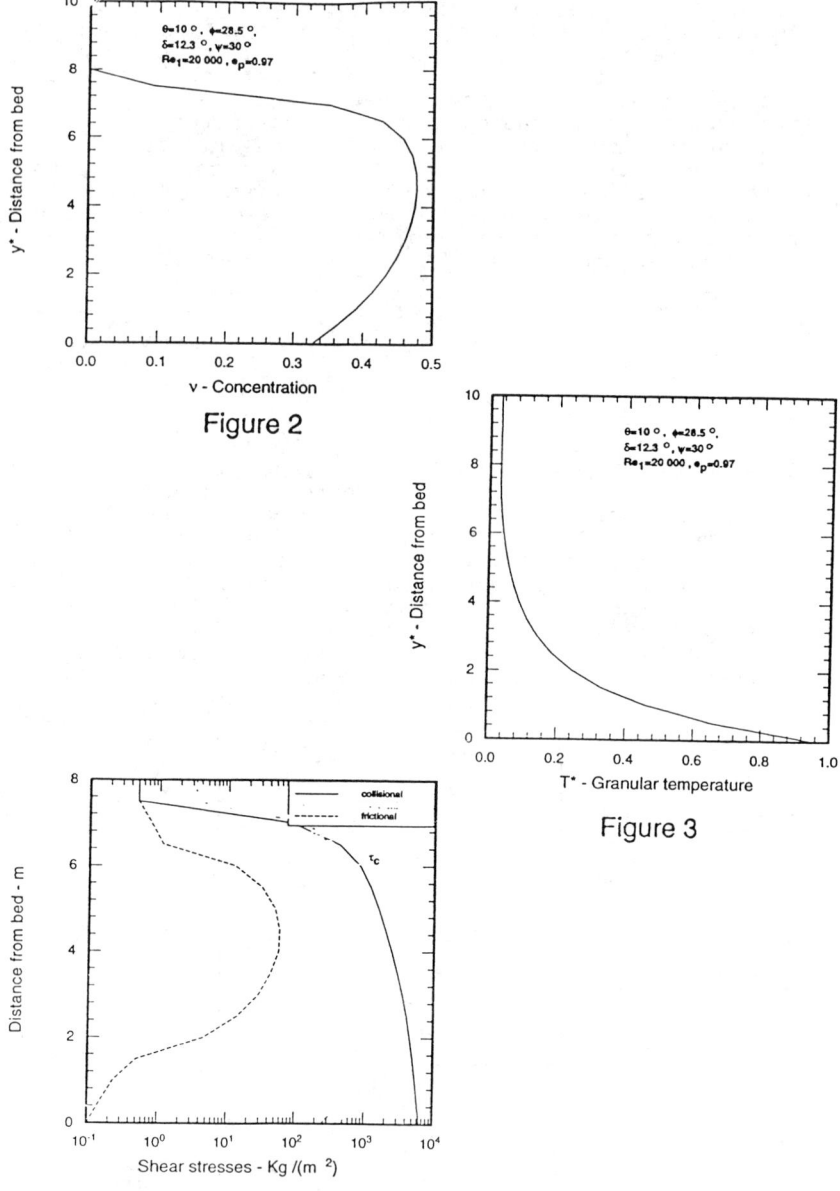

Figure 2

Figure 3

Figure 4

REFERENCES

BAGNOLD, R.A. 1954 Experiments on a gravity-free dispersion of large solid spheres in a Newtonian fluid under shear. Proc. Royal Soc. London **A225**, 49-63.

BATCHELOR, G.K. 1972 The determination of the bulk stress in a suspension of spherical particles to order c^2. J. Fluid Mech. **56**, part 3.

FRYE CASA, A. 1987 Caracteristicas de los depositos del Nevado del Ruiz. Internal report Tolima University, Colombia.

GARCIA, J.A. 1992 Experimental and Analytical investigations of granular-fluid mixtures down inclines. Ph. D. thesis, Department of Civil Engineering and Applied Mechanics, McGill University.

JENKINS, J.T. and SAVAGE, S.B. 1983 A theory for the rapid flow of identical, smooth, nearly elastic particles. J. Fluid Mech. **130**, 187-202.

LUN, C.K.K, SAVAGE, S.B, JEFFREY, D.J and CHEPURNIY, N. 1984 Kinetic theories for granular flow: inelastic particles in Couette flow and slightly inelastic particles in a general flowfield. J. Fluid Mech. **140**, 223-256.

O'BRIAN, J.S. and JULIEN, P.Y. 1984 Physical properties and mechanics of hyperconcentrated sediment flows. **Proc. of the Conf. on Delineation of Landslides, Flash Flows and Debris Flows Hazards in Utah.** Logan, Utah.

PIERSON, T., JANDA, R., THOURET, J.C AND BORRERO, C. 1991 Perturbation and melting of snow and ice by the 13 November 1985 eruption of Nevado del Ruiz, Colombia, and subsequent mobilization, flow and deposition of lahars. J. Volcanology and Geothermal Research **41**, 17-66.

Continuum-Mechanics-Based Rheological Formulation for Debris Flow

Cheng-lung Chen,[1] Member, ASCE, and Chi-Hai Ling[1]

Abstract

This paper aims to assess the validity of the generalized viscoplastic fluid (GVF) model in the light of both the classical relative-viscosity versus concentration relation and the dimensionless stress versus shear-rate squared relations based on kinetic theory, thereby addressing how to evaluate the rheological parameters of the GVF model using Bagnold's data. In the GVF model, most material constants are supposedly measurable or known, but the rheological parameters, such as the flow behavior index (η_1) and the consistency index (μ_1), have yet to be determined from experimental data. A literature survey reveals that the value of η_1 (i.e., the exponent of the shear rate, du/dz) in the shear-stress (τ) versus $(du/dz)^{\eta_1}$ relation for simple shear flows is presumed to be either 1 or 2 by most researchers, and that μ_1 [which relates τ proportionally to $(du/dz)^{\eta_1}$] varies strongly with the solids volume concentration (C), but weakly with du/dz. By and large there are two groups of researchers that differ as to how to express μ_1, whether in terms of C or du/dz, or both. In one group advocating $\eta_1 = 1$, the dimensionless μ_1 (often called the relative viscosity or dimensionless τ versus du/dz relation) expressed in terms of C has been extensively studied for very small particles (micron-size) since Einstein first derived the relative viscosity theoretically for very low C of rigid uniform spheres in a suspension. It was later discovered by some scientists that the dimensionless μ_1 versus C relation for $\eta_1 = 1$ also varies weakly with du/dz. In the other group advocating $\eta_1 = 2$, the dimensionless τ versus $(du/dz)^2$ relation was formulated first empirically by Bagnold based on the rotating-drum testing results of neutrally buoyant spherical grains of 1.32 mm in diameter, and later theoretically by many fluid-mechanicians based on kinetic theory. Analogous to the weak variation of μ_1 with du/dz for $\eta_1 = 1$, the dimensionless τ versus $(du/dz)^2$ relation for $\eta_1 = 2$ was also found to be a weak function of du/dz. This paper examines why the dimensionless ratio of τ over $(du/dz)^{\eta_1}$ versus C relation for $\eta_1 = 1$ or 2 varies with du/dz, thereby addressing how its dependency on du/dz can be minimized by finding an optimum value of η_1 from Bagnold's data.

---

[1]Hydrologist, U.S. Geological Survey, Water Resources Division, WR, 345 Middlefield Road, MS-496, Menlo Park, California 94025.

RHEOMETRY OF NATURAL SEDIMENT SLURRIES

JON J. MAJOR[1]

ABSTRACT

Recent experimental analyses of natural sediment slurries yield diverse results yet exhibit broad commonality of rheological responses under a range of conditions and shear rates. Results show that the relation between shear stress and shear rate is primarily nonlinear, that the relation can display marked hysteresis, that minimum shear stress can occur following yield, that physical properties of slurries are extremely sensitive to sediment concentration, and that the concept of slurry yield strength is still debated.

New rheometric analyses have probed viscoelastic behavior of sediment slurries. Results show that slurries composed of particles ≤ 125 μm exhibit viscoelastic responses, and that shear stresses are relaxed over a range of time scales rather than by a single response time.

INTRODUCTION

Rheometric analyses of highly concentrated suspensions by the geophysical and engineering communities, in particular analyses of suspensions of natural sediments, have escalated in the past several years as a consequence of expanded research on mass transport of sediment. Most of the research studies utilized various configurations of rotational rheometers, but there also have been noteworthy flume studies and measurements using capillary viscometers. Despite differences in the size and kind of rheometer used, in the types and concentrations of sediment used, and in the range of conditions and shear rates employed, the results of these studies, although diverse, exhibit broad commonality. The purpose of this paper is to review the results of several experiments, to discuss the variety of interpretation of those results, and to suggest where future rheometric investigations might profitably be focused.

---

[1]U.S. Geological Survey, 5400 MacArthur Blvd, Vancouver, WA 98661

EXPERIMENTAL CONDITIONS AND RESULTS

Experimental investigations of natural sediment slurries can be broadly divided into two categories: those that analyzed slurries composed of mixed populations of granular materials (e.g. Fairchild, 1985; O'Brien and Julien, 1988; Holmes et al., 1990; Zhao, 1990; Phillips and Davies, 1991; Major and Pierson, 1992) and those that analyzed finer slurries composed of particles < 63 μm in diameter (e.g. Wan, 1982; Wright and Krone, 1987; Murray, 1990; Rheological Comparative Experiment Group, 1990; Coussot et al., 1992; Dade, 1992). Studies that used granular materials commonly employed sediment concentrations and shear rates thought to be reasonably characteristic of debris flows and kindred high-concentration slurries (Figure 1), and included particles as large as 0.25 mm (Fairchild, 1985), 0.5 mm (O'Brien and Julien, 1988), 2 mm (Major and Pierson, 1992), and 120 mm (Phillips and Davies, 1991). Sediment concentrations in those experiments ranged from about 10 to 66 percent by volume, whereas sediment concentrations for some of the fine-slurry experiments were less than 10 percent by volume. Sample volumes used in experiments spanned several decades, ranging from a few hundredths of a liter to nearly 1000 liters.

A focus of several investigations was to examine the influence of sediment concentration and particle size distribution on the rheological behavior of the slurries. Many experiments found that rheological properties are extremely sensitive to changes in sediment concentration. Changes in sediment concentration of as little as a few percent can produce order-of-magnitude variation in rheological properties (Major and Pierson, 1992; O'Brien and Julien, 1988; Phillips and Davies, 1991; Murray, 1990; Wan, 1982). An empirically determined exponential model commonly describes well the dependence of rheological properties on sediment concentration. Dade (1992), however, presents scaling arguments using a power law that relates yield strength of cohesive muds to solids concentration. The effect of particle-size distribution shows greater diversity than sediment concentration and is discussed below.

Results of several steady shear analyses can be interpreted in terms of generalized flow models. Major and Pierson (1992) showed that the behavior of slurries they examined was best represented by a scalar power-law model combined with a yield strength:

$$\tau = \psi + b\dot{\gamma}^n \quad (1)$$

where τ is the applied shear stress, ψ is yield strength, $\dot{\gamma}$ is shear rate, and b and n are constants. This generalized model can be reduced to more specific models. For a Bingham material, $\psi \neq 0$, $n=1$, and b becomes the plastic viscosity, η_p. If $\psi=0$ and $n \neq 1$ this expression reduces to a simple power-law model. Murray (1990) interpreted her results following the generalized Cross model (Barnes et al., 1989):

$$\eta = \frac{\eta_0 - \eta_\infty}{1 + (k\dot{\gamma})^m} + \eta_\infty \quad (2)$$

where η_0 and η_∞ are viscosities at very low and very high shear rates, respectively, and k and m are constants. This model represents a fluid that is Newtonian at the limits of high and low shear rate, but is nonlinear at intermediate shear rates. For $\eta \ll \eta_0$ and $m=1$, this model is equivalent to the Bingham model. For $\eta_0 \gg \eta \gg \eta_\infty$ the equation reduces to a power-law model. If η_0 is very high (simulating yield strength) and $m \neq 1$ the Cross model is similar to (1) with $n \neq 1$, that is, it represents a fluid exhibiting yield strength and nonlinear flow behavior.

Several experimental results indicate that the rheological behavior of natural sediment slurries over a range of compositions, sediment concentrations, and shear rates is complex, even without the presence of particles > 2mm. The deformation response is primarily nonlinear. Slurries commonly exhibited shear-thinning behavior (viscosity decrease with increased shear rate) although some appear to have responded in a weakly shear-thickening manner (viscosity increase with increased shear rate). Despite the complexities observed, several of the experimental data can be approximated by a linear deformation response particularly at shear rates beyond about 5 to 10 s^{-1}.

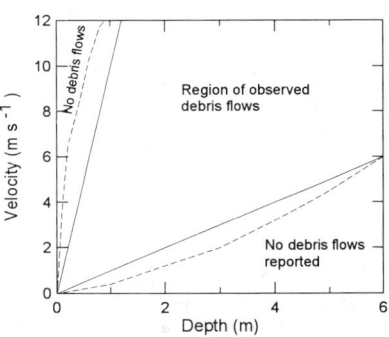

Figure 1. Field of mean shear rates (s^{-1}) for debris flows (modified from Phillips and Davies, 1991). The mean shear rate is approximated by ratio of mean flow velocity to flow depth. Solid lines define mean shear rates of 1 and 10 s^{-1}. The dashed lines envelope field data reported from several debris flows.

The effect of particle-size distribution on the rheometric response of natural sediment slurries is ambiguous. Major and Pierson (1992) found no systematic variation in the overall rheological behavior of slurries as sand content increased systematically, but they also report that the influence of granular particles appears to be rate dependent. At shear rates below about 5 s^{-1}, and at sand concentrations that exceed 20 percent, behavior of slurries differed significantly from behavior of slurries having lower sand concentrations or slurries sheared at higher rates. Murray (1990) comments that a most striking result of her experiments is the similarity in form of the deformation response for both clay- and silt-rich sediments. Results from other experiments also show no substantial variation of rheological response over a range of sediment mixtures (Phillips and Davies, 1991; O'Brien and Julien, 1988). Fairchild (1985), however, reports a systematic increase of the power law exponent in the general flow law (1), from a value $n < 1$ to $n = 2$, as sand concentration increased.

Whether or not natural sediment slurries possess yield strength remains subject to debate (e.g. Barnes and Walters, 1985). Results of several experiments suggest that both granular and fine (<63 μm) slurries possess yield strength, but its existence is based upon extrapolation of rheometric behavior to zero shear rate (e.g. Phillips and Davies, 1991; Holmes et al., 1990; O'Brien and Julien, 1988; Wan, 1982; Rheological Comparative Experiment Group, 1990). Other investigations appear to have measured yield strength directly by gradually increasing shear stress until the onset of flow was detected (Major and Pierson, 1992; Coussot et al., 1992; Dade, 1992; Wright and Krone, 1987), but rates of stress application possibly affect these results. Recent studies, performed using a variety of rheometric methods, strongly suggest that several dense suspensions not composed of natural sediments possess yield strength (e.g. Franck, 1988). Murray (1990), however, reports that clay- and silt-rich slurries behave nonlinearly but show no indication of having yield strength.

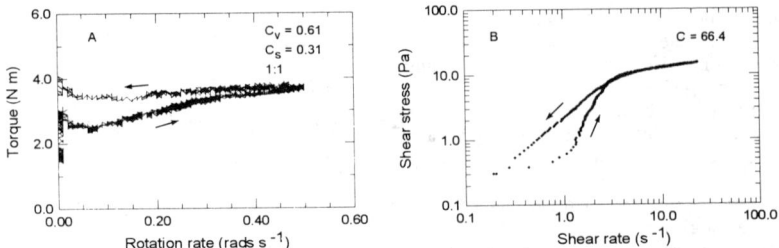

Figure 2. *Deformation responses exhibiting hysteresis. A. Granular sediment (from Major and Pierson, 1992); C_v is sediment concentration of slurry by volume, C_s is sand concentration by volume. Ratio is proportion of fines (<63μm) to sand (63μm-2mm). B. Silt-rich sediment (from Murray, 1990); C is sediment concentration by weight.*

Marked hysteresis of the rheological response of natural sediment suspensions to increasing and decreasing shear rate is reported in several experiments (e.g. Major and Pierson, 1992; Murray, 1990; Zhao, 1990; Wan, 1982), and is similar to results reported for other dense suspensions (Cheng and Richmond, 1978; Kirby, 1988). In each case, the shear stress measured as shear rate decreased was larger than the shear stress measured as shear rate increased (Figure 2), except perhaps at the very lowest shear rates. This hysteresis is opposite that commonly attributed to thixotropic behavior (Barnes et al., 1989). Major and Pierson (1992) attributed hysteresis, which they observed to be more pronounced as sand concentration increased, to enduring contact between grains and formation of "grain clusters" as shear rate gradually declined, and they speculated that such "friction-dominated" behavior may be characteristic of granular slurries. Murray (1990), however, has documented similar behavior in fine slurries. She attributes hysteresis to the interplay of orientation processes such as particle interaction, which act as shear stresses are increasing, and disorientation processes such as settling, mechanical interactions with pore fluid, and Brownian motion, which exert more influence when shear stresses are declining. Though the mechanism producing hysteresis remains debated, such a response, measured in a variety of sediment slurries, indicates clearly that deformation history affects rheological response and is an important aspect of slurry behavior.

Deformation responses suggesting that shear stress passes through a minimum value following yield (Figure 3) have been reported for granular slurries containing more than 20 percent sand by volume (Major and Pierson, 1992), for concrete slurries containing particles as large as 16 mm (Ukraincik, 1980), and for bentonite slurries (Wan, 1982; Wright and Krone, 1987; Coussot et al., 1992). Such behavior has been proposed as a general deformation response for debris flows at low to moderate shear rates (Davies, 1986; Phillips and Davies, 1991). Coussot et al. (1992) have shown such behavior to be a possible solution to a set of equations that describe the transition from solid-like to fluid behavior of clay slurries undergoing simple shear. Davies et al. (1991) have proposed that rheological behavior similar to that illustrated in Figure 3 might explain the sudden appearance and translation of large waves in debris flows. As a debris flow progresses down valley the shear stress driving the flow decreases. When the stress drops below some limiting value (τ_B in Figure 3) shear rate drops to zero, and the flow stops. If material is gradually and continuously supplied from upstream, the applied shear stress can gradually increase until it equals the yield stress needed to remobilize the mass. Davies et al. (1991) infer that the shear stress needed to remobilize the debris mass is the same as that needed to initially mobilize the saturated mass, namely τ_A. For stress-driven flow, meeting or exceeding the yield stress results in a rapid escalation of shear rate from zero to some larger value (point D in Figure 3), which Davies et al. (1991) infer to be manifest in the spontaneous appearance and growth of transient waves in stationary debris. This hypothesis requires the existence of both a stopping (τ_B) and a starting (τ_A) yield stress, and it further requires that $\tau_A > \tau_B$. The relatively simple depiction of rheological behavior illustrated in Figure 3 may, however, be complicated by hysteresis (Figure 2), a complication that may become more pronounced as the concentration of coarse particles increases (Major and Pierson, 1992). Exactly how hysteresis affects the hypothesis of Davies et al. (1991) remains uncertain owing to the presently limited understanding of the hysteresis phenomenon.

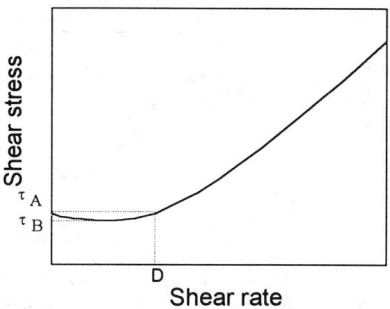

Figure 3. Schematic relation between shear stress and shear rate. See text for parameter definitions.

Investigations to date have focused primarily on steady shear analyses; other rheometric techniques provide further insight into sediment slurry behavior. Recently it has been recognized that fine slurries composed of clay or mixtures of silt and clay can exhibit both elastic solid- and fluid-like behavior characteristic of so-called viscoelastic material (Dade, 1992; Coussot et al., 1992). Experiments using mud from the Amazon shelf indicate that not only does the mud behave viscoelastically under small strains, but that shear stress exerted on the mud is relaxed over a broad distribution of response times (Dade, 1992). Following application of a small-amplitude strain that is held constant, shear stress gradually relaxes. With time, shear stress asymptotically approaches zero or some finite value for fluid- or solid-like behavior, respectively. From such a test a spectrum of relaxation times can be deduced. If the slurry is conceptualized as a field of Maxwell elements (springs connected in series with dashpots), then each element has a characteristic time in which it relaxes stress. If the number of elements is infinite, the relaxation spectrum is continuous; if the number of elements is finite, the relaxation spectrum is composed of discrete modes (Barnes et al., 1989). Dade (1992) has shown that the measured relaxation modulus for Amazon mud can be characterized by a six-component, eleven-term generalized Maxwell model. I recently conducted similar step-strain experiments using sediments collected from debris flow deposits in Owens Valley, CA. Slurries containing particles as large as 125 μm exhibit a similar broad spectral response of relaxation times (Figure 4). Dade (1992) interprets the broad spectral response of Amazon muds as a disruption of microstructures over a range of spatial scales. Disruption of microstructures may also affect stress relaxation in more granular materials. An additional interpretation is that the broad spectral response may reflect prolonged dissipation of pore pressures. Owing to the distribution of grain sizes in these more granular slurries, and to the heterogeneity of permeability, pore pressures might be expected to relax at different time scales as water seeps through a variety of spatial-scale microstructures.

Conclusions

Sufficient experimental work has now been conducted on the rheometric behavior of natural sediment slurries that certain universal trends are emerging. It is clear that the

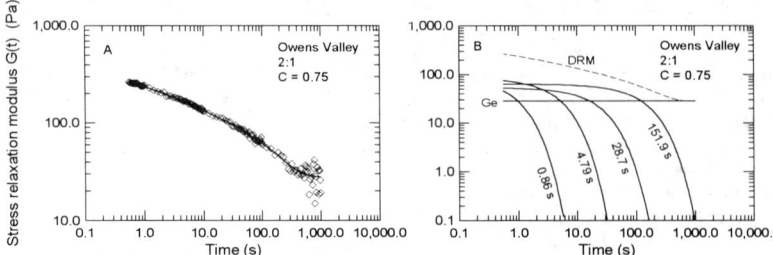

Figure 4. Step-strain shear-stress relaxation analysis (2% strain). A. Discrete Maxwell model approximation of relaxation modulus (solid line) fit to experimental data. B. Component elements of discrete Maxwell model. The discrete relaxation model (DRM) equals the sum of discrete decaying exponential elements. Characteristic time for each Maxwell element is the time needed for shear stress to relax to e^{-1} of its value following instantaneous strain; Ge is the equilibrium modulus at infinite time (Barnes et al., 1989; Dade, 1992). Ratio (2:1) is proportion of fines ($<63\mu m$) to sand ($63\mu m$-$125\mu m$); C is sediment concentration by weight.

rheometric response of virtually all natural sediment slurries is primarily nonlinear. The degree of nonlinearity varies, however, and experimental results suggest that, within measurement error, a linear response is a reasonable first-order approximation at shear rates beyond about 5 to 10 s^{-1}. At lower shear rates, however, experiments demonstrate that the rheometric response deviates from linearity. This result is important particularly in view of observations that suggest that many debris flows move at mean shear rates commonly ≤ 10 s^{-1}. Hysteresis of rheometric response has now been demonstrated to be characteristic of slurries having a wide variety of compositions. The cause of this behavior remains debated. The significance of hysteresis, however, is the demonstration that deformation history clearly affects sediment slurry behavior.

Despite the commonality of results exhibited by various investigations, there are still significant differences. Future work could be profitably aimed at systematically resolving the apparent discrepancies among reported results, and characterizing those differences that are dominantly experimental variations and those that reflect fundamental variations in sediment slurry behavior. Recognition of viscoelastic behavior of high-concentration slurries provides new avenues of exploration. Deeper analysis of that behavior is required, particularly examining its cause.

With one exception, all reported investigations have been restricted to slurries having constituent particle sizes ≤ 2 mm. Clear and unambiguous methods of measurement are required for analyzing high-concentration slurries that incorporate coarser components of debris flow sediments to determine whether, and to what degree, the rheological behavior of debris flows is influenced by those coarser particles.

APPENDIX 1. REFERENCES

Barnes, H.A., and K. Walters, 1985. The yield stress myth? *Rheologica Acta*, 24: 323-326.

Barnes, H.A., J.F. Hutton, and K. Walters, 1989. An Introduction to Rheology. Elsevier, Amsterdam, 199 pp.

Cheng, D.C.-H., and R.A. Richmond, 1978. Some observations on the rheological behavior of dense suspensions. *Rheologica Acta*, 17: 446-453.

Coussot, P., A.I. Leonov, and J-M. Piau, 1992. Rheological modelling and peculiar properties of some debris flows. Erosion, Debris Flows and Environment in Mountain Regions, *Proc. Chengdu Symposium, IAHS Publ. 209*: 207-216.

Dade, W.B., 1992. Studies on boundary conditions for fine-sediment transport. Ph.D. thesis, University of Washington, Seattle, WA, unpublished.

Davies, T.R.H., 1986. Large debris flows: A macro-viscous phenomenon. *Acta Mechanica*, 63: 161-178.

Davies, T.R.H., C.J. Phillips, A.J. Pearce, and Z.X. Bao, 1991. New aspects of debris flow behavior. *Proc. Japan-U.S. Symposium on Snow Avalanche, Landslide, Debris Flow Prediction and Control*: 443-451.

Fairchild, L.H., 1985. Lahars at Mount St. Helens, Washington. Ph.D. thesis, University of Washington, Seattle, WA, unpublished.

Franck, A.J.P., 1988. Rheological characterization of suspensions - comparison of steady and dynamic techniques. *Tenth Intl. Congr. Rheology*: 327-329.

Holmes, R.R., J.A. Westphal, and H.E. Jobson, 1990. Mudflow rheology in a vertically rotating flume. *Hydraulics/Hydrology of Arid Lands, Proc. ASCE 1990 Intl. Symp.*: 212-217.

Kirby, J.M., 1988. Rheological characteristics of sewage sludge: a granuloviscous material. *Rheologica Acta*, 27: 326-334.

Major, J.J., and T.C. Pierson, 1992. Debris flow rheology: experimental analysis of fine-grained slurries. *Water Resources Research*, 28: 841-857.

Murray, T., 1990. Deformable glacier beds: measurement and modelling. Ph.D. thesis, University of Wales, U.K., unpublished.

O'Brien, J.S., and P.Y. Julien, 1988. Laboratory analysis of mudflow properties. *ASCE J. Hydraul. Eng.*, 114(8): 877-887.

Phillips, C.J., and T.R.H. Davies, 1991. Determining rheological parameters of debris flow material. *Geomorphology*, 4: 101-110.

Rheological Comparative Experiment Group, 1990. A comparative experiment for soil and water mixture on several capillary viscometers. *Mountain Res.*, 8(3): 137-146 (in Chinese).

Takahashi, T., 1981. Debris flow. *Ann. Rev. Fluid Mech.*, 13: 57-77.

Ukraincik, V., 1980. Study on fresh concrete flow curves. *Cement and Concrete Res.*, 10: 203-212.

Wan, Z., 1982. Bed material movement in hyperconcentrated flow. Institute of Hydrodynamics and Hydraulic Engineering, Technical University of Denmark, Series paper 31: 79 p.

Wright, V.G., and R.B. Krone, 1987. Laboratory study of mud flows. *Hydraulic Engineering, Proc. ASCE 1987 Conf. Hydraul. Eng.*: 237-242.

Zhao, H., 1990. Experimental study on the rheologic behavior of the slurry of viscous debris flow. *Hydraulics/Hydrology of Arid Lands, Proc. ASCE 1990 Intl. Symp.*: 244-249.

SALT TRANSPORT IN A TIDAL CANAL, WEST NECK CREEK, VIRGINIA

By Jerad D. Bales [1], Member ASCE, and Stanley C. Skrobialowski[1]

ABSTRACT

Flow and salinity were monitored during 1989-92 in West Neck Creek, Virginia, which provides a direct hydraulic connection between the saline waters of Chesapeake Bay and the relatively fresh waters of Currituck Sound, North Carolina. Flow in the tidal creek was to the south 64 percent of the time, but 80 percent of the southward flows were less than 40 cubic feet per second. The highest flows were associated with rain storms. Salinity ranged from 0.1 parts per thousand to 24.5 parts per thousand, and the highest salinities were observed during periods of sustained, strong northerly winds. Salt loads ranged from 302 tons per day to the north to 4,500 tons per day to the south.

INTRODUCTION

West Neck Creek and its northward extension, London Bridge Creek, provide a hydraulic connection between the saline waters of Chesapeake Bay and the relatively fresh waters of the North Landing River and northern Currituck Sound (fig. 1). West Neck Creek and London Bridge Creek are also collectively known as Canal Number Two.

A 2.6-mile- (mi) long bypass canal was constructed east of Canal Number Two in 1989 to reduce potential flood damages. Although Canal Number Two has provided a direct connection between Chesapeake Bay and Currituck Sound for several years, construction of this bypass canal renewed concern about the effects of Canal Number Two and the new bypass canal on water quality in Currituck Sound, particularly on the movement of saltwater into Currituck Sound from Chesapeake Bay.

Consequently, the U.S. Geological Survey (USGS), in cooperation with the North Carolina Department of Environment, Health, and Natural Resources, conducted an investigation during 1989-92 of flow and salinity in West Neck Creek south of the new bypass canal. Subsequently, during 1991-92, salinity was monitored in the North Landing River near the North Carolina-Virginia State line for comparison with data collected in West Neck Creek. The purpose of this paper is to describe the flow and salinity regime in a segment of West Neck Creek, and to characterize the transport of salt in the North Landing River.

STUDY AREA AND DATA COLLECTION

Canal Number Two is located entirely within the city of Virginia Beach. Although an approximately 37 square-mile (mi^2) area originally drained to Canal Number Two (U.S. Army Corps of Engineers, 1980), development and associated changes in natural drainage patterns might have altered the size of the basin.

---
[1]Hydrologists, U.S. Geological Survey, 3916 Sunset Ridge Road, Raleigh, NC 27607.

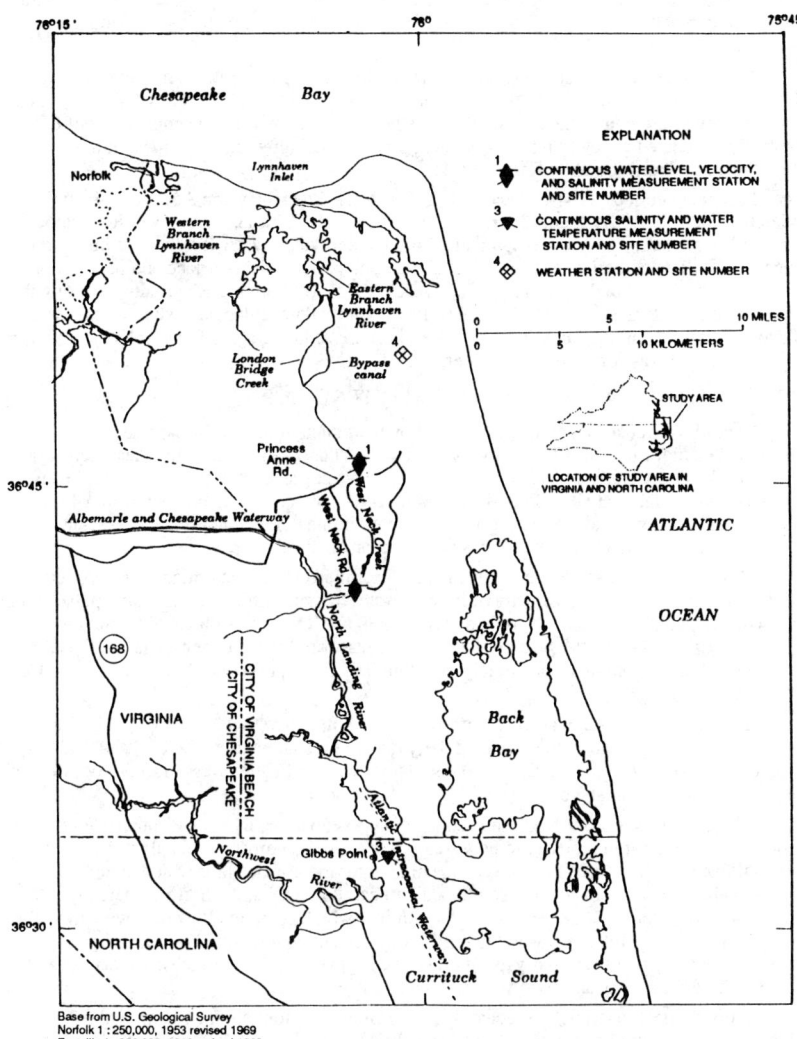

Figure 1.--Location of study area and data-collection sites.

The upstream boundary of the West Neck Creek study reach (site 1) is 3.9 mi downstream from the southern confluence of the bypass canal with London Bridge Creek. The distance between sites 1 and 2 (fig. 1) is about 4.8 mi. Canal Number Two is channelized from the Eastern Branch Lynnhaven River to a point 2.6 mi downstream from site 1. The channel width in the study reach varies from about 45 feet (ft) at site 1 to about 210 ft at site 2, and the average water depth is 6 ft.

The confluence of West Neck Creek with North Landing River is 1.8 mi downstream from site 2. The total drainage area of the North Landing River in Virginia is 116.6 mi^2. A lock at the western end of the Albemarle and Chesapeake Waterway prevents inflow of Elizabeth River (located about 13 mi west of site 1) water, containing high salinities, into the canal and, thus, into North Landing River (fig. 1).

Data were collected at two locations in West Neck Creek between August 1989 and March 1992 (fig. 1). At the northern West Neck Creek site (site 1), water level, flow velocity, and mid-depth salinity were measured at 15-minute intervals. Water level and salinity were measured at the southern site (site 2) during the same period. Near-surface and near-bottom salinities were measured at 15-minute intervals in the North Landing River (site 3) during the period January 1991-July 1992. Water-level and salinity data collected at sites 1 and 2 were fairly continuous throughout the study period; because of instrumentation problems, however, the flow record was somewhat intermittent.

FLOW CHARACTERISTICS

Water levels at sites 1 and 2 are affected by tidal fluctuations in Chesapeake Bay. Although semidiurnal tidal fluctuations were smaller at site 2 than site 1, the mean daily water level for the entire study period was about 0.5 ft above sea level at both sites. Water levels were typically higher at site 1 than site 2. Water levels at site 1 fluctuated between 4.45 ft and -1.18 ft and were generally lower during the winter months. Precipitation and associated runoff have a greater effect on water levels at site 1 than at site 2.

Velocity was measured at site 1 using an ultrasonic velocity metering (UVM) system. UVM systems measure the velocity of flowing water by transmitting an ultrasonic pulse along an acoustic path, which is at an angle of between 30 and 45 degrees diagonal to the flow (Laenen, 1985). The UVM system is capable of measuring low velocities, such as those present in West Neck Creek, accurately to within 0.05 foot per second (ft/s) (Laenen and Curtis, 1989).

The velocity measured at site 1 ranged from 1.48 ft/s (flow to the south) to -0.29 ft/s (flow to the north). The mean of the observed daily maximum southward velocities was 0.43 ft/s, and the mean of the observed daily maximum northward velocities was 0.12 ft/s. For the 308 days of record, the mean velocity was 0.15 ft/s to the south.

The maximum observed instantaneous flow was 356 cubic feet per second (ft^3/s) (to the south) and the minimum observed flow was -50 ft^3/s (to the north). The daily mean flow ranged from -39 ft^3/s to 214 ft^3/s. The mean flow for the 308 days for which complete data were available was 13 ft^3/s to the south. Daily mean flow was to the north 36 percent of the time, meaning that on 36 percent of the days during which complete flow data were available, the net movement of water was to the north. Daily mean flow was to the south 64 percent of the time. Nearly half of the daily mean flows were between -10 and 20 ft^3/s. Eighty percent of the southward daily mean flows were less than 40 ft^3/s.

Flow at site 1 is strongly affected by runoff from rain storms. For example, the 3.57 inches (in.) of rainfall on August 7, 1991, resulted in a peak flow of 268 ft^3/s at site 1 (fig. 2). Moreover, the large difference in water levels between sites 1 and 2 generated a strong sustained flow to the south; flow at site 1 was affected by the storm for a period of about 60 hours. In fact, each of the six recorded occurrences of a daily mean flow in excess of 100 ft^3/s was associated with storms having rainfall amounts greater than 2 in.

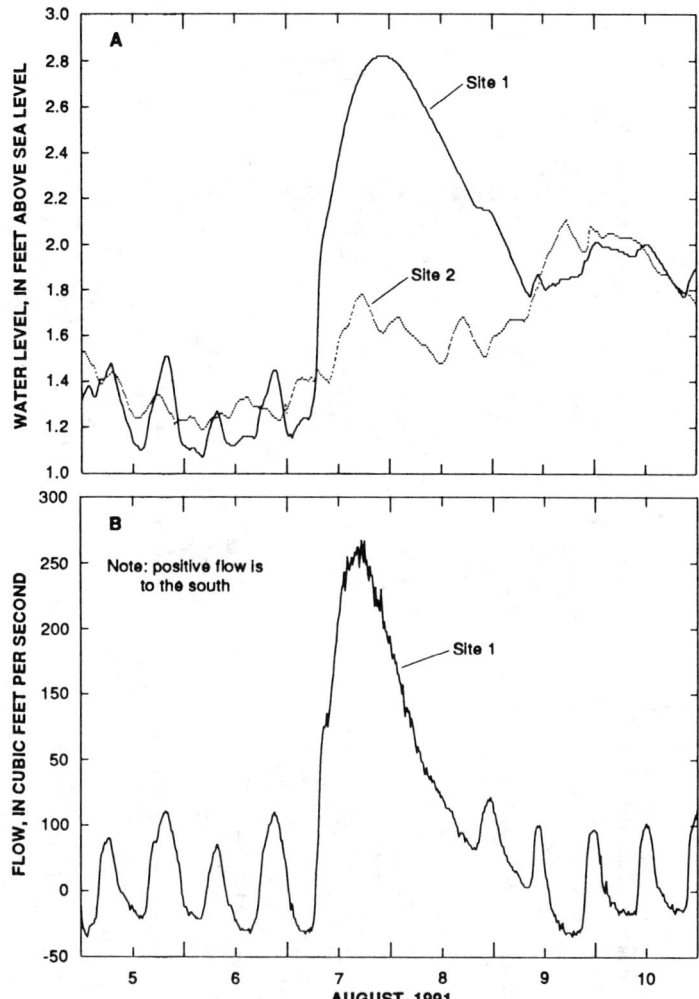

Figure 2.--(A) Water level at sites 1 and 2 and (B) flow at site 1 for the period August 5-10, in response to 3.57 inches of rainfall on August 7, 1991.

Wind also indirectly affects flow in West Neck Creek The highest sustained winds observed during the investigation occurred during late October and early November 1991 (fig. 3). High winds from the north and northeast resulted in a continuously southward flow from October 28 through November 2 (fig. 3). Nevertheless, flows, as well as water levels, during this period of extremely high winds were less than those observed during periods of high rainfall.

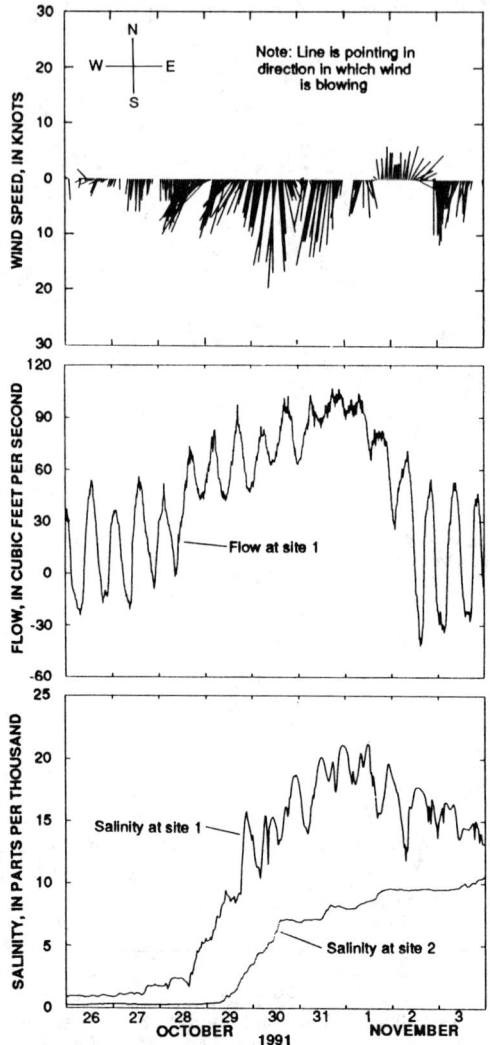

Figure 3.--Influence of wind on salinity of West Neck Creek: (A) wind speed and direction at weather station (site 4) (fig. 1), (B) flow at site 1, and (C) salinity at sites 1 and 2 for the period October 26-November 3, 1991.

SALINITY REGIME

Instantaneous observations of salinity at site 1 ranged between 0.1 parts per thousand (ppt) and 24.5 ppt. The daily mean salinity at site 1 was less than or equal to 1 ppt 55 percent of the time. At site 2, the salinity ranged from less than 0.1 ppt to 14.5 ppt. The daily mean salinity at site 2 was less than 1 ppt 58 percent of the time. Although the highest flows in the study reach were associated with rain storms, the highest salinities occurred during periods of sustained north to northeasterly winds, such as during late October and early November 1991 (fig. 3).

Daily salt loads were computed for site 1 as the daily sum of the products of the 15-minute interval observations of salinity and the 15-minute interval observations of flow. For the 294 days for which salinity and flow data were available for load computations at site 1, the net salt load was 34,510 tons to the south. The mean daily salt load was 117 tons to the south, and the median daily transport was 10 tons to the south. Observed daily salt load to the south ranged between 0.3 tons on August 23, 1991, and 4,500 tons on October 31, 1991. Observed daily northward salt load ranged from 0.2 tons on October 2, 1991, to 302 tons on November 14, 1991.

Salinity was also measured in the North Landing River near the North Carolina-Virginia State line (site 3). During the period January 1991-July 1992, near-surface and near-bottom salinities seldom differed by more than 0.2 ppt, and salt appeared to be uniformly distributed throughout the river cross section. There was little diurnal variation in salinity at the site.

Prior to November 1991, the salinity at site 3 was generally less than 0.8 ppt. Between April and July 1992, salinity at site 3 increased from about 1.3 ppt to about 2.5 ppt. It is not known whether this increase in salinity at site 3 resulted from southward transport of salt through West Neck Creek or northward transport through Currituck Sound, because salinity was not measured in West Neck Creek after March 1992.

SUMMARY

Flow and salinity were monitored during 1989-92 in West Neck Creek, Virginia, which provides a direct hydraulic connection between the saline waters of Chesapeake Bay and the relatively fresh waters of Currituck Sound, North Carolina. A complex system for measuring flow in tidal streams was used to obtain a record of flow in West Neck Creek. Flow in the tidal creek was to the south 64 percent of the time, but 80 percent of the southward flows were less than 40 ft^3/s. The highest flows were associated with storms having rainfall amounts in excess of 2 in. Sustained periods of strong northerly winds indirectly affected flows in West Neck Creek, but the effect was not as pronounced as runoff from high rainfall. Salinity ranged from 0.1 ppt to 24.5 ppt, and the highest salinities were observed during periods of sustained, strong northerly winds. Instantaneous salt loads ranged from 302 tons per day to the north to 4,500 tons per day to the south. Salinity in North Landing River increased from 0.8 ppt in November 1991, to 2.5 ppt in July 1992, but it is unknown whether this increase resulted from southward transport of salt through West Neck Creek or northward transport through Currituck Sound.

REFERENCES

Laenen, A., 1985, Acoustic velocity meter systems: Techniques of Water-Resources Investigations of the United States Geological Survey, Book 3, Chap. A17, 38 p.

Laenen, A., and Curtis, R.E., Jr., 1989, Accuracy of acoustic velocity metering systems for measurement of low velocity in open channels: U.S. Geological Survey Water-Resources Investigations Report 89-4090, 15 p.

U.S. Army Corps of Engineers, 1980, Virginia Beach streams, Canal Number Two general design memorandum phase 1, Virginia Beach, Virginia: Norfolk, Norfolk District, 75 p.

HYDRODYNAMIC AND TOXIC CONTAMINANT DISPERSION IN THE LOWER ST. MARYS RIVER

E.M. Yuen,[1] Associate Member, ASCE, and
J.A. McCorquodale,[2] Member, ASCE

Abstract

A numerical model that simulates the hydrodynamics and the transport of toxic contaminants in the Lower St. Marys River is being calibrated and verified using field data from the U.S. Army Corps of Engineers and the Ontario Ministry of Environment. The model solves the equations for continuity, momentum, turbulent kinetic energy κ, dissipation of turbulence energy ε, and contaminant dispersion concentration. It can treat multiple outfalls, islands, confluences, diversions and rapid mixing. Upon verification of the model, it was used to investigate waste load allocations in the river to meet specified water quality objectives. Isoconcentration maps were obtained with results from the model to determine Impact Zones downstream of the outfall.

Introduction

St. Marys River is a vital navigational channel link between Lake Superior and Lakes Huron and Michigan. It is one of the areas of concern in the International Joint Commission (IJC) report on water quality in the Great Lakes. The Lower St. Marys River is simulated by the model KETOX (McCorquodale and Yuen, 1987). This model has been successfully applied to the St. Clair, Detroit and Niagara Rivers.

---

[1] Assoc. Prof., Dept. of Civ. Engrg., Lawrence Tech. Univ., Southfield, MI 48075.

[2] Prof., Dept. of Civ. and Envir. Engrg., Univ. of Windsor, Windsor, Ontario, Canada N9B 3P4.

The Lower St. Marys River is discretized into a system of reaches, as shown in Fig. 1, in order to handle the occurrence of splitting of flows by the presence of islands, the confluence of flow at the end of islands and the occurrence of natural river curvature in the study area. It should be pointed out that each reach is further subdivided at the users discretion, into additional cross-sections with each section consisting of 15 grid points.

Model Description

A simplified hydrodynamic sub-model based on the work of McCorquodale et al. (1983) was used for KETOX. The model assumes that the lateral depth profiles and river flow rates are available and can be used to obtain velocity profiles along the longitudinal flow direction. The values for the Manning's bed roughness and the river bed slope are adjusted to best fit the U.S. Army Corps of Engineer (U.S. Army COE, 1984) field velociy data. The continuity equation to be satisfied during this adjustment is integrated numerically to ensure that mass is conserved from section to section along the river.
The pollutant dispersion sub-model is based on the work of Lau and Krishnappan (1981). This model uses stream function coordinates and solves the transverse mixing of a pollutant discharged into a river. KETOX has a sub-model that calculates the variation of the lateral dispersion coefficients at grid locations across each cross-section of the river using the equations for turbulent kinetic energy, κ, and dissipation of turbulence energy, ϵ (Rastogi and Rodi, 1978). There are sub-models to 'split' the flow at diversions or islands and to 'combine' the flow at confluences.

Model Calibration, Verification and Results

Field measurements for phenol concentration levels in the river were conducted in 1974 and 1983 (Hamdy et al., 1978; Ontario MOE, 1986). The hydrodynamic component of KETOX was calibrated using U.S. Army Corps of Engineer (U.S. Army COE, 1984) field data based on current meter measurements and drogue surveys. Since all current measurements and drogue survey data were used for calibrating the model, the hydrodynamic model cannot be considered to be verified at the moment. Model verification for the hydrodynamic submodel will only be possible if new field data are available in the near future.
The pollutant dispersion sub-model of KETOX was calibrated using the 1974 Ontario MOE (Hamdy et al., 1978) information on phenol loadings, ambient and other sample measurements along the river including some transects across the river samples. In order to represent the

variation in phenol loadings, a series of computer runs were made with KETOX to establish the upper and lower bounds on the predictions. A comparison of the 1974 measured values and the predicted phenol concentrations along the Canadian shoreline starting from the Algoma Terminal Basin outfall is shown in Fig. 2. It showed that most of the predicted values fall within the 95% confidence band indicating sufficient calibration effort on the KETOX sub-model. On the same plot were field measurements taken in 1983 (Ontario MOE, 1986). The 1983 data were used to verify the model. The model prediction indicated acceptable agreement thus meeting our verification objectives.

Discussion and Applications

KETOX is used to investigate Impact Zones for pollutants under different loading scenarios. For the Lower St. Marys River, the Impact Zone investigation was conducted to assess the problem of transboundary pollution between Canada and the United States. A volatile chemical, referred to as Y is used to show the application of the method to an Impact Zone study for different loading scenarios at the Algoma diffuser. The Provincial Water Quality Objectives (PWQO) for chemical Y is 1 ug/L. For steady state loading, isoconcentration maps can be developed with longitudinal resolution of the order of 15 m and lateral resolution as low as 1 percent of the flow in the reach. This permits a reasonably accurate Impact Zone to be defined so that various loading scenarios can be compared and evaluated. Figure 3 shows the results for a typical summer impact when the initial concentrations at the Algoma diffuser are 20 kg/d and 100 kg/d. The Algoma diffuser discharge was 4.2 m^3/s with an average summer flow of 2,464 m^3/s. The Impact Zone for 100 kg/d extends from the Algoma terminal basin to the easterly Sewage Treatment Plant (STP) in the North Channel. The extent of the Impact Zone is shortened to the near field zone of the diffuser when the load is reduced to 20 kg/d. The 1 ug/L isoconcentration line does not cross the International Boundary for either load.

A mixing zone (MZ) is defined by the Ontario Ministry of Environment as "an area of water contiguous to a point source where the water quality does not comply with the PWQO". This zone is to be minimized in order to protect aquatic life as well as other users. Limitations on mixing zones are established on a "case by case" basis with the size of a MZ limited by the nearest downstream user. The MZ should not be a "barrier to the migration of fish and aquatic life". The MZ is used in the allocation of loads for a particular outfall. The load is set so that the applicable guideline is satisfied at an acceptable compliance level at the edge of the MZ. Outside of this MZ there should be no long term (chronic) effects on the aquatic life; and within the mixing zone, the conditions should not be rapidly lethal to aquatic life. This is accomplished by using the model to establish a system response curve for the edge of the MZ, i.e. a concentration versus load curve as shown in Fig. 4. Figure 4 shows the most probable relationship between load and concentration at a distance of 300 m from the outfall for a contaminant Y with PQWO of 1 ug/L. Entering this curve with the PWQO yields the median load that would satisfy the objective 50% of the time (50% compliance). The median monthly load for 50% compliance in this example is 22 kg/day. The statistical nature of the controlling variables (river flow, contaminant load and calibration errors of the model) can be used to estimate the load that would satisfy the objective for a specified percent of time. The 95% compliance response curve which yields a 95% compliance median load of 10.6 kg/day is also shown in Fig. 4.

Conclusion

KETOX provides a means for the environmental managers to screen management alternatives. For instance, the model can be used to predict the relative differences in water quality to be expected from wastewater management changes such as consolidating individual wastewater treatment plants into regional facility, relocating existing wastewater discharges, or incrementally reducing wastewater loads. The effects of nonpoint source load reductions can also be evaluated.

Appendix I. References

Hamdy, Y.S., Kinkead, J.D. and Griffiths, M. 1978. St. Marys River Water Quality. Investigations, 1973-1974. Great Lakes Surveys Unit, Water Resources Branch, Ontario Ministry of the Environment, March.
Lau, Y.L. and Krishnappan, B.G. 1981. Modeling transverse mixing in natural streams. J. of Hyd. Div., ASCE Vol. 107.
McCorquodale, J.A. and Yuen, E.M., 1987. St. Marys River Hydrodynamic and Dispersion Study, Report to Ontario Ministry of Environment, IRI 18-61, University of Windsor, Windsor, Ontario.
Ontario Ministry of Environment, 1986. St. Marys River Water Quality Survey, 1983.
Rastogi, A.K. and Rodi, W. 1978. Predictions of heat and mass transfer in open channels. J. of Hyd. Div., ASCE Vol. 104.
U.S. Army Corps of Engineers, Detroit District, Great Lakes Hydraulics and Hydrology Branch, 1984. St. Marys River oil/toxic substance spill study current velocities and directions 1980-1983, December.

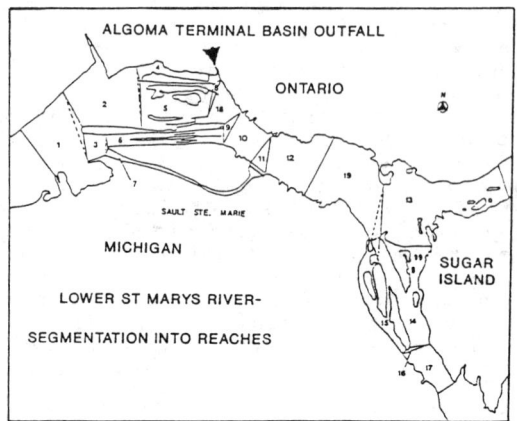

Figure 1. Definition of reaches in the Lower St. Marys River for the KETOX model.

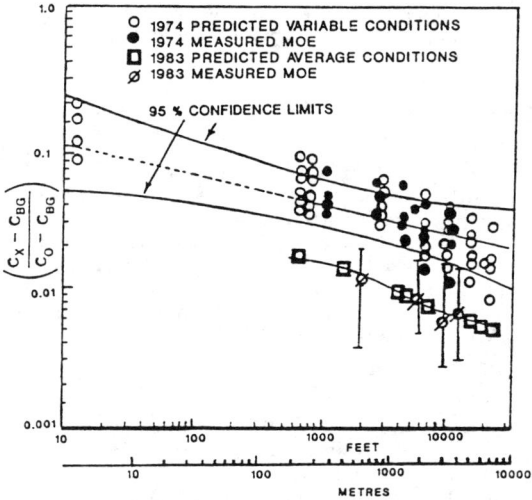

Figure 2. KETOX predicted and Ontario MOE measured concentrations of phenol along the Canadian shoreline of the Lower St. Marys River for 1974 and 1983.

CONTAMINANT DISPERSION

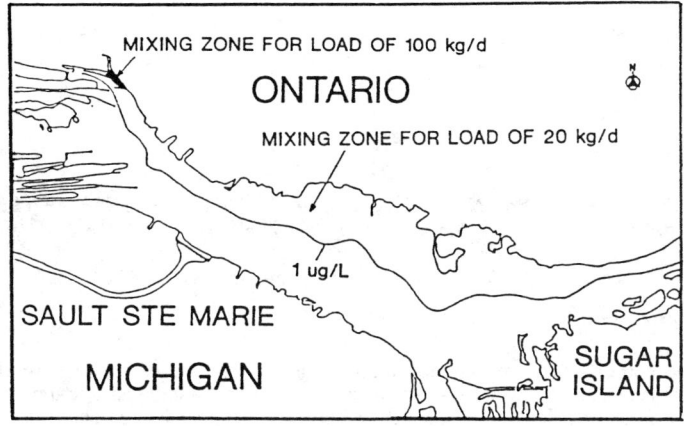

Figure 3. Load response curve for chemical Y at the boundary of a mixing zone.

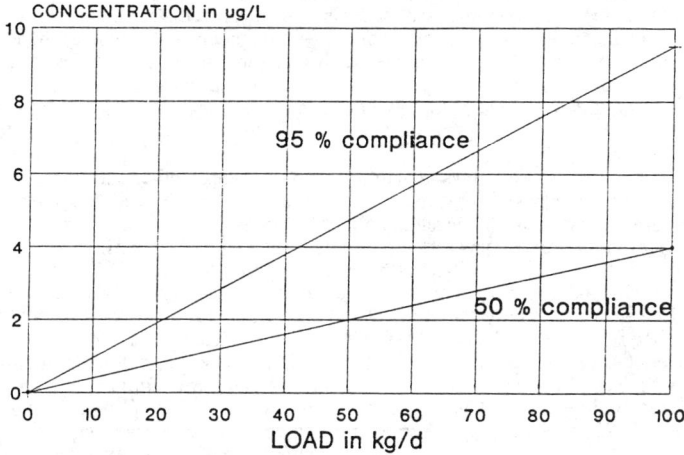

Summer Conditions- 300 m downstream of Outfall

Figure 4. Impact zones for chemical for the Lower St. Marys River.

MODELING LOW FLOW TRANSPORT OF NONCONSERVATIVE POLLUTANTS IN STREAMS

Il Won Seo, [1] Member, ASCE, and Dae Young Yu [2]

ABSTRACT

The complex nature of low flow transport and transformation of nonconservative pollutants in natural streams has been investigated using a numerical solution of a proposed mathematical model that is based on a set of mass balance equations describing the advection, dispersion, decay and mass exchange mechanisms in streams and in storage zones. In the present study, a mathematical model (named "Storage-Transformation Model") has been developed to predict adequately the non-Fickian nature of mixing and transformation mechanisms for decaying substances in natural streams under low flow conditions. Comparisons between the concentration-time curves predicted using the proposed model and the measured stream data shows that the Storage-Transformation Model yields better agreements in the general shape, peak concentration and time to peak than the 1-D dispersion model. The proposed model shows significant improvement over the conventional 1-D dispersion model in predicting natural mixing and storage processes in streams through pools and riffles.

INTRODUCTION

Characteristics of low flows in natural streams are substantially different from those observed at bank-full or flood stages. Under low flow conditions, pollution problems are most acute. The water quality of streams receiving municipal, industrial, and agricultural return flow is further degraded when natural streamflow is low. Dilution of contaminants decreases as streamflow decreases; thus the hazard associated with an accidental spill may be much greater at low flow than at a higher flow. Variations in bed geometry such as pool and riffle structure, dominant channel features during low flow (Leopold et al., 1964), play their strongest role in affecting mixing characteristics of polluted releases in the channel. In recent years, the investigation of the low flow condition has become important as a means of determining critical levels of water pollution, aquatic habitat and instream flow needs. Seo (1990) and Seo and Maxwell (1992) have conducted important research on the transport and mixing characteristics for pollutants discharged into natural streams with pools and riffles. They showed that in natural channels under low flow conditions, the effect of storage induced by the pool-riffle sequences should be considered adequately in the modeling

---

[1] Asst. Prof., Dept. of Civ. Engrg., Seoul Nat. Univ., Gwanak-Ku, Seoul 151-742, Korea

[2] Res. Asst., Dept. of Civ. Engrg., Seoul Nat. Univ., Gwanak-Ku, Seoul 151-742, Korea

of transport and mixing of conservative solutes. Knowledge of transport and mixing characteristics for nonconservative pollutants as well as conservative pollutants are required to establish sound water pollution control and water resources management programs.

The one-dimensional (1-D) Fickian-type dispersion equation derived by Taylor (1954) has been widely used to give a reasonable estimate of the rate of longitudinal dispersion. The 1-D Fickian dispersion equation for nonconservative pollutants in which the decay term is modeled as a first-order function is

$$\frac{\partial C}{\partial t} + U \frac{\partial C}{\partial x} = K \frac{\partial^2 C}{\partial x^2} - k_r C \qquad (1)$$

in which C = cross-sectional average concentration; U = cross-sectional average velocity; K = dispersion coefficient; k_r = decay coefficient; t = time; and x = longitudinal distance.

The analytical solutions of Eq. 1 for limited period of injection of nonconservative pollutants can be derived following Rose (1977) and Thomann and Mueller (1987). Boundary and initial conditions are as:

$$C(t, 0) = C_0, \qquad 0 < t < \tau$$
$$C(t, 0) = 0, \qquad t > \tau$$
$$C(0, x) = 0, \qquad x > 0, \qquad (2)$$

in which C_0 is the initial concentration injected. The solution of Eq. 1 for condition 2 is:

$$C(t, x) = \frac{C_0}{2} \exp\left(-\frac{k_r x}{U}\right) [\, G(t) \, \mathrm{erfc}\left(\frac{x - Ut(1+H)}{\sqrt{4Kt}}\right)$$
$$- G(t-\tau) \, \mathrm{erfc}\left(\frac{x - U(t-\tau)(1+H)}{\sqrt{4K(t-\tau)}}\right)] \qquad (3)$$

in which $H = 2k_r K/U^2$ and erfc(Z) is the complimentary error function which is defined by erfc(Z) = 1 - erf(Z). G(z) is the unit step-function which takes the values:
G(z) = 1, z > 0;   G(z) = 0,   z < 0.

An immediate limitation is that the Fickian dispersion model cannot be applied until after the initial period, i.e., the model should be limited to locations far downstream from the source at which the balance between advection and diffusion assumed by Taylor is reached (Fischer et al., 1979). Literature describing the field studies including Zand et al. (1976) and Legrand-Marcq and Laudelot (1985) has indicated that concentration distribution data collected in natural streams seem to indicate non-linear behavior of the variance for times even beyond the initial period. Furthermore, most experimental studies in natural streams have produced concentration-time curves which are significantly more skewed than the concentration distribution predicted by the solution of the 1-D Fickian dispersion equation. These show that water and dye are retained in the regions having storage effects along the channel bed and banks and then released slowly after the main cloud has passed. Several researchers including Seo

(1990), and Seo and Maxwell (1992) have suggested that a complete analysis must include the effect of channel storage zones.

The objective of the present study was to develop a mathematical model to predict adequately complex mixing characteristics of nonconservative pollutants in natural streams under low flow conditions. The predicted concentration-time curves were compared to the measured stream data.

MATHEMATICAL MODEL

The boundary geometry of natural streams is not smooth and regular. Under low flow conditions, irregularities and unevenness along the streams caused by pools and riffles can create storage zones that have significant storage effects. In this model a typical cross section is considered to consist of two distinct zones, a flow zone and a storage zone. In the flow zone, the dominant mass transport mechanisms are longitudinal advection and dispersion. The storage zones are considered as regions having vortex or recirculating flow and having mass interchange with the main flow across the interface between the flow and the storage zones. The storage zones serve to retain part of the solute as the main cloud passes, and the solute is then slowly released back into the flow zone. It is assumed that mass is decaying in both zones. Among several conceptually different physical models of the transient storage of mass in the storage zone (Jackman et al. 1984), the exchange model assumes a different uniform concentration in each zone. Mass transfer at the interface between the zones is considered to be proportional to the difference in the average concentrations.

Governing Equations

The equations describing the Storage-Transformation Model are derived using conservation of mass. The mass balance equation in the flow zone for steady flow is

$$A_f \frac{\partial C}{\partial t} = - U_f A_f \frac{\partial C}{\partial x} + \frac{\partial}{\partial x} (K A_f \frac{\partial C}{\partial x}) + k P (S-C) - A_f k_r C \qquad (4)$$

in which A_f = cross-sectional area of the flow zone; U_f = flow zone velocity; P = wetted contact length between the flow zone and the storage zone in the transverse or vertical direction; k = mass exchange coefficient; and S = the concentration of mass in the storage zone. A mass balance equation describing S as a function of longitudinal position (x) and time (t) is

$$A_s \frac{\partial S}{\partial t} = - k P (S-C) - A_s k_r S \qquad (5)$$

in which A_s = cross-sectional area of the storage zone perpendicular to the general flow direction.

Numerical Modeling

An analytical solution of the given set of governing equations (Eqs. 4-5) corresponding to the initial and boundary conditions (Eq. 2) was not available because of the non-uniform parameters and the existence of the mass exchange terms in each equation. Therefore, numerical techniques were applied to solve the given set of governing equations. Based on the preliminary numerical investigation (Seo 1990), among the various types of numerical schemes tested, the finite difference method (FDM) developed by Stone and Brian (1963) was selected to solve the given set of governing equations. This method, based on the Crank-Nicholson implicit approach, was considered to have no stability limitations as in the cases of other implicit schemes. The truncation error involved in this scheme was considered to be $O(\Delta t^2 + \Delta x^2)$, as in other Crank-Nicholson implicit approaches with the central difference approximation for space discretization, which is a higher-order than that of a fully implicit approach, $O(\Delta t + \Delta x^2)$. Δt is the time increment and Δx is the distance step.

The time derivative $\partial C/\partial t$ of the flow zone equation was represented by the spread form backward time difference approximation. The advective term was discretized by using the Crank-Nicholson approach, in which $\partial C/\partial x$ was centrally differenced. The dispersive term was also discretized by using the Crank-Nicholson type implicit method. Substituting each term into Eq. 4, and expanding the resulting equation for all the nodal points along the x axis, a set of simultaneous linear algebraic equations, of which the coefficient matrix is tridiagonal, can be obtained. The resulting system of algebraic equations was solved by using the Thomas algorithm, a variation of Gaussian elimination. The storage zone equation was also discretized by using the FDM developed by Stone and Brian similar to the flow zone equation. Because of the mass exchange terms in the mass balance equation for both flow and storage zones, another unknown term arises at the right hand side of the resulting system of algebraic equations for both flow and storage zones. So, additional iteration work was needed.

MODEL PREDICTIONS
Stream Data

The Storage-Transformation Model developed in this study was tested by using field data measured by Zand et al. (1976). This data was also used by Bencala (1983) for his solute transport model. Zand et al. described the dispersion study of nonconservative tracers in a small stream, Uvas Creek in California, U.S.A. The channel is highly irregular. It is composed of alternating pools and riffles and pool frequency ranges mostly 6 to 7 channel widths which falls into the range of that of the natural pool-riffle sequences studied by other investigators. The experiment was conducted in late summer during a period of low flow ($Q = 0.0125$ m^3/s). The strontium tracer, as a nonconservative pollutant, was injected at a constant rate for three hours and reached a maximum concentration of 1.73 mg/l a short distance below the injection point. Background concentration was measured to be 0.13 mg/l.

Simulation Results

In the numerical model, the simplified geometric and hydraulic characteristics of the pool-riffle sequences were used. The nonuniform hydraulic parameters, such as the flow depth and the storage zone area ratio, were considered to have single constant values at the pool and riffle, and then to vary linearly through the transition between the pool and the riffle. The mass exchange coefficient was also considered to follow the above assumption, but the dispersion and decay coefficients were assumed to be constant through the whole reach of pool-riffle sequences. The model parameters used for simulation are presented in Table 1.

TABLE 1.-Summary of the Model Parameters Used in the Simulation

A_s/A		Depth (m)		K	k (m/s)		k_r
Pool	Riffle	Pool	Riffle	(m^2/s)	Pool	Riffle	(1/s)
0.35	0.25	0.200	0.033	0.20	$0.6*10^{-3}$	$0.2*10^{-3}$	$0.47*10^{-4}$

Comparisons of the concentration-time curves of the model simulations with those obtained in the stream experiments are depicted in Fig. 1. In general, in overall shape, the concentration-time curves given by the storage zone model better fit the measured concentration-time curves than those given by the 1-D dispersion model. The tails of the concentration-time curves by the storage zone model are quite close to those of the measured concentration-time curves, whereas those by the 1-D dispersion model fail to fit. The peak concentrations predicted by the Storage-Transformation Model are quite close to the experimental data. However, the 1-D dispersion model underestimted peak concentrations. More important, the times to peak concentration predicted by the 1-D dispersion model are inaccurate, whereas the storage zone model predicts the elapsed times to peak concentration very accurately.

CONCLUSIONS

The comparison between measured and predicted concentration curves by the Storage-Transformation Model shows that there is a good level of agreement in the general shape, peak concentration and time to peak. The proposed model shows significant improvement over the conventional 1-D dispersion model in predicting natural mixing processes in natural channels with pools and riffles.

ACKNOWLEDGMENTS

This research work was partially supported by the Korean Science Foundation.

REFERENCES

Bencala, K. E. (1983). "Simulation of Solute Transport in a Mountain Pool-and-Riffle Stream With a Kinetic Mass Transfer Model for Sorption." Water Resour. Res., 19(3), 732-738.

Fischer, H. B., List, E. J., Koh, R. C. Y., Imberger, J., Brooks, N. H. (1979) Mixing in Inland and Coastal Waters. Academic Press, New York, N.Y.

Jackman, A. P., Walters, R. A., and Kennedy, V. C. (1984). "Low-flow Transport Models for Conservative and Sorbed Solute-Uvas Creek Studies." Water Resour. Investigation Reports, No. 84-4041, U.S. Geol. Surv., Menlo Park, Calif.

Legrand-Marcq, C., and Laudelot, H. (1985). "Longitudinal Dispersion in a Forest Stream." J. Hydro., 78, 317-324.

Leopold, L. B., Wolman, M. G., and Miller, J. P. (1964). Fluvial Processes in Geomorphology. W. H. Freeman, San Francisco, Calif.

Rose, D. A. (1977). Dilution and Decay of Aquatic Herbicides in Flowing Channels-Comments, J. Hydro. 32, 399-400.

Seo, I. W. (1990). "Low Flow Mixing in Open Channels," PhD thesis, Univ. of Illinois at Urbana-Champaign, Urbana, Ill.

Seo, I. W., and Maxwell, W. H. C. (1992). "Modeling Low Flow Mixing through Pools and Riffles." J. Hydr. Engr., ASCE, 118(10), 1406-1423.

Stone, H. L., and Brian, P. T. (1963). "Numerical Solution of Convective Transport Problems." AIChE J., 9(5), 681-688.

Taylor, G. I. (1954). "The Dispersion of Matter in Turbulent Flow through a Pipe." Proc. Royal Society of London, Ser. A, 223, 446-468.

Thomann, R. V., and Mueller, J. A. (1987). Principles of Surface Water Quality Modeling and Control. Harper & Row, New York, N.Y.

Zand, S. M. , Kennedy, G. W., Zellweger, G. W., and Avanzino, R. J. (1976). "Solute Transport and Modeling of Water Quality in a Small Stream." J. Research, U.S.Geol. Suv. 4(2), 233-240.

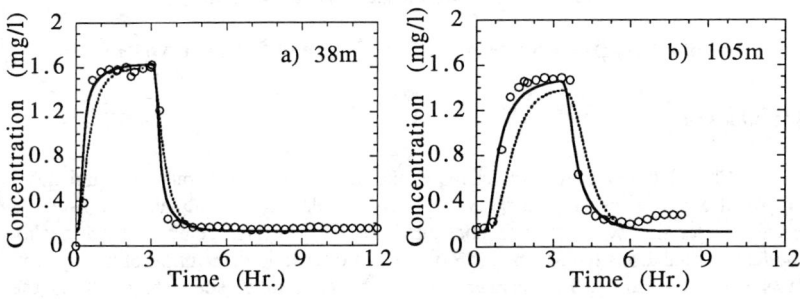

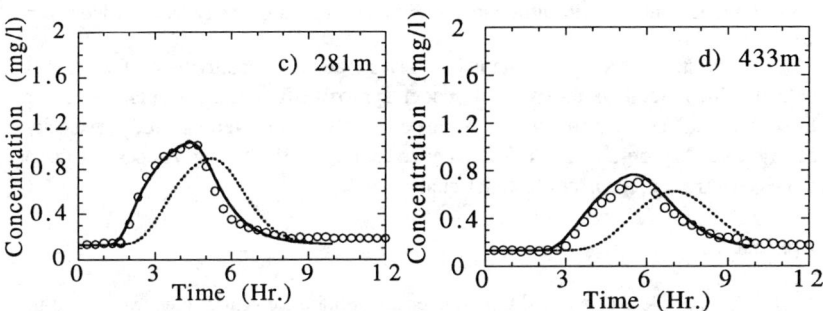

FIG. 1.-Concentration-Time Distribution of Observed Dispersion Data and Distributions Fitted by 1-D Dispersion and Storage-Transformation Models; —o— Observed Data; ----- 1-D Dispersion Model; ——— Storage-Transformation Model

Wind Induced Circulation in Shallow Lakes

Ian P. King[1], Parmeshwar L. Shrestha[2], and Gerald T. Orlob[3]

Introduction

Clear Lake is a relatively shallow lake in Northern California. Natural flows are small but at times they do transport heavy metals into the the lake. In order to predict water quality constituent concentrations in the water column, water quality modeling simulations require an appropriate hydrodynamic description of the system. It is imperative that a hydrodynamic model be chosenthat adequately reflects the hydromechanical behavior of the water body under the imposed stresses. The hydrodynamics of the water body influences transport parameters such as advection, dispersion, sedimentation, and other physico-chemical processes. Wind-induced circulation is one of the primary factors influencing circulation especially where the water body is confined with little streamflow as is typical of many lakes in Northern California.

This study has been designed to investigate the sensitivity of the overall velocity distribution to the type of model approximation (i.e., whether two- or three-dimensional) and to model parameters such as the vertical eddy viscosity coefficient. The objective is a better understanding of the appropriateness of these models for use as precursors to water quality models.

Approach

State of the art two- and three-dimensional finite element models were used to perform steady state hydrodynamic simulations for Clear Lake. The models were used

---

[1] Professor, Department of Civil and Environmental Engineering, University of California, Davis, CA 95616.
[2] Assistant Professor, Department of Civil Engineering, Virginia Polytechnic Institute and State University, Blacksburg, VA 24061.
[3] Professor Emeritus, Department of Civil and Environmental Engineering, University of California, Davis, CA 95616.

to investigate the hydraulic response of the lake as a result of wind stresses imposed on the lake surface. Output from the two models were depicted in the form of velocity vector plots and compared to assess the differences in velocity distribution and circulation patterns arising from use of the two different types of approximation. To determine the sensitivity of the circulation to the magnitude of the vertical eddy viscosity coefficient, further simulations were carried out using the three-dimensional model.

Hydrodynamic Modeling

Steady state simulations have been carried out using the two-and three-dimensional models, RMA-2V and RMA-10 .

RMA-2V uses the finite element method to solve the two-dimensional depth averaged equations of continuity and momentum to provide spatial and temporal (if required) descriptions of the velocity field and flow depths over the water body. The network structure may consist of curved triangles and quadrilaterals. The model has been extensively applied to streams and bays and estuaries such as, the San Francisco Bay and Delta system. It has been found to be reliable model for hydrodynamic simulation of relatively shallow, vertically mixed systems. Norton et al. (1973) and King (1990) have documented this model.

RMA-10 is a finite element model which solves the three-dimensional form of the shallow water equations for momentum and continuity. It is capable of predicting water surface elevations and the vertical distribution of the currents in the three Cartesian directions. In its most general form it is designed to simultaneously compute the advection and diffusion of salinity/temperature and to compute circulation in density stratified flow systems. For the purposes of these simulations Clear Lake has been assumed to be fully mixed. An essential additional input to this model is a description of the vertical eddy coefficients. These "not well known" coefficients control the shape of the vertical velocity distribution. As part of this study the sensitivity of the overall velocity distribution to the magnitude of these coefficients was investigated. The entire model is more fully described in King (1993).

Maximum water depths in Clear Lake vary from generally 10 meters west of the Narrows to about 15 meters near Sulphur Bank Mine. Inflows into the lake are from Rodman Slough, Rumsey Slough, Manning Creek, Adobe Creek, and Kelsey Creek. In addition, runoff from Sulphur Bank Mine carries mercury into the lake. Lake outflow is through Cache Creek.

A two-dimensional finite element grid was generated using RMAGEN (King, 1991) from digitized data on the bathymetry of Clear Lake. The network in plan view is comprised of 408 elements and 1433 nodes as shown in Figure 1 below.

The initial water surface elevation in Clear Lake was specified at 2.55 meters. Since the study was designed to investigate the hydrodynamic behavior of Clear Lake under stresses induced by wind, hydrologic boundary conditions due to inflows to the lake were not included in the simulation. However, at the downstream end, i.e., at Cache Creek, a steady water surface elevation boundary condition of 2.55 meters was

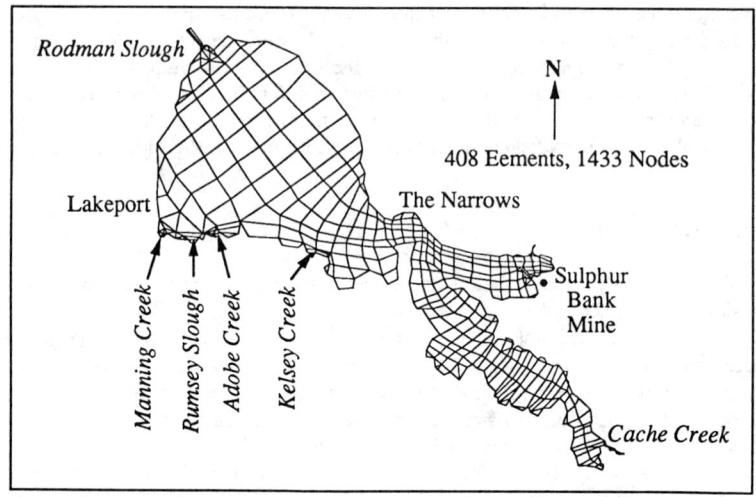

Figure 1 Finite Element Network for Clear Lake, California

specified. A South East wind of 16 kilometers per hour was imposed on the lake surface. The above initial and boundary conditions together with the network geometry were input first to RMA-2V and then to RMA-10. For the case of RMA-10 additonal data values were input to select a 2nd order vertical distribution. For the RMA-10 application there were approximately 3200 nodes. Simulations were carried out until steady state conditions were achieved. For these inital cases a "best estimate" of the vertical eddy viscosity was selected

Hydrodynamic Model Results

Results of the simulation were depicted as velocity vector plots using RMAVPLT (King, 1991b). In order to make a comparison meaningful the vertical distribution of velocities from RMA-10 was integrated. Velocities from the RMA-2V model results (Figure 2) were compared with these depth-integrated velocities obtained from the RMA-10 simulations (Figure 3). The velocities from the three-dimensional simulation were found to vary significantly in magnitude when compared to those from the depth-averaged two-dimensional simulation. For the simulations performed, the velocities predicted by the two-dimensional model were higher than the velocities derived from the three-dimensional model by a factor of about 5, although the same velocity pattern was apparent The following conclusions were deduced.
1. In RMA-2V, when wind is the primary driving horizontal force, wind set up is created that then drives flow along the deeper main axis of the lake against the prevailing shear stress. Note that because the water is deeper energy losses

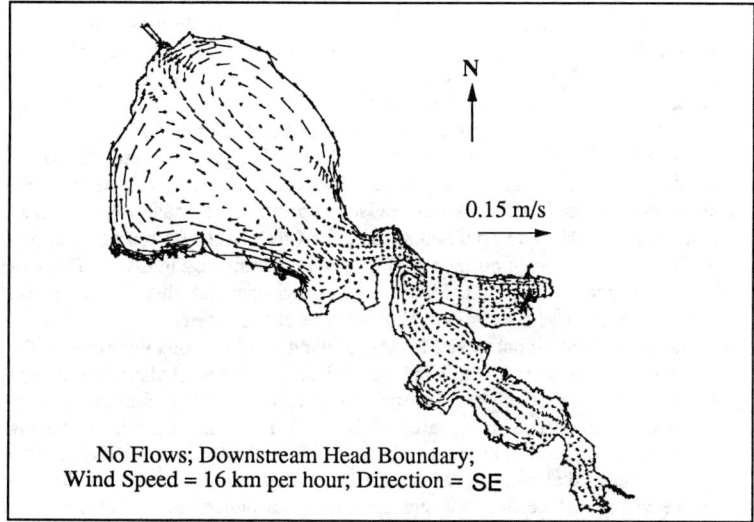

Figure 2 Velocity Vector Plot, RMA-2V Steady State Simulation

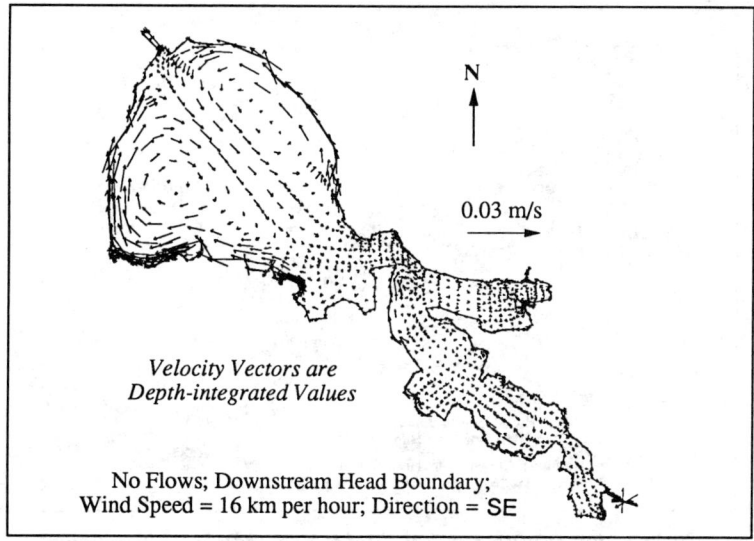

Figure 3 Velocity Vector Plot, RMA-10 Steady State Simulation

due to friction are smaller. Closer to the shore, in the shallow area, the applied wind stresses distribute over a smaller depth and serve to overcome the friction losses and the adverse water surface gradient. Circulation is thus achieved with downwind flow in the shallows and return flow in the deeper section. Since the two-dimensional model does not incorporate the velocity variations in the vertical direction, wind driven currents in the horizontal plane are larger in magnitude. This circulation is mostly caused by the irregular bathymetry. If the system is modeled with a uniform depth of about 15 meters RMA-2V predicts a much reduced magnitude of circulation RMA-2V thus causes horizontal circulation that is not realistic especially for aquatic environments that are generally more confined such as for Clear Lake. As can be seen from the figures, the average horizontal circulation computed by RMA-2V is greater than that computed by RMA-10. For environments that are deep and confined, horizontal circulation will tend to be exaggerated.

2. In the three-dimensional model RMA-10, wind stresses generally drive surface water in the direction of the wind. See Figure 4. Part of the wind stress is absorbed in overcoming the internal friction due to eddy viscosity and the remainder is absorbed in a water surface setup. This results in an adverse water surface elevation gradient. Differential depths account for the variation in magnitude. Water at lower levels is driven generally in the reverse direction by the water surface gradient and appears as an underflowing bottom return current. The overall circulation obtained for the RMA-10 simulation (shown in Figure 3) is therefore more realistic.

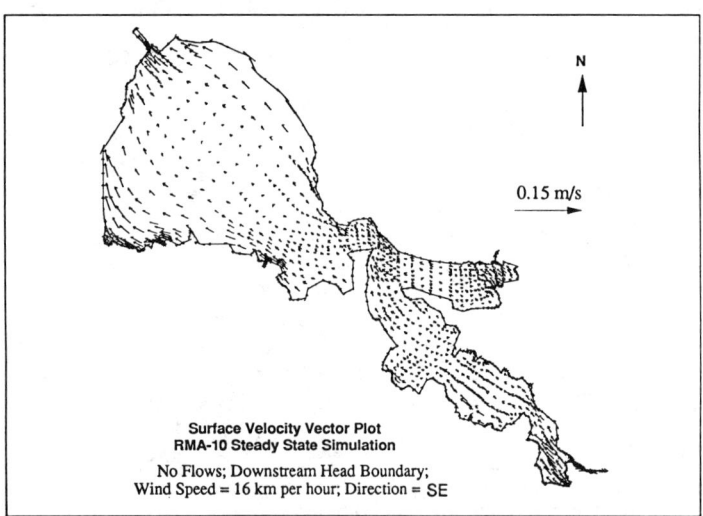

Figure 4   Surface Velocity Vector Plot, RMA-10 Steady State Simulation

To study the effect of eddy viscosity in the vertical direction, simulations were carried out using RMA-10 for varying the magnitude of the eddy viscosity. Inflows and outflows were not included. Wind conditions were input in a similar fashion as described for the above simulations. It was found that when the eddy viscosity coefficient was decreased, underflow increased and the resulting depth-integrated velocities decreased. and there was less horizontal circulation. When the coefficients were increased the flow distribution became more uniform vertically and the overall and depth integrated circulation more closely matched the two-dimensional depth averaged results.

Concluding Remarks

If velocities derived from the two hydrodynamic models were to be used to drive water quality models in a shallow lake subject to wind stress, results obtained would depend on the method used to obtain current circulation. Two dimensional depth averaged simulations could lead wrong direction transport in surface layers and to overestimates of the transport of contaminants in the lower levels of the lake. Overall circulation in these types of lakes are very sensitive to the assigned magnitude of vertical eddy coeeficients.

References

King, I.P., 1990. Program Documentation RMA2 - A Two-dimensional Finite Element Model for Flow in Estuaries and Streams, Version 4.3, Resource Management Associates, Lafayette, CA 94549.
King, I.P., 1991a. RMAGEN - A Program for Generation of Finite Element Networks - User Instructions, Version 2.0, Resource Management Associates, Lafayette, CA 94549.
King, I.P., 1991b. RMAVPLT - A Program for Generation of Velocity Vector Plots - User Instructions, Version 1.0, Resource Management Associates, Lafayette, CA 94549.
King, Ian P., 1993 RMA-10, A Finite Element Model for Three-Dimensional Density Stratified Flow, Report prepared for the New South Wales Environmental Protection Agency, Sydney, Australia.
Norton, W.R., I.P. King, and G.T. Orlob, 1973. "A Finite Element Model for Lower Granite Reservoir", Water Resources Engineers, Walnut Creek, CA.

Prediction of Filtrate Turbidity
by Parameter Estimation

Sadataka Shiba,[1] Member, ASCE

Abstract

In order to operate the rapid sand filter in response to the time variation of effluent quality, a model and a procedure to predict the filtrate quality of rapid sand filters have been developed and they have been verified by the simulation using the experimental results of Eliassen. The simulated results prove that the procedure using a simple model may be practical for operation of the facilities.

Introduction

The removal of suspended solids or turbidity from water by rapid sand filtration is an essential process in the purification of waters. More than 80 years have passed since rapid sand filters in water treatment plants began to be utilized in Japan. Nowadays the rapid sand filter has attained to a position of indispensable facilities in water treatment plants. Due to lack of knowledge of the filtration mechanism, however, the mathematical model which predicts the filtrate quality precisely has not been yet established.

On the other hand, in order to recycle the water resouces the need for the precise prediction of the turbidity of the filtrate is increasing more and more, because the filtrate quality should be controlled severely to meet the various extent of the treatment in the recycling. Therefore, the prediction must be as accurate as possible, even if the mathematical model is incomplete.

Construction of Rapid Sand Filter Model

---

[1] Res. Assoc., Dept. of Chem. Engrg., Osaka Univ., Toyonaka, Osaka 560, Japan

A large number of variables are involved in rapid sand filtration such as media characteristics, suspended solid characteristics and operating characteristics. Therefore, it is very difficult to develop a complete mathematical filtration model incorporating the unsteady variation of suspension concentration with time and there has been no adequate mathematical model expressing filter performance. The accurate prediction of filtrate quality, however, is feasible by making frequent estimation of the parameters in the model though the model is incomplete.

On this account the mathematical model used here is a simple linear one, which is composed of the two conservation equations for the volumetric concentration of filtrate quality C and the detention volume V as follows:

$$\frac{d(VC)}{dt} = Q_{IN} C_{IN} - (Q_{OUT} + k S)C \qquad (1)$$

$$\frac{dV}{dt} = Q_{IN} - Q_{OUT} \qquad (2)$$

where t = time; Q = flow rate; k = mass transfer rate; S = effective interface area between suspension and filter material; and subscripts IN and OUT = influent and effluent, respectively. The detention volume V is equal to the integrated pore volume over the whole bed, i.e., from the filter surface to the bottom as

$$V(t) = A \int_0^L [\varepsilon_0 - \sigma(t, z)] dz = AL\varepsilon_0 \left[1 - \frac{\overline{\sigma(t)}}{\varepsilon_0}\right] \qquad (3)$$

where A = cross sectional area of the filter; L = filter length; ε_0 = initial porosity of clean filter; $\sigma(t, z)$ = specific deposit, i.e., the ratio of the deposit volume to the total volume of pores, deposit and filter medium; z = vertical coordinate measured from bed surface; and $\overline{\sigma}(t)$ = filter-length-averaged specific deposit as

$$\overline{\sigma}(t) = \frac{1}{L} \int_0^L \sigma(t, z) dz \qquad (4)$$

Since the pore geometry of bed during filtration changes due to the suspended particle deposit inside the filter layer, that is, the bed becomes clogged, the filtrate steadily deteriorates, though it is at first improved. As the filter run proceeds the clogging gradually penetrates down through the bed. Therefore, the clogging of the filter is worst at the surface of the filter. The relationship between specific deposit σ and filter depth z in cm at the end of 20 hours' run, which has correlated by the author using the arrangement of the Eliassen's experimental data (Camp 1964), is given by

$$\sigma(20, z) = 0.185 \exp(-0.252z) \quad \text{at } t = 20 \text{ hours} \qquad (5)$$

By Eq. (5) $\overline{\sigma}$ is calculated as 0.0123 when L = 59.7 cm. This value is only 3% of ε_0 (= 0.408), though 28% for the upper layer (L = 4.26 cm). Anyway, for the usual filter length and run time $\varepsilon_0 \gg \overline{\sigma(t)}$. Then, from Eq. (3) V can be approximated to $AL\varepsilon_0$. This means that as a whole the decrease in pore volume of filter bed can be neglected to assume V is constant. Hence, from Eq. (2) $dV/dt = 0$ and $Q_{IN} = Q_{OUT} = Q(t)$. Consequently, Eq. (2) can be dropped from the governing equations.

For convenience let introduce the dimensionless quantities as follows:

$$\hat{t} = \frac{t}{T}; \quad \hat{C} = \frac{C}{C_M}; \quad \hat{C}_{IN} = \frac{C_{IN}}{C_M}; \quad \hat{Q} = \frac{Q}{Q_M}; \quad I = \frac{kS}{Q_M} \qquad (6, 7, 8, 9, 10)$$

where subscript M refers to the average value in an appropriate filtration time; and T = theoretical detention time (= V_M/Q_M). As V is approximated to be the constant value V_M, the governing equations are simplified as

$$\frac{d\hat{C}}{d\hat{t}} = -(\hat{Q} + I)\hat{C} + \hat{Q}\,\hat{C}_{IN} \tag{11}$$

Parameter I has the following meanings as: $I > 0$ -- tendency for deposition; $I = 0$ -- balance between deposition and scouring; and $I < 0$ -- tendency for scouring.

Parameter Estimation and Filtrate Prediction

If the parameter I is known, the effluent concentration $\hat{C}$ can be predicted by Eq. (11). Let define a vector variable $\pmb{x}$ and a vector function $\pmb{f}(\pmb{x})$ as

$$\pmb{x}^T = [\,\hat{C},\ I\,] \tag{12}$$

$$\pmb{f}^T(\pmb{x}) = [\,-(\hat{Q} + I)\hat{C} + \hat{Q}\,\hat{C}_{IN},\ 0\,] \tag{13}$$

If the estimation and prediction is made at appropriately short intervals, I can be assumed to be time-invariant in the interval, though it is globally time-varying. In that case the governing equation can be expressed as

$$\frac{d\pmb{x}}{d\hat{t}} = \pmb{f}(\pmb{x}) \tag{14}$$

Since $\pmb{f}(\pmb{x})$ is nonlinear with respect to $\pmb{x}$, it is difficult to obtain the general solution of Eq. (14). Using standard linearizing method (Bellman 1965), a linear differential equation for the (n+1)-th approximated solution $\pmb{x}^{(n+1)}$ is induced as follows:

$$\frac{d\pmb{x}^{(n+1)}}{d\hat{t}} = \pmb{J}^{(n)} \cdot (\pmb{x}^{(n+1)} - \pmb{x}^{(n)}) + \pmb{f}^{(n)}(\pmb{x}^{(n)}) \tag{15}$$

where $\pmb{J}^{(n)}$ is Jacobian matrix as

$$J^{(n)} = \begin{bmatrix} \partial f_1/\partial x_1 & \partial f_1/\partial x_2 \\ \partial f_2/\partial x_1 & \partial f_2/\partial x_2 \end{bmatrix} = \begin{bmatrix} -(\hat{Q}+I) & -\hat{C} \\ 0 & 0 \end{bmatrix} \quad (16)$$

The general solution of Eq. (15) is represented as

$$\boldsymbol{x}^{(n+1)}(\hat{t}) = \Phi^{(n+1)}(\hat{t}) \cdot \boldsymbol{x}^{(n+1)}(\hat{t}_0) + \boldsymbol{P}^{(n+1)}(\hat{t}) \quad (17)$$

where $\Phi^{(n+1)}(\hat{t})$ is the transition matrix given by the next equation

$$\frac{d\Phi^{(n+1)}}{d\hat{t}} = J^{(n)} \cdot \Phi^{(n+1)} \quad (18)$$

$$\Phi^{(n+1)}(\hat{t}_0) = I \quad (I : \text{unit matrix}) \quad (19)$$

and $\boldsymbol{P}^{(n+1)}(\hat{t})$ is the particular solution of Eq. (15) as given by

$$\frac{d\boldsymbol{P}^{(n+1)}}{d\hat{t}} = J^{(n)} \cdot (\boldsymbol{P}^{(n+1)} - \boldsymbol{P}^{(n)}) + \boldsymbol{f}^{(n)}(\boldsymbol{x}^{(n)}) \quad (20)$$

$$\boldsymbol{P}^{(n+1)}(\hat{t}_0) = \boldsymbol{0} \quad (\boldsymbol{0} : \text{zero vector}) \quad (21)$$

Computing the general solution of Eq. (17), the estimation of parameter I and the prediction of filtrate $\hat{C}$ are realized. For the sake of simplicity the superscripts (n) and (n+1) will be omitted hereafter.

$\boldsymbol{x}$ is obtained so as to minimize the objective function F as follows:

$$\frac{\partial F}{\partial \boldsymbol{x}(\hat{t}_0)} = 0 \quad (22)$$

$$F = \sum_{j=1}^{m} |X(\hat{t}_j) - \boldsymbol{x}(\hat{t}_j)|^2 \cdot W(\hat{t}_j) = \sum_{j=1}^{m} |X(\hat{t}_j) - [\Phi(\hat{t}_j) \cdot \boldsymbol{x}(\hat{t}_0) + \boldsymbol{P}(\hat{t}_j)]| \cdot W(\hat{t}_j) \quad (23)$$

where X = observed value of $\boldsymbol{x}$; and W = weighting matrix. From Eq. (22) $\boldsymbol{x}(\hat{t}_0)$ is given by

$$\boldsymbol{x}(\hat{t}_0) = [\sum_{j=1}^{m} \Phi(\hat{t}_j)^T \cdot W(\hat{t}_j)]^{-1} \cdot \{\sum_{j=1}^{m} \Phi(\hat{t}_j) \cdot W(\hat{t}_j) \cdot [X(\hat{t}_j) - \boldsymbol{P}(\hat{t}_j)]\} \quad (24)$$

Usually it can be assumed that W is unit matrix. As I is piecewise constant in each local interval, $I = x_2(\hat{t}_j) = x_2(\hat{t}_0)$. Then, I is obtained as

$$I = x_2(\hat{t}_0) = \frac{\sum_{j=1}^{m} \phi_{12}(\hat{t}_j) \cdot [X_1(\hat{t}_j) - P_1(\hat{t}_j)] - [\sum_{j=1}^{m} \phi_{11}(\hat{t}_j) \cdot \phi_{12}(\hat{t}_j)] \cdot x_1(\hat{t}_0)}{\sum_{j=1}^{m} \phi_{12}(\hat{t}_j) \cdot \phi_{12}(\hat{t}_j)} \quad (25)$$

where

$$\phi_{11}(\hat{t}_j) = g(\hat{t}_j) \tag{26}$$

$$\phi_{12}(\hat{t}_j) = -g(\hat{t}_j)\int_{\hat{t}_0}^{\hat{t}_j}\hat{C}\,g(u)^{-1}du \tag{27}$$

$$P_1(\hat{t}_j) = g(\hat{t}_j)\int_{\hat{t}_0}^{\hat{t}_j}(I\,\hat{C} + \hat{C}_{1N}\,\hat{Q})g(u)^{-1}du \tag{28}$$

$$g(\hat{t}_j) = \exp[-(\hat{Q} + I)(\hat{t}_j - \hat{t}_0)] \tag{29}$$

Since the right hand sides of Eqs. (25)-(29) contain unknown I, I may be computed iteratively starting with the assumed I for the right hand sides. The effluent concentration $\hat{C}$ can be predicted by inserting thus obtained I into Eq. (17).

Making use of the experimental data of Eliassen the estimation of parameter I (Fig. 1) and prediction of filtrate quality $\hat{C}$ (Fig. 2) have been carried out to verify the validity of the model developed herein and that of the estimation and prediction method. Estimated I is discrete-valued and should be represented by a sequence of short horizontal line, however, in the figure I is represented by a curve for the sake of simplicity. The interval of prediction is very small compared with that of the experimental data plotted. It is seen that the filtrate turbidity can be predicted fairly well.

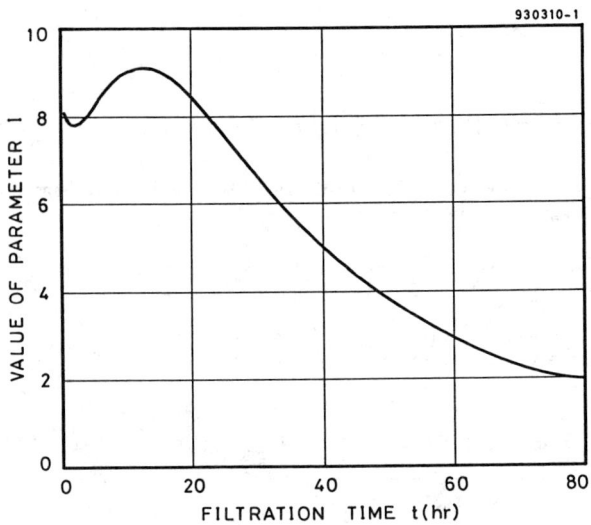

FIG. 1. TIME VARIATION OF ESTIMATED PARAMETER I

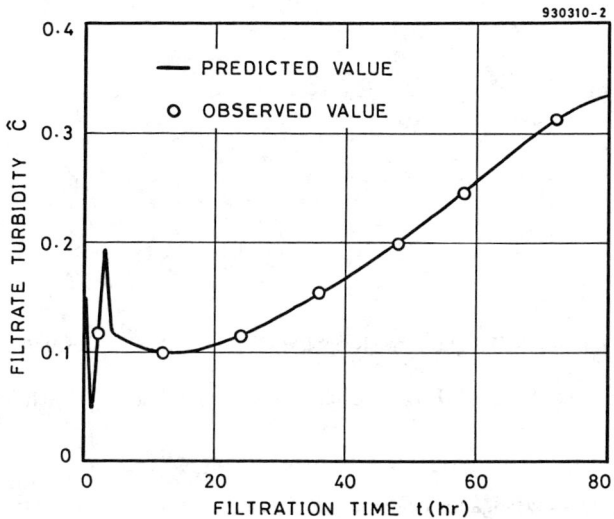

FIG. 2. TIME VARIATION OF FILTRATE TURBIDITY $\hat{C}$

As is seen from Figs. 1 and 2, the more the parameter I decreases in its value, the more the filtrate is deteriorated, that is, the higher the filtrate concentration $\hat{C}$ becomes. This relationship between I and $\hat{C}$ accords with the physical meaning of I which is considered in the model construction.

Conclusions

The conclusions drawn from this study are as follows:
(1) To predict the effluent concentration of rapid sand filters in water treatment plants a model and a method have been developed.
(2) The method of the prediction of filtrate concentration by the estimation of model parameter has been confirmed with the experimental data acquired by Eliassen.
(3) The model parameter can represent a kind of filter performance, i.e., ability of water pulification.

Appendix. References

Bellman, R. E., and Kalaba, R. E. (1965). *Quasilinearization and Nonlinear Boundary Value Problems*. American Elsevier, New York, N. Y., 76-84.
Camp, T. R. (1964). "Theory of Water Filtration." *J. Sanitary Engineering Division*, ASCE, 90(SA4), 1-30.

Repeatability and Oblique Flow Response Characteristics of Current Meters

Janice M. Fulford[1], Kirk G. Thibodeaux[1], and William R. Kaehrle[1]

Abstract

Laboratory investigations into the precision and accuracy of various mechanical-current meters are presented. Horizontal-axis and vertical-axis meters that are used for the measurement of point velocities in streams and rivers were tested. Meters were tested for repeatability and response to oblique flows. Both horizontal- and vertical-axis meters were found to under- and over-register oblique flows with errors generally increasing as the velocity and angle of the flow increased. For the oblique flow tests, magnitude of errors were smallest for horizontal-axis meters. Repeatability of all meters tested was good, with the horizontal- and vertical-axis meters performing similarly.

Introduction

Mechanical-current meters measure velocity by translating linear motion into angular motion through either a vertical or horizontal axis. Studies of current-meter accuracy and precision have been done by many investigators (Thibodeaux, 1992), but stream gagers debate as to which type is the better, horizontal- or vertical-axis meters. Most of these meter studies were published prior to 1960. Yarnell and Nagler's (1931) study on current meters, which is probably the most quoted, includes results of testing meters for response to oblique flows. The meters tested by Yarnell and Nagler have been modified in the subsequent years. This paper presents results of recent testing of mechanical meters that are currently used by stream gagers throughout the world.

Descriptions of test facility and meters tested

All tests were conducted in the submerged-jet tank at the U.S. Geological Survey's Hydraulic Laboratory at Stennis Space Center in Mississippi. The jet tank

---

[1]Hydrologist, U.S. Geological Survey, Stennis Space Center, MS 39529

Table 1. Characteristics of horizontal-axis meters. [mm millimeters, m meters]

Meter	impeller
Ott C-31, standard	125 mm dia., 0.25 m pitch, brass
Ott C-31, plastic	125 mm dia., 0.25 m pitch, plastic
Ott C-31, A	100 mm dia., 0.125 m pitch, brass
Ott C-31, R	100 mm dia., 0.25 m pitch, brass
Valeport BFM001	125 mm dia., 0.27 m pitch, plastic
PRC LS25-3A, metal	120 mm dia., 0.20 m pitch, aluminum
PRC LS25-3A, plastic	120 mm dia., 0.20 m pitch, plastic

is 120 feet long with a 12 X 12 feet cross section and can provide constant "live" velocities from 0.25 to 10 feet per second (ft/s). Velocity is determined by timed volumetric measurement and the area of the jet orifice. Meter angular velocity in revolutions per second (rev/sec) is determined by counting and timing revolutions using an electronic-counter timer.

Results are presented for five vertical-axis meters, Price[2] type-AA, optic Price type-AA, Price pygmy, winter Price type-AA, and winter Price type-AA with polymer rotor, and seven horizontal-axis meters, Ott[3] C-31 with metal, plastic, A, and R impellers, Valeport BFM001, People's Republic of China (PRC) LS25-3A with metal and plastic impellers. The vertical-axis meters have six conical cups that rotate about a shaft. The type-AA meters have a rotor 2 inches in height and 5 inches in diameter. The pygmy meter is two-fifths as large as the type-AAs. Price meters use a mechanical-contact switch, except for the optic meter, which uses a smaller optical switch. The winter meters are equipped with a modified type-AA yoke that facilitates its use under ice. The type-AA metal rotor has open cups. The polymer rotor has solid cups of the same dimensions as the type-AA metal rotor.

The horizontal-axis meters use a screw type impeller that rotates about a horizontal shaft. The PRC and Ott meters have a similar ball bearing and shaft assembly that requires oil lubrication. The Valeport's bearing surface is inside the impeller nose and uses water as the lubricant. Two Ott impellers, the A and R, are designed to measure the cosine component of oblique flows accurately. Other characteristics of the horizontal-axis meters tested are listed in table 1.

Repeatability Testing

The repeatability test is a measure of meter precision and measures how repeatable or consistently a meter measures velocity. For each meter ten measurements were made at each of five velocities, 0.25, 0.8, 1.5, 5 and 8 ft/s. Meters were

---

[2]Price type-AA meters and Price pygmy meters are U.S. Geological Survey meters.

[3]The use of trade names in this report is for identification purposes only and does not constitute endorsement by the U.S. Geological Survey.

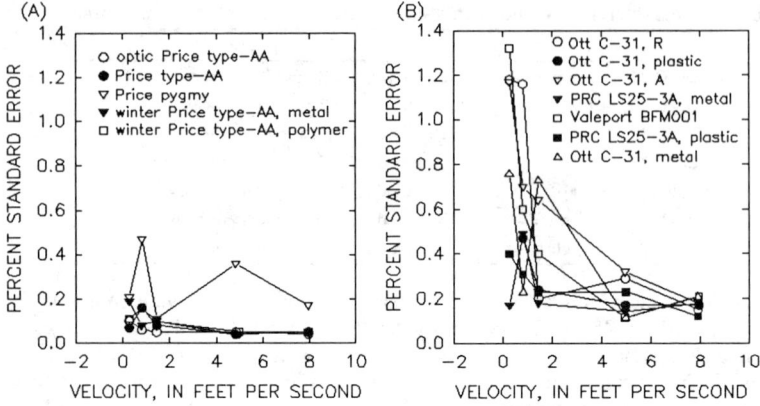

Figure 1. Percent standard error computed by velocity for (A) vertical-axis meters and for (B) horizontal axis meters.

positioned pointing into the flow at the center of the jet. For each of the five velocities, standard errors were computed. Percent standard errors were computed by dividing the standard errors by the jet velocity and multiplying by 100. In fig. 1 are plots of percent standard error versus jet velocity. For all meters, percent standard errors decrease with increasing velocity. The vertical-axis meters have a more consistent response than do the horizontal-axis meters. For all velocities tested the vertical-axis meters repeat with percent standard errors less than 0.5%. The Price-pygmy meter has the largest percent standard errors of the vertical-axis meters. Horizontal-axis meters repeat with percent standard errors of less than 1% for most velocities. At the lowest test velocity (0.25 ft/s) the Ott A, Ott R, and Valeport had percent standard errors from 1.2 to 1.3%. For the horizontal-axis meters, the PRC meters, and the plastic impeller Ott had the smallest percent standard errors.

Repeatability data were also fitted by a straight line using linear regression to determine their linear response. For each meter, a total of 50 measurements were regressed against the measured jet velocity. The computed regression coefficients and root mean squared errors (RMS) are listed by meter in table 2. Both horizontal- and vertical-axis meters generally have a similar range of RMS, and thus a similar linear response. The two PRC meters and the optic Price type-AA have the smallest RMS of the tested meters. The largest RMS is for the Ott with the A impeller. All meters had good linear responses (small RMS and percent standard errors).

Oblique flow response testing

The oblique flow response test is a measure of how accurately a meter measures the appropriate vector component of the flow. This test is also known as

Table 2.-Root mean squared errors and regression coefficients for repeatability data.

Meter	RMS error	slope	intercept	axis type
optic Price type-AA	0.0172	2.211	-0.014	vertical
Price type-AA	0.0350	2.263	-0.019	vertical
winter Price type-AA/metal	0.0204	2.294	-0.004	vertical
winter Price type-AA/polymer	0.0563	2.592	0.056	vertical
Price pygmy	0.0536	1.051	-0.055	vertical
PRC LS25-3A /metal	0.0186	0.654	0.009	horizontal
PRC LS25-3A /plastic	0.0173	0.653	-0.002	horizontal
Valeport BFM001	0.0405	0.870	-0.002	horizontal
Ott C-31 /metal	0.0507	0.840	0.087	horizontal
Ott C-31 /plastic	0.0450	0.835	0.007	horizontal
Ott C-31 /R impeller	0.0460	0.823	0.076	horizontal
Ott C-31 /A impeller	0.0656	0.421	0.127	horizontal

the cosine response because an ideal meter would register the cosine component of the angled flow. Each meter was tested at velocities of 0.25, 0.8, 1.5, 5, and 8 ft/s and at flow angles ranging from 90° to -90° in increments of 10°. Positive angles were flows directed downward onto the vertical-axis meters or from center to right side for horizontal-axis meters. At each combination of velocity and angle, two velocity measurements were made with each meter. Because only the meter and not the actual flow could be angled, the vertical-axis meters were positioned with the axis perpendicular to the force of gravity when testing for response to vertical angles of flow. Percent error for all types of meters was computed as $100 \times [\text{revs/sec}_\alpha \div (\cos\alpha \times \text{revs/sec}_0) - 1]$ where the subscripts α denotes the angle of flow and 0 straight flow. Vertical-axis meters were tested for response to vertical and horizontal angles of flow. Due to limited length of this paper, only the results of vertical-angle testing are shown. Stream gagers are unable to correct for errors due to vertically angled flow during field use of meters.

Both the vertical- and horizontal-axis meters under-register and over-register velocity depending on the angle and velocity of flow. In fig. 2 are plots of percent error averaged over the five test velocities versus angle. Most vertical-axis meters over-register for positive angles and under-register for angles between -40° and 0° the flow velocity. The winter Price type-AA meter with polymer rotor under-registered for all angles tested. Horizontal-axis meters except for the Ott with either the A or R impeller tended to under-register the velocity for most angles. For flow angles greater than 70°, all horizontal meters stalled. At angles between ±10° the vertical-axis meter errors range from -3.30% to -0.17% for the optic Price type-AA and from -7.87% to 8.92% for the Price pygmy. Errors for horizontal-axis meters range from 0.58% to 0.91% for the Ott with plastic impeller and from -2.02% to -3.77% for the PRC meter with plastic impeller. At angles of ±30° the vertical-axis meter errors range from -6.71% to 1.01% for the winter Price type-AA with metal rotor and from −31.83% to -33.97% for the winter Price type-AA with polymer rotor. For the horizontal-axis meters the errors range from -0.68% to 2.95% for the

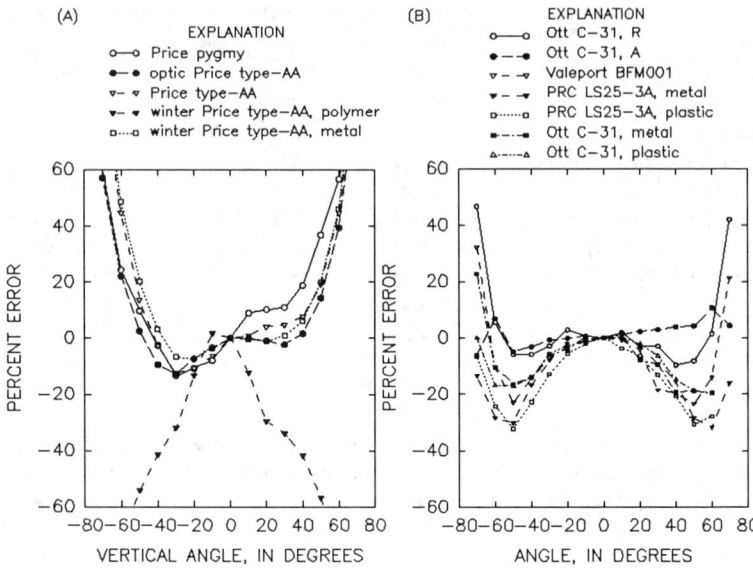

Figure 2. Average response for oblique flows for (A) vertical-axis meters and for (B) horizontal-axis meters.

2.95% for the Ott C-31 with A impeller and from -12.87% to -13.19% for the PRC meter with plastic impeller.

Error due to oblique flows increases slightly with velocity, except at the lowest velocity tested. At the lowest velocity tested, errors are larger than or nearly equal to the errors found for the highest velocity tested. In fig. 3 are plots of percent error versus jet velocity for 10° and 30° flows. Because the horizontal-axis meters stalled at low velocities in oblique flows, test results for 0.25 ft/s are omitted from the plot. Both groups of meters have larger errors at larger angles of flow. The Ott meters equipped with component impellers, A and R, have the smallest errors and the Price pygmy and the winter Price with polymer cup have the largest errors due to oblique flow.

Summary

Seven horizontal-axis meters and five vertical-axis meters were tested for linearity of response, repeatability, and response to oblique flows. Both horizontal- and vertical-axis meters tested had good repeatability (small RMS) and a similar linear response. Two horizontal-axis meters, the Ott meter with A and R impellers, had the smallest error in oblique flows. Except for the winter Price type-AA with polymer rotor, the vertical-axis meters over- and under-register oblique flows with

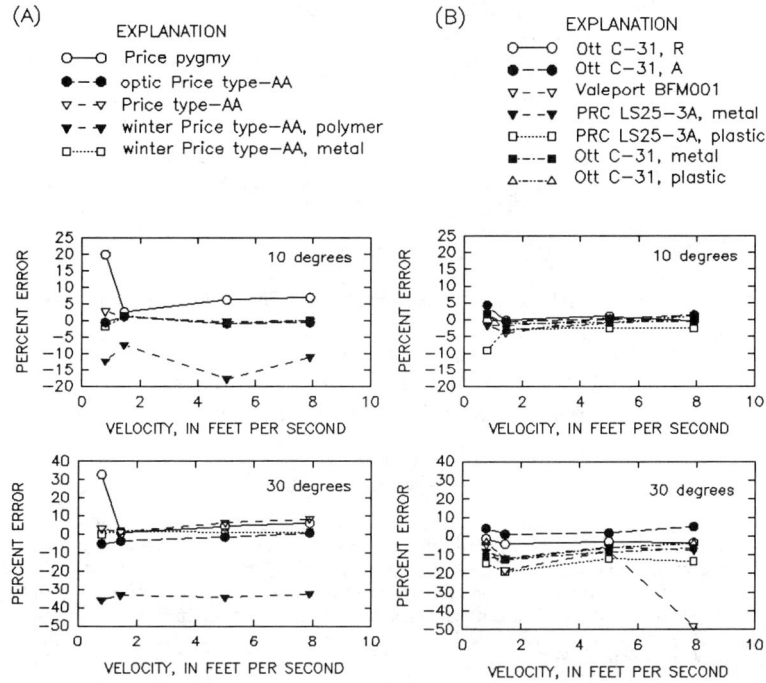

Figure 3. Response for 10° and 30° oblique flows for (A) vertical-axis meters and for (B) horizontal-axis meters.

angles between ±40°. Horizontal-axis meters tended to under-register oblique flows with angles between ±40°. The magnitude of error for horizontal-axis meters is usually smaller than those for vertical-axis meters in oblique flows. Test results were similar to those found by Yarnell and Nagler. However, because the range of oblique flows that occur in streams is unknown, it is impossible to determine whether horizontal-axis meters are better than vertical-axis meters for practical stream gaging.

References

Thibodeaux, Kirk G. (1992). "A brief literature review of open-channel current meter testing," Proc. of the Hydraulic Engineering sessions at Water Forum '92", ASCE, New York, NY, 458-463.

Yarnell, David L. and Nagler, Floyd A. (1931). "Effect of turbulence on the registration of current meters." Trans. of the ASCE, 766-860.

BED SHEAR STRESS IN UNSTEADY OPEN-CHANNEL FLOWS

Iehisa NEZU[1], Member, ASCE, Hiroji NAKAGAWA[1],
Yoshihiro ISHIDA[2] and Akihiro KADOTA[1]

ABSTRACT

In this study, velocity distributions over smooth and rough beds were accurately measured with an LDA in unsteady open-channel flows. The log-law distributions yielded well in the wall region for both the rising and falling stages of flood. The bed shear stress was evaluated reasonably from the log-law. These variations against the flood time coincided well with those evaluated from the momentum equation and also with the Reynolds stress measured directly. Of particular significance is the findings of counterclockwise loop of bed shear stress against the flow depth in unsteady flows.

INTRODUCTION

Many researches on turbulent structures and the associated sediment transport in *steady* open-channel flows have been conducted intensively in the last 20 years by making use of hydrogen-bubble techniques, hot-film anemometers and laser-Doppler anemometer (LDA). The comprehensive knowledge on these topics is reviewed in the IAHR-monograph written by Nezu and Nakagawa (1993a). Nezu *et al.*(1993b) have recently verified that accurate LDA data obtained in laboratory flumes can also be applied to flow structures of actual rivers and estuaries at high Reynolds numbers, including the complex features of secondary currents and turbulence. This suggests strongly that the accurate laboratory data of open-channel flows are very contributory to river engineering in steady field flows.

In contrast, rivers in flood are characterized by unsteadiness. It is often observed in rivers that the relation between the flow depth h(t) and the discharge Q(t) does not show a single curve, but a loop curve in the stronger *unsteady* flow. The significant difference of flow and the associated sediment transport between the rising and falling stages of flood has been pointed out by many researchers. Hayashi *et al.*(1988) have first conducted turbulence measurements in unsteady open channel flows by making use of hot-film anemometers; their unsteady equipment was controlled by a mechanical valve. They suggested that turbulence might be stronger in the rising stage than in the falling one. Tu and Graf(1992) used three micropropellers to measure velocities over gravel bed in unsteady open-channel flows. They found the noticeable feature that the shear stress became considerably larger in the rising stage than in the falling one.

---

1) Dept. of Civil and Global Environment Eng., Kyoto University, Kyoto 606, Japan.
2) Hankyu Railway Inc., Osaka 530, Japan.

Nezu and Nakagawa(1991) have made a new hydraulic circulation system of water discharge that was automatically controlled by a microcomputer; this system can reproduce any hydrograph of flood in *unsteady* open-channel flows. They have successfully conducted then accurate measurements of velocity components by making use of two-component LDA system; no probe (without laser light) in an LDA is especially suitable to measure unsteady flows because its device never disturbs the flow. Nezu and Nakagawa found the loop properties of the h-Q curve and some turbulence characteristics in unsteady open-channel flow over smooth bed.

In this study, mean velocity distributions over smooth and rough beds were accurately measured with an LDA in *unsteady* open-channel flows. The log-law distributions yielded well in the wall region for both the rising and falling stages of flood. The bed shear stress was evaluated reasonably from the log-law. These variations against the flood time coincided well with those evaluated from the momentum equation and also with the Reynolds shear stress measured directly.

THEORY OF BED SHEAR STRESS τ_b

The momentum equation of unsteady open-channel flows is obtained from the N-S equation or the Reynolds equation, (e.g., see IAHR-monograph of Nezu and Nakagawa 1993), as follows:

$$\frac{\partial U}{\partial t} + \frac{\partial U^2}{\partial x} + \frac{\partial UV}{\partial y} = gS_s + \frac{\partial}{\partial y}(\frac{\tau}{\rho}) \quad (1) \qquad S_s \equiv \sin\theta - \cos\theta \frac{\partial h}{\partial x} \quad (2)$$

U and V are the mean velocity components, u and v the turbulence components in the streamwise and vertical directions, respectively. $S_b = \sin\theta$ is the bed slope, and h is the flow depth. Because $|\partial U/\partial t| \gg |\partial U^2/\partial x|, |\partial UV/\partial y|$ in strong unsteady flow, the total shear stress τ can be given by integrating (1) from y to the free surface, i.e., y=h.

$$\frac{\tau}{\rho} = gS_s(h-y) + \int_h^y \frac{\partial U}{\partial t} dy \quad (3) \qquad \therefore \frac{\tau_b}{\rho} = gS_s h - \int_0^h \frac{\partial U}{\partial t} dy \approx gS_s R - \frac{\partial Q}{B \partial t} \quad (4)$$

The hydraulic radius R is adopted for the side-wall effects. Since the depth gradient $\partial h/\partial x$ is approximated by $-(1/c)\partial h/\partial t$, the surface slope S_s is given by $S_b + (1/c)\partial h/\partial t$. c is the celerity of flood wave; $c=(gh)^{1/2}$ for long wave. Therefore, S_s is larger in the rising stage of flood, whereas it is smaller in the falling stage.

UNSTEADINESS PARAMETER α

It is necessary to define a reasonable unsteadiness parameter that characterizes the effects on velocity profiles and turbulence in flood flows. Nezu et al.(1993b) have proposed the following new parameter α that correlates with the pressure gradient dP/dx.

Assuming the hydrostatic pressure $P=\rho g(h-y)$ in unsteady flows, one can obtain

$$\alpha \equiv \frac{1}{U_c} \frac{h_p - h_b}{T_d} \approx \frac{1}{\rho g} \frac{dP}{dx} \text{ for rising stage (5), since } \frac{1}{\rho g}\frac{dP}{dx} = \frac{1}{U}(1-\frac{U}{c})\frac{\partial h}{\partial t} \approx \frac{1}{U}\frac{\partial h}{\partial t} \quad (6)$$

where, T_d is the duration time from the base flow depth h_b to the peak depth h_p. U_c is the convection velocity of turbulent eddies; it is roughly equal to $(U_b+U_p)/2$. The parameter α is the ratio of the rising speed $V_s \equiv (h_p-h_b)/T_d$ of flood water surface to the convection velocity U_c. α also implies that the adverse pressure gradient is negative in the rising stage of flood, which corresponds to the deceleration period of unsteady pipe flow.

ANALYTIC METHOD OF UNSTEADY FLOWS

One of the most difficult aspects to investigate turbulent structures in unsteady flows is how to separate the mean-velocity component $U(t)$ and the turbulence component $u(t)$ from the instantaneous velocity, $U(t)+u(t)$. Three kinds of methods to define mean velocity are considered as follows: (a) *the ensemble average method*, (b) *the moving time-average method*, and (c) *the Fourier-component method*. The ensemble average method is the best one theoretically, in which the same phase of flow must be used as a trigger of sampling. This method is often used in periodic pipe flows and oscillating boundary layers, e.g., see a comprehensive review by Carr(1981).

Nezu and Nakagawa(1991) have examined the applicability of these three methods to unsteady open-channel flows and concluded that the Fourier-component method was the most suitable for flood surface flows because it was very difficult to determine accurately the same phase of flood waves from the wave gauges. The cutoff frequency of Fourier components for mean-velocity component was reasonably chosen to be much smaller by order of 10^{-3} than the bursting frequency. The present study adopted the same size of Fourier components, m=7, for mean velocity as used in Nezu and Nakagawa(1991).

EXPERIMENTAL EQUIPMENT AND HYDRAULIC CONDITIONS

The experiments were conducted in a 10m long, 40cm wide and 50cm deep tilting flume, which was the same as used in Nezu and Nakagawa(1991). In this water flume, the flood discharge can be automatically contolled by a microcomputer in which the rotation speed of a water pump motor involving a transistor inverter is controlled by the feedback from the signals of a electromagnetic flow meter. If any hydrograph of flood flow is input through a disc or a keyboard into the computer, the corresponding discharge can be reproduced accurately in this circulation system.

Two components of velocities (u, v) were measured with a three-beam LDA system operated in the forward-scattering differential mode (DANTEC-made). The flow depth $h(t)$ was measured by two sets of condensor-type water-wave gauges. The slope S_s of water surface in flood could be evaluated from simultaneous measurements of water depth using these two gauges; the distance between the two gauges was set 1.5m. The typical flood waves were given in Nezu et al.(1993a). All of the output signals of LDA and gauges were recorded in digital form with sampling frequency f=200-400Hz and sample size N=10000-50000, depending on hydraulic conditions. Statistical analyses were conducted then using a large digital computer in the Data Processing Center.

4 runs of experiments were systematically conducted in the duration time of flood, T_d= 30s, 60s, 90s, and 120s, for each smooth bed ($S_b=10^{-4}$) and rough bed($S_b=10^{-3}$). The depth h_b and bulk velocity U_b of the base flow were kept constant, i.e., h_b=4.0cm, U_b=15.6cm/s. In each case, the ratio of the peak to base discharges, Q_p/Q_b, was changed as Q_p/Q_b=3, 4, and 5. From these systematical experiments, the unsteadiness parameter α varied as $(0.68\text{-}2.25)\times 10^{-3}$. More detailed information is given in Nezu et al.(1993a).

RESULTS AND DISCUSSION

Velocity Distributions

Fig.1 shows an example of velocity distributions over smooth bed in the case of the strongest unsteadiness, i.e., α=2.25×10^{-3}. The figures on the left-hand side indicate the velocity profiles in the rising stage, i.e., 0≤T≤1, where T≡ t/T_d is the normalized flood time, and the figures on the right-hand side indicate those in the falling one, i.e., 1≤T≤2. The measured values of mean velocity U(y: T) in the wall region (y/h<0.2) obeyed very well the log-law distribution, irrespective of the rising and falling stages for all runs of the present experiments. Therefore, the friction velocity U∗ can be evaluated accurately from the log-law, which is described by Nezu and Rodi(1986), as follows:

$$\frac{U}{U_*} = \frac{1}{\kappa} \ln(y^+) + A + w\left(\frac{y}{h}\right) \quad (7) \qquad w\left(\frac{y}{h}\right) = \frac{2\Pi}{\kappa} \sin^2\left(\frac{\pi y}{2h}\right) \quad (8)$$

where, $U^+ \equiv U/U_*$, $y^+ \equiv yU_*/\nu$. κ is the von Karman constant and A is the integration constant; $\kappa=0.41$ and $A=5.0$-5.3. $w(y/h)$ is the Coles' wake function which should be applied to the outer region $(y/h>0.2)$. The curve of (7) is indicated in all figures of Fig.1; each figure is shifted by 10 units in U^+ to avoid confusion. The wake parameter Π became zero in the steady base flow (T<0) because this Reynolds number was retatively low, i.e., Re=6250; see Nezu and Rodi(1986). Nevertheless, the value of Π increases with T and attains a maximum ($\Pi \approx 0.3$) before the flow depth attains a peak depth h_p. On the other hand, it decreases in the falling stage and comes back to zero at T=2. This noticeable feature was also observed over rough beds by Nezu et al. (1993a), but more detailed discussion is not made here because this study focuses on the wall region and especially on the bed shear stress. The velocity in the very near wall ($y^+<30$) deviates below from the low-law due to viscosity in the beginning and ending of flood, but the log-law tends to extend toward the wall as the flow depth h(t) increases. This suggests strongly that the viscous sublayer in *unsteady* open-channel flow may be smaller than that of steady base flow due to a strong acceleration effect, i.e. Eq.(5).

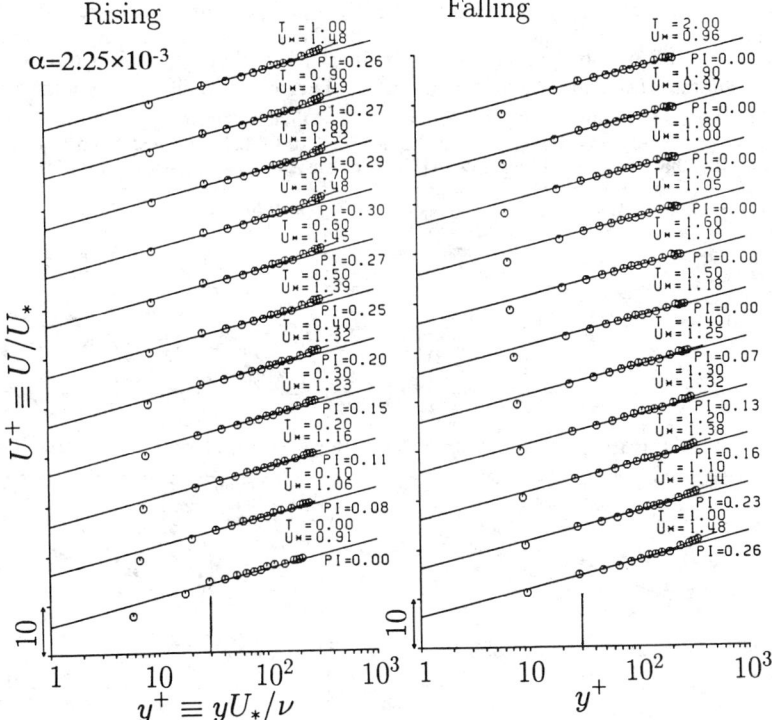

Fig.1 Mean Velocity Distributions over Smooth Bed in Unsteady Open-Channel Fow.

Variations of Bed Shear Stress

Fig.2 (a) and (b) show the variations of bed shear stress τ_b against the flood time $T \equiv t/T_d$, for smooth and rough beds, respectively. These values are normalized by its time-average value $\overline{\tau_b}$. The value of $\tau_b/\overline{\tau_b}$ increases with the time T and attains a maximum before the peak depth appears. On the other hand, it decreases monotonically with T in the falling stage. Of particular significance is that the maximum value of τ_b is larger and also the time lag between the peaks of the shear stress τ_b and the depth h is larger, as the unsteadiness parameter α is stronger. For example, the peak value of τ_b is about $2\overline{\tau_b}$ and $1.8\overline{\tau_b}$ over smooth and rough beds, respectively, for $\alpha = 2 \times 10^{-3}$; these maximum values attain about 4 times as large as that of base flow (T<0). These findings suggest strongly that various sediment transports including both bed load and suspended load in unsteady flood flows become stronger in the rising stage than the corresponding falling stage; especially, the bed shear stress and friction velocity attain peaks before the peak depth.

Loop Property of Bed Shear Stress v.s. Flow Depth

Fig.3 (a) and (b) show the variations of bed shear stress τ_b v.s. the change of flow depth $\Delta h \equiv (h-h_b)$ for smooth and rough beds, respectively. These values are normalized by the corresponding peak values. The counterclockwise loops are clearly seen due to the time lag between the peaks of τ_b and Δh. These loops of bed shear stress $\tau_b(t)$ are similar to those of discharge Q(t) observed in flood rivers. Of particular significance is the difference of loop property between the rising and falling stages due to the unsteadiness. It should be noted that the effect of unsteadiness on τ_b/τ_{bmax} at the same value of $\Delta h/\Delta h_p$ is much stronger in the falling stage than in the rising stage, irrespective of wall roughness. This implies that the decrease of bed shear stress becomes much faster than that of the subsiding flow depth as the unsteadiness of flow is larger.

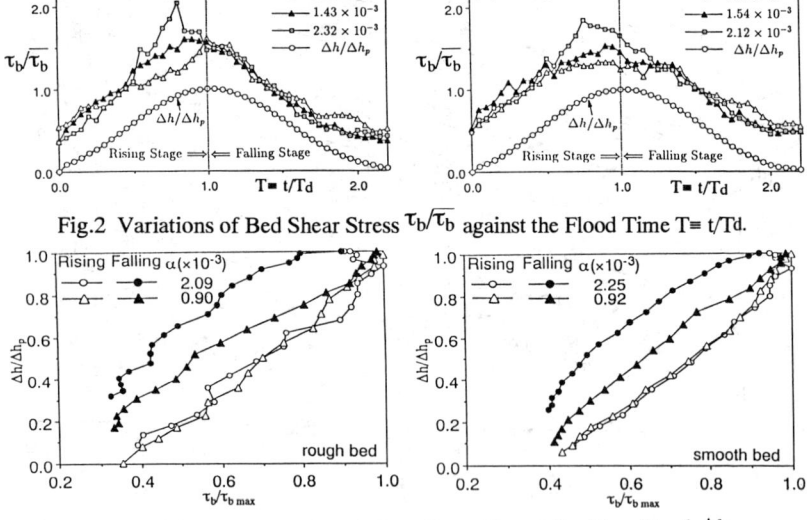

Fig.2 Variations of Bed Shear Stress $\tau_b/\overline{\tau_b}$ against the Flood Time $T \equiv t/T_d$.

Fig.3. Loop Property of the Bed Shear Stress τ_b v.s. the Flow Depth Δh.

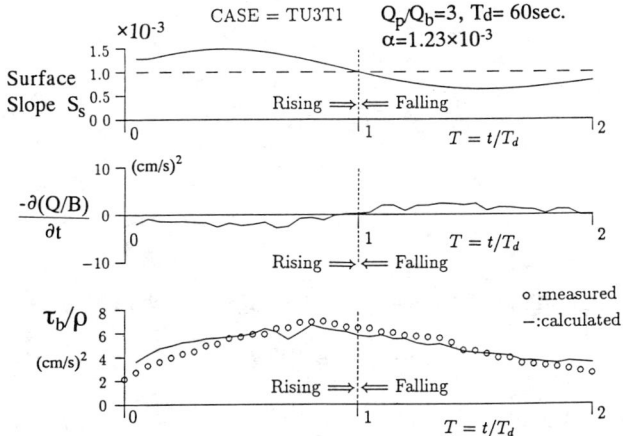

Fig.4. Comparison between the observed values of τ_b/ρ and the theoretical ones.

COMPARISON BETWEEN THEORY AND EXPERIMENTS

Fig.4 shows a comparison between the observed values τ_b and the theoretical ones. The surface slope S_s was calculated from $S_b + (1/c)\partial h/\partial t$, because it was less accurate to obtain $\partial h/\partial x$ from the two gauges, see Nezu et al.(1993a). The observed values are in a good agreement with the theoretical ones calculated from (4). The unsteady term $-\partial Q/\partial t$ is negative contribution to τ_b, at most 30% of τ_b. The directly measured data of Reynolds stress $-\overline{uv}$ also coincided well with the momentum Eq.(3). The shear stress increases in the rising stage, whereas it decreases in the falling one. The turbulence structures in *unsteady* open-channel flows will be discussed in Nezu & Nakagawa(1993b).

CONCLUSION

In this study, velocity distributions over smooth and rough beds were accurately measured with an LDA in *unsteady* open-channel flows. The log-law distributions yielded well in the wall region for both the rising and falling stages of flood. The variations of bed shear stress coincided well with those calculated from the momentum equation and also with the Reynolds stress measured directly. Of particular significance is the findings of counterclockwise loop of bed shear stress against the flow depth.

REFERENCES

1) Carr, L.W.(1981): A review of unsteady turbulent boundary-layer experiments, *Unsteady Turbulent Shear Flows* (eds. R. Michel et al.), Springer-Verlag, pp.3-34.
2) Hayashi, T., Ohashi, M. & Oshima, M.(1988): Unsteadiness and turbulence structure of a flood wave, *20th Symp. on turbulence*, pp.154-159, (in Japanese).
3) Nezu, I. & Nakagawa, H.(1991): *Int. Symp. on Transport of Suspended sediments and Its Mathematical Modelling*, IAHR, Florence, Italy, pp.165-190.
4) Nezu, I. & Nakagawa, H.(1993a): *Turbulence in Open-Channel Flows*, IAHR-Monograph, Balkema.
5) Nezu, I. & Nakagawa, H.(1993b); *9th Int. Symp. on Turbulent Shear Flows*, Kyoto, Japan.
6) Nezu, I. & Nakagawa, H., Ishida, Y. & Fujimoto, H.(1993a); 25th IAHR Congress, Tokyo, Japan, (to be published).
7) Nezu, I. & Rodi, W.(1986);*J. Hydraulic Eng.*, ASCE, vol.112, pp.335-355.
8) Nezu, I., Tominaga, A. & Nakagawa, H.(1993b); *J. Hydraulic Eng.*, ASCE, No.5.(in printing).
9) Tu, H. & Graf, W.H.(1992): Vertical distribution of shear stress in unsteady open-channel flow, *Proc. Inst. Civ. Engrs Wat., Marit. & Energy*, vol.96, pp.63-69.

Limitations of Reduced Scale Testing on Parshall Flumes

Steven J. Wright[1]

Abstract

A model for the boundary layer development in the entrance section of a Parshall flume is developed to account for the effects of fluid viscosity on flume calibration equations. Data for a 30.5 cm (1 ft) throat width flume and a one-third scale model are analyzed to demonstrate that Froude scaling relations can be used to predict calibrations from reduced scale models only after boundary layer corrections are applied. These corrections are most significant at low discharges and recommendations are made for flow conditions above which viscous effects are negligible.

Introduction

During calibration of a 1/3 scale model of a one foot (0.305 m) Parshall flume, difficulties were encountered in reproducing the standard rating equation proposed by Parshall (1926). Since it appeared that the discrepancies might be due to viscous effects that are not properly accounted for in a Froude scaled model, Wright and Taheri (1990) attempted to develop a boundary layer correction that would allow scaling of the data from the 1/3 scale flume to compare with those from a full size flume. This was based on the approach by Keller (1984) in which an effective depth was determined by reducing the actual flow depth by a displacement thickness computed from the flat plate boundary layer solution. However, the use of the flat plate, zero-pressure gradient boundary layer solution in the converging section of the Parshall flume led to unsatisfactory results. A more detailed analysis was developed by Wright, et al (1993) that accounts for the influence of the flume geometry on the boundary layer development. The model is applied in this paper to assess the validity of using reduced scale models for prediction of calibration equations for Parshall flumes.

Background

The Parshall flume is basically a critical depth meter which creates a unique depth-discharge relation at the measurement section so long as unsubmerged flow occurs in the throat. A theoretical expression relating the discharge to the water depth at the measurement location upstream

---

Prof. of Civil and Envir. Eng., Univ. of Michigan, Ann Arbor, MI 48109.

from the throat (assuming that critical flow occurs at the entrance to the throat) is developed as:

$$Q = B_T \sqrt{g} \left[\frac{2}{3}\left(H_a + \frac{V_a^2}{2g}\right)\right]^{3/2} \qquad (1)$$

Here Q is the total discharge, H_a is the water surface elevation (relative to the floor elevation at the throat entrance) at the upstream measurement section a, $V_a^2/2g$ is the velocity head at the upstream location and B_T is the throat width. One can define a discharge coefficient C_D by dividing the discharge by the right hand side of Eq. 1; a plot of C_D versus H_a/B_T for the experimental data reported by Wright and Taheri (1991) is presented in Fig. 1. The plot indicates a decline in C_D with decreasing head, a result which might be due to the effect of fluid viscosity since viscous effects would be more important at low heads and reduce the discharge. There is some additional factor other than viscosity present in these results since C_D values are greater than 1.0 at large heads. It is presumed that this deviation from theory is due to a nonhydrostatic pressure variation at the throat section. The drop in the flume floor creates pressures less than hydrostatic, resulting in discharges greater than predicted by Eq. 1.

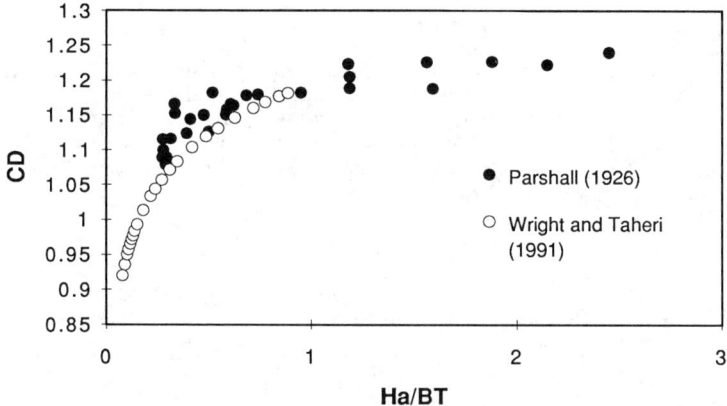

Figure 1. Discharge Coefficient for One Foot Parshall Flume.

Model Development

The theoretical model is based upon two concepts; the conservation of energy in the main flow and the development of a boundary layer along the bottom and sides of the Parshall flume. An effective flow depth y is defined from continuity considerations as y = Q/(BV) with Q the discharge, B the local width and V the local uniform velocity outside the boundary layer. The actual depth is computed by adding the displacement thickness δ^*:

$$y_a = y + \delta^* \qquad (2)$$

with
$$\delta^* = \int_0^\delta \left(1 - \frac{u}{V}\right) dz \tag{3}$$

where u is the local velocity within the boundary layer and z is elevation above the flume floor. The flow width B is also reduced by $2\delta^*$ from its nominal value to account for boundary layer development along both side walls of the flume. The specific energy E of the flow is given by

$$E = y_a + \frac{V^2}{2g} = \frac{Q}{BV} + \delta^* + \frac{V^2}{2g} \tag{4}$$

Assuming that E is conserved in the entrance region of the flume, a differential equation for the variation of V is developed from Eq. 4:

$$\frac{dV}{dx} = \frac{QgV\frac{dB}{dx}}{B(V^3B - gQ)} - \frac{d\delta^*}{dx} \frac{gBV^2}{V^3B - gQ} \tag{5}$$

where dB/dx is determined from the geometry of the flume approach section. As implied by Fig. 1, it is inappropriate to assume that E is equal to the energy at the critical flow condition; rather it is assumed that

$$E = C_p E_{critical} = C_p \left(\frac{Q^2}{gB_T^2}\right)^{1/3} \tag{6}$$

with C_p a so-called pressure correction factor with a value less than one to account for the nonhydrostatic pressure distribution at the entrance to the throat section. A relation for C_p was determined from experimental data and is reported by Wright, et al (1993). The boundary layer displacement thickness δ^* is computed from the integral momentum equation for the boundary layer in a nonuniform flow:

$$\tau_0 = \mu \frac{du}{dy}\bigg|_{y=0} = \rho \frac{d(V^2\delta_m)}{dx} + \rho V \frac{dV}{dx} \delta^* \tag{7}$$

with the momentum thickness δ_m defined by

$$\delta_m = \int_0^\delta \left(1 - \frac{u}{V}\right) \frac{u}{V} dz \tag{8}$$

The boundary layer is assumed to be laminar as the Reynolds number based upon distance from the entrance to the flume is less than 500,000 for the flume sizes and discharges considered in this study. In order to avoid a computational singularity in the equations at x=0 where $\delta^* = 0$, the differential equation is developed in terms of δ^{*2}:

$$2\delta^* \frac{d\delta^*}{dx} = \frac{d\delta^{*2}}{dx} = \frac{2\delta^{*2}}{\delta_m} \left(\frac{\mu \frac{du}{dy}\big|_{y=0}}{\rho V^2} - \frac{1}{V} \frac{dV}{dx} (2\delta_m + \delta^*)\right) \tag{9}$$

Eq. 9 is simplified by using the Blasius solution for flat plate boundary layers to provide the self-similar velocity profile:

$$\frac{d\delta^{*2}}{dx} = \frac{C_1 \mu}{\rho V} - \frac{C_2 \frac{dV}{dx} \delta^{*2}}{V} \qquad (10)$$

with $C_1 = 2.96$ and $C_2 = 9.183$.

Solutions of Eqs. 5 and 10 were implemented utilizing a standard routine for the numerical integration of nonlinear ordinary differential equations. The computation procedure consisted of the following steps:
1. Compute the energy in the entrance section of the flume for a specified Q using Eq. 6. The depth at the flume entrance (the beginning of the converging section, $x = 0$) was computed as the subcritical alternate depth at that energy level;
2. Numerically integrate Eqs. 5 and 10 downstream to the throat and compute an effective throat width by reducing the nominal throat width by $2\delta^*$;
3. Using the reduced throat width, repeat step 1 and integrate Eqs. 5 and 10 downstream to compute δ^* at the measurement section. The actual depth y_a at that section is determined from Eq. 2 and, assuming a level flume floor, is set equal to H_a;

Experiments

Experimental data collected in the 1 foot (0.305 m) Parshall flume were reported previously by Wright and Taheri (1991). Data were also available from tests on a 1/3 scale model of this flume. Standard dimensions presented by Chow (1959) were reproduced in the construction of the painted plywood flumes with extreme care in the construction of the throat dimensions. The as-built throat widths were measured and the actual throat width was used in the data analysis.

A critical measurement at low discharges is the determination of H_a. From the theoretical consideration in the development of Eq. 1, H_a should be the elevation difference between the water surface at the measurement section *a* and the floor of the flume section where critical flow occurs, so it is important to install the flume as level as possible. The critical flow section was assumed to be at the entrance to the throat. The zero level for the measurement of H_a was determined by blocking the downstream end of the flume and filling with an arbitrary volume of water. Water depths at intervals across the throat entrance section were measured with a point gage (variations were on the order of 0.001 ft) and an average depth computed. The zero level for all subsequent measurements of H_a was computed by subtracting this depth from the point gage reading measured in the stilling well at the measurement section. Until this procedure was developed, it was not possible to precisely replicate experimental results if the flume was moved or re-leveled.

Discharges were determined from calibrated venturi meters installed in the constant head supply lines. Measurements were made only for flow rates less than about 25 percent of the maximum discharge recommended for this flume size.

Experimental Results

A dimensionless comparison of the data for the one foot (0.305 m) Parshall flume and for the 1/3 scale model in Fig. 2 indicates that Froude scale modeling would be inappropriate for determining the head-discharge relation for Parshall flumes at the low discharges discharges. At a dimensionless head H_a/B_T of 0.5, for example, the discharge coefficient is more than 5 percent greater for the full scale flume than for the reduced scale model. The difference between the two is still around 2-3 percent at $H_a/B_T = 1.0$ where δ^* is only about one percent of the flow depth in the reduced scale model. A method for scaling the reduced scale model to the prototype size is as follows:

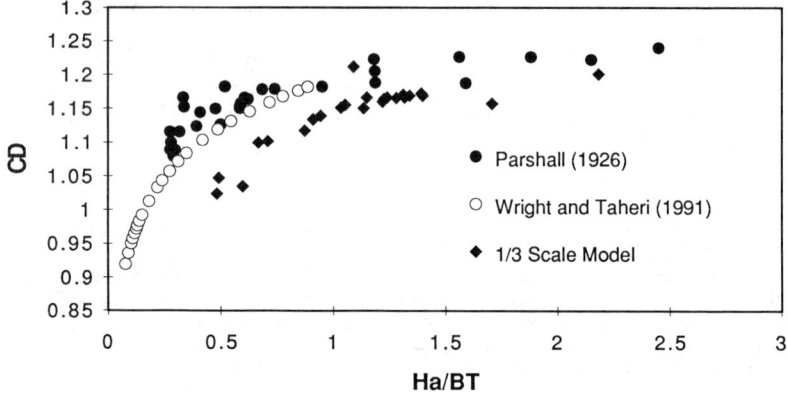

Figure 2. Experimental Data for 1/3 Scale and Prototype Parshall Flume.

1.) Reduce the head H_a in the scale model measurements by the displacement thickness predicted by the numerical model. This effective depth can then be scaled to the prototype size by the geometric scale;

2.) The computed prototype effective depth is increased by δ^* predicted by the numerical model at the prototype condition;

3.) The discharge is scaled by conventional Froude scale relations with an additional correction accounting for the reduction in effective width at the throat by the side wall boundary layer development.

Fig. 3 presents the results of applying this procedure to the data in Fig. 2. It is seen that the agreement between the two data sets is very good once the boundary layer corrections have been made. Since the boundary layer corrections are made to both side walls and the depth, it is not possible to generalize the findings to different flume sizes since the width correction will be less important for larger flume sizes.

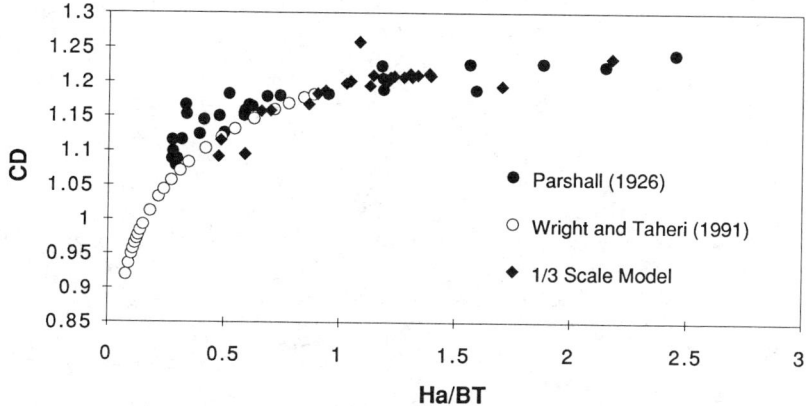

Figure 3. Results of Scaling the Reduced Scale Model to Prototype Size.

Summary and Conclusions

A comparison of experimental data for a 1/3 scale and a full size one foot Parshall flume indicates that viscous effects influence the flume calibrations for discharges less than 20 percent of the recommended maximum capacity. Reduced scale models should not be used to determine the flume calibrations for these conditions unless an attempt to correct for the viscous effects is made. The model for boundary layer development presented by Wright, et al (1993) is capable of accounting for the boundary layer development and allows for a proper scaling of the reduced scale model results to the prototype flume.

APPENDIX I - REFERENCES

Chow, V.T., (1959) *Open Channel Hydraulics*, McGraw Hill Book Co.

Keller, R.J., (1984) *Cut-Throat Flume Characteristics*, Journal of Hydraulic Engineering, Vol. 110, No. 9, Sept. 1984, pp. 1248-1263.

Parshall, R.L., (1926) *The Improved Venturi Flume*, Transactions ASCE, Vol. 89, pp. 841-851.

Wright, S.J. and Taheri, B. (1990), *Limitations to Standard Parshall Flume Calibrations*, Proceedings National Conference on Hydraulic Engineering, San Diego, Ca., pp. 915-920.

Wright, S.J. and Taheri, B. (1991) *Correction to Parshall Flume Calibrations at Low Discharges*, Journal of Irrigation and Drainage Engineering, Vol. 117, No. 5, pp. 800-804.

Wright, S.J., Tullis, B.P. and Long, T.M. (1993) *Recalibration of Parshall Flumes at Low Discharges*, in press, Journal of Irrigation and Drainage Engineering

Urban Storm Water Instrumentation :
A Field Observation

R. Brad Jennings, A.M. ASCE[1]

ABSTRACT

The Dallas-Fort Worth, Texas metroplex has 7 cities with populations of at least 100,000 that are required to meet National Pollutant Discharge Elimination System (NPDES) permitting requirements outlined in section 402(p) of the Water Quality Act of 1987, (USEPA, 1990). The United States Geological Survey (USGS), Texas District, in association with the North Central Texas Council of Governments (NCTCOG), has established a regional network of storm water runoff sampling gages in the metroplex. The gaged network is designed to collect and transmit water level, flow, rainfall and sampling data to a central location from remote gaged sampling stations. A pilot program was initiated to test and evaluate three different automated storm water monitoring and sampling systems at the studies' gaged sites.

INTRODUCTION

Storm water runoff in urban and industrialized areas carries a considerable amount of pollutants to nearby lakes and streams. In an effort to control this source of pollution, the Environmental Protection Agency (USEPA) issued regulations in November, 1990, requiring cities and incorporated areas with populations 100,000 or greater to undertake a process to obtain permits that monitor and control the quality of their storm runoff. The Dallas-Fort Worth metroplex contains 7 cities which

---

[1]Environmental Engineer, USEPA, Dallas, TX, formerly with USGS, Fort Worth, TX

must meet these regulations and obtain discharge permits for their municipal storm sewer systems.

The purpose of this paper is to present the general findings of the testing and evaluation for the three different automated storm water monitoring and sampling systems that are in use in the Dallas-Fort Worth, Texas metroplex NPDES storm water quality study.

SCOPE OF THE FIELD STUDY

The Dallas-Fort Worth metroplex is a growing urban center in north central Texas with an estimated 1990 population of 4.1 million. There are 7 cities with populations of at least 100,000 covering an area of over 1000 square miles.

The USGS is assisting the NCTCOG, representing these metroplex cities, in meeting the water science requirements outlined in permitting regulations. Specifically, the USGS is undertaking the establishment and monitoring of a 30 gage regional network to collect data from over 200 storm events. This data will provide information to characterize storm water quality for the entire metroplex and to satisfy a portion of Part 2 of the NPDES storm water discharge permitting process.

The field study for storm water quality sampling is designed to collect accurate hydrologic information to quantify discharges and to collect flow-weighted composite samples for the duration of the storm event.

Site selection and gage establishment for the regional network was implemented in a three-phased approach with the larger cities' (Dallas and Fort Worth) sites being brought on-line first. Because of the size of the study area and the spatial extent to which the gages are placed, an automated system was chosen to collect the samples during the storm event. This was done to reduce the manpower that would be required if the samples were collected manually.

SITE INSTRUMENTATION

The storm water quality sampling systems for the gaged sites consist of many integral parts. The stated task of the system is to collect information for both the quantity and the quality of the storm water discharge. In order to determine the quantity of runoff for a particular storm, a flow measuring device is placed near the outfall of the flow conveyance. In this case, the flow is measured in either circular pipes, non-circular pipes, or concrete-lined open channels. A flume (Kilpatrick and others, 1985) for circular concrete pipes, or a weir for non-circular pipes and concrete-lined open channels, is used to measure the discharge.

Samples are collected using a sampler (ISCO Model 3700) connected to a controller/data recorder. The samples are collected at variable time intervals corresponding to a specific detected volume of water flowing by the sampling station. The controller/data logger system and its associated software are the actual "brains" of the storm water sampling system. Three different systems are being used at various sites in the Dallas-Fort Worth study. The systems are provided by SUTRON, ISCO and CAMPBELL SCIENTIFIC, respectively. Costs for the three systems range from approximately $8000.00 to $10,000.00. System costs are for instrumentation and equipment and do not include labor costs for installation and maintenance.

SUTRON STORM WATER RUNOFF SAMPLING SYSTEM

NPDES site CRA1, located in north central Fort Worth, was selected as a pilot site to test the Sutron storm water runoff sampling system. Equipment and instrumentation at the site includes the SUTRON Model 8200 controller/data recorder, the ISCO sampler, 2 stage-sensing pressure transducers fitted to a conoflow gas pressure system, a modified Palmer-Bowlus flume mounted in the 60 inch circular concrete pipe and a tipping bucket raingage. Because of its remote location, this site was equipped with a cellular telephone.

The system consists of a field station for remote sampling and a PC Base Station to provide central monitoring. The base PC station software is based in the Microsoft Windows(tm) environment. A threshold precipitation and stage value are the triggers for allowing the SUTRON system to initiate sampling. These threshold values, once reached, "awaken" the system which, in turn, alerts the user, via a voice synthesized phone message, of the impending site storm event.

As a data recorder the SUTRON system will store over 64,000 readings in a battery backed-up RAM. The data may be accessed in one of two ways, on-site and remotely via the PC Base Station modem. At the site, the data may be transferred from the data recorder to a portable PC. This allows the user to view current as well as archived sensor data and enable/disable the system. Additionally, the user may view desired sensor data in real time from a small LED panel on the front of the controller/data recorder. The user also has the ability to interrogate the system, arm/disarm the site, and acknowledge system alarms via any touch-tone phone.

USGS Texas District experience with the SUTRON storm water runoff system overall has been positive. The system is presented as a very integrated package, but does require a fairly sophisticated PC to run the Base Station software.

ISCO STORM WATER MONITORING SYSTEM

NPDES site BEL1, located in central Fort Worth, was selected as a pilot site to test the ISCO storm water runoff sampling system. Equipment and instrumentation at the site includes the ISCO Model 3230 Bubbler Flow Meter with modem, the ISCO sampler, one stage-sensing pressure transducer, a 120 degree V-notch, sharp crested weir mounted at the outlet of the 4'X 8' concrete box culvert and a tipping bucket raingage. A hardwire telephone system is used as the primary method of communications.

This system is designed to transmit level, flow, rainfall and sampling data from the internal modem across a hardwire telephone line to a central PC base station. The flow meter monitors signals from the raingage and flow/level in the culvert. The flow meter activates the sampler when a threshold amount of precipitation is detected in a certain duration and when the discharge stage reaches a user defined pre-programmed level. The flow meter then sends signals to the sampler to collect the flow-weighted composite samples. The flow rate is plotted on the internal strip chart along with sample data and hourly rainfall amounts. The rainfall, level, flow and sample data are stored in the internal memory of the flow meter. The data may then be transmitted to the base station where the ISCO FLOWLINK(tm) software is used to generate graphs and reports. Additionally, the user may download archived data to a portable PC at the site.

USGS Texas District experience with the ISCO storm water monitoring system has been quite good. The ISCO system has been in use for approximately one year with no significant downtime has recorded in the 12 month period. The flow meter's internal strip chart makes needed data available to the on-site user without the use of a portable PC. The system is presented as a very integrated package that interfaces directly with the ISCO sampler.

CAMPBELL SCIENTIFIC STORM WATER DISCHARGE AND SAMPLER CONTROL SYSTEM

NPDES site 325, located in west Dallas, was selected as a pilot site to test the CSI storm water discharge and sampler control system. Equipment and instrumentation at the site includes the CSI Model CR-10 datalogger/controller along with accompanying phone modem and storage modules, the ISCO sampler, 3 stage-sensing pressure transducers fitted to a conoflow gas pressure system, a modified Palmer-Bowlus flume mounted in the 72 inch circular concrete pipe, and a tipping bucket raingage. A hardwire telephone system is used for remote communication with the base station.

The CSI system, unlike the other two systems, is a fairly non-integrated package. The CR-10 datalogger with its modem and storage module come as separate components. In the case of the Dallas-Fort Worth study, a person skilled in electronics was needed to connect the components and install them in a weather-proof enclosure. The manufacturerer states the reason for this is to allow the user to tailor the system to their specific needs. As with the other systems, a pre-programmed threshold precipitation and stage value trigger the system to start recording data and taking samples. The flow data and rainfall are stored in final memory over the course of the rainfall event. Cumulative flows are totalized and the CR10 triggers the sampler to take flow proportional samples and records the time and a verification of successful samples. Storm runoff data is accessed via a PC modem over the phone lines. Unlike the other two systems, a portable PC is the only means available for the user to access information while at the site.

USGS Texas District experience with the CSI system has been good. Although District experience with the system was limited, CSI's technical staff assisted greatly in creating a system to suit the study's needs.

Although the cost of this system was the lowest of the three systems under consideration, many man-hours were spent in the assembly, integration and testing of the system components. This associated cost must be considered when determining which system will best suit the technical needs of a particular study.

SUMMARY

All of the three storm water monitoring and sampling systems under consideration worked well in the field. Two of the three systems have been in continuous use on-site for more than a year with good results. After the pilot study, the third system, on loan to the Texas district, was returned to the manufacturer. Each of the systems are able to measure flow and initiate the automatic sampler. Data interrogation for the systems and the systems' user alert capabilities are two of the differences noted among the three systems.

All of the three systems tested offer a mechanism for storm water samples to be taken at a remote location. Data from these remote stations may then be transmitted to a central location for processing and analysis. For sampling sites that are located far from the central location, remote sampling gages may be essential in obtaining the required data for NPDES storm water projects, as well as other urban storm water quality projects.

SELECTED REFERENCES

Jennings, R. Brad, Raines, Tim H. et al., 1992, USGS Urban Stormwater Investigations in the Dallas-Fort Worth, Texas Metroplex, Proceedings of the 1992 ASCE WaterForum '92 Conference, 7 p.

Kilpatrick, F.A., Kaerhle, W.R., et al., 1985, Development and testing of highway storm-sewer flow measurement and recording system: U.S. Geological Survey Water Resources Investigations Report 85-4111, 98 p.

U.S. Environmental Protection Agency, 1990, National Pollutant Discharge Elimination system permit application regulations for stormwater discharges; final rule: U.S. Federal Register, Vol. 55, No. 222, pp 47989-48091.

Water-Level, Velocity, and Dye Measurements in the Chicago Tunnels

by

K. A. Oberg and A. R. Schmidt[1]

Abstract

On April 13, 1992, a section of a 100-year-old underground freight tunnel in downtown Chicago, Illinois, was breached where the tunnel crosses under the Chicago River, about 15 meters below land surface. The breach allowed water from the Chicago River to flow into the freight tunnels and into buildings connected to the tunnels. As a result, utility services to more than 100 buildings in downtown Chicago were lost, several hundred thousand workers were sent home, and the entire subway system and a major expressway in the Loop were shut down. The breach in the tunnel was sealed and the tunnel dewatered by the U.S. Army Corps of Engineers (Corps) and its contractors. The U.S. Geological Survey (USGS) assisted the Corps in their efforts to plug and dewater the freight tunnels and connected buildings. This assistance included the installation and operation of telemetered gages for monitoring water levels in the tunnel system and velocity measurements made in the vicinity of the tunnel breach. A fluorescent dye tracer was used to check for leaks in the plugs, which isolated the damaged portion of the Chicago freight tunnel from the remainder of the tunnel system.

Introduction

A network of tunnels more than 80 kilometers (km) long were constructed beneath the part of downtown Chicago known as the Loop from 1899 to 1909. The purpose of the tunnel system was to haul freight between buildings in the Loop in order to reduce traffic on the streets above (Moffat, 1982). Miniature electric locomotives running on narrow-gauge rail were used to transport goods to and from Loop buildings. Rail cars could be brought to the surface by means of elevators for the purpose of loading or unloading. Several additions to the tunnel system were constructed from 1909 to 1915 and portions of the tunnel were removed during construction of the Loop subway from 1940 to 1959. The total length of the tunnel system varied from 95.1 km in 1916 to 75.9 km in 1959.

---

[1]Hydrologists, U.S. Geological Survey, 102 E. Main Street, 4th Floor Urbana, IL 61801

The system was an outstanding engineering accomplishment in its day, costing approximately $30 million to construct (Chicago Tunnel Terminal Corporation, 1928). The tunnel was hand-excavated and lined by dry-packing unreinforced concrete, 0.3 meter (m) thick, behind wooden forms (Moffat, 1982). The tunnel invert varies from 12 to 15 m below land surface. The invert is deepest where the tunnel crosses under the Chicago River. Most tunnels were approximately 1.8 m wide and 2.3 m high; however, in some locations, the tunnel was wider and higher in order to accommodate a second or third set of rails.

Initially the tunnel system was used for the transport of coal, freight, cinders, and mail throughout the Loop. In later years most of the material transported was waste, especially coal cinders. The rail system was abandoned in 1959 because of a lack of funds to repair equipment and declining demand for service. Since that time the tunnel system has been used to carry power and fiber-optic cables beneath the Loop. Over the years, bulkheads were installed at various locations. However, most of these were not designed to be water-tight.

A breach in the freight tunnels beneath the Chicago River in April 1992 allowed river water to enter the tunnel system and connected buildings. This report describes water level, velocity, and dye measurements made in the tunnels and how the measurements were used in the efforts to plug the breach and dewater the tunnel system.

The Great Chicago Flood of 1992

On April 13, 1992, a section of the freight tunnel crossing under the Chicago River was breached. The breach was first reported at 5:57 a.m. after a Merchandise Mart employee noticed water rushing into the third (lowest) basement of the building. Within an hour two basements of the Merchandise Mart were filled with water. Water rose in the basements at a rate of 0.6 m per hour, despite efforts to pump it out. At about 8:00 a.m., a large whirlpool was reported near pilings in the Chicago River on the southeast side of the Kinzie Street bridge (Fig. 1). Initial attempts to plug the breach by dumping debris and quick-setting concrete into the whirlpool were unsuccessful. Sonar readings in the vicinity of the breach indicated that a car-sized hole was located near recently installed pilings at the Kinzie Street bridge.

The breach allowed approximately 950,000 cubic meters of water from the Chicago River to flow into the freight tunnels and into buildings in the Loop having access to the tunnels. By noon on the April 13, utility services were lost for many buildings in the Loop causing several hundred thousand workers to be sent home. In the end, utility services for more than 100 buildings were lost, the entire subway system in the Loop was shut down, and a major expressway in the Loop was closed because of the flooding. Recent estimates of the damages resulting from the flood exceed $1 billion.

The Corps and its contractors sealed the breach in the tunnel by constructing plugs in the tunnel on either side of the Chicago River near the breach. The construction of the plugs was complicated because the tunnel on the west side of the river splits into two branches, one running under Canal Street and the other under West Kinzie Street (Fig. 1). Access shafts ranging in diameter from 1.52 to 2.13 m were drilled from street level down into the tunnel. Initially, two access shafts were drilled at each branch of the tunnel on both sides of the Kinzie Street bridge. For each branch, sandbags and stone were lowered into one of the two access shafts to slow the flow of water through the tunnel. Then concrete plugs were placed in the second access shaft, with a single pour, to at least 5 feet above the crown of the tunnel. In addition, the area between the two shafts and the bottom of each concrete plug was grouted using a combination of microfine cement, sodium silica grout, and cement grout (Inouye and Jacobazzi, 1992). As an added safety measure, the Corps installed additional plugs on each branch of the tunnel near the breach in order to allow greater head differential during dewatering of the tunnels.

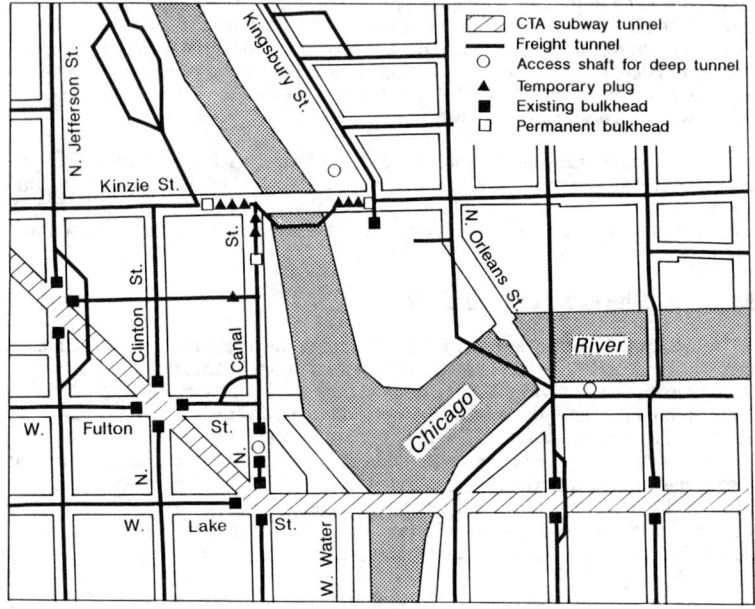

Figure 1.-- Chicago freight tunnels, plugs, and access shafts in the vicinity of the Kinzie Street breach. [Modified from Ferenczy, 1992, p. 9]

In order to dewater the tunnel system, the Corps initially established pumping stations at three locations in the Loop. The rate of drawdown in the tunnels was carefully controlled at about 5 to 8 centimeters per hour in order to minimize the risk of collapsing another part of the tunnel system. Seven other pumping stations in the Loop were eventually established in order to dewater the entire tunnel system. The dewatering was completed on May 22, 1992. After this, permanent bulkheads were constructed in the tunnels near Kinzie Street.

Water-level Measurements

Gages for monitoring water levels in the tunnels were installed by the USGS at seven locations throughout the Loop. The purpose of the gages was two-fold. The gages were used by the Corps to monitor the dewatering of the tunnel system and serve as an early warning system should another breach occur during the dewatering operation. Gages were installed at locations relatively near where the tunnel crossed under the Chicago River. Additional monitoring sites were desired, but were not feasible because of limitations in drilling access shafts and the possibility of destroying fiber-optic cables running through portions of the tunnel.

The first four gages were installed and operating within 24 hours of the initial request for assistance from the Corps. Each gage was equipped with an electronic datalogger, a stage sensor, and telephone telemetry. The electronic dataloggers were programmed to record water-level readings obtained from the stage sensor at 5 minute intervals. The stage sensor used was a potentiometer geared to a float and pulley system. The pulley turned in response to changes in water levels, causing the output voltage of the potentiometer to change. In addition, an auto-dial, auto-answer modem was installed with the datalogger. This allowed the Corps to interrogate each gage every 15 minutes, or at any desired time, via computer. All gages were removed by May 22, 1992, after all of the tunnel system, except the river crossings, had been dewatered.

Velocity Measurements

In order to evaluate the effectiveness of the temporary plugs installed on either side of the Kinzie Street bridge, the Corps requested that the USGS measure velocities in the tunnel on the landward side of the first temporary plugs. Velocities were measured in shafts drilled down into the tunnel from the top of Kinzie Street. The USGS proposed using a low-velocity, Price AA current meter and a Neil Brown Acoustic Velocity Meter[1] (hereafter referred to as Neil Brown). The advantage of the Neil Brown was the ability to measure low velocities and to determine flow direction. The latter was critical, because flow direction using Price AA current meters could only be estimated by noting the direction of flow at the surface. Water levels in these shafts were well above the tunnel, approximately equal to the water-surface elevation of the Chicago River. The shafts acted as stilling wells, making it difficult to determine the direction of flow by observing the orientation of the Price AA meter at the water surface. The Price AA meter was nevertheless used in some of the shafts so that any sudden changes in velocities could be observed.

There were several drawbacks to relying only on velocity meters to monitor leakage through the temporary plugs. These included (1) the resolution of the velocity meters, and (2) the possibility that velocities measured by the meters could be caused by pumping rather than a leak in the temporary plugs. The minimum resolution of the meters ranged from 0.9 centimeters per second (cm/s) for the Neil Brown meter to 3.0 cm/s for the Price AA meter. A measured velocity of 0.9 cm/s might therefore be interpreted as meaning there was negligible flow through the plug. However, a velocity of 0.9 cm/s would be equivalent to a leak of 25 liters per second (L/s) through the temporary plugs, assuming a cross-sectional area of 3.72 square meters.

---

[1] Use of brand names in this report is for identification purposes only and does not constitute endorsement by the U.S. Geological Survey

Velocity measurements were made using both meters in shafts on the west side of the Kinzie Street bridge (West Kinzie) every 15 minutes, 24-hours a day from April 18-20 and at periodic intervals thereafter until the second set of temporary plugs were installed. Velocity measurements using a Price AA meter also were made in a shaft on the east side of the Kinzie Street bridge (East Kinzie) 24-hours a day from April 18 to May 9. Unusual changes in water levels or velocities were immediately reported to the Corps. Velocities measured using the two meters were always at or below the resolution of the meters, except for a few short periods when velocities measured in a shaft on West Kinzie Street ranged from 1.5 to 3.0 cm/s. Even though little or no flow could be measured using the meters, the data were useful to the Corps in evaluating the effectiveness of the plugs and providing a safety measure, should one of the plugs being to leak.

Dye Injection and Sampling

For the reasons mentioned above, a fluorescent dye tracer was used to more precisely determine whether the plugs isolating the breached section of the Chicago freight tunnel from the remainder of the tunnel system were leaking. In order to accomplish this, dye would have to be injected on the riverward side of the plugs and then water samples would have to be taken from the landward side of the plugs and analyzed for the presence and concentration of dye. The dye injection was planned to allow detection of leaks as small as 0.06 L/s. The dye injection and sampling was planned initially to indicate only whether water was leaking through the plugs and not to determine leakage rates.

The dye injection was planned based on results from computer models of dispersion and advection in the tunnels. These models were developed to assure that the quantity of dye injected would be sufficient to detect any leaks in the plugs. Although the dye injection was not planned to determine accurate leakage rates, the two models did allow gross estimates of the leakage rate to be computed by comparing model results with measured dye concentrations. One model was written to simulate the mixing of the dye in the isolated portion of the tunnel. This model was necessary because the concentration of dye immediately adjacent to the plugs changed with time as the dye mixed throughout the isolated part of the tunnel. This model estimated the time until dye reached the plug, simulated the change in concentration over time at the plugs, and simulated the effect of injecting dye at three different locations in the isolated part of the tunnel.

A second model simulated the advection and dispersion of dye that had leaked through the plugs. This model was necessary because the sampling points were not immediately adjacent to the plugs and because the dye concentration changed over time because of mixing, transport away from the plugs, changes in the concentration of dye leaking through the plugs, and changes in the leakage rate.

Based on the initial model runs, 10 liters (L) of rhodamine WT dye were injected into the isolated part of the tunnel system; 5 L into a shaft approximately 24 m riverward from the plug on East Kinzie Street, and 2.5 L into each of two shafts approximately 6 m riverward from plugs in West Kinzie Street and Canal Street, respectively (Fig.1). All three dye-injection shafts were cased holes, 10 centimeters (cm) in diameter, drilled into the tunnel. The water levels in these shafts were approximately 6 m above the top of the tunnel. The dye was injected through 0.9 cm diameter tubing that was lowered to the approximate center of the tunnel. This was to assure that the dye was mixing in the tunnel rather than remaining above the tunnel in the injection shaft.

Samples were collected from the tunnel from shafts drilled landward from the plugs and analyzed for dye concentrations. Samples were analyzed in a laboratory van using a fluorometer calibrated to standard solutions with concentrations of 0.0, 1.0, 10, 50, 100, and 200 parts per billion. The background fluorescence of the water in the tunnels was determined from water samples collected prior to dye injection from the sampling shafts and from the Chicago River at the Kinzie Street bridge. Samples were collected from the dye injection shafts until May 2, 1992, to determine the mixing of dye in the isolated part of the tunnel.

Concentrations of dye detected in samples collected from a shaft on East Kinzie Street indicated that water was leaking through the plug in the tunnel beneath East Kinzie Street. Preliminary estimates of the leakage rate, based on comparing these concentrations with those from model simulations used to plan the dye injection, were that the rate of leakage was between 0.3 and 0.9 L/s. The models were revised later to better simulate some of the unique aspects of this system, including the initial mixing of dye in the tunnel and the effect of grouting on the leakage rate during the time being simulated. Results from the revised models indicated that the actual leakage rate through the plug beneath East Kinzie Street was between 0.1 and 0.2 L/s.

Summary

Water-level, velocity, and dye measurements were made in the Chicago freight tunnels from April 18 to May 22, 1992 by the USGS. Information from these measurements assisted the Corps in plugging and dewatering the tunnel system following the tunnel breach at Kinzie Street on April 13, 1992. Data from seven gages installed throughout the Loop were used by the Corps to monitor the dewatering effort. Velocity and dye measurements were made in the freight tunnels near Kinzie Street in order to evaluate the effectiveness of the tunnel plugs. Velocity meter measurements were useful in providing an early warning in the event major leaks in the tunnel plugs occurred. Dye injection and sampling allowed for a more accurate assessment of the integrity of the tunnel plugs than could be made using velocity meters alone.

References

Chicago Tunnel Terminal Corporation. (1928). City of Chicago Tunnel System. Chicago Tunnel Terminal Corporation, Chicago, Illinois, 34 pages.
Ferenczy, M. (1992). "Corps of Engineers Uses Computer Mapping to Drain the Chicago Freight Tunnels," Illinois GIS and Mapnotes. Center for Governmental Studies, Northern Illinois University, DeKalb, Illinois, 11(1), p. 7-9.
Inouye, R. R. and Jacobazzi, J. E. (1992). "The Great Chicago Flood of 1992." Civil Engineering. 62(11), p. 52-55.
Moffat, B. 1982. Forty Feet Below--The Story of Chicago's Freight Tunnels. Interurban Press, Glendale, California, 84 pages.

A Simple and Reliable Method For Analysis of Water Supply Distribution

T. H. Kuan   Member

Abstract

During the last decades, much mathematical computation has been done and studied concerning the characteristics of water supply distribution systems, and many of them have produced great excitement in their successful completion and utilization. However, the proposed analysis, which is presented in this paper, may be regarded as a bridge between the practical sample analysis and the optimization model development. From the analysis and observation of water distribution problems, a critical path of water supply, which is the key route to meet the applied demands at least energy cost, can be found; then the pattern of water distribution is uniquely defined. Several examples have been selected from the published literature to demonstrate the differences between the previous methods and the "critical path method". Finally, a complex network problem is presented to show the advantages of the critical path method.

In March 1992, the world's largest and most advanced water management(MGT) system (the London Water Ring Main), will be capable of supplying 340 MGD to 5.5 mil. people in the capital of England. The advanced automated control system and water supply strategy are very impressive. It calls us to develop a highly effective water MGT system to prevent the water shortages in Southern California.
        The system analysis for water network problems at

Assist. staff engineer, Public Works, City of Los Angeles, 200 Main Street, CHE. Rm.1545, Los Angeles, CA 90012

steady flow conditions has been studied since 1936 when the Hardy Cross method of successive approximations was developed to calculate the flow in each circuit of a network, and to bring the circuit into close balance so that continuity is satisfied at every junction. Head loss must also be zero for any closed circuit (Streeter 1966). After digital programs were developed for network analysis, Prof. Wood and his partners applied the Newton-Raphson procedure to linearize nonlinear equations, and solve the resulting system of linear simultaneous equations (Wood 1972, and Boulos 1992). Currently, this method is widely accepted in practice (Goulter 1992).

As further progress is made in modeling for water networks, it could be anticipated that computer based optimization design models would be in widespread use in offices. However, in reality it is quite different. The optimization models are showing no signs of succeeding in water supply system design (Goulter 1992).

In each of these three methods, the fundamental concepts used for solving the problem are same. They are (1) mass balance, and (2) energy conservation. In fact, the previous illustration of conservation-of-energy in flow around any loop usually is not appropriate. For example, the resistance to flow of conduit AB (fig 1), whether the flow is under the effect of gravity or against the effect of gravity, is not the same because the velocity distribution across the section is not the same (Rouse 1946). This difference, however, is usually negligible in energy analyses of water network problems, but could be the reason for the observed nonconvergence in linearized simultaneous equations. Perhaps, instead of making the algebraic sum of head losses in any loop be zero, the minimization of energy consumption in the water supply network should be considered. This is done not only to avoid the use of complicated and indeterminate formulas, but also to gain a simple and reliable strategy for solving the problems. Therefore, Dr. Albertson and I published a paper for the National Hydraulic Conference in 1990 to suggest this simple hydraulic analysis (Kuan 1990).

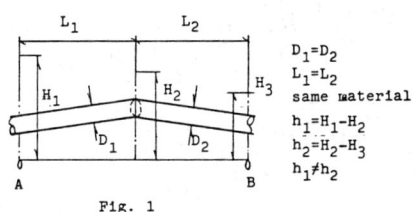

Fig. 1

$D_1 = D_2$
$L_1 = L_2$
same material
$h_1 = H_1 - H_2$
$h_2 = H_2 - H_3$
$h_1 \neq h_2$

The proposed technique for solving water supply distribution systems is based on the concept described in the following paragraphs:

(1) The calculation of energy:
With regard to the evaluation of the friction calculation, the Darcy-Weisbach formula is used because it permits the proper determination of friction factor (f) instead of (Ch) in the Hazen-Willian formula. The

quadratic Darcy-Weisbach formula can be easily converted to the linear items to be solved (Pipeline Design for Water and Wastewater provided by ASCE Committee 1975).

(2) The Continuity of mass:
The principle of mass balance is that the inflow must be equal to the outflow at each node. The results of that system analysis must show a positive value. That means that the original assumption of flow direction is correct, otherwise flow direction must be changed.

(3) The critical path:
For each particular project, there is a critical path or circuit, which can be found to assess the minimization of energy consumption for operation of the system. Hence, the objective function of optimization analysis is the minimization of head loss. The constraints are:
1. Mass balance at each node
2. Energy balance in the critical path
3. Necessary operation restraints.

The critical path for each particular project can be found by answering the following questions:
1. Is it the necessary path or circuit required to deliver the water to the whole project, and
2. Is it the path or circuit that uses the least energy?

Two examples have been selected from previous literature to demonstrate this method.

The first example is from D.T.Wood's "Algorithm for Pipe Network Analysis and Their Reliability " (Res. Rep.No. 127, University of Kentucky Water Resources Research Institute, Lexington, KY 1981).

To find the distribution of water flow in the problem presented in the following figure:

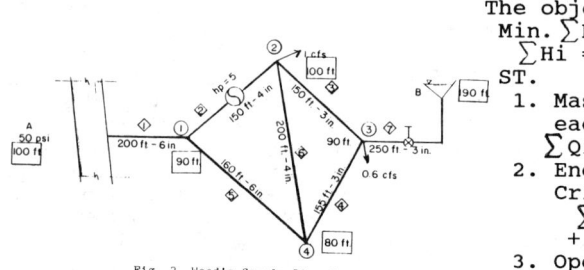

Fig. 2 Wood's Sample Pipe System

The objective:
Min. $\sum$ Head Loss
$\sum H_i = \sum K_i * Q_i^2$
ST.
1. Mass balance at each node,
$\sum Q_{in} - \sum Q_{out} = 0$
2. Energy balance at Critical path,
$\sum (G_i * Q_i - H_i) + \Delta E = 0$
3. Operational restrictions,
$Q_i > 0$

In this, the equation of energy balance is the linearization of the energy term of the approximate value of grade (Wood 1981). The quadratic forms can be solved by Linear Programming(Schrage 1989). The comparison of result of the two methods are:

Method	Q1	Q2	Q3	Q4	Q5	Q6	Q7	hi
Wood's	1.73	1.37	0.37	0.36	0.36	0.001	0.131	37.24
CPM	1.43	1.29	0.28	0.14	0.14	0.0	0.171	23.21

The second example is selected from M. B. McPherson's "Generalized Distribution Network Head Loss Characteristics"(Journal of Hydraulic Div. ASCE Vol.86 No. HY1 Jan. 1960). In this paper, the author presented a real system operation located at the Belmont Gravity District in Philadelphia.

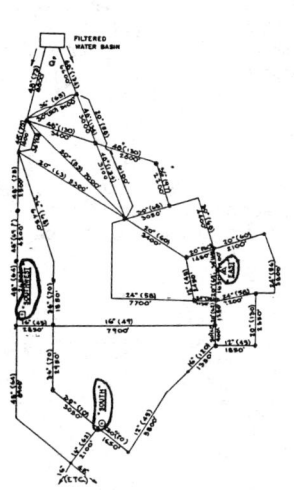

FIG. 3.—BELMONT GRAVITY DISTRICT, ARTERIAL PIPING NETWORK FOR "EXISTING" STUDY. NOT TO SCALE, SEE FIG. 4

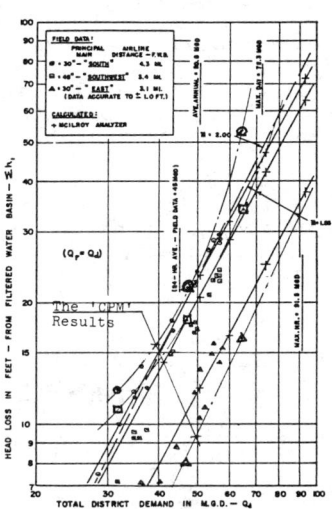

FIG. 4.—SYSTEM HEAD LOSSES, BELMONT GRAVITY DISTRICT, "EXISTING" NETWORK, SEE FIG. 3

The plot of data, which is Fig 4, showed the analysis and relationship between head loss and total demand. The results from the critical path method(CPM) are plotted in this chart to verify that the critical path method can be employed to model the real case.(see fig. 3 and fig. 4)

Consequently, the approach developed in this paper provided a simple and reliable method for determining the design, operation and upgrade of water supply distribution systems particularly for principal feeders. For instance, any change of hydraulic characteristics in the critical path will mainly influence the water supply system. This technique can be applied for efficient upgrading of deteriorated systems.

The nonconvergence of linearized simultaneous equations is the significant consideration for use of the Newton-Raphson method, which was gaining popularly. The advantage of CPM is that it always results in convergence because the optimization search analysis of minimum energy spent satisfies the Kuhn-Tucker condition (Wismer 1978). The method of searching for the critical path can be found in the optimization technique both in mathematic and graphical form (Bazaraa 1977). The use of the optimization algorithm to perform network analysis including determination of hydrant pressure requirements, and operation of elevated storage tanks will not be discussed in this paper because those features are related to water resources MGT techniques.

Conclusion: the CPM provides the methods for solving water supply distribution systems, while providing a number of advantages over previous techniques:
  (1) Elimination of tedious theoretical requirements,
  (2) Making the analysis more practical, and
  (3) Solving the problem quickly and reliably.
Indeed, it is reliable, but it is simple.

Acknowledgement

The literature has been reviewed by Mr. John Mandry, who is Control System engineer of Wastewater Treatment Engineering Division, Public Works, City of L. A. I appreciate his contribution.

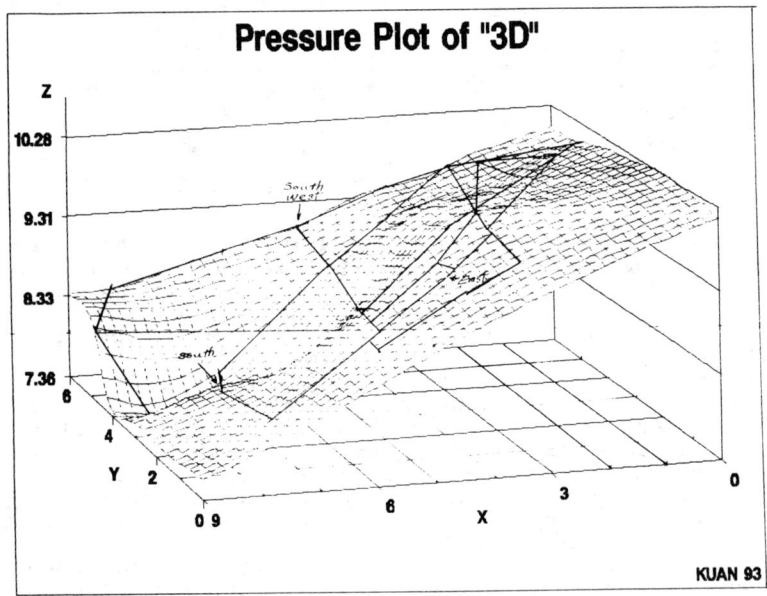

References :
Bararaa, M. S.,1977 "Linear Programming and Network Flow"

Boulos, P. F. & Wood, D. J. "Explicit Calculation of Pipe-Network Parameters", Journal of Hydraulic Div. Nov. 1990.

Committee on Pipeline Planning, Pipeline Div., ASCE, 1975 "Pipeline Design for Water and Wastewater".

Goulter, I. C. "Systems Analysis in Water-Distribution Network Design: From Theory to Practice", Journal of W.R.R.M. ASCE, May/June. 1992.

Kuan, T. H. & Albertson, M. L. "Analysis and Application of Linear Programming to solve Water Resources Problems", Proceeding of the 1990 National Conference of Hydraulics.

Rouse, Hunter, 1946 "Elementary Mechanics of Fluids".

Streeter, V. L., 1966 "Fluid Mechanics".

Schrage, Linus, 1989 "LINDO" (User's Manual)

Wood, D. J. & Charles, O. A. "Hydraulic Network Analysis Using Linear Theory", Journal of Hydraulic Div. ASCE, July 1972.

Wood, D. J. "Algorithm for Pipe Network Analysis and Their Reliability", 1981 (Res. Rep. No. 127, University of Kentucky, Water Resources Research Institute).

Wismer, D. A., 1978 "Introduction to Nonlinear Optimization".

Friction Loss Equations for Hydropower Facilities

Ed A. Toms, P.E.[1] ; E. June Busse, P.E.[2] ; Curtis A. Thompson, P.E.[3]

Abstract Many water suppliers are pursuing economic benefits from water supply and distribution systems through development of hydropower facilities on pipelines. Excessive pressure head that used to be dissipated by pressure relief valves or other energy dissipation devices is now analyzed for potential hydropower development. Feasibility assessment of these potential developments is based on benefit to cost economic analyses that are dependent on hydraulic friction loss calculations and pipeline roughness assumptions that can vary significantly. This paper presents sensitivity analyses that compare the effect of widely used pipe friction loss equations on hydropower feasibility analyses and facility sizing.

Introduction Many water suppliers have discovered an energy source that was previously been neglected or considered a nuisance. This new source is electricity generated by hydroelectric power plants installed on water pipelines. Closed conduit water systems that traverse significant elevation changes can develop excessive hydraulic pressure that requires energy dissipation to allow safe water delivery to the customer. This dissipated energy used to be considered detrimental to the delivery system, but is now recognized as an efficient, environmentally benign, economical energy source.

The City of Boulder, Colorado is a leader in developing alternative/renewable energy sources to keep costs (and water rates) low. The City has installed five hydropower units on water supply pipelines and annually generates four megawatts of energy that was previously neglected. A sixth hydropower development project being studied by the City is the Lakewood Pipeline, the subject of this study.

The Lakewood Pipeline is approximately 55,000 feet long and traverses an 1,800 feet drop in elevation. The pipeline's hydraulic profile includes open channel and pressure flow, and the fluctuating flow regime results in air entrainment that disrupts the water treatment plant operations. By replacing the pipeline with an all-pressure flow pipeline, the City can improve water treatment operations and develop hydropower to reduce the cost of the new pipeline.

This paper studies variations within widely used hydraulic friction loss equations and related surface roughness estimations and compares their effects on hydraulic and

---

[1]Civil/Hydraulic Engineer, Engineering Consultants Inc., 5660 Greenwood Plaza Boulevard, Suite 500, Englewood, Colorado 80111.

[2]Utilities Project Manager, City of Boulder, 1739 Broadway, Boulder, Colorado 80306.

[3]Senior Water Resources Engineer, Engineering Consultants Inc.

economic analyses for penstock, turbine, and generator sizing and design. The study considers four equations: Darcy-Weisbach (Colebrook-White equation and Swamee and Jain approximation), Scobey, Manning's, and Hazen-Williams. Minor losses were assumed to be negligible for all calculations. The assumed system efficiency is 82%[4], and the flow-duration curve procedure (Corps, 1985) was used.

Pipe Friction Headloss Equations The four friction loss equations are described below.

Darcy-Weisbach Equation - The classical by Darcy-Weisbach equation for friction loss of flow in a pipeline is:

$$H_L = f \frac{L}{D} \frac{V^2}{2g}$$

Where:
H_L is the head loss in feet.
L is the length of pipe in feet.
D is the average inside diameter in feet.
V is the average velocity.
f is friction factor.

The Darcy friction coefficient, f, may be calculated by the Colebrook-White equation for flow velocities normally encountered in water transmission:

$$\frac{1}{\sqrt{f}} = -2 \log \left(\frac{\epsilon}{3.7D} + \frac{2.51}{N_r \sqrt{f}} \right)$$

Where:
f is the friction factor.
ϵ is absolute surface roughness in feet.
D is pipe inside diameter in feet.
ϵ/D is relative roughness.
N_r is Reynolds number.

The Colebrook-White equation requires a trial and error solution for determining the friction factor, f, for a given roughness, ϵ, and Reynolds number, N_r. Use of the Moody diagram reduces the trial and error solution for determining the friction factor, f. Design values for absolute surface roughness, ϵ, range from 5 to 20 x 10^{-6} feet (Corps, 1985) and 35 to 40 x 10^{-6} (AWWA, 1989). For this study, a value of 30 x 10^{-6} was used as a conservative value. Also, water viscosity of 1.41 x 10^{-5} ft^2/sec was used in the analysis. The ϵ/D values ranged between 0.000086 to 0.00012 for the assumed ϵ value of 0.0003.

Swamee and Jain (Swamee and Jain, 1976) developed an explicit equation for the friction factor, f, in the transition zone of turbulent pipe flow. This explicit equation was based on curve fitting the solutions of the Colebrook-White equation for the smooth-turbulent and rough-turbulent flow through different size pipes. The resulting equation is:

$$f = \frac{0.25}{\left[\log \left(\frac{\epsilon}{3.7D} + \frac{5.74}{N_r^{0.9}} \right) \right]^2}$$

Figure 1 compares the Colebrook-White (Moody diagram) and Swamee and Jain equations for computing the friction factor, f.

---

[4] 80% to 85% system efficiency is used for preliminary hydropower studies as indicated in reference (Corps, 1985).

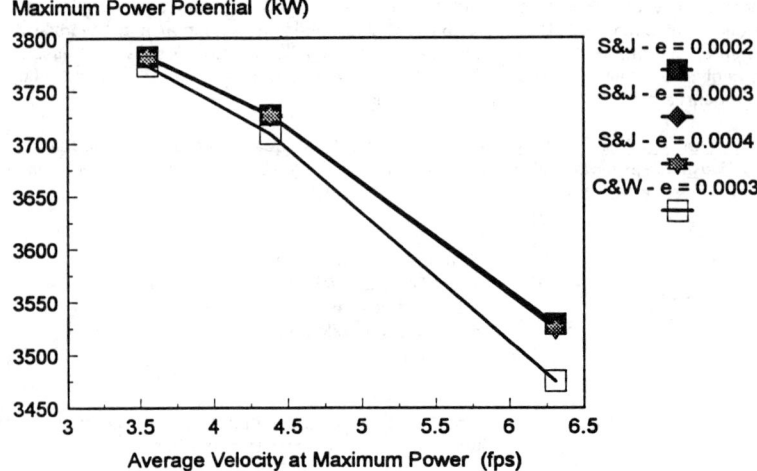

Figure 1 - Colebrook-White and Swamee-Jain friction factor compared.

Scobey Equation - The Scobey equation for steel pipe is the following:

$$V = \frac{D^{0.58} H_L^{0.526}}{K_s^{0.526}}$$

For computing headloss through a steel pipe:

$$H_L = K_s \left(\frac{V^{1.9}}{D^{1.1}} \right)$$

Where:
H_L is the headloss due to pipe friction in feet per thousand feet;
D is the pipe diameter in feet.
V is the average pipe velocity in feet per second.
K_s is a friction loss coefficient.

The CORPS (Corps, 1985) recommend a K_s value of 0.34 for design. The AWWA (AWWA, 1989) recommend a value of 0.36. The average pipe velocity for this project ranges from 0 to 10 fps. The calculated relationship between maximum power potential and average pipe velocity (Figure 2) is very close in the range of 0 to 4.5 fps. After 4.5 fps, the lines start to diverge more rapidly.

Manning Equation - The Manning equation is :

$$H_L = 2.87 \, n^2 \left(\frac{LV^2}{D^{1.33}} \right)$$

Where:
H_L is the headloss due to pipe friction in feet.
L is the length of pipe in feet.
V is the average velocity in the pipe in feet per second.
D is the inside pipe diameter in feet.
n is Manning's coefficient.

FRICTION LOSS EQUATIONS 1491

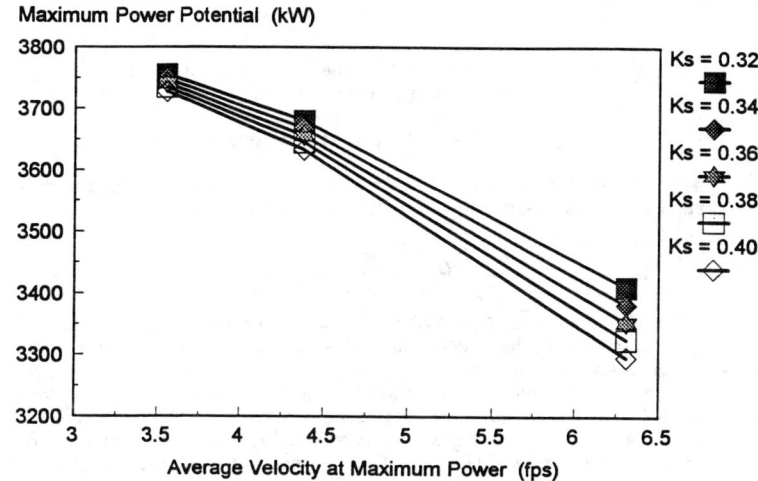

Figure 2 - Scobey equation coefficient compared.

The AWWA (AWWA, 1989) recommends a Manning's n value of 0.011 for steel pipe with linings conforming to its standards. A range of Manning's n values are evaluated from 0.011 to 0.040 as shown in Figure 3.

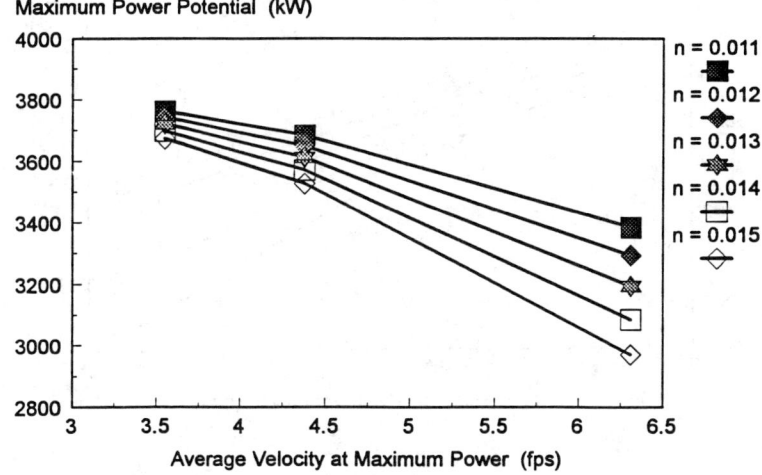

Figure 3 - Manning equation coefficients compared.

Hazen-Williams Equation -
by the following equation:

Pipe headloss is computed by the Hazen-Williams equation

$$H_L = \frac{4.72 \, Q^{1.852} \, L}{C^{1.852} \, D^{4.87}}$$

Where:
H_L is the headloss due to pipe friction in feet.
Q is the pipe discharge in cubic feet per second.
L is the pipe length in feet.
C is Hazen-Williams roughness coefficient.

AWWA (AWWA, 1989) states that for consideration of long-term lining deterioration a value can be obtained from the following equation:

$$C = 130 + 0.16D$$

A range of coefficients from 140 to 100 are compared in Figure 4.

<u>Summary</u> Figure 5 compares AWWA and CORPS recommended coefficients for the different friction loss equations related to maximum power potential and average pipe velocity. Evaluating a recommended range of absolute roughness values (Swamee and Jain, 1976)(Chow, 1959), indicates that using the Swamee & Jain approximation estimates the highest power potential for a hydropower site compared to using the Colebrook-White equation (Moody diagram). A comparative trend line can be computed with an average value, for the absolute roughness, of 0.0003 used in the Colebrook-White equation (Figure 5). The Scobey equations with a coefficient of 0.34 gives a conservative maximum power potential with respect to optimization of the penstock and hydropower unit size.

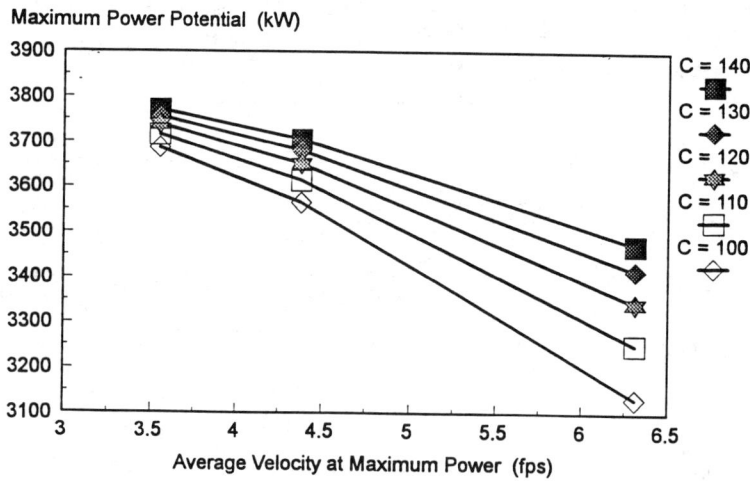

Figure 4 - Hazen-Williams equation coefficients compared.

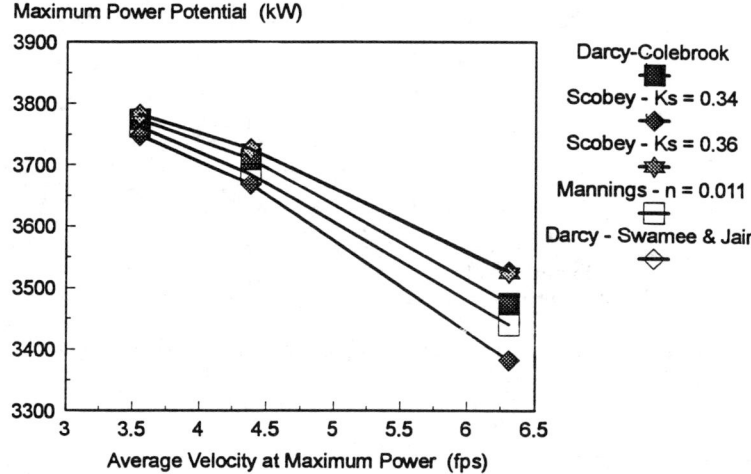

Figure 5 - Friction equation summary.

Conclusions Technical analyses of the origins and background of the four analyzed friction loss equations indicates that Manning equation is least qualified for hydropower hydraulic analyses and facility sizing. Considering that Manning's n value is only appropriate for a limited range of absolute roughness and it makes no allowance for Reynold's number effects, Manning equation could logically be expected to result in the most skewed effects for a hydropower hydraulic analysis. The irony of mother nature is the only possible cause of the unexpected comparison in Figure 5 that implies Manning equation to be the best compromise of the extremes. The lesson is to be ever conscious of the limitations of human endeavors to simulate nature. That awareness should lead to a comprehensive and deliberate approach to hydraulic analyses.

References
AWWA Manual 11 Steel Pipe - A Guide for Design and Installation, American Water Works Association (AWWA), 1989.
Engineering and Design Hydropower, Department of the Army Corps of Engineers, December 31, 1985.
Explicit Equations for Pipe-Flow Problems, Swamee and Jain, Journal of the Hydraulics Division, May 1976.
Fundamentals of Hydraulic Engineering Systems, Hwang and Hita, Prentice-Hall, Inc., 1987. **Open-Channel Hydraulics**, Chow, McGraw-Hill Book Company, 1959.

Development of Enhanced Tools for the Integrated
Analysis of Reservoir and Power System Operations

Barbara Miller[1], Pete Ostrowski, Jr.[1], and Vahid Alavian[1]

INTRODUCTION

The Tennessee Valley Authority (TVA) and the Electric Power Research Institute (EPRI) have initiated a collaborative project - the INTEGRAL Project - to develop improved tools for coordinating power generation and multipurpose reservoir operations. A Decision Support System (DSS) will be developed to aid in the collection, analysis, and display of reservoir and power system data. An Integrated Modeling System will also be developed to support the scheduling, forecasting, and planning functions of the DSS. The decision support and modeling systems will be generalized for use by a variety of electric utilities and water resource agencies. An overview of major INTEGRAL Project components is presented.

THE DECISION SUPPORT SYSTEM

A Decision Support System is an integrated suite of hardware and software organized around a common data architecture. DSS software can consist of models, data, analysis tools, and visualization aids which help the user identify a problem, evaluate options, arrive at a solution, and communicate results to other locations. The DSS provides a consistent set of displays and graphical controls to facilitate the decision making process (see Figure 1). TVA is working with the Center for Advanced Decision Support for Water and Environmental Systems (CADSWES) to plan, design, and implement the DSS (CADSWES, 1993).

The DSS, as tailored to the TVA system, is the TVA Environment and River Resource Aid (TERRA). TERRA will run on individual

---

[1]Senior Engineer, TVA Engineering Laboratory, Norris, TN 37828

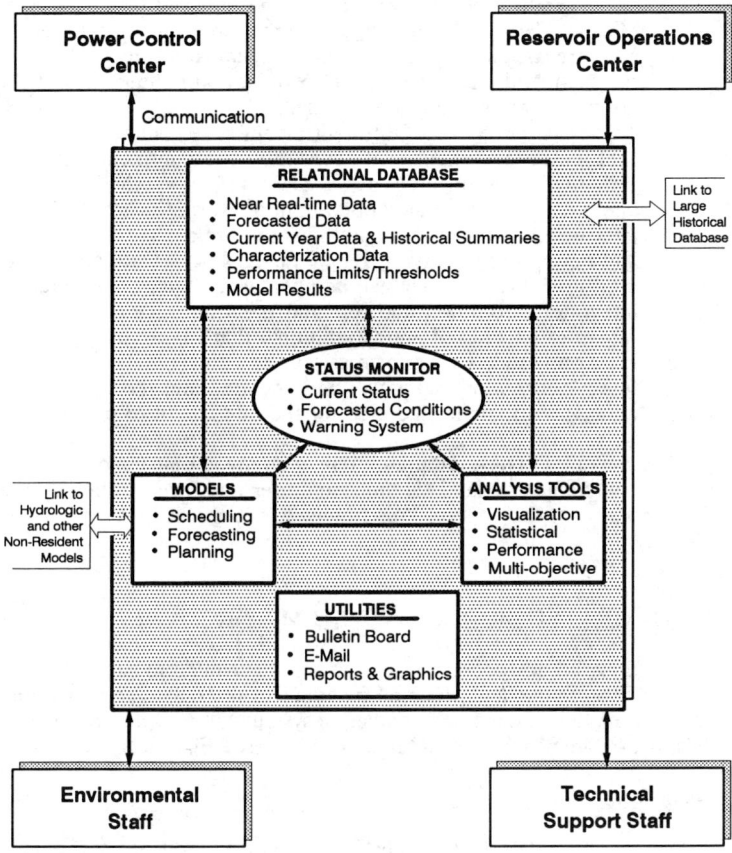

Figure 1. Schematic of the INTEGRAL Decision Support System

workstations networked over a corporate-wide area network. These workstations will be located at decision making centers to support communication and data sharing between TVA's power control center, power supply staff, reservoir operations center, water quality staff, and technical support staff. TERRA will serve five primary functions:

1. Manage System Constraints -- Important power and reservoir system constraints, commitments, and guidelines that affect operations will be updated, stored, and reported.
2. Manage Current System Data -- Current year, near-real-time, and forecast data will be automatically updated, stored, reported, and integrated into DSS operation. Summaries of historical data will also be included.
3. Track Current System Status -- Current system status will be checked for compliance with constraints, commitments, and guidelines. This includes actual measured data, scheduled operation, and forecasted operation. A warning system will indicate when thresholds and limits are broken.
4. Aid Power and Reservoir System Scheduling -- Data and models used in daily scheduling will be updated, stored, reported, and integrated into power and reservoir operations.
5. Integrate Forecasted and Operational Planning Information -- Data and models used for reservoir and power system forecasting and planning will be integrated into the DSS to anticipate problems and plan for the best use of resources.

The overall TERRA database architecture, communication framework, model access, analysis tools, and user interface will be designed to allow tailoring of the general DSS framework to the needs of other utilities.

THE INTEGRATED MODELING SYSTEM

An Integrated Modeling System will be developed to support the scheduling, forecasting, and planning functions of the DSS. To address a spectrum of modeling needs, the modeling system will include: resident models, non-resident models, and analysis tools (see Figure 2).

Resident Models

'Resident Models' will be fully integrated into the DSS, always loaded when the DSS is active and readily accessible by any DSS user. There will be two types of resident models: a Power and Reservoir System model and site-specific models.

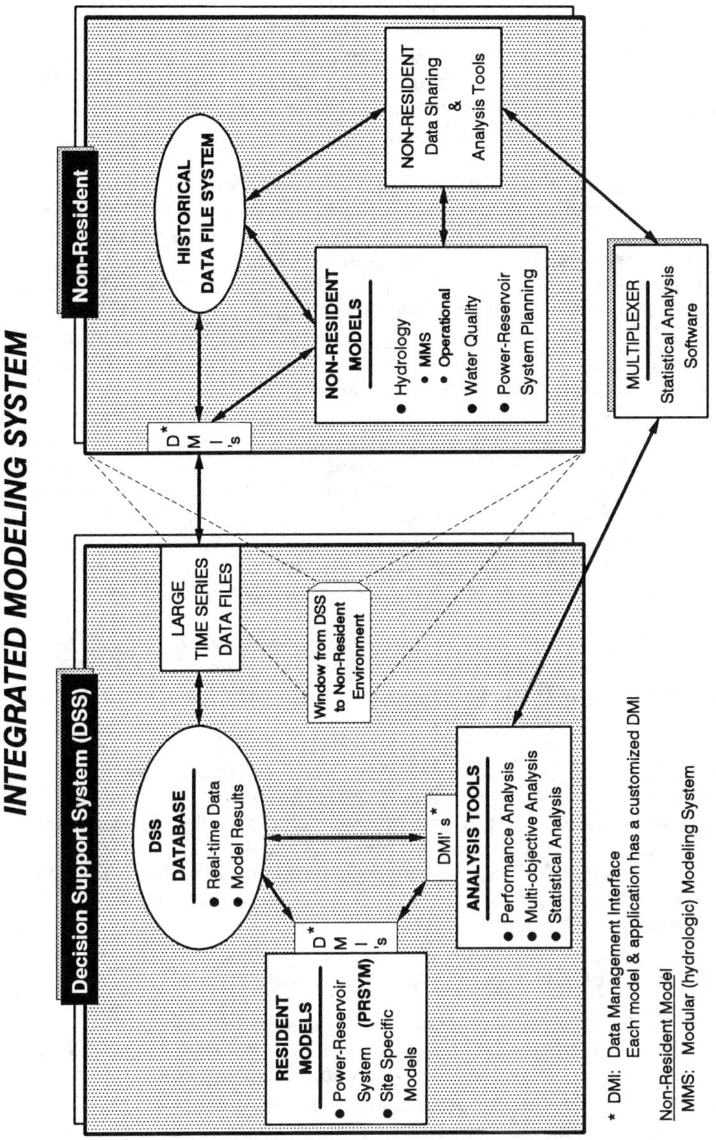

DEVELOPMENT OF ENHANCED TOOLS

Figure 2. Overview of the Integrated Modeling System

Power and Reservoir System Model. A major project task is the development of a generic Power and Reservoir System Model (PRSYM) that can integrate multipurpose reservoir operations with power system economics (Miller et al., 1993). PRSYM will be used to develop near-term and seasonal reservoir schedules based on current operating conditions, as well as to support multi-year evaluations for planning purposes. PRSYM will initially include a simulation module, but will be designed to include an optimization module in the future.

PRSYM must simulate the behavior of a large, complex, multi-purpose reservoir system subject to system constraints and objectives. The model will include capabilities for: routing, a graduated time step (daily, weekly, monthly), sub-system analysis, computation of hydropower generation at varying operating modes, and bulk water quality (temperature and DO) modeling. PRSYM will include a method to express, modify, add/delete, and prioritize constraints and operating policy without code modifications. PRSYM must also include a procedure for estimating the value of water in storage, including the value of hydropower over time in relationship to the overall power system.

Once system constraints have been satisfied, there may still exist a range of feasible solutions. PRSYM will include a simplified method for selecting an 'operationally acceptable' schedule within the available zone of flexibility. For example, the model could minimize deviations from guide curves or, to address power issues, could attempt to meet economic hydro generation targets which reflect forecasted power system conditions.

The PRSYM model also must be easily ported from one river basin to another and easily modified to accommodate utility-specific issues. The object-oriented software/modeling approach supports this need for a data-centered, modular, and extensible model.

Site-Specific Models. The DSS will also include site-specific models to address operational issues at individual power plants or river reaches. A TVA example includes a diffuser mixing model at a nuclear power plant. Other site-specific models could include water quality models at critical river reaches or detailed routing models at particular locations. The DSS will allow for the incorporation of site-specific models relevant to individual utilities or water resource agencies.

Non-Resident Models

Utility staff and reservoir operators expressed the need to use existing utility-specific models in conjunction with the DSS. Examples of such models include hydrologic and streamflow forecasting models, reservoir and power system operational and planning models, and detailed water quality models. To preserve the generic nature of the DSS, while facilitating access to the results of such models, these model will not be directly integrated into the DSS but will be considered 'non-resident' models. Links (referred to as Data Management Interfaces) will be developed to enable sharing of data between the non-resident models, as well as large historical databases, and the DSS.

In addition to utility-specific non-resident models, a generic framework for hydrologic modeling will be accessible through a window from the DSS. The Modular Modeling System, developed by CADSWES and the U.S. Geological Survey (Leavesley et al., 1992), will provide this general methodology for hydrologic modeling.

Analysis Tools

The DSS will provide a common set of tools to analyze the results from resident and non-resident models. These tools will allow for visualization of data in pre-defined formats as well as offer the flexibility to customize analysis of results. The analysis tools will include capabilities to: summarize model results; evaluate system behavior with respect to performance criteria; track model performance; portray multi-objective tradeoffs; and provide general statistical analysis, graphics, and reporting utilities. The system will be designed to accommodate other types of analytical procedures, such as risk and uncertainty analysis, at a future date.

PROJECT SCHEDULE AND FUTURE WORK

The INTEGRAL Project was initiated in 1992. The DSS with supporting models should be operational for the TVA system by the end of 1995. TVA and CADSWES will work with a second utility to demonstrate the generic nature of the DSS framework.

REFERENCES

Leavesley, G.H., P. Restrepo, L.G. Stannard, and M. Dixon, 1992. "The Modular Hydrologic Modeling System - MHMS." Proceedings of the AWRA 28th Annual Conference & Symposium, Managing Water Resources During Climate Change, Reno, Nevada, November.

Miller, B.A., V. Alavian, J.E. Giles, and J.A. Parsly, 1993. Requirements for an Integrated Modeling System to Support Water Resources and Power Operations.

CADSWES, 1993. "TERRA Phase 2 Detailed Design for the TVA/EPRI INTEGRAL Project Decision Support System," University of Colorado, Boulder, CO (in preparation).

Hydraulic Modeling of
High Unit Discharge Energy Dissipators

David R. Moore[1,2], Donald H. Burn[1] A.M. ASCE, Philip D. Wang[3], and Dennis E. Lemke[3]

Abstract

The incorporation of an efficient and economical energy dissipation device is critical with respect to the safe operation of a high unit discharge spillway. The development and design of a high unit discharge energy dissipator usually involves physical hydraulic modeling. This enables the designer to assess the operational viability of the proposed design while analyzing its performance with respect to the hydraulic interaction of the dissipator with the surrounding structure. To fully assess a high unit discharge energy dissipator, it is necessary to examine several variables including dynamic pressure fluctuations, cavitation likelihood, specific energy dissipation efficiency, sweep-out resistance capability, the velocity distribution downstream of the dissipator, and the relationship of the hydraulic jump height curve to the tailwater depth curve. This paper describes a methodology by which one may acquire and manage data with regards to a physical Froude model analysis of a high unit discharge energy dissipation device.

INTRODUCTION

As water travels over a spillway, its potential energy is rapidly converted, via the law of energy conservation, to kinetic energy. This energy must be dissipated so as to reduce the likelihood of tailrace scour and hence prevent the possible catastrophic destabilization of the foundation of the hydraulic structure. This is especially true if the unit discharge is high, say in excess of 75 cubic meters per second per meter of

---

[1] Department of Civil Engineering, University of Manitoba, Winnipeg, Manitoba, Canada, R3T 2N2
[2] Now with Freshwater Institute, Department of Fisheries and Oceans, 501 University Crescent, Winnipeg, Manitoba, Canada, R3T 2N6
[3] Hydro Development, Manitoba Hydro, P.O. Box 815, Winnipeg, Manitoba, Canada, R3C 2P4

spillway width. It is thus necessary to convert a portion of the kinetic energy to potential energy via friction and turbulence. This conversion and/or dissipation process is usually accomplished by a hydraulic energy dissipator and hence the avoidance of catastrophic structural damage may be directly related to the effectiveness of the design process for the hydraulic energy dissipator.

The design and development of a hydraulic energy dissipator usually involves physical hydraulic modeling. In order to effectively model an energy dissipator, it is imperative that the mechanics and/or dynamics involved in its operation be correctly scaled with respect to the Froudian laws of geometric, kinematic, and dynamic similarity. Thus, one must ensure that the important physical processes of the dissipator are capable of being simulated at the model scale used without being significantly distorted.

MODEL EVALUATION METHODOLOGY

In order to fully assess the operational viability of a proposed high unit discharge energy dissipator, it is necessary to examine several variables. Several authors (Toso and Bowers, 1987; Lopardo et al., 1987; Moore, 1992) note the importance of analyzing dynamic pressure fluctuations that are associated with the operation of a high unit discharge energy dissipator. An examination of the statistical nature of the fluctuations and the primary frequency distribution should be conducted so as to assist in the quantification of the macroturbulence phenomena. Such analysis is important for evaluating the likelihood of structural damage to dissipator components (for example, drains). It is very important that any structure(s) in the vicinity of the point of maximum turbulence (for example, the baffle blocks) be analyzed with respect to the longitudinal and lateral distribution of the fluctuations. It should be noted that a small model scale of say 1:70 may under-predict the actual magnitude of the fluctuations by 30%. In addition, the statistical distribution of the fluctuations may not be Gaussian throughout the zone of dissipation (Moore, 1992) and thus the prediction of extreme pressure values should not rely on this distribution.

A second evaluation variable which should be examined is the relative specific energy dissipation efficiency. By computing the quantity of specific energy dissipated by the respective energy dissipation process and dividing this quantity by the specific energy dissipated by a classical hydraulic jump (i.e., under identical hydraulic conditions), it is possible to quantify the efficiency of the dissipation process with respect to a known physical process.

Cavitation is a primary concern in the design and operation of a high unit discharge energy dissipator. Even though cavitation cannot be properly represented with a Froude model, it is possible to evaluate the likelihood of cavitation. This may be done by evaluating the coefficient of variation of a specific pressure fluctuation distribution (i.e., the standard deviation divided by the mean). It is known that cavitation occurs in low pressure zones, hence if the mean pressure is low and the standard deviation is high, there is a high likelihood of cavitation. Therefore the coefficient of variation of the pressure fluctuations should ideally be positive and as low as possible.

Another method of evaluation is the capability of the dissipator to resist hydraulic 'sweep-out' (commonly referred to as basin sweep-out or full sweep-out). Under sweep-out conditions, the hydraulic jump occurs downstream of the energy dissipation structure. Thus, the tailwater level should be lowered below design level during testing so as to assess the capability of the dissipator to resist the downstream movement of the zone of energy dissipation. A useful design tool and method of evaluation is the relationship of the jump height to the tailwater depth, as discussed by Elevatorski (1959). The graphical technique is intended for hydraulic jump energy dissipators, and should be used as a preliminary evaluation method with respect to tailwater depth requirements.

The lateral and longitudinal distribution of velocity downstream of the proposed energy dissipator should be fully examined. An 'unbalanced' velocity distribution may indicate that the design of the dissipator is inappropriate and may subsequently cause erosion within and/or downstream of the dissipator.

EVALUATION CRITERIA RANKING

The design of a high unit discharge energy dissipator usually involves the examination of numerous design variations. A comparative analysis of the resultant data for the aforementioned evaluation criteria permits one to rank each of the designs and hence allows the designer to choose the most effective design. Even though each of the evaluation criteria is important with respect to defining the performance of several proposed designs, some of the criteria are more important to a designer than others. For example, a dissipator which resists both cavitation and sweep-out may be a poor design if it is not an efficient energy dissipator. With this in mind, it is possible to rank the evaluation criteria with respect to their importance vis-a-vis the efficient and economical hydraulic performance of an energy dissipator. Thus, it is suggested that the hydraulic evaluation criteria in order of importance are: energy dissipation efficiency, sweep-out resistance capability, likelihood of cavitation, dynamic pressure fluctuations, velocity distribution downstream of the dissipator, and the relationship of the tailwater depth to the jump height (assuming the dissipator uses a hydraulic jump). A comparative analysis (i.e., one design versus another) of the ranked evaluation criteria, in conjunction with economic and other relevant analyses, will enable the designer to select the most desirable dissipator design.

DATA ACQUISITION AND MANAGEMENT

In general, four types of data must be acquired during the hydraulic modeling analysis of a high unit discharge energy dissipator. Requirements for data collection include dynamic pressure fluctuation data, depth of flow data, flow velocity data, and visual (photographic) data. Due to the dynamic nature of the physical processes associated with the operation of a high unit discharge energy dissipator, it is important that the acquisition process be conducted using a range of flow conditions and for a sufficient period of time so as to record the physical processes and phenomena associated with the operation of the respective dissipator. With this in mind, a computerized data acquisition and storage system should be employed.

Dynamic pressure fluctuations may be efficiently analyzed by employing an acquisition system similar to that used by Toso (1986). Pressure transducers connected in series to amplifiers, filters, and an oscilloscope provide an efficient means by which one can rapidly analyze pressure fluctuations. Several oscilloscopes have software packages which may be used in conjunction with a personal computer so as to convert 'raw' machine language data into an ASCII format. The resultant ASCII files may then be subsequently imported into a spreadsheet for rapid analysis.

Flow depth and velocity data should be measured several times per hour and may be entered either manually or electronically into the aforementioned spreadsheet. By employing a simple macro one can automate the computerized acquisition and storage process, hence increasing the speed of the modeling and design process. Photographs, which should be taken at speeds not less than 1/500 of a second, may also be included (in a digital form) in the data file. Figure 1 shows an example of a data acquisition and storage system.

CONCLUSIONS

A Froude model may be used to examine the operational viability of a proposed high unit discharge energy dissipator. By employing a computerized data acquisition and storage system, it is possible to manage the data from numerous hydraulic model tests more efficiently. The examination of dynamic pressure fluctuations, cavitation likelihood, specific energy dissipation efficiency, sweep-out resistance capability, the velocity distribution downstream of the dissipator, and the relationship of the hydraulic jump height to the tailwater depth permits the designer to assess and rank the performance of the respective dissipators and hence greatly assist in the selection of the most desirable design.

Acknowledgements

The financial support and technical assistance from Manitoba Hydro towards the completion of the research described in this paper is gratefully acknowledged.

Appendix References

Elevatorski, E. A. (1959). Hydraulic Energy Dissipators, McGraw-Hill Book Company, Inc., Toronto.
Lopardo, R.A., De Lio, J.C. and Vernet, G.F. (1987). "The Role of Pressure Fluctuations in the Design of Hydraulic Structures", Proceedings of International Symposium on Design of Hydraulic Structures, Colorado State University, Fort Collins.
Moore, D.R. (1992). "An Analysis of High Unit Discharge Energy Dissipator Options", M.Sc. Thesis, University of Manitoba, Winnipeg, Manitoba.
Toso, J.W. (1986). "The Magnitude and Extent of Extreme Pressure Fluctuations in The Hydraulic Jump", Ph.D. Thesis, University of Minnesota, Minneapolis, Minn.
Toso, J.W. and Bowers, E. (1987). "Design Considerations For Hydraulic Jump Structures", Proceedings of 1987 National Conference on Hydraulic Engineering, Williamsburg, Virginia.

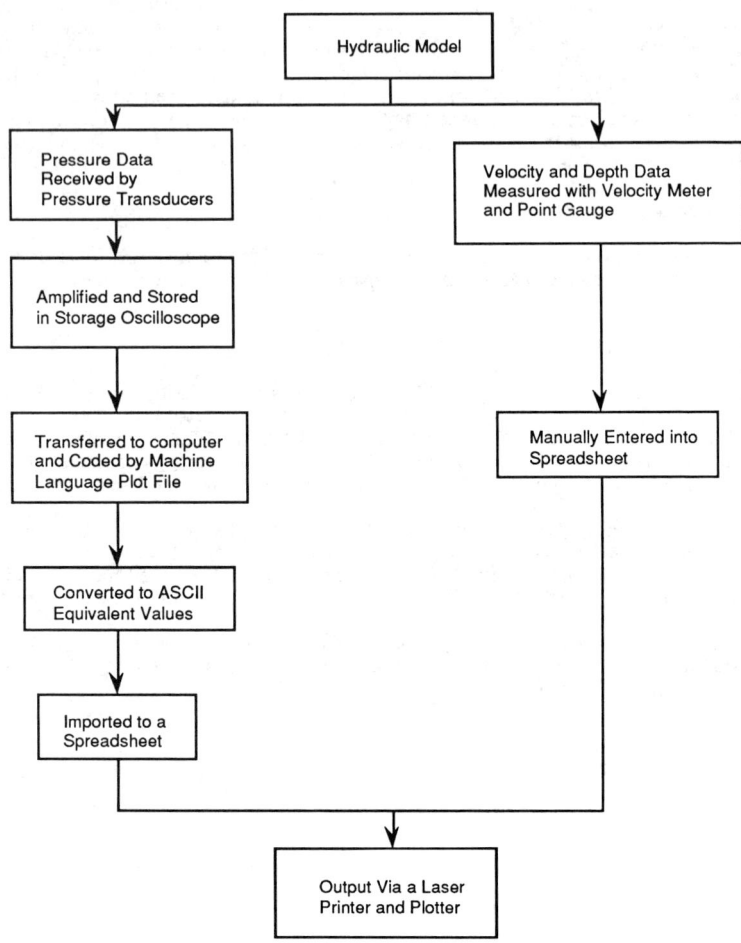

Figure 1 Data Acquisition and Storage System

Model Study of Center Hill Fuse Plug Spillway

Bobby P. Fletcher[1] and Paul A. Gilbert[2]

Abstract

Model tests of a fuseplug spillway mounted on the fixed crest of an embankment dam were conducted to develop a satisfactory design that would fail in a orderly fashion at a prescribed rate. Model materials were processed and manufactured to simulate characteristics of the prototype soil so that behavior in the model would simulate behavior in the prototype. The impervious core in the model consisted of prototype clay taken from the embankment site. Three physical models were used to screen and investigate various designs and to determine if scale effects would significantly influence hydraulic performance of the models. Test results indicated that scale effects were insignificant. A fuseplug design was developed in which the 600-ft-long spillway would fail at a constant rate in approximately 29 min.

Introduction

Because of topographical constraints around the Center Hill Dam in northern middle Tennessee, the US Army Engineer District, Nashville, determined that a fuseplug structure is a reasonable and feasible means of providing emergency spillway capacity. Since the location and timing of a fuseplug washout must be controlled and predictable, it was decided to conduct physical model tests to investigate design, performance, and expected behavior characteristics in the fuseplug structure.

---

[1]Research Hydraulic Engineer, Hydraulics Laboratory, US Army Engineer Waterways Experiment Station, 3909 Halls Ferry Road, Vicksburg, MS 39180-6199.

[2]Research Geotechnical Engineer, Geotechnical Laboratory, US Army Engineer Waterways Experiment Station, 3909 Halls Ferry Road, Vicksburg, MS 39180-6199

The Prototype

The Center Hill Dam is located on the Caney Fork River, 55 miles east of Nashville and 19 miles southwest of Cookeville.

The existing dam includes a concrete gravity dam and a rolled filled embankment with a top elevation of 696.0 (Figure 1).

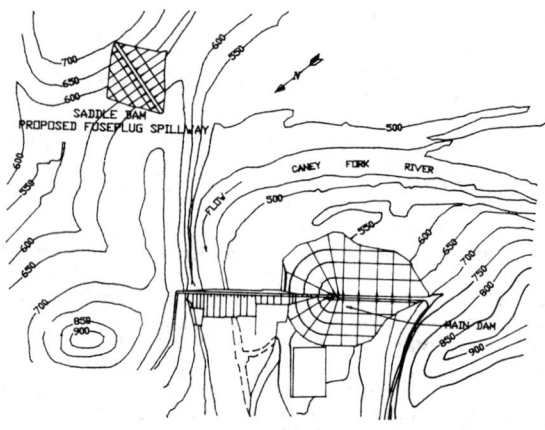

Figure 1. Location Map

About 1,800 ft upstream from the dam on the right reservoir rim is a rolled fill dam (saddle dam) 775 ft long with a crest elevation of 696.0 (Figure 1).

The 34.4-ft-high fuseplug spillway was designed to be mounted on a concrete-lined fixed spillway with a crest elevation of 658.0 and invert length of 600 ft (Figure 2). The fuseplug spillway was designed to be a stable structure except when overtopped. When overtopped, it should erode from the surface of the fixed crest in 30 min leaving a concrete spillway with a 600-ft-long crest. To initiate erosion of the fuseplug, a pilot channel will be located in the center of the fuseplug spillway. Wave barriers will be installed to prevent premature failure induced by over-topping waves.

The Models

Three physical models (Models A, B, and C) were used to investigate the proposed fuseplug spillway. These three models complemented each other and permitted the evaluation of scale effects.

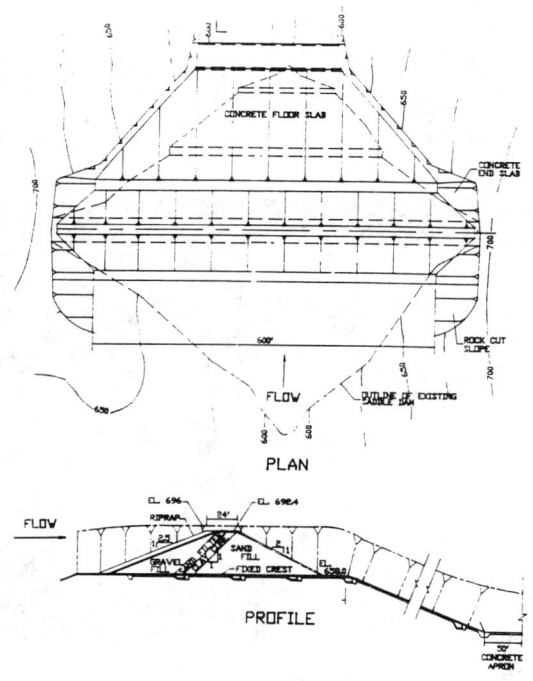

Figure 2. Plan View and Typical Section

Model A was constructed to a 1:40 scale and simulated a 40-ft-long section of the fixed crest and fuseplug spillway in a 1-ft-wide flume (Figure 3). Model A was used as a two-dimensional screening tool for qualitative guidance in operation of the larger models (Models B and C).

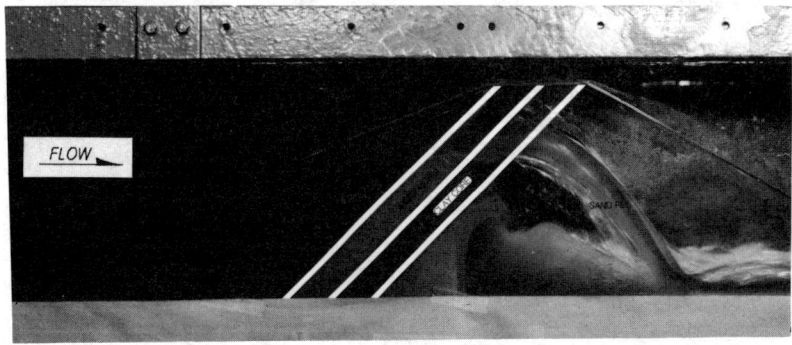

Figure 3. Model A

Model B was constructed to a 1:40 scale in a 12-ft-wide flume (Figure 4). Model B simulated the right abutment (80-ft length), the pilot channel, a 400-ft length of the fixed crest and fuseplug spillways, and 800- and 600-ft lengths of the approach and exit channels, respectively. Model B was used to determine the vertical and lateral erosion rates of the most favorable designs developed in Model A.

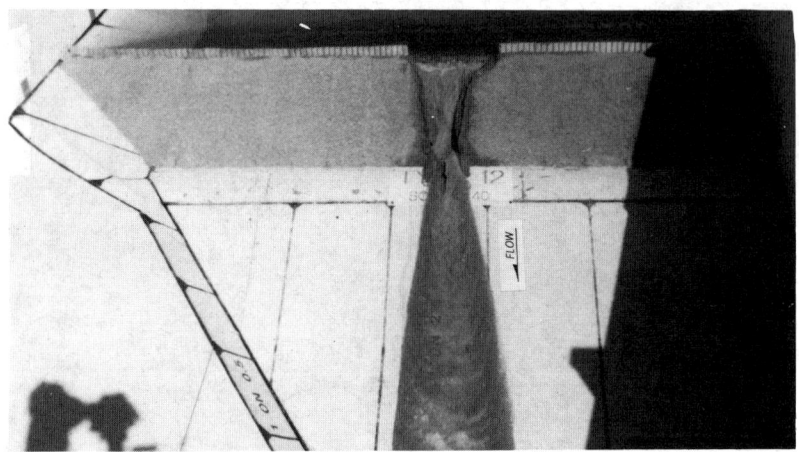

Figure 4. Model B Initial Overtopping

Model C was constructed at a 1:20 scale in the 12-ft-wide flume. The model simulated the center 240 ft of the fixed and fuseplug spillways, the pilot channel, and 400- and 300-ft lengths of the approach and exit channels, respectively. The primary purpose of Model C was to determine if scale effects influenced the results obtained from Model B.

Model Material

Tests were performed by the Geotechnical Laboratory to characterize the soils of which the prototype is to be constructed and to produce appropriate materials for use in the models.

The models were constructed primarily of two granular soils: (a) a manufactured material consisting primarily of a commercially available Ottawa silica sand and a concrete sand and (b) a limestone gravel to serve as the upstream gravel for the fuseplug structure.

It must be stated that if micro dimensions within the prototype (e.g., the dimensions of soil particles) were scaled down exactly along with macro dimensions, then model soil particles would be so small that the dimensionless ratio of inertia force to viscous force (i.e., Reynolds numbers) would be severely distorted between model and prototype. In such a case, it can readily be seen that the resulting soil particles would be so small and drag forces so disproportionately large that particle settlement velocities in the model would be much too small. Therefore, certain adjustments to soil grain-size distribution must be made in classical Froude modeling to better match viscous force similitude between model and prototype. One way to achieve this similitude is to scale settlement velocities directly between model and prototype.

A significant element in the zoned embankment being modeled in this investigation is the clay core, which serves as a impervious barrier to water seepage through the structure. Clay is characterized by cohesive strength; one of the weaknesses of Froude modeling is that cohesive strength does not model well. It may be shown from analysis of the governing equations that cohesive strength will be too large in the models. To achieve similitude of cohesive strength, the thickness of the clay core was investigated in the models at thicknesses equal to one-third its correctly scaled value.

Model Tests

A typical test for the two-dimensional Hydraulics Laboratory model (Model A) consisted of raising the water surface upstream of the spillway to el 692.4 to permit an initial 0.4-ft depth of flow over the crest of the pilot channel. As the spillway eroded, the upstream water surface was held constant at el 692.4 by manually increasing the model inflow to compensate for the gradually increasing flow through the expanding breech. Models B and C (three-dimensional models) were operated similar to Model A except after breaching, the lateral erosion rate was measured using timing lines located on the downstream slope of the spillway. Model tests were recorded with high resolution motion picture photography.

Two-dimensional tests conducted at a 1:40 scale in the 1-ft-wide section model (Model A) permitted a qualitative evaluation of vertical failure characteristics of the fuse-plug spillway. Tests in Model A were directed primarily toward developing a design that would readily permit initiation of failure by overtopping.

Three-dimensional tests conducted at a 1:40 scale in the 12-ft-wide flume (Model B) revealed the need for lowering the invert of the 50-ft-long pilot channel by 0.5 ft to permit an initial depth of flow of 0.9 ft rather than 0.4 ft.

About 7 min (prototype) was required for flow through the pilot channel to erode vertically down to the surface of the fixed crest and open a 75-ft-long breach (measured at the surface of the fixed crest). Lateral erosion on each side of the breach occurred at a rate of 12.5 ft per minute (prototype). Breaching and failure of the 600-ft-long spillway required 28 min.

Three-dimensional tests were conducted in the 12-ft-wide flume (Model C) at a 1:20 scale to determine if scale effects had a significant influence on hydraulic performance. The spillway failed in 29.8 min in Model C. The failure times of the 600-ft-long spillway computed from results obtained from Models B and C are within 6.5 percent of each other. This is good correlation and indicates that scale effects did not significantly affect model results.

Acknowledgement

The tests described and data presented, unless otherwise noted, were obtained from Research conducted for the US Army Engineer District, Nashville, by the US Army Engineer Waterways Experiment Station, Vicksburg, MS. Permission was granted by the Chief of Engineers to publish this information.

Appendix I. Conversion Factors, Non-SI to SI Units

Non-SI units of measurements used in this paper can be converted to SI units as follows:

To Convert	To	Multiply By
cubic foot per second (ft^3/sec)	cubic meter per second (m^3/s)	0.28
foot	meter (m)	0.3048
gallon (US liquid)	cubic meter (m^3)	0.004
mile	kilometer (km)	1.61

Appendix II. Reference

Fletcher, B. P. and Gilbert, P. A. (1992). "Center Hill Fuseplug Spillway Caney Fork River, Tennessee; Hydraulic Model Investigation." Technical Report No. HL-92-15, U.S. Army Engineer Waterways Experiment Station, Vicksburg, MS.

AN APPLICATION OF NONHOMOGENEOUS POISSON PROCESS IN SEDIMENT INFILTRATION INTO GRAVEL BED

Fu-Chun Wu[1] and Hsieh Wen Shen, Member ASCE[2]

Abstract

The compound nonhomogeneous Poisson process (CNHPP) model (Shen & Todorovic, 1971) is found applicable in sediment infiltration into gravel bed due to the unsteady and nonuniform characteristics of the sediment infiltration. In order to apply this model, we require the two model parameters: i) intensity function in time domain $\lambda_1(t)$, and ii) intensity function in space domain $\lambda_2(x)$. This paper provides a way to determine these two parameters from the experiments.

Introduction

It is reported that the intrusion of sediment particles into the gravel bed substrate is of important engineering concern to the following aspects:

1. Fisheries: spawning streambeds (Lisle, 1989)
2. River bed contamination (Cerling et al., 1990)
3. Groundwater recharge (Schuh, 1991)

Since the promising general nonhomogeneous Poisson process model was proposed in 1971, there was hardly any application of this model to sediment transport problem in the past two decades because it is very hard for one to imagine a case in which the step length distribution varies with the distance but not with the time while the rest period distribution varies with the time but not with the distance, and also the two intensity functions $\lambda_1(t)$ and $\lambda_2(x)$ required in this model are not easy to determine from the experiment (Hung & Shen, 1971).

---

[1] Graduate student, Dept. of Civil Engineering, Univ. of California at Berkeley

[2] Professor, Dept. of Civil Engineering, Univ. of California at Berkeley

Observations in the laboratory indicate that the sand infiltration into the gravel bed substrate is an unsteady and nonuniform process, therefore the application of the CNHPP model in this case seems reasonable. In order to apply this model, we require the two intensity functions which satisfy the restrictions stated above. A first-order finite-difference approach to evaluate these two model parameters is presented in the following section.

Determination of Intensity Functions

According to the CNHPP model (Shen & Todorovic, 1971), the cumulative probability distribution function (CDF) $F_t(x)$ is

$$F_t(x) = P\{X_t \leq x\}$$

$$= \exp\left[-\int_{t_0}^{t} \lambda_1(s)ds\right] \exp\left[-\int_{x_0}^{x} \lambda_2(s)ds\right] \sum_{n=0}^{\infty} \sum_{j=n}^{\infty} \frac{\left[\int_{t_0}^{t} \lambda_1(s)ds\right]^n \left[\int_{x_0}^{x} \lambda_2(s)ds\right]^j}{n! j!} \quad (1)$$

$$= \exp[-\Lambda_1(t)] \exp[-\Lambda_2(x)] \sum_{n=0}^{\infty} \sum_{j=n}^{\infty} \frac{[\Lambda_1(t)]^n [\Lambda_2(x)]^j}{n! j!}$$

where X_t = x-direction displacement of the particle at time t
$\lambda_1(t)$ = intensity function in time domain with unit [1/time]
$\lambda_2(x)$ = intensity function in space domain with unit [1/length]

By Taylor's expansion, for small Δt and Δx,

$$\Lambda_1(t+\Delta t) = \Lambda_1(t) + (\Delta t)\Lambda_1'(t) + O[(\Delta t)^2] \approx \Lambda_1(t) + (\Delta t)\Lambda_1'(t) \quad (2)$$

$$\Lambda_2(x+\Delta x) = \Lambda_2(x) + (\Delta x)\Lambda_2'(x) + O[(\Delta x)^2] \approx \Lambda_2(x) + (\Delta x)\Lambda_2'(x) \quad (3)$$

From Equation (1), we know

$$\Lambda_1'(t) = \lambda_1(t) \quad (4)$$

$$\Lambda_2'(x) = \lambda_2(x) \quad (5)$$

From Equations (2), (3), (4), and (5), we have

$$\Lambda_1(t+\Delta t) = \Lambda_1(t) + (\Delta t)\lambda_1(t) \quad (6)$$

$$\Lambda_2(x+\Delta x) = \Lambda_2(x)+(\Delta x)\lambda_2(x) \tag{7}$$

Also,

$$\begin{aligned}
\left[\Lambda_1(t+\Delta t)\right]^n &= \left[\Lambda_1(t)+(\Delta t)\lambda_1(t)\right]^n \\
&= \left[\Lambda_1(t)\right]^n + n\left[\Lambda_1(t)\right]^{n-1}(\Delta t)\lambda_1(t) + O\!\left[(\Delta t)^2\right] \\
&\approx \left[\Lambda_1(t)\right]^n + n\left[\Lambda_1(t)\right]^{n-1}(\Delta t)\lambda_1(t)
\end{aligned} \tag{8}$$

$$\begin{aligned}
\left[\Lambda_2(x+\Delta x)\right]^j &= \left[\Lambda_2(x)+(\Delta x)\lambda_2(x)\right]^j \\
&= \left[\Lambda_2(x)\right]^j + j\left[\Lambda_2(x)\right]^{j-1}(\Delta x)\lambda_2(x) + O\!\left[(\Delta x)^2\right] \\
&\approx \left[\Lambda_2(x)\right]^j + j\left[\Lambda_2(x)\right]^{j-1}(\Delta x)\lambda_2(x)
\end{aligned} \tag{9}$$

Part I. Determination of $\lambda_2(x)$

By the definitions given in the CNHPP model (Shen & Todorovic, 1971), and the equations listed above, we can simplify Equation (1) as

$$F_t(x+\Delta x) = \exp\!\left[-(\Delta x)\lambda_2(x)\right] \Big\{ F_t(x) + \\
\left[(\Delta x)\lambda_2(x)\right] \cdot \left[P\!\left(E_0^{t \triangleright t}\right) + \sum_{n=1}^{\infty} P(X_{n-1} \le x) \cdot P\!\left(E_n^{t \triangleright t}\right)\right] \Big\} \tag{10}$$

Then, from Equations (1) and (10)

$$\frac{F_t(x+\Delta x)}{F_t(x)} = \exp\!\left[-(\Delta x)\lambda_2(x)\right] \Big\{ 1 + \\
\left[(\Delta x)\lambda_2(x)\right] \cdot \left[\frac{P\!\left(E_0^{t \triangleright t}\right)}{P(X_t \le x)} + \frac{\sum_{n=1}^{\infty} P(X_{n-1} \le x) \cdot P\!\left(E_n^{t \triangleright t}\right)}{P(X_t \le x)}\right] \Big\} \tag{11}$$

Let $\dfrac{\sum_{n=1}^{\infty} P(X_{n-1} \le x) \cdot P(E_n^{t_0,t})}{P(X_t \le x)} = a(x)$ at a fixed time t, and $(\Delta x)\lambda_2(x) = k(x)$,

Equation (11) becomes

$$\dfrac{F_t(x+\Delta x)}{F_t(x)} = \exp[-k(x)] \cdot \left\{ 1 + k(x) \cdot \left[\dfrac{P(E_0^{t_0,t})}{P(X_t \le x)} + a(x) \right] \right\} \quad (12)$$

Consider the experimental setup in Figure 1, at time t_0, total amount M_t of sand was put onto the surface of the clean gravel bed, and then at time t, sand distribution within the gravel bed is measured, where m_0 is the amount of sand stays on the bed surface, m_1 is the amount of sand deposit in the first layer, etc. Each layer is of equal thickness Δx, and thus $x_j = x_0 + j\Delta x$ for $j = 0, 1, \ldots, N$. Obviously,

$$F_t(x_j) = P\{X_t \le x_j\} = \dfrac{\left(\sum_{i=0}^{j} m_i\right)}{M_t} = \dfrac{M_j}{M_t} \quad (13)$$

$$P\left\{ E_0^{t_0,t} \right\} = \dfrac{m_0}{M_t} \quad (14)$$

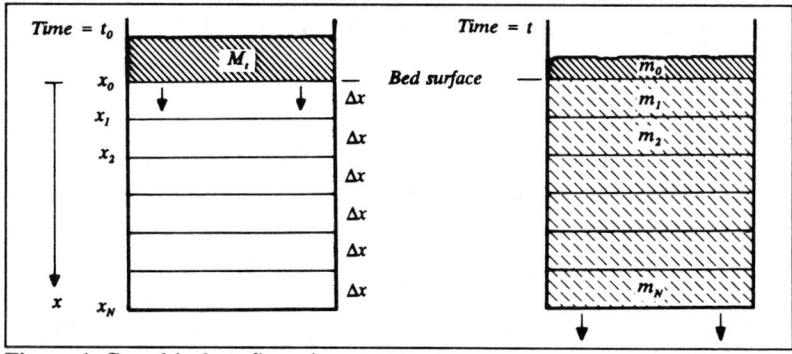

Figure 1. Gravel bed configuration

From Equations (12), (13), and (14), we obtain

$$\dfrac{F_t(x_{j+1})}{F_t(x_j)} = \dfrac{M_{j+1}}{M_j} = \exp(-k_j) \cdot \left[1 + k_j \left(\dfrac{m_0}{M_j} + a_j \right) \right] \quad (15)$$

where $k_j = k(x_j)$ and $a_j = a(x_j)$.

Similarly,

$$\frac{F_t(x_{j-1})}{F_t(x_j)} = \frac{M_{j-1}}{M_j} = \exp(k_j) \cdot \left[1 - k_j \left(\frac{m_0}{M_j} + a_j \right) \right] \quad (16)$$

To solve k_j and a_j, use the system of Equations (15) and (16) for $j = 1, 2, \ldots, N-1$. Then the intensity function $\lambda_2(x_j) = k_j /\Delta x$ can be calculated at each point.

Part II. Determination of $\lambda_1(t)$

In the same manner, $\lambda_1(t)$ can be evaluated as follows:

$$\frac{F_{t+\Delta t}(x_j)}{F_t(x_j)} = \frac{M_j(t+\Delta t)}{M_j(t)} = \exp[-k(t)] \cdot [1 + k(t) \cdot a(t)] \quad (17)$$

$$\frac{F_{t-\Delta t}(x_j)}{F_t(x_j)} = \frac{M_j(t-\Delta t)}{M_j(t)} = \exp[k(t)] \cdot [1 - k(t) \cdot a(t)] \quad (18)$$

where $\dfrac{\sum_{m=0}^{\infty} P(X_{m+1} \le x) \cdot P(E_m^{t_0,t})}{P(X_t \le x)} = a(t)$ at a fixed location $x = x_j$, $(\Delta t)\lambda_1(t) = k(t)$,

and $M_j(t) = \sum_{i=0}^{j} m_i$ at time $= t$, $M_j(t \pm \Delta t) = \sum_{i=0}^{j} m_i$ at time $= t \pm \Delta t$.

Use the system of Equations (17) and (18) to solve $k(t)$ and $a(t)$, and then calculate the intensity function in time domain $\lambda_1(t) = k(t)/\Delta t$.

Results

To investigate the intensity functions $\lambda_1(t)$ and $\lambda_2(x)$, so far we have performed three series of sediment infiltration experiments to measure the sand distribution (i.e. m_0, m_1, m_2, ..., m_N) in each layer of the gravel bed at different times. The intensity functions $\lambda_1(t)$ and $\lambda_2(x)$ calculated by solving the systems of nonlinear Equations (17) and (18), (15) and (16) are shown in Figures 3 and 2 respectively. From these results, we find
i) The intensity function $\lambda_2(x)$ only varies with the bed depth (Figure 2).
ii) The intensity function $\lambda_1(t)$ only varies with the time (Figure 3).
iii) From the curve obtained under different sediment input and different seepage rate conditions, it seems that the intensity functions $\lambda_1(t)$ and $\lambda_2(x)$ are not functions of the sediment input and the seepage rate. They are the properties corresponding to the ratio (gravel size/sand size).
iv) More experimental data are needed to establish the functional forms of the intensity functions $\lambda_1(t)$ and $\lambda_2(x)$ with the time and the depth.

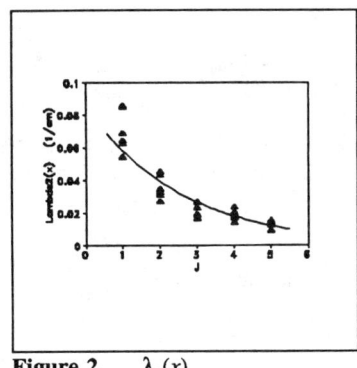

Figure 2. $\lambda_2(x)$

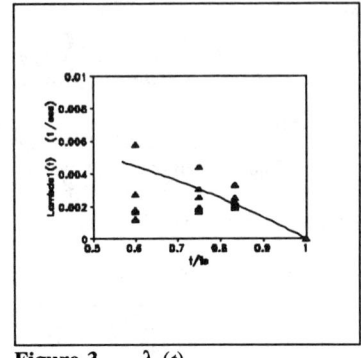

Figure 3. $\lambda_1(t)$

Conclusion

1. A first-order approximation to evaluate the intensity functions $\lambda_1(t)$ and $\lambda_2(x)$ is introduced.
2. We have proved that the CNHPP model can be applied to the case of sediment infiltration into the gravel bed because the intensity functions $\lambda_1(t)$ and $\lambda_2(x)$ satisfy the restrictions of this model.
3. The intensity functions $\lambda_1(t)$ and $\lambda_2(x)$ are the properties corresponding to the ratio of the gravel size relative to the sand size.

References

1. Cerling, T.E., Morrison, S.J., Sobocinski, R.W., and Larsen, I.L., "Sediment-Water Interaction in a Small Stream: Adsorption of ^{137}Cs by Bed Load Sediments", Water Res. Res., Vol. 26, No. 6, 1990.
2. Hung, C.S., and Shen, H.W., "Research in Stochastic Models for Bed-Load Transport", IASPS Symposium of Hydrology, Tucson, Arizona, Aug. 29 - Sept. 2, 1971.
3. Lisle, T.E., "Sediment Transport and Resulting Deposition in Spawning Gravels, North Coastal California", Water Res. Res., Vol. 25, No. 6, 1989.
4. Schuh, W.M., "Effects of an Organic Mat Filter on Artificial Recharge with Turbid Water", Water Res. Res., Vol. 27, No. 6, 1991.
5. Shen, H.W., and Todorovic, P.N., "A General Stochastic Model for the Transport of Sediment Bed Material", Stochastic Hydraulics, C.L. Chiu ed., University of Pittsburgh, Pittsburgh, Pa., 1971.

Variation of Froude Number with Discharge for Large-Gradient Streams

By Kenneth L. Wahl[1], Member, ASCE

Abstract

Under channel-control conditions, the Froude number (F) for a cross section can be approximated as a function of the ratio $R^{2/3}/d^{1/2}$, where R is the hydraulic radius and d is the average depth. For cross sections where this ratio increases with increasing depth, F can also increase with depth. Current-meter measurement data for 433 streamflow gaging stations in Colorado were reviewed, and 62 stations were identified at which F increases with depth of flow. Data for four streamflow gaging stations are presented. In some cases, F approaches 1 as the discharge approaches the magnitude of the median annual peak discharge. The data also indicate that few actual current meter measurements have been made at the large discharges where velocities can be supercritical.

Introduction

A controversy persists over the behavior of flow velocity in the channels of natural rivers. Some investigators believe that upper regime (supercritical) flows cannot be sustained in natural channels (Jarrett, 1984, 1987; Trieste, 1992). They cite the lack of actual observations of supercritical flow as evidence in support of this theory. Yet hydraulic relations indicate that, under certain conditions, the Froude number for cross sections having particular geometric properties will increase with increasing flow depth (stage). Under those conditions, supercritical flow can occur at large discharges if flow approaches critical depth at moderate discharges.

The data needed to settle this controversy have not been forthcoming for several reasons. Velocity is known to positively correlate to stream slope, but most streamflow data are collected on streams of only moderate slope. Also, actual velocity measurements of large floods are difficult to obtain; such floods are commonly of short duration, and their measurement is not without personal risk. Consequently,

---

1. Hydrologist, U.S. Geological Survey, WRD, Lakewood, Colorado

most measurements of major floods for large-gradient streams are done by indirect methods after the flood waters have receded.

The U.S. Geological Survey (USGS) collects streamflow data at approximately 7,000 continuous-record streamflow gaging stations. Collectively, thousands of current-meter measurements of discharge are made each year at these gaging stations. Summary data from most measurements made since about 1985 and for selected measurements made before that time are in computer files of the USGS. The author is presently examining these data to identify streams that may provide information about flow regimes. This paper reports the results of initial phases of that examination.

Review of Hydraulic Relations

The average velocity, V, of a river can be described by the Manning equation if channel control prevails. Although the Manning equation is empirical, the applicability of the relation is widely accepted for steady flow conditions. One form of that equation is:

$$V = \frac{C_0}{n} R^{2/3} S^{1/2}, \tag{1}$$

where C_0 is a constant (1.0 for metric units and 1.49 for English units), n = Manning roughness coefficient, R = hydraulic radius (cross-sectional area divided by wetted perimeter), and S = friction slope.

Similarly, the flow regime of a river can be described by the Froude number, F. One form of that equation is:

$$F = \frac{\sqrt{\alpha} V}{\sqrt{g d}}, \tag{2}$$

where α = the kinetic-energy correction factor, g = acceleration due to gravity, and d = the average depth.

Recognizing that $C_0/\sqrt{g}$ is a constant, C_1, equations 1 and 2 can be combined and simplified to yield:

$$F = \sqrt{\alpha} C_1 \frac{S^{1/2}}{n} \frac{R^{2/3}}{d^{1/2}}. \tag{3}$$

If the behavior of the individual terms is known, the behavior of F can be predicted. Although α is often treated as equal to 1 for a prismatic channel cross section, α is actually greater than 1 unless velocities at all points of the section are equal. However, unless cross-section shape changes significantly as depth increases, α will remain relatively constant or will perhaps increase slightly with increasing depth.

Studies in the United States (Barnes, 1967; Limerinos, 1970; Jarrett, 1984) and

New Zealand (Hicks and Mason, 1991) have shown that, in general, n decreases and friction slope increases with increasing depth. Roughness elements are submerged as the water depth increases, and channel control becomes fully developed. Bank effects that increase with stream depth, such as heavy or overhanging vegetation, will alter this relation. Where the relation between $S^{1/2}/n$ and discharge is either constant or increasing, equation 3 shows that F will increase if $R^{2/3}$ increases at a greater rate than $d^{1/2}$. This condition is a function of the shape of the channel cross section, and will almost always occur on very wide and shallow channels because R approaches d. The ratio $S^{1/2}/n$ was determined for data collected by Jarrett (1984) for Colorado streams with gradients of more than 0.002. That ratio plotted against a dimensionless discharge (Figure 1) shows that $S^{1/2}/n$ generally increases with discharge. Individual relations are shown in Figure 1 only for Jarrett's (1984) sites numbers 1 and 17; the relations for each of the other sites followed those general patterns. Overhanging vegetation affected n values at three sites reported by Jarrett (1984); data for those sites are not included in Figure 1.

In this paper, the measured discharges (Q) have been divided by the discharge of the median annual peak discharge (Qm) to produce a dimensionless discharge. No significance is attached to the median annual peak discharge; it simply serves as a convenient and easily defined index discharge that provides perspective about the relative magnitudes of the measured discharges and facilitates comparison between streams.

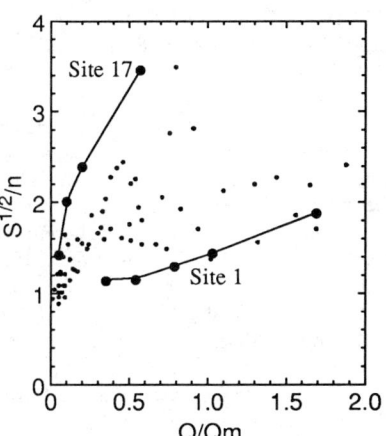

Figure 1. Relation between $S^{1/2}/n$ and Q/Qm for data collected by Jarrett (1984) from 18 Colorado streams.

Data Analysis

Because of the criteria used to locate USGS gaging sites, current-meter measurements at a specific cross section might not reflect hydraulic conditions over a long reach of channel; however, measuring sections are generally located where flow is as uniform as possible. Measurements at low to moderate stages are generally made by wading and are not always made at exactly the same location along the stream. When stages are too high for wading, however, measurements are generally made at a fixed location from a bridge or cableway.

The data analysis used computer files of summary data from current-meter measurements made in Colorado. The measurement data in the computer files includes Q, V, the top width (W), and the cross-sectional area (A). Because $d = A/W$, these data are sufficient to compute F for the measurements, if α is assumed equal to 1. Because velocity is nonuniform in the cross section, α is known to be greater than 1; therefore, the true Froude number for the measurements will be at least as great as the calculated value. Data were retrieved for more than 29,000 individual discharge measurements from 433 gaging stations in Colorado. Additional data are available, but for the present the search was limited to the data in the computer files.

Stations were identified where Froude numbers generally increased with depth and where at least one measurement had a computed Froude number of 0.8 or greater. A total of 62 stations met these criteria. Data from four representative stations are presented in Figure 2. The USGS station numbers and names, drainage areas, and the record lengths for the four gaging stations are: 07105800 Fountain Creek at Security, 1280 km^2, 27 years; 09046600 Blue River near Dillon, 308 km^2, 34 years; 09074800 Castle Creek near Aspen, 83.4 km^2, 22 years; and 09241000 Elk River at Clark, 559 km^2, 73 years.

The four streams used as examples in figure 2 represent a diverse set of channel conditions. Fountain Creek has a sand bed that is constrained from further downcutting by the underlying shale. The slope of Fountain Creek is about 0.011, and the median diameter (d_{50}) of the bed material is 2.33 mm (Von Guerard, 1989). At Q_m, $d=0.9$ m and $W/d=50$. The remaining three sites have cobble and boulder streambeds. For the Blue River, $S=0.013$, $d_{50}=110$ mm (Marchand and others, 1984), and, at Q_m, $d=0.5$ m and $W/d=30$. For Castle Creek, $S=0.021$ and $d_{50}=91$ mm (Andrews, 1984); at Q_m, $d=0.5$ m and $W/d=30$. For Elk River, $S=0.006$ and $d_{50}=210$ mm (Jarrett, 1984); at Q_m, d=1.2 m and $W/d=20$.

High-water measurements for Castle Creek are made from a concrete bridge. The bridge does not contract the flow at the stages that have been measured. High-water measurements for the other three gaging stations are made from cableways.

The data for these four streams are typical of data for many gaging stations examined, in that few current-meter measurements are available for discharges greater than the median annual peak discharge. Although the relations suggest that the Froude numbers will equal or exceed 1 for very large floods, data are not available to support extrapolation to larger discharges.

Chow (1959) noted that α has been found to vary from about 1.03 to 1.36 for fairly straight prismatic channels. A value for α was computed from the actual current-meter measurements for the four largest measurements made at Blue River near Dillon. Those values of α ranged from 1.13 to 1.17 and averaged 1.15. If F were adjusted for the influence of α using an average α of 1.15, the true F values would be about 7 percent larger than the values shown in Figure 2. Given the possible range in α, one can reasonably assume that the actual values of F are about 10 percent larger than the values computed from measured average velocities and shown in this paper.

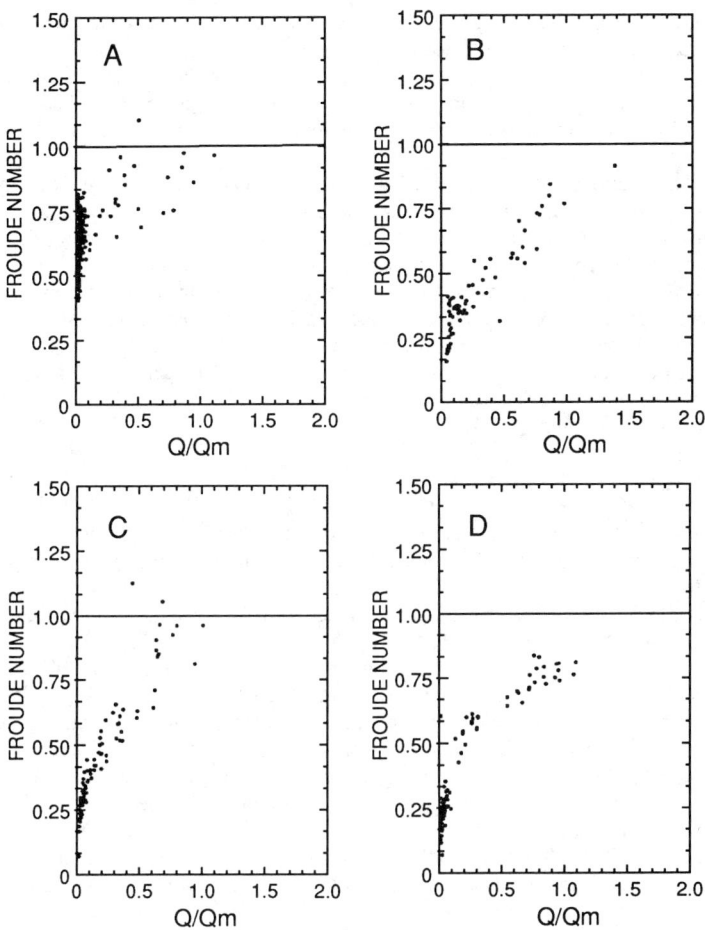

Figure 2.--Relation between Froude number and discharge for (A) Fountain Creek at Security, CO; (B) Blue River near Dillon, CO; (C) Castle Creek near Aspen, CO; and (D) Elk River at Clark, CO.

Discussion

Some investigators believe that streams adjust their channels to achieve subcritical flow. The data shown here do not contradict that. However, those adjustments require that flows be sustained for a sufficient time for the channel to reach equilibrium and that the channel has complete freedom to adjust. Unless both conditions are satisfied, equilibrium might not be reached. Thus, the mechanisms can exist to produce supercritical velocities during large discharges of short duration. The scatter shown in Figure 2 suggests that the channels in these example stations do change, but the data also show that Froude numbers greater than 1, indicating supercritical flow, might occur for the infrequent large discharges.

The relations shown in Figure 2 are not isolated cases. There are more than 60 active gaging stations in Colorado where current-meter measurement data show Froude numbers increasing with discharge and, thus, a potential for supercritical flow at large discharges. Because measuring sections are generally located in reaches where flows are as uniform as possible, the flow regime in the measurement cross section might be representative of a longer reach. Without additional data, however, nothing is known about the length of reach that exhibits the large Froude numbers. The results of the data analyses now being conducted should serve to identify gaging sites where additional data can be collected. Such data could provide answers to the questions being raised.

References

Andrews, E. D., 1984, Bed-material entrainment and hydraulic geometry of gravel-bed rivers in Colorado: Geological Society of America Bulletin, Vol. 95, p. 371-378.

Barnes, H.H., 1967, Roughness characteristics of natural channels: U.S. Geological Survey Water-Supply Paper 1849, 213 p.

Chow, V.T., 1959, Open channel hydraulics: McGraw-Hill, New York, 680 p.

Hicks, D.M., and Mason, P.D., 1991, Roughness Characteristics of New Zealand Rivers: Water Resources Survey, Kilbirnie, Wellington, New Zealand, 329 p.

Jarrett, R.D., 1984, Hydraulics of high-gradient streams: American Society of Civil Engineers Journal of Hydraulic Engineering, Vol. 110, No. 11, pp. 1519-1539.

_____, 1987, Errors in slope-area computations of peak discharges in mountain streams: Journal of Hydrology, Vol. 96, pp. 53-67.

Limerinos, J.T., 1970, Determination of the Manning coefficient from measured bed roughness in natural channels: U.S. Geological Survey Water-Supply Paper 1898-B, 47 p.

Marchand, J.P., Jarrett, R.D., and Jones, L.L., 1984, Velocity profile, water-slope, and bed-material size for selected streams in Colorado: U.S. Geological Survey Open-File Report 84-733, 82 p.

Trieste, D.J., 1992, Evaluation of supercritical/subcritical flows in high-gradient channel: American Society of Civil Engineers Journal of Hydraulic Engineering, Vol. 118, No. 8, pp. 1107-1118.

Von Guerard, P.B., 1989, Sediment transport characteristics and effects of sediment transport on benthic invertebrates in the Fountain Creek drainage basin upstream from Widefield, southeastern Colorado, 1985-88: U.S. Geological Survey Water-Resources Investigations Report 89-4161, 133 p.

Prediction of Gravel Transport Using Parker's Algorithm

David R. Dawdy[1], A.M., ASCE
Wen C. Wang[2], M., ASCE

Abstract

This paper demonstrates that sediment discharge in gravel streams can be reasonably estimated by calibrating the Parker and Klingeman procedure. This paper also identifies the need for a protocol for data collection for bedload prediction using the Parker and Klingeman procedure.

Introduction

One of the major points of technical argument raised by the Forest Service in a reserved water rights case in Water District 1 in Colorado concerned sediment transport in high mountain streams. In order to maintain "favorable conditions of flow", the Forest Service asserted that the channels had to be maintained by the flowing stream. That required that sufficient flow remain in the stream so that the sediment delivered to the stream system could be moved through the system, rather than depositing and clogging the channel. In order to determine that the sediment could be moved through the system, the sediment transport had to be predicted for the range of discharges for which the Forest Service claimed reserved rights.

Field Data and Calibration

Several data sets were collected to demonstrate that sediment discharge in gravel streams could be predicted.

---
[1]Consultant, San Francisco, California
[2]Multech Engineering Consultants, San Jose, California

The data were analyzed using the Parker and Klingeman procedure, which includes the effect of a "hiding factor" (Parker and Klingeman, 1982). That procedure includes a physically-based but semi-empirical engineering equation which contains three parameters. Those three parameters are first, a reference critical shear value, TRS50, which is the shear at which the median diameter bed material moves. Second is an exponent which relates the shear value required to move any other size present in the bed to TRS50. The third and final parameter is a scaling parameter, W_r^*, which determines the amount of bedload movement for a given shear in excess of the critical shear for a given size of material. The prediction of the size distribution of the bedload is based on the size distribution of a parent distribution. That parent distribution may be either for the pavement on the bed or the sub-pavement under the pavement. The sub-pavement was used for all calculations shown in this paper.

One set of data chosen for prediction in the Water District 1 case was for Coon Creek, an experimental watershed in the Medicine Bow National Forest in Wyoming. The site is described in a report by Marc Wilcox (1989). The Forest Service traps bedload in a structure in the bed of Coon Creek. Four years of data on total load and flow duration were available. Various estimates of bedload were made with the Coon Creek 1988 data for a cross-section 100 feet upstream of the sediment weir, at the site of the discharge measurement section. The reference Shields stress (TRS50, bed shear) and the exponent for the hiding factor for the subpavement calculations were varied. Results were:

	Shear (TRS50)	Exponent	D50 mm	Bedload tons/day
Measured Bedload			2.	48.16
Parker Values	0.0876	0.982	32.	51.7
Fitted Value	0.0876	0.95	13.5	59.5
Fitted Value	0.0876	0.9	4.	80.4
Fitted Value	0.09	0.9	3.75	71.7
Fitted Value	0.1045	0.87	2.09	48.3

Raising the reference bed shear, TRS50, causes the predicted load to decrease. Lowering the exponent causes the predicted median diameter to decrease. Where bedload is measured and the bedload size is determined, the two parameters can be calibrated. Calibration depends on the proper determination of the pavement and sub-pavement size distributions. Under what is termed equal mobility, the coarser half of the bedload must move in the same amount as the finer half. This holds whether the calculation is based on the subpavement or the pavement. However, the exponents and the reference stresses used for estimation using the distributions of pavement and sub-pavement must be different. That this is so is shown by Parker and Klingeman's Figure 5 in their paper, where the slope for the subpavement is -0.982 (their equation 21), whereas the slope for the pavement is about -0.885 (estimated value not given in Parker and Klingeman).

True "equal mobility" a la Parker and Klingeman cannot exist because the exponents for the pavement and sub-pavement size distributions are different. If all sizes move together, and the movement is proportional to the size in the parent distribution, then the resulting size of the bedload moving must be similarly proportional to the parent distribution. This cannot hold true for both the pavement and sub-pavement size distributions. The bedload size distribution is smaller than the sub-pavement, which is smaller than the pavement. Thus, the Parker and Klingeman algorithm really predicts "almost equal mobility", or "relative mobility".

A "Meyer-Peter and Mueller mobility" has a Parker and Klingeman exponent of zero, and gives too small a bedload size distribution when used to predict gravel movement. Therefore, the "true mobility" must be somewhere in between the Meyer-Peter and Mueller approach and equal mobility. If Parker and Klingeman's algorithm is used, the exponent in their Equation 21 must be different from 1.0 (Parker and Klingeman use 0.982 when the sub-pavement distribution is used for prediction). That exponent determines how the size distribution of the bedload is related to the size distribution of the parent material. The size distributions of the two potential parent distributions (the pavement and the sub-pavement) are of different sizes. That is why it follows that the exponent should be different for the prediction equations based on the two bed material distributions. The pavement exponent should be smaller than the sub-pavement exponent because the bedload is more different from the pavement than from the sub-pavement. The farther the exponent differs from one, the more the size of the bedload differs from the parent distribution, either pavement or sub-pavement.

The exponent in Parker and Klingeman's equation 21 and reference Shields stress in their equations 22 and 27 are related. They also are related to the W_r^*, for which Parker and Klingeman choose 0.002 (p. 1412). The value of 0.002 is a "small but measurable bedload movement" used to determine the reference shear stress, TRS50. If the size is determined by a proper choice of exponent, the volume transported can be fixed by a proper choice of TRS50. If there were a bias in prediction of volume of bedload, then any bias in prediction could be accounted for by proper choice of W_r^*. This means that with a good set of data the Parker and Klingeman empirical approach can be calibrated for a wide set of conditions. The factors which influence the three parameters need to be determined, as well as a proper choice of parameters. The third parameter, W_r^*, is an arbitrary level for identification of the other parameters by Parker and Klingeman, and perhaps should be left fixed. Thus, there would be two parameters to fit and two factors to fit to, bedload and size.

Determination of a proper "calibrated" reference shear stress depends upon the determination of energy slope. The energy slope is the local slope at the point of prediction of bedload transport, but how local is local? Figure 1 shows the relation of fitted values of Parker and Klingeman's TRS50, the reference shear stress for D50, to assumed energy slope as the slope is arbitrarily varied. Lower Trap Creek is one of the Forest Service measurement sites in Colorado for which data were collected specifically for the court case. The accuracy of prediction of the various sizes of sediment is almost identical over the range of assumed slopes. Thus, slope must be known in order for TRS50 to be used to predict the bedload without calibration. Variation in prediction of the subsurface reference size, $D50_s$, has a similar variation. If predicted $D50_s$ is increased, a change in reference shear, TRS50, will result in prediction of equal accuracy. Thus, subsurface size distribution and energy slope both must be properly determined in order to use the Parker and Klingeman method. There is no protocol at present for the determination of either.

Figure 2 shows that the determination of slope is not straightforward in high mountain streams. For Lower Trap Creek the slope is variable over short distances. The determination of slope and the location of the measurement section should be related. There should be a protocol for selection of the cross-section for bedload measurement, and it should be based on the local energy slope.

The sieve screens for determination of surface, subsurface, and bedload size distributions should be the same. This makes analysis much easier, consistent, and comparable.

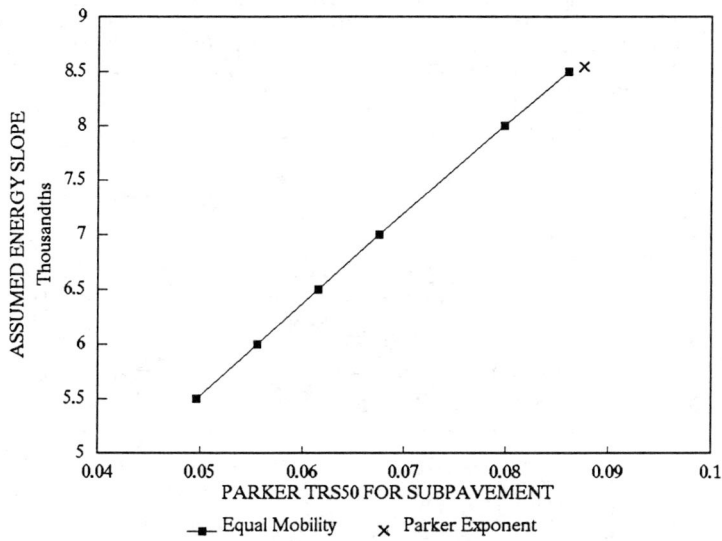

Figure 1 Variation of TRS50 with Assumed Slope for Lower Trap Creek

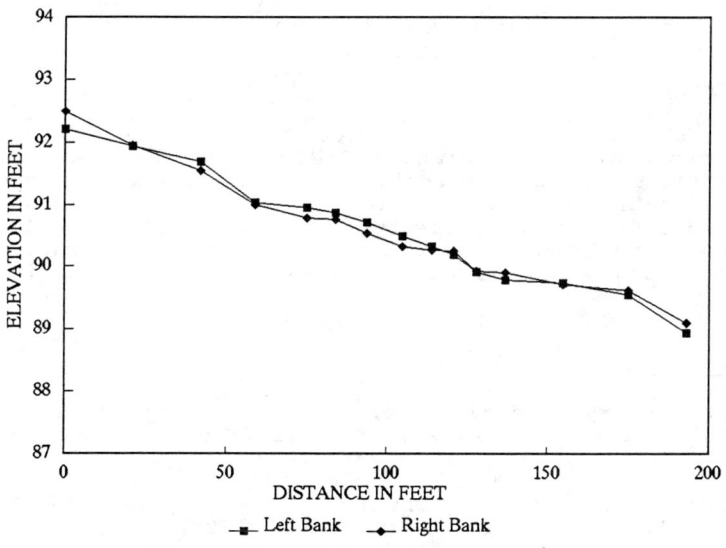

Figure 2 Water Surface Profiles for Lower Trap Creek

Figure 3 shows the resulting fit of measured to predicted bedload for Lower Trap Creek with a calibrated set of parameter values. The square symbols are for individual size breaks for the bed material, whereas the crosses are for prediction of total bedload for a complete measurement.

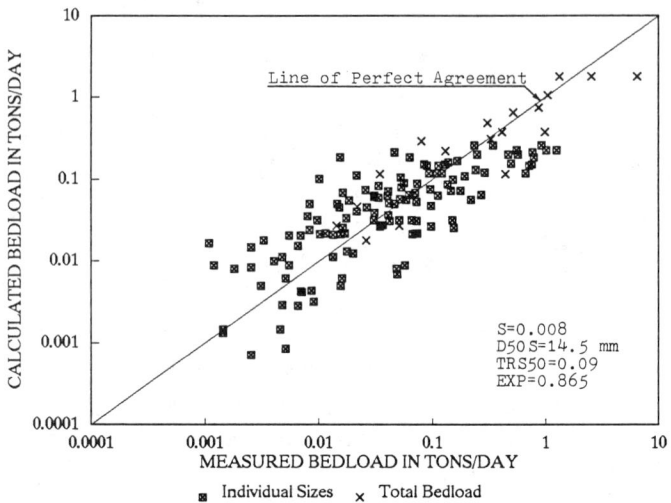

Figure 3 Comparison of Calculated and Measured Bedload for Lower Trap Creek

Summary

In summary, there is a need for a protocol for: a) determination of surface and subsurface bed material distributions, b) determination of local energy slope, c) location of cross-sections for bedload measurement, and d) data to be collected and its format. In addition, there should be a study to compare the accuracy of prediction using surface and subsurface size distributions.

REFERENCES

Parker, Gary, and Klingeman, P. C., On Why Gravel Bed Streams are Paved, Water Resources Research, Vol. 18, No. 5, October 1982.

Wilcox, Marc, Coon Creek Data Report, Medicine Bow National Forest, October 13, 1989.

EMERGENT TECHNIQUES IN SCOUR MONITORING DEVICES

J. R. Richardson[1] AM. ASCE and J. Price[2]

Abstract

This paper presents a brief status report of a research and development project funded by the National Cooperative Highway Research Program (NCHRP) to develop and test scour monitoring devices. At this time the project has completed Phase 1 and is currently working on Phase 2. The results of this project has led to the development of several instruments which may soon, or in the not too distant future, be available for monitoring scour at bridge piers and abutments. This paper discusses these emergent instruments.

Introduction

Scour in the form of erosion of the soil in and around bridge piers and abutments has and continues to be a concern for the nations highway agencies. This concern has manifested in Federal Highway Administration (FHWA) requiring the States to implement evaluation, screening and remedial programs to protect the nations bridges from scour. In an appraisal of the nations 484,000 bridges, which is approximately 95% complete, it was found that approximately 1,464 of these bridges are scour critical, and approximately 108,000 bridges are scour susceptible.

To protect the nations investment in highway infrastructure, and the traveling public, remedial measures (countermeasures) are needed for the approximately 109,464 bridges in the nation which are either scour critical or scour susceptible. Countermeasures which are acceptable to the FHWA for the mitigation of scour problems range from bridge closure and bridge replacement, to construction of scour protection such as spurs, guide banks (spur dikes), grade control structures, rehabilitation of foundations and other physical construction works.

Although it is desirable to replace bridges or construct physical countermeasures for all of these scour critical and susceptible bridges, the cost would be prohibitive given the number of bridges requiring countermeasures. Realizing that the immediate rehabilitation of all the nation's bridges is an unobtainable goal, coupled with the conflicting need to insure the safety of the nations bridges, the FHWA has also recognized monitoring of scour as a countermeasure until such time that the bridge can be repaired or replaced. Therefore, there is a pressing need for instrumentation which can be utilized to measure and monitor scour. Scour monitoring and measuring instrumentation would greatly extend the time frame for physical remedial action for all the nation's bridges which are either scour critical or scour susceptible.

---

[1] Senior Engineer, Resource Consultants and Engineers, Inc., Ft. Collins, CO
[2] President ETI Inc., Ft. Collins, CO

This paper presents the results and status of an ongoing research project funded by NCHRP which was initiated in 1990 to develop and test scour monitoring and measuring devices. From the results of this research project, emergent techniques and equipment for measuring and monitoring scour at bridge piers and abutments are being developed. While it is too soon to claim that there are new instruments available, these emergent instruments may become available in the near future.

Brief History Of The Project

In Phase 1 of the NCHRP project, an in-depth literature review was conducted to identify the most promising techniques which could be adapted or developed to measure and monitor scour at bridge piers and abutments. Mandatory and desirable criteria were also developed during the early stages of Phase 1. A synopsis of these activities is presented by Lagasse et al. (1991a & 1991b). From this review, four of the most promising techniques, namely low cost sonar, sounding rod devices, buried rod devices and other buried devices were singled out for further study.

Laboratory testing and limited field testing was conducted to test instruments from these four categories. Laboratory testing was conducted at the Hydraulics Laboratory of Colorado State University. These test were conducted at both small and near-prototype scale. Limited field testing was conducted to obtain needed experience in actual installation techniques and problem which could be expected in the field. A Phase 1 final report was submitted to NCHRP documenting the results of this Phase of the Project. This Report has subsequently been edited for publication by NCHRP in Research Results Digest, and should be available in early to mid 1993.

The results of the laboratory testing indicated that two types of instruments corresponding to a buried rod with magnetic trip switches and/or piezoelectric film sensors, and a buried rod with a sliding collar showed promise as potential instruments for monitoring and measuring scour. It was also determined during Phase 1 that there is no single class of instruments which can be relied on to meet scour monitoring needs for all of the various bridge and stream characteristics typically encountered in the field. Rather it was determined that a "tool-box" approach with a variety of instrument types would be needed.

Laboratory testing of low-cost sonic fathometers (fish finders) indicated that these instruments are sensitive to high concentrations of entrained air in the water. Because of this, the applicability of the sonic fathometer, based on the laboratory results, was considered questionable. However the problems encountered with this instrument were considered to be more specific to the conditions imposed by the laboratory testing rather than conditions which would be expected to occur in the field. Consequently, the low cost sonic fathometer, was also recommended for further study during Phase 2. Additionally, a commercially available sounding rod device, which was loaned to the research team by the supplier was also recommended for further field testing during Phase 2.

Phase 2 focused on further testing the sonic fathometer and sounding rod device in the field, and pursuing development of the buried rod devices so that they could be deployed in the field for testing. Ultimately it was planned to deploy the most promising devices at selected locations throughout the United States for further testing and evaluation. Once this task was accomplished the devices could potentially be offered as instrumentation to meet the needs for scour monitoring and measuring devices.

As an initial step of Phase 2, a sounding rod device and sonic fathometer were tested on a bridge spanning the Platte river, near the town of Orchard Colorado during the summer of 1992. The sonic sounder was specially designed so that the transducer head which must be mounted on a bridge pier or abutment, could be removed and service from the top of the bridge deck. This "bridge deck serviceable" design has greatly enhanced the ease of use of this particular instrument. Furthermore, testing during the summer of 1992 has indicated that this instrument has performed well, contrary to the findings of the laboratory study in Phase 1.

At the beginning of Phase 2, the Research Team was also directed to investigate the possibility of developing and fielding a simple, low cost device which could be manually operated. The interest in developing a simple manual device acknowledged the pressing need of the States for scour monitoring instrumentation. It was reasoned that a simple device might be designed, tested, evaluated and refined more quickly than more sophisticated devices. This would provide for the immediate needs of the States, while more sophisticated systems were being developed.

The development process currently being implemented in Phase 2 is to move each type of instrument through development, testing and refinement which would culminate in final development of devices suitable for monitoring and measuring scour at bridges. This process involves; (1) performing additional research and development necessary to fabricate field prototypes, (2) to install, test and evaluate field prototypes; (3) to refine the design and field test the devices at other sites; and (4) provide additional instruments for evaluation and testing by other agencies not associated with the Research Team. This last stage of development is referred to as "wider deployment" and is intended to be the last stage of development and testing before a device can be considered to be applicable for general usage as a scour monitoring device.

The above process provides a logical framework whereby scour monitoring instruments can be brought through research and development. At this time, the research team has different instruments at various stages in this development. The most promising of these instruments, and their current level of development is discussed in the next section of this text.

Emergent Scour Monitoring Instrumentation

This section describes the most promising instruments which are being developed by the Research Team. The order in which these instruments are discussed is from the least advanced to the most advanced instrument in terms of development. For example, instruments such as the buried rod with piezoelectric sensors are at a lower level of research and development than instruments such as the low cost sonar and simple magnetic sliding collar which have been advance to wider deployment status, and are as a consequence of this nearing the point where the can be made available to the highway community for general usage as a countermeasure.

Buried Rod With Piezoelectric Sensors

This instrument consists of a series piezoelectric sensors attached to a vertical support (rod or pipe) which can be either driven, augured or otherwise buried in the stream bed at a location where scour is anticipated to occur. Scour is indicated whenever a sensor is exposed to the flow as a result of scour in the form of a voltage output which is recorded by a data logger. Advantages of this instrument are that the instrument can be automated with data logging and telemetry equipment to provide a time history of scour. Another advantages of this instrument is that the instrument is capable of recording the scour and fill cycle which typically occurs at most bridge piers and abutments.

This instrument is currently in the research and development stage prior to initial field prototype deployment. Although laboratory tests have clearly indicated that the instrument is viable, additional research and testing to determine long term durability and weatherability of the sensors, methods to install the instrument, and optimum shape and orientation of the sensors is necessary. Two to three prototype instruments will need to be built and carefully field tested before this instrument can be introduced to the highway community. At this time, no field prototype has been fabricated or installed for testing in the field. It is anticipated that a field prototype may be deployed for the summer of 1994, with wider deployment and final development of this instrument approximately two years away.

Bridge Deck Serviceable Sonic Fathometer

This instrument utilizes an off-the-shelf sonic fathometer typically sold for locating fish to monitor and measure scour. The transducer for this device is mounted in a housing which is designed to slide freely inside a conduit mounted to the bridge pier or abutment and aimed at the location where scour is anticipated to occur. The transducer is slid down the inside of the conduit and snapped into place from the bridge deck. This design allows for the servicing or replacement of the transducer without the need for scaffolds, inspection cranes or other specialized equipment.

Scour data is obtained and recorded by a specially designed data logger connected to the sonic fathometer. In this way, a time stamped record of the scour and fill process can be recorded, stored and retrieved for future analysis. This system also lends itself to remote telemetry of scour data from a remote site.

Results of field tests conducted in the summer of 1992 at the Orchard field site in eastern Colorado has shown that this instrument has performs well, documenting the scour and fill process. Additional development of this instrument include improving the data logging, filtering and storage hardware and software, and packaging the instrument in a more ergonomic modular package. Currently it is planned to install a second instrument at a field test site on the Rio Grande river in New Mexico.

At this time this instrument has progressed to the point where it can be tested by State Agencies willing to cooperate with the Research team under the wider deployment stage of development discussed previously. At this time there has been interest by at lest one state.

Simple Magnetic Sliding Collar

The Research Team has developed a simple magnetic collar device on the direction of the NCHRP advisory panel to develop a simple device which might progress through the development stages quickly. This device consists of an open architecture collar equipped with magnets which is free to slide over a small diameter pipe which can be driven into the streambed at a location where scour is expected to occur.

The design of the collar is identical to the collar which was successful tested during near-prototype testing during Phase 1. The main difference between the simple sliding collar which was developed and the device used during laboratory testing was the usage of a small diameter, non-ferrous support rod which can be made and shipped in sections and driven into the bed using either manual, pneumatic or other mechanical methods. Scour can be determined by inserting a magnetic sensor mounted at the end of a graduated cable through the annulus of the support pipe from the bridge deck. When the sensor is in proximity to the collar, and audible signal can be detected at the bridge deck.

A field prototype of this instrument was fabricated and manually installed at the Orchard site in the fall of 1992. A second instrument has been fabricated and supplied to the Michigan Department of Transportation for wider deployment testing.

As anticipated this instrument has progressed rapidly through the design testing and refinement process. As such it is expected that this instrument will be the first instrument to emerge from the design, testing and development process and be made available to the highway community. This development ha been so rapid that patent rights for this instrument have been applied for.

The design of this instrument is simple, elegant, easy to install and operate, and can be upgraded to provide for automated data logging. The device can be fabricated in a variety of lengths, and installed using bends to route the extension conduit from the top of the buried rod to the bridge deck so that the instrument can be attached to most bridge piers and abutments to protect the instrument from debris and ice impact. In sand bed streams the instrument can be driven into the bed using a modified fence post type driver. In more densely packed streams the instrument can be installed by using a pneumatic post driver. At this time only manual determination of the scour depth can be made using the probe and buzzer. However, a multiple magnetic sensor array which can be used to upgrade existing field sites has been designed will be fabricated and tested in the near future. This multiple sensor array will allow for linkage of the instrument to a data logger so that the time history of scour can be automatically logged.

Conclusions

The results of the research being conducted by the Research Team have determined that there is a need for a collection of differing instruments with varying capabilities to meet the needs for scour monitoring at the wide variety of river bridge and flow conditions which are found throughout the United States.

As a result of this research project, there are emergent scour monitoring and measuring instruments which either in the near future will be available for general usage. With this assortment of instruments, the nations investment in bridge infrastructure, and the traveling public ca be better protected, until such time that all of the nation's bridges which are susceptible to scour can be either repaired, replaced or rehabilitated.

References

Lagasse, P.F., and Nordin, C.F. (1991a) Scour Measuring and Monitoring Equipment for Bridges, (1991) Proceedings, 1991 National Hydraulic Engineering Conference, Hyd. Div. ASCE, Nashville TN.

Lagasse, P.F., Nordin, C.F., Schall, J.D., and Sabol, G.V. Scour Monitoring Devices for Bridges, (1991b) Transportation Research Record 1290, vol. 2, Third Bridge Engineering Conference, Transportation Research Board, National Research Council, Washington, D.C.

Riprap Design at Bridges - Factor of Safety Approach

by George K. Cotton[1], Member

ABSTRACT

The factor of safety method offers a physically based methodology for design of riprap that relates the applied shear stress on the channel boundary to the stability of the riprap layer. This conceptual framework allows the method to be applied to a wide variety of river conditions provided that the boundary shear stress can be estimated. The particular advantages of the method at bridges is reviewed based on case studies for Arizona bridges.

INTRODUCTION

Arizona's rivers have presented transportation system designers with many difficult hydraulic problems. The combination of human activities in what are normally dry stream channels, and the natural behavior of sand and gravel channel has prompted the Arizona Department of Transportation to review many aspects of bridge design. This paper discussed specific research conducted by ADOT (Cotton, et. al. 1987) on riprap failures in Arizona and the riprap design procedure that was adopted for water control structures at bridges.

ARIZONA CASE HISTORIES

Actual bridge sites often involve a combination of flow conditions. Further complications include: changes in channel pattern over time and changes resulting from man's activity. Over the roughly fifty year design-life of a bridge, stream conditions often evolve to be very different from those present during the original design work. Similarly, man's activity in the form of in-stream mining, flood control improvements, or channel alterations can alter the stream channel near a bridge. While river conditions often change significantly, the water control structures and bridge protection systems remain unchanged and are expected to survive the changing stream environment. The following three case histories show clearly the challenges that confront bridge designers in dynamic alluvial systems.

---

[1] Senior Water Resources Engineer, CRSS Civil Engineers, Inc., Phoenix, Arizona

I-19 Santa Cruz River Bridge: This structure is located southwest of Tucson, Arizona, and consists of dual bridges with a 5-span northbound bridge of 157 m, and a 4-span southbound bridge of 125 m. The Santa Cruz River has a contributing watershed area of 5360 sq. km. at the site. The north abutments of the structures are located on the outside of a channel bend. Constructed in 1967, the bridge sustained minor damage in 1968 at a discharge of 160 cms and was repaired with sheet piling and cyclopian concrete. During floods in 1977-78 (620 and 380 cms, respectively) there was lateral migrations of the channel that changed the bank alignment at both the north and south abutments. River degradation was estimated at 1.5 m. Damaged banks were extensively rebuilt as a riprap system in 1978, replacing the 1968 bank protection. Riprap design was based on mean hydraulic conditions for the design flood without special consideration for river migration or the channel bend.

In February 1983 a small flood (140 cms) damaged 45 m of the north bank protection. This missing protection was replace with heavy rubble. In October 1983, the flood of record occurred at the bridge site with a discharge of 1270 cms (about 35 percent larger than the design flood). Extensive damage occurred to the north bank approaching the bridge and to the south bank downstream. The north span of the northbound structure collapsed, and there was extensive erosion of the riprap in the vicinity. Lateral migration of the Santa Cruz River also damaged the west road approach of the San Xavier Mission Road bridge.

Key factors in the riprap failure at this site were:

1. Underestimation of boundary shear stress at the channel bank, particularly the effect of the channel bend.
2. Limited knowledge of on-going lateral migration over a range of flood magnitudes.

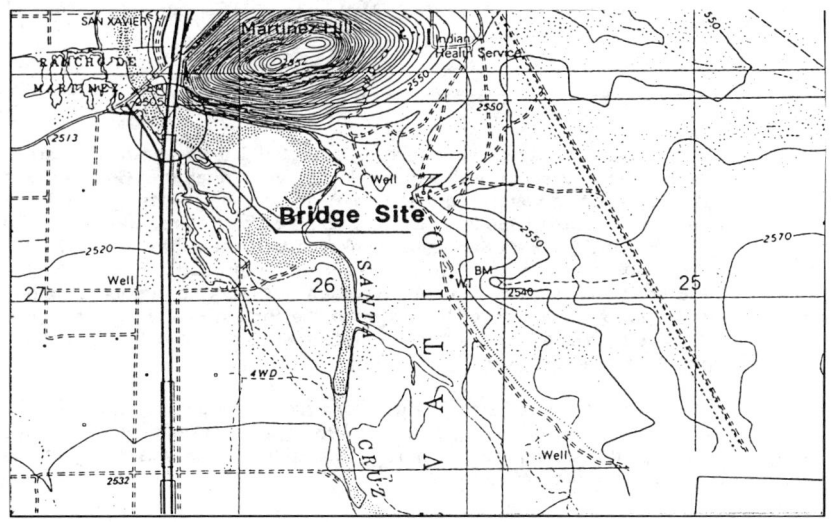

I-10 Rillito Creek Bridge: This structure is located 16 km north of Tucson, Arizona, and consists of dual 11-span bridges, 104 m in length,. Rillito Creek has a contributing watershed area of 1500 sq. km. at the site. The bridges are located 120 m downstream of the east frontage road and railroad bridges. When originally constucted the frontage road and railroad bridges were 150 m downstream of a major channel bend. By 1983, lateral migration had moved the channel parallel to the railroad embankment, creating an abrupt bend at the railroad bridge. In December 1978 flooding of 450 cms caused significant channel migration. There was a total loss of riprap along the north bank between the east frontage road and the I-10 bridge, and loss of 50 feet of the north approach of the frontage road. The riprap was replaced with a rock and rail type bank protection system that in October 1983 withstood a flood of 840 cms.

Key factors in the riprap failure at this site were:

1. Extensive lateral migration.
2. Underestimation of channel bend stresses.

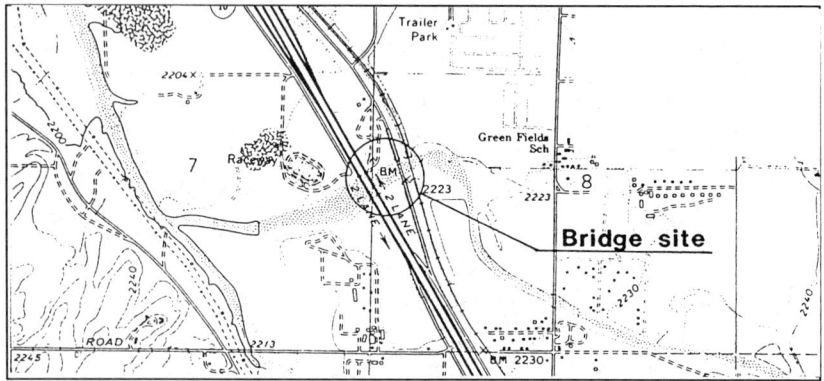

U.S. 93 Granite Creek Bridge: This structure is located 1.6 km north of Prescott, Arizona, and consists of dual 3-span bridges, 64 m in length. Granite Creek has a contributing watershed area of 101 sq. km. at the site. The bridges were constructed in 1956 and had riprap bank protection for about 80 to 90 m on the north and south channel banks, respectively. Through 1976 the bridge and its bank protection were report to be in good condition. In the 1978 bridge inspection, the bank protection was reported in good condition but 1.5 m of general scour had exposed the footing of the southern pier. Extensive sand and gravel mining both upstream and downstream of ADOT right-of-way was noted.

In 1980 a grade control structure was constructed using rock and rail for the sill and riprap for the stilling basin. In 1981 a flood of unknown magnitude destroyed the stilling basin. In 1982 the stilling basin was enlarged and constructed of grouted riprap.

Key factors in the riprap failure at this site were:

1. Degradation due to in-stream mining on land adjacent to highway right-of-way.
2. Underestimation of riprap size for the stilling basin plunge pool.

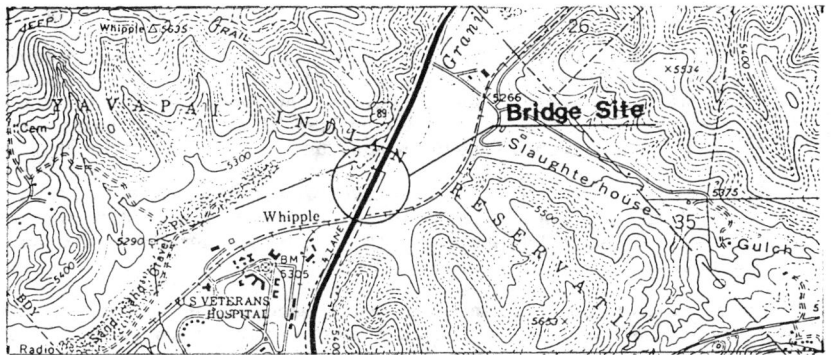

FACTOR OF SAFETY DESIGN APPROACH

As can be seen from the previous case studies, boundary shear stress is typically underestimated at a bridge site. This is particularly true at complex sites where two or more hydraulic conditions combine to produce a locally high boundary shear stress. Previous design relied on empirical approaches that consider only a few select cases, making it difficult for the designer to evaluate complex site conditions. Similarly, empirical methods are one-dimensional approaches derived from flume data or simple river channels without a floodplain. Application of such approaches to flows that are two-dimensional is problematic.

The factor of safety approach provides a clear conceptual framework for riprap design. To provide a stable riprap layer, the design shear stress for the riprap must exceed the boundary shear stress, as follows:

$$\tau_d > \tau_b \tag{1}$$

where: τ_d is the design shear stress, and τ_b is the boundary shear stress.

The right side of the inequality addresses the flow condition in terms of the local boundary shear condition at the design location. The left represents mainly the riprap stability at a particular location in the flow field.

The design shear stress is the maximum shear stress that a stone can safely withstand and still provide an acceptable factor of safety against failure. It can be defined in relation to the critical shear stress for a characteristic stone size as follows:

$$\tau_d = Ca * \tau_c \tag{2}$$

where τ_c is the critical shear stress for the characteristic stone size, and Ca is a dimensionless adjustment coefficient. The design shear stress is effected by factors such as: the stone density, the angle of repose, the bank slope, and the flow angle relative to the bank. Based on these factors the adjustment coefficient can be defined as follows:

$$Ca = \left(\frac{1}{SF}\right) Cl\; Cz\; Cw\; (1 - Cr\; Cb\; SF) \quad\quad\quad (3)$$

The term Cl accounts for flow impinging on the bank, term Cz accounts for the affect of bank slope, term Cw accounts for stone density, term Cr accounts for stone bearing angle relative to bank slope, term Cb accounts for the effect of hydraulic forces (lift and drag) relative to stone weight, and SF is the safety factor. (See Appendix I for an expansion of the terms in equation 3.)

Equation 3 fully addresses what can be termed the geotechnical aspects of riprap design. With the exception of the terms that are effected by the orientation of the flow field, Cl and Cb, the adjustment coefficient is simply a function of riprap and slope attributes.

Typically, standard backwater calculations will provide only an average value of boundary shear stress for an entire channel cross section. Determination of the local boundary shear stress for equation 1 requires an estimate of the boundary shear distribution. For a one-dimensional analysis, the average boundary shear stress or so-called tractive force, is given as follows:

$$\tau_o = \gamma\; da\; Se \quad\quad\quad (4)$$

where τ_o is the tractive force, and Se is the slope of the energy grade line. The local boundary shear stress can be defined in terms of the tractive force as follows:

$$\tau_b = Ka * \tau_o \quad\quad\quad (5)$$

where Ka is the local boundary shear stress adjustment factor. Typical boundary shear stress adjustment factors are summarized in the following table:

Boundary Shear Adjustment Factor	Description of Shear Distribution
Kbs	Channel bed in a straight, symmetric reach
Kbn	Channel bed in a straight, non-symmetric reach
Kbc	Channel bed at a contraction (bridge section)
Kss	Channel side slope
Ksf	Channel side slope, high Froude number conditions
Kas	Abutments and spurs
Kr	Channel bend
Kp	Channel bed at a bridge pier

In most cases, Ka will be the product of adjustment factors. For example a bridge pier in a contracted channel section would have Ka = Kbc * Kp.

These adjustment factors are available from a number of reports. Anderson et. al (1970) provide charts for the bank and bed coefficients, Kbs and Kss. The Corps of Engineers manual EM 1601, provides a chart for the channel bend coefficient, Kr. The author (Cotton, 1989) adapted charts by Lewis (Richardson, 1975) to provide coefficient, Kas, for abutment and spur dikes. The distribution of shear stresses in channels with high Froude numbers was measured by Davidian and Cahl (1963). An approximation of shear at a pier can be made from potential theory, where shear stress near the pier should be proportional to twice the approach velocity squared, giving Kp = 4.0.

For non-symmetric cross-sections, particularly where there is a complex interaction between the floodplain and main channel, the boundary shear stress distribution is difficult to estimate. Brown and Blodgett (1987) developed an approximate method for this case. However, it may be more reliable in the more complicated cases to simply use two-dimensional modeling tools.

The the more general formula for determination local boundary shear stress is computed as follows:

$$\tau_b = \rho \left(\frac{Us}{cf}\right)^2 \quad \text{..} (6)$$

where Us is the local streamwise velocity, cf is the local flow resistance coefficient, and ρ is water density. For the Manning equation the resistance coefficient is given as follows:

$$cf = \frac{da^{1/6}}{n\sqrt{g}} \quad \text{..} (7)$$

where da is the local average flow depth, n is Manning coefficient, and g is the gravitational acceleration.

APPENDIX I - Critical Shear Stress Adjustment Factor Terms

$Cl = \frac{2}{(1+\sin(Al+Ab))}$ $\quad Cz = \cos(Az) \quad\quad Cb = \cos(Ab)$

$Cw = \frac{(S-1)}{(1.4)}$ $\quad\quad Cr = \frac{\tan(Az)}{\tan(Ar)} \quad Ab = \text{atan}\left(\frac{\cos(Al)}{(E+\sin(Al))}\right)$

where:
$E = 2 * \left(\frac{\tau c}{\tau b}\right) Cz \; Cr$ $\quad$ Ab is the particle $\quad\quad$ S is the stone specific
$\quad\quad\quad\quad\quad\quad\quad\quad\quad\quad$ movement angle $\quad\quad\quad$ gravity
Az is the side-slope angle $\quad$ Ar is the bearing angle $\quad$ Al is the flow angle relative to the bank

APPENDIX II - References

Anderson, A.G., A.S. Paintal, and J.T. Davenport, 1970, "Tentative design of riprap-lined channels," NCHRP Report 108, Highway Research Board, Washington, D.C.

Annon., 1970, "Hydraulic design of flood control channels," Department of the Army Corps of Engineers, EM 1110-2-1601.

Brown, S.A. and J.C. Blodgett, 1987, "Application of natural stream characteristics to riprap design," TRB 67th Annual Meeting.

Cotton, G.K., R.M. Li, 1989, "Sizing riprap for the protection of approach embankments and spur dikes and limiting the depth of scour at bridge piers and abutments," Arizona Department of Transportation, FHWA-AZ89-260 vol. 1 & 2.

Davidian, J. and D.I. Cahal, 1963, "Distribution of shear in rectangular channels," USGS Prof. Paper, Article 113, 475-C, C206-C208, Washington, D.C.

Richardson, E.V., D.B. Simons, S. Karaki, K. Mahmood, and M.A. Stevens, 1975, "Highways in the river environment - hydraulic and environmental design considerations," FHWA.

RIPRAP COVERAGE AROUND BRIDGE PIERS

James F. Ruff,[1] F. ASCE and James R. Nickelson [2]

Abstract

Experiments were conducted to examine the effect of riprap size and coverage on reducing local scour at bridge piers. Pier diameter, bed material size, riprap size and diameter were varied during the study. Scour depths with no riprap coverage were compared to scour depths when a single layer riprap mat with 100% and 50% areal coverage was in place. It was found that scour depths could be reduced by placing a riprap mat around the pier that is five to eight times the diameter of the pier. Scour depths at bridge piers with riprap protected mats were reduced by 40% to 99% compared to scour at unprotected piers.

Introduction

Local scour around bridge piers is a major cause of bridge failures. Bridge collapses such as the Schoharie Creek Bridge in New York in 1987 and the Hatchie River Bridge in Tennessee in 1989 have been attributed to local scour. These failures caused the loss of human life and cost millions of dollars to rebuild or repair.

In attempting to prevent local scour around bridge piers, investigators have studied numerous methods. These include structural additions to piers, various mats placed at the base of the pier, and rock riprap placed around the pier. Riprap is the most often used technique for scour prevention. Riprap has some distinct advantages over other techniques. Rock riprap in sufficient size and quantity is readily available in most areas, and rock can easily be placed around a pier during new construction and with reasonable care during low water at existing structures. Once installed, riprap provides good scour protection as long as it remains in place. However, because riprap is generally obscured by the flowing water, inspection and

---

[1] Professor of Civil Engineering, Engineering Research Center, Colorado State University, Fort Collins, CO 80523.

[2] Engineer, WGM Group, P.O. Box 3418, Missoula, Montana 59806

maintenance are often lacking. Some of the individual rocks in the riprap mat can be moved or lost with time and flood flows. This loss of protective coverage increases the potential depth of scour. The objective of the study was to examine the depth of local scour at a bridge pier resulting from different percentages of areal coverage of the channel bed using a single layer of riprap.

Test Program

A testing program was conducted at the Engineering Research Center Hydraulics Laboratory of Colorado State University using a 61 x 1.2 x 2.48 m flume. Tests were conducted using three different circular bridge piers fabricated from acrylic plastic cylinders having diameters of 15, 23, and 30 cm. The piers were placed in the movable bed portion of the flume and scour measurements were taken at the conclusion of each test run. A single-layer thick blanket of uniformly graded rocks was placed around the pier before each test run to simulate riprap. Four uniformly graded rock sizes with d_{50} of 1.6, 3.2, 5.1, and 6.3 cm were tested. Two graded bed materials, one sand with d_{50} of 0.61 mm and one gravel with d_{50} of 3.3 mm, were used for the bed during the tests. Additional details of the test facilities can be found in Nickelson (1992).

A series of 29 tests were performed. The test runs for the gravel bed condition were run at a discharge of 0.96 cms (34 cfs) and a depth of 0.46 m (1.5 ft). The test runs for the sand bed condition were run at a discharge of 0.88 cmp (31 cfs) and a depth of 0.46 m. The flume was kept at a constant slope of 0.0015. The first test run for each pier was conducted with a flat bed and no riprap protection in place to get a baseline scour depth. Subsequent tests were run with uniformly graded riprap of varying sizes placed around the pier. The riprap was placed in a circular mat around the pier. The riprap mat varied in diameter from 4 to 8 times the diameter of the pier. After each run scour depths were taken and recorded.

For the gravel bed condition, the discharge of 0.96 cms was set right at the point of incipient motion of the bed material. By adjusting the flow it
was determined that flows of 0.96 cms allowed limited motion of the gravel bed without causing general scour of the gravel bed. Because the sand bed was inserted into the gravel bed, the discharge needed to be lowered so that the gravel bed did not move into the sand bed portion of the model. A discharge of 0.88 cms (31 cfs) met this condition.

A transparent pier was used so that the scour depth could be monitored during the testing. Scour depth measurements were made by placing a mirror inside the pier and viewing the bed in front of the pier. It was observed that in the first 10 minutes of the tests about 70% of the maximum scour depth was reached; for all practical purposes, the maximum scour depth developed in about two hours. The tests were continued for a period of three hours to ensure that the maximum scour depth was reached.

Following the baseline scour tests, the riprap coverage tests were conducted. Uniform sized rock was placed around the pier in the area affected by the local scour at the level of the bed. The riprap was placed in a circle around the pier that varied in diameter from 4 to 5 times the diameter of the pier for the gravel bed condition, and from 5 to 6 times the pier for the sand bed condition. The amount of rock placed on the bed around the pier comprised either 50% coverage or 100% coverage. The 100% coverage consisted of a single layer of rock tightly placed around the pier. The 50% coverage consisted of placing the 100% coverage and then removing every other rock in a grid pattern. Test were conducted for 3 hours, and scour depth measurements were then made. The variables that were changed for each pier and bed condition included the rock size and the coverage of the bed by the rock. Scour depths were taken in a grid area centered around the pier. Depending on the size of the scour hole this grid area was changed from test to test.

Experimental Observations

Observations made during the testing runs included watching the scour action take place and timing the scour depth. Leaching of the bed material through the riprap could be noted in all runs. The leaching activity was highest during the first portion of the test runs. The scour action appeared to start very rapidly and diminish with time. Observations made after the testing consisted of examining the location of the riprap and noting any movement. In all tests the riprap at the nose of the pier stayed in front of the pier. The rocks at the nose of the pier moved downward and towards the pier as the bed material was leached from between and underneath the rocks. In some of the tests the riprap behind and to the sides of the pier moved toward the center of the flume directly behind the pier.

In three of the 29 tests, scour depth development were observed through the inside of the cylinders. The results of the scour depth with respect to time are shown in Figure 1. The ratio of scour depth $d_s(t)$ to the maximum scour depth $d_{s,m}$, recorded in each run is plotted versus the time of the reading, t. The graph shows that the maximum scour depth, for practical purposes, occurs within one and one half hours in the tests that were run. The depth recorded approaches a limit asymptotically with time. From this data it is shown that about 70% of the maximum depth occurs in the first 10 minutes of the testing period.

The leaching action was viewed through a mirror placed inside the cylinder. Viewing the leaching action provides some qualitative information on the scour action. The leaching action began as the discharge was slowly increased and the depth was slowly decreased. The first area to begin to erode was directly in front of the pier.

At first, the erosion process was uniform, with the bed material slowly eroding from between the rocks. The eroded material becomes suspended in the flow and travels just above and between the rock riprap toward the side and rear of the pier. As the discharge increased, the erosion process increased in intensity. Small vortices formed in the voids between the rocks and the bed material between the rocks would then be carried into the flow, forming a hole. The bed material underneath the rocks would fall into this hole when the hole became large. The slope fails when the slope of the bed material reaches a critical point, the angle of repose. When there is no longer enough material under the rock to support the rock, the rock becomes suspended in the flow for a brief period. As the rock slowly falls towards the bed, large quantities of the bed material are carried into the

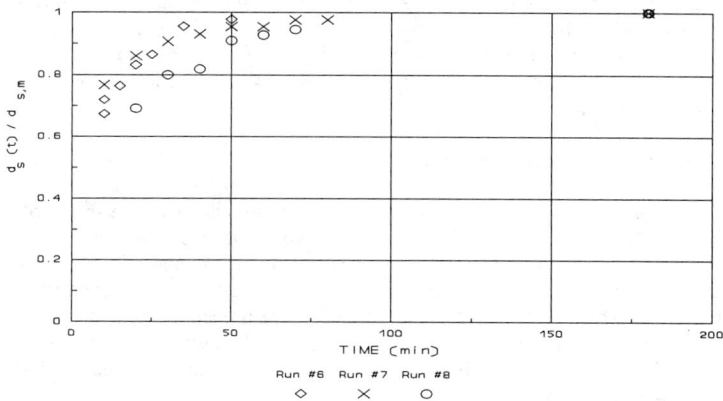

Figure 1. Scour ratio vs. time.

flow. When the rock comes to rest, the bed is once again even and a hole begins to form at locations where voids between the rocks occur.

The process described above, continues in a cyclic fashion with small amounts of material being constantly eroded away and rapid erosion occurring when the rocks are undermined. Rocks that are on the slope of the scour hole slide toward the nose of the pier as the scour hole becomes larger. When the scour hole becomes deeper, the scour action slows down, but the cycle continues. When the holes between the rocks form, larger size bed particles swirl around in the hole, but do not become entrained into the flow. These large particles eventually are leached away when the rocks slide or roll and the rapid erosion process takes place. Stability eventually occurs when the scour hole becomes large enough to absorb most of the energy of the horseshoe vortex and the small vortices formed between rocks diminish in strength. In some tests which had pea gravel as the bed material, large particles would become concentrated in the voids between rocks, causing the scour action to slow and eventually stop. The effect of the large particles slowing the scour action is similar to the armoring effect seen in river beds where large particles provide a filter system to protect against scour.

The riprap movement during the tests was observed at the end of each test run. In all cases the riprap in front of the pier at the beginning of the run was still in front of the pier at the end of the run. Riprap in front of the pier would move into the scour hole but did not move laterally or move beyond the front of the pier. In some cases the riprap behind and to the sides of the pier would move towards the center of the flume directly behind the pier. This appears to be the effect of the wake vortex system.

The general shape of the scour hole with or without riprap appears to be about the same. Generally the shape of the scour hole is conical in nature. The scour hole in front of the pier is a smooth, almost uniform half cone. The use of riprap tends to make the scour hole smaller in width and depth. The maximum scour depth occurs directly in front of the pier. The scour hole behind the pier is still generally a conical shape but it is much shallower and wider than the portion in front of the pier.

Scour Depth Versus Riprap Coverage

The riprap coverage, C, in percentage is plotted versus the scour ratio for three pier sizes in Figure 2. With zero riprap coverage, the scour depth equals the maximum scour depth. The data indicates that as the amount of riprap increases the scour ratio decreases. The amount of riprap coverage is an important parameter in predicting scour depths for the conditions examined in this study. Scour essentially ceased when the riprap which slid or rolled into the scour hole formed a mat of approximately 100% riprap coverage in the scour hole when it started at 50% coverage on the bed.

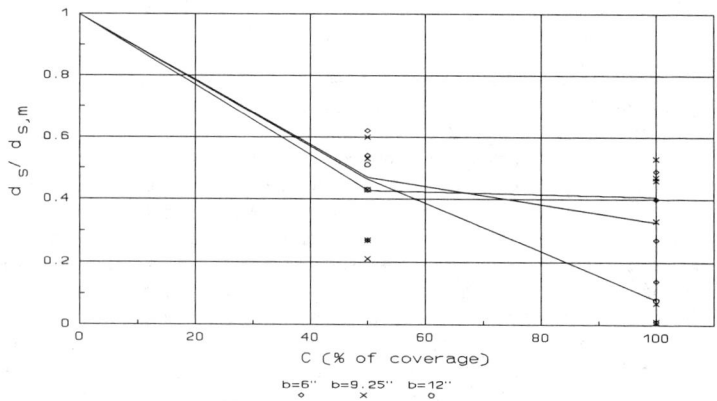

Figure 2 Scour ratio vs. percent of riprap coverage.

Riprap Displacement and Movement

Riprap movement was observed in two areas, in front of the pier and to the side and behind the pier. Displacement of the riprap occurred in front of the pier when the scour hole was formed. Riprap in front of the pier moved closer to the pier, but did not move beyond the front of the pier. This movement occurred in each experiment. The movement appears to be related to the scour action and not to the riprap size. If no undermining scour occurred, the riprap in this area would not move.

The second area in which riprap movement occurred was behind and to the sides of the pier. The riprap would move from the sides and behind the pier and migrate to an area directly behind the pier. The movement is caused by the wake vortex system dislodging the rocks and depositing them in the wake zone behind the pier. This type of movement occurred in six of the 29 riprap tests performed. In all such cases, sand was used as the bed material and a 50% coverage configuration was in place. In all tests with such conditions, the riprap moved. The reason behind this appears to be two fold. The reason the high coverage riprap does not move is thought to be the result of some interlocking of the riprap. When one rock is in contact with another rock, it is more difficult to dislodge. The reason that the sand bed allowed riprap movement appears to be that sand was more easily eroded and removed from the voids between the rocks than was the gravel. The rocks would begin to slide or roll into the eroded areas and once in motion are more susceptible to being lifted or rolled downstream.

Summary and Conclusions

This study points out the need for periodic inspections of bridge pier riprap protection at regular intervals. Time and floods can remove individual rocks in a riprap mat and reduce the coverage. Some protection of the piers may still be provided by the riprap, but it may not be sufficient to protect the structure. This study investigated the reduction in local scour around bridge piers resulting from different riprap coverage in a single layer mat of uniformly sized rocks. If the riprap coverage is reduced significantly (less than 50 %), the scour may undermine footings or piers.

As the coverage by riprap increases, the scour depth decreases. Scour depths with 100% coverage varied from one to 48% of the scour depth observed with no riprap protection. Scour depths for 50% coverage varied from 21 to 62% of the maximum scour depth. The reduction in scour depth was determined to be a function of the amount of riprap used, the ratio of the pier diameter to the bed material diameter, and the ratio of the riprap diameter to the bed material diameter.

The above conclusions are based on a hydraulic model using steady uniform flow. In order to extrapolate the results of the study to prototype size and field type conditions, larger scale studies should be conducted. However, the qualitative trends discovered in the study should hold true for field conditions.

Bibliography

Nickelson, J.R., 1992, "Experimental study of riprap around bridge piers," M.S. Thesis, Department of Civil Engineering, Colorado State University.

PREDICTING CRITICAL SCOUR STAGE AT BRIDGES

John N. Paine[1], M. ASCE, Darrell Kim Beatley[1], M. ASCE, and James N. Wigfield[2]

Abstract

After an initial Level I screening process, most Level II bridge scour studies result in recommendations for costly scour countermeasures. Typically, existing transportation budgets cannot accommodate the total demand for funds. Transportation officials are now charged with the task of scheduling the improvements while safeguarding the public. The Virginia Department of Transportation follows a procedure in which monitoring is the initial countermeasure. An engineering analysis establishes a water surface elevation for bridge closure. This water surface elevation is called the "critical scour stage." If the critical scour stage is met or exceeded, the bridge will be closed to traffic until a scour inspection can be performed. The procedure described in this paper is based on equations prescribed in HEC-18 (Richardson et al., 1991).

Introduction

One approach to determine critical scour stage could be to make a number of hydraulic model runs, in a trial-and-error manner, to find a discharge that produces the maximum allowable scour depth under a given set of boundary conditions. There are two significant drawbacks to this approach: scores of computer runs could be required, and significant time can be wasted computing resultant scour depths. A more direct approach is needed to save time and limit the feasible solution range. Such an approach can be

---

[1]ESPEY, HUSTON & ASSOCIATES, INC., 460 McLaws Circle, Williamsburg, VA 23185.

[2]VIRGINIA DEPARTMENT OF TRANSPORTATION, 1401 East Broad Street, Richmond, VA 23219.

employed by back-solving the scour equations. The result is a trial-and-error approach that methodically leads to a "reasonable" solution within a minimum number of trials. The technique incorporates the interdependent relationships between depth and velocity under a variety of discharge and hydraulic boundary conditions. The governing scour equations are used to reduce the possible combinations of hydraulic parameters to a single measurable flow depth that can be used to justify closing the bridge for scour inspection.

A procedure for estimating critical scour stage, based on back-solved scour equations, has been described previously (Paine et al., 1992). The procedure described here employs an improved approximation of abutment scour parameters, and includes contraction scour in the basic formulation. The ultimate goal of the procedure is to develop a single, measurable water surface elevation, y_1, that can be used to close the bridge for scour inspection.

Development of Equations and Procedure

According to Paine et al. (1992), if contraction scour is negligible, the depth just upstream of a pier that produces scour to the bottom of a pier footing can be computed as:

$$y_1 = \left(\frac{2.11 \, y_{SP}}{2.0 \, K_{1p} K_{2p} \cdot \alpha^{0.65} V_1^{0.43}} \right)^{7.4} \quad (1)$$

in which: y_1 = flow depth just upstream of the pier; y_{SP} = pier scour depth; K_{1p} = correction factor for pier nose shape; K_{2p} = correction factor for angle of attack; α = pier width; Fr_1 = Froude number just upstream of the pier; V_1 = velocity just upstream of the pier. For a given pier configuration, the flow depth upstream of the pier that produces a given pier scour depth, y_{SP}, is a function of velocity just upstream of the pier. Eq. (1) has no singular solution; it does however describe a solution range. Specifically, it describes a relationship between flow depths and velocities just upstream of the pier that will produce a pier scour depth equal to y_{SP}.

Live-bed scour at abutments can be computed, according to Froehlich (1989), as:

$$\frac{y_{SA}}{y_a} = 2.27 \, K_{as} K_{aa} \left(\frac{a'}{y_a} \right)^{0.43} Fr_e^{0.61} + 1 \quad (2)$$

in which: y_{SA} = abutment scour depth; y_a = depth of flood plain flow at the abutment; K_{as} = abutment shape coefficient; K_{aa} = coefficient for angle of embankment to flow; a' = length of abutment projected normal to the flow; and Fr_e = Froude number of approach flow upstream of the abutment. Note that a' = A_e/y_a, where A_e = flow area of the approach cross-section obstructed by the embankment. Also $Fr_e = V_e/(gy_a)^{0.5}$, where: V_e = cross-sectional velocity of the flow upstream of the abutment; and g = gravitational acceleration. A_e can often be approximated by the shaded area in Fig. 1, which is valid for many types of abutments under low to moderate flow conditions (not just spill-through abutments). Note that each abutment is considered individually.

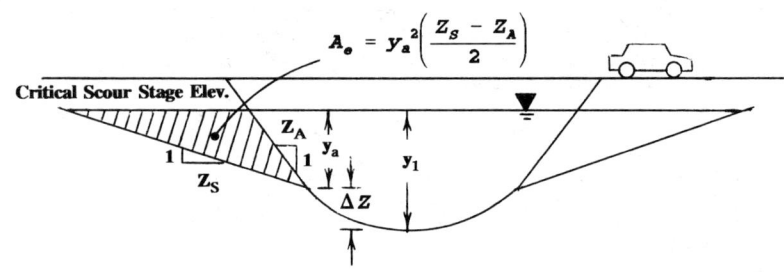

Figure 1. An Approximation of A_e

In Fig. 1, Z_S is a linear approximation of the overbank slope at the approach section, and Z_A represents the abutment slope under the bridge, horizontal distance per unit distance vertical. Then, $A_e = 0.5y_a^2(Z_S - Z_A)$. By definition (Richardson et al., 1991), a'= A_e/y_a. Eq. (2) can then be rearranged as a function in terms of y_a:

$$F(y_a, y_{SA}, V_e, K_{as}, K_{aa}, Z_S, Z_A) = \\ 2.27\, K_{as} K_{aa}\, y_a \left(\frac{Z_S - Z_A}{2}\right)^{0.43} \left(\frac{V_e}{(g\, y_a)^{0.5}}\right)^{0.61} + y_a - y_{SA} \quad (3)$$

Eq. (3) is implicit in terms of y_a and thus requires solution by numerical procedures. According to Richardson et al. (1991), contraction scour can be written as:

$$\frac{y_2}{y_1} = \left(\frac{Q_{mc2}}{Q_{mc1}}\right)^{\frac{6}{7}} \left(\frac{W_{c1}}{W_{c2}}\right)^{K_{1c}} \left(\frac{n_2}{n_1}\right)^{K_{2c}} \quad (4)$$

where: y_1 = average depth in the main channel; y_2 = average depth in the contracted section; Q_{mc1} = flow in the approach channel that is transporting sediment; Q_{mc2} = flow in the contracted channel; W_{c1} = bottom width of the main channel; W_{c2} = bottom width of the bridge opening; n_1 = Manning's n for main channel; n_2 = Manning's n for contracted section; and K_{1c} and K_{2c} are exponents determined by procedures described by Richardson et al. (1991). Also, $y_{sc} = y_2 - y_1$, where y_{sc} is the contraction scour depth. In most applications, according to Richardson et al. (1991), $n_1 = n_2$, which eliminates the ratio n_2/n_1, as well as K_{2c}. On bridges in Eastern Virginia, K_{1c} is typically equal to 0.69 (almost always). The procedure for determining K_{1c} is described by Richardson et al. (1991). Total scour y_T, for the purposes of this paper, is the sum of local and contraction scour. If contraction scour is significant at a pier, a function can be constructed to solve for y_1:

$$F(y_1, y_T, K_{1p}, K_{2p}, \alpha, V_1, Q_{mc1}, Q_{mc2}, W_{c1}, W_{c2}) =$$

$$y_1 2.0 K_{1p} K_{2p} \left(\frac{\alpha}{y_1}\right)^{0.65} \left(\frac{V_1}{(g y_1)^{0.5}}\right)^{0.43} + y_1 \left(\frac{Q_{mc2}}{Q_{mc1}}\right)^{\frac{6}{7}} \left(\frac{W_{c1}}{W_{c2}}\right)^{0.69} - y_1 - y_T \quad (5)$$

Referring to Fig. 1, y_a can be approximated by $y_a = y_1 - \Delta Z$. If contraction scour is significant at an abutment, then the solution function becomes:

$$F(y_1, y_T, K_{as}, K_{aa}, \Delta Z, Z_S, Z_A, V_e, Q_{mc1}, Q_{mc2}, W_{c1}, W_{c2}) =$$

$$2.27 K_{as} K_{aa} (y_1 - \Delta Z) \left(\frac{Z_S - Z_A}{2}\right)^{0.43} \left(\frac{V_e}{[g(y_1 - \Delta Z)]^{0.5}}\right)^{0.61}$$

$$\Delta Z + y_1 \left(\frac{Q_{mc2}}{Q_{mc1}}\right)^{\frac{6}{7}} \left(\frac{W_{c1}}{W_{c2}}\right)^{0.69} - y_T \quad (6)$$

Equations (5) and (6) are implicit, and can be easily solved using a numerical method such as the bisection algorithm. Note that the variables V_1 or V_e, Q_{mc1}, and Q_{mc2} are related, and vary over the range of depths and discharges considered; all other independent variables are constants. The velocity and discharge variables must be given in (5) and (6) in combinations that occur, or will occur, at the bridge site. That is, several hydraulic model runs should be made to determine the relationship between V_1 or V_e, Q_{mc1}, and Q_{mc2}. These variables are interdependent, and cannot be entered in random fashion.

For a given set of soil conditions and channel bottom geometry, scour becomes critical when a scour hole reaches the bottom of a footing. The minimum discharge and boundary conditions required to produce scour to the bottom of a footing establish the critical scour stage. Several WSPRO (or HEC-2) runs are made using a range of discharges and appropriate hydraulic boundary conditions. A plot of depth versus velocity is plotted as shown in Fig. 2. Then the appropriate back-solved scour equation, (1), (3), (5), or (6) is solved for a range of V_1 or V_e, Q_{mc1}, and Q_{mc2} values. The resulting values of depth versus velocity are plotted as shown in Fig. 2. The intersection of the two plots is the desired solution point; that is, a depth that can be reasonably expected to occur which would produce just enough scour to reach the bottom of a structural footing. Spreadsheet software is useful in finding the solution point. Simultaneous graphing aids in isolating the solution.

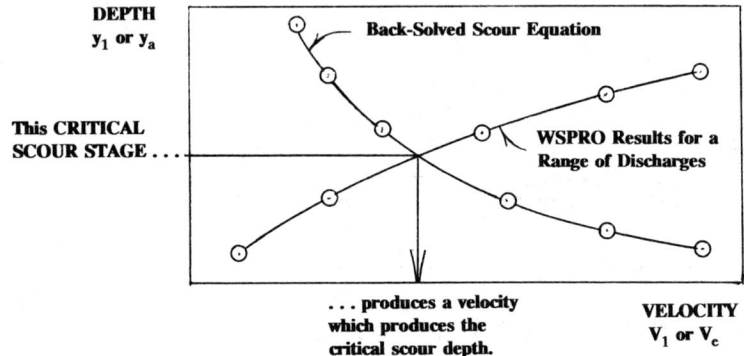

Figure 2. Determination of Critical Scour Stage

When the critical scour stage is determined, an estimate of its corresponding return period can be obtained by plotting the associated discharge on probability paper with the discharges used in the standard scour analysis. The return period is helpful in determining the frequency at which the bridge will be closed to traffic.

Conclusion

A procedure has been developed that considerably reduces the trial-and-error effort required to determine a critical scour stage at a bridge site. The approach incorporates the interdependent relationships between depth and velocity under a variety of discharges and boundary conditions. Governing scour equations are used to reduce the possible combinations of hydraulic parameters to a single measurable flow depth that can be used to justify closing the bridge for scour inspection.

The equations derived in this paper are not universally applicable, although the conceptual approach can be used successfully if the back-solved scour equations are appropriate for the site being analyzed. In order to validate the results, the scour equations should be solved in a straightforward manner to confirm that the selected stage and discharge do indeed produce critical scour. All physical data should be evaluated for sensitivity, and the final estimate of critical scour stage should be based on sound judgement. It should be acknowledged that the limiting assumptions applied when computing scour depths are often tenuous. For example, in the absence of reliable soils data, assumptions about stratigraphy of the soil layers can greatly affect scour calculations. These assumptions, if not carefully considered, can completely negate the validity of the entire study, including the estimate of the critical scour stage.

Appendix I - System of Units

The equations in this paper are based on equations published in HEC-18 (Richardson et al., 1991). U.S. Customary units (length in feet, discharge in cubic feet per second) were used in HEC-18. The metric equivalents of these equations are not available through simple conversion because of the imbedded dimensions in the constants and exponents. Versions of these equations in SI units have not yet been published.

Appendix II - References

Froehlich, D.C. (1989). "Abutment Scour Prediction," paper presented at the 68 TRB Annual Meeting, Washington, D.C.

Richardson, E.V., Harrison, L.J., and Davis, S.R. (1991). *Evaluating Scour at Bridges*. (1991). Pub. No. FHWA-IP-90-017, HEC 18, U.S. Dept. of Transportation, Federal Highway Administration, National Highway Institute, McLean, Va.

Paine, J.N., Leedy, R.J., Jr., and Wigfield, J.N. (1992). "Addressing Bridge Scour When Funding Falls Short," Proceedings of the Hydraulic Engineering Sessions at Water Forum '92, ASCE Hydraulics Division, New York, N.Y., pp. 204-209.

Riprap Incipient Motion and Shields' Parameter

by:

Roger T. Kilgore[1], Member ASCE, and
G. Kenneth Young[2], Member ASCE

Abstract

Unobstructed incipient motion data from a variety of sources are compiled to re-examine current thinking regarding riprap stability. Shields' framework is investigated and relationships are proposed between Froude number and Shields' parameter. Such relationships suggest that current methodologies result in over-sized riprap when flow fields are at or above Froude number of 0.8.

Introduction

The objective of this analysis is to re-evaluate riprap incipient motion using the framework developed by Shields (1936). This will enable better prediction of incipient motion of riprap and alluvial material over a wide range of particle size and turbulent flow conditions. The approach is to collect riprap data from published sources and develop a model that is applicable to designing riprap counter measures.

Revisiting Shear Velocity and Shields' Parameter

Implicit in Shields' work on the incipient motion of particles is that once wholly turbulent flow over a rough surface is achieved, the ratio of bed shear stress to the ability of a particle to resist it becomes a constant. However, Shields' data only include Reynolds' numbers less than 400. Nikuradse (1933) reported similar findings relating friction (stress) and turbulence. Experimental data with higher Reynolds' numbers collected subsequent to Shields' work (GKY & Assoc.,1993; Parola, 1991; Neill, 1967; Wang and Shen, 1985) do not support this hypothesis. However, Shields' hypothesis can be extended with an adjustment for Froude number.

Figure 1 depicts the data in Shields' framework. D_{50} values shown in Figure 1 are : 0.41, 0.57, 0.62, 0.80, 0.85, 1.06, 1.11, 2.00, 2.38, 2.91, 7.57, 9.65, 10.70, and 40.0 cm. The latter four values are field data from prototype waterways. As presented, these data do not suggest a constant Shields' parameter at Reynolds' numbers greater than 400. Wang and Shen (1985), when analyzing the data of Wang and Li (shown in the figure), suggested that the drag coefficient drops at high Reynolds' numbers. However, the data show variation in Shields' parameter at much lower Reynolds' numbers as well.

---

[1] Vice President and [2] President, GKY & Assoc., 5411-E Backlick Road, Springfield, VA 22151

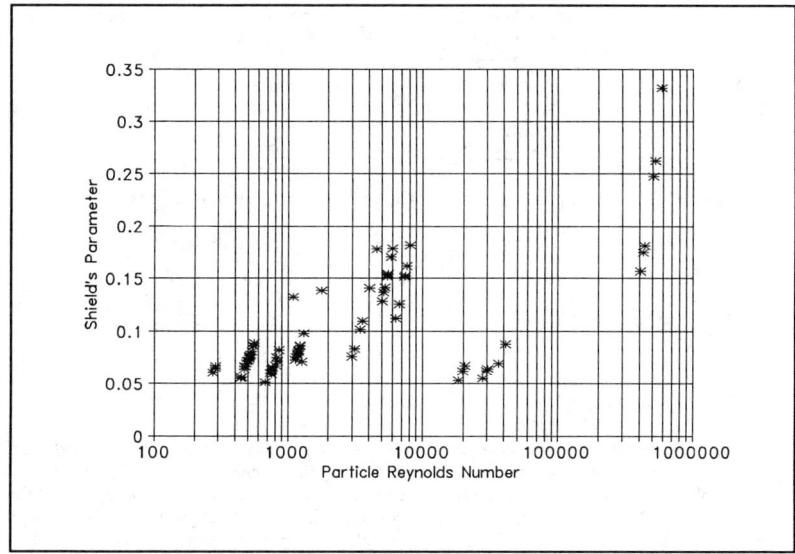

Figure 1
Shields Diagram for Riprap Data

It is helpful to re-examine the definition of Shields' parameter. At critical conditions, that is, at the threshold point where motion of a particle is imminent, the drag moment pulling the particle out of position is equal to the net gravitational moment (weight-buoyancy) holding the particle in position. The ratio of these two moments would be equal to one at incipient motion and may be defined as follows:

$$\frac{M_D}{M_g} = \frac{r_D\,F_D}{r_g\,F_g} \rightarrow \frac{r_D\,\rho\,C_D\,U_{*,c}^2\,C_1\,D_{50}^2}{r_g\,\gamma\,(SG-1)\,C_2\,D_{50}^3} \tag{1}$$

where:

M_D	=	drag moment, N · m;
M_g	=	net gravitational moment, N · m;
r_D	=	moment arm for the drag force, m;
r_g	=	moment arm for the net gravitational force, m;
F_D	=	drag force, N;
F_g	=	net gravitational force, N;
ρ	=	density at water, kg/m^3;
C_D	=	drag coefficient on the particle, dimensionless;
$U_{*,c}$	=	critical bed shear velocity, m/s;

$C_1 D_{50}^2$	=	surface area of particle to which bed shear stress is applied, m² (C_1 is a dimensionless particle-dependent constant);
γ	=	specific gravity of water, N/m³;
SG	=	specific gravity of the particle, dimensionless;
$C_2 D_{50}^3$	=	volume of particle, m³ (C_2 is a dimensionless particle dependent constant); and
D_{50}	=	median particle diameter, m.

The equation can be rearranged to show the Shields' parameter as follows:

$$SP = \frac{C_2}{C_1} \cdot \frac{r_g}{r_D} \cdot \frac{1}{C_D} \rightarrow \frac{(U_{*,c})^2}{g(SG-1)D_{50}} \qquad (2)$$

where:

SP = Shields' parameter, dimensionless.

For a given particle, it can be argued that the constants comprising Shields' parameter are, in fact, constant over a wide range of hydraulic conditions. For a sphere, $C_1 = \pi/4$ and $C_2 = \pi/6$. The ratio of r_g/r_d is probably related to the manner of placement as well as the angularity of the particle; and the drag coefficient, C_D, is also probably related to the angularity and manner of placement. It may be argued that for particles of different sizes, the Shields' parameter will be constant if the shape of the particle is not varied greatly.

On the other side of the equation, SG, D_{50}, and $U_{*,c}$ are the only variables. SG and D_{50} are constant for a given particle and $U_{*,c}$ is a parameter difficult to define or measure. If this side of the equation is also constant the question is why are varying Shields' parameter values observed. A particle will move only when the drag moment overcomes the net gravitational moment.

Model Paradigms

Two paradigms are proposed here for understanding this phenomenon. Both are based on determining the cause of the variation and finding an analytical framework for explaining it. After several trials, it was discovered that Shields' parameter for these data yielded a strong correlation with Froude number. Figure 2 summarizes the data along with an equation representing a relationship between Shields' parameter and Froude number as follows:

$$SP = 0.052 * Fr^{2.7} + 0.05 \qquad (3)$$

where: SP = Shields' parameter
Fr = Froude number $(V/(gy)^{0.5})$

Equation (3) suggests that as Froude number decreases, Shields' parameter approaches 0.05. Some of Shields' data exhibit values in the vicinity of 0.034 with unknown Froude numbers. However, it is unclear that these data represent turbulent flow.

Referring to equation (2), the variation suggested in equation (3) suggests that C_1, C_2, r_g, r_D, and/or C_D vary with Froude number. There is little reason to believe that C_2 (volume constant) or the moment arms would change significantly with Froude number or any other parameter. Drag coefficient (C_D) can change with varying hydraulic conditions and may explain the variations shown in Figure 2. Also, C_1, the effective surface area of the particle subjected to the shear stress generated by the flowing water, may contribute to the variation.

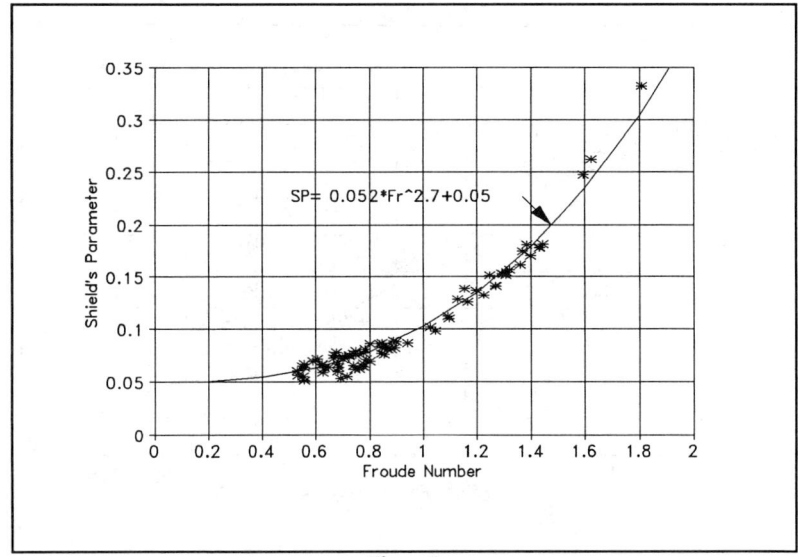

Figure 2
Shields Parameter versus Froude Number

A second paradigm for examining this phenomenon is to consider the resultant angle of the forces, holding the particle in place: net gravitational force (vertical) and the force generated by contact with other particles downstream of the given particle (horizontal). Since the net gravitational force is a constant for a given particle, when the other force is varied with flow conditions the angle of the resultant resistance force varies as is shown in Figure 3. The resistance angle is calculated as arctan(1/SP). The equation for the relationship is:

$$\Theta = 88.05 - 4.37 * Fr^2 \qquad (4)$$

where: Θ = resistance angle (degrees)
        Fr = Froude number

Spurious Self Correlation

Evaluation of the relationships between parameters created by multiplying and dividing a set of independent variables must consider whether correlations identified are merely spurious self correlations (Mahmood and Siddiqui, 1980; Kenny, 1982). In this analysis, the depth-averaged field velocity and the depth of flow appear in both parameters analyzed: Shields' parameter and Froude number. Mahmood and Siddiqui (1980) suggest that "this concern can be alleviated if methods like tests of hypothesis and computation of confidence intervals can be incorporated in data analysis."

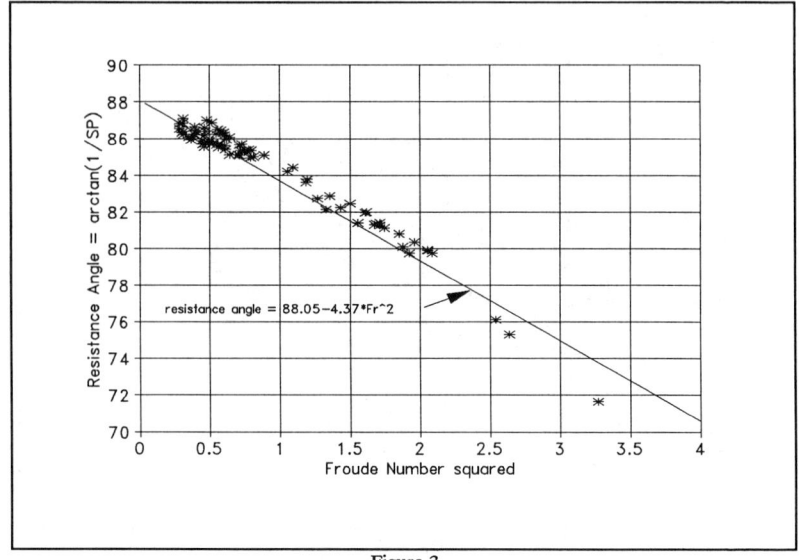

Figure 3
Resistance Angle

A Monte Carlo analysis was performed by generating 173 replicates of data sets containing 80 observations each (the number of observations in the composite data set). Normal distributions were developed based on the means and standard deviations of velocity, depth, D_{50}, and specific gravity in the composite data sets. (Uniform distributions were also tested with no change in conclusion.) The resulting correlation analyses on these random replicates produced an R^2 of 0.32 plus or minus 0.05 at the 95 percent confidence limit. The R^2 of the data shown in Figure 2 is 0.96 for 80 observations. Since 0.96 is much greater than 0.32 it is concluded that, although self correlation exists, the relationships reported above are far from spurious. In fact, the effective R^2, removing the effect of self correlation, can be estimated as follows:

$$R^2_{eff} = \frac{0.96-0.32}{1.00-0.32} = \frac{0.64}{0.68} = 0.94 \tag{5}$$

In other words, the percentage of the variation in Shields' parameter above that caused by common factors explained by the Froude number is 94 percent.

Conclusions

Shields suggests that for particle Reynolds' numbers greater than 1000, a constant ratio of moments will exist and Shield's parameter will be approximately equal to 0.06. This observation is in agreement with the values observed for Froude numbers up to 0.8. The above analysis of Shields' parameter suggests that the use of a constant Shields' parameter in flow fields with a Froude number greater than 0.8 is misleading. Assuming a constant Shield's parameter for higher Froude numbers results in conservative, in some cases very conservative, sizes in the design of riprap and other procedures outlined in Richardson, et al., (1991). Evaluation of the observations presented in this paper will result in significant cost savings in riprap protection. These concepts may also be applied to alluvial conditions (Young, et al., 1993) by considering the data of Shields' which includes Shields' parameters of 0.034 at low (6-10) Reynolds' numbers.

References

GKY & Associates, "Lab Report on Scour Protection Alternatives at Bridge Piers," FHWA, 1993 (in progress).

Kenny, Bernard C., "Beware of Spurious Self Correlation!" Water Resources Research, August, 1982.

Mahmood, Khalid and M.M. Siddiqui, "Spurious Correlation in Dimensional Analysis," ASCE Journal of the Engineering Mechanics Division, February, 1980.

Neill, C.R., "Mean Velocity Criterion for Scour of Coarse Uniform Bed Material," Journal of the International Association for Hydraulic Research, 1967.

Nikuradse, J., "Laws of Flow in Rough Pipes," National Advisory Committee for Aeronautics, Technical Memorandum 1292, July/August 1933.

Parola, Arthur C., "The Stability of Riprap Used to Protect Bridge Piers," Federal Highway Administration Report No. FHWA-RD-91-063, July 1991.

Richardson, E.V., Harrison, L.J., and Davis, S.R., "Evaluating Scour at Bridges," FHWA-IP-90-017, HEC-18, Federal Highway Administration, McLean, Virginia, 1991.

Shields, I.A. "Application of the Theory of Similarity and Turbulence Research to Bed Load Movement," (English Translation), 1936.

Wang, Sany-yi and Hsieh Wen Shen, "Incipient Sediment Motion and Riprap Design," ASCE Journal of Hydraulic Engineering, March, 1985.

Young, G.K., M. Palaviccini, and R. Kilgore, "Scour Prediction at Bridge Abutments," Proceedings of the ASCE Hydraulic Division Conference, San Francisco, CA, July 25-30, 1993.

Scour Retrofit Case Studies for Arizona

by George K. Cotton[1] and Jiri Vitek[2], Members

ABSTRACT

Scour retrofit involves a coordinated analysis and joint determination of action by several engineering disciplines: typically, geotechnical, hydraulic and structural engineering. The solution to scour-critical conditions at a bridge also involves a shared committment to project goals on the part of the owner, design team, and contractor. A cooperative approach based on the strategy of partnering offers a means of successfully reducing retrofit costs, litigation and stress.

INTRODUCTION

Scour retrofit of existing bridges presents a number of difficult problems that are not encountered in the design of a new structure. The investigation of the structure, together with the design of countermeasures, requires a multi-disciplined design team. Of equal importance is a shared commitment by the owner, design team and contractor to the goals of the retrofit. In this paper, two case studies are presented that contrast two bridge retrofit designs that were recently developed in Arizona by CRSS . The first case study discusses the retrofit design for a 1905 railroad bridge over the Salt River in Tempe, Arizona conducted as part of the Rio Salado improvements to the Salt River. The second case study discusses the improvements made to a scour-critical bridge on U.S. 93 that crosses Big Wash about 39 km north of Kingman, Arizona. The paper closes with a brief summary of partnering methodology as promoted by the Arizona Department of Transportation.

SOUTHERN PACIFIC RAILROAD SALT RIVER BRIDGE

In the Fall of 1988, CRSS was contracted by the City of Tempe to undertake a series of preliminary design studies for the improvements to the Salt River through Tempe. The following year design work was begun for river improvements and stabilization measures. During the study phase, CRSS assessed scour problems at all bridges and utilities across the Salt River in the Rio Salado corridor. We found that many facilities in the corridor had problems and would need to be improved to prevent scour damage. To accomplish this we installed countermeasures at two existing bridges of recent contruction, relocated a major water line and sewer trunkline below design scour depths, and protected other utilities by incorporating them within stablized channel banks. Initial investigation of the Southern Pacific Railroad bridge uncovered a more difficult set of problems.

---

[1] Senior Water Resources Engineer, CRSS Civil Engineers, Inc., Phoenix, Arizona
[2] Senior Structures Manager, CRSS Civil Engineers, Inc., Phoenix, Arizona

Hydraulic analysis for the existing river channel provided estimates of bridge scour that exceeded caisson depth for the bridge foundation. Two-dimensional modeling of the flow pattern near the railroad bridge showed the formation of eddies and a flow contraction along the south bank of the Salt River near the bridge. Bridge plans for the 1905 structure showed that while the northern piers of the bridge were founded on a shallow bedrock outcrop, the strata dipped abruptly to the south; as a result, the southern piers consisted of caissons placed on driven timber piles. Confirmation of the vulnerability of these conditions was provided by Southern Pacific maintenance staff who had repaired a settlement of one bridge pier near the south bank following major floods on the Salt River in February 1980.

Based on this analysis, the CRSS proceeded with a retrofit design of the railroad bridge. At Southern Pacific Railroad's request, testing of the strength of the concrete in the caissons was conducted for piers on timber piles. This data was important because of the crude nature of the foundation on this turn-of-the-century bridge. The piers consisted of steel cylinders filled with a cyclopian concrete. If the concrete did not have sufficient strength, the load would not transfer when scour undercut the caisson, and the pier would settle. Original construction methods for the bridge probably involved mostly human labor for excavation of the caissons and placement of the concrete. A series of three corings taken at each pier showed a wide variation in test results. At one pier, Pier No. 8, no cores were taken that were in good enough condition to test. Interestingly, it was Pier No. 7, located nearer the south bank, that had been repaired following the previous major flood while Pier No. 8, that had been exposed to more severe hydraulic conditions, had no reported problems.

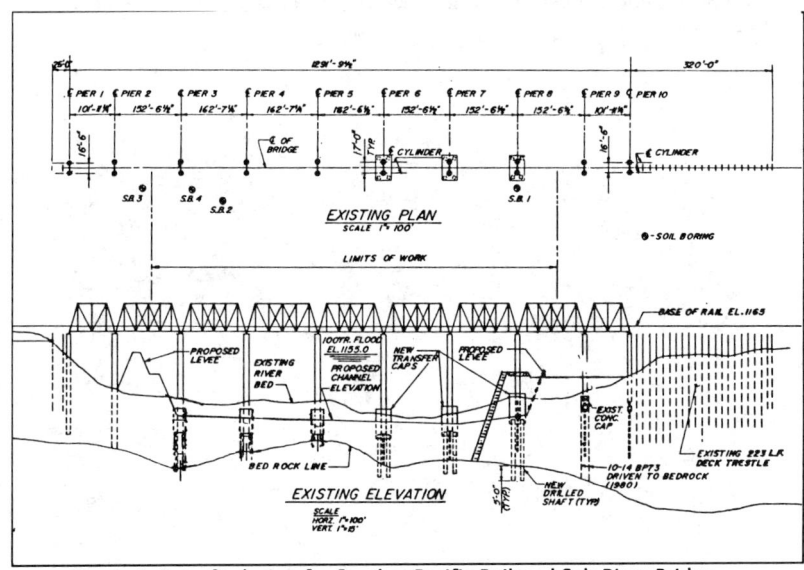

Figure 1. Retrofit design for Souther Pacific Railroad Salt River Bridge

A dilemma that was never fully resolved was created at this point in the retrofit design. The design team found that Pier No. 8 had failed; that is, the pier had little chance of surviving another severe flood event *with or without* planned changes to the Salt River channel. However, current operations of the railroad during low flows and dry weather were completely acceptable to the owner. A method of allocating retrofit cost that accounted for the existing risk was never accomplished since the owner perceived no risk. From the standpoint of the party responsible for conducting the retrofit, this was viewed as simply passing on existing risk cost to their project. For the design team, this resulted in the pursuit of alternatives that solved the existing deficiencies at a cost commensurate with the owner's perception of the risk. In reality, this simply passed much of the risk to the contractor.

The method chosen for stabilizing Pier No. 8 relied upon extensive grouting of loose material within the caisson and soil at the piling and caisson interface. The pier would then be upgraded with new pilings and a transfer cap attached to the old caisson. In the field, the critical element of soil grouting proved unworkable. The excavation required for other components of the retrofit posed an immediate risk to the structure. Limited excavation found rotting pilings beneath the caisson that emphasized further the existing deficiencies of the foundation.

Under the circumstances, the essential relationships between owner, retrofit party, design team and contractor were strained. To reach a workable solution, changes were made to the channel bank design. The pier was encased within the bank protection system and all excavation remained a safe distance away for the pier. This construction was not preferred treatment since it created an abrupt encroachment into the main channel. However, it was an acceptable middle ground between the owner and retrofit party.

Figure 2. Completed Southern Pacific Railroad Salt River Bridge retrofit

U.S. 93 BIG WASH BRIDGE

The U.S. 93 Big Wash Bridge is located 39 km northwest of Kingman, Arizona along the southwestern edge of Detrital Valley. Big Wash originates in the Cerbat Mountains and crosses U.S. 93 on the Cerbat piedmont. The piedmont consists of a 120 to 240 m thickness of alluvial sediments eroded from the Cerbat Range. The surficial layer extends about 7.5 to 10 m below existing grade and consists of non-cohesive sands and gravels with cobbles and boulders. The upper 10 to 15 feet of this layer is loose and uncemented, gradually increasing to a very high degree of cementation.

Typical of many piedmont channels in the Basin and Range physiographic province of the Southwestern United States, the channel is wide and braided in form and has a steep gradient (about 2.5 percent). Approaching U.S. 93, Big Wash has an active channel width of about 210 m and an average depth of less than 0.6 m. The existing U.S. 93 bridge spans Big Wash in 49.4 m at a 15-degree skew. The bridge is supported by four pier bents with 0.75 m circular piers. All piers are on spread footings embedded 3.0 or 4.3 m below the channel bed. Planned widening of the highway adds a new structure of similar size and alignment, about 18 m upstream of the existing bridge.

A scour assessment was conducted for the existing bridge using the 1987 version of the FHWA Technical Advisory on Scour, together with a geomorphic analysis of long-term scour. In accordance with this advisory, a multi-disciplined team that included structural, hydraulic, and geotechnical engineers was involved in the scour assessment. CRSS decided to conduct the scour analysis as an integrated, multi-disciplined task. One of the first issues that this team confronted was the fact that each discipline referred scour processes using different terms. For example, in Arizona a structural engineer familiar with computing stream forces uses the term "general scour" to refer to long-term scour, while the hydraulic engineer uses "general scour" or "natural scour" to apply to the *combination of long-term scour plus contraction scour*. Geotechnical engineers view scour potential in terms of the *perceived erodibility* of various stata and generally do not distinguish among scour processes in assessing scour potential. The initial findings of each discipline tended to contradict each other. Hydraulic based scour estimates were greater than the maximum depth estimated based on interpretation of borings by the geotechnical engineer. Neither estimate was provided to the structural engineer in terms that were consistent with his understanding of scour processes.

To resolve this interdisciplinary communication problem, lead engineers from CRSS' structural and hydraulic groups developed a Technical Note that cross-referenced the scour terminology given by FHWA with AASHTO requirements for analysis of stream forces. It was also decided that a geotechnical opinion on the erodibility of material could not be based solely on an interpretation of boring logs. Further tests of durability and erodibility of the underlying strata would need to be conducted before the computed scour depths could be reduced.

This resulted in a consistent framework for preparation of the bridge scour evaluation. The new FHWA terminology for scour processes was accepted and used in the bridge scour evaluation report. Tables in the report provided scour estimates for each specific AASHTO Loading Case so that the structural engineer did not need to independently interpret the report findings. A simple test of the underlying cemented material was performed which showed that, in a saturated condition, the material had little durability.

Following design, ADOT structured the contract as a Partnered Agreement. Partnering creates an owner-contractor relationship that promotes the achievement of mutually beneficial goals. It involves an agreement in principle to share the risks involved in completing the project, and to establish and promote a nurturing partnership environment. Partnering is not a contractual agreement, however, nor does it create any legally enforceable rights or duties. Rather, Partnering seeks to create a new *cooperative attitude* in completing contracts. To create this attitude, each party must seek to understand the goals, objectives, and needs of the other -- their "win" situation -- and seek ways that these objectives can overlap.

In this case, two specific partnering sessions were conducted: one regarding the Big Wash bridge retrofit, and one regarding the water control structures for the new bridge. As in the design phase, communication was the essential element. One session discussed design intent. This gave the contractor greater insight into the purpose of the channel improvements and structures. The second session addressed difficulties that the contractor had in excavating in hard caliche and the issue of durability of the underlying stata. The simple test of the caliche in a wet condition showed that the material had no durability. The contractor completed the excavation of bank and grade control structure embedment to the full design depth. Because of the Partnering agreement, issues were resolved quickly in the field with members of the design team and ADOT's resident engineering staff, dealing with the equivalent level of contractor's representative.

Figure 3. Complete U.S. 93 Big Wash Bridge Scour retrofit

CONCLUSIONS

Two different case histories of scour retrofit projects in the State of Arizona have been presented. These projects had similar technical requirements, but differ from each other primarily in the methods goals and objectives were communicated. The following list gives a summary of important points that are generally applicable to other bridge retrofit projects.

1. Risk cannot be transferred without the consent of all parties. In other words, if one party is unwilling or unable to assume his risk cost, that cost still exists. As a result certain parties in a retrofit may benefit because of the work of the other parties and this occurrence must be recognized as a reasonable outcome.

2. All parties to a scour retrofit must agree on a mutually acceptable statement of the level of risk. It is essential to achieve a clear understanding of what contribution, if any, each of the parties can make to reducing the existing risk.

3. The various disciplines within the design team for scour retrofit must use the same terms for scour processes. Each major discipline involved in scour retrofit has a different technical perspective that has historically used different terminology. At a minimum, a cross reference between each of the existing technical standards should be implemented at the beginning of the project.

4. The technical findings of each discipline should, to the extent practicable, be backed up with tests or calculations and not rely solely on the opinion or judgement of one discipline. Simple tests, a field investigation, or a enhanced model should not be overlooked in the retrofit design.

5. Partnering provides a process where the owner, designer and contractor can communicate clearly. We have found that Partnering promotes a better understanding of design intent, an environment where value engineering is encouraged, and methods of resolving disputes or difficulties at the appropriate level.

REFERENCES

Richardson, E.V., Harrison, L.J., and Davis, S.R., 1991, "Evaluating Scour at Bridges", Hydraulic Engineering Circular No. 18, FHWA-IP-90-017.

Lagasse, P.F., Schall, J.D., Johnson, F., Richardson, E.V., Richardson, J.R., and Chang, F., 1991, "Stream Stability at Highway Structures", Hydraulic Engineering Circular No. 20, FHWA-IP-90-014.

AASHTO, 1989, "Standard Specifications for Highway Bridges", Fourteenth Edition, 1989, Interims to date.

Vitek, J, 1990, "Structure Selection Report for Big Wash Bridge", prepared for the Arizona Department of Transportation, CRSS Civil Engineers, Inc.

Applicability of Two Simplified Flood Routing Methods: Level-Pool and Muskingum-Cunge

by

D.L. Fread and K.S. Hsu[1]

Abstract. Simplified flood routing models for unsteady flow simulation in reservoirs and rivers have advantages of relatively small computing requirements when compared to dynamic routing models based on the complete Saint-Venant equations of unsteady flow. The range of applicability as governed by accuracy is investigated for two simplified routing models, a level-pool reservoir routing model and a Muskingum-Cunge river routing model. The routing error for each simplified model is determined by systematic comparison with an accurate dynamic routing model (DAMBRK). Error properties of each simplified model are presented graphically as functions of dominant channel and flood hydrograph parameters.

Introduction

Within the National Weather Service (NWS) hydrology program for river and water resource forecasting services, sophisticated dynamic routing models such as DAMBRK, DWOPER, and FLDWAV (Fread, 1985; Chow et al., 1988), based on the complete one-dimensional Saint-Venant equations of unsteady flow, are being implemented, as resources are available, for particularly complex routing applications such as dam-break floods, major river systems subject to backwater effects, and tidal estuaries. This paper presents guidance for selecting which routing applications should be considered for the dynamic routing models and for two very simple routing models, i.e., (1) a level-pool reservoir routing model and (2) a diffusion-type Muskingum-Cunge river routing model. Suitability of the simplified models is assessed on the basis of the routing error as determined by the deviation of computed flows between the simplified models and the dynamic model (DAMBRK).

Level-Pool Reservoir Routing

Usually unsteady flow routing in reservoirs is approximated by a simple level-pool routing technique which is based on the principle of conservation of mass, i.e.,

---

[1] Director and Research Hydrologist, National Weather Service, Hydrologic Res. Lab., 1325 East-West Highway, Silver Spring, MD 20910.

$$I(t) - Q(t) = dS/dt \tag{1}$$

in which inflow (I) and outflow (Q) are functions of time (t), and the storage (S) is a function of the water-surface elevation (h) which changes with time (t). The reservoir is assumed always to have a horizontal water surface throughout its length, hence level-pool, and Q is assumed to be a function only of h(t). Eq. (1), an ordinary differential equation, can be solved by an iterative trapezoidal integration method. Using average values for I(t) and Q(t) over a Δt interval, and expressing dS/dt as the product of reservoir surface area (Sa, a known tabular function of h) and change of water-surface elevation (h) over the jth time step (Δt^j), Eq. (1) becomes:

$$0.5(I^j + I^{j+1}) - 0.5(Q^j + Q^{j+1}) - 0.5(Sa^j + Sa^{j+1})(h^{j+1} - h^j)/\Delta t^j = 0 \tag{2}$$

The inflows (I) at times j and j+1 are known from the specified inflow hydrograph, the outflow (Q) at time j can be computed from the known water-surface elevation (h^j) and an appropriate spillway discharge equation. The surface area (Sa^j) can be determined from the known value of h^j. The unknowns in the equation consist of h^{j+1}, Q^{j+1}, Sa^{j+1}; the latter two are known nonlinear functions of h^{j+1}. Hence, Eq. (2) can be solved for h^{j+1} by an iterative method such as Newton-Raphson, in which

$$h_{k+1}^{j+1} = h_k^{j+1} - f(h_k^{j+1})/f'(h_k^{j+1}) \tag{3}$$

and k is an iteration counter; $f(h_k^{j+1})$ is the left-hand side of Eq. (2) evaluated with the first estimate for h_k^{j+1} which for k=1 is either h^j (must be known at t=0) or a linear extrapolated estimate of h^{j+1}; and $f'(h_k^{j+1})$ is the derivative of Eq. (2) with respect to h^{j+1}. It can be approximated with a numerical derivative, i.e. $f'(h_k^j+1) \simeq [f(h_k^{j+1}+\varepsilon) - f(h_k^{j+1}+\varepsilon)] / [(h_k^{j+1}+\varepsilon) - (h_k^{j+1}-\varepsilon)]$ where ε is a small value, say 0.1 ft (0.03 m). One or two iterations usually solve Eq. (2) for h^{j+1}. Once h^{j+1} is obtained, Q^{j+1} is computed from the spillway discharge equation.

<u>Accuracy of Reservoir Routing Models.</u> The accuracy of level-pool routing models relative to the more accurate distributed dynamic routing models such as DAMBRK (Fread, 1985; Chow, et al., 1988) is shown in Fig. 1. The error (E_q, in percent) of the rising limb of the outflow hydrograph, as normalized by the peak outflow, is:

$$E_q = 100/Q_{D_p} \cdot [\sum_{i=1}^{N} (Q_{L_i} - Q_{D_i})^2/N]^{1/2} \tag{4}$$

in which Q_{L_i} is the level-pool routed flow; Q_{D_i} is the dynamic routed flow, Q_{D_p} is the dynamic routed flow peak, and N is the number of computed discharges comprising the rising limb of the routed hydrograph. The error (E_q) increases as (1) reservoir mean depth (D_r) decreases, (2) reservoir length (L_r) increases, (3) time of rise (T_r) of inflow hydrograph decreases, and (4) inflow hydrograph volume decreases. These effects can be represented by three dimensionless parameters, σ_ℓ, σ_t, and σ_v; where, $\sigma_\ell = D_r/L_r$, $\sigma_t = L_r/[3600\,T_r\,(gD)^{1/2}]$ in which g is the gravity acceleration constant and T_r is the time (hrs) from beginning of rise until the peak of the hydrograph, and σ_v = hydrograph volume/reservoir volume. As shown in Fig. 1, E_q increases as σ_t increases and as σ_ℓ and σ_v decrease; also the influence of σ_v

increases as σ_ℓ decreases. An analysis of Fig. 1 indicates that E_q exceeds 10% for (a) most reservoirs subjected to rapidly rising unsteady flows in which T_r is less than 1 hr such as dam-break floods or intermittent turbine releases, and (b) very long reservoirs ($L_r > 50$ mi) subjected to flash floods with T_r less than 18 hrs.

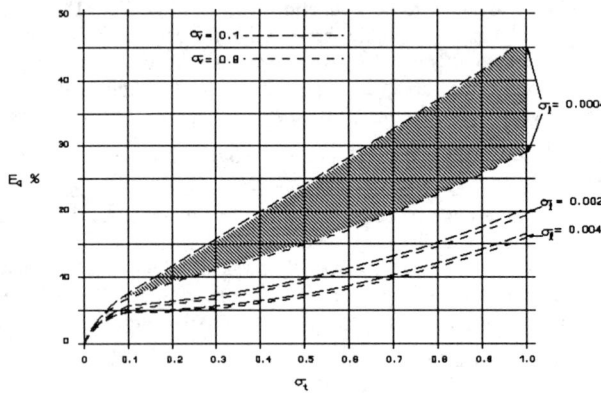

Fig. 1 Level-pool routing error (E_q) as a function of σ_t, σ_ℓ and σ_v (dimensionless parameters)

Muskingum-Cunge Method

The popular Muskingum method, described in standard reference books, e.g. Chow, et al., 1988, can be modified by computing the routing coefficients in a particular way as shown by Cunge (1969); this changes the kinematic-based Muskingum method to one based on the diffusion analogy which is capable of predicting hydrograph attenuation. This modified Muskingum method (known as the Muskingum-Cunge method) is most effectively used as a distributed flow routing technique. The recursive equation applicable to each Δx_i subreach for each Δt^j time step is:

$$Q_{i+1}^{j+1} = C_1 Q_i^{j+1} + C_2 Q_i^j + C_3 Q_{i+1}^j + C_4 \tag{5}$$

The coefficients C_1, C_2, and C_3 are positive values whose sum must equal unity, and the last term (C_4) accounts for the effect of lateral inflow ($\bar{q}_i$) along the Δx_i subreach:

$$C_1 = (\Delta t - 2KX)/[2K(1-X) + \Delta t] \tag{6}$$

$$C_2 = (\Delta t + 2KX)/[2K(1-X) + \Delta t] \tag{7}$$

$$C_3 = [2K(1-X) - \Delta t]/[2K(1-X) + \Delta t] \tag{8}$$

$$C_4 = \bar{q}_i \Delta x \Delta t/[2K(1-X) + \Delta t] \tag{9}$$

in which K is a storage constant and X is a weighting factor. It can be shown (Cunge, 1969) that Eq. (5) is a finite-difference form of the classical kinematic wave equation; however, if X is expressed as a particular function of the flow properties, Eq. (5) is able to account for wave attenuation but not for reverse (negative) flows or backwater effects. In this method, K and X are computed as follows:

$$K = \Delta x / \bar{c} \tag{10}$$

$$X = 0.5 \left[1 - \bar{Q}/(\bar{c}\ \bar{B}\ S\ \Delta x)\right] \tag{11}$$

in which $\bar{c}$ is the kinematic wave celerity, $\bar{Q}$ is discharge, $\bar{B}$ is cross-sectional top-width associated with $\bar{Q}$, and S is the energy slope approximated by the water surface slope as computed from the backwater solution of the initial steady flow condition to properly approximate the energy slope for channels with irregular and even adverse channel bottom slopes. The bar (-) indicates the variable is averaged over Δx and Δt. The coefficients, defined by Eqs. (6-9), are functions of Δx and Δt (the independent parameters), and K and X, which are functions of Q (the dependent variable) and its corresponding water-surface elevation (h) that may be obtained from a steady, uniform flow formula such as the Manning equation. Using a nonlinear solution procedure, the coefficients (K and X) are computed from the known flow properties, i.e., Q_i^j, Q_i^{j+1}, Q_{i+1}^j, h_i^j, h_i^{j+1}, h_{i+1}^j, and estimated values of the unknown flow (Q_{i+1}^{j+1}) and its h value. The estimated values are determined by extrapolation from previously computed values. The solution procedure is iterative and converges when computed and estimated values of h agree within a suitably small tolerance, 0.01 ft (0.003 m).

<u>Error Properties of Muskingum-Cunge Routing.</u> In order to assess the magnitude of errors associated with the nonlinear Muskingum-Cunge routing algorithm, a number of routing applications having a range of hydrographs and channel bottom slopes were simulated with the Muskingum-Cunge method as well as with a highly accurate implicit dynamic routing algorithm within the DAMBRK model (Fread, 1985; Chow, et al., 1988). The Muskingum-Cunge algorithm's peak discharges and corresponding water-surface elevations were compared with those computed by the dynamic routing algorithm. The difference between the two was taken as the magnitude of the error associated with the Muskingum-Cunge algorithm. This error (ε) was defined as $\varepsilon = (\varepsilon_Q^2 + \varepsilon_h^2)^{1/2}$, where ε_Q, and ε_h are the peak error in the discharge and depth, normalized about each peak. The results of this empirical error analysis via comparative routings through 10-, 20-, 50-, and 100-mile channel reaches (L) are shown in Fig. 2. The lines representing a constant value of ε are plotted against the dominant hydrograph property, T_r (the time of rise in hrs) along the vertical axis, and the energy slope, S (which is approximated by the average channel bottom slope, ft/ft). The Muskingum-Cunge algorithm is shown for 10% ε curves for the various L routing reaches. The shaded area below each curve represents all conditions of S and T_r that cause the error (ε) to exceed 10% while the unshaded areas are ε values less than 10%. The family of curves is represented by the following expression:

$$T_{r_{min}} = 0.0024\ S^{-1.03}\ (L/20)^{0.13}\ S^{-0.18} \tag{12}$$

These curves show that as S increases, there is a gradual nonlinear decrease in the minimum T_r value that can be accommodated by the algorithm for a given ε value. In general, Fig. 2 indicates that the Muskingum-Cunge algorithm incurs errors less than 10% when applied to rapidly rising hydrographs ($T_r > 1$ hr) for channels with

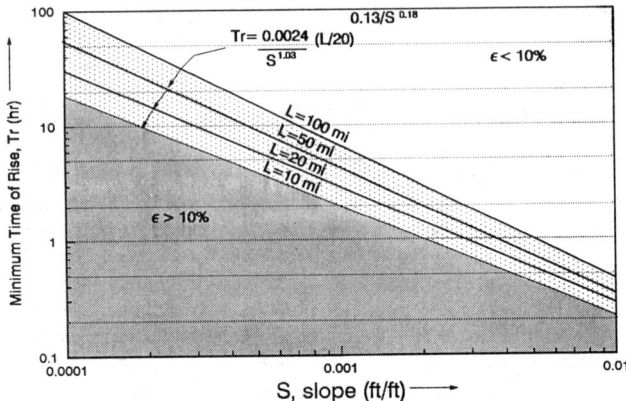

Fig. 2 Minimum allowable time of rise (Tr) which restricts errors (ϵ) in Muskingum-Cunge routing applications to less than 10%

$S > 0.002$ ft/ft (10 ft/mi). Also, the minimum allowable T_r gradually increases to 20-100 hrs (depending on the reach length) as the slope decreases to 0.0001 (0.5 ft/mi).

Conclusions

The level-pool reservoir routing model is shown empirically to incur errors exceeding 10% for (a) most reservoirs subjected to rapidly rising unsteady flows in which the time of rise of the hydrograph is less than 1.0 hour such as dam-break floods or intermittent turbine releases and (b) very long reservoirs (length > 50 mi) subjected to flash floods with times of rise less than 18 hours. The Muskingum-Cunge river routing model's error properties, excluding those due to neglecting backwater effects, increase with (a) decrease in the time of rise of the hydrograph, (b) decrease in the channel bottom slope, and (c) increase in the length of the routing reach. In general, the Muskingum-Cunge method incurs errors exceeding 10% when applied to hydrographs having a time of rise (hrs) smaller than that given by Eq. (12).

References

1. Chow, V.T., Maidment, D.R., and Mays, L.W. (1988). Applied Hydrology, McGraw-Hill, NY.

2. Cunge, J.A. (1969). 'On the subject of a flood propagation computation method (Muskingum method),' J. Hydraul. Res., Vol. 7, No. 2, pp. 205-230.

3. Fread, D.L. (1985). 'Channel routing,' Hydrological Forecasting, (Eds: M.G. Anderson and T.P. Burt), John Wiley and Sons, NY, Chapter 14, pp. 437-503.

Selection of Δx and Δt Computational Steps for Four-Point Implicit Nonlinear Dynamic Routing Models

by

D.L. Fread and J.M. Lewis[1]

Abstract. A major step in a successful application of unsteady flow models based on the numerical (four-point implicit, nonlinear finite-difference) solution of the complete one-dimensional Saint-Venant equations, is the selection of the magnitudes of the computational distance step (Δx) and time step (Δt) used in the numerical solution technique. A theoretical explanation is presented for the basis of the empirical selection criteria used rather successfully for a number of years; also, an enhanced time step selection criterion is presented. The suitability of the selection criteria is demonstrated using a numerical convergence testing technique for a wide spectrum of unsteady flow applications ranging from rapidly to slowly rising hydrographs in very flat to very steep sloping channels.

Introduction

Four-point implicit nonlinear finite-difference approximation equations of the complete Saint-Venant unsteady flow equations constitute the most extensively used basis of implicit dynamic routing models such as the NWS DAMBRK (Dam-Break), DWOPER (Dynamic Wave Operational) and FLDWAV (Flood Wave) (Fread, 1985, 1988). It is most important that appropriate computational distance (Δx) and time step (Δt) parameters be used in the application of these routing models. If the selected values are too small, the computations are inefficient, sometimes to the extent of making the application too expensive or time consuming and therefore infeasible; however, if the values are too large, the resulting truncation error (the difference between the true solution of the partial differential Saint-Venant equations and the approximate solution of the four-point implicit finite-difference approximations of the Saint-Venant equations) can cause significant errors in the computed discharges and corresponding water-surface elevations; and the errors may be so large as to make the computations totally unrealistic. Unrealistic solutions can cause the computer program

---

[1] Director and Research Hydrologist, National Weather Service, Hydrologic Res. Lab., 1325 East-West Highway, Silver Spring, MD 20910.

to abort when computed elevations result in negative depths; also, unrealistic solutions can result in significant irregularities (spurious spikes) in the computed hydrographs.

Over several years of experience with the selection of Δx and Δt values for the NWS implicit dynamic routing models in numerous applications, the following empirical selection criteria evolved:

$$\Delta x \leq c\, T_r/20 \tag{1}$$

and, $\quad \Delta t \simeq T_r/20 \tag{2}$

where T_r is the hydrograph's time of rise (time from the significant beginning of increased discharge to the peak of the discharge hydrograph), in hours; and c is the bulk wave speed (the celerity associated with an essential characteristic of the unsteady flow such as the peak or center of gravity of the hydrograph), in miles/hour; Δt is the computational time step size, in hours; and Δx is the computational distance step size, in miles. In most applications, the bulk wave speed is well approximated as a kinematic wave. Since c can vary along the waterway (channel, river, reservoir, estuary), Δx may not be constant. The kinematic wave celerity is approximated as:

$$c \simeq k' V \tag{3}$$

in which k' is the kinematic wave ratio having values ranging from $4/3 \leq k' \leq 5/3$ ($k' \simeq 3/2$ for most natural channels), and V is the flow velocity.

Herein, a theoretical explanation for the Δx and Δt empirical selection criteria is presented. The suitability of the selection criteria is demonstrated using numerical convergence testing for a wide spectrum of unsteady flow applications.

Theoretical Derivation of Δx and Δt Criteria

Theoretical wave damping (attenuation) and celerity (velocity) error (e)-diagrams were obtained previously by Fread (1974) using a Fourier technique to analyze linearized Saint-Venant equations. The e-diagrams showed convergence ratios (ratio of implicit finite-difference solution of linearized equations to their analytical solution) for wave damping and celerity plotted against D_L (wave discretization numbers) for a range of D_C (Courant numbers) values and (D_f) dimensionless friction numbers. Recently, a relationship between D_L and D_C was found for error values in the range of 0 to 5 percent; i.e.,

$$D_L \geq \eta\, D_C \tag{4}$$

where η is approximately 12 for $e \simeq 2$ percent, and $\eta \simeq 7$ for $e \simeq 5$ percent. The wave discretization number (D_L) is defined as:

$$D_L = L_w/\Delta x \tag{5}$$

where L_w is the wave length and Δx is the computational distance step. However,

$$L_w = c\, T \simeq c\, 3\, T_r \tag{6}$$

where c is the kinematic wave celerity, T is the wave period of the unsteady disturbance (wave), and T_r is the time of rise of the wave or hydrograph. Substituting Eq. (6) into Eq. (5) yields:

$$D_L = 3\, c\, T_r/\Delta x \tag{7}$$

The Courant number (D_C) is defined as:

$$D_C = c' \, \Delta t / \Delta x \tag{8}$$

where

$$c' = V + \sqrt{gD} \tag{9}$$

in which c' is the dynamic wave celerity, V is flow velocity, g is the gravity acceleration constant, and D is the hydraulic depth of flow.

<u>Δx Selection Criteria</u>. Substituting Eq. (7) into Eq. (4) yields:

$$3 c T_r / \Delta x \geq \eta D_C \tag{10}$$

which can be rearranged to give:

$$\Delta x \leq \frac{c T_r}{\eta D_C / 3} \tag{11}$$

If η is replaced with the conservative value of 12, i.e., a 2 percent level of truncation error is tolerated, and if $D_C \geq 5$, then

$$\Delta x \leq c T_r / 20 \tag{12}$$

which is identical with the empirical formula for Δx selection, i.e, Eq. (1).

The Δx selection criterion, Eq. (12), is based on the linearized form of the Saint-Venant equations; however, the complete Saint-Venant equations used in the NWS implicit routing models are nonlinear. The nonlinear terms can interact with highly nonlinear data, e.g. channel properties, so as to require even smaller Δx computational distance steps (Fread, 1988) than specified by Eq. (12).

<u>Δt Selection Criterion</u>. In order to find an expression for the selection of Δt, Eq. (7) and Eq. (8) are substituted into Eq. (4). This gives:

$$3 c T_r / \Delta x \geq \eta c' \, \Delta t / \Delta x \tag{13}$$

which can be rearranged to give:

$$\Delta t \leq T_r / M \tag{14}$$

where

$$M = \eta c' / (3 c) \tag{15}$$

Replacing η with the conservative value of 12 which allows a 2 percent level of truncation error, and substituting Eq. (3) and Eq. (9) into Eq. (15) yields:

$$M = 4 (V + \sqrt{gD}) / (1.5V) \tag{16}$$

Now, the Manning equation is used for V, i.e.,

$$V = \mu' D^{2/3} S_o^{1/2} / n \tag{17}$$

in which μ' is 1.49 (1.0 if SI units), D is hydraulic depth, S_o is bottom slope, and n is the Manning roughness coefficient. Substituting this in Eq. (16) gives:

$$M = 2.67 \left[1 + \tilde{\mu} \, n^{0.9} / (q^{0.1} S_o^{0.45}) \right] \tag{18}$$

in which $\tilde{\mu}$ is 3.97 (US units) and 3.13 (SI units), and q is the average unit width discharge along the routing reach. Using typical values for S_o, n, and q provides a range of M values generally not exceeding $6 \leq M \leq 30$. Thus, Eq. (14) with an M value of 20 is the same as Eq. (2). Unlike Eq. (2), Δt is variable in Eq. (18).

Convergence Testing to Validate Δx and Δt Selection Criteria

Numerical convergence testing is a technique wherein sensitivity tests are performed for a given problem to see if a sequence of unsteady numerical solutions with increasingly refined computational distance (Δx) and time (Δt) steps approach a fixed value, i.e., the numerical solution has converged if further refinement of Δx and Δt produces insignificant change in the solution.

Convergence testing is applied to three cases spanning a wide spectrum of unsteady flow applications ranging from rapidly to slowly rising hydrographs in very flat to steep sloping channels. In each case, the channel is 100 ft (30 m) wide with a constant Manning n of 0.045, an initial flow of 2000 cfs (56.6 cms), and a single-peaked inflow hydrograph with Q_{peak} = 20000 cfs (566.3 cms). Case (1) has a flat channel slope of 0.0002 and an inflow hydrograph with T_r = 1.0 hr; the routing reach is 20 miles long; convergence testing is at mile 10. Case (2) is identical to case (1) except the channel slope is steep (0.01). Case (3) has a very flat channel slope of 0.0001 and an inflow hydrograph with T_r = 72 hr; the routing reach is 100 miles; convergence testing is at mile 50.

Case (1) (the flat channel and rapidly rising hydrograph) convergence testing results are shown in Fig. 1a for Δt of 0.05 hr and Δx distance steps of 5, 2.5, 2.0, 1.0, 0.5, and 0.25 miles. Erroneous leading waves appear for all distance steps, $\Delta x \geq 1.0$ mile, and the hydrograph peak ceases to vary by less than 0.1 percent for the 0.5 mile distance step. Eq. (12), using a wave speed (c) of 7.0 mi/hr, gives a Δx value of 0.36 mile which is near that (0.5 mile) obtained from the convergence testing. As indicated in Fig. 1b, convergence seems to be is reached with the time step of about 0.05 hr. This is in agreement with Eq. (14) which also yields a value of 0.05 hr using a computed value of 21 for M from Eq. (18).

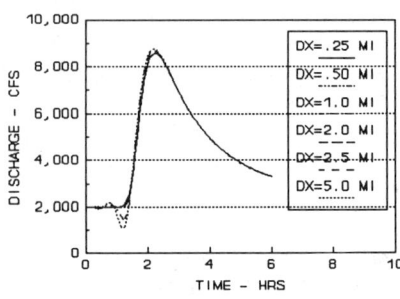
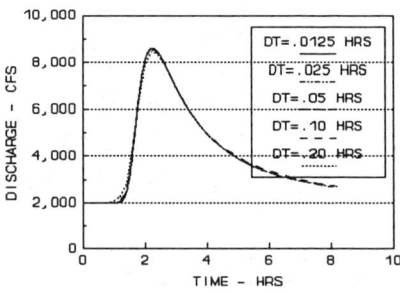

Fig. 1a Convergence testing for Δx (Case 1)

Fig. 1b Convergence testing for Δt (Case 1)

Case (2) (the steep channel and rapidly rising hydrograph) convergence testing results are shown in Fig. 2 for a Δt of 0.10 hr. Convergence appears to be reached with a distance step Δx ≤ 2.0 miles compared with a value of 1.67 miles provided by Eq. (12) using a wave celerity (c) of 33.3 mi/hr.

Case (3) (the very flat channel and slow rising hydrograph) convergence testing results are shown in Fig. 3 for a Δt of 3 hr. Convergence appears to be attained for Δx ≤ 25 miles compared to a computed value of 15 miles provided by Eq. (12) using a wave celerity of 4.2 mi/hr.

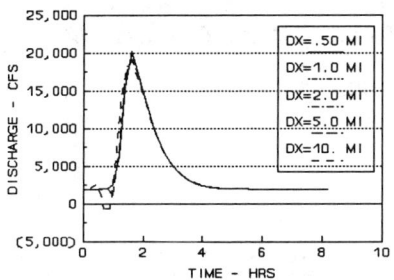

Fig. 2 Convergence Testing for Δx (Case 2)

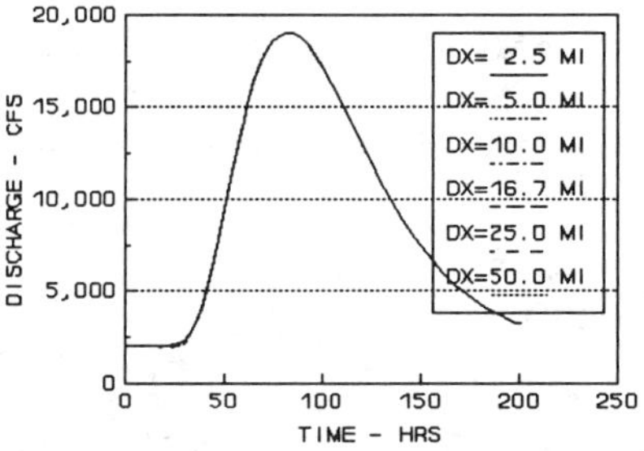

Fig. 3 Convergence Testing for Δx (Case 3)

Conclusions

Critical to applications of Saint-Venant based implicit dynamic routing models is the selection of the computational distance steps (Δx) and time steps (Δt). A theoretical derivation is given for the Δx and Δt selection criteria. These criteria not only explain the utility of the previous empirical formulae in producing acceptable computational results, but are also capable of yielding appropriate Δx and Δt values for routing applications significantly differing from past experience. The validity of the selection criteria is demonstrated through convergence testing.

References

1. Fread, D.L. (1974). <u>Numerical Properties of Implicit Four-Point Finite Difference Equations of Unsteady Flow</u>, HRL-45, NOAA Tech. Memo NWS HYDRO-18, Hydrologic Res. Lab., National Weather Service, Silver Spring, Md., 38 pp.

2. Fread, D.L. (1985). 'Channel routing,' <u>Hydrological Forecasting</u>, (Eds: M.G. Anderson and T.P. Burt), John Wiley and Sons, NY, Chapter 14, pp. 437-503.

3. Fread, D.L. (1988). <u>The NWS DAMBRK Model: Theoretical Background/User Documentation</u>, HRL-256, Hydrologic Res. Lab., National Weather Service, Silver Spring, Md., 315 pp.

ADVANCED HEC-2 MODELING ON FORESTER CREEK

Lisa T.M. Vomero, ASCE Affiliate[1]

Abstract

The goal of this paper is to share information regarding advanced modelling techniques used to solve complex hydraulic flow issues using the HEC-2 Computer Model (1990). There are several hydraulically significant aspects of this otherwise routine HEC-2 study. The problems encountered included: split flow, both returning and non-returning to the main channel stem as well as changing flow regimes in the channel. The channel's flow regime changed from supercritical at the start of the study to subcritical flow within the first several cross sections; however, the flow regime remained subcritical for the remainder of the reach length.

Introduction

The purpose of the study was to delineate the floodplain/floodway boundaries and to produce a Flood Boundary Map as well as a Flood Insurance Rate Map. This work was performed under contract with the Federal Emergency Management Agency (FEMA). As such, work was coordinated with the local jurisdictions. In this case there are two: San Diego County Department of Public Works and the City of El Cajon. Forester Creek is a predominantly natural channel located in southern San Diego County, California. The area of study is bounded by its natural drainage divides to the north, east, and south. The western limit corresponds to the City of El Cajon's eastern boundary. Please refer to Figure 1.

---

[1]Hydrogeologist and Director of Marketing, WEST Consultants, Inc., 2111 Palomar Airport Road, Suite 180, Carlsbad, CA 92009-1419; (619) 431-8113 Office; (619) 431-8220 FAX

FIGURE 1. FORESTER CREEK

(SHEET 1 OF 4 TOTAL)

PRELIMINARY MAP RELEASE SUBJECT TO CHANGE/REVISION

Map Not to Scale

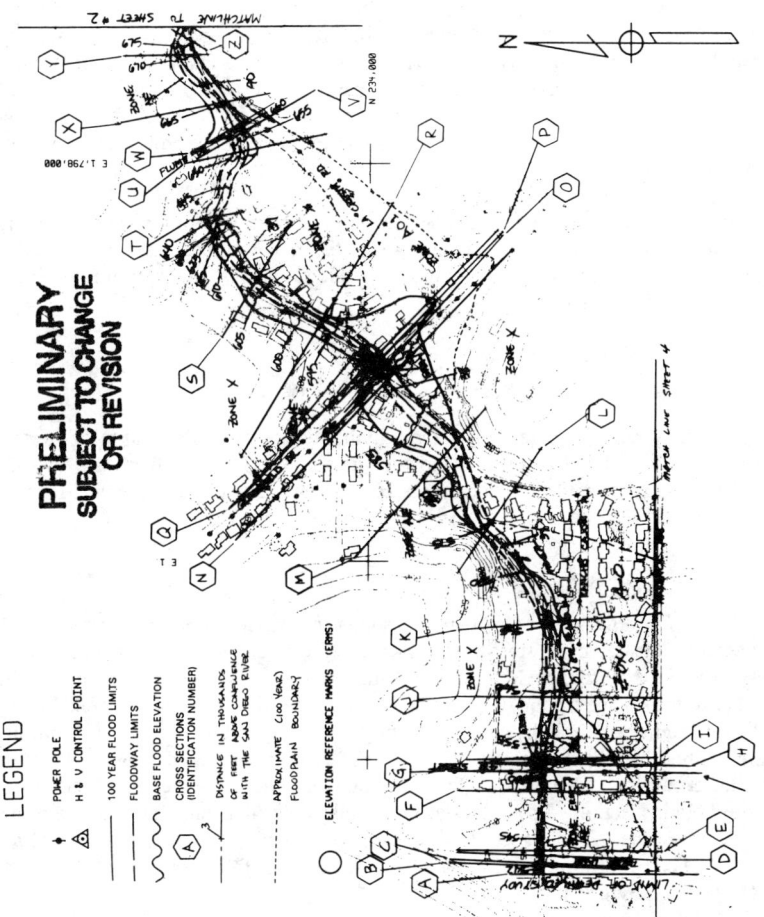

Methodology and Limitations

The hydrology for the basin was calculated using the County of San Diego's Hydrology Manual, dated 1985. Peak discharge values used in the study were calculated as follows:

Concentration Point	Area (sq.mi.)	Q 10 (cfs)	Q100 (cfs)
1	1.44	280	740
2	3.12	480	1420

The stream hydraulics were modeled using the one-dimensional HEC-2 flow model, entitled: "Water Surface Profiles", dated 1990. This model does have the capability to handle split flow conditions and this program option was employed in our study. It should also be noted, that there exists a separate document for which this type of modeling is very useful. It is entitled: "Application of the HEC-2 Split Flow Option" and it is available from the Hydrologic Engineering Center at the Corps of Engineers.

The HEC-2 model, however, does not have the capability to run the subcritical and supercritical flow conditions together. As a result, it was necessary to break the stream reach into two (2) separate computer models. This is discussed further in the latter portion of this paper.

Split Flows

As previously stated, there are several hydraulically significant aspects of this otherwise routine HEC-2 study. The first item is the fact that there are two (2) separate areas were split flow conditions are present. There are however different, in that one has flow returning to the channel and the other one does not.

Split One

The first split flow area occurs in the southern, downstream area of the reach immediately upstream of 4th Street. Please refer to Figure 1. It is modelled in cross sections A through G. This first area of split flow does not return to the channel in the course of our study reach length. Split flow one was modelled using both weir and normal depth methods of the HEC-2 split flow option. The normal depth method of split flow calculation was preferred as it resulted in what is a more realistic occurrence. The actual flow conditions in the field were determined from on-site field inspection. From the flat topography as well as undefined nature of the channel it is expected that sheet flow would result after the water breaks out of the channel. The water after leaving the channel, flows south across Naranca Avenue, please refer to Figure 1.

Split One (Continued)

The split flow out of the area overflows into an adjacent southern drainage basin. The result is that a separate study including the addition of more mapping and digitized cross sections had to be obtained. Additional peak flow volumes for the 100 year event also had to be calculating using the County of San Diego's Hydrology Manual (1985).

It is also interesting to note that the adjacent area to the south were the overflow goes was no longer in the county's jurisdiction and there was no topography available for it; since, it was outside our original study limits. This adjacent drainage basin corresponded to the San Diego County and the City of El Cajon boundaries, respectively. As a result, there were also political ramifications which had to be communicated and addressed by all persons involved with regards to the actual hydraulics of the stream.

Split Two

The second split flow area occurred upstream of the first at the intersection of the creek with La Cresta Road. This location is characterized by a culvert crossing consisting of three concrete pipes. This crossing was modelled using the (relatively new) special culvert routine of the program. The culvert crossing is located upstream of a super elevated curve near the crest of a small hill.

The culverts have no headwalls and do not have capacity for the 100-year flow. This was determined by employing the "J3" record, codes 101 and 105 in the HEC-2 model. Code 101 prints hydraulic calculations for culverts only and Code 105 provides cross section output at both special bridges and culverts. From the analyses of the output is was determined that there was weir flow over the road. Then, care was taken to see were that water would naturally flow. Since the road is super elevated and near the crest of a hill, the water flows down the hill and away from the channel instead of flowing across the road back into the original channel on the other side of the road. As a result, the water is carried in the road which acts as a channel during the 100-year storm event. Once the water flows into the road, it becomes temporarily lost from the system. Therefore, the "QT" record in the model was adjusted to reflect this loss.

Next, the area of the split flow out had to be assessed and a flood hazard zone designated. Since the peak discharge and longitudinal slope of the road was known, Manning's equation was used to calculate normal depth. The normal depth calculations were performed using the geometry of the typical street section. Based upon this depth, the Flood Zones were assigned to this outflow area. The lost discharge was subsequently added back into the model further downstream at Greenfield Drive by again adjusting the "QT" record.

Flow Regime Change

As previously mentioned, this model also involved two separate flow regimes, super- and subcritical. The location were this occurred was in the downstream portion of the study reach, near the beginning of the model, in the vicinity of 4th Street, please refer to Figure 1. At 4th Street there is a box culvert. Downstream of this box culvert the channel is improved, upstream it is natural. In the downstream area, the channel is characterized by concrete lining on both the bottom and the side slopes.

This resulted in a change of flow regime from subcritical under natural conditions to a supercritical flow regime in the improved channel. Hydraulically the end result is that the model had to be broken into two separate flow regime models, one supercritical and the other subcritical. The flow regime boundary was conveniently located at the box culvert.

The supercritical flow first modelled and run as subcritical had to be reversed and rerun. It would not have been prudent to model the supercritical portion as subcritical as one might think; even though, the sub-critical run results in a higher, more conservative water surface elevation. The actual channel improvement was apparently designed specifically for a super critical flow condition. This was not known at the beginning of the study.

Therefore, if not modelled as such, the water would be shown as overflowing its bank carrying capacity and will flood into the adjacent overbank areas, which is not what has been historically observed. These conditions of nature, design and the model must be assessed before any assumptions are made for input into the model. It is important to verify that the results given will accurately reflect what is actually happening in the field, especially for cases like this!

Summary

In summary, it is hoped that you have gained insight regarding advanced modelling techniques available to solve complex hydraulic flow issues using the HEC-2 Computer Model (1990). The problems that can be modelled effectively are: split flow, both returning and non-returning to the main channel stem as well as changing flow regimes in the channel. As a final note, it is important for engineers to have a good working knowledge of the input needed to create a good hydraulic model. It is equally important to understand what the output means and to verify its accuracy with what is actually observed in the field.

References

County of San Diego, Department of Public Works Flood Control Division: Hydrology Manual, 1985.

Federal Emergency Management Agency (FEMA), Flood Insurance Administration: Flood Insurance Study, City of El Cajon, California, San Diego County, Community Number 060289, 1982.

Federal Emergency Management Agency (FEMA), National Flood Insurance Program: Floodway Flood Boundary and Floodway Map City of El Cajon, California San Diego County, Community Number 060289 0005, 1982.

Federal Emergency Management Agency (FEMA): Flood Insurance Study, Guidelines for Study Contractors, FEMA (Document) 37, 1991.

U.S. Army Corps of Engineers, Hydraulic Engineering Center: Application of the HEC-2 Split Flow Option, Training Document No. 18, 1982.

U.S. Army Corps of Engineers, Hydraulic Engineering Center: COED Corps of Engineers Editor, 1987.

U.S. Army Corps of Engineers, Hydraulic Engineering Center: Computing Water Surface Profiles with HEC-2 on a Personal Computer, Training Document No. 26., 1990.

U.S. Army Corps of Engineers, Hydraulic Engineering Center: HEC-2 Water Surface Profiles, 1990.

U.S. Department of the Interior Geological Survey, Alpine Quadrangle, 7.5 Minute Series (Topographic), 1975.

U.S. Department of the Interior Geological Survey, El Cajon Quadrangle, 7.5 Minute Series (Topographic), 1975.

Modeling Critical Depth in Open Channels

Robert G. Traver, M. ASCE[1]

Abstract
 This paper explores multiple critical depths. Reasons behind and characteristics of their occurrence are reviewed. A method to determine multiple critical depths is proposed for use in unsteady or steady surface water profile models.

Problem Statement
 The purpose of this work is to study the occurrences of multiple critical depths, and determine means of identification and employment for hydraulic models.

Introduction
 Most water resource planning incorporates a mathematical model of the system under analysis. This model is usually exercised to evaluate design changes to the river system under study. Therefore, the model being used must be reliable. One present complication that can cause a lack of accuracy is the occurrence of multiple critical depths (Blalock and Sturm 1981).
 Determination of critical depth is necessary for both steady and unsteady water surface profile modeling. From a purely hydraulic viewpoint, it represents the change between supercritical and subcritical flow. When viewed as a component of a numerical model, it may represent a boundary assumption, a point of decision, or a limit to the solution. An inaccurate depiction can lead to a false solution, or nonconvergence, both of which are unacceptable when modeling a stream system.

Governing Equations
 Blalock and Sturm (1981), and Chaudry and Bhallamudi (1989) have reported the occurrence of multiple critical depths. These authors conclude that the energy correc-

---

[1] Assistant Professor, Department of Civil Engineering; Villanova University, Villanova, PA 19085.

tion coefficient (α) and its derivative with respect to
depth are required when representing specific energy (Se)
(Eq. 1) or critical depth (yc) (Eq. 2):

$$Se = y + \frac{\alpha Q^2}{2gA^2} \qquad (1)$$

$$\frac{dSe}{dy} = 0 = 1 - \frac{\alpha Q^2}{gA^3}\frac{dA}{dy} + \frac{Q^2}{2gA^2}\frac{d\alpha}{dy} \qquad (2)$$

with A representing the channel area, g the gravitational
coefficient, y the depth and Q the flow. α is expressed
as a function of depth as defined by Henderson (1966).

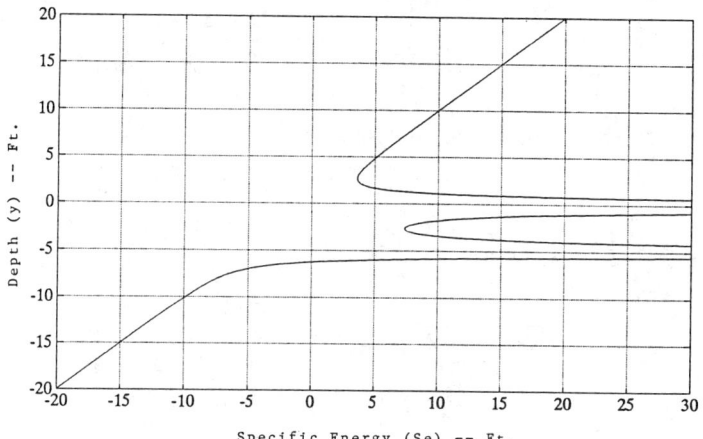

Figure 1. Trapezoidal Channel, Depth - Specific Energy.

Roots of the Equations
When examining any equation with multiple solutions,
it is necessary to determine the number of possible
positive and negative roots. As only positive depths are
valid, this defines the maximum number of solutions. As
shown by Fig. 1, five roots can occur for trapezoidal
channels, as the specific energy term includes a depth to
the fifth power. Assuming α is nondimensional, Descartes's rule of signs (Southworth and Deleeuw 1965) indicates that two positive and three negative roots exist.
The same procedures recognize two positive and one negative roots for rectangular channels. Critical depth is
therefore a multiple positive root. As most models use a
X-Y coordinate structure which mimics incremental trapezoidal or rectangular shapes, it is tempting to state
that only one critical depth is possible. This however

is incorrect as multiple critical depths exist for compound channels (Blalock and Sturm 1985).

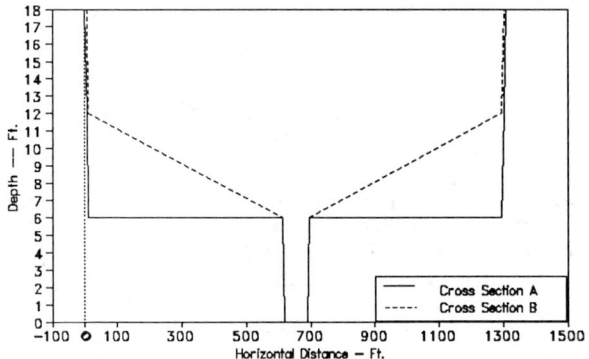

Figure 2. Compound Channels A and B.

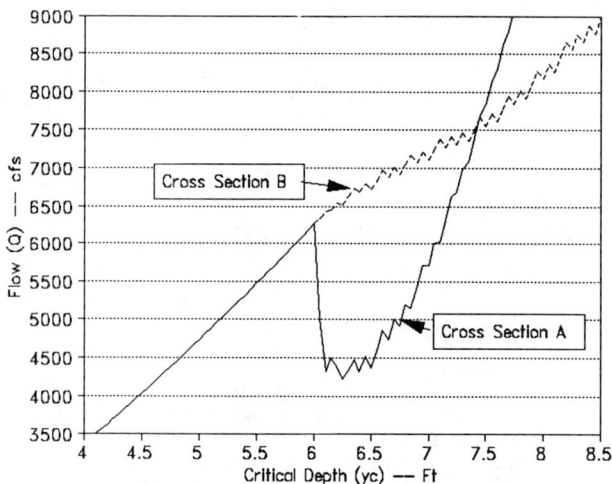

Figure #3. Flow versus Critical Depth.

Sample Channels

Two example compound channels flowing at 5000 cfs have been used in previous articles (Blalock and Sturm (1981), and Chaudry and Bhallamudi (1989)) to investigate multiple critical depths (Fig. 2). This previous work was expanded by modeling a range of flows for these channels. Equation 3 was derived from critical depth, to focus on the influence of channel geometry.

$$Q = \left(\frac{\alpha}{gA^3} \frac{dA}{dy} - \frac{1}{2gA^2} \frac{d\alpha}{dy} \right)^{-.5} \quad (3)$$

Plotting critical depth as a function of flow (Fig. 3) provides a new perspective. For each cross section, the relationship is stable, and can be developed from channel geometry. Fig. 3 shows that it is possible to have a second critical depth, adding two positive roots to the specific energy plot. Using channel A as an example, it is possible for some flows to have a single specific energy represented by four separate depths. This would easily confuse any flow model, causing non-convergence, or a false solution. For 5500 cfs, critical depth ranges from 5.5 to 6.8 feet, a difference of +/- 1.3 feet. This is clearly a significant source of error.

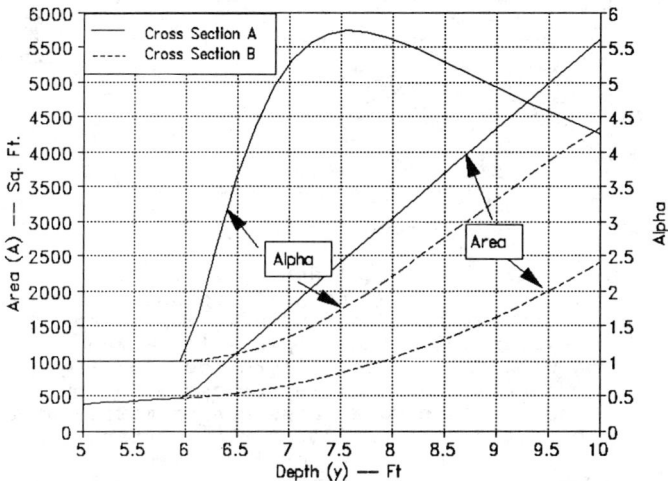

Figure 4. Area and α versus depth.

Source of Multiple Critical Depths

It is evident, that multiple critical depths begin as the flow goes out of bank and end after a depth of around a foot is reached in the overbank. From Fig. 3, a flow of 5500 cfs produces three solutions to the critical depth equation (additional solutions apparently created by the jagged portions of the plot are due to the numeric approximation of the derivatives used in this study, and should be ignored). The middle point represents a local specific energy maxima and is therefore not interpreted as a critical depth by this researcher. By examining compound channels with and without the α term, it can be shown that the α term is not the source and its use in fact eliminates some occurrences. Therefore, it can be assumed that the substantial change in area as the flow goes out of bank is the cause of the multiple roots. Figure 4 depicts this change for both area and α.

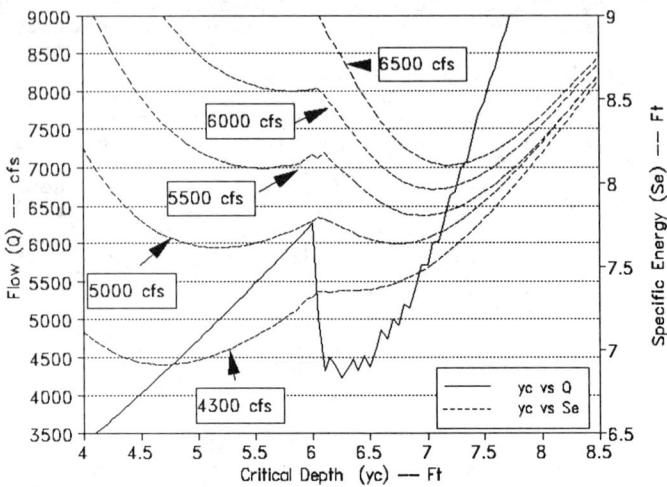

Figure 5.   Channel A - Specific Energy.

Before modeling critical depth, it is necessary to determine whether this is a natural hydraulic phenomena or simply an aberration of our numeric representation. Fig. 5 displays specific energy plots for a range of flows, along with the critical depth equation (Eq. 3). For one dimensional modeling, we are assuming that both the channel and the overbank flows are primarily at a 90 degree angle to the orientation of the overbank, and have the same friction slope and water surface elevation. This researcher believes this occurrence of multiple critical depths is an aberration of this depiction. If we consider a channel depth of 6.1 feet, or a depth of .1 feet in the overbank areas, we are assuming the relatively low velocity overbank flows have the same Energy gradient as the high velocity channel flow. Naturally this is untrue as one dimensional flow is not a valid assumption in this region and our model basically states that our numerical depiction is invalid. At this low depth the overbank flow velocity is negligible, and can be assumed to be directed into or out of the main channel, depending on whether the water surface is rising or falling. This is reinforced as the local specific energy maxima always occurs at the data point immediately above bank full conditions. Modeling of momentum produced similar results.

Modeling and Multiple Critical Depths

There is some good news for the hydraulic modeler. First, it is difficult to create a situation where there are multiple critical depth as long as α is correctly employed. Second, this researcher was unable to create any situation where more than two critical depths occurred. Third, as the relationship is stable and dependent upon channel geometry, it is relatively easy to determine whether multiple critical depths occur before applying a water surface profile model. Fourth, when multiple critical depths occur, one is always located within the main channel, and one is located within the overbank. This allows us to split the channel into two regions, one above and one below bank full conditions. As this point is constant for all flows for a given cross section it becomes a boundary. It is therefore possible to use root finding methods such as a combined Bisection - Newton Raphson scheme (Traver 1988) to determine critical depths for both channel and overbank regions.

Treatment of multiple critical depths within steady and unsteady flow models is currently under study by the author. The best solution would be to develop a revised mathematical depiction, to more correctly model this area of extreme change. A depth of .25 feet within the overbank of cross section A, effectively doubles the velocity head by doubling the α term. It takes over 1.5 feet to produce the same result for cross section B. This is clearly unrealistic. In the meantime, various methods including linkage of the critical depths and simply ignoring this region are being investigated.

Bibliography

Blalock, M.E., and Sturm, T.W., 'Minimum Specific Energy in Compound Open Channel,' *Journal of the Hydraulics Division*, ASCE, Vol. 107, June 1981, pp. 699-717.

Chaudry, M.H., and Bhallamudi, S.M., 'Computation of Critical Depth in Symmetrical Open Channels.' *Journal of Hydraulic Research*, Vol. 26, No. 4, 1988, pp. 377-396.

Henderson, F.M., *Open Channel Flow*, Macmillan Publishing, NY, 1966.

Southworth, R.W., and Deleeuw, S.L., *Digital Computation and Numerical Methods*, McGraw-Hill Book Co., NY, 1965.

Traver, R.G., *Transition Modeling of Unsteady One Dimensional Open Channel Flow Through the Subcritical-Supercritical Interface*, Ph. D., Civil Engineering, Pennsylvania State University, 1988.

A Numerical Model for Learning Concepts of Streamflow Simulation

L. L. DeLong[1]

Abstract

Numerical models are useful for demonstrating principles of open-channel flow. Such models can allow experimentation with cause-and-effect relations, testing concepts of physics and numerical techniques. FOURPT is a numerical model written primarily as a teaching supplement for a course in one-dimensional streamflow modeling. FOURPT options particularly useful in training include selection of governing equations, boundary-value perturbation, and user-programmable constraint equations. The model can simulate non-trivial concepts such as flow in complex interconnected channel networks, meandering channels with variable effective flow lengths, hydraulic structures defined by unique three-parameter relations, and density-driven flow. The model is coded in FORTRAN 77, and data encapsulation is used extensively to simplify maintenance and modification and to enhance the use of FOURPT modules by other programs and programmers.

Introduction

The computer program FOURPT is a one-dimensional, unsteady, open-channel streamflow model written primarily as a teaching supplement for a course, "One-Dimensional Streamflow Modeling Concepts," that is part of the U. S. Geological Survey, Water Resources Division, training program. Although many numerical models exist for practical use in simulating unsteady streamflow, data preparation and model use are generally dominated by requirements specific to individual models. Ideally, students should concentrate their efforts on learning principles of flow computation, not on learning the eccentricities of a specific model. FOURPT has been written to simplify data requirements and yet provide the capability to demonstrate non-trivial concepts that are encountered in actual field applications.

Governing Equations and Numerical Formulation

The governing mass- and momentum-conservation equations solved by FOURPT

---

[1]Hydrologist, U.S. Geological Survey, Building 2101, Stennis Space Center, MS 39529.

are
$$\frac{\partial}{\partial t}(\rho M_a A) + \frac{\partial}{\partial x}(\rho Q) - \rho_\ell q = 0, \qquad (1)$$
and
$$\frac{\partial}{\partial t}(\rho M_q Q) + \frac{\partial}{\partial x}\left(\beta\rho\frac{Q^2}{A}\right) + gA\left(\rho\frac{\partial Z}{\partial x} + \rho\frac{Q|Q|}{K^2} + \frac{\partial \rho}{\partial x}\bar{z}\right) = 0, \qquad (2)$$

where t is time, ρ is density, A is cross-sectional area, M_a is an area-weighted sinuosity coefficient, x is downstream reference distance, Q is volumetric discharge, ρ_ℓ is density of lateral inflow, q is lateral inflow, M_q is a flow-weighted sinuosity coefficient, β is the momentum coefficient, g is acceleration due to gravity, K is channel conveyance, Z is distance of the water surface above a common datum, and $\bar{z}$ is the distance of the cross-section centroid from the water surface. Equations 1 and 2 are similar to those presented by DeLong (1986) with the exception that density is allowed to vary with stream distance and time. The area-weighted sinuosity coefficient M_a (DeLong 1986; DeLong 1989) and flow-weighted sinuosity coefficient M_q (Froehlich 1990) can vary with depth of flow and distance.

Numerical Formulation

General analytical solutions to the unsteady one-dimensional open-channel flow equations do not exist. Governing equations are solved numerically in FOURPT using a four-point-implicit method. This method was selected because it is easy to present in a training environment and is a method used in many practical flow models. The simultaneous-equation solver used in FOURPT is capable of operating on large and complex networks of interconnected channels. It is efficient in terms of memory and computational speed and has been successfully applied to a network of greater than 400 channels.

Model Capabilities

FOURPT has many capabilities particularly useful for training. Selection of governing equations (kinematic-, diffusion-, or dynamic-wave equations) while simulating flows of practical interest helps students develop an understanding of the importance of each of the terms in the momentum equation under a variety of circumstances. This knowledge is useful for understanding model limitations and selecting models for actual field applications.

The representation of a channel by a sufficient number of descriptive cross sections and the subsequent use of an adequate number and spacing of computational cross sections in the numerical solution are separate and independent notions. Each can significantly effect model results. Descriptive channel cross sections should be determined with the sole thought of adequately and efficiently representing channel geometry and hydraulic properties. Spacing of computational cross sections is chosen to ensure numerical convergence, which depends on the numerical scheme employed and the specific properties of the channel and flow simulated. FOURPT helps to enforce these concepts by treating channel description and spacing of computational cross sections separately in data input and program execution. This simplifies and encourages testing for adequate channel representation and numerical convergence with respect to spatial discretization — spacing between computational cross sections can be reduced independently of descriptive cross sections.

FOURPT can be linked with alternative channel-properties modules that assume the channel to be rectangular, trapezoidal, or irregular in cross section. Simple rectangular- or trapezoidal-geometry routines require only a single line of data per channel and are very convenient for use in training exercises or network prototyping. Complex, irregular-geometry routines require tabulated values of geometric and hydraulic properties at multiple depths for each cross section. The irregular geometry routines allow values of momentum coefficients and effective flow lengths to vary with flow depths. This adds the capability to simulate flow in channels of compound section (Chow 1959) and flow in meandering channels — including inundation of adjacent flood plains (DeLong 1986).

For convenience, time-dependent boundary conditions can be computed from harmonic equations. Optionally, parameters in the equations can be perturbed by user-defined normal distributions, using pseudo-random number generators, to simulate uncertainty in known boundary conditions. Analysis of a sequence of model runs provides estimates of the relation of uncertainty in results to uncertainty in boundary conditions. This is a useful exercise for students and for persons prototyping potential field applications.

Although density is allowed to vary with time and distance along a channel FOURPT does not at present (January 1993) include transport computations and is not coupled with a transport model. A user may optionally force density to be constant or supply known time series of density values at the extremities of each channel as was done for the example in fig. 1. Until transport computations are included or the model is coupled with an appropriate transport model, density is assumed to vary linearly between values supplied at channel extremities.

Any hydraulic structure (such as a culvert, bridge opening, or gate) that can be described by a unique three-parameter relation can be simulated by FOURPT. The effects of such structures are enforced implicitly through equations constraining dependent variables at channel junctions. Three-parameter relations are described by tables of points with coordinates corresponding to water-surface elevations immediately upstream and downstream of the structure and flow through the structure.

For special applications requiring boundary conditions not already incorporated in FOURPT, a facility for user-programmed constraints exists. The facility requires only minimal knowledge of the FOURPT computer code. The desired constraint is programmed independently of most FOURPT code — required flow variables

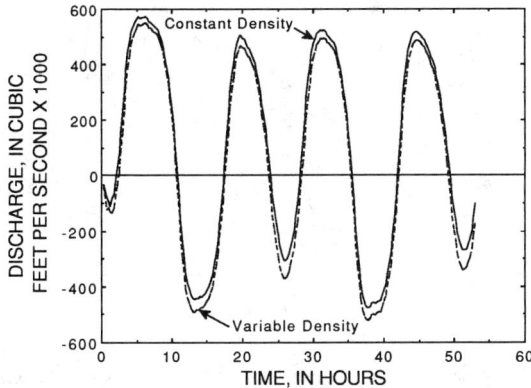

Figure 1: Effect of variable density on flow.

are automatically set and only two FOURPT functions require minor modification to initialize and invoke the added constraint. Constraining routines enforcing logarithmic and semi-logarithmic stage-discharge relations and a routine enforcing dam-break functions (expanding trapezoidal breach, flow over dam, flow over spillway, flow through gates, and flow through turbines) patterned after Fread (1984) are included in the FOURPT code as working examples.

The Computer Code

FOURPT is written in FORTRAN 77 (American National Standards Institute 1978) and has been used on a variety of minicomputers, workstations, and microcomputers without modification. Probably the greatest departure from traditional FORTRAN programming in FOURPT is in the extensive use of data encapsulation. The code itself has instructional value for persons concerned with modifying or writing models or auxiliary FORTRAN programs. It provides a rare, practical example of data encapsulation in FORTRAN.

FOURPT is constructed of FORTRAN 77 modules (DeLong et al. 1992; Thompson et al. 1992). Herein, module is defined (Stroustruap 1988) as a programming construct composed of a set of procedures combined with the data that they manipulate. Data encapsulation refers to the technique of encapsulating data with functions that are used by client routines to access and manipulate data indirectly. Direct use of data is restricted to the few functions comprising the module. This technique helps to reduce overall program dependence on the form of the data. For example, FOURPT uses a channel-properties module to supply geometric or hydraulic properties needed throughout the program. Cross-sectional

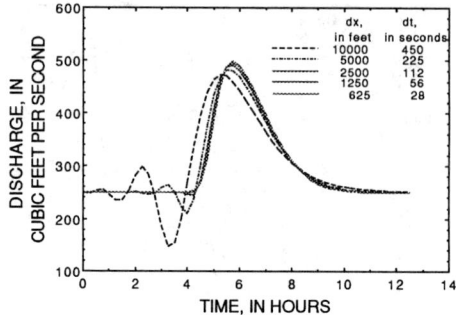

Figure 2: Spatial and temporal convergence in a prismatic channel.

area at x distance downstream and h depth of flow is obtained from a function invocation such as
$$area = CxArea(\,x,\,h\,).$$
The arguments x and h are all that are required by $CxArea$ (or any of the properties functions) and would logically be known in routines requiring cross-sectional area. No knowledge is required concerning the shape of the channel, number of data points representing the channel, or any detail specific to the computation of cross-sectional area. Consequently, routines invoking $CxArea$ are not dependent on the form of the data describing the channel, and a variety of channel descriptions may be substituted without modification to the client code. Other data fundamental to the simulation of unsteady open-channel flow are grouped according to type and encapsulated with functions in a manner similar to channel properties. Modules and submodules used in FOURPT include channel properties, flow status, channel schematic, equation solver, program control, general constraints, three-parameter ratings, logarithmic ratings, semi-logarithmic ratings, file names and units, space- and time-series output, volume and mass balance, and fluid density. Because modules originally written for FOURPT encapsulate data that are fundamental to the general field of streamflow computation, some of them could be of use to other models.

Computer Exercises

Computer exercises with FOURPT provide first-hand demonstration and experimentation with basic concepts. Spatial convergence (independent of temporal convergence) and the need for proper schematization can be demonstrated with the simulation of steady flow through an expanding reach. The effect of spatial and temporal convergence, relations between spatial and temporal discretization, and phase errors can be demonstrated (fig. 2) through the simulation of selected hydrographs. By calibrating a model applied to a hypothetical data set with known stage boundaries, the effects of timing and datum errors can become readily apparent. Considerable effort is sometimes required to synthesize exercises that demonstrate salient points, common misapplications, and appropriate corrective actions without causing model failure in the process. Successful exercises can be bundled with source code for distribution to persons not attending formal training, thereby providing tests for successfully installing the model as well as basic tutorials.

References

American National Standards Institute (1978). *American National Standard Programming Language FORTRAN.* New York, N. Y.: American National Standards Institute.

Chow, V. T. (1959). *Open-channel hydraulics.* N. Y.: McGraw-Hill.

DeLong, L. L. (1986). Extension of the unsteady one-dimensional open-channel flow equations for flow simulation in meandering channels with flood plains. In S. Subitzky (Ed.), *Selected papers in the hydrol. sci.*, pp. 101–105. U.S. Geolog. Water-Supply Paper 2290.

DeLong, L. L. (1989). Mass conservation: 1-d open channel flow equations. *J. Hydr. Div., ASCE 115*(2), 263–269.

DeLong, L. L., D. B. Thompson, and J. M. Fulford (1992). Data encapsulation using fortran 77 modules. *Fortran Forum 11*(3), 11–19.

Fread, D. L. (1984). *DAMBRK: The NWS dam break forecasting model.* Silver Spring, Md.: Ofc. of Hydrol., National Weather Service.

Froehlich, D. C. (1990, Jul.). Hermite-Galerkin model for dam-break floods. In *Proc. 1990 Nat. Conf. on Hydr. Engrg.*, pp. 557–562. ASCE.

Stroustruap, B. (1988, May). What is object-oriented programming? *Software Magazine, IEEE*, 10–20.

Thompson, D. B., L. L. DeLong, and J. M. Fulford (1992). *Data Encapsulation Using Fortran-77 Modules — a First Step Toward Object-Oriented Programming.* U.S. Geolog. Survey Water Resources Investigations Report 92-4123.

A COMPARISON BETWEEN TWO KINEMATIC WAVE SOLUTIONS FOR MOVEMENT OF DEBRIS FLOWS

By M. Arattano[1], and W.Z. Savage[2]

Abstract

We compare an asymptotic kinematic wave solution for movement of debris flows with a solution obtained by the method of characteristics. The differences between the two solutions in the field of applicability of the theory are negligible for wide rectangular channels but increase for narrow channels. The two solutions converge for large elapsed times after debris-flow initiation.

Introduction

Arattano and Savage (1992) show that kinematic wave theory can be used for debris-flow modeling. Takahashi (1991) suggests using kinematic wave theory for modeling short duration debris flows resulting from natural dam failure and Weir (1982) uses kinematic wave theory to model lahars.

The continuity equation, the basis of kinematic wave theory, can be integrated by the method of characteristics (Hunt, 1982; Arattano and Savage, 1992) or by an asymptotic method proposed by Whitham (1979). We will show in this paper that Whitham's solution can be derived from the solution obtained by Arattano and Savage (1992) by making appropriate approximations.

Whitham's solution for the continuity equation

One-dimensional, unsteady flow in a rectangular channel with a fixed slope is described by the momentum equation,

---

[1]CNR IRPI Str. delle Cacce, 73, 10135 Torino, Italy
[2]U.S. Geological Survey, Box 25046, MS 966, Denver Federal Center, Denver, CO 80225

$$u\frac{\partial u}{\partial x}+\frac{\partial u}{\partial t}+g\frac{\partial h}{\partial x} = gi-gi_f \qquad (1)$$

and the continuity equation (Abbott, 1966),

$$\frac{\partial h}{\partial t}+u\frac{\partial h}{\partial x}+h\frac{\partial u}{\partial x} = 0 \qquad (2).$$

Here, t is time, g is gravitational acceleration, u is velocity, x is distance along the channel, i is channel slope, h is flow height, and i_f is the bed-resistance term. Kinematic wave theory requires the derivative terms in equation (1) to be neglected (Lighthill and Whitham, 1955). Equation (1) then becomes simply $i = i_f$. The bed-resistance term is approximated (Hunt, 1982; Arattano and Savage, 1992) by

$$i_f = n^2 u^2/R^{\frac{4}{3}} \qquad (3).$$

Here n is the Manning coefficient (Chow, 1959) and R is the hydraulic radius defined as

$$R = ah^{k_1} \qquad (4)$$

(Arattano and Savage, 1992) where a is a coefficient and k_1 is a parameter that accounts for changes of hydraulic radius with flow depth. This expression for R is used in equation (3) to give the mean fluid velocity

$$u = Ch^k i^{\frac{1}{2}} \qquad (5)$$

with $C = a^{2/3}/n$ and $k = 2k_1/3$. The parameter k_1 ranges between 0 and 1 as the ratio w/h ranges between 0 and ∞ and allows us to account for flow in narrow debris flow channels (Arattano and Savage, 1992).

Substituting equation (5) in equation (2) and using the dimensionless variables $h_* = h/H$, $x_* = x/L$, $u_* = u/U$, and $t_* = Ut/L$ where L is a typical length, H is a typical flow height, and U is a typical velocity in the flow defined by $U = CH^k i^{1/2}$, the continuity equation becomes (Arattano and Savage, 1992)

$$\frac{\partial h_*}{\partial t_*}+v_*(h_*)\frac{\partial h_*}{\partial x_*} = 0 \qquad (6)$$

where $v_*(h_*(x_*,t_*)) = (k + 1)h_*^k$ is the propagation speed (Whitham, 1979). Multiplying by $v'_*(h_*)$ equation 6 becomes

$$\frac{\partial V_*}{\partial t_*} + V_* \frac{\partial V_*}{\partial x_*} = 0 \qquad (7)$$

where $V_*(x_*,t_*) = v_*(h_*(x_*,t_*))$.

Whitham (1979) proposed the asymptotic solution of equation (7),

$$V_* = v_{*0}, \quad x_* \leq v_0^* t_* \qquad (8a)$$

$$V_* = \frac{x_*}{t_*}, \quad v_{*0} t_* \leq x_* \leq x_{*s}(t_*) \qquad (8b)$$

$$V_* = v_{*0} \quad x_* > x_{*s}(t_*) \qquad (8c)$$

where $x_{*s}(t_*)$ which is still to be determined, is the position of the shock front. Recalling that $V_*(x_*,t_*) = v_*(h_*(x_*,t_*)) = (k+1)h_*^k$, assuming $v_{*0} = 0$, which is equivalent to having zero fluid flow in the channel ahead of the kinematic shock and behind the point of inception of the debris flow, and assuming that k is constant, we write equations (8) as

$$h_* = 0, \quad x_* \leq 0 \qquad (9a)$$

$$x_* = (k+1)h_*^k t_*, \quad 0 \leq x_* \leq x_{*s}(t_*) \qquad (9b)$$

$$h_* = 0 \quad x_* > x_{*s}(t_*) \qquad (9c).$$

The shock-front position is located by requiring that the total debris-flow volume remains constant. This leads (Hunt, 1982; Weir, 1982; Arattano and Savage, 1992) to the parametric equations of the shock front

$$t_{*s} = \frac{A_*}{k h_{*s}^{k+1}} \qquad (10a)$$

$$x_{*s} = \frac{A_*(k+1)}{k h_{*s}} \qquad (10b)$$

or by eliminating the parameter h_{*s}, the equation of the shock front,

$$x_{*s} = (k+1)(\frac{A_*}{k})^{\frac{k}{k+1}} t_{*s}^{\frac{1}{k+1}} \qquad (11),$$

where $A_* = A/HL$ is the non-dimensional cross-sectional area of the flow. The solution (dashed lines in Fig. 1) in the region between the shock front and the t_*-axis is given by equation (9b), whereas outside of this region the solution is given by $h_* = 0$.

Weir's (1982) solution for the position of the shock front in the non-dimensional x_*, t_* plane for a rectangular channel of constant slope is

$$x_{*s} = \kappa \left(\frac{A_*}{(\kappa-1)}\right)^{\frac{\kappa-1}{\kappa}} t_{*s}^{\frac{1}{\kappa}} \qquad (12).$$

This is the same form as equation (11). However κ in equation (12) is the exponent of the flow height in the relationship between discharge per unit channel width and flow height and k in equation (11) is the exponent of the flow height in the flow height and velocity relation (equation 5).

Comparison with the solution obtained by the method of characteristics

The solution to the governing equations of kinematic wave theory obtained by the method of characteristics is (Arattano and Savage, 1992)

$$h_* = 0, \quad x_* \leq 0 \qquad (12a)$$

$$x_* = (k+1)h_*^k t_* + h_* \qquad (12b)$$

$$h_* = 0 \quad x_* > x_{*s}(t_*) \qquad (12c)$$

These equations, based on equation (5), hold when a triangular form is assumed for the initial shape of the debris mass (Arattano and Savage, 1992).

The parametric equations of the shock front from this solution are (Arattano and Savage, 1992),

$$t_{*s} = \frac{A_*(1-h_{*s}^2)}{kh_{*s}^{k+1}} \qquad (13a)$$

and

$$x_{*s} = \frac{A_*(k+1)}{kh_{*s}} - \frac{A_*(1-k)}{k}h_{*s} \qquad (13b)$$

In this case, an explicit equation for the shock-front position such as equation (11) is not easily found.

Equation (12b) can be rewritten as

$$1 - \frac{h_*}{x_*} = \frac{(k+1)h_*^k t_*}{x_*} \qquad (14).$$

For large x_*, h_*/x_* is small and equation (14) can be approximated by Whitham's solution (equation 9b). Equation (9b) represents a family of straight lines with constant values of h_* originating at the origin of the axes in the x_*,t_* plane (dashed lines in Fig. 1).

The method of characteristics (equation 12b) also gives straight lines which carry constant flow height values. However these curves leave the x_*-axis along a segment of unit length in the x_*, t_* plane (solid lines in Fig. 1). The curves from the two solutions for the same flow-height value, h_*, are parallel straight lines that become closer for decreasing values of h_*. Because the straight lines with larger values of h_* rapidly meet the shock front and decay, we see (Fig. 1) that, for large times the two solutions tend to coincide. Also, for large x_*, the shock height $h_{*s} << 1$ (Hunt, 1982). Hence, equation (13a) can be approximated by equation (10a), equation (13b) can be approximated by equation (10b) and the shock curve given by Whitham's (1979) solution asymptotically converges to the shock curve obtained by the method of characteristics.

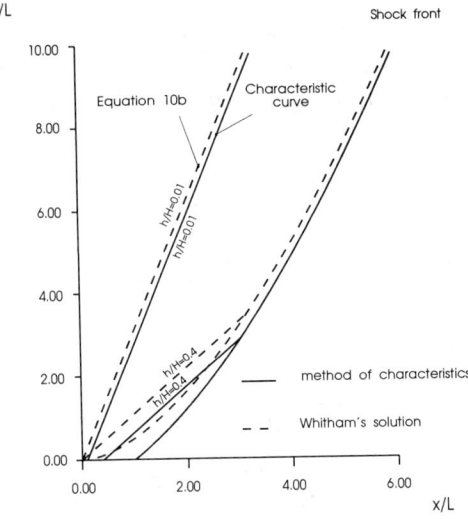

Figure 1 - Comparison between Whitham's (1979) solution and the solution obtained by the method of characteristics ($A_* = 1/2$ and $k = 2/3$)

The differences between the solutions also depend on the parameter k. For a rectangular channel (Fig. 1) with a high ratio w/h ($k = 0.66$) the differences between the shock-height values predicted by the two methods is 1.3 percent after the flow has traveled a distance $x/L = 5$. For smaller w/h ratios, the differences are larger. If $k = 0.2$, the difference is 20 percent for $x/L = 5$, and it reduces to 2.6 percent after the flow has traveled a distance $x/L = 15$.

Conclusions

The differences between Whitham's (1979) solution and the solution by the method of characteristics of the continuity equation depend on time and the width of the channel. For rectangular flow cross sections with a high ratio between width, w, and flow height, h, the two solutions differ little after the flow has traveled a short distance from the point of inception. For narrower channels, the differences remain significant for greater distances. However, both solutions asymptotically converge for large times after flow initiation.

References

Arattano, M., and Savage, W. Z. (1992). "A kinematic wave model for debris flows". *U.S. Geological Survey Open-File Report 92-290*, 39 p.

Abbott, M. B. (1966). *An introduction to the method of characteristics.* American Elsevier, New York.

Chow, V. T. (1959). *Open-channel hydraulics.* McGraw-Hill Book Co., New York.

Hunt, B. (1982). "Asymptotic solution for dam-break problem." American Society of Civil Engineers, vol. 108, no. HY1, p. 115-126.

Lighthill, M. J., and Whitham, G. B. (1955). "On kinematic waves," *I. Flood movement in long rivers.* Proceedings of the Royal Society of London, vol. A 229, p. 281-316.

Takahashi, T. (1991). *Debris flows, IAHR monograph.* A.A. Balkema, Rotterdam.

Weir, G. J. (1982). "Kinematic wave theory for Ruapehu lahars," *New Zealand Journal of Science*, vol. 25, p. 197-203.

Whitham, G. B. (1979). *Lectures on wave propagation.* Springer-Verlag, New York.

Roll Waves in Mud Flow

Chiu-on Ng[1] and Chiang C. Mei[2]

Abstract

Permanent roll waves on a shallow layer of fluid mud which is modeled as a power-law fluid is studied. Based on long-wave approximation, momentum integral method is applied to derive the averaged continuity and the momentum equations. Linearized instability analysis of a uniform flow is insufficient to suggest a preferred wavelength for the nonlinear roll waves, which are obtained as periodic shocks connected by smooth profiles. Among all wavelengths only those longer than certain threshold correspond to positive energy loss across the shock, suggesting that a roll wave developed spontaneously from infinitesimal disturbances should have the shortest wavelength corresponding to zero dissipation across the shock, though finite dissipation elsewhere. For longer roll waves, no roll wave exists if the uniform flow is linearly stable, when the fluid is slightly non-Newtonian. However when the fluid is highly non-Newtonian, very long roll waves may still exist even if the corresponding uniform flow is stable to infinitesimal disturbances.

Introduction

In muddy rivers intermittent flows of clay/water mixture are frequently observed, especially after a torrential rain. Mud flows resulting from volcano eruptions are also accompanied by intermittent waves. The waves feature bores sandwiched between long stretches of gentle profiles and are known as roll waves. In Jiang-jia Ravine of Southwest China, which is a gully about 20 m in width, the mudflow surges down during the wet seasons in groups of successive waves (Li et al., 1983). The maximum wave height once reached 4 m and the maximum wave velocity 13 m/s, while the discharge per wave was as much as 2420 m^3/s. The surge fronts splattered with so much force that even large stones were thrown into the air. The flow in the rear of the waves was however much shallower, slower and essentially laminar, and frequently stagnant before the next surge. Such intermittent mudflows are responsible for the severe erosion and deposition of bed materials in Jiang-jia Ravine.

---

[1]Research assistant, [2]Professor and MASCE, Department of Civil and Environmental Engineering, Massachusetts Institute of Technology, MA 02139.

The mathematical theory for roll waves of finite amplitude in turbulent clear water streamflow was first given by Dressler (1949) who fitted shocks between stretches of smooth profiles. Dressler found that roll waves will occur only if the bottom roughness of the channel is non-zero and less than a certain critical value, as required by the instability criterion for a linearized disturbance. Dressler's theory was extended to laminar water flows by Ishihara, Iwagaki, and Iwasa (1954), and Tamada and Tougou (1979). With or without bores, intermittent mud or debris flows can be laminar throughout because of the high viscosity. The problem of roll waves in laminar flow of non-Newtonian fluids has only been studied by Liu (1990) who examined the rapid flow of Bingham-plastic fluids down an incline. Taking an unstable infinitesimal progressive wave as the initial disturbance, Liu calculated numerically the transient development into periodic shocks in a thin layer of Bingham-plastic mudflow.

The Bingham model is however a mathematical idealization. At low shearing rates (10-100 s^{-1}), the rheological character of thick fluid mud is known to be shear-thinning and a more appropriate and convenient model is the power-law which is,

$$\tau = \mu_n \left(\frac{\partial u}{\partial z}\right)^n \quad (1)$$

where μ_n is the viscosity coefficient of dimension $[ML^{-1}T^{n-2}]$, and n is the flow index which is between 0 and 1 for a shear-thinning fluid. The special limit of $n = 1$ corresponds to a Newtonian fluid, and μ_1 is the ordinary dynamic viscosity. The parameters μ_n and n vary with the concentration, the particle size distribution and the chemical composition of the solid contents. The range $0.1 \leq n \leq 0.4$ has been observed for some fluid muds.

<u>Long Wave Approximation</u>

Consider a two-dimensional laminar flow of a thin layer of mud down a plane of inclination θ which can be any positive value not greater than 90°. An (x, z) coordinate system is defined with the x-axis along and the z-axis normal to the plane bed. The longitudinal velocity component is denoted by $u(x, z, t)$, and the depth normal to the bed by $h(x, t)$. The equations of motion are approximated by following standard arguments in long wave approximation. The pressure is hydrostatic and the limiting velocity profile for a steady uniform flow is

$$u(z) = \left(\frac{1+2n}{1+n}\right) \bar{u} \left[1 - \left(1 - \frac{z}{h}\right)^{\frac{1+n}{n}}\right] \quad (2)$$

where $\bar{u}$ is the depth averaged velocity. For long waves, we can apply the momentum integral method by assuming that (2) is also true for $u(x, z, t)$ in a transient and non-uniform flow, but with $\bar{u}$ and h dependent on x and t. The final equations, in normalized form, are

$$h_t + (\bar{u}h)_x = 0 \quad (3)$$

$$(\bar{u}h)_t + (\beta \bar{u}^2 h + \tfrac{1}{2}\alpha h^2)_x = h - \left(\tfrac{\bar{u}}{h}\right)^n \quad (4)$$

$$\tau_b = \left(\frac{\bar{u}}{h}\right)^n \quad (5)$$

where τ_b is the bottom stress, and $\beta = \frac{2(1+2n)}{(2+3n)}$ is the momentum flux factor, and

$$\alpha = \cot\theta \left(\frac{1+2n}{n}\right)^n Re^{-1} \qquad (6)$$

and Re is the Reynolds number of the flow $Re \equiv \rho \bar{u}_o^{2-n} h_o^n / \mu_n$. Equations (3) and (4) are the governing equations for $\bar{u}(x,t)$ and $h(x,t)$, with α and n as the parameters. Note all the subsequent variables are normalized with respect to the corresponding uniform flow quantities. For the flow to remain laminar, it is necessary that the Reynolds number be below certain threshold Re_c. We adopt the empirical results by Darby (1986) and obtain

$$Re < Re_c = 0.125 \left(\frac{1+3n}{2n}\right)^n [2100 + 875(1-n)] \qquad (7)$$

Linearized Instability of Uniform Flow

We perturb the uniform flow with small normal mode disturbances $\propto \exp\{i(kx - \omega t)\}$. By standard linear analysis, an eigenvalue condition is obtained

$$\omega^2 - (2\beta k - in)\omega - i(1+2n)k + (\beta - \alpha)k^2 = 0 \qquad (8)$$

From (8) we can show that, for $k \neq 0$, disturbances will die out and the flow will be stable if

$$\alpha > \frac{1+2n}{n^2} \equiv \alpha_s(n) \qquad (9)$$

and the neutral stability is given by $k = 0$ and $\alpha = \alpha_s$. It can also be shown that the growth rates of the unstable disturbances increase monotonically with the wavenumber k, and therefore the linear analysis is insufficient to predict the preferred wavelength of the nonlinear roll waves.

Permanent Roll Waves

Considering roll waves in a permanent form moving at a constant speed c, and a moving coordinate $\xi = x - ct$, (3) can be integrated to

$$h(c - \bar{u}) = q \qquad (10)$$

where q is a constant which is the discharge rate as seen in the moving system. With (10), (4) gives the profile equation:

$$\frac{dh}{d\xi} = \frac{h - (ch^{-1} - qh^{-2})^n}{(\beta-1)c^2 - \beta q^2 h^{-2} + \alpha h} \equiv \frac{A(h)}{B(h)} \qquad (11)$$

In accordance with all the reported observations, we seek a surface profile with its depth increasing monotonically towards the front. It follows that A and B must be of the same sign. By the method of characteristics we can show that there exists a critical depth h_c where

$$A(h_c) = h_c - \left(\frac{c}{h_c} - \frac{q}{h_c^2}\right)^n = 0 \qquad (12)$$

$$B(h_c) = (\beta - 1)c^2 - \frac{\beta q^2}{h_c^2} + \alpha h_c = 0 \qquad (13)$$

and both $A'(h_c)$ and $B'(h_c)$ are positive so that the profile is continuous between two successive shocks and $dh/d\xi > 0$. Numerical integration of (11) will only give a smooth monotonic wave profile $h(\xi)$. For finite result, it is necessary to match this profile with two shocks at the ends. Defining λ = wavelength, $h_{1,2}$ = depth immediately before, after a shock, $\langle h \rangle$ = average depth of the roll wave profile, we get the following relations

$$\lambda = \int_{h_1}^{h_2} \frac{B(h)}{A(h)} dh \qquad \langle h \rangle = \lambda^{-1} \int_{h_1}^{h_2} \frac{h\, B(h)}{A(h)} dh \qquad 1 = c\langle h \rangle - q \qquad (14.a, b, c)$$

where (14.c) is the conservation of mass in the moving system. The momentum conservation across a shock gives a further relation

$$\alpha h_1 h_2^2 + \alpha h_2 h_1^2 + 2(\beta - 1)c^2 h_1 h_2 - 2\beta q^2 = 0 \qquad (15)$$

Physically there must be a loss of mechanical energy across a jump. By considering a small control volume which just encloses the shock and moves at the same speed c, the rate of change of mechanical energy across a jump ($\dot{E}$) can be found. From the result we must conclude that a physically possible jump will satisfy

$$q > \left[\frac{3n}{2(3+4n)} \alpha c (h_2 + h_1) + \frac{(1+3n)n}{(2+3n)(3+4n)} c^3 \right] \times$$
$$\left\{ \frac{3(1+2n)}{4(3+4n)} \alpha \frac{(h_2 - h_1)^2}{h_2 h_1} + \frac{2n}{3+4n} \alpha + \frac{3(1+2n)n}{2(2+3n)(3+4n)} c^2 \frac{(h_2 + h_1)}{h_2 h_1} \right\}^{-1} (16)$$

Since the right side of (16) is positive, there is a minimum discharge in the moving coordinate system below which stationary roll waves cannot be maintained, and the shock speed c must be greater than the speed of the fluid particles. In contrast with the turbulent flow, the conditions $h_2 > h_1$ and $q > 0$ are no longer sufficient to guarantee that $\dot{E}$ to be negative. Numerical results show that the implication of (16) is to rule out roll waves with a small jump height. At the lower limit of the finite shock amplitude, $\dot{E} = 0$, i.e., there is no loss of energy. Mathematically higher order derivatives which are ignored here must be important near the lossless shock and smoothens the profile.

On specifying the three parameters: n, α, λ (or the wavenumber k), we may solve for the other variables: c, q, h_c, $\langle h \rangle$, h_1, h_2 from the equations (12), (13), (14.a,b,c) and (15). After completing the solution, the rate of energy dissipation across the shock is calculated according to (16) in order to eliminate physically unacceptable solution. Note that all solutions must satisfy two further conditions, which can be used to determine the admissible range of α and the corresponding range of h_c. Recall first that $A'(h_c) > 0$ which implies

$$G(h_c) \equiv \left(\frac{1+2n}{n^2}\right) h_c^{\frac{2+n}{n}} > \alpha \qquad (17)$$

Secondly the depth $h = 1$ must lie between $\langle h \rangle$ and h_c, which yields

$$F(h_c) \equiv \frac{-(\beta-1)h_c^{\frac{2(1+2n)}{n}} + \beta h_c^{\frac{2(1+n)}{n}} + 2(\beta-1)h_c^{\frac{1+2n}{n}} - 2\beta h_c^{\frac{1+n}{n}} + 1}{(h_c - 1)^2 h_c} > \alpha \tag{18}$$

According to these conditions, a solution exists for a h_c only if α is less than both $F(h_c)$ and $G(h_c)$. By examining the properties of F and G, we find that: (i) For $n \geq 1/\sqrt{2}$, no roll wave will exist if $\alpha > \alpha_s$ i.e., if the uniform flow is stable to infinitesimal disturbances. However for $n < 1/\sqrt{2}$, there exists roll wave solution even when the corresponding uniform flow is unstable! This result generalizes to power-law fluids the argument by Dressler and Pohle (1953) for clear water in open-channel flows. (ii) When $F(h_c) = \alpha$ and $h_c > 1$, $h_a \to h_1 \to \langle h \rangle \to 1 < h_c < h_2$. The corresponding wave must be infinitely long ($\lambda = \infty$), and may be called a *solitary roll wave*. For $n < 1/\sqrt{2}$ and $\alpha_{max} > \alpha > \alpha_s$, there are two h_c's satisfying $F(h_c) = \alpha$ and therefore two solitary waves are possible.

Numerical Results

Numerical results show that because of the energy criterion (16), roll waves exist only when k is below a threshold k_c and when c is above a threshold c_c. On the other hand, the linear theory predicts that the shorter the disturbance, the more likely it will reach the nonlinear state of roll waves. Therefore it is plausible that roll waves generated spontaneously from linearized disturbances should have the largest wavenumber allowed by the requirement that energy is not created across a shock. This means that the preferred wavenumber of such a roll wave is simply k_c, corresponding to the smallest wavespeed c_c and height H_c, and no energy loss at the jump. We shall call this the *minimum roll wave*. At the shock, the gradient is of course too large so that the long wave approximation must require local improvement, such as higher order or perhaps locally exact treatment, which we do not pursue here. The numerical results also show that the minimum roll waves have appreciably different depth and velocities profiles, but nearly the same bottom stress as the uniform flow with the same discharge.

Roll waves of longer wavelengths, corresponding to positive dissipation across the shock can still be generated, of course, by imposing initial or boundary conditions, for example by periodically varying the influx rate at far upstream at different periods where infinitesimal disturbances are unstable. Figure 1 shows the snapshot profiles for $h(x)$, $\bar{u}(x)$ and $\tau_b(x)$ for $k = 0.1$, $\alpha = 0, 1, 5$ and $n = 0.1$. From the velocity profiles, the flow is seen to have practically vanished when the depth is below a certain value. As a result, the fluid along more than half of the stretch of a wave appears stagnant, while the fluid in the wave front is moving with a tremendous speed. This behavior closely resembles the bursting mudflows characterized by a period of cessation of flow between successive waves, as observed by Li *et al.* (1983). These long roll waves are potentially more damaging than the minimum roll waves and the corresponding uniform flows and pose greater threats in bottom scour.

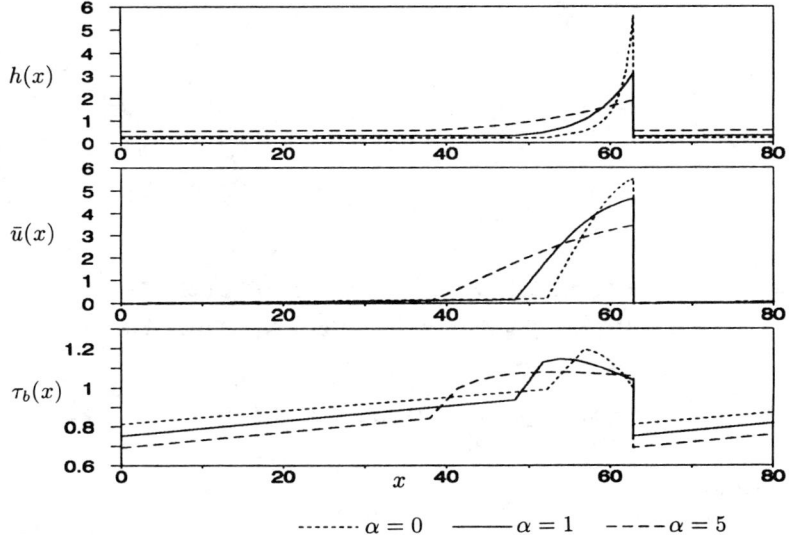

Figure 1. Snapshots of $h(x)$, $\bar{u}(x)$ and $\tau_b(x)$ for $k = 0.1$, $\alpha = 0, 1, 5$, $n = 0.1$

Acknowledgement

We thank the ONR, Ocean Engineering Division (Contract N00014-89-J-3128) and the NSF, Natural Hazards Program (Grant BCS-9112748).

References

Darby, R. (1986) Laminar and turbulent pipe flows of non-Newtonian fluids. *Encycl. Fluid Mech.*, Vol. 7, Cheremisinoff (ed.), Gulf Pub. Co., 19-53.

Dressler, R.F. (1949) Mathematical solution of the problem of roll-waves in inclined open channels. *Comm. Pure Appl. Math.* 2, 149-194.

Dressler, R.F. and Pohle, F.V. (1953) Resistance effects on hydraulic instability. *Comm. Pure Appl. Math.* 6, 93-96.

Ishihara, T., Iwagaki, Y. and Iwasa, Y. (1954) Theory of the roll wave train in laminar water flow on a steep slope surface. *Trans. JSCE* 19, 46-57.

Li, J., Yuan, J., Bi, C. and Luo, D. (1983) The main features of the mudflow in Jiang-jia Ravine. *Zeit. Geomorph.* 27(3), 325-341.

Liu, K.F. (1990) Dynamics of a shallow layer of fluid mud. Ph.D. Thesis, Civil Engineering Department, Mass. Inst. Tech.

Tamada, K. and Tougou, H. (1979) Stability of roll waves on thin laminar flow down an inclined plane wall. *J. Phy. Soc. Japan* 47(6), 1992-1998.

Friction in Debris Flows: Inferences from Large-scale Flume Experiments

Richard M. Iverson and Richard G. LaHusen[1]

Abstract

A recently constructed flume, 95 m long and 2 m wide, permits systematic experimentation with unsteady, nonuniform flows of poorly sorted geological debris. Preliminary experiments with water-saturated mixtures of sand and gravel show that they flow in a manner consistent with Coulomb frictional behavior. The Coulomb flow model of Savage and Hutter (1989, 1991), modified to include quasi-static pore-pressure effects, predicts flow-front velocities and flow depths reasonably well. Moreover, simple scaling analyses show that grain friction, rather than liquid viscosity or grain collisions, probably dominates shear resistance and momentum transport in the experimental flows. The same scaling indicates that grain friction is also important in many natural debris flows.

Introduction

Progress in predicting the behavior of debris flows has been hampered by a dearth of data suitable for testing hypotheses. The capricious timing, location, and magnitude of most debris flows make systematic field measurements both difficult and dangerous. Consequently, in 1991 and 1992 the U.S. Geological Survey constructed a flume to conduct controlled experiments on debris flows. Located on a hillside about 70 kilometers east of Eugene, Oregon, near the headquarters of the H.J. Andrews Experimental Forest, Willamette National Forest, the flume provides research opportunities available nowhere else. This paper describes the rationale for flume design and operation and presents some results and interpretations of preliminary experiments. The interpretations focus on a key question that must be addressed in debris-flow modeling: what are the relative contributions of liquid viscosity, grain collisions, and grain friction to shear resistance and momentum transport in debris flows?

---

[1]U.S. Geological Survey, Cascades Volcano Observatory, Vancouver, WA 98661

Flume Design and Operation

Three considerations guided the flume design. (1) We wished to study flow of finite debris masses from initiation to deposition. This dictated the choice of a non-recirculating flume with a sediment-loading area at its head and a run-out area at its toe. (2) We wished to study flow of realistic geological debris, which can include large particles with the potential for strong inertial interactions. (Miniature, laboratory debris flows provide inadequate similitude of coarse-grained, natural debris flows owing to the low inertia of small particles and the difficulty of scaling the peculiar physico-chemical properties of interstitial water and clay.) This dictated that the flume would be large. Cost restrictions consequently necessitated an outdoor location and a fixed bed slope. (3) We wished to study both water-saturated debris flows and relatively dry flows such as rock avalanches and pyroclastic flows. This dictated a flume slope steeper than 30° -- steep enough for some dry mixtures to flow. This slope also typifies steep hillsides where natural debris flows originate.

These considerations resulted in construction of a reinforced concrete flume 95 m long, 2 m wide, and 1.2 m deep that slopes 31° throughout its upper 88 m and flattens gradually to $2\frac{1}{2}°$ over the last 7 m (Figure 1a). Ten meters downslope from the flume head, steel gates attach with hinges to the flume walls. Up to 20 m^3 of sediment can be loaded behind the gates and saturated with water applied via subsurface channels and surface sprinklers. To initiate a debris flow, a hydraulic piston unlatches the gates, which swing fully open in about half a second. Alternatively, a sloping mass of sediment can be placed behind a low retaining wall and watered until slope failure occurs. In either case the ensuing debris flow descends the channel and forms a deposit on a concrete run-out pad at the flume base (Figure 1b). The deposited sediment can be studied and then recycled by excavating it with a loader and trucking it to the head of the flume.

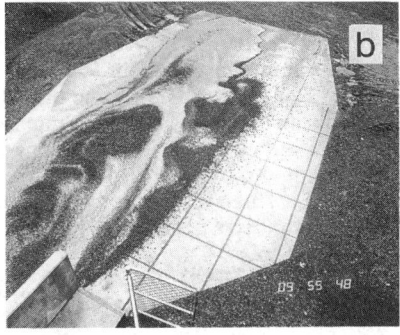

Figure 1. Photographs of the experimental debris flow of September 25, 1992.

Opening of the flume gate triggers a clock that synchronizes digital data acquisition. Collection of flow-dynamics data is focused at three cross sections, located 33 m and 67 m below the gate and 3 m above the gate. The cross sections include ports in the flume bed, glass windows and steel panels in the flume wall, and booms for suspending instruments above the flow. Load cells and piezoelectric sensors mounted in the ports measure normal stresses at the bed over areas of 1, 22, and 500 cm² and at frequencies up to 4000 Hz. Transducers that measure shear stresses or pore-fluid pressures also can be installed in the ports. Windows permit cross-sectional viewing and videotaping of flows as they pass, and steel panels provide a low-attenuation environment for making active and passive acoustic measurements. Booms facilitate placement of cameras and ultrasonic flow-depth meters directly above the flows.

Other data-collection efforts focus on refinement of an automated debris-flow detection system, quantitative interpretation of deposit morphology and sedimentology, and testing of formulae for estimation of flow velocities, discharges, and impact forces.

The Coulomb Flow Hypothesis: An Experimental Test

Preliminary experiments during the spring and summer of 1992 emphasized development and refinement of experimental techniques, but also yielded useful data. Simultaneous, high-speed measurements of flow depth and normal stress on the flume bed permitted calculation of the dynamic bulk density of a flowing sand-gravel-water mixture. The data (see Iverson et al., 1992, figure 4) yielded a dynamic bulk density close to 2000 kg/m³, indistinguishable from the static bulk density of the uncompacted material. This indicated that sediment grains in the flow may have contacted one another almost continuously. Accordingly, we hypothesized that solid grain friction dominated shear resistance and momentum transport in the experimental flow.

Solid friction is only one possible mechanism of shear resistance and momentum transport in debris flows. Liquid viscosity, the mechanism in viscoplastic or viscous models, and particle collisions, the mechanism in inertial grain-flow models, may also be important (Iverson and Denlinger, 1987). Evaluation of key dimensionless scaling parameters for open-channel flows of sediment-liquid mixtures helps quantify the relative importance of these three mechanisms. Three key scaling parameters may be labeled the Bagnold number, B, Savage number, S, and friction number, F, and defined as

$$B = \frac{\dot{\gamma} \rho_s \delta^2}{\mu} \lambda^{1/2} \qquad S = \frac{\dot{\gamma}^2 \delta^2}{v g h} \qquad F = \frac{\rho_s v g h}{\dot{\gamma} \mu}$$

where $\dot{\gamma}$ is a typical shear-strain rate, ρ_s is the sediment density, δ is a typical grain diameter, μ is the liquid viscosity, v is the mean volume fraction of the granular phase, h is a typical flow depth, and g is the gravitational acceleration.

The factor λ depends on v and the maximum (close-packed) grain volume fraction, v^*, in the manner $\lambda = v^{1/3}/(v^{*1/3} - v^{1/3})$. For a range of hypothetical debris flows, with $0.5 \leq v \leq 0.7$ and $0.7 \leq v^* \leq 0.9$, λ is typically of order 10.

The definitions of B and S derive from the work of Bagnold (1954) and Savage (1984) and represent ratios of the characteristic shear stresses due to grain collisions, $\dot{\gamma}^2 \rho_s \delta^2$, liquid viscosity, $\dot{\gamma} \mu$, and solid friction, $v \rho_s g h \tan\phi$. (To match the usage of Savage, the definitions of S and F omit $\tan\phi$, the solid bulk friction coefficient, which typically ranges only from 0.5 to 1.0.) Shear-cell experiments performed by Bagnold (1954) and subsequently replicated by others demonstrate that stresses due to grain collisions dominate viscous fluid stresses if $B > 450$, whereas viscous effects dominate if $B < 40$. Similarly, inferences drawn by Savage and Hutter (1989) from a variety of data indicate that grain collisions probably dominate grain friction if $S > 0.1$. Experimental data that define the value of F at which grain friction dominates liquid viscosity are not available. However, because $F = \lambda^{-1/2} B/S$, a plausible estimate is that friction dominates viscosity if $F > 10^{-1/2} \cdot (450 \div 0.1)$, that is, $F > 1400$.

Table 1 summarizes evaluation of B, S, and F for a water-saturated sand-and-gravel debris flow released in the flume on Sept. 25, 1992, and for larger, muddier, stonier, hypothetical debris flows. Values of B, S, and F for the experimental flow indicate that Coulomb friction probably dominated both grain collisions and liquid viscosity as a mode of shear resistance and momentum transport. This reflects the low viscosity of the liquid (water) and small size of the grains (median diameter ~ 0.007 m) in the flow. The values of B, S, and F for the hypothetical debris flows are more equivocal and indicate the importance of collisions and viscosity in addition to solid friction.

Table 1. Estimation of scaling parameters B, S, and F for debris flows.

	Flume flow, 9/25/92	Hypothetical flow 1	Hypothetical flow 2
h (m)	0.1	1	2
δ (m)	0.007 (median)	0.1 (cobbles)	1 (boulders)
$\dot{\gamma}$ (s^{-1})	20@	5*	5*
μ (Pa-s)	0.001 (water)	1 (dense slurry)#	0.1 (dilute slurry)#
ρ_s (kg/m^3)	2700	2700	2700
g (m/s^2)	9.8	9.8	9.8
v	0.6	0.6	0.6
v^*	0.75	0.75	0.75
B	10,000	500	500,000
S	0.03	0.04	2
F	80,000	3000	60,000

@Employs estimate (from videotapes) that bed slip accounted for 2/3 of mean motion.
*Typical of natural debris flows as inferred by Phillips and Davies (1991).
#Typical of clay-silt slurry tests by O'Brien & Julien (1988) and Major & Pierson (1992).

As a test of the Coulomb hypothesis, we used the model of Savage and Hutter (1989, 1991) to predict the velocity and depth of the Sept. 25 flume debris flow. The model employs depth-averaged equations of motion to simulate unsteady, nonuniform flow of a finite mass of Coulomb material. The flow begins and ends with the mass at rest. The model contains only two material parameters, an internal friction angle, ϕ_i, and a bed friction angle, ϕ_b. We independently measured quasi-static values of these parameters and obtained ϕ_i = 42° and ϕ_b = 27° for input to the model. To account for pore-water pressure effects, which are not included in B, S, and F as defined above, we added the term $-\overline{\partial p/\partial \eta}(\tan\phi_b)/\rho_s g$ to the right-hand side of Savage and Hutter's (1991) normalized momentum equation, 2.39. Here $\overline{\partial p/\partial \eta}$ is the depth-averaged pore-pressure gradient in the debris flow. In making model predictions, we estimated that $\overline{\partial p/\partial \eta} = -\rho_w g\cos\theta$; thus we assumed that the vertical pore-pressure gradient is hydrostatic and that the pore-water head gradient parallels the flume bed, which slopes at an angle θ. This represents the simplest, uncoupled, quasi-static pore-pressure effect. More complex, dynamic and coupled pore-pressure effects (e.g., Iverson and LaHusen, 1989) undoubtedly arise in debris flows and should be predicted, not estimated, as part of a more comprehensive model.

Figure 2 illustrates a comparison between model predictions and measured behavior of the experimental debris flow. The model over-predicts

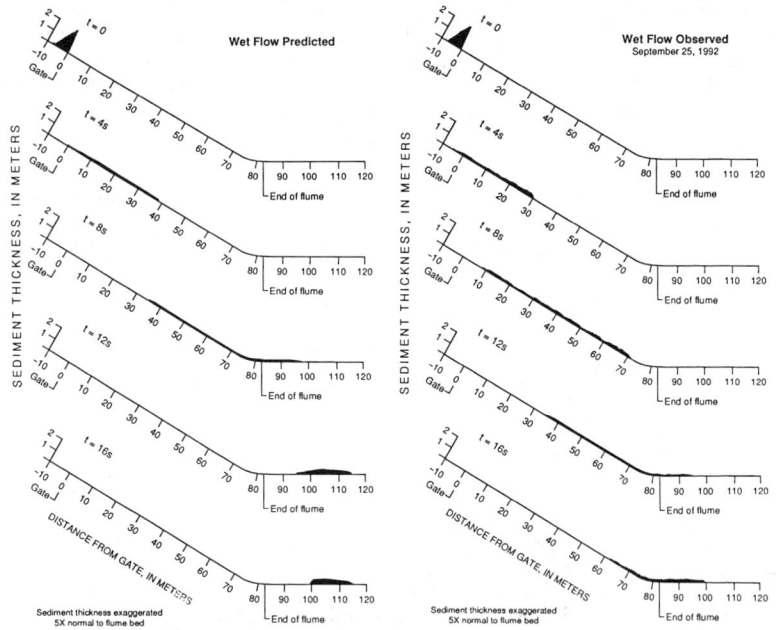

Figure 2. *Predicted and observed debris-flow motion from initiation to deposition.*

flow-front speed and flow depth, but typically only by a few tens of percent. The error can be reduced somewhat by accounting for sidewall friction, finite opening speed of the flume gate, and lateral spreading of sediment that occurs when the flow discharges from the flume. Moreover, if we calibrate the model by adjusting friction angles or the estimated pore-pressure gradient, the error can be can be made almost zero. However, we emphasize instead that the predictions of the *uncalibrated* model are quite good, given the complexity of debris flows and the simplicity of the model.

Conclusions

Results from preliminary experiments in a recently constructed hillside flume support the hypothesis that Coulomb friction dominates shear resistance and momentum transport in rapid debris flows that consist of water-saturated sand and gravel. Simple scaling analyses show that solid friction may also dominate in some natural debris flows, although grain collisions or liquid viscosity may dominate in others. Because pore-fluid pressures can strongly mediate both grain friction and collisions, improved models of debris flow should incorporate solid-fluid coupling that generates pore-pressure gradients.

Acknowledgments

We thank S.B. Savage and J.W. Vallance, who supplied a C code for implementation of the Lagrangian numerical scheme of Savage & Hutter (1991).

References

Bagnold, R.A. (1954). Experiments on a gravity free dispersion of large solid spheres in a Newtonian fluid under shear, *Proc. R. Soc. Lond.*, A225: 49-63.

Iverson, R.M., and Denlinger, R.P. (1987). The physics of debris flows -- a conceptual assessment, *Intl. Assoc. Hydrol. Sci. Publ. 165*: 155-165.

Iverson, R.M., and LaHusen, R.G. (1989). Dynamic pore-pressure fluctuations in rapidly shearing granular materials, *Science*, 246: 796-799.

Iverson, R.M., Costa, J.E., and LaHusen, R.G. (1992). Debris-flow flume at H.J. Andrews Experimental Forest, Oregon, *U.S. Geol. Surv. Open-file Report* 92-483: 2 p.

Major, J.J., and Pierson, T.C. (1992). Debris flow rheology: experimental analysis of fine-grained slurries, *Water Resources Research*, 28: 841-857.

O'Brien, J.S., and Julien, P.Y. (1988). Laboratory analysis of mudflow properties, *J. Hydraul. Eng.*, 114: 877-887.

Phillips, C.J., and Davies, T.R.H., (1991). Determining rheological parameters of debris flow material, *Geomorphology*, 4: 101-110.

Savage, S.B. (1984). The mechanics of rapid granular flows, *Adv. Applied Mech.*, 24: 289-366.

Savage, S.B., and Hutter, K. (1989). The motion of a finite mass of granular material down a rough incline, *J. Fluid Mech.*, 199: 177-215.

Savage, S.B., and Hutter, K. (1991). The dynamics of avalanches of granular materials from initiation to runout. Part I: Analysis, *Acta Mechanica*, 86: 201-223.

ANALYSIS OF PULSING PHENOMENON IN VISCOUS DEBRIS FLOW

Wan Zhaohui[1]   Hua Jingshen[2]

Abstract

Pulsing motion is the main pattern of viscous debris flow. Results of study on hyperconcentrated flow are briefly illustrated in this paper. By referring study results of hyperconcentrated flow, the pulsing phenomenon in viscous debris flow is analysed.

Introduction

According to its characteristics of grain size composition and its flow characteristics, debris flow can be classified into two categories, i.e. viscous debris flow and non−viscous debris flow. Debris flow with clay particles at high concentration behaves as a viscous one.

Based on long−term observation at Jiangjia Gully, pulsing motion is the main pattern of viscous debris flow. According to statistics, about 70 percent of time viscous debris flow moves in the form of pulsing motion(Wu et. al., 1990). Only if the discharge supply is abundant, it moves continuously. Pulsing motion has also been observed in Jiamajimei Gully, Tibet and some places else.

Pulsing viscous debris flow is a highly unsteady and non−uniform one. It consists of three parts, that is, snout, body and tail. The snout possesses a vertical or even forward inclined front, looking like a bore with violent turbulence, big noice and splashing drops. Sometimes smoke caused by strong colli-

---

[1] Professor − Senior Engineer
[2] Engineer, Institute of Water Conservancy and Hydroelectric Power Research, P.O. Box 366, Beijing, China

sion between stones can be observed. After the snout water depth as well as flow velocity drop dramatically, see Fig. 1. The body is obviously a laminar flow. One can see gravels and pebbles moving parallelly in the center with obvious shear near boundaries. Water depth and flow velocity further decrease in the tail part. Quite often that part stops moving and keeps stagnation for a while before the arrival of the next snout.

There is tremendous difficulty in the survey of debris flow. Due to the extremely high velocity of large stones in debris flow, direct measurement of velocity, concentration, etc. is almost impossible. The suddenness of its occurence increases the difficulty. As a result, very little survey data of debris flow are available. On the contrary, field data of hyperconcentrated flow are comparably rich, and there are also some laboratorial studies on hyperconcentrated flow. It might be worthwhile referring hyperconcentrated flow study in the analysis of debris flow.

Characteristics of hyperconcentrated flow

Based on abundant rheological measurement, hyperconcentrated flow can be described by a Bingham fluid (Wan and Wang, 1993). Both yield stress τ_B and rigidity η increase with increasing concentration. Yield stress τ_B is initiated due to the flocculation structure, which is fragile and prone to be destroyed by turbulence or shear on the one hand, and is liable to form again on the other hand. Under certain conditions a corresponding flocculation structure in dynamic equilibrium exists. Corresponding to that flocculation structure, a certain yield stress occurs. It means that for a certain sediment—water suspension yield stress varies with turbulent intensity and other flow conditions.

Yang et al. (1986) conducted a special experiment in an equipment like that used by H. Rouse in developing the diffusion theory. They measured the concentration profiles of sand in clay suspension under different clay concentration and different turbulence intensity and then deduced the variation of yield stress under different conditions, shown in Fig. 2. In the figure τ_B is the yield stress obtained through rheological measurement, τ_{BT} is the yield stress under turbulent conditions, deduced indirectly through the concentration profiles of sand. F is the oscillation frequency of grids and ε the momentum exchange coefficient. Both of them reflect the turbulence intensity.

Main conclusion deduced from it is as follows: turbulence has dieffects on flocculation. On the one hand, turbulence increases the oppoturnity of contact of flocs and makes yield stress increase when the clay concentration is low. On the other hand, turbulence breaks the flocculation structure and causes the decrease of yield stress. In most cases, the latter effect dominates.

Flow resistance of Bingham fluid can be described by f—Re correlationship, which is the same as that for clear water, only if yield stress τ_B has been taken into consideration in the composition of Reynolds number (Qian and Wan, 1986), see Fig. 3. Here Reynolds number for Bingham fluid takes the following form:

$$Re = \frac{4\rho_m RU}{\eta\left(1 + \dfrac{R\tau_B}{2\eta U}\right)}$$

and

$$f = \frac{8gRJ}{U^2}$$

In laminar flow region

$$f = \frac{96}{Re}$$

in which U and R are the average velocity and hydraulic radius of the flow, g gravity, J slope.

It is found that $S - U^3/(gH\omega)$ curve has a hook-like outline as shown in Fig. 4. Here S is concentration, H water depth of the flow, w fall velocity of a single sediment particle in water at rest at infinity. The hook-like outline means that in the region of high concentration more sediment can be conveyed under weaker flow intensity, or eroded bed deposit can be carried away easily.

"Clogging" and pulsing flow phenomenon associated with hyperconcentrated flow have been reproduced in laboratorial flumes (Wan et. al., 1979). Under the condition of constant low incoming discharge periodical oscillation of water stage, associated with periodical variation of flow velocity and longitudinal slope of water surface, was observed in laboratorial flumes. If the incoming discharge was further decreased, the whole hyperconcentrated muddy stopped moving for a while. As the incoming discharge at the flume inlet still continued, the water stage at the inlet rose gradually. As the water stage at the inlet reached a certain elevation and the longitudinal slope of water surface reached a certain magnitude correspondingly, the hyperconcentrated muddy started moving again. Starting at the inlet and propagating downstreamward, flow lasted for a while and the water stage lowered. Then flow stopped again and the next circulation started again. Further lowering the incoming discharge made the flow stop moving completely. The longitudinal profile of the stopped muddy had an outline as shown in Fig. 5. Boundary shear $\gamma_m RJ$ nearly kept a constant value close to τ_B, in which γ_m is the specific weight of the mud.

The relative variation of water depth in the middle of the flume (7#) $\triangle H/H$ can be used as a parameter reflecting the intensity of pulsing phenomenon, in which H is the water depth and $\triangle H$ is its variation at 7#. It is found that puling phenomenon occurred under the condition of $Re < 2000$, i.e. in laminar flow region, see Fig. 6.

Analysis on pulsing phenomenon of viscous debris flow

Based on wide rheological measurement, matrix of viscous debris flow can be described by Bingham model (Wan and Wang, 1993).

Due to Bingham yield stress a layer of viscous debris flow stagnates on bed while the flow stops. The thickness of the stagnant layer H is decided by the criterior $\tau_{Bt} = \gamma_m HJ$. Here τ_{Bt} is the yield stress of the viscous debris flow under the condition of turbulence.

Provided upstream discharge supply continues (due to upstream steeper slope, etc.), the thickness of stagnant layer increases gradually and the shear stress on boundary increases correspondingly. Once the shear stress on boundary reaches a certain value τ_{Bs}, which is the yield stress of the viscous debris flow under the condition of standstill and is larger than τ_{Bt}, the stagnant layer starts moving.

Once the stagnant layer starts moving, the flocculation structure will be broken and the yield stress resulting from the structure reduces correspondingly. Consequently, Reynolds number Re increases. And if the flow is in laminar flow region, the friction factor f reduces and the flow velocity increases, see Fig. 3. With increasing velocity the flocculation structure is further destroyed and yield stress is further decreases. Such a circulation might make flow velocity rather high. Flow with high velocity runs over the downstream stagnant layer and forms a snout with large local longitudinal slope and corresponding large local shear stress. The large shear stress might make the stagnant layer move and the originally standstill material join the flow. As a result, the water depth and the discharge increase. Quite often the flow at the snout turns to be a turbulent one. Provided upstream discharge supply is not enough, only a wave propagates downstreamwards and after the wave the water depth drops and the boundary shear drops consequently. If the shear stress is less than τ_{Bt}, the whole material stagnates again and the next circulation starts again. Because the wave has just passed and the flocculent structure has just been destroyed, τ_{Bt} should be used in this case.

Analysis of field data in Jiangjia Gully, Yunnan (Wu et. al., 1990) shows that pulsing flow happened under the condition of Re < 2000 and continuous flow occurred when Re > 2000. Field data agree with laboratorial result well.

References

Qian, N. (N. Chien) and Wan, Z., 1986, A critical review of the research on the hyperconcentrated flow in China, Series of publication IRTCES.

Wan, Z., and Wang, Z., 1993, Hyperconcentrated flow (to be published).

Wu, J., Kang, Z., Tian, L. and Zhuang, S., 1990, Observation and study of debris flow in Jiangjia Gully, Yunnan, Science Press.

Yang, M. and Qian N. (N. Chein), 1986, The effect of turbulence on the flocculent structure of the slurry of fine grains, Journal of Hydraulic Engineering, No. 8.

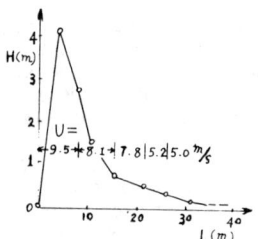

Fig. 1 Profile of a pulsing debris flow

Fig. 2 Variation of τ_{BT}

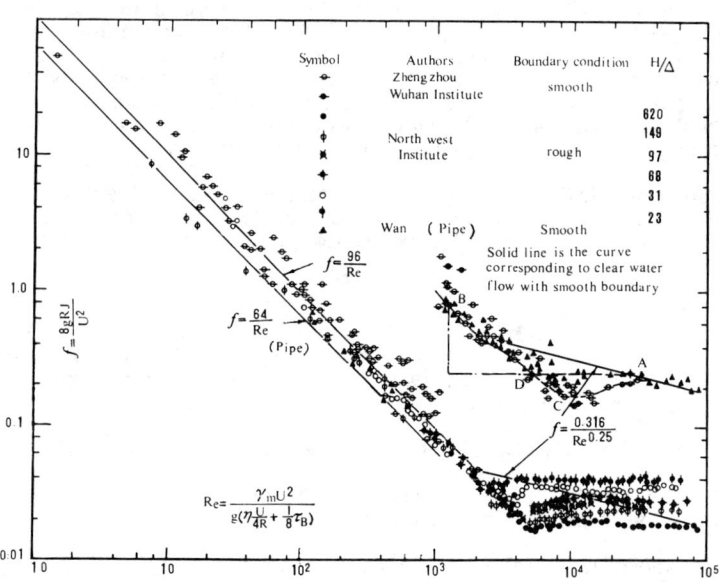

Fig. 3 f —— Re

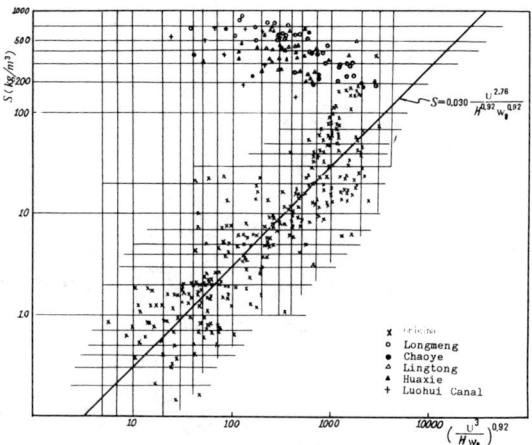

Fig. 4 Sediment carrying capacity

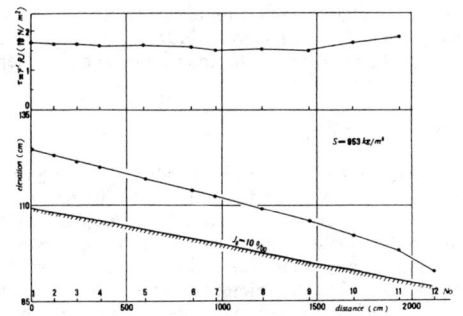

Fig. 5 Profile of "clogging"

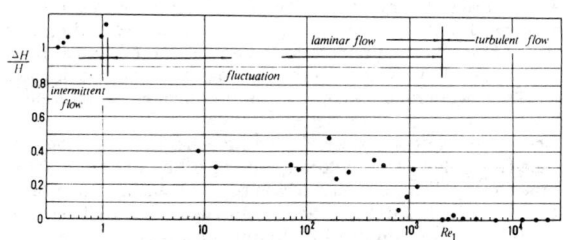

Fig. 6 $\Delta H / H$ --- Re

A STUDY ON DEBRIS FLOW SURGES

Zhaoyin Wang[1]

ABSTRACT

The mechanism of development of debris flow surges is studied theoretically and experimentally. It is proved that the development of viscous debris flow from continuous flow into a series of roll waves is attributed to the yield stress of the matrix. Slight fluctuation in depth grew up and developed into surges in flow of clay suspensions when the suspensions transformed from Newtonian into non-Newtonian fluid and exhibited yield stress. The greater the yield stress and the smaller the flow depth, the more liable the flow developes into surge waves.

1. INTRODUCTION

Debris flow is classified into thin debris flow, which is turbulent and continuous, and viscous debris flow, which is laminar and pulsing, by Chinese scientists. One of the most stricking and important characteristic of viscous debris flow is development from steady flow into successive pulses or surges. Usually, a low wave would appear following increase in solid concentration, much like a tidal bore, moving much faster than the current. Eddies would shrink and the agitated surface become smooth as turbulent flow yields to laminar flow owing to a greatly increase in viscosity. The low wave grows up and sometimes grows so large that the maximum discharge is more than double the average one and the residual mixture stops moving after the wave passes. Thus the debris flow develops into a series of surge waves. The time interval between the waves is from 20-300 seconds (Kang, 1985) to 10-20 minuts (Pierson, 1980) depending on the incoming discharge, the size of the channel and compositions of the mixtures. It seems that the phenomenon has a close relationship with the rheological properties of the matrix of the debris flow because it occurs only when the mixture changes from a Newtonian fluid into a non-Newtonian one.

Engelund and Wan (1984) studied the mechanism of the instability of hyperconcentrated flow. They used bentonite and kaolinite suspensions as the flow media and found that the flow of bentonite suspension became pulsing when the flow rate was small. They attributed the instability to thixotropy of the fluid. Thus a bentonite clay slurry, which exhibited thixotropy, developed into pulsing flow while a kaolinite slurry,

---

[1] Dr. Senior Engineer, Institiute of Water Conservancy and Hydroelectric Power Research, China; Research Fellow of the A.v.Humboldt Foundation, Guest Scientist, Inst. für Hydrologie und Wasserwirschaft, Kaiserstr. 12, 7500 Karlsruhe 1, Germany.

which exhibited no thixotropy, did not. They concluded that pulsing flow will occur only in liquid which has a static yield shear stress higher than the minimum shear stress during flowing. Davies (1988) studied the phenomenon, too. He used a moving bed channel for the study and concluded that the occurrence of pulsing debris flow results from the presence of high concentration of coarse grains shearing in dense, viscous slurry of fine grains in water.

It is noticed that the pulsing flow phenomenon occurs also in mud flow and hyperconcentrated flow. A series of waves developed in the Black River in North West China in July 1965 when the concentration of suspended sediment rose to 980 kg/m^3 (Qian, 1980). There were few coarse grains in the flow. On the other hand, flow of clay suspensions, which exhibited no thixotropy, could also develop into pulsing flow and regular surge waves (Wang et al., 1990). Hence a different mechanism is needed.

2. THEORY

In the steep, narrow debris flow gullies, there are many channel boundary nonuniformities able to disturb the uniform flow of a slurry, and any of these will be able to initiate the instability which give rises to incipient surge waves. How these surges can evolve into the much larger, more regular surge waves in the larger channels downstresm, such as described by Li et al.(1983)? A proper answer may be the follows:

Employing the one-dimensional continuity equation and equation of motion of unsteady open channel flow and using the method of characteristics, one get a group of differential equations that follows the C_1-family of characteristic curves:

$$\frac{dx}{dt} = U + \sqrt{gH} \quad (1)$$

$$\frac{d}{dt}(U + 2\sqrt{gH}) = gJ - \frac{\tau_0}{\rho_m H} \quad (2)$$

and another group follows the C_2-family of characteristic curves:

$$\frac{dx}{dt} = U - \sqrt{gH} \quad (3)$$

$$\frac{d}{dt} (U - 2\sqrt{gH}) = gJ - \frac{\tau_0}{\rho_m H} \quad (4)$$

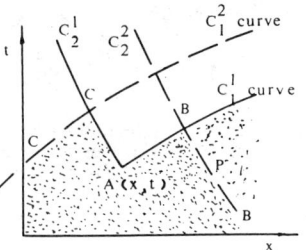

Fig.1 Characteristic curves on x-t plane

where τ_0 is the shear stress on the channel bed, J the energy slope. If a flow is steady and uniform, the velocity U and the depth H are constants, and the frictional resistance must then be equal to the tractive force, i.e., $gJ=\tau_0/\rho_m H$. If a disturbance induces increments ΔU, $\Delta(2\sqrt{gH})$ and $\Delta(\tau_0/\rho_m H)$, then Eq.(2) becomes

$$\frac{d}{dt}[(U + 2\sqrt{gH}) + \Delta U + \Delta(2\sqrt{gH})] = gJ - \frac{\tau_0}{\rho_m H} - \Delta(\frac{\tau_0}{\rho_m H}) \quad (5)$$

If one substracts Eq.(2) from Eq.(5), the result is

$$\frac{d}{dt}[\Delta U + \Delta(2\sqrt{gH})] = - \Delta(\frac{\tau_0}{\rho_m H}) \quad (6)$$

Eq.(6) is called disturbance equation along the C_1-family of characteristic curves. Similarly, the disturbance equation along the C_2-family of characteristic curves is

$$\frac{d}{dt}[\Delta U - \Delta(2\sqrt{gH})] = - \Delta(\frac{\tau_0}{\rho_m H}) \quad (7)$$

Viscous debris flow roughly follows the Bingham model, i.e.

$$\tau = \tau_B + \eta du/dy \tag{8}$$

where τ and τ_B are shear stress and yield shear stress, respectively, η rigidity coefficient. In laminar open channel flow of a Bingham fluid, the upper part is a flow plug in which the fluid flows at an uniform velocity u_p ($\approx U$). Only in the zone near the bed does the velocity vary, from zero to u_p. The thickness of the layer is assumed d. Observations and measurements in the experiments suggested that d vary little as a wave passes, and the velocity gradient in the lower zone varied with fluctuation of average velocity and depth. The velocity gradient is, thereofore, roughly U/d, and

$$\tau_o = \tau_B + \eta \frac{du}{dy} \approx \tau_B + \eta \frac{U}{d} \tag{9}$$

Substituting the equation into Eq.(6) one obtains

$$\frac{d}{dt}[(\Delta U + \Delta(2\sqrt{gH})] = -\Delta(\frac{\tau_B}{\rho_m H} + \eta \frac{U}{\rho_m H d}) \tag{10}$$

As shown in Fig.1, a disturbance that occurs at point A(x,t) in the x-t plane will propagate along the characteristic curves. For any point B on the C_1^1 characteristic curve that passes through the point A(x,t), there is a characteristic curve C_2^2 intersecting the C_1^1 curve at the point. The dotted area is undisturbed and initial disturbance has no effect in the area. The velocity U and depth H in it remain constant. Integration of Eq.(7) along the C_2^2 characteristic curve yields

$$\Delta U - \Delta(2\sqrt{gH}) = \int_{B'}^{B} -\Delta(\frac{\tau_o}{\rho_m H}) dt \tag{11}$$

Point P is always in the undisturbed area as it moves from B' to a point very near to B, and $\Delta(\tau_o/\rho_m H)$ is zero in the process of integration except at the point B. Since $\Delta(\tau_o/\rho_m H)$ is not infinite at B, Eq.(11) yields

$$\Delta U - \Delta(2\sqrt{gH}) = 0 \quad \text{or} \quad \Delta U = \Delta(2\sqrt{gH}) \tag{12}$$

For a low discharge of mud flow, if the yield stress of the mud is large, the second term on the right hand side of Eq.(10) is negligible, and Eq.(10) can be rewritten as

$$\frac{d}{dt}[\sqrt{\frac{g}{H}} \Delta H] = -\frac{1}{2}\Delta(\frac{\tau_B}{\rho_m H}) = \frac{1}{2}\frac{\tau_B}{\rho_m H^2}\Delta H \tag{13}$$

In which, Eq.(12) and the following formulas

$$\Delta(2\sqrt{gH}) = \Delta H \frac{d}{dH}(2\sqrt{gH}) = \sqrt{\frac{g}{H}} \Delta H \tag{14}$$

$$\Delta(\frac{\tau_B}{\rho_m H}) = \Delta H \frac{d}{dH}(\frac{\tau_B}{\rho_m H}) = -\frac{\tau_B}{\rho_m H^2}\Delta H \tag{15}$$

have been used. The integration of Eq.(13) yields

$$\frac{\Delta H}{\Delta H_o} = \exp[\frac{\tau_B}{2\sqrt{gH}H\rho_m}t] \tag{16}$$

where ΔH_o is the initial disturbance in depth.

Eq.(16) indicates that the initial disturbance ΔH_o will grow, and the larger the yield stress τ_B and the smaller the mud depth H, the faster the wave will grow. This trend is the reason that roll waves occurred only in the flow of non-Newtonian fluid of small discharge and small depth. After a disturbance develops into a roll wave, the continuities of velocity and depth no longer hold, and the wave stops growing when it reaches a certain amplitude.

On the other hand, if velocity is high and η is large, the first term on the right hand side of Eq.(10) is negligible compared with the second term, and Eq.(10) becomes

$$\frac{d}{dt}(\Delta U) = -\frac{1}{2}\Delta(\frac{\eta U}{\rho_m H d}) \qquad (17)$$

in which Eq.(12) has been employed. Since

$$\Delta(\frac{\eta U}{\rho_m H d}) = \frac{\partial}{\partial U}(\frac{\eta U}{\rho_m H d})\Delta U + \frac{\partial}{\partial H}(\frac{\eta U}{\rho_m H d})\Delta H$$

Eq.(17) can be rewritten as

$$\frac{d}{dt}(\Delta U) = -\frac{\eta}{2\rho_m H d}(1 - Fr)\Delta U \qquad (18)$$

or after integration

$$\frac{\Delta U}{\Delta U_o} = e^{-\frac{1-Fr}{2\rho_m H d}t} \qquad (19)$$

where ΔU_o is the initial disturbance in velocity, $Fr = U/\sqrt{gH}$ is the Froude number. Eq.(19) proves that at high flow rate, as long as Fr<1, the disturbance in velocity ΔU_o always decreases, hence, a large discharge of mud flow is stable.

Eqs. (16) and (19) give the results for the two extream cases. In the general case, both terms on the right hand side of Eq.(10) should be taken into account. Then we have

$$\frac{d}{dt}(\Delta U) = \frac{1}{2\rho_m H}[\frac{\tau_B}{\sqrt{gH}} - \frac{\eta}{d}(1 - Fr)]\Delta U$$

$$\frac{\Delta U}{\Delta U_o} = \exp\{\frac{t}{2\rho_m H}[\frac{\tau_B}{\sqrt{gH}} - \frac{\eta}{d}(1 - Fr)]\} \qquad (20)$$

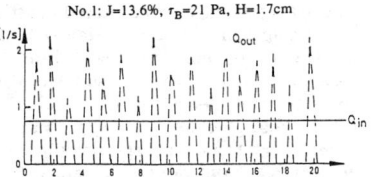

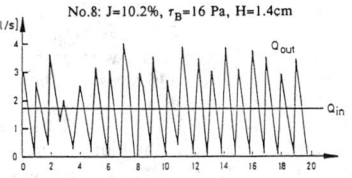

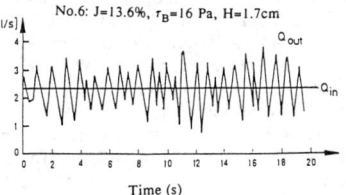

Fig.2 Hydrographs of Q_{in} and Q_{out}

If the fluid has large yield stress and small rigidity coefficient, a flow is unstable over a large range of discharges and is liable to develop into surge waves. Whereas if the fluid has small τ_B and great η, a flow is stable in most cases; only if the discharge and depth are very small is it possible to develop into pulsing flow.

3 EXPERIMENTS

Two experiments were conducted for investigation of debris flow surges in tilting flumes. Two kinds of clay mud, having simuilar rheological properties with debris flow matrix, were used to simulate debris flow mixture. As concentration of clay was higher than 50 kg/m^3, the sediment suspension was Bingham fluid. The yield stress and the rigidity coefficient increased with increasing concentration rapidly and no thixotropy was observed in the rheological measurements. In the experiment incoming discharge, Q_{in}, was recorded by means of a magnetic flow meter and the discharhge at the outlet end of the flume, Q_{out}, was measured by using a gauging tank and a fluviograph. Wave height at different distances from the entrance was recorded by using some scaling rods. All flows in the experiments were laminar.

The mud flow remained stable if the concentration was low and the yield stress was not large enough, or the incoming flow rate and mud depth were large, whereas the mud flow became unstable and a series of surges developed if the concentration was high and incoming flow rate and depth were small. Generally speaking, a slight fluctuation in velocity occurred first, and then some ripples appearred on the surface. The ripples grew into waves as they propageted downstream, and more other ripples formed at the same time. Sometimes the waves grew so large that their maximum discharges were more than double the incoming discharge, and the residual mud stopped moving after waves passed. This process was similar to the phenomenon "river clogging" in hyperconcentrated flow in the middle reach of the Yellow River and its tributaries (Qian, 1980). Roll waves first grew as they propagated downstream and stopped growing when they reached a certain amplitude. The limiting amplitude appeared to be directly related to the incoming flow rate and the clay concentration.

The first experiment was conducted in a 8.7 m long, 10 cm wide flume. Fig.2 shows the hydro-graphs of Q_{in} and Q_{out} of three runs in the experiment, where H is the mean depth. The yield stress of the muds in the three runs was 21 Pa (No.1) and 16 Pa (No.6 and 8). It shows that the mud flow at the outlet of the flume fluctuated and even developed into an intermittent flow (No.1), although the incoming discharge remained constant. The smaller the incoming discharge, the longer the time interval between the waves appeared to be. If Q_{in} was larger (No.6), the flow was continuous but fluctuate. The intermittent flow shown in Fig.2 is similar to viscous debris flow surges. In both cases, the slope is large, and the depth is small. The height of debris flow surge is one to several meters but the height of wave in the experiment was only 1-2 cm because the size of the flume was much smaller. The period of debris flow surges are longer than that in the experiment because the geometric scale of debris flow is much larger. Based on dimensional considerations, time scale λ_t has the following relation with gepmetric scale λ_L,

$$\lambda_t = \lambda_L^{0.5} \quad (21)$$

Geometric scale of debris flow is generally 100-1 000 times greater than the flow in the flume, so that the time period of surge waves are several tens times longer.

Another experiment was conducted in a 24 m long, 60 cm wide flume. It was

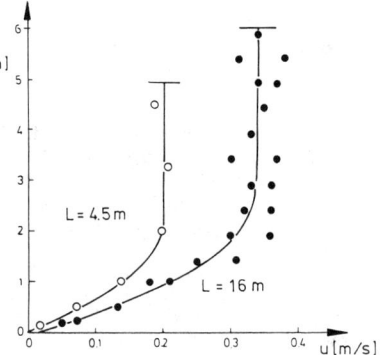

Fig.3 Velocity profile near the entrance and that of wave near the outlet

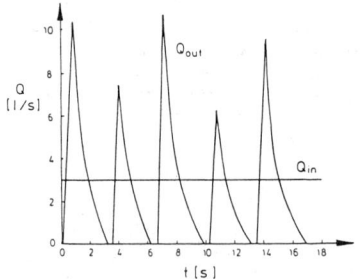

Fig.4 Hydrographs of Q_{in} and Q_{out} in the second experiment

found that at small bed slope (1.2%), a flow of clay suspension with concentration of 310 kg/m^3 (τ_B=5.3 Pa) at flow rate of 5.5 l/s was continuous and there was no obvious surge waves, but there appeared depth and velocity fluctuation. The mud flow near the entrance was stable but in the downstream section was fluctuate. The mud depth in the downstream section first rose gradually from 4.5 cm to 6.2 cm in 37 min. and then suddenly reduced to 3.6 cm in 6 min. The velocity distributions of the stable flow and fluctuate flow are shown in Fig.3. The latter was measured when the velocity reached its maximum and was meansured many times. One can see that the velocity distribution still follows the laminar velocity distribution of a Bingham fluid. At higher bed slope (4.8%) and higher concentration (456 kg/m^3, τ_B=19 Pa), the mud flow developed into surge waves if the flow rate was low (3 l/s). Fig.4 shows the hydrographes of Q_{in} and Q_{out} of the flow. The flow was intermittent with the maximum output discharge nearly three times of the incoming one. The height of the surge waves was about 3-5 cm and the period of the surge waves about 3.5 s, both values were greater than those in the first experiment because of the greater size of the flume. Fig.5 shows the growth of the wave height along the distance. The wave height, ΔH, grew faster and faster at first, then grew slowly, and kept constant when it reached a certain amplitude. The experimental results proved the conclusion from the theoretical analysis.

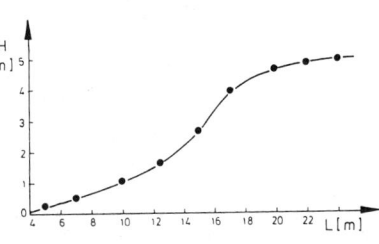

Fig.5 Growth of the wave height

CONCLUSION

Development of viscous debris flow from continuous flow into a series of roll waves is attributed to the yield stress of the matrix. Slight fluctuation in depth can grow up and developed into surges if the sediment mixture transformed from Newtonian into non-Newtonian fluid. The greater the yield stress and the smaller the flow depth, the more liable the flow developes into surge waves.

ACKNOWLEDGEMENTS

The author would like to acknowledge the financial support of the Chinese National Natural Science Foundation and the Alexander von Humboldt Foundation.

REFERENCES

Davies,T.R.H. (1988) "Debris flow surges - a laboratory investigation", Mitteilungen der Versuchsanstalt für Wasserbau, Hydrologie und Glaziologie, Nr. 96.

Englunde,F. and Wan,Z. (1984) " Instability of hyperconcentrated flow", J of Hyd. Engin., Proc., ASCE, Vol. 110, No. HY3, pp.219-232.

Kang, Z. (1985) "Characteristics of the flow patterns of debris flow at the Jiangjia Gully in Yunnan" Memoirs of Lanzhou Institute of Glaciology and Cryopedology, No.4.

Li,J., Yuan,J., Bi,C. and Luo, D. (1983) " The main featuresof the mud flow in the Jiangjia Ravine", Zeit. Geomorph., Vol.27, No.3, pp. 325-341.

Qian,N. (N. Chien) (1980) "Apreliminary study of the mechanism of hyperconcentrated flow in North-West China " Academic Report on Sediment Problems in the Yellow River, No.4 (in Chinese)

Wang,Z., Lin,B. and Zhang,X. (1990) "Instability of non-Newtonian open channel flow", Acta Mechanica Sinica, Vol.22, No.3, pp.266-275 (in Chinese).

Comparisons between experimental and numerical studies
on laminar flow with tracer transport

Christian Forkel, Helmut Daniels, Jens Birkhölzer and Gerhard Rouvé[1]

Abstract

In this paper we present an experimental and an accompanying numerical study of two- and three dimensional transient flow and tracer transport in complex flow systems at low Reynolds numbers. The physical experiment was carried out in an irregular wall-bounded domain with one inlet for water and tracer (Uranin) and one outlet. The tracer movement was taped by a CCD-camera, converted by a digital image processing technique and visualized with AVS. The numerical simulations were carried out with two-dimensional depth averaged and fully three-dimensional finite element codes for the incompressible Navier-Stokes equations (PASTIS-3D) and the transport equation. Comparisons between the experimental data and the numerical simulations were made. The results are in excellent agreement.

Introduction

Accidental contamination in natural flow systems are a challenge for alert systems in environmental hydraulics. Numerical codes may help to solve these problems. But still a major problem is the verification and the validation of such models. Of course models can be verified by analytical results or by simple experiments. But if the models should be used for simulations in natural flow systems this procedure might not suffice. In this case the model has to be tested in very complex geometries. Therefore in this paper not only two dimensional depth-averaged and fully three-dimensional finite element models for laminar flow with tracer transport will be presented. It will also be shown with comparisons between an experiment with irregular boundaries and numerical simulations that these models are capable of calculating this experiment very accurate. Another aim of this study is to show the proper use of scientific visualization techniques in the comparison between experiments and numerical calculations.

Mathematical background and numerical methods

For the calculation of the fully three-dimensional laminar flow field in a wall-bounded geometry we use the incompressible Navier-Stokes equations:

---

[1]Institut für Wasserbau, Aachen University of Technology, Mies-van-der-Rohe Str.1, W-5100 Aachen, Germany

$$\frac{\partial v_i}{\partial x_i} = 0 \tag{1}$$

$$\frac{\partial v_i}{\partial t} + v_j \frac{\partial v_i}{\partial x_j} = -\frac{\partial P}{\partial x_i} + v_0 \frac{\partial^2 v_i}{\partial x_j^2} \tag{2}$$

and for the tracer movement the ordinary transport equation:

$$\frac{\partial c}{\partial t} + \frac{\partial}{\partial x_i}(v_i c) - \frac{\partial}{\partial x_i} D_{ij} \frac{\partial c}{\partial x_j} = Q_c \tag{3}$$

D_{ij} is the diffusion tensor which is formed by the identity matrix multiplied with the molecular viscosity of the tracer in water. From these equations we derived the depth-averaged equations for the laminar flow by integrating over the vertical direction (bottom and top of modeled area are horizontal).

$$\frac{\partial q_i}{\partial x_i} = 0, \text{ with } q_i = v_i \cdot 2b \quad \text{(b: channel half width)} \tag{4}$$

$$\frac{\partial q_i}{\partial t} + q_j \frac{\partial}{\partial x_j}\left(\frac{q_i}{2b}\right) = -2b \frac{\partial P}{\partial x_i} + 2bv_0 \frac{\partial^2}{\partial x_j^2}\left(\frac{q_i}{2b}\right) - 3v_0 \frac{q_i}{b^2} \tag{5}$$

For the simulation of the tracer transport the upper symmetric half of the domain was split into six layers. In each layer the ordinary transport equation (3) was solved with:

$$v_i = \frac{3q_i}{4b}\left(1 - \frac{z^2}{b^2}\right) \tag{6}$$

The vertically averaged simulations were carried out with the code PASTIS-3D (DANIELS, 1990+1992) for the flow field and the code STRANK (FORKEL, 1990) for the tracer transport, in the three-dimensional simulations both flow and transport were calculated with PASTIS-3D.

PASTIS-3D solves the transient incompressible Navier-Stokes equations time-accurately in two and three space dimensions. Finite elements of arbitrary shape and with unstructured mesh topology are used to approximate spatial operators. The element types have mixed C^0 and C^{-1} approximation functions. In 3D the trilinear velocity/constant pressure hexahedron and in 2D the bilinear velocity/constant pressure quadrilateral isoparametric element are used. Because of the arbitrary shape of the elements all integrations on the element level are obtained by full 2 x 2 x 2 Gauss point integration in the general case. Time-accurate solutions are obtained by using the `projection 2´ algorithm recently described and analysed for stability and convergence behaviour by GRESHO (1990). Modifications to the original algorithm were made where it was appropriate.

The code contains within the framework of `projection 2´ an explicit (forward Euler) option as well as an implicit (backward Euler) option, and trapezoidal rule as well as several semi-implicit options to solve the momentum equations and transport equation. It allows subcycling of the pressure solution as well as monitoring the divergence, dissipation and some measures of local as well as temporal error. Also, the convective term can be approximated by honest Galerkin, element-averaged values or the product approximation. The two-dimensional Finite Element code STRANK uses biquadratic approximation functions from 9-node elements and contains also semi-implicit time integration methods. Further details of the numerical scheme are shown in FORKEL et al. (1992).

Model Experiment

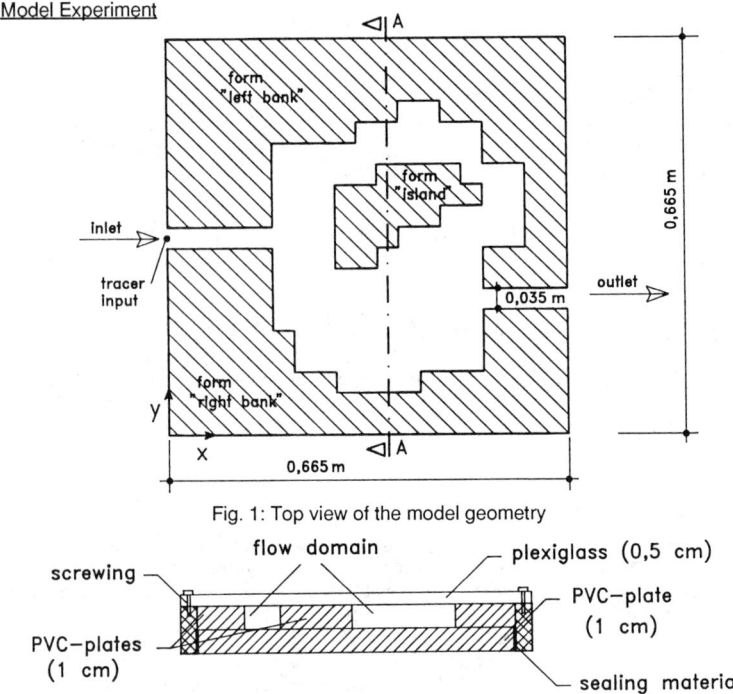

Fig. 1: Top view of the model geometry

Fig. 2: Side view (Cut A - A) of the model geometry (5 times increased in height)

The domain of the model experiment is shown in Fig. 1 and Fig. 2, details of the surrounding apparatus are given in BIRKHÖLZER (1991). The model experiment has an irregular geometry in the horizontal directions, a constant depth of 1 cm in the vertical direction and is totally wall-bounded except of one inlet for water and tracer and one outlet. The top of the domain consists of plexiglas so that the tracer movement can be taped by a CCD-Camera. The tracer Uranin is a fluorescent substance that lights when it is exposured to UV-light. The video-signal (brightness) is then converted into data by a digital image processing technique (Optimas) and can be visualized for example by AVS.

The flow rate is controlled by a rotameter. The tracer is mixed with water and is led in over the whole depth in the middle of the inlet. Fig. 3 shows the boundary conditions of the experiment. First we had a steady flow rate of 36,4 l/h, at time t = 0s we started a tracer input with an additional flow rate of 3,77 l/h and a concentration of 40 mg/l. This gave a tracer input of 150,8 mg/h. The tracer input was stopped at time t = 270s, after that time only fresh water with a flow rate of 36,4 l/h was led in and cleaned up the area until time t = 450 s. Therefore the boundary condition were piecewise constant but implied short transient flow fields at the times t = 0 and t = 270s. The water had a temperature of nearly 10° C, so the viscosity of the water was $\nu = 1,31 \cdot 10^{-6}$ m²/s. The maximum Reynolds number based on the channel half width b = 0,5 cm and the maximum velocity in the inlet was nearly Re ≈ 400.

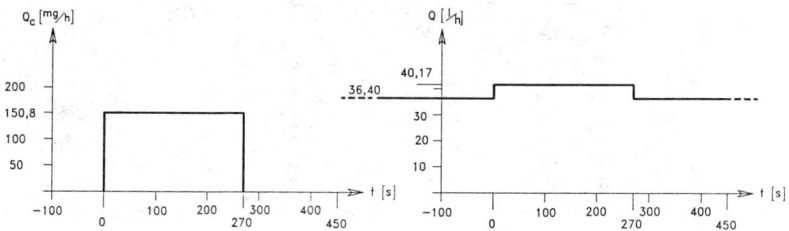

Fig. 3: Boundary Conditions: tracer input and flow rate

Numerical Simulation

The simulations were performed on a Siemens-Fujitsu VP 600 supercomputer and took about 15 CPU-hours for all calculations (the vector rates reached up to 98,5 %). The finite element mesh consists of 7424 rectangular elements (side length 0,4375 cm) in the horizontal plane so that there were 8 elements for the width of the inlet channel. For the three-dimensional simulations the vertical direction was discretized with 5 elements for one symmetric half of the domain by refining the mesh at the wall. The boundary conditions we used are: **a)** Dirichlet Boundary Conditions for the velocities at all walls($v_i = 0$), **b)** Dirichlet Boundary Conditions for the velocities at the inflow boundary ($v_x \neq 0$, $v_y = v_z = 0$), **c)** Neumann Boundary Conditions for the friction at the outflow boundary ($F_n = F_\tau = 0$), **d)** Dirichlet Boundary Conditions for the velocities in the vertical direction at the symmetric plane ($v_z = 0$), **e)** Neumann boundary conditions for the friction at the symmetric plane ($F\tau = 0$).

The x-velocities at the inflow boundary were calculated in a pre-processing program by simulating the inflow channel with pressure boundary conditions at the open boundary, taking the velocity distribution in the inlet and fitting it to the given flow rate of the experiment.

For the 2D-simulation of the tracer transport we used 1856 9-node elements with the same nodes as for the flow field. The boundary conditions were: **a)** no mass exchange through the walls and the symmetric plane ($Q_c = 0$) and **b)** Neumann Boundary Conditions at the inlet and an open boundary at outlet.

The diffusion coefficient of Uranin in water was determined to be $2 \cdot 10^{-7}$ m²/s (Pe ≈ 1000). A problem was the disturbance of the flow field because of the tracer injection. This effect could not be simulated with only 8 elements in the channel width so we had to increase the diffusion coefficient in the inlet channel with a factor of 100. The maximum Courant number is between Cu ~ 0,1 in the 2D simulations and Cu ~ 1,0 in the 3D simulations.

Comparison of the measured data with the results of the numerical simulation

Both the measured data and the results of the numerical simulations were visualized with AVS. Figure 4 shows a top view of the vertically averaged concentration fields. The bright areas are zones with higher concentrations, the darker areas mark the less contaminated zones.

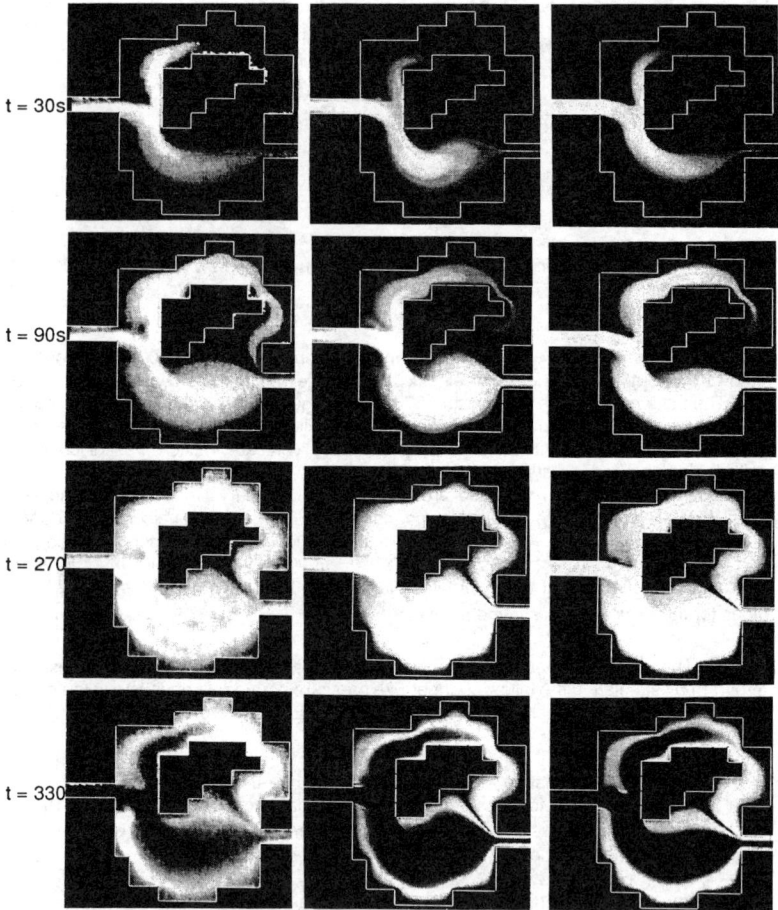

Fig. 4: Comparisons of measured data (left side) with the results of the numerical simulations (middle: 2D, right side: 3D) at various time steps

The calculated and the measured data are in very good agreement besides some little discrepancies in 2D: first the higher concentrations in the left stream in the experiment than in the calculation and second the greater expansion of the main stream in the calculation in front of the outlet. These discrepancies vanish in the views of the three-dimensional simulations. This is due to the fact that despite there is only a little extension in the vertical direction we have a three-dimensional roll just in front of the "island" which could not be simulated with the 2D-code but with the 3D-code. So the three-dimensional simulations are in excellent agreement with the experiment data.

Conclusions and outlook on future work

With the Finite Element Code PASTIS-3D we have a very efficient code for simulating laminar flow and tracer transport also in very complex geometries. This was proofed by the numerical simulation of a model experiment which gave excellent agreement between the measured and the calculated data. We intend to use this code with slight modifications (density dependent flow, surface turbulences) to simulate contaminant transport in reservoirs. The code will also be basis for a new Finite Element code to model turbulent structures with Large Eddy Simulations.

Acknowledgements

This work was supported by the German Research Foundation (DFG) under Grant Ro 365/44-1.

Literature

BIRKHÖLZER, J. (1991): *Stofftransport in Kluftgrundwasserleitern*; Mitteilungsheft 81 des Instituts für Wasserbau und Wasserwirtschaft "Vorträge Wasserbau-Seminar 1990/91: Schadstofftransport in Grund- und Oberflächengewässern", Hrsg.: Prof. Dr.-Ing. G. Rouvé, ISBN 0343 - 1045.

DANIELS, H. (1991): *Numerische Berechnung instationärer Strömungsvorgänge in Wärmespeichern*; Mitteilungsheft 77 des Instituts für Wasserbau und Wasserwirtschaft, Hrsg.: Prof. Dr. Ing. G. Rouvé, RWTH Aachen.

DANIELS, H.(1992): *PASTIS-3D Finite Element Projection Algorithm Solver for Transient Incompressible Flow Simulations*, UCRL-MA-111833, Lawrence Livermore National Laboratory.

FORKEL, C. (1990): *Einfluß der räumlichen Diskretisierung bei numerischen Berechnungen instabiler instationärer Dichteströmungen*; Diplomarbeit am Institut für Wasserbau, RWTH Aachen.

FORKEL, C., H. DANIELS AND G. ROUVÉ (1992): *Simulation instationärer Ausbreitungsprozesse in dreidimensionalen laminaren und turbulenten Strömungen*, 1. Arbeitsbericht zum DFG-Forschungsvorhaben, Kenn-Nr.: Ro 365/44-1, Institut für Wasserbau und Wasserwirtschaft, RWTH Aachen

GRESHO, P. (1990): *On the Theory of Semi-Implicit Projection Methods for Viscous Incompressible Flow and Its Implementation via a Finite Element Method that also Introduces a Nearly Consistent Mass Matrix*; Part 1 and 2, Int. J. Num. Meth. in Fluids, Vol. 11, Wiley and Sons Ltd, pp. 587 - 620 and 621 - 659.

Comparison of Advective Transport Algorithms with an Application in Suisun Bay, a Sub-Embayment of San Francisco Bay, California.

Jon R. Burau[1], Stephen Monismith[1], and Jeffrey Koseff[1]

Abstract

This paper examines the conservative properties of several numerical treatments of the salt conservation equation. Specifically, computationally efficient first and second order Eulerian-Lagrangian Methods (ELM) are compared against (1) fully conservative finite volume methods including upwind, LUICKEST, and QUICKEST. This paper shows that under the numerically rigorous requirements of long-term estuarine transport, that the non-conservative nature of ELM may be problematic. Moreover, solution of advective terms using any of these methods is shown to be a small percentage of the overall computational effort in hydrodynamic models: arguments over their relative computational efficiency is merely academic.

Introduction

The transport of salt and other conservative substances within natural systems is one of the most important and challenging problems in estuarine research. Although the tidal motions in estuaries are typically large, when averaged over a tidal cycle, they nearly cancel. The generally small net differences in the tidal motion, known as a residual, determines the ultimate fate of salt (and other conservative substances) within estuaries. Because the residuals are orders of magnitude smaller that the tidal motions, careful attention must be paid to the numerical treatment of the transport equations. This paper discusses the conservative nature and relative costs of several advection schemes applied to the salt conservation equation.

Depth Averaged Salt Conservation Equation

The depth averaged salt balance equation is well known (Pritchard,1984) and is given here in conservative form,

$$\frac{\partial(Hs)}{\partial t} + \frac{\partial(Hsu)}{\partial x} + \frac{\partial(Hsv)}{\partial y} = \frac{\partial}{\partial x}(D_x \frac{\partial Hs}{\partial x}) + \frac{\partial}{\partial y}(D_x \frac{\partial Hs}{\partial y}) \quad (1)$$

where s, u, and v represent depth averaged salinity, and velocity components in the x and y directions respectively. H represents the total depth, and D_x, and D_y are diffusion coefficients in the x and y directions. Eq.(1) is applicable to any conservative quantity and, thus, is equally valid for any generic conservative substance. Because salt transport in estuaries is advection dominated, we will limit our discussion to the case of pure advection.

[1] Environmental Fluid Mechanics Laboratory
Stanford University
Stanford, CA 94305-4020

Numerical modeling of sharp gradient, pure advective processes is inherently difficult. Low order schemes suffer from excessive numerical diffusion whereas high order schemes generally produce unwanted and occasionally catastrophic oscillations. The stability, accuracy, and consistency of all of the numerical schemes discussed in this paper have been previously addressed in the literature (see Baptista et.al.(1984) for ELM methods and Anderson et.al. (1984) for integral approaches). We begin with a discussion of the methods, then discuss their conservative attributes and relative computational costs.

Eulerian-Lagrangian Methods

Eulerian-Lagrangian Methods (ELM) are well documented in the literature (Cheng et.al. 1984, Baptista 1984) consequently we will only provide a brief description of the method here. Basically, ELM updates the salt field by using the velocity field at the previous time step to find the location of each parcel of water [point "P" in figure (1)] that will be advected (translated) to each grid point at the next time step. This traceback procedure to find "P" is known as the method of reverse streaklines and is illustrated in figure (1). Once we know the position and salinity at "P" we simply advect the parcel of water that "P" represents (Lagrangian viewpoint) to its fixed grid (Eulerian) location at the next time step. Point "P" almost never falls exactly on a grid point, therefore the salinity at point "P" must be estimated using some kind of interpolation based on the salinities at adjacent grid points before it can be advected to its new location by the method of reverse streaklines. In this paper we compare two different interpolation procedures: (1) bilinear, and (2) second order Lagrangian polynomials which we will refer to as quadratic interpolation henceforth.

Most explicit methods can not exceed a Courant-Friedrich-Lewy (CFL) number $CFL = \frac{\Delta t}{\Delta x}\sqrt{u^2 + v^2}$ of unity. The primary advantage of ELM is that it is not limited by the CFL constraint. In theory, ELM will work for any CFL number, provided accurate streaklines can be computed. Because ELM is not CFL condition limited it can be run using large time steps which translates into lower computational costs. A fundamental weakness of ELM, however, is that it is a non-conservative method. Therefore, from a practical standpoint, one needs to evaluate the degree to which the lack of a conservative property affects ELM generated solutions and decide whether the advantages of ELM outweigh its lack of conservation. We examine the non-conservative nature of ELM in the next section.

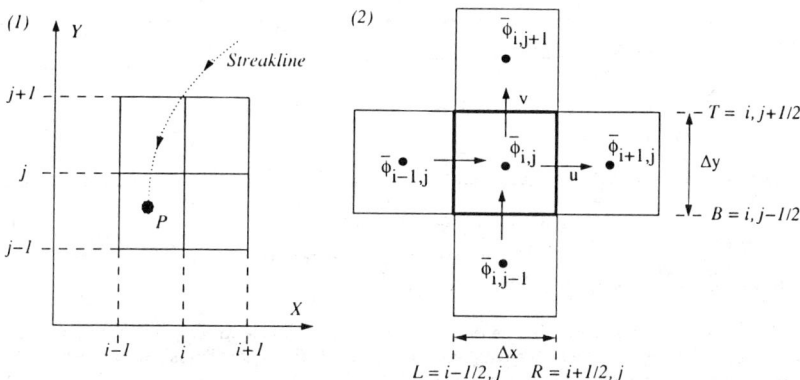

Figure (1) Eulerian–Lagrangian Method

Figure (2) TRIM numerical stencil.

Salt Conservation

For a conservative substance, like salt, the total change in salt within a system should just balance the flux of salt through the system boundaries and any internal sources or sinks. To evaluate the conservative property in ELM we ran several test cases, the most practical of which we present in this paper. Suisun Bay California has highly variable bottom topography (see figure 3) making it an excellent test case on which to evaluate the conservative properties of ELM. In this test case, sea level is forced at the open boundaries at Chipps Island and Carquinez Strait. We placed a non-dynamic tracer in the center of the bay and monitored the total amount of tracer in the bay over time. For a perfectly conservative method we expect no change in the total salt mass, since in our example we have no exchange through the boundaries and no internal sources or sinks of salt. The total salt mass in this problem, however, declines steadily to roughly 20 percent of its original concentration in 4 days as shown in figure (4). With this magnitude of mass loss, ELM is probably a poor choice for estuarine transport since most estuarine transport calculations of ecological significance are on the order of weeks. The methods described in the next sections are explicit alternatives to ELM that strictly conserve.

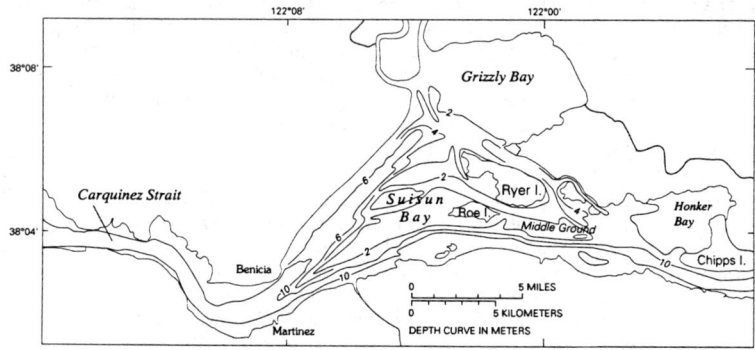

Figure (3) Suisun Bay Bathymetry

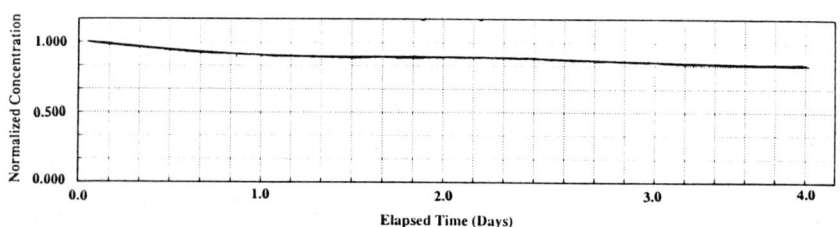

Figure (4) Percent change in total tacer mass using bilinear ELM

Finite Volume (integral) Methods

The finite volume method is one of many procedures (finite difference, finite element, finite analytic, etc.) used to conceptualize numerical solution algorithms that approximate some differential equation. Finite volume relations are derived by *equating* the flux of something (mass, etc.) through the sides of a finite control volume *with* the change in storage of that something (mass, etc.) within the control volume.

The quantity we wish to conserve in the depth-averaged salt equation is $\phi = Hs$, where ϕ is the conservative variable. Integrating Eq.(1) over an arbitrary area, A, enclosed by surface, S, and applying the divergence (Gauss) theorem we have

$$\int_t \frac{\partial \bar{\phi}}{\partial t} dt + \frac{1}{A} \int_t \oint_S \vec{F} \cdot d\vec{S} \, dt = 0 \tag{2}$$

where $\vec{F} = u\phi \vec{i}_x + v\phi \vec{i}_y$ and $(\vec{i}_x, \vec{i}_y)$ are unit vectors in the (x,y) directions and $d\vec{S}$ represents the local outward pointing normal to the surface whose magnitudes are equal to an incremental length along the surface, and $\bar{\phi}$ is the area average of ϕ. In this case our spatial integrations will conform to the staggared grid on which the hydrodynamics are solved as shown in figure (2). From this point forward, we drop the over-bar in Eq.(2) recognizing that ϕ represents the average over the area. Performing the surface integration in a piecewise fashion for a single computational cell, assuming a simple backward time integration for $\frac{\partial \phi}{\partial t}$, and using the following notation for the left, right, bottom, and top faces: L = $(i - 1/2, j)$, R = $(i + 1/2, j)$, B = $(i, j - 1/2)$, T = $(i, j + 1/2)$ our final finite volume estimate for ϕ at the new time level is

$$\phi_{i,j}^{n+1} = \phi_{i,j}^n - f_R + f_L - f_T + f_B \tag{3}$$

where the individual flux face contributions are

$$f_R = \frac{1}{\Delta x} \int_0^{\Delta t} \int_0^{\Delta y} (u\phi)_{i+1/2,y} \, dy \, dt, \quad f_L = \frac{1}{\Delta x} \int_0^{\Delta t} \int_0^{\Delta y} (u\phi)_{i-1/2,y} \, dy \, dt \tag{4}$$

$$f_T = \frac{1}{\Delta y} \int_0^{\Delta t} \int_0^{\Delta x} (v\phi)_{x,j+1/2} \, dx \, dt, \quad f_B = \frac{1}{\Delta y} \int_0^{\Delta t} \int_0^{\Delta x} (v\phi)_{x,j-1/2,y} \, dx \, dt \tag{5}$$

In our case, the hydrodynamic code solves for the velocities at the flux faces so we can use them directly as they are computed in Eqs.(4) and (5). The conservative quantity, ϕ, is computed at the centers of the cells and is not directly available at the flux faces. The way in which the ϕ's are interpolated [or, alternatively, the integrals are evaluated in Eqs.(4) and (5)] to the flux faces determines the method and its spatial accuracy. The next sections describe the various ways in which the flux face average ϕ's are computed.

QUICKEST

Our approach in the following sections will be to derive the highest order (3rd order) algorithm we will consider in this paper (QUICKEST), then show, without proof, that the lower order algorithms are a subset of the higher. Quadratic Upstream Interpolation of Convective Kinematics with Estimated Streaming Terms (QUICKEST), after Leonard, (1979), is an odd, third order scheme that has gained wide spread acceptance where sharp gradient solutions are desired. To arrive at the fully two-dimensional QUICKEST formulation we begin with a general two dimensional Gauss upwind quadratic interpolation formula for ϕ in non-dimensional coordinates ($\xi = \frac{x}{\Delta x}, \eta = \frac{y}{\Delta y}$) written at $x = i\Delta x, y = j\Delta y$ (Hildebrand, 1956):

$$\phi_{\xi,\eta} = [1 + \xi \nabla_\xi + \eta \nabla_\eta + \xi \eta \nabla_\xi \nabla_\eta + \frac{\xi(\xi + \beta_u)}{2} \delta_\xi^2 + \frac{\eta(\eta + \beta_v)}{2} \delta_\eta^2] \phi_{i,j} \tag{6}$$

where $\nabla_\xi = (1-\alpha_u)\nabla_\xi^b + \alpha_u \nabla_\xi^f$, and $\nabla_\eta = (1-\alpha_v)\nabla_\eta^b + \alpha_v \nabla_\eta^f$ are difference operators in the (ξ, η) directions, and $\nabla_\xi \nabla_\eta = \alpha_u \alpha_v \nabla_\xi^f \nabla_\eta^f + \alpha_u (1-\alpha_v) \nabla_\xi^f \nabla_\eta^b + \alpha_v (1-\alpha_u) \nabla_\xi^b \nabla_\eta^f + (1-\alpha_u)(1-

$\alpha_v)\nabla^b_\xi \nabla^b_\eta$ is a cross derivative operator. The 'f' superscript represents a forward operator: where $\nabla^f_\xi = \phi_{i+1,j} - \phi_{i,j}$ and $\nabla^f_\eta = \phi_{i,j+1} - \phi_{i,j}$. The '$b$' superscript represents a backward operator: where $\nabla^b_\xi = \phi_{i,j} - \phi_{i-1,j}$ and $\nabla^b_\eta = \phi_{i,j} - \phi_{i,j-1}$. The cross derivative operators are $\nabla^f_\xi \nabla^f_\eta = \phi_{i+1,j+1} - \phi_{i,j+1} - \phi_{i+1,j} + \phi_{i,j}$, $\nabla^f_\xi \nabla^b_\eta = \phi_{i+1,j} - \phi_{i,j} - \phi_{i+1,j-1} + \phi_{i,j-1}$, $\nabla^b_\xi \nabla^f_\eta = \phi_{i,j+1} - \phi_{i-1,j+1} - \phi_{i,j} + \phi_{i-1,j}$, $\nabla^b_\xi \nabla^b_\eta = \phi_{i,j} - \phi_{i-1,j} - \phi_{i,j-1} + \phi_{i-1,j-1}$, and finally $\delta^2_\xi = \phi_{i+1,j} - 2\phi_{i,j} + \phi_{i-1,j}$ and $\delta^2_\eta = \phi_{i,j+1} - 2\phi_{i,j} + \phi_{i,j-1}$ represent second order operators.

Obviously "Upstream" depends on the flow direction. Using standard forward and backward operators in the above relations requires selectively removing certain terms depending on the flow direction. To do this we introduce the following flags:

$$\begin{aligned}
\alpha_u &= 0, \quad for \quad u \geq 0 \quad and \quad \alpha_u = 1, \quad for \quad u < 0 \\
\alpha_v &= 0, \quad for \quad v \geq 0 \quad and \quad \alpha_v = 1, \quad for \quad v < 0 \\
\beta_u &= 1 - 2\alpha_u \quad and \quad \beta_v = 1 - 2\alpha_v
\end{aligned} \qquad (7)$$

To evaluate f_R and f_T in Eq.(3) one must integrate ϕ in both time and space by applying a Taylor series expansion about t=0 and using the advection equation to exchange spatial for temporal derivatives with the final result,

$$\begin{aligned}
f_R = \frac{C_{uR}}{2} \Big[&\; 2 + (1 - C_{uR})\nabla_\xi - C_{vR}\nabla_\eta + C_{vR}(\frac{C_{uR}}{3} - \frac{1}{2})\nabla_\xi \nabla_\eta \\
&+ \{\frac{1}{4} + C_{uR}(\frac{C_{uR}}{3} - \frac{1}{2}) + \frac{\beta_u}{2}(1 - C_{uR})\}\delta^2_\xi \\
&+ \{\frac{1}{12} - \frac{\beta_v}{2}C_{vR} + \frac{C^2_{vR}}{3}\}\delta^2_\eta \Big] \phi_{i,j}
\end{aligned} \qquad (8)$$

$$\begin{aligned}
f_T = \frac{C_{vT}}{2} \Big[&\; 2 + (1 - C_{vT})\nabla_\eta - C_{uT}\nabla_\xi + C_{uT}(\frac{C_{vT}}{3} - \frac{1}{2})\nabla_\xi \nabla_\eta \\
&+ \{\frac{1}{4} + C_{vT}(\frac{C_{vT}}{3} - \frac{1}{2}) + \frac{\beta_v}{2}(1 - C_{vT})\}\delta^2_\eta \\
&+ \{\frac{1}{12} - \frac{\beta_u}{2}C_{uT} + \frac{C^2_{uT}}{3}\}\delta^2_\xi \Big] \phi_{i,j}
\end{aligned} \qquad (9)$$

where $C_{uR} = u_R \frac{\Delta t}{\Delta x}$, $C_{vR} = v_R \frac{\Delta t}{\Delta y}$, $C_{uT} = u_T \frac{\Delta t}{\Delta x}$, and $C_{vT} = v_T \frac{\Delta t}{\Delta y}$ are the flux face Courant numbers.

LUICKEST

Linear Upstream Interpolation of Convective Kinematics (LUICKEST) is an even, second order operator which can be deduced from Eqs.(8-9) by neglecting all 2nd order terms, or

$$f_R = \frac{C_{uR}}{2}[2 + (1 - C_{uR})\nabla_\xi - C_{vR}\nabla_\eta - \frac{C_{vR}}{2}\nabla_\xi \nabla_\eta]\phi_{i,j} \qquad (10)$$

$$f_T = \frac{C_{vT}}{2}[2 + (1 - C_{vT})\nabla_\eta - C_{uT}\nabla_\xi - \frac{C_{uT}}{2}\nabla_\xi \nabla_\eta]\phi_{i,j} \qquad (11)$$

UPWIND

Finally by neglecting terms transverse to each flux face in Eqs.(10-11) we arrive at the relations, using our notation, for upwinding,

$$f_R = \frac{C_{uR}}{2}[2 + \nabla_\xi]\,\phi_{i,j}, \qquad f_T = \frac{C_{vT}}{2}[2 + \nabla_\eta]\,\phi_{i,j} \qquad (12)$$

Results and Conclusions

The increase in complexity that accompanies an increase in accuracy is clearly evident when Eq.(12) is compared with Eqs.(10-11) and Eqs.(8-9). Typical results from a test problem involving the advection of a Gauss hill down a (100x20km) channel oriented 45 degrees from the grid axes are given in figure (5) for the methods described in this paper. The relative cpu effort, normalized by the cpu effort of bilinear ELM, for each of these algorithms is plotted in figure (6). Figure (6) was constructed for simulations using a Courant number $\sim$ 1.0. For Courant numbers $\geq$ 1.0, the finite volume methods must be sub-cycled. Whereas the cost of ELM will remain roughly static over a wide range of Courant numbers (generally the reverse streakline calculations in ELM take slightly longer the larger the Courant number), the cost of the finite volume methods scale with the value of the Courant number (e.g. a Courant number of 5 implies a five fold increase in the cpu effort over what is given in figure 6). Fortunately, the advection calculations are typically a small part of a total hydrodynamic code. Generally, solving for the water surface in most hydrodynamic codes involves the inversion of a large matrix (for semi-implicit or implicit codes) which accounts for most of the cpu time. Therefore, if we use TRIM2D [Cheng et.al.,(1993)] as an example hydrodynamic code, and plot total cpu time for hydrodynamics and salt transport (see figure 7) we see that using a highly accurate fully conservative method like QUICKEST is competitive, particularly since it strictly conserves.

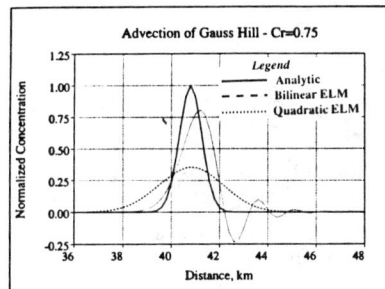
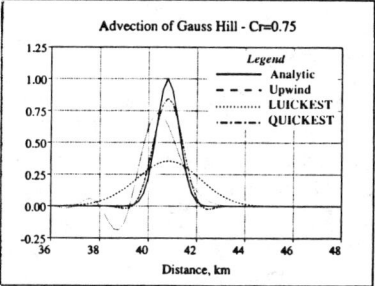

Figure (6) Numerical results from a problem of pure advection of a Gauss hill down a 100x20 km channel oriented 45 degrees to the grid axes.

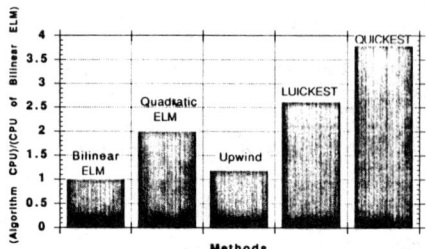

Figure (7) Comparison of methods based on algorithm cpu times.

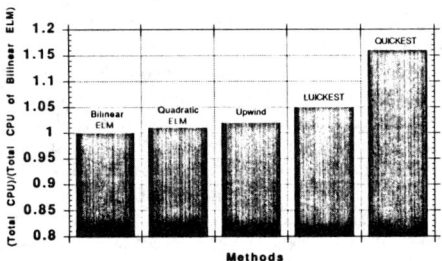

Figure (8) Comparison of methods when total cpu times (hydrodynamic plus salt) are considered.

References

Anderson, D.A., J.C. Tannehill, and R.H. Pletcher, 1984. Computational Fluid Mechanics and Heat Transfer, Mcgraw-Hill

Baptista A.M., E.E. Adams, and K.D. Stolzenbach, 1984. Eulerian-Lagrangian Analysis of Pollutant Transport in Shallow Water, Dept. of Civil Eng., MIT report No. 296

Cheng R.T., V. Casulli, and J.W. Gartner, 1993. "Tidal, Residual, Intertidal Mudflat (TRIM) Model and its Applications to San Francisco Bay, California." Estuarine, Coastal and Shelf Science, Vol. 36, In press.

Cheng R.T., V. Casulli, and S.N. Milford, 1984. "Eulerian-Lagrangian Solution of the Convective-Dispersion Equation in Natural Coordinates." Water Resources Research, Vol. 20, no. 7, pp. 944-952

Pritchard, D.W., 1982. in Estuarine modeling: An assessment, NTIS Rep. PB-206 807, Environ. Protect. Agency, Washington, D.C.

An Optimization Approach to River Adjustments

Robert Millar and Michael Quick,[1] Member, ASCE

Abstract
The adjustment of alluvial channels is presented here as a non-linear optimization problem in which the ideal channel will tend to adopt an optimum configuration subject to the imposed constraints. Bank stability is shown to play a key role, and increasing the bank stability causes the channel width to decrease, the depth to increase, and the slope to decrease. The theoretical predictions are tested on the gravel river data from Andrews (1984) and Hey and Thorne (1986).

Introduction
Alluvial rivers are generally believed to develop a hydraulic geometry which is in equilibrium with the prevailing hydrologic and sediment supply regime. River basin development often produces changes in the volume and timing of the water and sediment which can result in downstream changes as the channel adjusts to the new regime. Channel adjustments can also result from disturbance of the channel banks, particularly following the removal of the bank vegetation. It is often desirable from both economic and environmental standpoints to predict the nature of these alluvial adjustments.

The hydraulic geometry of an alluvial channel is defined by several dependent variables which include the channel width, depth, slope, roughness, and sinuosity, as well as other planform variables such as the meander wavelength, and the radius of curvature which are not considered in this paper. These variables are highly interdependent and a condition which forces a change in any one of these variables will result in a dynamic adjustment of the other dependent variables until equilibrium is attained. Therefore it is necessary to consider the alluvial channel as an integrated system, rather than by isolating variables and studying them independently.

The problem of modeling the adjustments of alluvial rivers is approached as a constrained optimization problem whereby the channel achieves an optimum configuration without violating any of the constraints. This approach is not new. As early as

---

[1]Respectively, Graduate Student and Professor, Department of Civil Engineering, University of British Columbia, Vancouver, British Columbia, Canada, V6T 1Z4.

1914 Gilbert (p. 135) demonstrated experimentally the existence of an "optimum form ratio". More recently Yang (1976), Chang (1979), and White *et al.* (1982) and others have developed optimization routines which predict the geometry of channels with relative success. These models consider only the bankfull or dominant discharge, and the corresponding sediment transport rate as constraints. These researchers recognized that the solution of the equations describing continuity, flow resistance, and sediment transport yielded a family of solutions because the number of unknown dependent variables exceeded the number of equations available for solution. A further relationship was required, this is commonly termed an "extremal hypothesis". White *et al.* use the maximum transport capacity hypothesis, Chang preferred the minimum stream power hypothesis, while Yang suggested a minimum unit stream power criterion. The extremal hypothesis corresponds to the objective function in an optimization formulation. In practice each of the hypotheses are generally equivalent, and represent alternate formulations of essentially the same multi-dimensional, non-linear optimization problem.

The procedure outlined in this paper was developed for gravel rivers with banks which are comprised of non-cohesive coarse sand or gravel. The constraints considered in this paper are sediment transport, flow resistance, continuity, the bank stability, and the valley slope. Basically for a channel to attain a stable, equilibrium geometry it must be able to transport the imposed sediment load without significant net deposition or erosion. Since water is the driving force, this sediment transport must be accomplished by the imposed flow regime. As with previous workers, the channel is modeled at the bankfull discharge, together with the corresponding sediment transport rate. Stable banks are an additional requirement for an equilibrium geometry. This does not preclude lateral migration; a channel is considered stable as long as there are no significant changes in the width, depth, or slope.

Over engineering time scales, less than 100 years or so, the valley slope can usually be considered to be constant. Any changes in the valley slope requires large volumes of deposition or erosion of sediment. Adjustments of channel slope are possible only through meandering. The valley slope is therefore a constraint on the channel geometry as the channel slope must be less than or equal to the valley slope.

For brevity only the bank stability constraint will be considered in more detail in this paper. A full description of the model can be found in Millar and Quick (1993).

Objective Function
In the present formulation there are four primary dependent variables: width, depth, channel slope, and bank angle. A solution is obtained whereby the stream power, γQS, is a minimum:

$$\min f(W, Y, S, \theta) = \gamma QS \quad \text{..}[1]$$

where W is the channel width, Y is the channel depth, S is the channel slope, θ is the bank angle, and γ is the specific weight of water. Since Q, the bankfull discharge, is an independent variable which is imposed on the channel, the channel develops a geometry where S is minimized.

Bank Stability Constraint

The bank stability constraint is comprised of two parts, evaluation of the mean bank shear stress, and assessment of the bank stability. The shear stress distribution is necessary for evaluation of the bank stability, as well as for bedload transport calculations. The mean bed shear stress can vary greatly from γYS which is commonly assumed, particularly for small aspect ratios. The method of Flintham and Carling (1988) is used to determine the distribution of the boundary shear stress. The method is based upon the distribution of the shear force, defined as the product of shear stress and wetted perimeter, across the channel. The equations were developed from data collected in roughened rectangular and trapezoidal laboratory channels.

The U.S.B.R. equation developed for the design of banks composed of non-cohesive gravel sediment (see Lane, 1955) has been modified in Millar and Quick (1993) to:

$$\frac{\tau_{bank}}{\gamma(s-1)D_{50bank}} = 0.067 \tan\phi' \sqrt{1 - \frac{\sin^2\theta}{\sin^2\phi'}} \quad\quad\quad [2]$$

where τ_{bank} is the mean bank shear stress, s is the specific gravity of the sediment, D_{50bank} is the median bank grain diameter, ϕ' is the modified friction angle of the bank sediment.

Unlike the mobile bed sediment, the gravel comprising the channel banks may be consolidated, cemented by fine silt and clay, and stabilized by roots which have penetrated the gravel banks. If the banks are composed of loose, unconsolidated gravel ϕ' reduces to the angle of repose. The effect of bank vegetation, cementing, and consolidation is to increase ϕ' which can take a maximum value of 90°. In equation [2], the critical dimensionless shear stress of the bank sediment is expressed as a function of ϕ'.

Model Verification

The published data of Andrews (1984) and Hey and Thorne (1986) were used to test the model. Hey and Thorne subdivided their data into 4 subsets from vegetation type I (grass) to type IV (> 50% trees and bushes). Similarly Andrews subdivided his data into *thin* and *thick* bank vegetation types.

Data pertaining to the nature of the bank sediments were absent in both papers. It was assumed that D_{50bank} was equal to D_{50}, the median grain diameter of the bed sediment. Furthermore ϕ' was also not given. Initially a value of $\phi'=40°$, which is close to the angle of repose for coarse gravel, was used as input. The initial model run resulted in large discrepancies between the modeled and observed geometries. The scatter was strongly asymmetric, with the modeled channels being consistently wider and shallower than the observed channels. The average values of the ratio of the observed width divided by the modeled width, W_{obs}/W_{mod}, for each bank vegetation type are presented in Table 1. Note that the modeled channels with the weakest bank vegetation agree well with the observed, and as the vegetation density increases the modeled width deviates systematically from the observed geometry.

Andrews (1984) and Hey and Thorne (1986) both performed empirical regime analyses to document the influence of the bank vegetation on the channel geometry. The channel width was found in both papers to be the most sensitive. Empirical width equations of the following form were developed:

$$W = aQ^{0.5} \quad \quad [3]$$

Q is the bankfull discharge (dimensionless in Andrews), a is a coefficient which varied with the bank vegetation density, becoming smaller as the bank vegetation density increased. The ratios of the observed channel widths divided by the unvegetated channel width, W_{obs}/W_{unveg}, was obtained by dividing the coefficient a obtained for each vegetation type, by the a value obtained for the channels with the weakest bank vegetation from each study. These results are also shown in Table 1. These ratios represent the influence of the bank vegetation on the channel width. For example the channels in the type III group of Hey and Thorne are on average 0.63 time their unvegetated counterparts.

Vegetation Type	W_{obs}/W_{mod} This Paper	W_{obs}/W_{unveg} Andrews(1984)	W_{obs}/W_{unveg} Hey and Thorne (1986)
I	0.94	-	1.0
II	0.79	-	0.77
III	0.68	-	0.63
IV	0.57	-	0.54
THIN	0.98	1.0	-
THICK	0.65	0.79	-

Table 1. Comparison of optimization and empirical regime analyses.

Note the good correspondence between W_{obs}/W_{mod} obtained from this study, and W_{obs}/W_{unveg} from the regime analyses of Andrews and Hey and Thorne which indicates that W_{mod} assuming $\phi'=40°$ is approximately equal to W_{unveg}. This implies that the bank vegetation increases ϕ', which in turn increases the bank stability resulting in narrower channels.

For the second stage ϕ' was determined analytically for each of the channels in the published data sets. For each channel a range of ϕ' values were input until the modeled channel width agreed with the observed width. An example is shown in Table 2. Reach 13 of Hey and Thorne was selected as it displayed the largest deviation from the observed geometry when $\phi'=40°$ was used. Note that with $\phi'=40°$

not only is the modeled width over 3 times greater than the observed, but that the modeled channel is much shallower and steeper than the observed. When ϕ' is increased to 74.1° near perfect agreement is noted between the observed and modeled channels. It is noteworthy that 85 of the 86 channels listed in the two data sets could be brought into agreement by varying only the single parameter, ϕ'.

	Width (meters)	Depth (meters)	Slope	γQS (N/sec)
Observed	18.4	1.07	0.0133	6525
$\phi'=40°$	62.0	0.53	0.0170	8340
$\phi'=74.1°$	18.4	1.10	0.0133	6525

Table 2. Effect of ϕ' on channel geometry (Reach 13 from Hey and Thorne, 1986).

A summary of the values of ϕ' obtained analytically is presented in Table 3. Despite the large variation within each vegetation type, it is clear that the mean values of ϕ' increase systematically with increasing bank vegetation density.

Vegetation Type	Min	Mean (degrees)	Max
I	28.4	43.7	61.1
II	34.9	51.6	67.5
III	40.9	60.0	83.3
IV	46.9	66.3	90.0
Thin	28.4	43.1	58.0
Thick	47.0	60.2	73.3

Table 3. Variation of ϕ' with bank vegetation type.

Discussion

The adjustment of alluvial channels is presented here as a non-linear optimization problem. The ideal channel will tend to adopt a minimum stream power configuration subject to the imposed constraints. When one of the constraints is modified the channel will adjust to a new optimum configuration. This is evident from results of the modeling presented in Table 2. By varying a single bank stability parameter (ϕ') not only did the width adjust, as one would expect, but the channel depth, and even the channel slope show a large adjustment. Thus the dependent variables which describe the hydraulic geometry are shown to be strongly interdependent, and will dynamically adjust to changes in any of the independent variables which influence the constraints. The solutions corresponding to the two ϕ' values have significantly different values of γQS. In both cases γQS is a minimum for the given constraints.

This optimization approach is of value because it permits each of the independent variables to be perturbed in isolation, and the influence on the channel to be assessed. Consider the effect of vegetation removal on the observed channel in Table 2 (Reach 13, Hey and Thorne). If it had been determined that most of the increase of the value of ϕ' was due to the bank vegetation, the effect of vegetation removal could be assessed by reducing ϕ' to close to the angle of repose of the bank sediment ($\approx 40°$). In this example the effect on the channel geometry is considerable. Similarly the impact of other activities which affect one or more of the constraints can be examined.

The valley slope can exert a large influence on the optimum solution. Again consider the calculated geometries from Table 2. If the valley slope were 0.016, the channel with $\phi'=74.1°$ would develop a stable, single thread, meandering morphology with a sinuosity of 1.2. The slope required for a channel with $\phi'=40°$ is greater than the valley slope, and is therefore in violation of the valley slope constraint. There is no feasible solution, and an unstable, aggrading, braided channel would result.

Despite encouraging results, significant features are still not being addressed by the above formulation. A good example in gravel rivers is the role of the coarse pavement or armor layer which is a key feature in controlling the sediment transport rates. The inclusion of the pavement layer influence, as well as other features, are currently being investigated as part of the first author's Ph.D. research.

Appendix I. References

Andrews, E.D., 1984: Bed material entrainment and the hydraulic geometry of gravel-bed rivers in Colorado. *Geol. Soc. Am. Bull.,* 95, 371-378.

Chang, H.H., 1979: Minimum stream power and river channel patterns. *J. Hydrol.,* 41, 303-327.

Flintham, T.P., and Carling, P.A., 1988: The predictions of mean bed and wall boundary shear in uniformly and compositely roughened channels. In *International Conference on River Regime*, W.P. White (Ed.). John Wiley and Sons, 267-287.

Gilbert, G.K., 1914: The transport of debris by running water. *USGS. Prof. Paper.* 86. 263 p.

Hey, R.D., and Thorne, C.R., 1986: Stable channels with mobile gravel beds. *J. Hydr. Eng., ASCE,* 112 (8), 671-689.

Lane, E.W., 1955: The design of stable channels. *Trans. ASCE,* 120, 1234-1279.

Millar, R.G., and Quick, M.C., 1993: The effect of bank stability on the geometry of alluvial gravel rivers. Submitted for review for publication in the *J. Hydr. Eng.*

White, W.R., Bettess, R., and Paris, E., 1982: An analytical approach to river regime. *J. Hydr. Eng., ASCE,* 108 (10), 1179-1193.

Yang, C.T., 1976: Minimum unit stream power and fluvial hydraulics. *J. Hydr. Eng., ASCE,* 102 (7), 919-934.

Gully Intrusion on Reclaimed Disposal Sites

Scott A. Hogan[1], AM. ASCE, Christopher J. Pauley[2], SM. ASCE,
Steven R. Abt[3], M. ASCE, Terry L. Johnson[4]

Abstract

A three-phase gully erosion study was initiated at the Engineering Research Center of Colorado State University, in cooperation with the U.S. Nuclear Regulatory Commission. The objective of this study is to investigate the gullying processes and attempt to develop an empirical model that could aid in the design of earthen soil covers by means of predicting the gully incision potential on reclaimed disposal sites in the semi-arid western United States.

Statistical analysis of results from several simulated gully flume tests combined with data retrieved from natural occurring gullies at reclaimed mine sites, has yielded a preliminary empirical model, which characterizes the maximum gully incision as a nonlinear process. In addition the locations of maximum depth have been observed to occur predominantly near the crest of the slope.

---

[1] Research Associate, Department of Civil Engineering, Colorado State University, Fort Collins, Colorado, 80523.

[2] Graduate Research Assistant, Department of Civil Engineering, Colorado State University, Fort Collins, Colorado, 80523.

[3] Professor and Director, Hydraulics Laboratory, Department of Civil Engineering, Colorado State University, Fort Collins, Colorado, 80523.

[4] Senior Hydraulic Engineer, U.S. Nuclear Regulatory Commission, Washington, D.C., 20555.

Introduction

Increased environmental concern throughout the United States has renewed the importance of mined land reclamation and disposal of wastes associated with the excavation and processing of natural resources such as coal, uranium, and other minerals. The restoration of mined land and long term stabilization of waste disposal areas is of priority to protect public safety and conserve other natural resources, such as soil and water. Long term stabilization of waste requires the waste to remain undisturbed, usually for periods of two hundred to one thousand years.

Current stabilization methods for waste disposal require that the waste material be concentrated at or below the ground surface. An earthen cap, or cover, is then placed over the material with side slopes ranging from 1:3 to 1:10. Numerous materials, such as rip rap, synthetic fabrics, and artificial barriers, have been used to prevent erosion of the slopes. Where large areas of erosion protection are required, the costs encountered in using most of the available protection measures are extensive. Therefore, the earthen covers are generally vegetated and left unprotected. In unprotected conditions the safeguarded material is susceptible to exposure resulting from gully erosion.

In cooperation with the U. S. Nuclear Regulatory Commission, efforts are underway at Colorado State University to investigate the gully incision potential. A laboratory testing program and field data collection program are currently in progress. The investigation is being conducted in three phases; field data collection, outdoor flume testing, and indoor flume testing. To determine the maximum depth of gully incision the results from the three phases are being combined and analyzed to derive an empirical model that could be used to aid in the design of earthen soil covers for reclaimed disposal sites.

Phase I - Field Data Collection

The first phase of this investigation involved the characterization and measurement of eleven gullies at sites located across Colorado and Wyoming. The parameters determined to be most pertinent to the gully growth process included; maximum gully depth, gully width, slope, area tributary to the gully, percentage of vegetal cover, runoff estimation, soil densities, and soil characteristics. Profile, cross-section, and drainage area measurements were made with a theodolite, Philadelphia rod, and standard stadia reduction calculations. The soil samples collected were analyzed for composition, gradation, plasticity index, and compaction potential. Runoff at each site was estimated by gathering precipitation data from the nearest recording station to each site, and applying the SCS Runoff Curve Number method (Soil Conservation Service, 1986). The soil classification and curve number for each site were determined from county soil surveys. A partial summary of the field data collected is presented in Table 1. Data not reported in Table 1. and their respective ranges include; vegetation cover (5-60 percent), plasticity index (0-21.5 percent), median grain size, d_{50} (0.29-4.0 mm), uniformity coefficient (2.7-14.7), and soil density (785-1586 kg/m^3).

Table 1. Summary of Field Data

Gully No.	Age (yrs)	Drainage Area (m²)	Initial Slope	Pile Height (m)	Gully Length (m)	Max. Gully Depth (m)	Width at D_{max} (m)	Runoff Volume (m³)
1	12	3116	0.16	8.84	61.87	2.19	4.57	199.3
2	30	3764	0.38	22.56	102.41	4.11	5.03	458.3
3	12	182554	0.03	1.52	89.00	0.73	0.67	105424
4	5.5	1464	0.24	8.23	50.29	2.23	5.49	179.2
5	4.5	1485	0.10	1.83	30.78	0.82	2.90	174.2
6	4.5	366	0.13	3.05	18.29	0.37	0.91	49.0
7	8	17813	0.11	6.10	65.84	0.46	0.91	834.3
8	8	4753	0.22	7.62	57.91	0.82	1.07	222.6
9	5	60386	0.46	7.01	33.53	1.98	1.52	872.7
10	14	4312	0.45	53.34	131.06	2.65	3.20	1140.5
11	19	479	0.70	16.15	57.30	0.85	1.83	32.7

Phase II - Outdoor Laboratory Modeling

To study gullying processes under controlled conditions an outdoor laboratory modeling program was initiated. In order to correlate the effects of soil materials, embankment geometries and hydrologic events with gully growth, a rainfall/runoff facility was constructed. The facility is capable of simulating hydrologic activity on a modeled disposal embankment.

The outdoor facility is housed in a 30.5 meter long, 6.1 meter wide and 2.4 meter deep concrete flume. The flume was equipped with a rainfall/runoff simulator and the appropriate measurement instrumentation. The facility is capable of simulating precipitation hyetographs with intensities as high as 250 mm per hour, and additional surface runoff from a header box up to 0.05 m³/s. Point gage instrumentation was installed to contour the embankment surface at periodic intervals during each test. The embankments were constructed 2.4 meters high and 6.1 meters wide. The 9.1 meter long top of the pile was sloped at 2 percent to insure runoff moved to the middle of the crest. At the crest, the embankment transitioned to slopes of 10-20 percent. The following soil parameters and ranges are characteristic of the soils tested; plasticity index (0-12.5 percent), uniformity coefficient (10-200), median grain size d_{50} (0.002-8.0 mm). Table 2. summarizes the embankment characteristics and data from the five outdoor flume tests completed to date.

Table 2. Summary of Outdoor Modeling Data

Test No.	Soil Type (USC)	Initial Slope	Compaction Standard Proctor (%)	Max. Gully Depth (m)	Width at D_{max} (m)	Runoff Volume (m^3)
O1	CL	0.20	86	0.18	0.55	2927.5
O2	CL	0.20	79	1.10	2.15	3066.8
O3	SM	0.19	80	1.26	2.88	2618.1
O4	SM	0.19	90	1.07	0.80	3313.8
O5	SM	0.12	80	1.05	2.65	3313.8

Phase III - Indoor Laboratory Modeling

An indoor flume was utilized to allow small scale testing during the winter months. Indoor test modifications and embankment constructions require significantly less effort than those of the outdoor flume. The steep flume is 9.8 meters in length, 1.2 meters wide, and 0.6 meters deep. Only runoff from a simulated tributary area was incorporated into the testing. The embankments were constructed 0.6 meters high and 1.2 meters wide. The 2.3 meter long top of the pile was sloped at 2 percent to insure runoff moved to the middle of the crest. At the crest, the embankment transitioned to slopes of 10-20 percent. The parameters and ranges characteristic of the outdoor phase soils are also characteristic of the indoor phase soils. Table 3. summarizes the embankment characteristics and data from the six indoor flume tests completed to date. The I4 test was interrupted and not reported.

Table 3. Summary of Indoor Modeling Data

Test No.	Soil Type (USC)	Initial Slope	Compaction Standard Proctor (%)	Max. Gully Depth (m)	Width at D_{max} (m)	Runoff Volume (m^3)
I1	CL	0.22	76	0.10	0.16	48.8
I2	CL	0.23	74	0.09	0.24	153.3
I3	SM	0.25	84	0.51	0.27	153.3
I5	SM	0.20	90	0.49	0.20	165.7
I6	SM	0.13	90	0.47	0.12	75.9
I7	SM	0.13	80	0.40	0.12	45.9

Results

The data recorded to date from all three phases was compiled into a single data base for analysis. Previous researchers, Beer & Johnson (1963), Thompson (1964) and Seginer (1966), have utilized linear and logarithmic regression techniques to analyze various aspects of gully growth. While these studies indentified prominent variables and processes, they were unable to fully represent the non-linearity of gully development. An empirical nonlinear equation based on a general exponential growth model is proposed to predict the deepest point of a gully on a reclaimed slope. The model has a general form of: $y = \beta - 1/(1+\alpha e^{\phi x})$ where y is the cumulative effect at a time x and β represents the maximum growth value. α and ϕ are constants determined by regression. The second term is a S-curve growth function, which exhibits slow growth initially, an accelerated rate in the middle with a gradual decay in the later stages of life. Secondary parameters, such as soil characteristics, were incorporated into the general model to allow for their effect on the growth rate. With a limited data set of only 22 individual gullies, results have proved promising. Regression coefficients in excess of 0.85 were computed for several variations of the base model. Equation (1) presents a preliminary model that predicts the depth of maximum gully incision.

$$D_{max} = 1.4H^{0.2} - \frac{1}{1+1.7e^{-4.3R}} + 2.2(CU^{-2.5} LL^{1.1} S^{1.6} C^{0.3}) \qquad (1)$$

where: D_{max} = maximum gully depth (m)
 H = total height of embankment or pile (m)
 R = total runoff (m^3)
 CU = soil uniformity coefficient
 LL = soil liquid limit (% moisture content)
 S = slope of embankment
 C = compaction of embankment (% standard proctor)

Figure 1. displays the predicted versus the observed depths, for the model in Equation 1. The correlation coefficient for the model is $R^2 = 0.87$. Preliminary findings indicate that the deepest point occurs 75 percent of the time within 5 Dmax's upslope to 15 Dmax's downslope of the embankment crest. The overall confidence in the developed equation is limited due to the small size of the data set. Research is continuing to enlarge the data set and refine the preliminary equations.

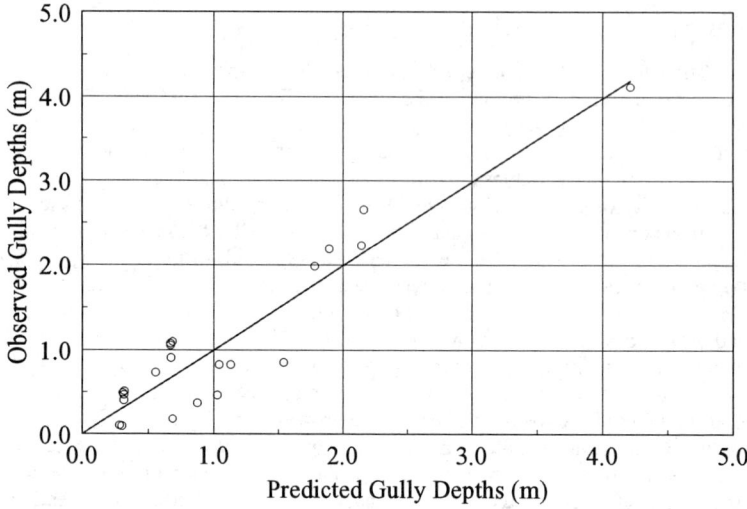

Figure 1. Observed versus Predicted Gully Depths

Summary

The major objective of this study is to define an empirical relationship to predict the maximum depth that a gully on reclaimed tailings can reach under conditions found in the semi-arid western United States. Efforts to date have resulted in a preliminary empirical equation based on twenty-two observed gullies. The location of the maximum depths has occurred most often near the slope crest. The ultimate findings will assist in the safe and economical design of earthen covers for mine tailings.

References

Beer, C.E. and Johnson, H.P., 1963. 'Factors in gully growth in the deep loess area of western Iowa', Transactions of the ASAE., Vol. 6, No. 1, pp. 237-240.

Seginer, I., 1966. 'Gully Development and Sediment Yield', Journal Hydrology (Amsterdam), Vol. 4, pp. 236-253.

Soil Conservation Service, 1986. 'Urban Hydrology for Small Watersheds', United States Department of Agriculture, Engineering Division, Technical Release 55, 2nd ed.

Thompson, J.R., 1964. 'Quantitative effects of watershed variables on rate of gully-head advancement', Transactions of the ASAE, Vol 7 No. 1, pp. 54-55.

SEDIMENT DEPOSITION IN JENNINGS RANDOLPH RESERVOIR, MARYLAND AND WEST VIRGINIA

By Margaret M. Burns[1] and Robert C. MacArthur[2], M. ASCE

ABSTRACT: The watershed of the Jennings Randolph Reservoir covers 263 square miles of mountainous terrain in western Maryland and West Virginia. Sedimentation studies performed prior to impoundment predicted a long-term average sedimentation rate of 20 acre-feet per year, or 0.08 acre-feet per square mile per year. After three years of operation, large volumes of sediment were observed in the headwaters of the reservoir. Field measurements showed that sediment was accumulating at a rate of three to four times that anticipated. Additional field measurements after the flood of record in 1985 showed that this one event had contributed approximately 600 acre-feet of sediment, or approximately 30 years of predicted sediment inflow. This study was initiated in response to the observed high rates of sedimentation in the reservoir, with the goal of predicting the average annual sediment yield to the reservoir. Several different sediment prediction techniques were used in combination to come up with a range of predicted annual average sediment yields of 0.23 to 0.50 acre-feet per square mile per year, or from three to five times the original estimate.

Introduction

The Jennings Randolph Reservoir is located in the Appalachian highlands on the North Branch of the Potomac River (NBPR) on the state line between western Maryland and northeastern West Virginia. Portions of the watershed lie in Garrett County, Maryland and Grant and Mineral Counties, West Virginia. The watershed above the reservoir drains an area of approximately 263 square miles. The main channel is relatively steep with a representative channel slope of approximately 0.3 to 0.8 percent. The hills and valleys bordering the NBPR are steep, narrow and heavily wooded. Much of the lands around the project area have been mined for coal, with some clear-cutting for hardwood products also occurring in the basin. Generally the water draining from the hillsides is heavily polluted

---

[1]Hydraulic Engineer, Hydrology-Hydraulics Section, Baltimore District, U.S. Army Corps of Engineers, P.O. Box 1715, Baltimore, MD 21203
[2]Principal, Resource Consultants and Engineers, Inc., 1477 Drew Avenue, Suite 107, Davis, CA, 95617

with acid mine wastes, thus affecting the water quality of the river and reservoir. Over 60 percent of the NBPR basin is covered by forest. Farming in the project area is limited due to the steep terrain and poor soils.

Based on several sediment studies conducted prior to impoundment (as well as extensive measured data at Kitzmiller and other locations in the Potomac River Basin), the annual sediment yield to Jennings Randolph Reservoir was estimated to be approximately 20 acre-feet per year. 2065 acre-feet of reservoir storage was allocated to sediment for the 100-year life of the reservoir. The dam was completed in May 1981, and the conservation pool was filled in May 1982. A large amount of deposited sediment was noticed in the headwaters of the reservoir in the fall of 1984, after the pool was drawn down. Reconnaissance sediment surveys performed by District personnel in November 1984 and January 1986 (after the flood of record, Tropical Storm Juan) yielded computed depositions of 270 and 900 acre-feet respectively (total deposition since impoundment). Since the measured sediment inflow thus far seems to greatly exceed the initial estimate, there has been a continued interest in accurately assessing the amount of sediment deposition and, if necessary, computing a revised annual average sediment yield to the reservoir.

Sediment Transport Mechanisms, Sources and Sinks

From field inspections and subsequent review of available reports and documents, it can be concluded that the amount of sediment in the streams in the basin varies according to its source, the local streamflow characteristics, the antecedent watershed conditions and the seasons of the year. Streams draining the heavily forested lands in the headwater areas generally have low sediment production and delivery to the main river channel, except during a severe event. Normal (during low or medium flows) basin sediment yield is primarily limited to suspended load and wash load materials that enter the NBPR from surface erosion, road cuts, rilling and gullying processes. Sediment loading to the main channel is supply-limited during low and medium flow periods and the river is capable of carrying the fine sands and silts supplied to it by naturally occurring and man-induced activities throughout the basin. Areas where agriculture, active mining, clear cutting and heavy industry occur have higher sediment loads. The availability of transportable material is a major factor in explaining the variations of sediment discharge within the basin. Sediment delivery to the main channel is also affected by sediment trapping and local storage that occurs along the way to the main river. The sediment delivery ratio for the NBPR has been estimated by the Soil Conservation Service to be approximately 6 to 8 percent for average annual sediment production. Therefore, far more sediment materials are trapped and go into storage throughout the system (approx. 92-94%) than go through the system on an average annual basis. Steep pool-riffle and boulder-step channels (typical of the tributaries) are capable of storing and accumulating sediment materials over many years during periods of relatively low to medium runoff. Once the critical threshold of the system is exceeded during a large storm event, large volumes of colluvial and alluvial materials can flush from storage into the high transport capacity main channel. Fine to medium sized sediments that enter the main channel are easily transported through the system to the reservoir during high flows where the main channel velocities average from 10 to 17 feet per second.

Comparison of Sediment Data from Various Sources

The average annual sediment yield to Jennings Randolph Reservoir was determined after review of many methods and published reports. A total of ten different references were compared. Of these, six references were regional studies predicting sediment loads for either the Potomac River basin or the Appalachian region (items 1 through 6 in Table 1). Measured reservoir deposition data for both large and small drainage basins in the nearby area was included as items 7 and 8 in Table 1. The Baltimore District's computation of sediment deposition in Jennings Randolph Reservoir based on a comparison of the 1991 hydrographic surveys and the pre-impoundment aerial topographic mapping is included as item 9 in Table 1. A computation of sediment yield using the Pacific-Southwest Inter-Agency Committee (PSIAC) method is included as item 10 in Table 1. Figure 1 and Table 1 summarize estimated and published annual sediment yields from these numerous sources. Note the wide range of scatter and the width of the confidence band (approximately one log cycle). This is typical of these types of basins and is a direct result of the episodic processes described above.

In general, the information from the regional studies (items 1 through 6 in Table 1) agreed very closely with the Corps design sediment yield of 0.076 acre-feet per square mile. These regional studies were published during the 1960s. The Corps design value was based on suspended sediment measurements on the Potomac River at Kitzmiller, MD during 1961-1962. All these data were collected for river discharges less than 5000 cfs, which corresponds to a recurrence interval of approximately one year. Therefore, extrapolation of a sediment rating curve based on this data to higher events may not provide an accurate indication of sediment production and delivery processes in the basin during less frequent runoff events.

Regional published yields were generally in the low range (0.05 to 0.09 acre-feet per year per square mile). Measured single event sediment accumulations were generally much higher, ranging from 0.49 to 1.05 acre-feet per square mile. The PSIAC computation (item 10 in Table 1) produces estimated yields of approximately 0.33 acre-feet per square mile. According to the literature, the PSIAC method produces reasonable yield estimates for drainage basins of the size and character of the Jennings Randolph watershed. The PSIAC method is based on nine physically-based parameters which depict watershed characteristics.

A best fit through the data as shown in Figure 1 indicates that the approximate average annual sediment yield at the reservoir is 0.5 ac-ft/sq mi/yr. During very wet years with high runoff, the yield can be as high as 1.75 ac-ft/sq mi/yr. Conversely, during dry low flow periods the yield can reduce to 0.075 ac-ft/sq mi/yr or less. This variation is common. A basin will not have a constant sediment production or delivery year after year. The actual yield to the Jennings Randolph Reservoir is a function of event sequencing, basin characteristics,

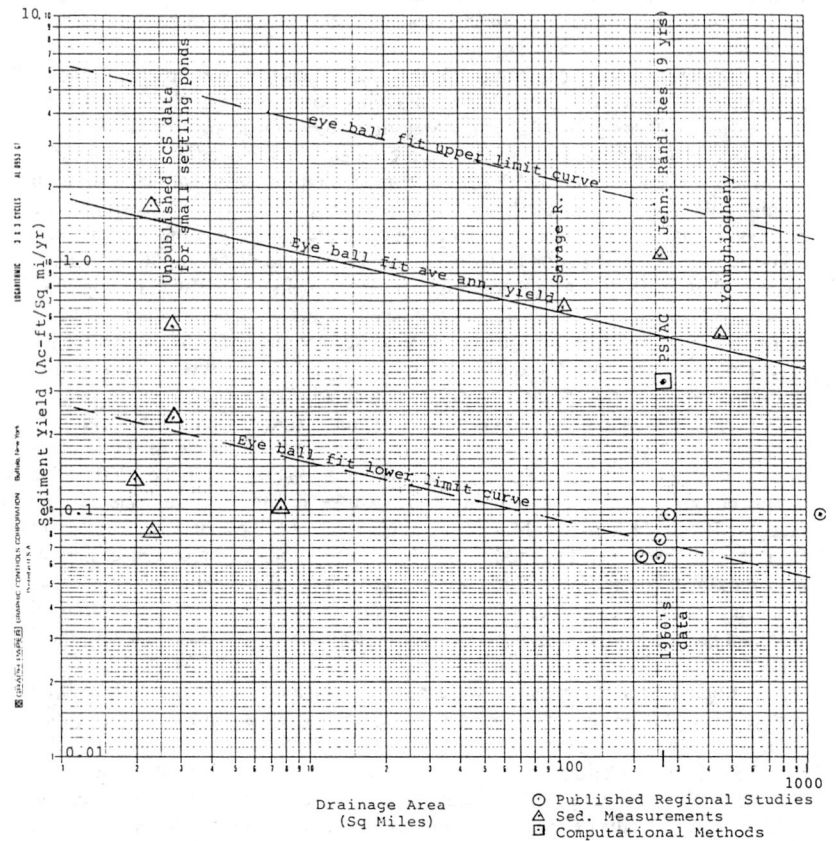

Sediment Yield Versus Drainage Area From Various Sources

Figure 1

Table 1. Summary of Methods Used and Yields.

No.	Reference	Drainage Area (sq.mi.)	Average Annual Sediment Yield (ac-ft/sq mi)	(tons/sq mi)
1.	Reconn. of Sed.& Chem Quality of Surface Water in the Potomac R. Basin USGS 1961 (p.47)	287	0.09	150
2.	Appendix Q, Erosion and Sed., North Atlantic Regional Water Res. Study Coord. Comm. (Table Q-8)	263	0.07	118
3.	Prelim. Study of Sed. Sources and Trans. in the Potomac R. Basin, Interstate Comm. on the Potomac River Basin, 1963	225	0.06	94
4.	Water Resources in the Appalachian Region, Pennsylvania to Alabama, Atlas HA-198, USGS, 1965	263	0.06	95
5.	Preliminary Appraisal of Stream Sedimentation in the Susquehanna River Basin, USGS, 1968	263	0.06	90
6.	Geomorphology, by Chorley et al, Methuen, 1984	1500 ave.	0.09	141
7.	Sediment Deposition in U.S. Reservoirs, ARS Misc. Pub. 1362, Feb. 1978 a. Savage River b. Youghiogheny River	105 428	0.64 0.49	840 516
8.	Small Reservoir Surveys, West Virginia SCS, 1985 (unpublished data)	2.3 2.8 2.8 7.7 2.3	1.68 0.55 0.23 0.10 0.08	2379 779 326 142 113
9.	Jennings Randolph Reservoir, Sediment Survey, COE, 1991	263	1.05	1714
10.	PSIAC (computed yield for the Jennings Randolph watershed)	263	0.33	539

sediment availability, transport capacity, and the specific magnitude and duration of single event storms.

Annual Sediment Yield to Jennings Randolph Reservoir

Depending on the analytical approach and the interpretation of the data, the annual sediment yield to Jennings Randolph Reservoir ranges from a low of 0.06 to a high of 1.75 acre-feet per square mile per year. Based on this study, a recommended long-term range is 0.23 to 0.50 acre-feet per square mile per year. A value of 0.35 acre-feet per square mile per year is a reasonable figure for planning purposes, and corresponds to an annual sediment yield to the reservoir of 92 acre-feet per year. This is much higher than the pre-impoundment estimate of 20 acre-feet per year.

As part of this study, it was verified that the original estimate of 20 acre-feet per year was based on valid data, was computed using accepted methods, and was in concurrence with numerous other studies performed for this basin and similar areas. However, the sediment transport in this mountainous watershed tends to be dominated by extreme events (flood flows). During normal and low flows, the streams in the watershed are supply-limited and carry low sediment loads. Flood flows, however, mobilize large volumes of stored sediment in both the main channel and its tributaries, and dislodge the upper layer of streambed armoring material to expose finer material underneath. In addition to the materials in the main stem of the NBPR becoming mobilized, another, perhaps more significant source of sediment materials comes from tributary flushing during high flows. Above a critical discharge the tributaries will flush (unload) significant volumes of materials that have been accumulating over many years of lesser flows. The lesser flows only have sufficient energy to get the materials to the tributaries where they are trapped until a large enough event can flush them through to the main channel. This is a natural process of storage and release that occurs in steep gradient, coarse bed and/or bed rock controlled systems. Watersheds such as these are high producers of sediment for very short periods of time (during major events), and low producers of sediment for long periods of time during normal flows. A single extreme event may produce one to two orders of magnitude more sediment than a typical two-year event. Because the nature of the sediment transport process changes after a certain threshold is reached, a sediment-discharge rating curve based on data up to the two-year event (as used in the pre-impoundment studies) cannot be accurately extrapolated to reflect the sediment discharge for major flood events. The traditional and accepted methods for computing sediment yield may greatly underestimate the sediment yield of a basin dominated by episodic and extreme events. This conclusion may be applicable to other reservoirs in similar terrain.

Laboratory Study of the Characteristics of Shallow Open Channel Flow using Fiber-optic Laser Doppler Velocimetry

Jau-Yau Lu[1] and Li-Chuan Chen[2]

Abstract

An experimental study was proposed to investigate the characteristics of shallow rain-impacted flow using a recirculating flume and an artificial rainfall simulator. The Fiber-optic Laser Doppler Velocimetry (FLDV) technique was used to measure both the mean velocity profiles and the distributions of longitudinal and vertical turbulence intensities for very shallow flow conditions. The results were compared with other experimental studies.

Introduction

The mechanics of soil erosion is related to the flow parameters of shallow rain-impacted flow. An understanding of the effect of rainfall on these parameters is needed as a basic step to quantitatively predict the erosion rates. The FLDV system has the following unique advantages: no contact, no calibration, high spatial and temporal resolution, and multi-component measurements.

---

[1] Professor, Department of Civil Engineering, National Chung-Hsing University, Taichung, Taiwan, R.O.C.
[2] Graduate Research Assistant, Department of Civil Engineering, National Chung-Hsing University, Taichung, Taiwan, R.O.C.

A 3-years laboratory study was initiated to investigate the flow characteristics of shallow rain-impacted flow. This paper describes part of the results for the first phase of the study, which includes the experimental study without rain, the construction of a rainfall simulator, and the measurements of raindrop characteristics.

Experimental Equipment and Procedure

The recirculating flume used in this study consisted of a 0.5 m wide, 0.5 m high, and 12 m long flume with 10 mm thick plate glass bottom and side walls. The slope of the flume can be adjusted up to 10%. The flume were very carefully designed and constructed to minimize possible disturbance in the shallow flow.

The velocity measurements were performed using a two-component (two-color, four-beam) TSI-made Fiber-optic Laser Doppler Velocimeter (FLDV) operated in the backscatter differential mode. The FLDV system consisted of six components: a 2W Argon-Ion Laser light source, colorlink (multicolor receiver), IFA-550 signal processors, and a personal computer with a system control software "FIND". Figure. 1 is a schematic diagram of the FLDV system. A specially designed 3-D traverse system were used to move the fiber-optic probe with an accuracy of 0.001 mm.

Based on the preliminary tests, titanium dioxide particles were used as seeding particles to improve the data quality. The mean size and specific gravity of titanium dioxide were 1 μm and 4.28, respectively. In addition, the frequency shift, data rate, sampling frequency, sample size and sampling time chose in the experiments were 200 KHz, 300 Hz, 50 Hz, 4096 and 82 sec, respectively. The measuring section and measuring vertical were selected with consideration of the fully development of the flow, the minimum side-wall effect, and the constraint of the optic system for shallow flow conditions.

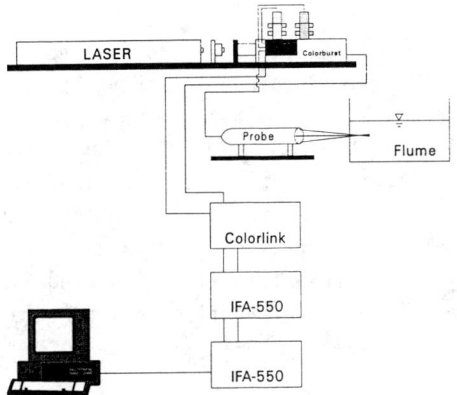

Figure 1 Schematic Diagram of FLDV System

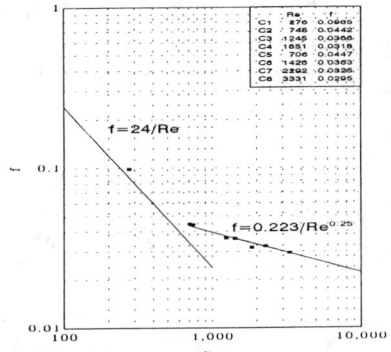

Figure 2 Variation of Darcy-Weisbach Friction versus Reynolds Number Factor

Case	S(%)	H(mm)	T(°C)	Q(cms)	U_m(m/s)	U_*(m/s)	Re	F
C1	0.1	4	30	0.000112	0.056	0.00622	276	0.28
C2	0.1	6	29	0.000305	0.102	0.00758	745	0.42
C3	0.1	8	30	0.000514	0.129	0.00872	1245	0.46
C4	0.1	10	30	0.000770	0.154	0.00971	1851	0.49
C5	0.3	4	30	0.000287	0.144	0.01076	706	0.72
C6	0.3	6	32	0.000584	0.195	0.01313	1426	0.80
C7	0.3	8	30	0.000946	0.237	0.01510	2292	0.84
C8	0.3	10	29	0.001386	0.277	0.01682	3331	0.88

Table 1 Experimental Conditions

Results

Table 1 shows the experimental conditions and the related flow parameters of the laboratory study, in which S=channel slope, H=flow depth, T=flow temperature, Q=flow discharge, Um= mean flow velocity, U_*=shear velocity, Re=Reynolds number, F= Froude number. Figure 2 shows the relationship between the Darcy-Weisbach friction factor and Reynolds number. Figure 2 also implies that Run C1 was in the laminar flow region, Runs C2 and C5 were in the transitional region, and Runs C3, C4, C6, C7, and C8 were in the turbulent flow region and agreed with the Blasius formula very well.

Both the values of von Karman constant k and integral constant A for the "log law" were evaluated with regression analysis. Figure 3 shows the relationship between von Karman constant k and Reynolds number for our experimental data. The mean and standard deviation of the von Karman constant k were 0.414 and 0.014, respectively. Figure 4 is a plot of integral constant A versus Reynolds number for our data and Tominaga & Nezu's (1992) data. The values of integral constant A decreased consistently as Reynolds number decreased for the shallow flow conditions. The mean velocity data fitted both the velocity-

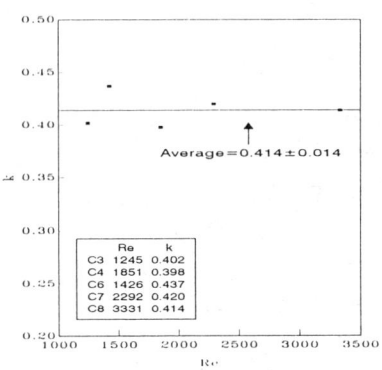 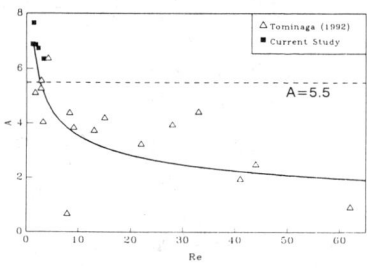

Figure 3 Variation of von Karman Constant Reynolds Number versus

Figure 4 Variation of Integral Constant Reynolds Number versus

defect law and one-seventh-power law reasonably well for the outer region, especially for the experimental runs with high Reynolds numbers, i.e. Runs C3, C4, C6, C7 and C8.

Figure 5 and 6 are the distributions of the normalized turbuleuce intensities for both the longitudinal (u'/U_*) and vertical directions (v'/U_*), where u' and v' are the root-mean-square values of the fluctuating components of the flow velocity. Data collected by McQuivey and Richardson (1969), and Nezu and Rodi (1986) are also included for comparizon. In gengeral, there was a peak for u'/U_* near the outer edge of the viscous sublayer for our data. Three data sets with subcritical flow were selected randomly for Nezu and Rodi's (1986) data (including high, medium and low Reynolds numbers). The results were reasonably close for different studies except those near the boundary. McQuivey & Richardson (1969) did not collect collect enough data near the boundary, which probably was the reason why their curve did not have a peak near the boundary.

Summary

This paper describes the results of a laboratory study conducted for the investigation of the characteristics of very

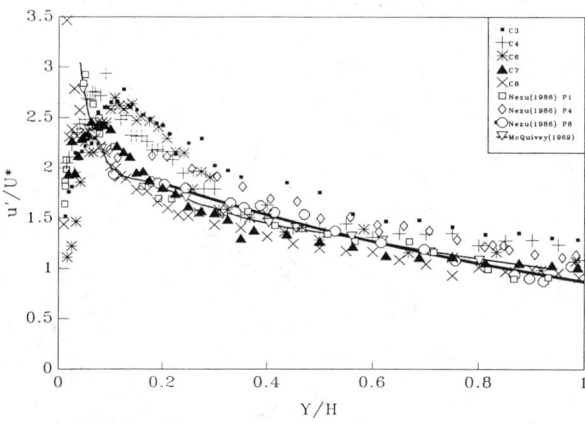

Figure 5 Comparison of Normalized Turbulence Itensity in Longitudinal Direction

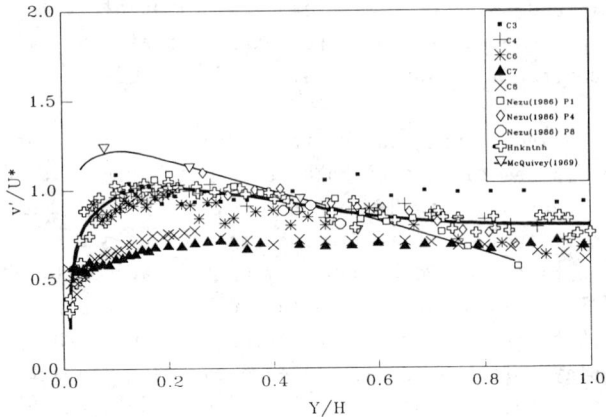

Figure 6 Comparison of Normalized Turbulence Intensity in Vertical Direction

shallow (4~10 mm) open channel flow using a 2-D FLDV system. Both the mean flow and the turbulence intensity were measured and compared with related theories and data. The von Karman constant k in the log law was nearly a constant of 0.41. The integral constant A in the log law, however, decreased consistently as Reynolds number decreased for the shallow flow conditions. The distributions of the turbulence intensity were reasonably close for different studies, though deviations did exist for the data near the boundary. A 1 m × 4 m oscillating type rainfall simulator has been constructed. This study will be extended to the cases with rain in the near future.

Appendix I -References

McQuivey, R. S., and Richardson, E. V. 1969. "Some Turbulence Measurements in Open Channel Flow."J. Hydr. Div., ASCE, 95(1), P.209-223.

Nezu, I., and Rodi, W. 1986. "Open-channel Flow Measurements with a Laser Doppler Anemometer." J. Hydr. Engrg., ASCE, 112(5), P.335-355.

Tominaga, A., and Nezu, I. 1992. "Velocity Profiles in Steep Open-channel Flows." J. Hydr. Engrg., ASCE, 118(1), P.73-90.

DISTORTED PHYSICAL MODELS FOR MIXING STUDIES
Jianlu Xu[1] and Hsieh Wen Shen[2] M. ASCE

Abstract: The frictional resistance and mixing process affected by roughness bars are investigated. Specific combinations of the size and layout of the bars for various flow conditions are suggested through analyzing the scaling of friction factor, transverse mixing coefficient and longitudinal dispersion coefficient. The result agrees reasonably well with the San Francisco Bay-Delta Model.

Introduction
Most physical hydraulic models are constructed with distorted scales due to space shortage. The impacts of such distortion on the simulation of flow and dispersion have received close scrutiny by many investigators and still remained a serious concern.

Fisher (1971) analyzed various factors controlling the dispersion in prototype and in a distorted model without roughness bars and concluded that this type of model was unlikely to imitate the dispersion. Through a comparison of data taken in a distorted model also without the bars and data from the prototype for a power station site near Heysham, England, Crickmore (1972) showed that the model exaggerated the rate of transverse turbulent mixing and reduced the rate of vertical turbulent mixing. However, most distorted models are implemented with roughness bars which dominate the frictional resistance. In such flow, the rate of transverse mixing depended mainly on the size and layout of the bars, as shown by Fisher(1975) in an experiment.

This paper discusses the possible size and layout of roughness bars such that a model could meet the dynamic similarity for both flow and mixing process.

Review of similarity laws for a distorted model
The basic requirement for a physical model is to satisfy dynamic similarity, that is, the driving mechanisms of water motion and mixing in model are similar to those in prototype. Therefore, the velocity scale, time scale, etc. must be related to horizontal and vertical scales by similarity laws. Hudson (1979) analyzed the similarity laws for a distorted model. The velocity scale, N_u, and

---

[1] Graduate student, Dept. of Civil Eng., UC. Berkeley
[2] Professor, Dept. of Civil Eng., UC. Berkeley

Darcy-Weibach's friction factor scale, N_f, are expressed by:

$$N_u = N_h^{1/2} \tag{1}$$

$$N_f = \frac{N_h}{N_x} = D_t \tag{2}$$

and the scale of horizontal mixing coefficient, N_e, is:

$$N_e = N_h^{1/2} * N_x = D_t^{1/2} * N_x^{3/2} \tag{3}$$

in which N_x is the horizontal scale, N_h the vertical scale, and D_t the distortion ratio. In deriving above formulae, the acceleration of gravity, fluid density and salinity concentration are assumed identical in model and prototype, as they are in common practice.

Eqs.(2) and (3) indicate that both frictional resistance and horizontal mixing must be augmented with the distortion ratio. In a model of a distortion ratio, say 10, the friction factor must be 10 times as large as the value in prototype. Such effect is rather difficult to achieve through only roughening the model bed. Hence, roughness bars are generally used in severely distorted models.

Frictional resistance from roughness bars

The force transmitted to the flow from a single bar may be approximated by

$$F = \frac{1}{2}\rho C_d A U^2 \tag{4}$$

in which ρ is the water density, C_d the drag coefficient, U the cross-sectional mean velocity, and A the projecting area of the bar on a plane normal to the flow direction, equal to $a \cdot h \cdot \cos(\alpha)$, where a is the width of the bar, h the water depth, and α the angle between flow direction and the shortest axis of the bar. The drag coefficient depends on the local flow conditions and the shape of the bar, with an approximate magnitude 2 for typical flow conditions in model (the Reynolds number around 10^4). The transmitted force is assumed uniformly distributed over an "effective" area, as the dotted zone shown in Fig.1.

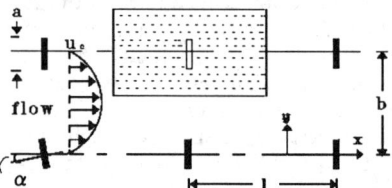

Fig.1 Sketch of the layout of roughness bars

In this manner, a nominal stress can be obtained as:

$$\tau_m = \frac{F}{b*l} = \rho \frac{a*h}{b*l} U^2 * \cos\alpha \tag{5}$$

Where b and l are the distances at which bars are separated laterally and in the direction of flow, respectively. To compare the nominal stress with the bottom shear stress, Eq.(5) may be rewritten in Darcy-Weisbach's form, and this gives a nominal friction factor f_b:

$$f_b = 8 * \frac{a*h}{b*l} * \cos\alpha \tag{6}$$

Vertical and transverse turbulent mixing

Vertical diffusion in model is affected by the bars only to a minor extent since overall vertical structure of flow is little changed. Moreover, the vertical mixing is insignificant compared to transverse mixing for wide and shallow rivers as far as large scale dispersion is concerned. Thus, no further consideration about the vertical mixing will be given.

On the contrary, the velocity distribution in transverse direction is modified completely and quite different from that in prototype, so is the transverse turbulent structure; consequently, the mechanisms of the transverse mixing in model would differ significantly from those in prototype.

Based on the assumptions that transverse mixing coefficient for mass is the same as that for momentum, that roughness strips provide all the resistance, and that velocity is uniform over depth, Fisher (1975) obtained an analytical expression for transverse mixing coefficient as

$$\epsilon_t = \frac{C_d a b U}{24 l} * (\frac{U}{U - u_c}) \tag{7}$$

in which u_c is the velocity along the axis of symmetry through the bars. For this study, the approximations of $C_d = 2$ and $u_c = 0$ are adequate. To check the analytical result, Fisher (1975) conducted an experiment in a flume with bars equi-distantly arranged, and found $\epsilon_t = (0.07-0.14)aU$. When b=1, Eq.(7) reduces to $0.083 \cdot a \cdot U$. In reality u_c will have some magnitude and vary with x between the bars, so the proportionality would be a little higher than 0.08. This experiment well verified the analytical result when b=1.

The above analyses assumed $\alpha = 0$. It is not always the case in reality. The transverse mixing coefficient for a general situation can be extended from Eq.(7) as follows:

$$\epsilon_t = \frac{b a U \cos\alpha}{12 l} \tag{8}$$

This is valid only for small α, since the effects of roughness bars decrease as α increases, and eventually become negligible when α approaches 90°, which would invalidate the assumptions used in deriving Eq.(7).

Longitudinal dispersion

As in natural rivers, the longitudinal dispersion in model is mainly due to the non-uniform lateral velocity distribution. Under this circumstance, the longitudinal dispersion coefficient may be defined as

$$K = -\frac{1}{A} \int_0^b h u' \int_0^y \frac{1}{h \epsilon_t} \int_0^y h u' dy dy dy \tag{9}$$

where A is the area of the cross section, and u' the velocity deviation from cross-sectional mean velocity. Based on the same assumptions for deriving the transverse mixing coefficient, the lateral velocity distribution is

$$u = 6*U*(1-\frac{y}{b})\frac{y}{b} \tag{10}$$

and the velocity deviation is:

$$u' = 6*U*(\frac{y}{b}-\frac{y^2}{b^2}-\frac{1}{6}) \tag{11}$$

Substituting u' in Eq.(9) and taking integration give:

$$K = \frac{U^2 b^2}{210 \, \epsilon_t} \tag{12}$$

In deriving this formula, the transverse mixing coefficient and water depth are assumed constant over the whole cross section.

The size and layout of the bars

The above analyses show that the resistance and mixing provided by bars depend on the size and layout of the bars. For dynamic similarities, the scales of friction factor and dispersion coefficient must equal the respective values from Eqs.(2) and (3). This gives some requirements on the size and layout of the bars.

The total friction comes from the bottom friction and the resistance transmitted from the bars. If the friction factor of bottom shear stress is f_m and the friction factor in prototype is f_p, the resultant friction factor in model, according to Eq.(2), should equal:

$$f_b + 8*\frac{a*h}{b*l}\cos\alpha = D_t f_p \tag{13}$$

In the model the Reynolds number will be much less than it is in the prototype, but the bottom surface roughness will also be much less. Hence to a reasonable approximation the friction factors in model and prototype will be the same. This leads to:

$$8*\frac{ah}{bl}\cos\alpha = (D_t - 1)f_p \tag{14}$$

To calculate the scale of transverse mixing coefficient, its value in prototype needs to be estimated. An approximate estimate is suggested by Fisher (1979):

$$\epsilon_{t,p} = 0.60 h_p u_{*p} \tag{15}$$

in which u_{p*} is the frictional velocity, h_p the water depth, with the subscript p denoting the prototype. Ratio of Eq.(8) over Eq.(15) gives the scale of transverse mixing coefficient in model:

$$N_\epsilon = \frac{\epsilon_{t,m}}{\epsilon_{t,p}} = \frac{ba \, U_m \cos\alpha}{12l \, 0.6 \, h_p \, u_{p*}} \approx 0.14 \cos\alpha \frac{U_m}{U_{p*}} * \frac{a}{h_p} * \frac{b}{l} \tag{16}$$

Assuming $u_{p*} = 0.1 \, U_p$, U_p being the cross sectional mean velocity in prototype, and recognizing that $U_p = U_m/N_h^{1/2}$ and that $h_p = h_m/N_h$, the above expression can be rewritten as

$$N_e = 1.4 N_h^{3/2} \frac{a}{h} \frac{b}{l} \cos\alpha \tag{17}$$

Let this scale equal Eq.(3), it gives:

$$\frac{a}{h_m} \frac{b}{l} \cos\alpha = \frac{1}{1.4 D_t} \tag{18}$$

Similarly, the longitudinal dispersion coefficient in prototype should also be estimated, and it may be formulated by (Fisher, 1979):

$$K_p = \frac{IU_p^{\prime 2} L_p^2}{\epsilon_{t,p}} \tag{19}$$

in which L_p is the characteristic length of transverse mixing, usually 0.5 W_p, half the river width, I a dimensionless coefficient measuring the irregularity of velocity distribution over the cross section, and $u_p^{\prime 2}$ the mean square of the velocity deviation. If using $I=0.07$, $u_p^{\prime 2}=0.2 U_p^2$ (Fisher, 1979, p.136), and comparing Eq.(12) to Eq.(19), the scale of the longitudinal dispersion coefficient is:

$$N_k = \frac{U_m^2 b_m^2}{0.003 * U_p^2 W_p^2} \frac{\epsilon_{t,p}}{210 \epsilon_{t,m}} = 1.6 \, (N_h)^{\frac{1}{2}} N_x \, (\frac{W_m}{b})^2 \tag{20}$$

in which W_m is the width of a river in model. Again let the scale equal Eq.(3), it yields

$$(W_m / b)^2 = 1.6 \tag{21}$$

This implies that only one bar is needed for each cross section from the viewpoint of longitudinal dispersion and that the bar should not locate in the middle of the river. Solving from Eq.(14) and Eq.(18), two ratios are obtained:

$$\frac{a}{l} = (\frac{(D_t-1) f_p}{11.2 D_t \cos^2\alpha})^{1/2} \tag{22}$$

$$\frac{b}{h} = (\frac{5.7}{(D_t-1) D_t f_p})^{1/2} \tag{23}$$

The above equations together with Eq.(21) govern the size and layout of the bars. b could satisfy both Eq.(21) and Eq.(23) only when the ratio of width to depth assumes a very special value. For a prototype width-to-depth ratio 50 and a model distortion ratio 10, b satisfying Eq.(23) would lead the model to underachieve the longitudinal dispersion coefficient about 50%. This may be not unreasonable because the predictions by Eq.(19) agree with observations within a factor of four or so (Fisher, 1979, p.136).

Comparison with San Francisco Bay-Delta Model (SFBDM).

Sizes of 20 bars and distances between the bars were measured in each of 3 rivers in San Francisco Bay-Delta Model which is scaled by 1/1000 horizontally, and 1/100 vertically. This model was reasonably well calibrated. The bars are generally placed normal to the flow direction. Table 1 shows the mean size and space interval of the 20 bars along with the analytical results when $\alpha=0$.

Table 1 Comparison between analytical results and SFBDM

River	f_p	h cm	W_m cm	a cm	l cm	b cm	a/l mea.	a/l anl.	b/h mea.	b/h anl.
Sacramento	0.06	4.8	16.6	1.0	23.6	8.3	0.042	0.069	1.73	1.05
Sutter Slough	0.057	3.6	5.0	0.8	30.1	2.5	0.027	0.068	0.69	1.04
San Joaquin	0.028	5.8	22.4	2.0	33.8	11.2	0.059	0.047	1.93	1.52

In the above table, the friction factor was estimated by $8g/C^2$, where C is prototype Chezy coefficient which was calculated with the water depth and Manning roughness used in Fisher's Delta Model from Hugo B. Fisher Inc.(1991). Water depths in the above table were from the same source as Chezy Coefficient but scaled by 1/100. For simplicity, b was taken as half of the river width measured in model. The agreement between the layout of the roughness bars in SFBDM and that from this study is reasonably good.

Conclusions

This paper has investigated the effects of roughness bars on the flow and mixing process in model. In a hydraulic model in which the primary resistance to flow is by vertical bars, the horizontal mixing coefficients depend on the arrangement of bars, the width of bars, and flow velocity. A certain combination of the size and layout of the bars is required to obtain proper scaling for each of the friction factor, transverse mixing coefficient and longitudinal dispersion coefficient. These particular combinations are valid only for a special river width-to-depth ratio. Therefore in the layout of the bars, compromises among the frictional resistance, transverse mixing and longitudinal dispersion should be taken, and an optimum procedure is desirable. Eq.(22) and Eq.(23) may be used for the initial setup of a model, but careful calibration should be conducted.

Acknowledgement:

This study was sponsored by the U.S. Army Corps of Engineers, San Francisco District.

References:

Crickmore, M.J., "Tracer Tests of Eddy Dispersion in Field and Model", J. Hydraulic Div. ASCE. Vol.98 No.HY.10, 1972, pp.1737-1752

Fisher, H.B. and Holly, E.R.,"Analysis of the Use of the Distorted Hydraulic Models for Dispersion Studies." Water Resour. Res.,7, 1971, pp.46-51.

Fisher, H.B. and Hanamura T., "The Effects of Roughness Strips on Transverse Mixing in Hydraulic Models," Water Resour. Res. 11, 1975, pp. 362-264.

Fisher, H.B., List, B.J., Koh, R.C.Y., Imberger, J. and Brooks, N.H, "Mixing in Inland and Coastal Waters." Academic Press Inc. 1979.

Hudson,R.Y.,Hermann,F.A.,Sager,R.A.,Whalin,R.W.,Keulgan,G.H,Chatham,C.E. and Hales,L.Z.,"Coastal Hydraulic Models", CERC, Special Rep.No.5., 1979.

Hugo B. Fisher,Inc. "Documentation for Version 8 of the Enhanced Fisher Delta Model", 1991.

INVESTIGATION OF SALTATING PARTICLE MOTIONS USING FLOW VISUALIZATION TECHNIQUE

By Hong-Yuan Lee[1], Associate Member, and In-Song Hsu[2]

ABSTRACT: A real-time flow visualization technigue is applied in this study to investigate these phenomena. This technique is able to measure the particle trajectories and particle velocities without interfering the flow field. Several cases with combinations of different water depths, channel slopes, particle sizes and channel bed conditions were tested. Relations between the average saltation height, length and velocity were obtained.

INTRODUCTION

A number of researchers have worked on various aspects of grain saltation. Basing on his flume study, Einstein (1950) found the thickness of the bed load movement was about two times particle size and the saltation length was about 100 times particle size. The saltation length was a function of particle size, shape and hydraulic characteristics. Bagnold (1966) found the bed load movement was dominated by the gravity force and not affected by the turbulent eddies. Fernandez Luque and Van Beek (1976), Abbott and Francis (1977) and Murphy and Hooshiari (1982) have investigated saltation in water using high- speed photography and determined the saltation height and length and the average particle velocity. Using the similar technique, Withe and Schultz (1977) measured the saltation trajectory in wind and investigated the Magnus effect due to particle spinning. A real-time flow visualization technique is applied in this study to investigate the particle saltation motion near the channel bed. This technique is able to measure the particle trajectories and velocities without interfering the flow field.

EXPERIMENTAL SETUP

1. Experiment conditions

The experiments were conducted in a 12 m long, 0.3 m wide and slope-adjustable recirculating flume. Several cases with combinations of different water depths, channel slopes, channel bed conditions and particle sizes were tested. The range of the water depth is between 3.71 cm and 12.08 cm and that of the slope is between 0.002 and 0.023. Two bed conditions, smooth and rough,

---
[1]Prof., Dept. Civil Engrg. and Hydr. Res. Lab., National Taiwan Univ., Taipei, Taiwan, R.O.C.
[2]Grad. Res. Asst., Dept. Civil Engrg., National Taiwan Univ., Taipei, Taiwan, R.O.C.

were tested. The effective roughness of the rough surface is about 1.36 mm. Two particle sizes, with specific gravity equals to 2.64 and mean diameters equal to 1.36 mm and 2.47 mm respectively were used in the experiments. The flow conditions are turbulent flow. The range of the Reynolds number is between 21,000 and 73,000 and the particle Reynolds number varies between 50.3 and 224.6. The value of the parameter U_*/W_f was kept 1 in the experiments, hence it was in a bed load dominate mode. The parameter W_f is the particle fall velocity. The important hydraulic and sediment transport characteristics are summarized in Table 1.

2. Experiment procedures

In order to measure the instantaneous particle saltation trajectories accurately and efficiently, a real-time flow visualization technique is used in this study. The general configuration of the experimental setup are shown in Fig.s. 1, and 2 respectively. The light source is generated by a 4 watt argon ion laser. It passes a beam steering device and three mirrors to reach top of the flume, and then use another beam steering device and a cylindrical mirror to create a light sheet passing through the center plane of the flume. The thickness of the light sheet is adjustable. It is between 1 mm and 2 mm. The length of the light sheet has to be longer than that of the working section which is about 15 cm . During the experimental process, sediment particles were released at about 2 m upstream of the working section. As the salting particles passed through the light sheet, laser light was reflected and recorded by a CCD camera (charge couple device camera).

The image obtained from the CCD was first processed and stored by an image processing board installed by an IBM PC 486-50 computer. The image stored was then digitized by an input look-up table (ILUT). Different gray levels were assigned to the images collected at different time. The digitized images were then processed by an arithmetic–logic unit (ALU) and stored in a frame buffer zone. After accumulating several hundred images, it was then processed by an output look-up table (OLUT). Different colors were assigned to different gray levels. The final colored particle trajectory was thus obtained and shown in the display monitor. Each color represents different time interval, and length of the color segment represents the distance travelled during the time interval. The instantaneous velocity vector can thus be obtained. In this experiment, the time interval chosen was 1/30 second, i.e. 30 pictures were taken in 1 second, and seven different colors were assigned to represent different time intervals.

3. Experimental results

The experimental results are summarized in Table 2, where $Y = \tau_o/(r_s - r_w)D_m$ and $Y_c = \tau_c/(r_s - r_w)D_m$ are the dimensionless shear stress, and critical shear stress, τ_o and τ_c are the shear stress and critical shear stress, D_m is the mean particle size and r_s and r_w are specific gravity of sediment and water respectively. L_b, H_b and U_b are the average saltation length, height and velocity respectively and U_* is the shear velocity. The range of the dimensionless shear stress is between 0.06 and 0.46. For the case of the smooth bed, the range of the dimensionless saltation length, L_b/D_m, is betweern 14.8 and 106.8, that of the dimensionless saltation height, H_b/D_m, is between 2.3 and 9.3 and that of the dimensionless saltation velocity, U_b/U_*, is between 6.5 and 9.9. For the case of the rough bed, the range of L_b/D_m is between 14.5 and 105.0, that of H_b/D_m is between 2.2 and 9.2 and that of U_b/U_* is between 6.4 and 9.9. The saltation length, height and velocity increase as the shear stress increases. It also indicates that as Y increases, U_*/W_f increases accordingly and thus particle velcity approaches to the flow velocity. Sediment transport shifts to a suspended mode.

The saltation characteristics for both smooth and rough beds are similar. The saltation length, height and velocity of the smooth bed conditions are slightly larger than those of the rough bed. This is due to influence of the flow resistance. The bottom velocity is smaller for the case of the rough bed and hence smaller amount of energy is available to transport the sediment particles.

The average particle velocity is slow at the rising limb of the trajectory, reaches a maximum value at the central portion and then slows down at the falling limb. This can be explained as follows. In the rising limb, the sediment particle is picked up by water and transported to a higher velocity zone. Hence, it is accelerating. Up to certain elevation, the relative velocity decreases and thus the lift force decreases accordingly. The particle starts to settle down due to gravity, and decelerates. However due to the inertial effect the velocity of the falling limb is larger than that of the rising limb.

Regression equations for the dimensionless saltation length, height and velocity are obtained. They are expressed as:

$$\frac{L_b}{D_m} = 196.3[Y - Y_c]^{0.788} \tag{1}$$

$$\frac{H_b}{D_m} = 14.27[Y - Y_C]^{0.575} \tag{2}$$

$$U_b - U_* = 11.53[Y - Y_c]^{0.174} \tag{3}$$

The corresponding figures are shown in Fig 3.

REFERENCE

1. Abbott, J.E., and Francis, J. R. D., "Saltation and Suspension Trajectories of Solid Grains in a Water Stream," Proc. Royal Soc., London, England, Vol. 284, A 1321, 1977.

2. Bagnold, R.A., "An approach to the Sediment Transport Problem from General Physics," Geological Survey Prof. Paper 422-I, Wash., 1966.

3. Bagnold, R.A., "The Nature of Saltation and of Bed-Load Transport in Water," Proc. Royal Soc., London, England, A 332, 1973.

4. Fernandez Luque, R., and Beek, R. van, "Erosion and Transport of Bed-load Sediment," Journal of Hydraulic Research, Vol. 14, No. 2, 1976.

5. Murphy, P.J., Hooshiar, Hamid, "Saltation in Water Dynamics," J. Hydr. Engr., ASCE, Vol. 108, No. 11, 1982, pp.1251.

6. White, B.R., and Schultz, J.C., "Magnus Effect in Saltation," Journal of Fluid Mechanics, Vol. 81, 1977, pp. 507.

Table 1 Experiment arrangement

Bed Condition (1)	Particle Properties		Hydraulic Characteristics						
	Run (2)	D_m(mm) (3)	H(cm) (4)	slope (5)	U_* (m/s) (6)	U_*/W_f (7)	R_e (8)	R_* (9)	$T(°C)$ (10)
Smoth	S1	1.36	12.08	0.002	0.036	0.14	42,000	50.3	21.0
	S2	1.36	3.71	0.006	0.042	0.17	21,000	58.0	20.9
	S3	1.36	4.94	0.006	0.047	0.19	24,000	64.9	21.0
	S4	2.47	4.64	0.016	0.075	0.22	40,000	187.9	21.2
	S5	2.47	6.55	0.016	0.085	0.25	64,000	213.1	21.0
	S6	2.47	7.64	0.016	0.089	0.26	73,000	224.6	21.1
	S7	1.36	4.54	0.016	0.074	0.29	39,000	102.6	21.0
	S8	1.36	5.80	0.016	0.081	0.32	54,000	112.4	21.0
	S9	1.36	7.23	0.016	0.088	0.35	70,000	121.4	21.1
	S10	1.36	6.35	0.020	0.094	0.37	67,000	129.8	21.1
	S11	1.36	6.15	0.023	0.099	0.39	66,000	137.6	21.0
	S12	1.36	6.35	0.023	0.105	0.41	69,000	145.1	21.0
rough	R1	1.36	12.08	0.002	0.036	0.14	42,000	3.7	21.2
	R2	1.36	3.71	0.006	0.042	0.17	21,000	4.3	21.0
	R3	1.36	4.94	0.006	0.047	0.19	24,000	4.8	21.1
	R4	2.47	4.64	0.016	0.075	0.22	40,000	7.7	20.8
	R5	2.47	6.55	0.016	0.085	0.25	64,000	8.7	21.0
	R6	2.47	7.64	0.016	0.089	0.26	73,000	9.1	21.1
	R7	1.36	4.54	0.016	0.074	0.29	39,000	7.6	21.0
	R8	1.36	5.80	0.016	0.081	0.32	54,000	8.3	21.0
	R9	1.36	7.23	0.016	0.088	0.35	70,000	9.0	21.2
	R10	1.36	6.35	0.020	0.094	0.37	67,000	9.6	21.0
	R11	1.36	6.15	0.023	0.099	0.39	66,000	10.1	20.9
	R12	1.36	6.35	0.023	0.105	0.41	69,000	10.7	21.0

*D_m-particle size, G_s-specific gravity, W_f-fall velocity, H-water depth, U_*-shear velocity, U_{*c}-critical shear velocity

Table 2 Average saltation length, height and velocities under different flow conditions

Bed Condition (1)	Dimensionless Shear Y (2)	Dimensionless Critical Shear Y_c (3)	Average Saltation Length $\overline{L}_b/D_m$ (4)	Average Saltation Hight $\overline{H}_b/D_m$ (5)	Average Saltation Velocity $\overline{U}_b/U_*$ (6)
Smooth	0.06	0.026	14.8	2.3	6.5
	0.08	0.026	21.1	2.6	6.9
	0.10	0.026	25.2	3.2	7.5
	0.14	0.031	34.0	3.9	7.7
	0.18	0.031	41.6	4.9	8.1
	0.20	0.031	49.3	5.2	8.4
	0.25	0.026	62.0	6.0	9.1
	0.30	0.026	71.0	6.9	9.4
	0.35	0.026	86.6	7.4	9.6
	0.40	0.026	95.0	8.3	9.7
	0.45	0.026	103.2	9.0	9.8
	0.46	0.026	106.8	9.3	9.9
Rought	0.06	0.026	14.5	2.2	6.4
	0.08	0.026	20.0	2.6	6.9
	0.10	0.026	24.6	3.1	7.5
	0.14	0.031	32.4	3.8	7.7
	0.18	0.031	40.0	4.7	8.1
	0.20	0.031	48.2	5.1	8.4
	0.25	0.026	59.1	5.9	9.1
	0.30	0.026	69.0	6.7	9.4
	0.35	0.026	85.3	7.4	9.6
	0.40	0.026	93.2	8.2	9.7
	0.45	0.026	101.0	9.0	9.8
	0.46	0.026	105.0	9.2	9.9

*$Y = \gamma_o/(\tau_s - \tau_w)D_m, Y_c = \gamma_c/(\tau_s - \tau_w)D_m$

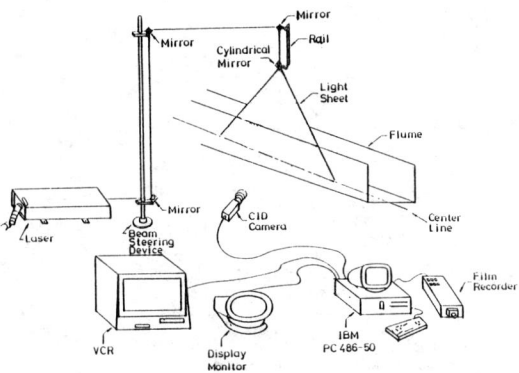

Fig. 1 General configuration of the experimental setup

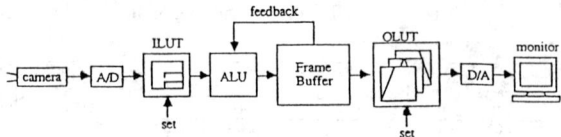

Fig. 2 Flow chart of the image processing system

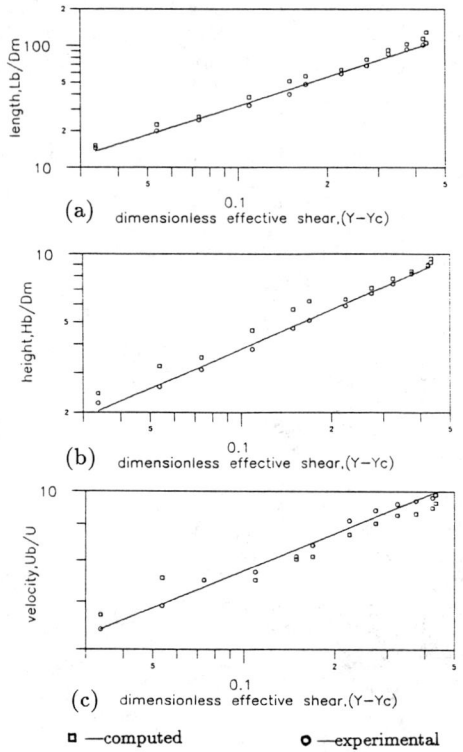

Fig. 3 Regression for dimensionless saltation length, height, and velocity.

ACOUSTIC DOPPLER CURRENT PROFILER
FOR INLET CURRENT MEASUREMENTS

By Yen-hsi Chu[1], Michael Metcalf[2], J. Bailey Smith[3],
Trap Puckette[3], and Gilbert K. Nersesian[4]

ABSTRACT: A BroadBand Acoustic Doppler Current Profiler (BroadBand ADCP) was used to survey tidal currents at Shinnecock and Moriches Inlets, Long Island, NY. Measurements were made remotely from a survey vessel, at regular intervals throughout the water column. The "continuous" data collected along the measured transects provides the cross-sectional velocity distributions and reveals large scale eddies occuring at channel sides of the inlet near jetty heads during the flood phase of tides.

INTRODUCTION

Shinnecock and Moriches Inlets, 24 km apart, are the two easternmost inlets on the south shore of Long Island, New York. Both inlets provide vessel passages from the Atlantic Ocean to an interconnected tidal bay system to the north (Figure 1). Rock jetties were built in 1952-53 at both inlets to stabilize channels and improve navigation conditions. Prior to the construction of jetties, both inlets were migrating slowly westward and experiencing classic natural opening and closure cycles (Czerniak, 1977; Kassner and Black, 1982).

Since the jetty construction, two large scour holes have developed at the tips of the two west jetties (Chu and Nersesian, 1992). These scour holes reach 15 m and 24 m in depth at Moriches and Shinnecock Inlets, respectively as compared to the average inlet depth of 6 m. These scour holes undermine the jetties, thus causing the seaward ends of the jetties to collapse into the scour holes.

A field monitoring program was initiated in 1990 to study the inlet hydrodynamic processes in the vicinity of the jetty scour holes. Current measurements were conducted to determine current patterns at the inlet entrances that may be responsible for the formation of the scour holes. Due to the swift currents and large wave actions

---

[1]Supervisory Hydraulic Engineer, U.S. Army Engineer Waterways Experiment Station, 3909 Halls Ferry Rd., Vicksburg, MS 39180-6199
[2]Technical Marketing Specialist, RD Instruments, 9855 Businesspark Ave., San Diego, CA, 92131
[3]Physical Scientist, U.S. Army Engineer Waterways Experiment Station, 3909 Halls Ferry Road, Vicksburg, MS 39180-6199
[4]Senior Coastal Engineer, U.S. Army Engineer District, New York, 26 Federal Plaza, New York, NY 10278-0090

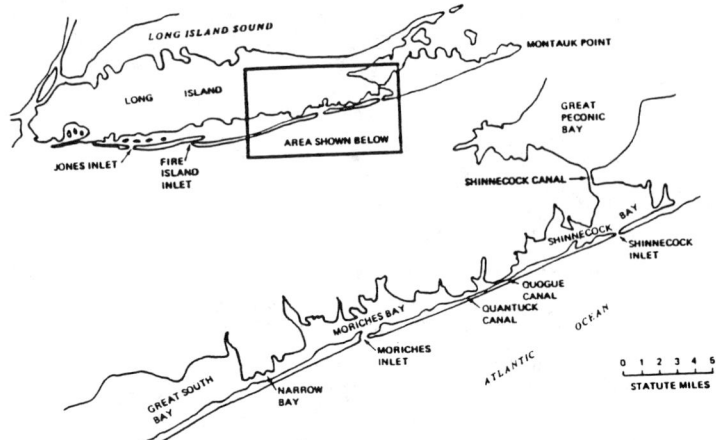

Figure 1. Moriches and Shinnecock Inlets, Long Island, NY

at both entrances, conventional current meters were not able to obtain sufficient data across the tidal inlet to determine current patterns in the vicinity of the scour hole. This paper discusses the experience of using an Acoustic Doppler Current Profiler (ADCP) to measure tidal currents at Moriches and Shinnecock Inlets.

ACOUSTIC DOPPLER CURRENT PROFILER

The ADCP operates by sending out a series of acoustic pulses into the water and listening to the backscattered signals. These signals are processed to measure the rate of phase change with time to determine any change in frequency from the transmitted pulse - a Doppler shift. The Doppler shift is caused by the relative motion between the ADCP and scattering material floating with the current. Since the equation for relating Doppler shift to relative velocity is well known, the velocity of the current can be readily determined.

The unique feature of the ADCP is its ability to perform a series of current measurements made at regular intevals throughout the water column. The ADCP receives a continuous backscatter echo record from the water column and the receivers are range-gated (i.e., turned on and off in time) to assign regular intervals of the echo record to different depths. In each of these regular intervals, called a depth cell, three velocity components are measured. The use of an internal fluxgate compass in the ADCP allowed the data to be transformed to an earth-referenced coordinate system. For this study, 60 depth cells (0.5 m each cell) were used to accommodate a maximum possible profiling range of 30 m.

The instrument used for present measurements was a BroadBand Acoustic Doppler Current Profiler (BroadBand ADCP) manufactured by RD Instruments (RDI) in San Diego, CA. It is a new generation instrument which is more accurate than the previous model, boasting a lower standard deviation for a velocity measurement and improved spatial resolution of water-current patterns.

APPLICATION OF BROADBAND ADCP AT LONG ISLAND INLET

One of the advantages of the BroadBand ADCP is the relatively simple deployment and operation of the system. The system consists of the instrument itself, an aluminum pressure housing approximately 80 cm in length and 22 cm in diameter with acoustic transducers mounted on one end, a small deck box for connecting the instrument to the computer, a computer (a 386 or faster IBM-PC type), and connecting cables. For the deployment at Shinnecock and Moriches Inlets on 21-23 July 1992, the instrument was strapped to an aluminum pipe which was mounted on the side of a survey boat (length of 7.5 m) (Figure 2). The transducer head of the BroadBand ADCP was set at a depth of approximately 80 cm. A communication cable was connected to the top on the instrument by a water tight connector and led to the deck box which was stationed in the cabin of the boat. The deck box was in turn connected to a laptop computer which ran the data collection software "TRANSECT". The entire system was powered by a small portable 120-volt generator on the boat.

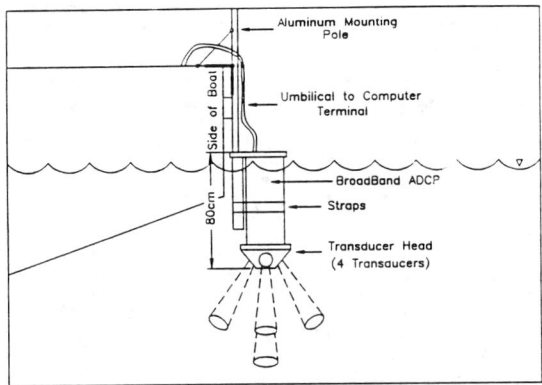

Figure 2. Deployment of BroadBand ADCP

During the field investigation, current measurements were run at each inlet in the vicinity of scour holes both across the inlet and out the inlet seaward from the tip of the jetties. Additional transects were run to measure overall tidal current patterns at landward portions of the inlet, and across both the east and west back bay channels. The BroadBand ADCP was able to collect tidal current data across the inlet along specified transects at a much faster rate and with great improvement in spatial resolution. During the 1992 field study, 50 to 64 profiles were measured in one tidal cycle comparing to approximately 15 to 20 profiles (about 1 to 4 vertical measurements per profile) conducted during the 1991 field study (Chu and Nersesian, 1992) using an electromagnetic current meter.

The software running the instrument allows for operator input of a variety of parameters making it possible to customize the data collection to different locations and conditions. The operator can adjust the bin size, depth range, pings per ensemble (average time), positioning method (GPS, LORAN-C, or Microfix), and other parameters pertaining to the data collection and storage.

Current data were collected by making transects across the inlets, starting and stopping approximately 5 m from the jetty on either side. The collection of data closer to the jetties is ineffectual as the jetties would begin to interfere with the beam patterns of the transducers. Boat speed averaged approximately 0.75 m/s. Associated instantaneous wave orbital velocities affecting currents could not be measured as a 7-second sampling interval was utilized. Data were stored on the hard drive of the computer and later downloaded to floppy disks. The data could then be reviewed using the same software or transformed into data files for uses in other data processing programs.

DISCUSSION OF RESULTS

Recorded current data at Shinnecock and Moriches Inlets were analyzed to determine the north-south, east-west, and vertical components of currents through the entire water column at each transect location. Additional analyses of current data included the degree of backscatter (suspended matter), resultant vectors of currents at depths throughout the water column, tabular format listing current velocity in two directions, and discharge across the transect. Vessel heading, speed, pitch and roll, and transect length and time were also calculated.

Figures 3 and 4 illustrate typical current profiles across Shinnecock Inlet entrance between the jetties. The horizontal axis plots time, and the vertical axis plots depth (m). Current velocity (cm/s) and direction are indicated by the scale to the right where "positive" currents are directed toward the north or east and "negative" currents are directed toward the south or west. Figures 3 and 4 illustrate the north/south and the east/west components, respectively, of the currents at Shinnecock Inlet during the mid-flood tide stage. Currents to the north and east (of up to 110 cm/s) flood the inlet. Note that currents to the south (of up to 60 cm/s) exit the inlet on the sides of the inlet adjacent to the jetties during the flood phase of the tide (currents to the north). This ebb-type current measured during the flood tidal phase clearly shows the presence of the separation eddy at the jettied entrance as discussed by Chu and Nersesian (1992).

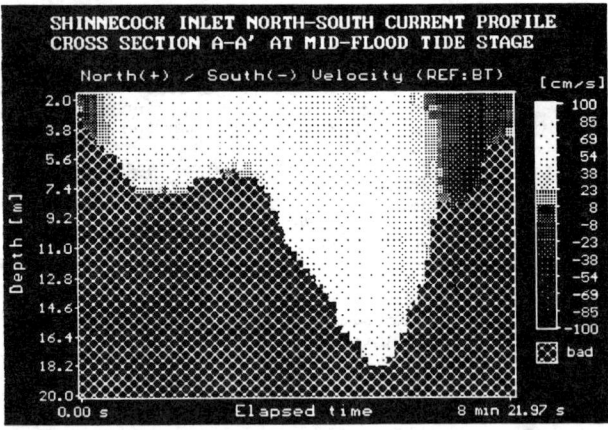

Figure 3. North/South Current Components During the Mid-Flood Tide

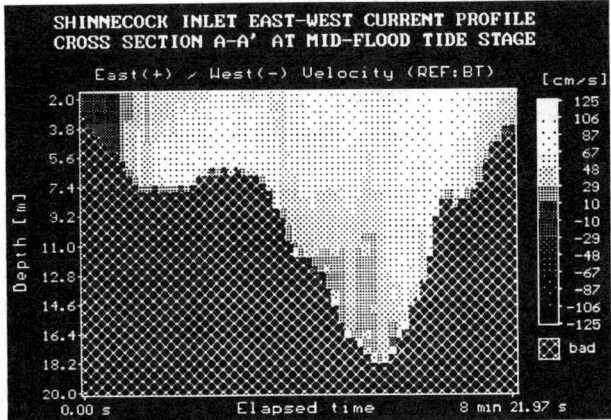

Figure 4. East/West Current Components During Mid-Flood Tide

Figure 5 shows the contours of the north/south current components. This figure supports the earlier measurements by conventional current meters that the currents are relatively uniform across the inlet. Inside the eddy, steep velocity gradients (or high shear zone) were measured at Ensembles 10-12 (Figure 5). It is suspected that the flow turbulance generated in this high shear zone is the major energy source scouring the inlet bottom.

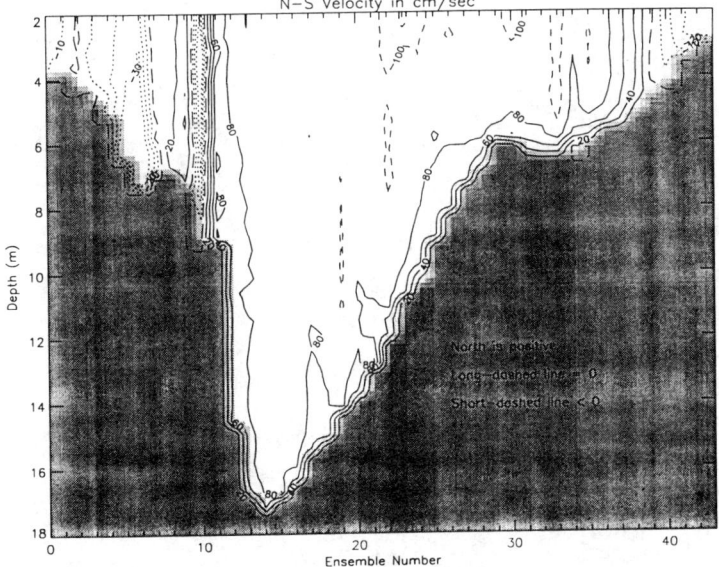

Figure 5. Contours of North/south Currents at Shinnecock Inlet

CONCLUSIONS AND RECOMMENDATIONS

The ADCP technology provided the present study with an effective and powerful tool to map currents at the two tidal inlets. This new type of instrument has significantly improved the current measurement technique with the remote sensing ability (the ADCP is not required to be positioned at the point of measurement), the high rate of data return, and the enhanced spatial resolution. With these three distinct capabilities, the ADCP is extremely useful for measuring time varying flows. Beacuse of its ability to collect large volumes of data the ADCP requires additional software for data reduction and analysis.

ACKNOWLEDGEMENT

The present study is sponsored by the U.S. Army Engineer District, New York. Permission to publish this paper is granted by the Chief of Engineers.

REFERENCES

Chu, Y.H. and Nersesian, G. K., 1992, "Scour Hole Development and Stabilization at Shinnecock and Moriches Inlets, New York," Proceedings, Coastal Engineering Practice '92, ASCE, Long Beach, CA.

Czerniak, M. T., 1977, "Inlet Interaction and Stability Theory Verification," Procedings of Coastal Sediments '77, ASCE, Charleston, SC.

Kassner, J. and Black, J. A.,1982, "Efforts to Stabilize a Coastal Inlet: a Case Study of Moriches Inlet, New York," **Shore and Beach**, pp. 21-29.

TURBULENT VELOCITY FLUCTUATIONS IN NATURAL RIVERS

Nani G. Bhowmik,[1] F. ASCE, and Renjie Xia[2]
Illinois State Water Survey
Office of Hydraulics and River Mechanics
Champaign, IL 61820-7495

ABSTRACT

The longitudinal and transverse components of flow velocity, as well as the normal and tangential stresses, must be expected to fluctuate with time and space when flow at a high Reynolds number moves between fixed boundaries. Although these nonperiodic fluctuations are generally secondary in magnitude compared to the mean motion, they have profound effects on properties of the primary mean motion. Scientists from the Illinois State Water Survey are involved in collecting and analyzing detailed velocity data from the Illinois and Mississippi Rivers using 2-D electromagnetic current meters. The goal of the present research is to understand and evaluate the turbulent structure in natural river systems, especially near the channel border areas.

The fluctuation characteristics of flow velocity were analyzed systematically. Flow velocities were measured with time at six different lateral locations and at three different vertical elevations. Analyses of the velocity data included the cross-sectional and vertical distributions of longitudinal and transverse velocity components (u, v), the fluctuating velocity components (u', v') and their frequency-distribution curves, turbulent intensities (σ_x, σ_y), turbulent shear stress ($-\rho u'v'$), and turbulent kinetic energy.

It appears that velocity, velocity fluctuation, turbulent intensity, turbulent shear stress, and turbulent kinetic energy are all relatively high within the main

---

[1]Principal Scientist and Acting Head of Hydrology Division, Illinois State Water Survey, 2204 Griffith Drive, Champaign, IL 61820-7595

[2]Associate Professional Scientist, Office of Hydraulics and River Mechanics, Illinois State Water Survey, 2204 Griffith Drive, Champaign, IL 61820-7595

channel. Moreover, all of these flow parameters decrease in the channel border areas. Thus the main channel area above the riverbed appears to be the most active zone in the river. The study shows that the values of σ_x, σ_y, τ_{xy}, and average KE are high at elevations of 10 to 20 percent above the bed.

INTRODUCTION AND DATA COLLECTION

A site-specific field study was conducted to investigate changes of velocity and turbulent shear stress in the Illinois River, a large natural river. Figure 1 shows a schematic diagram of the field set-up on the Illinois River at Kampsville. This study is a component of a major research undertaking focusing on changes in hydrodynamic characteristics on the Upper Mississippi River System (UMRS) due to navigation traffic (Mazumder et al., 1991; Bhowmik, 1992; and Soong et al., 1990).

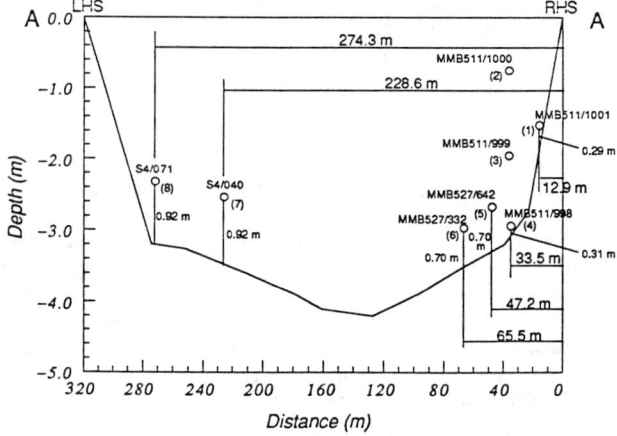

Figure 1. Schematic diagram showing the data collection set-up on the Illinois River at Kampsville.

Velocity data were collected using eight 2-D electromagnetic current meters (six Marsh McBirney, Inc., models MMB511 and MMB527 and two Inter Ocean model S4). The relative locations of these meters are also shown in Figure 1. The signal processing unit in these meters provides visual readings for the x and y components of flow velocity. The x-axis is the longitudinal direction of river flow; the y-axis is perpendicular to the sailing line, with the positive direction toward the left-hand side (LHS) of the river; and the z-axis is normal to the xy-plane.

The x and y components of flow velocity (u, v) were continuously measured with time at six different lateral locations at distances of 12.9, 33.5,

47.2, 65.5, 228.6 and 274.3 meters from the right-hand side (RHS) of the river. Three MMB511 current meters were also mounted at three vertical heights of 0.31, 1.22, and 2.44 meters above the river bottom, at a lateral distance of 33.5 meters from the RHS. Velocity data were collected at the rate of one sample per second to study the turbulent velocity fluctuations.

ANALYSIS OF VELOCITY DATA

The longitudinal and transverse components of flow velocity (u, v), turbulent intensities (σ_x, σ_y), turbulent shear stress ($-\rho u',v'$), and turbulent kinetic energy have been computed for a better understanding of turbulent fluctuations in natural rivers.

The temporal variations of longitudinal and lateral velocities and turbulent shear stress at any particular point in the river are of equal importance to those variations in speed. The variation of velocities with time indicates acceleration and the variation of turbulent shear stress at any particular point in the river are of equal importance to those variations in space. At the same time the variation of turbulent shear stress with time indicates the transfer of momentum. These changes generate secondary currents and eddies. Instantaneous flow velocity were denoted into two parts such as u' and v', and the three components of the velocities were denoted as u, v, w.

The time series of plots of flow velocity for the u and v components and their respective fluctuations u' and v' for a period of 60 minutes for the October 15, 1990, data set are shown in Figure 2. A comparison of the magnitude of u with that of v shows that the transverse velocity component v is much less than the longitudinal velocity component u. These differences are approximately one order in magnitude. This is normally expected in open channel flow where the overall motion of the flow is in the direction of the longitudinal x-axis.

VARIABILITY OF TURBULENT
FLUCTUATIONS ACROSS THE RIVER WIDTH

Fluctuations of flow velocity at any fixed location can be studied by developing a frequency-distribution curve of the differences between the instantaneous velocity and the mean velocity (Rouse, 1959). This analysis, when applied to each measured point in a river cross section, can show the strong or weak turbulent fluctuations of the flow velocity at different lateral and vertical locations. Figure 3 shows such a frequency distribution of the longitudinal velocity fluctuation, u', across the width of the river (RM 35.2) at six locations and over the depth at three elevations (refer to Figure 1) for the October 15 data set. When the value of u' (u' = u - ū) is small, its frequency distribution is clustered close to the zero on the x-axis, and this indicates that the fluctuation is weak. Conversely, when the value of u' is large, its frequency distribution will spread out, indicating the strong fluctuation. Thus, based on Figure 3, the velocity fluctuation is relatively weak near the shoreline and quite strong near the main channel area of the river.

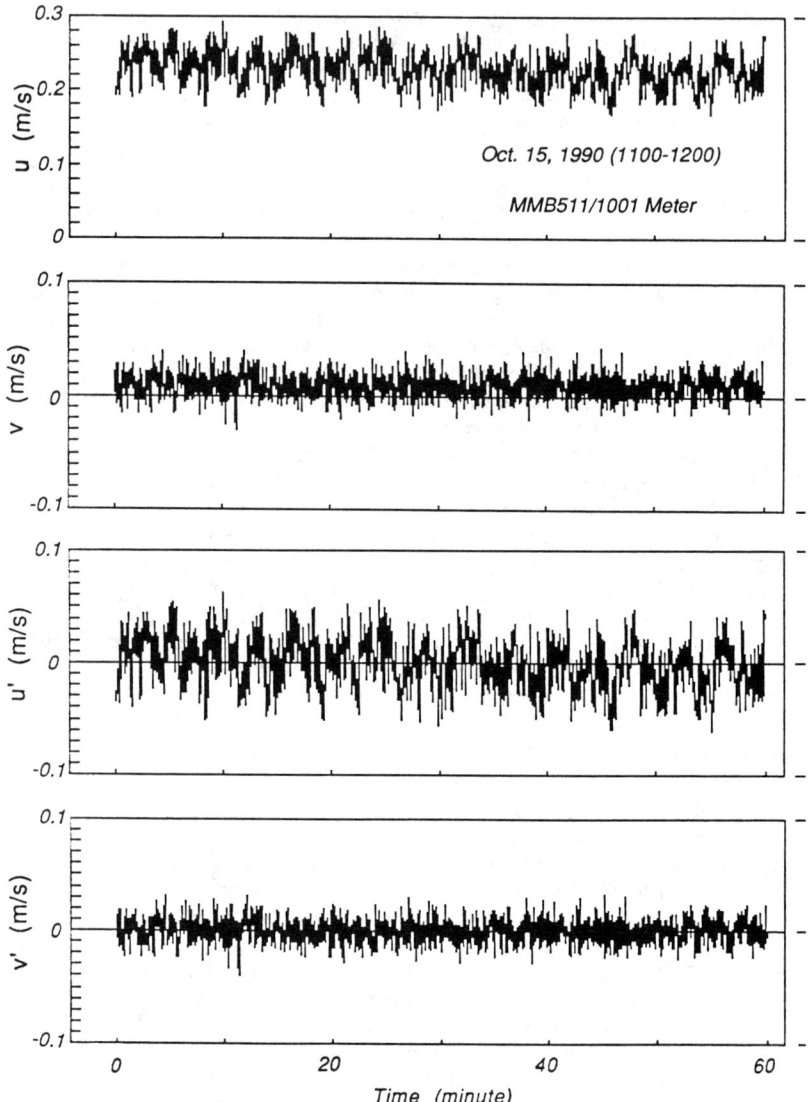

Figure 2. Variations of measured u and v and u' and v' with time.

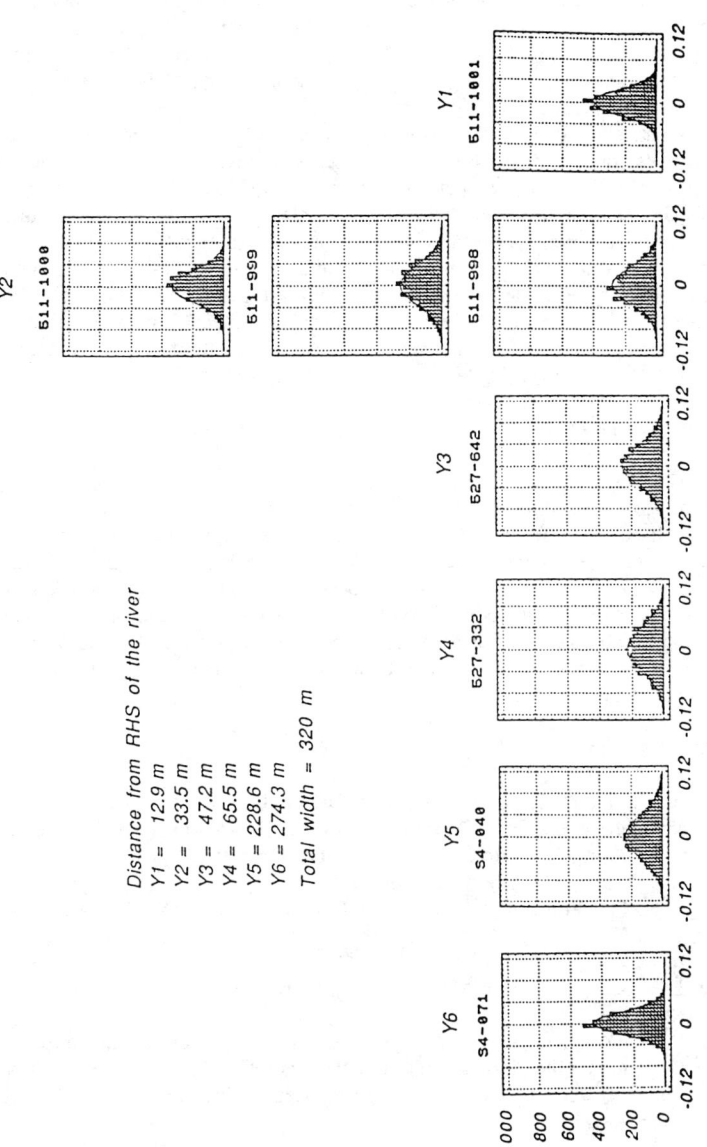

Figure 3. Distribution of longitudinal turbulent velocity fluctuation, u', in the river cross-section.

Analyses of the variability of the frequency distribution curve for the lateral velocity fluctuation, v', for the same eight points as in Figure 3 indicated similar variations. The lateral velocity fluctuation is quite small near the channel border areas compared to that near the main channel area. This analyses indicates that turbulence is strong in the main channel near the river bottom and quite weak near the channel border areas.

Statistical analysis of the frequency-distribution plots for the longitudinal and lateral velocity fluctuations, u' and v', across the width of the river and in the vertical direction indicated that the velocity fluctuations are well fitted by a normal (Gaussian) distribution, and that the selection of the 60-minute time interval is correct since the mean values of u' and v' are almost always equal to zero when $T = 60$ minutes. Other parameters that were evaluated included turbulent intensity, turbulent shear stress, and turbulent kinetic energy. The variability is similar to those observed for turbulent intensity with the highest values occurring within the main channel and also at a depth of 20 percent above the bed.

ACKNOWLEDGEMENTS

This research was done as part of a project supported by the U.S. Army Corps of Engineers through the Environmental Management Program for the Upper Mississippi River system. Funding was provided through the Environmental Management Technical Center, Onalaska, Wisconsin, with Ken Lubinski as the project manager. David Soong, Bill Bogner, and Jim Slowikowski of the Water Survey and Rodger Adams, a previous staff member of the Water Survey, were also involved extensively in this research on the physical changes in large rivers associated with river traffic. The camera ready copy of the paper was prepared by Lori Nappe.

REFERENCES CITED

Bhowmik, N.G. 1992. Lateral Velocity Distribution in Large Rivers Due to Navigation Traffic. Proceedings of the Mississippi River Research Consortium, Inc. Vol. 24 April 30-May 1, pp. 13-14.
Mazumder, B.S., N.G. Bhowmik, and T.W. Soong. 1991. Turbulence and Reynolds Stress Distribution in a Natural River. ASCE Proceedings, 1991 National Conference on Hydraulic Engineering, Nashville, TN, July 29-August 2, pp. 906-911.
Rouse, H. 1959. Advanced Mechanics of Fluids. John Wiley and Sons, Inc., New York.
Soong, T.W., W.C. Bogner, and W. F. Reichett. 1990. Data Acquisition for Determining the Physical Impact of Navigation. ASCE Proceedings, 1990 National Conference on Hydraulic Engineering, July 30-August 3, Vol. 1, pp. 616-621.

Determination of Pulsating Pressures for
Baldhill Dam, N.D. Spillway

Bobby P. Fletcher[1] and Gregory W. Eggers[2]

Abstract

Model tests of the existing Baldhill Spillway were conducted with symmetrical and asymmetrical gate openings to determine for various flow conditions the characteristics of the hydraulic pulsating pressures acting on the surface of the spillway chute and stilling basin apron. The 1:30-scale model simulated the entire width and length of the spillway crest, gates, chute, and stilling basin. The model was capable of simulating various upper and lower pool elevations and discharges as high as 65,000 cfs. Pulsating pressures were simultaneously measured with 30 surface-mounted transducers located in turbulent areas. Data were collected with a data acquisition system capable of collecting data for various lengths of time and sampling at various rates. The test results provided sufficient information of the potential pulsation pressure acting on the surface of the spillway chute and stilling basin apron to enable design decision to insure the integrity of the structure.

Introduction

Hydraulic loads on stilling basin floor slabs include not only static loads but also loads resulting from the dynamic forces due to fluctuating (pulsating) pressures within the hydraulic jump. These fluctuations about the mean pressure can be extreme under certain conditions and can act as both positive (maximum) and negative (minimum)

---

[1]Research Hydraulic Engineer, Waterways Experiment Station, Corps of Engineers, 3909 Halls Ferry Road, Vicksburg, MS 39180-6199.

[2]Hydraulic Engineer, Department of the Army, Corps of Engineers, 180 Kellogg Blvd. E., St. Paul, MN 55101.

in conjunction with the static loading condition. Pulsating pressures can act to rapidly increase or decrease hydrostatic loads in stilling basins. The minimum pulses can abruptly decrease static pressure below the mean pressure on the surface of the stilling basin slabs, creating a dynamic loading condition on the floor slab. The maximum pulses can rapidly increase local pressures in stilling basins.

Past Research

The pulsating pressure phenomenon in the hydraulic jump has been researched extensively in recent years. Extreme pressures were measured (Bowers and Toso 1988) in a study conducted at the St. Anthony Falls Hydraulic Laboratory at the University of Minnesota. Stilling basin uplift forces were measured (Fletcher 1988) on the Old River Control Structure to determine the effects these dynamic loads impart on spillway structures. Fluctuating uplift was also measured (Fiorotto and Rinaldo 1992) to determine the increased loads associated with pulsating pressures in stilling basins.

The pressure measured during Toso's tests indicate that the magnitude of these pressure fluctuations can be extreme, but they do not provide an adequate feel for the areal extent or duration of the pulses. The longitudinal and lateral dimensions were estimated to be 8 and 13 times the entering flow depth (d_1), and the magnitudes of peak instantaneous pressures approached full velocity head for the incident incoming velocity (V_1). The use of these data in development of dynamic loads would result in extremely large loadings which were not corroborated by research conducted at the U.S. Army Engineer Waterways Experiment Station (WES) on a model of the Old River Control Structure. This model study measured the total uplift force on a stilling basin resulting from the dynamic forces on the surface of the stilling basin slab. The research by Fiorotto et al. measured pressures both on the surface as well as underslab pressures and on the global behavior of floor linings in the presence of pulsating uplift.

The location, magnitude, duration, and scale of these pressure pulses is of particular concern for the Baldhill Dam spillway (Figure 1). The new design discharge for the spillway (65,000 cfs) results in the stilling basin being surcharged by a unit discharge 50 percent greater than it was originally designed for. Also, higher tailwaters will force the stilling action out of the stilling basin and up on a spillway chute. Uplift beneath the slabs will be increased due to higher tailwater elevations and the thin chute slabs will be subjected to stilling action on the surface. It is the ability of these thin slabs to carry

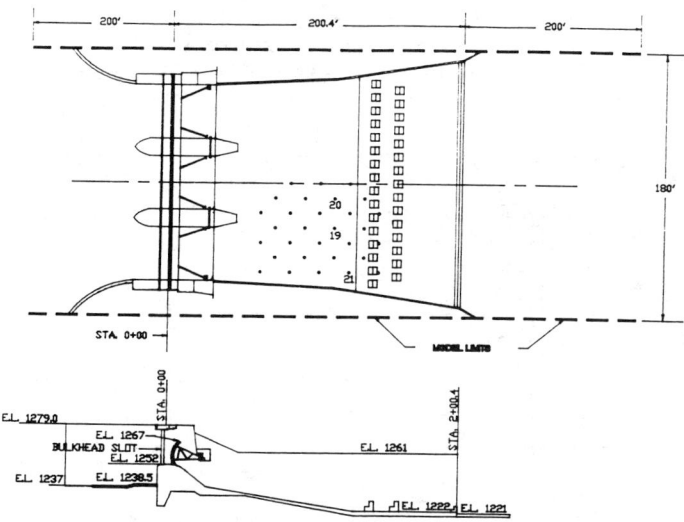

Figure 1. Baldhill Spillway Plan and Profile and Type 1 Transducer Pattern

large dynamic loads that precipitated the need for further investigation of the pulsating pressure phenomenon. Additional physical model testing would help to define the magnitude, scale, and duration of the pressure pulses which cannot be reliably estimated based on the previously mentioned research.

Hydraulic Model Study

The hydraulic model study was conducted at WES using a 1:30-scale model of the Baldhill spillway and stilling basin. Pulsating pressures were simultaneously measured using 30 surface-mounted pressure transducers for specified lengths of time and at various sampling rates. Sensitivity of sampling rates was tested to determine the optimum rate to use without losing resolution on data gathered. A rate of 9.13 samples/sec was determined to be optimum, which allowed for a 5.48-hr run in prototype time scale (3,600 sec model time scale) with the data acquisition equipment used. The transducers were located in the stilling basin in a configuration as shown in Figure 1. It was assumed that the flow patterns either side of the spillway center line were symmetric, which was based on observations from the previous general model. Initial tests with this configuration revealed that a closer transducer spacing was required to achieve the resolution necessary to define the areal extent of the pressure pulses. Figure 2 displays the revised transducer network which isolates an area around transducers 20 and 21 which consistently provided some of the higher readings in early runs.

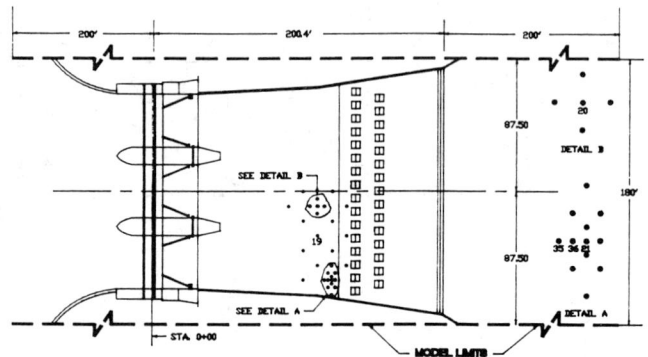

Figure 2.   Type 2 Transducer Pattern

Time history plots for the 65,000 cfs run for a series of three transducers are shown in Figure 3 for the 65,000 cfs run. Figure 3a shows a typical section of the plot and Figure 3b windows in on one of the critical pressure pulses to provide better definition of the scale and duration of the pulse.

The most extreme positive pulse recorded was about 25 ft above the mean as opposed to in excess of 40 ft below the mean for the minimum pulse displayed in Figure 3b.

The transducer pattern shown in Figure 2 facilitated fairly good resolution of individual events around transducer No. 21, whereby contour plots could be developed for any instant in time for some of the more critical loading cases. Contour plots of one of the larger scale events which occurred for the 65,000 cfs test condition are shown in Figure 4. The two frames show contour plots for pressure measurements recorded about 1/10 second apart. The contour interval is 5 ft of water. These plots indicate that the scale of these pulses is much smaller than previous research suggested, possibly related to this particular spillway geometry and head/discharge relationship.

Conclusions

The Baldhill model study generated a tremendous amount of data for various discharge relationships for this low head baffled spillway. The data will allow the engineer to

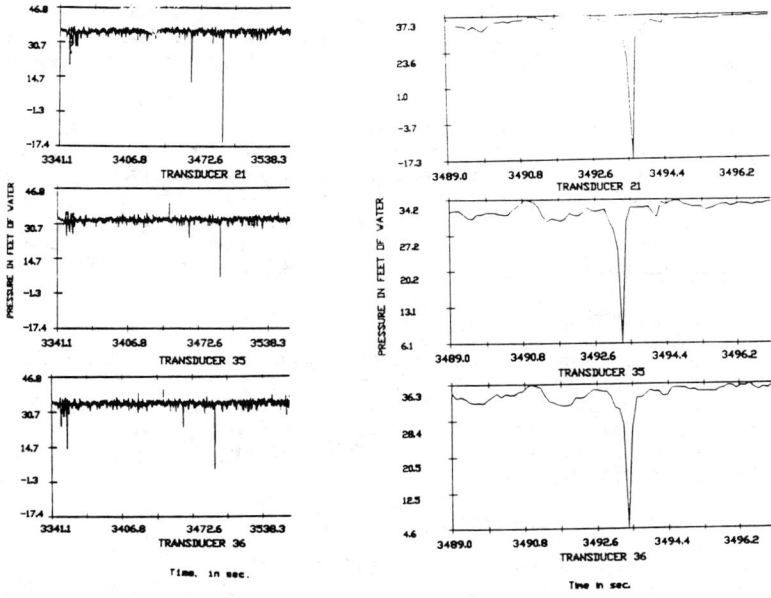

a. 3341.1 to 3538.3 sec   b. 3489.0 to 3496.2 sec

Figure 3. Time History Plots

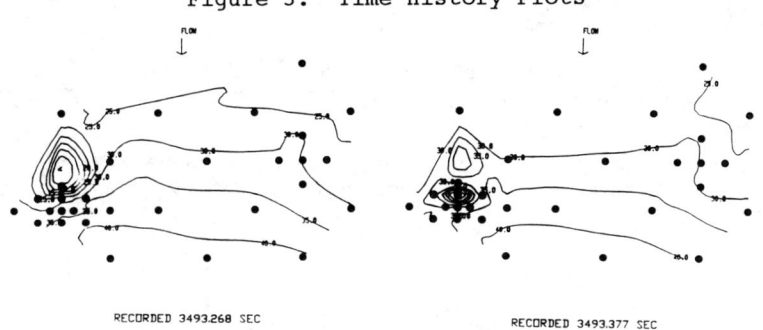

Figure 4. Pressure Contour Plots

interpret the dynamic loads for the critical design conditions for the rehabilitation of the Baldhill Dam spillway. The data will be summarized in a WES technical report to be released in March 1993. The data indicate that pressure fluctuations about the mean can approach magnitudes nearly as great as the incoming velocity head to spillways. The areal extent of the pressure pulses is on the order of one times the incident flow depth (d_1), considerably smaller in scale than that which is suggested in Bowers and Toso (1988).

The scope of this model study was limited to the spillway geometry and discharge characteristics at Baldhill Dam. The tests show that adequate scale and duration exist to make pulsating pressure a valid design concern. Further pulsating pressure research is required to better define relationships of scale and duration for various Froude relationships, design heads, and spillway geometries.

Acknowledgement

The tests described and data presented, unless otherwise noted, were obtained from research conducted for the U.S. Army Engineer District, St. Paul, by the U.S. Army Engineer Waterways Experiment Station, Vicksburg, MS. Permission was granted by the Chief of Engineers to publish this information.

APPENDIX I. References

Toso, J. and Bowers, C. E. (1988). "Extreme pressures in hydraulic jump stilling basin." J. Hyd. Engr., ASCE, 114(8), pp. 829-843.

Fletcher, Bobby P. (1988). Technical Report. HL-88-6, "Old River Low Sill Control Structure." U.S. Army Engineer Waterways Experiment Station.

Fiorotto, V. and Rinaldo, A. (1992). "Fluctuating uplift and lining design in Spillway Stilling Basins." J. Hyd. Engr., ASCE, 118(4), pp. 578.

Fletcher, Bobby P. (To Be Published). Technical Report. "Baldhill Spillway" U.S. Army Engineer Waterways Experiment Station.

HYDRAULIC MODEL STUDIES OF Y-BRANCH

Baum K. Lee,[1] M. ASCE, H.W. Coleman,[2] M. ASCE,
J.H. Kim,[3] H.I. Kwon[4]

Abstract

Hydraulic model studies were made for the Y-branch steel penstock located in a vertical bend. The model tests were performed for various modes of pumping and generating operation.

The prime objectives of the model studies were to determine if the location of Y-branch in the bend is acceptable, to determine if head loss is sensitive to geometry of sickle plate, and to determine if pressure pulsations generated by the Y-branch and/or sickle plate would be significant to the structural design of the Y-branch.

The model test indicated that head loss is not sensitive to the sickle plate size and any of the three sickle plates was acceptable from standpoint of head loss, and the decision for sickle plate geometry was based solely on structural consideration. The test also demonstrated that relocation of the Y-branch was not necessary in view of acceptable head losses and pressure fluctuations at the bend.

Background

The Muju Pumped-Storage Project is a hydroelectric power-plant project which will have a total installed capacity of 600 MW split between two units. The project is located in the extreme southern portion of the Keum River Basin, Republic of Korea. The project features include a 61-m high upper dam, a 47-m high lower dam, and power tunnels. The gross head is approximately 590 m. The project is owned by Korea Electric Power Corporation.

---

[1] Chief Hydrologist, Harza Engineering Company, Sears Tower, 233 S. Wacker Drive, Chicago, Illinois 60606-6392.
[2] Head, Hydraulic Section, Harza Engineering Company, Sears Tower, 233 S. Wacker Drive, Chicago, Illinois 60606-6392.
[3] Director of Design Group II, Korea Heavy Industries & Construction Co., Changwon, Korea.
[4] Senior Engineer, Industrial Equipment Design Department, Korea Heavy Industries & Construction Co., Changwon, Korea.

The power tunnel consists of a concrete lined upper tunnel, a steel lined inclined shaft, and a steel lined lower tunnel. Of particular interest in this model study was the Y-branch in the lower power tunnel. The Y-branch not only splits into two branches with a diameter reduction from 4.0 m to 2.8 m, but simultaneously serves as a 48° vertical bend from the inclined tunnel to the horizontal branches. In addition, the Y-branch structural design includes an internal rib or sickle plate the width of which may affect the head losses in the Y-branch. Uncertainties about the hydraulic performance of the Y-branch as affected by the vertical bend and the sickle plate motivated a hydraulic model study.

Hydraulic Model Study

The hydraulic model study for the Y-branch was performed at the Georgia Institute of Technology in Atlanta, Georgia during the period January-May, 1992. Scale of the model was 1:13.7. The model included a portion of the 4.0 m diameter lower power tunnel (hereafter simply called main tunnel), the Y-branch, and a portion of each of the 2.8 m diameter branch tunnels. Figure 1 shows the Y-branch.

The objectives of the hydraulic model study included the following:

- To determine if the Y-branch located in a vertical bend produces acceptable flow conditions, or if a relocation was required.

- To determine sensitivity of head loss to sickle plate size geometry. Three sickle plate geometries were tested.

- To assess if pressure pulsations generated by the Y-branch or sickle plate would affect the structural design of the Y-branch and pump-turbine operation.

The model study included the measurement of the hydraulic grade lines in the main tunnel and in the branch tunnels for several specified pumping and generating discharges, some flow visualization of the major flow patterns, and dynamic pressure measurements at critical locations in the Y-branch. The entire measurement program was repeated for three sickle plates, each with a different width extending into the Y-branch at the centerline. Ten separate tests for different pumping and generating modes were conducted for each of the sickle plates for a total of 30 tests.

The model length of the main tunnel was approximately 15 diameters of which 10 diameters were clear acrylic pipe. The branch pipes were 10 diameters long and were also clear acrylic pipe. The entrance to the main tunnel contained an expanded-metal screen for equalizing the approach velocity distribution and a flow straightener which removed any swirl in the approach flow.

Of particular interest for the hydraulic model study was the width of the sickle plate at the centerline (743 mm for Sickle Plate 1), because this represents the maximum distance that the plate extends into the interior of the Y-branch from the crotch. All three sickle plates were fabricated from clear acrylic plastic with a 90° bevel along the inside edge of the plate and were all 150 mm (prototype) thick. The upper halves of the three sickle plates which were tested are shown in Figure 2 for comparison of their widths and shapes. The prototype plate widths were 0.743 m, 0.947 m, and 1.150 m for Sickle Plates 1, 2, and 3, respectively.

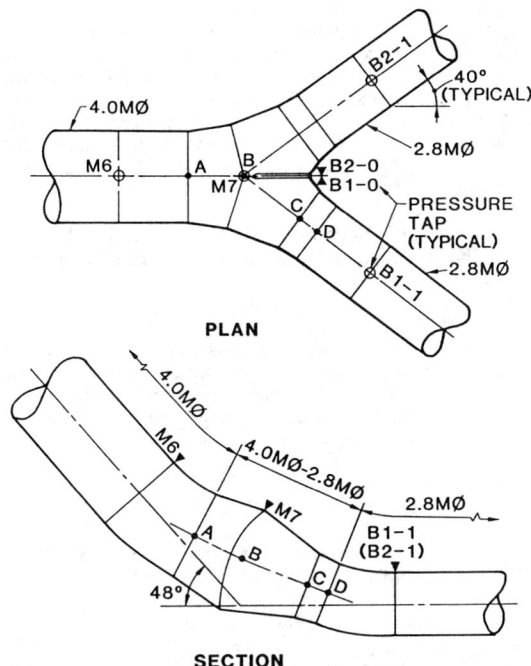

FIGURE 1. GEOMETRY OF Y-BRANCH

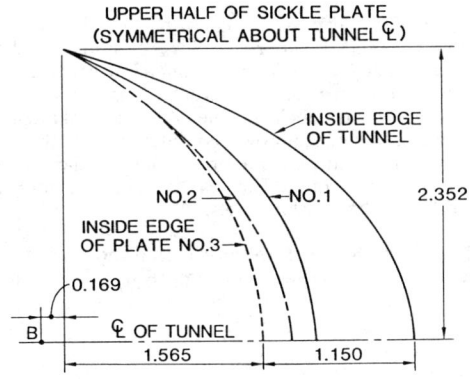

FIGURE 2. SICKLE PLATES SHAPES

Study Results

Location of Y-Branch. Model tests indicated that relocation of the Y-branch in the vertical bend was not necessary. The test demonstrated that head loss factors were not excessive and were in the expected range, which means that head losses used for turbine specifications were acceptable. Flow conditions indicated some separation and vorticity generated by the expansion from 4.0 m diameter to 4.6 m diameter. However, those conditions did not warrant a relocation. In addition, pressure pulsations measured were not considered sufficient to affect the design of the Y-branch or sickle plate.

Head Loss. The determination of head loss factors for the Y-branch was the primary effort of the model study. Head loss factor, K, is defined as:

$$K = h_L/h_v$$

in which h_L is energy loss excluding friction losses, and h_v is velocity head in operating branch. Head loss, for purpose of this study is defined as the energy loss between main tunnel (4.0 m diameter), and each operating branch (2.8 m diameter) for various modes of operation, pumping and generating. Typical plots of energy grade lines for Sickle Plate 1 are shown on Figure 3.

Prior to the tests, it was expected that head loss would increase with size of sickle plate. However, the test results indicated that:

- Within experimental accuracy of the model, there was not significant difference in head loss for the three sickle plates tested.

- The head loss factors were in the expected range, 0.2 to 0.3, for symmetrical generating or pumping. For asymmetric pumping or generating K factors in the range 0.3 to 0.4 were found, which resulted from the greater flow separation and vorticity generated for these infrequent operating modes.

Therefore, any of the three sickle plates tested were acceptable from the standpoint of head loss, and the decision for sickle plate size could be based on structural considerations only.

An unusual aspect of the head-loss results was the sensitivity of the relative head losses in the two branches to the approach flow condition. Preliminary tests for Sickle Plate 1 showed that for maximum symmetrical generating flow, the head losses to the two branches were vastly different. This condition was found to be largely related to the approach velocity profile. After several attempts to straighten the approach flow, the profile was improved considerably, and the resulting K factors were more similar. This sensitivity to approach flow in the model seemed to be related to the generating mode only. The head losses for pumping mode were very nearly symmetrical for all three sickle plates.

Flow Conditions. Flow directions were visualized and documented by photography. Limited use was made of injection of compressed air into the model.

HYDRAULIC MODEL STUDIES

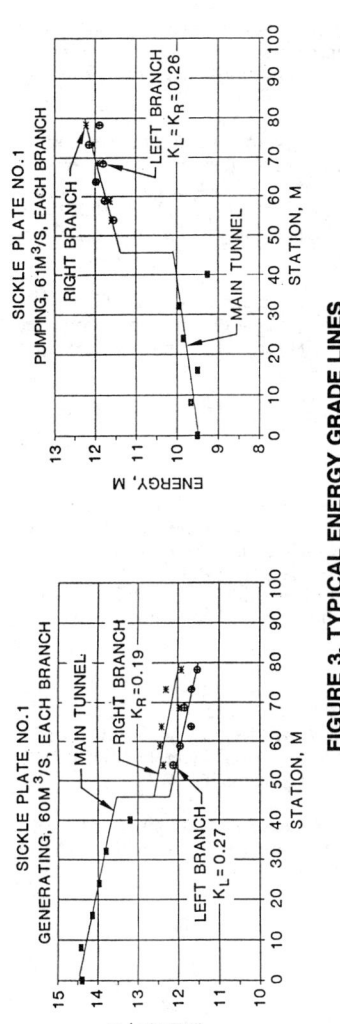

FIGURE 3. TYPICAL ENERGY GRADE LINES

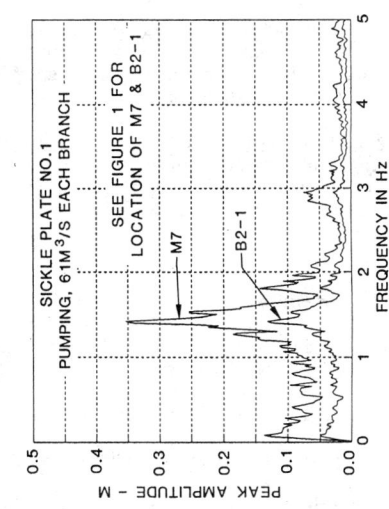

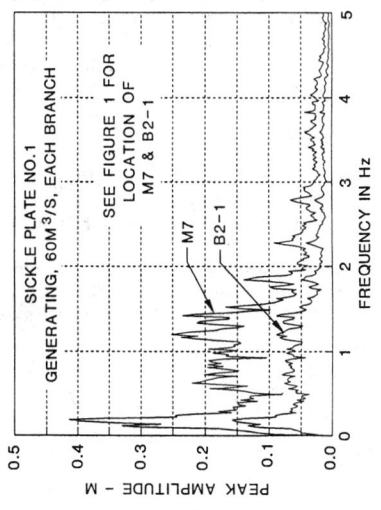

FIGURE 4. TYPICAL DYNAMIC PRESSURE SPECTRA

For symmetrical flow, the only noticeable separation and vorticity was related to the expansion from the 4.0 m diameter to the 4.6 m diameter, prior to the bifurcation under generating mode. This effect is accentuated somewhat by the vertical curvature of the top of the penstock just upstream.

For asymmetrical flow, and particularly for one unit shut down, separation and vorticity were observed. These conditions are clearly unavoidable for such operating conditions. However, no obvious unacceptable flow conditions were noted for these asymmetrical operating conditions.

Structural Aspects of Pressure Fluctuations. Pressure fluctuations were measured by Kulite piezo-resistive pressure transducers at four locations; on either side of the widest part of the sickle plate, at the top end of the sickle plate, and at the end of the Y-branch in the top of the branch tunnel. It was considered that these locations would include the areas of greatest flow instability from separation and vorticity.

The pressure data were generated as dynamic pressure spectra which show amplitude and frequency at the measurement points for all the flow conditions, for all three sickle plates. Figure 4 shows typical results for Sickle Plate 1. Based on these plots, it was concluded that:

- The amplitudes of pressure fluctuations are small, less than 0.5 m.
- The dominant frequencies are under 2 Hz.

From the structural design standpoint, the frequencies are the governing factor. If the frequencies predicted by the model are near the natural frequency of vibration of the sickle plate, then resonance can occur. In such a case, the magnitude of the pressure fluctuation would be important to evaluate danger of fatigue. As the sickle plate is very rigid, its natural frequency is much higher than the frequency of the pressure pulsations. Therefore, the potential for flow induced vibration of the sickle plate is minimal.

References

Georgia Institute of Technology, June 1992, Final Report, Hydraulic Model Study of Penstock Y-Branch, Muju Pumped Storage Project, by Sturm, T.W. and Martin, C. Samuel for Harza Engineering Company.

Harza Engineering Company International L.P., June 1992, Evaluation of Test Results, Hydraulic Model Studies, Steel Penstock Y-Bifurcation, Muju Pumped Storage Power Plant Project.

Model Study of
Rio Hondo Flood Control Channel
Los Angeles, California

John E. Hite, Jr.,[1] Scott E. Stonestreet,[2]
and Michael E. Mulvihill,[2] Members, ASCE

Abstract

The Rio Hondo Flood Control Channel located within the Los Angeles County Drainage Area (LACDA) is part of a comprehensive flood control system which includes 20 dams, 129 debris basins, and 386 km (240 miles) of improved flood control channels. The U.S. Army Engineer District, Los Angeles, plans to increase the level of flood protection provided by the existing system. Increasing the capacity of the Rio Hondo Flood Control Channel is included in this plan and consists of installing parapet walls on top of existing channel levees and modifying bridges which constrict the flow. A physical model study is currently being conducted by the Hydraulic Structures Division of the U.S. Army Engineer Waterways Experiment Station for the Los Angeles District to assist in the final design for the Rio Hondo Channel. The model results to date are presented along with the modeling techniques used for the study.

Purpose of Model Investigation

The existing concrete and grouted stone lined Rio Hondo Channel was designed for a peak discharge varying from 1,133 m^3/s (40,000 cfs) to 1,203 m^3/s (42,500 cfs). Due to changes in hydrology and more available statistical

---

[1]Research Hydraulic Engineer, Hydraulics Laboratory, US Army Engineer Waterways Experiment Station, 3909 Halls Ferry Road, Vicksburg, MS 39180-6199

[2]Hydraulic Engineer, Los Angeles District, US Army Corps of Engineers, P.O. Box 2711, Los Angeles, CA 90053-2325.

data, the new design flow for this channel is 1,283 m^3/s (45,300 cfs) upstream from Whittier Blvd. bridge and increases to 1,424 m^3/s (50,300 cfs) at the lower portion of the channel. The additional discharge enters the channel through side inflows. Feasibility level analyses indicated that flooding problems in highly urbanized areas will result from the higher design discharge.

A physical model study was considered beneficial to eliminate the uncertainties involved in the numerical methods used to predict water depths and energy losses in the vicinity of bridges with highly three-dimensional near critical flows. The purpose of the study, therefore, was to determine the flow depths in the existing channel for the new design discharge. Also, the model would be used to develop designs to increase the flow capacity of existing bridges and to determine the portions of channel that need parapet walls. Finally, the model would be used to verify that the recommended flood control plan will provide the desired level of protection.

Model Description

The 192-m-long (630 ft), 1:50-scale model reproduces the channel from upstream of Whittier Blvd. bridge (channel sta 389+00) to sta 74+00 downstream of Firestone Blvd. bridge, approximately 9.66 km (6 miles) of the prototype channel. The model also reproduces 8 vehicle bridges, 4 railroad bridges, and 2 pedestrian bridges. The model contains 5 curves, numerous channel invert transitions, bike paths, vehicle access ramps, and equestrian ramps. Three large side inflows which introduce additional discharge to the model are reproduced along with the correct geometry for several more inflow drains.

The channel was constructed of plastic-coated plywood and very smooth concrete and the bridges were made from plastic. A model slope adjustment was used to account for the difference in roughness between the model and prototype channel surfaces. Since portions of the Rio Hondo side slopes were grouted stone, a series of tests were conducted in a separate laboratory flume to verify the model Manning n values used for the study. It was found that 0.635 cm (1/4 in.) wire screen reproduced the appropriate composite roughness.

Model Testing Program

Prior to model testing, the Los Angeles District developed a schedule, shown in Figure 1, outlining a scheme for testing each bridge modification. The District requested that bridge modifications be tested starting with the least expensive alternative and continuing with the more expensive alternatives. Generally, the most expensive alternative was to raise or build a new bridge.

ALTERNATIVE BRIDGE MATRIX FOR RIO HONDO CHANNEL

BRIDGE	Existing Conditions	Pier Extensions	Streamline Soffit	Smooth Sides	Adjust Slope	Transition w/in channel	Transition w/in R/W	Raise Bridge	New Bridge	Diversion Channel
Whittier	●	●	●		●	●	●	●	●	●
UPRR	●	●			●					●
Washington	●	●	●		●					●
AT & SFRR	●	●	●	●	●	●	●	●	●	●
Slauson	●	●	●	●	●	●	●	●	●	●
SPRR (U/S)	●	●		●	●	●	●	●	●	●
Telegraph	●									
Santa Ana Fwy	●									
Suva	●	●	●	●	●	●		●	●	
Florence	●	●	●	●	●	●		●	●	
SPRR (D/S)	●	●		●	●	●		●	●	
Firestone	●	●	●	●	●	●		●	●	

Figure 1. Alternative Bridge Matrix

The initial tests were conducted with the entire channel reproducing a concrete lined channel. Since the wire screen had to be added to give the correct composite roughness, it was decided to obtain data with a smooth channel first which could later be compared to the data obtained after the wire was in place. This comparison would show the effect of paving over and smoothing the grouted stone side slopes.

The testing program consisted of obtaining water-surface profiles for the following conditions:

   a. A smooth channel and no debris on the pier noses.

   b. A smooth channel and debris on the pier noses.

c. The existing channel (with grouted stone side slopes) with debris on the pier noses.

d. The improved channel, that is, the condition with modifications and channel improvements, including debris on the pier noses.

The standard debris loading for this study was 0.61 m (2 ft) of debris projecting out from each side of the pier nose and to a depth 1.83 m (6 ft) below the water surface for the design discharge. The Los Angeles District determined this loading from physical evidence left from previous flows in prototype channels located in the Los Angeles area.

Results to Date

The preliminary model results indicate that none of the vehicle bridges have to be raised and the modifications suggested in Table 1 may be effective in increasing the capacity of the bridges. The pier extension has proven to be beneficial in reducing the depth of flow underneath several bridges to an acceptable level. Flow conditions at Whittier Blvd. bridge for the existing channel conditions and the design flow of 1,283 m^3/s (45,300 cfs) are shown in Figure 2. The desired clearance between the low steel of the bridge and the water-surface did not occur for these conditions. A pier extension was designed and installed at Whittier Blvd. bridge in an effort to reduce the flow depth at the bridge. Figure 3 shows the flow conditions with the new pier extension and indicates an acceptable clearance. Further modifications such as streamlining the existing pier underneath the bridge and adding a pier tail to the downstream side will be tested at Whittier bridge in an effort to reduce wave heights downstream of the bridge.

The current testing program is approximately 50 percent complete and plans are to finish all tests by July 1993.

Acknowledgements

The tests described and the resulting data presented herein, unless otherwise noted, were obtained from research sponsored by the U.S. Army Engineer District, Los Angeles, and the Headquarters, U.S. Army Corps of Engineers (USACE). The tests were conducted by the U.S. Army Engineer Waterways Experiment Station. Permission was granted by the Chief of Engineers to publish this information.

TABLE 1. Rio Hondo Bridge Modifications (Preliminary)

Bridge (1)	Station (2)	Type Modifications (3)
Whittier	378+50.99	U/S pier extension Streamline existing pier D/S tapered pier tail
UPRR	369+03.79	U/S pier extension (pier 4) U/S pier extensions (piers 4 & 5) Raise low steel 6 ft
Washington	308+43.86	U/S pier extension Left training wall Modify left bike path D/S tapered pier tail
ATSFRR	268+33.74	Modify left bike path D/S tapered pier tail Raise, modify D/S concrete bridge
Slauson	243+91.25	No change
SPRR (U/S)	235+51.90	D/S tapered pier tail
Telegraph	232+62.24	U/S pier extensions
Santa Ana	225+65.38	No change
Suva	180+00.44	U/S pier extensions Remove left U/S equestrian ramp
Florence	150+29.57	U/S pier extensions Remove left U/S equestrian ramp
SPRR (D/S)	94+95.56	D/S tapered pier tail
Firestone	81+54.92	U/S pier extensions and align with flow

Figure 2. Flow Conditions at Whittier Blvd.

Figure 3. Flow Conditions at Whittier Blvd. with pier extension

CONFIRMATION OF SEISMIC SAFETY AT STAFFORD DAM

Chris D. DeGabriele, Member ASCE[1]
Glen A. Roycroft, Member ASCE[2]

Abstract

A critical aspect of the North Marin Water District's mission to maintain reliable water supply facilities and water service is periodic assessment and re-evaluation of the seismic safety and performance of Stafford Dam. This paper chronicles the District's program as it keeps pace with advances in analytical methods and exploration techniques. Investigations of the seismic safety of Stafford Dam including response to the 1989 Loma Prieta earthquake are described.

Overview of Stafford Dam

Stafford Dam is a 79-foot high compacted earthfill dam founded on streambed alluvium about 40-feet thick. It is located on Novato Creek approximately 4 miles west of the City of Novato in northern Marin County California. This is a highly seismic area subject to potentially severe ground shaking from events on the nearby San Andreas and Hayward faults.

The dam was designed by the Engineering Office of Clyde C. Kennedy and constructed in 1951. The dam impounds Stafford Lake, a water supply reservoir with a capacity of approximately 4,300 acre feet. A large portion of the downstream shell was removed and reconstructed in 1984 to replace the toe drain. The dam was raised 8 feet in 1985 in conjunction with installation of a new spillway to accommodate the probable maximum flood. This provided increased flood protection for Novato Creek which flows through the City of Novato. An inundation study performed in 1973 estimated that approximately 10,000 people downstream of the dam might be affected by catastrophic failure of the structure.

---

[1]Chief Engineer, North Marin Water District, P.O. Box 146, Novato, CA 94948

[2]Senior Engineer, Miller Pacific Engineering Group, 165 N. Redwood Dr., San Rafael, CA 94903

A significant amount of information including cyclic triaxial laboratory test data on the embankment and foundation materials is available from engineering studies of the dam (Lee & Praszker, 1978 & 1986; Harlan Miller Tait, 1985a, 1985b & 1986; Miller Pacific Engineering Group 1992). The embankment and foundation consists of four main soil types as shown on the idealized cross-section of the dam in Figure 1:

(1) A relatively homogenous embankment of compacted fill, including old and new portions (zones 1, 2 and 3). The fill is primarily a Clayey Sand (SC) with gravel. The percent of materials passing the No. 200 sieve averages 32%. Typical liquid limits are between 36 and 42, and the plasticity index is about 18.

(2) Brown sandy clay and clayey sands directly underlying the dam (zone 4).

(3) Blue-grey clay, silty clay and clayey silt (zone 5) underlie the brown sandy clay and clayey sand. This clay has between 70% and 98% passing the No. 200 sieve, an average liquid limit of 39, and an average plasticity index of about 19.

(4) A discontinuous zone of interlayered sands, silts and clays (zone 6) directly overlying Franciscan bedrock (zone 7) and underlying the blue clay. These interlayered soils contain up to 33% gravel-size particles.

Instrumentation for the dam includes piezometers and survey monuments. Piezometer readings are recorded weekly. Dam crest monuments are surveyed annually. The District measures the turbidity daily and the rate of flow at the toe drain outlet weekly.

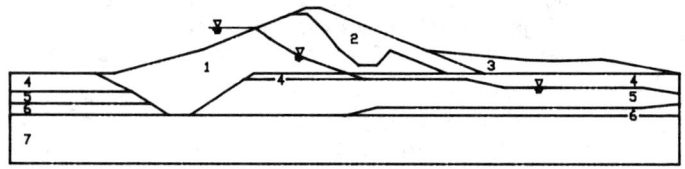

IDEALIZED CROSS SECTION OF DAM FOR SEISMIC ANALYSIS

LEGEND:
1) EMBANKMENT: ORIGINAL MAIN PORTION
2) EMBANKMENT: TOE DRAIN REPLACEMENT AND RAISED PORTION
3) EMBANKMENT: TOE FILL
4) UPPER ALLUVIUM: BROWN CLAYEY SAND, SANDY CLAY
5) LOWER ALLUVIUM: BLUE-GREY CLAYEY SILT, SILTY CLAY
6) ALLUVIUM OVER BEDROCK: INTERLAYERED SAND, SILT AND CLAY
7) BEDROCK: FRANCISCAN FORMATION
▼) IDEALIZED PHREATIC SURFACE FOR ANALYSIS

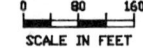
SCALE IN FEET

FIGURE 1

Previous Seismic Investigations

At the time of the design of Stafford Dam, design philosophy considered that every dam was designed to make it completely safe and the likelihood or consequences of failure were not necessarily examined. A fundamental change in the philosophy of dam design occurred during the 1960's when engineers and owners recognized that dam failures do occur and that no structure is completely safe. Risk and the consequences of failure became concepts receiving greater consideration by engineers and decision makers. The District realizes its responsibility to public safety and has similarly kept pace with the times by using the best available analytical techniques in evaluating dam performance and when considering rehabilitation or expansion projects.

Four evaluations of seismic safety have been conducted thus far in the life of the dam. The original design of Stafford Dam relied primarily on engineering judgement based on past experience to minimize the danger of embankment failure during earthquakes. At that time, engineering judgement was confirmed through the use of pseudostatic analytical methods of slope stability. This was the state of practice at the time of design of Stafford Dam and remained as the primary analytical method through the 1960's. Each of the subsequent seismic evaluations took advantage of advances in geotechnical exploration and testing techniques, and analytical methods.

The first major seismic evaluation was conducted in 1978 when the District retained a geotechnical consultant to evaluate the seismic stability and provide a qualitative evaluation of the possibility of increasing the height of the existing dam (Lee & Praszker, 1978). This comprehensive study included: review of the limited existing information on the original dam design and construction; subsurface exploration including six test borings, installation of 4 piezometers, and geophysical testing to determine shear wave velocities of the embankment and foundation materials; laboratory testing of soil samples obtained from the test borings including cyclic triaxial tests; evaluation of soil liquefaction potential; two dimensional static and dynamic finite element analyses; and stability analyses. The study concluded that the then existing dam could withstand a magnitude 8.25 earthquake on the San Andreas Fault with only minor permanent deformation resulting. As a follow-up to this study, the District installed accelerographs in the hope of recording ground motion for use in future analyses.

The second seismic evaluation commenced in 1985 when the District replaced the toe drain and the following year, reconstructed the spillway and raised the dam 8 feet to its present height. Additional static and pseudostatic stability analyses of the enlarged embankment were conducted by District consultants as part of this effort (Harlan Miller Tait,

1985 a,b). Additional test borings were drilled on the downstream face of the dam and laboratory testing performed on the samples obtained. These studies again confirmed the seismic stability of Stafford Dam.

In 1986, a dynamic response analysis of the enlarged embankment was performed (Lee & Praszker, 1986) for a magnitude 8.25 event on the San Andreas Fault. Pseudostatic slope stability analyses were then used to estimate yield accelerations and potential sliding masses in the embankment and foundation materials. The classic Newmark (1965) approach was used to estimate deformations due to seismic loading. The study reconfirmed seismic safety and concluded that deformation of up to 0.1 feet on the upstream face and 0.4 feet on the downstream face could occur.

Recent Investigations

The dam was shaken by the October 17, 1989 Loma Prieta Earthquake which was centered approximately 80 miles south of the dam. Dynamic response of the dam as a result of this event was recorded by the accelerographs installed at the dam crest and on rock at the right abutment approximately ¼ mile downstream of the dam crest.

In 1992, the District retained Woodward-Clyde Consultants (WCC) to conduct the fourth seismic evaluation. In the state of practice today, engineers usually estimate dynamic response with total stress, equivalent linear two-dimensional finite element methods. Thus, a numeric model of Stafford Dam based on the computer code FLUSH was developed to represent embankment and foundation geometry, static and dynamic material properties, and analyze dynamic response.

Dynamic Properties by CPT Characterization of material type and dynamic properties were developed based on prior investigations and recent field explorations consisting of Cone Penetration Tests (CPT) conducted by Miller Pacific Engineering Group (1992). The use of the CPT represented a significant advance in evaluation of dynamic properties for Stafford Dam. Past practice to evaluate dynamic properties in-situ has been primarily based on Standard Penetration Test (SPT) data simply because this constituted the most voluminous data base available. However, engineers are beginning to emphasize the CPT as the preferred in-situ testing technique because it is unequalled in delineation of soil stratigraphy, is repeatable and accurate, and provides continuous measurement of subsurface parameters. Moreover, advances in the interpretation of CPT data by various researches (Robertson and Campanella, 1983; Olsen, 1984; Seed and DeAlba, 1986) allow the practical application of this technique.

Calibration of Model The recorded Loma Prieta ground motions were used to calibrate the numeric model of the dam. The input motion for the model was taken as the recorded ground motion on rock. To account for the difference in motions between a rock outcrop and bedrock underlying the dam embankment, the recorded outcrop motion was adjusted by using several "deconvolution" analyses (Mejia and Boulanger, 1992). The deconvolution analyses were performed using one-dimensional wave propagation analyses by the computer program SHAKE. The ground motions adjusted by deconvolution were used as input motions for the dynamic response model. Model parameters (i.e., shear wave velocity, Poisson's ratio, unit weight, shear modulus and damping relationships) were varied and sensitivity analyses were performed in order to provide a good match to the recorded response. The dams response to recorded Loma Prieta earthquake motions shows significantly stiffer soil properties than those indicated by field measurement and used in the previous seismic studies.

Design Earthquake Motions The design earthquake, a magnitude 8.25 event on the San Andreas Fault at a distance of about 10 miles from the site, was selected for the seismic evaluation. This is considered to be the maximum credible earthquake for this portion of the San Andreas Fault. A peak ground acceleration (PGA) on rock for this event was calculated to be 0.48g based on available attenuation relationships. Considering the severe consequences of dam failure, the maximum credible earthquake and high level of PGA was judged to be appropriate for use in the recent seismic study. Seismologists have not recorded ground motion for earthquakes of magnitude 8+ as yet. Therefore, the Seed and Idriss (1969) synthetic ground motion as modified by Seed and Sun (1989) was used as the design earthquake for input into the dynamic response model. This ground motion record was scaled to the PGA of 0.48g.

Response to Design Earthquake The WCC (1992) dynamic response model predicts amplification of ground motion during the design earthquake. Peak accelerations ranging from 0.49g to 0.72g are predicted at the dam crest. Conversely, the 1978 Lee & Praszker studies had predicted attenuation of earthquake motions from bedrock to the dam crest. Both studies concluded that because the embankment and foundation are composed primarily of well compacted, stiff to dense clayey sands, earthquake loading is not expected to result in significant strength loss and/or liquefaction of soils. Analyses using CPT data for the zone 6 sands indicate limited liquefaction under the design level seismic shaking.

Seismic Stability Using the results of the dynamic response model for a PGA of 0.72g, the seismic stability of Stafford Dam was evaluated in terms of potential deformations under the design earthquake. The CPT data provided useful information on the extent and residual strength of the potentially liquefiable (zone 6) sands. The seismic deformations were estimated using the methods outlined by Makdisi and Seed (1978). Yield accelerations were computed using limit equilibrium slope stability analyses for non-circular, sliding-block failure mechanisms. Deformations were conservatively estimated assuming that liquefaction occurs early in the duration of shaking. The potential deformations were estimated to be less than 1 foot.

Conclusions

The four seismic studies analyzing Stafford Dam were conducted to evaluate seismic response and stability and all concluded that the design earthquake is not expected to cause significant damage to the dam. Each study used the most appropriate and up-to-date technology available at that time. The ultimate test of our analytical procedures is the comparison of computed results with observed behavior. The recent studies by the District accomplished this goal by calibrating a dynamic model of Stafford Dam with Loma Prieta ground motions. Using the calibrated dynamic model to estimate the Dam's response to the design earthquake is the best logical step available at this time. Our evaluations will be updated when new response data is recorded and better analytical techniques become available.

References

Harlan Miler Tait Associates, "Construction Report Stafford Dam Toe Drain Replacement Report," January 9, 1985.

Harlan Miller Tait Associates, "Design Report Stafford Flood Control and Spillway Project," April 8, 1985.

Harlan Miller Tait Associates, "Construction Report, Stafford Flood Control and Spillway Project," February 7, 1986.

Lee & Praszker, "Stafford Dam Seismic Study and Comments on its Enlargement," April 1978.

Lee & Praszker, "Seismic Study of the Enlarged Stafford Dam," June 1986.

Makdisi, F., and Seed, H.B. (1978), "Simplified Procedures for Estimating Dam and Embankment Earthquake-Induced Deformations," Journal of the Geotechnical Engineering Division, ASCE, Vol. 104, No. GT7, July 1978.

Mejia, L.H. and Boulanger, R.W., "Calibrated Dynamic Response Analysis of Stafford Dam," submitted for publication in the Proceedings of the Third International Conference on Case Histories in Geotechnical Engineering, University of Missouri-Rolla, June 1993.

Miller Pacific Engineering Group, "North Marin Water District Stafford Dam, Seismic Stability CPT Exploration Program," February 20, 1992.

Newmark, N.M. (1965), "Effects of Earthquakes on Dams and Embankments," Geotechnique, Vol, 15, No. 2, pp. 139-160.

Olsen, R.S. (1984), "Liquefaction Analysis Using the Cone Penetrometer Test (CPT)," Proceedings, Eight World Conference on Earthquake Engineering, San Francisco, Vol. 3, pp. 247-254.

Robertson, P.K., and Campanella, R.G. (1983), "Evaluation of Liquefaction Potential Using the Cone Penetration Test," Soil Mechanics Series Report No. 64, Department of Civil Engineering, University of British Columbia, Vancouver.

Seed, H.B. and Idriss, I.M. (1969), "Rock Motion Accelerograms for High Magnitude Earthquakes," Report No. UCB/EERC-69/7, Earthquake Engineering Research Center, University of California, April 1969.

Seed, H.B., and DeAlba, P.M. (1986), "Use of SPT and CPT Tests for Evaluating the Liquefaction Resistance of Sands," Proceedings, INSITU 86, ASCE Specialty Conference, Blacksburg, Virginia, pp. 281-302.

Seed, H.B. and Sun, J.I., (1989), "Implications of Site Effects in the Mexico City Earthquake of September 19, 1985 for Earthquake-Resistant Design Criteria in the San Francisco Bay Area of California," Report No. EERC-89/03, University of California, March 1989.

Woodward-Clyde Consultants, "Stafford Dam, Analysis of Recorded Earthquake: Ground Motions and Evaluation of Seismic Stability," April 1992.

An Improved Method for Measuring System Performance of Hydraulic Infrastructure Systems

Sue-Jen Wu[1] and Ru-Lin Hsu[2]

Abstract

Due to the public awareness of the importance of functioning of hydraulic infrastructure systems and the serious consequences caused by their failure, many researches were undertaken to evaluate the performance of these systems. However, most of the work was concentrated on evaluating system performance based only on simple connectivity concept. This measure is not realistic and is ineffective in capturing system behavior since it does not address both the capacity and pressure requirements. Although some work has been done to incorporate the capacity effect into the performance measure, current methodologies are all in the class of NP-hard. This paper presents an improved method for measuring system performance of hydraulic infrastructure systems. Large systems can now be analyzed in reasonable computation time by the presented algorithm. This method provides us with quantitative measurements which can be utilized as the basis for developing sound maintenance strategies.

Introduction

A hydraulic infrastructure system is the key to the wealth and quality of life of a society. As the system deteriorates, civilization unravels and public health is threatened due to its inability to transport clean water.

---

[1] Engineer, Dept. of Hydraulic Engineering, Sinotech Engineering Consultants, Inc., 171 Nanking E. Rd. Sec. 5, Taipei, Taiwan, R.O.C.
[2] Chief Engineer, Sinotech Engineering Consultants, Inc., 171 Nanking E. Rd. Sec. 5, Taipei, Taiwan, R.O.C.

However, the system has finite life. Pipes, valves, tanks and reservoirs have rapidly degraded because of age, leaks, breaks, and excessive demands. In order to reduce the hazard caused by a deteriorated hydraulic infrastructure system, a reliable system is needed. The basis for maintaining a reliable system depends on sound maintenance policies, which require quantitative measurements for making decision. Theory of reliability analysis has been proved to be able to assess system performance successfully (Wu, 1992). Based on this information, optimal maintenance decision criteria can be set up.

In real-life systems, the adequacy of water supply requires that the source not only be connected to the demand point but that the needed amount of water must be available at a prescribed pressure(Shamsi and Quimpo,1988). The performance measure which only considers simple connectivity is therefore not realistic. Since some developed algorithms which incorporate capacity effect into this measure are NP-hard, this study will attempt to overcome the NP-hard difficulty to develop an improved method for measuring system performance.

A Polynomial Time Algorithm

"NP-hard" is the term originally introduced in the theory of computational complexity and NP-completeness. Any algorithm which is referred to be NP-hard indicates that its running time exponentially increases with system size and can not be bounded. The problem becomes intractable when the analyzed system is large. It is generally accepted that a problem has not been "well-solved" until a polynomial time algorithm is known for it(Garey and Johnson,1979). The polynomial time algorithm presented in this paper is named as the Reliability Index Algorithm, which applies the concept of reliability reductions (Satyanarayana and Wood, 1982) to reduce the size of network in analysis. Three reductions are used in this algorithm : series reduction, parallel reduction, and Type 1 polygon-to-chain reduction. Together with the factoring theorem, these reductions are applied recursively in the algorithm. The value obtained from this algorithm is a reliability index which reflects system performance. The network reliability index is determined by multiplying the reliability index of the reduced graph before factoring by its reliability index correction factor Ω . These three reductions are shown in Figure 1.

In order to incorporate the capacity effect, the Reliability Index Algorithm follows Aggarwal's concept(Aggarwal, 1985,1988) that each probability term is multiplied by a normalized weight. Detailed derivation of reliability indices and capacities for the reduced links of the above three reductions and the factoring theorem is given by Wu (1992). They are briefly introduced as follows.

(1) Series Reduction

A series reduction replaces two adjacent arcs e_1 and e_2 with a single link e_r. The reliability index correction factor Ω is 1 for this reduction and the reliability index for the reduced link is

$$\frac{\min(C_1, C_2)}{C_s} p_1 p_2 \tag{1}$$

where C_1 and C_2 are the capacities of the two original arcs. C_s represents the required system capacity at the demand point. The capacity for the reduced link is $\min(C_1, C_2)$.

(2) Parallel Reduction

A parallel reduction replaces a pair of arcs e_1 and e_2 with a single link e_r. The reliability index correction factor Ω is 1 for this reduction and the reliability index for the reduced link is

$$\frac{\min(C_s, C_1 + C_2)}{C_s} p_1 p_2 + \frac{\min(C_s, C_1)}{C_s} p_1 q_2 + \frac{\min(C_s, C_2)}{C_s} q_1 p_2 \tag{2}$$

The capacity for the reduced link is $C_1 + C_2$.

(3) Type 1 Polygon-to-Chain Reduction

The reliability index correction factor and the reliability indices for the reduced chain is

$$RI_r = \frac{\frac{X_s}{X_t}\delta}{\alpha + \frac{X_s}{X_t}\delta}, \quad RI_s = \frac{\frac{X_r}{X_t}\delta}{\beta + \frac{X_r}{X_t}\delta}, \quad \Omega = \frac{(\alpha + \frac{X_s}{X_t}\delta)(\beta + \frac{X_r}{X_t}\delta) X_t}{X_r X_s \delta}$$

where X's are the ratios of arc capacities to the system capa-

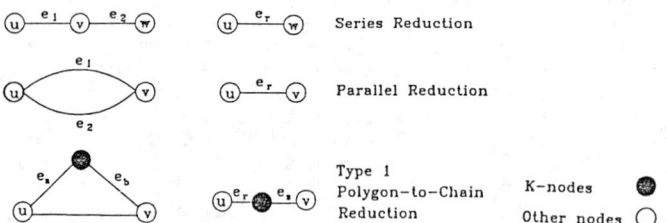

Figure 1. Series, Parallel, and Type 1 Polygon-to-Chain Reductions

city. α, β and δ are the summed state reliability indices of the polygon. The capacities for the reduced chain are

$$C_{e_1} = C_a + q_a p_b p_c \min(C_b, C_c) \quad (3)$$
$$C_{e_2} = C_b + q_b p_a p_c \min(C_a, C_c) \quad (4)$$

(4) Factoring Theorem

The factoring theorem is applied whenever the network being analyzed can not be further reduced by any of the above three reductions. In applying this theorem, the key point is to select an arc e_i which makes further reductions possible. By factoring this arc, two subnetworks are generated : one with arc e_i being contracted, the other with arc e_i being deleted. The equation used for this theorem is

$$RI(G_K) = p_i RI(G_{K'} * e_i) + q_i RI(G_K - e_i) \quad (5)$$

Case Study

One of the major advantages of the Reliability Index Algorithm is that it can analyze large systems. Results of a large-sized system which is analyzed by this algorithm are therefore presented in this paper. The hydraulic infrastructure system investigated is the Norwich system. The layout plan of this system is shown in Figure 2 and the corresponding network representation is shown in Figure 3. There are 150 arcs and 127 nodes in this network. Water is supplied to the system from more than one source. A 1.5 MG (5678.1 m³), in the west side of the system, supplies water to the network. In the east side, a combination of supplies, including two surface storage facilities and two wells, supplies water to the system. For simplicity, the east side supplies are treated as a subsystem. Together with the west

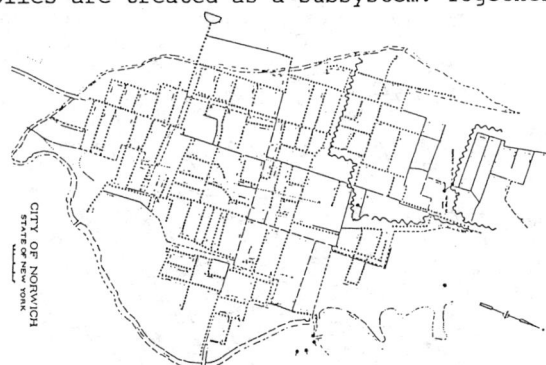

Figure 2. Layout Plan of the Norwich Hydraulic Infrastructure System

side supply, a dual supply sources is formed. An artificial node (node 1) is therefore introduced to transform this system into a standard single source system.

Results from the Reliability Index Algorithm are shown in Figure 4, a reliability surface plot. A reliability surface plot provides us with an overall idea about the conditions of system performance. Areas with low reliability indices can easily be identified from this plot. It can also be used to determine where in the system maintenance and rehabilitation will improve the supply reliability most (Shamsi and Quimpo, 1989).

Following traditional non-polynomial time algorithms, the full expansion of minimal paths or cut sets of an undirected system with 150 arcs may take $2^{150} = 1.427247691 \times 10^{45}$ terms. Current computers' capabilities can not handle this huge amount of computations. For example, one of the fastest CRAY supercomputers has the computational capability about 10^{10} floating point operations (FLOPS) per second. If it is estimated that each term takes 1 FLOP, a total of $1.427247691 \times 10^{45}$ FLOPS are necessary. Using this CRAY supercomputer, the estimated computation time to calculate the reliability for each sink node would be $1.427247691 \times 10^{35}$ sec. This is equivalent to 4.5×10^{27} years. Even the expansion of minimal paths or cut sets is reduced to $2^{60} = 1.15 \times 10^{18}$ terms. Using the same CRAY supercomputer, the computation time still needs 1.15×10^{18} sec, which is equal to 133 days. Hence, it is very difficult to analyze this system by the non-polynomial time algorithm.

The average time needed to obtain reliability index for each sink in this network is about 2 minutes by using the presented algorithm in a 486DX-33 personal computer. From the above comparison, one can see how much improvement has been achieved by the presented method.

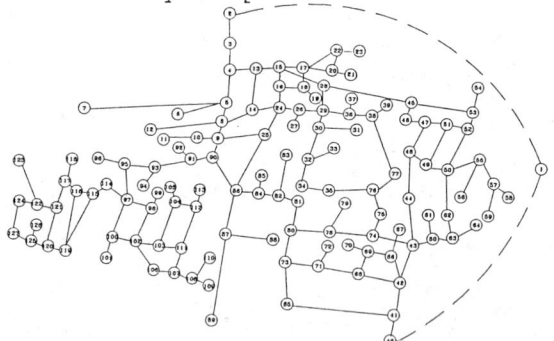

Figure 3. Standard Network Representation for the Norwich System

Conclusion

The presented Reliability Index Algorithm provides us with an important tool for evaluating the performance of large-sized hydraulic infrastructure systems, which was intractable before. Regions with low reliability indices should be assigned a higher priority for maintenance. Simulations can easily be performed by applying the presented algorithm. Important information can be obtained from these simulation results. Maintenance policies can then be set up to keep the hydraulic infrastructure system more reliable.

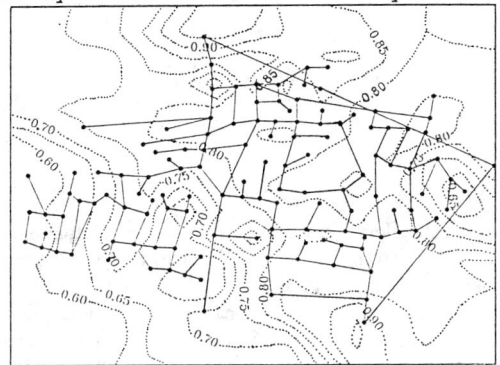

Figure 4. Reliability Surface for the Norwich System

References

Aggarwal, K.K. (1985). "Integration of Reliability and Capacity in Performance Measure of a Telecommunication Network", IEEE Trans. on Reliability, Vol. R-34, No.2.

Aggarwal, K.K.(1988). "A Fast Algorithm for the Performance Index of a Telecommunication Network", IEEE Trans. on Reliability, Vol. 37, No. 1.

Garey, Michael R. and David S. Johnson(1979). "Computers and Intractability, A guide to the Theory of NP-Completeness", Bell Telephone Laboratories, Inc.

Satyanarayana, A., and R. Kevin Wood (1982). "Polygon-to-Chain reductions and Network Reliability", ORC 82-4, Operation Research Center, Univ. of California, Berkeley, CA.

Shamsi, U.M. and Rafael G. Quimpo (1988). "Reliability Approach to Water Supply Infrastructure Maintenance", National Science Foundation Grant No. ECE-8601169.

Wu, Sue-Jen (1992). "Predictive Model and Reliability Analysis for Water Distribution Systems", Ph.D. Dissertation, University of Pittsburgh, Pittsburgh, PA.

Bridge-Scour Analysis using the
Water-Surface Profile (WSPRO) Model

David S. Mueller[1], A.M. ASCE

Abstract

   A program was developed to extract hydraulic information, required for bridge-scour computations, from the Water-Surface Profile computation model (WSPRO). The program is written in compiled BASIC and is menu driven. Using only ground points, the program can compute average ground elevation, cross-sectional area below a specified datum, or create a Drawing Exchange Format[2] (DXF) file of a cross section. Using both ground points and hydraulic information from the equal-conveyance tubes computed by WSPRO, the program can compute hydraulic parameters at a user-specified station or in a user-specified subsection of the cross section. The program can identify the maximum velocity in a cross section and the velocity and depth at a user-specified station. The program also can identify the maximum velocity in the cross section and the average velocity, average depth, average ground elevation, width perpendicular to the flow, cross-sectional area of flow, and discharge in a subsection of the cross section. This program does not include any help or suggestions as to what data should be extracted; therefore, the user must understand the scour equations and associated variables to be able to extract the proper information from the WSPRO output.

Introduction

   Bridge-scour calculations require knowledge of specific hydraulic conditions during floods for which such calculations are to be performed. Measured hydraulic data are not generally available because of the extreme nature of the floods of

---

[1]Hydrologist, U.S. Geological Survey, Water Resources Division, 2301 Bradley Avenue, Louisville, KY 40217

[2]Drawing Exchange Format is a trademark of AutoDesk, Inc. Use of trade, product, or firm names is for descriptive purposes only and does not constitute endorsement by the U.S. Government.

BRIDGE-SCOUR ANALYSIS

interest. Therefore, the hydraulic conditions must be estimated from hydraulic or numerical models. The Water-Surface PROfile computation model (WSPRO) was developed by the U.S. Geological Survey in cooperation with the Federal Highway Administration (FHWA) and includes special routines to compute the hydraulic conditions in the vicinity of bridges (Shearman, 1990; Shearman and others, 1986). In addition, the model approximates the 1-dimensional velocity distribution by dividing the flow through each cross section into 20 equal-conveyance tubes. Contraction- and abutment-scour equations require hydraulic information associated with various subsections of the cross sections at the bridge and approach, and pier-scour equations require hydraulic information at specific locations or points along the bridge cross section. Although WSPRO has very flexible output tables, there is no option to allow output of user-defined subsection properties without specifying geometric or roughness break points. Although the necessary information can be extracted from the WSPRO output by simple calculations or visual interpolation, the Bridge Scour Analysis using WSPRO (BSAW) program was developed to improve the precision and efficiency of extracting this information. The accuracy of the BSAW program has been verified by manual computations for more than 25 WSPRO output files containing different geometric and hydraulic data. This paper describes of the user interface, input, special routines, and output of the BSAW program.

Data input requirements

Input for BSAW is the standard output file generated by WSPRO. No special WSPRO user-defined tables are required, but velocity distributions must be requested for all cross sections for which hydraulic information is required. The BSAW program scans the WSPRO output file for particular headings such as "VELOCITY DISTRIBUTION" and "*** START PROCESSING CROSS SECTION" and reads the appropriate data that follows. The only other data required by BSAW is supplied by the user through the menus and prompts.

Computations using only ground points

If only ground points are available in the WSPRO output file, BSAW can compute geometric properties for subsections of the cross section. BSAW will display all cross sections for which there are ground points in the WSPRO output file. After selection of the desired cross section, BSAW will display the maximum and minimum stations for that cross section and a menu to allow the user to select 'Compute Average Ground Elevation' or 'Compute Subsection Properties'. If the user selects 'Compute Average Ground Elevation' the program will prompt for the beginning and ending stations of the desired portion of the cross section. The program computes the average ground elevation by integrating the area between the ground and a reference plane at zero elevation and then dividing this area by the distance between the beginning and ending station (width). If the user-specified beginning and ending stations do not correspond directly to stations read from the WSPRO output file, the elevations at the user-specified stations are linearly interpolated from the adjacent stations. The width perpendicular to the flow is computed by multiplying the width by the cosine of the angle of skew for the cross section. Upon completion of the calculations the program will display the specified stations, the width perpendicular to flow, and the average elevation of the subsection (Figure 1).

```
MINIMUM STATION =           0.0
MAXIMUM STATION =        4100.0

ENTER BEGINNING STATION => 2000
ENTER ENDING STATION     => 3000

BEGINNING STATION         2000.0
ENDING STATION            3000.0
WIDTH PERP. TO FLOW       1000.0
AVERAGE ELEVATION          456.7

STRIKE ANY KEY TO CONTINUE
```

Figure 1. Output screen for average ground elevation option.

The user can also compute the geometric properties of a subsection by selecting 'Compute Subsection Properties' and entering the beginning and ending stations and a reference elevation. The cross-sectional area is computed by integrating the area between the ground and the user-specified reference plane. The user-specified plane is expected to represent the water surface, so the area computed will represent the cross-sectional area of flow. It is common for the bed elevation within a specified subsection to be above the water surface at the waters edge, at piers that have been coded into the ground points, and at bars or islands in the stream. Therefore, routines were developed to check for situations where the bed elevation is above the reference elevation. If the bed elevation is above the reference elevation, the depth of flow is forced to zero, the location at which the bed elevation equals the reference elevation is computed, and integration continues with no contribution from the portion of the cross section above the specified reference plane. After the gross cross-sectional area is computed, it is corrected for skew by multiplying the area by the cosine of the angle of skew of the cross section. The average depth of flow is computed by dividing the corrected cross-sectional area by the width perpendicular to the flow. Upon completion of the calculations, the program will display the specified stations, the width perpendicular to flow, the average elevation, the cross-sectional area of flow, and the average depth (Figure 2).

Using only ground points, the user can also generate a Drawing Exchange Format[2] (DXF) file of the desired cross section. The DXF file format is a common format which can be read by most computer-aided drafting and geographical information systems.

```
MINIMUM STATION =           0.0
MAXIMUM STATION =        4100.0

ENTER BEGINNING STATION => 2000 ENTER
ENDING STATION      => 3000

ENTER WATER-SURFACE ELEVATION => 460

BEGINNING STATION         2000.0
ENDING STATION            3000.0
WIDTH PERP. TO FLOW       1000.0
AVERAGE ELEVATION          456.7
X-SECTION AREA            5428.0
AVERAGE DEPTH                5.4

STRIKE ANY KEY TO CONTINUE
```

Figure 2. Output screen for geometric properties computations.

Computation of point hydraulic parameters

The option to compute point hydraulic parameters was included to support pier-scour computations. The user must select the desired cross section and flow conditions from the menus, and 'Pier Hydraulic Parameters' from the hydraulics menu. The program will prompt for the station of the pier. The cross-section identification, angle of skew, total discharge, water-surface elevation, specified pier station, depth of flow, and velocity are then displayed (Figure 3).

Several special routines have been included to estimate the depth at or near the specified station. If the elevation of the specified station is below the water-surface elevation, the program will compute the depth at the specified station. The maximum depth in the cross section and the maximum depth near the pier (within a distance equal to the depth of flow on either side of the specified station) also are computed and displayed (Figure 3). The specified station could be at a vertical wall (a station that has two elevations associated with it), such as at a vertical abutment or at the edge of a pier coded in the ground points. The program checks for vertical walls and uses the lowest elevation for that station to compute the depth at the station. If the specified station is near the centerline of a pier that has been coded in the ground points, the elevation at that station may be above the elevation of the water surface. If the elevation at the station is above the water-surface elevation, the program will print a message stating that the ground elevation is above the water-surface elevation. The program will then display the depth and location of the nearest station having a positive depth, the maximum depth near the pier, and the maximum depth in the cross section.

The velocity at the specified station is determined using the hydraulic properties associated with each of the 20 equal-conveyance tubes computed by

```
      MINIMUM STATION =       0.0        LEFT WATERS EDGE =      168.2
      MAXIMUM STATION =    4100.0        RIGHT WATERS EDGE =    3245.0
                       X-SECTION NAME            COMP
                       SKEW ANGLE                 0.0
                       TOTAL DISCHARGE          14190
                       WATER SURFACE ELEV.      459.0
                       PIER STATION             200.0

      LOCATION         STATION           BED ELEV.               DEPTH
      AT PIER           200.0              457.6                   1.4
      NEAR PIER         300.0              453.2                   5.8
      CROSS-SECTION    3157.0              420.7                  38.3

      PIER             STATION           VELOCITY              TUBE NO.
                                            0.00                   0
                        168.2
      200.0  ------------------------->     0.39                   1
                        519.4
                                            0.49                   2

      MAXIMUM FOR CROSS SECTION             1.28                  18
                       STRIKE ANY KEY TO CONTINUE
```

Figure 3. Output screen for pier hydraulic computation.

WSPRO. The velocity, boundaries, and number of the equal-conveyance tube containing the specified station is presented. In addition, the velocity and number of the adjacent equal-conveyance tubes and the equal-conveyance tube containing the maximum velocity for the cross section are also presented (Figure 3). In certain situations, the user may want to adjust the velocity at a specified station. For example, if the velocity in the equal-conveyance tube containing the bridge pier is less than the velocities in the adjacent equal-conveyance tubes it may be reasonable to adjust the velocity at the specified station based on the velocities in the adjacent equal-conveyance tubes. When the specified station is very close to the boundary of two adjacent equal-conveyance tubes the value of the adjacent equal-conveyance tube may need to be considered in the determination of the velocity of the specified station. For design purposes, especially in streams with frequent changes in thalweg location, the maximum velocity in the cross section may be used for all of the piers in the main channel.

Computation of subsection hydraulic properties

Abutment- and contraction-scour computations require knowledge of the hydraulic parameters associated with a specific subsection of a cross section. WSPRO provides subsection parameters at specified break points. However, break points must only be added to represent a change in roughness or geometry. If break points are added without meeting these criteria, an artificial increase in the kinetic energy correction factor may result. Therefore, addition of break points to compute subsection parameters is not desirable. If parameters of a subsection are desired and the subsection does not begin and end at a break point, the parameters must be computed from the equal-conveyance-tube information.

The BSAW program will compute the hydraulic parameters associated with a user-specified subsection using the ground points and equal-conveyance-tube information. The user must select the desired cross section and flow condition from the menus, and 'Compute Hydraulic Parameters' from the hydraulics menu. The program will prompt the user for the beginning and ending stations of the desired subsection. The maximum and minimum stations in the cross section and the stations for each water edge are displayed. The program will not allow station input outside the edge of water. The program will display the cross-section identification, total discharge, water-surface elevation, and beginning and ending stations of the subsection. The program will compute and display the average ground elevation, width perpendicular to flow, average depth, cross-sectional area of flow, discharge, and average velocity of the specified subsection (Figure 4). The average ground elevation, width perpendicular to the flow, and average depth are computed in the same manner as described previously in this paper. The cross-sectional area for the specified subsection is computed by summing the cross-sectional areas (computed by WSPRO) of all the equal-conveyance tubes which are completely contained in the specified subsection and adding the area for the portions of the subsection which split equal-conveyance tubes. The areas for the parts of the subsection that split equal-

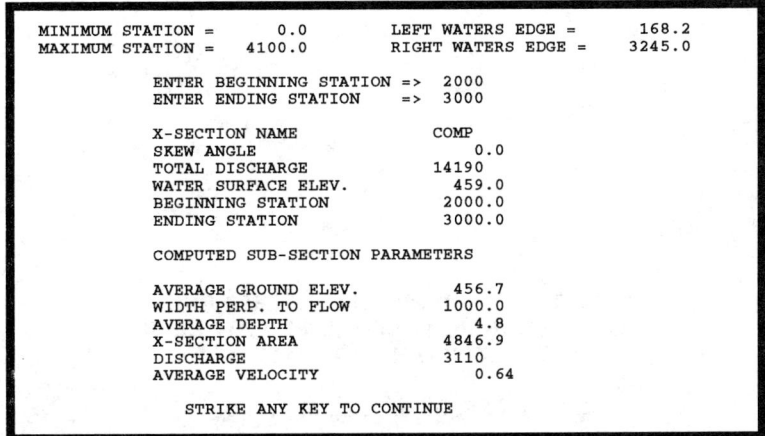

Figure 4. Output screen for hydraulic parameters computations.

conveyance tubes are computed in the same manner as described previously in this paper with the reference elevation set to the water-surface elevation. The discharge is computed by summing the product of the applicable area of each equal-conveyance tube and the velocity of that tube. The average velocity in the subsection is computed as the discharge of the subsection divided by the cross-sectional flow area of the subsection.

Summary

The BSAW program is an accurate and efficient means for extracting hydraulic and geometric information from WSPRO output. The options and computations used to extract the information are specifically tailored to variables necessary for bridge scour computations. The program provides a menu-driven interface and sufficient information to allow the user to adjust the computed values, if necessary.

References

Shearman, J. O., 1990, User's manual for WSPRO--a computer model for water surface profile computations: Federal Highway Administration Publication FHWA-IP-89-027, 112 p.

Shearman, J. O., Kirby, W. H., Schneider, V. R., and Flippo, H. N., 1986, Bridge waterways analysis model-research report: Federal Highway Administration Publication FHWA-RD-86-108.

INNOVATIONS AND PRACTICAL PROCEDURES FOR HYDRAULIC MODEL APPLICATIONS IN BRIDGE SCOUR EVALUATIONS

Jeffrey S. Glenn[1], P.E., M. ASCE

Abstract

Procedures for data conversion, and establishment of downstream boundary conditions for a hydraulic computer model for estimation of riverbed scour at bridges are presented. The data conversion procedure involves an improved method from the current practice of taking the results of a field survey of the river reach and converting them for input into a hydraulic model. Field survey data are entered into coordinate geometry and computer aided drafting and design programs, cross sections are then generated, and converted into data files. A computer program called XY2WSPRO reads the data file and converts the X,Y data points into the "GR card" format of the hydraulic computer model, WSPRO. The simplified technique for establishing downstream boundary conditions for the hydraulic model is for use in the absence of known calibration data. The trial-and-error method can be used to achieve consistent results from the hydraulic model at the point of interest, in this case, the bridge.

Introduction

Evaluating a bridge spanning a waterway for its susceptibility to bridge scour is a complex task. There are many engineering investigations and analyses that need to be completed before the magnitude of riverbed scour that could occur at the bridge can be computed. A hydraulic model of a river reach is often needed to accurately describe the flow characteristics at a bridge in preparation for calculating the potential riverbed scour. Estimating riverbed scour requires hydraulic information such as discharge, velocity, and depth; structural properties such as the location and dimensions of the footings, piers, abutments, and superstructure; as well as the

---

[1]Water Resources Engineer, Whitman & Howard, Inc., 143 North Main Street, Concord, New Hampshire 03301-5089

geometry and bed material of the waterway. As shown in Figure 1, this information is obtained by a preliminary review of the bridge and waterway, topographic survey of the waterway cross sections, development and calibration of hydrologic and hydraulic models, and finally calculation of the riverbed scour at the bridge.

Preliminary Review

A review of existing information pertaining to the bridge including construction plans, inspection reports, underwater inspection reports, boring logs, photographs, channel surveys, and engineering studies, is conducted to acquaint the engineer with the history of the bridge. A field review of the bridge and waterway is then conducted to gather information regarding the structural and hydraulic vulnerability to scour. The field review would include a check for visible signs of scour, obtaining samples of the channel and flood plain soils, photographs of the structure and waterway, assessment of hydraulic and geomorphic characteristics of the waterway, and review of the hydrologic characteristics of the drainage area upstream of the bridge.

Field Survey

The engineer determines where cross sections for the hydraulic model will need to be surveyed. A topographic survey is then conducted of the bridge, channel, and flood plain to establish the geometric data required for the hydraulic model.

The field survey data is downloaded into a coordinate geometry (COGO) program, AdCADD (Softdesk, 1992) and a computer aided drafting and design (CADD) program, AutoCAD (AutoDesk, 1992) on a personal computer, where it is converted into three-dimensional X,Y,Z data points. Cross sections are generated and plotted, along with a plan-view and profile of the surveyed river reach. The coordinates of each cross section are written as two-dimensional X and Y pairs to data files whose names contain the bridge and cross section identification numbers.

XY2WSPRO

XY2WSPRO (Glenn, 1992) is a computer program which reads the waterway cross section data files and arranges the cross sections in the proper order, and converts the X,Y data points into the "GR card" format of the Federal Highway Administration's (FHWA) HY-7 WSPRO (Shearman, 1988) computer program. It then includes the bridge and cross section identification numbers, computation parameters required by WSPRO, and any predetermined WSPRO output options. XY2WSPRO allows the user to raise or lower the cross sections by adding a constant to the Y-value of each data point, shift the cross sections to the left or right by adding a constant to the X-value of each data point, and make the resulting cross sections mirror images of the original field-surveyed cross sections. To complete the resulting WSPRO data file, the user must develop and insert the remaining input

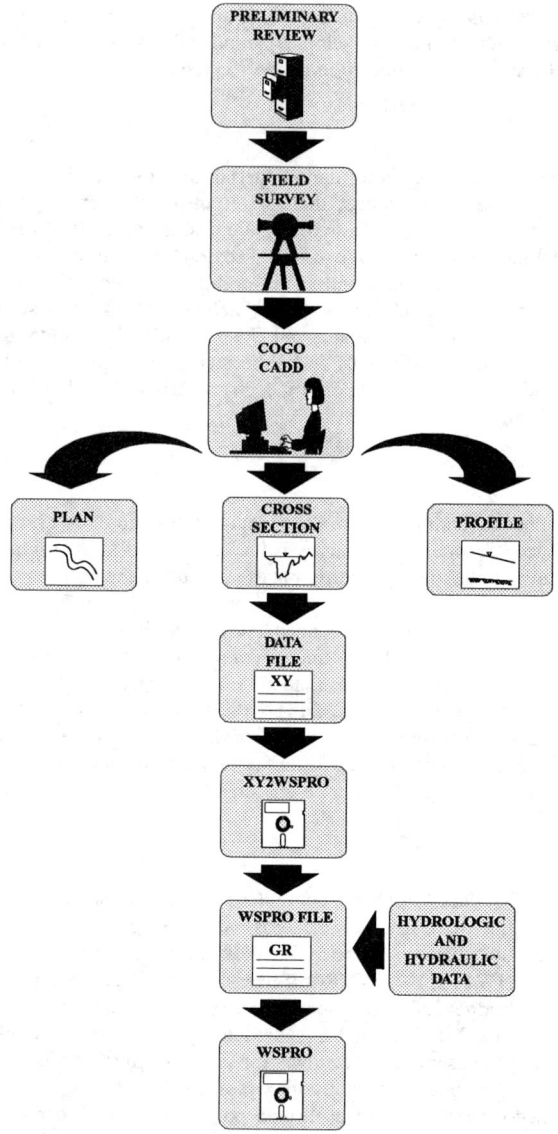

Figure 1. Data Collection, Conversion, and Modeling Procedure.

parameters, such as channel and flood plain roughness, reach length, bridge data, and flow. By using XY2WSPRO to convert surveyed cross sections into the WSPRO input format, a considerable amount of time can be saved, and the potential for data entry errors is reduced.

Calibration

Calibration of the hydraulic model (e.g. WSPRO) can be achieved by matching computed water surface elevations with high-water marks from distinct flood events, or river gauging station information. In the absence of observed data for calibration, a trial-and-error method can be used to establish an appropriate downstream boundary condition for the hydraulic model (see Figure 2).

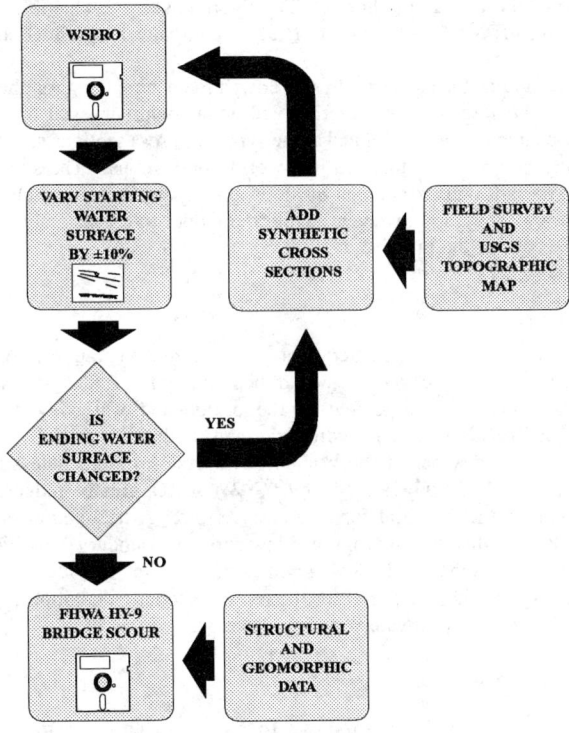

Figure 2. Boundary Condition Procedure.

In the case of a subcritical flow regime, the assumed water surface elevation of the cross section furthest downstream is both raised and lowered by 10% of the depth, and the modeled water surface elevation at the bridge is observed. If it changes, the downstream boundary is too near the bridge, resulting in an unstable model, and a cross section (or cross sections) further downstream is (are) required. If additional surveyed cross sections are not available, synthetic cross sections can be created partly from USGS topographic sheets and partly from the surveyed cross section furthest downstream. The surveyed cross section can provide the basic shape of the main channel, while the USGS sheet can provide the basic shape of the flood plain and the slope of the channel, as well as the type of ground cover for determining Manning's n-values. The synthetic cross section is then placed in the data file, and WSPRO is re-run. The assumed downstream starting water surface is, again, both raised and lowered by 10% of the depth, and the modeled water surface elevation at the bridge is observed. When this process results in no-change of the modeled water surface elevation at the bridge, then the model is stable; and if it otherwise properly depicts the river hydraulics, can provide the hydraulic characteristics of the flow at the bridge. The riverbed scour evaluation can then be completed, using FHWA HY-9, Scour at Bridges computer program (Fraher, 1992).

This technique for establishing the downstream boundary for the hydraulic model has the advantage of not requiring additional field surveyed cross sections. The hydraulic characteristics calculated at the synthetic cross sections should be given careful scrutiny before using them in real-world applications. The synthetic cross sections are, after all, approximations of the waterway geometry and characteristics at their locations. They do, however, enable the model to consistently estimate the hydraulic properties at the bridge.

Summary

Calculating riverbed scour requires structural and hydraulic information, as well as knowledge of the geometry and bed material of the waterway. This information is obtained by a review of the bridge and waterway, field survey, development and calibration of hydrologic and hydraulic models, and finally calculation of riverbed scour at the bridge. Conversion of the field survey cross section data into a WSPRO data file by XY2WSPRO greatly reduces the time required, as well as the potential for data entry errors. A trial-and-error technique can be used to establish an appropriate downstream boundary condition for the hydraulic model in subcritical flow regimes, so as to provide the hydraulic characteristics of the bridge required for the riverbed scour calculations, without the need for additional surveyed waterway cross sections.

Appendix I. References

AutoDesk, Inc. (1992) "AutoCAD Release 12 Reference Manual", Sausalito, CA.

Fraher, M. (1992) "HY-9, Scour at Bridges", Federal Highway Administration, McLean, VA.

Glenn, J. S. (1992) "XY2WSPRO", Whitman & Howard, Inc., Concord, NH.

Shearman, J. O. (1988) "WSPRO, Water Surface Profile", Federal Highway Administration, McLean, VA.

Softdesk, Inc. (1992) "AdCADD Civil/Survey COGO Reference Manual", Henniker, NH.

Appendix II. Additional References

Lagasse, P. F., et. al. (1991) "Stream Stability at Highway Structures", FHWA-IP-90-014 (HEC-20) Federal Highway Administration, McLean, VA.

Richardson, E. V., L. J. Harrison and S. R. Davis (1991) "Evaluating Scour at Bridges", FHWA-IP-90-017 (HEC-18) Federal Highway Administration, McLean, VA.

Richardson, E. V., D. B. Simons and P. Y. Julien (1990) "Highways in the River Environment", FHWA-HI-90-016, Federal Highway Administration, McLean, VA.

Shearman, J. O. (1990) "User's Manual for WSPRO--A computer Model for Water Surface Profile Computation (Hydraulic Computer Program HY-7)", FHWA-IP-89-027, Federal Highway Administration, McLean, VA.

Shearman, J. O. (1986) "Bridge Waterways Analysis Model: Research Report", FHWA/RD-86/108, Federal Highway Administration, McLean, VA.

BRI-STARS MODEL FOR ALLUVIAL RIVER SIMULATION

Albert Molinas[1]

ABSTRACT

The BRIdge StreamTube Model for Alluvial River Simulation (BRI-STARS) is a semi-two-dimensional model capable of computing alluvial scour/deposition through subcritical, supercritical, and a combination of both flow conditions involving hydraulic jumps. The use of streamtubes allow the modeling of scour/deposition for applications involving bridges not only along a study reach but also across alluvial channels. In this paper, the basic formulations used in the BRI-STARS model are presented and various hydraulic, sediment transport, local scour, and stream power minimization computations are described. The BRI-STARS model runs on IBM PC-AT or compatibles.

MODEL DESCRIPTION

The BRIdge StreamTube Model for Alluvial River Simulation (BRI-STARS) which was developed under the National Cooperative Highway Research Program (NCHRP) is a semi-two-dimensional model capable of computing alluvial scour/deposition through subcritical, supercritical, and a combination of both flow conditions involving hydraulic jumps. This model, unlike the conventional water and sediment routing computer models, is capable of simulating channel widening/narrowing phenomenon as well as local scour due to highway encroachments. The BRI-STARS model contains specialized routines for the computation of bridge hydraulics and local scour. Channel widening/narrowing is accomplished by coupling the stream tube computer model with a decision-making algorithm using rate of energy dissipation or total stream power minimization.

The BRI-STARS model was developed to simulate long-term streambed variations in rivers for which sediment and hydraulic data is limited. The computer model using stream tubes can be applied to a variety of river problems. It can be

---

[1]President, Hydrau-Tech, Inc., 333 West Drake Road, Suite 40, Fort Collins, CO 80526

used as a fixed-bed model to compute water surface profiles for subcritical, supercritical or the combination of both flow conditions involving hydraulic jumps. As a movable-bed model, the computer program can be applied to route water and sediment through natural river channels. The use of stream tubes allows the variation of hydraulic conditions and sediment activity not only in the longitudinal, but also in the lateral direction. With the selection of a single stream tube, the model becomes one-dimensional. Average channel response to changes in certain river flow or sediment conditions can be studied. With the selection of multiple stream tubes, the model becomes semi-two-dimensional. With this selection, changes in the cross section geometries in the lateral direction can be simulated. Since the bed-elevation changes are not averaged over the entire active channel widths as in one-dimensional models, more realistic channel erosion or aggradation can be simulated. The use of streamtubes provides valuable information in model applications where the variation of flow variables across the channel is needed. The armouring process provided in the program allows the study of river sedimentation problems for longer periods of time.

At each time step, the Standard Step Method of water surface profile computations are first carried out for the entire reach treating the channel as a single tube. Secondly, using the computed water surface elevations, lateral locations of stream tubes at each cross section are determined. Treating each stream tube as independent channels, new water surface profiles and hydraulic variables along them are computed. Third, sediment is routed through each stream tube, satisfying the sediment continuity equation. At the end of these computations, bed material compositions are revised and channel bed elevations are updated. An armouring procedure is incorporated into the sediment routing computations. Computations are proceeded in time through defined water and sediment discharge hydrographs.

The model has four major components:

- Hydraulic modeling;
- Sediment transport modeling;
- Stream power minimization; and
- Scour components

The first two components simulate the scouring/deposition process taking place in the vertical direction across the channel. The total stream power minimization component determines whether at a given cross-section the channel adjustments due to scour/deposition should take place in the lateral or vertical direction. It is this component that allows the lateral changes in channel geometries. Finally, the scour component allows the computation of the scour due to highway encroachments.

HYDRAULIC COMPUTATIONS

Backwater Computations

Water discharge hydrographs are appproximated by bursts of constant discharges. During each constant discharge time block, backwater computations are carried out without interruption for subcritical, supercritical, or a combination of both flow conditions modeling hydraulic jumps. The details of these computations are presented in several publications (Molinas and Yang, 1985; Molinas, 1983). These uninterrupted water surface profile computations are one of the most significant features of the Stream Tube Computer Model. It is this unique component that makes the model applicable to water and sediment routing computations through complex flow conditions in bridge waterways.

Streamtube Computations

Stream tubes are imaginary tubes bounded by stream lines. Since the discharge between stream lines is constant, each stream tube carries a constant discharge along its length. For steady, incompressible flows, it is possible to write

$$\frac{P}{\gamma} + \frac{V^2}{2g} + z = H_t \tag{1}$$

where H_t is a constant along the stream tube. When applied to real fluids, the total head is not a constant. Due to friction and other local losses, it is reduced in the direciton fo flow. Along a river it is possible to determine the variation of this quantity. This is the basic assumption in the Stream Tube computer Model.

The use of stream tubes in routing water and sediment through alluvial channels is another unique feature of the BRI-STARS model. In the model, the total discharge carried through the channel is distributed equally among the preselected number of stream tubes. Along each stream tube the water discharge remains constant. Nonlateral inflow into individual stream tubes fron neighboring tubes is allowed. Due to the assumptions involved, at a given station water surface elevation across the channel should remain constant. Under these circumstances, the equal discharge location, and therefore lateral stream tube locations, correspond to equal channel conveyances. Following the initial backwater computations at each station, stream tube location across the channel which satisfy equal conveyance requirements are determined. Next, since total energy along each stream tube should remain constant, backwater computations are performed to establish the hydraulic conditions along individual tubes.

Bridge Computations

The BRI-STARS model contains the bridge hydraulics routines utilized in the FHWA's WSPRO water surface profile computation program. The details of these computations are described in detail in the WSPRO User's Manual. A second option which utilizes user supplied local loss coefficients at the bridge sections is also available. In this "simple bridge" approach, the bridge loss coefficients can be calibrated with measured/computed water surface profiles for closer agreement with more accurate methods of computations. The "simple bridge" approach is applicable only for free surface bridge hydraulics computations since it treats the bridge piers as a part of the channel geometry.

SEDIMENT TRANSPORT MODELING

In BRI-STARS, sediment routing computations are carried out along each stream tube by satisfying the sediment continuity equation, which is given as:

$$\frac{\partial Q_s}{\partial x} + \eta \frac{\partial A_d}{\partial t} = 0$$

where η is the volume of sediment in a unit bed layer volume or one minus porosity; A_d is the volume of sediment deposition/scour per unit length; Q_s is the volumetric sediment discharge. The sediment continuity equation is discretized as follows:

$$\eta \frac{\partial A_d}{\partial t} = \eta \ (2W_i + W_{i+1} + W_{i-1}) \frac{\Delta Z_i}{4 \ \Delta t}$$

and

$$\frac{\partial Q_s}{\partial x} = 2 \frac{Q_{s\,i} - Q_{s\,i-1}}{\Delta x_i + \Delta x_{i-1}}$$

where W is the width of channel; ΔZ is the change in bed elevation and i is the cross-sectional index. The change in bed elevation, ΔZ_i, is obtained by satisfying the sediment continuity equation for each size class.

Sediment routing procedure is followed in the downstream direction. Starting from an upstream control station where the input sediment discharge hydrograph for different size classes is known, sediment discharge per size class, ($Q_{si,k}$), is predicted using the selected sediment transport equation. The change in bed elevation for each size class ($\Delta Z_{i,k}$) is then computed and summed.

BRI-STARS uses an armouring procedure by Bennett and Nordin (1977). This procedure divides the bed into conceptual layers of different thicknesses termed the *active* and *inactive layers*. The active layer is the layer on the surface of the bed which can be sorted through within a single time simulation time step. Due to the use of several stream tubes across the channel, the Bennett and Nordin method has been modified to accomodate stream tubes. The locations of stream tubes change with changing flow conditions and channel geometry. To account for bed material composition, active and inactive bed layer thicknesses for the stream tubes are transferred into point values across the channel. The sediment transport capacity computations in the present mathematical model can be carried out by the use of: (1) Yang's 1973 (sand) and 1984 (gravel) equations; (2) Ackers and White's equation; (3) Engelund and Hansen's equation; (4) Meyer-Peter and Muller's equation; and (5) User supplied generic equation.

STREAM POWER MINIMIZATION

The third major component of the BRI-STARS is that of making decisions as to whether the channel adjustments taking place at a given cross section due to scouring/deposition should advance in the lateral or vertical directions. The basic tool for this decision-making component is the "Minimum Rate of Energy Dissipation Theory" developed by Yang and Song and this general theory's special case "Minimum Unit Stream Power Theory" by Yang.

The integrated total unit stream power for a given river reach is defined as:

$$\Phi_t = \sum_{i=1}^{N-1} \frac{\gamma}{2}(V_i S_i A_i + V_{i+1} S_{i+1} A_{i+1}) \Delta X_i$$

where ΔX_i is the reach length, or distance between stations i and $i+1$; V_i, S_i and A_i are velocity, slope, and area at station i, respectively.

In the BRI-STARS, selecting directions for channel adjustmenst is accomplished by minimizing the integral expression for total stream power given above at different stations along the reach. At a given time step, if altering the channel width at a given cross section results in lower total stream power than raising/lowering the channel, channel adjustments are progressed in the lateral direction. For the opposite case, the adjustments are made in the vertical direction.

At cross sections where the sediment routing algorithm predicts erosion, channel adjustments can be made either by deepening or widening the existing channel. If both channel widening and channel deepening will reduce the total stream power for the study reach, the selected mode of channel adjustment in the computer model is the one which results in the minimum total stream power for the reach.

Similarly, at cross sections where sediment accumulation is predicted by the sediment routing algorithm, channel adjustments can proceed either by raising the channel bed or by narrowing the channel along the banks. In this case, the selected mode of channel adjustment in BRI-STARS will also be the one resulting in the minimum total stream power for the reach.

In both the aggrading and degrading channel cases, the sediment load is treated as a constraint in the minimization. In cases where geological/manmade restrictions are applied to the channel deepening/widening, computations are performed to accommodate these constraints.

The amount of channel width adjustment during a time step is determined by the sediment continuity equation. Channel widening/narrowing computations are carried out only in the stream tubes neighboring the channel banks. In the stream tubes which are not adjacent to the banks, sediment routing computations are carried out by satisfying the sediment continuity equation. In these tubes the channel adjustments can proceed only in the vertical direction. The amount of sediment eroded from the banks in a given reach is treated as part of the sediment influx for the downstream cross sections.

SCOUR COMPONENTS

The aggradation/degradation taking place in a river reach can be classified as:

- General aggradation/degradation;
- Contraction scour; and
- Local scour

General aggradation/degradation is simulated through the use of the sediment continuity equation. Similarly, contraction scour is computed through the use of the sediment surplus/deficit concept and the associated armouring algorithms. However, BRI-STARS has Laursen's contraction scour equations as an option for comparison to the sediment routing results.

The local scour due to bridge piers in the BRI-STARS model is computed utilizing the following methods: (1) Colorado State University equation; (2) Jain and Fisher equation; (3) Laursen equation; (4) Froelich equations; and (5) User supplied generic equations. The local scour due to abutments is calculated using the Froehlich or user-supplied generic equations. The model computes and lists the local scour at bridge piers and abutments separately from the computed general stream aggradation/degradation values.

SUMMARY AND CONCLUSIONS

In this paper the BRI-STARS model which was developed under the National Cooperative Highway Research Program (NCHRP) Project No. 15-11 was presented. This model utilizes an integrated approach in solving alluvial channel changes for highway applications. The specialized hydraulic and local scour algorithms for bridge structures (piers, abutments, constrictions) are designed to provide highway engineers a computer modeling tool in their analysis and design of highway structures in river environments. The BRI-STARS model provides the engineer a number of computational options to accomodate differences in individual case studies. The model has been designed to allow for the implementation of new technology in scour and sediment transport equations as it becomes available.

ACKNOWLEDGEMENTS

The development of the BRI-STARS model was funded through the Transportation Research Board, National Cooperative Highway Research Program's Project No. 15-11 and 15-11A.

PRACTICAL COMPARISON OF ONE-DIMENSIONAL AND TWO-DIMENSIONAL HYDRAULIC ANALYSES FOR BRIDGE SCOUR

BY

M.A. PORTS[1], F. ASCE, T.G. TURNER[2], M. ASCE, D.C. FROEHLICH[3], M. ASCE

Introduction

The Kentucky Transportation Cabinet, Department of Highways is working with the Indiana Department of Transportation and the Federal Highway Administration to plan, design, and construct a new bridge over the Ohio River near Owensboro, Kentucky. The proposed project involves the design and construction of a new four lane bridge and roadway approaches over the Ohio River between Owensboro in Davies County, Kentucky and Rockport in Spencer County, Indiana. As shown in Figure 1, the proposed project crosses the Ohio River floodplain at one of its widest points. The width of the floodplain exceeds four miles. The overbank areas are nearly level and generally characterized by numerous swells and swales and ephemeral streams aligned parallel to the main river channel. During the conceptual and early preliminary phases, the potential scour was estimated based upon a conventional hydraulic evaluation using WSPRO. The Kentucky Department of Highways determined, however, that a two-dimensional hydraulic analysis, using FESWMS, was required for the successful design of the proposed bridge and approach roadways constricting the natural flood plain.

This paper compares the hydraulic analysis of the proposed bridge over the Ohio River and approach roadways across the flood plain using one-dimensional model, WSPRO, to the two-dimensional model, FESWMS. The discussion concentrates on the advantages of using FESWMS to model complex flow situations. A comparison of bridge scour estimates for this project based on results from both one-dimensional models and a two-dimensional model is also presented. The two-dimensional results are shown to improve estimates of scour for complex flow conditions and multiple bridge openings. The advantages of using the two-

---

[1]Principal Water Resources Engineer, Parsons Brinckerhoff, 301 North Charles Street, Baltimore, Maryland 21201

[2]Civil Engineer, Parsons Brinckerhoff, 301 North Charles Street, Baltimore, Maryland 21201

[3]Assistant Professor, Department of Civil Engineering, University of Kentucky, Lexington, Kentucky 40506

dimensional model results are discussed for each component of scour, i.e., contraction, pier, and abutment scour. Estimates of scour are based on the procedures presented in HEC-18.

The original hydraulic analysis, using WSPRO, relied on the conventional assumptions of one-dimensional flow. The WSPRO model was calibrated to the official 100-year flood profile published by the Federal Emergency Management Agency. The official 100-year flood profile is based upon water surface profiles prepared by the U. S. Army Corps of Engineers using HEC-2. As shown on Figure 1, the proposed bridge crosses the Ohio River approximately midway around a 90 degree bend. Further complicating the hydraulic analysis is the fact that the proposed crossing lies immediately upstream from a natural channel contraction of approximately 35 percent. The contraction is caused by the rock bluff upon which the City of Rockport is located. Under baseline, or existing, conditions, the contraction causes significant flow to spill out of the channel onto the adjacent Kentucky floodplain. The proposed project further complicates the rather complex flow conditions at the site. The main span is approximately 4,500 feet in length and completely spans the main river channel. Therefore, the main bridge span has little hydraulic impact upon flows wholly within the main river channel, except for the relatively minor impact from the proposed piers. However, the approximately 13,080 feet of approach roadway across the Kentucky floodplain is elevated above the 100-year water surface elevation. Thus, the approach embankment acts as a dam across the Kentucky floodplain. In order to alleviate the total redistribution of flow from the floodplain into the main channel, five bridges are proposed for the approach embankment. Two bridges are 300 feet in length, one bridge is 400 feet in length, and two bridges are 500 feet in length. However, the total bridge opening of 2,000 feet represents a significant constriction of the 2.5 mile wide Kentucky floodplain.

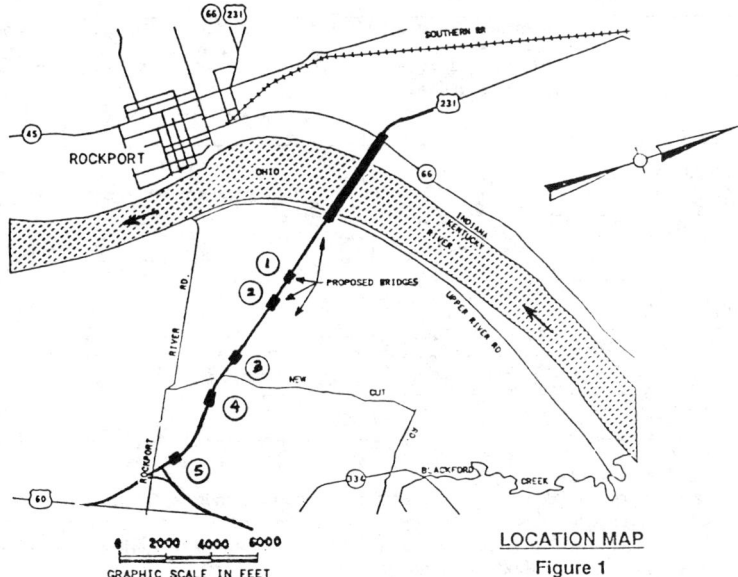

LOCATION MAP
Figure 1

The results from the WSPRO analysis were then utilized to estimate the depth, velocity, and direction of the flow at the main bridge span and the five overflow bridges. The hydraulic parameters were then used to estimate the potential contraction scour, abutment scour, and pier scour for each of the six bridge structures independently. The subsequent hydraulic analysis, using FESWMS, relied on the assumptions of depth-averaged, two-dimensional flow. The FESWMS model was calibrated to the observed high water marks reported by the U. S. Army Corps of Engineers. Also, the calibrated FESWMS model agrees closely with their HEC-2 water surface profile. The results from the FESWMS analysis were then utilized to estimate the depth, velocity, and direction of the flow at the main bridge span and the five overflow bridges. The hydraulic parameters were then used to estimate the potential contraction scour, abutment scour, pier scour for each of the six bridge structures. The resulting scour estimates are summarized and compared as follows.

Contraction Scour

Laursen's live-bed contraction scour equation is used to estimate the average contraction scour for the main span structure. And, Laursen's clear-water equation is used to estimate the average contraction scour for each of the five overflow bridges. The choice of contraction scour equations reflects the assumption that the bed load is carried by the flows within the main river channel, but that relatively little bed load is carried by the flows over the Kentucky floodplain. A comparison of the contraction scour estimates is summarized in Table 1.

TABLE 1

COMPARISON OF CONTRACTION SCOUR ESTIMATES

	Main Span Sta 875+00 to 920+00	Bridge 1 Sta 842+87 to 846+87	Bridge 2 Sta 829+45 to 834+45	Bridge 3 Sta 800+59 to 803+59	Bridge 4 Sta 778+70 to 783+70	Bridge 5 Sta 748+94 to 751+94
FESWMS RESULTS:						
Net Opening, ft.	2,496	372	466	282 ft.	466	280
Flow Depth, ft.	55	22	21	16	12	14
Flow, cfs	704,00	32,600	43,000	20,900	33,000	18,200
Scour Depth, ft.	3	57	61	52	54	47
WSPRO RESULTS:						
Net Opening, ft.	2,491	359	449	269	449	269
Flow Depth, ft.	55	22	21	16	12	14
Flow, cfs	789,000	22,800	29,600	13,900	19,200	10,400
Scour Depth, ft.	3	38	41	34	31	25

For the live-bed condition within the main river channel, there is no difference in the contraction scour estimates, even though the estimates of flow in the contracted section differ by as much as 12 percent. Because relatively no contraction is caused by the main span structure, both methods estimate merely 3 feet of contraction scour. However, significant differences are apparent between the scour estimates for the five overflow bridges. For the clear-water condition, the contraction scour is a function of the width of the contracted section, the flow through the contracted section, the depth of flow, and the median diameter of the bed material. The depth of flow on the floodplain upstream from the relief bridges and the bed material properties are the same for both the WSPRO and FESWMS analyses. For the WSPRO analysis, the direction of the flow approaching the relief structure is assumed. The width of the contracted section is computed to be the width of the bridge opening less the effective pier widths. For the FESWMS analysis, the direction of the flow approaching the structure is computed directly. Thus, differences in the net opening width, reflect differences in the direction of the approach flow. For the WSPRO analysis, flows through the bridge structures are apportioned based upon conveyance. While, for the FESWMS analysis, flows through the bridge structures are computed directly. As a result, differences in the direction and magnitude of the flows through each of the structures account for differences in contraction scour estimates of 49 to 88 percent.

Abutment Scour

Froehlich's live-bed equation is used to estimate scour at the abutments of the structures. A comparison of the abutment scour estimates for the five overflow bridges is summarized in Table 2.

TABLE 2
COMPARISON OF ABUTMENT SCOUR ESTIMATES

	Bridge 1		Bridge 2		Bridge 3		Bridge 4		Bridge 5	
	North 846+187	South 842+87	North 834+45	South 829+45	North 803+59	South 800+59	North 783+70	South 778+70	North 751+94	South 748+94
FESWMS RESULTS										
Flow Depth, ft	20.5	19.5	21.5	21.5	13.5	12.5	12.5	11.5	14	14
Abutment Length, ft	1615	350	490	2200	385	1050	640	1925	750	500
Abutment Angle, °	90	140	90	90	90	90	90	90	90	90
Froude Number	0.09	0.06	0.04	0.06	0.07	0.06	0.07	0.08	0.08	0.07
Scour Depth, ft	58	34	36	57	28	31	29	40	35	30
WSPRO RESULTS										
Flow Depth, ft	23.4	23.4	23.5	23.5	20.5	20.5	18.5	18.5	14.5	14.5
Abutment Length, ft	87	219	394	1328	120	94	475	870	577	76
Abutment Angle, °	90	90	90	90	90	90	90	90	90	90
Froude Number	0.07	0.07	0.02	0.02	0.02	0.02	0.03	0.03	0.04	0.04
Scour Depth, ft	34	39	33	40	25	24	29	32	28	20

The Froehlich equation is an empirical relationship based upon 170 live-bed scour measurements taken in laboratory flumes. For each of the 170 cases, the abutments were symmetrical and the depth of floodplain flow was constant across the approach cross-section obstructed by the embankment. However, for wide natural floodplains, the length of the abutment projected normal to the flow, the flow area of the approach cross-section obstructed by the embankment, the flow obstructed by the abutment and embankment, and the depth of floodplain flow approaching the abutment are all difficult parameters to estimate based solely upon a one-dimensional hydraulic analysis. However, the estimation of these parameters is facilitated by a two-dimensional hydraulic analysis. For example, from the WSPRO analysis, the approach flow is assumed to be normal to the obstructing embankment in all cases. However, the FESWMS analysis demonstrates that the flow approaches the south abutment of Bridge 1 at an angle of 140 degrees. In addition, the FESWMS analysis more directly computes the flow distributions across the wide Kentucky floodplain. The estimates of abutment scour based on the FESWMS analysis vary from -13 percent to as much as 70 percent of the estimates based on the WSPRO analysis.

Pier Scour

The Colorado State University equation is used to estimate pier scour for the piers of all of the bridge structures. A comparison of the local scour estimates for the two southernmost piers of the main span are summarized in Table 3.

TABLE 3

COMPARISON OF PIER SCOUR ESTIMATES FOR MAIN SPAN

	Pier 1 Station 876+05		Pier 2 Station 877+30	
FESWMS RESULTS	ACTUAL		ACTUAL	
Skew Angle	35°		35°	
Pier Width, ft.	31		31	
Pier Length, ft.	51.8		51.8	
Velocity, fps.	2.9		3.3	
Flow Depth, ft.	9		9	
Froude Number,	0.17		0.19	
Scour Depth, ft.	20		21	
WSPRO RESULTS	MAXIMUM	MINIMUM	MAXIMUM	MINIMUM
Skew Angle	15°	10°	15°	10°
Pier Width, ft.	15.9	11.8	15.9	11.8
Pier Length, ft.	51.8	51.8	51.8	51.8
Velocity, fps.	2.9	2.9	2.9	2.9
Flow Depth, ft.	9	9	9	9
Froude Number	0.17	0.17	0.17	0.17
Scour Depth, ft.	13	11	13	11

Because, for the WSPRO analysis, the angle at which the flow approaches the pier is estimated, two estimates of the angle are assumed. Thus, pier scour is estimated for both a maximum and minimum approach angle. The maximum and minimum approach angles are estimated based on the results from the WSPRO analysis. However, the results form the FESWMS analysis directly compute the angle of approach. The approach angle of the flow directly affects the projected width of the pier, thus directly affecting the scour depth. While there was no significant difference between the approach angles for the WSPRO and FESWMS analyses for 14 of the main span piers, significant differences were found for the southernmost two piers. The FESWMS model computed an approach angle of 35 degrees, more than twice the maximum estimate from the WSPRO analysis. Thus, based on the FESWMS analysis, the pier scour is increased from 11 feet to as much as 21 feet.

In summary, the two-dimensional hydraulic analysis provides significantly improved estimates of velocity magnitude, direction of flow, and flow distribution for complex flow conditions around bends, through contractions, and through multiple openings. As discussed, these flow characteristics are essential for the estimation of potential bridge scour.

REFERENCES

1. Arcement, Jr., G. J. and V.R. Schneider, 1984. *Guide for Selecting Manning's Roughness Coefficients for Natural Channels and Flood Plains.* Office of Implementation, Federal Highway Administration, McLean, Virginia.

2. Bradkley, J.N., 1978. *Hydraulics of Bridge Waterways.* Hydraulic Design Series No. 1, Federal Highway Administration, Washington, D.C.

3. Froehlich, D.C., 1990. *Finite Element Surface Water Modeling System: Two-Dimensional Flow in a Horizontal Plane.* User's Manual, Department of Civil Engineering, University of Kentucky, Lexington, Kentucky.

4. Lagasse, P.F., J.D. Schall, F. Johnson, E.V. Richardson, J.R. Richardson, and F. Chang, 1991. *Stream Stability at Highway Structures.* Hydraulic Engineering Circular No. 20, Office of Implementation, Federal Highway Administration, McLean, Virginia.

5. Laursen, E.M., 1963. *An Analysis of Relief Bridge Scour.* Journal of Hydraulic Engineering, Volume 89, Number 3, American Society of Civil Engineers, New York, New York.

6. Lee, J.K., D.C. Froehlich, J.J. Gilbert, and G.J. Gregg, 1982. *Two-Dimensional Analysis of Bridge Backwater.* Proceedings of the Conference Applying Research to Hydraulic Practice, American Society of Civil Engineers, New York, New York.

7. Lee J.K., D.C. Froehlich, J.J. Gilbert and G.J. Wiche, 1983. *A Two-Dimensional Finite Element Model Study of Backwater and Flow Distribution at the I-10 Crossing of the Pearl River Near Slidell, Louisiana.* Water Resources Investigations Report 82-4119, U.S. Geological Survey, NSTL Station, Mississippi.

8. Ports, M.A., T.G. Turner, and D.C. Froehlich, 1992. *Two-Dimensional Hydraulic Analysis of the Owensboro Bridge and Approaches.* Water Resources Planning and Management: Saving a Threatened Resource - In Search of Solutions, Proceedings of the Water Resources Sessions at Water Forum '92, American Society of Civil Engineers, New York, New York.

9. Richardson, E.V., D.B. Simons, and P.Y. Julien, 1990. *Highways in the River Environment: Participant Notebook.* NHI Course No. 13010, National Highway Institute, Federal Highway Administration, McLean Virginia.

10. Richardson, E.V., 1991. *Evaluating Scour at Bridges.* Hydraulic Engineering Circular No. 18, Office of Research and Development, Federal Highway Administration, McLean, Virginia.

11. Shen, H.W., V.R. Schneider, and S. Karaki, 1969. *Local Scour Around Bridge Piers.* Journal of Hydraulic Engineering, Volume 95, Number 6, American Society of Civil Engineers, New York, New York.

12. Wiche, G.J., J.J. Gilbert, and J.K. Lee, 1982. *Analysis of Alternatives for Reducing Bridge Backwater.* Proceedings of the Conference Applying Research to Hydraulic Practice, American Society of Civil Engineers, New York, New York.

Model Study of Local Scour Downstream Bridge Piers

Laila Abed[1], M.M. Gasser[2]

Abstract: A deep scour hole downstream of a large circular pier has been developed at Imbaba bridge, which is considered one of the major bridges across the Nile River in Cairo, Egypt. At that location the Nile River is considered almost clear water river with fine sand bed. An undistorted mobile bed model, with a scale 1:60, was constructed at the Hydraulics and Sediment Research Institute (HSRI), Cairo. A series of clear water scour tests were performed to investigate the causes of the local scour downstream the circular pier. It was found that the large scour hole downstream the circular pier was produced by the conflicting velocity fields at the intersection of the wake vortex streams from adjacent piers, and increased by the confluence flow. Based upon the results of this investigation, an empirical formula was developed to predict the wake and confluence maximum local scour depth downstream of a circular pier for a clear water condition.

Introduction:

One of the major consideration in the design and construction of a bridge is the local scour around its foundation. Local scour occurs where water has been accelerated as it moves along the obstruction and where large vortex systems (horse shoe, wake and trailing vortex systems) are generated as the flow separates around the obstruction, (Laursen, 1980, 1987). At most bridges, wake vortex scour is insignificant and confluence scour does not exist. In clear water rivers flowing on fine sand, these two forms of local scour can be very large, (Stevens, Gasser and Saad, 1991). The HSRI has been monitoring the behavior of particular deep scour hole 32 m downstream (DS) the centerline of Imbaba bridge and 97 m from the left bank. There is a concern that the hole may enlarge and/or move

---

[1] Senior Researcher, Hydraulics and Sediment Research Institute, Delta Barrage, Cairo, Egypt

[2] Director of the Hydraulics and Sediment Research Institute, Delta Barrage, Cairo, Egypt

upstream (US) endangering the pier and the bridge. At this location the Nile River is considered nearly clear water river, (the bridge is located in the backwater curve of the Delta Barrages and the sediment transport is very low), with fine sand bed (d_{50} = 0.115 mm). A construction drawing of the bridge was found and shows that there is a clay layer around the main circular pier of at least 2 m thickness at level -3 m, covered with a pure sand layer up to zero level. It shows also that the bed around the first three piers to the left bank was scoured due to the confluence flow to zero level before the bridge construction. The survey of 1992 clearly defines the confluence scour as well as the wake vortex scour DS the large circular pier. It defines also that the bed was scoured to a level -3 m US and around the first three piers to the left and ceased since the survey of 1981. Approaching the location of the bridge, at Imbaba, the main channel is flat (bed elevation = +10 m). The side channel has a U-shape with its bottom at elevation +10 m. The two channels intersect at a certain angle and the confluence scour depth was approximately 10 m before the bridge construction. The confluence scour hole is aligned with the bisector of the intersection angle. The wake scour hole was at elevation -8.30 m, about 18.30 m below the adjacent bed level to the east. The depth of scour due to the piers and confluence after the bridge construction is 18.30 - 10 = 8.30 m.

Laboratory Tests:

Fourteen experiments were conducted in an undistorted physical model, scale 1:60, 20 m long, 11 m wide and 0.45 m deep. Each test was run for a minimum of five hours to allow maximum scour and stabilization of the system. The model depends on discharges which give mean velocities less than or equal to the mean critical velocity of particle incipient motion. The model was operated according to the Froude simulation. The bridge was simulated by its seven piers and fenders (the circular pier is the second one to the left). The bed was configured such that the bed level was zero in the vicinity of the first three piers to the left, see Figure 1. According to the market facilities, the sand available was with mean diameter d_{50} = 0.268 mm. Therefore, it was necessary to exaggerate the velocity to compensate the effect of the bed material size difference in both model and prototype in such a way that:

$$n_{usc} = \frac{\bar{U}_p}{\bar{U}_m} = \frac{\bar{U}_{pcr}}{\bar{U}_{mcr}} \tag{1}$$

Where $\bar{U}$ is the average velocity, n_{usc} is scour velocity scale ratio, p is prototype and m is model. Using Shields method, the average critical velocities $\bar{U}_{cr}$ for model and prototype were calculated and that gives:

$$n_{usc} = \frac{\bar{U}_{crd_{50}=0.115mm}}{\bar{U}_{crd_{50}=0.268mm}} = \frac{0.432}{0.24} = 1.8 \tag{2}$$

$$n_{qsc} = F * n_q = (\frac{n_{usc}}{n_u}) \, n_q = 0.232 \, n_q \qquad (3)$$

Where n_{qsc} is the scour discharge scale ratio, and F is a correction factor less than unity to compensate for the effect of the velocity exaggeration.

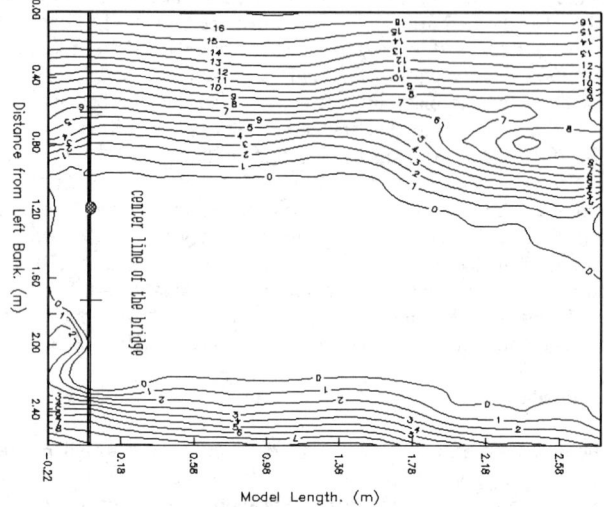

Figure 1. Bed contours for the model before any run according to the prototype at the time of bridge construction

Discussion of the results

The scour downstream the large circular pier at Imbaba bridge was produced by the conflicting velocity fields at the intersection of the wake vortex streams from the adjacent pier.

The set of variables governing the downstream local scour are: 1) the bridge piers; 2) the flow; 3) the bed material; 4) the vortex shedding frequency; and 5) the fluid properties. A dimensional analysis (Streeter and Wylie, 1979) leads to the following functional relation :

$$\frac{Y_s}{Y_1} = f\left(F_r, \, \frac{a}{Y_1}, \, R_p, \, S, \, C_D, \, \frac{Y_1}{d_{50}} \right) \qquad (4)$$

Where R_p is the Reynolds number = $U_1 a/\nu$, based on approach flow velocity U_1, and pier width a, F_r is the Froude number of the approach flow = $U_1/\sqrt{gY_1}$, S is the Strouhal number = na/U_1, n is the frequency of shedding of vortices from one side of the pier in Hertz which is given in terms of Strouhal number S, and this in turn is a function of pier Reynolds number. C_D is the coefficient of drag due to the pressure variation around the pier and it is a function of the pier Reynolds number, the shape, and streamlining of the pier, it is constant at high pier Reynolds number, d_{50} is the bed size which is constant and can be eliminated.

The relation given in Equation (4) is reduced to:

$$\frac{Y_s}{Y_1} = f\left(F_r, \frac{a}{Y_1}, S, C_D\right) \tag{5}$$

A multiple linear regression analysis was used to develop a prediction equation for the downstream local clear water scour, based on the functional relation given in equation (5). Y_s/Y_1 was regressed against every possible combination of the independent variables. The equation providing the largest coefficient of determination, r^2, is:

$$\frac{Y_s}{Y_1} = -2.6\,F_r - 0.119\,\frac{a}{Y_1} + 404.34\,S^2\,F_r^2 + 0.28 \tag{6}$$

In the case of two intersecting wake vortex streams and for design purposes the following relation is suggested:

$$\left(\frac{Y_s}{Y_1}\right)_{design} = K\left(\frac{Y_s}{Y_1}\right)_{computed} \tag{7}$$

Where $(Y_s/Y_1)_{computed}$ is the calculated value using Equation (6), the factor K can be predicted from the factor relations of vortices in the wake with or without intersection, a value of 1.3 was estimated. It was found that the measured velocities U at the intersection of the wake is almost equal to 1.5 the approach velocities U_1. Table 1 shows the predicted downstream scour depths from Equations (6), (7) and the total scour depths measured in the prototype in different years. The table shows that the scour downstream of the bridge was took place at high discharges in the time of the High Aswan Dam construction then the scour hole depth started to decrease with the decrease in the discharge until it became stable from 1986 up to now. It shows that equation (6) could be a good guide in predicting the scour depth DS the circular pier due to the wake vortex with the existence of confluence flow in clear water condition. At confluence, interface of the two merging streams creates vortices and turbulence, which scours the river bed under the interface (Stevens, Gasser and Saad, 1991). In the model, it was difficult to simulate the very high velocities that caused the large confluence scour in the past. Therefore it was decided to simulate the bed configuration as it was at the time of Imbaba bridge construction. At this time the confluence maximum scour was at level zero at the vicinity of the large circular pier. The circular pier sits directly in the confluence scour hole and is protected naturally by a 2 m pure clay layer, which was exposed as the fine sand layer was being washed away after the bridge construction. Therefore, there was no any further US scour around the pier due to the horse shoe vortex. Most wake vortex scour was likely caused during the construction of the High Aswan Dam. In 1964, the first poundage was created at the dam. The flood peak passing through the diversion tunnels was approximately 10,000 m³/s, but most of the sediment that had previously accompanied this water was left in the pond. From 1965 to 1968 the peaks were much less. In 1968, the closure was completed and the maximum yearly release has been only a quarter of the former flood peaks. This maximum yearly release which reaches the Imbaba area is about 1700 or 1655 m³/s, with sediment in suspension which is only 15-30 mg/L (Gasser, 1987). The scour hole has been filling slightly between 1968 and 1981, and became stable since 1986. Equation (6) and (7) show that with a discharges of 1700 and 1655 m³/s, the scour depth is very small and the deepest point in the scour hole should not be more than the level -0.101 m which means that there is no scour since this discharges started to flow in the Nile. These two discharges are the maximum discharges around the year and any small deposition in the scour hole due to smaller discharges are scoured by the maximum ones. This could explain the equilibrium happened in this scour hole from 1986 to 1992 in the prototype. The measured scour in the model was due to the wake vortex with the existence of confluence flow. Then by adding confluence scour occurred before the bridge construction which was equal to 10 m (average bed level in the area minus average bed level at the confluence, +10 - 0 = 10 m) the total scour depth due to the confluence and wake vortex can be obtained.

Table 2 Scour depths predicted from the predicted equations and the ones measured in the prototype

date	max.Q in m³/s	Y_1 in m	U_1 in m/s	F_r	measured values from prototype		predicted values from equations (6) and (7)	
					level of deepest point in the scour hole m	total scour depth m	level of deepest point in the scour hole m	total scour depth m
1964	10000	9.0	2.48	0.264	n.m.	n.m.	-14.08	24.08
1965 1968	7000	7.0	2.187	0.269	n.m.	n.m.	-11.19	21.19
1981 1983	1700 or 1655	6.13	0.516 or 0.502	0.066 or 0.065	-11.0	21.0	-0.101	10.101
1986 1992	1700 or 1655	6.13	0.516 or 0.502	0.066 or 0.065	-8.3	18.3	-0.101	10.101

n.m. = not measured

It was found that the velocity increased about 16 % at the confluence and

33 % in the wake scour hole downstream the bridge.

Conclusions

The deep scour hole downstream the large circular pier at Imbaba Bridge over the River Nile in Cairo was produced by the conflicting velocity fields at the intersection of the wake vortex streams from adjacent pier. This scour hole was increased by the clear water confluence scour. The wake vortex scour does not threaten the piers that produced the vortices. The existing discharges during the year at present time are very low and they could not cause any further scour. The existing scour hole is stable since 1986, and it is downstream, away from the pier. Equation (6) can be used to predict the wake vortex scour downstream of a circular pier in clear water condition, where a confluence flow exist. Local scour depth at the intersection of two wakes can be predicted by using both equations (6) and (7). The velocity at the wake is 1.5 the approach velocity, and at confluence, the velocity is 1.2 the approach velocity.

Acknowledgment

This paper is a part of a research program in scour around bridges in the River Nile. Thanks are due to the HSRI for financing this research. Thanks are also due to Dr. Mahmoud Samy Ain Shams University for providing valuable suggestion.

Appendix References

Gasser, M.M. (1987), "Hydrological aspects of the Nile River at Shoubra Elkheima Power Plant." The Hydr. and Sediment Res. Inst., Delta Barrage,Egypt.
Laursen, E.M., (1980), "Predicting Scour at Bridge Piers and Abutments", General Report No. 3, University of Arizona.
Laursen, E.M., (1987), "Prediction of Scour in River Engineering", River Engineering Book, Appendix A.
Streeter, V.L., and Wylie, E.B., (1979), "Fluid Mechanics" 17th ed., McGraw-Hill, NewYork, 562p.
Stevens M.A., Gasser, and Saad, (1991), "Wake Vortex Scour at Bridge Piers" Journal of Hydr. Eng., Vol. 117, No. 7, July.

SELECTING SEDIMENT TRANSPORT EQUATION FOR SCOUR SIMULATION AT BRIDGE CROSSING

By Howard H. Chang[1], M.ASCE, Carroll Harris[2],
Bill Lindsay[3], Steve S. Nakao[4], Ray Kia[5], M.ASCE

ABSTRACT: Gravel transport formulas were tested in order to select a formula for simulating stream channel changes of Stony Creek in northern California. At a stream reach undergoing scour, the amount of scour will be overpredicted by a sediment formula giving too high values of transport, and vice versa. If a formula overpredicts the transport rate, it will indicate a refill in gravel pits greater than the measured value. On the other hand, a formula under- predicting the transport rate will show less refill than the measured amount.

For the Stony Creek study, the following formulas were used separately for test and calibration: Meyer-Peter Muller (MPM) formula, Yang formula for gravel, Parker formula, and Engelund-Hansen Formula. It should be noted that the Engelund-Hansen formula was not developed for gravel bed streams. The deliveries obtained from Yang's formula and Parker's formula are more or less similar. The Engelund-Hansen formula overpredicts the transport rate while the Meyer-Peter Muller formula underpredicts the transport rate. Erosional changes at the Highway 32 bridge crossing and depositional changes in the gravel pits are most significant. Simulated changes at this bridge crossing compare favorably with the measurement for all the sediment formulas used, but certain formulas are apparently better than others on the basis of the comparison.

INTRODUCTION

The fluvial processes of Stony Creek in Glen County, California have been influenced by flood control projects and instream gravel and sand extraction in

---

[1] Professor of Civil Engrg., San Diego State Univ., San Diego, CA 92182
[2] Chief of Planning and Liaison Branch, Division of Structures, Caltrans, Sacramento, CA
[3] Chief of Hydraulics and Hydrology Section, Caltrans, Sacramento, CA
[4] Hydraulics Specialist, Caltrans, Santa Ana, CA
[5] Division of Structures, Caltrans, Sacramento, CA

recent years. As a result of these activities, channel-bed degradation and undermining of bridge piers have occurred. Such changes are the most severe at the Highway 32 bridge (see Fig. 1), where the bed has degraded up to 16 feet, exposing certain footings and support pilings.

The mathematical approach was used to quantify the problems and to quantify the mining impacts, based on the FLUVIAL-12 model (Chang, 1988), which has been formulated and developed since 1972 for water and sediment routing in alluvial streams. The combined effects of flow hydraulics, sediment transport and stream channel changes are simulated for a given flow period. Test and calibration for the FLUVIAL-12 model were made using data from Stony Creek in order to assess the applicability of the model for the Creek and to select an appropriate sediment transport formula.

TEST AND CALIBRATION OF FLUVIAL-12 MODEL

Test and calibration are important steps to be taken for more effective use of a model. Because of the difference in sensitivity of simulated results to each relation or empirical coefficient, more attention needs to be paid to those that generate sensitive results. Major items that require calibration include the roughness coefficient, sediment transport equation, and so on. Field data are generally required for test and calibration of a model. Generally, the channel configuration before and after the changes, a flow record, sediment characteristics are required.

Test and calibration of the FLUVIAL-12 are for the following major objectives: (1) To select an appropriate sediment transport formula for Stony Creek, (2) to assess the applicability of the FLUVIAL-12 model for simulating river hydraulics, sediment transport and stream channel changes for Stony Creek, and (3) to evaluate the applicability of the model and sediment transport formula for simulating channel changes at the bridge crossings as affected by sand and gravel mining. The study reach is in the vicinity of the Highway 32 bridge and major gravel pits. Because of sand and gravel extraction, this stream reach has experienced significant changes, thus providing an opportunity for calibration study of the model.

DATA BASE FOR TEST AND CALIBRATION

The data base for the calibration study includes cross-sectional data, flood hydrograph and sediment data. The cross-sectional data were generated from aerial photographs for aerial surveys of the stream made on February 26, 1977 and on March 24, 1978. Stony Creek is an ephemeral stream; flood flow is limited to short durations. The January-February 1978 flood is assumed to be responsible for most of the changes in the stream channel between the two aerial surveys. It is, however, understood that mining activities also contributed to the stream channel changes during this period. This flood has an estimated peak discharge of 14,000 cfs, a significant figure when compared with the 100-yr flood of about 24,000 cfs. The effective flow periods from January 9 to January 23 and from February 4 to February 21 are used in the simulation study and the small flow from January 23 to February 4 is neglected.

TESTING OF SEDIMENT TRANSPORT FORMULAS

Several sediment transport formulas have been developed for stream channels with gravel as the principal bed material. Most of the bed material in Stony Creek is in the gravel-size range, with median diameters ranging from 5 mm to 9 mm. The following formulas are used separately for test and calibration: Meyer-Peter Muller (MPM) formula (1948), Yang formula (1984), Parker formula (1990), and Engelund-Hansen formula (1967). Among these formulas, the MPM formula is a bedload formula and it is widely used for gravel-bed streams in Europe. The Yang formula consists of a component on sand transport and another component on gravel transport. The Parker formula is developed primarily for gravel transport in the form of bedload. The Engelund-Hansen formula is primarily for sand beds. It is still included in this study partly because one also likes to see the differences in results due to the selection of a sediment formula.

The selection of a gravel transport formula, ideally, should be based on field measurements of gravel transport. In an ephemeral stream, such as Stony Creek, measurements may only be made during flood events when there is substantial movement of the bed material. Due to the lack of measured sediment transport data for Stony Creek, selection of a sediment transport formula must follow a different approach. In the study, the sediment transport formulas are tested based on the formula's capability in simulating stream channel changes. This approach becomes feasible since channel geometric data bracking the 1978 flood event have been identified. During this flood event, most significant changes in the stream channel were induced by mining and therefore occurred in the vicinity of the gravel pits. Refill of the gravel pits is related to the sediment formula used. If a formula overpredicts the transport rate, it will indicate a refill greater than the measured value. On the other hand, a formula underpredicting the transport rate will show less refill than the measured amount. At a section undergoing scour, the amount of scour will be overpredicted by a sediment formula giving too high values of transport, and vice versa.

Stream channel changes in Stony Creek during the 1978 flood are noted to have the following features: (1) channel bed scour at the Highway-32 crossing, and (2) refill of the gravel pits downstream of the highway. These changes are significant and thus useful for assessing the predictability of each sediment transport formula.

CHANGES IN CHANNEL GEOMETRY

Changes in longitudinal and cross-sectional profiles were simulated for the study reach. Measured channel-bed profiles taken from the 1978 survey are also plotted for comparison with the respective simulated cross-sectional profiles.

Simulated cross-sectional changes at the Highway 32 crossing using different sediment formulas are compared with the 1978 measurement in order to assess the predictability of these sediment formulas. The cross-sectional changes simulated using the MPM formulas are somewhat less than the measurement. This is related to the smaller sediment transport rates computed by the formula. The simulated results based on the Engelund-Hansen formula as shown in Fig. 2 depict greater scour than the measurement. The overpredicted scour is related to the overprediction of the sediment transport rates. Both Yang's formula and Parker's formula generated cross-sectional changes at this section closely similar to the measurement, as exemplified in Fig. 3. The simulated results and measurement indicate the general trend of streambed scour at the bridge crossing. Since gravel pits are located in the vicinity of the bridge crossing, the general results concur with the speculation that gravel mining and export generally induce streambed scour.

Cross-sectional changes in the gravel pits are characterized by a distinct deposition pattern. Simulated changes at these cross sections using the four respective formulas are also depositional, but with various degrees of correlation with measurement. The deposition predicted by the MPM formula is generally less the measured value, while Engelund-Hansen formula overpredicts deposition. Patterns of deposition simulated using the Parker formula and the Yang formula are generally similar. On the basis of this testing, the Parker formula and the Yang formula were selected for further simulation applications.

ACKNOWLEDGEMENTS

The study was financially supported by District 3 of California Department of Transportation. Hydrology and geomorphology information were provided by M. Swanson and M. Kondolf.

REFERENCES

ASCE, *Sedimentation Engineering*, Manuals and Reports on Engineering Practice, No. 54, Vito A. Vanoni, ed., 1975.

Borah, D. K., Alonso, C. V., and Prasad, S. N., "Routing Graded Sediments in Streams: Formations," *J. Hydraul. Div.*, ASCE, 108(HY12), pp. 1486-1503, December 1982.

Chang, H. H., *Fluvial Processes in River Engineering*, John Wiley & Sons, 432 pp., 1988.

Engelund, F. and Hansen, E., "A Monograph on Sediment Transport in Alluvial Streams", Teknisk Vorlag, Copenhagen, Denmark, 1967.

Meyer-Peter, E. and Muller, R., "Formulas for Bed-Load Transport," Paper No. 2, Proceedings of the Second Meeting, IAHR, pp. 39-64, 1948.

Parker, G., "Surface-Based Bedload Transport Relation for Gravel Rivers", *Journal of Hydraulic Research*, 20(4), 417-436, 1990.

Yang, C. T., "Unit Stream Power Equation for Gravel", *J. Hydraul. Eng.*, ASCE, 110(HY12), pp. 1783-1798, December 1984.

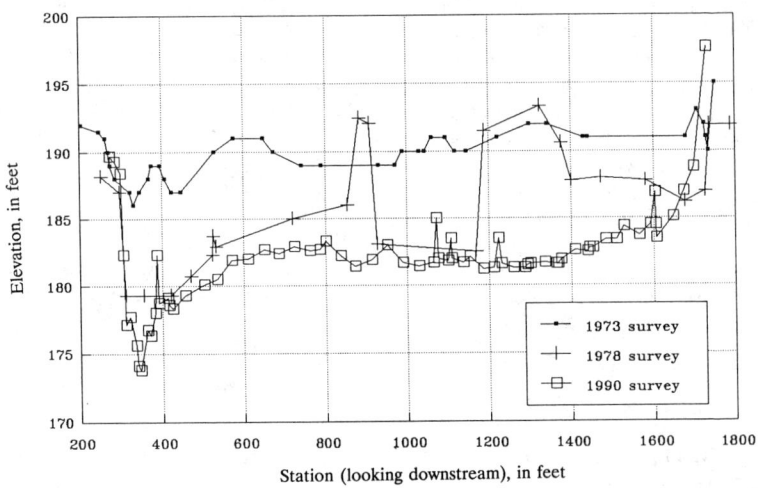

Fig. 1. Measured cross-sectional changes at the Highway 32 stream crossing

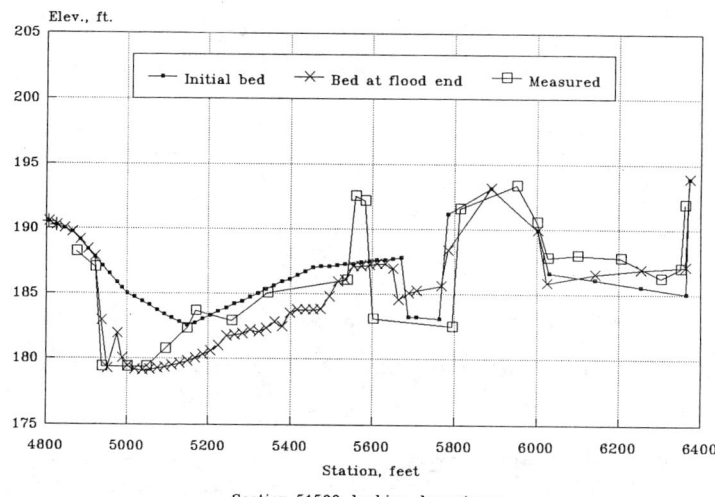

Fig. 2. Simulated and measured cross-sectional changes at Highway 32 crossing based on the Engelund-Hansen formula

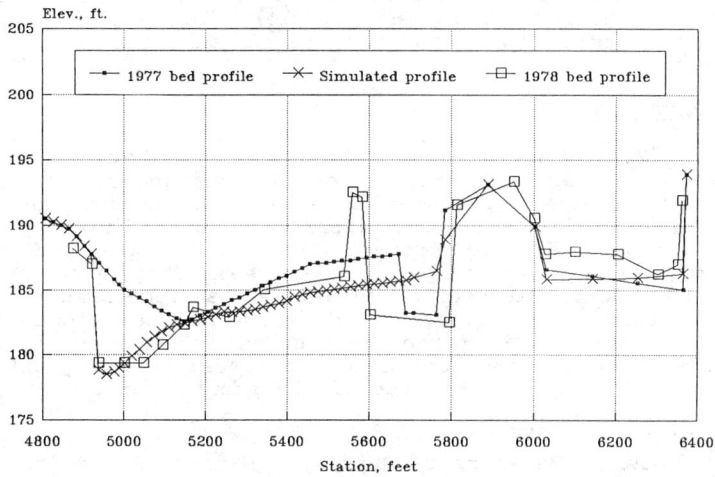

Fig. 3. Simulated and measured cross-sectional changes at Highway 32 crossing based on the Parker formula

Effect of Fenders on Local Pier Scour

L. M. Abed[1] and E.V. Richardson[2] F. ASCE

ABSTRACT

Physical model experiments were performed in the Hydraulics and Sediment Research Institute, Delta Barrage, Egypt, to determine the effect of fenders on the scour depth around circular bridge piers. The experiments determined the size of the scour hole around a circular pier with and without a floating box shape (open from the bottom) fender. The fender was placed at different submerged depths. It was found that fenders decrease the depth of local scour around circular piers and the scour depths decreased with an increase of submergence of the fender.

INTRODUCTION

Many bridges along the River Nile have circular piers which are protected from navigation impact with fenders. Fenders are open steel structures composed of piles and trusses which are attached to but **floating** around the bridge piers to protect them from impact of the navigation units. In the last few years deep scour holes has been identified near circular piers at some bridges. It was believed that fenders could be one of the reasons for the increased depth of the scour holes. Therefore, it was decided to use a physical model to investigate the effect of these fenders on the size of the scour hole.

---

[1]Senior Researcher, Hydraulics and Sedimentation Research Institute, Water Research Center, Ministry of Public Works and Water Resources, Cairo, Egypt.

[2]Senior Associate, Resource Consultants & Engineers, Inc. P.O. Box 270460 and Professor of Civil Engineering, Colorado State University, Ft Collins, CO

The shape used in simulating fenders in the model was selected to test the worst condition which is a closed structure, (box shape, open from the bottom).

Nine experiments were conducted in the Hydraulics and Sediment Research Institute. Two experiments were carried out without fenders and the rest with fenders.

EXPERIMENTAL SETUP

A physical undistorted model, with a scale 1:75, was constructed. The pier model was circular with diameter = 0.20 m. The wood fender model was sharp-nosed box shape, open from the bottom, with 2.28 m length and 0.28 m width, see Figure 1. The discharge was 90 l/s, the depth of flow was 0.08 m, the mean velocity was 0.375 m/s and the average velocity in the vicinity of the pier was 0.70 m/s. The sediment size was 0.396 mm with a standard deviation of 1.6. The model was placed in a recirculating sediment flume to simulate live bed scour. The model was calibrated to obtain equilibrium sediment movement and constant flow depth. The submerged depths of the fender tested were 1, 2, 4, and 6 cm. Submerged depth was measured from the water surface. The model was kept running for 3 to 4 hours to reach equilibrium scour depth. Measurements of the bed were taken by forming of contour lines around the pier, and upstream and downstream of the fender. The experiments were conducted at two different locations along the flume to assure the results.

Summary of the results for local pier scour are given in Table 1. And for the scour upstream and downstream of the fenders are given in Table 2.

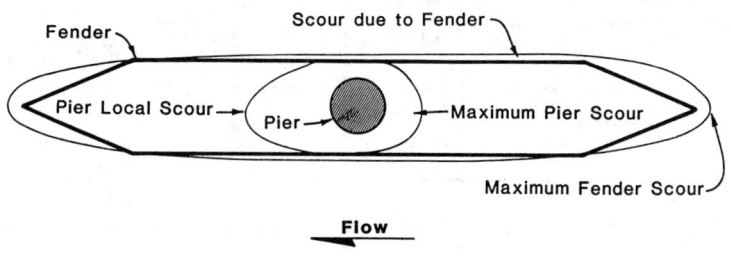

Figure 1. Location and Shape of Fender and Pier Scour.

Table 1. Results of local scour around the pier.

pier without fender		pier with fender		
experiment number	scour depth cm.	fender submerged depth cm.	scour depth cm.	scour width cm.
(1)	12	1	10	40 circular
		2	10	40 circular
		4	8	35 circular
(2)	8	1	6	50 circular
		2	6	50x45 oval
		4	4	40 circular
		6	4	32x40 oval

Table 2. Results of scour upstream and downstream the fender (experiment 2).

just upstream the fender			downstream the fender		
fender submerged depth cm.	scour depth cm.	scour width cm.	scour depth cm.	scour width cm.	distance from C.L. of bridge cm.
1	no scour	no scour	2	20x10 oval	85
2	no scour	no scour	4	23x23 oval	80
4	4	40 circular	4	20x24 oval	75
6	4	25x50 oval	6	13x27 oval	73

LOCAL PIER SCOUR DEPTHS

The effect of the fender on local pier scour is given in Table 1 and illustrated in Figure 2. The location and general shape of the pier scour hole is shown in Figure 1. In both experiments local scour depth decreases with the presence of the fender and scour depth decreases with an increase in submerged depth of the fender. In experiment 1 the scour depth without fenders was decrease by 33 percent when the fender was submerged 50 percent. In experiment 2 the scour depth decreased 50 percent when the fender was submerged 50 percent.

SCOUR DEPTHS UPSTREAM AND DOWNSTREAM OF THE FENDER

The maximum scour depths upstream and downstream of the fender are given in Table 2 for experiment 2. The shape and location of the scour holes is given in Figure 1. In Figure 3 the ratio of the maximum downstream scour depth to the flow depth is given as a function of the ratio of the fender submerged depth to the flow depth.

Table 2 and Figure 3 show that at the downstream end of the fender the scour depth increases with the increase of fender submerged depth. Table 2 also shows that scour occurs at the upstream of the fender at higher fender submergence depth only.

CONCLUSIONS

It was found that a floating sharp-nosed box fender decreased the local scour depth at a circular pier from 17 to 50 percent depending on the depth the fender was submerged under the water surface. In the two experiments the scour depth at the circular pier was decrease from 17 to 25 percent when the fender was submerged 1 cm. And was decreased 33 to 50 percent when the fender was submerged 4 cm. Depth of flow in the two experiments was 8 cm. Flow conditions and bed material in the two experiments were the same. Only the location of the pier and fender in the flume was different.

Scour holes occurred at the upstream and downstream ends of the fender. Although no scour occurred at the upstream end of the fender for fender submergence of 1 and 2 cm. Scour depths were the same at the upstream and downstream ends of the fender at larger submergence.

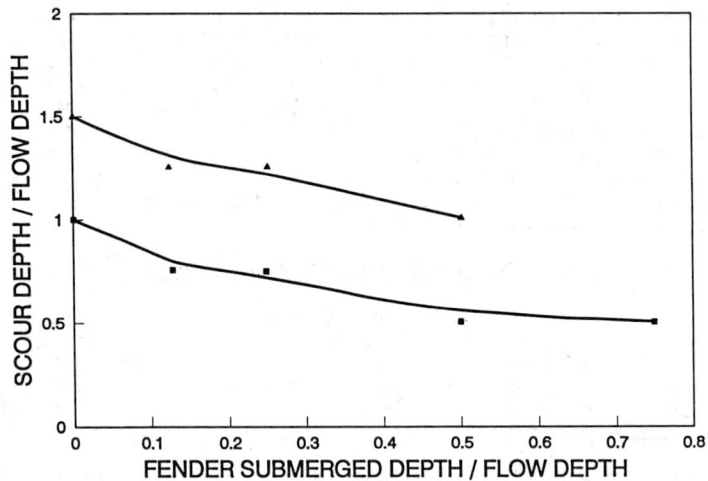

Figure 2. Relative Scour Depth at the Pier vs. Relative Fender Submerged Depth.

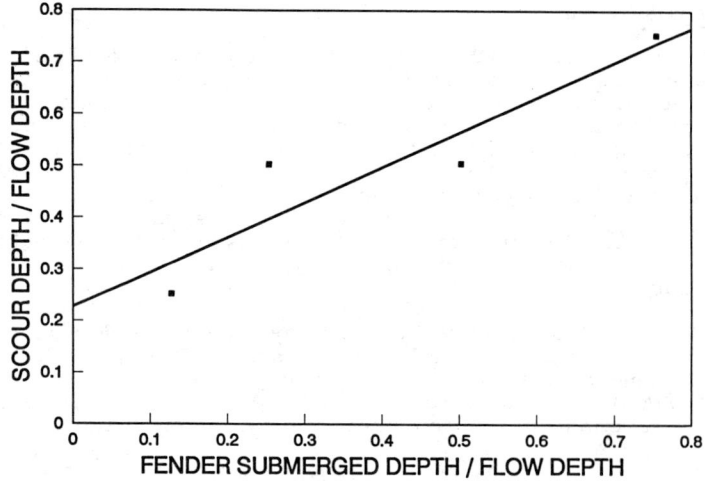

Figure 3. Maximum Scour Depths Downstream of the Fender.

ACKNOWLEDGEMENTS

Thanks are due to Dr. Gasser director of the Hydraulics and Sediment Research Institute for providing the financial assistance and the opportunity for conducting this study. And also to engineer Mahmoud Kamel for his help in conducting the tests and collecting the data. Thanks are also due to all personnel in constructing and maintenance of the testing facility.

APPENDIX I: REFERENCES

Abed, L.M., (1991), "Local Scour Around Bridge Piers in Pressure Flow", Ph. D. Dissertation, Civil Eng. Dept., Colorado State University, Ft.Collins, CO

Karaki, S. S. and R.M. Haynie, (1963), "Mechanics of Local Scour Discussion and Bibliography", Part I and II, Civil Engineering Department, Colorado State University, Fort Collins, Colorado, November.

Laursen, E.M. and A. Toch, (1956), "Scour Around Bridge Piers and Abutments", Bulletin No 4, Iowa Highway Research Board.

Laursen, E.M., (1980), "Predicting Scour at Bridge Piers and Abutments", General Report No. 3, University of Arizona, Tucson, AZ

Melville, B.W. and A.J. Raudkivi, (1977), "Flow Characteristics in Local Scour at Bridge Piers", J. Hydr. Div., ASCE, 15(4), 373-380, New York, NY

Richardson, E.V., Simons, D. and P.Y. Julien, (1990), "Highways in the River Environment", FHWA PUB. No. FHWA-HI-90-016, U.S. Depart. of Trans., Washington, D.C.

Richardson, E.V., Harrison, L.J. and Davis, S.R., (1991), "Evaluating Scour at Bridges,"FHWA HEC 18, FHWA PUB. No. FHWA-IP-90-017, U.S. Depart. of Trans. Washington, D.C.

Stevens, M.A., (1987), "Scour at Bridge Piers," Lecture Notes, Boulder, Colorado.

DEBRIS FLOW INITIATION AND TERMINATION IN A GULLY

Tamotsu Takahashi [1]

1. INTRODUCTION

A debris flow may be produced when a mass of earth and stone which starts to move on very steep channel contains water more than to saturate the voids among particles. Three major processes in such phenomena have been pointed out. The first one is due to the erosion and entrainment of bed materials (saturated or unsaturated) by appearance of the surface water flow. The second is due to failure of the natural dam and the third is continuity of motion of a saturated earth mass originated from a landslide. Majority of natural dams fail by overtopping of water, so that the debris flow genetic mechanism in the second category may be similar to the first. The third mechanism would be the main cause of the viscous type debris flows since the cohesive bed may hinder the erosion and thereby the development of debris flows.

This paper focuses on development and deposition of the stony-inertial type debris flow which is generated and developed by the bed erosion due to appearance of surface water flow in a very steep gully or on the downstream slope of a natural dam.

2. THE EQUATIONS DESCRIBING THE DEVELOPMENT AND DEPOSITION OF A DEBRIS FLOW

2.1 Equation of motion

The one dimensional equation of conservation of momentum is

$$\frac{\partial q}{\partial t} + \beta \frac{\partial (uq)}{\partial x} = gh\sin\theta_{bo} - gh\cos\theta_{bo}\frac{\partial (z_b + h)}{\partial x} - \frac{\tau_b}{\rho_T} \quad (1)$$

where $q(=uh)$ = discharge of flow in the unit width; u = mean velocity; h = depth of flow; z_b = erosion or deposition thickness; θ_{bo} = inclination of the original bed surface; τ_b = shear stress at the bottom of flow; ρ_T = specific density of the flow [$\rho_T = c_L(\sigma - \rho_m) + \rho_m$]; c_L = mean volume concentration of the large solids fraction throughout the entire flow depth; σ = density of the solids fraction; ρ_m = density of the interstitial fluid containing the fine solids fraction; β = momentum correction coefficient; g = gravitational acceleration; t = time and x = flow direction.

The resistance at the bottom τ_b in (1) in the stony-inertial debris flow; i.e., particles are distributing in the whole depth and $h/d_L \leq 30$ (d_L: mean particle diameter of the large solids fraction) is

$$\tau_b = (\sigma/8)[(c_{*L}/c_L)^{1/3} - 1]^{-2}(d_L/h)^2 u^2 \quad (2)$$

where c_{*L} = volume concentration of the large solids when packed. At an early stage

---

[1] Disaster Prevention Research Institute, Kyoto University, Gokasho, Uji, Kyoto 611, Japan

of development, or when the channel slope is flatter than the critical value, c_L is less than that is enough to distribute in the entire flow depth and the mixture layer appears only in the lower part of the flow. This is called the immature debris flow, for which the following empirical formula is obtained.

$$\tau_b = (\rho_T/0.49)(d_L/h)^2 u^2 \tag{3}$$

2.2 Equations of continuity and variation of the bed

The continuity equation for total volume is

$$\frac{\partial h}{\partial t} + \frac{\partial q}{\partial x} = i[c_b + (1-c_b)s_b] + r \tag{4}$$

where i = erosion velocity (> 0) or the deposition velocity (< 0); c_b = volume concentration of the total solids in the static bed; s_b = degree of saturation of the bed (applicable only for erosion; when deposition takes place, assigning $s_b = 1$); r = water supply per unit length from the sides of the channel.

The continuity equation for coarse solids fraction which is sustained in the flow by the effect of particle encounters is

$$\frac{\partial(c_L h)}{\partial t} + \frac{\partial(c_L q)}{\partial x} = ic_{bL}; \quad i \geq 0 \quad \text{or} \quad ic_{dL}; \quad i < 0 \tag{5}$$

where c_{bL} = volume concentration of the large solids fraction in the bed; c_{dL} = volume concentration of the large solids fraction when deposited.

The continuity equation for the fine solids fraction which is suspended in the interstitial fluid is

$$\frac{\partial[(1-c_L)c_F h]}{\partial t} + \frac{\partial[(1-c_L)c_F q]}{\partial x} = ic_{bF}; i \geq 0 \quad \text{or} \quad ic_{dF}; \quad i < 0 \tag{6}$$

where c_F = volume concentration of the fine solids fraction in the interstitial fluid of the flow [$\rho_m = c_F(\sigma - \rho_w) + \rho_w$; ρ_w = density of plane water]; c_{bF} = volume concentration of the fine solids in the bed; c_{dF} = volume concentration of the fine solids when deposited.

Equation for the variation of bed surface elevation is given by

$$(\partial z_b/\partial t) + i = 0 \tag{7}$$

where z_b = erosion or deposition thickness of the bed.

2.3 Equation of bed erosion velocity

The bed is assumed to be eroded by the shear stress on the bed assigned by the interstitial fluid of the overlying sediment-laden flow. One may approximate this shear stress, τ_f, by the difference between the applied shearing and resisting stresses as follows:

$$\tau_f = [(\sigma - \rho_m)c_L + \rho_m]gh\sin\theta - (\sigma - \rho_m)ghc_L\cos\theta R \tag{8}$$

where θ = energy gradient [$\tan\theta = \tau_b/(\rho_T gh)$]; R = dynamic friction coefficient of moving particles on the bed. This equation shows the larger c_L becomes by the erosion of the bed, the smaller τ_f becomes, and when c_L attains an equilibrium value, c_{eq}, no more erosion takes place and τ_f is equal to τ_{fe}. So far investigation reveals that the equilibrium concentration for the stony-inertial debris flow is

$$c_{eq} \equiv c_\infty = \frac{\rho_m \tan\theta}{(\sigma - \rho_m)(\tan\phi - \tan\theta)} \tag{9}$$

and the one for the immature debris flow is

$$c_{eq} \equiv c_{s\infty} = 6.7c_\infty^2 \tag{10}$$

where ϕ = internal friction angle of the bed and $\tan\phi$ is a little larger than R. It should be noted here that (10) is applicable only when $c_{s\infty}$ has a value less than c_∞ calculated by (9).

The asymptotic behavior of bed erosion process may be explained by

$$i = K(\tau_{*f} - \tau_{*fe})\sqrt{\tau_f/\rho_m} \tag{11}$$

where K = a numerical constant; $\tau_{*f}[= \tau_f/(\sigma - \rho_m)gd_L]$ and $\tau_{*fe}[= \tau_{fe}/(\sigma - \rho_m)gd_L]$ = respectively, the nondimensional shear stress and nondimensional equilibrium shear stress produced by the interstitial fluid of the flow.

Substitution of (8) and (9) or (10) into (11) obtains

$$\frac{i}{\sqrt{gh}} = K \sin^{3/2}\theta \frac{\rho_m}{\sigma - \rho_m} c_{L\infty}^{1/2}(c_{L\infty} - c_L)^{1/2}(c_{eq} - c_L)\frac{h}{d_L} \qquad (12)$$

Note that for a bed steeper than the value that satisfies

$$\tan\theta = \frac{c_{*L}(\sigma - \rho_m)}{c_{*L}(\sigma - \rho_m) + \rho_m}\tan\phi \qquad (13)$$

the value of c_∞ from (9) exceeds c_{*L}. Because no flow is possible at such a high concentration, c in (12) must be replaced by the maximum possible flow concentration, which experimentally is about $0.9c_{*L}$; consequently, $i = 0$.

When the bed is saturated, the shearing stress as well as the pressure in the sediment-laden flow on the surface are continuously transmitted into the bed, and above certain depth in the bed, the applied shearing stress becomes larger than the resisting stress. Due to this imbalance between the two stresses this part of the bed becomes unstable and easily entrained to flow. The shearing stress at depth a_s in the bed is given by

$$\tau = g\sin\theta[c_L(\sigma - \rho_m)h + \rho_m h + c_b(\sigma - \rho_w)a_s + \rho_w a_s] \qquad (14)$$

and the resistance at the same depth is given by

$$\tau_L = g\cos\theta[c_L(\sigma - \rho_m)h + c_b(\sigma - \rho_w)a_s]\tan\phi \qquad (15)$$

The unstable bed thickness is obtained by equating these two stresses. The unstable part, however, does not immediately move but there is a delay before completion of erosion. If this delay is measured by the time necessary for the debris flow to travel a certain length, $(d_L/u)/\delta_e$, one can express the erosion velocity as

$$i = \delta_e\left(\frac{\sigma - \rho_m}{\sigma - \rho_w}\right)\left(\frac{c_{L\infty} - c_L}{c_b - c_{T\infty}}\right)\frac{q}{d_L} \qquad (16)$$

where $\delta_e =$ a numerical constant and

$$c_{T\infty} = \rho_w\tan\theta/[(\sigma - \rho_w)(\tan\phi - \tan\theta)] \qquad (17)$$

Note that this description applies only to a bed flatter than the value that satisfies (13), because with a steeper slope the entire bed will slide before it becomes saturated.

The experimental data for the erosion velocity of a saturated bed, in which the bed is composed of only the coarse materials whose mean diameter is $d_L = 1.88$mm and the slope is between 12° and 15°, are compared in Fig.1 with (16) in which δ_e is set equal to 0.003.

Judgement of degree of saturation in the bed is rather difficult, and one may prefer to use (12) than (16) regardless the bed is saturated or not. The same data used in Fig.1 are used to correlate with (12) in Fig.2, and $K = 2.3$ is found appropriate. The other experimental data for the erosion of unsaturated bed, however, suggest $K \simeq 0.06$. This may imply that development of debris flow on saturated bed is much faster than on unsaturated bed.

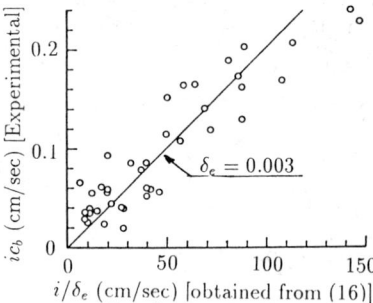

Fig. 1. Erosion Velocity of the Saturated Bed; Eq.(16)

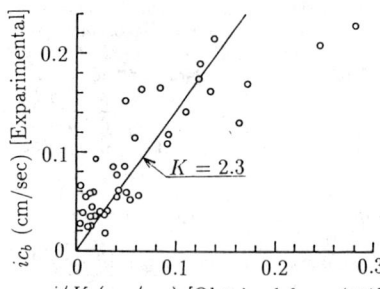

Fig. 2. Erosion Velocity of the Saturated Bed: Eq.(12)

2.4 Equation of lateral erosion velocity

Lateral erosion velocity is assumed as a function of the shear stress on the side wall assigned by the interstitial fluid of the sediment-laden flow, τ_{sf}. The value of τ_{sf} is then assumed as half of the bed shear stress, τ_f. One may write the recession velocity of the wetted side wall under the surface of the flow, i_s, as

$$\frac{i_s}{\sqrt{gh}} = \left(\frac{1}{2}\right)^{3/2} K_s \sin^{3/2}\theta \left[1 - \frac{\sigma - \rho_m}{\rho_m} c_L \left(\frac{\tan\phi}{\tan\theta} - 1\right)\right]^{1/2} \left(\frac{\tan\phi}{\tan\theta} - 1\right)(c_{eq} - c_L)\frac{h}{d_L} \quad (18)$$

where K_s = a numerical constant. By recession of the wetted side wall, the part of the wall above the surface of the flow may lose its stability and fall into the flow. If one assumes the whole side wall is kept vertical and the recession is parallel, the mean recession velocity of the side walls would be

$$i_{sml} = i_s h/(l_l + h), \quad i_{smr} = i_s h/(l_r + h) \quad (19)$$

where i_{sml} and i_{smr} = mean recession velocity at the left and right banks, respectively; l_l and l_r = parts of the left and right banks above the flow surface, respectively.

2.5 Equation of deposition velocity

The velocity u_e that is needed for a debris flow to transport the load neither with erosion nor deposition is given by the equilibrium steady uniform velocity equation obtained from (1) and (2)

$$u_e = \frac{2}{5d}\left[\frac{g\sin\theta_e}{0.02}\frac{\rho_T}{\sigma}\right]^{1/2}\left[\left(\frac{c_*L}{c_L}\right)^{1/3} - 1\right]h^{3/2} \quad (20)$$

Similarly, for the case of the immature debris flow

$$u_e = 0.7 g \sin\theta_e h^{3/2}/d \quad (21)$$

where θ_e = channel slope in which concentration c_L is in equilibrium; therefore, for the case of the debris flow $\tan\theta_e$ is given by substituting c_L in (9)

$$\tan\theta_e = [c_L(\sigma - \rho_m)\tan\phi]/[c_L(\sigma - \rho_m) + \rho_m] \quad (22)$$

and for the immature debris flow from (10)

$$\tan\theta_e = [\sqrt{c_L}(\sigma - \rho_m)\tan\phi]/[\sqrt{c_L}(\sigma - \rho_m) + 2.6\rho_m] \quad (23)$$

If a debris flow is decelerated to a velocity less than u_e by some measures on a slope steeper than or equal to θ_e, or if the debris flow comes to a flat area, the flow is now over-freighted and it tends to deposit. But experiments reveals deposition starts only from the point where the velocity becomes less than pu_e, in which $p(< 1)$ is a numerical constant. The amount of excess particles deposited is $h(c_L - c_{eq})$ per unit bed area. If one measures the time necessary to deposit this amount by $h/(\delta_d u_e)$, the deposition rate, i, may be given by

$$i = \delta_d\left(1 - \frac{u}{pu_e}\right)\frac{c_{eq} - c_L}{c_{dL}}u_e \quad (24)$$

where δ_d = a numerical constant. In the reach where $u > pu_e$, one assumes, $i = 0$, no matter what i is calculated from (24).

3. NUMERICAL SIMULATION OF DEVELOPMENT AND DEPOSITION OF A DEBRIS FLOW

3.1 Division into the coarse and fine fractions

If the shear velocity, u_*, assigned by the interstitial fluid on the bed, which is given from (8) by

$$u_*^2 = gh\sin\theta\left[1 - \frac{c_L}{c_{L\infty}}\frac{R - \tan\theta}{\tan\phi - \tan\theta}\right] \quad (25)$$

exceeds the settling velocity of the particle, the particle may be suspended in the interstitial fluid by the effect of turbulence; thereby increases the density of the fluid, ρ_m. To check this conjecture two kinds of experiments were conducted. In the first experiment a constant slope (18°) bed composed of uniformmaterial ($d_L = 1.07$mm) and saturated by pure water was eroded by a constant rate of muddy flow ($q = 53.3 \text{cm}^2/\text{sec}$), which contain particles of 0.084mm in diameter with 35% in volume concentration. In the second experiment the same bed was eroded by the muddy water with the same rate and concentration but the particle diameter in the flow was 0.2mm. Fig.3 compares the longitudinal changes of experimentally obtained solids concentrations in the first experiment to the results of calculation using the system of said equations. Both of the concentrations of coarse and fine fraction are explained by this model. Therefore, it may be concluded that the particles of 0.084mm in diameter is suspended in the debris flow. In this case u_{*f} well exceeded

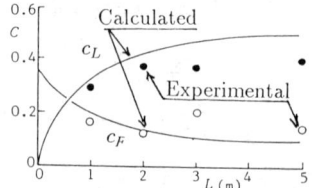

Fig. 3. Variation of Concentration with Distance

the settling velocity. In the second experiment, however, the coarse as well as fine solids concentrations in the debris flow could not be explained by considering the 0.2mm particle was suspended. In this case, on the other hand, the total solids concentration was obtained from the calculation considering both particle fractions were large and the debris flow developed by the action of pure interstitial water. The value of u_{*f} in this case was about the same to the settling velocity of the fine particles. Although this examination proves the validity of the conjecture, how one can divide particles into the two categories in case if the materials are well graded is the theme for further research.

3.2 Formation of a debris flow in an incised channel

A numerical simulation for the case corresponding the experiment is accomplished using the above system of equations. In the experiment a 5cm wide, 2cm deep and 3m long channel was incised in the bed adjacent to the transparent flume wall. The bed was composed of the unsaturated sediment mixture whose mean diameter was 1.5mm. The total width of the bed was 40cm and the slope was 13°. A steady water flow of 150cm³/sec was introduced from the upstream end and changes in vertical and lateral erosion rates with time were measured. The sediment runoff from the channel was also measured by separating sediment from the sediment-laden flow at the downstream end of the flume. Comparisons of the experimental values of the erosion velocities to those of calculation are given in Fig.4. In the calculation following values are used: $K = 0.06, K_s = 1.0, \delta_d = 1.0, p = 2/3, c_{bL} = c_{dL} = c_{*L} = 0.655, c_{bF} = c_{dF} = 0$. Lateral erosion did not necessarily procede uniformly along the channel, but sporadic sloughing created discountinuous widenings. The general tendency of widening and vertical erosion of the channel are, however, rather well explained in Fig.4. Fig.5 shows the comparison of the sediment discharges at the downstream end of the channel. The experimental sediment discharge fluctuates presumably depending on the local sloughing and other irregularities, but the characteristic sediment yield such as large flow rate

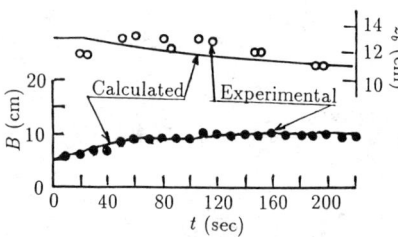

Fig. 4. Variation of Channel Width and Bed Elevation

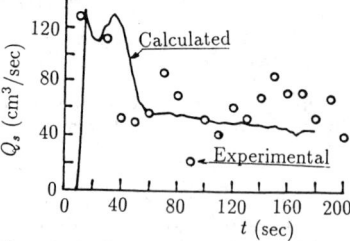

Fig. 5. Sediment Discharge at the Channel Outlet

at early stage and gradual diminution is explained.

3.3 Deposition of a debris flow at the channel slope change; formation of a debris fan

The process of development of a debris fan at the debouchment of a gully where the longitudinal slope changed abruptly was simulated by using the two-dimensional version of the system of equations appeared above and compared with the experimental data. Detailes of the experiment and the equations that takes into account the particle segregation process during flowing in the upstream channel can be found elsewhere (Takahashi et al. 1992). Fig.6 shows the time-varying shapes and thickness distributions of the debris fan. One may recognize the shape as well as the thickness are well reproduced by the numerical simulation. One point to be emphasized is that, as is clearly seen in the figure, the maximum deposit thickness in many experiments appears not at the outlet of the channel but downstream from there, and this characteristics can only be reproduced if one uses the deposition velocity equation which takes account the inertial motion downstream from the outlet as (24) considers.

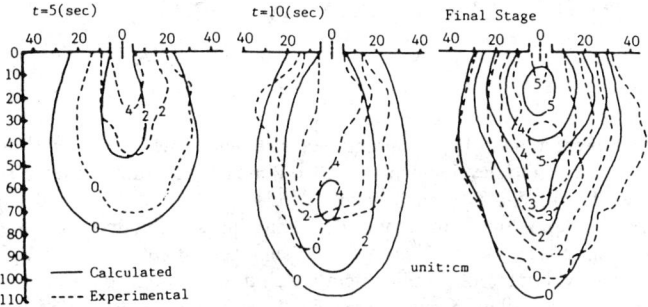

Fig. 6. Time-Varying Shapes and Elevations of Deposited Debris- Flow Surface in Debris Fan

4. CONCLUSION

A comprehensive simulation model predicting the debris-flow discharge hydrograph, solids concentration in it and the process of variation of the channel due to vertical as well as lateral erosion or deposition was formulated and then verified by flume experiments. This model is applicable to developing processes of both the erosion type debris-flow and the majority of natural dam failure type debris-flow and to the deposition process of the stony type debris flow regardless the cause of it generation. In this model the solids composition in the flow entrained from the bed is divided into two categories; the fine fraction that is suspended in the fluid and thereby effective to increase the competence to transport coarse particles, and the coarse fraction that is sustained in the flow by the effect of particle collisions. A criterion for dividing those fractions is given but further research is needed for the case of well graded materials. Determination of the appropriate values of many of the model's parameters require further investigation.

Appendix   REFERENCE

Takahashi, T., Nakagawa, H., Harada, T. and Yamashoki, Y. (1992). "Routing Debris Flows with Particle Segregation." *J.Hydr.Eng.*, ASCE, 118(11), 1490-1507.

Hydraulic Modeling and Mapping of Mud and Debris Flows

J. S. O'Brien[1], Member, ASCE

Abstract

The prediction of hyperconcentrated sediment flows on alluvial fans was advanced through the development of a two-dimensional, finite difference flood routing model FLO-2D. The model simulates clear water flood hazards, mudflows and debris flows on urbanized alluvial fans and floodplains. Interactive flood and mudflow routing between the channel, streets and floodplain is accomplished with a uniform grid system which describes the complex floodplain topography. A mudflow event created by the eruption of Mount St. Helen is simulated with FLO-2D.

The FLO-2D finite difference grid system is established using computer-aided design (CAD) graphics on a digitized map. A file with FLO-2D predicted flood depths can be imported into the digitized base map drawing and depth contours can then be plotted with the CAD software thus automating the flood hazard delineation process.

Introduction

The two-dimensional finite difference, flood routing model FLO-2D was created to simulate non-Newtonian flows on alluvial fans. The model estimates flow hydraulics (velocity and depth), predicts a justifiable area of inundation and simulates flow cessation. This versatile model routes channel flow with variable area cross sections, predicts overbank discharge and simulates overland flow over complex topography in eight directions. Urban flooding on developed fans and floodplains is analyzed with components which simulate street flow and account for flow path obstructions and flood storage loss due to walls or buildings. The model is described in more detail in O'Brien, et. al., (1993).

---

[1]Prin., FLO Engrg., Inc. P.O. Box 1659, Breckenridge, CO 80424

Hydraulic Modeling

Flood routing is accomplished with a finite difference algorithm which solves the continuity equation and the diffusive wave approximation to the momentum equation. Hyperconcentrated sediment flow, including mud floods, mud and debris flows are analyzed by a quadratic rheologic model initially described by O'Brien and Julien (1985):

$$\tau = \tau_y + \eta \left[\frac{dv}{dy}\right] + C \left[\frac{dv}{dy}\right]^2 \tag{1}$$

where (dv/dy) is the shear rate; η is the dynamic viscosity; τ_y is the yield strength; and C denotes the inertial shear stress coefficient. The quadratic model includes terms to define both the viscous and inertial flow regimes. The sum of the yield and viscous stresses comprise the viscous flow regime while the turbulent and dispersive stresses dominate in the inertial regime. With a approximate solution for the sum of the inertial and viscous terms using empirical equations to represent friction slope terms, flow hydraulics can be computed for the continuum of flow regime from viscous to inertial (O'Brien, et. al., 1993). Accordingly, the friction slope components can be written as:

$$S_f = \frac{\tau_y}{\gamma_m h} + \frac{K\eta V}{8\gamma_m h^2} + \frac{n^2 V^2}{h^{4/3}} \tag{2}$$

The total friction slope S_f is the sum of the yield stress slope, viscous slope and the turbulent-dispersive slope which is written in terms of the flow depth h, depth-averaged velocity V and the specific weight of the fluid matrix γ_m. Reasonable values of resistance parameters K and Manning n can be assumed for channel and overland flow roughness. The yield stress and the viscosity vary principally with sediment concentration:

$$\eta = \alpha_1 e^{\beta_1 C_v} \tag{3}$$

and

$$\tau_y = \alpha_2 e^{\beta_2 C_v} \tag{4}$$

where α_i and β_i are empirical coefficients defined by a rheologic analysis of the material (O'Brien and Julien, 1988).

A number of mudflow hazard delineation projects have been completed using the FLO-2D model (O'Brien, et. al., 1993). Recently the USGS Cascades Volcanic Observatory provided an opportunity to simulate the Pine Creek mudflow initiated by the eruption of Mount St. Helens. The mudflow traveled 22.5 km in 20 ± 3 minutes before entering Swift Reservoir where the mudflow volume and the peak discharge was estimated by response of the stage recorder at Swift Dam (Pierson, 1985). The USGS provided cross sections of Pine Creek surveyed after the mudflow.

The simulation of the Pine Creek mudflow was accomplished as follows:

- A 15 minute topo map was digitized and a uniform grid system of 500 ft square elements was established over the potential flow area.

- A (CAD) program with a digital terrain model was used to export the grid element coordinates and elevations to a FLO-2D file.

- Rheological parameters for the mudflow were selected from the Major and Pierson (1992) paper on Mount St. Helen's mudflows assuming a silt-and-clay to sand ratio of 1:1.

- The channel geometry data for 12 cross sections were reduced and prepared in a data input file. Selected Manning's n values ranged from 0.03 to 0.1.

- The inflow hydrograph was estimated at the first cross section to reproduce the peak discharge at cross section P2. The first grid element was located several thousand feet upstream of the first cross section.

FLO-2D was run several times to replicate the following conditions: Estimated peak discharge at cross section P2 of 28,600 m^3; estimated peak discharge at the reservoir of 7,500 m^3; timing of the peak discharge arrival at the reservoir; and estimated volumetric inflow to the reservoir. When these conditions were satisfactorily met, the computed flow hydraulics were compared with those estimated by Pierson (1985) at the 12 cross sections.

Preparing for the simulation, it was realized the magnitude of the mudflow event could tax the capability of the model. The peak discharge was in excess of 1 million cfs (28,000 m^3) and estimated velocities exceeded 20 m/s (65 fps). The discharge was two orders of magnitude greater than any previous FLO-2D simulations and required very small timesteps (0.01 sec.) for the steep rising limb of the hydrograph. Computational time, however, was reasonable (25 minutes).

One modeling concern was whether the estimated sediment volumetric concentration and associated the rheologic parameters of viscosity and yield stress for mudflows could replicate the large flow hydraulics. Based on an appropriate definition of the fluid matrix, viscous mudflows occur in the range of 45 to 50% sediment concentration by volume (O'Brien and Julien, 1988). A comparison of computed mudflow viscosity and yield stress revealed that the equivalent range of sediment concentration for the Major and Pierson (1992) samples was 60 to 65%. They reported concentrations up to 61% by volume in their rheological analysis. The difference in concentration and rheologic properties may have arisen from the definition of the flow matrix, the type and quantity of clay and the large quantities of clastic material in the samples.

Pierson (1985) reported concentrations of 64 to 70% by volume in the post-event analysis of the Pine Creek mudflow. Volumetric concentrations of 63 to 66% were used FLO-2D simulation knowing that comparable viscosity and yield stress values as those reported by O'Brien and Julien, 1988 would be computed. It was reported that only minor amounts of deposition took place in the Pine Creek (Pierson, 1985). In the simulation, an increase of 2% sediment concentration by volume resulted in flow cessation on the falling limb of the hydrograph which assisted in defining the limits of the sediment concentration. The question remained, however, whether the viscous mudflow rheologic parameters would allow computation of such high velocities and replicate the reservoir inflow hydrograph. The following FLO-2D results were obtained:

- **Hydrograph Timing: Arrival of the Peak Discharge in Swift Reservoir**

    Pierson Estimate: 20 ± 3 min   FLO-2D Simulated: 20.4 min

- **Volume: Total Inflow Volume to Swift Reservoir**

    Pierson Estimate: 13,431,000 m^3 FLO-2D Simulated: 13,490,000 m^3

- **Peak Discharge: Peak Discharge into Swift Reservoir**

    Pierson Estimate: 7,500 m^3 FLO-2D Simulated: 11,750 m^3

Several runs were required to satisfactorily replicate the reservoir inflow hydrograph timing and volume. These were the most accurate data based on the response of the Swift Reservoir recording gage. The reservoir inflow peak discharge should be verified by reservoir routing and may be underestimated.

Reasonably good correlation was obtained between the estimated flow hydraulics from field data and the FLO-2D mudflow simulation as shown in Table 1. It should be noted that the FLO-2D predicted maximum velocities and flow depths at the cross sections do not necessarily occur at the same instant. Also, the cross sections for the FLO-2D simulation were only approximately located. Pierson's (1985) peak discharge was crudely estimated from assumptions of superelevation in bends and potential flow depths. The peak discharge was assumed to correspond to his reported depth and velocity estimates.

Table 1. Comparison of Estimated and FLO-2D Predicted Flow Hydraulics in Pine Creek

Xsection - Grid	Peak Discharge (m³/s)		Max. Velocity (m/s)		Max. Flow Depth (m)	
	Pierson[1]	FLO-2D	Pierson	FLO-2D	Pierson	FLO-2D
P1 - 1166	17,100	28,300	23.5	22.1	9.8	20.6
P2 - 1043	28,600	27,200	17.7	20.9	15.2	19.2
P2.1 - 1009	25,900	27,000	20.8	21.8	12.6	14.0
P3 - 942	28,200	26,200	13.1	15.6	14.5	16.3
P4 - 915	21,700	25,000	12.4	19.4	14.9	18.2
P5 - 856	19,900	24,200	10.9	15.8	14.8	18.7
P6 - 672	21,000	21,700	14.2	14.0	13.9	20.2
P7 - 571	19,200	19,200	21.1	12.8	10.7	19.4
P8 - 432	16,600	18,000	15.3	12.7	9.4	13.0
P9 - 415	6,250	13,500	9.3	11.7	9.3	19.9
P10 - 372	8,930	12,500	11.0	9.6	9.0	14.6
P11 - 196	7,320	12,000	12.0	20.9	6.0	6.4

[1] Table 1 Average Flow Hydraulics from Pierson, GSA Bulletin, 1985, Vol. 96, p. 1064

Mapping Mudflow Hazard

Applying FLO-2D with a CAD digital terrain model enables the flood contours (or velocity contours) to be plotted directly on the digitized base map. Similarly, temporally varied flow depths can be plotted to visualize the progression of the floodwave (O'Brien, et. al., 1993). This is accomplished as follows: The potential flow surface on a base map is digitized with CAD and a triangulation irregular network (TIN) is created. Next, a grid system is overlain. A data file containing the grid element node number, coordinates, elevation and a descriptor (Manning's n) is exported from the CAD program. This file is then reformatted with a preprocessor program and becomes a FLO-2D data input file.

To plot the flood contours, the FLO-2D predicted maximum flow depths and supplemental grid element coordinate data are written to file. This file is then imported to the CAD base map. A new TIN is run and contours representing the maximum flow depths are plotted over the base map. This essentially automates the hazard delineation process. Results of this mapping procedure are shown in a conference companion paper.

The procedural limitations include constraints of the modeling system. The simulation detail is limited by the grid element size which is represented by a single elevation. The channels may require breaklines in the CAD drawing to assist the creation of the contour lines. In addition, the maximum flood contour location is interpolated between the grid element nodes. These contours may have to be adjusted to reflect rigid boundaries such as flood containment walls.

When conducting a FLO-2D flood simulation, several hazard scenarios should be investigated and the final flood hazard maps should be a compilation of the worst case scenarios. These flood scenarios may include: various return period flood events, mudflows, loss of channel conveyance capacity, culvert or bridge plugging, or redirection of the fan flow paths. The selection of various flood scenarios should be coordinated with the local flood control agency.

Summary

FLO-2D is a flood routing model designed to simulate both water floods and mudflows in channels, over alluvial fans and on urban floodplains. The Pine Creek mudflow initiated with the eruption of Mount St. Helens was simulated with the FLO-2D model. The relatively good correlation of the FLO-2D simulated mudflow results with the estimated flow hydraulics from field data required appropriate estimates of the sediment concentration and mudflow rheologic parameters of viscosity and yield stress.

References

O'Brien, J. S., Julien, P. Y., and Fullerton, W. T. (1993). "Two-dimensional water flood and mudflow simulation." J. Hydr. Engrg., ASCE 119(2), 244-261.

O'Brien, J. S. and Julien, P. Y. (1988). "Laboratory Analysis of Mudflow Properties." J. Hydr. Div., ASCE, 114(8), pp. 877-887.

O'Brien, J. S. and Julien, P. Y. (1985). "Physical properties and mechanics of hyperconcentrated sediment flows." ASCE Specialty Conf. on the Delineation of Landslides, Flash Floods and Debris Flow Hazards in Utah, Utah Water Research Lab, Univ. of Utah, Logan, Utah, pp. 260-279.

Pierson, T. C. (1985). "Initiation and flow behavior of the 1980 Pine Creek and Muddy River lahars, Mount St. Helens, Washington." Geological Society of America Bulletin, V. 96, 1056-1069.

Major, J. M. and Pierson, T. C. (1992). "Debris flow rheology: Experimental analysis of fine-grained slurries," Water Resources Rsrch., 28(3), 841-857.

AN EMPIRICAL MODEL FOR THE VOLUME-CHANGE BEHAVIOR OF DEBRIS FLOWS

By S.H. Cannon[1]

Abstract

The potential travel distance of debris flows can be estimated by considering the volume-change behavior of flows as they travel down hillsides; movement stops where the volume of actively flowing debris becomes negligible. The average change in volume over distance for 26 recent debris flows in the Honolulu area was assumed to be a function of the slope over which the debris flow traveled, the degree of flow confinement by the channel, and an assigned value for the type of vegetation through which the debris flow traveled. Analysis of the data yielded a relation that can be incorporated into digital elevation models to characterize debris-flow travel on Oahu.

Introduction

Residential development in the Honolulu District of Oahu, Hawaii, has expanded into a complex and steep landscape, and fast-moving debris flows, which originate on steep valley walls and in headwaters, have caused recurring problems (Torikai and Wilson, 1992). Most debris flows in the area stop on steep hillslopes, but some flow down aprons and channels into populated areas (Ellen et al., 1991). Because only some debris flows on Oahu travel far enough to pose a hazard to life and property, an understanding of the physical controls of debris-flow travel distance is critical to adequately define the hazard.

Work by Cannon (1989) and Cannon and Savage (1988) suggests that the travel-distance potential of debris flows can be estimated using an approach based on the mass-change rate of the flow, by which an initial mass, or as in this paper, volume, of material is modified by loss of material as it travels downslope. Travel stops where the volume of the actively flowing debris becomes negligible. Cannon (1989) describes the primary controls on

---

[1]U.S. Geological Survey, Box 25046, MS 966, Denver Federal Center, Denver, CO 80225

the mass-change rate as the gradient of the ground surface over which the debris flow is traveling, the channel morphology, and the strength of the moving slurry. Field observations on Oahu suggest that vegetation also exerts a considerable influence on volume-change behavior.

The work presented here evaluates the influence of slope, vegetation type, and channel morphology on the average volume-change behavior of debris flows on Oahu. (Physical properties of the moving slurry have not been incorporated into the analysis because they are difficult to measure. These properties were assumed to be constant in the parts of the Honolulu District evaluated during this study where relatively homogeneous rock and soil types make up the hillslopes). A statistical model for volume-change behavior is developed that can be combined with digital elevation models to characterize debris-flow travel. An example of this application is described. The model was used to map the hazard from debris flows in the Honolulu District (Ellen and Mark, this volume).

Approach

The average volume-change behavior of debris flows on Oahu was characterized by dividing the volume of material that mobilized from the landslide at the head of each debris-flow path by the distance traveled down the hillside, measured from the downslope margin of the landslide scar to the path terminus (Figure 1). The total scar volume was assumed to be the initial volume of the debris flow, and the volume was considered to diminish linearly with travel.

Note that this approach does not allow for the incorporation of substantial material into the debris flow through erosion of the channel base and sides, a process that is considered to be significant in other areas. Close examination of recent debris flows on Oahu indicates that erosion is not a dominant process for the majority of flows.

The average change in volume over distance was calculated only for debris flows that traveled over what appeared on maps of topography and vegetation to be nearly constant slopes and through nearly uniform channels and vegetation types. Hence, one representative value for each of the independent variables, (slope angle, θ; degree of channel confinement, R; and vegetation type V_t), was determined for each flow. By restricting the sample to debris flows with fixed variables, a model for volume change could be developed that can be applied on a cell-by-cell basis in digital elevation models, where the variables in each cell are likewise assumed to remain at constant average values.

Data Collection

The average change in volume over distance and the three independent

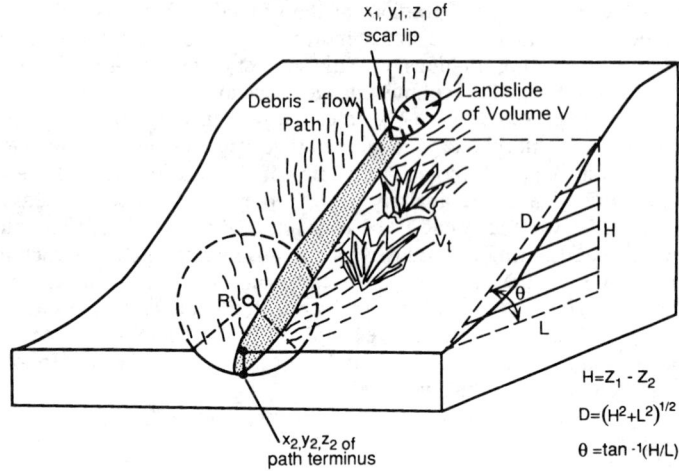

Figure 1. Three-dimensional view of a debris flow. H, vertical drop of path (m); L, map length of path measured from scar lip to terminus of deposits (m); D, length of path along hillside (m); θ, gradient of path (degrees); R, transverse radius of curvature of channel (m); V_t, vegetation type.

variables of slope, confinement, and vegetation were measured for 26 debris flows selected from 72 flows mapped at 1:12,000 from aerial photographs taken following the New Year's Eve storm of 1987-1988 or from field observations in the three years following the storm (S.D. Ellen, written communication, 1991).

The volumes of debris-flow scars included in the sample were determined from direct measurement in the field where access was possible. When the scar was inaccessible, volumes were estimated from measurements of scar area made remotely in the field or from aerial photographs, combined with a relation between scar area and volume developed from scars measured directly. Scar volumes of the debris flows included in the data set varied from between 25 and 938 m³, with an average of 194 m³ and a standard deviation of 183 m³.

The travel distances of the debris flows included in the data set were determined by digitizing the locations of the downslope margin of the scar (the starting position of the flow) and the terminus of the path. The digital x, y, and z coordinates of these locations, determined from a 10-m digital elevation grid, were used to calculate the parameters H, L, and D, as shown in Figure 1. Path lengths measured over the hillside (D) varied between 19 and 220 m, with an average of 85 m, and a standard deviation of 55 m.

The gradient, (θ), of the hillside over which the debris flows traveled was calculated as shown in Figure 1. Slopes varied from 26 to 47 degrees, with an average of 33 degrees and a standard deviation of 5.7 degrees. A simple relation between slope and average volume-change behavior is expected; in general, steeper slopes result in less deposition than do gentle slopes (Cannon, 1989).

The influence of vegetation type on volume-change behavior was evaluated using the vegetation units mapped by M.A. McKittrick (written communication, 1991) which were designed to reflect the general sizes of vegetation in the area, and consist of both dominant and subordinate vegetation types. Based on the simplified assumption that the size of vegetation directly affects the volume-change behavior, a value of 10 through 60, in increments of 10 and with 60 being trees, was assigned to each dominant unit. A value of 1 through 6, with 6 being trees, was added for the first subordinate vegetation type included in the unit. This ranking system gives a semblance of continuity, a prerequisite for regression analysis, to the vegetation-type variable. Assigned vegetation rankings ranged from 25 to 60, with an average of 49 and a standard deviation of 7.

The degree of confinement provided by a channel is thought to affect how much material is deposited, and consequently how far a debris flow will travel (Cannon, 1989). Tighter, more confining channels will result in less deposition than wide, relatively unconfined channels (Cannon, 1989). The radius of curvature of channels in a vertical plane, calculated from a 10-m digital elevation grid, was used as a measure of the degree of confinement; smaller radii indicate tighter, more confining channels (Figure 1). A value of 2,000 m was assigned for planar and convex hillsides to provide a finite upper limit for the radius of curvature of Oahu hillsides. Curvature radii varied from 14 to 2,000 m, with an average of 442 m and a standard deviation of 688 m.

Statistical model

To evaluate the influence of the independent variables on the average volume-change behavior, a step-wise multiple regression, following the procedure described by Draper and Smith (1981, p. 309-310), was carried out. In this procedure, the effect of each independent variable on the dependent variable is evaluated in a step-wise order, while the effects of the remaining variables are held constant.

The significance of the equation generated in each step was evaluated using the overall F-statistic, and the criterion that the new independent variable provides a significant decrease in the residual sum of squares, i.e., significantly improves the equation, was evaluated using the smallest partial F-value for each step (Draper and Smith, 1981). In addition, the mean square about the regression provides an estimate of the variance about the regression, which represents a measure of the error with which any observed value of the dependent variable could be predicted from the equation.

By transforming the data to a normal distribution, which is a necessary prerequisite for normal conditional distributions, the basic assumption in regression analysis that the conditional distributions of the residuals are normal (Johnston, 1980) was adhered to. The log, natural log, square, and square root transformations of each variable were examined to determine the best approximation of a normal distribution. A number of different data transformations resulted in normal distributions of the variables, and thus a number of regressions were performed to find the best possible model.

The regression procedure indicated that the average volume-change behavior can best be represented by the following relation:

$$\log((V_i - V_f)/D) = 0.14 \log R - 1.40 \log \theta + 2.16, \qquad (1)$$

where θ and R were entered into the equation in that order, and the numbers used to characterize vegetation did not significantly improve the equation. A plot of the residuals reveals satisfactory behavior, and the mean square about the regression indicates up to a 26% error in estimating the dependent variable from the regression equation.

Note that the signs of the coefficients in the model agree with the physical understanding of likely relations; volume change is promoted by large values of R (lack of confinement) and by gentle gradients.

Model application

The model developed above can be used to simulate the volume-change behavior of debris flows routed through a digital landscape composed of 10x10-m cells (Ellen and Mark, this volume). The initial volume of material in the flow is assumed to equal the volume of the landslide scar at the head of the path. This volume of material is then progressively reduced, on a cell-by-cell basis, as the flow progresses through the digital landscape. The volume-change behavior for the cell-by-cell analysis is represented by equation 1 where V_i is the volume of material entering the cell, V_f is the volume of material that leaves the cell, and D is the length of the path segment through the cell.

Summary

The potential travel distance of debris flows is estimated by considering the volume-change behavior of flows as they travel down hillsides; movement stops where the volume of actively flowing debris becomes negligible. The average change in volume over distance for 26 recent debris flows in the Honolulu area was characterized by dividing the volume of material that mobilized from the landslide at the head of the flow (V) by the distance traveled over the hillside (D) deposit. The volume-change behavior was assumed to be a function of the slope over which debris flow traveled (θ), the degree of flow confinement by the channel (R)), and an assigned value for the type of vegetation (V_t) through which the debris flow traveled. A step-wise multiple regression analysis of the effect of the independent variables on the change in volume over distance yielded a relation that can be used to simulate the volume-change behavior of debris flows when they are routed through a digital landscape. The initial volume of material in the flow is progressively reduced, on a cell-by-cell basis, as the flow progresses downslope.

References

Cannon, S. H. (1989). "An evaluation of the travel-distance potential of debris flows". *Utah Geological and Mineral Survey Miscellaneous Publication 89-2*, 35 p.

Cannon, S. H., and Savage, W. Z. (1988). "A mass-change model for debris flow". *Journal of Geology*, vol. 96, p. 221-227.

Draper, N. R., and Smith, H. (1981). *Applied regression analysis*, 2d edition. John Wiley and Sons, New York.

Ellen, S. D., Iverson, R. M., and Pierson, T. C. (1991). "Map showing the distribution of debris flows during the New Year's Eve storm of 1987-1988 in southeastern Oahu, Hawaii". *U.S. Geological Survey Open-File Report 91-129*, scale 1:20,000, 1 plate.

Johnston, R. J. (1980). *Multivariate statistical analysis in geography*. Longman, London and New York.

Torikai, J. D., and Wilson, R. C. (1992). "Hourly rainfall and reported debris flows for selected storm periods, 1935-91, in and near the Honolulu District, Hawaii." *U.S. Geological Survey Open-File Report 92-486*, 76 p.

Mapping Debris-Flow Hazard in Honolulu Using a DEM

By Stephen D. Ellen and Robert K. Mark[1]

Abstract

A method for mapping hazard posed by debris flows has been developed and applied to an area near Honolulu, Hawaii. The method uses studies of past debris flows to characterize sites of initiation, volume at initiation, and volume-change behavior during flow. Digital simulations of debris flows based on these characteristics are then routed through a digital elevation model (DEM) to estimate degree of hazard over the area.

Introduction

We have developed a semi-automated method for mapping debris-flow hazard and applied it to a 180-km^2 area near Honolulu, Hawaii (fig. 1). The resulting hazard map, along with more complete description of the method, is presented elsewhere (Ellen and others, 1993). The method uses information from geomorphic analysis and historical debris flows to calibrate digital simulations of debris flows that are routed through a digital elevation model (DEM) of the study area. The DEM is used because it provides a convenient numerical representation of topography, which commonly exerts a strong influence on spatial distribution of debris flows, as on other forms of slope instability (Wadge, 1988; McEwen and Malin, 1989).

The debris flows evaluated by this method include slope movements that have been variously termed soil avalanches (Wentworth, 1943), mud flows, debris slides, or debris avalanches (Varnes, 1978). They are a rapid and destructive form of slope movement that in the Honolulu area typically begin when intense rainfall triggers shallow slope failures on steep hillsides (fig. 2). The sliding masses of soil, weathered bedrock, and vegetation generally become fluid enough to flow rapidly down hillslopes and channels as debris flows. The New Year's Eve storm of 1987-1988 triggered more than 400 debris flows in the area (Ellen and others, 1991). Several of these struck homes and others contributed debris that diverted floodwaters to cause additional damage (State of Hawaii, 1988; Interagency Flood Hazard Mitigation Team, 1988; Dracup and others, 1991).

---

[1]Geologist and Physical Scientist, U.S. Geological Survey, 345 Middlefield Road MS-975, Menlo Park, CA 94025

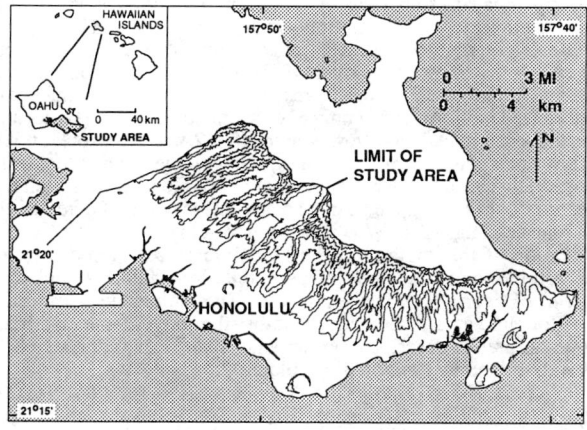

Figure 1: Study area, Honolulu District of Oahu, Hawaii. Contour interval 122 m (400 ft).

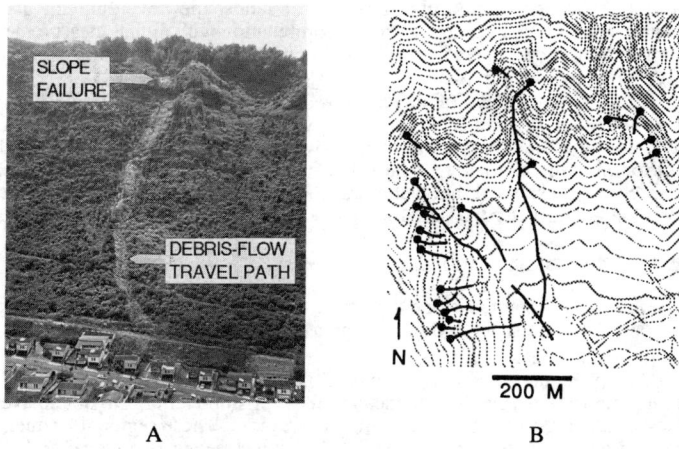

Figure 2: Debris flows near Honolulu triggered by New Year's Eve storm of 1987-1988. A, Debris flow that damaged two houses in Kuliouou Valley. B, Distribution of debris flows in part of Hahaione Valley. Contour interval 12.2 m (40 ft).

Method

We mapped hazard from debris flows digitally by systematically simulating the initiation and movement of debris flows in a DEM (Ellen and Mark, 1988). The DEM is a numerical representation of topography that consists of elevations interpolated at 10-m X,Y spacing from 12.2-m (40-ft) elevation contours. The method is illustrated in figure 2 of Wieczorek (this volume). Topography is central to the method because land form commonly exerts a major influence on location of the slope failures that constitute debris-flow sources (fig. 2A of Wieczorek, this volume), location of flow paths (fig. 2B of Wieczorek, this volume), and extent of flow (fig. 2C of Wieczorek, this volume). Once potential debris flows were simulated, hazard from all potential flows was mapped using an automated count of potential flow paths through each map cell (fig. 2D of Wieczorek, this volume).

Application of this method in the Honolulu area required information on distribution, volume, and frequency of potential slope failures, and on volume-change behavior of debris flows during travel. These elements were estimated by studies in the Honolulu area. Our analysis of distribution, volume, and frequency of potential slope failures is described below; analysis of travel-distance behavior of debris flows is described by Cannon (this volume).

Distribution of slope failures

We restricted analysis to the majority of the Honolulu District that is underlain by the Koolau Basalt, a homogeneous sequence of basaltic lava flows. Within this unit, regional distribution of slope failures and resulting debris flows appears controlled by weathering, which produces the loose materials susceptible to slope failure (Scott and Street, 1976). Intensity of weathering varies greatly across the study area because of an extreme gradient in annual precipitation. We estimated rate of weathering through geomorphic analysis of landscape development, using a known age for the Koolau volcanic dome in combination with digital measurement of the volume removed during erosion. The analysis resulted in a map showing regional rates of generation of materials susceptible to slope failure.

Within these regions of differing weathering rate, we estimated relative probability of slope failures at individual map cells from the distribution of 1,512 historical slope failures mapped by Peterson and others (in press) among cells of different topographic character. Analysis of DEM-computed terrain measures by logistical regression showed that historical slope failures are significantly related to steepness, but not curvature, of hillslopes.

Volume and frequency of slope failures

Volumes of potential slope failures were estimated from measurements of historical slope-failure scars in the study area (fig. 3). The volumes are distributed lognormally with a mean of 120 m^3.

To map frequency of debris-flow initiation, the mapped values of regional weathering rate were divided by the mean volume of slope failures to obtain average frequency of slope failure at map cells in each region. These regional frequencies, combined with the relation between mapped hillslope steepness and local probability of slope failure, resulted in a map showing for each cell the long-term average frequency of slope failure.

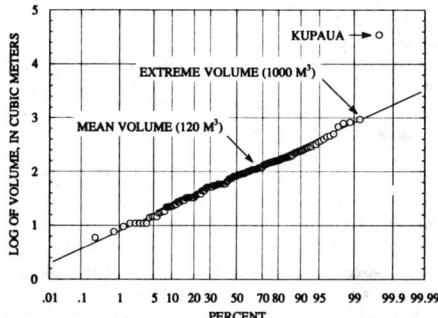

Figure 3: Probability plot showing approximately lognormal distribution of volumes of 201 historical slope failures in and near study area, measured or estimated during this study and by Scott (1969). Mean volume was used in simulations for high, moderate, and low hazard from hillslopes; extreme volume was used in simulations for maximum likely extent of hazard from hillslopes. Outlier labeled "Kupaua" is deep landslide source of anomalously large 1987-1988 debris flow in Kupaua Valley; hazard from such debris flows was mapped separately as "hazard along drainages from headwaters."

Mapping hazard

Simulated debris flows were routed through the DEM from random weighted samples of starting points taken from the map of slope-failure frequency. The simulated flows started with the initial volumes described in figure 3, and traveled according to the volume-change rates described by Cannon (this volume), to produce travel paths like those in figure 4.

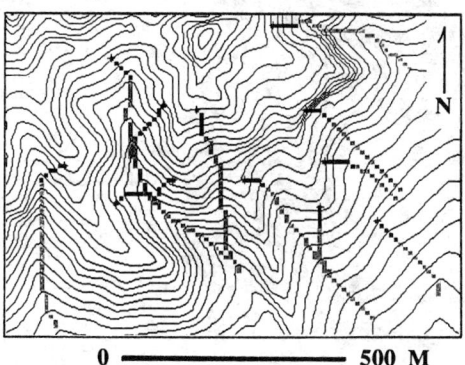

Figure 4: Map showing simulated debris-flow travel paths in part of study area. Black cells, travel paths for mean initial volume; gray cells, additional travel resulting from extreme initial volume (fig. 3). Contour interval 12.2 m (40 ft).

The simulations were compiled on a 10-m grid of the study area by counting the number of times each map cell is entered by simulated debris flows. Counts of simulated debris flows that started with the mean slope-failure volume (fig. 3) were translated into categories of high, moderate, and low hazard calibrated by approximate return period. Approximately 800,000 simulations, representing a period of 10,000 years, were run to create these categories of the hazard map. Strikes by simulated debris flows that started with the extreme initial volume (fig. 3) were translated into the category of maximum likely extent of hazard. These several hazard zones were then combined with drainages leading from hazardous hillslopes and from steep headwaters to produce the map of debris-flow hazard, a preliminary version of which is illustrated in figure 5.

The preliminary form of the hazard map agrees closely with historical observations. More than 99.6 percent of cells struck by mapped historical debris flows occur in areas shown as hazardous, and 96 percent occur in hazard zones other than "maximum likely extent." For the final map, historical debris flows are superimposed on the hazard zones to illustrate this accordance as well as the topographic control and erratic distribution characteristic of historical debris flows.

Acknowledgments

This study was supported in part by the City and County of Honolulu.

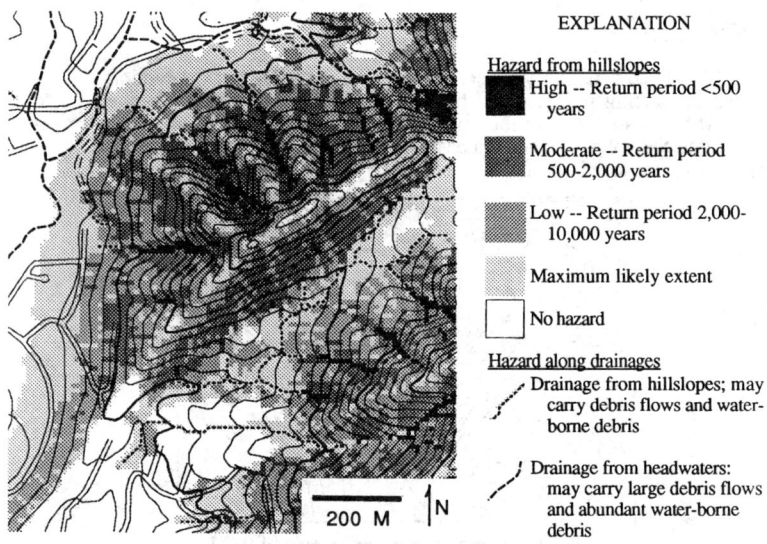

Figure 5: Preliminary version of map of debris-flow hazard in part of study area. Return periods are measured on 10-m cells; larger cells would result in smaller return periods.

References cited

Cannon, S.H. (this volume). "An empirical model for the volume-change behavior of debris flows and its application in estimating travel distances in the Honolulu area of Oahu, Hawaii."

Dracup, J.A., Cheng, E.D.H., Nigg, J.M., and Schroeder, T.A. (1991). "The New Year's Eve flood on Oahu, Hawaii, December 31, 1987 - January 1, 1988." National Research Council, *Natural Disaster Studies*, 1.

Ellen, S.D., and Mark, R.K. (1988). "Automated modeling of debris-flow hazard using digital elevation models (abs.)." *Eos*, 69(16), 347.

Ellen, S.D., Iverson, R.M., and Pierson, T.C. (1991). "Map showing the distribution of debris flows during the New Year's Eve storm of 1987-1988 in southeastern Oahu, Hawaii." U.S. Geological Survey *Open-File Report* 91-129, scale 1:20,000.

Ellen, S.D., Mark, R.K., Cannon, S.H., and Knifong, D.L. (1993). "Map of debris-flow hazard in the Honolulu District of Oahu, Hawaii." U.S. Geological Survey *Open-File Report* 93-213, scale 1:30,000.

Interagency Flood Hazard Mitigation Team (1988). "Interagency flood hazard mitigation report, in response to the January 8, 1988, disaster declaration FEMA-808-DR-HI." Federal Emergency Management Agency.

McEwen, A.S., and Malin, M.C. (1989). "Dynamics of Mount St. Helens' 1980 pyroclastic flows, rockslide-avalanche, lahars, and blast." *J. Volcanology and Geothermal Research*, 37, 205-231.

Peterson, D.M., Ellen, S.D., and Knifong, D.L. (in press). "Distribution of past debris flows and other rapid slope movements in the Honolulu District of Oahu, Hawaii." U.S. Geological Survey *Open-File Report*.

Scott, G.A.J. (1969). "Relationships between vegetation and soil avalanching in the high rainfall areas of Oahu, Hawaii," thesis presented to the University of Hawaii at Honolulu, Hawaii, in partial fulfillment of the requirements for the degree of Master of Arts.

Scott, G.A.J., and Street, J.M. (1976). "The role of chemical weathering in the formation of Hawaiian amphitheatre-headed valleys." *Zeitschrift für Geomorphologie*, 20(2), 171-189.

State of Hawaii (1988). "Post flood report, New Year's Eve storm, December 31, 1987 - January 1, 1988, windward and leeward east Oahu." State of Hawaii, Department of Land and Natural Resources, *Circular* C119.

Varnes, D.J. (1978). "Slope movement types and processes," chap. 2 of *Landslides, analysis and control*. U.S. National Academy of Sciences, *Transportation Research Board Special Report* 176, 11-33.

Wadge, G. (1988). "The potential of GIS modeling of gravity flows and slope instabilities." *International J. Geographical Information Systems*, 2(2), 143-152.

Wentworth, C.K. (1943). "Soil avalanches on Oahu, Hawaii." *Geological Society of America Bulletin*, 54, 53-64.

Wieczorek, G.F. (this volume). "Assessment and prediction of debris-flow hazards."

Prediction of Occurrence and Runoff Analysis of Debris Flow

Muneo Hirano[1] and Toshiyuki Moriyama[2]

Abstract

A debris flow will occur on a slope when the rainfall exceeds a certain value determined by the hydraulic, geological and topographic properties of the slope. To estimate this critical value, the system analysis technique appears to be useful. In this study, a neural-networks model was used to predict the occurrence of debris flow. Reliability of the method was verified by applying to the debris flows in Unzen Volcano which recently had begun to erupt. Equation of debris flow runoff was derived by solving the basic equations using the kinematic wave theory. It is found from the derived equation that the runoff intensity of debris flow is in proportion to the rainfall intensity and cumulative rainfall, jointly. This gives a theoretical basis to the conventional method which has been widely used.

Introduction

A debris flow will occur on a slope when the slope is saturated and surface flow appears due to heavy rainfall. This suggests that the occurrence criteria of debris flow may be obtained from measured values of the hydraulic conductivity, porosity, depth of the deposits, the length and gradient of the slope, etc. But the errors in the measurements of these factors would make the estimated critical value unreliable. Moreover, the field conditions of a slope are not always stationary. Therefore, the system analysis technique may be desired to identify the critical value. In this study, a neural-networks model is used as a trial to predict the occurrence of debris flow.

---

[1] Professor, Dept. of Civil Eng., Kyushu University, Fukuoka 812, Japan
[2] Research Associate, Dept. of Civil Eng., Kyushu University, Fukuoka 812, Japan

A runoff model of debris flow is obtained by use of the kinematic wave theory in the similar way as the rainfall-runoff model.

To check the reliability, these models are applied to the debris flows in Unzen Volcano which recently began to erupt.

The Critical Rainfall for Occurrence of Debris Flow

On a slope of deposits, the critical occurrence condition of debris flow is that the shear stress excesses the resisting stress. But according to the experiments by Hirano et al.(1976), debris flow occurs when surface flow appears on a slope due to the heavy rainfall. The criteria for occurrence of the surface flow is given by solving the momentum and continuity equations of subsurface flow as

$$l \geq kT \sin\theta/\lambda \quad \text{and} \quad \lambda D \geq \int_0^T r \cos\theta \, dt \tag{1}$$

where, l is the length of the slope, k is the hydraulic conductivity, T is the time of concentration, θ is the angle of the slope, λ is the porosity, D is the depth of the deposit, and r is the rainfall intensity. Assuming that debris flow occurs when surface flow appears on a slope, the occurrence criteria of debris flow is derived from Eq.(1) as

$$r_T = \frac{1}{T}\int_0^T r \, dt \geq \frac{Dk}{l}\tan\theta \quad \text{or} \quad R(t,T) = \int_{t-T}^t r(\tau) \, d\tau \geq \frac{TDk}{l}\tan\theta = R_c \tag{2}$$

where, t is the time, and R_C is the critical rainfall.

Equation (2) indicates that debris flow will occur when cumulative rainfall within the time of concentration exceeds a certain value related to the properties of the slope. The time of concentration T should be estimated to obtain this critical value. It may be possible to estimate the value of R_C by measuring the hydraulic conductivity, the depth, the length and the gradient of the slope, but the obtained value will not be accurate enough for practical use due to the errors in the measurements. This is the reason why the system analysis technique is desired to identify the parameters.

To estimate the time of concentration, we plotted the maximum values of cumulative rainfall for each duration against the duration. If there is no error in the data and in the theory, the plotted line when debris flow occurred should exceed a certain point, and the line without debris flow should not as schematically illustrated in Fig. 1(a). This point corresponds to the time of concentration and the critical rainfall R_C. But because of the errors in the data and the unsteady field conditions, the upper limit of occurrence and the lower limit of non-occurrence will be like two curves as illustrated in Fig. 1(b). The point where the difference between two curves is minimum is assumed to correspond to the time of concentration.

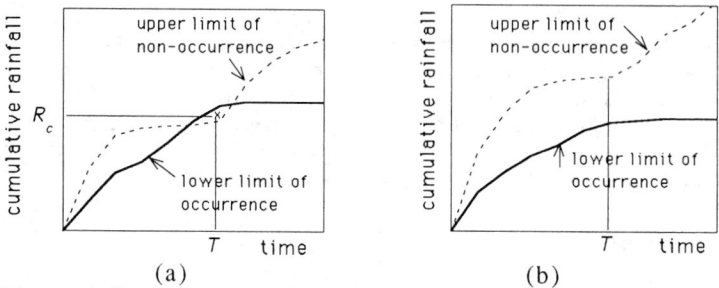

Figure 1 Schematic illustration of the upper limit of non-occurrence and the lower limit of occurrence.

The maximum cumulative rainfalls of the cumulative durations were calculated using the rainfall data collected by the Unzen Meteorological Observatory, both when debris flows had occurred and when they had not. Using the calculated values in both cases, the lower limit with debris flows and the upper limit without are plotted in Fig. 2.

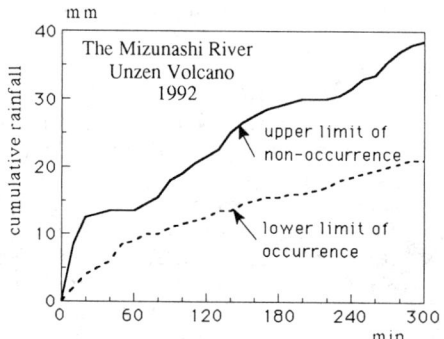

Figure 2 The upper limit of non-occurrence and lower limit of occurrence of debris flow at the Mizunashi River(1992).

From Fig. 2, the following are confirmed: 1) the time of concentration is estimated to be 70 minutes; 2) the occurrence of debris flows is possible when the amount of rainfall per 70 minutes rises over the limit of 10 mm; and 3) debris flows will definitely occur when this amount rises over the limit of 14 mm.

Prediction of Debris Flow by the use of Neural Networks

Neural networks with three layes were used to predict the occurrence of debris flow at Unzen Volcano. Cumulative rainfall of every ten minutes from 10 to 300 minutes were fed to the input layer as the input data. The hidden layer has twelve cells and the output layer has one. The error back propagation method was introduced to find the optimum networks by using the learning data. The rainfall data in May, 1991 were used for the learning data with teacher signals of 0.01 for non-occurrence and 0.99 for occurrence data. After learning these data, the neural networks model predicted the occurrence of debris flows by using the rainfall data from June, 1991 to September, 1992.

Results of the prediction are as follows: In 1991, thirteen debris flows had occurred and all were predicted by the neural networks. Ten of 23 rainfalls without occurrence of debris flow were predicted to occur. In 1992, debris flows occurred seventeen times which were all correctly predicted, and by seven rainfalls no debris flow occurred contrary to the prediction. In both years, no debris flow occurred by the rainfall which the neural networks judged as non-occurrence.

Runoff Analysis of Debris Flow

The equation of continuity in a stream is given by

$$\frac{\partial A_s}{\partial t} + \frac{\partial Q}{\partial x} = q_s \quad (3)$$

where, A_S is the cross sectional area of the stream, Q is the discharge of the flow, and q_S is the lateral inflow rate. Assuming A_S to be a function of Q, the above equation is solved by use of the characteristic curve as follows:

$$\frac{dx}{dt} = \frac{dQ}{dA_s} \quad \text{and} \quad Q(t) = \int_0^L q_s \, dx = \int_0^L f r l \cos\theta \, dx \quad (4)$$

where, L is the length of the stream and f is the runoff coefficient which is given from the continuity condition on a slope as

$$f = (1 - \lambda)/(1 - \lambda - C) \quad (5)$$

where, C is the concentration of debris flow. It is seen in the above equation that the range of f is from unity to infinitive. According to the experiments(Hirano et al.(1976)), $l = 0.54$, $C=0.50$ and $f =18$ for debris flows.

On a slope where the conditions given by Eq.(1) is satisfied, a debris flow will occur. When all slopes in a watershed are shorter and the deposits are deeper than the critical values given by Eq.(1), water flows into a stream, but no debris flow occurs in the watershed. In this case, Eq.(4) can be rewritten as

$$Q(t) = A \int_0^\infty \int_0^\infty f\, r\, \phi(\eta, l)\, dl\, d\eta = A \int_0^\infty f\, r_T\, \Phi(T)\, dT = A \int_0^\infty f\, r(t - \tau)\, u(\tau)d\tau \quad (6)$$

where,

$$r_T = \frac{1}{T}\int_0^T r\, dt\, , \quad \Phi(T) = \int_0^\infty \phi(\eta, l)\, d\eta \frac{dl}{dT} \quad \text{and} \quad u(\tau) = \int_0^\infty \frac{1}{T}\Phi(T)\, dT \quad (7)$$

Equation (6) leads to instantaneous unit hydrograph(IUH) when no debris flow occurs. It is clarified that IUH is a function of the time of concentration(Hirano(1988))).

Flow rate of debris flow, Q_S, is obtained by applying Eq.(2) to Eq.(6) as

$$Q_s(t) = \sum_{t_o=-\infty}^t fr\, \phi(\eta, l)\Delta\eta\Delta l = fr\, \{ \int_{kt\sin\theta/\lambda}^\infty \phi(\eta_o, l) dl\, \Delta\eta_o + \sum_{t_o=0}^t \phi(\eta, l)\Delta\eta\Delta l \} \quad (8)$$

where,

$$\eta_o = \int_0^t r\cos\theta\, dt \quad \text{and} \quad \eta = \int_{t_o}^t r\cos\theta\, dt \quad (9)$$

Assuming that debris on a slope outflows in a short period of time Δt, one obtains

$$\Delta t = Dl / \{(f_s - 1)\, r\cos\theta)\} \quad (10)$$

Substitution of Eqs.(9)-(11) into (8) yields

$$Q_s(t) = A\, r(t)\frac{f_s}{(f_s-1)\lambda}\{\eta_o \int_{kt\sin\theta/\lambda}^\infty \phi(\eta_o, l)\, dl + \frac{k\sin\theta}{\lambda}\int_0^t \eta\, \phi(\eta, \frac{k(t-t_o)\sin\theta}{\lambda})\, dt_o\} \quad (11)$$

Since the first term of the right hand side of the equation is dominant compared with the second one, Eq.(11) can be simplified as

$$\frac{Q_s(t)}{A} \approx M\, r(t) \int_0^t r\cos\theta\, dt\, , \quad M = \frac{f_s}{f_s-1}\frac{1}{\lambda}\int_{kt\sin\theta/\lambda}^\infty \varphi(\eta_o, l)\, dl \quad (12)$$

Equation(12) indicates that the runoff intensity of debris flow is in proportion to the rainfall intensity at the time, $r(t)$, and the cumulative rainfall up to that time, $R(t)$, jointly. This means that a constant value of $Q_S/(AM)$ is shown as a hyperbolic curve on a $[r(t), R(t)]$ plane. An empirical method by use of hyperbola-like curve(s) on $[r(t), R(t)]$ plane has been widely used to forecast the occurrence of debris disasters. Equation (12) gives a theoretical basis to this conventional method.

If we assume that λD and l are independent each other, Eq.(12) reduces to a parametric model with four parameters, the means and standard deviations of λD and l. In Fig. 3, the computed hydrographs are compared with the observed ones of the Mizunashi River in Unzen Volcano and the Nojiri River in Sakurajima Volcano. In the computations, λD and l are assumed to distribute normally and log-normally, respectively. The standard deviation and the mean of $l/(k\sin\theta/\lambda)$ were put as log 2 and 3.5 hr, respectively, for both rivers. The mean of $\lambda D=30$cm is used for the Mizunashi River and that of 8cm is adopted for the Nojiri River. The standard deviation of λD are put to be 1/4 of its mean value for the both rivers. The computed hydrographs show close agreement with the observed ones.

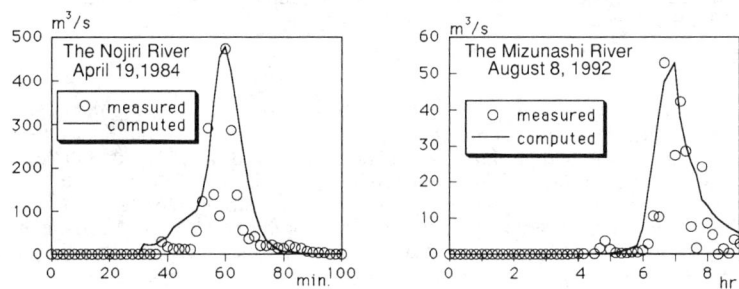

Fig. 3 Comparison between the computed and observed hydrographs of debris flows

Appendix, References

Hirano, M., M. Iwamoto and T. Harada (1976). "Study on the mechanism of occurrence of debris flow by artificial rainfall", *Preprints of the Annual meeting of JSCE*, pp.299-301, (in Japanese).

Hirano, M.(1983). "Modeling of runoff process in a first- order basin", *Journal of Hydroscience and Hydraulic Engineering*, Vol. 1, No. 2, pp.113-123.

Sediment Deposition from Debris Flow on a Gentle and Wide Slope

Haruyuki Hashimoto[1] and Muneo Hirano[2]

Abstract

Motion of sediment grains in the deposition process from debris flow on a gentle and wide slope was examined. A model of the deposition was presented on the basis of an analysis of grain motion. With this model bed change due to the deposition was calculated. It was found that a concave bed configuration is produced along the main flow on the slope.

Introduction

There occurs rapid deposition of sediment from debris flows at the mouth of a valley because of a decrease in its slope and an increase in its width. Tsubaki and Hashimoto (1984), and Hashimoto and Hirano (1990) dealt with a two-dimensional case of such deposition, which was caused by a rapid decrease in channel slope. They investigated the motion of sediment grains in the deposition theoretically and experimentally. From this result they derived an equation for bed change due to the deposition. However they did not discuss more natural deposition, that is a three-dimensional case of deposition.

In the present work we deal with the deposition on a gentle and wide slope like a fan or a plain at the mouth of a valley. First, we experimentally examine motion of grains during deposition. Second, we analyze the grain motion theoretically. Finally, we calculate bed deposition profile on the gentle slope.

Experiments of Grain Deposition from Debris Flows

We used a flume 9.5 m long and 10 cm wide and a slope 195 cm long and 123 cm wide. The flume was made steep and the slope gentle; these are connected as shown in Fig.1. Debris flow occurs in the steep flume and deposits grains on the gentle slope.

---

[1] Associate Professor, Department of Civil Engineering (Suiko), Kyushu University, Fukuoka 812, Japan.
[2] Professor, Department of Civil Engineering (Suiko), Kyushu University, Fukuoka 812, Japan.

SEDIMENT DEPOSITION

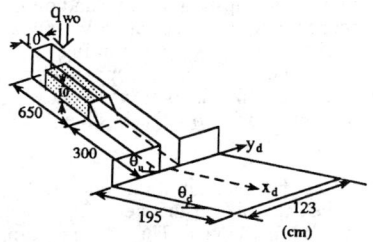

Fig. 1. Steep flume and the gentle, wide slope

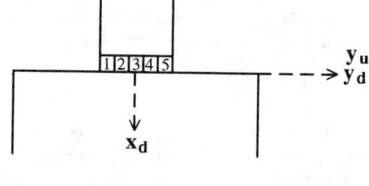

Fig. 2. Apparatus for introducing colored grains on flow surface

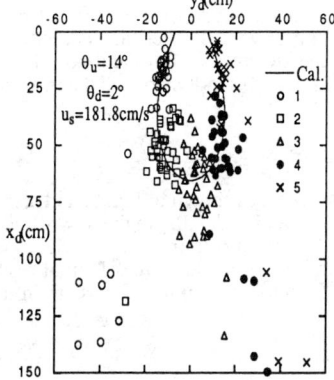

Fig. 3. Rest positions of colored grains

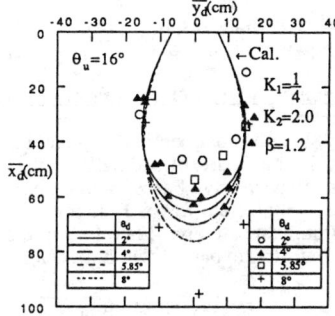

Fig. 4. Mean rest positions of colored grains in each subdivision

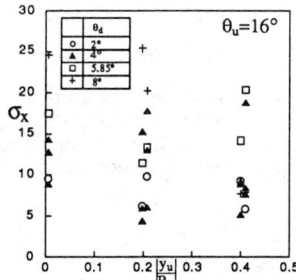

Fig. 5. Standard deviation of x_d component of rest positions of the colored grains in each subdivision

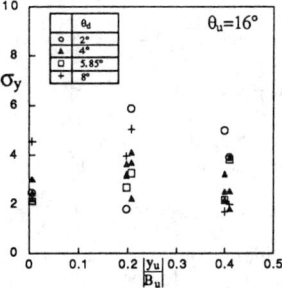

Fig. 6. Standard deviation of y_d component of rest positions

A coarse sand with a diameter of 1.9 mm and a specific gravity of 2.58 was used for the movable bed in the upstream flume. The fixed bed in the upstream flume and on the downstream slope, on the other hand, was roughened with the same sand. We put an apparatus above the bed at the downstream end of the flume, as shown in Fig. 2, in order to introduce colored sand grains on the surface of debris flow. This apparatus was divided into five subdivisions in which the grains of five different colors were put. The bottom of the apparatus can open automatically when debris flow reaches the downstream flume end.

Water was supplied at a rate of $Q=10^3$ cm^3/s at the upstream flume end. Debris flow occurred on the movable bed and then moved rapidly on the steep fixed bed. Debris flowing onto the gentle slope lasted for about ten seconds. The colored grains were dropped on the flow surface at the mouth, deposited and then buried; they were found out with a brush. We measured the positions (x_d, y_d) of rest of the colored grains. The result is shown in Fig. 3. Here, θ_u denotes a sloping angle of the upstream flume and θ_d indicates an angle of the downstream slope. It is found that the colored grains are deposited in an U-shaped curve. The colored grains in subdivision 3 travel a longer distance, while the ones in the subdivisions 1 and 5 travel a shorter distance. Means $(\bar{x}_d, \bar{y}_d)$, standard deviations (σ_x, σ_y), variation coefficient $\alpha_x = \sigma_x / \bar{x}_d$ and correlation coefficient of the rest positions (x_d, y_d) of the grains in each subdivision are determined, as shown in Figs. 4, 5, 6 and 7.

Motion of Grains in Debris Flow on the Gentle Slope

We consider debris flow reaching the gentle slope, as shown in Fig. 8, and also a grain in small parts of width dB_u in the upstream flow and width dB_d in the downstream flow, respectively. Forces acting on the grain in the flow direction are gravity component and the fluid forces F^f that cause motion, and intergranular forces F^g that resist motion, respectively. Therefore, the equation of motion of a grain of mass m in the small upstream part of width dB_u can be written as

$$0 = -F^g_{ux} + mg \sin\theta_u + F^f_{ux} \qquad (1)$$

For a grain just in the small downstream part of width dB_d, on the other hand, we can have

$$x_d : m(1 + \frac{\rho}{2\sigma}) \frac{d^2x}{dt^2} = -F^g_{dx} + F^f_{dx} + mg \sin\theta_d \qquad (2)$$

$$y_d : m(1 + \frac{\rho}{2\sigma}) \frac{d^2y}{dt^2} = -F^g_{dy} + F^f_{dy} \qquad (3)$$

where subscript u and d denote the upstream steep flume and the downstream gentle slope, respectively. Grains change their velocity and position from $(u(z_u), 0, 0)$ and $(0, y_u, z_u)$ in the steep flume to $(u_{d0}, v_{d0}, 0)$ and $(0, y_{d0}, z_{d0})$ on the gentle slope, respectively. Here u_{d0} is approximated by $u_{d0} = u(z_u) \cos(\theta_u - \theta_d)$, and hence z_{d0} can be written as

$$z_{d0} = z_u \bigg/ \left[a_0 \cos(\theta_u - \theta_d) \frac{dB_d}{dB_u} \right]$$

from a continuity equation, where $a_0 = \sqrt{1 + (v_{d0}/u_{d0})^2}$. Therefore, from the expression for intergranular forces (Hashimoto and Tsubaki, 1983) we found that

$$F^g_{dx} = a_0 F^g_{ux} (\frac{dy_d}{dy_u})^3 \cos^5(\theta_u - \theta_d) \quad (4), \qquad F^g_{dy} = (\frac{v_{d0}}{u_{d0}}) F^g_{dx} \qquad (5)$$

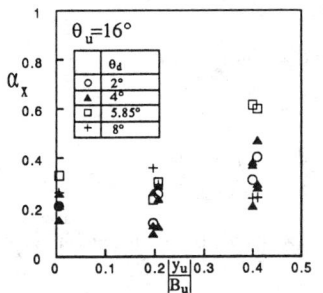

Fig. 7. Variation coefficient of x_d component of rest positions

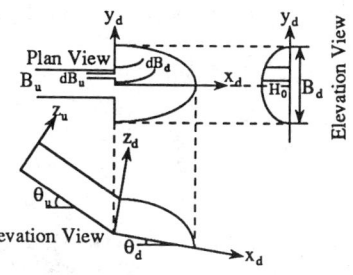

Fig. 8. Schematic figure of debris flowing onto the slope

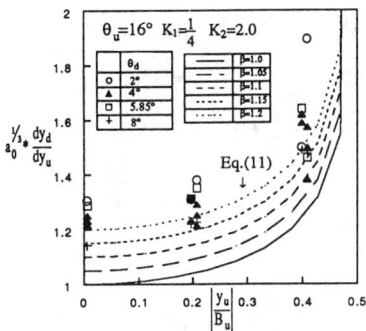

Fig. 9. Rate of increase in flow width

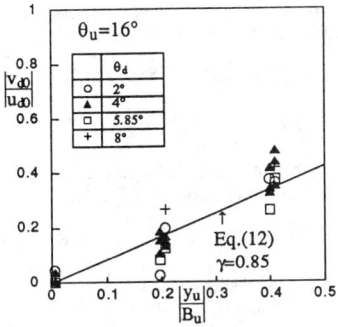

Fig. 10. Ratio of initial velocity components of grains in the y_d-direction and in the x_d-direction

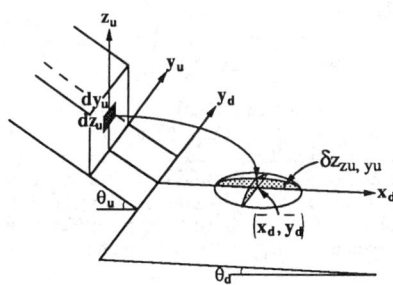

Fig. 11. Model for the deposition of grains

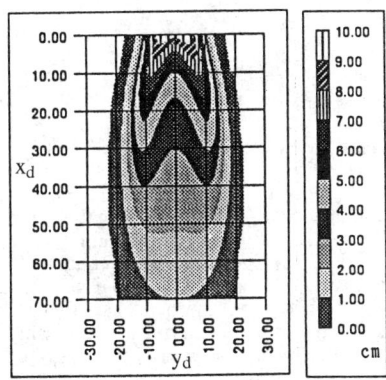

Fig. 12. Plan view of bed configuration calculated by Eq. (14) ($\theta_u=16°$, $\theta_d=4°$, t=3.6 sec)

, where dy_d/dy_u represents the rate of increase in flow width and is equal to $a_0 \, dB_d/dB_u$. Substituting Eqs. (1), (4), and (5) into Eqs. (2) and (3) yields

$$\frac{d^2x}{dt^2} = -\frac{g}{1+\frac{\rho}{2\sigma}} \{a_0(\frac{dy_d}{dy_u})^3 \sin\theta_u \cos^5(\theta_u - \theta_d) - \sin\theta_d\} \equiv -G_x \tag{6}$$

$$\frac{d^2y}{dt^2} = -\frac{g}{1+\frac{\rho}{2\sigma}} \{a_0(\frac{dy_d}{dy_u})^3 \sin\theta_u \cos^5(\theta_u - \theta_d) \frac{v_{d0}}{u_{d0}}\} \equiv -G_y \tag{7}$$

In the deposition due to a rapid decrease in flume slope, that is two-dimensional deposition, dy_d/dy_u becomes 1 and v_{d0} becomes 0.

Solving Eqs. (6) and (7) under the initial conditions $x = 0$, $dx/dt = u_{d0}$, $y = y_{d0}$, $dy/dt = v_{d0}$, we can have equations for grain velocity and trajectories. From these equations we can obtain the period (t_x, t_y) and distance (x_{sp}, y_{sp}) from the passage of grains past the z_d-axis to their rest:

$$t_x = u(z_u) \cos(\theta_u - \theta_d)/G_x \tag{8}, \quad t_y = v_{d0}/G_y \tag{8'}$$

$$x_{sp} = u^2(z_u) \cos^2(\theta_u - \theta_d)/(2G_x) \tag{9}, \quad y_{sp} = v_{d0}^2/(2G_y) + y_{d0} \tag{9'}$$

In the above equations there remain dy_d/dy_u and v_{d0}/u_{d0} undetermined. They can be obtained in the following way:

First, assuming the transverse profile of flow depth as

$$\frac{H}{H_0} = \left\{ 1 - \left(\frac{2y_{d0}}{B_d}\right)^{K_2} \right\}^{K_1} \tag{10}$$

and using a continuity equation, we can have

$$\frac{dy_d}{dy_u} = \beta \left(1 - \left(\frac{2y_{d0}}{B_d}\right)^{K_2}\right)^{-K_1}, \quad \beta = \frac{h}{H_0 \cos(\theta_u - \theta_d)} \tag{11}$$

where $\quad \frac{y_u}{B_u} = \frac{1}{2} \int_0^{2y_{d0}/B_d} (1 - Y_d^{K_2})^{K_1} dY_d \Big/ \int_0^1 (1 - Y_d^{K_2})^{K_1} dY_d \tag{11'}$

The increase rate of flow width dy_d/dy_u is found to be a function of y_u.

Second, we use the means ($\bar{x}_d$, $\bar{y}_d$) of rest position of the colored grains for the estimate of v_{d0}/u_{d0}. Substituting their experimental values into Eqs. (9) and (9'), we can determine $a_0^{1/3}(dy_d/dy_u)$ in G_x and v_{d0}/u_{d0}, respectively. The results are plotted in Figs. 9 and 10. a_0 is approximately equal to 1 because of $v_{d0}/u_{d0} < 1$ and hence the value of $a_0^{1/3}(dy_d/dy_u)$ can be compared with Eq.(11). The values of coefficients $K_1=1/4$, $K_2=2.0$ and $\beta=1.2$ in Eq. (11) are found suitable for $\theta_u=16°$. In addition from Fig. 10 it is found that there is a linear relationship between v_{d0}/u_{d0} and y_u/B_u. Hence we can obtain the equation

$$\frac{v_{d0}}{u_{d0}} = \gamma \frac{y_u}{B_u} \tag{12}$$

, where γ is a constant of proportionality. By using these constant and coefficients we can calculate Eqs. (9) and (9'). The results are shown by some lines in Figs. 3 and 4.

Calculation of Bed Elevation on the Gentle Slope

In a similar way to that presented by Tsubaki and Hashimoto (1984) the deposition of grains can be modeled as shown in Fig. 11. Grains with velocity $u(z_u)$ and concentration C coming through small section of area $dz_u \cdot dy_u$ from the upstream steep flume settle somewhere about an average position on the downstream gentle slope. Rest positions of grains with the same velocity are regarded as a random variable and their mean ($\bar{x}_d$, $\bar{y}_d$) can be given by Eqs. (9) and (9').

Denoting the probability density function of the random variable, bed elevation of deposited grains originating in position (z_u, y_u) and volumetric concentration of grains in bed by $p_{zu,yu}(x_d, y_d)$, $\delta z_{zu, yu}$ and C_* respectively, we can express the mass conservation of deposited grains as

$$C_* \frac{\partial}{\partial t} (\delta z_{zu, yu}) \, dx_d \, dy_d = u(z_u) \, C \, dz_u \, dy_u \, p_{zu, yu}(x_d, y_d) \, dx_d \, dy_d \qquad (13)$$

$p_{zu, yu}(x_d, y_d)$ is assumed to be a normal density function. Here, the experimental values of the variation coefficient α_x, standard deviation σ_y, and correlation coefficient are used. Integrating Eq. (13), we obtain

$$z = \int_{d/2}^{h} \int_{-B_u/2}^{B_u/2} \frac{C}{C_*} u(z_u) \, (t - t_{sp}) \, p_{zu, yu} \, (x_d, y_d) \, dz_u \, dy_u \qquad (14)$$

, where t_{sp} indicates time of the beginning of grain deposition. Since t_{sp} represents the period of grain motion and t_x is larger than t_y, we can have the relation $t_{sp} = t_x$. Fig. 12 shows a plan view of the result calculated by Eq. (14). It is found that a concave bed configuration is produced along the x_d - axis.

Conclusions

Grains on the surface of debris flow are deposited in an U-shaped curve on the gentle and wide slope. Means, standard deviations, variation coefficient, correlation coefficient of rest positions of grains were determined. The equations for estimating average positions were obtained. From these results and the law of mass conservation bed change due to the deposition was calculated. It was found that there occurs a concave bed configuration along the main flow on the slope.

References

Hashimoto, H., and Tsubaki, T. (1983). " Reverse grading in debris flow." Proc. JSCE, 336, 75-84 (in Japanese).
Hashimoto, H., and Hirano, M. (1990). " Model of deposition of grains from debris flow." Hydraulics/Hydrology of Arid Lands, Proc. Int. Sym., ASCE, 537-542.
Tsubaki, T., and Hshimoto, H. (1984). " Deposition of debris flow due to abrupt change of bed slope." Proc. 28th Japanese Conf. Hydraulics, JSCE, 711-716 (in Japanese).

INTEGRATION OF ENVIRONMENTAL MANAGEMENT WITH RESERVOIR AND POWER SYSTEM OPERATIONS

Vahid Alavian[1]

Abstract

Two examples related to the Tennessee Valley Authority's reservoir and power systems operation with environmental performance and management as major goals are given in this paper. One describes a policy study examining the long-term strategy for river system operation and the other describes the development of advanced tools and methodologies to assist in the daily decision making process regarding timely and effective operation of the system.

Introduction

The need to integrate environmental concerns and issues in the management of water resources and power systems has become increasingly apparent in recent years. In most large systems incorporation of environmental factors creates conflict with the traditional operation rules and policies. In some cases, this conflict has turned controversial and confrontational resulting in no action or implementation of expensive, unsatisfactory alternatives.

Studies to examine policies and strategies for integrated operation of water, power, and ecological resources must consider central issues related to the environmental and ecological variables that have become significant. One of the most important issues is the relative weight of the uses of the resource. For example, should water quality and recreation be given the same footing as flood control, navigation, and power generation? If so, what would be the cost and who should pay?

---

[1]Senior Engineer, TVA Engineering Laboratory, Norris, TN 37828

The decision process to evaluate the environmental issues to be integrated into an existing water management system must be based on a clear understanding of the values, goals, and constraints involved. Open public participation coupled by sound technical analysis is the approach taken by TVA in its attempt to integrate environmental management into its operation. Examples given below illustrate the point.

The Lake Improvement Plan

The Tennessee Valley Authority (TVA), a federal agency created in 1933, is responsible for a wide variety of programs within the Tennessee River Basin. These programs include water resources management; power production; social and economic development; and natural resources conservation. Currently, TVA operates some 42 major multi-purpose dams, 11 coal-fired plants, and 2 nuclear plants. Figure 1 shows a diagram of the TVA reservoir and power system.

In accordance with the TVA Act of 1933, the reservoir system was developed and operated primarily for the multi-objective purposes of flood control, navigation, and power production. Over the past 50 years, however, the region's water resources needs have changed significantly. Public demand for clean water, high quality recreation, and general ecological health of the river system now competes with the traditional demands. In response to these changes, TVA conducted an environmental impact study and arrived at an alternative operation policy which attempts to explicitly incorporate environmental considerations into the operating objectives (TVA, 1990).

The plan, known as the Lake Improvement Plan, resulted from an in-depth, interdisciplinary study of TVA's river management policies. The process was aimed at providing the decision makers with sound basis, supported by public review and technical analysis, for changing or maintaining the existing policies. To achieve this goal, two basic rules were followed: 1) all parties involved have equal position in the process regardless of their legislative standing; 2) all analyses, qualitative or quantitative, are subject to peer review and challenge in an open review process. A flow chart of the process is shown in Figure 2.

Public meetings were held to identify the key issues related to use of the river-based resources by the citizens. Integrated modelling of the river system was used to find feasible and economic alternatives for incorporating the environmental constraints. The operational solutions include maintaining the summer lake levels longer and increasing minimum releases from projects to enhance the ecological health of the system and the region's economy.

Where operational modifications appeared insufficient to meet environmental objectives (e.g., dissolved oxygen), technological solutions were

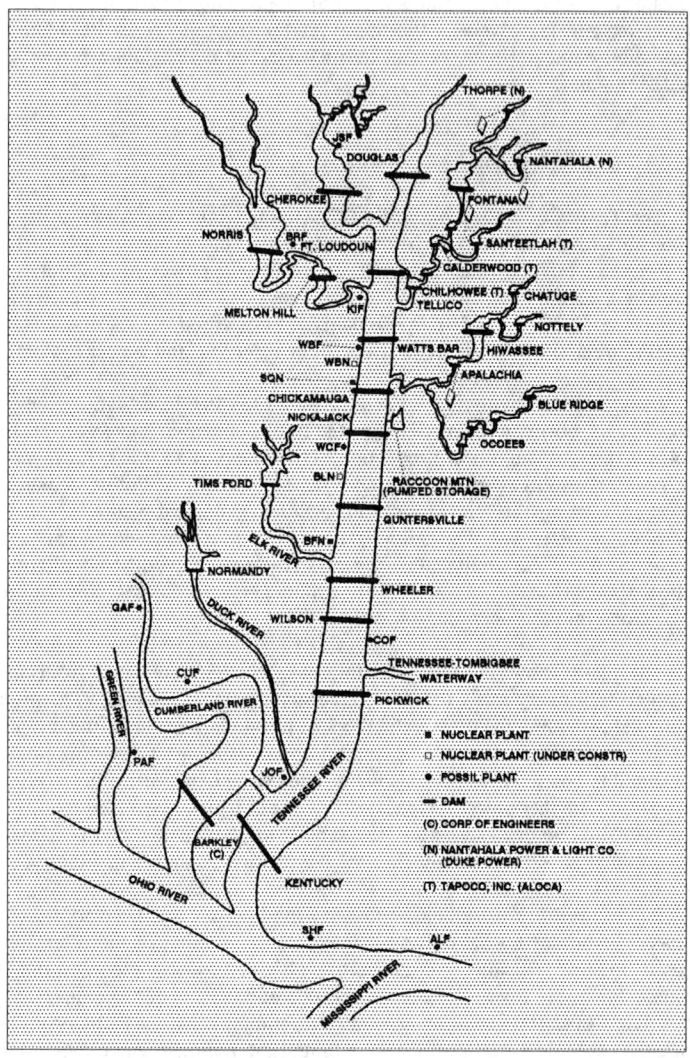

Figure 1. TVA Reservoir/Power System

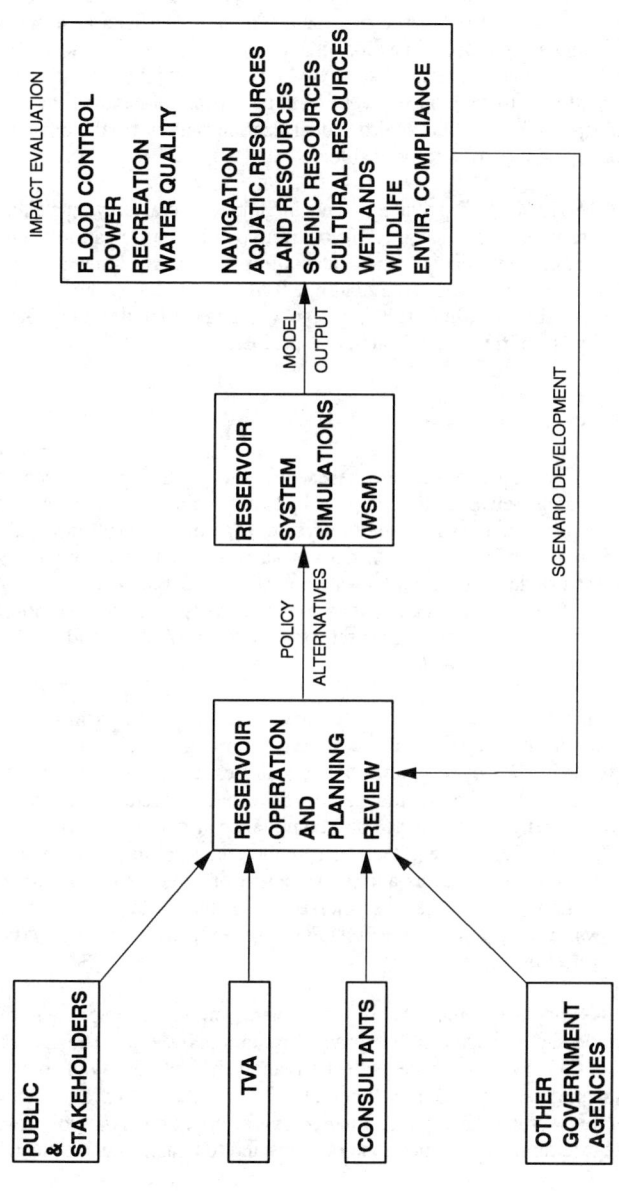

Figure 2. Development and Evaluation of Policy Alternatives

devised and implemented. Low head, long-crested weirs have been installed as an environmentally appropriate solution to maintain aquatic habitat and to improve dissolved oxygen levels in the tailrace of tributary dams. Artificial aerators and surface water pumps have been installed in the forebay of some reservoirs to increase dissolved oxygen levels. Research in autoventing hydro turbines is underway in an attempt to develop self-venting units capable of aerating the water during normal operation. TVA's hydro modernization plan will take advantage of this technology, if it proves to be viable.

The TVA Lake Improvement Plan includes several key sustainable components: response to changing priorities; public participation; multi-disciplinary systems approach; and adaptation of environmentally and technologically sound solutions. In a continued effort to redefine the Agency's stewardship responsibilities, its strategic plan explicitly includes the goal of leading the nation's utilities in environmental performance.

Daily Operations

TVA's daily operation of the reservoir and power systems requires continuous balancing between the region's hydrology and demands on water resources. Reservoir operators regulate the system for flood control; navigation; water quality; and recreation based on current storage and forecast hydrology. Power schedulers use the available releases from each project to generate peaking hydro power. These two organizations interact daily, sometimes hourly, according to a set of multiple objectives set forth by the TVA ACT and the TVA Board Policy.

Environmental issues are explicitly taken into consideration in the daily decisions through limits and standards set by regulatory agencies and/or releases and lake levels established by long-term policy analyses (i.e., Lake Improvement Plan). A set of analytical tools and models have been developed in order to assist with environmental performance and compliance at various projects. An EPRI-TVA Collaborative Project has been undertaken to develop enhanced methods for the integrated analysis and management of water resources systems and power operations. The project is referred to as the INTEGRAL Project. Figure 3 shows a diagram of the INTEGRAL Project indicating various components of the system.

The two major components of the project are the development of a Decision Support System and an Integrated Modeling System. The Decision Support System (DSS) will aid in the collection, analysis, and display of reservoir and power system data. The primary functions of the DSS are: to provide power system, reservoir system, and water quality characterization data; to provide information regarding near real-time system status and forecasted conditions with

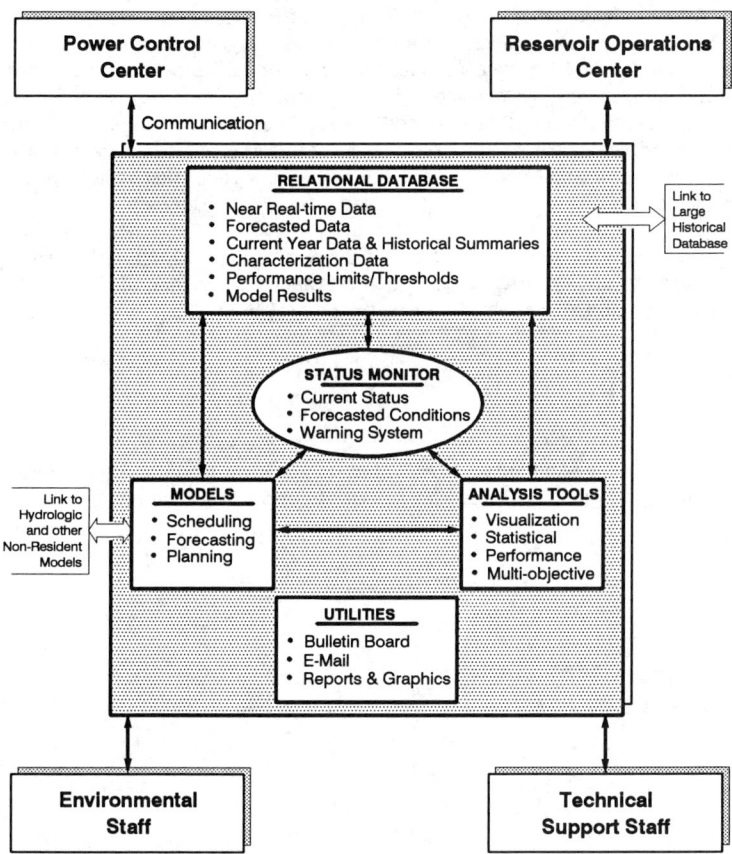

Figure 3. Schematic of the INTEGRAL Decision Support System

respect to power operations, reservoir operations, and environmental indicators; to aid in reservoir and power system scheduling; and to provide useful information for forecasting and planning for integrated multipurpose reservoir and power system operations.

An Integrated Modeling System will support the scheduling, forecasting, and planning functions of the DSS. The types of models to be incorporated into the modeling system include: hydrologic models to generate reservoir system inflows and account for meteorological and climatic influences on basin runoff; power and reservoir system scheduling and planning models; river system temperature and dissolved oxygen models to address thermal compliance, water quality, and ecological health issues; and site-specific hydro and power plant models where detailed operation scheduling and environmental issues must be considered.

An early version of the Decision Support System including the existing mathematical models will be on line for testing by summer of 1993. The complete system is expected to become operational in the TVA system by early 1994.

References

Tennessee Valley Authority, 1990. Tennessee River and Reservoir System Operation and Planning Review, TVA/RDG/EQA-91/1, Knoxville, TN.

DECISION SUPPORT FOR PREDICTING THE HYDRAULIC AND WATER QUALITY CHARACTERISTICS OF NATURAL RIVERS

Christian Jokiel[1], Peter Ruland[2], Gerhard Rouvé[3]

Abstract

This paper describes a decision support system (DSS) based on a geographic information system (GIS). It combines simulation models for river flow and water quality with tools for the management of topographically related data. Its application is demonstrated by a case study of the river system "Jüchener Bach/Nordkanal" in Germany.

Introduction

Water quality management in Germany becomes increasingly more important because most water bodies suffer severe pollution due to increasing waste water discharges, intensive industrial water use and direct runoff. A second major ecologic concern is river restoration. It includes "close-to-nature" orientated reconstruction of watercourses and the re-establishment of flood-plains. Water quality management and river restoration are both complex processes that must be modeled to study their possible impacts. Modelling begins with the data storage and handling of geographical, hydrological and water quality boundary conditions. This is followed by the determination of the hydrological and hydrodynamical characteristics and the simulation of the water quality to find and finally present the best ecologic and economic solution and its impacts on the environment.

To allow the user to perform all these necessary steps within one DSS a GIS (Smallworld) was chosen as the host environment. Two separated but interacting simulation models were integrated into this framework: The first is a one-dimensional water surface model (ESNA) based on the Darcy-Weisbach law considering all flow resistance factors in natural rivers. The dynamic water quality model is derived from the "Water Quality for River Reservoir Systems (WQRRSQ)" (HEC/1978).

---

[1] Research Engineer, Institute of Hydraulic Engineering and Water Resources Management (IWW) Aachen, University of Technology (RWTH), 5100 Aachen, Germany
[2] Director of Research, same Institution
[3] Professor of Civil Engineering and Institute Director, same Institution

The following sections will first give an overview of the approaches and basic concepts used in the two simulation models. Afterwards the background and key features of Smallworld GIS and how it can be custumised to the specific user needs will be explained. In addition the advantages of using GIS as a powerful management tool for controlling the whole model from the pre-processing to the presentation of simulation results will be illustrated within a case study.

The water surface model for natural rivers, ESNA

ESNA is a tool for one-dimensional water surface modelling and was developed at the IWW. It is based on many years of research studies of flow phenomena in a straight compound channel with vegetated flood plains. In natural rivers the main characteristics regarding the hydrodynamic aspects are:

- compound cross-section with extreme lateral variation of flow depth
- bank and flood plain vegetation
- varying bed roughness
- meanders

These characteristics evoke various effects such as extreme transverse velocity gradients, momentum exchange due to high turbulent shear stresses in the transition zone and additional flow resistance. For reliable water surface prediction these effects have to be considered. Therefore each cross-section is subdivided into subsections (Fig. 1).

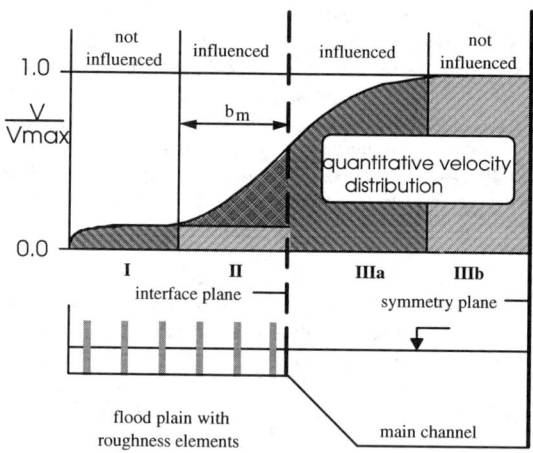

Fig.1: Subdivision of compound channel and transverse velocity distribution

In each subsection the velocity is computed using the general Darcy-Weisbach formula (1) with the non-dimensional resistance factor λ.

$$v_i = \sqrt{\frac{8g}{1/u_i \cdot \sum_k \lambda_{k,i} \cdot u_{k,i}}} \cdot r_{hy,i} \cdot I_e \qquad (1)$$

where:

v_i	velocity in subsection i	$u_{k,i}$	wetted parameter to $\lambda_{k,i}$
g	gravity acceleration	$r_{hy,i}$	hydraulic radius
u_i	wetted parameter	I_e	slope of energy line
$\lambda_{k,i}$	friction coefficient of influence k		

Compared to conventional formulas the main advantage of this approach is the non-dimensional friction coefficient λ. That allows the superposition of the different flow resistances due to bed roughness, main channel-flood plain interaction, varying vegetation, etc..

The formula for water surface computation (2) was derived from the differential equation of St. Venant. Starting with the known water surface at any section downstream as boundary condition, the upstream water surface profile is computed stepwise in an implicit difference process.

$$y_u = y_d + \beta[\alpha(y_d) - \alpha(y_u)]\frac{Q^2}{2 \cdot g} + \frac{\Delta x}{2}[I_{e,d} + I_{e,u}] \qquad (2)$$

where:

y	water elevation		β	momentum coefficient
I_e	slope of energy line		Q	discharge
g	gravity acceleration		Δx	distance between section
d	denoting downstream		u	denoting upstream
α	coefficient for non-uniform velocity distribution			

For more detailed information about the governing equations for determining the different flow resistance factors, the data preparation and the use of the model see SCHRÖDER/1990.

The ecological water quality model

The fundamental principle of conservation of heat and mass is used to derive the following differential equation model for the dynamics of heat and biotic and abiotic material.

$$V\frac{\partial C}{\partial t} = Q_i \cdot C_i - Q_0 \cdot C \pm V \cdot S \qquad (3)$$

where:

V	volume of fluid element		t	time coordinates
Q_i	inflow		Q_0	outflow
C	thermal energy or concentration		S	all source and sinks

C_i inflow thermal energy or concentration

In the water quality module three categories of constituents are considered: conservative pollutants where the source and sink term S is equal to zero, abiotic components (temperature, dissolved oxygen, biochemical oxygen demand, forms of nitrogen and phosphorus), and a variety of biological constituents like phytoplankton, zooplankton, etc..

To solve the general mass balance Equation (3) it is rewritten in a finite difference scheme for each element and integrated with respect to time.

$$V_i \cdot \dot{c}_i = c_{i-1} \cdot s_{i-1} - c_i \cdot s_i + c_{i+1} \cdot s_{i+1} + p_i \qquad (4)$$

where:

V_i volume of element i $\qquad\qquad$ c_i concentration of c in element i

s_i advection and diffusion term for element i $\qquad$ p_i constant term for each element (sources and sinks)

$\dot{c}_i$ time rate of change of concentration c in element i

The Gaussian reduction scheme is used to solve this set of simultaneous equations beginning with the least dynamic constituent and regressing to the most dynamic.

Smallworld GIS

Smallworld GIS is based on an object oriented language called Magik and runs on Unix - workstations. Its Graphics is based on Xwindow libraries such as OpenLook or Motiv. For data storage and management it uses a relational database. Included are objects which operate the window system such as the class "frame" and "canvac" forming together a drawable window on the screen.

The complete GIS is implemented by means of object classes and methods in Magik. The user can change these objects or add new objects to adopt the system to specific needs. For example so called "real world object" (rwo) include geometric and non-geometric information stored in the database and methods processing these information. These methods can be inheritated from upper classes. Figure 1 shows the layout of the rwo "teatment plant".

object: treatment plant		inherits methods from higher classes
geometric attributes	non geometric attributes	behavior
- area - label	- river mile location - discharge - quality -	method to compute the output

Fig. 2: Layout of the real world object "treatment plant"

Integration of the two simulation models into the GIS

To customise the GIS appropriate rwo´s had to be created. Among these are rwo´s like "cross_section" to administer the data for the water surface model and "river_section" to visualise the results of water quality computations. The creation of specific rwo´s for the simulation models is combined with definition of tables in the database in order to store the results of different simulation runs.

RWO "cross_section": As mentioned before the water surface model divides the river reaches into cross-sections. The objects are used to store the input data such as geometry of a cross-section, slope, discharge or roughness elements. The rwo contains additional functions that visualise the cross-section in a sub window in form of an x-y-graph and that allow the interactive graphical change of the cross-sections geometry.

RWO "river_section": This objects stores parts of the river being investigated. It contains the river mile location and nodes of the elements for the water quality model. Besides being a data carrier for the water quality simulation it is used to visualise simulation results. Data pertaining to specific geographic sites can be accessed by picking these sites.

Furthermore the standard graphical user interface of the GIS is changed. Buttons were added to the main menu and sub menus were created that invoke additional functions necessary to operate the simulation models within the GIS.

The functions of the customised DSS include:

- storage of all input data including boundary conditions
- easy entering, editing and retrieval of these data through interactive computer graphics
- controling of the simulation runs
- visualisation of input data, boundary conditions and simulation results as

x-y graph, diagram or geographically referenced map
- tools for analysing and comparing results

Fig. 3 shows a screen shot of the system together with some sub windows containing information of a water quality run.

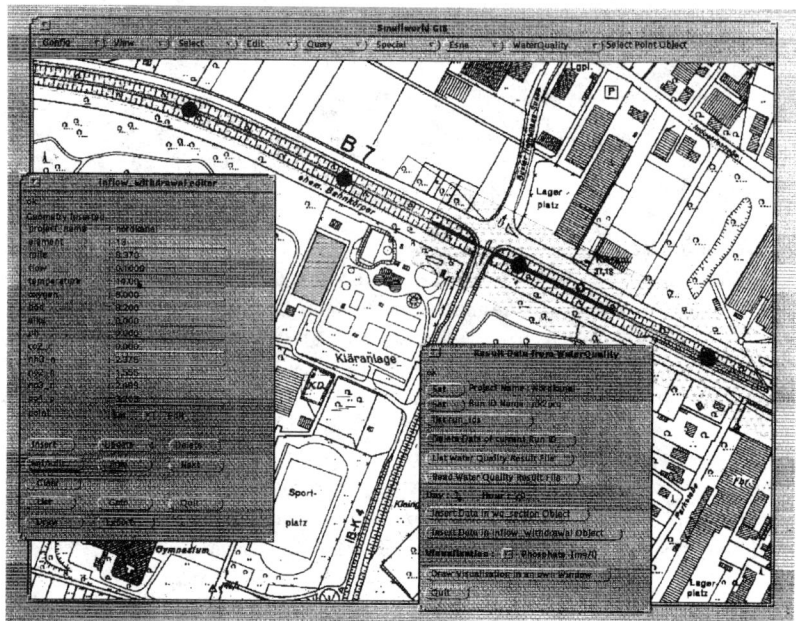

Fig. 3: Screen shot 1 of the DSS

River system "Jüchener Bach/Nordkanal"

The major characteristics of the river-system "Jüchener Bach/Nordkanal", situated in North-Rhine-Westphalia / Germany are:

- 25 river miles with natural characteristics subdivided into 150 cross-sections
- Infiltration from the river bed into the ground water along the whole river
- Discharge of coal mining drainage water to guarantee a minimum water level
- Intensive pollution due to inflows by two sewage treatment plants

The water quality is governed by the discharge of the coal mining drainage water and of the two treatment plants. The major aim is to achieve a river water quality of 2 ("moderate polluted") and to find the best restoration plan for this river system. By means of the created information system the various impacts, advantages and disadvantages of candidate plans are illustrated at once and analysed, supporting the decision process (fig.4).

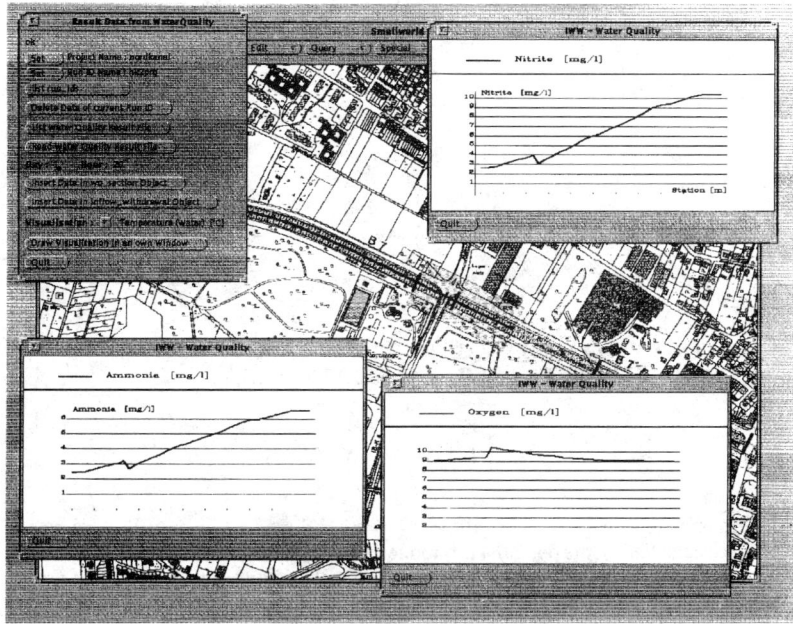

Fig. 4: Screen shot 2 of the DSS

Conclusion

GISs are perfecly suitable for the integration of simulation models. They provide many functions for managing, analysing and visualising topographic data, which otherwise must be progammed by the user itself.

The use of a GIS with an underlying object-oriented programming language permits considerable flexibility in the developement of an interactive interface facilitating the use of such a complex system. Thus it becomes easier to use.

References

Schröder, M.: "Optimierung eines mathematischen Modells zur
   Staulinienberechnung", IWW, RWTH Aachen, 1990

QUANTITY AND QUALITY OF DRY WEATHER FLOW IN LAS VEGAS VALLEY, NEVADA
by
Steve A. Mizell[1],
Richard H French[2], M. ASCE

Abstract

In recognition of the potential for dry weather flows to convey pollutants from the Las Vegas Valley, the potential economic and resource value of these flows within the Valley, and the need to characterize dry weather flows under the stormwater discharge permit program, the quantity and quality of dry weather flow in Flamingo Wash was investigated during the period of May 1991 to April 1993. Discharge, temperature, pH, and electrical conductivity were measured and samples were collected for analysis. Dry weather discharge averages about 4.6 cfs. The quality of this flow, with the exception of moderately high TDS and bacteria, is adequate for selected uses.

Introduction

Historically, washes in Las Vegas Valley have been ephemeral. Although some washes also carried natural spring discharge, this flow was generally curtailed as the result of ground water development. As development within the Valley has increased, washes in the Valley have begun to carry perennial flow. This flow generally originates in the western half of the metropolitan area and is conveyed eastward across the Valley to Las Vegas Wash which conveys the water to Lake Mead. Sources of this modern perennial flow likely include: runoff from over–irrigation of ornamental landscaping, de–watering in areas where the ground water table is high, and natural discharge of high ground water to the washes.

Flamingo Wash was selected for this pilot study because the tributary watershed encompasses typical land uses (light industrial, commercial, resort hotels, low– and high–density residential, parks and golf courses, and undeveloped land); and because the wash incorporates natural, channelized, and lined reaches, which are typical of washes in the metropolitan area.

----

[1]Assistant Research Professor, Water Resources Center, Desert Research Institute, P.O. Box 19040, Las Vegas, Nevada 89132–0040.
[2]Research Professor, Water Resources Center, Desert Research Institute, P.O. Box 19040, Las Vegas, Nevada 89132–0040.

Flamingo Wash originates in the mountains west of the Las Vegas Valley and crosses the southern half of the metropolitan area of Las Vegas (Figure 1). In the developed area, the topographic basin encompasses approximately 110 square miles along a reach which is more than 12 miles long. Additionally, 74 square miles contribute to the Wash through storm drain diversions.

Quantity of Flow

Flamingo Wash may be divided into two reaches based on the nature of flow. To the west of Las Vegas Blvd., the flow is characterized by channel reaches where surface water flow is present and reaches with no surface flow but where there is presumably ground water flow below the channel bed. Reaches containing flow are generally associated with outfalls from lined drains. East of Las Vegas Blvd., Flamingo Wash conveys a continuous flow of surface water to Las Vegas Wash.

Just above the confluence with Las Vegas Wash, dry weather flow in Flamingo Wash is diverted to a 30–inch pipe that conveys the water beneath a golf course. Flood flows are transmitted through the golf course in a grassed floodway. In May 1991, an electromagnetic flow meter was installed to monitor dry weather flow. From June 1991 through December 1992 (568 days of record), the daily average flow was 4.6 cubic feet per second (cfs) (Figure 2) with a standard deviation of 2.4 cfs. Comparison of monthly average discharge and monthly precipitation suggests correlation but one which is unable to account for all variations in observed monthly discharge. Preliminary analysis of periodicity in hourly discharge values suggests a strong 24–hour cycle during periods unaffected by flood flows (summer months). Daily minimum flows occur between 6 pm and 8 pm and maximum flows occur about 12 hours later.

At a daily flow rate of 4.6 cfs, Flamingo Wash conveys approximately 3300 acre–feet of water per year out of the Valley. In comparison, the average daily flow of Las Vegas Wash below the wastewater treatment plants and tributaries was 170 cfs for Water Year 1990, or approximately 123,000 acre–feet per year.

Quality of Flow

Investigation of dry weather flow quality included: regular measurement of field indicator parameters and seasonal sample collection for analytical testing. Field parameters (temperature, pH, electrical conductivity (EC)) were measured at about 20 locations in the Wash and in outfalls to the Wash. Initially, these measurements were made weekly; this routine was later reduced to an alternate week schedule. Temperature of the Wash waters closely reflected air temperature, exhibiting minimums in winter and maximums in summer. pH values were highly variable within a range of 7.4 to 9.2 but averaged about 8.2. Conductivity increased from the "headwaters" to the confluence with Las Vegas Wash. This pattern is represented by the average value of EC at four stations listed in order of increasing distance downstream: Rochelle Street, 1270 micro–mhos; Industrial Road, 1640 micro–mhos; Cambridge Street, 2990 micro–mhos; and Nellis Blvd., 3880 micro–mhos. In comparison, the average EC for Las Vegas Wash below the wastewater plants and natural tributaries for Water Year 1990 was 2440 micro–mhos.

Samples were collected for chemistry and biologic analysis at four locations along Flamingo Wash at six times (each February, May and September) during the term of this study. Analyses of these samples included determination of major ion,

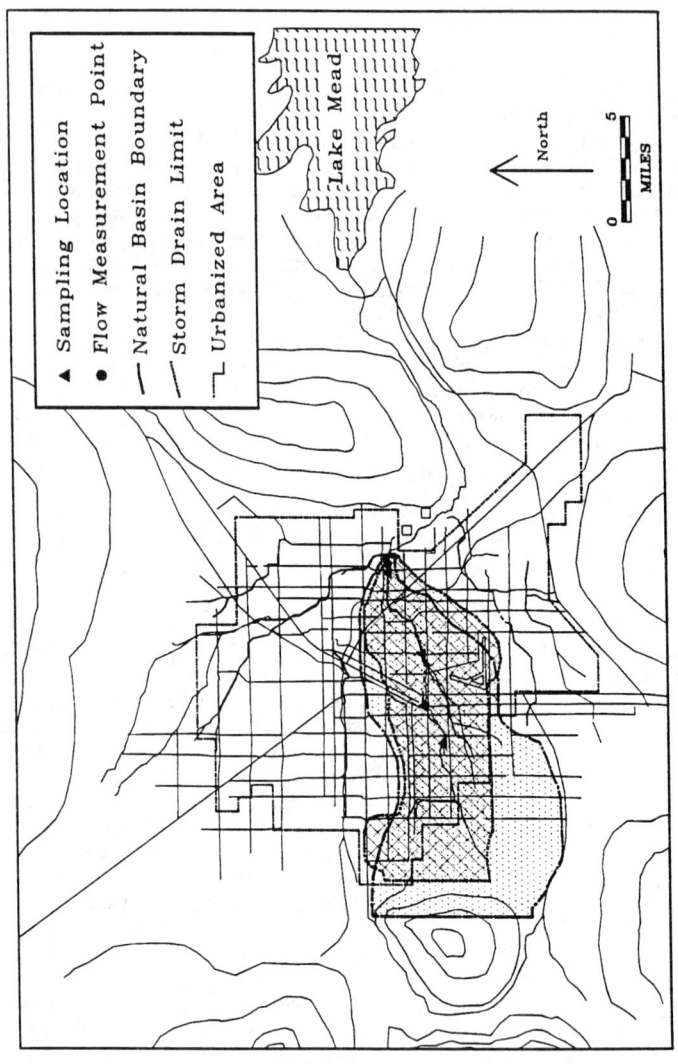

Figure 1. Flamingo Wash drainage, Las Vegas Valley, Nevada.

DRY WEATHER FLOW

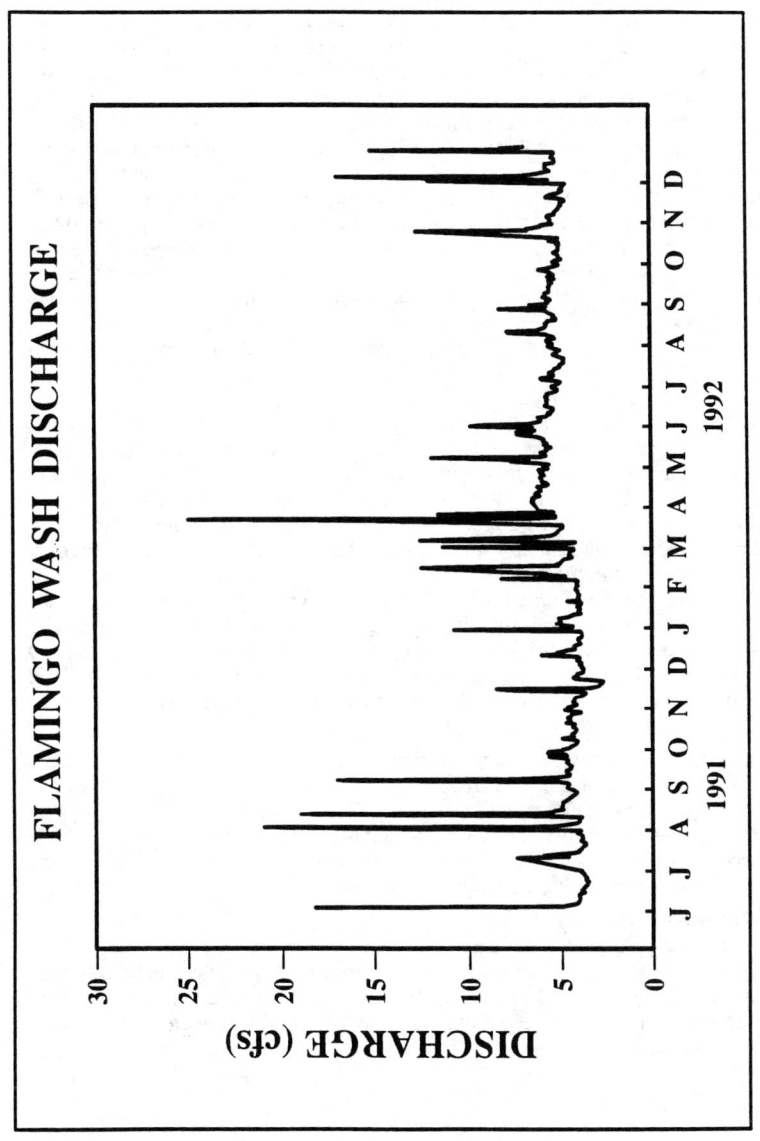

Figure 2. Discharge hydrograph for Flamingo Wash.

selected trace element, nutrient, volatile and semi-volatile compound concentrations, bacteria counts and evaluation of BOD.

In general, the analytical results indicated only minor quality problems in terms of moderately high salinity levels and high bacteria counts. Major ion concentration ratios remained relatively constant at each of the sample sites. The only trace elements detected were antimony, arsenic, boron, barium, copper, fluoride, iron, manganese, selenium, and zinc. None of the detected trace elements were present in significant concentrations. Phosphate compounds exhibited higher concentrations at the upstream sampling points while nitrogen compounds had higher concentrations at the downstream sampling points. Nitrate-N dominated the nutrient compounds showing concentrations of up to 9 mg/l. Only two organic compounds were detected. Di-n-butylphalate was observed at all four sample locations in September 1991; these concentrations were below health concern levels and the compound was not detected in any other sample. Bis (2-ethylhexyl) phthalate was detected in the preliminary sampling at one site but has never reappeared. BOD analysis has generally resulted in reported values below 10 mg/l. Total coliform counts have ranged about 2 orders of magnitude, from 2,300 MPN/100ml to 240,000 MPN/100ml.

The major ion data may give insight to source water influences along the Wash. Figure 3 shows major ion ratios for Flamingo Wash waters and for potential sources of dry weather flow. At the Rochelle sampling point, the major ion chemistry is very similar to that of surface waters imported from Lake Mead. At the Cambridge and Nellis sampling points, major chemistry of the Wash water is more like that of the shallow ground water. This suggests that runoff of potable water from excess landscape irrigation and shallow ground water discharge to the Wash are important sources of dry weather flow.

Dry weather flow in Flamingo Wash contributes only minor amounts of the chemical loading discharged from the Las Vegas Valley. Dissolved nitrogen species in Wash waters produce less than one percent of the total nitrogen load discharged from Valley. Additionally, using the EC data as an indicator, it appears that Flamingo Wash contributes less than 5 percent of the salinity load discharged from the Valley.

Conclusions

Available data suggest: dry weather flow in Flamingo Wash is a minor contributor to salinity and nutrient loads discharged from the Valley; and the quantity and quality of dry weather flow in Flamingo Wash is such that, with appropriate treatment, retention and use of this water within the Valley is not unreasonable. While these conclusions imply the technical feasibility of using this water, it should be noted that there may be policy issues that must also be resolved.

Acknowledgements

Research presented in this paper was sponsored by the Las Vegas Valley Water District, the Clark County Department of Comprehensive Planning, and the Nevada Division of Environmental Protection. However, the opinions and interpretations discussed are not necessarily those of the sponsors. Biologic parameter analyses were performed by the Clark County Sanitation District.

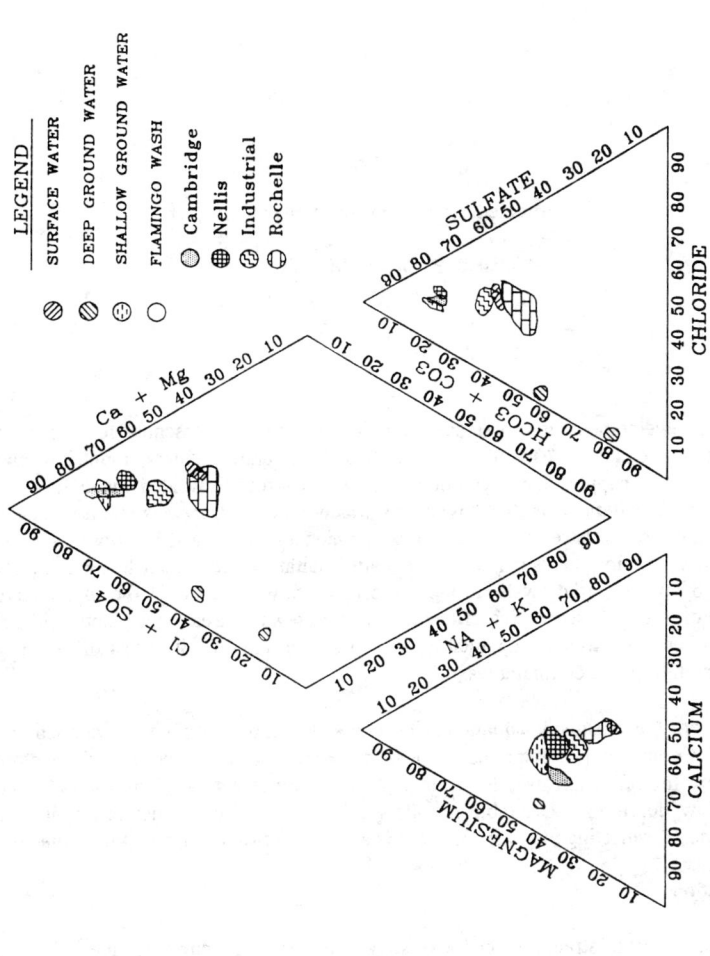

Figure 3. Piper diagram showing major ion chemistries for Flamingo Wash waters and potential source waters.

Protected Streamflow and Water Uses

Krishan P. Singh[1], M. ASCE

Abstract

The development of protected-streamflow standards is essential for the most desirable and equitable use of stream waters for aquatic habitats, municipal and industrial water supply, and recreation. Protected streamflow at a given location along a stream is defined as that flow below which water withdrawals will not usually be permitted for offstream uses such as municipal and industrial water supply. Protected streamflow helps maintain 1) aquatic habitats without their being seriously affected by water withdrawals during critical low-flow periods, 2) the assimilative capacity of a stream to receive effluents from wastewater treatment plants without adverse effects on streamwater quality, and 3) stream integrity in terms of diversity and strength of biotic communities.

A versatile, basinwide aquatic habitat assessment model has been developed and tested. Computer programs have been developed to calculate the daily flow availability for offstream uses for four protected flow levels based on monthly and yearly flow durations. Desirable low-flow releases from in-channel reservoirs are investigated for meeting instream flow needs and extra costs for meeting these needs.

Introduction

Availability of stream water for offstream uses such as municipal and industrial water supply may vary from day to day, month to month, and year to year, and regulating structures such as in-channel dams and reservoirs may affect the natural riverine environment adversely. Thus, protected flows are needed to minimize the adverse impacts of flow reductions on aquatic habitats and river ecology. Seasonal

---

[1]Principal Scientist and Director, Office of Surface Water Resources and Systems Analysis, Illinois State Water Survey, Champaign, IL 61820.

protected flows provide reasonable protection to indigenous fish species at various life stages in terms of maintaining suitable weighted usable areas (WUA). The physical habitat simulation model PHABSIM (Milhous et al., 1984) of the U.S. Fish and Wildlife Service (USFWS), together with fish suitability or preference curves, are used to define the seasonal WUA requirements. PHABSIM has two submodels: a hydraulic simulation model and a habitat suitability model. Singh et al. (1986) and Singh and Broeren (1989) developed a basinwide flow model to replace the hydraulic simulation model and create better aquatic habitat assessments.

Singh and Ramamurthy (1987) developed procedures and computer programs to calculate the daily flow availability from a stream over a period of record, considering various levels of protected flows. These flows can be used to determine the level to which various offstream uses can be met at different times; for optimal reallocation of streamflow for instream and offstream uses, depending on local conditions; to estimate the offstream storage requirements for a water supply system; or to plan for conjunctive use of ground water.

Protected Streamflow

Streams are used for various purposes, such as municipal and industrial water supply, recreation, waste assimilation, navigation, and aquatic habitat. As communities and industries expand, so does the quantity of effluents discharged to a stream and the demand for water withdrawals and streamflow regulation. Water regulation and withdrawals alter the stream regimen and may severely impair the stream environment if a sufficient quantity of water is not retained in the channel to support fish and wildlife and to maintain stream ecology and water quality.

Instream Flow Incremental Methodology—Instream flow needs for defining protected flow levels are investigated by relating the amount of suitable aquatic habitat to the quantity of flow. Comparisons of the quantity of suitable habitat provided at various protected flow levels can be used to assess the impact of water withdrawals on the stream environment. The Instream Flow Group (IFG) of the USFWS has been actively involved in developing the instream flow incremental methodology (IFIM), which relates habitat suitability to measurable streamflow parameters such as width, velocity, depth of flow, and substrate. IFIM has been computerized as a physical habitat simulation model or PHABSIM with hydraulic simulation and habitat suitability submodels. The hydraulic simulation is based on assumptions that preclude extrapolation to higher or lower flows than the flows and field data used for calibration. Results are reach- and discharge-specific and lack spatial transferability. Illinois studies (Herricks et al., 1983) indicate inability to achieve hydraulic calibration in many reaches of a river basin, serious calibration problems in some reaches, and more troubles under low-flow conditions. Miller and Wenzel (1984) investigated hydraulic modeling of low flows and found it to be futile in case of alluvial channels.

Singh and Broeren Method—Natural streams have riffles and pools -- the riffles act as dams holding water in the pools behind them during low flows. As streamflow decreases, the relative difference in depth and velocities in riffles and pools increases. These differences afford suitable habitat for various fish species at various life stages. Leopold and Maddock (1953) have shown that stream characteristics of width, depth, and velocity are related to discharge as simple power functions up to the point of bankful discharges. Stall and Fok (1968) extended the concept by incorporating flow duration and drainage area in these relationships:

$$\ln (\text{parameter}) = a + b\,F + c \ln (A_d) \tag{1}$$

in which a, b, and c are coefficients, F is the decimal flow duration, and A_d is the drainage area.

Singh and Broeren (1989) developed a basinwide flow model to replace the hydraulic simulation submodel of the PHABSIM. Evaluation of the WUA of a habitat suitable for a particular fish species requires knowledge of the local variations in depth and velocity. The model integrates flows of various durations and drainage area relationships, stream hydraulic geometry relations developed from U.S. Geological Survey flow measurement data, statistical joint distributions of depths and velocities in riffle-pool sequences, and adjustment factors for modifying hydraulic geometry parameters to those derived from field studies.

The first step is the development of relationships between discharges corresponding to various flow durations and watershed parameters such as drainage area and stream length, but mostly drainage area, A_d:.

$$\log Q(F) = A(F) + B(F) \log A_d \tag{2}$$

Values of A(F) and B(F) depend on flow duration, F. The second step initially entails development of at-station stream hydraulic geometry relations for in-channel flows:

$$\log (\text{VAR}) = a_0 + a_1 (\log Q) + a_2 (\log Q)^2 + \ldots \tag{3}$$
VAR = velocity V, depth D, or width W

and then deriving the most significant regression equations using calculated values of variables at specified flow durations for all stations within a hydrologically homogeneous region:

$$\log (\text{VAR}) = a + b\,F + (c + d\,F) \log A_d \tag{4}$$

The third step involves selection of representative study reaches in streams of various orders. Each reach should have a minimum of two pools and three riffles. Five equally spaced transects are placed between riffles for a minimum of 13 transects in each selected reach. Six measurements of velocities and depths are taken along each transect. From the set of 78 measures for velocity and depth, the statistical joint distribution of depths and velocities is determined.

The fourth step deals with the development of adjustment factors applicable to stream hydraulic geometry parameters on the basis of values calculated from field studies. The details of these steps are given in detail by Singh and Broeren (1989) and Singh et al. (1986). The integration of this model with the habitat model of the PHABSIM allows basinwide habitat evaluations and simulations for any flow scenarios.

Monthly vs. Yearly Flow Duration

Some states have defined protected flow (e.g., Q_y) as the flow corresponding to 90, 85, or 80% flow duration based on the entire daily flow record. Such flows based on monthly flow durations (e.g., Q_m) vary from month to month. Q_y (90) exceeds Q_m(90) for the months of August through December, and Q_y (85), Q_y (80), and Q_y (75), do so for the months of August through January for Shoal Creek near Breese in south-central Illinois. For all intents and purposes, the flows corresponding to 90 to 75% yearly flow durations are higher than those corresponding to monthly flow durations for the months of August through December, and they allow for higher protected flows. However, during June and July, the monthly flow-duration values are much higher than those for the yearly flow duration, and use of the yearly flow duration considerably lowers the protected flow. If the protected flow is to correspond to some flow-duration value, the interaction between monthly and yearly flow-duration values must be recognized in determining desirable protected flows for each month, season, or full year. During low streamflows most fish are found in the pools of the riffle-pool sequences throughout the stream length. The average depth of the pool below the riffle bed increases with drainage area, although the relation needs validation for regional application.

About 90% of the water withdrawn from a stream for municipal water supply is returned to the stream from the wastewater treatment plant at some location downstream of the intake. Thus the low flow in the stream is decreased in the stream reach between the intake and the wastewater plant outfall. This distance can be reduced to minimize the affected length of the stream. In the Midwest, water for irrigation is usually needed from mid-June to mid-August, and only a very small portion of this water reaches the stream in the form of baseflow during dry summer months. Although some water may be available for irrigation purposes in June and July, it may not be sufficient to meet expanding irrigation needs. Some on-farm storage ponds may be filled with water withdrawn from the stream during the relatively high-flow months of April and May.

Flow Availability for Other Uses and Protected Flow

Flow availability statistics basically consist of month-by-month analyses of daily flows above a selected protected flow level for that month. The statistics provide information on average deficit or surplus flow available and the percentage of years and days during which these deficits and surpluses occur. The information is necessary for the optimal design of a side-channel reservoir where the need for offstream water use has been established. The nature of surpluses and deficits provides the data for optimizing reservoir storage. Adoption of a protected flow level for a stream is largely guided by environmental considerations, instream flow needs, offstream water uses, and economics. A computer program was developed to read the daily flows of record at a gaging station and to compute and print the relevant information (Singh and Ramamurthy, 1987).

Instream Reservoirs and Protected Flows

Modification of river flow resulting from construction and operation of a dam and reservoir has been identified as a significant factor causing water quality and aquatic habitat problems. However, if the flow releases to meet instream flow needs are considered in the hydrologic design of the dam and reservoir, adverse impacts can be mitigated to a large degree. A study was conducted by Singh and Ramamurthy (1981) to investigate the increase in the project cost for the increase in storage required for the protected flow releases. The information developed in this study can be used to make rational decisions about the choice of protected flow levels, not only to ensure satisfactory maintenance of aquatic habitats and stream ecology, but also to keep the project cost increases to a minimum. The mandatory low-flow releases during droughts are usually smaller than available during average and wet years. The percent of incremental project cost increases with 1) the decrease in the drainage area of the stream, 2) the decrease in water withdrawals for water supply, and 3) increase in flow variability of the stream under consideration. For a water supply reservoir, the increased reservoir storage may double the cost of raw water withdrawals. However, the raw water cost is only about 10 to 30% of the cost of potable water after treatment. Thus a significant increase in raw water cost may increase the cost of water supply by a similar amount.

It is imperative to follow reservoir storage conservation policies by minimizing the sediment input to the reservoir, as well as by venting most of the incoming sediment downstream in the interest of reservoir water quality, long-term storage utility, and reduction in downstream bed degradation. The venting of sediments can be achieved by releasing excess flows through bottom sluices and by installation of automatic siphons.

Conclusion

The adoption of a protected-flow standard involves consideration of conflicting goals and needs. Both tangible and intangible benefits are associated with a protected-flow level, and they vary with the level of protection. There is also an associated cost for adopting and maintaining a protected-flow level in a stream. A cost-benefit approach provides a framework for analyzing the economics of adopting a particular protected-flow level and selecting the best level to meet various needs.

Appendix-1. References

Leopold, L.B., and Maddock, T. (1953). "The hydraulic geometry of stream channels and some physiographic implications." U.S. Geological Survey Professional Paper 252, Washington, DC.

Herricks, E.E., Eheart, J.W., Stall, J.B., and Gantzer, C.J. (1983). "Instream flow needs analysis of the Sangamon River basin." University of Illinois *UIUC-ENG-83-2017*, Champaign, Illinois.

Milhous, R.T., Wegner, D.L., and Waddle, T. (1984). "*User's guide to the physical habitat simulation system (PHABSIM).*" Instream Flow Information Paper 11, U.S. Fish and Wildlife Service, Ft. Collins, Colorado.

Miller, B.A., and Wenzel, H.G. (1984). "Low-flow hydraulics in alluvial channels." Univ. of Illinois Water Res. Center *Research Report 192*, Urbana, Illinois.

Singh, K.P., and Ramamurthy, G.S. (1981). "Desirable low flow releases from impounding reservoirs: fish habitat and reservoir cost." Illinois State Water Survey *Contract Report 273*, Champaign, Illinois.

Singh, K.P., Broeren, S.M., and King, R.B. (1986). "Interactive basinwide model for instream flow and aquatic habitat assessment." Illinois State Water Survey *Contract Report 394*, Champaign, Illinois.

Singh, K.P., and Ramamurthy, G.S. (1987). "Pertinent considerations in the development of protected-streamflow criteria for Illinois streams." Illinois State Water Survey *Contract Report 431*, Champaign, Illinois.

Singh, K.P., and Broeren, S.M. (1989). "Hydraulic geometry of streams and stream habitat assessment." *J. Water Resources Planning and Management*, ASCE, 115(5), 583-597.

Stall, J.B., and Fok, Y.S. (1968). "Hydraulic geometry of Illinois streams." Univ. of Illinois Water Res. Center *Research Report 15*, Urbana, Illinois.

Application of Environmental Regulations on Design of Hydraulic Structures for Open Cooling Water System

Jagdish K. Virmani[1] M. ASCE, Mahmood Naghash[2] M. ASCE, and Adnan M. Alsaffar[3] F. ASCE

Abstract Design of intake and discharge structures for an open cooling water system for a thermal power plant requires that the design of hydraulic structures meets the federal, state, and local environmental regulations. The pertaining regulations evolve from Section 316(a) and 316(b) of the Federal Water Pollution Control Act where best available technology must be applied for the specific site.

For a site with a limited river front available within the plant property along with large variation in river water level and flow, detailed optimization studies are performed to meet these requirements. The intake was designed to minimize impact on fish entrapment and impingement and to meet the Section 316(b) permit requirement and to reduce warm water recirculation so that plant efficiency and the size of the thermal plume will not be affected. The design of the discharge structure had to meet the thermal discharge criteria governed by the 316(b) permit requirement, and minimize warm water recirculation into the intake. Low river flow and associated velocities had contributed to the low levels of dissolved oxygen (DO) in the river water. Therefore; provisions were also made to aerate the heated effluent before discharging into the river.

Introduction

The circulating water system of a power plant is an integral part of the steam generating cycle. Its conceptual development design and operation encompass several environmental, engineering, economic, and operation and maintenance considerations. The environmental regulations play a major role in the design of cooling system. The major factors affecting the design are availability of the cooling water, the impact of water withdrawal on fish and larvae entrainment and impingement, and the impact of plant heated effluent on the aquatic habitat. Most frequently, detailed studies are performed in support of the design, and to comply with the federal, state, and local 316(a) and 316(b) permit requirements. State and Federal Government regulations with respect to water withdrawal, type of intake and acceptable thermal discharge constraints are identified in developing the design concept.

---

[1] Engineering Specialist
[2] Senior Engineer
[3] Chief Hydraulic Engineer
Geotechnical & Hydraulic Engineering Services, Bechtel Corporation, Gaithersburg, MD

Thermal Discharge Criteria Section 316(a) of the Federal Water Pollution Control Act [1], provides that any permitted thermal effluent be within the range of the specific state water quality standards in question. The applicable thermal criteria for this project required that, "Temperature rise shall be limited to no more than 2.8 °C above natural temperature, not to exceed 31 °C at any time during months of May through November and not to exceed 23 °C at any time during the months of December through April. During any months of year, heat should not be added to a stream in excess of the amount that will raise the temperature of water more than 2.8 °C above natural temperature." It should be noted that these requirements may vary from stream to stream and state to state. Frequently for open cooling water systems, in order to meet the thermal discharge criteria a mixing zone is required. The mixing zone is defined as the area in which the temperature is greater than the thermal discharge criteria.

Intake Withdrawal Criteria Section 316(b) of the Federal Water Pollution Act [1], provides: "Any standard established pursuant to Section 301 or Section 306 of the Act and applicable to a point source shall require that the location, design, construction, and capacity of cooling water intake structures reflect the best available technology for minimizing adverse environmental impact." To meet this requirement, the effect of intake facility on the fish should be minimized by proper selection and design of intake [2,3].

Cooling Water System Design

A particular power plant was designed with a once-through cooling water system. The pump intake withdraws the cooling water from a river under a skimmer wall and discharges the heated water back to the same river via a concrete seal weir aeration/discharge structure. This particular design created a vertical and horizontal separation between the intake and discharge flows; thus, minimizing the recirculation. The water system schematic is shown on Figure 1.

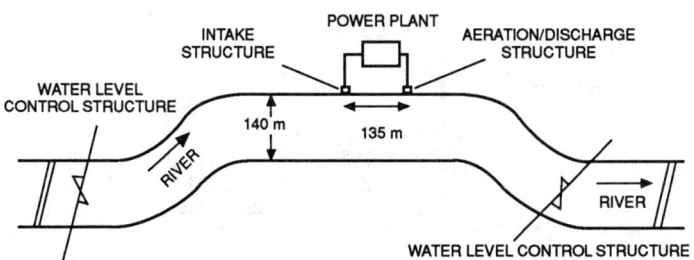

Figure 1. SCHEMATIC OF THE OPEN COOLING WATER SYSTEM

The total piping length was about 250 m. The circulating water cooling system consists of the circulating water pumps, main condenser, pipes, and control valves. The system was designed to provide a circulating water flow of 3.4 m³/s with a design temperature rise of 8 °C above the ambient.

River Conditions

The river at the plant site was 140 m wide with an average normal depth of 4 m. The average river flow area in the vicinity of the plant at normal pool level is about 420 m². The average river velocity during an average annual flow of 130 m³/s is about 0.15 m/s for ordinary high water level of 8 m. For a low flow of 10 m³/s the average river flow velocity at a normal low water level of 4 m was about 0.03 m/s. The river monthly average water temperature varies from 2 °C in winter to 24 °C in summer.

Intake Structure

The shoreline river intake structure for such an application included a skimmer wall, trash rack, travelling screen, stoplog slots and circulating water and service water pumps.

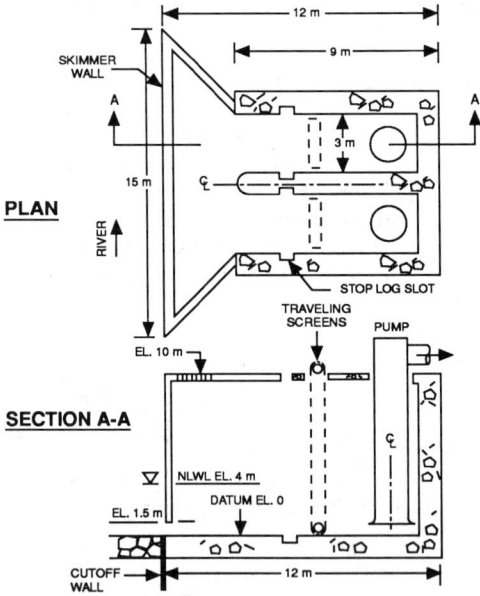

Figure 2. INTAKE STRUCTURE CONCEPT

The skimmer wall was provided to facilitate the withdrawal of cold water from near the bottom. The height of the opening for the skimmer wall was based on the selective withdrawal criteria considering the densimetric Froude Number of the flow, with consideration of the effect of the surface water discharge on the surface water temperature.

In developing the intake structure concept, the river hydrographic data and the thermal plume analysis of the thermal discharge including the recirculation effect were used. In the design, primary consideration was given to fish protection using the best available technology and to minimize the impact on the river during the construction and operation phases. The techniques applied for fish protection included: the skimmer wall which was aligned flush with the shoreline for preventing the formation of quiescent areas that attract the fish, low flow velocity of 0.15 m/s under skimmer wall, the utilization of through flow vertical traveling screens, and the 0.15 m/s approach velocity to the screens at normal low water level of 4 m. The typical details of the intake structure are shown on Figure 2.

Discharge Structure

The function of discharge structure for the plant is to convey the plant heated water to the river and to meet the thermal discharge criteria. In selecting the location and features of the discharge structure, the aforementioned factors are considered. The outfall alternatives considered included: multiport submerged diffuser, high velocity near-surface discharge, and the low velocity near-surface discharge. To prevent interference with the navigation during the construction and operation phases, the offshore discharge scheme was eliminated. The shoreline near-surface discharge scheme with low velocity would have resulted in a large surface plume (large mixing zone) compared to the high velocity near-surface discharge concept; therefore, it was deleted. Consequently, the high velocity near-surface shoreline discharge was selected for the design.

To mitigate the potential for warm water recirculation into the intake during periods of low or negligible river flow, the distance between intake and discharge structures was maximized by locating these structures at opposite ends of the property boundary. Furthermore, the use of skimmer wall at the intake structure provided the vertical separation between the intake and discharge flows. The discharge outfall consisted of a 1.5-m diameter horizontal pipe located on the shoreline and discharged perpendicular to the river flow stream. The top of the pipe was set about 0.5 m below the river normal low water level of 4 m. Using this concept, the average discharge pipe velocity is about 1.9 m/s and the thermal plume criteria of 2.8 °C above the ambient temperature is met at a distance of about 30 m from the point of discharge.

At ordinary high water levels, the discharge pipe experiences higher submergence leading to a smaller mixing zone. The typical details of the discharge structure are shown on Figure 3.

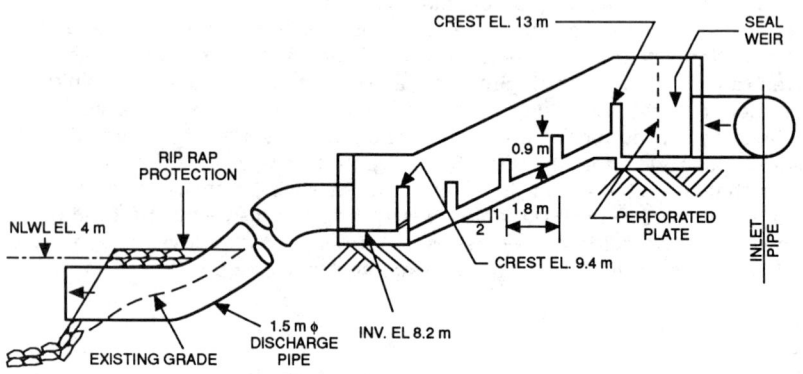

Figure 3. AERATION/DISCHARGE STRUCTURE

Aeration Structure

The existence of control structures along the river and low river flow velocities, contributed to the reduction of river dissolved oxygen (DO) levels. Furthermore, for this once-through cooling water system, the condenser sub-atmospheric pressure of about 0.3 bar absolute and condenser temperature rise of 8 °C created a super-saturation condition resulting in air release [4]. Consequently, the cooling water on the downstream side of the condenser at the seal weir was under-saturated and was capable of absorbing air. For this purpose an aeration structure was added at the outfall to increase the dissolved air content of the thermal discharge and to provide a mechanism for air entrainment in assisting the aeration of river water at the point of discharge.

Applying the available gravity force as the main source of input energy, a free surface aeration structure was designed to provide sufficient interfacial area and turbulence necessary for an accelerated gas transfer process. On a general slope of 2 horizontal to 1 vertical, a cascade of sharp crested weirs were utilized as the design concept [5], (Figure 3). The structure was a rectangular chute open to the atmosphere, 3 m wide and 12 m long including the seal weir. A total of 5 sharp crested weirs each 0.9 m high with 1.8 m horizontal spacing were utilized.

The free fall of water over the weirs and subsequent plunge pools are the mechanism for providing the surface transfer area, the shearing and impact forces, and the turbulence intensity, all necessary parameters for an accelerated gas transfer process. The cascade of plunge pools in the aeration structure in conjunction with high flow velocities allow a large amount of free air to be entrained in the discharge water and be carried to the river for further aeration.

Conclusions

The potential environmental impacts associated with the construction and operation of the once-through cooling water intake and discharge structures were controlled through the use of state-of-the-art techniques and use of best available technology. In addition, by use of an aeration structure as an integral part of the discharge structure, the DO levels in the heated discharge and the river water were improved.

References

1. U. S. Environmental Protection Agency, "Guidance for Evaluating the Adverse Impact of Cooling Water Intake Structures on the Aquatic Environment: Section 316(b) P. L. 92-500," Office of water Enforcement, Washington, D. C., 1977.

2. U. S. Environmental Protection Agency, "A State of the Art Report on Intake Technologies," 1982.

3. Design of Water Intake Structures For Fish Protection, ASCE Hydraulics Division, 1982.

4. Naghash, M, 1990, "Steady Turbulent Gas Desorption in Surface Condenser Tubes", Air-Water Mass Transfer, Proceedings of Symposium on Gas Transfer at Water Surfaces, Minneapolis, MN, September 11-14.

5. Albrecht, Detlef, 1969, "Schätzung der Sauerstoffzufuhr durch Wehre und Kaskaden", Die Wasserwirtschaft 11.

The Concept of "Local Euler Number" as an
Aid for Sizing Pitot Tubes

José Roberto Bonilha[1]

Abstract

Correlations with drag coefficients enable determining the relations between the form of Pitot tubes and their velocity coefficients. A numerical example is presented to clear the use of a methodology of analysis which can constitute a helpful tool for designing instruments or checking calibration results.

Introduction

In spite of the existence of up to date types of high technology velocity measurement instruments, the Pitot tube, one of the most simple, ancient and reliable hydraulic devices has never lost its importance. The following theoretical formula, which enables the calculation of the velocity v at one point, in function of Δh, the difference of readings at two piezometer tubes, is supplied at initial lessons on Fluid Mechanics:
$$v = \sqrt{2 g \Delta h} \qquad (1)$$
Further references, commonly found only at Handbooks and Catalogues of Fabricators notice that for each different tip, specific experimental velocity coefficient is proposed. Thus, formula (1) becomes:
$$v = c \sqrt{2 g \Delta h} \qquad (2)$$
As a rule, no further information about the theoretical or physical meaning of this coefficient, treated as a function of velocity, is issued.

This author intends to point out that a methodology based on the Euler number, allows the analysis of tips in

---
[1] Professor, Escola Politécnica da USP, UNICAMP Universidade de Campinas, Consultant Engineer, CTH Centro Tecnológico de Hidráulica, Av. Prof. Lúcio M. Rodrigues, 120, Brazil

order to identify relations between their forms and their respective coefficients c. In certain cases, the numerical value of c can be analytically calculated as well. In order to justify the fundamentals of this analysis, a brief summary of the concerned theoretical concepts will be presented, before discussing the numerical example.

Theoretical concepts

Euler number and velocity coefficient

As long as only the pressure gradient, the density and the acceleration are considered, the assumption of steady, irrotational flow yields a unique solution for the distribution of velocity and pressure around any given form of boundary and so, a particular value of the dimensionless ratio $E = \Delta p /(\rho v^2/2)$ characterizes the flow at each point, regardless of the absolute magnitudes of the pressure p, the velocity v, and the boundary scale. Any parameter of this form may be called an Euler number.

$$E = f(\text{form}) \qquad (3)$$

As a consequence of this definition, the feasibility of determining an one-valued function $c = f(E)$, constitutes a prove of the dependance of c on the form of the tip of the instrument. The function $c = f(E)$ can be determined as follows:

Considering a tip as a solid body immersed in a moving fluid and supposing known Euler numbers E_1 and E_2 at two different locations, 1 and 2, upon its surface, it is possible to write the following equations, provided the flow is steady and irrotational:

$$\frac{p_1 - p}{\rho v^2/2} - \frac{p_2 - p}{\rho v^2/2} \left[v = \frac{1}{\sqrt{E_1 - E_2}} \sqrt{2 g \Delta H} \right.$$

Thus: $c = 1 / \sqrt{E_1 - E_2}$ (4)

For instance, in the case of the Prandtl Pitot $E_1 = 1$ at the stagnation point and $E_2 = 0$, because 2 is situated downstream where the flow is practically undisturbed. Thus: $c = 1 / \sqrt{E_1 - E_2} = 1 / \sqrt{1 - 0} = 1$

This result complies with formulas (1) and (2).

The drag coefficient Cd

This dimensionless number enables the calculation of the longitudinal force exerted upon a body immersed in a moving fluid.

Existence of viscous forces is a necessary condition for existing surface resistance and so:

$$Cd = \frac{F/A}{\rho v^2 / 2} = E'_1 - E'_2 = f(form, R) \quad (5)$$

A is the aspect area.

Correlations between c e Cd

As these two coefficients are functions, whose arguments are differences of Euler numbers measured upon the same immersed body, correlations between certain aspects concerned to both can be expected.

Geometrical prototype.

First imagined as simple tubes during the XVIII century, the Pitot instruments more recently appeared with various different shapes to face specific needs of engineering applications. Amongst these types, the Pitot Cole was designed by the Cole Pitometer & Co. to measure the rate of flow in pipes. It can be imagined as two pieces of bent pipes of small diameter which can be turned around in order to facilitate their introduction into pipes through taps. This tip was taken as geometrical prototype to exemplify numerically the methodology treated herein.

Proposed methodology for analyzing and sizing tips. Pitot Cole as example.

Idealization of a geometrical model

As a first approach it can be assumed that the Pitot Cole tip is simply a circular disk. According to this assumption, the front face of the disk is represented by the cut cross section of the upstream small pipe. Its rear face is represented by the cut cross section of the downstream small pipe.

Rouse [2] supplied values of Euler numbers for circular disks, according to a chart whose readings are: 0,1;.286,.992;.543,.918;.718,.814;.901,.596;.956,.407;1,0

At downstream, zone of separated flow, the Euler number has a negative and constant value $E_2 = -.440$

The following expression was used to calculate the average value of E_1 related to one of several tips calibrated in Sao Paulo Hydraulic Laboratory, Brazil, [tip 8AB], with D = 4.65mm and relation $r/r_o = .6323$:

$$E_1 = \frac{\int_0^{r/r_o} 2\Pi (r/r_o) \cdot E(r/r_o) \cdot d(r/r_o)}{\Pi (r/r_o)^2} \quad (6)$$

This integration between the limits $r/r_o = 0$ and $r/r_o = .6323$ led to the value $E_1 = .948$.

Finally the value of the coefficient c of the tip [8AB] can be calculated as follows:

$$c = 1/\sqrt{E_1 - E_2} = 1/\sqrt{.948 - (-.440)} = .849$$

Numerical comparison between c and Cd

For the disk imagined as geometrical model to represent the tip [8AB], the drag coefficient would be calculated as follows: $Cd = .760 - (-.440) = 1.2$. The value .760 was calculated using the same equation (6) but considering different limits of integration: $r/r_o = 0$ and $r/r_o = 1$. E'_2 has the same constant value $-.440$.

These numerical calculations clarify questions related to correlations between c and Cd. Considering formulas (4) and (5) it can be noted that if Cd increases c decreases and vice versa.

Available experimental data for comparisons

Original data of Cole Pitometer & Co

The following expression can summarize the data of a table of coefficients considering extended range of velocities from .2 m/s till 10.3 m/s [1].

$$c = .98299604 \, R^{-.01404503} \quad [r^2 = .99446]$$

Data referring to the tip [8AB]

This tip, a copy of an American prototype, was calibrated in São Paulo, Brazil, on a revolving boom device and supplied data which were plotted on the Chart .

Result of comparisons

According to Rouse [2] the coefficient Cd related to the disk is a constant for R higher than 2000 but the measured values of c indicate that only for R higher than 30000 this experimental coefficient could be taken as a constant equal to .849. Another geometrical model had to be adopted to represent the tip within the range 2000<R<30000.

New geometrical model

As result of other tests it can be concluded that: For sake of analysis of the velocity coefficient this tip can be idealized as a tandem formed by a circular disk and two ellipsoids. Considering this new model it was possible to explain the fact that the function c = f(R) has points of maximum at values of R near 4000. In fact the curve Cd=f(R) related to a ellipsoid 3:4 has points of minimum within this same range of Reynolds numbers as can be observed on the Chart.

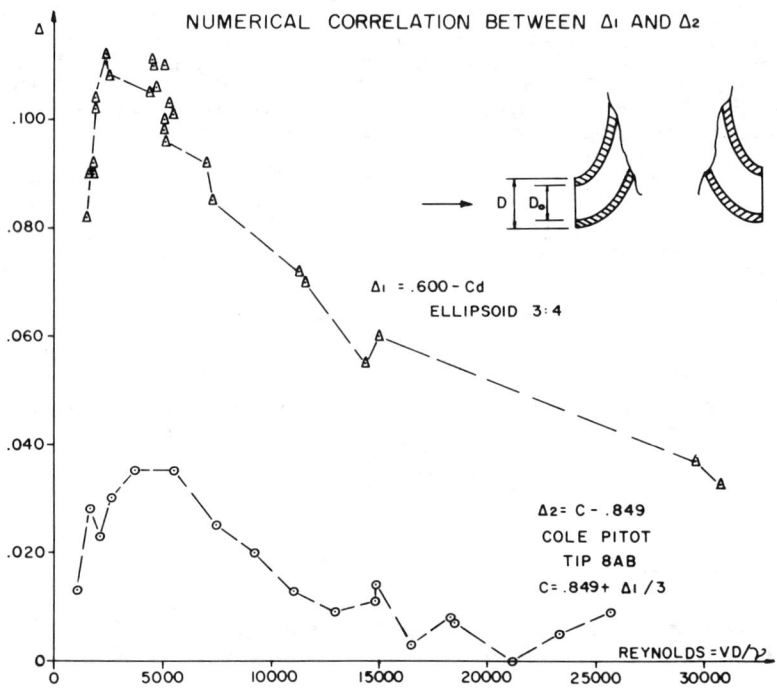

Drag coefficients for ellipsoids

Lazzari [1] supplied values of Cd for ellipsoids with various relations of D/d, within an extended range of Reynolds numbers, 2000<R<50000. According to Rouse [2] the value of Cd can be taken as a constant Cd = .600 for an extended range of Reynolds numbers at the vicinity of R = 100000. The values of Cd for an ellipsoid 3:4 were used to calculate the values of the negative increment Δ_1 which are plotted on the Chart.

Analytical expression of c

To prove that the new geometrical model can adequately simulate the Pitot Cole tip, it was assumed that the velocity coefficient c is a constant value c=.849 plus an increment, function of Reynolds number. Identically the drag coefficient of the ellipsoid 3:4 was assumed as a constant Cd=.600 minus an increment, also a function of Reynolds number. Thus:

$Cd = .600 - \Delta_1 (R)$ $c = .849 + \Delta_2 (R)$

The functions $\Delta_1 (R)$ and $\Delta_2 (R)$ representing these positive and negative increments were plotted on the Chart. Visually it can be noted that both have the same trends of variations and points of maximum at the same range of Reynolds numbers. 3 can be taken as a factor of proportionality between both functions, i.e.: $\Delta_1 (R) = 3 \cdot \Delta_2 (R)$

This conclusion proved that the idealized geometrical model, a tandem formed by a circular disk and two ellipsoids, is adequate to represent tips of Pitot Cole within the range 2000<R<30000.

General conclusion

As it has been pointed out, within several fields of practical utilization of Fluid Mechanics, the application of the theoretical concept of Euler number can generalize conclusions based upon tests conduct on an individual device. This fact, is due mainly to the possibility of determining correlations with other well known fundamental parameters of this Science.

In addition, when designing or refurbishing Pitot devices the correlations between c and Cd can constitute a sound basis for the better understanding of the related flow phenomena and the consequent choice of adequate forms for each type of instrument.

References

[1] PEDRAZZI, José Afonso (1992) Critérios de projeto para um novo tip do Pitot-Cole. (Design criteria for refurbishing the Cole Pitot tip).Thesis - USP - São Paulo, EPUSP.

[2] ROUSE, Hunter (1957) Elementary mechanics of fluids. N.York, John Wiley, 1957.

METHODS FOR PREDICTION OF MAXIMUM SCOUR AT COASTAL STRUCTURES

Jimmy E. Fowler[1]

Abstract

The most common coastal scour-related problems are toe scour at rubble mound structures and vertical seawalls, and scour at the base of piles and horizontal pipelines. Existing scour prediction methods for these problems vary from rules of thumb, to empirically derived equations, to theoretically derived relationships. Recent studies at the U. S. Army Waterways Experiment Station's (WES) Coastal Engineering Research Center (CERC) indicate that sufficient design guidance exists for vertical walls, pipelines, and vertical piles, however, additional research is still needed for rubble-mound structures.

Introduction

Scour at coastal structures is a serious problem which causes damage to structures. The consequences of scour at coastal structures have long been known, and elaborate and expensive toe protection schemes have often been implemented. Among the most common coastal scour-related problems are toe scour at rubble mound structures and vertical seawalls, and scour at the base of piles and horizontal pipelines. Existing prediction methods for these types of scour problems vary from rules of thumb, to empirically derived equations to theoretically derived relationships. When existing prediction methods are felt to be insufficient, physical model studies often are performed.

Scour at Vertical Seawalls

Laboratory studies conducted at WES by Fowler (1992) lead to the following relationship for predicting maximum scour at the toe of seawalls:

$$\frac{S_{max}}{H_o} = \sqrt{22.72\ h_w/L_o + .25} \qquad (1)$$

The range of conditions tested in the laboratory study included $-0.011 \leq h_w/L_o \leq 0.05$ and $0.015 \leq H_o/L_o \leq 0.04$. For the above equation, h_w is water depth at the seawall, L_o is deep water wave length, H_o is deep water wave height, and S_{max} is maximum scour depth, seaward of vertical seawalls. For cases not covered by equation 1, the rule of thumb, $S_{max}/H_o \leq 1$, should be used.

---

[1]Research Hydraulic Engineer, Coastal Engineering Research Center, U.S. Army Engineer Waterways Experiment Station, Vicksburg, MS.

Scour at Rubble Mound Structures

Little guidance is available for prediction of scour depths at the toe of rubble-mound structures. The majority of efforts in this area have focused on predicting size and amount of toe protection which should be used to avoid wave induced instability of toe armor layers. Depth of scour at the toe of rubble-mound structures is extremely difficult to isolate and measure. In light of this, very few researchers have attempted to develop prediction equations for depth of scour at toes of rubble mound structures.

For small rubble-mound structures, such as revetments, the $S_{max}/H_o \leq 1$ rule of thumb, developed for vertical seawalls, should be adequate for determining maximum scour depth. For larger rubble-mound structures constructed in deeper water, two-dimensional studies conducted at WES during 1992 have shown that toe scour may be a more pronounced problem when there is wave action in the presence of strong currents. Prediction techniques for these structures are presently insufficient and are currently being studied at WES. The type and amount of toe scour protection is often given in terms of the mean weight of the individual stable armor unit, W_a, lb_f. W_a is determined by various empirically derived equations, the most common of which is by Hudson (Shore Protection Manual, 1984):

$$W_a = \frac{\gamma_r H_D^3}{K_D (S_r - 1)^3 \cot \theta} \qquad (2)$$

For the above, γ_r is the specific weight of armor unit, lb_f/ft^3, H_D is the design wave height at the structure site, ft, S_r is the specific gravity of the armor unit relative to the water at the structure given by $S_r = \gamma_r/\gamma$, where γ is the specific weight of the water, lb_f/ft^3. θ is the angle in degrees of the structure slope measured relative to the horizontal plane, and experimentally determined K_D is the stability coefficient which varies with the type of armor unit as well as other parameters. Consult the Shore Protection Manual (1984) for additional information regarding use of equation 2.

Hales (1980) conducted a survey of scour protection practices in the U.S. and found that a rule of thumb for minimum toe scour protection is a toe apron measuring 2.0 to 3.0 ft thick and about 5 ft wide. A study by Eckert (1983) indicated toe scour protection should be designed to accommodate the maximum scouring force which exists where wave downrush on the structure face extends to the toe. According to Eckert (1983), the rule of thumb for minimum toe scour protection found by Hales will be inadequate if the following conditions exist:

1) Water depth at toe of the structure is less than twice the height of the maximum unbroken wave height that can exist in that water depth.

2) Wave reflection coefficient in excess of 0.25, which is generally true for structures with slopes steeper than 1 v on 3 h.

Movable bed model tests conducted by Lee (1970; 1972) on a quarrystone-armored jetty with slope of 1 vertical on 1.25 horizontal indicated that a double layer of rock having mean weight, W_{apron}, where

$$W_{apron} = W_a/30 \qquad (3)$$

provided suitable toe scour protection. Lee's tests also showed that the width of toe protection should be equal to the width of 4 - 6 of the stones having mean weight given by equation 3, estimated by:

$$B_{apron} = n_s \, k_\Delta \left(\frac{W_{apron}}{\gamma_r} \right)^{1/3} \qquad (4)$$

In the above, B_{apron} is the apron width in ft, n_s is the number of stones, and k_Δ is a layer coefficient (for stone only) varying between 0.94 and 1.15, dependent upon armor type, shape and construction method as detailed in the Shore Protection Manual (1984), and γ_r is unit weight of armor stone, lb_f/ft^3. The above rules of thumb for rubble-mound toe protection are not valid for concrete armor units and should only be used for toe protection consisting of stones.

Markle (1989) conducted laboratory studies to address sizing of toe berm and toe buttressing stone in breaking wave environments (see figures 1 and 2). Guidance is given in terms of the stability number, N_s, defined by

$$N_s = \left(\frac{\gamma_{rb}}{W_{50}} \right)^{1/3} \frac{H_D}{(S_r - 1)} \qquad (5)$$

with W_{50}, the median weight of individual berm stone in lb_f, as defined previously in equation 2. In addition,

γ_{rb} = specific weight of berm stone, lb_f/ft^3
S_r = specific gravity of berm stone relative to the water in which the structure resides, i.e., $S_r = \gamma_{rb}/\gamma$
H_D = design wave height, ft
γ = specific weight of water in which structure resides, lb_f/ft^3

Markle's conclusion for *toe berm stone* states that "unless site-specific model tests are conducted to justify higher values of N_s, stability number should be selected based on the lower limit curves presented in figures contained in Markle (1989), and individual toe berm armor stone weights should range from a maximum 1.3 W_{50} to a minimum of 0.7 W_{50}." For *toe buttressing stone*, limited 2-D stability tests for toe buttressing a one-layer uniformly placed tribar structure, a stability number N_s equal to 1.5 should be used in a breaking wave environment.

Scour Prediction at Piles or Other Vertical Supports

Based on a 2-D laboratory study conducted to examine effects of waves and currents, Herbich et al (1984) developed a procedure for estimating scour at the base of vertical piles. The procedure provides order of magnitude estimates for both *local scour* and *total scour* as described below.

For *local scour*, S_1, in ft, which occurs in the immediate vicinity of the obstruction causing the scour:

$$\log_{10}\left(\frac{S_1}{h}\right) = -1.2935 + 0.1917 \log_{10} \beta \qquad (6)$$

$$\beta = \frac{H_o^2 \, L_o \, u_b^3 \, D_p \, [u_b + (1/T - u_b/L_o) \, H_o L_o/2h]^2}{[(\rho_s - \rho)/\rho] \, v \, g^2 \, h^4 \, d_g} \qquad (7)$$

In the above equations,

L_o = deep water wave length, ft
H_o = deep water wave height, ft
d_g = median sediment diameter, ft
ρ_s = sediment density, lb/ft^3
u_b = near bottom velocity, ft/s

D_p = pile diameter, ft
T = wave period, sec
h = depth, ft
ρ = fluid density, lb/ft^3
ν = kinematic viscosity, ft^2/s

For *total scour depth*, S_t, in ft, which occurs over a much larger area and includes local and general scour:

$$\log_{10}\left(\frac{S_t}{h}\right) = -1.4071 + 0.2667 \log_{10}\beta \tag{8}$$

In addition to the above equations, an additional parameter α, which can be used to determine whether general scour will occur:

$$\alpha = \frac{H_o^2 L_o u_b^2 [u_b + (1/T - u_b/L_o) H_o L_o/2h]^2}{[(\rho_s - \rho)/\rho] g^2 h^4 d_g} \tag{9}$$

According to the method, general scour will not occur when $\alpha < 0.02$ and total scour will be limited to that associated with local scour. The above relations were not verified using prototype data. A useful relationship between α and β is

$$\beta = \frac{u_b D}{\nu}\alpha \tag{10}$$

Scour Prediction at Submerged Pipelines

Chao and Hennessy (1972) developed a method for estimating maximum scour depth under offshore pipelines. Based primarily on 2-D flow theory, the method uses certain reasonable assumptions to obtain:

$$q_s = u_o \left(H_p - \frac{R_p^2}{2H_p - R_p}\right) \quad \text{for } H_p \geq R_p \tag{11}$$

$$\frac{q_s}{(H_p - R_p)} = u_o \left(\frac{2(H_p/R_p)^2 - (H_p/R_p) - 1}{2(H_p/R_p)^2 - 3(H_p/R_p) + 1}\right) \quad \text{for } H_p \geq R_p \tag{12}$$

$$\tau_b = \frac{f_f \rho (u_{avg})^2}{8} \tag{13}$$

In the above, q_s is the discharge through the scour hole in ft^3/sec, u_{avg} is the average velocity through the scour hole in ft/sec, u_o is the undisturbed velocity at the top of the pipe, ft/sec. H_p is the sum of the pipe radius, R_p, and maximum scour depth, both in feet. Friction factor, f_f, is obtained using a figure similar to that contained in Hales (1980) where f_f is plotted versus Reynolds Number, R, defined by

$$R = \frac{R_h \, \overline{u}}{\nu} \qquad (14)$$

In the above equation, R_h is the hydraulic radius approximated by H_p-R_p, ft, and $\overline{u}$ is mean current velocity, ft/sec. Using Chao and Hennessy's method, Herbich (1981) produced a series of charts to estimate bottom scour for combinations of sediment size, bottom current velocity, and pipeline diameter. Consult Herbich (1981) for these charts and details concerning their proper usage.

Summary and Conclusions

This paper has briefly discussed existing scour prediction techniques for various coastal scour situations. In general, prediction techniques for scour at structure toes are either rule-of-thumb methods or semi-empirical equations based on limited laboratory and field studies. Table 1 contains a summary of suggested methods for estimating maximum scour depths for coastal structure designs. Probably sufficient guidance exists for vertical walls, piles, and pipelines. Additional research is needed in the area of scour prediction methods for rubble-mound structures.

Acknowledgement

Research presented herein was sponsored by the Coastal Structure Evaluation and Design Work Program of the U. S. Army Engineer Waterways Experiment Station. The Office, Chief of Engineers, is acknowledged for authorizing publication of this paper.

References

Chao,, J. L., and P. V. Hennesy, 1972. "Local Scour Under Ocean Outfall Pipelines," Journal of Water Pollution Control Fed., Vol. 44, no 7, pp 1443-1447.

Eckert, J. W. 1983. "Design of Toe Protection for Coastal Structures," Proc. of Coastal Structures '83, ASCE, Alexandria, Va. pp. 331-341.

Fowler, J. E., 1992. "Scour Problems and Methods for Prediction of Maximum Scour at Vertical Seawalls," In publication. Technical Report CERC 90- . US Army Engineer Waterways Experiment Station, Coastal Engineering Research Center, Vicksburg, MS.

Hales, Lyndall Z., 1980. "Erosion Control of Scour During Construction," HL-80-3, Report 2, U.S. Army Engineer Waterways Experiment Station, Vicksburg, MS.

Herbich, John B., 1981. "Offshore Pipeline Design Elements," Marine Technology Society.

Herbich, John B., Schiller, Robert E., Jr., Dunlap, Wayne A. and Watanabe, Ronald K. 1984. "Seafloor Scour; Design Guidelines for Ocean-Founded Structures," Marine Technology Society.

Kraus, N. C. 1988. "The Effects of Seawalls on the Beach: An Extended Literature Review," Journal of Coastal Research, Special Issue No. 4, N. C. Krause and O. H. Pilkey, Eds., pp 1-28.

Markle, Dennis G. 1986. "Stability of Rubble-Mound Breakwater and Jetty Toes: Survey of Field Experience," Technical Report REMR-CO-1, U. S. Army Engineer Waterways Experiment Station, Vicksburg, MS.

Markle, Dennis G. 1989. "Stability of Toe Berm Armor Stone and Toe Buttressing Stone on Rubble-Mound Breakwaters and Jetties," Technical Report REMR-CO-12, U. S. Army Engineer Waterways Experiment Station, Vicksburg, MS.

Shore Protection Manual. 1984. 4th ed., 2 Vols, U.S. Army Engineer Waterways Experiment Station, Coastal Engineering Research Center, U.S. Government Printing Office, Washington, DC.

Table 1. Scour Prediction Methods For Various Scour Modes

Scour Mode	Method	Appropriate Equation(s)	Remarks
Vertical Piles	Herbich et al (1984)	$\log_{10}\left(\dfrac{S_l}{h}\right) = -1.2935 + 0.1917 \log_{10}\beta$ $\beta = \dfrac{H_o^2 L_o u_b^3 D_P [u_b + (1/T - u_b/L_o)\, H_o L_o/2h]^2}{[(\rho_s - \rho)/\rho]\, \nu\, g^2\, h^4\, d_\pi}$	Provides method for estimating local and general scour
Submerged Pipelines	Chao and Hennessy (1972)	$q_s = u_o \left(H_p - \dfrac{R_p^2}{2H_p - R_p} \right)$ $\dfrac{q_s}{(H_p - R_p)} = u_o \left(\dfrac{2(H_p/R_p)^2 - (H_p/R_p) - 1}{2(H_p/R_p)^2 - 3(H_p/R_p) + 1} \right)$	Provides order of magnitude estimates only; valid for $H_p \geq R_p$.
Vertical Seawalls	SPM (1984)	$\dfrac{S_{max}}{H_o} \leq 1$	Some laboratory results using regular waves have exceeded this rule.
Vertical Seawalls	Fowler (1992)	$\dfrac{S_{max}}{H_o} = \sqrt{22.72\, h_w/L_o} + .25$	Valid for cases where $-0.011 \leq h_w/L_o \leq 0.05$ and $0.015 \leq H_o/L_o \leq 0.04$
Smaller Rubble Mound Structures in Shallow Water	Rule of thumb SPM (1984)	$\dfrac{S_{max}}{H_o} \leq 1$	This has not been proven for rubble mound structures, but should be sufficient for revetments and shallow water rubble mound groins
Rubble Mound Structures in Deep Water	Markle (1989)	Stability number guidance is given for toe berm and toe buttressing stone design for wave stability only.	No guidance for scour depth estimation or protection is yet available

Design of a Curved Baffle Energy Dissipation Structure

R.Joseph Bergquist, P.E., M.ASCE[1] and
Charles C. Hutton, P.E., M.ASCE[2]

Abstract

A curved baffle energy dissipation structure was designed and constructed based on the results of hydraulic and analytical models. Field tests of the prototype demonstrate close comparison between predicted and actual hydraulic and structural responses. The curved impact baffle is located downstream of two parallel 1067.5 mm (42 inch) fixed cone values discharging into a delivery canal where quiescent conditions are required to prevent excessive superelevation of water surface through a nearby curve and for measurements of flow with a broad-crested weir. The results of the hydraulic and analytical models used to design the energy dissipation structure are described. Analytical investigations were performed using a three-dimensional finite element model and a time-history dynamic analysis to predict structure vibrations. Field measurements performed on the structure included the use of accelerometers and pressure transducers. Model and field measurements include flows, water levels, wave heights, velocities, impact pressures, and air entrainment. Comparison of models and prototype are presented and discussed. The design effort resulted in a very practical and economic structure capable of dissipating energy for discharges up to 34 cms (1200 cfs) under 48.8 m (160 feet) gross head.

Introduction

Two 2133.6 mm (84 inch) diameter pipes were constructed to convey water from an existing canal down a hillside to a new canal about 122 m (400 feet) away. The discharge through the pipes is controlled by two 1067.5 mm (42 inch) diameter and

---

[1] Senior Water Resource Engineer, ECI, a Division of Frederic R. Harris, 5660 Greenwood Plaza Boulevard, Suite 500, Englewood, Colorado 80111, Phone (303) 773-3788

[2] Vice President-Design, ECI, a Division of Frederic R. Harris, 5660 Greenwood Plaza Boulevard, Suite 500, Englewood, Colorado 80111, Phone (303) 773-3788

one 304.8 mm (12 inch) hooded fixed-cone valves located in a regulating structure situated at the upstream end of the lower canal. The two 1067.5 mm (42 inch) valves each discharge through 2.95 m (9.67 foot) diameter cylindrical steel lined concrete hoods. The discharged waters enter the concrete lined canal in the plunge pool area about 183 m (600 feet) upstream of the curve in the canal and 457 m (1500 feet) upstream of the broadcrested measuring weir. The released waters must pass through the curve and over the weir in a tranquil manner so the design flow does not overtop the canal lining and accurate flow measurements can be made at the weir.

During initial testing it was found that the cylindrical hoods did not adequately dissipate the energy created by the operation of the valves. The discharging jet of water, estimated from field data to have a velocity of 20 to 23 mps (60 to 70 feet per second), swept the plunge pool waters downstream and caused supercritical flow conditions in the canal. The hydraulic jump that resulted was unsteady. A significant amount of turbulence developed in the canal before the curve. Waves rose up the sloping canal side on the outside of the curve, overtopped the lining and eroded the embankment material above the lining.

The lack of adequate energy dissipation necessitated the design and construction of a stilling basin energy dissipator to dissipate the energy contained in the discharge of the valve jets. The energy reduction had to be great enough to achieve: 1) tranquil flow in the canal; 2) passage of design flow without overtopping of the canal lining; and 3) provision of flow conditions whereby accurate flow measurements can be made at the downstream measuring weir.

Model Study Preparation

A thorough literature search was made to determine the types of structures that should be considered for dissipating efficiently the energy in the water jet discharge. The literature search identified a number of alternative stilling basins and energy dissipators that were studied in more detail for adaptation to the existing facility. Preliminary hydraulic designs were prepared for U.S. Bureau of Reclamation (USBR) Type II, Type VI and Type VIII stilling basins, a box-type structure, and impact-type energy dissipators. Also investigated was a combination stilling basin and orifice outlet control structure.

The most promising design, however, was the Enders Dam Outlet Works Stilling Basin which was model studied and constructed by the USBR in 1948. This structure is a combination of a curved wall impact-type energy dissipator and a stilling basin. Preliminary evaluation revealed that its principal advantage is that it achieves good energy dissipation for the jet discharge from a valve with a structure that is shorter and shallower than a conventional stilling basin. These aspects translate into worthwhile cost savings.

More comprehensive comparative calculations of impact alternatives provided a beginning point for physical modelling. Field testing of the fixed-cone valves with cylindrical hoods was performed to obtain data on discharge jets velocities, forces and trajectories.

Model Study

The model study was conducted to determine the most appropriate impact basin between the vertical and curved wall impact-type of energy dissipators and establish the required dimensions to achieve the desired flow regime. Model studies were undertaken at a scale of 1:7 of the existing conditions as well as with energy dissipation structures investigated. Each fixed-cone valve in the model was preceded by a pipe 20 diameters in length to establish a flow with uniform velocity. Scoping tests performed with existing conditions assured similarities between model and prototype. The variables monitored for existing condition testing included air entrainment; velocities and flow patterns; and water jet pressure profiles, spectra and impact points.

Upon completion of the scoping tests a portion of the canal in the model was removed and the various perspex energy dissipation basin were installed and studied. The first basin investigated was a vertical impact wall concept, the second was the curved impact wall concept. Two different positions of the vertical baffle were tested, but this type of structure did not prove satisfactory due to excessive splashback, vibrations, and high water levels in the chamber. The initial tests of the second concept confirmed that the same problems did not exist with the curved baffle. This basin type was studied in more detail and optimized to minimize downstream wave action while ensuring that water levels in the basin did not submerge the valves.

Modifications to the curve baffle dissipator during the optimization process raised the floor elevation by 0.9 m (3 feet); set the width to 8.9 m (29.3 feet); extended the basin length to 13.4 m (44 feet); and required a fillet under the hoods to improve the flow regime by reducing the size and intensity of turbulent eddies. Moving the baffle blocks downstream, adding a vertical end wall and lengthening the basin to 16.7 m (55 feet) reduced canal wave amplitudes to ±152 mm (6 inches).

Model study results demonstrated that:

- The rate of air entrainment increased up to discharge velocities of 10 mps (30 fps) and then leveled off. The data obtained at 17 cms (600 cfs) discharge indicated the air flow rate ranges from 15.6 to 19.8 cms (550 to 700 cfs) for approximately 100 percent entrainment by volume. On the basis of this data it was expected that the prototype discharge would entrain about 10 to 20 percent more air than the model. Therefore, the air entrainment at 48.8 m (160 feet) of head and 17 cms (600 cfs) was expected to range in the prototype from 18.7 to 24.1 cms (660 to 850 cfs) per valve. Since the average air concentrations in the flow below the baffle were generally in the range of 15 to 30 percent by volume, about 70 to 80 percent of the entrained air escaped prior to passing underneath the baffle. The average prototype velocities underneath the baffle for 34 cms (1200 cfs) was expected to range from 4.5 to 5.2 mps (13.6 to 15.7 fps). Thus air entrainment was not considered a problem.

- The cumulative measured force applied to the baffle by both jets at 34 cms (1200 cfs) was 90 to 100 kips. When discharges were reduced to 17 cms (600 psf) the cumulative measured force applied to the baffle was 65 kips. The force varied directly with total discharge regardless of distribution between the valves.

- Mean recorded impact pressures on the baffle were 89.6 kp (13.0 psi) for 17 cms (600 cfs), 71.7 kp (10.4 psi) for 11.3 cms (400 cfs), and 32.8 kp (4.7 psi) for 5.7 cms (200 cfs). The relatively low mean pressures are indicative of the high air content of the jet. No pressures fell below atmospheric.

Analytical Studies

Final dimensions and configuration of the curved impact wall were based on the results of the model study and analytical studies including stability, structural and vibration analysis. The stability studies investigated flotation, overturning, sliding and bearing pressures on the foundation material. The structural studies determined the concrete thicknesses and reinforcement required to resist the imposed design forces. A three-dimensional finite element analysis was performed to determine the moments and forces in walls and slabs.

Vibration studies of dynamic loadings were conducted to determine if the stresses and motions induced in the structure by the pulsating valve jet cause excessive deflection which could result in structural cracking or failure due to overstressing or resonance. Data was collected during the model study on time variation of the magnitude of the valve jet force for free discharge and for response of the model wall. These data were used to determine the potential deflections and stresses in the structure due to the impact from the valve jet. Simplified dynamic studies were completed first to develop an understanding of the basic behavior of the curved baffle wall and for comparison with more detailed analysis. A three-dimensional finite element analysis of the isolated curved baffle included spectral analysis of hydraulic model transducer force measurements, modal analysis, spectrum analyses, time-history analysis, and a harmonic response analysis.

Results of the vibration studies include:

- The first three natural frequencies for a pinned support condition and the first three natural frequencies for fixed supports are shown in Table 1.

- Maximum displacement of 0.28 mm (0.0110 inches) for maximum discharge loading.

- Vertical displacement versus time for the time history analysis are negative for the entire loading.

An additional three-dimensional finite element analysis of the full structure with curved baffle, walls, and slabs was completed to verify modal frequencies and stresses. The natural frequencies in the baffle obtained with this analysis are also shown in Table 1.

These results indicated that vibrations should not be a problem because the jet pressures are always positive and are not large enough to create stresses to overcome those due to gravity.

Recommendation

The recommended dissipator dimensions based on the hydraulic model and analytical studies are shown in Figure 1. The basin has a width of 9.7 m (30 feet). This structure resulted in a net energy loss of approximately 68.7 m (225.3 feet) of head per pound of liquid. It is slightly wider and longer than recommended by the model study to incorporate the 304.8 mm (12 inch) valve and further reduce wave amplitudes in the canal

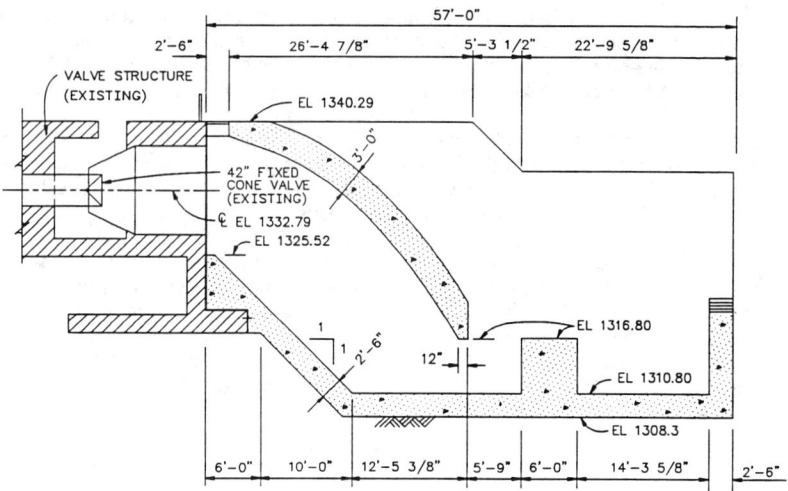

Figure 1: Curved Baffle Dissipator

Prototype Measurements

During startup testing of the curved impact dissipator a series of measurements were made to compare the model and analytical results with the prototype. These included vibration monitoring and modal vibration tests. The vibration monitoring tests included time history and frequency spectra of baffle vibrations during delivery water tests. Modal vibration tests were performed on the baffle during dry conditions to determine its natural resonant frequencies.

The maximum recorded vertical vibration amplitudes were about 0.028mm (0.0011 inches) of peak to peak displacement for maximum loading conditions. The maximum recorded horizontal vibrations were approximately the amplitudes. The general trend was for vibration amplitude to increase with delivery flow. The time history records indicated that typical peak to peak displacements are approximately 1/2 of the maximum displacements. Modal vibration tests were performed using an impulse sledgehammer to excite the baffle. The responses to the impacts were sensed by seismic accelerometers to be at frequencies shown in Table 1.

Comparison

As demonstrated by the following tabulated data there was a good comparison between model and prototype.

Table 1: Comparison of Modal Frequencies				
Mode Number	3-D Finite Element Models			Prototype (Hz)
	Walls & Baffle (Hz)	Isolated Baffle		
		Fixed (Hz)	Pinned (Hz)	
1	26.0	33.7	19.8	31.0
2	30.0	36.3	23.8	32.1
3	37.1	53.2	43.5	56.7

Summary

An exhaustive literature search, model study, and analytical analysis resulted in an efficient curved baffle energy dissipator. Excellent comparison was found between the hydraulic model analytical studies and prototype behavior and measurements. The structure functions as predicted and meets all the stated goals for an energy dissipator for this situation. Net energy loss approximated 68.7 m of head leaving only 2.3 m of head (2.1 m water depth + 0.2 m velocity head) at the start of the canal. Wave amplitudes in the canal were ±152 mm. Air entrainment proved not to be a problem.

References

1) Department of Interior, United States Bureau of Reclamation, Hydraulic Model Studies of the Enders Dam Spillway and Outlet Works - Missouri Basin Project - Hydraulic Laboratory Report No. Hyd. 252, Dec. 1948.

2) Department of Interior, United States Bureau of Reclamation, Hydraulic Model Study of Fixed-Cone Valve Energy Dissipator - Central Arizona Project Bypass Structure, Waddell Pumping-Generating Plant - Denver, Colorado, Aug. 1986.

3) Department of Interior, United States Bureau of Reclamation, Hydraulic Design of Stilling Basin for Pipe or Channel Outlet - Research Report No. 24.

Simulation of Rapid Reservoir Drawdown
for Flood Control, Cowlitz Falls Project

by
Robert H.A. Janssen[1], Associate Member, ASCE,
and Frederick A. Locher[2], Member, ASCE

Abstract

The Cowlitz Falls Dam is a run-of-river project, currently under construction on the Cowlitz River in Washington State, 23 Km downstream from the town of Randle. The requirements for reservoir drawdown during floods were that flood levels at Randle should not exceed those that would have occurred before construction of the dam, and that the rate of rise in water surface elevation downstream from the dam should not endanger anyone fishing or camping along the river banks. Initial steady state and sedimentation studies were used to determine the necessary water surface elevations for drawdown. The results of the reservoir drawdown simulation using the DAMBRK and UNET programs, and the drawdown procedure developed, are presented in this paper.

1. Introduction

Cowlitz Falls Dam is currently under construction at river kilometer 142.4 on the Cowlitz River in southwestern Washington State. As shown in Figure 1, the town of Randle is located about 23 Km upstream from the dam site, and Riffe Lake, a reservoir formed by Mossyrock Dam, is located about 5 Km downstream from the dam site. The total drainage area

---

[1]Hydraulic Engineer, Geotechnical and Hydraulic Engineering Services, Bechtel Corporation, San Francisco, CA 94119

[2]Chief Hydraulic Engineer, Geotechnical and Hydraulic Engineering Services, Bechtel Corporation, San Francisco, CA 94119

above the dam is about 2667 km², and the long term average annual flow at the damsite is 132 m³/s. The Cispus River, a major tributary with a drainage area of 850 Km², joins the Cowlitz River approximately 2 Km upstream from the dam site.

The dam is a run-of-river project for power generation, and will be a 213 m long, 44.2 m high concrete gravity structure, with the crest elevation of 268.8 m, impounding 12.3 million cubic meters. In order to pass sediment during floods, the dam incorporates two low-level sluices, with inverts at El. 234.7 m. Four spillway bays will pass major flood flows. The crests of Bays 1 and 4 will be at El. 251.8 m, and the crests of Bays 2 and 3 will be at El. 253.0 m.

The construction license for the project required that the flood levels in the Randle area should not exceed the flood levels that would have occurred prior to construction of the dam for all floods up to the 100-year flood. This means that the reservoir must be drawn down during floods to minimize backwater effects and sediment deposition.

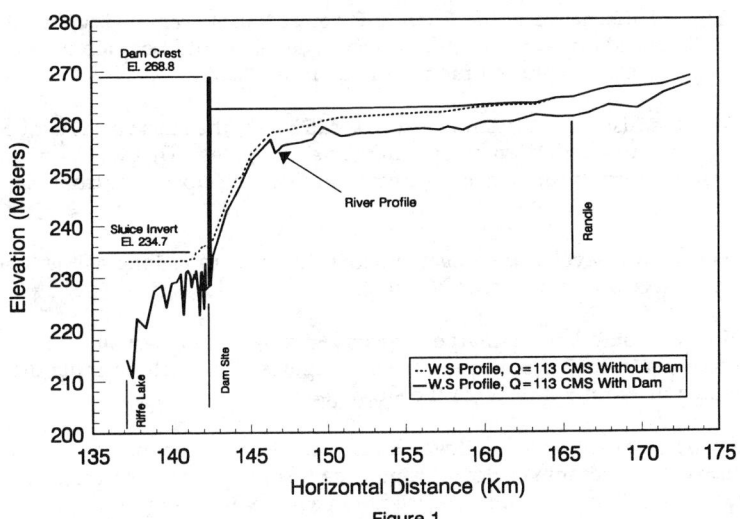

Figure 1
River and Water Surface Profiles

2. Initial Studies

Flow profiles upstream from the damsite were calculated using the U.S. Army Corps of Engineers computer program, HEC-2. These profiles showed that with the reservoir at the normal operating level of 262.7 m, the backwater effects will extend upstream from Km 151. However, if the reservoir level is at El. 259.7 m, then the Cowlitz River will flow uncontrolled upstream from Km 151.

A separate study of sediment transport in the river, using the U.S. Army Corps of Engineers computer program, HEC-6, showed that drawing the reservoir down below El. 259.7 m during large floods will increase the volume of sediment bypassed through the dam. However, the turbines cannot continue to operate if the reservoir level is below the cooling water intakes at El. 256.6 m. Sediment bypass, while still generating power, is therefore maximized if the reservoir is drawn down to El. 256.6 m.

Based on the results of the above studies, the following guidelines were initially developed for reservoir drawdown for all floods up to the 100-year flood. The turbine units were assumed to be operating throughout this procedure. Should this not be the case, then spillway Bay 4 will be partially opened to make up for the turbine flow of 283 m^3/s.

- While the discharge in the Cowlitz River at Randle remains within the bank full capacity of 425 m^3/s, the reservoir level should be maintained within the normal operating range of 262.1 m to 262.7 m.

- When the discharge at Randle exceeds 425 m^3/s, the reservoir must be drawn down to El. 259.7 m by opening the sluice gates. This will prevent backwater effects from causing incremental flooding upstream from Km 151.

- The reservoir level will be maintained at El. 259.7 m as long as the total dam discharge does not exceed 765 m3/s.

- When the total dam discharge exceeds 765 m3/s, the reservoir will be drawn down to El. 256.6 m using the sluice gates. This will promote the bypassing of sediment through the sluice gates.

- As the flood progresses, spillway Bays 1 and 4 will be opened as necessary. The combined flow through the turbines, the sluice gates, and spillway Bays 1 and 4, will be sufficient to pass any floods up to the 100-year flood.

3. Reservoir Drawdown Simulation

Two computer models were used to further develop the reservoir drawdown procedure, according to the guidelines stated previously. The first was DAMBRK, developed by the U.S. National Weather Service, and the second was UNET, developed by the U.S. Army Corps of Engineers. A 36 Km reach of the Cowlitz River was modelled, from Randle to Riffe Lake. For both models, the Cispus River was treated as a simple side inflow hydrograph. The inflow hydrographs at the upstream end of the Cowlitz River, and for the Cispus River, were derived from data for the December 1977 flood of record. The peak of this flood was about equal to the 100-year peak flow of 2590 m^3/s.

The purpose of these simulations was to develop a drawdown procedure that could be performed for all floods up to the 100-year flood. This involved demonstrating that the post-project flood levels the in the Cowlitz River, upstream from Km 151, will not exceed the pre-project levels, and ensuring that the rate of rise of the water surface elevation downstream from the dam will not endanger anyone fishing or camping along the river banks. Our initial simulations yielded the following observations:

- The sluice gates have sufficient capacity to lower the reservoir to the prescribed levels well before the flood wave can travel from Randle to the dam.

- The rate of rise of water levels downstream from the dam can be controlled by the time taken to open the sluice gates.

- Adjustment of the flow criterion for drawdown from El. 259.7 m to El. 256.6 m was necessary to draw the reservoir down fully to bypass sediment.

4. Discussion of Final Results

Based on the results of the initial simulations, a drawdown procedure was developed for all floods up to the 100-year flood. During the initial drawdown stage, the sluice gates were gradually opened over 3 hours in order to limit the rate of rise of the downstream water surface elevation. The final drawdown stage was initiated when the total dam discharge reached 623 m^3/s in order to ensure that the reservoir could be drawn down before the arrival of the incoming flood wave. The reservoir level and dam discharge for the simulations of this drawdown procedure are plotted in Figures 2 and 3 respectively.

Initial drawdown started at T = 4 hours, and the sluice gates were opened in 3 hours. The reservoir was drawn down from the normal operating level at El. 262.7 m, to El. 259.7 m, in about 4 hours. During this time, the water surface level downstream from the dam rose almost 2.7 m, at a rate of 0.9 meters per hour.

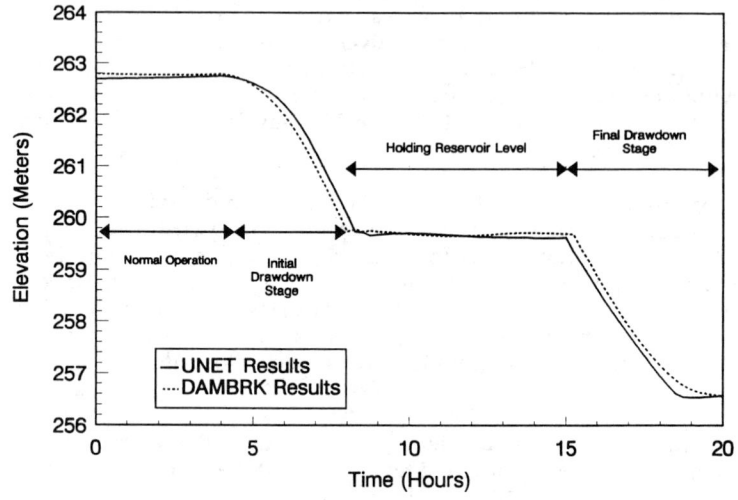

Figure 2
Reservoir Water Surface Elevations During Drawdown

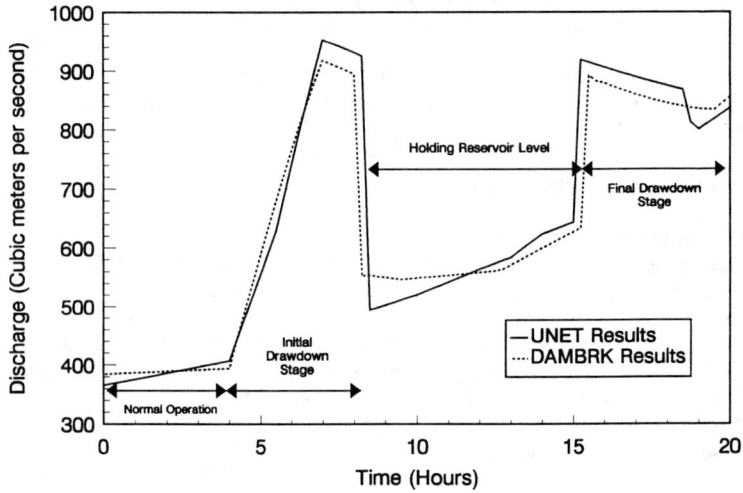

Figure 3
Total Discharge From Dam During Drawdown

As the reservoir level reached 259.7 m, the sluice gates were partially closed in order to maintain this level until the total dam discharge reached 623 m³/s at T = 15 hours. By fully opening the sluice gates at this time, the reservoir level was drawn down to El. 256.6 in about 3.5 hours. The reservoir level was held at El. 259.7 m for about 8 hours, which lend some flexibility to the drawdown procedure. It would be possible to either extend the initial drawdown stage, or start the final drawdown stage earlier.

The results presented here are for drawdown during a 100-year flood. For smaller floods, once the discharge at Randle drops below bank full capacity, all gates at the dam will be closed in order to allow the reservoir to fill up again. The drawdown procedure presented here is therefore applicable to all floods up to the 100-year flood.

It was not possible to accurately model the sluice gate rating curves in the DAMBRK model. This is reflected by the difference between the total dam discharges calculated by UNET and by DAMBRK, as shown in Figure 3. However, the results in Figure 2 shown that this did not make a significant difference to the reservoir levels, or to the timing of the drawdown procedure. Overall, the UNET program was easier to use, especially with its graphical presentation of results.

5. Conclusions

1. Drawdown of the Cowlitz Falls reservoir, for all floods up to the 100-year flood, can be performed using a single drawdown procedure.

2. The reservoir can be drawn down using only the sluice gates.

3. The flood levels at Randle, 23 Km upstream from the dam site, will not exceed those that would have occurred prior to construction of the dam.

4. The rate of rise in water surface elevation downstream from the dam, is controlled by the opening rate of the sluice gates. Hence, a rate of rise criterion and feedback control of the gates is not necessary.

PIER SCOUR ON THE SOUTH SASKATCHEWAN RIVER

Ferdous Ahmed[1], M. A. Sabur[2], and D. D. Andres[2]

INTRODUCTION

A continuing program to measure river bed scour near bridge piers and abutments on Alberta rivers has been on going for over thirty years. The objectives are to verify the existing formulae or develop new design criteria for scour prediction on both sand and gravel bed rivers. A number of priority bridge sites throughout the province have been selected on the basis of river characteristics, bridge configuration, and accessibility; and when a flood is forecast every effort is made to undertake surveys to quantify the flood hydraulics and measure the pertinent scour characteristics.

In 1990 a significant flood was forecast for the lower South Saskatchewan River in southeastern Alberta. A field crew was dispatches to the Hwy 41 bridge site, located downstream of Medicine Hat and about 15 km upstream of Alberta/Saskatchewan border. Three separate surveys at various discharges over a duration of the flood were carried out to document the channel characteristics, the bed forms, the water surface slopes, and the scour characteristics. This paper presents the results of the surveys, documents the hydraulic characteristics, and compares the measured pier scour depths with those calculated using some of the more contemporary formulae.

---

[1]Graduate Student, Department of Civil Engineering, University of Alberta, Edmonton, Canada T6G 2G7; on leave from the Institute of Flood Control and Drainage Research, Bangladesh University of Engineering and Technology, Dhaka 1000, Bangladesh.

[2]Alberta Research Council, Edmonton, Canada T6H 5X2.

BACKGROUND

A 6.4 km long study reach was established on the South Saskatchewan River at the Hwy 41 bridge site located downstream of Medicine Hat. The drainage area at this location is about 66,000 km^2, although there is substantial storage of both hydropower and irrigation in the upstream part of the basin. A Water Survey of Canada gauge has been located at this site since 1966, to enhance the flow records that has been collected at Medicine Hat since 1912. There is no difference in the flood characteristics between the two sites, with the mean annual flood, the ten year flood, the twenty year flood, and the one hundred year flood estimated to 1000, 2750, 3500 and 5250 m^3/s respectively (Alberta Environment; 1986).

The river valley is located within an open plain. The valley is stream cut with frequent slumps and no evidence of a valley flat. Two levels of fragmentary terraces are apparent. The channel is obviously entrenched, and is relatively stable in the lateral direction although it exhibits both point bars and occasional islands. The sinuosity is 1.01. The channel banks are composed of gravel overlain by silts and easily erodible shales. The channel bed is composed of sands and gravels with a D_{50}, D_{65}, and D_{90} of 0.25, 0.42, and 3.0 mm respectively. This material extends down to at least 10 m below the existing bed level (Kellerhals, Neill, and Bray; 1972).

The existing four-span bridge was constructed in 1967. The length of the bridge is 24 m and its three rectangular piers are situated within the low flow waterway. Both abutments are protected by rock-faced guidebanks. At bankful discharge there is no constriction of the waterway due to the guidebanks. The piers are aligned approximately parallel to the flow, although the bridge deck is skewed 30 degrees to the flow. The width and length of these piers are 2.6 and 21 m respectively. The piers are situated on piles which extend down to an elevation of about 562.7 m or about 20 m below the nominal existing bed level. The elevation of the bottom of the pile caps is 576.1 m.

RIVER SURVEYS

Three complete river surveys were undertaken during the course of the flood. The first survey was made at a flow of 1280 m^3/s on June 3/4, 1990. This was within a few hours of the peak discharge of 1310 m^3/m. The second survey was done three days later at a discharge of 1230 m^3/s followed by a low water survey on July 17 at a discharge of 210 m^3/s. Each survey documented the bathemerty of nine channel cross sections which were established at salient locations throughout the reach. The water surface slope was measured, along with bed topography from long lines established at quarter points across the channel along the whole reach. Finally, bed material samples were taken to confirm the bed characteristics and scour depths were measured around the piers and guidebanks. All the hydrographic measurements were taken using a boat and echo-sounder. In addition, Water Survey of Canada carried out a discharge measurement on June 6, thereby providing accurate estimates of the discharge over the course of the flood.

HYDRAULIC CHARACTERISTICS

The reach-averaged hydraulic characteristics for each of the surveys are summarized in Table 1. A more detailed description can be found in Ahmed, Sabur, and Andres (1993), however some of the salient features are discussed herein. The water surface slope remained essentially constant for all three surveys and was measured to be 0.00021. At the peak measured discharge or at the flow responsible for the bulk of the scour around the piers, the flow was approximately bankful. The mean velocity was 1.68 m/s, the mean flow depth was 3.3 m, and the bed shear stress was 6.8 Pa. For a characteristic bed material size of 0.4 mm the critical shear stress is in the order of 2 Pa; therefore one would expect to see a significant amount of bed material transport. The literature suggests the dunes would be the predominant bed form and this was confirmed from the long line surveys. Measurements showed that about 80% of the bed within the study reach exhibited dunes with typical lengths of 30 m and amplitudes of about one meter.

Table 1 Reach-averaged hydraulic parameters

Survey date >>>	June 3/4, 1990	June 7, 1990	July 17, 1990
Discharge, Q (m^3/s)	1280	1230	207
Water surface slope, S	0.00021	0.00021	0.00021
Mean depth, H (m)	3.3	3.3	1.6
Top width, W (m)	235	235	190
Water elevation (m)	587.578	587.516	585.109
Cross-sectional area (m^2)	760	750	350
Maximum depth (m)	5.0	5.0	2.8
Mean Velocity, V (m/s)	1.68	1.64	0.60
Bed shear, τ_0 (N/m^2)	6.8	6.8	3.3
Froude number, Fr	0.30	0.3	0.15
Manning's n	0.019	0.020	0.033

PIER SCOUR

Scour measurements along both sides and at the nose of all three piers were taken on June 3, June 7 and July 17, 1990. It was observed that the scour depth was essentially constant for all piers for all the surveys. The local hydraulic characteristics for peak discharge of 1310 m^3/s and a measured water elevation of 587.9 m at each of the piers is shown in Table 2. The scour depths at the three pier locations were computed by different methods using the local hydraulic conditions prevailing at each of them. This procedure is expected to give better estimates than those obtained using hydraulic parameters averaged over the entire cross section.

A number of equations are available to compute equilibrium scour depth, d_{se}, at bridge piers. No attempt is made here to describe them since a number of good reviews are already available (Breusers et al. 1977; Dargahi 1982; Raudkivi 1991). Only the more recent and those frequently used approaches were evaluated herein. These include: Laursen and Toch (1956), Laursen (1960), Shen et al. (1966), Carstens (1966), Shen et al. (1969), Coleman (1971), Hancu (1971), Breusers et al. (1977), Jain and Fischer (1980), Qadar (1981), Jain (1981), and Melville and Sutherland (1988) as described in the above-mentioned reviews. It should be noted that for the design method of Melville and Sutherland (1988), a D_{90} of 3.0 mm was taken as the armor-forming size. For rectangular pier with L/B ratio of 8, K_S was taken as 1.11.

For the methods of Hancu (1971), Breusers et al. (1977), Jain and Fisher (1980) and Jain (1981), the mean velocity at incipient sediment motion was estimated using Sheilds' diagram and the conventional relationship between the shear velocity and the mean velocity. For a D_{50} of 0.42 mm this procedure gave a critical shear velocity of 0.014 m/s and critical mean velocity of 0.225, 0.340 and 0.247 m/s for for the left, center and right pier respectively.

Table 2   Measured pier scour characteristics

	Left	Center	Right
Ambient bed elevation (m)	584.5	584.5	584.8
Maximum scoured elevation (m)	582.8	583.1	583.6
Maximum scour depth (m)	1.7	1.4	1.2
Nominal approach depth (m)	3.0	3.0	2.7
Local bed elevation (m)	583.24	581.78	584.76
Local depth (m)	4.26	5.72	2.74
Local velocity (m/s)	1.32	1.99	1.45
Local discharge intensity (m^3/s/m)	5.62	11.39	3.97
Local Froude Number	0.204	0.265	0.280
Dimensionless scour depth (d_S/B)	0.65	0.54	0.46

CONCLUSIONS

The equilibrium scour depths as predicted by different methods are shown in Figure 1. Two formulae, namely Hancu (1971) and Shen et al. (1966), gave values considerably lower than the measured value. Carsten's (1966) method gave good results on average but it underestimated the scour at the left pier. Most of the other methods gave scour depths much higher than the measured depth. The best result, in this case, was that given by Jain's (1960) method, which gave estimates that are around 25% greater than what was measured.

This paper summarizes the measured scour levels characteristics around piers for an unconstricted bridge across a mobile sand bed river. Most of the conventional formulae overpredicted the expected scour, some by as much as 500%. The main reason of discrepancy appears to be as follows.

The scour depth increases with time until an equilibrium is achieved which all the formulas are meant to compute. Laboratory investigations suggest that the time required to develop equilibrium scour is generally much greater than the duration of a typical flood. Therefore, the equations tend to overestimate scour depth.

ACKNOWLEDGMENTS

This paper is part of a continuing research program to study river processes in Alberta, under the auspices of the Alberta Co-operative Research Program in Surface Water Engineering. Participants of this program are the Alberta research Council, Alberta Transportation and Utilities, and the Civil Engineering Department of the University of Alberta. The authors would like to acknowledge J. Thompson, B. Trover, P. Mostert, and S. Procyk for carrying out the field surveys.

REFERENCES

Ahmed, F., Sabur, M. A., and Andres, D. D. (1993): *Evaluation of Bed Scour and Related Flow Processes of the South Saskatchewan River near Sandy Point, Alberta,* Report of Alberta Research Council, Edmonton, Canada. (to be published)

Breusers, H. N. C. et al. (1977): *IAHR J. Hyd. Res.* **15**(3): 211-252.

Carstens, M. R. (1966): *ASCE J. Hyd. Div.* **92**(HY3): 13-36.

Coleman, N. L. (1971): *Proc. 14th IAHR Congress* **3**: paper C37.

Dargahi, B. (1982): *Local Scour at Bridge Piers - A Review of Theory & Practice,* Bulletin No. TRITA-VBI-114, Royal Institute of Technology, Stockholm.

Hancu, S. (1971): *Proc. 14th IAHR Congress* **3**: 299-313.

Jain, S. C. (1981): *ASCE J. Hyd. Div.* **107**(HY5): 611-625.

Jain, S. C., and Fischer, E. E. (1980): *ASCE J. Hyd. Div.* **106**(HY11): 1827-1842.

Kellerhals, R., Neill, C. R., and Bray, D. I. (1972): *Hydraulic and geomorphic characteristics of rivers in Alberta,* Report No. 72/1, Alberta Research Council, Edmonton.

Laursen, E. M. (1960): *ASCE J. Hyd. Div.* **86**(HY3): 39-54.

Laursen, E. M., and Toch, A. (1956):*Scour around Bridge Piers and Abutments,* Bulletin No. 4, Iowa Highway Research Board.

Melville, B. W., and Suthreland, A. J. (1988): *ASCE J. Hyd. Engrg.* **114**(10): 1210-1226.

Neill, C. R. (1964): *River-Bed Scour: A Review for Bridge Engineers,* Technical Publication No. 23, Canadian Good Road Association, Ottawa.

Qadar, A. (1981): *Proc. Inst. Civil Engnr.* **71**(2): 739-757.

Raudkivi, A. J. (1991): chapter 5 in H. N. C. Breusers and A. J. Raudkivi, *Scouring: IAHR Hydraulic Structures Design Manual 2,* Balkema, Rotterdam.

Shen, H. W, et al. (1966): *Mechanics of Local Scour,* Report No. CER66HWS22, Colorado State University.

Shen, H. W., et al. (1969): *ASCE J. Hyd. Div.* **95**(HY6): 1919-1940.

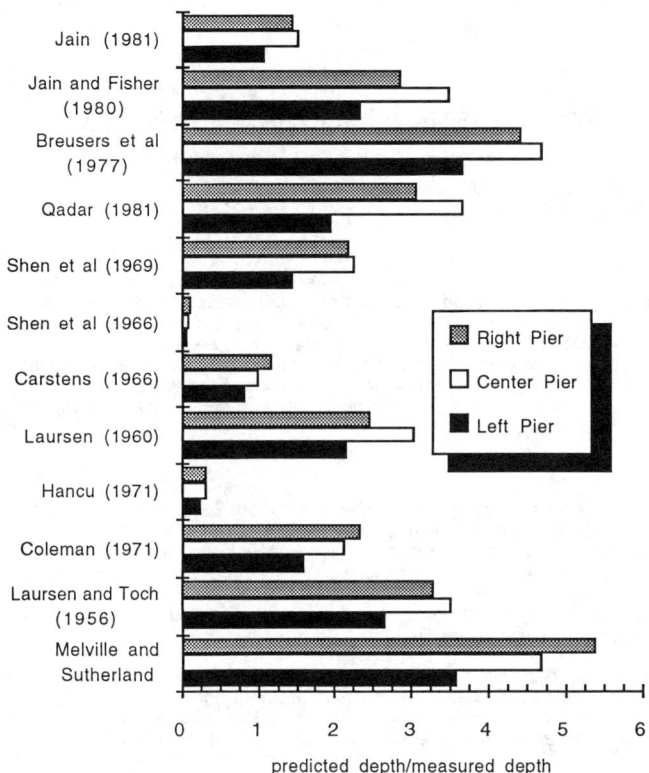

Figure 1   Comparison of predicted scour depths

SCOUR AT A BRIDGE OVER THE WELDON RIVER, IOWA

By Edward E. Fischer[1], Member, ASCE

ABSTRACT

Contraction scour at the State Highway 2 bridge over the Weldon River in south-central Iowa was caused by a flood of record proportions on September 14 and 15, 1992. The peak discharge was 1,930 cubic meters per second, which was 4 times the probable 100-year flood used to design the bridge, and resulted in road overflow. Contraction scour exposed the pier footings, but a subsurface layer of glacial clay apparently resisted additional vertical scour and caused the scouring process to move laterally. The embankment at the left abutment was eroded away, exposing 3 m of vertical abutment piling.

INTRODUCTION

Intense thunderstorms in south-central Iowa on September 14 and 15, 1992, produced floods of record proportions that caused considerable damage. More than 75 mm of rain fell over a large area (fig. 1). Floodwaters closed many highways and local roads, including the Interstate 35 crossing of the Thompson River in southern Decatur County (fig. 2). The magnitudes of the floods were such that two USGS streamflow-gaging station shelters were inundated, disabling the data-collection and satellite-telemetry equipment inside.

The flooding damaged many bridges throughout the area. For example, the earth embankments at the right-bank abutments of the dual Interstate 35 bridges over the Thompson River were eroded to a nearly 1:1 slope (Darrell Coy, Iowa Dept. of Transportation, oral commun., December 1992), and the State Highway 2 bridge abutments at the Thompson River and the Weldon River sustained considerable scour damage. Highway 2 subsequently was closed to traffic at the Weldon River bridge because the embankment at the left abutment was eroded away, exposing 3 m of vertical abutment piling.

This paper presents a detailed account of the flooding and scour that occurred at the Highway 2 bridge over the Weldon River.

STORM DESCRIPTION

South-central Iowa was already very wet on September 14 when the first rainfall was recorded during the morning hours (Harry Hillaker, State Climatologist, Iowa Dept. of Agriculture & Land Stewardship, Des Moines, Iowa, oral and written commun., January 1993). Then,

---

[1] Hydrologist, U.S. Geological Survey, P.O. Box 1230, Iowa City, IA 52244.

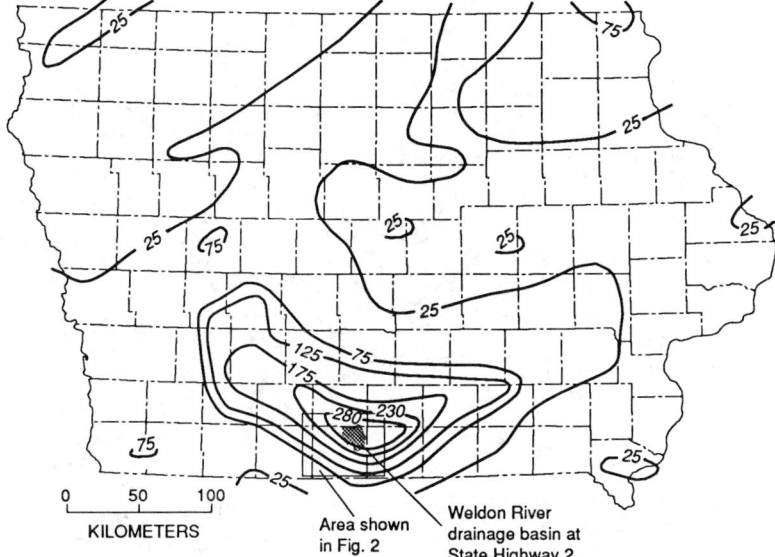

Figure 1. Areal distribution of rainfall from intense thunderstorms in Iowa, September 14-15, 1992. (Lines of equal rainfall, in mm, variable intervals, from Hillaker, 1992, p. 3.)

Towards evening, the combination of strong warm air advection on a strong southerly low level jet, deep tropical moisture, thickness diffluence, and a stalled cold frontal boundary resulted in a classic back-developing Mesoscale Convective System. As the low level jet collided with an old thunderstorm outflow boundary, deep convection began. The thunderstorms formed and moved eastward, training over the same areas most of the night. The rain tapered off by mid-morning on the 15th. (Miles Schumacher, Meteorologist, National Weather Service, Des Moines, Iowa, written commun., January 1992.)

The area of greatest rainfall was centered over Clarke, Decatur, Lucas, and Wayne Counties (see figures 1 and 2). The highest unofficial reported rainfall total for the storm was 400 mm at Van Wert, in northern Decatur County. The highest official rainfall total was 292 mm at Derby in Lucas County. A total of 236 mm of rain fell at Derby between 23:00 hrs September 14 and 05:30 hrs the next morning (fig. 3). The average rainfall intensity for this 6.5-hour period was 36.3 mm/hr. The maximum intensity was 30.5 mm in 15 minutes, which occurred between 23:00 hrs and 23:15 hrs (Harry Hillaker, written commun., January 1993).

DRAINAGE BASIN CHARACTERISTICS

A major part of the Weldon River drainage basin upstream from State Highway 2 was in the area of greatest rainfall (fig. 1). The drainage area at Highway 2 is 188 km^2 (Larimer 1957) and the drainage-basin perimeter is about 63 km. A major tributary, Jonathon Creek (drainage area, 63 km^2), drains into the Weldon River about 30 m upstream of the bridge. Stream channelization in the lower part of the basin is apparent on topographic maps.

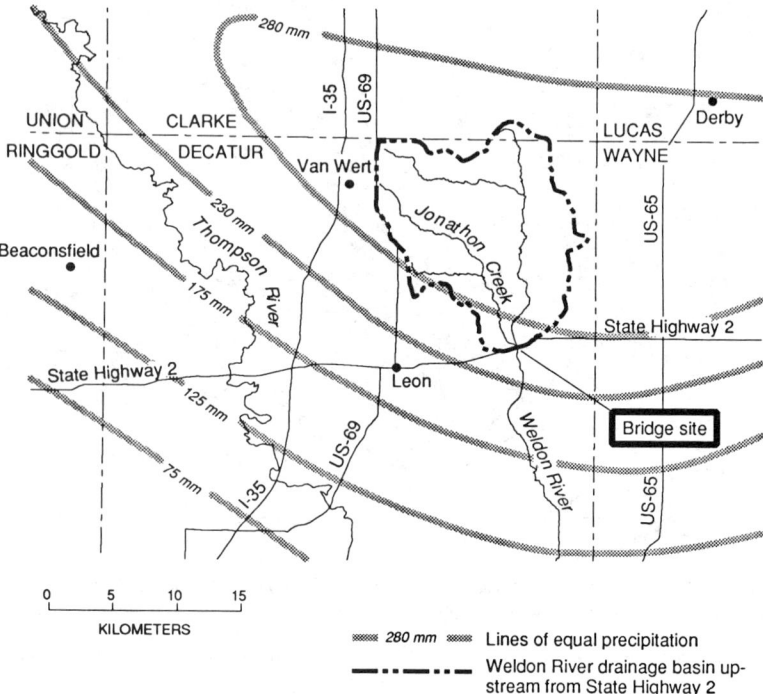

Figure 2. Location of State Highway 2 bridge over the Weldon River.

The topography consists of rolling hills that surround a wide river valley. Maximum relief, the difference between the highest point in the drainage basin and the elevation at the basin outlet, is approximately 70 m as determined from USGS 1:24,000-scale topographic maps. The width of the floodplain at the bridge is about 670 m.

Land use throughout the basin is primarily agricultural. Both Weldon River and Jonathon Creek are lined with narrow bands of trees in the vicinity of the bridge. The trees separate the stream channels from adjacent corn and soybean fields.

SITE DESCRIPTION

State Highway 2 crosses the Weldon River on a 68.0-meter concrete beam bridge supported by two monolithic piers and concrete abutments. The piers and abutments are supported by steel pilings driven into the underlying glacial clay. The bridge opening is characterized by spill-through earth embankments. The bridge replaced a truss bridge at the same location and was completed in 1985. The old piers and abutments were truncated and buried at the time of construction. The roadway is on an embankment that is approximately 4.5 m above the flood plain and is perpendicular to the principal axis of the valley. The flood plain upstream and downstream of the highway was planted in corn and soybeans.

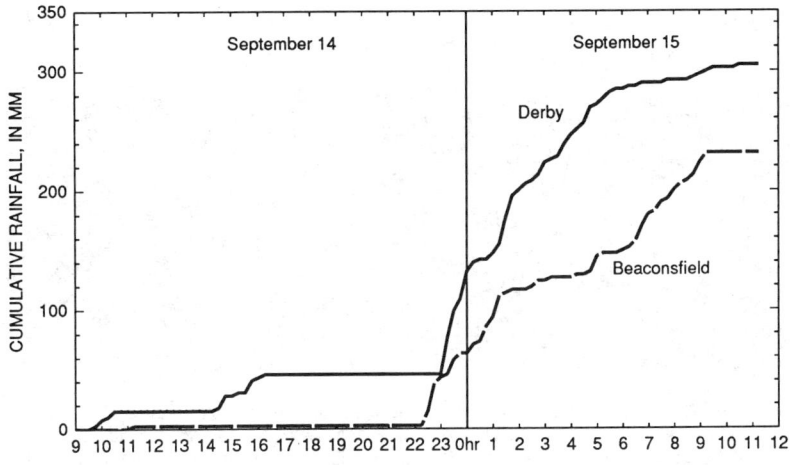

Figure 3. Cummulative rainfall for September 14-15, 1992, at Derby and Beaconsfield, Iowa. (Data provided by Harry Hillaker, State Climatologist, Des Moines, Iowa, written commun., January 1993.)

Design-flood criteria shown on the bridge plans are: 50-year flood = 370 m^3/s, 100-year flood = 480 m^3/s, and 500-year flood = 690 m^3/s (Iowa Department of Transportation-Highway Division, 1984). The prevalent methods for determining 50-year and 100-year flood values at the time the bridge was designed are described by Lara (1973) (Brad Barrett, Iowa Dept. of Transportation, oral commun., January 1993).

FLOOD DESCRIPTION

The peak discharge of the September 14-15 flood for the Weldon River at Highway 2 was 1,930 m^3/s. It was determined on the basis of a combined contracted-opening and road-overflow measurement of peak discharge (Matthai 1968, Hulsing 1968). The peak discharge was about 4 times the 100-year design flood.

The peak of the flood is estimated to have occurred between 06:00 and 07:00 hrs on September 15. Photographs taken between 07:15 and 7:30 hrs by Richard Bettin of Leon, Iowa, show the river as a large lake covering the flood plain (fig. 4). Floodwaters covered the road and bridge deck for several hours. The fall of the water surface at the highway was 1.45 m. This was the difference between the upstream and downstream high water levels determined from high water marks surveyed after the flood. The maximum water depth over the road was about 0.8 m.

SCOUR DESCRIPTION

Bed profiles measured after the flood at both the upstream and downstream sides of the bridge are shown in figure 5. Contraction scour exposed the pier footings, the entire streamward face of the left abutment, and about 3 m of the left abutment pilings. It also exposed the old abutment and pier that were covered over during construction of the bridge. According to the bridge plans, the footings were placed on glacial clay,

Figure 4. Flood of September 14-15, 1992, Weldon River at State Highway 2, Decatur County, Iowa. Photograph by Richard Bettin, Leon, Iowa. Photograph was taken looking west-southwest about 07:30 hrs on September 15. Direction of flow is from right to left.

capping 11-meter steel pilings driven into the clay. No evidence was found that any pier-footing piling was exposed. It is surmised that the clay layer was resistant to vertical scour and forced the scouring process to erode the channel at the left abutment.

The average velocity through the bridge opening during the flood peak was computed to be 3.02 m/s. Velocities were likely greater when the entire discharge flowed through the contracted bridge opening prior to overtopping the road, resulting in pressure flow conditions as the water surface came into contact with the bridge structure. Evidence that a strong jet of water issued from the contracted opening was apparent by the general devastation downstream--eroded banks, downed trees, sand deposition on the flood plain, and cobble deposition in the channel. This was in stark contrast to the situation upstream where the trees and channel appeared undamaged.

Despite the erosion at the abutment, the bridge sustained no apparent structural damage. It was closed to traffic to prevent collapse of the roadway at the left abutment.

SUMMARY

Intense thunderstorms in south-central Iowa on September 14 and 15, 1992, produced floods of record proportions throughout the area. The peak discharge of the flood in the Weldon River at the State Highway 2 bridge was 1,930 m^3/s. The peak stage was about 0.8 m above the roadway. Contraction scour exposed the pier footings down to a layer of glacial clay and eroded the channel at the left abutment, exposing 3 m of abutment piling. No evidence was found that any pier-footing piles were exposed. The bridge subsequently was closed to traffic until the scour damage at the left abutment could be repaired.

ACKNOWLEDGMENTS

Special acknowledgment is given to Brad Barrett and Darrell Coy of the Iowa Department of Transportation, Preliminary Bridge Design Section, Ames, Iowa, for providing information relating to the bridge; and to Harry Hillaker, State Climatologist, Iowa Department of Agriculture & Land Stewardship, Des Moines, and Miles Schumacher, Meteorologist, National Weather Service, Des Moines, for providing the storm data and description.

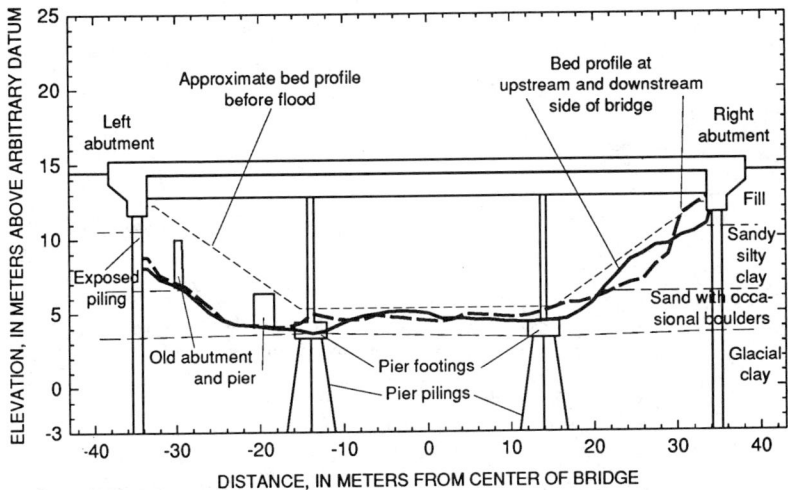

Figure 5. Channel cross section at State Highway 2 bridge over the Weldon River after the flood of September 14-15, 1992. View is looking downstream. (Bridge dimensions, soil-boring data, and approximate bed profile before flood from Iowa Department of Transportation-Highway Division, 1984.)

REFERENCES

Hillaker, H. (1992). "Rainfall totals for 60 hours ending 7 a.m. CDT September 16, 1992." *Iowa Climate Review*, Iowa Dept. of Agriculture and Land Stewardship, Des Moines, Iowa, 6(9), 3.
Hulsing, H. (1968). "Measurement of peak discharge at dams by indirect methods." Techniques of Water-Resources Investigations, Book 3, Chp. A5, USGS, Denver Federal Center, Denver, Colo.
Iowa Department of Transportation-Highway Division (1984). "Design for 0-degree skew 223-foot x 40-foot pretensioned prestressed concrete beam bridge, Decatur County." *File No. 26716, Design No. 482*. Iowa Dept. of Transportation, Ames, Iowa.
Lara, O.G. (1973). "Floods in Iowa: Technical manual for estimating their magnitude and frequency." *Bulletin No. 11*, Iowa Natural Resources Council, Des Moines, Iowa.
Larimer, O.J. (1957). "Drainage areas of Iowa streams." *Bulletin No. 7*, Iowa Highway Research Board, Ames, Iowa.
Matthai, H.F. (1968). "Measurement of peak discharge at width contractions by indirect methods." *Techniques of Water-Resources Investigations*, Book 3, Chp. A4, USGS, Denver Federal Center, Denver, Colo.

Measurement of Bridge Scour at the SR-32 Crossing of the
Sacramento River at Hamilton City, California, 1987-92

by J.C. Blodgett[1] and Carroll D. Harris[2]

Abstract

A study of the State Route 32 crossing of the Sacramento River near Hamilton City, California, is being made to determine those channel and bridge factors that contribute to scour at the site. Three types of scour data have been measured--channel bed (natural) scour, constriction (general) scour, and local (bridge-pier induced) scour. During the years 1979-93, a maximum of 3.4 ft of channel bed scour, with a mean of 1.4 ft, has been measured. Constriction scour, which may include channel bed scour, has been measured at the site nine times during the years 1987-92. The calculated amount of constriction scour ranged from 0.2 to 3.0 ft, assuming the reference is the mean bed elevation. Local scour was measured four times at the site in 1991 and 1992 and ranged from -2.1 (fill) to 11.6 ft, with the calculated amounts dependent on the bed reference elevation and method of computation used. Surveys of the channel bed near the bridge piers indicate the horizontal location of lowest bed elevation (maximum depth of scour) may vary at least 17 ft between different surveys at the same pier and most frequently is located downstream from the upstream face of the pier.

Introduction

The California Department of Transportation (Caltrans) constructed a new bridge on State Route 32 across the Sacramento River at Hamilton City in 1987 (Figure 1). An old bridge constructed in 1908 (Brice and Blodgett, 1978) was removed, including piers, in 1988. The new bridge is 842 ft long and supported by six round-nose piers that are 7 ft wide and about 11 ft long. Channel width at the bridge is about 650 ft and the skew angle of the bridge to flow is about 17°. Piers in the waterway cause a flow constriction at the bridge for most flows occurring at the site.

Types of scour that occur at this site include channel-bed scour, constriction or general scour, and local scour (Blodgett, 1989). Channel-bed scour is a naturally occurring phenomenon in alluvial channels (Figure 2). Channel-bed scour varies, depending on the

---

[1]Hydrologist, U.S. Geological Survey, 2800 Cottage Way, Room W-2233, Sacramento, California 95825
[2]Chief of Planning and liaison for Division of Structures, California Department of Transportation, 1801 30th St., Sacramento, California 95816

MEASUREMENT OF BRIDGE SCOUR 1861

Figure 1. State Route 32 crossing of Sacramento River at Hamilton City, California. Photographed by California Department of Transportation, March 1993

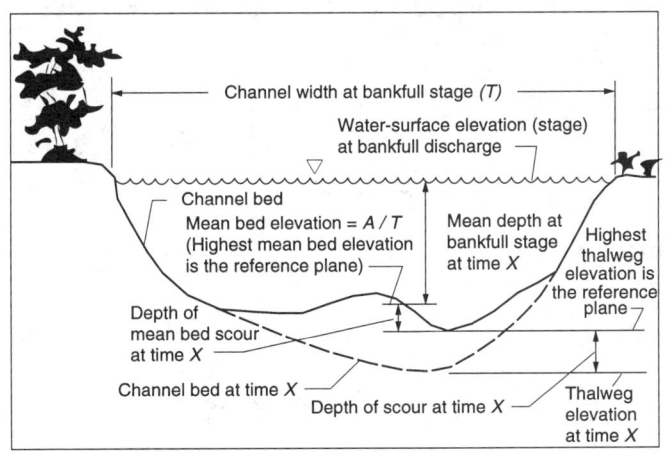

Figure 2. Sketch of channel bed scour and mean depth

magnitude and velocity of flow and the size and availability of the bed material, as described by Blodgett (1986). Constriction (general) scour is defined by Brice and Blodgett (1978) as scour that generally affects part of the channel bed at a bridge contraction and is not localized at a pier or other obstruction. Local scour (Brice and Blodgett, 1978) refers to scour that is localized at a pier, abutment, or other obstruction to flow. With adequate constriction and local scour data, the reliability of estimates of scour depth obtained by new equations, or those by Laursen and Toch (1956) and Copp and others (1988), can be compared.

The purpose of this study is to measure the various types of scour and their temporal characteristics that occur at the Sacramento River at State Route 32 bridge site near Hamilton City. Results obtained using various methods of establishing the reference plane (bed) elevation (Figure 2) from which scour is measured are presented. Where channel bed or constriction scour occurs at the bridge, a projected ambient bed elevation was estimated. The highest projected bed elevation for the sample of data is used as the reference plane from which constriction and local scour were calculated.

This study was done by the U.S. Geological Survey in cooperation with the California Department of Transportation (Caltrans), Division of Structures. The study is part of an effort to develop better methods for determining the potential for scour around bridge piers.

Results of Surveys

Hydrographic surveys were made at the new bridge site from 1987 to 1993. Water-surface and channel thalweg (lowest point in a cross section) profiles at this site have been surveyed since 1979. After demolition of the old bridge and piers, located about 100 ft upstream from the new bridge, the profile of the channel bed has changed with a scour area now located at the new bridge (Figure 3). A projected thalweg elevation at the downstream side of the new bridge (section DSN) on January 11, 1993, indicates about 0.7 ft of scour (Figure 3) caused by the bridge constriction. The thalweg and water-surface profile of the river at this site indicates also that a natural channel flow contraction, located over 1,000 ft downstream from the bridge (Figure 1), affects the channel bed elevation in the vicinity of the bridge (Figure 3).

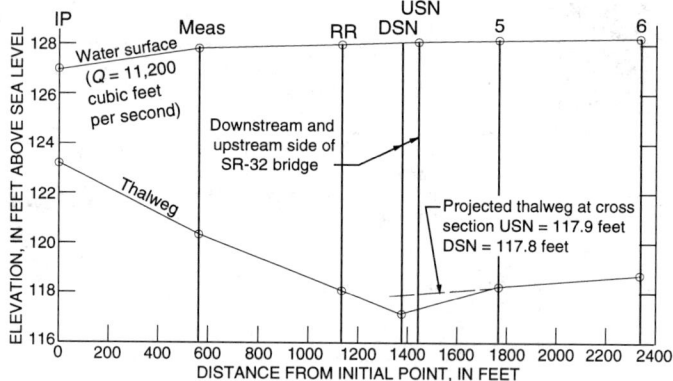

Figure 3. Profile of water surface and thalweg in study reach on January 11, 1993, Sacramento River near Hamilton City, California

Channel Bed Scour

The thalweg at the approach section (cross section 6, Figure 1) is unaffected by channel changes downstream at the bridge. All calculations of hydraulic properties at this cross section are based on a bank-full stage of 143.5 ft. This stage was used because the channel banks are not vertical, and the calculated mean bed elevation (cross-sectional area (A) divided by channel width (T), Figure 2) can vary, depending on the stage. The area of this cross section has changed less than 6% for the 11 surveys and indicates the channel size has remained in dynamic equilibrium. During 1979-93, the range of scour was 0.4 to 3.4 ft and averaged 1.4 ft.

One reference for measuring scour at a cross section is the elevation of the thalweg. Depth of scour is then calculated as the depth below the highest thalweg elevation (reference plane, Figure 2) for the period of record. Maximum scour at the thalweg for the 14-year period at cross section 6 was 1.9 ft (Table 1). These data indicate that several measurements of channel bed scour are needed over a range of streamflow conditions in order to obtain data that will most likely define average and maximum amounts of scour.

Table 1. Selected channel-bed scour data at cross sections 6 (approach section) and constriction scour at cross sections USN (upstream side of bridge), 1979-93

[--, no data]

Date	Discharge, in cubic feet per second	Channel Bed Scour			
		Mean Bed[1], in feet		Thalweg[1], in feet	
		Elevation	Scour	Elevation	Scour
Cross Section 6					
8-29-79	--	121.4	1.0	117.0	1.6
10-25-79	--	121.0	1.4	117.3	1.3
11-4-80	--	121.4	1.0	117.6	1.0
10-5-88	--	122.0	0.4	117.6	1.0
7-13-89	--	121.8	.6	118.1	0.5
10-26-89	--	119.5	2.9	116.7	1.9
1-18-90	--	122.4	--	117.5	1.1
3-27-91	--	119.0	3.4	116.8	1.8
10-1-91	--	121.6	.8	117.5	1.1
10-30-92	--	120.6	1.8	118.3	.3
1-12-93	--	121.2	1.2	118.6	--
Mean		121.1	1.4	117.5	1.16
Cross Section USN[2]					
6-24-87	12,700	123.0	0.4	115.5	2.4
10-6-88	7,170	123.1	.3	115.2	2.7
4-07-89	9,460	122.8	.6	117.7	.2
7-13-89	7,500	122.9	.5	117.5	.4
1-18-90	9,150	120.4	3.0	116.5	1.4
3-27-91	23,800	121.7	1.7	116.0	1.9
10-1-91	4,000	123.2	.2	117.5	.4
5-12-92	6,500	122.6	.8	118.0	-.1
10-30-92	5,050	123.4	--	118.4	-.5

[1]Calculated assuming bank-full conditions, stage 143.5 ft.
[2]Thalweg reference datum (117.9 ft) from projected thalweg profile of 1-11-93 (117.9 ft).

Constriction Scour

Scour caused by the constriction of flow at cross section USN (at the upstream side of the bridge, Figure 1) was measured nine times between 1987 and 1992 (Table 1). The bed elevation at the bridge was estimated on the basis of the projected thalweg profile for the 1993 survey (Figure 3). Using this reference elevation (117.9 ft), the depth of scour at the constriction ranged from -0.1 to 2.7 ft.

Local Scour

Local scour around bridge piers 3, 4, 5, and 6 (Figure 1) was measured four times in 1991-92. The depths of scour vary, depending on the reference elevation selected and method of calculation used (Table 2). Methods of establishing reference elevations in Table 2 include: A, Reference projected thalweg elevation of 117.9 ft from 1-11-93 surveys. B, Reference mean bed elevation at cross section USN of 123.4 ft from 1991-92 surveys. This method used the highest historical mean-bed elevation at cross section USN. C, Reference is concurrent surrounding stream bed elevations: 5-29-91, 120.0 ft; 10-1-91, 119.7 ft; 5-12-92, 120.5 ft; and 10-30-92, 119.8 ft (method described by Norman, 1975). D, Reference historical highest bed elevations: Pier 3, 122.7 ft; Pier 4, 120.7 ft; Pier 5, 121.3 ft; and Pier 6, 121.6 ft (method described by Jarrett and Boyle, 1986)

Method A (reference elevation is the projected thalweg) calculates the least amount of scour at the four piers. Method B (reference elevation is mean bed elevation near cross section USN) gives the greatest amount of scour. Method C (reference elevation is the highest concurrent surrounding streambed elevation) calculates local scour depths that are generally in the middle range of the results using the four methods. Method D (using the highest bed elevations near the piers) calculates depth of scour that ranges from 8 to 71% of those using Method B. The minimum and maximum local scour near the piers at the site is -2.1 ft (fill) and 10.0 ft, depending on the method of calculation used. The lowest channel bed elevation measured in 1991 and 1992 in the vicinity of the piers was 111.8 ft (10-1-91) which, using Method B, indicates a scour depth of 11.6 ft. Methods A, B, and D may include channel bed or constriction scour because these methods do not isolate depths caused by local scour.

Table 2. Variation in estimates of local scour depths at SR-32 bridge across Sacramento River near Hamilton City, California

Date	Method	Depth of scour at piers, in feet				Date	Method	Depth of scour at piers, in feet			
		Pier 3	Pier 4	Pier 5	Pier 6			Pier 3	Pier 4	Pier 5	Pier 6
5-29-91	A	0.8	0.3	-0.7	-0.3	5-12-92	A	-.8	-.8	-2.1	1.9
	B	6.3	5.8	4.8	5.2		B	4.7	4.7	3.4	7.4
	C	2.9	2.4	1.4	1.8		C	1.8	1.8	.5	4.5
	D	5.6	3.1	2.7	3.4		D	4.0	2.0	1.3	5.6
10-1-91	A	4.5	-1.7	1.0	2.9	10-30-92	A	3.3	.8	0	3.7
	B	10.0	3.8	6.5	8.4		B	8.8	6.3	5.5	9.2
	C	6.3	.1	2.8	4.7		C	5.2	2.7	1.9	5.6
	D	9.3	1.1	4.4	6.6		D	8.1	3.6	3.4	7.4

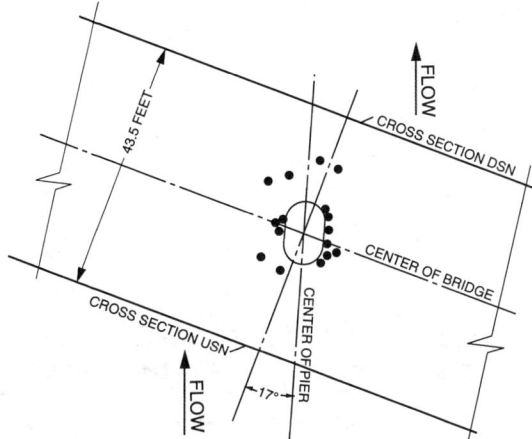

Figure 4. Composite location of points of lowest channel bed elevations in vicinity of bridge piers during years 1991-92 at SR-32 bridge across Sacramento River near Hamilton City. Note: Pier size is 7 × 11 feet

Methods A-D all reference the depth of scour to locations at the upstream side of the bridge or in front of the piers. The lowest bed elevation (or maximum depth of scour) most often occurs at locations downstream from the upstream side of the pier (Figure 4). The location of the lowest bed elevation in the vicinity of a pier is not consistent, and a maximum horizontal movement of 17 ft between these points at a given pier has been measured between 1991 and 1992.

References

Blodgett, J. C. (1986). "Rock riprap design for protection of stream channels near highway structures, Volume 1--Hydraulic characteristics of open channels," *U.S. Geological Survey Water-Resources Investigations Report* 86-4127, 33 p.

Blodgett, J. C. (1989). "Monitoring scour at the State Route 32 bridge across the Sacramento River at Hamilton city, California," in Proceedings of the Bridge Scour Symposium, October 17-19, McLean, Virginia, *Federal Highway Administration Report* No. RD-90-035, p. 211-226.

Brice, J. C., and Blodgett, J.C. (1978). "Countermeasures for hydraulic problems at bridges--Volume 1, Analysis and assessment," *Federal Highway Administration Report* No. FHWA-RD-78-162, 169 p.

Copp, H. D., Johnson, J. P., and McIntosh, J. L. (1988). "Prediction methods for local scour at intermediate bridge piers," *Transportation Research Board*, 67 Annual Meeting, January 11-14, Paper 870-1461, 23 p.

Jarrett, R. D., and Boyle, J. M. (1986). "Pilot study for collection of bridge-scour data," *U.S. Geological Survey Water-Resources Investigations Report* 86-4030, 46 p.

Laursen E. M., and Toch, A. (1956). "Scour around bridge piers and abutments," *Iowa Highway Research Board*, Bulletin 4, 54 p.

Norman, V. W. (1975). "Scour at selected bridge sites in Alaska," *U.S. Geological Survey Water-Resources Investigations Report* 32-75, 160 p.

ESTIMATING BRIDGE SCOUR IN NEW YORK FROM HISTORICAL U.S. GEOLOGICAL SURVEY STREAMFLOW MEASUREMENTS

Gerard K. Butch [1]

Abstract

Historical streamflow measurements by the U.S. Geological Survey and bridge-inspection reports by the New York State Department of Transportation are being used to estimate scour at 31 bridges in New York State. Streamflow measurements that were made before, during, or after high flows are used to estimate scour and to define hydraulic properties associated with floods. Clear-water scour is common at most sites; local scour holes that formed during high flows did not refill after subsequent high flows. The 31 streambeds are armored by gravel; median particle size ranges from 22 to 68 millimeters. Streambed elevations measured after a high flow are assumed to represent the elevations during peak flow.

Measurements at several bridges indicate scour by multiple high flows, severe floods, and debris. Three high flows at State Route 23 over the Otselic River in Cortland County produced 6.1 feet of local scour and partly exposed concrete pilings below the footing. Although the recurrence interval of each flow was less than 10 years, a 30-degree angle between the flow and the pier increased the tendency of the streambed to scour. State Route 427 over the Chemung River in Chemung County survived the 1972 flood (recurrence interval greater than 100 years) because pilings supported the undermined piers. The maximum local scour during the 1972 flood was estimated to be 5.4 feet. A local-scour hole, 2.4 feet deep before the flood, was deepened to 7.8 feet. Average contraction scour was estimated to be an additional 4.1 feet. Measurements at Bailey road over the Genesee River in Wyoming County indicate 3.2 feet of local scour. In 1989, 3.8 feet of contraction scour occurred when debris blocked a third of the bridge opening. Hydraulic analysis of the 1989 flow with debris blockage indicated an average flow velocity comparable to that of the 1972 flood (recurrence interval greater than 100 years), which destroyed the previous bridge.

---

[1] Hydrologist, U.S. Geological Survey, P.O. Box 1669, Albany, New York 12201

Introduction

In 1988, the U.S. Geological Survey (USGS) in cooperation with the New York State Department of Transportation (NYSDOT) began a 6-year study of bridge scour. Sites were selected near USGS stations along streams that contain erodible bed material or that appeared unstable from review of USGS rating curves and NYSDOT files (Butch, 1991). Locations of 31 study sites are shown in figure 1.

Historical USGS streamflow measurements are used to estimate scour because annual peak discharges at many sites have been less than mean-annual peak discharges. This paper summarizes streamflow measurements at several bridges that indicate scour by multiple high flows, severe floods, and debris. The data are assumed to represent streambed elevations during peak flow.

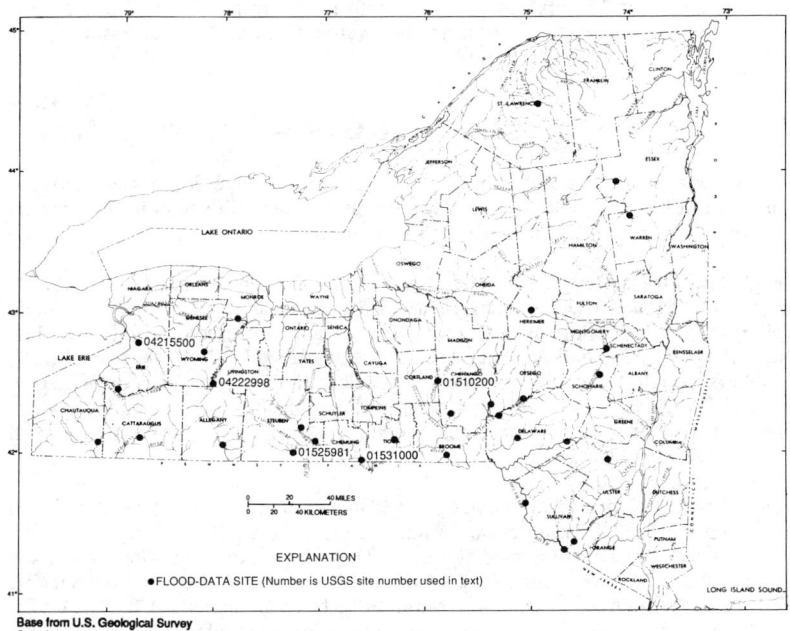

Figure 1. Locations of flood-data sites.

Methods

Data-collection methods are similar to the "limited approach" developed by the USGS in cooperative studies with the Federal Highway Administration, whereby discharge, velocity, streambed elevation, and bed-material data are collected with equipment and procedures compatible with the USGS stream-gaging program (Jarret and Boyle, 1986). Long-term data are needed to document the relation of scour to high flows. Field measurements are essential because scale-model laboratory measurements are uncertain. Streamflow measurements that were made before, during, or after high flows at 31 flood-data sites are being analyzed to estimate scour and define hydraulic properties associated with high flows. Local and contraction scour are estimated from elevations of the streambed along the upstream side of each bridge. General scour can be estimated from changes in the stage-to-discharge relation at nearby USGS streamflow-gaging stations. Scour that cannot be correlated with a specific flood is assumed to have occurred during the highest discharge since bridge construction.

Historical USGS streamflow measurements and NYSDOT bridge-inspection reports are used in the analysis because annual peak discharges at many sites have been less than the mean-annual peak discharges during the study. The historical data are assumed to represent streambed elevations during peak flows because clear-water scour is common; streambeds are armored by gravel, and the median particle size ranges from 22 to 68 mm. Local-scour depths ranged from 0.0 ft to 7.8 ft. Spread footings that were buried during construction were exposed at 18 sites, and subsequent high flows did not refill these holes.

Scour at Selected Sites

Measurements of scour caused by multiple high flows, severe floods, and debris are included in the study. Streambed elevations at five sites with significant scour are shown in figures 2A-2E (locations are shown in figure 1).

High flows at State Route 23 over the Otselic River in Cortland County (site 01510200) in 1983, 1988, and 1990 produced 6.1 ft of local scour and partly exposed concrete pilings below the footing (fig. 2A). Streambed elevations measured during a high flow on October 24, 1990 indicate that the scour hole has deepened 2.1 ft since October 17, 1990. Although the recurrence interval of each flow was less than 10 years, a 30-degree angle between the flow and the pier increased the tendency of the streambed to scour and extended the scour hole along the entire length of the pier. The hole did not refill after subsequent high flows.

State Route 427 over the Chemung River in Chemung County (site 01531000) survived the 1972 flood (recurrence interval greater than 100 years) because the undermined piers were supported by pilings. The estimated maximum local scour was 5.4 ft at pier 1. A local scour hole, 2.4 ft deep before the flood, deepened to 7.8 ft (fig. 2B). Average contraction scour was estimated to be an additional 4.1 ft. The USGS stage-to-discharge relation indicates that the streambed aggraded about 1 ft after the 1972 flood but was gradually scoured to preflood elevations. High flows in 1975 and 1979 eroded fill material that crews had added after the 1972 flood (fig. 2B, inset). Riprap was placed in the scour hole at pier 2 in 1988.

A NYSDOT inspection at Bailey road over the Genesee River in Wyoming County (site 04222998) in 1988 indicates 3.2 ft of local scour at pier 1. The hole is about 5 ft downstream from the nose of the pier, possibly as a result of turbulence induced by the large, sharp-nosed pier. In 1989, 3.8 ft of contraction scour occurred when debris blocked a third of the bridge opening (fig. 2C). Hydraulic analysis of the 1989 flow with debris blockage estimated an average flow velocity greater than 11 ft per second, comparable to that of the 1972 flood (recurrence interval greater than 100 years), which destroyed the previous bridge. Through 1991, flows with recurrence intervals of about 2 years caused the streambed to aggrade about 5 ft in a few locations, possibly from deposition of bed material that was mobilized in 1989. In 1992, a discharge with a recurrence interval of less than 2 years scoured the streambed about 2 ft in a few locations (fig. 2C).

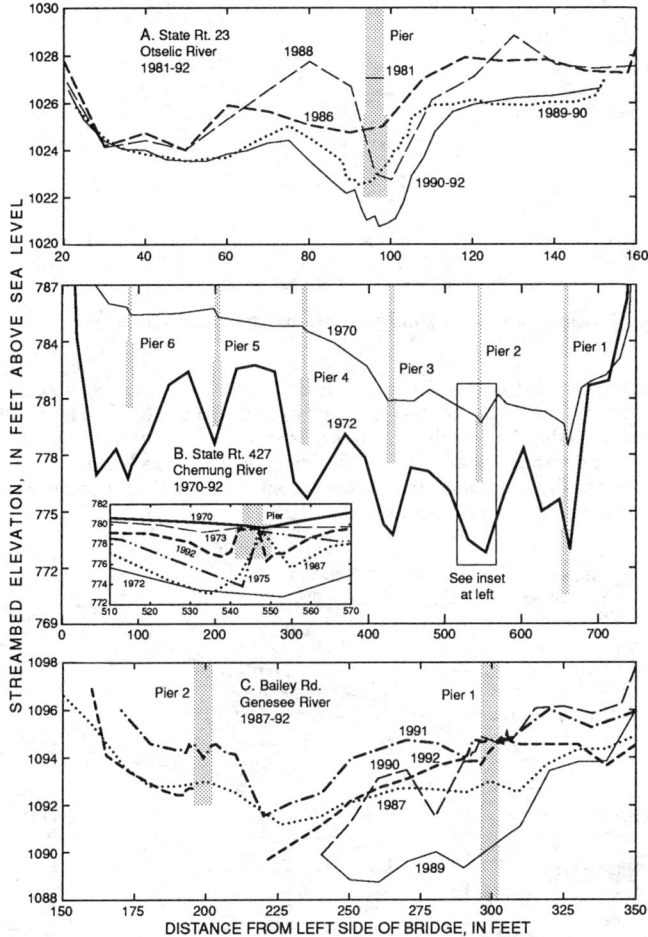

Figure 2. Streambed elevations at selected bridge sites.

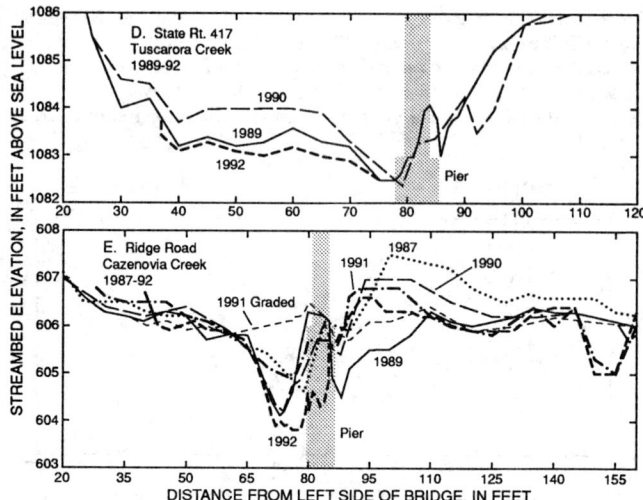

Figure 2 (continued). Streambed elevations at selected bridge sites.

The streambed mobility at State Route 417 over Tuscarora Creek in Steuben County (site 01525981) is shown by the stage-to-discharge relations in fig. 3. The effect that nearby gravel mining has on the streambed is unknown. Flows with recurrence intervals of less than 5 years produced 0.5 ft of aggradation in 1988, and 1.5 ft of scour during 1989-90. Cross-section measurements confirm streambed mobility and also indicate that the scour hole did not refill (fig. 2D).

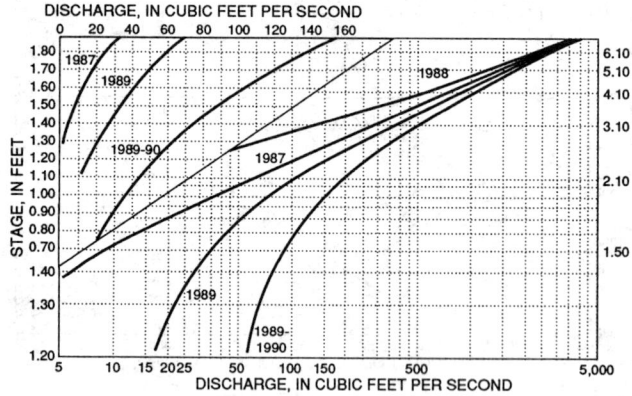

Figure 3. Stage-to-discharge relation at Tuscarora Creek above South Addison, 1987-90

The stage-to-discharge relation at Ridge road over Cazenovia Creek in Erie County (site 04215500) indicates streambed mobility. Although cross sections at the bridge confirm bed movement, the lowest elevation of the scour hole remains nearly constant (fig. 2E). A large debris pile increased local scour in 1989, but crews removed the debris in 1990 and graded the streambed in 1991. In 1992, a flow that was less than half the mean-annual peak discharge produced a scour hole similar to the previous hole.

Summary

Historical USGS streamflow measurements from 31 gravel-bed streams in New York are being used to define hydraulic properties associated with high flows and to calculate streambed elevations to estimate scour. The maximum estimated local scour is 7.8 ft. Measurements at several bridges indicate scour by multiple high flows, severe floods, and debris. Clear-water scour is common; local scour holes that formed during high flows did not refill after subsequent high flows, even at sites with mobile streambeds, which were identified by changes in streambed elevation and USGS stage-to-discharge relations.

References

Butch, G.K., 1991, Measurement of bridge scour at selected sites in New York, excluding Long Island: U.S. Geological Survey Water-Resources Investigations Report 91-4083, 17 p.

Jarrett, R.D., and Boyle, J.M., 1986, Pilot study for collection of bridge-scour data: U.S. Geological Survey Water-Resources Investigations Report 86-4030, 89 p.

RELATION OF LOCAL SCOUR TO HYDRAULIC PROPERTIES AT SELECTED BRIDGES IN NEW YORK

Gerard K. Butch [1]

Abstract

Hydraulic properties, bridge geometry, and basin characteristics at 31 bridges in New York are being investigated to identify factors that affect local scour. Streambed elevations measured by the U.S. Geological Survey and New York State Department of Transportation were used to estimate local-scour depth. Data that show zero or minor scour were included in the analysis to decrease bias and to estimate hydraulic properties related to local scour.

The maximum measured local scour at the 31 bridges for a single peak flow was 5.4 feet, but the deepening of scour holes at two sites to 6.1 feet and 7.8 feet by multiple peak flows could indicate that the number or duration of high flows is a factor. Local scour at a pier generally increased as the recurrence interval (magnitude) of the discharge increased, but the correlation between local-scour depth and recurrence interval was inconsistent among study sites. For example, flows with a 2-year recurrence interval produced 2 feet of local scour at two sites, whereas a flow with a recurrence interval of 50 years produced only 0.5 feet of local scour at another site. Local-scour depth increased with water depth, stream velocity, and Reynolds number but did not correlate well with bed-material size, Froude number, pier geometry, friction slope, or several other hydraulic and basin characteristics.

Introduction

Streambed elevations measured by the U.S. Geological Survey (USGS) and the New York State Department of Transportation (NYSDOT) are being used to estimate local scour at 31 bridges in New York (Butch, 1991). Drainage areas of the sites ranged from 39 mi^2 to 3,070 mi^2. The median particle size, d50, of the 31 streambeds, all armored by gravel, ranges from 22 mm to 68 mm.

---

[1] Hydrologist, U.S. Geological Survey, P.O. Box 1669, Albany, New York 12201

Hydraulic properties, bridge geometry, and basin characteristics are being investigated to identify factors that affect local scour. This paper describes the relation of local-scour depth to water depth, stream velocity, and Reynolds number.

Methods

Bridge-geometry data were obtained from plans and site inspections. Samples of the streambed were collected at each site for particle-size analysis. Streamflow data from nearby USGS streamflow-gaging stations are being used to calculate peak discharges. If scour holes developed before streambed elevations were measured, the largest discharge since bridge construction is used.

A step-backwater model was used to estimate stream velocity and water depth near each pier during a peak flow (Shearman, 1990). The model was calibrated to discharge measurements made before, during, or after high flows.

Relation of Local Scour to Hydraulic Properties

Analysis of 45 peak flows at 59 piers resulted in 90 local-scour depths. The median local-scour depth is 1.2 ft, but depths range from zero to 5.4 ft (table 1). Estimates of velocity, water depth, and Reynolds number near each pier range from 1.6 ft/s to 16.0 ft/s, 1.5 ft to 31.9 ft, and 0.000003 to 0.000379. Discharges represent recurrence intervals from less than 2 years to more than 100 years and range from 2,080 ft^3/s to 233,000 ft^3/s.

Local scour at a pier generally increases with the recurrence interval (magnitude) of the discharge, but the correlation between local-scour depth and recurrence interval are inconsistent among bridge sites. For example, flows having a 2-year recurrence interval produced 2 ft of local scour at two sites, whereas, a flow with a 50-year recurrence interval produced only 0.5 ft of local scour at another site. Velocity and water depth exceeded 10.8 ft/s and 16.2 ft (75th percentile, from table 1) at 7 of 10 local-scour depths greater than 3 ft. The increase in depth of two local-scour holes to 6.1 ft and 7.8 ft by multiple peak flows could indicate that the number or duration of high flows is a factor. Correlation of local-scour depth with water depth resulted in a

Table 1.—Local-scour depth factors and statistics

Factor	Range				Percentile	
	Median	Min.	Max.	Standard deviation	25th	75th
Local-scour depth (ft)	1.2	0.0	5.4	1.2	0.4	2.1
Velocity (ft/s)	8.7	1.6	15.0	3.3	6.8	10.8
Water depth (ft)	12.1	1.5	31.9	6.5	7.9	16.2
Reynolds number (x 10^6)	74.0	3.0	379.0	88.0	41.0	127.0
Discharge (ft^3/s)	32,300	2,080	23,300	51,900	12,200	65,000
Bed material, median (mm)	34.0	22.0	68.0	11.0	27.0	43.0

coefficient (r) of 0.74 (fig. 1A). Local-scour depth increased with water depth but was less than 3 ft if water depth was less than 10 ft. The correlation between local-scour depth and stream velocity was 0.54 (fig. 1B). Local-scour depth increased with velocity, and was less than 3 ft if velocity was less than 6 ft/s. The correlation between local-scour depth and Reynolds number was 0.73 (fig. 1C). Local-scour depth increased with Reynolds number and was less than 3 ft if the Reynolds number was less than 0.00005. Measurements at sites with zero scour are included in the analysis to decrease bias and to estimate hydraulic properties for local scour. The correlations of local-scour depth with water depth, velocity, and Reynolds number decreased to 0.69,

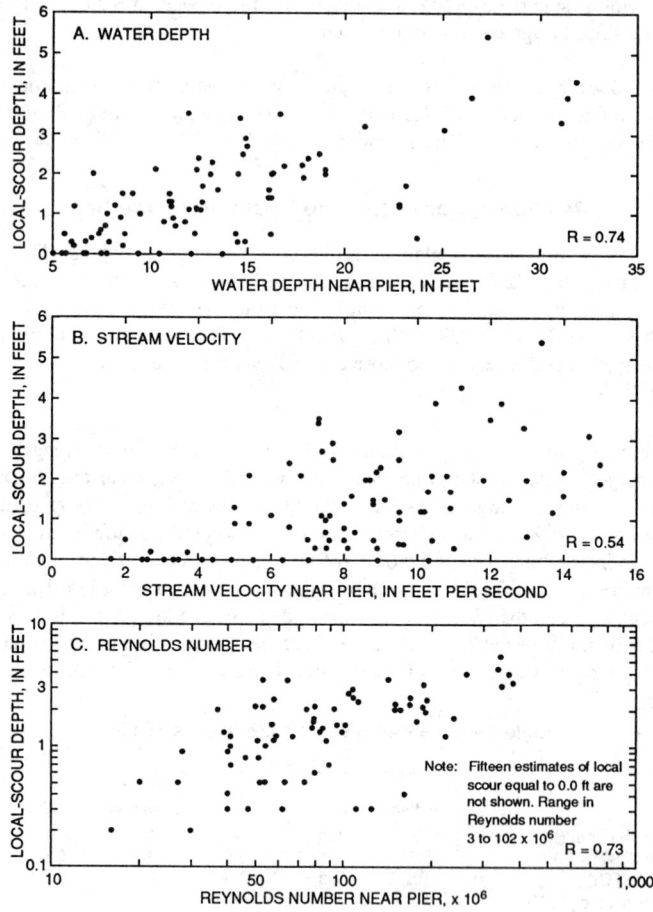

Figure 1. Relation of local-scour depth to: A. Water depth. B. Stream velocity. C. Reynolds number.

0.40, and 0.69 when measurements of zero scour were excluded, indicating that these measurements could include important factors that define hydraulic properties for local scour. The amount of time during which a factor exceeded a given value was not analyzed. Bed material, Froude number, and pier geometry, which are significant explanatory variables in other equations (Richardson and others, 1988), did not correlate well with local-scour depth, nor did friction slope, stream power, and several other hydraulic and basin characteristics. The medians and ranges of water depth, velocity, and Reynolds number for the ranges of local scour are given in table 2.

Table 2.—Median and range of local-scour depth and selected hydraulic factors

Range of local-scour depth (ft)	Number of data points	Local scour (ft)	Water depth (ft)		Velocity (ft/s)		Reynolds number (x 10^6)	
		median	median	range	median	range	median	range
0.0	15	0.0	6.6	1.5-13.7	4.4	1.6-10.3	18	3-102
0.0-0.9	37	0.3	7.8	1.5-23.7	7.0	1.6-13.0	40	3-161
1.0-2.9	43	1.7	12.9	6.1-23.1	9.0	5.0-15.0	86	37-238
3.0-5.4	10	3.5	25.8	12.0-31.9	11.6	7.3-14.7	300	53-379
0.0-5.4	90	1.2	12.1	1.5-31.9	8.7	1.6-15.0	74	3-379

Summary

Hydraulic properties, bridge geometry, and basin characteristics are being investigated at 31 gravel-bedded sites to identify factors that affect local scour at bridges. Measurements of zero scour are included in the analysis to decrease bias and to estimate hydraulic properties of local scour. The maximum measured local scour for a single peak flow was 5.4 ft, but multiple peak flows deepened two holes to 6.1 ft and 7.8 ft. Local scour generally increases with velocity, water depth, and Reynolds number but did not correlate with bed material, Froude number, pier geometry, or several other hydraulic and basin characteristics. Further study is needed to document the relation of local scour to hydraulic factors and the number or duration of high flows.

References

Butch, G.K., 1991, Measurement of bridge scour at selected sites in New York, excluding Long Island: U.S. Geological Survey Water-Resources Investigations Report 91-4083, 17 p.

Richardson, E.V., Simons, D.B., and Julien, P.Y., 1988, Highways in the river environment: Fort Collins, Colorado State University, p. V106-114.

Shearman, J.O., 1990, User's manual for WSPRO, a computer model for water-surface profile computations: Federal Highway Administration Publication FHWA-IP-89-027, 177 p.

Overmining Causes Undermining
(It's a Mad Mad River)

Catherine M. Crossett, P.E. [1]
Associate Member, ASCE

Instream aggregate mining has caused significant problems and expense to California bridges. Riverbed degradation on the order of 3-5 m (9-15 ft) beneath waterway crossings within actively mined streams is not uncommon. It is nearly impossible to solve the riverbed degradation and protect bridge foundations with engineering measures alone. "Fixing" bridges in actively mined rivers requires an integrated process including engineering, regulation and management. The Mad River exemplifies the struggle between competing riverine interests and this integrated process.

Background

Instream aggregate mining involves harvesting sand and gravel from riverbeds (river-run aggregate). The material removed from streams is economically extracted, quality grade aggregate, and provides revenue to the local economy. Extracting material from aggrading streams (rising streambed elevation) can increase the river's flood control capacity, however, it can induce riverbed degradation which can damage infrastructure by exposing bridge foundations, buried pipelines and municipal water intake systems if extraction exceeds the river's replenishment rate (Collins & Dunne, 1990).

Aggregates mined from streams are primarily used for construction — predominantly as road bases and concrete components. The California Department of Transportation (Caltrans) consumed about 15% of all aggregate in California, utilized approximately 18 million metric tons (20 million tons) of sand and gravel in 1991 for road and bridge construction.

Mining can damage bridges located both upstream and downstream from an operation. Galay (1983) explains the phenomenon, "the removal of part or all of the bed material results in a lower transport rate and therefore in slope flattening below the point of sediment removal . . . Upstream from this point there will also be extensive upstream progressing degradation because the flattening of the downstream slope requires a lowering of the river bed." The resulting degradation can undermine bridge foundations, threatening the structure's integrity. In addition, lateral channel migration

---

[1] Associate Hydraulic Engineer, Structure Hydraulics, California Department of Transportation, P.O. Box 942874 Sacramento, CA 94274-0001

triggered by instream mining can damage bridge approaches and wash away adjacent river-front property.

Controlling aggregate mining in streams to protect infrastructure requires a balance between engineering, regulation and management. Engineering can protect existing infrastructure from the effects of degradation, however, engineering alone cannot stop long-term riverbed degradation caused by overmining. The extraction operations need to be managed to extract material on a "safe yield"[2]. Regulation alone is not economically efficient (Davies, 1975). Once a management plan is developed, operators should be regulated to insure the management plan is implemented and there is a balance maintained between the competing riverine interests.

Engineering

Because it is beyond the scope of this paper to evaluate all the impacts of riverbed mining, this paper focuses on the impacts to the Route 299 bridge. Caltrans addresses instream mining through the Federal Highway Administration's (FHWA) Scour Critical Bridge Program (Technical Advisory 5140.23). The scour critical program was developed to prevent bridge scour failures and associated loss of life which recently occurred in other parts of the country. Caltrans identifies bridges which require mitigation measures and schedules the work through the State Transportation Replacement and Improvement Needs (STRAIN) program. The scour critical program has identified over 100 bridges near instream mining operations out of 4500 State bridges.

Mad River Gravel has been extracted from the Mad River for over 30 years. Riverine infrastructure has been adversely impacted by the degradation including:

Route 101 Bridges	4.5 m (15 ft) degradation since 1929 construction.
NCRA[3] Railroad Trestle	Piers on unknown foundations undermined since 1930 construction.
Route 299 Bridges	7.5 m (25 ft) degradation since 1941 construction.
Blue Lake Blvd. Bridge	1.5 m (5 ft) degradation since 1982 construction.
HBMWD[4] Ranney Gages	2.5-3 m (8- 10 ft) degradation.

The eastbound Route 299 bridge is the most scour critical State Highway bridge crossing the lower reaches of the Mad River. As shown in Figure 1, the riverbed has degraded continuously since the bridge was constructed in 1941. Caltrans utilized a multidisciplinary team of engineers to evaluate and design countermeasures for the Route 299 bridge. The existing structure is founded on .9 m (3 ft) spread footings on 4 m (13 ft) unreinforced tremie seals. During the underwater bridge inspection, the divers were able to probe .3 m (12 in) beneath the upstream face of the pier 4 footing. Following the underwater investigation, geotechnical engineers investigated the

---
[2] Extraction rate equals replenishment rate. Also referred to as gravel harvesting.
[3] North Coast Railroad Authority
[4] Humboldt Bay Municipal Water District

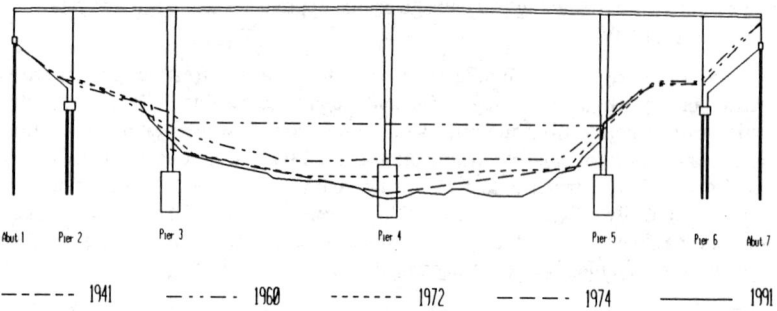

Figure 1. Eastbound Route 299 bridge over Mad River

substructure by drilling through and adjacent to the piers. The geotechnical investigation revealed the tremie seals are founded on shale which is characterized as easily erodible weak rock. The riverbed had degraded to within 1 m (3.5 ft) of the bottom of the spread footings.

As the substructure data was collected and analyzed, designers investigated potential structural and hydraulic countermeasures. Hydraulic countermeasures control scour — 1) rip-rapping piers to inhibit local scour and 2) constructing drop structures to control channel degradation. Structural countermeasures modify the bridge itself to withstand channel degradation and local scour — 1) sheet piling the piers 2) strengthening the foundations 3) reconstructing the foundations 4) replacing the piers or 5) replacing the structure. Structural calculations revealed the Route 299 bridge is now incapable of withstanding substantial lateral loads. The preferred retrofit scheme proposes installing supplemental piles adjacent to the existing foundation to handle the lateral loads. The piles will be placed into the scour resistant rock to protect the structure should local scour or additional channel degradation undermine the existing piers.

Regulation

California regulates mining through the Surface Mining and Reclamation Act (SMARA) of 1975 (Mining & Geology Board, 1979). SMARA delegates implementation to 117 local lead agencies — 57 Counties and 50 Cities. SMARA (section 2770.5) requires the lead agency to notify Caltrans of proposed instream mining activities within one mile upstream or downstream of a state highway bridge to allow Caltrans to comment on the mining's potential impact to the bridge.

SMARA fails to effectively regulate mining to mitigate Caltrans' concerns. SMARA's drawbacks include:

1) As a commenting agency, Caltrans relies on the lead agency to incorporate concerns into permit conditions.

2) Local agency bridges (70% of California bridges) are not addressed by the current regulation. The FHWA Highway Bridge Rehabilitation and Replacement Program (HBRR) requires the local agency to pay only 20% of the total cost to rehabilitate or replace the bridge.

3) Vested rights only require operators extracting after 1975 to obtain a permit from the lead agency. Humboldt County has taken the position that they have no ability to place operating restrictions on vested operations.

4) SMARA only addresses individual project impacts and contains no mechanism to address cumulative impacts.

State agencies recommended Humboldt County conduct a comprehensive analysis of the environmental impacts of gravel mining and prepared a plan to control gravel mining on a sustained yield basis in an environmentally sound manner. In order to allow limited gravel extraction in the 1992 season, the State prepared a Memorandum of Agreement (MOA) between the County, Resources Agency, Mining and Geology Board, Department of Conservation, Department of Fish and Game, State Lands Commission, and Mad River gravel operators (Resources Agency, 1992). The MOA formed a scientific committee which established the operational limits (depth, cross section and volume). In addition, the MOA mandated preparing a Program Environmental Impact Report (EIR) on the Mad River. This EIR was funded by the 4 gravel operators in the Mad River and scheduled for completion May 1, 1993.

Management

Education When engineering and regulatory measures are unsuccessful, State Agencies must convince the lead agency to manage their resources to insure the river system's long-term health. Caltrans' primary focus has been increasing awareness of instream mining's potential negative impacts on bridges. Caltrans presents the potential problems of instream mining to in-house staff, local lead agencies, Boards of Supervisors, County Planning Commissions and other State Agencies. Since the local lead agency administers SMARA and Caltrans is only a commenting agency, Caltrans must inform the lead agency of the potential adverse impacts of instream operations in order to allow the lead agency to make an informed decision.

Caltrans staff also provide technical assistance to lead agencies, for example, assisting in drafting the Mad River MOA. Caltrans helped form the Mad River State Agency Cooperation Committee to assist Humboldt County comply with the ambitious EIR schedule outlined in the MOA. By coordinating with other agencies, the State can present a united front, share information, and help the County manage their resources. Caltrans staff currently serve on the Statewide Instream Mining Task Force which is streamlining the permit process, dealing with the problems created by instream mining, and reducing duplication by different State Agencies. Caltrans is meeting with aggregate industry representatives to formulate the Department's instream mining policy including streamlining permitting, developing mitigation alternatives, establishing monitoring standards and determining potentially deleterious locations.

Economics The most recent change to SMARA prohibits all State Agencies from purchasing aggregate from non-permitted sources after July 1, 1993 (SMARA, 1993). Caltrans could further limit aggregate sources to those which do not contribute to unmitigated riverbed degradation and bridge foundation undermining[5]. Determining which operations are contributing to the bridge undermining is complex, requiring detailed sediment transport and geomorphological studies. These studies are expensive: one Caltrans District recently funded a $300,000 study of a 14 KM (9 mi) reach of Stony Creek in Glenn County to determine the progression of the riverbed degradation. The study concluded that the adjacent instream mining caused over 5 m (16 ft) of degradation beneath the Route 32 bridge (Swanson & Kondolf, 1991). Repairs have cost over $1 million since the bridge was constructed in 1975. The sources causing the degradation had obtained the necessary permits—and fully comply with SMARA. If mining continues to fully permitted amounts the riverbed will degrade an additional 4.5 m (15 ft) beneath the Route 32 bridge. (Chang, 1992)

Gravel Management Sonoma and Lake Counties are the only two agencies who have comprehensive plans for managing their aggregate resources. Sonoma County is currently updating their 1980 Aggregate Resources Management (ARM) Plan which identifies all existing and potential future aggregate sources in the county including hard rock quarries, off channel pits and instream sources. The plan estimates the 20 year demand for aggregate and identifies the proposed source to satisfy that demand. The management plan is also a Program EIR, which allows the County "to consider broad policy alternatives for aggregate management and programwide mitigation measures . . . [and] to consider cumulative impacts that could be slighted on a project-by-project basis" (Sonoma, 1992).

Caltrans encourages all lead agencies to establish operational mining limits on instream sources, initiate bi-annual monitoring programs to detect degradation and institute programs to trigger mitigation if degradation negatively impacts bridges. The safe limits for mining restrict or prohibit mining if the riverbed degraded below a specified elevation or "redline". Ventura County has successfully implemented a redline on the Santa Clara River. This redline allowed Caltrans to determine realistic elevations for the new Route 118 bridge foundations without fearing significant degradation would undermine the bridge or placing foundations significantly deeper than necessary.

---

[5] One individual commented, "If someone keeps steeling your hub-caps, should you continue to buy them back?"

Conclusion

Natural riverbed degradation can be difficult or impossible to curb, but degradation caused by in-channel mining can be controlled. The primary impediment and solution to controlling instream mining lies with the local lead agency. Local agencies are loath to place restrictions on an industry which provide jobs and tax revenue to the local economy but have the jurisdiction to efficiently regulate the mining. SMARA does not currently provide effective mechanisms to help the lead agency overcome this political roadblock. Humboldt County resisted addressing the problem created by instream mining until encouraged to by the State.

While far from ideal, the existing mining regulations do provide a framework in which to work. This framework should be used to develop measures to encourage lead agencies to manage their resources -- not to add more layers to an already bureaucratic system. The solution exists in working with the local agencies, aggregate industry and State Agencies to find ways to manage the extractions without burdening the industry with unnecessary regulations. Any aggregate cost increase will significantly effect Caltrans which consumes 15% of California's aggregate.

References

California State Mining and Geology Board, 1979, "California Surface Mining and Reclamation Policies and Procedures", Spec. Pub. 51, California Division of Mines and Geology.

California State Mining and Geology Board, 1993, "California Surface Mining and Reclamation Act", 1993, California Division of Mines and Geology.

Chang, Howard H., Swanson, Mitchell L., & Kondolf, G. Mathias, "An Investigation of the Causes of Accelerated Channel Erosion and Development of Countermeasures for Bridge Stabilization on Stony Creek", 1991, Executive Summary Prepared for California Department of Transportation.

Collins, Brian and Dunne, Thomas, "Fluvial Geomorphology and River -Gravel mining", California Department of Conservation, Division of Mines and Geology. Special Publication 98, 1990.

Davies, J. Clarence and Barbara S., "Federal Standards and Enforcement" in Economics of the Environment. W.W. Norton & Co., New York, Inc. 1977.

Galay, V.J., 1983, Causes of river bed degradation: Water Resources Research, v. 19, p. 1057-1090.

Resources Agency of California, "Memorandum of Agreement concerning Gravel Extraction Operations on the Mad River in Humboldt County During 1992."

Sonoma County Planning Department, 1992, "Aggregate Resources Management Plan: Draft Update Environmental Impact Report", Sonoma County Planning Department, Santa Rosa, CA.

Swanson, Mitchell L., and Kondolf, G. Mathias, "Geomorphic Study of Bed Degradation in Stony Creek, Glenn County, California" June 1991.

BRIDGE SCOUR PREDICTION METHODS
APPLICABLE TO STREAMS IN PENNSYLVANIA

By Dennis Johnson and Arthur C. Miller[1]

Abstract

The Pennsylvania Department of Transportation, PennDOT, currently owns or maintains in excess of 28,000 bridges over 8 feet in length. The most common type of damage sustained by these bridges is pier and/or abutment scour that is caused by floods or extreme flow situations. The Pennsylvania Department of Transportation in conjunction with Penn State University undertook a research project to evaluate the applicability of existing scour prediction equations to streams in Pennsylvania.

Introduction

The Federal Highway Administration, FHWA, has mandated that all State Department of Transportation identify bridges that are susceptible to scour. To accomplish this task, PennDOT developed a methodology to rank the bridges most prone scour, which necessitated predicting scour depths at these sites, as well as predict scour depths at future bridge sites.

Some of the difficulties encountered in the research included site identification, definitions of scour parameters, and actual measurements of scour depths. Initially, PennDOT utilized a cataloging system, the Bridge Management System (BMS), to identify "scour critical" bridge sites. Eventually, the most efficient way to identify scour prone bridges was by interacting with the maintenance and bridge staffs of the individual PennDot district offices. The maintenance foramens provided some of the best insight on specific bridge problems.

1. Ph.D. Candidate and Professor of Civil Engineering, respectively at Penn State, University Park, PA.

Ultimately 15 sites were chosen (7 for the abutment study and 8 for the pier study). An extensive literature search was performed and a large number of scour equations were investigated. The parameters that were thought to be relevant for scour, both pier and abutment, are summarized in Table 1.

Parameter Estimation

Data were collected for each bridge site via actual field visits, hydraulic and hydrologic reports, design drawings, bridge inspection reports, and information provided by local residents.
Survey data were taken for all locations as well as soil samples in and around the bridge site as well as upstream of the bridge. A grid system was used to collect the soil samples and the samples were later analyzed to determine particle size distribution. The sites were hydraulically modelled using the U.S. Army Corps of Engineers - Water Surface Profile computer program - HEC-2. It was necessary to estimate the actual maximum scour depth which occurred at the study sites, as well as the associated flow.

Table 1 - Hydraulic parameters influencing bridge scour.

No.	DESCRIPTION
1	Approach flow depth and velocity
2	Angle of attack
3	Constricted flow, area, depth, and topwidth
4	Pier width and shape
5	Foundation material and grain size distribution

The flows were estimated for the sites in a number of ways. In most cases, high water marks were known and the flow was determined by comparing the water surface elevations predicted by the HEC-2 computer program with the known high water elevations. Information gathered from local residents proved quite valuable in this stage of the research. In other cases, the flood return periods were estimated from rainfall estimates. In the complete absence of hydraulic and hydrologic information, the flood of record was assumed to be the 100 year event.

Given two hydraulically similar situations, the critical factor in formation of the maximum scour depth is the stream bed's grain size distribution. It is difficult if not impossible to represent the grain size distribution with a single parametric term. For this study it was decided to use the geometric standard deviation of the grain size distribution to represent the actual bed sediment configuration. The sediment samples were obtained from each site by shoveling the top layer (approximately 1 foot) of the stream bed. It was realized and considered that this was not necessarily most scientific approach but in discussion with PennDOT geotechnical staff there appeared to be no better alternative for collecting river bed samples. Split spoon samples were also taken for many of the sites.

Estimated scour depths were obtained from actual field measurements, inspection reports, and in two cases, measurements after the bridge had failed. In order to be conservative, the estimated maximum scour depths were increased by a foot or more. Also, the scour was assumed to be a product of local scour only, rather than a combination of local and contraction scour, which in all likelihood is the more realistic. Table 2 summarizes the scour parameters that were measured or computed and Table 3 summarizes the estimated parameters for the development of the abutment scour prediction equation.

TABLE 2 - Parameter descriptions for development of abutment scour equation.

Parameter	Description
K_1	Coefficient of abutment shape
θ	Angle of embankment to flow
$QC_{left/right}$	Flow constricted by the embankment
$AC_{left/right}$	Flow area constricted by the embankment
$TW_{left/right}$	Topwidth of the flow constricted by the embankment
$YC_{left/right}$	Depth of flow constricted by the embankment = $AC_{left/right}/TW_{left/right}$
VC	Velocity at the upstream side of bridge
G	Geometric standard deviation = $(D_{84}/D_{16})_{0.5}$

The data were rearranged in dimensionless form and a regression analysis was performed. The resulting equation had an R^2 value of 0.9682. Note the equation is also in dimensionless form.

$$\frac{YS}{YC} = 265 K_1 \left(\frac{\theta}{90^\circ}\right)^{0.13} \left(\frac{TW}{YC}\right)^{0.375} Fr^{1.28} G^{-2.34} \qquad (1)$$

$$\frac{YS}{YC} = 265 K_1 \left(\frac{\theta}{90°}\right)^{0.13} \left(\frac{TW}{YC}\right)^{0.375} Fr^{1.28} G^{-2.34} \qquad (1)$$

TABLE 3 - Hydraulic parameters for development of abutment scour equation for seven bridges - Case 1 (BMS# = identification no.)

BMS#	K_1	θ	QC	AC	VC	TW	YC	F_r	G
61/0027/0390/0000	1.0	140	950	410	2.32	115	3.56	0.22	6.36
61/4004/0020/2955	.82	105	120	32	3.75	32	1.0	0.66	6.36
64/4053/0010/0440	.82	85	238	116	2.05	62	1.87	0.26	5.50
46/0063/0390/0894	.82	60	900	225	4	25	9	.235	8.73
55/3003/0140/0000	1.0	45	120	73	1.64	41	1.78	.217	6.36
58/4039/0090/0487	.82	135	875	505	1.73	290	1.73	.232	6.73
14/1002/0390/0068	.82	105	1100	554	2.0	182	3.04	.202	7.8

For comparative purposes, the Penn State equation was compared to Froehlich's equation (as recommended by the Federal Highway Administration in the HEC-18 manual (1). Table 4 illustrates the actual estimated scour depth, the predicted scour depth via the Penn State equation, and the predicted scour depth via Froehlich's equation.

$$\frac{y_s}{y_a} = 2.27 K_1 \left(\frac{\theta}{90°}\right)^{0.13} \left(\frac{a'}{y_a}\right)^{0.43} Fr^{0.61} + 1 \qquad (2)$$

Table 4 - Comparison of actual and computed scour depths

BMS#	Measured scour	PSU Eq.1	Froehlich
61/0027/0390/0000	6.0	10.5	18.7
61/4004/0020/2955	6.0	7.3	7.5
64/4053/0010/0440	5.0	6.8	8.7
46/0063/0390/0894	3.0	11.6	19.2
55/3003/0140/0000	3.5	4.4	7.4
58/4039/0090/0487	3.5	6.5	14.3
14/1002/0390/0068	3.5	6.3	15.7

A similar approach was taken to arrive at the pier scour prediction equation. The equation utilized the geometric standard deviation of the sediment distribution and the correction coefficients for angle of attack and pier shape, as developed by Richardson, et al. (2). Tables 5 and 6 illustrate the parameter definitions and the actual estimated parameters, respectively.

Table - 5 Parameter description for development of pier scour equation

Parameter	Description
K_1	Correction coefficient for pier nose shape
θ	Angle of attack
PW	Pier width
y_a	Approach flow just upstream of the bridge
v_a	Approach velocity
D_{16}, D_{50}, D_{84}	The respective sediment grain sizes of which X% of the streambed is smaller than, where X = 16%, 50%, and 84% respectively.

Table 6 - Hydraulic parameters used in development of pier scour equation

BMS#	K_1	θ	PL	PW	K_2	y_2	Fr	G
14/1002/0350/0068	1.0	75	44	3.0	2.375	8.38	0.74	7.8
08/0220/0420/0000	1.1	90	N/A	16.0	1.0	16.8	0.48	4.32
49/4007/0010/0000	1.0	90	N/A	3.0	1.0	13.1	0.57	5.4
46/0063/0590/0894	1.0	65	44	3.5	3.3	13.8	0.56	8.73
46/4031/0060/0016	0.9	90	N/A	7.0	1.0	16.5	0.28	10.8
07/1014/0010/0156	1.1	90	N/A	6.0	1.0	16.3	0.83	6.1
11/0053/0490/0000	1.0	45	3.8	3.5	4.0	10.8	0.3	6.1
03/1021/0010/1398	1.1	90	N/A	7.0	1.0	10.7	0.52	6.1

The data was rearranged in dimensionless form and a regression analysis was performed. The resulting equation had an R^2 value of 0.255, confirming rather poor agreement. Note equation (3) is in dimensionless form.

$$\frac{y_s}{y_a} = 0.45 K_1 K_2 \left(\frac{PW}{y_a}\right)^{-0.48} Fr^{0.10} G^{-0.45} \qquad (3)$$

Comparison of Pier Results

Table 7 summarizes the depth of scour as computed by PSU equation 3 and the CSU equation (4) (as recommended by the Federal Highway Administration in the HEC-18 manual (1).

Table 7 - Comparison of pier scour depths - measured and computed- in feet

BMS#	Measured Scour	PSU Eq. 3	CSU Eq. 4
14/1002/0350/0068	2.5	4.95	18.9
08/0220/0420/0000	4.0	4.1	26.1
49/4007/0010/0000	5.0	5.3	7.9
46/0063/0590/0894	2.5	14.2	27.1
46/4031/0060/0016	2.5	3.1	16.5
07/1014/0010/0156	3.0	3.52	17.3
11/0053/0490/0000	2.0	13.2	24.5
03/1021/0010/1398	1.0	2.7	13.4

Over 26 different equations were utilized for the 8 bridges used in the pier analysis. Actual measured scour depth to pier width ratios were compared. The Penn State equation and Washington-DOT equation, which is actually a modified University of Aukland compared the best to actual measured scour.

Conclusions

In excess of 25 equations for both abutment and pier scour were tested for the applicability to Pennsylvania streams. None of the established equations predicted very well. The present study was limited in the data base used; however, the results of the study do indicate that the current scour predictive equations are not applicable to streams in Pennsylvania.

References

1. Richardson, E.V. and Richardson, J.R., "Bridge Scour", Bridge Scour Symposium, Cosponsored by FHWA and USGS, October 17-19, 1989.

2. U.S. Federal Highway Administration Technical Report, Procedures for Evaluating Scour at Bridges, (HEC-19), 1991.

SCOUR INSPECTION USING GROUND PENETRATING RADAR

William A. Horne, Associate Member ASCE[1]

ABSTRACT

This paper examines the capabilities of Ground Penetrating Radar (GPR) for river bed scour inspection based on a study carried out by Clarkson University, Potsdam, New York in association with the New York State Thruway Authority (NYSTA). With funding provided by the New York State Science and Technology Foundation (NYSSTF), researchers at Clarkson used GPR equipment to investigate a number of bridge piers and abutments with optimal water and soil conditions. The 300 Mhz antenna was most successful penetrating silty, granular material in less than two (2) meters of water. These bridges were part of a detailed scour inspection program scheduled by the NYSTA for 1990 and 1991 in which all Thruway bridges were analyzed for maximum scour potential and hydraulic adequacy (Horne, 1992). Clarkson's findings with GPR equipment were compared with actual results of the NYSTA's full scale geotechnical investigation. Soil boring information from the latter provided ground-truth measurements necessary for calculating the depths of scour holes and scour interfaces. More importantly, the soil samples were used to calibrate the natural material properties, such as density and gradation, with the radar wave travel time for penetrating different channel bottom layers. Comparing the recovered soil samples with both the color output of the radar equipment and the amplitude of the returning signal displayed on an oscilloscope verified the existing scour conditions.

INTRODUCTION

Bridge Scour is an increasingly important, but often difficult problem to define in terms of understanding and quantifying the scour process (Horne, 1992). Quantifying scour conditions by assuring the structural integrity of any bridge crossing a waterway requires examination of three separate areas of civil engineering: structural, geotechnical, and hydraulic/hydrologic. Defining these areas individually is not a difficult problem, but establishing the interaction of all three in designing for flood

---

[1] Structural Engineer, Clough, Harbour & Associates (Engineers, Surveyors, Planners & Landscape Architects), III Winners Circle, Albany, New York, 12205-5269

events and remediating deficient structures is much more complex (Horne, 1992). Determining both the structural adequacy of the bridge components and soil/structure interaction of the foundation system, can become very difficult and extremely expensive. A complete scour analysis quantifies the performance of the bridge foundation soils during fluctuating hydraulic turbulence patterns. This may be accomplished through computer modeling, such as the Hydraulic Engineering Circular, HEC 18 and HEC 20, which predicts a foundation's performance during adverse conditions. It is very difficult to obtain real time data from a particular flood event primarily due to practical and physical inspection limitations.

While the basic NBIS bridge inspection processes are critically important for determining the presence of structural defects, other methods, which may be quicker, easier, and more cost-effective, can often supply some of the required information needed to determine a structure's scour susceptibility. One such geophysical method, Ground Penetrating Radar (GPR), could be considered in place of traditional investigations, provided the GPR scour survey produces definitive results. Geophysical scour inspections can provide additional information at sites where NBIS procedures were only partially successful and a comprehensive scour evaluation using HEC 18 and HEC 20 may not be warranted. If the scour susceptibility must be quantified and results from other inspection techniques, including a diving inspection and soil borings, remain inconclusive, geophysical methods may be the only economical alternative that bridge owners have remaining to fully determine a structure's scour susceptibility, and predict its performance during a flood event.

BACKGROUND

Pulsed electromagnetic energy was first used in 1926 by Hulsenbeck to detect buried objects; this method was primitive compared to today's technology (Horne, 1992). Advancement of the technology was slow until the late 1960's, when more sophisticated applications were needed. While the NASA lunar investigations resulted in some progress, the real push for technological advancements came from the military during the Vietnam War (Compressed Air, 1990). The U.S. Army developed "Combat GPR" to help locate enemy mines, tunnels and bunkers. The new technological advancements resulted in the successful differentiation of layered media such as air, water and varying soil conditions. This opened up many commercial opportunities in the 1970's for GPR and other non-destructive testing (NDT).

Subsurface Interface Radar (SIR) or Impulse Radar (IR) are terms frequently used to describe the GPR process. The radar antenna transmits a pulse of electromagnetic energy into the subsurface at a given frequency. Upon reaching an interface or buried object, a portion of the energy is reflected back to the surface, while the main pulse continues deeper (Horne, 1992). The amount of energy that is reflected back to the receiver, and the returning angle, depends on the dielectric properties of the material that the pulse travels through. The process is analogous to Snell's Law of Reflection and Refraction of light waves through different media (Horne, 1992). When an interface or change in material is detected, the direction at which the energy is reflected, and the rate at which it is absorbed can be related to the inherent properties of the material, such as its chemical composition, density, and water content. (GSSI, 1990) Analyzing the output signal in both the oscilloscope and the color monitor is an essential part of the GPR inspection process. Furthermore, the ability to recognize the presence of a scour condition and reference its exact location, are the most challenging tasks.

The GPR system's output measures the energy's two-way travel time from the antenna to the interface and back to the receiver. The depth and thickness of the interface can be calculated knowing the natural dielectric constants or by ground truth measurements such as soil borings (Horne, 1992). The dielectric permitivity can be calculated by the following relationship:

$$E_R = (0.5\ t)^2\ (c/.3048d)^2 \qquad \text{EQUATION 1 (Horne, 1992)}$$

E_R = relative dielectric permitivity
t = two-way travel time, in seconds
c = speed of light in free space (approx. 3.0×10^8 m/sec)
d = depth to the reflected interface, in meters

The ground truth measurement provides a calibration standard to accurately calculate E_R for different layers. Once values for E_R are established, all other GPR data for that location is considered valid, provided the river sediment is uniform (Horne, 1992). Approximate values of E_R can be found in various reference sources on geophysics, but they are listed primarily for ideal materials such as pure fresh water, saturated sand, or average soil. These ideal materials are rarely encountered in the field, particularly at bridge sites. Generally, a natural silty sand and turbid fresh water with many organics may be common to sites in a particular geographic area. Water quality characteristics can significantly inhibit the GPR survey's success rate. Presently, conductivity meters can measure resistivity in the field and initially predict the GPR's effectiveness at a particular site. In order to establish E_R for many natural sediments, a correlation must be established between the reflected energy pulse data and the absorption coefficient or conductivity (energy loss) for a wide variety of materials. This is essentially a process of setting up a sequential reflection configuration for each medium using the strength of the returning signal from both natural and ideal radar absorption coefficients. Finding this relationship may help to accurately predict the Radar's conductivity in natural sediments and unique water conditions.

PLANNING

In late 1989, the New York State Thruway Authority (NYSTA) commenced a special investigative program to reassess the hydraulic adequacy of all Thruway bridges crossing rivers and streams (Horne, 1992). The 559-mile Thruway was divided into three sections: southern, central, and western. The central section (Albany to Syracuse) was awarded to the consulting firm of Edwards and Kelcey Inc. (E&K), and the soil boring subcontract was awarded to Atlantic Testing Laboratories, Limited (ATL). Clarkson University worked in conjunction with NYSTA, Edwards and Kelcey and ATL, with funding from NYSSTF, to develop valuable information on the capabilities of GPR as a tool for scour assessment (Horne, 1992).

A number of GPR surveys were conducted prior to mobilization of the soil boring subcontractor. This had three advantages (Horne, 1992): First, the scheduled soil boring would provide a ground truth measurement necessary for calculating the scour depths detected by the radar equipment. Second, the drilling subcontractor (ATL) agreed to perform additional soil borings and sampling at locations where the GPR survey indicated a scour interface. Finally, soil samples from the ground truth measurements were used to calibrate the material properties with the radar wave travel time for penetrating specific layers of natural sediment.

RESULTS: THRUWAY BRIDGE OVER ONEIDA CREEK

The GPR investigation at the Oneida Creek Bridge found what appeared to be a scour interface along the west side of the east pier, as suggested by a significant increase in the strength of the returning signal at a particular depth. ATL performed two additional soil borings at this three-span structure. The split-spoon sampling from the first boring revealed a sand interface located between two silty-clay layers at an approximate depth of 1.5 meters. The material below the sand seam was a stiffer silty clay, followed by hard glacial till. Interestingly, turbulent eddies were noticed on the water surface above the sand interface. The likely conclusion is that the degradation and aggradation processes of a previous flood had deposited the sand interface in an area where changing hydraulic conditions were evident, and roughly 1.5 meters of scour infilling had occurred (Horne, 1992). While the structural integrity of the spread footing was not in question, the radar results that produced a continuous subsurface profile did identify the presence of a scour interface that might not have been otherwise detected.

Table 1 summarizes the laboratory test program performed on the ground truth soil samples obtained at the Oneida Creek site. The most significant results are the gradation (% passing) changes for the sand interface, as well as the lowest value (2.680) for the specific gravity of this material when compared to the other samples (Horne, 1992). The second boring attempted to differentiate between the infilled sediment and the stiffer clay. The two materials appeared to be homogenous and could represent primary scour infilling between the sand interface and the glacial till, and secondary scour infilling between the sand interface and the stream channel bottom (Horne, 1992). It is suspected that all infilling occurred prior to 1987 when the NYSTA placed angular stone around this pier as part of their continual scour maintenance program (Horne, 1992). The presence of this stone was confirmed with the borings. Based on the color output of the radar equipment, the amplitudes of the returning signal, the laboratory test results and the natural hydraulic turbulence patterns observed directly above this particular site, it is the writer's opinion that the pier at this site obstructs the natural flow of Oneida Creek and has contributed to past scour conditions. However, no flood events to date have affected the structural stability of this pier. The approximate elevation of the primary scour interface is 394.0', which is at the top of the 1.2 meter thick, concrete reinforced spread footing. Furthermore, the foundation soil interaction area around this pier's spread footing has never been disturbed due to turbulence patterns associated with the hydraulic configuration of the structure.

Description	Depth (m)	Boring Number	Sample Number	SPT Blow Count (Avg/6")	Mechanical Analysis Percent Passing Sieve Sizes					Spec. Grav.	Atterberg Limits		
					1"	1/2"	#4	#40	#200	.005 mm			
Angular Stone & Gravel	0-0.61	B-1	S-1	NX CORE	100	73	40	2	1	---	---	---	---
Silt & Clay	0.61-1.22	B-1	S-2	3			100	100	100	40	2.726	27	13	14
Silt & Clay	1.22-1.83	B-1	S-3A	6				100	100	54	2.713	36	14	22
Sand Interface	1.55	B-1	S-3B	7			100	95	56	---	2.680	---	---	---
Weathered Till	1.83-2.44	B-1	S-4	12	89	59	34	19	15	4	2.693	14	9	5
Glacial Till	2.44-3.05	B-1	S-5	30		98	88	66	47	20	2.740	16	10	6
Secondary Scour Infilling	0.91-1.52	B-2	S-1	3				100	100	59	2.723	33	15	18
Primary Scour Infilling	1.52-2.13	B-2	S-2	6				100	99	57	2.729	34	14	20

TABLE 1
LABORATORY SOIL TESTING PROGRAM (SUMMARY)

CONCLUSIONS

It is difficult to predict the extent that GPR might be used in the future for determining scour susceptibility. One area that justifies evaluation is a variable frequency antenna combined with more signal processing advancements. Variable frequency antennae would allow the operator to successfully penetrate a wider variety of natural river sediments at different depths, thus producing clear and continuous output over the entire site. A remote control moveable antenna/boat would also allow inspections during peak flow situations, trying to obtain real time scour data for a particular flood event. These are all related to the technology itself. It is the writer's opinion that the system must be more user friendly and reliable, in order for GPR to gain acceptance throughout the engineering industry (Horne, 1992). Successful GPR scour detection is dependent upon the gathering of well defined data from the returning radar pulses. The amount of quantifiable information is directly related to the clarity of the output signal. Although unclear results sometimes lead to the dismissal of this advancing technology, as more successes are realized, more advancements can be made to improve the reliability of the technology and, therefore, further expand the use of GPR for bridge scour inspection programs. While GPR can offer promising advantages, adopting this technology for scour inspections will continue to be a difficult task until dependable definitive information can be assured. Presently there are no geophysical standards for scour inspections. The market for GPR is rapidly expanding as a result of significant technological advancements, the increasing success rate for GPR scour inspections, and the demand for cost effective, non-destructive test (NDT) methods. However, to gain wide acceptance with engineers, the hardware and software for GPR equipment must continue to improve. Finally, it will be critical to set standards for the accurate calibration of radar travel time through natural sediments and water conditions. GPR has already shown the capability of being an effective tool for scour detection. Further refinements through the resolution of the aforementioned issues will provide bridge owners and inspectors with an even more accurate and cost-effective system.

ACKNOWLEDGEMENTS

The financial support from the New York State Science and Technology Foundation (NYSSTF) to Clarkson University made this investigation possible. The cooperative efforts of the NYSTA, Edwards & Kelcey, and Atlantic Testing Laboratories (ATL) were greatly appreciated. I would like to sincerely thank Robert Donnaruma and Mark Hixson of the NYS Thruway Authority, Spencer Thew of ATL, Frank Huber from Edwards & Kelcey, and Tom Fenner of GSSI for their time and effort.

REFERENCES
- Horne, W.A., 1992, Bridge Scour Detection using Ground Penetrating Radar Clarkson University, Potsdam, New York.
- Ground Penetrating Radar, Compressed Air, 1990, Volume 95, No. 12, pp 12-19.
- Geophysical Survey Systems Incorporated, personal correspondence and Equipment Manual, Model SR80, Hudson, New Hampshire, 1989-1992.

Debris Flow & Hyperconcentrated Flow: A UK Perspective

PA Carling[1]

Abstract

Debris-flow activity within the UK is largely confined to upland regions and certain coastal locations. Within the uplands, debris-flow and hyperconcentrated flood waters pose a local but significant threat to life and infrastructure. Research effort has been fragmented between disciplines, but this paper reviews the contributions of a variety of disciplines within a context of prediction and control. This latter requirement has received greater impetus following debate concerning regional climatic change. This debate has coincided with a decade (1980's) of apparent increase in catastrophic upland flooding and geomorphic change.

Introduction

Debris-flow (sensu Takahashi, 1981; Postma, 1986) or hyper-concentrated flow (sensu Carling, 1987) generally is not recognized within the UK as a significant risk to life and property because flows are small in scale and are generally located in remote unpopulated regions. However, this does not reflect the true state of affairs. It is not possible to accurately assess the economic costs, because damage is almost always attributed to non-specific 'floods' or rarely 'landslide', 'avalanche' or 'mudslide'. However, Addison (1987) describes a debris-flow, of a typical magnitude and economic effect, which caused some £146,000 of damage. Each year there have been at least 10 similar events in recent years, so annual direct costs must be in excess of £1.5 million. Other

---

[1] Principal Scientific Officer, Institute of Freshwater Ecology, Windermere Laboratory, Ambleside, Cumbria LA22 0LP, UK.

economic costs which are less obvious include disruption of traffic flow, agriculture and fisheries. As the British uplands are increasingly being used for water supply purposes and recreation it is imperative to correctly identify economically disruptive processes and to propose counter-measures.

Source Areas & Material Characteristics

Three basic parent materials contribute to debris-flows.These are (i) peat released from raised mires or banket bogs, (ii) colluvium or diamicton from slope failure and (iii) channel bed materials.

(i) Occasionally raised bogs on level surfaces rupture but generally failures are associated with peat masses within shallow bedrock hollows on hill-sides. Slopes may be as gentle as 7^0; volumes of material in large debris flows from blanket peat are of the order of <80000 m^3 (Carling, 1986) although volumes from raised bogs can be greater (<1 x 10^6 m^3). Source material characteristics have been detailed by Hobbs (1986).

(ii) Addison (1987) and Ballantyne (1991) argue that source areas have common geological and site characteristics. They occur on impermeable concave rock slopes ($35-38^0$) smoothed and over-steepened by late-Devensian glacial erosion. Source material is glacigenic, colluvium and talus. The tills have point load strengths in the range 0.23 to 1.68 MN m^{-2}, silt-clay contents of 28 to 36% and low void ratios and so are subject to fluidization and sustained flow through matrix forces (Addison, 1987). Flow volumes may exceed 400 m^3 although most flows are <20 m^3 with a maximum volume evacuated by a total of 114 flows from one hillside >4000 m^3 (Innes, 1985). In contrast valley-confined flows may consist of much larger volumes of material (Brazier & Ballantyne, 1989; Wells & Harvey, 1987).

(iii) A debris flow (Carling,1987) was initiated by failure of peat in a bedrock hollow, but downstream channel bed sediments were locally evacuated so that the deposits were locally and alternatively representative of debris-flow and clear-water processes. The development of debris flows from mobilization of channel bed sediments is not uncommon but is poorly described in the literature compared to bedrock hollows (Wasson, 1977). Takahashi (1981) argues that the channel bed is the most common source of debris flow material but this is yet to be verified in the UK.

Frequency

There are few data on frequency of debris-flow events from individual source areas as little is known about the rate of accumulation of debris in hollows or river channels once material has been evacuated. As the volume of a flow is limited by the regolith thickness there must be an upper limit to the magnitude-frequency relationship determined by 'pedogenic', slope-wash processes and talus accumulation rates (Innes, 1985). Carling (1986) suggests that there is a recorded peat slides every 6 years on average in Ireland and every 36 years on the UK mainland but this is not from the same basins. Innes (1985) determined that in the Scottish Highlands during the last 500 years, low magnitude events have occurred with a greater frequency than high magnitude ones. However, the relative importance of large versus small events in transporting debris varied between sites with some sites being dominated by frequent small movements (<20 m^3) and other sites by occasional large movements (~130 m^3). However there is some difficulty in identifying individual events, as multiple flow surges often characterize individual storms (Baird & Lewis, 1957; Innes, 1985; Luckman,1992). Nevertheless individual sites were reactivated every ten years indicating rapid recharge of the source areas. At one active site the annual accretion rate during the last thee hundred years to the debris cone at the terminus is 50 to 60 m^3 (Ballantyne, 1991); implying similar accumulation rates in the source area.

Climatic & Meteorological Control

The most impressive debris-flow deposits (15000-20000 m^3) in the UK are well-vegetated and show little sign of recent activity. Some of these deposits accumulated under para-glacial conditions in the early Holocene (<10000 yr BP), but others, in Scotland especially, accumulated under Loch Lomond Stadial periglacial conditions. Activity seems to have ceased after 4000 yr BP owing to the exhaustion of available debris, but recommenced ca. 550 yr BP, when human interference (burning and grazing) as well as climatic deterioration destabilized the vegetation and soil masses (Innes, 1983; Brazier et al, 1988; Brazier & Ballantyne, 1989; Harvey et al, 1981). Dating relatively recent events is dependent on lichenometry and can be challenged on the basis of the inability to date older deposits and the possibility that older deposits are reworked or obscured (Ballantyne, 1991; Luckman, 1992). However, the evidence still seems to point to increased activity in the last few centuries, because (a) if the rate had been sustained through the post-glacial period much greater volumes of debris would have accumulated and (b) there is evidence of: modern debris-flow activity eroding relict talus (Statham, 1976); recent deposition on relict early Holocene debris-cone surfaces (Harvey 1987; Wells & Harvey, 1987); and debris-flow activity in fluvial channel systems where no previous debris-flow evidence exists (Carling, 1987).

Recent debris-flow activity is associated primarily with high-intensity short-duration rainfall following which slope failures occur or channelized flows become super-charged with debris. A perusal of the UK literature indicates that typically debris-flow activity is reported when total rainfall exceeds 100mm and where storm durations of ca. 2-3 hrs have intensities in the range 30 to 50mm hr^{-1}. Caine (1980) provides a relationship (much quoted in the UK) between rainfall duration (D) and intensity (I) sufficient to cause slope failure; $I = 14.82D^{-0.39}$, which for a two hour event would indicate that slopes could fail with an intensity of 11mm hr^{-1}. Available evidence does not support use of this equation in the UK uplands. Wells & Harvey (1987) suggest a threshold of 50mm hr^{-1} would be appropriate in the north of England; elsewhere 30mm hr^{-1} would seem sufficient. Such intensities represent return periods of 10 to 100 years depending on location, for not only does total rainfall vary primarily with altitude and with a traverse across the UK from west to east but also the synoptic origin of rainfall varies. For example in the west of Scotland, fronts, warm sectors and maritime polar air dominate, whilst in the east other categories contribute increased and variable parts to total rainfall (Smithson, 1970). If rainfall intensity and duration vary spatially it might be inferred that debris-flow initiation and character could be regionally variable. Support for this assertion comes from the application of a rainfall-infiltration model (Brooks et al, 1993) to a podzol soil profile. Statistically significant differences in rainfall distribution in time when comparing cyclonic and anticyclonic rainstorms result in different soil water response, which has implications for the likelihood and timing of debris-flow initiation and the depth of the failure plane (Brooks et al, 1993).

Form & Process

To the authors knowledge UK debris-flows have not been scientifically observed nor modelled. The nature of the flow processes consequently have been inferred from the form and sedimentology of deposits. Detailed sedimentological descriptions are provided by Wells & Harvey (1987) and Carling (1987) and process-deductions by Addison (1987) and Ballantyne (1992). Most hill-slope

flow lobes are thin (<0.5m; Innes, 1983) although Wells
& Harvey (1987) and Carling (1987) describe thicker
deposits. In the latter cases, the resultant sedimentary
features consist of a mixture of debris-flow lobes and
sheets and alluvial deposits overlaying fans at the base
of gullies, up-slope of which are levées sub-parallel to
the debris-flow track. In both latter examples, the
authors attribute the deposits to low-viscosity non-
Newtonian flow, transitional between debris-flow and
stream-flow. Morphological and sedimentological evidence
of pulsing (Addison, 1987; Carling, 1987) in channel-
confined flows indicate that the dynamic behaviour
probably varied in time and space as the flow progressed.
The sedimentological models advanced by Wells & Harvey
and Carling have not been widely tested. Alternative
models might apply in certain circumstances. For example,
in the case of a lobe 20m thick, Ballantyne (1992)
concluded that a high-velocity viscous flow-model
applied.

Conclusions

Although there is now a substantial body of literature
concerning the history of UK debris flow, a unified model
of form and process has yet to be advanced. Such a model
would require support from laboratory experiments and
theoretical considerations which are currently absent.
For practical consideration of prediction and control
emphasis needs placing on identifying regolith
characteristics which favour failures; the specific
rainfall characteristics promoting failure and; the run-
out characteristics of debris-flow which dictate the
likelihood that structures will be over-whelmed.

Appendix - References

Addison, K. (1987) Debris flow during intense rainfall in
Snowdonia, North Wales: A preliminary survey. *Earth
Surface Processes & Landforms*, **12**, 561-566.
Baird, PD. & Lewis, WV. (1957) The Cairngorm floods,
1956: summer solifluction and distributary formation.
Scottish Geographical Magazine, **73**, 91-100.
Ballantyne, CK. (1991) Holocene geomorphic activity in
the Scottish Highlands. *Scottish Geographical Magazine*,
107, 84-98.
Ballantyne, CK. (1992) Rock slope failure and debris
flow, Gleann na Guisrein, Knoydart: Comment. *Scottish J.
Geol.* **28**, 77-80.
Brazier, V., Whittington, GW. & Ballantyne, CK (1988)
Holocene debris cone evolution in Glen Etive, western
Grampian Highlands, Scotland. *Earth Surface Processes &
Landforms*, 13, 525-31.
Brazier, V. & Ballantyne, CK. (1989) Late Holocene debris

cone evolution in Glen Feshie, western Cairngorm Mountains, Scotland. *Trans. Royal Society of Edinburgh: Earth Science*, **80**, 17-24.

Brooks, SM., Richards,KS. & Anderson, MG. (1993) Shallow failure mechanisms during the Holocene: Utilisation of a coupled slope hydrology-slope stability model. In: Thomas, D.S.G. & Allison, R.J. (Eds.) *Landscape Sensitivity*, Wiley, Chichester, 149-175.

Caine, TN (1980) The rainfall intensity-duration control of shallow landslides and debris flows. *Geografiska Annaler*, **62A**, 23-27.

Carling, PA. (1986) Peat slides in Teesdale & Weardale, Northern Pennines, July 1983: Description and failure mechanisms. *Earth Surface Processes* **11**, 193-206.

Carling, PA. (1987) A terminal debris-flow lobe in the northern Pennines, United Kingdom. *Trans. Royal Soc. Edinburgh: Earth Sciences*, **78**, 169-176.

Harvey, AM. (1987) Seasonality of processes on eroding gullies: a twelve year record of erosion rates. In: Godard, A. & Rapp, A. (Eds) *Processus et Mesure de l'érosion*. CNRS, Paris, 439-54.

Harvey, AM., Oldfield, F., Baron, AF. & Pearson, GW. (1981) Dating of postglacial landforms in the central Howgills. *Earth Surface Processes & Landforms*, **6**, 401-12.

Hobbs, NB. (1986) Mire morphology and the properties and behaviour of some British and foreign peats. *Quart. J. Engineering Geol.London*, **19**, 7-80.

Innes, JL. (1983) Lichenometric dating of debris-flow deposits in the Scottish Highlands. *Earth Surface Processes & Landforms*, **8**, 579-588.

Innes, JL. (1985) Magnitude-frequency relations of debris flows in NW Europe. *Geografiska Annaler*, **67A**, 23-32.

Luckman, BH. (1992) Debris flows and snow avalanche landforms in the Lairig Ghru, Cairngorm Mountains, Scotland. *Geografiska Annaler*, **74A**, 109-121.

Postma, G. (1986) Classification for sediment gravity-flow deposits based on flow conditions during sedimentation. *Geology*, **14**, 291-94.

Smithson, PA. (1970) Regional variations in the synoptic origin of rainfall across Scotland. *Scot. Geog. Mag*, **86**, 182-196.

Statham, I. (1976) Debris flows on vegetated screes in the Black Mountains, Carmarthenshire. *Earth Surface Processes*, **1**, 173-180.

Takahashi, T. (1981) Debris Flow. *Ann Rev Fluid Mech*, **13**, 57-77.

Wasson, RJ (1977) Last-glacial alluvial fan sedimentation in the Lower Derwent Valley, Tasmania. *Sedimentology*, **24**, 781-99.

Wells, SG. & Harvey, AM. (1987) Sedimentological and geomorphic variations in storm-generated alluvial fans, Howgill Fells. *Bull. Geol. Soc. Am.*, **98**, 182-98.

Differentiation of Debris-Flow and Flash-Flood Deposits:
Implications for Paleoflood Investigations

Christopher F. Waythomas[1] and Robert D. Jarrett[2]

Abstract

Debris flows and flash floods are common geomorphic processes in the Colorado Rocky Mountain Front Range and foothills. Usually, debris flows and flash floods are associated with excess summer rainfall or snowmelt, in areas were unconsolidated surficial deposits are relatively thick and slopes are steep. In the Front Range and foothills, flash flooding is limited to areas below about 2300 m whereas, debris flow activity is common throughout the foothill and alpine zones and is not necessarily elevation limited. Because flash floods and debris flows transport large quantities of bouldery sediment, the resulting deposits appear somewhat similar even though such deposits were produced by different processes. Discharge estimates based on debris-flow deposits interpreted as flash-flood deposits have large errors because techniques for discharge retrodiction were developed for water floods with negligible sediment concentrations.

Criteria for differentiating between debris-flow and flash-flood deposits are most useful for deposits that are fresh and well-exposed. However, with the passage of time, both debris-flow- and flash-flood deposits become modified by the combined effects of weathering, colluviation, changes in surface morphology, and in some instances removal of interstitial sediment. As a result, some of the physical characteristics of the deposits become more alike. Criteria especially applicable to older deposits are needed. We differentiated flash-flood from debris-flow and other deposits using clast fabric measurements

---

[1]Hydrologist, U.S. Geological Survey, 4230 University Dr., Suite 201, Anchorage, AK 99508

[2]Hydrologist, U.S. Geological Survey, Box 25046, MS418, Denver, CO 80225

and other morphologic and sedimentologic techniques (e.g., deposit morphology, clast lithology, particle size and shape, geomorphic setting).

Clast fabric data were obtained by measuring the azimuthal orientation and dip of the A-B plane (A-axis = long axis, B-axis = intermediate axis) of disc-shaped cobbles and boulders. Clast orientation data were analyzed and compared using an eigenvalue method. Our analysis indicates that flash-flood deposits have larger angles of clast imbrication (mode about 50-55°) and exhibit a much greater degree of preferred orientation than do debris-flow deposits (mode about 15-20°). The fabric-analysis technique may provide a method for investigating the origin of coarse-grained surficial deposits (flood deposits, debris-flow deposits, colluvium, till) whose mode of formation is unknown or uncertain and will improve subsequent estimates of discharge for floods, paleofloods, and debris flows.

TIME-DEPENDENT LANDSLIDE PROBABILITY MAPPING

Russell H. Campbell[1] and Richard L. Bernknopf[1]

Abstract

Case studies where time of failure is known for rainfall-triggered debris flows can be used to estimate the parameters of a hazard model in which the probability of failure is a function of time. As an example, a time-dependent function for the conditional probability of a soil slip is estimated from independent variables representing hillside morphology, approximations of material properties, and the duration and rate of rainfall. If probabilities are calculated in a GIS (geographic information system) environment, the spatial distribution of the result for any given hour can be displayed on a map. Although the probability levels in this example are uncalibrated, the method offers a potential for evaluating different physical models and different earth-science variables by comparing the map distribution of predicted probabilities with inventory maps for different areas and different storms. If linked with spatial and temporal socio-economic variables, this method could be used for short-term risk assessment.

Introduction

Landslide risk can be expressed in terms of expected losses (safety and property damage) and applied in economic analyses of decisions about land use (e.g., mitigation) and hazard response (e.g., warning). This is a progress report on a multidisciplinary study to characterize the conditional probability of rainfall-triggered debris flows that can result from soil slips during rainstorms. Estimating time-dependent soil-slip probability is an initial step in characterizing debris-flow risk.

We specify the conditional probability of a soil slip during a rainstorm, P, as the probability of failure at a given time, t and place, k, given that no landslide occurred there earlier in the storm. This specifies P as a stochastic process that has a discrete state space (slide at k, $s=\{0,1\}$) and a continuous parameter space (time, $t=\{0,1,...,T\}$) (Bhat, 1984). In this kind of probability model, the rainfall causes a

---

[1] U.S. Geological Survey, 922 National Center, Reston, VA 22092

time-dependent change in the probability that the site will survive successive time increments of the storm. We expect that as rain persists at a rate exceeding some minimum, the conditional probability of a soil slip will rise, and a change in state (from s=0 to s=1) can be viewed as the consequence of time-dependent reductions in stability at k. The model is used to test the null hypothesis, H_0, in equation 1.

$$H_0: P_s^k(T) = P_s^k(0); \quad where \ t=0,1,...,T; \quad k=1,...,K \quad (1)$$

To prepare an example, we used data available for the exceptional storm of January 3-5, 1982, in the San Francisco Bay region, when times of failure were observed for sites of several debris flows (Ellen and Wieczorek, 1988), and applied the model to an area of the Oakland hills (Figure 1).

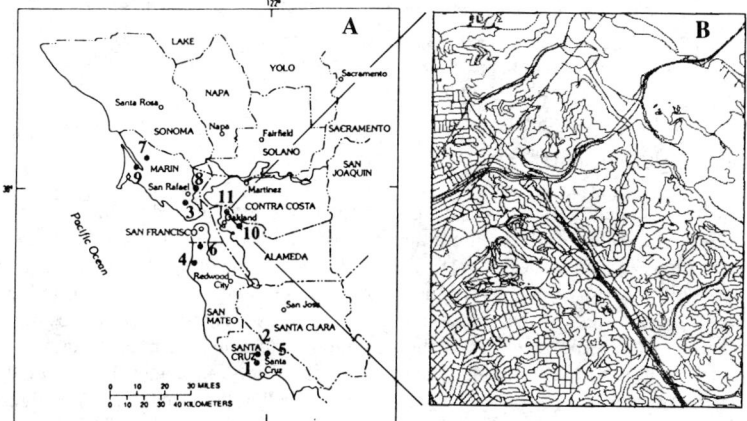

Figure 1. Index maps showing: (A) San Francisco Bay region, numbers 1-11 identify sites of 1982 rainstorm-triggered debris flows having known times of failure; and (B) roads and streams in the Oakland hills area in northwest quarter of the Oakland East 7.5' quadrangle.

Probability model

The conditional probability of soil slips is derived from a cumulative probability distribution of duration, $F(t)=P_s(T\leq t)$, of the current physical state, s=0, for which the survivor function is $S(t)=1-F(t)=P_s(T>t)$ (Kiefer, 1988; Lancaster, 1990). The model assumes that survival decreases with time at an increasing rate as high-intensity rainfall continues. We chose the Weibull distribution to model duration data because it has the assumed property. Substituting the Weibull distribution into the survivor function, we obtain equation 2.

$$S(t) = \exp(-(\lambda t)^p) \ ; \ with \ \lambda = e^{-(\beta_1 x_1 + \beta_2 x_2 + ... \beta_{n_s} x_{n_s})} \quad (2)$$

where x_i are independent variables (Greene, 1991). The coefficients, β, and parameter, p, determined in the regression, control the shape and scale of all the related functions. (See Figure 2.)

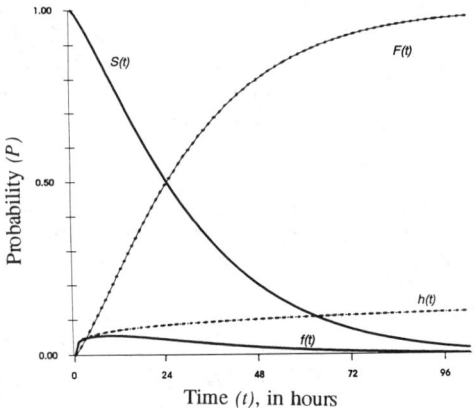

Figure 2. Weibull functions for the sample, extrapolated to 102 hours using mean values for x_i.
$S(t)$, survivor function;
$F(t)=1-S(t)$, cumulative probability distribution function;
$f(t)=dF(t)/dt,$
$=\lambda p(\lambda t)^{p-1} exp(-(\lambda t)^p),$
probability density function;
$h(t)=f(t)/S(t)=\lambda p(\lambda t)^{p-1},$
hazard function.

Specifying the variables

The dependent variable is a function of time, which requires defining (1) a time of origin, (2) a scale for measuring time, and (3) a failure time that occurs only once for each site (Cox and Oakes, 1984). The functions assume that, given enough time, all sites will fail. Any that survive the period of observation are considered to have failure times in excess of that period and are defined as "censored." In the example reported here, the time of origin is the beginning of continuous rainfall having an intensity greater than 0.25 mm/h at the nearest recording rain gage; the end of the observation period is defined by the end of that rainfall at the same gage (see Figure 3). The duration of survival, therefore, extends to either the time of failure (for sites where failure occurred) or beyond the censoring time at the end of storm rainfall (for sites that survived). A number of censored sites in the vicinity of each observed soil-slip site was estimated by using data reported in Ellen and Wieczorek (1988).

The independent variables reflect (1) the prestorm stability at each site, as characterized by prestorm hillside characteristics, and (2) the destabilizing effect of rainfall, as characterized by rain gage records. To use hillside characteristics in a probability model, they must be represented by numbers (e.g., slope in degrees or percent). Several hillside characteristics that vary spatially may be used singly or in functional combinations. In this example, we used a *stability index (SI)* that is the ratio of the tangent of the slope angle to the tangent of an angle of shear resistance estimated for the hillside material. Consistent measures of slope can be derived from 1:24,000-scale contour maps or from digital elevation models, and consistent preliminary estimates of broad categories of shear resistance in slope materials can be made by using descriptions in case studies or regional geologic and soils maps.

A separate independent variable represents the cumulative effect of rainfall at time, T, as a function of rainfall intensity and duration. It may be further combined with hillside characteristics if the independence of the variables is not compromised. The *cumulative rainfall index* (CRI_T) is a convenient way to characterize a cumulative effect at time (T) for bursts of rainfall rates (I) in time interval (t) that exceed selected minimum rates (I_0) (Figure 3). It is calculated as:

$$CRI_T = CRI_{T-1} + (I_t - I_0)t; \quad Subject\ to:\ CRI_T \geq 0 \tag{3}$$

The I_0's used in this exercise are 6.9 mm/h and 4.6 mm/h, where gages are in areas having mean annual precipitation greater or less than 660 mm, respectively; these minima were chosen because they were exceeded at all gages near failure sites at the times of failure. For regression and computation in the present example, we combine the (CRI_T) with the thickness of colluvium (estimated from case studies and soils maps) in a ratio M so that the product, MSI, is a time-varying fraction of SI.

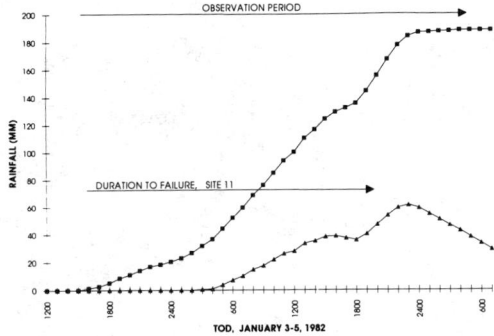

Figure 3. Gage record A-5, showing cumulative storm rainfall curve (boxes), cumulative rainfall index curve (triangles), period of observation, and time to failure at nearby site number 11.

Regression

The table below lists sites of rain-triggered debris flows in the San Francisco Bay region during the storm of January 3-5, 1982 (Fig. 1), their observed time of

Site No.	$I_{(0)}$ (mm/h)	Time of failure	Time to fail (h)	Slope	Shear resistance	Thickness (m)	Number censored
1	6.9	1930	18.5	30°	30°	4.5	13
2	6.9	1310	12.0	38°	35°	4.3	9
3	6.9	2115	17.0	30°	40°	1.8	13
4	4.6	2310	4.0	26°	40°	3.9	8
5	6.9	1900	18.0	31°	40°	7.7	8
6	4.6	2100	2.0	26°	40°	1.0	9
7	4.6	1200	7.0	20°	40°	2.0	3
8	6.9	1400	10.0	26°	40°	2.0	5
9	6.9	1030	5.0	23°	40°	2.0	4
10	4.6	1234	3.5	26°	40°	0.5	11
11	4.6	2000	17.5	17°	30°	1.5	16

occurrence on January 4, thresholds for rainfall rate, I_0, at the nearest recording raingages, the hillside characteristics at those sites, and an estimate of the number of unfailed (censored) sites in the vicinity of each site of failure.

Regression on equation 2 yields the following coefficients and other statistics. Both variables are significant, and the coefficients have the expected signs.

Variable	Coefficient (β)	t-ratio	Mean of x	Std. Dev. of x
SI	2.29	42.5	1.62	0.30
MSI	-1.80	-5.4	0.125	0.23

Log-Likelihood=-495.2; p=1.25; λ=.03; r for MSI and SI=0.04; Median time to failure=24.4 h

The coefficients and parameters were used to calculate probability estimates for each cell (k) in the Oakland hills study area for each hour of the storm. Maps for two selected hours are shown in Figure 4.

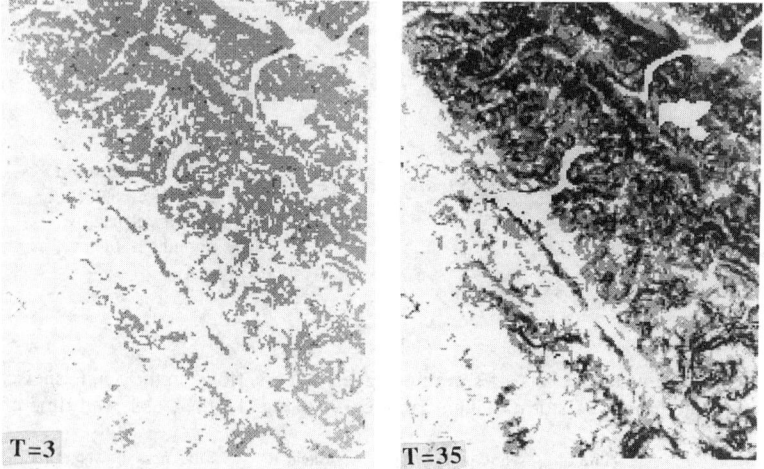

Figure 4. Maps of the Oakland hills study area (Figure 1B) showing spatial distribution of different levels of probability for soil-slip failure F(t,k) at two hours (T=3, T=35) during the storm of January 3-5, 1982. Darker gray shading indicates higher probabilities.

Discussion

The procedure illustrated by this example yields the conditional probability that a soil slip will occur in a map cell at a time T during a storm if (1) no failure has occurred in that cell before T and (2) rainfall continues to T. At present, they are uncalibrated by comprehensive inventories, and probabilities near the end of the storm seem high compared to the poststorm proportion of failed to unfailed slopes, where inventories provide that information. As relative probability levels, however, they indicate that the null hypothesis of equation 1 should be rejected. Clearly, these predicted probabilities are not equivalent to a deterministic prediction that those sites will fail; however, the model may be modified to improve correlation between predicted probability and observed failure frequency. The shape and scale of the functions (Figure 2) are sensitive to (1) variations in the proportion of censored to exited (failed) cells having the same attributes, (2) the definition of the observation period, (3) the geomorphic and geologic properties selected for regression, and (4) the I_0 selected for minimum rate of rainfall. The model in this example is derived from a sample that is probably not fully representative, and probably should contain a much higher proportion of censored sites. The sample also contains a higher proportion of sites where mean annual rainfall was higher and 1982 storm rainfall was more intense than in the Oakland hills area. In a better calibrated form, a model of this kind could be linked with spatial and temporal socio-economic variables to identify and delineate short-term increases in risk.

Acknowledgements

Special thanks to S. D. Ellen, G. F. Wieczorek, S. H. Cannon, and R. K. Mark for providing data and helpful discussions about the 1982 storm; to D. R. Soller for help with GIS programs; to W. Moy, J. Spears, M. Falkenstein, and G. Oppenheimer for assistance in preparing data and illustrations; and to L. K. Peng for hardware and software systems support. We appreciate reviews by G. F. Wieczorek, D. M. Perkins and R. W. Jibson. All these colleagues are from the U.S. Geological Survey.

Appendix A. References

Bhat, U. N. (1984). *Elements of applied stochastic processes*, 2nd ed.: John Wiley & Sons, New York

Cox, D. R. and Oakes, D. (1984). *Analysis of survival data*: Chapman and Hall, London

Ellen, S. D., and Wieczorek, G. F., eds. (1988). *Landslides, floods, and marine effects of the storm of January 3-5, 1982, in the San Francisco Bay region, California*: U.S. Geol. Survey Prof. Paper 1434

Greene, W. H. (1991). *LIMDEP user's manual and reference guide, version 6.0*: Econometric Software, Inc., Bellport, NY

Kiefer, N. M. (1988). "Economic duration data and hazard functions": *J. Econ. Lit.*, v 26, p. 646-679

Lancaster, T. (1990). *The econometric analysis of transition data*: Cambridge Univ. Press, New York

Operation of a Real-Time Warning System for Debris Flows in the San Francisco Bay Area, California

Raymond C. Wilson,[1] Robert K. Mark,[1] and Gary Barbato[2]

Abstract

The United States Geological Survey (USGS) and the National Weather Service (NWS) have developed an operational warning system for debris flows during severe rainstorms in the San Francisco Bay region. The NWS makes quantitative forecasts of precipitation from storm systems approaching the Bay area and coordinates a regional network of radio-telemetered rain gages. The USGS has formulated thresholds for the intensity and duration of rainfall required to initiate debris flows. The first successful public warnings were issued during a severe storm sequence in February 1986. Continued operation of the warning system since 1986 has provided valuable working experience in rainfall forecasting and monitoring, refined rainfall thresholds, and streamlined procedures for issuing public warnings. Advisory statements issued since 1986 are summarized.

Introduction

Rapidly moving debris flows, triggered by severe rainstorms, are among the most numerous and dangerous types of landslides in California. Debris flows can begin suddenly, accelerate quickly, reach high velocities (up to 60 km/hr), and flow down streams or other channels for distances of several kilometers. They can smash homes and other structures, wash out roads and bridges, sweep away cars, knock down trees, and, finally, lay down thick deposits of mud, rock, and other debris where they come to rest, obstructing drainages and roadways. Because debris flows occur abruptly and move swiftly, public warnings, to be effective, must be provided in "real time," as a major storm develops or approaches.

A catastrophic rainstorm over the San Francisco Bay area in January 1982, deposited nearly half of the normal annual rainfall in 32 hours and triggered more than 18,000 landslides, principally debris flows, that caused 25 fatalities and $66 million in property damage (Ellen and Wieczorek, 1988). Although the NWS had forecast heavy rainfall and issued several special weather statements, the destructiveness of the debris flows and other landslides triggered by the storm were unexpected. Following

---

[1] U. S. Geological Survey, 345 Middlefield Rd., Menlo Park, CA 94025
[2] National Weather Service, 660 Price Ave., Redwood City, CA 94063

this disaster, the USGS, in cooperation with the NWS, began to develop a system for warning the public when rainfall conditions reach or approach critical levels for triggering debris flows.

Previously, Campbell (1975) had proposed a debris-flow warning system for the Santa Monica mountains in southern California, based on threshold levels of rainfall intensity and pre-storm seasonal rainfall. Caine (1980) developed a threshold for the intensity and duration of heavy rainfall needed to trigger debris flows, using a worldwide data set. New thresholds were developed for the San Francisco Bay region, using data from the 1982 storm (Cannon and Ellen, 1985), and for a small (12 sq. km), but highly vulnerable, study area near La Honda, San Mateo County, using data from several storms (Wieczorek, 1987). Meanwhile, the NWS developed procedures for issuing quantitative precipitation forecasts throughout northern and central California and coordinated the development of the Automated Local Evaluation in Real Time (ALERT) system, a network of radio-telemetered rain gages across the San Francisco Bay region (Fig. 1). Lastly, the USGS installed an ALERT rain gage and a network of shallow (30 - 140 cm) piezometers on a hillslope in the La Honda study area that had produced debris flows in several storms.

These developments formed the basis for a debris-flow warning system that was initiated formally in February 1986, when public warnings were issued during a sequence of severe storms that triggered hundreds of debris flows across the San Francisco Bay region (Keefer et al. 1988). The two warnings, issued through the NWS radio broadcast system, were the first successful public warnings of debris-flow hazards issued in the United States.

Present Operation of the Warning System

When a storm approaches the San Francisco Bay region, the NWS Forecast Office in Redwood City attempts to make a quantitative forecast of the storm rainfall; the USGS Landslide Initiation and Warning Project in Menlo Park compares the observed and forecast rainfall to thresholds for debris-flow initiation; and both groups work together to assess the probable hazard from debris flows, so that appropriate public statements may be issued.

Forecasting the Rainfall: Most of the rainfall in the San Francisco Bay region is produced by weather systems that originate either in the Gulf of Alaska or in subtropical latitudes near Hawaii. In the January 1982 storm, extremely heavy rainfall resulted from a collision of air masses from both regions. The principal tool for tracking storms is the imagery from weather satellites, currently the GOES-7 (Geostationary Operational Environmental Satellite), launched in February 1987. Every 30 minutes, GOES-7 transmits an image of cloud cover across the northeastern Pacific from the Gulf of Alaska to Hawaii to the west coast of North America. The spatial patterns of the clouds and their movements, revealed by time-lapse sequencing of several images, allow the estimation of speed, direction, and intensity of large storm systems. Imagery in the infrared spectrum also indicates the temperatures of the cloud tops and provides important inferences about the expected intensity of rainfall.

Surface and upper-air weather observations, including barometric pressure, wind velocity, temperature, and precipitation, from a network of land-based weather stations, combined with reports from aircraft and ships, also furnish important data on approaching storm systems. Additional forecast guidance is provided by computer

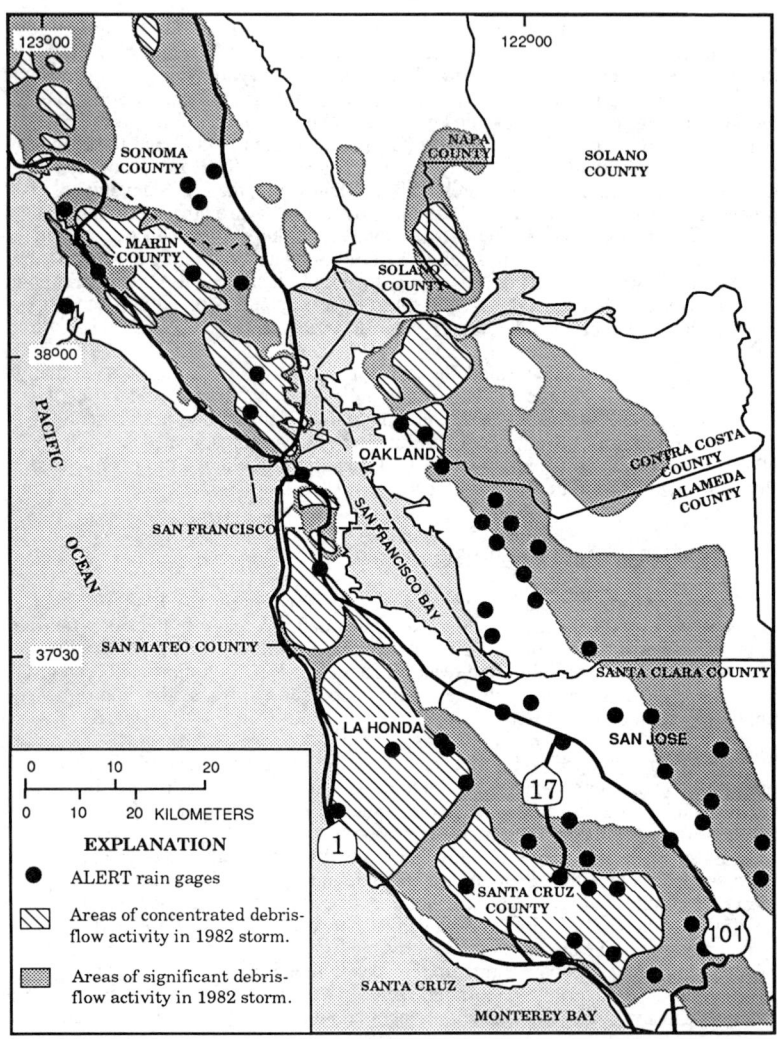

Figure 1. Map of the San Francisco Bay region, California, with ALERT Rain Gage Locations and Areas of Debris-Flow Activity During the Jan. 1982, Storm.

simulations of long-term weather trends from the NWS National Meteorological Center in Camp Springs, Maryland. These computer simulations, based on models of global atmospheric circulation, are updated frequently with surface and upper-air observations from throughout the northern hemisphere.

At the NWS Forecast Office in Redwood City, the lead forecaster compiles this information and prepares the Quantitative Precipitation Forecast (QPF). The QPF, issued twice daily, estimates the amount of rainfall expected in each of four 6-hour periods, for the following 24 hours, at 17 reference points throughout northern and central California, including a point near the southern tip of San Francisco Bay. The QPF is expressed in terms of rainy-day-ratios (RDR), because orographic effects lead to wide variations in local precipitation amounts. The RDR for a given location is defined as the mean rainfall for the month divided by the mean number of rainy days (days with > 1 mm rainfall) in that month (Barbato, 1987).

Rainfall Thresholds for Debris-Flow Initiation: The warning system must consider two complementary thresholds that relate to different time scales (Campbell, 1975; Cannon and Ellen, 1985): 1) an antecedent threshold, requiring an accumulation of a certain amount of rainfall during the season, and 2) a storm threshold, requiring that a critical combination of rainfall intensity and duration must be exceeded during the course of the storm.

The antecedent rainfall threshold exists because the hillside soils of the San Francisco Bay region become dehydrated during the course of the long summer dry season (late April through early October). A certain amount of rainfall is necessary to replenish soil moisture to a level that rehydrates the clay minerals, fills the capillary porosity, and reduces the soil suctions to levels where gravitational drainage can take place. Until the soil moisture is restored, the positive pore pressures necessary for slope failure cannot form and debris flows are unlikely, even in heavy rainfall. The USGS uses two methods to determine whether soil moisture is up to the antecedent threshold: 1) Seasonal rainfall totals are tracked for ALERT gages across the Bay area; 250 - 400 mm (8 - 16 in) are required, depending on soil type and thickness (Keefer et al. 1987). 2) Shallow piezometers are monitored at the La Honda study area, which serves as a benchmark site for the region. The soil moisture is considered to have reached the antecedent level when the piezometers first respond strongly to storm rainfall. This generally occurs somewhat before the seasonal rainfall total at La Honda reaches the antecedent value, 280 mm (11 in), empirically determined by Wieczorek (1987).

Once the antecedent rainfall threshold has been exceeded, approaching storms are evaluated to see if the intensity and duration of the expected rainfall are sufficient to trigger debris flows. The 1986 debris-flow warnings were based on empirical rainfall thresholds which vary over a significant range of values and predict different degrees of debris flow activity (Keefer et al. 1987). Since 1986, these empirical thresholds have been consolidated into a pair of relationships between cumulative rainfall and duration that outline a continuous spectrum of size and frequency of debris flows (Fig. 2). The lower, "safety" threshold is adapted from Wieczorek's (1987) threshold for the initiation of individual debris flows in the La Honda study area, and represents a rainfall level below which significant debris flow hazards are considered unlikely. The upper, "danger" threshold, is adapted from the threshold of Cannon and Ellen (1985), based on data from the January 1982 storm, and represents a rainfall level above which abundant debris flows large enough to destroy structures are likely to occur across broad areas.

A further modification of the threshold levels is to incorporate the effects of orographic variations in precipitation across the region by expressing the threshold levels in terms of RDR's. This also facilitates the evaluation of forecast rainfall, already expressed in RDR's. In order to compare these threshold levels with ALERT observations, these RDR thresholds must be converted to absolute rainfall values, using the rainy-day normal for the ALERT station. Such a conversion, for La Honda, is shown by the scale on the right side of Fig. 2.

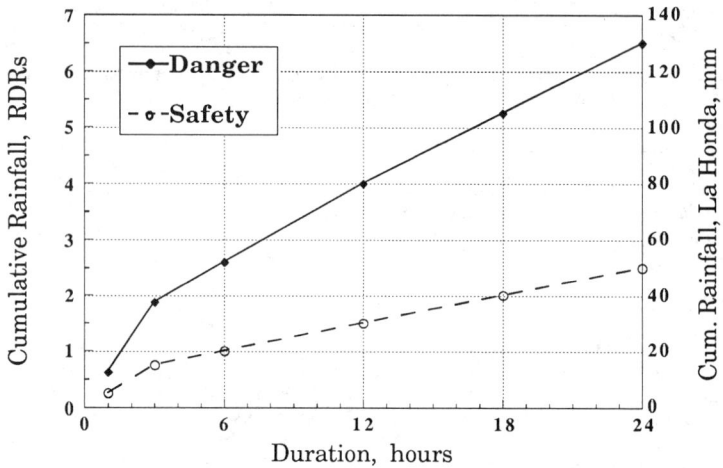

Figure 2. Rainfall Thresholds for Triggering Debris Flows, Expressed as Cumulative Rainfall (RDR or mm) Versus Duration for Peak-Rainfall Periods.

Issuing Debris-Flow Hazard Statements: As a storm begins to make landfall, the ALERT network of radio-telemetered rain gages can be used to monitor the rainfall intensities and estimate the speed of advance of the storm front. Observed rainfall amounts, combined with QPF estimates, are compared to the warning thresholds (Fig. 2) to determine the level of hazard and the type of public statement to be issued. Both the NWS and the USGS participate in this phase of the operation.

Storms with peak rainfall periods that fall below the lower threshold are considered unlikely to trigger hazardous debris flows and generally require no statements. For storms with rainfall levels just above the lower threshold, a brief statement may be added to a NWS "Urban and Small Streams Flood Advisory," warning motorists that roadways may be obstructed by rock falls or debris flows. If rainfall is forecast to approach the upper threshold, a Flash-Flood/Debris Flow Watch is issued, advising people living on or below step hillsides, or near creeks, to stay alert and be prepared to evacuate, as debris flows are a strong possibility during the watch period. Storms that exceed the upper threshold could trigger numerous, massive debris flows leading to loss of life and substantial property damage. Therefore, when rainfall is observed to exceed the upper threshold, or if reports of significant debris flow activity are received, the strongest statement--a Flash-Flood/Debris Flow Warning--is issued. Sample texts for these debris-flow statements have been prepared, with wording agreed upon by both the USGS and NWS, so that timely, informative advisories with complete, relevant information can be issued with a minimum of preparation time.

Summary of Debris-Flow Statements Issued Since 1986

Although rainfall was below average in the San Francisco Bay region from late 1986 through late 1992, the debris-flow warning system issued advisory statements in response to several unusual events: (1) A Flash-Flood/Debris Flow Warning was issued on March 24, 1991, when a brief, but intense, rainstorm triggered a large rock fall that closed Highway 17 in the Santa Cruz mountains. (2) A special warning threshold was devised for the area stripped of vegetation by the Oakland firestorm (10/20/91), with zero antecedent rainfall and reduced intensity-duration levels (USGS Press Release, 10/28/91). Relatively mild rainstorms and a successful re-seeding effort prevented significant debris-flow problems in the burn area. 3) Based on a forecast for heavy rainfall, a Flash-Flood/Debris Flow Watch was issued on 2/11/92 for the Oakland burn area. While rainfall was only moderate in Oakland, it reached levels of 80 to 120 per cent of the danger threshold in southwestern San Mateo County, triggering numerous, small debris flows. 4) Flash-Flood/Debris Flow Watches were issued on 1/13/93 and 1/15/93, during a closely spaced sequence of intense rainstorms. A post-storm field reconnaissance located a number of small, widely scattered debris flows on roadways and natural slopes in Marin, San Mateo, Alameda, Santa Clara, and Santa Cruz Counties (S. Ellen, pers. com.).

Appendix I. References

Barbato, G. E. (1987). *Quantitative precipitation forecasting in northern and central California by the National Weather Service*, National Weather Service, Redwood City, CA.
Caine, N. (1980). "The rainfall intensity-duration control of shallow landslides and debris flows," *Geografiska Annaler*, 62A, 23-27.
Campbell, R.H. (1975). "Soil slips, debris flows, and rainstorms in the Santa Monica Mountains and vicinity, Southern California," *USGS Professional Paper no. 851*, United States Geological Survey, Washington D. C., p. 31.
Cannon, S.H., and Ellen, S. (1985). "Rainfall conditions for abundant debris avalanches, San Francisco Bay region, California," *California Geology*, 38(12), 267-272.
Ellen, S.D., and Wieczorek, G.F. (1988). "Landslides, floods, and marine effects of the storm of January 3-5, 1982, in the San Francisco Bay region, California," *USGS Professional Paper no. 1434*, United States Geological Survey, Washington D. C..
Keefer, D.K., Wilson, R.C., Mark, R.K., Brabb, E.E., Brown, W.M., Ellen, S.D., Harp, E.L., Wieczorek, G.F., Alger, C.S., and Zatkin, R.S. (1987). "Real-time landslide warning during heavy rainfall," *Science*, 238, 921-925.
Wieczorek, G.F. (1987). "Effect of rainfall intensity and duration on debris flows on central Santa Cruz Mountains, California," *Debris Flows/Avalanches: process, recognition, and mitigation: Reviews in Engineering Geology, v. VII*, Geological Society of America, Boulder, CO, 23-104.

STRUCTURAL AND NON-STRUCTURAL DEBRIS-FLOW COUNTERMEASURES

Takahisa Mizuyama[1]

ABSTRACT

Debris-flow can potentially occur for almost any torrents whose gradient is steeper than a given critical value and the frequency at which the debris-flow occurs depends on geological and meteorological conditions. Current countermeasures against these debris flow hazards are either structural and non-structural in natures. The present situation surrounding these countermeasures is reviewed and discussed mainly based on Japanese methods.

INTRODUCTION

The mechanisms of debris-flow occurrence and movement have gradually been made clear. We can determine which torrents are debris-flow prone, where debris-flow deposit areas are, and which rainfalls may generate debris-flows to some extent. The dimensions of debris-flows, their peak discharges and magnitudes, however, can not be predicted accurately. Unfortunately the countermeasures against debris- flows are influenced greatly by the accuracy of these predictions. Under these circumstances, structural measures and/or non-structural measures are introduced. The structural measures consist of check dams, levees and channels. Check dams are also called debris basins, sabo dams or retarding basins. Non-structural measures include warning systems, evacuating systems, reinforced buildings and better land use. In this paper, the warning and evacuation system and structural measure systems adapted in Japan are introduced and discussed.

---

1 Department of Forestry, Kyoto University, Kyoto, 606-01 Japan

WARNING AND EVACUATION SYSTEMS

Debris-flow fans are potentially debris-flow prone areas. It is recommended that people do not to live near the mouths of a torrent. Land use control, however, is not easy in most countries. It is also difficult to persuade people who already live in debris-flow prone areas to move. The second alternative is warning and evacuation systems. Warning systems can be divided into two types. One is an indirect method. Rainfall data is used for this method in most cases. By analyzing data on former rainfall which cause debris-flows or do not cause debris-flows, the critical conditions can be found. When rainfall is likely to enter the critical domain, a warning is issued. A short time rainfall intensity forecast, one to three hours in advance is useful and necessary for the warning system. This system has not been used in Japan, though. The second type of warning system relies on a direct warning method. When debris-flow occurs, wires installed across a torrent are cut by the debris-flow or a vibration sensor is activated by the movement of debris flow. These signals are transmitted to a control device and a warning is issued automatically via a siren or a turning light. These direct sensor methods give only a few minutes to evacuate. The wire sensors are usually used in the torrents where debris flow occurs with relatively low frequency or where scientific observation is carried out since they are accurate but difficult to reset. Vibration sensors are more suited to high frequency debris-flow torrents such as the torrents of active volcanoes.

FUNDAMENTAL INFORMATION FOR STRUCTURAL MEASURES

Structural measures are planned differently depending on what is to be protected from the debris-flows, i.e., roads, houses and so on. The measures taken to protect houses from debris-flows are discussed hereinafter. The dimensions of anticipated debris-flows must be determined when making planning structural measures and when designing debris-flow control structures. We have not, however, had enough knowledge to predict these dimensions accurately.

Total Sediment Discharge

According to the Japanese technical standard prepared by the author, for the Ministry of Construction (Sabo Division, 1988), the planned total sediment discharge is the volume of sediment transportable by the debris flow for a given estimated scale or the movable sediment volume within the catchment area, whichever is smaller.

The movable sediment volume within the catchment area is calculated as follows:

Ve = Ae x Le

where, Ve; the movable sediment volume, Ae; the mean cross-sectional areas of deposits on torrent bed (= Be x De , where Be; the mean width of the torrent bed where erosion is predicted when debris-flow occurs and De; the mean depth of deposits of torrent bed, usually 1.2 meters to 2.0 meters) and Le; the distance measured along the torrent from the outlet of the valley to the furthermost point of the catchment area.

The volume of sediment transportable by the debris-flow for an estimated scale (Vec) is calculated by multiplying the sediment concentration of debris-flow (Cd, Takahashi, 1977) by the total volume of water. This volume of water is obtained by multiplying the catchment area (A km2) by the total rainfall for the estimated scale (R). The runoff correction rate (fr) is also considered; it changes with catchment area and is obtained empirically. Therefore,

Vec = ((1000xRxA)/(1-λ))((Cd/(1-Cd))fr

Cd = $\rho \tan\theta$ /(($\sigma - \rho$)($\tan\phi - \tan\theta$))

where λ : the void ratio, σ: the density of sediment, ρ: the density of water, ϕ : the angle of internal friction of deposited sediment, θ : torrent bed gradient.

Peak Debris-flow Discharge

The possible process by which a debris-flow can occur complex. Some debris-flows are initiated directly by small shallow landslides and some are created by the forming and breaking of small landslide dams. This complexity makes it difficult to predict the peak discharge of debris-flows. The following equation shown (Takahashi, 1977) is often used to estimate debris-flow peak discharge (Qdp).

Qdp = $(C_*/(C_*-Cd))$Qp

where C_*: volumetric sediment concentration of deposited sediment, usually about 0.6 and Qp: the discharge of water alone calculated hydrologic methods.

Other empirical methods, hydrologic methods, and

hydraulic methods are also studied. One of the empirical methods is based on the close relationship between the peak debris-flow discharge and the total debris-flow volume or magnitude (Mizuyama et al., 1992).

STRUCTURAL MEASURES PLAN

The detailed dimensions of debris-flow, the peak discharge for instance, cannot be predicted accurately. The total volume of debris-flow is, however, predicted more accurately and more easily. In cases of debris-flow hazard mitigation plans, safety is given the first priority. The structural measures are planned to check the entire volume of debris-flow. Figure 1 shows some typical plans schematically. It would be considered beneficial if we the occurrence of debris-flow could be contoled. This, however, is impossible at the present. In most cases, we make efforts to prepare sabo dams or check dams whose sediment trap capacity is equivalent to the planed total volume of debris-flow.

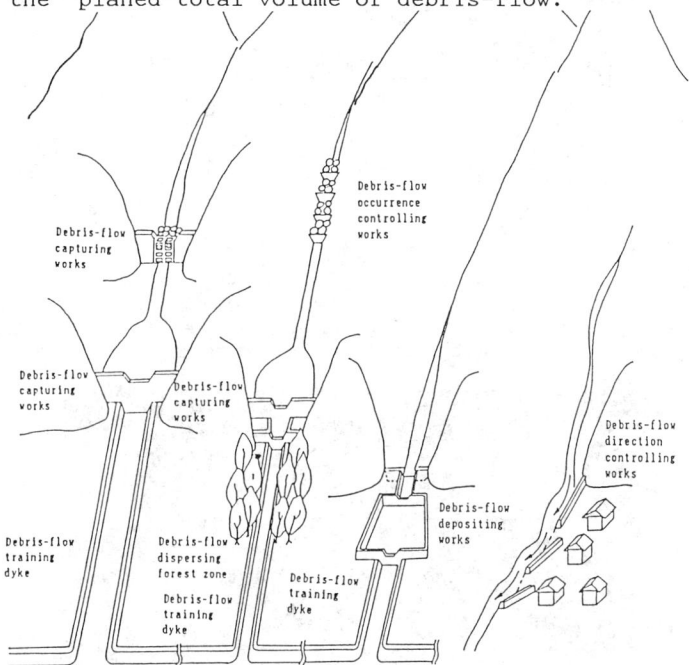

Figure 1 Typical examples of debris-flow control structure arrangements

OPEN-TYPE SABO DAMS

Ordinary sabo dams are gravity dams made of massive concrete. They have the disadvantage that they are filled with sediment by small discharge before debris-flow occurs. To make up for this disadvantage, various types of open-type sabo dams have been proposed and adopted. Concrete made slit sabo dams and steel pipe screens or grid dams are the major ones in use. According to recent studies (Mizuyama et al., 1988), it has been proven that slit sabo dams make the debris flow peak discharge small, but that almost all sediment is then washed away, mainly by successive flow. They are not always safe. Steel pipe open dams are better, although it is not easy to determine its opening space. Figure 2 shows how debris-flow peak discharge varies with the ratio between slit space and the maximum grain size (Watanabe et. al, 1977). Photo 1 is an example of a steel pipe grid sabo dam. Open-type sabo dams are expected to become a major debris-flow control structure.

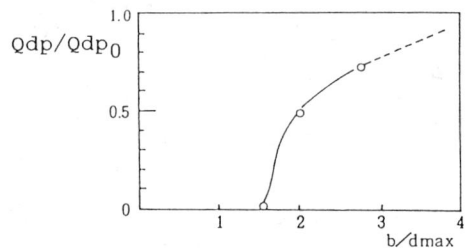

Figure 2 Experiment on the rate of decrease of peak discharge of sediment (Qdp/Qdp_0), where b: the spacing of grids and d_{max}: the maximum grain diameter of debris flow.

Photo 1 A steel pipe grid sabo dam

NEW TREND OF CONTROL STRUCTURES

A debris-flow breaker is a lateral screen set on torrent beds to remove water from debris-flow. It has been attempted. It worked beautifully, even though the effectiveness on the entire debris-flow was not so large. Sabo dams with gates and other steel pipe sabo dams are under development aimed at a more effective control structure.

CONCLUSIONS

Through reviewing debris-flow countermeasures, I recognize that there has not been much progresses in this field in the last fifteen years. It can interpreted from this that research work should contribute to further development of countermeasures. It also can be inferred that field engineers should make efforts to apply the latest research results to practical situations.

REFERENCES

Mizuyama, T., S.Abe and K.Ido (1988)" Sediment control by sabo dams with slits and/or large drainage conduits" 6th Congress Asian and Pacific Regional Division, IAHR, Kyoto, Japan

Mizuyama, T., S.Kobashi and OU, G.(1992)" Prediction of debris flow peak discharge" Interpraevent 1992, Bern
Sabo Division, Public Works Research Institute (1988)" Technical standard for the measures against debris flow (draft)" Technical Memorandum of PWRI, No.2632

Takahashi, T.(1977)" A mechanism of occurrence of mud-debris flows and their characteristics in motion" Kyoto Univ. Disaster Prevention Research Institute Annuals, Vol.20, B-2

Watanabe, M., T.Mizuyama and S.Uehara(1977)" Review of debris-flow countermeasure facilities" Journal of Japan Erosion Control Engineering Society, No.115

EVALUATION OF HYDRAULIC STRUCTURE RELIABILITY
CONSIDERING UNCERTAINTIES IN HYDROLOGIC MODELS

Yeou-Koung Tung[1], Bing Zhao[2] and Jinn-Chuang Yang[3]

Abstract

Uncertainties in a hydrologic model give rise to uncertainty in model outputs which, in turn, can be used for reliability-based design and/or evaluation of hydraulic structures. This paper assessed uncertainties associated with the well-known unit hydrograph (UH) model and incorporate them into the evaluation of overtopping risk and other related measures of a flood detention reservoir. Constrained simulation is employ-ed so that the generated UHs satisfy not only their statistical properties but also its physical definition.

Introduction

Hydrologic modelling often constitutes an important aspect of many hydrological investigations. The hydrological model provides essential information for designing hydraulic structures and other decision-makings. However, due to the presence of uncertainties in modelling process, parameters, and inputs, hydrologic model outputs are also infested with uncertainty. Consequently, the performance of hydraulic designs or decision-makings based on hydrologic model outputs cannot be assessed with certainty.

Although uncertainties in hydrologic and hydraulic

---

[1]Assoc. Prof., Wyoming Water Resources Center, University of Wyoming, Laramie, Wyoming 82071

[2] Graduate Student, Department of Civil Engineering, Arizona State University, Tempe, AZ 85280

[3] Department of Civil Engineering, National Chiao-Tung University, Hsinchu, Taiwan, Republic of China

investigation cannot be avoided, it is important to understand the extent of uncertainties involved in the planning, design, and evaluation. Such information provides decision-makers and engineers with full spectrum of possible outcomes as affected by any decision and design. Knowledge such as this is extremely valuable for selecting appropriate course of action.

In this paper, uncertainties associated with the discrete UH of a specified duration from complex storms in the multi-storm analysis are assessed. Information is further incorporated to evaluate the overtopping probability of a hypothetical detention reservoir.

Assessment of Uncertainty of a Discrete UH

Assume that there are R storms available. The discrete convolution relationship between effective rainfall hyetograph (ERH), direct runoff hydrograph (DRH), and UH can be written, in matrix form, as

$$\begin{bmatrix} q_1 \\ q_2 \\ \cdot \\ \cdot \\ \cdot \\ q_R \end{bmatrix} = \begin{bmatrix} P_1 \\ P_2 \\ \cdot \\ \cdot \\ \cdot \\ P_R \end{bmatrix} u \qquad (1)$$

in which P_i is the ERH matrix of i-th storm, q_i is the DRH vector of the i-th storm, and u is the vector containing the unknown UH ordinates that characterizes the watershed.

Many methods have been developed to solve Eq. 1 for u. Literature also has shown that the multi-storm analysis yields a more desirable UH than that from a single-storm analysis. Zhao (1992) recently provides two theorems supporting the use of the multi-storm analysis for deriving a basin-wide representative UH. Although the multi-storm analysis provides a unique UH, this does not imply that it is free of uncertainty.

To assess the uncertainty associated with the UH from the multi-storm analysis, this paper applies bootstrap resampling technique (Efron, 1982). The techni-que treats the R storm events considered in the multi-storm analysis as random samples from which R of them are randomly selected, with replacement, to form a set of synthesized events. Then, the multi-storm analysis is applied to the R synthesized storm events to derive a UH. The process is repeated for a large number of times to produce many multi-storm UHs from which the uncertainty features of the derived UH are derived.

Incorporating UH Uncertainty in Reliability Analysis

In many hydraulic designs and analyses, the selected design ERH (P_{dsgn}) is convolved with a UH to obtain a design DRH (q_{dsgn}) as

$$q_{dsgn} = P_{dsgn}\, u \qquad (2)$$

based on which the evaluation and design of a hydraulic system are performed. Due to the uncertainties in UH, the design DRH will also subject to uncertainty. In many evaluation and design of a hydraulic structure, the design DRH is routed through the structure from which the structural performance is examined. Under such circumstance, even for a very simple routing procedure, analytical function of outflow hydrograph generally cannot be obtained. In this study, Monte Carlo simulation is applied for reliability analysis.

In Monte Carlo simulation, the values of stochastic input parameters are generated according to their distributional properties. Although the method has been criticized in the past for being computationally intensive, its ability to account for many complex features of real-life systems along with the advent of computing power would make the method more and more attractive in future engineering applications.

Using Monte Carlo simulation for reliability analysis of hydraulic structures considering UH uncertainty, two approaches can be taken. One approach is to first derive the statistical properties of q_{dsgn} based on those of u and Eq. 2. Then, the ordinates in q_{dsgn} are treated as multivariate random variables and q_{dsgn} can be generated based on an assumed probability distribution. The second approach is to generate UH ordinates based on the statistical properties of u which convolved with design ERH to obtain q_{dsgn}. The most commonly used distribution is the multivariate normal distribution.

It should be kept in mind that the generated DRHs or UHs must satisfy the continuity constraint that the volume of each generated DRH equals to design ERH or UH equals to unity. Excellent discussions about the Gaussian unconditional simulation and conditional simulation were given by Borgman (1990). Borgman and Faucette (1993) also proposed a practical method to convert constrained simulation into conditional simulation.

An Application

The uncertainty analysis of multi-storm UH and incorporation of such uncertainty in reliability analysis are applied to Tong-Tou Watershed (820 sq. km.) in Taiwan using 9 storm events. Fig. 1 shows some uncertainty features of 1-hr UH.

For illustrating reliability analysis, a hypothetical detention reservoir is assumed in existence at the mouth of the watershed and one is interested in estimating the probability that water would overtops the reservoir. By constrained simulation, a large number of generated DRHs, plus the baseflow, are routed through the reservoir. The percent by which the water overtops the reservoir is the failure probability of interest. Fig. 2 shows 20 UHs generated by constrained Gaussian Monte Carlo simulation.

In the simulation, the design ERH used was made of the largest ordinates of the observed ERHs at a given time. Fig. 3 shows the design ERH used in simulation. Detail descriptions about the physical characteristics of the reservoir are given by Zhao (1992). The overtopping probability based on 2000 simulations is 0.081. The statistical properties of overtopping duration, peak discharge, and volume are given in Table 1.

Conclusions

Uncertainty analysis of hydrologic model provides useful information with regard to the uncertainty features of model outputs. Results from uncertainty analysis can be incorporated further to evaluate the reliability of existing hydraulic structures or to perform reliability-based designs.

Acknowledgement

The study is sponsored by the Council of Agriculture, Ministry of Economic Affairs, Taiwan, Republic of China. We are thankful to Mr. Wen-Jung Hu and Ms. Yue-

Chuan Huang for their supports.

References

Borgman, L.E. (1990). Irregular ocean waves : kinematics and forces. Ocean Engineering Science. Ed. by Mehaute, B.L. & Hanes, D.M., John Wiley & Sons, Inc.

Borgman, L.E. & Faucette, R.C. Frequency-domain simulation and stochastic interpolation of random vectors in multidimensional space. Computational Stochastic Mechanics. Ed. by Cheng, H-D. & Yang, C.Y. (sched. for pub. in 1993).

Efron, B. (1982). The Jackknife, the Bootstrap and Other Resampling Plans, CBMS 38, SIAM.

Zhao, B. (1992) Determination of a Unit Hydrograph and Its Uncertainty Applied to Reliability Analysis of Hydraulic Structures. M.S. Thesis, Department of Statistics, University of Wyoming, Laramie, Wyoming. July 1992. 352pp.

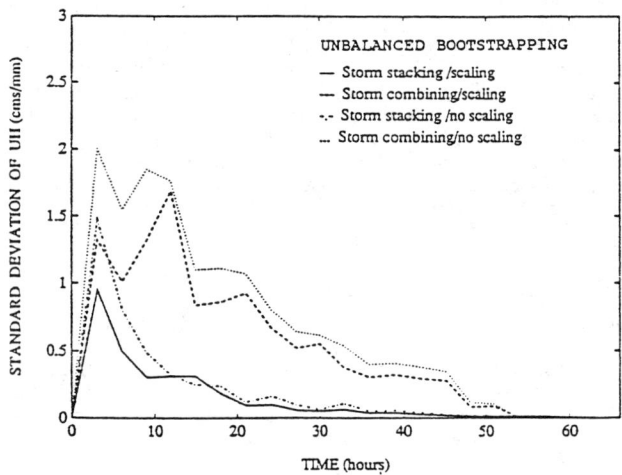

Figure 1. Standard Deviations of Derived Multi-storm UH by Different Methods

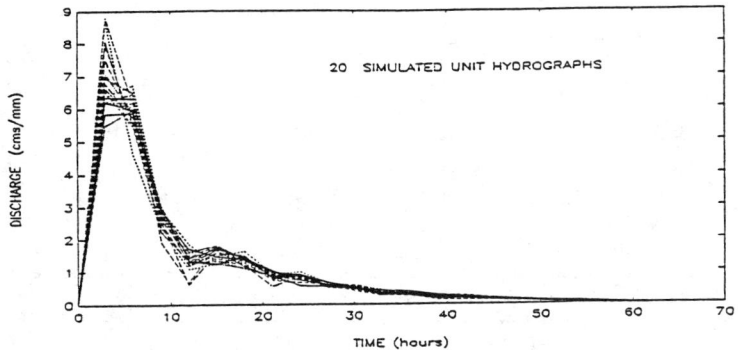

Figure 2. Sample UHs Generated by Constrained Simulation

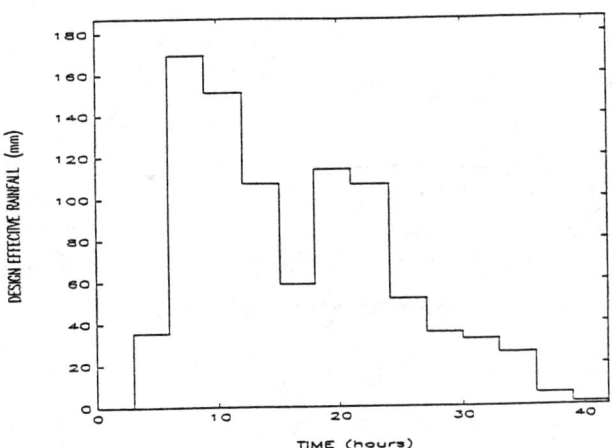

Figure 3. Design ERH Used in Reliability Analysis

Table 1. Statistical Properties of Overtopping Parameters

(a) Means, Standard Deviations, and Skew Coefficients

	Unbalanced bootstrap		
	Duration (hours)	Peak (cms)	Volume (cubic meter)
Mean	4.500e+00	4.444e-01	6.686e+03
STD	2.439e+00	3.801e-01	8.163e+03
CV	1.844e+00	1.169e+00	8.191e-01
Skewness	1.157e+00	1.165e+00	2.093e+00

(b) Correlation Coefficients

	Overflow Duration	Overflow Duration
Overflow Peak	0.8809	
Overflow Volume	0.8555	0.9666

Analyzing Uncertainty of IUH of Nash

J.C. Yang[1], S.Y. Tarng[1], and Y.K. Tung[2]

Abstract

Uncertainties in hydrologic modeling are the primary factors affecting the reliability of hydraulic structures. Among various hydrologic rainfall-runoff models for engineering applications, conceptual instantaneous unit hydrograph (IUH) models are frequently applied for computing design discharge hydrograph for designing hydraulic structrues or evaluating their performances. In this paper, Nash's IUH model is selected for demonstrating the methodological framework to analyze its uncertainty.

Introduction

Modelling of rainfall-runoff process is an important task in hydrologic investigation. Several hydrologic rainfall-runoff models of various degrees complexity are developed. However, the model is only an idealization of the reality and involves uncertainties in its parameters, inputs, and model itself. This paper briefly describes an approach to quantify uncertainties associated with the two parameters in Nash's IUH model which are further used to assess the uncertainty of the IUH ordinates.

Assessing Parameter Uncertainty of Nash IUH Model

Nash's IUH model is one of the frequently used

---

[1] Department of Civil Engineering, National Chiao-tung University, Hsinchu, Taiwan, Republic of China

[2] Wyoming Water Resources Center, University of Wyoming, Laramie, Wyoming 82071

conceptual models for rainfall-runoff study which has the form as

$$u(t) = \frac{1}{K\Gamma(N)} e^{-t/K} \left(\frac{t}{K}\right)^{N-1}, \quad t \geq 0 \tag{1}$$

where u(t) is the ordinate of IUH at time t, K and N are its two parameters. Based on the effective rainfall hyetograph (ERH) and direct runoff hydrograph (DRH) of an observed storm event, method of moments is commonely used to determine the values of K and N (Bras, 1987).

This paper is concerned mainly with the multi-storm setting in which data from several storms are available. In a multi-storm problem, it is not uncommon to obtain different parameter values for different individual storms regardless the approach used to determine the model paramters. Variabilities or uncertainties of the model parameters arise from inadequate model and data errors including recording and processing. When the uncertainties of parameters are large, the use of single-valued representation such as their respective mean values is not adequate.

One simple way to quantify the uncertainties associated with the model parameters is to compute the statistical properties of model parameter values on the basis of analyzing individual storms. However, it is long being recognized that the unit hydrograph (UH) from the single-storm analysis is not stable and less representative as compared to the one from multi-storm analysis in which observed storm data are considered simultaneously in deriving the UH. Although Nash's IUH has no problem with the noise fluctution in its ordinates, its parameters from single-storm analysis could also suffer from lack of representativeness. As the result, the statistical properties of model parameters from the single-storm analysis may also lack representativeness of the true uncertainty feature.

In the multi-storm analysis, the objective is to determine its K and N values which are considered to be the most representative. It should be pointed out that the conventional method of moments is not applicable to multi-storm analysis in which storm data are stacked. This paper adopts the downhill simplex algorithm for determining K and N that minimizes the total squared deviations between the observed and computed direct runoff ordinates for all storms simultaneously. Although, the model parameters obtained are more representative through the multi-storm analysis, it does not mean that they are free of uncertainty.

To assess uncertainty associated with the parameters in Nash's IUH model using multi-storm analysis, a statistical resampling technique, called bootstrap method (Efron, 1982), was used in this study. Assume that rainfall-runoff data about M storms are available. The bootstrap method treats each observed storm event containing the pair of ERH and DRH as a random sample from which storms are randomly selected with replacement to form a data set of M storms. Then, multi-storm analysis is applied to the generated storm set for determining the values of K and N. The process is repeated for a large number of times from which the corresponding model parameter values are computed. With many pairs of (K,N) computed from the boostrapped samples, the statistical properties such as the means, standard deviations, confidence intervals of K and N, and correlation between K and N can be obtained.

As an example, Figure 1(a) and 1(b), respectively, show the histograms of optimal K and N for Lan-yang watershed (821 km^2) in Taiwan, Republic of China. The results are obtained on the basis of 7 storm events from a 2000 bootstrap samples. As can be seen, the optimal N for the watershed appears to follow a shifted exponential distribution whereas optimal K fullows a normal distribution. The mean, standard deviation, and skew coefficient of K and N are shown in the figures. The correlation coefficient between K and N by unbalanced bootstrp is -0.59.

Assessing Uncertainty of Nash IUH

Due to the presence of uncertainties in model parameters, the ordinates of Nash's IUH at different times are also subject to uncertainty. Serrano and Serrano (1990) quantify the uncertainty of u(t) in Eq. 1 using stochastic differential equation without considering the uncertainty of N. To account for the uncertainties in both K and N and their correlation, a computationally simple method, called Rosenblueth point estimation (PE) method, is used herein.

The Rosenblueth PE method replaces the distribution of each involved random variables by concentrating its probability at two discrete points. The locations of the two points and their associated probability masses are determined in such a way that the first three moments of the random variable are preserved (Rosenblueth, 1981).

Furthermore, correlation between K and N can be incorporated into the assessment of uncertainty of u(t).

Figure 2 shows the uncertainty of u(t), in terms of standard deviation, by considering the total and partial effects of uncertainties in K and N. As can be seen, without considering the full effect of uncertainties in K and N, the estimated uncertainty in IUH ordinates could be quite inaccurate for some parts of the UH.

Summary and Conclusions

Uncertainty analysis of the well-known Nash IUH model is presented in the paper. The first part describes the bootstrap resampling technique along with the multi-storm analysis to quantify the uncertainties associated with the two parameters in the model. Once the uncertainties of K and N are computed, they are used to assess the uncertainty associated with the ordinates of the IUH. Specifically, Rosenblueth point estimation technique is used. Results indicate that to assess the uncertainty of u(t) correctly, it is necessary to account fully the uncertainty features of the two parameters in the model.

Acknowledgement

This study was funded by the Council of Agriculture, Ministry of Economic Affairs, Taiwan, Republic of China. The authors wish to thank Mr. Wen-Jung Hu of the Council of Agriculture and Ms. Yeu-Chuan Huang of Provincial Water Conservancy Bureau for their supports.

References

Bras, R. (1990) *Hydrology: An Introduction to Hydrologic Science*, Addison-Wesley Publishing Company, Reading, MA.

Efron, B. (1982) *The Jackknife, the Bootstrap and Other Resampling Plans*, CBMS 38, SIAM.

Rosenblueth, E. (1981) "Two-point estimates in probabilities," *Applied Mathematical Modelling*, 5: 329-335.

Serrano, S. and Serrano, S.E. (1990) "Development of the instantaneous unit hydrograph using stochastic differential equations," *J. of Stochastic Hydrology/ Hydraulics*, 4: 151-160.

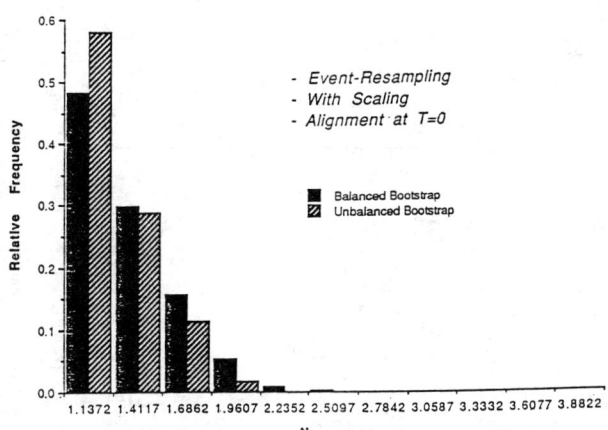

Histogram of N Values By SIMPLEX Method For Lan-Yang Watershed

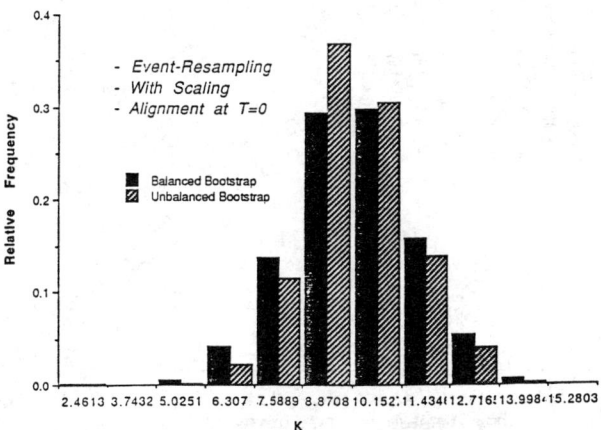

Histogram of K Values By SIMPLEX Method For Lan-Yang Watershed

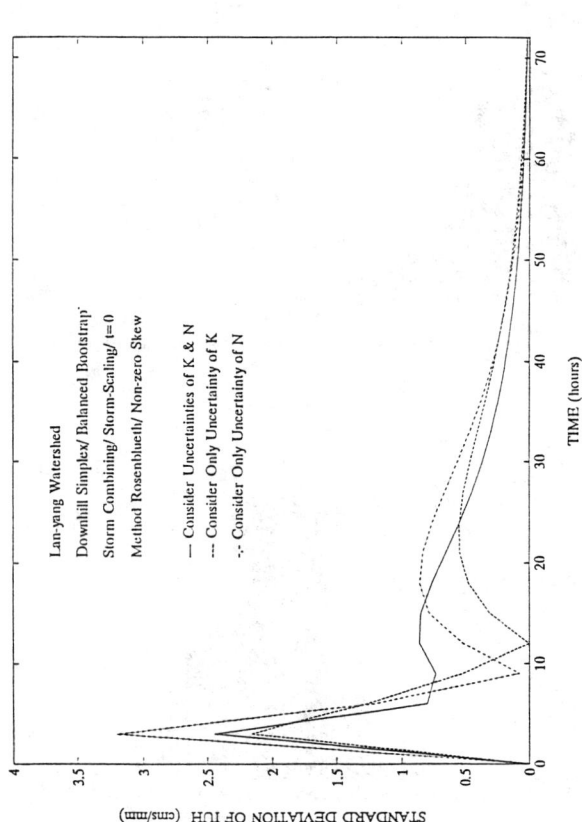

Figure 2 Effect of Considering Complete and Partial Parameter Uncertainty On Nash's IUH Uncertainty by Rosenblueth's Method for Lan-yang Watershed.

Hydrologic Design: Extending Traditional Methods

Delbert D. Franz, M.[1]

Abstract

A review of traditional hydrologic design shows that it is inadequate. The continued adaptation of manual techniques into computer software without a major change in fundamental assumptions is unable to meet the requirements for hydrologic analysis now or in the future. Hydrologists must seek methods that use the power of the computer to yield improved analyses; not just do analyses from the days of manual computation faster. New methods that exploit the ever increasing power of the digital computer and that are also a logical extension of traditional methods are outlined.

Introduction

Traditional hydrologic design has three major components: selection of a synthetic rainfall event, transformation of this synthetic rainfall event into a flow or a hydrograph, and an assignment of a probability of exceedance to the synthetic flow hydrograph. This process is followed whether the rational formula, HEC-1, or the SCS methods are used. Focusing on the basic assumptions or presuppositions for each of these components will reveal weaknesses in these methods.

Selection of Rainfall Event

The rainfall event, or design storm, is selected from rainfall-frequency analyses. Such an analysis assumes that it is meaningful to categorize rainfall events based on duration. By limiting attention to the total rainfall amount for a given duration, the hydrologist can rank the events from largest to smallest as required by frequency analysis. The loss of rainfall intensity pattern that this entails is serious because runoff intensity is dependent on the rainfall intensity. Some design-storm methods partially recover this information by using a standard distribution for the rainfall. This *ad hoc* adjustment presumably leaves the frequency of the event unchanged.

---

[1] Vice President, Linsley, Kraeger Associates, Ltd., 752 Ormonde Drive, Mountain View, CA 94043

Selecting the duration to use for a particular case assumes that there is a single critical duration for a watershed for a given return period. Generally a time of concentration or a time to equilibrium approximates this critical duration. If significant detention storage is present, the critical duration may have to be estimated by trial and error.

Transformation of the Rainfall Event

Once the design storm is selected, it must be transformed into a hydrologic event: a flow rate or a hydrograph. The key assumption used in transforming a design storm to a hydrologic event is the belief that the initial conditions can be set *a priori*. All rainfall-runoff models have initial conditions of some kind that influence how much runoff is produced from a design storm. The values for these initial conditions are stochastic because they depend on the antecedent rainfall as well as other meteorological factors. Clearly it is impossible to set the initial conditions *a priori* meaningfully in all but the simplest cases. Even a parking lot can have significant depression storage as evidenced by the scattered pattern of parked cars whose drivers try to avoid stepping into a puddle when exiting their automobiles. In practice these initial conditions are assigned following some rule or standard.

Assignment of Frequency

The hydrologic model transforms the design storm into a peak flow, a hydrograph, or elevations for reservoirs. These results are the hydrologic response of the model system to the synthetic rainfall event when the assumed initial conditions prevail. To assign a frequency to these results the hydrologist invokes the principle of conservation of rainfall frequency. The principle of conservation of rainfall frequency assumes that the outcome of the rainfall transformation will have the same frequency as the design storm. Furthermore, all parts of the outcome will be given this frequency. That is, the hydrograph produced by the model will be given the same frequency as the rainfall. This assignment is made even though such an assignment does not make sense. In order to assign a frequency to an event, that event must be such that it can be ranked in comparison to other events. But hydrographs cannot be ranked and therefore they cannot have a probability of exceedance and therefore they cannot have a return period. Only single-valued outcomes like peak flows or hydrograph volumes can be ranked and assigned a probability of exceedance. The return period for the peak flow could differ from the return period of the hydrograph volume. Furthermore, neither are likely to be the same as the return period of the design storm used to define them.

At face value the principle of conservation of rainfall frequency is false. The same flood peak could have resulted from a variety of combinations of antecedent conditions and rainfall events with return periods differing from the design return period. The principle of conservation of rainfall frequency fails because it does not integrate the probabilities of outcomes over the space of possible combinations of rainfall intensities, durations, and antecedent conditions.

Need for an Alternative

The major short comings of the traditional methodology are the artificial character of the design storm, the nearly arbitrary definition of the initial conditions in the stream system, and the invocation of the fallacious principle of conservation of rainfall frequency. These shortcomings are of increasing significance as the profession moves away from conveyance-only solutions to stormwater management problems and moves toward the combination of stormwater detention and conveyance. The errors inherent in the traditional methodology are more acceptable in the design of simple conveyance-only systems. However they are unacceptable in complex systems that embody significant amounts of detention storage to ameliorate the effects of watershed changes.

The crux of the problem is the *a priori* selection of the proper design event for the complex storm-water management systems being constructed in our urbanizing watersheds. These systems often include a plethora of structures that affect the flow: culverts, bridges, offline and on line storages, storm sewer discharges, *etc*. The proper selection of a design rainfall that will lead to a reasonable result is nearly impossible. The principle of conservation of rainfall frequency is clearly invalid in these cases. A sequence of short-return-period rainfall events can fill the detention storages because the time available to evacuate them between storms, especially in the case of offline storages, may be too short. Each of the rainfall events, taken in isolation, is too small to cause any problems. Their effect, in sequence is, however, significant.

No single event or small series of events will suffice to define the behavior of these systems. What is needed is a sequence of rainfall events that will reflect what the system is likely to encounter in the future. The use of a single design storm was necessary when most computations were done manually. We can do much better today. The power of the current desk top systems is such that much of that power can be directed to doing a better analysis and not just doing the old analyses faster.

An Alternative

One alternative approach that has great promise is the generation of a long sequence of synthetic storms using stochastic simulation methods. In these methods, a long series of observed rainfall at one or more gages is analyzed to estimate the parameters needed by the stochastic model. The goal is to produce a series of numbers that look like rainfall and have the salient characteristics of the observed rainfalls. This is not different in principle than the methods used for a design storm. The design storm is synthetic: no such storm may have ever occurred. However, based on the observed rainfall, and our statistical analyses, we conclude that it could occur with the given frequency. The stochastic data generation approach takes this process further by creating not one storm but many storms. Such a generator should produce all levels of rainfall as well as the dry periods between rainfall events to be used in the next step.

Just as in the traditional approach, the next step is the transformation of the synthetic rainfall into estimates of synthetic streamflow. However, in this case the transformation model to use is one that simulates the land-phase of the hydrologic cycle, such as the Stanford Watershed Model and its many variations. These models explicitly represent the major features of the surface and subsurface hydrologic processes such as infiltration, soil moisture storage, evapotranspiration, percolation, interflow, ground water discharge, etc.. They produce a record of synthetic streamflow at one or more points in a stream system that is continuous in time. This avoids the problem of selection of initial conditions. The computations can be started in a period between storms, and sufficient time can be provided so that the effect of assumed initial conditions dissipate. As a consequence the problem of choice of initial conditions is essentially eliminated.

Furthermore, the assignment of frequency to the outcomes is no longer a problem. The values produced by the rainfall-runoff model, be they water surface elevations or flow rates, can be analyzed statistically to derive the return periods of interest. Thus the three major problems with the basic assumptions of traditional methods are reduced or eliminated. No design event need be selected, the initial conditions are computed as part of the process, and the assignment of frequency follows naturally and rigorously from the generation process.

This combination of stochastic generation of synthetic rainfall data and rainfall-runoff modeling outlined here is not really new. Work has been done at Stanford University on this approach(Pattison, 1966; Franz, 1971; Ott, 1972; Nasseri, 1976). Bras, et. al.(1985) used a similar approach to study a stream system with eight flood-control reservoirs. Franz, et. al.(1989) reported applications to three streams, including one with two major reservoirs and having a network of nine hourly rainfall stations. The National Research Council(1988) encouraged further research and development on stochastic rainfall models and on the alternative presented here.

Technological Objections

With such a long history why has this combination method not been used more widely? A reason of some validity is that the technology for generating synthetic rainfall is just emerging. Some very general but not site-specific models and some site-specific models have appeared over the past 25 years. What is needed in the near term is a stochastic rainfall model that can use existing sequences of both daily and hourly rainfall in its parameter estimation process so that it can be region specific but simultaneously retain as much of the current knowledge of rainfall processes as possible. Any such model must have a structure that makes it possible to regionalize the parameter estimation process. Such a model does not now exist but the seminal ideas for such a model are available and are under study.

Other technological reasons involve the requirement to extrapolate in both the rainfall model and the rainfall-runoff model to values larger than any observed. However, this objection applies with equal or greater force to the traditional methods as well. All of our design methods extrapolate to conditions that have never been seen before. Extrapolation to rainfall amounts never experienced before is as much a problem for the traditional methods as it is for the combination methods. The same can be said for the transformation process from synthetic rainfall to synthetic streamflow.

Legal-Administrative Objections

Another rich source of objections to any new method in hydrology is the combination of legal and administrative concerns. Our legal system evaluates current practice by past practice. Stated negatively, the courts protect you if you practice mediocrity among the mediocre, and stated positively they protect you if you practice excellence among the excellent! The courts refuse to evaluate technology and leave it to the practice of the profession to determine what is accepted. Carried to it logical extreme this policy prevents the acceptance of any new methods.

The concern for administrative simplicity also favors traditional methods because they require much less skill to use. Much of hydrologic design is done by individuals not trained in hydrology. Much of what is done must be reviewed by one or more governmental agencies. Thus simple traditional methods, even when their technological shortcomings are obvious, may be preferred for these reasons.

Clearly some standards must exist. Unfortunately, there is the universal tendency for standards to become substitutes for thought. Too much of current hydrologic practice is the production of results according to regulation and not according to reality. Regulations, if properly formulated and used, can have a salutary effect on practice. However, they become substitutes for a cogent analysis when applied blindly.

Financial Constraints

A third source of objections falls under the heading of financial constraints. The combination method clearly exchanges simplifying assumptions for computational complexity. But that is exactly the reason we have digital computers: to help us with computational complexity. The current generation of microcomputers and work stations is more than powerful enough to implement this emerging technology. After nearly two decades of using computers to do analyses faster perhaps it is time to try to do them better as well. The radical changes wrought by urbanization will eventually require some form of the combination method. The traditional methods are already inadequate for many basin-wide problems. The result is costly projects that do not function properly.

Conclusions

An examination of the assumptions basic to traditional hydrologic design reveals that there are significant weaknesses. These weaknesses will become far more acute in their effect as the complexity of watersheds increases with continuing development. Alternative methods for analysis must be developed. An emerging method is the combination of stochastic generation of rainfall data with the deterministic simulation of streamflow. Research needs of immediate importance involve a better understanding of the aspects of rainfall that significantly affect flood frequency, and the development of stochastic rainfall models that can improve on those already available. Clearly the rapid increase in computational power offered by the computer revolution of the past several years makes it possible for hydrologists to improve on traditional methods.

References

Bras, R. L., Gaboury,D. G., Grossman, D. S., and Vicens, G. J.(1985), "Spatially Varying Rainfall and Floodrisk Analysis," *J. Hydr. Engrg.*, ASCE, 111(5), 754–773.

Franz, D. D.(1971), "Hourly Rainfall Synthesis for a Network of Stations," *J. Hydr. Div.*, ASCE, 77(9), 1349–1366.

Franz, D. D., Kraeger, B. A., and Linsley, R. K.(1989), 'A System for Generating Long Streamflow Records for Study of Floods of Long Return Period", NUREG/CR-4496, Vol. 2, Nuclear Regulatory Commission, Washington, D. C.

Nasseri,I.(1976), "Regional Flow Frequency Analysis Using Multi-station Stochastic and Deterministic Models," Ph. D. dissertation, Civil Engineering, Stanford University.

National Research Council(1988), *Estimating Probabilities of Extreme Floods: Methods and Recommended Research*, National Academy Press, Washington, D. C.

Ott, R. F.(1972),"Streamflow Frequency Using Stochastically Generated Hourly Rainfall", Tech. Rep. No. 151, Civil Engineering, Stanford University.

Pattison, A.(1966), "Synthesis of Hourly Rainfall Data", *Water Resources Research*, V. 1, 489-498.

IMPORTANCE OF HYDRAULIC-MODEL UNCERTAINTY IN FLOOD-STAGE ESTIMATION

Satvinder Singh,[1] Associate Member, ASCE, and
Charles S. Melching,[2] Member, ASCE

Abstract

The delineation of the regulatory floodplain and determination of the respective flood stage is subject to hydraulic-model uncertainty in transforming the design discharge into a flood stage, in addition to the hydrologic uncertainty in the design discharge. The hydraulic-model uncertainty includes uncertainties in the model formulation used to approximate the dynamic movement of the flood wave, the channel-geometry definition, and the model parameters. An HEC-2 model analysis was performed for the South Branch of the Raritan River near Stanton, New Jersey. The advanced first-order, second-moment method of reliability analysis was used to determine the exceedance probability for flood stage, subject to uncertainties in (1) model formulation and channel geometry, (2) Manning's n of the main channel and overbank, and (3) design discharge. The model formulation was assumed to have a coefficient of variation of prediction of 10 percent, including uncertainty in channel-geometry definition. The coefficients of variation for Manning's n for the overbank and main-channel conditions were 30 and 27.5 percent, respectively. The 100-year flood stage estimated by traditional methods, that is, with only the design discharge uncertain, has an exceedance probability of once in 57 years, considering all significant uncertainties. The combined model-formulation and channel-geometry uncertainties contributed much more to the prediction uncertainty than did Manning's n. Accurate estimation of the flood stage with an exceedance probability of once in 100 years is aided by the use of reliability analysis considering all significant sources of uncertainty. A more theoretically correct model aids in the reliable delineation of a regulatory floodplain when traditional frequency analysis procedures are used.

Introduction

Since the turn of the century, hydraulic engineers have used probabilistic approaches to estimate the magnitude of floods used in hydraulic design such that a specified level of safety is provided. The level of safety is set by statute or regulation as an allowable probability of discharges exceeding the capacity of the structure or flood-plain boundaries. Depending on the data available, the design discharge is determined either by frequency analysis of flows on the stream, regional-frequency analysis, transposition of frequency relations from hydrologically similar streams, or by rainfall-runoff modeling using a storm of the specified probability. The designed structural geometry or the flood-plain boundaries are then calculated from the design discharge using a hydraulic model. The calculation is performed without considering the uncertainties in the hydraulic modeling, which include uncertainties in (1) the model formulation, (2) the geometry of the channel or structure, and (3) the model parameters. Without considering these uncertainties, the level of safety provided by the structure or flood-plain boundary could be far less than is desireable.

---

[1] Sen. Env. Eng., New Jersey Dept. of Envir. Protection and Energy, 401 E. State St., Trenton, NJ 08625
[2] Hyd. Eng., Water Resour. Div., U.S. Geological Survey, 102 E. Main, 4th Floor, Urbana, IL 61801

This paper illustrates the effects of the sources of hydraulic-modeling uncertainty on the delineation of the regulatory flood-plain and, in particular, the importance of model formulation uncertainty. Several researchers have studied the effects of hydraulic-modeling uncertainty on flood-plain delineation. Burges (1979) was the first to consider the effects of such uncertainty. However, he incorrectly equated the variance of the design discharge with the variance of the channel capacity leading to questionable results (Chadderton and Miller 1980; Oegema and McBean 1987). McBean et al. (1984) estimated one standard-deviation confidence bounds on the flood-plain width for the 100-year flood by using the mean-value, first-order, reliability-analysis method (MFORM) to combine the uncertainties in the estimated 100-year flood discharge and Manning's roughness coefficient, n. Oegema and McBean (1987) used MFORM to combine the hydraulic uncertainties in the friction slope, channel geometry, and Manning's n into a probability distribution of hydraulic capacity. This distribution was integrated with the probability distribution that describes the uncertainty in the estimate of the 100-year flood discharge to determine one standard-deviation confidence bounds on flood-plain width. The Hydrologic Engineering Center (1986) used Monte Carlo simulation to determine the mean and maximum absolute errors in water-surface profiles considering uncertainty in Manning's n and hydraulic geometry as determined from aerial spot elevations or topographic maps.

Although each of the cited research efforts provides valuable information on the importance of hydraulic uncertainties, this research does not directly address the question of how the hydraulic uncertainties can be included in the flood-plain delineation such that the specified level of safety is achieved. Cesare (1991) used the advanced, first-order, reliability-analysis method (AFORM) to combine hydraulic uncertainties with the frequency analysis of flood discharges and determined the flood stage and flood-plain width corresponding to various levels of exceedance probability. He demonstrated this approach for a hypothetical trapezoidal stream. This paper extends the approach used by Cesare to a real case of flood-plain delineation using the HEC-2 Water Surface Profiles computer program (Hydrologic Engineering Center, 1990) that uses the standard-step backwater approach to consider one-dimensional, steady, gradually varying flows in open channels.

Reliability-Analysis Method

The reliability, R_L, of any engineering system can be defined as the probability that the capacity of the system, C, is greater than or equal to the load, L, placed on the system; that is,

$$R_L = 1 - P_e = \Pr(C \geq L), \tag{1}$$

where P_e is the exceedance probability and $\Pr(x)$ is the probability of event x. A convenient way to express the reliability of the system is to use a performance function, $Z = C - L$, such that

$$R_L = \Pr(Z \geq 0). \tag{2}$$

In AFORM, a first-order Taylor series expansion is used to approximate Z at the failure surface (i.e., where Z = 0). The relibility of the system can then be described in terms of a reliability index, β, which for the case of independent variables is defined as

$$\beta = E[Z]/\sigma_Z, \tag{3}$$

where $E[Z] \approx g(\mathbf{X}^*) + \sum_{i=1}^{m} C_i (\overline{x}_i - x_i^*)$, (4)

$$\sigma_Z \approx [\sum_{i=1}^{m} (C_i \sigma_i)^2]^{1/2},$$ (5)

and g( ) is the functional form of Z, **X** is the vector of basic variables (e.g., model parameters), * denotes the failure surface, C_i is $\partial g/\partial x_i$ evaluated at $\mathbf{X}^*$ (in this study numerical differentiation is used with $\Delta x_i = 0.04 x_i$), $\overline{x}_i$ is the mean value of basic variable i, σ_i is the standard deviation of basic variable i, and m is the number of basic variables. For convex failure surfaces where the basic variables are normally distributed or can be approximated as normally distributed using the appropriate transformation (Rackwitz, 1976), β can be considered a standard-normal variate. Thus, the reliability can be approximated as

$$R_L = 1 - P_e \approx \Phi(\beta),$$ (6)

where $\Phi( )$ is the standard normal integral. In this study, the point on the failure surface was determined by a modified Rackwitz (1976) iterative algorithm. The iterative algorithm determines the most likely combination of uncertain basic variables that results in a given probability of exceedance. Information on AFORM can be found in Ang and Tang (1984, pp. 340-383).

Example Application

Study Site

A 5,100 ft reach of the South Branch of the Raritan River near Stanton, New Jersey, straddling the State Highway Route 31 bridge (3,800 ft downstream and 1,300 ft upstream) was studied. The survey data for cross-section geometry and reach lengths are from the New Jersey Department of Environmental Protection and Energy's (NJDEPE) 1973 study for this river. Eighty-seven years of historical flow record (1904-90) are available from a U.S. Geological Survey streamflow gage about 11,000 ft upstream from the bridge. The drainage area at the gage is 147 square miles. Because of the large drainage area and the short distance from the gage to the study site, no areal correction was applied when transferring the flood-frequency distribution from the gage to the study site.

Problem Formulation

For the analysis of the uncertainty in the computed flood stage, the performance function was formulated as

$$Z = g(\mathbf{X}) = T_H - \lambda f(n_b, n_c, Q, C_e, C_c, C_{eb}, C_{cb}, C_o, C_w, K),$$ (7)

where T_H is a target level of stage which is varied to produce a curve of exceedance probabilities, λ is a model-correction factor, f( ) represents the HEC-2 model operations that estimate flow depth, n_b is Manning's n for overbank areas, n_c is Manning's n for the main channel, C_e is the coefficient of expansion for river reaches, C_c is the coefficient of contraction for river reaches, C_{eb} is the coefficient of expansion through the bridge, C_{cb} is the coefficient of contraction through the bridge, C_o is a coefficient to be used in the orifice equation for pressurized flow through the bridge, C_w is the coefficient to be used in the weir-flow equation for flow over the bridge, and K is the bridge-pier shape coefficient. The comparison of the estimated flow stage to the target level was performed at a location 770 ft upstream from the bridge. This location is at the upstream end of a 500-ft section of slightly adverse slope, and so it was highly sensitive to backwater effects.

Uncertainties in the Basic Variables

The probability distribution of flood-peak discharge was assumed to be lognormal and the method of moments was used to determine the parameters of the distribution from the 87-year record. The mean and standard deviation of the logarithms of the annual-maximum discharges in cfs were 8.441 and 0.5172, respectively.

The mean values of all other model parameters in f( ) in Eq. 7 were taken as the selected best values from the 1973 NJDEPE study of this site. All of these model parameters are assumed to be normally distributed for ease of illustration and lack of better information. The discussion below describes the estimation of the coefficient of variation (COV = $\sigma_i/\overline{x}_i$) values for the model parameters.

The COV of Manning's n for the main channel and overbank areas was estimated using the procedure developed by the Hydrologic Engineering Center (1986). Hydrologic Engineering Center (HEC) staff and other experienced hydraulic engineers were asked to estimate Manning's n corresponding to the 100-year flood discharge for 10 widely differing stream reaches. By analyzing the range of estimated Manning's n values for the streams, HEC developed a graph for estimating the COV of Manning's n estimates that were obtained by engineering judgement from site inspection, without making discharge measurements or preliminary calibrations to high-water marks. From this graph the main-channel n value of 0.04 has a COV of 27.5 percent and the overbank n value of 0.10 has a COV of 30.0 percent. These COVs are nearly twice those used by McBean et al. (1984), Oegema and McBean (1987), and Cesare (1991).

For the expansion, contraction, and bridge coefficients a range of COV values between 0.05 and 0.25 was examined. The flood-stage results were insensitive to the uncertainty in these coefficients. This insensitivity is due to the fact that the channel is fairly prismatic and the bridge does not significantly constrict the flow at stages above the one-percent exceedance probability level.

The mean value of the model-correction coefficient was assumed to be 1 (i.e., the model is unbiased) and the COV was assumed to be 0.10. These values were considered reasonable for the standard-step backwater method using Manning's equation to estimate the friction slope. In this study, the uncertainties in channel geometry are not considered directly; however, the channel-geometry effects are included in the uncertainty in λ.

Discussion of Results

The exceedance probability in terms of the return period ($=1/P_e$) as a function of stage for both traditional frequency analysis and AFORM considering all significant sources of uncertainty is shown in Fig. 1. As calculated by traditional frequency analysis the stage that is exceeded once, on average, every 100 years is 15.02 ft. This stage is exceeded once every 56.6 years if all significant sources of uncertainty are considered. Considering all significant sources of uncertainty, the stage that is exceeded once, on average, every 100 years is 15.89 ft.

Four combinations of sources of uncertainty were considered: (1)

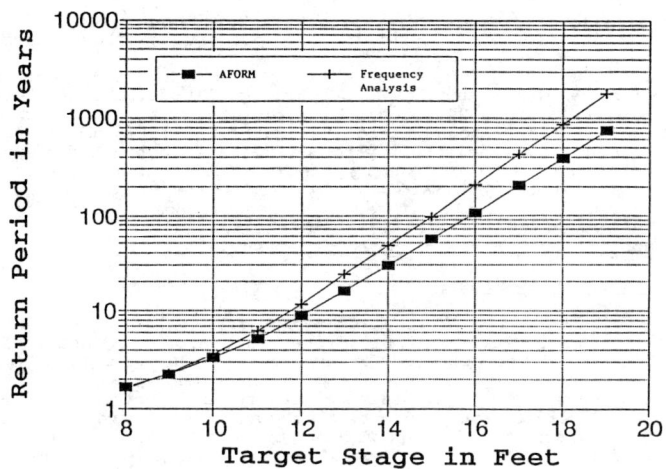

Figure 1. Flow-stage exceedance probability in terms of return period calculated using AFORM considering all significant sources of uncertainty and using traditional frequency analysis.

all sources in Eq. 7; (2) uncertainty in Manning's n, model-correction factor, and discharge only; (3) uncertainty in Manning's n and discharge only; and (4) uncertainty in discharge only (traditional frequency analysis). The exceedance probability in terms of return period for selected flow stages is compared in Table 1. The expansion, contraction, and bridge coefficients have negligible effect on uncertainty in stage. By ignoring uncertainty in the model-correction factor, the uncertainty in stage greatly decreases and the exceedance probability is closer to that calculated from traditional frequency analysis. Thus, the uncertainty in the model-correction factor has a greater influence on uncertainty in stage than does the uncertainty in Manning's n. This contrasts with the findings of Cesare (1991), who found that for a hypothetical trapezoidal channel, uncertainty in flood stage was slightly more influenced by uncertainty in Manning's n than by uncertainty in the model-correction coefficient for equal values of COV for both sources. In this study, the COV for Manning's n is three times that for the model-correction coefficient. A partial explanation for this discrepancy is that the stage in

Table 1. Exceedance probability of flow stage in the form of a return period for cases considering selected sources of uncertainty: (1) all sources in Eq. 7; (2) model-correction factor, λ, Manning's n, and discharge, Q, only; (3) Manning's n and discharge only; and (4) discharge only.

Flow Stage (feet)	Return Period (years) for Sources of Uncertainty			
	All	λ, n, Q	n, Q	Q
10	3.32	3.32	3.49	3.66
12	8.84	8.84	10.5	11.7
14	29.3	29.3	41.1	47.7
16	106	105	171	206
18	390	389	719	862

Cesare's study exceeded once, on average, in 100 years was 11.2 ft and the standard deviation of λ is directly proportional to this stage.

In this study, the uncertainty in the model-correction coefficient includes uncertainty in the model formulation and the channel geometry. The Hydrologic Engineering Center (1986) and Oegema and McBean (1987) found that the effects of channel-geometry uncertainties on the uncertainty in the computed water-surface profile are smaller than those of Manning's n (one fourth or less) unless the geometry is crudely estimated (e.g., from topographic maps with 10-foot contour intervals). Thus, the channel-geometry uncertainties probably amount to a small part of the uncertainty in λ and the consequent effect on the computed water-surface profile.

Summary and Conclusions

A realistic example of flood-stage estimation using HEC-2 has shown that consideration of flood frequency alone can underestimate the river stage exceeded once, on average, in 100 years by as much as a foot in comparison to an analysis considering all significant sources of uncertainty. In this example, the underestimation leads to a level of safety 80 percent less than is desired. Accurate estimation of the flood stage with an exceedance probability of once in one hundred years is aided by reliability analysis considering all significant sources of uncertainty. This example also illustrated that model formulation uncertainty is important in the uncertainty of the estimated flood stage. A theoretically more correct model aids in reliable delineation of a regulatory floodplain when traditional frequency analysis procedures are used.

References

Ang, A. H.-S. and Tang, W.H. (1984). Probability concepts in engineering planning and design Volume II: Decision, risk, and reliability. John Wiley and Sons, New York.

Burges, S. J. (1979). "Analysis of uncertainty in flood plain mapping." Water Resour. Bull., 15(1), 227-243.

Cesare, M. A. (1991). "First-order analysis of open-channel flow." J. Hydr. Eng., ASCE, 117(2), 242-247.

Chadderton, R. A. and Miller, A. C. (1980). "Discussion: Analysis of uncertainty in flood plain mapping, by S. J. Burges." Water Resour. Bull., 16(4), 752-754.

Hydrologic Engineering Center. (1986). "Accuracy of computed water surface profiles." Research Document 26, U.S. Army Corps of Engineers, Davis, CA.

Hydrologic Engineering Center. (1990). "HEC-2, Water surface profiles, User's manual." U.S. Army Corps of Engineers, Davis, CA.

McBean, E., Penel, J., and Siu, K.-L. (1984). "Uncertainty analysis of a delineated floodplain." Can. J. Civ. Eng., 11, 387-395.

Oegema, B. W. and McBean, E. A. (1987). "Uncertainties in flood plain mapping." in Application of Frequency and Risk in Water Resources, V. P. Singh, ed., D. Reidel Publishing Company, Dordrecht, The Netherlands, 293-303.

Rackwitz, R. (1976). "Practical probabilistic approach to design." Bulletin 112, Comite European du Beton, Paris, France.

RAINFALL DEPTH-DURATION-FREQUENCY ANALYSIS FOR MAJOR CITIES IN TAIWAN

Baolin Wu[1], Ming-hsi Hsu[2], Victor Yih[3], Shaohsing King[4], Chian-min Wu[5],

Abstract

The annual series of extreme rainfall amounts for various durations are compiled. The statistics such as mean, standard deviation, coefficient of variation, coefficient of skewness, and coefficient of kurtosis are calculated. The durations selected in this study range from 5 minutes to 72 hours. The average record length is 49.0 years. The computed coefficients of skewness and coefficients of kurtosis of each series are plotted in the skewness-kurtosis diagram to determine the best fit distribution. The probability distributions investigated in this study are the log-normal, Pearson Type III, Gumbel, Weibull, and Burr Type III distributions. The results reveal that the overall best fit distribution is the Pearson Type III Distribution. The rainfall amounts for each duration and each return period are calculated by the frequency factor method. The return periods used in this study range from 2 to 1000 years. Finally, the rainfall depth-duration-frequency curves are constructed for each city and the rainfall intensity-duration-frequency curves can be converted from the corresponding depth curves.

The resulting rainfall depth-duration-frequency and rainfall intensity-duration-frequency curves can be used for various hydrologic designs, analyses, and studies in Taiwan.

---

1. Senior Hydrologist, Contra Costa County Flood Control District, Martinez, California 94553

2. Professor and Director, Hydraulic Research Laboratory, National Taiwan University, Taipei, ROC

3. Professor, Department of Agricultural Engineering, National Taiwan University, Taipei, Taiwan, ROC

4. Chief, Information Systems Division, Water Resources Planning Commission, Taipei, Taiwan, ROC

5. Chairman, Water Resources Planning Commission, Taipei, Taiwan, ROC

1. Introduction

The island of Taiwan is located about 160 kilometers (100 miles) off the southeast coast of the mainland China. Latitudes range from 21.9° N to 25.3° N; longitudes range from 120° E to 122° E. The area is 36,000 square kilometers (13,900 square miles). The Center Mountain Range runs from north to south with its highest peak at 3,997 meters (13,113 feet).

The Climate of the island is subtropical with hot, wet summers and mild winters. The mean annual rainfall ia 2,500 millimeters (98.6 inches). From May through October, typhoons and thunderstorms bring in abundant rainfall equal to approximately 78 percent of the annual total. During the winter months, the northeast monsoon prevails and brings in moderate rainfall to the north.

2. Basic Data

The rainfall data used for frequency analysis was compiled chronologically in an annual series, which contains the extreme values of various durations for each year and for each raingage station. In this study, the selected rainfall durations are 5-, 10-, 15-, 30-, and 60-minute and 2-, 3-, 6-, 12-, 24-, 48-, and 72-hour. Most of this data was collected by the Central Weather Bureau.

The cities selected in this study are Keelung, Taipei, Hsinchu, Taichung, Chiayi, Tainan, and Kaohsiung. A total of eighty-four annual series were assembled. The record length ranged from 23 years to 87 years with an average of 49.0 years and a standard deviation of 19.5 years.

3. Method of Analysis

For a set of n discrete data, $x_1, x_2, \ldots, x_n$, with a mean of $\bar{x}$ the unbiased estimates of the first four central moments can be expressed as follows:

$$M_1' = \frac{1}{n} \sum_{i=1}^{n} (x_i - \bar{x}) = 0 \quad \quad (1)$$

$$M_2' = \frac{1}{n-1} \sum_{i=1}^{n} (x_i - \bar{x})^2 \quad \quad (2)$$

$$M_3' = \frac{1}{(n-1)(n-2)} \sum_{i=1}^{n} (x_i - \bar{x})^3 \quad \quad (3)$$

$$M_4' = \frac{1}{(n-1)(n-2)(n-3)} \left[n(n+1) \sum_{i=1}^{n} (x_i - \bar{x})^4 - 3(n-1) \left(\sum_{i=1}^{n} (x_i - \bar{x})^2 \right)^2 \right] \quad \quad (4)$$

Where $\bar{x}$ is the sample mean which can be computed by the following equation.

$$\bar{x} = \frac{1}{n-1} \sum_{i=1}^{n} x_i \quad \quad (5)$$

The unbiased sample standard deviation, $\hat{s}$, is one of the best descriptions of dispersion for the study of many hydrologic problems. It is equal to the square root of M_2' such as,

$$\hat{s} = (M_2')^{1/2} \quad \text{(6)}$$

The unbiased sample coefficient of variation, $\hat{g}_0$, the dimensionless dispersion parameter, is the ratio of the unbiased sample standard deviation to the sample mean,

$$\hat{g}_0 = \hat{s}/\bar{x} \quad \text{(7)}$$

The unbiased sample coefficient of skewness of the observed data, $\hat{g}_1$, can be estimated as

$$\hat{g}_1 = M_3'/(M_2')^{3/2} \quad \text{(8)}$$

The coefficient of skewness is the description of asymmetry. For a symmetrical distribution, the coefficient of skewness is equal to zero. A distribution with a negative skewness is said to be skewed to the left while a distribution with a positive skewness is said to be skewed to the right. In other words, a positively skewed distribution has a long tail on the right side. Rainfall is limited by zero at the left and virtually unlimited to the right. It is, therefore, positively skewed.

The unbiased sample coefficient of kurtosis, $\hat{g}_2$, can be estimated by the following equation.

$$\hat{g}_2 = M_4'/(M_2')^2 \quad \text{(9)}$$

This is the descriptor of peakedness or flatness of a distribution. For a normal distribution, the coefficient of kurtosis is equal to zero. A negative kurtosis indicates that a distribution is plateau-like near the center with relatively high density tails as compared to the normal distribution. A positive kurtosis indicates that a distribution has a relatively high peak and low density tails.

The statistics of each annual series, such as mean, standard deviation, coefficient of variation, coefficient of skewness and coefficient of kurtosis were calculated by the above equations.

4. Selection of Probability Distribution

A number of probability distributions are available for rainfall frequency analysis. They include: type I extremal distribution (Gumbel distribution), type III extremal distribution (Weibull distribution), log-normal distribution, and Pearson type III distribution. The relationship among those distributions is shown in Figure 1. Normal, Gumbel and exponential distributions are represented by single points indicating that those distributions have unique values of g_1 and g_2. On the other hand, log-normal, Pearson type III and Weibull distributions are represented by curves. These distributions have a wide range of g_1 and g_2.

It is interesting to note that the normal distribution is the special case of log-normal and Pearson type III distributions, the exponential distribution is the special case of Pearson type III and Weibull distributions, and the Gumbel distribution, practically, is a special case of the log-normal distribution when $g_1 = 1.13$. The diagram in Figure 1 can be used to select a probability distribution if the computed g_1 and g_2 for that data series closely match those of one particular distribution.

For each selected duration, the average values of the computed g_1 and g_2 are plotted in Figure 1. The distribution with the best overall fit was found to be the Pearson type III distribution. Therefore, Pearson type III distribution is employed for frequency analysis in this study.

5. Rainfall Depth-Duration-Frequency Curves

For each station and each duration, the rainfall amounts of various recurrence interval or return period, T, can be calculated by the following equation.

$$X_T = \bar{x} + s\, K_T \quad\quad\quad\quad (10)$$

Where x is the sample mean of the annual series, s is the unbiased estimate of the standard deviation and K_T is the percentile or frequency factor with a return period equal to T years.

Frequency factors of the Pearson type III distribution were prepared by Harter. Frequency factors of the log-normal, Weibull, Gumbel, exponential and normal distribution were prepared by Wu.

The return periods selected in this study are 2-, 5-, 10-, 25-, 50-, 100-, 500-, and 1000-year. For each return period, the computed rainfall depths (amounts) were plotted against their durations, then a smooth curve was fit through those points. The rainfall depth-duration-frequency curves established for the city of Taipei are shown in Figure 2.

6. Rainfall Intensity-Duration-Frequency Curves

For the short duration portion of the rainfall depth-duration-frequency curves, the rainfall depths can be converted into rainfall intensities. Thus, rainfall intensity-duration-frequency curves can be obtained from the corresponding rainfall-depth-duration-frequency curves.

7. Discussion

Based on the computed g_1 and g_2 of the annual series compiled for these major cities in Taiwan, it appears that the Pearson type III distribution is the best distribution for frequency analysis.

The rainfall depth-duration-frequency curves and the corresponding intensity-duration-frequency curves established for these cities can be used for various hydrologic designs and studies.

Reference

1. Chow, V. T., A General Formula For Hydrologic Frequency Analysis, Trans, Am. Geophys. Union, Vol. 32, pp. 231-237, 1951

2. Harter, H.L., A New Table of Percentage Points of the Pearson Type III Distribution, CTU Tech. Release No. 38, Soil Conservation Service, 1968

3. Wu, B. and J. D. Goodridge, On the Selection of Probability Distributions, AGU Fall Meeting, 1974

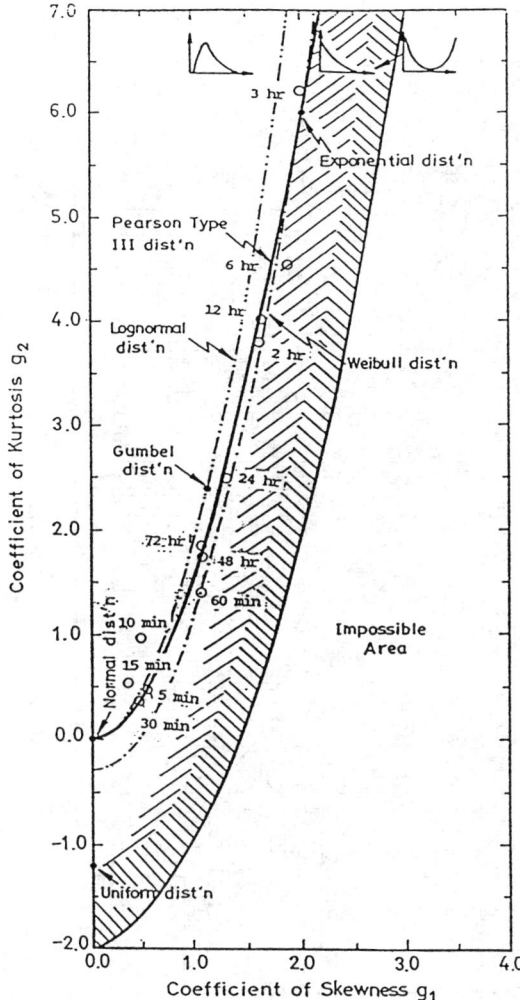

Figure 1. Overall Fittness between Coefficients of Skewness and Kurtosis of the Observed Data and Those of Various Distributions

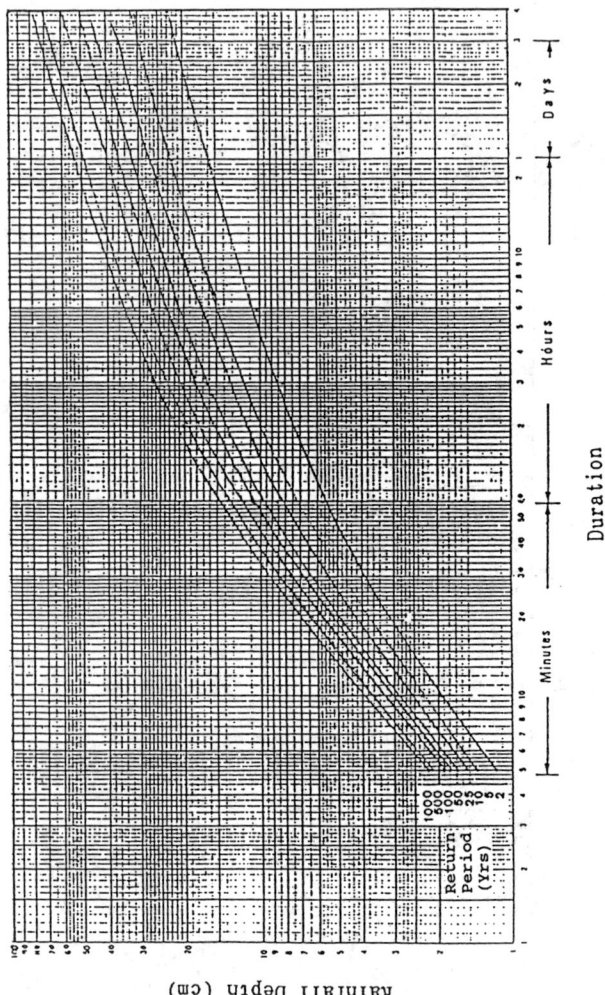

Figure 2. Rainfall Depth-Duration-Frequency Curves for the City of Taipei

LARGE-SCALE EMBANKMENT OVERTOPPING PROTECTION TESTS

Kathleen H. Frizell[1] and James F. Ruff,[2] F. ASCE

Abstract

The U.S. Bureau of Reclamation (Reclamation), Colorado State University (CSU), and the Electric Power Research Institute (EPRI) have an ongoing cooperative research effort to determine low-cost, feasible methods for providing overtopping protection for embankment dams. Investigations have progressed to testing an overlapping tapered concrete block shape developed from Reclamation's laboratory flume tests and installing the blocks over gravel filter material in a near-prototype size facility. The overlapping portion of the block produces an offset, or step, where drains, located through the blocks, provide relief of uplift pressure in the underlying filter. The test program in the large facility closely matched that of the laboratory. The stability of the overlapping tapered block system has been confirmed by the large scale tests.

Purpose

The purpose of conducting large scale tests of overtopping protection methods is to confirm Froude scaling relationships or develop other relationships between laboratory data and the near prototype size facility. Should the block system developed from the laboratory data (Frizell, 1992) show stability, then the results may be comfortably extended to any size actual embankment dam.

---

[1] Hydraulic Engineer, U.S. Bureau of Reclamation, P.O. Box 25007, Denver, CO 80225

[2] Prof. of Civil Engineering, Colorado State University, Engineering Research Center, Fort Collins, CO 80523.

Large-scale Facility and Initial Tests

The outdoor overtopping facility is near-prototype size, with a height of 15.24 m (50 ft) and is located at CSU, in Fort Collins, Colorado. The facility, shown in Figure 1, consists of a concrete headbox, chute, tailbox, and sump with a pump. The concrete chute on a 2:1 (H:V) slope, has a maximum width of 3 m (10 ft) with a removable wall installed to reduce the chute width to 1.52 m (5 ft) for the current testing program. Water is supplied through a 0.91 m (3-ft) pipe from Horsetooth Reservoir. A portion of the flow can be recirculated by pumping back from the tailbox to increase the total discharge through the facility. Unit discharges up to about 2.94 m^3/m/s (31.6 ft^3/ft/s) have been tested.

Figure 1. Fifty foot high flume facility used to test overlapping blocks for embankment dam protection.

Tests are currently being conducted on overlapping tapered concrete blocks design by Reclamation. The blocks are placed over 15.2 cm (6-in) of free-draining, angular, well graded, gravel filter material. The gravel is placed on the concrete floor with 10.1 cm (4-in) angle iron (with a gap above the floor to allow free discharge) placed every 1.81 m (6-ft) up the slope to prevent sliding. A wooden strip was installed along each wall to easily screen the gravel bedding and to prevent failure along the wall contact during operation.

The blocks, shown in Figure 2, are 0.37 m (1.23 ft) long and 63.5 mm (0.21 ft) high with a maximum thickness of 0.11 m (0.375 ft). The blocks are fabricated 0.61 m (2 ft) and/or 0.31 m (1 ft) wide with drains located through the block from the rise of the step to the underside. Two 0.61 m wide blocks and one 0.31 m wide block comprise each row in the facility. The blocks are installed shingle-fashion from the toe and are alternated so that there are no continuous seams in the flow direction except along the walls.

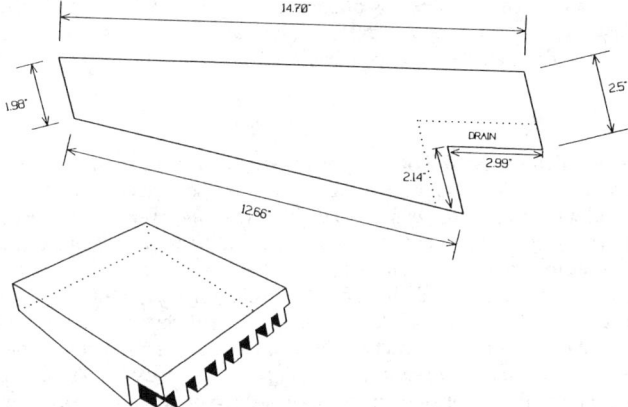

Figure 2. Wedge block dimensions.

At the crest of the structure, a small concrete cap was placed to transition from the flat approach to the first row of blocks. At the toe of the concrete slope is a fixed concrete end block to support the blocks up the slope. A row of blocks are tied down to the angle iron on the floor at about third points up the slope. Where the blocks will be under the tailwater at the toe of the slope, the blocks are pinned together longitudinally through the overlapping area parallel to the slope.

Test procedure

The laboratory tests performed by Reclamation in 1990 and 1991 are being repeated in the large scale facility. The initial tests, under flows similar to scaled laboratory flows, were conducted to obtain pressure data for block stability analyzes. The instrumented blocks (Fig. 2), and accompanying piezometer blocks buried in the gravel bedding, were installed in five locations down the slope of the facility. Pressures measured on the block faces and in the gravel bedding are used to determine the stability of the hydraulically designed block shape.

Flow description

During initial startup of the flume, under a very low discharge, the fines and dirt were flushed from the bedding material. Flushing lasted a very short time and was observed by the brief coloring of the water. After shutting off the water, slight settling of the blocks was apparent; however, there was no sliding or noticeable trend to the settling. Throughout the testing no further noticeable settling of the blocks occurred. The maximum settlement was about 2 to 3 cm (0.79 to 6.11 in).

The many discharges tested in the flume produced varied flow conditions over the blocks. The very small flows were almost entirely broken up by the block shape leaving no noticeable thickness of solid water. As the discharge increased, the boundary layer took longer to develop, eventually developing for the largest flow one third to one half the distance down the slope.

Stability

The question of stability of the protective system is the most critical for an embankment dam. Any failure or instability in the system could cause a catastrophic failure of the entire dam during an overtopping event. Laboratory data shows that the ability of the blocks to relieve the uplift pressure, combined with the impact of the water on the block surface, make the blocks inherently stable. The near-prototype tests, completed thus far to a unit discharge of 2.94 m^3/m/s (31.6 ft^3/ft/s), indicate that the blocks are stable and will perform satisfactorily.

The stability of the block system has been analyzed as a function of the total forces acting on individual blocks down the slope. The block weight and impact pressure act on the block and slope in a downward (positive) direction to keep the blocks on the slope (Fig. 3). The uplift pressure in the bedding material underneath the block and the low pressure zone created by the block offset act in an upward (negative) direction tending to lift the blocks from the embankment surface. In the analysis, a net positive force indicates a stable block.

Pressure data were gathered to compute the magnitude of the forces acting on the block surfaces. In general, the pressures in the impact zone on the block increased with discharge and remained the same or decreased slightly with distance down the slope. Decreasing pressure magnitudes with distance down the slope are, most likely, a function of flow aeration. Of course, the weight of the block is constant. In general, the pressure in the offset area of the block decreases with discharge and distance down the slope. Between step 44 and step 74 down the embankment the pressures in the offset area became negative.

The uplift pressures were measured by using piezometer blocks buried in the gravel bedding at about the same locations down the slope as the instrumented blocks where surface pressures were measured.

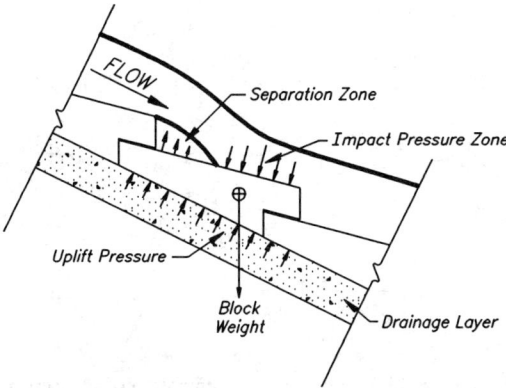

Figure 3. Forces acting on wedge block.

The underdrain pressures were assumed to be linear between the measurement locations. The underdrain pressures show a gradual increase over about the first 45 steps, as would be expected from the pressure data and low flow velocities. These data confirm the hypothesis that flow would be forced into the bedding near the top of the slope. At about 50 steps down the slope the pressures begin quickly decreasing to the fixed toe of the slope where the pressure increases slightly to about 0.15 m (0.5 ft) of positive pressure for all flow rates.

Conclusions

The overall stability of the block system down the slope is given in Figure 4. The resultant vertical force on the block at various locations down the slope is the sum of all the measured pressures integrated over the appropriate areas. These data show that the block system is stable at all locations down the slope and for all flow rates tested, with the exception of slight instability at the toe for the smallest unit discharge. In general, there is from 30 to 170 pounds of force per foot of width in the downward direction holding the blocks on the slope. In this initial analysis consideration was given to the additional benefit of block overlap. The overlap forces would further enhance the block system stability.

These initial calculations on the block stability confirm analytically the visual observation that the block system is inherently stable. This conclusion will be further investigated by more clearly defining the underdrain pressures with more measurement locations.

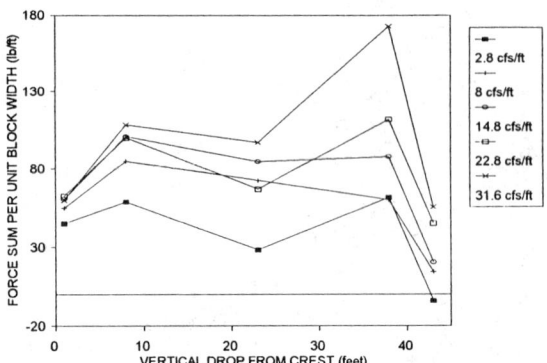

Figure 4. Block stability indicated by summation of pressure forces acting on wedge blocks at locations down the slope.

Future tests

Full model-prototype comparisons will be made at the completion of the tests in the spring of 1993. The remaining tests will primarily measure velocity, air concentration, additional pressures in the bedding material, and block stability under tailwater conditions. The final tests will address stability of the block system after satisfactorily initiating weaknesses in the block system.

Initial model/prototype comparisons show favorable scaling of the pressure field. Relationships and effect of air concentration on the velocity or pressure scaling have yet to be determined.

Upon completion of tests with the block system, the facility will be used to test large size riprap. This will allow confirmation of the numerous laboratory studies with riprap and determine the limits where riprap may be used to protect steep slopes during small overtopping events.

References

1. Frizell, K.H., 1992, "Hydraulics of stepped spillways for RCC dams and dam rehabilitations," Proceedings, 1992 ASCE Roller Compacted Concrete III Conference, pp. 423-439, San Diego, CA, February 2-5.

PROGRESSIVE FAILURE OF AN OVERTOPPED EMBANKMENT

Ghassan AlQaser [1] and James F. Ruff, [2] F. ASCE

Abstract

The failure of an overtopped embankment is a problem of a very complicated nature. Concerned organizations are interested in information that can be used in future planning and development of areas downstream from an embankment of a potential failure. A study of the progressive failure of an overtopped embankment was designed and implemented. The eventual use of the test results will be in modifying the hypotheses used so far in dam break models.

The test results showed completely different mechanics for the development and the progression of the breach than the mechanics adopted in the available dam break models. Based on the experimental results, the erosion processes of an overtopped embankment will involve tractive shear erosion, headward cutting, and structural failures. It is found that, depending on the size of the flood, the failure of an overtopped embankment may be minimal, partial, or complete.

Introduction

The shortage of the spillway capacity in maintaining a successful passage of the design flood is one of the principle reasons for a dam to be overtopped and to fail. Rehabilitation of dams, and contingency plans for specific locations require techniques that are able to assess the consequence of damage. These techniques are available, but only to limited degree of reliability. The lack of adequate information led engineers to resort to some substitutions which may not efficiently represent some of the substantial elements in a dam break model.

---

[1] Post Doctoral Fellow, Engineering Research Center, Colorado State University, Fort Collins, CO 80523.

[2] Professor of Civil Engineering, Engineering Research Center, Colorado State University, Fort Collins, CO 80523.

Overtopping any embankment will eventually result in damage. The extent of damage depends on the specific material properties of the dam fill, the geometrical design, the reservoir storage, the magnitude and the volume of the inflow hydrograph, and the geometrical properties of the valley (NRC, 1983). The interrelation of the variables listed above and the error involved in the estimation of each of them will produce uncertainty in the predictions of any dam break model. However, these predictions should still be within the reliable limits for practical use. A study of the progression of failure of an overtopped embankment was designed and implemented to better understand the processes involved.

Dam Break Modelling

Generally, the computation of the breach outflow hydrograph is one of the major tasks in dam break modelling. Engineers attempt to predict the breach characteristics and the outflow emanating from the breach based on physical principles and embankment material properties. Records show that gradual failures are more likely to happen which is the basic assumption of this approach (Froleich, 1990). The resulting model is called a physically-based mathematical model.

Presently, the major deficiencies in dam break models are the use of hypothetical geometrical relationships for simulating the progression of the failure, and the adaptation of an erosion model solely based on tractive shear theory (Fread, 1984). Another deficiency is the difficulty of modelling the erosion of compacted cohesive soils (Singh, 1989).

Experimental work and test procedure

Two embankments were built and tested in an outdoor flume facility. Locally available cohesive material was used to build the embankment models. The clay was mixed with graded sand comprising a sandy clay soil which is classified as SC soil type in accordance with the unified soil classification system. The embankments were 3.66 m long, 1.22 m high, and had a crest 0.3 m wide. The upstream and downstream slopes were 2.75:1 and 2.5:1, respectively. An average dry soil density was found to be 19.5×10^3 N/m^3 (124 lb/ft^3).

Testing the two models consisted of a series of overtopping runs for each of them. Overtopping was initiated by releasing water into the upstream reservoir. The water level was allowed to rise above the dam crest initiating an overtopping to the embankment crest. The overtopping was temporarily ceased when it was felt that there was a significant change in the geometry of the developing breach and data collection was necessary.

Analysis

Degradation of the downstream slope begins when the hydrodynamic forces exerted by the flowing water (tractive shear stress) exceed the resistive forces of the material comprising the surface (critical shear stress). The tendency of the flow to reach a state of equilibrium along the downstream slope results in the development of accelerating flow with a nonuniform distribution of the shear stress. This will result in the formation of a number of overfalls. A schematic presentation for the development of these overfalls is given in Figure 1. The erosion process will continue until the flow pattern changes into a free surface flow where the water jet will separate from the subreach body producing a free fall jet. This agrees with Ralston's description of the breach development (Ralston, 1987).

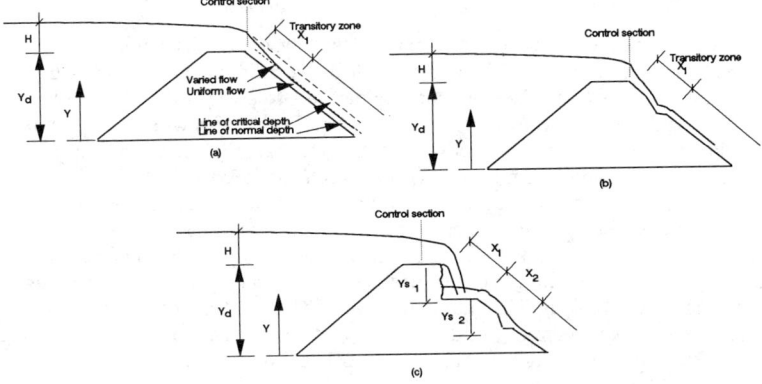

Figure 1. Progressive failure of an overtopped embankment.

The location of the control section changes with time as the crest erodes both headward and downward (see Figure 2). The high stress concentration at the lower edge of the crest will trigger a headward cutting that will move faster than the headward movement of the lower overfalls until it reaches the upstream slope.

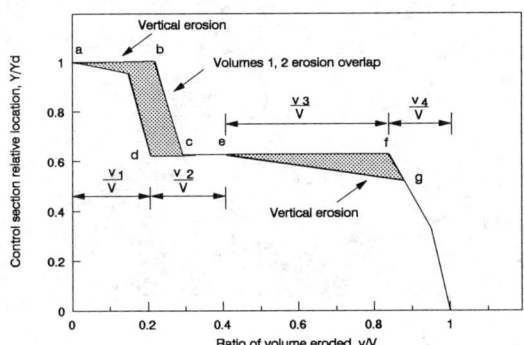

Figure 2. Location of the control section of an overtopped embankment.

The concentration of the flow due to the sag in the embankment crest or the presence of a section that was poorly compacted, will result in initiating a breach through the crest of the embankment. At this point, the breach will progress both laterally and vertically. This process will continue until the headward cutting of the next overfall takes over the crest. The longitudinal channel length of the bottom of the breach will get longer as the crest degrades downward and the side of the breach gets higher producing a narrow flow channel compared to the total width of the embankment.

The progressive headward cutting of the lower overfalls continues until the development of merely one big overfall with almost a flat top and a narrow breach channel. This also was observed in the failure of Black Creek Spillway (Temple, 1989). At this stage, headward cutting and structural failures will be the dominant erosion processes. The only tractive shear erosion that might be taking place is the downward degradation of the crest on the breach channel bed. The impact of the falling jet, and the turbulence associated with it, cause the base of the overfall to deepen and widen, and the face and sides of the overfall to be undermined. If the face and sides of the overfall are high enough, the cohesive strength of the soil can no longer support the weight of that portion with the undermined base resulting in cracking and falling of big chunks of soil. This process continues until the embankment is completely eroded out or stopped because no more water is flowing out. The final shape of the breach for model 1 is shown in Figure 3.

Figure 3.　　Progressive failure of an overtopped embankment, Model 1, Run 7, Time = 731 minutes.

Results

Based on the theoretical considerations, and experimental observations, results, and analysis, a procedure is developed to provide a basis for future dam break models. The proposed procedure was tested on a hypothetical overtopping failure of a 4.6 m high embankment. Typical reservoir characteristics were used. The computations were carried out for many combinations of breach widths and inflow hydrographs, see Figure 4 and Figure 5.

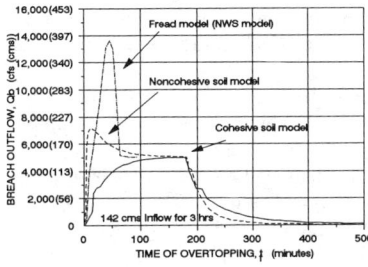 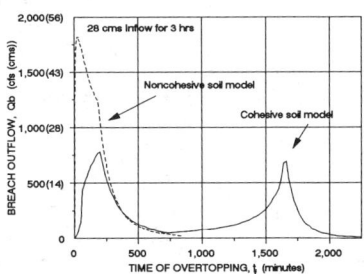

Figure 4. Breach outflow hydrograph.   Figure 5. Breach outflow hydrograph.

A constant inflow of 141.6 m^3/s was assumed to last three hours, four hours, and six hours after the initial overtopping had started. The six hours inflow resulted in a complete failure, whereas, the three hours inflow resulted in a partial failure of the embankment. The predicted breach outflow hydrographs were compared to predictions made by using a noncohesive soil erosion relationship with a corresponding geometrical model (Simmler, 1982) and to predictions by the NWS model, see Figure 4 and Figure 5. Both approaches predicted the same peak outflow with a complete failure for the three cases.

Conclusions

The probability of occurrence of adverse events in nature put reliable dam break models in great need. Professional engineers are reluctant to make definitive statements regarding the ability of a particular embankment to withstand overtopping. The test results from this study showed completely different mechanics and erosion processes for the development and the progression of the breach than what has been adopted in the available dam break models. The erosion processes will involve headward cutting, tractive shear erosion, and structural failures. The selection of an independent erosion model for the assessment of each of the processes involved is a necessity for realistic modelling of the problem.

Compacted cohesive soils have shown unsteady erosion rates when the exerted hydrodynamic forces remain unchanged. Therefore, the use of the classical (present) sediment transport models for the assessment of the eroded volumes will lead to over estimating the outflow hydrograph, and result in a relatively faster failure of the embankment.

References

1. AlQaser, G., (1991), "Progressive failure of an overtopped embankment," Ph.D. Dissertation submitted to Colorado State University, Ft. Collins, CO.
2. Dodge, R., (1988), "Overtopping flow on low embankment dams, summary report of model tests," Report No. Rec-ERC-88-3, U.S. Bureau of Reclamation.
3. Fread, D., (1984), "DAMBRK: THE NWS dam break flood forecasting model," National Weather Service, Hydro-Technical Note No. 4, Hydraulic Research Laboratory.
4. Froleich, D., (1990), "Dam BREAK flood modelling," Ph.D. Dissertation submitted to Colorado State University, Fort Collins, CO.
5. NRC, (1983), "Safety of existing dams, evaluation and improvement."
6. Ralston, D., (1987), "Mechanics of embankment erosion during overflow," ASCE, Hydraulic Engineering Conference.
7. Simmler, H. and Smater, L., (1982), "Dam failure from overtopping studied on a hydraulic model," ICOLD, 14th Congress, Vol. I, Q. 52, R. 26.
8. Singh, V. and Scarloctos, P., (1989), "Breach erosion of earth fill dams and flood routing (beed) model," Military Hydrology Report No. 14, U.S. Army Corps of Engineers.
9. Temple, D., (1989), "Mechanics of an earth spillway failure," ASAE, Paper No. SWR 89-2000.

A New Type of Concrete Trapezoid Dam

Yui Tsang Chan[1], M.ASCE

Abstract

A new type of 128.0 m height concrete trapezoid dam (t-d) has been built for the Wunan Town Hydroelectric Power station (WTHPS). The paper presents its optimum section design, several provisions for the monoliths on the river bank and rectifying the effect of longitudinal joints on the stresses at the dam heel, and a lot of the structural model tests and stress analyses by the finite element method (f-m), on which the design is based.

Introduction

The WTHPS in Zhejiang Province in China is composed of a surface power house, a 1.0 km length round 6.0 m dia. tunnel, a raft way, and the t-d dam of 128.0 m max. height and 430.0 m crest length to provide a yearly regulating reservoir whose capacity is $2100.0*10^6$ m^3.

To select a proper type of dam for this project was a dominant factor. A total concrete volume of a mass concrete gravity dam scheme was estimated to be $2.15*10^6$ m^3, but a concrete buttress dam (b-d) scheme was about $1.0*10^6$. The main problem of b-d is to be easy to crack at the upstream face of the buttress head, because such a great difference of the section width between the head and web exists, that it causes the tensile stress in the upstream part of the buttress head during the water pressure applied, especially in the case of the large temperature change, as induced by the climate and the hydration heat of the cement, its effect worsens the tensile zone, resulting in crack developed both along the upstream face and toward the buttress body. That once occurred at the b-d of the Heng Ren HPS in the north of china, and another one at the Zhe Xi HPS in China also.

---
[1]Engineer, Civil Engineering Department, China Light & Power Co. Ltd., 215 Fuk Wah Street, Hong Kong.

Although a double buttress dam is able to overcome this major disadvantage of the single buttress dam, it brings about not only more formwork work, but also other complications of the concreting in the more narrow construction blocks. These lead to a slow construction rate and cost raised. For greatly improving b-d, based upon the considerable analyses with the computer programs and model tests, we creatively designed the new type of concrete gravity dam with an enlarged head in the upstream side, called the t-d because of its plane shape being a trapezoid. In theory, there was only a connecting point between 2 monoliths. In engineering practice, this point was replaced by a transverse joint (t-j) for placing the water stop and the well for drainage and inspection, as shown in fig.1 & 2.

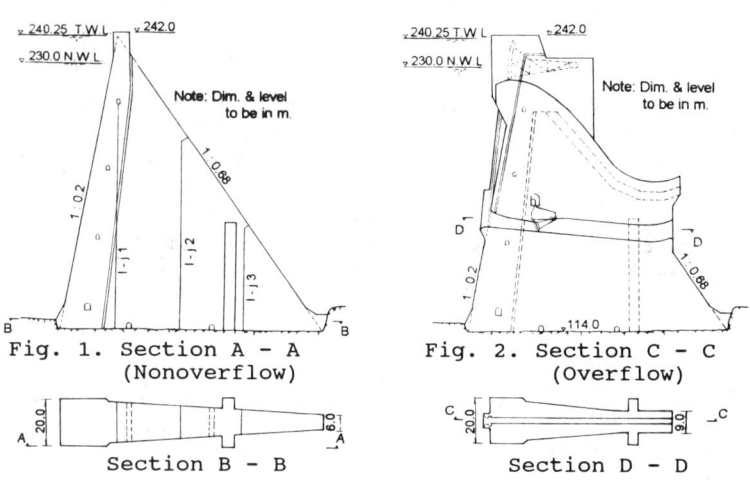

Fig. 1. Section A - A (Nonoverflow)

Fig. 2. Section C - C (Overflow)

Section B - B

Section D - D

The practice in the WTHPS has proved the following advantages of the trapezoid dam:

- cutting down the concrete volume of $0.87*10^6$ m^3, as compared with the mass concrete gravity dam (g-d);
- preventing the concrete dam from such cracks at the upstream face as that ones in the buttress dam;
- reducing uplift force, produced by seepage and floatage, in virtue of the existence of the relieving surface between monoliths;
- being of advantage to send out hydration heat as a result of more lateral surface of dam, so that saving cooling pipe system, etc.;
- being no need for slantingly suspended formwork, used for concreting of buttress head, and a great convenience of construction to use the large size formwork to speed up the construction rate.

Optimum Monolith Design

To minimize a total concrete volume for the t-d, we had to find out the optimum parameters for the monolith according with the following criteria for the reservoir full at the top water lever(T.W.L.), including the uplift unless otherwise indicated:

- 1.05 factor of safety against sliding at the dam base, depending on the friction only;
- σ_u, min. principal stress by the method of Strength of Materials (s-m) to be compressive at the dam base, and not less then 0.25p in the dam body (p = water head on the section in consideration, and the seepage excluded).

That was a complicated optimum design task with 4 parameters, i.e.the upstream and downstream slopes (n,m), and the upstream and downstream widths (T,t). Firstly, T1 & t1 were assumed. Analysing the curves, representing $n = f_1(\sigma_u)$ and $n = f_2(m)$, we got the optimum m_{T1} & n_{T1}, corresponding to T1 & t1. Proceeding with the same way,we had another m_{T1} & n_{T1} of T1 & t2, and then m_{T1} & n_{T1} of T1 & t3, etc. Based on the above results easily we got the optimum m_{T1}, n_{T1} & t_{T1}. corresponding to T1. The same processes were repeated, the m_{T2}, n_{T2} & t_{T2} of T2, etc. were found. After the huge detailed calculations for the WTHPS, we have obtained the optimum m, n, T & t, corresponding to the shape of the river valley and the friction coefficient. It should be pointed that as considering a deep outlet arranged inside the dam body, a 9.0 m overflow monolith width at the downstream face was adopted instead of 6.0 m for the nonoverflow monolith. The above results are shown in Fig.1 & 2.

Dam on River bank

Both river banks are steeper, especially in the left bank, having the angle of gradient between $30°$ and $40°$. The terraces were required to be excavated for the monoliths on the banks, but some still have been found on 2, even 3 terraces. therefor a problem of the lateral stability against sliding had to be dealt with. Its analyses were a greatly difficult task.

To solve the above problem, a 5.0 m thickness vertical concrete stiffener walls, separated by the t-js without key, have been added near the downstream side, as shown on Fig.1 & 2. These lower parts of the t-js, i.e. about 20.0 m height above the base, were required to be grouted after the reservoir impounding water to a certain level. The monolith's t-js were treated in the same way, so that these parts of dam body would work as a monolithic layer,being impossible to the lateral sliding.

As the t-js without any key were grouted only within the less areas near the base after a certain water pressure applied, the water pressures transferred from the monoliths in the middle of the valley to these ones on the banks, were not taken into consideration.

Stress Analyses by Strength of Material

To determine the monolith section to be needed, the conventional method of s-m was still employed. The dam for the WTHPS was divided into 22 monoliths. Every monolith was considered to be a vertical cantilever element. Except the foregoing stress criteria for σ_u, principal stress at the downstream face σ_d should not exceed the permissible compressive stress in concrete for the reservoir full, and 1.0 kg/cm^2 permissible tensile stress for the reservoir empty. The results of min. σ_u at the base, controlling the design, together with the results for the other dams in China for comparison are listed in Tab.1.

Table 1

Name of HPS	Dam	Height (m)	n	Load	σ_u(kg/cm^2) (1)	(2)
Wunan Town	t-d	128.0	1:0.2	T.W.L.	2.8	47.60 (0.380p)
Zhe Xi	b-d	104.0	1:0.45	T.W.L.	28.3	48.00 (0.470p)
Feng Shu	g-d	95.1	1:0.3	T.W.L.	-6.6	28.18 (0.486p)

Notes: $\sigma_{u(1)}$ -- uplift, taken into account;
$\sigma_{u(2)}$ -- uplift, not taken into account;
(-) -- tensile stress;
σ_u in the round bracket, represented by p.

Compressive Stress Loss at Dam Heel

Considering the effect of the hydration heat and contraction, and the concreting equipment capacity, the monolith was divided into 4 construction blocks by vertical longitudinal joints (l-j), as shown on Fig.1. The l-js were provided with about 2.0 m height keys and grouting system buried. Owing to the l-js grouted after about 44.0 m height construction blocks raised, the max. vertical compressive stress $\sigma_{u,v}$ at the heel due to the own weight of dam was decreased by 4.0 and 5.0 kg/cm^2, calculated with s-m and f-m respectively to consider the effect of the gap width of the l-j without grouting.

To rectify the stress loss, a 6.0 kg/cm^2 higher grout pressure at grout pipe outlet was adopted only for the second grout blanket of l-j1. the grouting to the first blanket of l-j1 right on the dam foundation progressed after the second blanket. This rectification work has created a 6.0 kg/cm^2 max. $\sigma_{u,v}$ at the heel, measured with the the strain gauge buried in the dam. This provision has been successful instead of a scheme of adding concrete to the upstream surface of the monolith.

Tests and Finite Element Method

As this is a new type of dam, a lot of the photoelastic and gypsum structural model tests (p-m & g-m), and numerous stress analyses by f-m were carried out. From studying these, we have concluded the followings[1]:

- stresses, obtained by the p-m & g-m, basically being uniform along the section width to let us simplify a 3-dimensional problem to a 2-dimensional one;
- stress distribution at the section more (1/6 - 1/7)H from the base by s-m, where H denotes height of dam, being similar to these by p-m & g-m;
- being no concentration of the max.σ_d at the toe, i.e. the ratio of 51.48 kg/cm^2 by s-m to these values of 45.0 to 48.1 kg/cm^2 by p-m & f-m to be approximately 1.0, in contrast with g-d of 1.5 to 3.0 times its values by s-m because of the more stiffness;
- principal tensile zone of 6.5 m height 13.5 m width at the dam heel due to corner effect, starting at the intersection of the base and upstream slope of dam.

The dam is founded on the rhyolite. The site investigation proved the rhyolite to be suitable for the t-d foundation. In the middle part of the valley, a cube crushing strength 2900 kg/cm^2 hard rock was encountered. There has been a more safety factor of bearing capacity.

In regard of the upstream tensile zone in the dam, there were not the criteria for that. The other similar dams, having been in operation, were analysed for comparison. These tensile zone (t-z) are listed in Tab.2.

Table 2

Name of HPS	Country	Dam	Height (m)	m-b (m)	t-z (H*B in m)	r	E_d/E_f
Wunan Town	China	t-d	128.00	110.8	6.5*13.5	0.12	1.33
SFJ	China	b-d	103.00	97.0	7.6*13.6	0.14	1.00
Norfork	USA	g-d	73.33	66.67	2.94*10.0	0.15	0.40
Dworshak	USA	g-d	211.3	161.0	1.3*37.0	0.23	1.05

Notes: SFJ -- Sun Fung Jiang HPS;
m-b -- monolith thickness at base;
r -- B/(m-b);
E_d -- modulus of elasticity of dam concrete;
E_f -- modulus of elasticity of rock foundation;
tensile zones, corresponding to load case of own weight of dam and normal water level.

In fact, working condition of dam depends not only on own weight of dam and water head, but also uplift, temperature change effect during construction and operation period, construction blockment and procedure, moduli of elasticity of dam and foundation, corner effect, expansion of concrete soaked in water, and concrete creep, etc. For the t-d, the above effects were analysed by the f-m. The principal tensile stress at the dam heel amounted to such a figure, that the concrete would crack. We assumed that the crack of about 15.0 m length developed toward the original joint of the rhyolite with much steeper dip, i.e. nearly vertical, the above mentioned tensile zone disappeared because of the stress redistribution. There were the compressive stresses at the end of this crack, indicating the crack being in the stable condition.

The principal stresses near the heel are the compressive, measured on the full scale dam, such as on the Sunanjian Concrete Gravity Dam in China, and Dworshak dam in U.S.A. At the 125.0 m height concrete gravity dam for the Bratsk HPS in C.O.I.S. the cracks of 0.1-0.15 mm gap width in the rock within about 9-11.0 m depth below the dam heel are found out after the reservoir impounding water$^{(2)}$. These phenomena coincide with the above analyses.

Conclusions

. The reservoir for the WTHPS stored up water in 1979, the trapezoid dam experienced several floods, especially that one in 1991. The concrete trapezoid dam has been in a more satisfactory working condition;
. The stress analyses for the section near the base by the conventional method of Strength of Materials greatly differ from these by f-m and models test. For further studying its real working condition, the the full-size observation is very important.

Reference

1. Y.T. Chan, The design and stress analyses for a new type of concrete trapezoid dam in the Wunan Town Hydroelectric Power Station, East China Institute of Exploration & Design of Electricity, 1974.
2. Y.T. Chan, Stress analyses of concrete trapezoid dam, Proc. of Hydro-Electricity, 1975.

Assessing the True Value of Flood Control Reservoirs: The Experience of Folsom Dam in the February 1986 Flood

Philip B. Williams, MASCE[1]

Abstract

Over the last 40 years considerable public investments have been made in large flood control storage reservoirs throughout the U.S. and internationally. These projects were typically designed to control floods in excess of the 100-year event. Although there have been only a few decades of operating experience for these projects, there is now sufficient evidence to show that the actual flood benefits achieved can be considerably smaller than promised in their design. The operating performance of Folsom Dam on the American River above Sacramento is illustrative. In the February, 1986 flood—variously estimated to be between the 40- and 70-year event—the reservoir was completely filled forcing releases higher than the designated floodway capacity and use of surcharge storage space. Possible catastrophe to Sacramento was averted by the early cessation of rainfall.

This paper describes two different assessments of this event, both of which have profound implications for flood control policy in the U.S. One carried out by the Corps of Engineers significantly downgrades the capability of Folsom, thereby justifying the need for a new flood control reservoir upstream at Auburn. The second, carried out by the author, focusses on the role of Folsom as part of a larger flood management system emphasizing the importance of efficient flood control operation.

Operating Reservoirs for Flood Control

Most major flood control reservoirs protecting urban areas are designed to store flood flows larger than a designated floodway channel capacity downstream. The objective is to reduce downstream flood stages to "non-damaging" levels during extremely large floods. The flood storage, the outlet capacity, and operational regime are determined by simulations of the operation during historical and synthesized flood hydrographs. These simulations form the basis of both the prescribed operating regulations for the reservoir and the calculation of the non-reimbursable public flood protection benefits.

---

[1]President, Philip Williams & Assoc. Ltd., Pier 35, The Embarcadero, San Francisco, CA 94133.

Because many flood control reservoirs are also multi-purpose reservoirs there can be competition for the use of storage between irrigation, power generation, recreation and flood control. In part because it recognized this inherent conflict, the US Congress' 1944 Flood Control Act gave the Corps of Engineers responsibility for developing flood control operational plans and regulating their implementation to ensure flood benefits are realized in Federal multipurpose reservoirs.

To carry out this responsibility the Corps developed guidelines and procedures for reservoir operation in the 1950's when many large flood control reservoirs were being designed. These procedures emphasized the importance of proper operational management in achieving flood benefits. For example:

"The temptation to infringe on flood-control space is sometimes strong, because usually losses to other functions are obvious, and losses to flood control (although usually much greater) may not occur or indeed probably will not occur in any particular case. Consequently, a very rigid attitude against infringement on flood-control space must be maintained at all times."

The Design Operation for Folsom Dam

The operational plan for the 1.2 billion m^3 multipurpose Folsom Reservoir was formulated in the 1950's and reserved 490 million m^3 of storage in the winter period to store flood flows in excess of the 115,000 cfs designated floodway capacity of the lower American River downstream. Folsom was originally intended to control the record historic flood of 1862, and was sized to control what was then thought to be the Standard Project Flood (SPF) with a peak flow of 9,620 m^3/s and storage requirement above the 3,250 m^3/s maximum release of about 490 million m^3. While the dam was under construction revised hydrologic analysis indicated that the SPF had been underestimated and that actually this flood had a return frequency of about 125 years, and accordingly it was redesignated the Reservoir Design Flood.

In establishing the flood control operation for Folsom the Corps emphasized the importance of following the prescribed operating criteria:

"Flood control releases from Folsom Reservoir will be very large in comparison with the space reserved for flood control, and the adequacy of this reservoir for flood control is dependent upon the full and prompt use of the release capacity. For example, a 12-hour delay in the initiation of flood control releases would have the same effect on flood control as would diminishing the flood control space by 100,000 acre-feet. Consequently, it is essential that the need for releases be determined at the earliest possible time and that action to initiate releases be taken as soon as possible."

The Operation of Folsom Dam in the 1986 Flood

The first part of the winter of 1985-86 was dry and attention was mainly focussed on possible shortfalls in irrigation water deliveries. On February 12, the reservoir was being maintained at a storage level of 876 million m^3, leaving 370 million out of the 490 million m^3 of flood storage reservation available. On this day weather forecasts were broadcast predicting the "worst storms of the season" and "record low pressure systems" advancing into California from the Pacific.

By February 14 major flooding was occurring on the Russian and Napa Rivers that drain the coastal range, and heavy rain was falling in the American River watershed. However, no attempt was made to draw down the reservoir and releases were maintained at about 570 m³/s until 1400 hrs on the 16th. During this time inflow increased to about 60,000 cfs 1,700 m³/s and the reservoir rose to 968 million m³. Although increased releases were made later on the 16th they lagged the increased inflow by about 12 hours (Figure 1) and it was not until 0200 on the 18th that the designated release of 3,250 m³/s was made. By this time the reservoir storage level had reached 1.1 billion m³ and two more days of heavy rain were predicted.

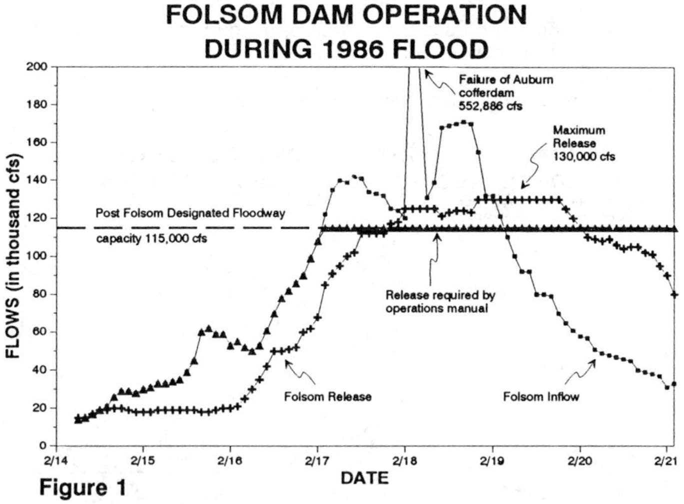

FOLSOM DAM OPERATION DURING 1986 FLOOD

Figure 1

At this point a temporary cofferdam built across the American River above Folsom Dam started to overtop. This cofferdam, constructed 10 years before as part of the now suspended Auburn Dam, had been designed to overtop and washout in the event of a 10- to 30-year flood, releasing about 120 million m³ of stored water. On the 18th, apparently to take this inflow into account, releases were increased to 3,540 m³/s. At about 1400 the cofferdam collapsed and the reservoir was almost completely filled at a time when the rainflood inflow was rising to a peak of 4,920 m³/s.

Fortunately for the City of Sacramento downstream, at this point the rain stopped and the flood was controlled by storing water in the emergency surcharge space above the level of the flood control reservation, and by making releases of 3,680 m³/s. Although severe levee erosion occurred at a few locations downstream there was adequate freeboard. However record flood stages occurred at the confluence of the American and Sacramento Rivers, in part because the delay in releases at Folsom caused the flood peaks to coincide.

Analysis of Folsom Dam's Operation

Although the 1986 flood produced record flood volumes in the Sacramento Basin, the size of flood on the American River as it affected the flood operation of Folsom, was not exceptional and smaller than the design flood (Figure 2). The one day unimpaired peak flow is between about the 30- and 70-year event depending on which flood frequency methodology is selected. The actual total flood inflow volume above

3,250 m³/s was about 350 million m3, somewhat lower than the unimpaired flow due to the storage effect of upstream hydro reservoirs.

The actual performance of Folsom Dam during the 1986 flood was significantly less effective than the operation on which its flood benefits were based, and which was required in the Corps' Reservoir Regulation Manual. The most significant deviations were:
- storage of water in the flood reservation prior to the flood
- significant delays in making required releases ·
- use of surcharge storage
- making releases higher than designated floodway capacity

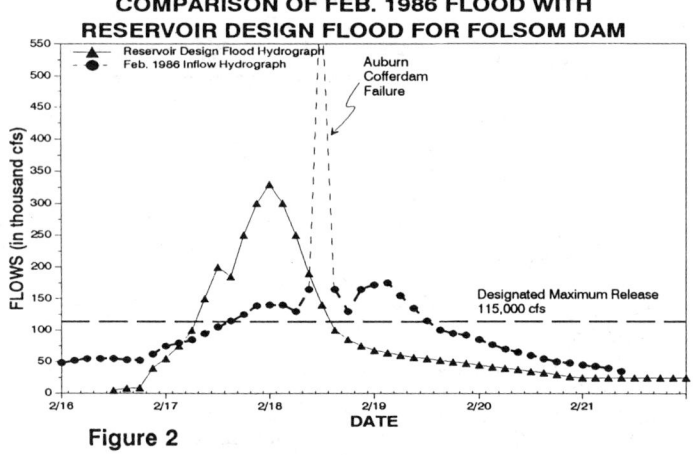

Figure 2

The Corps Response to the 1986 Flood

The narrow escape of the City of Sacramento from a major flooding catastrophe initiated a series of major studies by the Corps of Engineers on the adequacy of flood protection that concluded:
- The 1986 flood was a record flood with a maximum one day flow of 5,770 m³/s.
- The size of the 100-year and 200-year floods had been underestimated.
- It was impossible to operate Folsom as effectively as it had been planned in the 1950s and that it was now only able to control the 70-year flood.
- The most cost effective means of providing 200-year protection is the construction of a major new flood control dam at Auburn.

To reflect its new view of the downgraded operation of Folsom, the Corps made substantial revisions to its operating manual in 1987. These revisions de-emphasize the need of the dam operators to follow strict operating instructions and allow for significant delays in releases. In effect the new operating manual ratified retroactively the operational deviations that occurred in 1986.

An Alternative Evaluation of the 1986 Flood

Because of the environmental impacts, the costs, and the long time required for implementing the proposed Auburn Dam, a number of citizens organizations and agencies were interested in alternatives to the Corps proposal. These studies had very different conclusions that derived from a different conception of flood control. These studies examined flood control reservoir storage as a component of the larger flood management system. They then identified the most cost effective steps for incremental improvements in flood damage reduction within the existing system. This approach is to be contrasted with the Corps emphasis on structural components of the flood management system, its assumption that flood control reservoirs provide the most reliable means of protection, and their focus on providing protection for a specific size of flood (e.g. 200-year).

The alternative studies concluded:
- The Corps had exaggerated the size of the 1986 flood, particularly as it affected the operation of Folsom Dam. After correcting for computational errors the maximum one day unimpaired flow is 4,882 m^3/s, smaller than three historic floods.
- If the Dam had been operated as required by the Corps at that time, there would have been no need for the emergency measures taken and there would have been about 120 million m^3 of flood storage available on February 19 to control additional inflow if it had occurred as forecasted.
- Although there was a small increase in the estimated size of the 100-year flood, the construction and operation of a number of large hydro dams caused a more significant reduction in flood peaks. This beneficial effect could be recognized and incorporated in the flood management system.
- The most cost effective flood protection measure was taking steps to ensure that Folsom is operated the way it was intended. These steps would include training of operators, ensuring adequate oversight and accountability for operational decisions, lifting of perceived restrictions on timely releases such as removal of portable restrooms in the floodway, reinstatement of unambiguous operating instructions.
- The operational effectiveness of Folsom can be improved by making operational decisions based on real time monitoring of stream flow and precipitation.
- The reliability of flood protection can be significantly improved by recognizing the importance of ensuring the integrity of the downstream levee system as the primary flood defense. This requires repair of existing levees and upgrading the maintenance and inspection system. An independent study has shown that the lower American River levees have deteriorated significantly and require immediate remedial action.
- The floodway capacity can be increased at the same time as repairing the existing levees.
- The spillway of Folsom can be modified to allow higher releases at low reservoir levels and make more effective use of its flood storage.
- The risk of coincident flood peaks on the Sacramento and American Rivers can be reduced by operating Folsom as originally intended.

Implications for National Flood Management Policy

Congress has now deferred action on the proposal to construct a new Auburn Dam and has instructed the Corps to engage in a new evaluation that takes into account these alternative perspectives on flood management. As part of this study the National Academy of Engineering has been asked to advise on methodology. This new evaluation has profound implications for flood management and the operation of flood control reservoirs throughout the U.S.

If the Corps is correct in its assessment of Folsom's operation it means that protection provided by many other flood control reservoirs throughout the country may be substantially less reliable than previously assumed, requiring the downgrading of the effectiveness of similar projects that were planned up until 1987. This substantial loss of public flood protection benefits could then lead to the need for massive increases in flood control expenses as new dams and levees are proposed to correct the perceived deficiencies of existing dams.

If the flood management systems approach is correct it signals the need for a substantial revision and rethinking of the Corps' new approach to reservoir operational management. It also points the way to a number of inexpensive means of providing more cost effective improvements in flood protection by ensuring that our existing flood control investment is used in the most efficient way.

References

U.S. Army COE. Folsom Dam & Reservoir Reoperation, California: Operation Plan and Environmental Impact Statement (Draft Report). Sacramento: March 1992.

U.S. Army COE. Folsom Dam and Lake, American River, CA, Water Control Manual. Sacramento: December 1987.

U.S. Army COE and The State of California Reclamation Board. American River Watershed Investigation California Feasibility Report. Sacramento: December 1991.

U.S. Army COE. Reservoir Regulation Manual for Flood Control: Folsom Dam and Reservoir, American River, CA. Sacramento: October 1956; Rev., March 1959.

U.S. Army COE. Reservoir Operation Criteria for Flood Control: Sacramento-San Joaquin Valley, CA. Sacramento: October 1959.

Williams, Philip B. and Julie Galton. Analysis of the 100-Year Flood into Folsom Reservoir. San Francisco: March 27, 1987.

Williams, Philip B. Analysis of the February 1986 Flood on the Lower American River. San Francisco: February 25, 1986.

WRC Environmental and Mitchell Swanson and Associates. A Review of Key Issues from the American River Watershed Investigation Feasibility Study and an Analysis of Alternatives for Increasing Flood Protection in Sacramento. Sacramento: March 16, 1992.

CHATUGE HYDROPROJECT AERATING INFUSER PHYSICAL MODEL STUDY

TONY A. RIZK* and GARY E. HAUSER*

ABSTRACT

The Tennessee Valley Authority is investigating the feasibility of a dual purpose reregulation and aeration weir to increase the dissolved oxygen in water releases and to provide continuous minimum flow downstream of the Chatuge hydroproject, on the Hiwassee River in North Carolina. The new structure is dubbed the infuser. The infuser weir consists of a reregulation weir to which a metal grating is attached. Turbulent water jets fall through the grating openings producing a frothy tailwater with excellent aeration performance. The full-scale testing issues included infuser aeration performance, hydraulic safety considerations, and an examination of fish and trash passage. The infuser weir will maintain minimum flow and mitigate seasonal low dissolved oxygen water releases. The tailwater was found to be free of hydraulic jumps. The infuser aeration performance is affected by the sizes and spacings of the blockages. An optimum infuser blockages configuration yielded an infuser aeration efficiency exceeding 70 percent.

INTRODUCTION

Historically, weirs have been used to regulate water supplies. Water flowing over a weir entrains air in its wake as it plunges into the pool downstream. Gameson (1957) reported water falling over weirs may gain 1 mg/L for every foot of fall height. With interest in aerating weirs on the rise, major contributions were made by Avery and Novak (1978); Markovsky and Kobus (1978); Nakasone (1987); among others. Gameson et al., (1958) reported that weir aeration was a function of temperature according to the relation

$$R_{(t=15°C)} = 1 + \left[\frac{R_t - 1}{1 + 0.027 * (t - 15)} \right] \quad (1)$$

where R is the deficit ratio defined as

---

*Tennessee Valley Authority, Engineering Laboratory, Norris, TN

$$R = \frac{C_{up} - C_{sat}}{C_{down} - C_{sat}},\qquad(2)$$

and t is the temperature in degrees Celsius. The deficit removal efficiency is

$$E = 1 - \left(\frac{1}{R}\right)\qquad(3)$$

Avery and Novak (1978) published data relating the improvement of weir aeration to increases in salinity. The salinity data was correlated by the authors to yield

$$R_{(s=0)} = 1 + (R_s - 1)\,e^{(0.73s)}\qquad(4)$$

Nakasone (1987) showed that optimum weir aeration performance is obtained at a straight weir specific discharge of about 0.67 cfs/ft, with a nappe drop height of 4 feet in a plunge pool depth exceeding 4 feet. Bin (1988) correlated the volume of air entrained by laminar jets to Froude, Reynolds, and Weber numbers, and geometric aspect ratios. His results indicated the jet entrainment velocity is mainly a function of jet length. Van De Sande and Smith (1975) demonstrated that jet aeration is a linear function of the air/water volume fraction. Their experiments demonstrated that the aeration of vertical jets are up to 30 percent higher than equivalent jets inclined at 30 degrees to the horizontal. Ervine, McKeogh and Elsawy (1980) and McKeogh and Elsawy (1980) observed that aeration performance of jets improved for highly turbulent, broken up jets. Leutheusser and Birk (1991) reported that life threatening hydraulic jump conditions occur at specific discharges exceeding 2.0 cfs/ft. Independently, Hauser (1991) observed that safe weir tailwater hydraulic conditions are exhibited at discharges below 1.5 cfs/ft.

To summarize, the weir and jet aeration literature have suggested that optimum aeration may be achieved for highly turbulent, mildly broken-up, vertical jets, with 0.67 cfs/ft specific discharge. The infuser weir (Figure 1) is comprised of a water retaining wall and a deck overlain with grating through which water jets fall into the plunge pool and entrain air bubbles in the process. Air access is provided through blockages located under the grate in a nonuniform fashion. Air is entrained in the water jets falling through the openings between the blockages and into the plunge pool. Plunge pool air and water mixture moves downstream. The plunge pool air bubbles are reentrained by other water curtains. This compounding effect produces an extremely frothy pool with excellent aerating characteristics. Aeration is further improved by reentraining the plunge pool air bubbles.

The principal objective of the TVA infuser testing program was to design an infuser with optimum aeration. The infuser testing program considered flat, step-down, and inclined infusers; infusers with longitudinal (parallel to flow) blockages and transverse (normal to flow) blockages; and infusers with different configurations of waterfall blockages and openings.

Experimental Test Setup

The Chatuge tailwater flow rate during normal generation is about 1300 cfs corresponding to about 13 cfs/ft for a 100-ft wide infuser deck. An infuser height exceeding 8 ft is required to maintain a water storage capacity of 30 dsf

(2.6 million cubic feet) to provide a minimum flow of 60 cfs for at least 12 hours. At normal turbine generation flow rates, the height of the head water above the weir crest is about 2.5 ft. Upstream head water elevation at the infuser weir would exceed 10.5 ft above channel bottom. A near full-scale model was built at the TVA Engineering Laboratory flume and shown in Figures 2 and 3.

The flume was instrumented with dissolved oxygen monitors, pressure transducers in the flume bottom, and thermocouples. Ambient temperature, pressure, and humidity were recorded. An impact gage was designed, built, and installed at the TVA EL to measure the impact of falling water on the channel bottom.

The infuser test setup was a near full-scale model of the proposed prototype. The infuser height, the tailwater height, and the free fall height (all of which are critical parameters in aeration performance) were less than the prototype dimensions. Thus, the aeration observed in the test facility was considered to be a conservative estimate of the expected oxygen transfer rate in the prototype.

The flume sump water was stripped using sodium sulfite with cobalt chloride to a steady-state DO content of 1 to 3 mg/L. Depending on the pumping rate, each test period lasted from 30 minutes to one hour, with several minutes in which the dissolved oxygen content in the upstream water measured in the range of 2 mg/L. The concentration of total dissolved solids in the water was monitored closely. The total dissolved solids were kept low by periodically flushing and refilling the sump water.

RESULTS

Table 1 summarizes the relevant infuser designs tested in the period from August 1991 to the end of December 1991. Included in Table 1 are the infuser length, infuser blockages layout, type of grating used, flow rate, headwater and tailwater elevations, length of wet infuser section, and minimum upstream and downstream dissolved oxygen. The resulting aeration deficits and efficiencies are normalized to a 15 degree Celsius water temperature and zero salinity according to Equations (1) and (4). The test results are discussed below.

The test results indicated that the infuser aeration performance is dependent on the water free fall, plunge pool depth, width of the waterfall blockages, open slot pattern, and type of grate used. Optimum aeration was obtained at a mean horizontal velocity (q/Lwet) of 0.65 ft/s as shown in Figure 3. Changing from metal grate to fiberglass grate (while maintaining a similar infuser deck hydraulic profile) reduced the infuser aerating efficiency from 58 to 48 percent (Tests 1.16 and 3.2). Increasing the plunge drop height from 3 ft to 4 ft increased the aeration efficiency by 12 percent (Tests 3.1 and 3.2). Aeration efficiency was reduced from 56 to 55 percent by allowing for a 1.5 cfs/ft overfall at the end of the infuser as shown from Tests 4.1 and 13.1 with the effective length of the infuser reduced by 4 feet.

Safety concerns included potentially dangerous hydraulic jump conditions downstream of the infuser and under the infuser deck, as well as danger of slipping and falling while walking on the infuser deck. To resolve these concerns, the principal author entered the laboratory flume and examined first hand the hydraulic conditions under the infuser deck and downstream of the infuser deck at flow rates increasing from about 3 cfs/ft up to 10.7 cfs/ft. Safe hydraulic conditions exist downstream of and under the infuser deck.

In the laboratory, the high velocity of the water was observed to drive trash and fish over and off the structure. Injury rate of fish passing over the structure was studied using a large sample of fish (Yeager and Tomlanjovitch, 1992). Passing over the structure did not cause any significant injury to fish.

CHATUGE INFUSER RECOMMENDATIONS

The Chatuge infuser weir crest elevation is about 9.5 ft above channel bottom. The specific infuser discharge is about 13.1 cfs/ft at normal turbine flow rate. The average infuser water jets free fall height is about 4 ft. The upstream to downstream infuser deck length is 14 ft. The water curtains are ventilated via aerating chimneys. The infuser grating of choice is a standard 1 inch by 4 inch metal safety grating with the 4-inch dimension oriented transverse to flow. Additional infuser design issues addressed by the TVA EL team included height and location, testing and design of the reregulation weir portion, minimum flow and regulating valves, aeration during minimum flow, infuser deck ventilation chimneys, infuser hydraulics during the 100-year flood and associated load requirements, infuser hydraulics during construction phases, and the infuser weir maintenance plan. The final recommended infuser design provided a platform from which the TVA designers developed the final infuser construction drawings. The final design allowed for a ready removal or modification of the blockages. To prevent potential safety related problems, access to the infuser plunge pool was controlled. A steel cage with a restricted access gate was constructed at the downstream end of the infuser. Preliminary prototype measurements indicated an aeration efficiency about 73 percent.

REFERENCES

Avery, S.T., and P. Novak, 1978. "Oxygen Transfer at Hydraulic Structures," J. Hydr. Div., ASCE, 104(HY11):1521-1540.

Bin, A.K., 1988. "Minimum Air Entrainment Velocity of Vertical Plunging Liquid Jets," Chemical Engg. Sci., 43(2):379-389.

Davis, J., 1987. "Improving Reservoir Releases," TVA Report TVA/ONRED/AWR 87/33.

Ervine, D.A., 1976. "The Entrainment of Air in Water," Water Power and Dam Construction, 28(12):27-30.

Ervine, D.A., E. McKeogh, and E.M. Elsawy, 1980. "Effect of Turbulence Intensity on the Rate of Air Entrainment by Plunging Water Jets," I. Inst Civ. Engrs, Part 2, Vol. 69, pp. 425-445.

Gameson, A.L.H., 1957. "Weirs and the Aeration of Rivers," J. Inst. of Water Engineers, 11:477-490.

Gameson, A.L.H., K.G. VanDyke, and C.G. Ogden, 1958. "The Effect of Temperature on Aeration at Weirs," Water and Water Engineering, pp. 489-492.

Hauser, G., 1991. "Full Scale Physical Modeling of Plunge Pool Hydraulics Downstream of a Vertical Weir," Tennessee Valley Authority, Engineering Laboratory Report WR28-1-590-153.

Hauser, G., W. Proctor, and T. Rizk, 1992. "Development of an Aerating Reregulation Weir for the Chatuge Tailwater," Tennessee Valley Authority, Engineering Laboratory, in preparation.

Leutheusser, H.J., and W.M. Birk, 1991. "Downproofing of Low Overflow Structures," J. Hydr. Engg., 117(2):205-213.

Markovsky, M., and H. Kobus, 1978. "Unified Presentation of Weir Aeration Data," J. Hydr. Div., ASCE, 104(4):562-568.

Maxwell, W., C. Hall, and R. Weggel, 1969. "Surface Tension in Froude Models," J. Hydr. Div., 95(2):677-701.

McKeogh, E.J., and E.M. Elsawy, 1980. "Air Retained in Pool by Plunging Water Jet," J. Hydr. Div., 106(10):1577-1593.

Nakasone, H., 1987. "Study of Aeration of Weirs and Cascades," J. Env. Eng., 113(1):64-81.

Rizk, T.A., and G. Hauser, 1993. "Chatuge Hydroproject Aerating Infuser Physical Model Study," Tennessee Valley Authority, Engineering Laboratory Report No. WR28-1-17-101.

Sene K.J., N.H. Thomas, and B.T. Goldring, 1989. "Planar Plunge-Zone Flow Patterns and Entrained Bubble Transport," J. Hydr. Res., 27(3):363-382.

Van De Sande, E., and J.M. Smith, 1975. "Mass Transfer from Plunging Water Jets," Chemical Engg. Journal, 10(3):225-233.

Yeager, B., and D. Tomljanovich, 1992. Research Communication, 1992.

TABLE 1

Large Scale Test Results

Test #	Grate Type	Wet Inf Length (ft)	Openings Blockage Size, (# Open - # Slots)	Flow (cfs/ft)	HW (ft)	TW (ft)	R	E
1.13	SB	23	1', (1-2, 3-4, 6-6, rest 10)	10.8	7.17	3	2.05	51.3
1.14	SB	20	1', (1-2, 3-4, 6-6, rest 10)	8.9	7.05	3	2.10	52.4
1.15	SB	16	1', (1-2, 3-4, 6-6, rest 10)	6.7	6.64	3	2.02	50.5
1.16	ST	23	1', (1-2, 3-4, 6-6, rest 10)	10.8	7.13	3	2.36	57.5
2.1	ST	31	1', (4-1, 8-2, 12-3, 4-4)	10.8	7.13	3	1.62	38.3
2.3	ST	28	1', (4-1, 8-2, 12-3, 4-4)	8.9	7.05	3	1.51	33.9
2.2	ST	24	1', (4-1, 8-2, 12-3, 4-4)	6.7	6.63	3	1.58	36.8
3.1	FT	20	1', (1-2, 2-3, 1-4, 1-5, 4-7)	10.8	7.13	4	1.55	35.7
3.2	FT	20	1', (1-2, 2-3, 1-4, 1-5, 4-7)	10.8	7.13	3	1.92	47.8
4.1	ST	18	6", (1-2, 3-4, 6-6, rest 10)	10.8	7.08	3	2.25	55.7
4.2	ST	18	6", (1-2, 3-4, 6-6, rest 10)	10.8	7.08	4	1.74	42.5
4.3	ST	12	6", (1-2, 3-4, 6-6, rest 10)	8.9	6.90	3	2.31	56.6
5.1	ST	12	0", (All Open)	10.8	7.11	3	2.02	50.5
6.1	FT	12	0", (All Open)	10.8	7.11	3	2.00	50.0
7.1	FT	12.5	6", (1-2, 3-4)	10.8	7.07	3	1.92	48.0
8.1	ST	12	6", (1-2, 3-4)	10.8	7.10	3	1.76	43.1
11.1	ST	14	6", (1-2, 3-4, 6-6, 2-8, 1-9)	10.8	7.09	3	2.12	52.8
12.1	ST	14	6", (1-2, 3-4, 2-6, 2-8, 1-9, 3-10)	10.8	7.10	3	2.14	53.3
13.1	ST	14	6", (1-2, 3-4, 6-6, 4-10)	10.8	7.10	3	2.21	54.7
14.1	ST	14	6", (1-2, 3-4, 2-6, 2-10, 1-9, 2-10)	10.8	7.10	3	1.83	45.3

NOTES:
   GRATE TYPES: SB = steel with bars on bottom, 1" x 4" openings
                         ST = steel with bars on top, 1" x 4" openings
                         FT = fiberglass with roughness on top, 1" x 1" openings
   INF HEIGHT = 5 ft for all cases studied

PHYSICAL MODEL STUDY

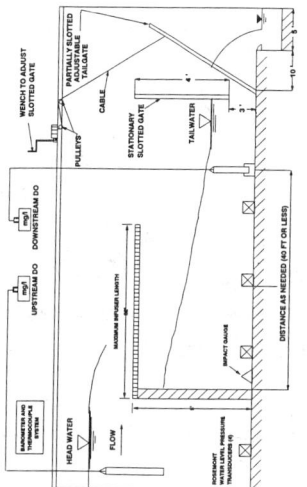

Figure 2. Large-Scale Diffuser Test Layout, Vertical View

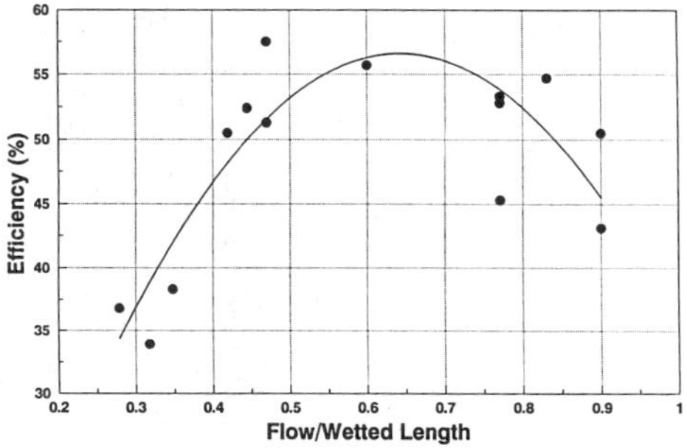

Figure 4. Aeration Efficiency vs. "Deck Flux"

Figure 1. Chatuge Diffuser

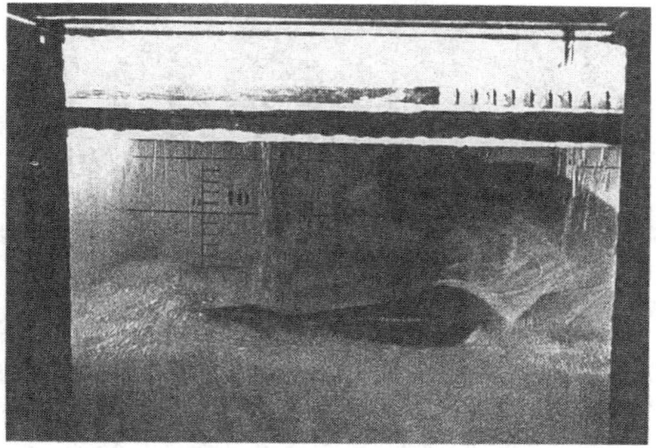

Figure 3. Large-Scale Infuser Physical Model 10.8 cfs/ft Specific Discharge

Recent Developments in Three-dimensional Numerical Estuarine Models

Ralph T. Cheng and Peter E. Smith
U. S. Geological Survey

Vincenzo Casulli
University of Trento, Italy

Abstract

For a fixed cost, computing power increases 5 to 10 times every five years. The readily available computing resources have inspired new model formulations and innovative model applications. Significant progress has been advanced in three-dimensional numerical estuarine modeling within the past three or four years. This paper attempts to review and summarize properties of new 3-D estuarine hydrodynamic models. The emphasis of the review is placed on the formulation, numerical methods, spatial and temporal resolution, computational efficiency, and turbulence closure of new models. Recent research has provided guidelines for the proper use of 3-D models involving the σ-transformation. Other models resort to a fixed level discretization in the vertical. The semi-implicit treatment in time-stepping models appears to have gained momentum. Future research in three-dimensional numerical modeling remains to be on computational efficiency and turbulent closure.

Introduction

Readily available computing resources have inspired new model formulations and creative model applications. A survey of three-dimensional numerical estuarine models was given by Cheng and Smith (1989) about four years ago. The scope of this paper complements the previous survey (Cheng and Smith, 1989) by placing the emphasis on new 3-D estuarine hydrodynamic model formulation and numerical methods. No attempt is made to discuss any specific model in detail, nor is there any effort to list all published models since 1989. In the following sections, the governing equations, assumptions, and approximations are summarized. Recent research on the effects of the σ-transformation in 3-D models is discussed. Models using fixed vertical levels, and semi-implicit time-stepping integration will be reviewed. In conclusion, the priorities for future research on three-dimensional numerical modeling is discussed.

Governing equations and approximations

The governing system of equations for flows and transport of scalar variables in an estuary includes the conservation equations of mass, momentum, conservative scalar variables, an equation of state, and a kinematic condition at air-water interface. The estuarine system is assumed to be large enough so that the Coriolis acceleration must be included, yet small enough so that the Coriolis acceleration can be treated as a constant. The governing equations which are usually written for turbulent mean variables are derived from time averaging over a period much longer than turbulent characteristic time. Additional turbulent correlations result from the averaging, and they represent transport of turbulent fluxes often referred to as turbulent mixing or dispersion. These turbulent fluxes must be dealt with using turbulence closure "sub-models".

The dependent variables are the velocity components u, v, w in the x, y, z directions, respectively, the hydrostatic pressure, p, the water surface elevation, ζ, density, ρ, salinity, s, and temperature, T. Density is related to s and T through an equation of state. The system of equations is not closed until the turbulent correlations are specified, and the appropriate initial and boundary conditions are given. No analytical solution for the nonlinear governing system of equations is known to exist, and the numerical solutions for transient and stratified tidal circulation in estuaries are extremely complex. Additional computational difficulty can result from the usually complex bathymetry of tidal basins. Before attempting a numerical solution of such a problem, some simplifying assumptions and approximations are typically introduced.

Nearly all models assume that the water is incompressible, the Boussinesq approximation applies. The pressure is assumed to be hydrostatic, thus the z-momentum equation and the hydrostatic pressure can be eliminated leaving the pressure gradients in the x and y momentum equations given as the sum of a barotropic (water surface gradient) and a baroclinic (density gradient) pressure gradient terms. Because estuaries are characterized as regions that connect the river and the ocean, the water is mixed oceanic and fresh water, therefore, density stratification should be included in the model formulation.

Turbulent Closure Schemes and Boundary Conditions

The governing equations are not complete until a turbulent closure scheme is specified. Turbulence closure "sub-models" are meant to characterize the turbulent mixing and entrainment processes, and express these processes in terms of the dependent mean variables. There is not a generally agreed upon scheme for turbulent closure. Typically, the turbulent transport fluxes are written in an analogous form like diffusion, and leaving the eddy viscosity and diffusivity coefficients to be defined by semi-empirical equations or by solving additional turbulence transport equations.

In most estuaries, the ratio of vertical to horizontal length scales is less than 0.01 or 0.001, thus crude approximations for horizontal mixing will suffice. Most models treat the horizontal eddy viscosity and eddy diffusivity coefficients as constants.

The treatment of vertical mixing or vertical turbulent transport is extremely important in correctly representing the dynamics of partially mixed estuaries. Among many different types of eddy viscosity-diffusivity closure models, they fall in the general form of

$$Kv = q\ L\ F(Ri) \quad \text{and} \quad Nv = q\ L\ G(Ri)$$

where Kv is the vertical eddy viscosity;
Nv is the vertical eddy diffusivity;
q is turbulent velocity scale;
L is a turbulent length scale;

Ri is a gradient Richardson number defined as

$$Ri = -\frac{g}{\rho}\frac{\partial\rho/\partial z}{(\partial|U|/\partial z)^2}$$

F(Ri) and G(Ri) are referred to as stability functions including the effects of stratification. The intensity of turbulence is characterized by three variables, namely, turbulent kinetic energy, κ, turbulent dissipation, ϵ, and a mixing length, ℓ. The various turbulent closure schemes have been proposed to relate κ, ϵ and ℓ to q, L, F(Ri) and G(Ri).

The simplest approximations for Kv and Nv assume q and L as constants or as empirical functions. The stability functions, F(Ri) and G(Ri) are also given by empirical expressions (Rodi, 1984). Assuming that L is given and a solutions of κ with a specification of ϵ (given as a function of κ) is sought, this approach is known as a one-equation model. The next level of turbulence closure solves an additional equation which determines the distribution of a turbulence length scale, ℓ. This approach is referred to as a two-equation model of turbulence closure (Rodi, 1984).

A form of bottom stress is used to specify the rate of momentum loss at the water-sediment interface. The bottom stress which is given as a drag coefficient times a quadratic form of velocity, enters into the 3-D model through the boundary condition at the sediment-water interface. The drag coefficient, C_d, can be assigned as a constant, computed by a Manning-Chezy formula, or evaluated by fitting the bottom velocity profile to the classic turbulent boundary layer. Specification of shoreline boundary conditions is usually straightforward, while proper specification of boundary conditions at open boundaries remains an unresolved issue. The approximations introduced at open boundaries depend upon the specific applications.

Methods of Solution

Coordinate Transformations:

One of the common difficulties in 3-D estuarine hydrodynamic modeling is the high demand of computing resources. As discussed in Cheng and Smith (1989), different modeling strategies have been used to minimize computing demands. In attempts to improve computational efficiency, various coordinate transformations have been proposed. A σ-transformation, which maps the vertical dimension z to $\sigma \in [a,b]$, is introduced where a,b take on values such as -1, 0, or 1. The σ-transformation simplifies computational algorithm when large bathymetry variations are involved. The same number of vertical points is used throughout the model domain. Although this approach offers certain advantages due to the mapping, Haney (1991) has recently shown that serious numerical errors can result from computing the pressure gradient force near steep topography in σ-coordinate models. The errors are due to spatial truncation errors and a problem of 'hydrostatic inconsistency'. Deleersnijder and Beckers (1992) have noted that truncation errors in σ-models actually apply to spatial derivatives of any variable in the 3-D equations, not just pressure. Several studies have shown that the level of grid refinement needed to produce accurate results near highly irregular topography can be unreasonable with a σ-model. Johnson et. al. (1990) pointed out that in their 3-D model of Chesapeake Bay using σ-coordinates, salt was incorrectly advected from deep channels to shallow areas in the bay. They were forced to convert their model to level coordinates to obtain correct simulations. In estuaries, high vertical grid resolution is needed in deep channels, and only coarse vertical resolution is needed in well-mixed shallow areas. Using the same number of vertical points everywhere in a σ-model seems to be wasteful of computing resources. In consideration of both computational accuracy and efficiency, a fixed-level coordinate model may be better suited for modeling estuarine hydrodynamics.

Furthermore, in most σ-models, the transformation becomes singular when the water depth becomes zero. Thus the treatment of flooding and drying of tidal mudflats is more difficult in σ-models. The fixed-level type of model has been considered by several authors, (Duwe at. al., 1983; Leenderste, 1989; Casulli and Cheng, 1992).

Motivated by a reduction in the number of horizontal computing points, various transformations of the x,y variables have been proposed (Cheng and Smith, 1989). For simplicity, other models suggest the use of rectangular computational mesh (Casulli and Cheng, 1992). Computational efficiency can be improved by considering only the active points with the aid of careful structure of the model code.

Time Integration:

Time integration of the governing equations can be treated by solving for the dependent variables in the frequency domain. When the dependent variables are expressed in harmonic decompositions, the governing equations are reduced to an elliptic partial differential equation whose solution defines the spatial distribution of tidal harmonic constants for partial tides (Lynch and Werner, 1987; Walters and Foreman, 1992). Alternatively, the governing hyperbolic equations are integrated with respect to time using either implicit or explicit methods. Neither implicit nor explicit methods are computationally efficient, and are thus deemed not satisfactory.

Among the time-stepping methods, a commonly used strategy is a mode-splitting integration, (Simons, 1972), which splits the over-all 3-D computations into two stages. A set of depth averaged shallow water equations which is dominated by the propagation of the surface gravity wave (barotropic pressure gradients), are solved explicitly or semi-implicitly using a small Δt. The use of the hydrostatic approximation implies that the barotropic pressure gradients in the full 3-D system are the same barotropic pressure gradients in the 2-D shallow water system. Once the barotropic pressure gradients are solved in 2-D, the 3-D velocity field is solved locally and semi-implicitly (Blumberg and Mellor, 1987; Sheng, 1987; Hamrick, 1992). The governing equation for the external mode can be written as a wave equation, which is solved by semi-implicit time-stepping integration by Lynch and Werner (1992), the internal 3-D velocity field is solved locally and implicitly. When the mode-splitting is used, care must be exercised in treating the coupling between 2-D and 3-D solutions; because "consistency" of the 2-D and 3-D solutions is violated in the general case that the nonlinear convective terms are retained.

In the last several years, a semi-implicit method of solution has emerged and gained recognition, Casulli (1990). At the same time, improved ADI schemes have been proposed and tested (Stelling 1983; Leenderste, 1989; de Goede, 1991). Applications of semi-implicit methods of solution in estuaries can be found in Backhaus et al. (1985), Duwe et al. (1983), Casulli and Cheng (1992). In the semi-implicit formulation for three-dimensional estuarine flow, in order to maintain numerical stability and computational efficiency, the governing equations are finite-differenced by keeping the minimum number of terms implicit, and the remaining terms explicit. Following a stability analysis, the barotropic pressure gradients and the divergence in the depth integrated continuity equation must be kept implicit, Casulli and Cattani (1993). No coordinate transformation is introduced, the computational points are defined on rectangular mesh in the horizontal plane and fixed levels in the vertical. To maintain numerical stability, the vertical diffusion terms must also be treated implicitly. By substituting the momentum equations into the depth averaged continuity equation, the result is a five diagonal, positive definite matrix for the water surface elevation over the entire domain. The coefficients of the five-diagonal matrix are formed by first solving the tri-diagonal systems that relate the distribution of u and v velocities in the vertical by implicit finite differences. The tri-diagonal systems can be solved by direct elimination. The symmetric, positive definite, five-diagonal matrix can be solved efficiently by a pre-conditioned conjugate gradient method. The conservative form of the continuity equation is used, the presence of islands, shoreline boundaries, and tidal mudflats can be handled naturally without special treatment. In the solution algorithm, only wetted points need to be considered. For example, in an application to San Francisco Bay, with a 250 m horizontal grid-mesh (72,000 points) and 34 vertical layers, there are less than 19,000 active points on the horizontal plane, and the total active points are less than 50,000. High computational efficiency has been achieved by taking advantage of both the numerical properties of the semi-implicit method of solution as well as the fixed level grid structure without introducing any transformation (Casulli and Cheng, 1992).

Discussion

Numerical modeling of three-dimensional tidal circulation in partially mixed estuaries is very challenging indeed. The data requirements in 3-D modeling are an order of magnitude greater than its 2-D counterpart. Fortunately, some advances in instrumentation, such as in-situ acoustic Doppler current profiler (ADCP), appear to be promising for providing 3-D field data. Dramatic increases in computing power in commonly available workstations have made many 3-D estuarine model applications possible. Further improvements in computing speed and computer graphics can be anticipated for the future. As the 3-D circulation modeling becomes routine practice, suitable graphics need to be developed to assist interpretation of model results, and to aid model calibration and verification.

Although great advances have been made in several aspects of 3-D estuarine modeling, further research needs remain to be made in the areas of improving computational efficiency, critical evaluation of turbulent closure, and detailed model applications to complex, moderately to highly stratified estuaries. Several additional complications associated with stratified flows, apparently, do not have satisfactory solutions. For example, the specification of boundary conditions at open boundaries, the treatment for the propagation of internal waves require immediate attention. Most model applications use fairly coarse spatial grids, thus the truncation errors, which appear as numerical dispersion, are sufficiently large making any choice of turbulent closure irrelevant. More refined grid resolutions are needed to define the three-dimensional flow field. When the flow properties can be resolved on a fine spatial grid, the study of turbulent closure becomes more meaningful and challenging.

References

Backhaus, J. O., 1985, A three-dimensional model for the simulation of shelf sea dynamics, Dt. Hydrogr. Z., 38, 165-187.

Blumberg, A.F., and Mellor, G. L., 1987, A description of a three-dimensional coastal ocean circulation model, in Three-dimensional Coastal Ocean Models, N. S. Heaps, (Ed). AGU, 1-16.

Casulli, V., 1990, Semi-implicit finite difference methods for the two-dimensional shallow water equations, J. of Comp. Physics, 86, 56-74.

Casulli, V. and Cheng, R. T., 1992, Semi-implicit finite difference methods for three-dimensional shallow water flow, Inter. J. for Num. Methods in Fluids, 15, 629-648.

Casulli, V. and Cattani, E., 1993, Stability, accuracy and efficiency of a semi-implicit method for three-dimensional shallow water flow, to appear in Computers and Mathematics with Applications

Cheng, R. T., and Smith, P. E., 1989, A survey of three-dimensional numerical estuarine models, Estuarine and Coastal Modeling Proceedings, WW Div/ASCE, Newport, R. I., p. 1-15.

Deleersnijder, E. and Beckers, J.-M. 1992, On the use of the sigma-coordinate system in regions of large bathymetric variations, J. Mar. Syst., 3, 381-390.

de Goede, E. D., 1991, A time-splitting method for the three-dimensional shallow water equations, Inter. J. for Num. Methods in Fluids, 15, 629-648.

Duwe, K. C., Hewer, R. R., and Backhaus, J. O., 1983, Results of a semi-implicit two-step method for the simulation of markedly nonlinear flow in coastal seas, Continental Shelf Research, 2, 255-274.

Hamrick, J., 1992, A Three-dimensional Environmental Fluid Dynamics Computer Code: Theoretical and Computational Aspects, Rept. 317, Virginia Inst. of Marine Science, Gloucester Point, VA.

Haney, R.L., 1991, On the pressure gradient force over steep topography in sigma coordinate ocean models, J. Phys. Oceanogr., 21, 610-619.

Johnson, B.H., Kim, K.W., Sheng, Y.P., and Heath, R.E., 1990, Development of a three-dimensional hydrodynamic model of Chesapeake Bay, in Proceedings of Conf. on Estuarine and Coastal Modeling, Newport, RI, November 15-17, 1989, 162-171.

Leendertse, J. J., 1989, A new approach of a three-dimensional free-surface flow modeling, Rept. R-3712-NETH/RC, Rand Corporation, Santa Monica.

Leendertse, J. J., Alexander, R. C., and Liu, S. K., 1973, A three-dimensional model for estuaries and coastal seas, Volume I: Principles of computation, R-1417-OWRR, Rand Corp, Santa Monica.

Lynch, D.R. and Werner, F. E., Greenberg, D. A., and Loder, J. W., 1992, Diagnostic model for baroclinic, wind-driven and tidal circulation in shallow seas, Cont/ Shelf Res., Vol. 12, p.37-64.

Rodi, W., 1984, Turbulence models and their applications in hydraulics - a state of the art review, 2nd edition, IAHR, The Netherlands, pp.104.

Sheng, Y. P., 1987, On modeling three-dimensional estuarine and marine hydrodynamics, in <u>Three-dimensional models of marine and estuarine dynamics</u>, Edited by J.C.J. Nihoul and B.M. Jamart, Elsevier, 35-53.

Stelling, G. S., 1983, On the construction of computational methods for shallow water flow problems, Ph. D. Thesis, TU Delft.

Walters, R.A. and Foreman, M.G.G., 1992, A 3D, finite element model for baroclinic circulation on Vancouver Island continental shelf, J. Mar. Syst., 3, 507-518.

TRIM_3D: A Three-Dimensional Model for Accurate Simulation of Shallow Water Flow

Vincenzo Casulli[1]   *Enrico Bertolazzi*[1]
and
Ralph T. Cheng[2]

Abstract

A semi-implicit finite difference formulation for the numerical solution of three-dimensional tidal circulation is discussed. The governing equations are the three-dimensional Reynolds equations in which the pressure is assumed to be hydrostatic. A minimal degree of implicitness has been introduced in the finite difference formula so that the resulting algorithm permits the use of large time steps at a minimal computational cost. This formulation includes the simulation of flooding and drying of tidal flats, and is fully vectorizable for an efficient implementation on modern vector computers. The high computational efficiency of this method has made it possible to provide the fine details of circulation structure in complex regions that previous studies were unable to obtain. For proper interpretation of the model results suitable interactive graphics is also an essential tool.

1. Introduction

A characteristic analysis of the two-dimensional, vertically integrated shallow water equations has shown that the celerity term $\sqrt{gH}$ in the equation for the characteristic cone arises from the barotropic pressure gradient in the momentum equations and from the velocity derivatives in the free surface equation (Casulli, 1990). Results of this analysis have led to a practical semi-implicit method which has been proved to be unconditionally stable, and which has been proven to be very useful in several applications (see, e.g., Cheng et al., 1993; Signell and Butman, 1992).

Recently, the semi-implicit finite difference method for the two-dimensional shallow water equations has been extended to the three-dimensional shallow water equations (Casulli and Cheng, 1992). The Courant-Friedrich-Lewy (CFL) stability condition is not required by this method, because the barotropic pressure gradient in

---

[1] Dipartimento di Matematica, Universita' di Trento, 38050 Povo (TN), Italy
[2] US Geological Survey WRD, 345 Middlefield Rd MS 496, Menlo Park, Ca 94025

the momentum equations and the velocities in the vertically integrated continuity equation are finite-differenced implicitly.

Numerical experiments of the three-dimensional shallow water equations have shown that this algorithm is stable and is highly efficient. Moreover, when only one vertical layer is specified, this method reduces, as a special case, to the semi-implicit method for the two-dimensional vertically integrated shallow water equations as described by Casulli (1990). The resulting two- and three-dimensional methods, however, are only first order accurate in time, and introduce some artificial damping.

The stability, the accuracy and the efficiency of this three-dimensional algorithm has been studied by Casulli and Cattani (1993) who introduced an implicitness parameter θ. When θ=1, this method reverts to the semi-implicit scheme proposed by Casulli and Cheng. When θ=1/2 the pressure gradient in the momentum equations and the velocities in the free surface equation are evaluated as an average of their values at time levels n and n+1, so that this discretization is second order accurate in time. A rigorous stability analysis carried out by Casulli and Cattani (1993) has shown that, indeed, this algorithm is stable for 1/2≤θ≤1, and highest accuracy and efficiency is achieved when θ=1/2.

Computationally, the resulting algorithm is suitable for simulations of complex three-dimensional flows using fine spatial resolution and relatively large time steps. The present formulation is fully vectorizable and naturally allows for the simulation of flooding and drying of tidal flats.

2. Governing Equations

The governing three-dimensional, primitive variable equations describing constant density, free surface flows in estuarine embayments and coastal oceans can be derived from the Navier-Stokes equations after turbulent averaging and under the simplifying assumption that the pressure is hydrostatic (see, e.g., Casulli and Cheng, 1992). Such equations have the following form

(1) $$\frac{\partial u}{\partial t} + u\frac{\partial u}{\partial x} + v\frac{\partial u}{\partial y} + w\frac{\partial u}{\partial z} = -g\frac{\partial \eta}{\partial x} + v_h(\frac{\partial^2 u}{\partial x^2} + \frac{\partial^2 u}{\partial y^2}) + \frac{\partial}{\partial z}(v_v \frac{\partial u}{\partial z}) + fv$$

(2) $$\frac{\partial v}{\partial t} + u\frac{\partial v}{\partial x} + v\frac{\partial v}{\partial y} + w\frac{\partial v}{\partial z} = -g\frac{\partial \eta}{\partial y} + v_h(\frac{\partial^2 v}{\partial x^2} + \frac{\partial^2 v}{\partial y^2}) + \frac{\partial}{\partial z}(v_v \frac{\partial v}{\partial z}) - fu$$

(3) $$\frac{\partial u}{\partial x} + \frac{\partial v}{\partial y} + \frac{\partial w}{\partial z} = 0,$$

where $u(x,y,z,t)$, $v(x,y,z,t)$ and $w(x,y,z,t)$ are the velocity components in the horizontal x, y and vertical z-direction, t is the time, $\eta(x,y,t)$ is the water surface elevation measured from the undisturbed water surface, g is the gravitational acceleration, f is the Coriolis parameter, assumed to be constant, and v_h and v_v are the coefficients of horizontal and vertical eddy viscosity, respectively.

Integrating the continuity equation over the depth and using a kinematic condition at the free surface leads to the following free-surface equation

(4) $$\frac{\partial \eta}{\partial t} + \frac{\partial}{\partial x}\left[\int_{-h}^{\eta} u\, dz\right] + \frac{\partial}{\partial y}\left[\int_{-h}^{\eta} v\, dz\right] = 0,$$

where $h(x,y)$ is the water depth measured from the undisturbed water surface and $H(x,y,t)$ is the total water depth, $H(x,y,t) = h(x,y) + \eta(x,y,t)$.

The boundary conditions at the free surface are specified by the prescribed wind stresses, (τ_x^w, τ_y^w),

(5) $$\nu \frac{\partial u}{\partial z} = \tau_x^w, \qquad \nu \frac{\partial v}{\partial z} = \tau_y^w,$$

and the boundary conditions at the sediment-water interface are given by

(6) $$\nu \frac{\partial u}{\partial z} = \gamma u, \qquad \nu \frac{\partial v}{\partial z} = \gamma v,$$

where γ is a nonnegative bottom friction coefficient. Typically, γ can be given by the Manning-Chezy formula, or by fitting it to a turbulent boundary layer.

3. A three-dimensional semi-implicit numerical method

In order to derive a stable and efficient semi-implicit numerical method for equations (1)-(6) the gradient of surface elevation in the momentum equations (1) and (2), and the velocity in the free surface equation (4) are discretized by the θ-method. The convective, Coriolis and horizontal viscosity terms are discretized explicitly, and, in order to eliminate also the stability condition due to the vertical eddy viscosity, the vertical mixing terms are discretized implicitly.

The spatial mesh consists of rectangular boxes of length Δx, width Δy and height Δz_k. Each box is numbered at its center with indices i, j and k. The discrete u velocity is then defined at half integer i and integers j and k; v is defined at integers i, k and half integer j; w is defined at integers i, j and half integer k. Finally, η is defined at integer i, j. The water depth $h(x,y)$ is specified at the u and v horizontal points. Then, a parametrized semi-implicit discretization of the momentum equations (1) and (2), in compact vector notation, takes the following form

(7) $$\mathbf{A}^n_{i+1/2,j} \mathbf{U}^{n+1}_{i+1/2,j} = \mathbf{G}^n_{i+1/2,j} - g\theta \frac{\Delta t}{\Delta x} (\eta^{n+1}_{i+1,j} - \eta^{n+1}_{i,j}) \Delta \mathbf{Z}^n_{i+1/2,j}$$

(8) $$\mathbf{A}^n_{i,j+1/2} \mathbf{V}^{n+1}_{i,j+1/2} = \mathbf{G}^n_{i,j+1/2} - g\theta \frac{\Delta t}{\Delta y} (\eta^{n+1}_{i,j+1} - \eta^{n+1}_{i,j}) \Delta \mathbf{Z}^n_{i,j+1/2},$$

where $\mathbf{U}$ and $\mathbf{V}$ are vectors containing the horizontal velocity components u and v on a water column at the specified horizontal grid location; $\Delta \mathbf{Z}$ a vector containing the vertical space increments; G is a finite difference operator which includes the explicit Eulerian-Lagrangian discretization of the convective terms, the horizontal viscosity terms, and the explicit contribution of the barotropic pressure terms; finally, A is the tridiagonal matrix which results from the implicit discretization of the vertical viscosity term. Matrix A is symmetric and positive definite. Equations (7), (8) also include the appropriate semi-implicit discretizations of the boundary conditions at the free surface and at the sediment-water interface (5), (6). The details of these terms can be found in Casulli and Cattani (1993).

Equations (7) and (8) are linear tridiagonal systems which are coupled to the water surface elevation η^{n+1} at time t^{n+1}. In order to determine $\eta^{n+1}_{i,j}$, and for numerical stability, the new velocity field must satisfy, for each i,j, the discrete analogue of the free surface equation (4),

(9) $$\eta^{n+1}_{i,j} + \theta \frac{\Delta t}{\Delta x}\left[(\Delta Z_{i+1/2,j})^T \cdot U^{n+1}_{i+1/2,j} - (\Delta Z_{i-1/2,j})^T \cdot U^{n+1}_{i-1/2,j}\right]$$

$$+ \theta \frac{\Delta t}{\Delta y}\left[(\Delta Z_{i,j+1/2})^T \cdot V^{n+1}_{i,j+1/2} - (\Delta Z_{i-1/2,j})^T \cdot V^{n+1}_{i,j-1/2}\right] = \delta^n_{i,j}$$

where

(10) $$\delta^n_{i,j} = \eta^n_{i,j} - (1-\theta)\frac{\Delta t}{\Delta x}\left[(\Delta Z_{i+1/2,j})^T \cdot U^n_{i+1/2,j} - (\Delta Z_{i-1/2,j})^T \cdot U^n_{i-1/2,j}\right]$$

$$-(1-\theta)\frac{\Delta t}{\Delta y}\left[(\Delta Z_{i,j+1/2})^T \cdot V^n_{i,j+1/2} - (\Delta Z_{i-1/2,j})^T \cdot V^n_{i,j-1/2}\right].$$

Equations (7), (8) and (9) now constitute a closed linear system of equations with unknowns $u^{n+1}_{i+1/2,j,k}$, $v^{n+1}_{i,j+1/2,k}$ and $\eta^{n+1}_{i,j}$ over the entire computational mesh. For computational convenience this system is first reduced to a smaller system in which $\eta^{n+1}_{i,j}$ are the only unknowns. Specifically, formal substitution of the expressions for $U^{n+1}_{i\pm1/2,j}$ and $V^{n+1}_{i,j\pm1/2}$ from (7) and (8) into (9) yields

$$\eta^{n+1}_{i,j} - g\theta^2 \frac{\Delta t^2}{\Delta x^2}\left\{[(\Delta Z)^T A^{-1} \Delta Z]^n_{i+1/2,j}(\eta^{n+1}_{i+1,j}-\eta^{n+1}_{i,j}) - [(\Delta Z)^T A^{-1} \Delta Z]^n_{i-1/2,j}(\eta^{n+1}_{i,j}-\eta^{n+1}_{i-1,j})\right\}$$

$$- g\theta^2 \frac{\Delta t^2}{\Delta y^2}\left\{[(\Delta Z)^T A^{-1} \Delta Z]^n_{i,j+1/2}(\eta^{n+1}_{i,j+1}-\eta^{n+1}_{i,j}) - [(\Delta Z)^T A^{-1} \Delta Z]^n_{i,j-1/2}(\eta^{n+1}_{i,j}-\eta^{n+1}_{i,j-1})\right\}$$

(11) $$= \delta^n_{i,j} - \frac{\Delta t}{\Delta x}\left\{[(\Delta Z)^T A^{-1} G]^n_{i+1/2,j} - [(\Delta Z)^T A^{-1} G]^n_{i-1/2,j}\right\}$$

$$- \frac{\Delta t}{\Delta y}\left\{[(\Delta Z)^T A^{-1} G]^n_{i,j-1/2} - [(\Delta Z)^T A^{-1} G]^n_{i,j-1/2}\right\}$$

Since A is positive definite, A^{-1} is also positive definite and therefore $(\Delta Z)^T A^{-1} \Delta Z$ is a positive number. Hence equations (11) constitute a linear five-diagonal system of equations for $\eta^{n+1}_{i,j}$ which is symmetric and positive definite. Thus, it has a unique solution which can be determined very efficiently by a conjugate gradient method. It is important to emphasize that when $\theta=1/2$ is used, the off diagonal terms in system (11) are reduced by a factor $\theta^2=1/4$. Thus, when $\theta=1/2$ system (11) is better conditioned, and, accordingly, a faster convergence of the conjugate gradient method is achieved.

Once the new free surface location has been determined, equations (7) and (8) are used to evaluate the new velocity u and v. Finally, the vertical component of the velocity is determined by the discretizing the continuity equation.

4. Properties of the Model

Once the free surface (and hence the new water velocity) has been computed throughout the computational domain, the total water depth $H_{i+1/2,j}^{n+1}$ and $H_{i,j+1/2}^{n+1}$ at the side of each computational cell are updated to account for the new free surface location. Since a negative value for the water depth is physically meaningless, the new total depth H is defined to be H=max(0,h+η) and a resulting zero total depth simply means that no water can flow across this side of the cell. If the total depth is positive, then the corresponding cell side is wet. Moreover, when H is zero, the respective friction factor will be assumed to be infinity, and, accordingly, the corresponding velocities u or v across the side of the cell is forced to vanish. The resulting finite-difference equation for the water surface elevation, equation (11), correctly accounts for positive and for zero value of the total depth on each side of a computational stencil. The occurrence of zero value for the total depth H on one side of a cell implies zero velocity or zero mass flux until, at a later time, H becomes positive. A cell is considered a dry cell when the total water depths at all sides are zero. Accordingly, in a dry cell, equation (11) is reduced to $\eta_{i,j}^{n+1}=\eta_{i,j}^{n}$. Similarly, on a dry side of a cell, equations (7) or (8) will be replaced by $U_{i+1/2,j}^{n+1}=0$ or $V_{i,j+1/2}^{n+1}=0$. Hence, whether the flooding and drying of cells takes place or not, equation (11) can be applied to all points throughout the domain resulting in an algorithm that can be vectorized for efficient computations. The presence of islands, and other permanently dry areas, as well as tidal flats will be accounted for appropriately and automatically.

Another important property of the present formulation arises from the fact that when the vertical spacing Δz is taken to be large enough so that both the bottom and the free surface always fall within one vertical layer, then this algorithm reduces to a two-dimensional semi-implicit numerical method as described by Casulli (1990) and which is consistent with the two-dimensional, vertically integrated shallow water equations. This property of the algorithm leads to a computer code that can be used for both three-dimensional problems as well as two-dimensional problems. More importantly, when the three-dimensional model is applied to a typical coastal plain tidal embayment characterized by deep channels connected to large and flat shallow areas, a great saving in computing is achieved because the deep channels are correctly represented in three dimensions while the flat shallow areas are represented only in two dimensions.

Finally, the stability of this model (Casulli and Cattani, 1993) has been shown to be given by

$$(12) \quad \Delta t \leq \left[2\nu_h\left(\frac{1}{\Delta x^2}+\frac{1}{\Delta y^2}\right)\right]^{-1}.$$

Therefore, in absence of horizontal viscosity ($\nu_h=0$), the semi-implicit algorithm becomes unconditionally stable.

5. Applications

The present model has been tested on several sites, one of the more complex being the Lagoon of Venice whose area covers about 50 km^2 and consists of several interconnected narrow channels with a maximum width of 1 km, and up to 50 m deep encircling large and flat shallow areas. The Lagoon is connected to the Adriatic Sea through three narrow inlets, namely Lido, Malamocco, and Chioggia. The city of Venice is located in the Lagoon near Lido inlet. Considerable portion of the Lagoon of Venice consists of tidal flats, and proper treatment of flooding and drying is essential. The tidal amplitudes in the Adriatic Sea are about 0.5 m. Tides propagate from the Adriatic Sea into the Lagoon through the three inlets. The Lagoon has been covered with a 642 by 847 by 200 finite-difference mesh of $\Delta x=\Delta y=50$ m and with the maximum Δz being 0.25 m. Thus, the total number of grid points is therefore 108,754,800, but only 1,177,729 of these are active. This fine computational mesh allows for a very accurate description of the tree-like structure of the main channels. At the three inlets, an M_2 tide of 0.5 m amplitude and 12 lunar hour period has been specified. With an integration time step $\Delta t=15$ minutes this simulation runs about five times faster than real time on a 6 Megaflops workstation.

6. Conclusions

A semi-implicit finite-difference method for solving the three-dimensional shallow water equations has been outlined. The combination of the judicious selection of terms that are finite-differenced implicitly, and the use of an Eulerian-Lagrangian method for treating the convective terms makes this formulation fast, accurate and stable. This solution scheme solves a set of tridiagonal systems along the vertical layers, and one five-diagonal linear system defined throughout the horizontal flow field. Computationally, each tridiagonal system is solved by a direct method, while the numerical solution of the large five-diagonal system can be conveniently obtained by a conjugate gradient method. Furthermore, the structure of the solution algorithm leads to a computer code which is completely vectorizable. The performance of the present algorithm on the Lagoon of Venice, Italy has been discussed.

References

Casulli, V., 1990, Semi-Implicit Finite Difference Methods for the Two-Dimensional Shallow Water Equations, Jour. of Computational Physics, Vol. 86, No. 1, p. 56-74.

Casulli, V. and Cattani, E., 1993, Stability, Accuracy and Efficiency of a Semi-Implicit Method for Three-Dimensional Shallow Water Flow, Computers & Mathematics with Applications.

Casulli, V. and Cheng, R.T., 1992, Semi-Implicit Finite Difference Methods for Three-Dimensional Shallow Water Flow, Int. Jour. for Numerical Methods in Fluids, Vol.15, 629-648.

Cheng, R. T., Casulli, V. and Gartner, J.W., 1993, Tidal, Residual, Inter-tidal Mudflat (TRIM) Model with Applications to San Francisco Bay. Estuarine, Coastal Shelf Science, to appear.

Signell, R.P. and Butman, B, 1992, Modeling Tidal Exchange and Dispersion in Boston Harbor, Jour. of Geophysical Research, Vol.97, No. C10, 15591-15606.

Parameterization of Turbulence for Three-Dimensional Circulation Modeling

by

Y. Peter Sheng[1], M. ASCE

Introduction

Estuarine circulations are driven by the actions of tide, wind, and density gradients associated with saline ocean water and fresh riverine water, and are also influenced by the complex geometry and bathymetry of the estuaries. Estuaries in the U.S. (e.g., Tampa Bay and Chesapeake Bay) are generally quite shallow except within the navigation channels. In general, there is significant vertical and horizontal stratification in the channel. In the shallow shoals, there is often significant horizontal stratification, although it is generally well mixed in the vertical direction. Proper parameterization of turbulent mixing in the vertical and horizontal directions is essential to the accurate simulation of mean circulation and transport, and has been a challenge for estuarine modeling in the past decades.

A variety of schemes for parameterizing horizontal and vertical turbulent mixing have been developed and utilized in three-dimensional models of estuarine circulation and transport. These schemes often differ significantly in assumptions and levels of complexity, but have not been critically compared again st one another in terms of their robustness for realistic estuarine simulations. For example, some estuarine circulation models have very efficient numerical algorithm, but still uses constant turbulent eddy coefficients or simple mixing length theory to simulate stratified estuarine flow. As another example, some models use very elaborate 2-equation turbulence model to simulate horizontal turbulent mixing in estuaries, although the model uses very crude horizontal grid. Thus, one purpose of this paper is to compare the various turbulence schemes for estuarine applications.

---

[1]Professor, Coastal & Oceanographic Engineering Department, 336 Weil Hall, University of Florida, Gainesville, Florida 32611.

In general, the various turbulence models in estuarine circulation models assume that the water column is fully turbulent at all times with vertical scales ranging from a few millimeters to the water depth and horizontal scales ranging from a few millimeters to the size of the estuary. It is often assumed that the cascade of energy is similar to that in a normal turbulent shear flow. However, recent studies (e.g., Luketina and Imberger, 1987; Wolanski *et al.*, 1984) have shown that bottom friction and estuarine fronts prevent the cascade of energy and the development of full turbulent spectrum. Flow features in estuaries often resemble slow viscous flow or laminar flow rather than fully rough turbulent flow. Thus, it is necessary to revisit the conventional turbulence schemes used in all three-dimensional circulation models to examine their ability to simulate newly observed estuarine circulation features.

Approach

This paper first provides a brief generic review of turbulence schemes used to parameterize vertical turbulent mixing in most three-dimensional estuarine circulation models. In particular, we review turbulence models developed by several schools including the following: (i) the various second-order closure models developed by Sheng (1982), Sheng (1984) and Sheng and Villaret (1989), (ii) the various second-order closure models developed by Mellor *et al.* (e.g., Mellor and Yamada, 1982), and (iii) the k-e model developed by Rodi *et al.* (e.g., Rodi, 1980). These models are briefly described and compared in terms of their assumptions, capabilities, and limitations.

Following the above generic review, model results obtained with turbulence models of varying degrees of complexity (i.e., Reynolds stress closure, T.K.E. closure, and equilibrium closure) are compared against the same laboratory and field data to determine what is the most desirable level of complexity. Particular emphasis is placed on testing the models' ability to simulate intermittently turbulent flow situations in estuaries.

In addition to vertical turbulence schemes, various schemes (constant horizontal diffusivity vs. grid-dependent horizontal diffusivity) for parameterizing horizontal turbulence are also examined. All tests are conducted using an enhanced three-dimensional curvilinear-grid circulation model which is based on the original CH3D model (Sheng, 1989).

Results of model simulations of an estuarine front using various levels of turbulence model are analyzed and presented. Recommendation on how to improve existing turbulence models are provided.

References

Luketina, D.A. and J. Imberger, 1987: "Characteristics of a Surface Buoyant Jet," *J. Geophys. Res.*, **92**:5435-5447.

Mellor, G.L. and T. Yamada, 1982: "Development of a Turbulence Closure Model for Geophysical Fluid Problems," *Rev. Geophys. and Space Phys.*, **20**:851-875.

Rodi, W., 1980: "Turbulence Models and Their Application in Hydraulics," *IAHR*.

Sheng, Y.P., 1982: "Hydraulic Applications of a Turbulent Transport Model," *Proceedings 1982 ASCE Hydraulic Division Specialty Conference on Applying Research to Hydraulic Practice, American Society of Civil Engineers*, Jackson, MS, pp. 106-119.

Sheng, Y.P., 1984: "A Turbulent Transport Model of Coastal Processes," *Proceedings 19th International Conference on Coastal Engineering*, American Society of Civil Engineers, pp. 2380-2396.

Sheng, Y.P., 1989: "Evolution of a Three-Dimensional Curvilinear-Grid Hydrodynamic Model for Estuaries, Lakes and Coastal Waters: CH3D," *Estuarine and Coastal Modeling* (M.L. Spaulding, Ed.), ASCE, pp. 40-49.

Sheng, Y.P. and C. Villaret, 1989: "Modeling the Effect of Suspended Sediment Stratification on Bottom Exchange Processes," *Journal of Geophysical Research*, **94:C10**, pp. 14429-14444.

Wolanski, E., J. Imberger, and M. Heron, 1984: "Island Wakes in Coastal Waters," *J. Geophys. Res.* **89**:10553-10569.

Three-Dimensional Free Surface Modeling
(Is There A Best Approach?)

B. H. Johnson[1], H. L. Butler[2], R. C. Berger[3]

Abstract

With increases in computing power, researchers all over the world are developing and applying three-dimensional (3D) hydrodynamic models. The numerical techniques employed are usually determined by the developer's background in computational fluid dynamics. As a result, many different types of models exist. For example, a major distinction between models is the approach taken for the spatial integration of the governing equations. Various approaches that have been taken and opinions based upon experience gained from 3D modeling activities at the U.S. Army Engineer Waterways Experiment Station (WES) are presented. However, the basic intent of this paper is to generate discussion concerning the question, "Is there a best approach for computing the three-dimensional flow fields required in the solution of typical estuarine/coastal problems?"

Introduction

The governing equations of fluid motion can be cast into different forms, e.g., the wave equation approach. For the spatial integration of the governing equations, either the finite difference, finite volume, or finite element method is normally employed. All of these approaches can be traced to the concept of weighted

---

[1] Special Assistant for Numerical Modeling, Hydraulics Laboratory, US Army Engineer Waterways Experiment Station, 3909 Halls Ferry Road 39180-6199.
[2] Chief, Research Division, Coastal Engineering Research Center, US Army Engineer Waterways Experiment Station, Vicksburg, MS.
[3] Research Hydraulic Engineer, Hydraulics Laboratory, US Army Engineer Waterways Experiment Station, Vicksburg, MS.

residuals. Generally, finite differences are employed
for the time integration, regardless of the method used
for the spatial integration. With the finite difference
method, either Cartesian or boundary-fitted structured
grids are employed. Boundary-fitted grids can be either
generalized nonorthogonal curvilinear grids or orthogonal grids. The finite element is generally associated
with the Galerkin weighted residual method and is normally applied on irregular unstructured grids. The
finite volume method is derived by setting the weight
function in the method of weighted residuals approach to
unity. The finite volume and finite difference methods
applied on staggered, structured, curvilinear grids are
similar. Applied on an unstructured grid, theoretically
the finite volume method provides geometric flexibility
at a lower computational expense but with reduced accuracy compared to the finite element method.

At WES, 3D models that use all of these approaches
for the spatial integration of the governing equations
are being developed and applied to various estuarine and
coastal areas. Examples of studies at WES employing
each approach are presented with comments concerning the
relative advantages/disadvantages of each.

A Boundary-fitted Finite Difference Model-Chesapeake Bay

In this study, a 3D hydrodynamic model of
Chesapeake Bay was developed and verified to provide
flow fields to drive long-term (years) water quality
computations. The basic code was developed by Sheng
(1986) and extensively modified in its application to
Chesapeake Bay by Johnson et al. (1991). The model is
known as CH3D-WES (Curvilinear Hydrodynamics in Three
Dimensions), and as its name implies, makes hydrodynamic
computations on a curvilinear or boundary-fitted planform grid. Initially the vertical dimension was handled
through the use of what is commonly called a sigma
stretched grid. However, with the relatively coarse
grid resolution employed, stratification in the deep
channels could not be maintained during long-term simulations. As a result, the governing equations were
rederived for solution on the Cartesian or z-plane in
the vertical direction.

After verification, the 3D model was employed in a
production mode to yield year-long simulations of 1984-
1986 to provide flow fields to drive the long-term water
quality computations. Using a time-step of 5 minutes,
about 10 hours of CPU on a Cray-MP were required to
simulate one year of hydrodynamics on a grid composed of
approximately 4,000 cells.

The planform boundary-fitted coordinates of the model provide enhancement to fit the deep navigation channel and irregular shoreline of the bay. Thus, an obvious benefit of generating finite difference solutions on boundary-fitted grids is that geometric features are modeled more accurately than with Cartesian grid models. The structured nature of the grid allows for the employment of economical solution techniques, e.g., approximate factorization.

Primitive Equation Finite Element Model-Galveston Bay

The model selection for the Galveston Bay Study was guided by the need to be able to incorporate combinations of 16 to 18 small proposed disposal island sites and to resolve a narrow deep channel through a wide shallow bay. Therefore, geometric flexibility was critical. Thus, the RMA10-WES finite element model was used. These simulations have been produced for approximately 12 prototype years on a grid containing approximately 12,000 computational nodes with a time step that varied from 15 to 60 minutes.

RMA10-WES was developed by King (1993), and modified by the staff at WES. The model solves for primitive variables, employs Galerkin test functions, and uses mixed interpolation to eliminate spurious water surface elevations. The Newton-Raphson technique is used to solve the resulting system of nonlinear equations. With the ability to quickly generate unstructured grids and the model's geometric flexibility that allows for the transition between 1D, 2D, and fully 3D computations within a domain, an ideal project for RMA10-WES is one that requires many structural design modifications and contains only a few regions in which high resolution with complex geometry is needed.

A Wave Equation Finite Element Model

Early finite element shallow-water equation models were plagued with spurious mode problems. However, the introduction of the generalized wave-continuity equation (GWCE) formulation by Lynch and Gray (1979) led to the development of robust depth-integrated coastal circulation finite element models with excellent numerical amplitude and phase propagation characteristics and monotonic discipation which suppresses generation of 2 x oscillations. One example of a finite element GWCE-based model being developed for WES is that of Westerink et al. (1992) called the ADvanced CIRCulation model (ADCIRC). A time-differentiated form of the primitive continuity equation is combined with a spatially differentiated form of the primitive momentum equations. The resulting wave type equation is then solved along with the primitive momentum equations. The finite element method is implemented by expanding the variables over three node linear triangles. The discrete GWCE is decoupled from the discrete momentum equations, allowing for a sequential solution procedure. A major advantage is the time-independence of the GWCE coefficient matrix, allowing a one-time assembly and inversion. Even though the discrete momentum equations coefficient matrix is time-dependent, it is diagonal. The 3D version of ADCIRC continues to undergo testing and is not yet viewed as a production tool. Details are presented in Luettich, Westerink, and Scheffner (1992).

The pilot application of the depth-averaged version of ADCIRC presented in Westerink et al. (1992) for the computation of tides and storm surges was in the Gulf of Mexico on a grid of approximately 6,800 elements and 4,000 nodes. The maximum to minimum element area ratio was greater than 15,000 with a very smooth transition from large to small elements. Additional applications on grids exceeding 50,000 elements have been made in the western North Atlantic, Gulf of Mexico, and Caribbean Sea to develop a tidal constituent and storm response database.

A Finite Volume Model

The MAC3D computer code now under development at WES is a 3D finite-volume model for buoyant incompressible flow. However, computations are made on a structured rather than an unstructured grid. As with CH3D-WES, since the discrete equations are formulated in general curvilinear coordinates, the code accommodates any geometry that can be mapped into rectangular computational blocks. The present version is limited to single blocks, but future versions will incorporate multiblock grids. Plans call for the code to be used to predict flow conditions and test operational scenarios for the proposed McCook Reservoir near Chicago, IL.

Summary and Conclusions

Four numerical models that differ significantly in the approach taken for the spatial integration of the governing equations and/or the form of the equations solved have been presented. Major advantages of the structured grid models, i.e., the boundary-fitted finite difference model CH3D-WES and the finite volume MAC3D model, are that a certain degree of geometric flexibility is obtained while still retaining the ability to use efficient solution techniques to make long-term (years) simulations feasible. Other advantages are the ease of making programming changes and the capability to conserve mass on individual cells. A disadvantage is that the transformed equations contain many more terms. In addition, although these models possess substantial geometric flexibility, structured grid models cannot be easily adapted to some problems, e.g., the Galveston Bay Study.

The finite element method as normally applied does not conserve mass on the element level. Although this problem has been noted when attempting to apply flow fields generated by RMA10-WES type models to finite volume water quality transport codes, if the formulation of the constituent transport and hydrodynamic models is consistent, there should be no impact on long term constituent computations. The advantages of finite element models such as RMA10-WES that retain all nonlinearities, allow mixed element types, and retain the complete coefficient matrix are virtually unlimited geometric flexibility and increased theoretical accuracy. The major disadvantage is the increased computational cost per time-step. However, being fully implicit does allow for relatively large time-steps in many applications. The ADCIRC model applications to date demonstrate that the finite element method can be realistically implemented within the framework of GWCE formulations and efficiently applied in a depth-averaged mode on grids with highly variable nodal densities. Similar efficiency and performance of the 3D version of ADCIRC for typical estuarine and coastal problems remain to be demonstrated.

Individual models generally have peculiarities that often determine the best available model for a particular application. However, the major thrust of this paper has been to focus on the basic question of whether one approach for handling spatial dependency, e.g., boundary-fitted finite difference, unstructured finite volume, the finite element method applied to the primitive equations in a completely nonlinear and fully implicit form, or the finite element method applied within the GWCE framework with various approximations to increase the efficiency of the computations, can be considered "best." The authors do not claim to have the answer. However, it might be noted that the finite volume method applied on unstructured grids offers attractive features of both finite element and finite difference methods, namely, mass conservation over individual cells, essentially unlimited geometric flexibility, and computational expense that will be less than that of the finite element method. As a final note, researchers in other areas of computational fluid dynamics, e.g., aerodynamics, are combining structured and unstructured discretizations in hybrid computational schemes that may prove to be more efficient and versatile than one method used alone.

Acknowledgments

The models presented herein were developed and applied as a result of research conducted by the U.S. Army Engineer Waterways Experiment Station. Permission to publish this paper was granted by the Chief of Engineers.

APPENDIX I. REFERENCES

Johnson, B. H., Kim, K. W., Heath, R. E., Hsieh, B. B., and Butler, H. L. (1991). "Development and verification of a three dimensional numerical hydrodynamic, salinity, and temperature model of Chesapeake Bay," Technical Report HL-91-7, U.S. Army Engineer Waterways Experiment Station, Vicksburg, MS.

King, I. P. (1993). "RMA10, A finite element model for three-dimensional density-stratified flow" (in preparation), University of California, Davis, CA.

Luettich, R. A., Westerink, J. J., and Scheffner, N. W. (1992). "ADCIRC: an advanced three-dimensional circulation model for shelves, coasts, and estuaries; Report 1: theory and methodology of ADCIRC-2DDI and ADCIRC-3DL," Technical Report DRP-92-6, U.S. Army Engineer Waterways Experiment Station, Vicksburg, MS.

Lynch, D. R., and Gray, W. G. (1979). "A wave equation model for finite element tidal computations," Computation of Fluids, 7, 207-228.

Sheng, Y. P. (1986). "A three-dimensional mathematical model of coastal, estuarine and lake currents using boundary-fitted grid," Report No. 585, A.R.A.P Group of Titan Systems, New Jersey, Princeton, NJ.

Westerink, J. J., Luettich, R. A., Baptista, A. M., Scheffner, N. W., and Farrar, P. (1992). "Tide and storm surge predictions using finite element model," ASCE Journal of Hydraulic Engineering, 118 (10), 1373-1390.

Galveston Bay 3-D Model Study Channel Deepening
Lessons Learned in Management of a Large Modeling Study

J. H. Schmidt, R. T. McAdory, W. D. Martin, and R. C. Berger[1]

Abstract

The numerical modeling of Galveston Bay in 3-D for long time durations with many geometry and hydrologic scenarios, coupled with the need to store all data for subsequent use in a model to predict oyster populations, produced a challenge in data and resource management. There were roughly 12 years of simulation generating nominally 50 gigabytes of data. The data consisted of thousands of input, output, and hotstart files submitted to and generated by the Cray-YMP supercomputer. The computer's six processors enable simultaneous execution of multiple jobs. Processes for managing such a large volume of input data and multiple output files and maintaining quality control are presented. Backup, archiving and transfer of results is also addressed. Visualization of 3-D results in a meaningful fashion is discussed.

Introduction

The problem of generating long-duration 3-D numerical runs, is a likely task in the future. The Galveston project serves as an early 3-D example that could benefit other modeling efforts. The project had large required durations of about 1 year in length, multiple plans, and the requirement that the flow and salinity fields be saved so that subsequent biological models could estimate oyster production. This paper is divided into 5 sections dealing with:

1. Computational speed
2. Run philosophy

---

[1]USAE Waterways Experiment Station, 3909 Halls Ferry Road, Vicksburg, MS 39180.

3. Quality assurance
4. File and data handling
5. Visualization

Computational Speed

The complexity of the domain with the narrow ship channel through a wide shallow bay with multiple entrances and a host of unknown proposed dredge material disposal islands dictated geometrically flexible numerical model. In our case this is the RMA10-WES code which was originally developed by Dr. Ian King of Resource Management Associates and extensively modified by the staff of the Waterways Experiment Station (WES), Hydraulics Laboratory (HL). The basic philosophy of this model is to attain computational efficiency by tremendous flexibility. The code is finite element and implicit. It allows 1, 2 and 3-D elements and their transitions and variable time steps. The obvious cost for this flexibility is a large time expense per time step. This then was our first task, to address computational speed on the Cray-YMP. The code as originally configured took roughly 60 cpu seconds each iteration and required 16 megawords of storage.

Our tests showed that about 22.3 seconds or 37.5% of the time was spent integrating the equations and 30.5 seconds or 51.1% in solving the matrix using the Frontal method. It was obvious that the normal improvement we could expect from minor changes in the vectorization was not going to make a significant difference. The time spent integrating the 3-D equations was addressed by changing the order of the loops of a typical finite element program so that the innermost was now the loop on gauss points instead of being the outermost. This made the inner loop a 27 count instead of 20 but more importantly it made many calculations that had been vector times a scalar operation into a matrix times a vector operation. This matrix times a vector operation is an efficient Cray optimized routine. The time spent integrating the 3-D equations was cut from 14.8 to 5.6 seconds per iteration or about 2.6 times faster than the original.

In the case of the Front solver many manipulations were combined to improve efficiency. The Frontal method is a direct solver which tries to minimize the amount of storage required. As coefficients are eliminated and the upper triangular portion written off to storage the remaining matrix is shifted to fill the voids. This series of operations was changed to save up several of these shifts and perform all of them at one time. For example instead of making 20 shifts of 1 position we now would make 1 shift of 20 positions. This along with the use of Cray specific routines reduced the time in the solver per iteration from 30.5 to 14.6 seconds or 2.1 times the previous solver speed.

While we believe that other basic changes could make still more speed improvements, our time constraints would not allow more effort spent here.

Run Philosophy

Our method of running was to try to make good use of the time available on the Cray. By that we wanted to make sure that the runs continued throughout all times that the Cray was available. While our nominal time step size was 1 hr. frequently smaller time steps were required. For example, our early experience found that the winter months with the passage of strong "Northers" caused numerical stability difficulties which required smaller time steps. The appropriate time step for stability is difficult to determine a priori, so hot-start files are generated periodically. The convergence of the nonlinear iteration is monitored and when an instability occurs the model backs up in time to a previous hotstart file and runs again with shorter time steps. This can be repeated for particularly rough time periods. This philosophy of rerunning difficult regions is more computationally efficient than running the entire simulation at the minimum required time step. This also keeps the code running through these periods without being continuously monitored. Current work to predict an appropriate time step is in progress. With the number of runs that were required it made more sense to run separate jobs concurrently rather than looking at parallelization of a single run. This is definitely more efficient but requires more logistical and quality control.

Quality Assurance

Since our calculations took weeks, and sometimes months, to complete, we needed ongoing methods of determining the quality of the results soon after their computation. This was accomplished by visual inspection of the file sizes, a special results file that contained the calculated salinities, water velocities and water surface elevations of 39 selected nodes, and by displaying the results for the entire system using the Cray graphics package Multi-Purpose Graphics System (MPGS).

The special results files contained nodal results from nodes distributed throughout the computational mesh and included locations representing all portions of the system including the ocean and various prototype data collection sites. Time history plots of the numerical results for these nodes were then plotted to check behavior.

Use of MPGS allowed the entire three dimensional system to be viewed at once with animation of a set of time steps. This global view allowed us to examine gross trends of salinity and water velocity through multiple hour long time steps for intervals up to seven tidal cycles. A cursory look at the MPGS image allowed us to check, for instance, that freshwater inflows were placed correctly, that power plant inflows and outflows were transporting salinities as expected, and that the overall features expected in a tidal driven estuary such as Galveston Bay were present. The image could also be sliced to reveal interior results, inverted to examine the system's bottom, and contour plots of any of these views could be produced to indicate more quantitatively plan view or vertical slicing structure, particularly stratification.

File and Data Handling

The end result of the calculation is a set of files containing the nodal results for each hour. These time step results, along with the geometry information, are roughly 50 megabyte files, referred to here as "mondo" files, that typically contained information for about 100 time steps. These files had a total size exceeding 5 gigabytes for a one year run. The complete files are created by combining the individual run results files which may include fractional hour time steps and regions of overlap. The nonoverlapping, integral hour time steps are stripped out of the individual results files and combined to form a continuous record. This cleaned up time step file is then combined with a geometry file, saved, and archived. These files are averaged, filtered, and otherwise processed to obtained isohaline, circulation, time history and other plots of results as desired. They are also read out as ASCII files for incorporation into the oyster model.

Each of the various manipulations of the mondo files requires large blocks of memory that are on the order of that required to run the calculations to begin with. Since all of these processes could, in principle, have to be run simultaneously for one or more of the various cases, a careful balance of resources is required. The limits of available memory, especially since the results files and the created mondos are Cray binary files, are every bit as restrictive as the run time capabilities of the RMA10-WES code itself.

The tape backup system used is 8 mm tapes. While this is an economical arrangement, (each tape holds over 2 gigabytes of data) it is slow and current tape operations are not standard and sometimes fairly primitive. For example, thus far we are unable to append files to a partially filled tape; the writing must be one operation.

Visualization

The results calculated are 3-D, and so efforts were made to present this feature in the final results. Often bottom results were most meaningful and so 2-D isohalines and circulation plots were created. Also time dependent plots were made of specific locations to show long term trends. Vertical slices of various cuts across the system and along the channel provided contour plots of isohalines. These results revealed features such as stratification wedges and allowed for base versus plan comparisons of these intrinsically 3-D features.

The MPGS system mentioned earlier was used to make a video of the system in which the system is turned and rotated to reveal views form the side and bottom. Running the solution from one of these alternative views provides great insight into the workings of the model.

Conclusion

As long term numerical model simulations are becoming more common this poses significant computational, logistical and storage difficulties. If the model study is 3-D the these problems are tremendously amplified. In the case of Galveston future transport studies required that the flow fields be saved thus the data requirements were as challenging as the computational demands. The use of 8 mm tapes appears to be an economical path but read/write features for this medium are still primitive. With many existing 3-D codes simple vectorization may not produce significant speedup, but we have found that more basic algorithmic modifications do sometimes produce significantly faster codes without huge labor time.

Acknowledgement

These tests were funded by the U.S. Army Engineer District, Galveston. Permission to publish this paper was granted by the Chief of Engineers.

OXYGENATION EXPERIMENTS IN THE WAVE BREAKING ZONE
E. I. Daniil[1], A.M., ASCE and C. I. Moutzouris[2]

Abstract

Experiments on oxygenation due to waves breaking on a uniformly sloping beach were carried out at the Laboratory of Harbor Works, National Technical University of Athens. The influence of increasing wave frequency and wave height on the oxygen transfer coefficient is discussed. Results are compared to experimental data for non-breaking waves.

It is concluded that oxygen transfer coefficients vary linearly with the maximum vertical wave velocity at the water surface. The presence of breaking waves leads to much higher transfer coefficients than in the case of non-breaking waves. As a consequence, it is recommended that the wave breaking zone and its related parameters be determined in order to improve the design of sewage outfall and to avoid adverse environmental effects.

Introduction

Air/water gas transfer is controlled by the interaction of the air and water boundary layers and represents one of the main sources of dissolved oxygen (D.O.) for water bodies. D.O. is often used as a water quality index. Research on oxygenation started as early as 1925, when river pollution problems started to appear, and many equations relating the oxygen transfer coefficient to hydraulic parameters were developed. Gradually interest spread to other water bodies, i.e. lakes and oceans, as well. The wave field is a parameter that has not been included in the predictive equations so far, although its effect has often been noted qualitatively. Experiments on the effect of waves on gas transfer have been reported by Hosoi et al (1977), Hosoi and Murakami (1986), Jähne et al. (1984, 1985, 1987), Daniil and Gulliver (1991) and Wanninkhof and Bliven (1991).

The gas transfer coefficient, which is the most characteristic parameter of the phenomenon, cannot be measured directly. It can only be deduced

---

[1] Researcher, [2] Professor
Laboratory of Harbor Works, Civil Engineering Department, National Technical University of Athens , 5 Iroon Polytechniou, 15773 Zografou, Athens, GREECE, Tel. - 30 - 1 - 778 3866, FAX - 30 - 1 - 775 9565.

through the transport equation, if the rest of the terms in the equation can be estimated accurately. In the presence of waves the transport processes are becoming more complicated and determination of the transfer coefficient is more difficult.

In the present paper experimental data and a theoretical investigation on the oxygenation of the water in the presence of waves breaking on a uniformly sloping beach is presented. It is believed that the estimation of the oxygen transfer in the nearshore zone is very important for countries like Greece with long coastlines.

Experimental Procedure and Data Acquired

Oxygenation due to waves breaking on a uniformly sloping beach was studied experimentally in a wave flume of the Laboratory of Harbor Works, National Technical University of Athens (NTUA). The dimensions of the flume are 27.40 x 0.60 x 1.53 m (L, W, H). A beach with a slope of 1:2.3 (vertical: horizontal) was located at a distance of 15.40 m (12.50 m) from the wave maker for series A (B) of experiments. The depth of water was 0.60 m in all experiments. Three sampling locations were selected for each series of experiments. Locations 1, 2, 3 were situated at a distance of 16.3, 14.8, 12.2 m from the wave maker for series A and at a distance of 13.4, 12.2, 10.0 m for series B, as shown in Fig. 1. At location 1 samples were taken from the bottom of the beach, whereas at the two upstream locations (2, 3) samples were taken from 30 cm depth. Experimental conditions for all experiments are given in Table 1. Photographs were also taken to document the type of breaking.

The water was chemically deoxygenated and D.O. concentration was followed over time for all three locations. D.O. samples were taken frequently and analyzed with the azide modification of the Winkler titration method. Sampling started shortly after the water was deoxygenated and the waves were started, and continued for 2 to 4 hours, till the oxygen deficit was low.

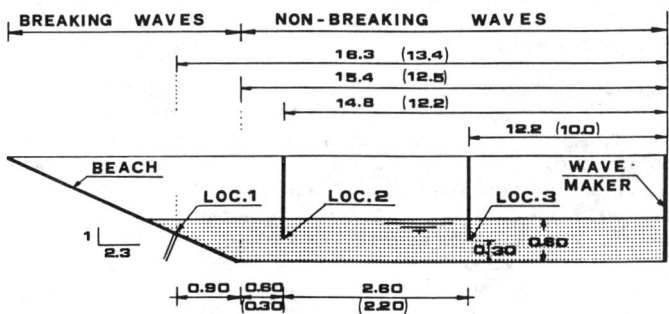

Fig. 1. Schematic diagram of sampling locations for series A (B) of experiments

TABLE 1. Experimental conditions and oxygen transfer coefficients

Exp. No.	Wave Period T (sec)	Wave Height H (cm)	Saturation Conc. C_s (mg/l)	Schmidt number Sc (-)	Transfer Coefficient $K_L \times 10^5$ (m/sec)	Transfer Coefficient $K_{L20} \times 10^5$ (m/sec)
A 3	1.60	9.9	8.4	458	2.9	2.7
A 2	1.45	7.6	8.3	446	5.5	5.0
B 7	1.45	9.1	11.0	918	6.4	8.3
B 6	1.45	12.0	11.2	967	11.3	15.0
B 9	1.45	14.9	11.0	912	17.0	22.0
A 4	0.75	5.6	8.7	506	5.3	5.1
A 5	0.75	7.1	8.4	455	9.9	9.0
B 8	0.75	7.3	12.0	1096	10.6	15.0

In Fig. 2 D.O. - time histories obtained from experiments B7, B6, B9 with waves of the same frequency but different wave height are compared. Curves are placed so that they have a common starting point of 3 mg/l. To achieve this the time axis is shifted by 30 minutes from one experiment to the other. It is concluded that the transfer velocity increases as the wave height increases. Moreover D.O. concentration differences between the three sampling locations decrease as the wave height increases. This is due to the fact that breaking becomes more intense as the wave height increases and, therefore, the horizontal dispersion increases as well.

In Fig. 3 D.O. - time histories from experiments A2, A5 with waves of almost the same height but different frequency are compared in a similar manner. The transfer velocity increases as the wave frequency increases.

From the above comparisons, it can be concluded that the oxygen transfer coefficient increases with wave height for a given wave frequency. It also increases with wave frequency for a given wave height. The determination of the oxygen transfer coefficients is described in the following section.

Data Analysis

The one-dimensional transport equation was used for the analysis of the data. For a complete analysis, the horizontal transport needs to be determined. As a first approximation it could be assumed that the composite effect of the horizontal transport is negligible, if data from the first sampling point upstream of the sloping beach (location 2) were used. In this case, the transfer coefficient could be determined from the following equation, using linear regression and the measured D.O. - time history:

$$ln(C_s - C) = -\frac{K_L A}{V} t + ln(C_s - C_o) \qquad (1)$$

where C, C_o, C_s are average, initial, saturation D.O. concentration respectively, K_L is the oxygen transfer coefficient, t is time, A is the projected

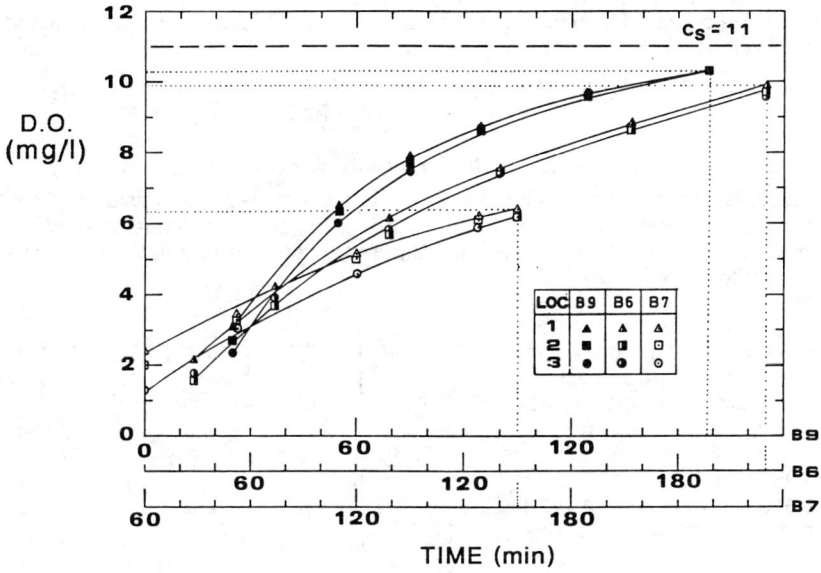

Fig. 2. Comparison of D.O. - time histories for waves with the same frequency and different wave heights

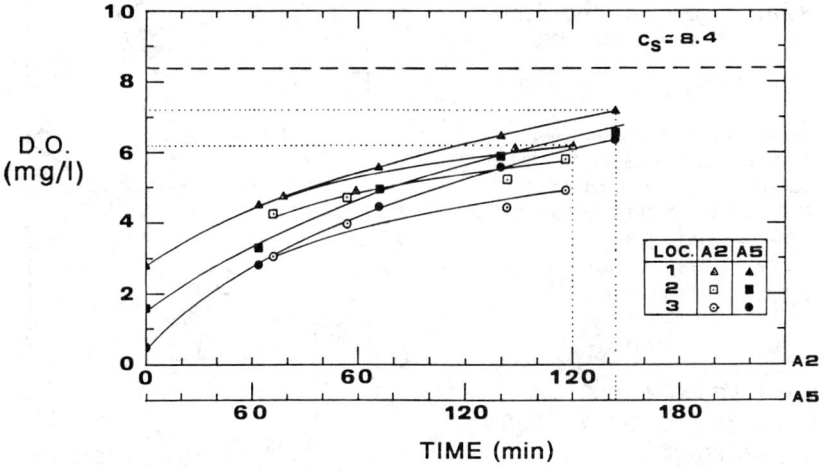

Fig. 2. Comparison of D.O. - time histories for waves with the same wave height and different wave frequencies

free surface area on the horizontal plane and V is the aerated volume extending from the free surface to the bottom of the channel.

The obtained coefficients were translated to 20 °C as follows (Daniil and Gulliver, 1988):

$$\frac{K_{L20}}{K_L} = \left[\frac{Sc}{Sc_{20}}\right]^{1/2} = \frac{v}{v_{20}}\left[\frac{293}{\theta+273}\right]^{1/2}\left[\frac{\rho}{\rho_{20}}\right]^{1/2} \quad (2)$$

where $Sc=v/D$ is the Schmidt number ($Sc_{20}=546$), v is the kinematic viscosity of the water, θ is the water temperature in degrees Celsius and ρ is the water density. Quantities without subscript refer to the temperature of the experiment, whereas the subscript 20 indicates their values at the reference temperature of 20°C.

The oxygen transfer coefficients, as determined from the above analysis, are compared to the relation obtained for non-breaking waves (Daniil and Gulliver, 1991), in which case the transfer coefficient is related to the vertical wave velocity at the water surface, in Fig. 4.

Daniil and Gulliver used a renewal model and expressed the average renewal rate as a function of a wave Reynolds number and derived the following equation:

$$K_L Sc^{1/2} = a_r \pi H f \quad (3)$$

where H is the average wave height and f the wave frequency. For the experiments with non-breaking waves they obtained $\pi a_r = 0.0159$.

For breaking waves an equation of the form:

$$K_L Sc^{1/2} = \alpha H f - \beta \quad (4)$$

was fitted to the data. Using linear regression $\alpha=0.066$ and $\beta=2.80\times10^{-3}$ m/s.

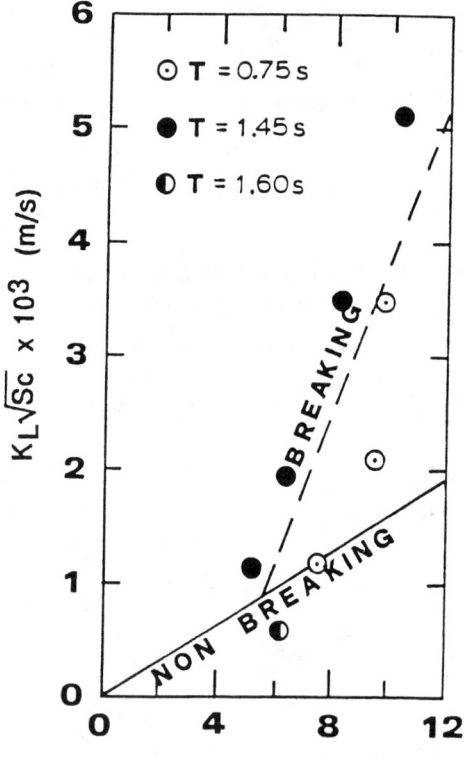

Fig. 4 Correlation of the oxygen transfer coefficients with the wave parameter H f.

Equation (4) holds for H f >0.056 m/s in the area where the resulting $K_L Sc^{1/2}$ values are higher than those for non-breaking waves for the same Hf values. Equation (4) expresses the general trend of the oxygen transfer coefficients for breaking waves, but it should be further expanded to include a parameter describing the non-linearity of the breaking.

Conclusions

The reported experiments demonstrate the importance of breaking waves in nearshore oxygenation and the need of including wave parameters in water quality models.

The oxygen transfer coefficient is described through a surface renewal model. The average surface renewal rate is correlated to the wave characteristics. The oxygen transfer coefficient is proportional to the maximum vertical wave velocity at the water surface. The correlation for the breaking wave data gives a much higher slope than that for non-breaking waves. The relation needs to be extended in order to incorporate breaking wave parameters.

Incorporating wave characteristics in water quality models can improve the design of sewage outfall near the shore, in order to avoid adverse environmental effects and to facilitate compliance with state or federal regulations, taking advantage of the "self-cleaning" capacity of the water body.

References

Daniil, E. I. and J. S. Gulliver, (1988), "Temperature Dependence of the Liquid Film Coefficient for Gas Transfer", *J. of Environ. Engrg.*, ASCE, 114(5), 1224-1229.

Daniil, E. I. and J. S. Gulliver (1991), "The Influence of Waves on Air-Water Gas Transfer" ASCE, *J. Environmental Engrg.* 117(5), pp. 522-540.

Hosoi, M., A. Ishida and K. Imoto, (1977), "Study on Reaeration by Waves," *Coastal Engineering in Japan*, 20, 121-127.

Hosoi, Y. and H. Murakami (1986) "Effect of waves on dissolved oxygen and organic matter", 20th Coastal Eng. Conference, Proc./CER Council ASCE/Taipei, Taiwan, Nov. 9-14, 1986. Chapter 184, 2498 - 2512.

Jähne, B., W. Huber, A. Dutzi, T. Wais and J. Imberger, (1984), "Wind/Wave-Tunnel Experiments on the Schmidt Number - and Wave Field Dependence of Air/ Water Gas Exchange," In: *Gas Transfer at Water Surfaces*, ed. W. Brutsaert and G. H. Jirka, Reidel Publishing Co., 303-309..

Jähne, B., T. Wais, L. Mémery, G. Caulliez, L. Merlivat, K. O. Münnich and M. Coantic, (1985), "He and Rn Gas Exchange Experiments in the Large Wind-Wave Facility of IMST," *J. of Geophys. Res.*, 90, C6, 11989-11997.

Jähne, B., K. O., Münnich, R. Bösinger, A. Dutzi, W. Huber, and P. Libner, (1987), "On the parameters influencing air-water gas exchange," *J. Geophys. Res.*, 92, C2, pp. 1937-1949.

Wanninkhof, R. H. and L.F. Bliven (1991) "Relationship between gas exchange, wind speed, and radar backscatter in a large wind-wave tank", *J. Geophys. Res.*, 96(2), 2785 - 2796.

Integrated Planning Analysis
for an Aggregate Mine

Edward Wallace, M.ASCE[1], and Robert MacArthur, M.ASCE[2]

ABSTRACT

Planning and design work for an aggregate mining facility within the Sutter Bypass, a key component of California's flood control system, is being performed by a multi-disciplinary team of engineers and resource scientists using state of the science hydrodynamic modeling techniques. A cooperative relationship between project planners and agency review staff from different disciplines is identifying opportunities for project features which combine mining, environmental, flood control, and agricultural objectives.

INTRODUCTION

Gravel extraction from rivers and flood plains can have detrimental effects on river morphology, sediment transport, flood control capabilities, water quality, and environmental resources if poorly planned and implemented (Collins and Dunne, 1990). This paper describes the role that hydrologic and hydraulic analysis plays in a comprehensive, multi-disciplinary planning study being conducted to assess the potential impacts of extraction of gravel from a floodway. The goal of the study is to develop a workable plan for extracting 50 to 60 years of aggregate resources from the heart of one of Northern California's most important flood control features, the Sutter Bypass. Figure 1 shows the 6,000 acre project site located near the confluence of the Sacramento and Feather Rivers, a location with diverse habitat, valuable environmental resources, and prime agricultural land as well as a large aggregate deposit. The challenge is to develop a plan that provides a method for extracting large volumes of aggregates located from 50 to 200 feet beneath the

---

[1]Principal, Meridian Consulting Engineers, 1215 119th Street, Sacramento, CA 95814
[2]Principal, Resource Consultants and Engineers, 1477 Drew Avenue, Davis, CA 95617

surface, while not adversely impacting flood control characteristics of the river and bypass system or the environment.

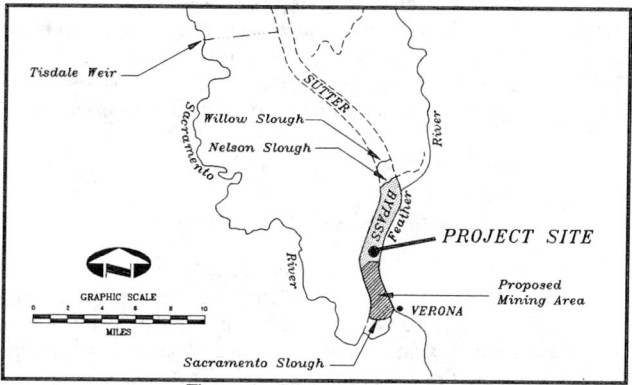

Figure 1. Project Location

APPROACH

A unique approach has been taken which solicits concerns and questions from a variety of agencies at an early stage. This has resulted in a candid and constructive working relationship with the technical staff and upper level managers of state and federal agencies responsible for flood control and natural resource protection. Environmental considerations such as preservation and restoration of riparian habitat have traditionally competed with the need to provide adequate conveyance capacity in constructed floodways and channels. The approach taken for this project emphasizes mutual understanding of flood control and environmental opportunities and constraints among project planners and review agencies. This cooperative effort will result in a balanced project plan with benefits to flood control, ecological resources, and sustained agricultural operations.

ISSUES AND CONCERNS OF REVIEWERS

The project will be subject to a complex review process involving The Reclamation Board, California Department of Water Resources, California Department of Fish and Game, U.S. Fish and Wildlife Service, National Marine Fisheries Service, U.S. Army Corps of Engineers (Regulatory and Operations Sections), California Regional Water Quality Control Board, Sutter County, and the State Lands Commission.

The Reclamation Board has the responsibility for controlling encroachments and ensuring the proper maintenance of federally funded levee and flood control projects(The Reclamation Board, 1987). Under this authority, a permit will be required for construction of the proposed project, and The Reclamation Board has taken the lead in review of the project's potential impacts on flood conveyance and flood control operations. The Reclamation Board will review the project with the assistance of the California Department of Water Resources and the U.S. Army Corps of Engineers Operations Section, and will involve various staff in hydrologic, hydraulic, sediment transport, legal, flood control operations and maintenance, geotechnical and environmental specializations. Areas of potential flood control impacts identified in planning meetings fall into three general cataegories:

a) Hydraulic impacts
b) Sediment transport impacts
c) River channel and levee stability impacts

Concepts for the project are being developed which involve a number of environmental and topographic changes which must be evaluated. Examples of potential project features include gravel extraction pits in the floodway, seasonal and permanent wetland creation along the pit shorelines, increased widths and densities of riparian forest along the river and sloughs, vegetative clearing and graded openings to enhance flood carrying capacity and flow distribution, bank stability improvements, wetland and other habitat restoration on the site but away from the active mining area, and changes in existing agricultural practices to integrate farming into long term environmental and flood control property management goals.

ENGINEERING AND SCIENTIFIC ANALYSES

The proposed aggregate mine site is characterized by a complex history of fluvial dynamics and river engineering works, both of which have been strongly impacted by hydraulic mining in the Bear, Yuba and Feather River watersheds during the late 1800's. A multi-disciplinary effort is required to develop a process-based understanding of fluvial dynamics, sediment transport and flood control operations in the project area. The following studies were initiated to address potential project impacts:

- geomorphologic assessment of the project area
- ecological/environmental resources assessment of the site
- development of hydrologic design criteria
- multi-dimensional hydraulic modeling
- sediment transport analyses
- river channel and levee stability analyses
- detailed geotechnical stability assessment, including seismic analyses
- evaluation of habitat restoration and reclamation opportunities
- river channel historical location investigation
- fisheries impact assessment
- flood frequency analyses

To date, work on the project has resulted in completion of several tasks which provide the necessary background for hydraulic modeling. Modeling work is in progress to evaluate potential impacts and identify opportunities to meet flood control, environmental, mining, and agricultural objectives. Completion of the studies, and assemblage of the information into a detailed Reclamation Board permit application is anticipated in the fall of 1993.

Hydrologic Criteria and Hydraulic Modeling - Development of the hydrologic design criteria for the site and creation of a two-dimensional hydraulic model form the basis for evaluation of a variety of project features and their interactions. The primary concern of the Reclamation Board and Corps of Engineers is that no project features adversely affect flood conveyance or water surface elevations in the flood control system. The "marsh elements" version of computer model RMA-2V (King and Roig,1988 and MacArthur et al,1991) was used to simulate the 1983 and 1986 high flow periods in the bypass for existing and project conditions. RMA-2V is a hydrodynamic model written in Fortran 77 for the implicit time integration and solution of the non-linear two-dimensional shallow water equations. It uses the finite element method to describe the rivers and bypass system as an assemblage of triangles and quadrilaterals. The elements are allowed to have variable shapes, depths, bottom roughnesses and sizes, thus permitting accurate representation of complex geometries and hydraulic systems. Element roughnesses (Manning n) were selected based on aerial photographs of land use. Flow and hydraulic head boundary conditions were established at exterior locations where flows enter or leave the bypass. Boundary conditions were established from measured flow and stage data at Nicholas, Sacramento River below Wilkins Slough, Sutter Bypass gage downstream from Tisdale Weir, Verona and the Fremont Weir. Topographic data based on 1987 photogrammetry of the site, supplemented by USGS quad sheet information and Corps of Engineers river cross-section data has been used to generate model geometry. Peak flows observed during the 1983 and 1986 flood events were used for calibration and validation of the 2-D model. These two events were selected based on the availability of gage data, magnitudes of flows near the flood control system design flows, and the differences for the two events in flow distribution between the Sacramento and Feather River systems. Volume continuity and computed stage at the external boundaries were calibrated to less than one percent for both events. No adjustment to the element parameters, including bottom roughness and eddy diffusivity, was required between the two events.

PROJECT ASSESSMENT AND DEVELOPMENT

The project assessment includes a step wise progression of work in a variety of disciplines on parallel tracks. Initial hydraulic modeling results have been used to identify constraints and opportunities for environmental enhancement, mining operations, and agriculture. Coordination between disciplines identifies conjunctive use opportunities for project features and minimizes adverse interactions. The project provides a unique opportunity, through the use of hydrodynamic modeling as a planning tool, to combine traditionally competitive uses in the floodplain into a mutually beneficial plan. Project scenarios for modeling were developed by a team of civil, mining and geotechnical engineers; geomorphologists; fisheries and wildlife biologists; ecologists; agricultural managers; and mining operations managers.

Evaluation of Flood Control Issues - After calibration and validation of RMA-2V with measured data, production simulations are being made to compare existing hydraulic conditions in the Bypass and river system with three different project scenarios. Project conditions include the physical features of the mined pit area, low berms, overburden stock piles, settling ponds, enhanced riparian forest components, flow transfer weirs and flow distribution improvements, habitat improvements, wetlands and other vegetative enhancements, and proposed changes in agricultural practices.

A systematic procedure for quantifying the cumulative effects of project features was developed. Computed water surface contours throughout the modeling area for various incremental installations of project features will be compared to water surface contours for the calibrated 1986 peak flow conditions. The same iterative comparisons for computed velocity (magnitude and direction) and bed shear stress will be made for the various project scenarios. Incremental changes in computed water surface elevation, flow velocity and bed shear can then be compiled for each major project modification expected to occur over time, starting with the installation of a small portion of the proposed mining pits. The mining area is proposed to increase slowly over time to a total acreage of 600 to 800 acres. The model provides a method for tracking incremental and cumulative effects on water surface, velocity and shear stress due to installation of major project features over the life of the project. Where measurable increases in water surface elevations are predicted by the model, project modifications will be recommended.

Sediment Transport Issues - Increased sediment loading into the Feather and Sacramento Rivers or into the Yolo Bypass as a result of the project is a concern to the flood control, water quality, and fisheries agencies. Future sediment transport assessments will compare sediment loading to the rivers and Yolo Bypass for existing and project conditions. Results from the two-dimensional hydrodynamic model will be used as the basis for sediment transport computations. Local scour and possible head cutting in the mined areas or near habitat enhancement sites will be examined separately.

Biological, Environmental and Habitat Issues - The project area presents unique opportunities for restoration of native central valley habitats. Many of these habitat types, which have been lost due to urban and agricultural development, are functionally related to river and floodplain hydrology. Therefore, flood control planning can also identify opportunities for habitat enhancements. Examples of this cooperative relationship include increased riparian forest bands for protection of levees against wind-generated wave erosion; stabilization of broad overflow channels with perennial grasses; creation of wetlands to replace brushy growth or to serve as energy dissipators against potentially erosive flows; selective clearing to establish larger trees rather than dense undergrowth; design of channel improvements to enhance migratory fish passage; construction of channels to distribute flows, control erosion, and minimize entrapment of fish during flood recession; and terracing of mine slopes to create habitat and improve pit slope stability.

The hydraulic model also provides a means to assess projected future conditions. For example, the hydraulic effects of vegetation growth under various management practices can be predicted. These types of maintenance and operations issues must be addressed to formulate permit conditions and management plans which are mutually acceptable to agencies with differing objectives.

CONCLUSIONS

1. A coordinated approach between resource scientists and engineers allows hydraulic and sediment transport assessments to be made comprehensively, resulting in mutual benefit to traditionally competing land uses in the floodplain.

2. A state of the science hydraulic modeling tool such as a two-dimensional finite element model is essential to assess the interactive effects of complex project features.

3. Planning and assessment procedures for projects which involve multiple uses and objectives in sensitive flood control areas must be as flexible and dynamic as the system being evaluated. Project assessments for flood control impacts must anticipate the the long term and phased effects of the proposed facilities.

References

Collins, Brian and Thomas Dunne (1990) "Fluvial Geomorphology and River Gravel Mining: A Guide for Planners, Case Studies Included, California Department of Conservation, Division of Mines aand Geology Special Publication 98.

King, I.P. and L.C. Roig (1988) "Recent Applications of RMA's Finite Element Models for Two Dimensional Hydrodynamics and Water Quality," Proc. 2nd. Int. Conf. on Finite Elements in Water Resources, Pentech Press, London, UK.

MacArthur, Robert C., James Pennaz, Gary E. Freeman, Lisa L. Weissinger and Ian P. King (1991) "Enhanced Multi-Dimensional Modeling of Marshes and Wetlands," Proc. of the Natl. Conf. on Irrigation and Drainage, ASCE, Honolulu, Hawaii, July 22-26, 1991.

The Reclamation Board (1987). "The Reclamation Board", Department of Water Resources, The Resources Agency, 1987.

BANK STABILIZATION WITH ENVIRONMENTAL FEATURES ON THE MIDDLE RIO GRANDE

Drew C. Baird (ASCE Member)[1], James P. Wilber[1], Richard W. Slater[1]

ABSTRACT

River Maintenance on the the Middle Rio Grande has changed significantly since channel work began in the mid 1950's. Past practices included pilot or relocation channels, vegetation control, and island and bar removal. Techniques which encorporate river geomorphology, fish habitat structures and vegetation that have more desireable environmental effects are now being used. Preliminary fish and invertebrate data show significant use of stabilized banks and other environmental features. Past river maintenance practices, river geomorphology, implementation of present techniques, construction of fish habitat structures, and results of aquatic resources monitoring are presented.

INTRODUCTION

Construction of river channel rectification works by the U.S. Bureau of Reclamation (Reclamation) on the Middle Rio Grande began in 1953, and consisted of floodway vegetation clearing, pilot channeling, island and bar removal and construction of Kelner jack fields. After the initial construction river channel maintenance concentrated heavily on pilot channeling, vegetation control, and removal of islands and bars. Researchers since 1953 have investigated the character and nature of river channels (Leopold and Wolman, 1957). These investigations have led to the development of new river maintenance techniques which have more desireable effects on the river geomorphology and environment (Nunnally and Keller, 1979). This paper focuses on the historical river maintenance practices, river morphology research, and the implementation and results of new methodologies.

---

[1] Bureau of Reclamation, 505 Marquette NW, Suite 1313, Albuquerque, NM 87102-2162

HISTORICAL BACKGROUND

Major channel rectification work started in the late 1950's and continued until the mid-1960's. River channel straightening, thereby shortening the length of the river was common. This was the approach of that era on many rivers throughout the world (Parker and Andres, 1976; and Winkley, 1976). The initial construction and subsequent river channel maintenance, in many cases, did not contribute to the long term stability of the river channel and resulted in constant maintenance. Regular pilot channel reconstruction was frequently required in many areas. Annual maintenance consisted of moving all bars and re-forming the straight alignments. This type of work was last done in 1985.

RIVER GEOMORPHOLOGY

Leopold and Wolman (1957) determined that rivers have three basic planform shapes: meanders, braides, and straight reaches. Straight reaches longer than 10 times the river width are rare in natural channels. Meandering rivers can accomodate a wide range of flows, and have riffles and pools which are important for the movement of sediment and for aquatic habitat. Rivers adopt these planforms to adjust their length, width, and depth, in relation to local geology, valley slope, bed and bank material composition, and sediment load.

River straightening or channelization as was done on the Middle Rio Grande has been shown to have adverse impacts (Parker and Andres, 1976; and Winkley, 1976). Nunnally and Keller (1979) identified several general problems with past channelization practices. Channelization can cause destruction of riparian vegetation, reduction or elimination of pools and riffles, and a reduction in aquatic habitat. Channelization offers no erosion protection from high flow velocities during bank-full discharges, and results in increased bank erosion from increased stream slope.

Nunnally and Keller (1979), and Vanoni (1975) advocated stabilizing the river banks and preserving existing river planform as much as possible in maintenance activities, while providing erosion protection for the outside of bends. Nunnally and Keller (1979) gave several advantages of this approach: 1) this minimizes the destruction of trees and other vegetation, 2) river banks are more stable because gradients are not increased and the river can accomodate wider ranges of flow without increased erosion, 3) stream reaches are almost never straightened or relocated, however there is sometimes the need to remove constrictions or bank obstructions and to narrow the stream where erosion has produced an overly wide channel, and 4) during construction there is minimal disturbance to the streambed or to vegetation except where heavy undergrowth is removed. Trees and saplings often are left undisturbed.

The final result using these procedures is a more stable stream with aesthetic value. Pool and riffle sequences associated with meanders recover quickly after high flow events and provide better habitat and higher aquatic and wildlife populations than on channelized streams (Nunnally and Keller, 1979). In addition there is improved water quality through reduced turbidity from bank erosion. Riprap stabilization and groins also provide additional aquatic habitat.

PRESENT RIVER MAINTENANCE APPROACH

After a reviewing Reclamation's practices in the mid 1980's it was determined this new approach should be pursued. We are now leaving most areas in the same or nearly the same planform as is possible. This preserves the geomorphology of the system, maintains pools and riffles, and minimizes disturbances to the river bottom and vegetation. River bars which are partially removed for fill material recover after the next spring runoff. This new approach has enabled more permanent site maintenance and has significantly reduced recurring maintenance and annual environmental disturbances. Not every reach of the Middle Rio Grande is in equilibrium. One reach is degrading and another is aggrading. However, general principles of geomorphology and sediment transport have been successfully used.

Vanoni (1975) indicated that the bend radius divided by the width (radius/width ratio) of 10 was best on some river stabilization projects. On the the Middle Rio Grande there has been success stabilizing curves with riprap revetments, or groins, at or near the natural radius/width ratio of about 2 or 3. This has significantly reduced changes to the river channel, the amount of required work, and site disturbance.

In many areas bars that were removed annually are left providing natural backwaters. Many of these bars are now re-vegetating. At one site a pond was constructed on the inside of a bend. The pond now contains fish and is frequented by numerous bird species.

In one area a geomorphological study was completed which showed that the river had reached an equilibrium length. The reach was subsequently stabilized in the equilibrium planform. Mitigation for this project included re-vegetation and construction of backwaters on downstream sides of point bars. River bar material was used to shape the banks prior to riprap placement. Bars were excavated in the center leaving a dike at waters edge. The dike was then removed minimizing the amount of time equipment was in the water reducing turbidity during construction. These bars reformed after the first post construction spring runoff.

In areas where sediment loads are significant the river channel has become entirely plugged with sediment deposits. In the past sediment plugs were removed with land based equipment. Currently amphibious excavators are used to construct small pilot cuts through sediment plugs. For example, a small 5,000 cubic yard (3,800 cubic meter) pilot cut was made through a sediment plug in 1991. Discharge from a thunderstorm flowed through the pilot cut and removed the entire sediment plug, about 300,000 cubic yards (229,000 cubic meters), in 12-14 days. Under past practices about 300,000 cubic yards (229,000 cubic meters) of material would have been excavated and deposited on the channel banks. The new technique eliminates the damage to riparian vegetation on the channel banks.

FISH HABITAT STRUCTURES

Construction of aquatic habitat structures is being incorporated in many river stabilization projects to diversify and enhance existing habitat. At one revetment site a middle bar created a side channel where several boulder groupings were placed to create habitat diversity. A riprap riffle was also constructed. At other sites, snags (i.e. dead trees) were buried in the bank and extended into the flow. Groins are also being used which create small side channels because of downstream sediment deposits. The groins also create areas of slow and fast water, and areas of slow and fast water enhancing aquatic habitat. Boulders along with snags cabled to the banks are placed to provide habitat diversity. Figure 1 shows some of the typical habitat structures.

AQUATIC RESOURCE MONITORING

The Middle Rio Grande contains both cool and warmwater aquatic habitat. The study area contains primarily cool water habitat with the associated aquatic communities. The fishery is composed of native riverine fish and introduced fish. The most abundant fish within this reach, are white sucker (Catostomus commersoni), longnose dace (Rhinichthys cataractae), and carp (Cyprinus carpio).

Fish were sampled by electrofishing and seine in 1991 and 1992, to document short and long term impacts of river maintenance activities and environmental features. Table 1 presents the percent composition of fish collected from nonstabilized banks, banks stabilized with riprap, and created backwaters. Preliminary data show that bank stabilization has a short term impact on fish through displacement and in some cases direct mortality. Recolonization of stabilized banks occurred relatively rapidly. Fish community structure was similar in stabilized and nonstabilized banks. Longnose dace increased in abundance and mosquitofish (Gambusia affinis) decreased in abundance

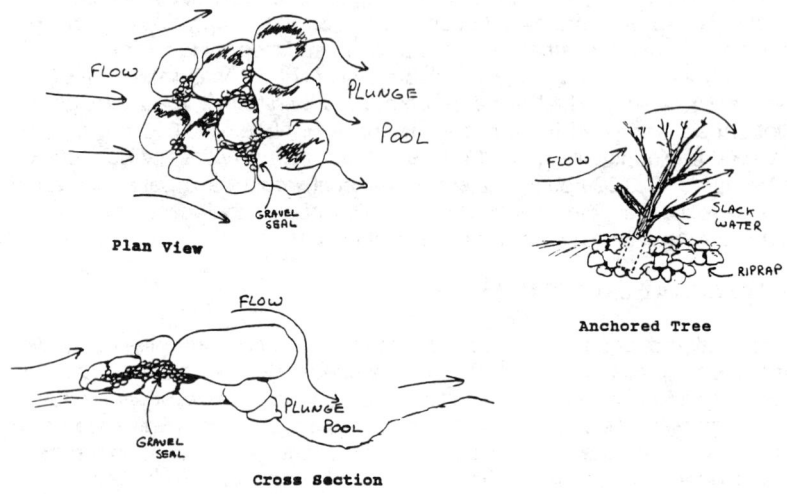

Figure 1. Typical Aquatic Habitat Structures

Table 1. Percent composition of fish collected from nonstabilized cutting banks, banks stabilized with riprap, and created backwaters.

Percent Composition

Species	Nonstabilized Banks	Stabilized Banks	Created Backwaters
White sucker	48	43	13
Longnose dace	18	41	0
Carp	9	7	2
Flathead chub	0	4	0
Channel catfish	0	2	<1
Rainbow trout	0	1	0
Brown trout	3	1	0
River carpsucker	3	<1	4
White crappie	0	<1	13
Mosquito fish	15	<1	57
Red shiner	3	0	7
TOTAL (N)	100% (33)	100% (303)	96% (260)

within stabilized banks. Electrofishing catch-per-unit-effort (CPUE) (# of fish/10 minute sampling effort) in 1992 was 14.6 for nonstabilized banks and 30.1 for stabilized banks. This difference was primarily due to a corresponding increase in the CPUE of longnose dace between nonstabilized and stabilized banks, 1.9 to 18.4, respectively.

Created backwaters showed the highest diversity of fish by species and life stage. Mosquitofish and white sucker were the dominant species in backwaters during 1992 and white crappie (Pomoxis annularis) was the dominant game fish. Diversity of macro invertebrates increased within riprap placed on stabilized banks, probably due to the stable cover and substrate provided by the rock, as opposed to a shifting, eroding bank.

CONCLUSIONS

In general the new overall approach is the best management practice. Alternatives are selected based on efficiency, cost, resource management, and endangered species to minimize short term impacts and maximize long term benefits and habitat improvements.

Preliminary aquatic resource monitoring data show significant use of stabilized banks and other environmental features. Additional research is needed to study species and life stage selectivity of environmental features along with the seasonality and long-term stability of the associated communities.

REFERENCES

Leopold, L.B., and Wolman, M.G., "River Channel Patterns: Braided, Meandering, and Straight," USGS Professional Paper 282-B, 1957.

Nunnally, N.R., and Keller, E., (1979) "Use of Fluvial Processes to Minimize Adverse Effects of Stream Channelization," Water Resources Research Institute, University of North CArolina, Report No. 144.

Parker, G., and Andres, D., (1976) "Detrimental Effects of River Channelization," Proceedings, Rivers'76 Conference, ASCE, August, pp. 1248-1266.

Winkley, B.R., (1976) "Response of the Mississippi River to the Cutoffs," Proceedings, Rivers '76 Conference, ASCE, August, pp. 1267-1284.

Two Options for Disposal of Desalination Reject Water

Louis J. Armstrong A.M. ASCE[1], Phillip R. Mineart A.M. ASCE[1], and Ralph H. Cross, III M. ASCE[2]

Abstract

A major problem associated with the use of desalination plants for the production of fresh water is the disposal of the reject water. This water is typically double the salinity of the intake water. The increase in salinity results in a discharge that is denser than the ambient water, which can significantly reduce the dilution achieved from an outfall or diffuser located on the sea floor. Two alternative disposal options were studied. The first option is to dispose of the reject water in an existing wastewater outfall and diffuser. This option was considered for the Central Marin Sanitation Agency's (CMSA) waste treatment plant outfall in San Rafael, California. The second option is to use a diffuser mounted just below the water surface and pointing downward. The second option was analyzed for a site in an enclosed bay. We present a comparison of the two disposal options and discuss the relevant parameters associated with each.

Introduction

Two studies were conducted on the disposal of desalination plant reject water. The first study evaluated the effects of using

---

[1] Woodward-Clyde Consultants, 500 12th Street, Suite 100, Oakland, California, 94607

[2] Consultant, 2365 Dapplegray Lane, Walnut Creek, California, 94596

an existing wastewater outfall to dispose of reject water from a proposed reverse osmosis desalination plant. The wastewater flow was 30.28 million liters per day (MLD) (8 MGD) and the proposed addition was 18.93 to 37.85 MLD (5 to 10 MGD). The mixture may be positively, neutrally or negatively buoyant in the bay water, depending on the CMSA treatment plant effluent and reject water flowrates. The buoyancy affects not only the manner in which the diffuser discharges into the bay, but also the subsequent mixing and dilution of the combined effluent with the bay waters. This study included both a dye study and a modeling study (Woodward-Clyde, 1991).

The second study was performed to evaluate discharging the reject water from a proposed reverse osmosis desalination plant via a diffuser mounted below the water surface and pointing downward. A 1.14 to 2.65 MLD (0.3 to 0.7 MGD) flow of reject water was proposed for the desalination plant. A hydraulic analysis was performed to develop a preliminary diffuser design. Dilution modeling was then performed to evaluate the effectiveness of the diffuser.

The goal of both studies was to estimate the dilution from each disposal system in order to better understand the environmental impacts of disposing of the desalination reject water.

Disposal Using an Existing Wastewater Outfall

When considering the use of an existing wastewater outfall for disposal of desalination reject water there are both hydraulic and environmental parameters to consider. Hydraulically, the existing outfall must be able to handle the increased flow. If the diffuser is operating below its design flow then the additional discharge may enhance mixing due to the increased effective length of the diffuser. This was the case for the CMSA diffuser. If the diffuser is operating near its design capacity the increased flow will cause the port velocities to increase. This may or may not increase dilution depending on the depth of the water body and whether or not the discharge can be characterized as a jet or a plume.

Another consideration when discharging desalination water into an existing wastewater outfall is the density of the combined effluent. Since the discharge from a wastewater treatment plant can be cyclical on a daily basis, the density of the combined discharge can also vary. These variations can cause the discharge to vary from positively to neutrally to negatively buoyant in a single day, as was the case with the CMSA discharge. As the buoyancy of the discharge decreases, the amount of dilution achieved from buoyant mixing also decreases. For the case of the CMSA diffuser, if the discharge became negatively buoyant the dilution from the diffuser was reduced by two-thirds over the wastewater discharge. If the discharge became close to neutrally buoyant the dilution was decreased by 1/2. If the discharge remained positively buoyant the dilution depended on the environmental conditions of the receiving water.

If the discharge remained positively buoyant and the receiving water was not stratified, the dilution from the diffuser was approximately the same as the existing wastewater dilution. This occurs because the decrease in buoyant mixing was compensated for by the increased efficiency of the diffuser. If the receiving water was stratified, the discharge would tend to trap, thus decreasing the distance over which buoyant mixing can occur. This trapping caused the dilution to decrease by approximately 1/2 over the wastewater discharge.

In addition to dilution from the diffuser, a conservative constituent may be diluted when the desalination reject water and the wastewater mix in the outfall pipe. The dilution achieved in the outfall pipe depends on the flowrate and concentration of both waste streams. If the wastewater discharge has a lower concentration of a particular constituent, the wastewater will help dilute the desalination reject water. The reverse is also true, the desalination reject water may help dilute a constituent in the wastewater. Therefore, this inpipe dilution may be beneficial to both waste stream dischargers.

Table 1 shows the predicted dilutions caused by mixing in the receiving water, ie. from the diffuser, and caused by mixing in the outfall pipe for the various discharges and ambient conditions discussed above. From the table, it can be shown that when both the inpipe dilution and dilution caused by mixing in the Bay are accounted for, the predicted dilution before and after the desalination reject water is added are approximately equal.

Disposal Using a Downward Pointed, Surface Mounted Diffuser

Another option for disposal of desalination reject water is to discharge it through a traditional diffuser that is mounted at the surface and is pointing downward. Since the desalination reject water is denser than the receiving water, it will be negatively buoyant and will sink. This can be viewed as the traditional diffuser problem turned upside down.

The advantage of this discharge set-up over using a pre-existing diffuser is that the diffuser can be designed for maximum dilution. As with a bottom diffuser and a positively buoyant discharge, there is a optimal diffuser design in terms of number of ports, port spacing and port diameter. Other parameters such as mounting depth and location are also important. The diffuser can be mounted underneath a pier or along side a ship dock and should be mounted below lower low water to avoid surface expressions. The mounting configuration and location will govern whether the diffuser ports are all on one side of the diffuser or are alternating.

An analysis of a desalination discharge into an enclosed bay was performed. The analysis was performed for two diffusers, one mounted from a pier with alternating ports, and one mounted from a ship dock with ports all on one side. The depth of the water was approximately 10.7 meters (35 feet). The flowrates analyzed were between 1.14 to 2.65 MLD (0.3 to 0.7 MGD). Stratified and non stratified bay conditions were considered. For the two discharges analyzed, the diffusers were designed to achieve at least 100:1 dilution. The proposed diffuser diameter was 30.5 centimeters (12 inches) and the proposed diffuser length was between 36.6 and 54.9 meters (120 and 180 feet).

Conclusion

In conclusion both disposal options can be useful for disposal of desalination reject water. The first option requires that a wastewater outfall exist and have sufficient capacity to handle the additional flow. However, the impacts to dilution should be studied very carefully to assure continued dilution of the wastewater effluent in the receiving water. The second option requires that a mounting structure exist and that a diffuser can be mounted on it. Beyond that, a diffuser can be designed that will achieve the desired dilution given the discharge volume and receiving water depth.

Acknowledgements

The authors would like to thank the Marin Municipal Water District for funding the evaluation of discharging desalination reject water into an existing wastewater outfall and the staff of the Central Marin Sanitation Agency Treatment Plant for their cooperation during the study.

References

Woodward-Clyde Consultants. 1991. Appendix C: Hydrological and
    Water Quality Evaluation of Discharging MMWD's Proposed
    Desalination R.O. Concentrate Via CMSA's Treated Effluented
    Outfall of the Marin Municipal Water Supply Project EIR

Scenario	Buoyancy	Discharge (Million Liters/Day)	Predicted Dilutions: Mixing in Bay	Predicted Dilutions: Mixing in Outfall	Predicted Minimum Dilution
Existing Wastewater, Uniform Density	Positive	30.28	30	1	30
Existing Wastewater, Stratified	Positive	30.28	15	1	15
18.93 MLD Addition, Uniform Density	Slightly Positive	49.21	29	1.625	47
18.93 MLD Addition, Stratified	Slightly Positive	49.21	10	1.625	16
37.85 MLD Addition	Negative	68.13	10	2.25	22
37.85 MLD Addition	Neutral	68.13	14	2.25	31

Table 1: Predicted Near Field Dilutions from CMSA Diffuser

Ballast Water Treatment Effluent Dispersion Studies

Peter A. Mangarella, Ph.D., P.E., Member[1]
Joseph M. Colonell, Ph.D., P.E., Fellow[1]

Abstract

Treated tanker ballast water is discharged from Alyeska Pipeline Service Company's ballast water treatment facility through a multiport diffuser into Port Valdez, Alaska. The discharge is required in the NPDES permit to comply with water quality regulations for aromatic hydrocarbons. This paper presents the methods, results, and insights on effluent plume behavior obtained from two dye dispersion studies conducted to help evaluate compliance with the water quality regulations.

Introduction

Alyeska's NPDES permit requires compliance with water quality standards for aromatic hydrocarbons (benzene, ethylene, toluene, and xylene) at the edge of a prescribed mixing zone. As part of the permit renewal process, EPA and the Alaska Department of Environmental Conservation (ADEC) required Alyeska to conduct dye dispersion and water quality testing to evaluate compliance with the NPDES permit. In response, Alyeska contracted with Woodward-Clyde to conduct two water quality and dye dilution studies in Port Valdez, one in the fall and one in the spring to characterize plume behavior under two water column stratification conditions.

This paper describes the methods and results of those studies with emphasis on the insights on plume behavior in a situation where effluent and ambient conditions are quite variable over seasonal and shorter (days to weeks) time scales. Previous papers have focused on how the near field dilution data compare with laboratory data and

---

[1] Senior Consultants, Woodward-Clyde, 500 12th Street, Oakland, California, 94606 and 701 Sesame St., Anchorage, Alaska, 99503, respectively.

predictions using the EPA UDKHDEN model (Wright et al, 1988) and coupling the UDKHDEN model with a Monte Carlo shell to statistically characterize plume behavior with effluent and ambient variability (Colonell et al, 1991).

Site Description

Ballast water from tankers that call at the Trans-Alaska Pipeline Terminal in Valdez, Alaska, is pumped ashore and treated in a ballast water treatment facility consisting of physical, chemical, and biological processes. After treatment, the ballast water is discharged through a 1.2 m (4 ft) diameter outfall into Port Valdez, a 220 m deep fjord in south central Alaska. Ambient stratification in Port Valdez varies seasonally between strongly stratified late summer and early fall conditions to a weakly stratified, nearly homogeneous conditions from late fall thorough early spring. Currents in the vicinity of the diffuser are induced by density effects, tides, and winds; but are generally low ranging between about 0.5-8 cm/sec, with a median speed of about 3.5 cm/sec.

The outfall includes a 61 m (200 ft) long diffuser consisting of 20 ports with 3 m (10 ft) spacing on alternate sides of the diffuser. The 10 inshore ports are 10 cm (4 in) in diameter and the 10 offshore ports 12.7 cm (5 in) in diameter. The ports are oriented 45 degrees to the plane of the diffuser. Video surveys of the diffuser indicate that the diffuser is intact although there is a "collar" of marine growth that has grown around each port. The diffuser is laid on a slope of 34 percent such that the water depth at mean lower low water is 62 m at the upstream end of the diffuser and 82 m at the downstream end.

Effluent flow and density vary in response to the amount and source of ballast water carried by the tanker fleet. The effluent flowrate tends to be highest during the autumn and winter months when the tankers' need for ballast is greatest. For example, the median flowrate for the months of Nov. through April is about 0.67 m^3/sec (15 MGD) whereas the corresponding flowrate during the rest of the year is about 0.5 m3/sec (11 MGD). Effluent density varies dramatically (16 -25 sigma-t units) from tanker to tanker depending on the tanker's point of origin and where it picked up ballast.

Study Design and Methods

Two field studies were conducted, one in October 18-22, 1985 and one in March 24-29, 1986 to examine plume behavior and dilution under strong and weak stratification, respectively. Rhodamine WT dye was injected onshore into the effluent outfall. As we had only limited control over the flowrate but wanted to maintain a relatively constant effluent dye concentration, we attempted (with only modest success) to maintain the dye injection rate proportional to the effluent flowrate.

The effluent plume was tracked with a profiling in situ fluorometer. Ambient stratification was monitored with a conductivity-temperature-depth (CTD) probe that was coupled with the fluorometer. The fluorometer and CTD were lowered at a rate to obtain data at about 2 m intervals. Water samples were obtained with Go-Flo bottles arranged in a rosette at the depth where the fluorometer indicated the highest dye concentration. This depth was determined through the real time graphical display of the dye concentration as we lowered the fluorometer. Having such real time display is essential! The water samples were analyzed for aromatic hydrocarbons and dye concentration (the latter as a check on the in situ fluorometer). During the 5 day October test 118 CTD/fluorometer casts were made; 276 casts were made during the 6 day March test which involved around-the-clock sampling. Current meters were moored near the diffuser to characterize currents for data interpretation and modeling.

Results

In October 1985, effluent densities ranged between 20.7 and 24.0 while the ambient density at the discharge level was approximately 24 sigma-t units. The vertical density gradient was about 0.03 sigma-t units per meter at diffuser depth. Ambient currents were typically small with a median value of about 3.5 cm/s. This combination of diffuser discharge and ambient conditions restricted plume rise to less than 20 meters and kept spreading layer thicknesses to less than 10 m. The minimum observed dilution near the diffuser was 30:1; the average of all observed minimum dilutions near the diffuser was 60:1. During one period the effluent density increased to the point where the effluent was almost neutrally buoyant (Figure 1) and the discharge behaved like a turbulent jet.

In March 1986, the weak density stratification allowed the plume to rise to the surface (a distance of 60 to 80 m) or very near so and spreading layers were generally 20-30 m which is large relative to October conditions. Minimum observed dilutions were about 80:1, while the average of all observed minimum dilutions was 200:1. During the March 1986 test snowmelt runoff increased the vertical density gradient from zero to about 0.001 sigma-t units per meter causing the plume to be trapped at depths of 20-40 meters, whereas the plume had surfaced only a few days before (Figure 2).

Acknowledgements

The writers gratefully acknowledge the cooperation and support of Alyeska Pipeline Service company in the conduct of this work and for permission to publish these results.

References

Colonell, J.M., P. Mangarella, and R. Schanz (1991). Discharge of Tanker Ballast Water: A Regulatory Compliance Analysis, Proceedings of the International Symposium on Environmental Hydraulics, Hong Kong, 16-18 December, J.H.W. Lee and Y.K. Cheung, Editors, pp. 287-293.

Wright, S.J.. P.A. Mangarella, and J.M. Colonell (1988). Outfall Plume Dilution in Stratified Fluids, Proc. National ASCE Conf. on Hydraulic Engrg., Colorado Springs, CO.

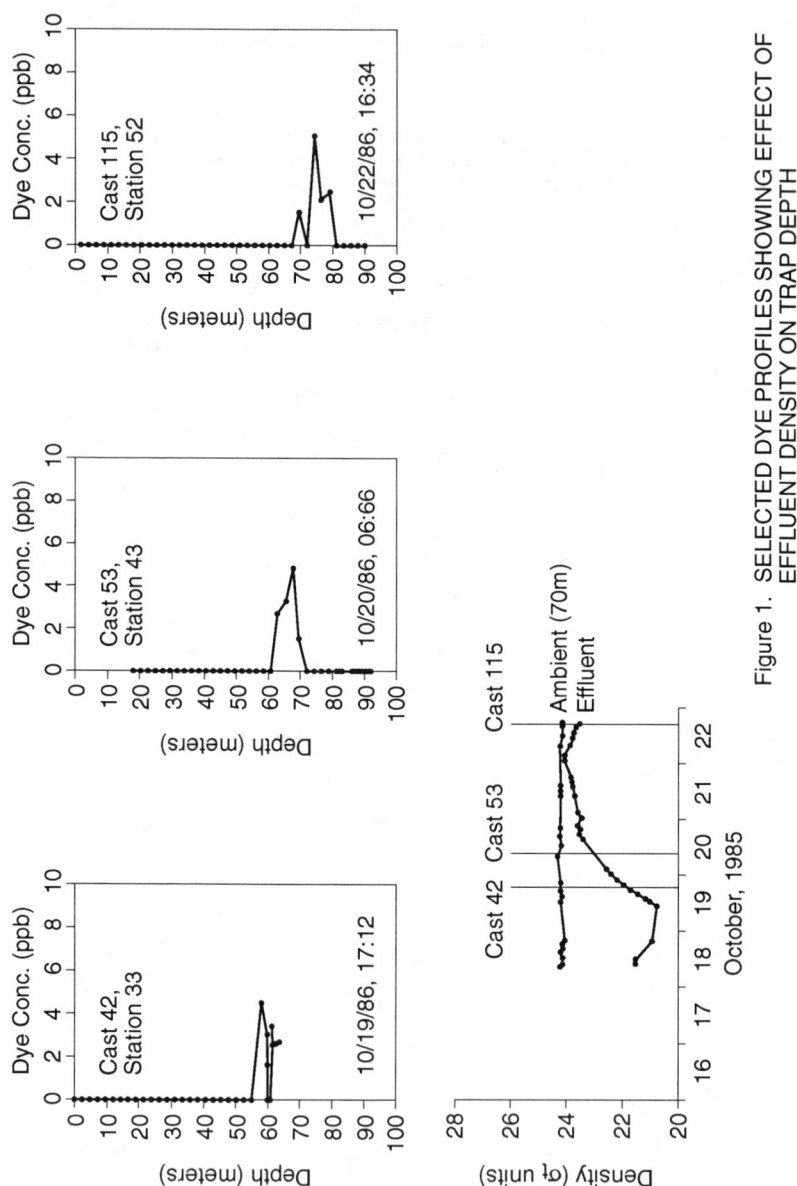

Figure 1. SELECTED DYE PROFILES SHOWING EFFECT OF EFFLUENT DENSITY ON TRAP DEPTH

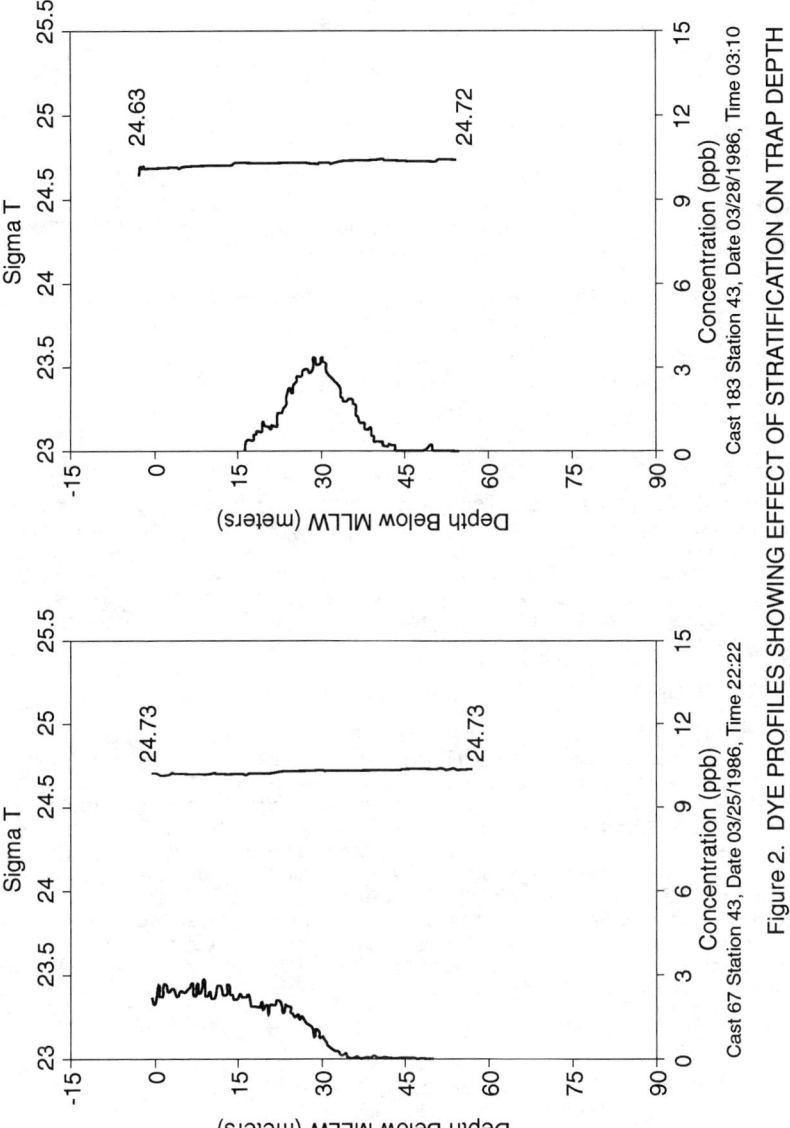

Figure 2. DYE PROFILES SHOWING EFFECT OF STRATIFICATION ON TRAP DEPTH

EVALUATION OF AN EXISTING SCOUR HOLE AT THE CASTLETON BRIDGE, A TIDAL CROSSING

Sufian A Khondker[1], Ph.D., P.E., M.ASCE and Mark A Hixson[2], P.E.

Abstract

Several fathometric surveys conducted in recent years showed a large depression or scour hole on the downstream southeast corner of the central pier of the Castleton Bridge over the Hudson River in New York. This paper attempts to evaluate the causes of the existing depression or scour hole as well as to ascertain the possibility of its enlargement or migration upstream.

Introduction

The Castleton Bridge over the Hudson River, approximately fifteen miles south of Albany, New York, is forty-three span structure and over a mile long. The three central main spans are steel thru-struss type supported on concrete piers (nos. 10 thru 13) founded either on bedrock or on piles driven to bedrock. All three piers are rectangular in cross section with rounded noses, approximately 19 feet in width and 98 feet in length. The central pier (Pier 12) has 60 feet wide wooden fender, which extends 60 feet upstream from the upstream face and 75 feet downstream from the downstream face of the pier. The pointed fender extends several feet below water and is supported by timber piles.

---

[1] Associate Consulting Engineer, Ebasco Infrastructure, New York, NY

[2] Managing Engineer, Structures Design, New York State Thruway Authority, Albany, NY

Field Investigations

In order to delineate the subsurface material surrounding the hole a unique sub-bottom profiler known as Chirp Acoustic Profiler was used. This technology uses a swept frequency acoustic sub-bottom profiler integrated with the navigation capabilities of an automated hydrographic survey system to provide precise sub-bottom profiles. The profile data thus obtained was calibrated against the borings taken later to ascertain its accuracy. In addition, divers were used to take samples of the bottom material at two locations in the hole and the velocity along with direction was also measured inside the hole.

The result of the hydrographic survey along with the three central piers are shown in Figure 1. The Hudson River navigation channel is located in the western half of the river between Piers 11 and 12. The eastern half of the river between piers 12 and 13 is relatively shallow.

Sub-bottom profiling lines were run at the upstream and downstream fascias of the bridge, along the transverse faces of the piers and on a grid in the area of the depression or scour hole southeast of Pier 12. A typical output depicting the scour hole is presented in Figure 2. Based on the color coordination, the current technology of the Chirp system cannot clearly delineate the type of the subsurface material, however, it can definitely define the compactness or relative density of the underlying material. The darker the color, the more dense is the material and hence less susceptible to scour. For evaluation of the scourability of the bed material the Chirp system can provide excellent data particularly when it is correlated with actual borings.

Figure 2 indicates a thin, soft layer of silt (indicated by blue) on top of a harder layer (black). Borings B-1 and B-2, located within the apparent scour hole, encountered hard shale at Elevation 53.5 ft. The location of highly reflective layer (black) from the output of the profiler (Figure 2) at the location of boring B-1 corresponds to El 54.0, which is very close to the elevation where the shale was found.

Current measurements were taken inside the hole and the divers also noted the direction of the current in the depressions (Figure 2).

Several bulk samples as well as split spoon samples from the borings were analyzed. The bed material primarily consists of fine sand with traces of silt with $D_{50}=0.25$mm.

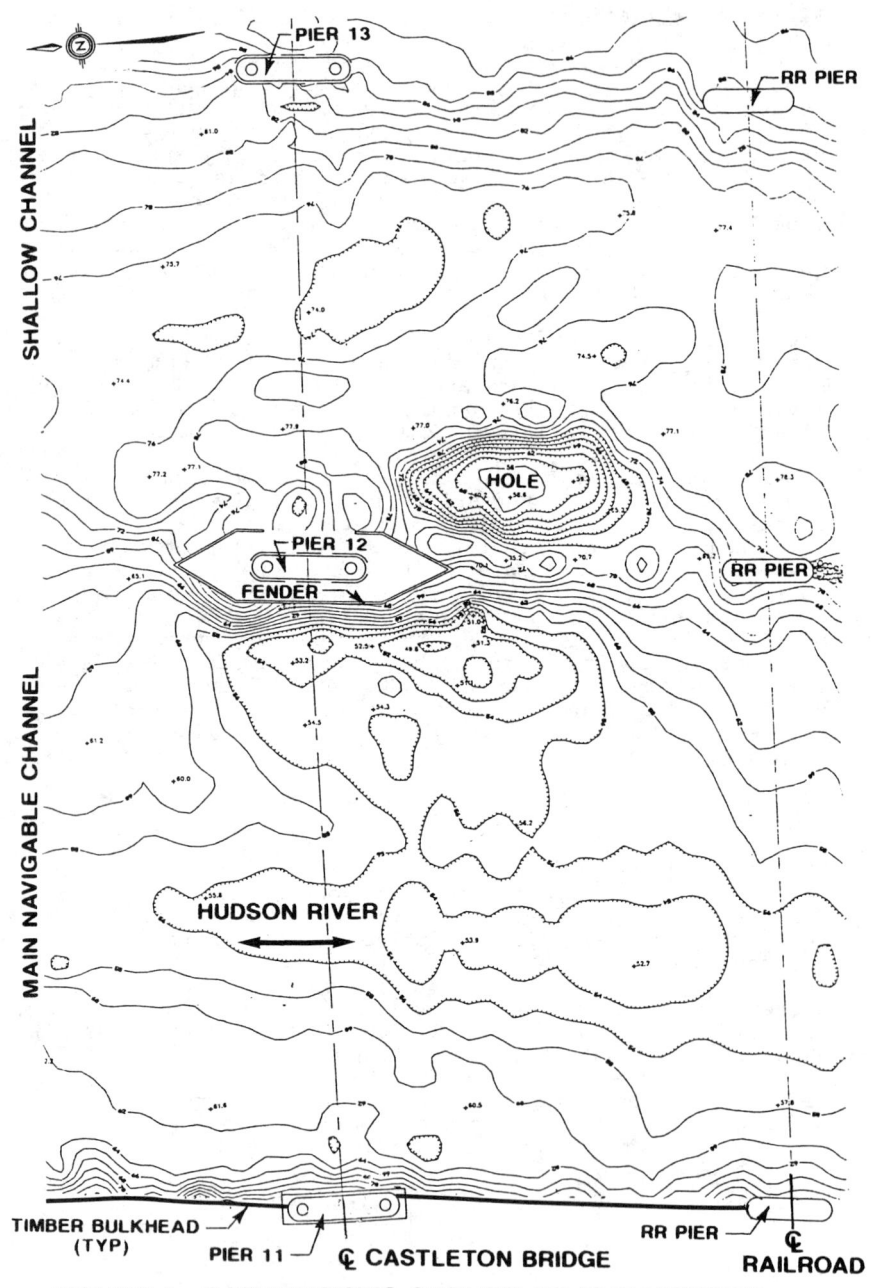

FIGURE 1 : BATHYMETRIC SURVEY OF THE VICINITY

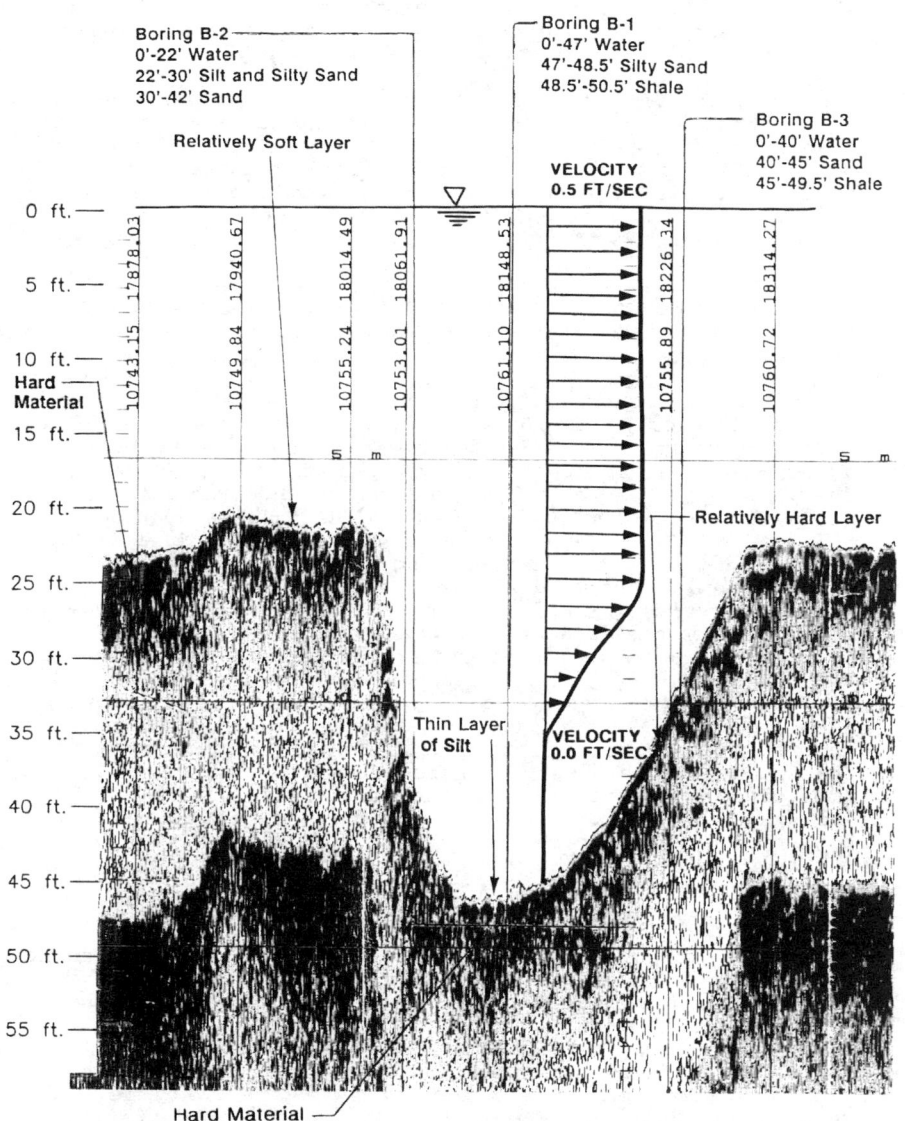

FIGURE 2 : TYPICAL OUTPUT FROM THE CHIRP PROFILER

Evaluation of the Scour Hole

There are several school of thoughts as to the presence of the apparent scour hole on the downstream southeast corner of Pier 12.

One postulation is that Army Corps of Engineers inadvertently dug this hole while dredging for the navigable channel on the western half between Pier 11 and 12. This postulation, probably, has the least validity because the hole, which is located at the lower reach of the Hudson River, where deposition is the primary function, will fill-up in no time. There must be a sustaining force to keep the hole open.

Second postulation, which is probably an accurate one, is that the apparent scour hole was created by wake vortex intensified by a stream of wake vortices from the adjacent pier No. 13 (Figure 3). Stevens, Gasser and Saad (1991) conducted a thorough study of similar scour holes downstream of two large circular piers at two bridges across the Nile River in Cairo. On a different occasion, the principal author had the opportunity to witness the model of one of the bridges at the Hydraulic and Sediment Research Institute in Egypt. According to Stevens et.al.(1991) the deep scour holes at the Tahrir and Imbaba Bridges over the Nile River were produced by the conflicting velocity fields at the intersection of the wake vortex streams from adjacent piers. At the Castleton Bridge, however, the stream of wake vortices from the adjacent piers are not obvious. The right pier (Pier 11) has bulkheads (training walls) both upstream and downstream which will preclude any flow separation and formation of vortices. The left pier (Pier 13) is located in a shallow water and too far apart (600 feet) from the central pier. Even if the wake vortices are generated at this pier, the chances of intersecting these vortices with the vortex generated at the central pier is probably less (Figure 3). According to Carstens and Sharma (1975) however, for wide piers conventional horseshoe vortex does not occur and only wake vortex can occur. In addition all piers at the Castleton Bridge have riprap protection to prevent horseshoe vortex scour, if it occurred.

The third postulation, which the authors favor, is the vortex formation due to the rejoining of the two separated water bodies having different velocities as shown in Figure 3. The main flow at higher velocity is at the west side of the central pier (Pier 12) whereas the shallower slower flow is on the east side of the pier. The pier, along with the elongated fender, acts as a buffer between the two differing velocities. When these two bodies of water with differing

EXISTING SCOUR HOLE

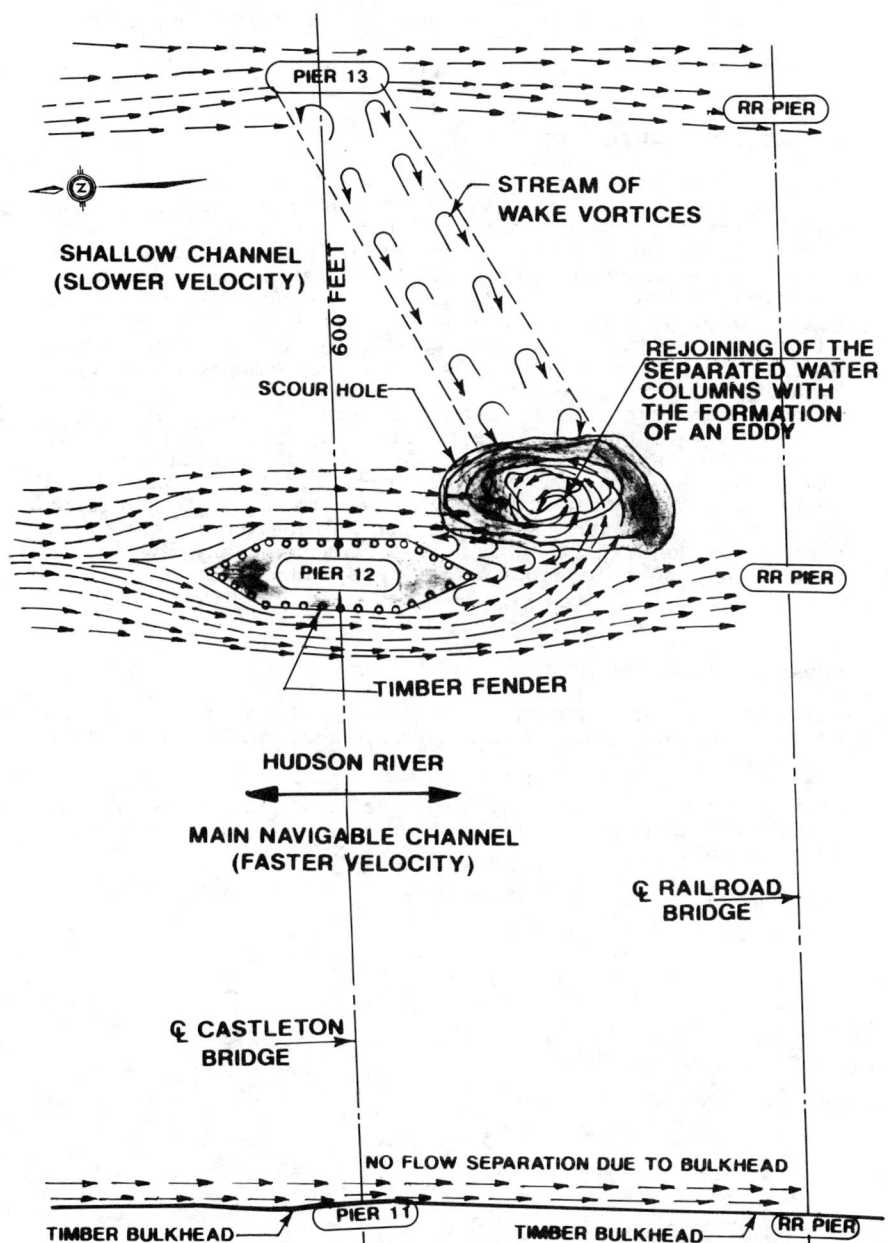

FIGURE 3 : TWO PROBABLE CAUSES OF THE SCOUR HOLE

velocities rejoins at the downstream end of the central pier an eddy is formed resulting in the scour hole (Figure 3). The measured velocities (Figure 2) inside the hole also confirms the formation of such eddy.

Summary and Conclusions

Based on the field investigations and above evaluations it is concluded that the presence of the apparent scour hole is not man-made. Most likely, it is the result of a wake vortex formed at the downstream end of the central pier and the fender supported on piles plays an important role in the creation of this vortex. Other predominant reason is the formation of a vortex due to the rejoining of the separated water columns having differing velocities. A model study may be needed to better understand the actual phenomena. Finally, whether the hole is created by wake vortex or vortex resulting from the differing velocities or a combination of the two, the possibility that the hole will migrate upstream is unlikely. As the wake vortex scour is always at the downstream of the pier, it has little or no destructive potential to undermine the pier foundation. Besides the profiler records of the depression and the river bottom surrounding it (Figure 2) do not show any soft zones into which the hole might easily propagate.

Appendix I: References

Carstens, T. and Sharma H.R. (1975), "Local Scour Around Large Obstructions." Proc. Int. Assoc. Hyd. Res., 16th Congress, IAHR, Delft, the Netherlands

Stevens, M.A., Gasser, M.M. and Saad M. (1991), "Wake Vortex Scour At Bridge Piers." ASCE Journal of Hydraulic Engineering, Vol. 117, No. 7, Paper No. 26006

Development of Bridge-Scour Instrumentation for
Inspection and Maintenance Personnel

David S. Mueller[1], A.M. ASCE and Mark N. Landers[2], M. ASCE

Abstract

Inspecting bridges and monitoring scour during high flow can improve public transportation safety by providing early identification of scour and stream stability problems at bridges. Most bridge-inspection data are collected during low flow, when scour holes may have refilled. More than 25 percent of the States that responded to a questionnaire identified lack of adequate methodology and/or equipment as reasons for not collecting scour data during high-flow conditions. Therefore, the U.S. Geological Survey (USGS), in cooperation with the Federal Highway Administration, has begun to develop instrumentation for measuring scour that could be used by inspection and maintenance personnel during high-flow conditions. A variety of instruments and techniques for measuring scour were tested and evaluated in real-time bridge-scour data-collection studies by the USGS. In the National Scour Study, fathometers were found to be superior to sounding weights and will be the primary bed-measuring instrument. The ability of low-cost fathometers and fish finders to locate the bed accurately is being evaluated. Simple and efficient methods for deploying the transducer during floods are also important for a successful measurement. This information and additional testing are being used to design new, portable scour-measuring systems.

---

[1]Hydrologist, U.S. Geological Survey, Water Resources Division, 2301 Bradley Avenue, Louisville, KY 40217

[2]Hydrologist, U.S. Geological Survey, Water Resources Division, 415 National Center, Reston, VA 22092

Introduction

Inspection of bridges for scour during floods is needed to identify scour and stream-stability problems that may not be apparent during low-water inspections. At present, most inspection data are collected during low-flow conditions. Although low-flow data collection is adequate for many aspects of bridge inspections, it may not be adequate for identifying scour and stream-stability problems. Low-flow inspections may fail to detect scour holes that were present during a previous flood but may have subsequently refilled. Safety considerations often prohibit the use of a manned boat to measure scour during high flow. The most common method for collecting scour data is to measure cross sections from the bridge deck using a sounding weight. However, this technique can only be used along the upstream and downstream edges of the bridge deck, and requires heavy weights to avoid excessive downstream drift of the weight during high flow. This paper describes instrumentation and procedures being developed to overcome limitations of the present practices and permit inspection and maintenance personnel to collect bridge-scour data during floods.

Survey of need and requirements

Questionnaires were sent to State departments of transportation (DOT) and the Federal Highway Administration (FHWA) office in each State to evaluate procedures and techniques currently in use and to determine the features and target cost of any new equipment. Responses to the questionnaires were very positive, indicating an overall interest in and awareness of the potential problems. Only seven States did not respond to the questionnaire and a total of 62 responses were received.

Responses indicate that no State has an organized program for collecting scour data during high flows. Some States are collecting data when possible and other States are monitoring scour at bridges with known scour problems. In addition, more than 20 percent of the States have cooperative study programs with the U.S. Geological Survey (USGS) to collect scour data during floods at selected sites. Numerous States indicated that lack of funds and personnel are the primary reasons for not implementing a program to collect scour data during floods. Other reasons listed by several States include:

1) Lack of adequate methodology and/or equipment (14 States)
2) Logistics - remote areas or short duration floods (4 States)
3) Perceived low-risk of bridge failure does not justify expense (3 States)

The responses indicated that bridge inspectors and maintenance personnel in most States are or will be responsible for collecting some real-time scour data. Most inspectors use a sounding weight to collect spot bed-elevations or cross-sections data. Some inspectors use a fathometer, but no State reported using a digital fathometer and automatic digital data storage. The responses were also used to identify characteristics that are important for new portable scour measuring equipment. Simplicity of operation is the most important characteristic, followed by personnel requirements, portability and size, durability, cost, and ability to record a permanent record. A target price of $1,000 for such a system was identified, although price is not a high priority.

Evaluation of existing methods and equipment

The USGS has numerous bridge scour data collection projects that involve collecting data during high-flow conditions. Several of the project chiefs and staff have developed instrumentation to assist in the data collection (Trent and Landers, 1991). Most of the USGS studies initially used a sounding weight to measure the cross section along the upstream and downstream face of the bridge. However, this method is slow and, during high flow, weights in excess of 100 lbs must be used to prevent excessive downstream drift of the weight. A bridge crane or truck-mounted boom is required to raise and lower these heavy weights. Complete documentation of important features such as maximum depth and width of the scour holes, is difficult because data can only be collected at discrete points using a sounding weight. Therefore, the use of fathometers was investigated.

Fathometers provide a continuous record of the cross section, thus eliminating any gaps in the data, except where obstructions prohibit an instrument from being lowered into the water. The accuracy of fathometers in deep (greater than 10 ft) slow moving (less than 5 ft/sec) flow conditions is good; however, less satisfactory results were obtained in shallow (6-8 feet), swift (6-15 feet/second) water (G. K. Butch, USGS, written, commun., 1992). The storage of sounding data by digital recorders greatly reduces the data analysis time by eliminating the need to digitize strip charts. Several attempts have been made to modify an inexpensive analog fathometer and interface it with a data logger to collect digital information, only limited success has been achieved to date. However, fathometers are superior to sounding weights in regards to the time required to make a measurement and in the amount of detail obtained.

The typical equipment configuration presently in use is a chart recording analog fathometer with the transducer mounted on the bottom of a sounding weight. The weight and transducer are lowered from the bridge then slowly moved along the bridge; the locations of piers and other fixed stations are marked on the chart to identify the horizontal position (Trent and Landers, 1991). Although this technique provides a rapid and continuous record of bed elevations along the edge of the bridge, its use poses several problems. During high flow, a 100- to 200-pound weight may be required to maintain the transducer in a stable position, thus requiring a truck-mounted boom to raise and lower the weight. Some bridges require specially constructed long booms in order to deploy the transducer over the side of the bridge (G. H. Carlson, USGS, written commun., 1991). Other bridges have security fences and structural supports that make deploying the transducer difficult or require the transducer to be removed from the water in order to move around a support. Detailed notes and marks on the chart are required to set the scales for digitizing the chart, if the transducer is frequently moved out of the water. This system requires both a signal cable to the transducer and a steel cable to hold the weight. In some situations, the management of these cables is a problem.

Floats have been used to help reduce the weight required to deploy the transducer, and to allow the transducer to be maneuvered beneath the bridge and along the sides of the piers. Various types of floats have been tried, including a spherical warning marker for power lines, 'hoppity hop' balls, a raft made from PVC pipe, and water skis. Spherical floats, warning markers and 'hoppity hop' balls, did not worked well due to substantial drag on the sphere when partly submerged and the resulting instability which causes the transducer to be raised and tilted out of the vertical position. A raft made of PVC pipe worked reasonably well (G. H. Carlson, USGS, written commun., 1991). Both the Texas and Arkansas DOT's had success using a water ski to deploy a transducer. The primary problems associated with the water skis are some air entrainment and instability during very high flow (Garland Land, Arkansas DOT, oral commun., 1992).

Instrumentation development plan

Portable scour-measuring systems consist of four components: (1) a sounding instrument, (2) a data storage device, (3) a deployment system, and (4) a method to identify and record the horizontal position of the sounding. Both 'off-the-shelf' and custom designed equipment are being evaluated for each of the components; however, 'off-the-shelf' components will be given preference due to their immediate availability and potentially lower cost.

Fathometers will be the primary sounding instrument, based on the evaluation described previously. In-dash style depth sonar and fish finders are low-cost alternatives to more expensive recording and survey-grade fathometers. The performance of low-cost fathometers and fish finders will be compared to survey-grade fathometers which use advanced signal processing to digitize the location of the stream bottom.

Data from fathometers can be stored by different techniques, each having unique advantages. The best technique depends on the purpose of the scour measurement. Fathometers with paper-chart recorders provide a permanent record that can be used for real-time annotations about the location of the transducer and other measurement conditions. Chart-recording fathometers typically cost from $500 to $1,000. However, if a digital record of a cross section is desired, the paper chart must be digitized, which requires tedious adjustments to obtain a uniform horizontal scale. Digital fathometers can be very expensive and do not provide annotation features; however, digital data are easily plotted or incorporated into a computer-aided design system or database. Survey-grade fathometers usually provide both a chart and a digital interface; however, these systems usually cost considerably more than $10,000. If the purpose of the scour measurement is to obtain data to describe scour processes, then digital storage of cross-section data is desirable. If the purpose of the scour measurement is inspection of the bed to determine support for the bridge foundation, then spot soundings recorded using a notepad and a simple in-dash style depth sonar with numerical readout may be sufficient. A plot of the cross section during the measurement is valuable because it allows the operator to determine where and how much data should be collected. The best alternative may be a fathometer with a numerical and graphical display and some method of permanently storing the data, either graphically on a chart or digitally. A low-cost ($500) video depth sounder has been identified that provides a color display of the bottom, a numerical readout of the depth, and provides digital output of the data for storage on an external computer. This device has been procured and an evaluation is in progress.

Three categories of deployment systems are being investigated: (1) hand-held systems with floatation device; (2) manually-operated mechanical systems, with and without floatation device; and (3) electric-powered vehicle-mounted systems, with and without floatation device. The hand-held deployment system with floatation device provides the maximum portability and should be the least expensive. The water ski concept used by the Texas and Arkansas DOT's appears to have the most promise. In an attempt to eliminate the stability problems experienced with the water ski, a knee board will be tested. The knee board is wider and will allow weight or a submerged wing to be added for additional stability. In addition, a single cable may be used to maneuver the knee board and provide electrical connection to the transducer. A manually-operated mechanical deployment system will be evaluated for situations that require a weight or weighted knee board that is too heavy for one person to deploy safely. For this system, small manual and power reels, fishing down-riggers, and bridge cranes used by the USGS are being investigated. The feasibility of using a Kevlar reinforced electrical cable to bear the weight and provide the electrical signal connection is also being investigated for use with mechanical and vehicle-mounted deployment systems. The vehicle-mounted deployment systems are the least attractive option, due to higher cost and decreased portability.

The final component of the portable scour-measuring system is a method for locating and recording the position of the transducer. A survey-quality positioning system would be too expensive and require too many people to satisfy the requirements of the inspection instrumentation system. Therefore, measuring wheels, tapes, and stationing marked directly on the bridge will be used to identify the transverse location of the transducer as it is moved across the stream. If a flotation device is not used, only the transverse position of the transducer will be recorded and the longitudinal position will be assumed to be along the edge of the bridge. If a deployment system with a flotation device is utilized, both the longitudinal and transverse position of the transducer must be recorded. The operator must be able to see under the bridge to estimate the longitudinal position of the transducer. A safety hazard could result from the operator leaning over the bridge railing to see beneath the bridge. Therefore, a system to hang convex mirrors from the side of the bridge is being developed to assist the operator in seeing beneath the bridge. It is anticipated that the positioning will be recorded using paper notes or a tape recorder. Notes could also be keyed directly into a portable computer, if desired.

Conclusions

Currently no State has an organized program for collecting scour data during floods. One obstacle in developing such a program is the lack of adequate methodology and equipment. The USGS, in cooperation with the FHWA, has begun to develop equipment that could be used by inspection and maintenance personnel to measure scour during high flows. The first system to be evaluated will be a hand-held deployment system using a knee board. In-dash depth sonars, low-cost chart recording fathometers, and a color video fathometer will be evaluated with this and other deployment systems. The estimated cost of this system is very close to the target cost of $1,000. The major components of this system have been procured and testing of the equipment is scheduled for 1993.

Acknowledgments

The authors would like to express their appreciation to many people in the USGS, the State DOT's, and the FHWA who took the time to complete the surveys and/or share their instrumentation ideas and experiences with the authors.

References

Trent, R. E. and Landers, M. N., 1991, Chasing floods and measuring scour: Transportation Research Board, Transportation Research Record 1290, Third Bridge Engineering Conference, v. 2, p. 235-244.

Using Geophysical Data to Assess Scour Development
Gary Placzek, F.Peter Haeni[1], and
Roy Trent[2], Member ASCE

Abstract
The development of scour holes in the Connecticut River near the new Baldwin Bridge has been documented by comparing geophysical records collected before (1989), during (1990), and after (1992) bridge construction. Eight piers that support the 579-m (meter) span over the Connecticut River were protected by 12-m wide cofferdams during construction. The maximum flow during the study period was equivalent to a 3-year recurrence-interval flood, indicating no significant floods.

Fathometer data indicate that deep scour holes, 1.5 to 6.4 m deep, developed north of piers 6, 7, and 8. Scour holes, less than 1.3 m-deep, developed south of these piers. The deepest scour hole was north of pier 7, where data show a flat river bottom in 1989, a scour hole 3.3-m deep in 1990, and a scour hole 6.4-m deep in 1992. Continuous seismic-profiling (CSP) data show that a 1.5-m deep scour hole north of pier 6 in 1990 was filled in with 1.5 m of material by 1992. No infilling was detected in the scour holes north of piers 7 and 8. Numerous subbottom reflectors from geologic layers, up to 7.6-m deep, were identified in the CSP records.

Introduction
Surface-geophysical methods have been used in several studies to evaluate existing and infilled scour holes at bridge piers (Gorin and Haeni, 1989; Haeni and Gorin, 1989; Crumrine, 1991; Haeni and Placzek, 1991). The U.S. Geological Survey (USGS), in cooperation with the Federal Highway Administration, used geophysical data to monitor the development of scour holes in the

---

[1] Electronics Eng., Hydrologist, U.S. Geological Survey, 450 Main St., Room 525, Hartford, CT 06103

[2] Research Hydraulics Eng., Federal Highway Admin., Structures Div., 6300 Georgetown Pke, McLean, VA 22101

Connecticut River at the new Baldwin Bridge. For this study, a fathometer was used to collect bathymetric data to determine the geometry of any existing scour holes, and a continuous seismic-profiling (CSP) system was used to detect infilled scour holes and subsurface geologic layers. This paper presents the results of the geophysical study at the new Baldwin Bridge.

The new Baldwin Bridge, 18.3 m (meters) south of the old Baldwin Bridge, carries Interstate 95 over the 579-m wide Connecticut River near Old Saybrook(fig. 1). The new bridge has eight 4.9-m wide rounded-nose piers. During construction, each pier was protected by a cofferdam that was 12 m wide and 44 m long. The flow of the river was perpendicular to the 12-m side of the cofferdam. After construction, each cofferdam was cut at the river bed level; however, where the top of the footing was above the river bed, the cofferdam was cut at the top of the footing.

Tidal fluctuations are about 1.2 m per day at the bridge. During the study period, the maximum discharge of the Connecticut River near Middletown (the closest USGS streamflow-gaging station) was 266,208 liters per second on March 22, 1990 (Cervione and others, 1991). This discharge represents a peak flow equivalent to a 3-year recurrence-interval flood; therefore, no significant floods created the scour holes discussed in this report.

Geophysical Methods and Data Collection

The fathometer used in this study transmits a 192-kHz (kilohertz) acoustic signal downward from the transducer with an 8-degree beam angle. This narrow beam angle allows the fathometer to accurately record bathymetric data with little interference from side echoes from the piers.

The CSP system used in this study transmits a seismic signal tuned between 3.5 and 14.0 kHz. The seismic signal can penetrate the river bottom and provides subsurface and bathymetric data. A 7-kHz signal radiates downward from the transducer with an 80-degree by 35-degree beam angle. The 3.5-kHz signal has a 120-degree by 55-degree beam angle. Side echoes from piers can interfere with CSP data because of these wide beam angles.

Fathometer and CSP data were collected on May 25, 1989, before construction; on October 31, 1990, during construction; and on July 22, 1992, after construction. Multiple fathometer and CSP profiles were collected in an east-west line to the north and south of the piers and in a north-south line alongside the piers. The data were collected using a 6.7-m boat with a 10-horsepower outboard motor. Before construction of the new bridge,

the data were collected along the proposed bridge
alignment, and the center-lines of the old bridge piers
were marked on the record. The new piers are located
15 m west of the corresponding old piers. During
construction, the locations of the cofferdams were
marked on the record; after construction the center-line
of the new piers were marked.

Analysis of Geophysical Data

The CSP data were replayed to adjust the gains and
the vertical time scale. Depth scales were corrected
for varying transducer depths and tidal stages.
Approximate horizontal scales were determined by
correlating visually estimated position data with bridge
plans. Multiple fathometer and CSP records were
compared to confirm bathymetric and subsurface
information. The CSP data also were compared with the
fathometer data to identify reflections in the CSP data
caused by side echoes from piers. The bridge plans were
used to help interpret the geophysical records. Scour
hole depths were referenced to the original river bed
level.

Fathometer data collected before construction (fig.
2) show an undulating river bottom with about 1.8 m of
relief, a flat river bottom at old pier 7, and a 12-m
deep navigation channel east of old pier 8. Fathometer
data collected 5 m north of the piers during
construction (fig. 3) show scour holes 1.5-m deep at
pier 6, 3.3-m deep at pier 7, and 2.1-m deep at pier 8.
Fathometer data collected 5 m north of the piers after
construction (fig. 4) show no scour hole at pier 6, and
scour holes 6.4-m deep at pier 7 and 4.6-m deep at pier
8. The top of the footings were seen in the records at
piers 7 and 8. The deepest scour hole south of the
piers was 1.2-m deep at pier 7.

Additional fathometer data were collected laterally
alongside of old and new pier 7 on December 15, 1992.
This line extended from 150 m north to 75 m south of the
new bridge, 3 m east of both piers. On the north side
of each pier, scour depths of 1.5 m and 6.1 m were
detected. Little or no scour was present on the south
side of the piers (fig. 5).

CSP records showed the same bathymetric data as the
fathometer records, but in addition, provided
information about the subsurface. CSP data collected
before construction (fig. 6) showed subbottom reflectors
up to 7.6 m deep. CSP data collected during
construction (fig. 7) showed that the scour holes north
of piers 6, 7, and 8 had no infilling. CSP data
collected after construction (fig. 8) showed that 1.5 m
of material was deposited in the scour hole on the north
side of pier 6, and that the scour holes at piers 7 and

8 increased in depth, but were not filled in. CSP data on the south side of the piers showed geologic layers in the subsurface, but no infilled scour holes.

Summary

CSP and fathometer data collected in the Connecticut River before, during, and after construction of the new Baldwin Bridge documented the formation of scour holes. The data showed scour holes up to 6.4-m deep on the north side of the piers and up to 1.2-m deep on the south side of the piers. The deepest holes were at pier 7 where no infilling was detected. Data collected before construction show an undulating river bottom with no scour holes. Data collected during construction, when the piers were protected by cofferdams, show scour holes up to 3.3-m deep on the north side of the piers. Data collected after construction on the north side of the piers show 1.5 m of infilled material in the scour hole at pier 6, and no infilled material in the scour holes 6.4-m deep at pier 7, and 4.6-m deep at pier 8. These holes were formed from May 1989 to July 1992 during which time no significant flood events occurred.

Appendix 1. References

Cervione, M.A., Jr., Davies, B.S. 3rd, Bohr, J.R., and Hunter, B.W., 1991, Water resources data, Connecticut, water year 1990: U.S. Geological Survey Water-Data Report CT-90-1, 274 p.

Crumrine, M.D., 1991, Results of a reconnaissance bridge-scour study at selected sites in Oregon using surface-geophysical methods, 1989: U.S. Geological Survey Water-Resources Investigations Report 90-4199, 44 p.

Gorin, S.R., and Haeni, F.P., 1989, Use of surface-geophysical methods to assess riverbed scour at bridge piers: U.S. Geological Survey Water-Resources Investigations Report 88-4212, 33 p.

Haeni, F.P., and Gorin, S.R., 1989, Post measurement of a refilled scour hole at the Bulkeley Bridge in Hartford, Connecticut, in Proceedings of the Bridge Scour Symposium: McLean, Virginia, Federal Highway Administration Report No. FHWA-RD-90-035, p. 147-174.

Haeni, F.P., and Placzek, Gary, 1991, Use of processed geophysical data to improve the delineation of infilled scour holes at bridge piers, in Expanded Abstracts with Biographies, SEG 61st Annual International Meeting, Houston, Texas, November 10-14, 1991: Houston, Texas; Society of Exploration Geophysicists, p. 549-552.

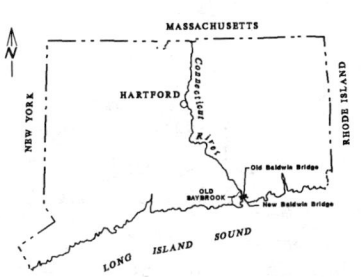

Figure 1. Site location map

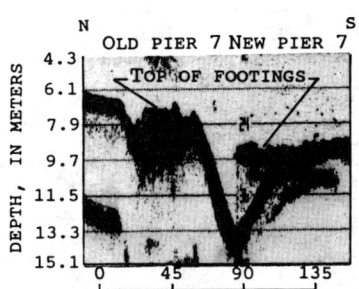

Figure 5. Lateral fathometer record, east of piers

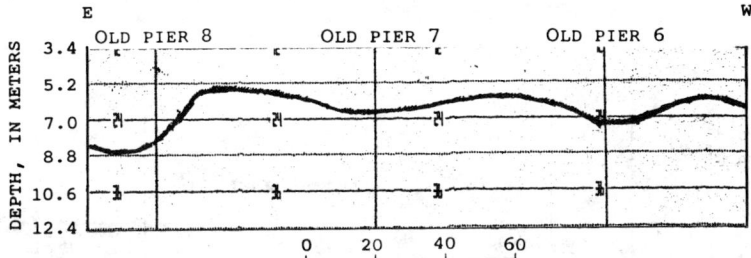

Figure 2. Fathometer record before bridge construction

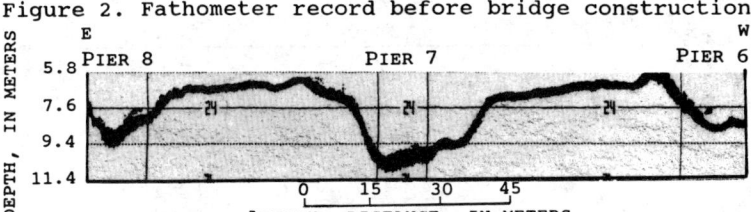

Figure 3. Fathometer record during bridge construction

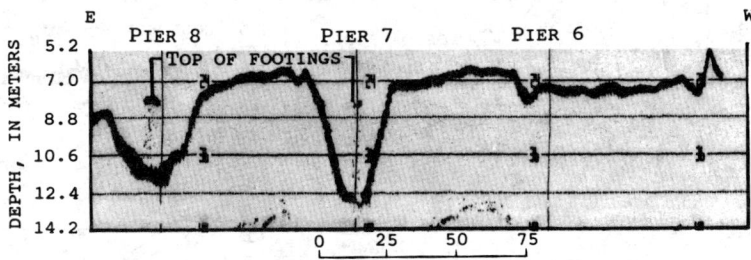

Figure 4. Fathometer record after bridge construction

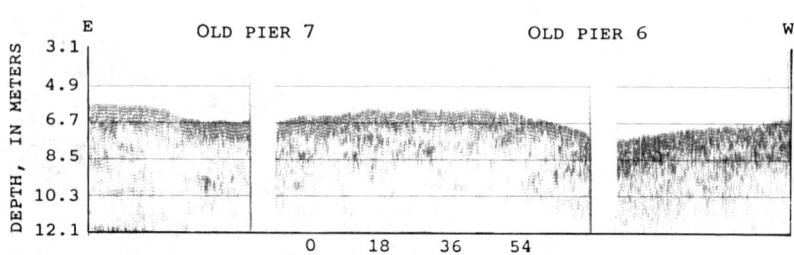
Figure 6. 7kHz CSP record before bridge construction

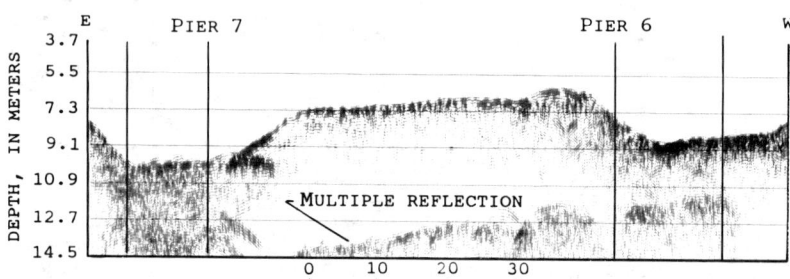
Figure 7. 7kHz CSP record during bridge construction

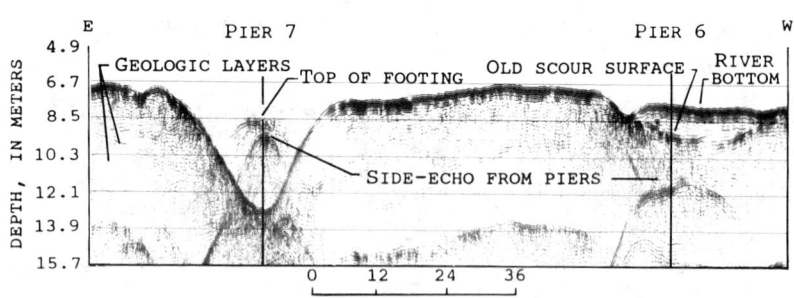
Figure 8. 3.5kHz CSP record after bridge construction

Local Scour Measurements At Bridge Piers In Alberta

Des Williamson[1]

Abstract

A summary of local pier scour data collected by Alberta Transportation and Utilities (AT&U) over the past 30 years is presented. Several methods used to gather this data are described. These measurements, along with other published field data, are compared to the results of some recent local scour prediction formulae based on laboratory data. In general, measured scour depths are less than predicted. Possible reasons for this discrepancy are discussed.

Introduction

Local scour at bridge piers can be defined as an "abrupt decrease in bed elevation near a pier due to the erosion of the bed material by the local flow structure induced by the pier" (Shen,1971). The basic mechanism of local scour is the large-scale eddy structure, or system of vortices, which develops near the pier. Many attempts have been made in the last 40 years to quantify the magnitude of local scour at bridge piers. Most of these methods have been based on the results of small-scale laboratory experiments.

AT&U has collected field scour measurements at bridges in Alberta for more than 30 years. Some of the problems encountered when collecting accurate field scour measurements are discussed. A selection of this collected data, as well as other published field data, are compared to the results of several of the most recent prediction equations based on laboratory data. This comparison identifies some factors that should be considered when using prediction equations based on laboratory results to design for local scour at piers.

Field Data Collected in Alberta

The terrain of the province of Alberta includes the eastern slopes of the Rockies, foothills, and prairies, resulting in many types of rivers. AT&U has been conducting field surveys at bridges and river protection works on many of these rivers for more than 30 years. These surveys are part of a monitoring system that has been developed

---

[1]Bridge Planning Engineer, Bridge Engineering Branch, Alberta Transportation and Utilities, 4999-98th Avenue, Edmonton, Alberta, T6B 2X3

to aid in the maintenance and the future design of bridges in the province. A considerable amount of data has been collected on local scour at bridge piers, abutments and river protection works, as well as general scour at bridge openings. Several methods have been used to collect this data.

For smaller bridge sites, a "handrail profile" can be used to obtain a channel cross-section at the upstream and downstream ends of the bridge. This procedure involves lowering a weighted tape or an echo sounder off the bridge handrail at several points across the bridge. A weighted tape is effective in shallow depths and low velocities, and the survey can be done by one person. In regions of higher velocities, the echo sounder attached to a float can provide accurate results by allowing it to travel in a straight line starting a short distance upstream of the point to be measured and then marking the chart when the transducer crosses the desired point. In areas of high turbulence, such as close to the nose of a pier, measurements may have to be repeated several times as the float and transducer can be tossed about, taking measurements that are not perpendicular to the streambed. The use of an echo sounder requires two people. Some bridge structures, such as trusses and bridges with narrow clear roadways, can make this procedure difficult and dangerous.

Point measurements over an area of the streambed can be taken using an extendable boom which can be rotated to various angles. Alberta Transportation and Utilities use a lightweight aluminum boom with 5 - 3.66m sections to position an echo sounder transducer over a certain point. Once several points have been collected, the results can be plotted as a contour plan of the streambed which will show any areas of local scour or deposition of bed material. The boom can be connected to a truck or bridge handrail, and therefore can be used to gather data at piers, abutments, spurs, and guidebanks.

For larger sites, the echo sounder is attached to a boat. Various surveying techniques are used to determine the position of the boat, and point measurements are taken. The resulting data can be plotted in the form of streambed contour plans and channel cross-sections. Problems often encountered with the use of a boat include locating an access point to the channel, running into gravel bars, and negotiating around piers during periods of high flow velocities. This procedure usually requires three people.

Other data that is often collected in the field includes highwater marks, the angle of attack of flow near piers, and flow velocity estimates. Table 1 presents a summary of local pier scour related data at several sites in Alberta. The local scour depths have been extracted from streambed contour plans. Approach flow depths and velocities have been estimated for peak flood conditions, and the median bed particle sizes have been obtained from surveys published by the Alberta Research Council (Kellerhals, Neill, and Bray, 1972).

Prediction Equations Based On Laboratory Data

Four of the most recent pier scour prediction methods have been selected to be compared with the collected field data as well as field data that has been previously published. These methods are : Roads and Transportation Association of Canada (RTAC, 1973), International Association for Hydraulic Research (IAHR) Task Force (Breusers, Nicollet, and Shen, 1977), Melville and Sutherland (1988), and Colorado

State University (CSU) as presented in the HEC-18 manual (US Army Corps of Engineers, 1990).

Table 1 - Pier Scour Measurements in Alberta

File No.	Location	b (m)	Pier Shape	d_s (m)	D_{50} (mm)	T (yrs)	Y (m)	V (m/s)	a (°)	d_s/b	Y/b	F
73949	Peace R, Dunvegan	11.4	C	1.7	53	50	7.9	2.7	0	0.1	0.7	0.31
71315	Athabasca R, Fort Ass.	6.4	R	4.6		*	5.5	1.6	0	0.7	0.9	0.22
1980	Red Deer R, Sundre	2.7	R	1.8	28	10	2.8	2.1	0	0.7	1.0	0.40
73949	Peace R, Dunvegan	11.4	C	3.4	53	50	12.7	2.7	0	0.3	1.1	0.24
74228	Wapiti R, Grande Prairie	4.9	C	1.8	48	*	6.5	1.8	0	0.4	1.3	0.23
71291	N. Sask. R, Pakan	2.4	C	2.1		*	3.4	1.2	0	0.9	1.4	0.21
74233	N. Sask. R, Redwater	2.9	C	2.7	31	5	4.6	1.8	0	0.9	1.6	0.27
74228	Wapiti R, Grande Prairie	4.9	C	2.1	48	*	8.3	1.8	0	0.4	1.7	0.20
286	St. Mary R, Kimball	1.6	R	1.7		70	2.8	3.4	0	1.1	1.8	0.65
73810	Athabasca R, Whitecrt.	1.2	C	1.5	52	*	2.2	1.8	0	1.3	1.8	0.39
1980	Red Deer R, Sundre	2.7	R	1.8	28	50	5.2	2.7	20	0.7	1.9	0.38
1293	Bow R, Gleichen	2.6	T	1.2		15	5.2	2.3	0	0.5	2.0	0.32
73802	S. Sask. R, Med. Hat	3.7	R	2.7		5	7.6	2.0	0	0.7	2.1	0.23
74236	N. Sask. R, Myrnam	2.6	C	1.2		5	5.4	1.6	0	0.5	2.1	0.22
74381	N. Sask. R, Dray. Valley	2.4	C	2.4		*	5.2	1.8	0	1.0	2.2	0.25
74236	N. Sask. R, Myrnam	2.6	C	1.5		5	6.1	1.6	0	0.6	2.3	0.21
8800	Red Deer R, Garrington	1.2	C	1.4		10	2.7	2.4	0	1.2	2.3	0.47
9847	N. Sask. R, Rk Mt. Hse.	3.0	T	2.0	63	*	7.1	1.8	0	0.7	2.4	0.22
75701	Red Deer R, Sharrow	1.5	C	2.4		15	3.6	1.5	0	1.6	2.4	0.25
74236	N. Sask. R, Myrnam	2.6	C	1.2		5	6.7	1.6	0	0.5	2.6	0.20
9590	Pembina R, Pibroch	2.0	T	1.5		10	5.3	1.5	15	0.8	2.7	0.21
8800	Red Deer R, Garrington	1.2	C	1.4		10	3.2	2.4	0	1.2	2.7	0.43
74381	N. Sask. R, Dray. Valley	2.4	C	1.8		*	6.7	2.0	0	0.8	2.8	0.25
71291	N. Sask. R, Pakan	2.6	C	1.0		*	7.5	1.1	0	0.4	2.9	0.13
71291	N. Sask. R, Pakan	2.6	C	1.4		*	7.5	1.2	0	0.5	2.9	0.14
74236	N. Sask. R, Myrnam	2.6	C	2.0		*	8.2	1.0	10	0.8	3.2	0.11
74236	N. Sask. R, Myrnam	2.6	C	2.1		*	8.2	1.2	10	0.8	3.2	0.13
315	St. Mary R, Cardston	1.2	C	1.2		100	3.9	3.2	35	1.0	3.3	0.52
315	St. Mary R, Cardston	1.2	C	2.0		100	3.9	3.2	35	1.7	3.3	0.52
13371	Pembina R, Jarvie	1.5	C	1.2	0.32	5	4.9	1.4	0	0.8	3.3	0.20
70509	Pembina R, Manola	1.4	T	2.1		5	4.6	1.2	0	1.5	3.3	0.18
71145	N. Sask. R, Lea Park	2.1	T	1.5	24	*	7.1	1.2	0	0.7	3.4	0.14
75315	N. Sask. R, Heinsburg	2.6	C	1.6		*	8.9	0.9	0	0.6	3.4	0.10
13742	N. Sask. R, Waskatenau	2.0	C	1.8		5	6.9	1.6	0	0.9	3.5	0.19
75315	N. Sask. R, Heinsburg	2.1	C	1.8		5	7.4	1.5	0	0.9	3.5	0.18
70509	Pembina R, Manola	1.4	T	1.8		5	5.4	1.3	0	1.3	3.9	0.18
74227	Peace R, Fort Vermilion	2.7	C	2.5	0.31	*	11.0	2.2	0	0.9	4.1	0.21
74141	Oldman R, Brocket	1.5	T	1.2	43	*	6.3	1.0	35	0.8	4.2	0.13
7461	Red Deer R, Duchess	1.2	C	1.2		15	5.3	1.4	0	1.0	4.4	0.19
70241	Smoky R, Watino	1.8	T	1.5	80	50	10.4	2.4	0	0.8	5.8	0.24
73275	Smoky R, Bezanson	2.2	T	1.7		50	12.8	1.8	20	0.8	5.8	0.16
70241	Smoky R, Watino	1.8	T	1.5	80	50	11.0	2.3	0	0.8	6.1	0.22

b = pier width
d_s = depth of local scour below stream bed
D_{50} = median bed particle size
T = return period of flood (* indicates < 5 years)
Y = depth of approach flow
V = mean velocity of approach flow
a = angle of attack (skew)
F = Froude number of approach flow
C = circular or rounded pier shape
R = rectangular pier shape
T = triangular pier shape

The RTAC method assumes a linear relationship between depth of scour and pier width, with multiplication factors to account for pier shape, depth of approach flow, and angle of attack (skew).

The IAHR method is a little more complex. It includes a term which varies linearly with flow velocity in the range of 0.5 to 1.0 times the competent velocity of the streambed, modelling the effect of velocity on pier scour in the clearwater (no general motion of the bed material) scour range. In addition, the hyperbolic tangent function is used to account for the flow depth to pier width ratio. Pier shape and angle of attack are accounted for by using similar multiplication factors to the RTAC method.

The method of Melville and Sutherland (1988) starts with a depth of scour value of 2.4 times the pier width and then allows reductions in this number to account for bed armouring, low velocities for the clearwater scour case, and shallow depths. Again, pier shape and angle of attack are accounted for by using multiplication factors.

The Colorado State University method relates the depth of scour to the pier width, depth of flow, and the Froude number of the flow. The inclusion of the Froude number implies that the depth of scour is affected by flow velocity in the live-bed (bed particles in motion) scour case. Although this is not consistent with the observations of some authors, such as Shen(1971), the equation does give comparable results to those of the other methods for smaller Froude numbers.

Comparison Between Prediction Equations And Field Data

The AT&U data, as well as published field data from Froehlich (1991) and the Alberta Research Council (ARC) (Malcovish and Andres, 1991), have been plotted in Figure 1 along with the predicted results of the RTAC, IAHR, Melville and Sutherland, and CSU methods.

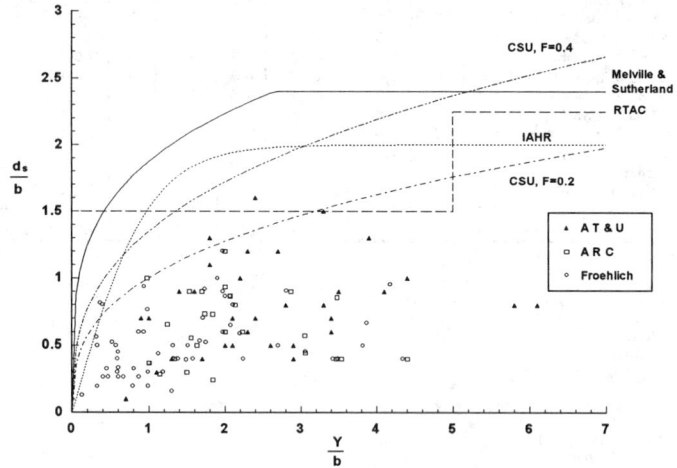

Figure 1 - Comparison Between Prediction Equations and Field Data

In this plot, d_s is the depth of local scour, Y is the depth of approach flow, and b is the pier width. The IAHR and Melville and Sutherland results assume no reduction due to velocity (live-bed case), and the CSU results are shown for Froude numbers of 0.2 and 0.4. All of the prediction methods assume no skew (0° angle of attack) and a rounded pier shape. To be consistent, only field data with an angle of attack less than 10° are shown on the plot. However, no pier shapes were excluded.

As can be seen on the plot, all of the prediction methods show reasonable consistency in their results. The equations appear to provide an upper envelope to the field data, with most of the field data showing considerably less scour than predicted. Some possible reasons for the discrepancy between the predictions based on laboratory data and the observed field results are as follows.

Backfilling - The maximum scour depth usually occurs close to the peak of a flood. It is therefore desirable to take scour measurements as close to the peak as possible. However, in many cases the scour survey cannot be made until several days after the peak. As a result, it is likely that some backfilling of scour holes will have occurred, and methods such as echo sounding will not be able to measure the maximum scour depth. Some modern equipment, such as ground penetrating radar (GPR) which can identify changes in density of certain materials, may solve this problem by being able to identify historic scour depths. If GPR proves to be a reliable method, the need to chase floods for scour information will be eliminated, and more reliable field measurements will result.

Time dependency - Most of the laboratory data upon which the prediction equations are based have been measured under steady state conditions where the discharge was kept constant until a maximum scour depth was observed. In the field, however, the discharge rises to the peak and then recedes back down to normal values during a flood. As a result, high discharges and velocities occur for a certain length of time, and the maximum (equilibrium) scour depth may not have been reached. The effect of unsteady flow on scour depths has been shown to be significant (Yanmaz and Altmbilek, 1990). It is therefore desirable to account for the time dependency of scour during design. Yanmaz and Altmbilek present some results which may be useful in determining if a reduction in design scour depth is warranted based on a given flood hydrograph. However, more work in this area is needed.

Combination of constriction and local scour - In laboratory experiments, the conditions can be controlled so that constriction scour will be negligible and only local scour will occur. However, at most bridge sites, there is a combination of the two types of scour occurring at the same time. This can make interpretation of scour survey results difficult when it comes to separating the effects of the two processes. In addition, Malcovish and Andres (1991) suggest that the two processes are not additive, and that if significant constriction scour occurs, negligible local scour will occur.

Location of footing - Jones, Kilgore, and Mistichelli (1990) performed a laboratory study to assess the effect of footing location on the depth of scour. They observed that local scour at piers can be significantly reduced by a footing located at or below the streambed, confirming previous observations. The amount of reduction depends on the distance the footing extends to the front and to the sides of the pier. However, they also noted that the scour depth can be increased by a footing if it is located above the streambed. In this case, the footing acts to increase the pier width.

Armouring of streambed - Many laboratory scour studies have used uniform bed material, which means bed armouring will not occur. For non-uniform beds, which are common in rivers, a layer of larger material will form at the top of the bed as the smaller particles are carried away, limiting the depth of scour. Some laboratory studies have examined this process, and the method of Melville and Sutherland (1988) allows for a reduction in scour depth due to armouring.

Accuracy of Field Measurements - Under laboratory conditions, the location of maximum local scour can be observed and measurements can be taken at that location. However, under field conditions, it is often impossible to see where the location of the maximum local scour is. Streambed contours plotted from spatial data points aid in locating the deepest scour hole, but it is still possible that the maximum depth of scour was missed because of the spacing between measurements.

Conclusions

In general, the four prediction methods discussed all give reasonable results and can be considered in the design of bridge piers. The discrepancies between the results of these methods and observed field data can be explained by factors such as backfilling, time dependency, combination of constriction and local scour, location of footing, armouring of the streambed, and accuracy of field measurements. Pier scour estimates will be improved by considering these factors. Further studies in these areas will lead to better predictions of local scour depths at piers. In addition, the accuracy of field data will likely be improved by the use of new methods such as ground penetrating radar.

References

Breusers, H.N.C., Nicollet, G., Shen, H.W., 1977. Local scour around cylindrical piers. Journal of Hydraulic Research.
Froehlich, D.C., 1991. Upper confidence limit of local pier-scour predictions. Transportation Research Record.
Jones, J.S., Kilgore, R.T., Mistichelli, M.P., 1990. Effects of footing location on bridge pier scour. Journal of Hydraulic Engineering, ASCE.
Kellerhals, R., Neill, C.R., Bray, D.I., 1972. Hydraulic and geomorphic characteristics of rivers in Alberta. Research Council of Alberta, River Engineering and Surface Hydrology Report 72-1.
Malcovish, C.D., Andres, D.D., 1991. Evaluation of bed scour and related flow processes along the North Saskatchewan River near Berrymoor, Alberta. Alberta Research Council Report No. SWE-91/01.
Melville, B.W., Sutherland, A.J., 1988. Design method for local scour at bridge piers. Journal of Hydraulic Engineering, ASCE.
Roads and Transportation Association of Canada, 1973. Guide to bridge hydraulics.
Shen, H.W., 1971. Scour near piers. River Mechanics, Volume II.
US Army Corps of Engineers, 1990. Scour at bridges. HEC - 18 Manual.
Yanmaz, A.M., Altmbilek, H.D., 1990. Study of time-dependent local scour around bridge piers. Journal of Hydraulic Engineering, ASCE.

INSTRUMENTATION FOR DETAILED BRIDGE-SCOUR MEASUREMENTS

by Mark N. Landers[1], M.ASCE, David S. Mueller[2], A.M. ASCE,
and Roy E. Trent[3], M. ASCE

Abstract

A portable instrumentation system is being developed to obtain channel bathymetry during floods for detailed bridge-scour measurements. Portable scour measuring systems have four components: sounding instrument, horizontal positioning instrument, deployment mechanism, and data storage device. The sounding instrument will be a digital fathometer. Horizontal position will be measured using a range-azimuth based hydrographic survey system. The deployment mechanism designed for this system is a remote-controlled boat using a small waterplane area, twin-hull design. An on-board computer and radio will monitor the vessel instrumentation, record measured data, and telemeter data to shore.

Introduction

Bridge scour is a long-standing transportation-engineering problem. Current understanding of and predictive methods for bridge scour are based primarily on laboratory investigations using scale models. However, scour predictions based on these investigations vary considerably, probably due to the typically limited and somewhat unique conditions modeled in each investigation. Scour predictions also differ from some scour observations at bridges, probably due to dynamic dissimilarity between field conditions and scale models and to the range of deterministic scour variables in the field that are difficult to design for or measure in the laboratory. Thus, the conclusion after numerous studies is that more field data are needed; however, resources for field data-collection efforts continue to be only a small fraction of those committed to scour evaluation, structure design, and countermeasures. Detailed scour data, beyond that which can be obtained by sounding from bridges, are especially needed. Detailed scour data sets include real-time measurements of hydraulic and channel geometry data at intervals over the duration of a flood hydrograph, through a channel reach extending from downstream of, to upstream of the hydraulic influence of the bridge. This type of data set is essential to permit distinction between the local, contraction, and general scour occurring at the bridge cross section. The development of appropriate instrumentation is

---

[1] Hydrologist, USGS, 415 National Center, Reston, VA 22092
[2] Hydrologist, USGS, 2301 Bradley Avenue, Louisville, KY 40217
[3] Research and Development, FHWA, 6300 Georgetown Pike, McLean VA, 22101

essential to obtaining detailed scour measurements; and that development is increasingly possible because of advances in hydrographic and surveying instrumentation.

A portable system to make detailed scour measurements is being developed by the U.S. Geological Survey in cooperation with the Federal Highway Administration. The basic function of this equipment is to measure and record detailed channel bathymetry through a bridge study reach during floods. Principal components of a portable scour measuring system are a sounding instrument, horizontal positioning instrument, deployment mechanism and data storage device. Instruments for each of these components have been evaluated according to design criteria selected for this application. This paper discusses the characteristics of each component and the integrated system.

Design Criteria

Design goals for each instrument component and the overall system include:
- operation in floods - the system must be field durable and operate well in high velocity and turbulence;
- accuracy - each component of the system should be designed to optimize overall measurement accuracy;
- portability - the system should be of a size and weight that can be commercially shipped;
- cost - cost should be minimized;
- ease of use - the system should be deployable and operable by two hydrographers;
- weight - weight should be minimized for each component to permit design flexibility of the remote-control boat deployment mechanism.

Design accuracy criteria for the integrated system was initially 15 cm (0.5 ft) or better in the vertical and in the horizontal, for the purpose of measuring bathymetric changes on the scale at which sediment transport and scour processes occur on most streams. However, at the time of these evaluations 15 cm (0.5 ft) accuracy was not achievable by any of the horizontal positioning systems within the cost constraints. Furthermore, vertical fluctuations or heave of the remote-control boat are difficult to measure and may substantially degrade the accuracy of the final measurement. Although multi-axis accelerometers may be able to measure vessel heave to within 5 cm (0.2 ft), these devices are currently beyond the budget of this investigation. Thus the system was optimized to provide the highest accuracy for the design cost. Digital output is required for all of the instruments because of the quantity and detail of data being collected; to provide real-time, remote monitoring of data collection; and in order to reduce the time required for data analysis.

Sounding Instruments

A digital fathometer is the preferred instrument for making detailed bathymetric surveys. A digital fathometer is composed of a transducer, the electronics to process and digitize the transducer signal, a digital interface to output the data, and an optional hardcopy recording device. Transducers are characterized by their frequency and cone angle. Higher frequency signals provide better resolution of the bottom echo but poorer penetration of deep or sediment-laden waters. For this application a typical high-frequency signal, in the range of 200kHz, provides a good balance between the resolution of the acoustic signal and the maximum sounding depth under potentially high sediment conditions. The angle of the cone within which acoustic signals propagate away from the center of the transducer is referred to as the cone angle. The cone angle and the water depth determine the footprint of the acoustic wave when it strikes the streambed. If the acoustic footprint is large relative to a scour hole, or if a pier intersects the acoustic cone (producing side echos), the recorded depth may not be accurate. A three degree transducer was selected for this application. Signal processing of the received acoustic echo was also evaluated. Most fathometers evaluated process signals using threshold detection. In threshold detection, the first received echo that exceeds a fixed threshold amplitude is recorded as the bottom echo. More accurate signal digitization is provided by peak detection. Peak detection processes the complete echo to locate the peak amplitude of the acoustic echo. The time between the signal transmission and this peak echo is multiplied by the speed of sound in water to obtain the water depth.

The interface between the digitizer and an external data logger is typically parallel Binary Code Decimal (BCD) or RS-232C serial communication, either of which can easily be interfaced with a measurement system data-logger. Survey-grade fathometers have a chart record in addition to the digital output. The digital record provides a single depth value for a sounding, but a paper chart record of the analog signal can indicate weak echos off debris, piers, or other objects, in addition to the bottom echo. The chart record is desirable for this valuable information and for it's immediate hard copy recording of results. The disadvantage of chart recording fathometers is their weight. Most digital and chart recording fathometers are developed for use in bathymetric surveys from manned boats where weight is not a major design factor.

Fathometers evaluated in this study were primarily of survey-grade and cost from $10,000 to $40,000. Fathometers in this cost range typically have the adjustable controls for reverb blanking distance, sensitivity (gain), draft, and speed of sound. The unit selected for this application has a 200 kHz, three degree transducer, peak detection, chart record, digital RS-232C output, and is the lightest of the survey-grade fathometers evaluated.

Horizontal Positioning Instruments

Horizontal positioning instruments are used to measure and record the horizontal position for which the depth is being measured. Vertical positions are also measured by some systems. Measurement of the physical position of an object on the earth's surface is a rapidly evolving field. In general, position is determined using microwave range-range measurements, range-azimuth measurements, or Global Positioning Systems (GPS). Stadia systems were not reviewed because the objective was to evaluate a system for operation during floods and for a remote control platform. All of the reviewed positioning systems are digital and integrate easily with microcomputers. All can be integrated with data from a digital fathometer on the microcomputer in real time; however this integration may be easier with the range-range or range-azimuth systems. Data can be processed and displayed in real time by all of the systems; however, for differential GPS this requires extra telemetry and post-processing may improve the accuracy of the data.

Range-range systems use the measured distance from the survey vessel to 3 or more transponders to triangulate the vessel position in real time. The systems measure the average time it takes for a known frequency signal to travel between the transmitters and the receiver, and solves for the survey vessel position. Microwave-based systems are typical, but radiowave, laser, and underwater-ascoustic systems were also considered. The cost of these systems vary, but $70,000 appears to be representative. Typical accuracy for these systems is one to three meters. The set-up time required to locate transponders over pre-surveyed positions and the weight of range-range systems are significantly greater than for range-azimuth or GPS systems.

The Global Positioning System (GPS) is a $10 billion satellite positioning and navigation network developed by the U.S. Department of Defense. GPS promises to provide an amazing degree of accuracy and range of application for navigation and position fixing. Development of GPS began in 1973 and the system will be fully operational in 1993 with 21 satellites and three operational spare satellites. There are already many manufacturers of GPS systems, who offer extensive and specialized systems and technical support. Because of it's rapid development, the actual accuracy and applicability of GPS for hydrographic surveying in a flooding river environment has been difficult to assess. According to some manufacturers and literature, carrier-phase, differential GPS in kinematic mode will provide better than decimeter (0.3 ft) accuracy for a moving antenna, as this application would have. Carrier-phase technology is not yet generally available and it's initial cost will probably be high. However, the principal disadvantage of kinematic differential GPS for this application is that four or more satellites must be kept 'locked' in view during the entire survey. This would be an unrealistic constraint when operating very close to and under bridges for scour measurements. The overall accuracy that this technology promises would enable more accurate bathymetric surveys than are now possible.

Range-azimuth systems operate like engineering survey total stations which combine an electronic distance measurer (EDM) with an electronic theodolite. The system sits at a known (horizontal and vertical) point and measures the distance, azimuth and vertical angles to a target and computes the target position. The thickness and width of the EDM beam is enlarged from that of the standard survey beam so that it can obtain a reading even when the target is not centered. For some systems, this beam is four milliradians for the vertical by seven milliradians for the horizontal, so that the system would read and record the position of the center of the beam, so long as the prism was anywhere within the beam area (0.4 m by 0.7 m at 100 m). Range-azimuth systems are available in several configurations with accuracies from meters to one centimeter and costs from $20,000 to about $150,000. Unlike range-range or GPS systems, range-azimuth systems require active-tracking of a reflector on the vessel. If the target could be tracked perfectly, the measurement accuracy would equal the instrument accuracy. However, operator tracking of a small, moving boat on a flooding river can be very difficult. The most expensive evaluated system provides automated tracking in the horizontal and vertical planes; however, the vertical tracking response is slower than the vertical motions anticipated for the remote control boat.

A range-azimuth hydrographic positioning system was selected for this project. The selected unit has a laser EDM modified to achieve 0.3 foot accuracy, mounted on a 5 second electronic theodolite. The selected EDM has distance gating functions and an adjustable scope to make manual tracking somewhat easier.

Deployment Mechanisms

A remote control boat was selected as the deployment mechanism for the detailed scour measurement system. This is a highly developmental part of this investigation. Skinner (1985) reviewed remote control platforms, but was unable to find a recommendable system. However, recreational remote control boat technology has advanced rapidly and is readily adaptable for scientific applications. Key design criteria for this component of the system are viability, stability, and operability in a flood environment. Also key to the design are total size and weight. All aspects of the design seek to make the boat and instrumentation operational and viable in the flood environment.

Roll and pitch of small boats is typically high, which would degrade the quality of a bathymetric survey. Stability of ships in high seas is critical for high-speed ferries, antisubmarine warfare, and other applications. In 1969 the U.S. Navy began development of a Small Waterplane Area, Twin-Hull (SWATH) ship concept to provide increased platform stability in high seas. The SWATH concept derives from the conventional catamaran hull and ocean oil-drilling platforms. A SWATH boat consists of two submerged pontoons that are attached to an above-water structural box by relatively thin struts on each side (Figure 1). A typical SWATH design would have only 20 percent of the waterplane area of a conventional monohull. The reduced waterplane area and redistribution of buoyant volume into submerged hulls reduce wave excitation forces and wave period to which the boat would normally respond. Consultants designing the remote-control boat for this project typically design full-scale SWATH boats for commercial applications. Models of full-scale designs are sometimes in the size range of the remote-control prototype. To the authors' knowledge this will be the first time the SWATH concept has been applied to a remote-control boat for use in the river environment.

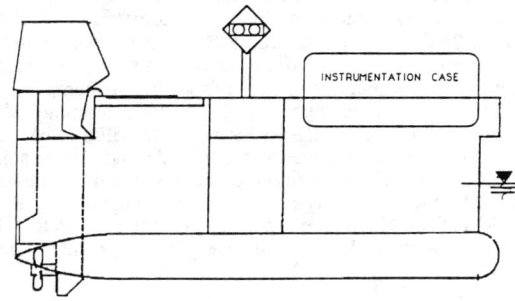

Profile View

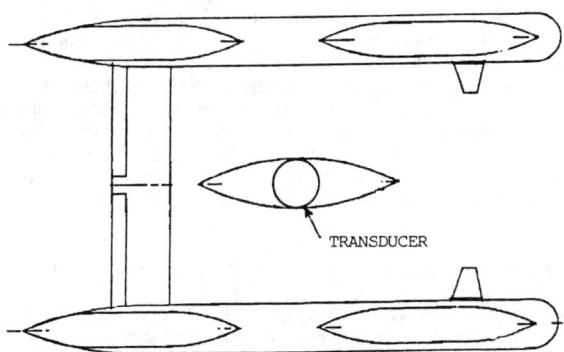

Top View of Section Below Waterline

Figure 1 - Remote control boat for deploying instrumentation for detailed bridge-scour measurements

The boat will be powered by a 9-horsepower gasoline outboard motor (with remote-controlled electric start) that can propel the boat at speeds of at least 10 knots for up to 1 hour. Electric motors, although preferred for their greater reliability, were not selected because they could not achieve the sustained speed needed to operate in floods without exceeding overall weight constraints. The fathometer transducer will be housed in a faired instrument pod located near the middle of the vessel. Other instrumentation on the vessel will be housed in waterproof enclosures. Hull cavities will be filled with foam, high-density plastic bumpers will be placed around the vessel perimeter, and an optional tether system may be used to improve security of the boat in the flood environment. The selected design has a draft of 46 cm (1.5 ft), length of 198 cm (6.5 ft), and a total dry weight of 400 pounds (including payload). The vessel design is complete, fabrication is underway, and laboratory testing will begin in April 1993.

Integrated System

The integrated system includes the remote control boat and the instrumentation it deploys, the positioning system, telemetry instrumentation, and personal computer. Roll and pitch of the boat will be measured by on-board digital inclinometers or accelerometers. Boat heading will be measured using a digital flux-gate compass. An onboard microprocessor will poll the onboard devices for readings and send a concatenated data string to a telemetry unit at a minimum rate of once per second. A companion, on-shore telemetry unit will read the real-time data into a rugged, portable computer which serves as the system data storage device. Horizontal positioning data will also be read into the computer in real time to complete the bathymetric data set. Attitude and heading data will be used to adjust the recorded transducer position to represent the true bottom position of the acoustic footprint for each sounding. The computer will integrate the data, compute, and display real-time plots of bed-elevations and horizontal positions so that the operator can concentrate data collection where the bed changes significantly and ensure adequate coverage of the entire reach.

Summary

Scour around bridges depends upon processes that extend from downstream of to upstream of the hydraulic and erosive influence of the bridge. Sediment transport into a reach can influence channel scour and fill to a degree that obscures the influence of bridge-related factors typically used to design for bridge scour. Measurements of scour made only at the bridge commonly indicate unsteady sediment transport to the bridge section, but do not provide adequate information to separate bridge-related from sediment-inflow related scour and fill. The detailed scour measurement instrumentation described in this paper is being developed by the U.S. Geological Survey in cooperation with the Federal Highway Administration to obtain detailed channel bathymetry measurements through a bridge-study reach so that scour processes can be better understood and so that better scour prediction methods can be developed.

References

Cohen, P. M., 1970, Bathymetric navigation and charting: U.S. Naval Inst., Annapolis, Maryland, 138 p

Fein, James A., Ochi, Margret D., and McCreight, Kathryn K., 1980, The seakeeping characteristics of a small waterplane area, twin-hull ship: Proceedings of the 13th ONR Symposium on Naval Hydrodynamics, Toyko, Japan, October 1980, 20p.

Skinner, J. V., 1985, Measurement of scour-depth near bridge piers: U.S. Geological Survey Water-Resources Investigations Report 85-4106, 33 p.

Trent, R. E. and Landers, M. N., 1991, Chasing floods and measuring scour: Transportation Research Record, no. 1290, vol. 2, p. 236-244.

LOCAL SCOUR AT BRIDGE PIERS IN ALBERTA - CASE HISTORY

Alan Humphries Ph.D., P.Eng[1]

Abstract

A two span steel truss bridge with a rectangular concrete central pier was built over the Oldman River near Brocket in south-west Alberta in 1954. Initially the bridge was built square to the river but the shifting river channel has progressively increased the flow skew at the bridge. In 1975 the Oldman River suffered an extreme flood event but no significant pier scour was observed. However by 1987 the skew angle was 45 degrees and a 3 m deep scour hole had formed at the pier. The maximum recorded flow between 1975 and 1987 had less than a 1 in 7 year return period. In 1988 the pier footing was protected with a heavy rock riprap apron.

Introduction

Local scour at bridge piers is well recognised as a major cause of bridge failures. Considerable research has been conducted into the phenomenon which has culminated in the publication of a number of prediction methods. For over 30 years the Bridge Engineering Branch of Alberta Transportation and Utilities, as part of the Cooperative Program in Surface Water Engineering with the Alberta Research Council and the University of Alberta, has been measuring scour at bridge sites throughout the province of Alberta in Canada. The branch has recently consolidated this data and has attempted to correlate the measurements with some of the prediction methods. The results of this work is presented by Williamson (1993) in a companion paper.

Detailed scour records exist for the Brocket bridge over the Oldman River in south-western Alberta which enable it to be used as a case study into the development of pier scour. The Brocket bridge consists of two 53.3 m steel through trusses with a mass concrete central pier and concrete box type abutments. The bridge was built in 1954 replacing old bridges located approximately 4 km upstream and downstream.

[1]Bridge Planning Engineer, Bridge Engineering Branch, Alberta Transportation and Utilities, 4999-98th Avenue, Edmonton, Alberta, Canada T6B 2X3

The pier has a rectangular footing 15.85 m long, 3.5 m wide and 1.5 m thick. The pier shaft is rectangular with 90 degree pointed ends. The shaft is 14.25 m long and 1.85 m wide at the base, tapering to 11.3 m long and 1.5 m wide at the top. The total height of the pier from the bottom of the footing is 9.55 m. When constructed in 1954 the base of the footing was 3.3 m below streambed. The footing is supported by forty-nine 2.6 m long steel H piles (fig 1).

The bridge is located on a local road 5 km north-west of Brocket in the eastern foothills of the Rocky Mountains. The Oldman River is an unstable gravel bed river and has gravel deposits over 11 m deep. The bed material has a D_{90} of 56 mm and a D_{50} of 12.5 mm. The bank material tends to be somewhat coarser gravel with a D_{50} of 52 mm. The river has an average stream slope at the bridge site of approximately 0.2% (Kellerhals,1972). The Water Survey of Canada has maintained a streamflow gauge at the bridge since 1966. The gauge datum corresponds to the approximate streambed elevation. The highest recorded instantaneous discharge of 1560 m^3/s occurred on June 20, 1975. The gauge height for this flood was 4.0 m and the estimated average velocity through the bridge was 3.9 m/s. The 1975 flood had an estimated return period of 75 years (Phinney,1982).

Scour History

The bridge was built in 1954 square to the river with the north abutment face set back from the river bank (fig 2). In 1966 an inspection reported that the north bank upstream was eroding rapidly. A site survey was completed in 1967 (fig 3). The survey found that the upstream north bank had moved beyond the abutment and a 2 m deep scour was measured by the upstream north bank. The pier was surrounded by a gravel bar with the river channel under the north span The river was skewed 17 degrees to the bridge. A 45 m long guidebank, protected with Class 2 rock riprap (0.8 m maximum diameter) was constructed in 1968 to protect the north abutment.

Immediately following the 1975 flood a detailed scour survey was completed (fig 3). A 4 m deep scour hole was measured by the upstream end of the guidebank. If this scour hole had occurred at the pier the bottom of the footing would have been exposed by approximately 0.5 m. Fortunately the pier was still at the south edge of the low water channel and no significant pier scour was observed. However the skew had increased to 25 degrees and had caused the flow to start to impinge on the pier. The north bank had continued to erode and the nose of the guidebank needed reinforcing with Class 3 rock riprap (1.1 m maximum diameter).

In July 1983 the river skew had increased to about 40 degrees and another scour survey was completed (fig 3). The survey found that the scour hole by the nose of the guidebank was now 4.6 m deep but more significantly a 1.2 m deep scour hole had formed around the nose and along the north side of the pier. The bottom of the pier scour hole was approximately 1.75 m above the bottom of the pier footing. The highest recorded discharge between 1975 and 1983 was 529 m^3/s, with a gauge height of 2.63 m and a mean velocity of approximately 2.5 m/s. This flow is equivalent to about a 1 in 7 year return period.

Between 1983 and 1987 the north upstream bank continued to erode and the skew angle increased to 45 degrees. On May 28, 1987 scour profiles of the guidebank and pier were taken (fig 1). The bottom of the scour hole along the north side of the pier was only 0.3 m above the bottom of the footing. The scour hole had deepened by 1.45 m within three years and was reaching a critical depth that would soon expose the bottom of the pier footing. The discharge on May 28, 1987 was 45.5 m^3/s, with a gauge height of 1.14 m and a mean flow velocity of approximately 0.6 m/s. The highest recorded discharge between July 1983 and May 1987 was 345 m^3/s, with a gauge height of 2.21 m and a mean velocity of approximately 2 m/s. This flow is equivalent to about the 1 in 3 or 4 year flood.

Pier Scour Protection Works

The pier scour hole measured in 1987 was becoming critical and some remedial action was required. The design was based on the known 1975 flood discharge. Based on the methods proposed by Neill (1975) the average general scour was estimated at 1.1 m, ie. 2.9 m above the bottom of the footing. Using the methods outlined by Melville (1975) the local pier scour depth was estimated at 3.9 m, giving a total scour elevation 1 m below the pier footing. The estimated total scour of 5 m compared favourably with the 4.6 m of scour measured adjacent to the guidebank.

A rock riprap apron was designed to protect the scour sensitive nose and north side of the pier. The apron design width was based on a 2:1 slope from 0.3 m above the bottom of the footing (the measured scour elevation) to the theoretical scour elevation. This gave a theoretical launched apron width of 2.9 m and using a 33% loss factor a total required apron width of 4 m. The final design specified an apron that extends the full length of the pier footing on the north side, around the front of the footing and along the upstream 4 m of the south side. The bottom of the rock was placed at the elevation of the bottom of the existing scour hole. Care was taken not to excavate below this elevation during installation to ensure the stability of the pier at all times. The apron thickness is 2.0 m around the nose of the footing and for the upstream 4.0 m against either side, it then reduces to 1.5 m for the remainder of the apron along the north side of the footing. The total volume of rock used in the apron was 160 m^3.

The rock met the Bridge Engineering Branch specification for Class 3 Heavy Rock Riprap. This specification gives the rock gradation as 100% smaller than 1100 mm or 1800 kg, at least 20% larger than 900 mm or 1100 kg, at least 50% larger than 800 mm or 700 kg and at least 80% larger than 500 mm or 200 kg. The percentages are quoted by mass. The sizes quoted are equivalent spherical diameters and are for guidance only. The specification also states that the rock shall be hard, durable and angular in shape, resistant to weathering and water action, free from overburden, spoil, shale or shale seams and organic material. The minimum dimension of any single rock shall not be less than one third of its maximum dimension. The minimum acceptable unit weight of the rock is 2.5 t/m³. Experience within Alberta has found that rock riprap meeting these specifications will offer protection against local velocities of up to 4.7 m/s.

Summary and Conclusions

The Oldman river has a mobile gravel bed and, over the years, progressive channel migration has increased the skew angle under the Brocket bridge. A combination of scour surveys, aerial photographs, stream flow data and inspection records dating back over the life of the bridge offers the opportunity to follow the development of a scour hole at the pier due to the increase in flow angle of attack.

In 1954 the bridge was built square to the river. The north bank upstream of the bridge is actively eroding redirecting the flow under the bridge. By 1992 the low flow was skewed by 50 degrees to the pier. A 45 m long guidebank was built in 1968 to protect the north abutment and in 1975 the Oldman river experienced an extreme flood. The flood flow generated a 4 m deep scour hole adjacent to the guidebank but no significant pier scour was measured. Between 1975 and 1983 the skew angle increased from 25 to 40 degrees and a 1.2 m deep scour hole developed by the nose and north side of the pier. In 1987 this hole was 3 m deep and the bottom of the hole was only 0.3 m above the bottom of the pier footing. The skew angle had increased to 45 degrees. In 1988 a Class 3 rock riprap apron 4 m wide and up to 2 m thick was placed around the pier footing to prevent further scour. An inspection in 1992 found the apron to be working well. Between 1975 and 1987 the Oldman river did not experience a flow any greater than a 1 in 7 year flood.

It is concluded that the deep scour hole at the pier was generated by the increased angle of attack of the flow at the pier. This demonstrates that when designing or assessing a bridge structure for local scour the potential for changes in stream flow skew angle must be considered. The case study also highlights how regular monitoring and consistent record keeping can identify developing problems allowing timely and cost-effective preventative maintenance to be done.

References

Kellerhals R.et.al.,"Hydraulic and Geomorphic Characteristics of Rivers in Alberta", Alta Coop Res Prog in Hwy and Riv Eng, Edmonton, Alberta, 1972.
Melville B.W.,"Local Scour at Bridge Sites",Eng Rep No.117, Univ Auckland,1975.
Neill C.R.(ed), "Guide to Bridge Hydraulics",R.T.A.C., Univ Toronto Press, 1975.
Phinney R.B.,"Flood of June 1975 in the Oldman River Basin, Alberta",Water Resources Branch, Env Canada, Calgary, Alberta, Canada, Oct 1982.
Williamson D., "Local Scour Measurements at Bridges Piers in Alberta", to be presented at ASCE Hyd Div Conf, San Francisco, July 1993.

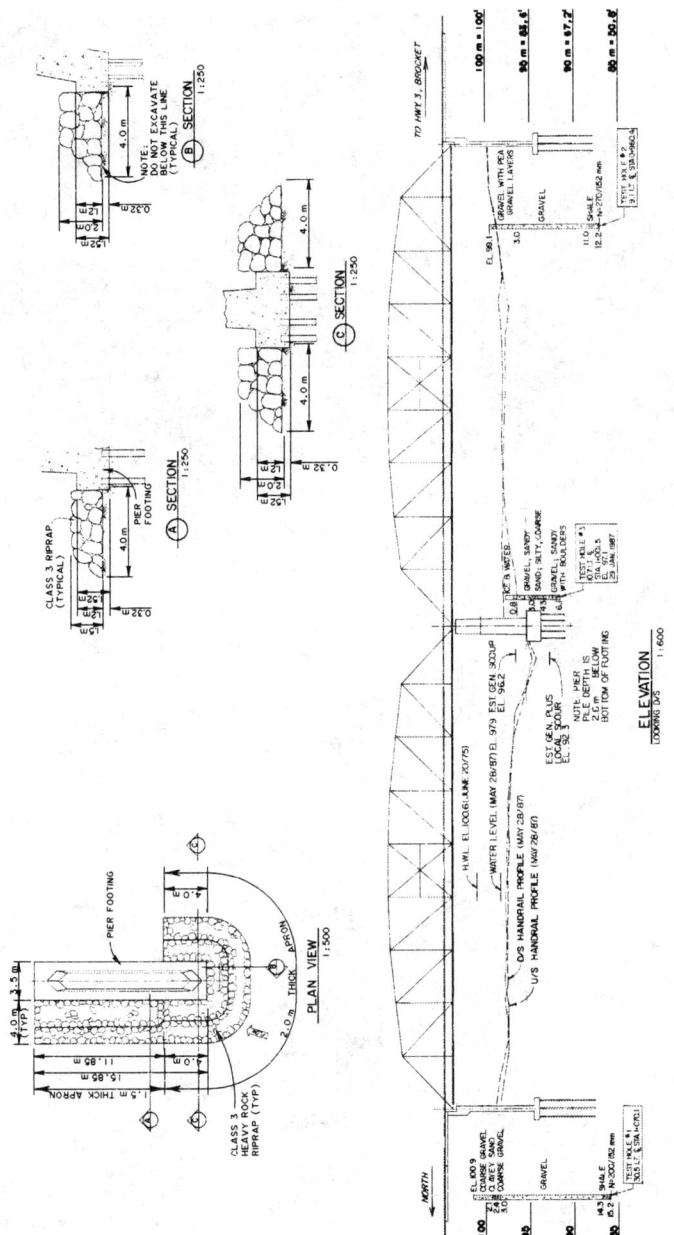

FIGURE 1 GENERAL LAYOUT

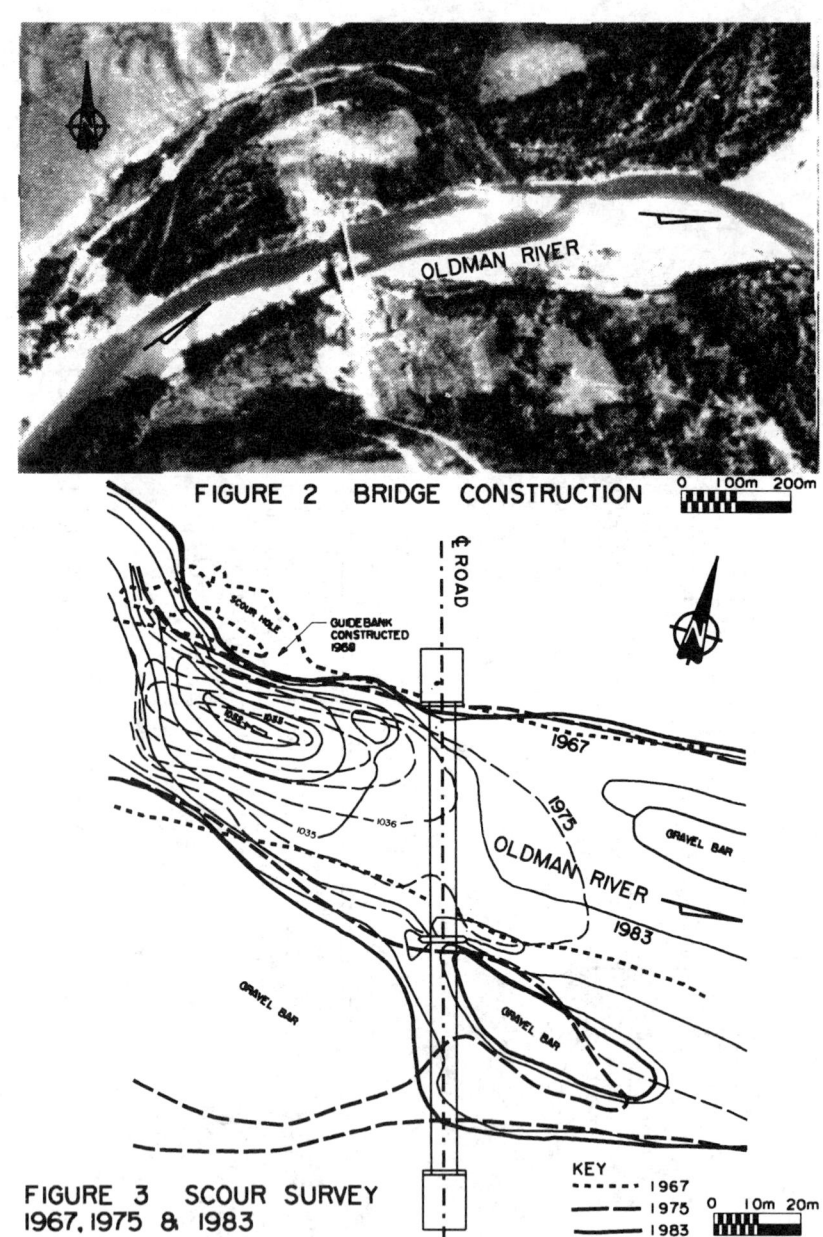

FIGURE 2 BRIDGE CONSTRUCTION

FIGURE 3 SCOUR SURVEY
1967, 1975 & 1983

KEY
------- 1967
— — — 1975
———— 1983

REFERENCE SURFACES FOR BRIDGE SCOUR DEPTHS

Mark N. Landers[1] M.ASCE and David S. Mueller[2] A.M.ASCE

Abstract
Depth of scour is measured as the vertical distance between scoured channel geometry and a measurement reference surface. A scour depth measurement can have a wide range depending on the method used to establish the reference surface. A consistent method to establish reference surfaces for bridge scour measurements is needed to facilitate transferability of scour data and scour analyses. This paper describes and evaluates techniques for establishing reference surfaces from which local and contraction scour are measured.

Introduction
Field measurements of channel scour at bridges are fundamental to a better understanding of scour processes and development of improved scour prediction methodologies. The complexity and natural variability of channel erosion in response to hydraulic, sediment, and bridge characteristics makes field measurements of scour at bridges particularly important. A complete bridge scour data set includes concurrent measurements of channel geometry, flow velocity, and flow depth during scouring conditions, as well as measurements of bed-material and bridge characteristics. The magnitude of scour for a scour data set is the vertical distance between the measured channel geometry and a surface, line, or point that represents the reference channel geometry for the base-line condition; i.e. for conditions in the absence of the bridge structure. The term reference surface is used here even when the three dimensional surface is represented by a line or a point. This reference surface has also been referred to as reference datum in the literature.

Several methods have been used to establish reference surfaces from which bridge scour depths are measured using channel geometry data. Reference surfaces in past bridge scour investigations have been based on mean bed elevation, concurrent ambient bed elevation, water-surface

---

[1] Hydrologist, U.S. Geological Survey, MS 415, Reston, VA 22092

[2] Hydrologist, U.S. Geological Survey, 2301 Bradley Ave. Louisville, KY

elevation, and maximum observed bed elevation. Reference surfaces defined by different methods, which may not isolate particular scour components, can result in measured scour depths which vary by as much as 100 percent for a given data set. An evaluation of methods to determine reference surfaces is needed to ensure the correct and consistent interpretation of bridge-scour data being collected in ongoing studies, which constitute the broadest and most intensive bridge-scour data-collection efforts to date in the United States. Reference surfaces should be selected so that the local, contraction, and general (long- and short-term changes unrelated to the effects of the bridge) components of total scour may be quantified separately. The reference surface for each component of scour will be unique; however, the technique for determining each reference surface requires consideration of the overall scour process at the bridge. This paper evaluates techniques for establishing reference surfaces from which local and contraction scour are measured using channel geometry data at bridges. Potential problems in applying these methods to field data are also discussed. This paper does not discuss reference surfaces for predicting minimum channel elevations from computed scour depths for bridge foundation design.

Local Scour

Local scour of the channel bed occurs where obstructions cause flow acceleration and the formation of vortices around the base of the obstruction, which locally increase the erosive capacity of the flow. Local scour typically occurs around bridge piers, abutments, spurs and embankments. The depth of local scour is the difference between the bed level with and without the flow obstruction present for a given flow condition. The maximum local scour from a measurement is the quantity usually reported, unless otherwise noted. Methods to determine reference surfaces from channel geometry are perhaps simplest to define for the case of local scour. This is also the case most commonly discussed in the literature.

Scour measurement data have usually been obtained from models in laboratory flumes where bed elevations outside of the scour hole usually do not change significantly. The reference surface is typically taken as the average of several points measured in the unscoured region around the obstruction after equilibrium bed conditions are established and after the model run is completed. This ambient or mean equilibrium (in the case of dunes) bed level has been used as the reference surface in most flume studies of local scour. Other reference surfaces used in flume studies include the initial condition bed level, pre-equilibrium bed level, and the water-surface. Field measurements of local scour have generally used concurrent ambient bed level as a reference (Harrington and McLean, 1984; Norman, 1975; Chang 1980). However, in field data studies by Inglis (1949) the water surface was used as the reference surface, and Jarrett and Boyle (1986) use the highest observed bed-elevation at the point where local scour is being measured. Results from Blodgett's (1989) comparison of reference surfaces for a scour measurement on Sacramento River are shown in Table 1. Blodgett considered the concurrent ambient bed as the best reference surface, given the available data.

Table 1.- Measured local scour using several reference surfaces

Method	Description of Reference Surface	Local Scour (feet)
1	Concurrent ambient bed level	4.3
2	Concurrent thalweg at upstream side of bridge	3.0
3	Projected upstream to downstream thalweg profile	4.5
4	Projected upstream to downstream mean bed elevation.	8.3
5	Highest bed elevation observed at same pier (Jarret and Boyle, 1986)	1.7

Local scour measurements using reference surfaces other than concurrent ambient bed level may include amounts of contraction or general scour, which would significantly reduce the value of these data. Such combined scour quantities cannot be effectively analyzed in relation to separate local, contraction, and sediment supply deterministic processes. Therefore, concurrent ambient bed level is the preferred reference surface for measurements of local scour depth from scoured channel geometry.

The concept and description of this preferred surface is simple. However, a representative, concurrent ambient bed level is not always apparent, given the range of channel geometry conditions and data limitations. Two examples are presented to illustrate the method for different scoured channel geometry measurements. Figure 1 shows data from a model of pressure flow for a rectangular pier in a sand bed from a flume with a sediment box; thus clear-water conditions. The reference surface is taken as the concurrent ambient bed level, illustrated by the heavy line. Figure 2 shows data from a scour measurement on South Altamaha River, at southbound Interstate Highway 95 near Brunswick, Georgia. The reference surface is represented by the sloping line, and the maximum vertical distance between this line and the locally scoured bed is the measured scour. The measured depth of scour at pier 13 for this measurement is taken as the maximum vertical distance between the reference surface and the locally scoured channel geometry. Associated

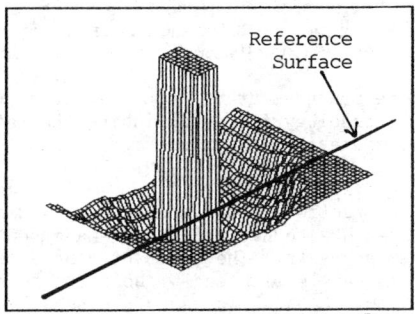

Figure 1. - Reference surface for local scour

data for this measurement are given in the following table.

Time (hours)	Scour Depth (feet)	Flow Depth (feet)	Velocity (feet per second)
13:00	3.9	26.4	1.7
17:05	5.2	23.1	2.2

Pier 13 is supported by three rows of one foot square concrete piles that extend from the bed to near the water surface. The effective pier width is 3 feet. The bed-material has a median diameter of 0.0033 ft. Establishment of local scour reference surfaces can be difficult and require much judgement for some complex cross-section geometries, as for the case in which the thalweg coincides with the local scour hole. Additional factors that must be evaluated in measuring local scour include presence of remnant scour holes, debris, scour countermeasures, time-rate of scour, and dune-bed forms.

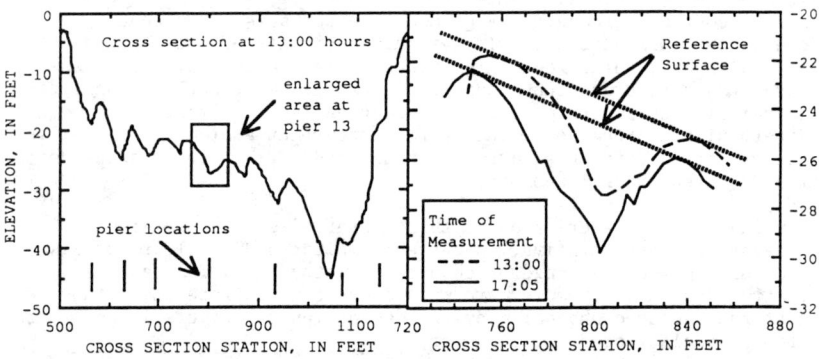

Figure 2.- Reference surface for local scour at pier 13 of the Interstate 95 crossing of So. Altamaha River, near Brunswick, GA

The measurement of scour requires judgment; however, consistent methodology for establishing the reference surface will facilitate transferability of scour data and scour analyses.

Contraction Scour

Contraction scour is caused by a decrease in the normal channel flow area due to natural or man-made contractions such as highway embankments. The depth of contraction scour is the difference between bed levels with and without the contraction in place for a particular flow. Contraction scour is usually defined as the difference between the average bed elevations of contracted and uncontracted sections.

Contraction scour is typically classified as live-bed or clear-water scour. Live-bed scour refers to scour conditions in which the uncontracted section is actively transporting bed load into the contracted section. Clear-water scour occurs when the uncontracted section does not transport bed load into the contracted section. Live-bed scour is typical of many main-channel sections while clear-water scour is typical at most overbank, flow-relief bridges.

Blodgett (1989) reported one of the few field investigations in which contraction scour was quantified separately from general long-term or short-term aggradation and degradation. Blodgett's contraction scour reference surface was represented by a straight line projected over the contracted section from the upstream to the downstream portions of a thalweg profile during moderate flow conditions. The contraction scour was reported as the difference between this reference surface and the thalweg (minimum elevation) of the contracted section. Blodgett stated that using the thalweg to measure contraction scour represented a worse case condition. However, differences between thalweg elevations may not be consistent with existing contraction scour equations that are based on average changes in the contracted section required to achieve equilibrium sediment transport (FHWA, 1991). The equations do not have the ability to compute the distribution of contraction scour; thus the location and magnitude of maximum contraction scour cannot be predicted. A sediment transport model could predict the spatial distribution of contraction scour.

The reference surface should characterize the mean bed elevation of an uncontracted section at the location of the contraction scour measurement. The reference surface can be established by passing a line through the average elevation of uncontracted sections, located upstream and downstream of the contracted section. Ideally, the contracted and uncontracted sections would be measured concurrently. The effects of local scour should be removed by excluding locally scoured areas when determining the average contracted bed-elevation. Cross sections that have some subareas with live-bed and others with clear-water bed-load transport conditions require separate analysis of those subareas. The reference surface and the contraction-scoured bed elevation would be the average elevation of the portion of the uncontracted and contracted sections, respectively, where the specific bed-load transport is occurring.

There are several potential problems with this ideal reference surface for live-bed conditions. Some of these problems are as follows. (1) It is often difficult to identify the bottom width over which active bed-load transport is occurring, because of irregular cross-section geometry. (2) Upstream and downstream cross-sections may be in natural contractions or expansions, due to channel bends or other factors, which make them unrepresentative of an uncontracted condition at the bridge. (3) Large dune bed forms can produce misleading results when dune crests or troughs predominate in one of the measured sections. (4) Slow, downstream migration of large sand and gravel bars can make a measurement nonrepresentative of equilibrium conditions. (5) Measured contraction scour may not represent equilibrium scour if the scour develops over many years due to the infrequency of channel-formative flows and the resistance of the bed to scour. (6) Most flood-flow scour measurements are made only from the bridge deck along the upstream and downstream sides of the bridge because boats are usually unavailable to obtain concurrent uncontracted channel geometry.

Pre- and(or) post-flood measurements of uncontracted sections are often used to establish a contraction scour reference surface. These

are usually obtained upstream and sometimes also downstream of the
hydraulic influence of the bridge contraction. Live-bed contraction
scour measurements using this reference surface may be useful, if
there are sufficient data to support the assumption that the geometry
of the approach and exit sections change relatively little during high
flow. The stability of the uncontracted sections and the accuracy of
the measurement should be assessed using comparisons of pre- and post-
flood measurements, and upstream and downstream uncontracted sections
or profiles through the study reach.

Clear-water contraction scour occurs when there is no sediment
transported into the scour hole so that the geometry of the
uncontracted section will remain the same after the flood has passed.
Therefore, post-flood surveys can be used to measure clear-water
contraction scour; although real-time flood measurements of hydraulic
characteristics are still desirable. Post-flood surveys should extend
downstream beyond the influence of deposited scour-hole material.

Summary
Depth of scour is measured as the vertical distance between scoured
channel geometry and a measurement reference surface. A scour depth
measurement can have a wide range depending on the method used to
establish the reference surface. Consistent methods to establish
reference surfaces that isolate the individual components of total
scour are needed to facilitate transferability of scour data and scour
analyses. A reference surface based on concurrent ambient bed level
can isolate the local scour component of total scour. A reference
surface that characterizes the mean bed elevation of an uncontracted
section at the location of the contraction scour measurement isolates
the contraction scour component of total scour.

References
Blodgett, J.C., 1989, Monitoring scour at the State Route 32 bridge
across the Sacramento River at Hamilton City, California: Proceedings
of the Bridge Scour Symposium, October 1989. FHWA-RD-90-035: 211-226

Chang, F.M., 1980, Scour at bridge piers - field data from Louisiana
files: Federal Highway Administration Report no. FHWA-RD-79-105, 34p

Federal Highway Administration (FHWA), 1991, Evaluating Scour at
Bridges, U.S. Dept. of Transportation FHWA-IP-90-017 HEC-18, 191 p.

Harrington, R. A., and Mclean, D. G., 1984, Field observations of
river bed scour on the Peace River near Fort Vermilion, Alberta:
Canadian Journal of Civil Engineering, V.11, p782-797

Inglis, C.C., 1949, The behavior and control of rivers and canals:
Government of India, C.W.I.& N. Research Publication 13, Poona,India

Jarrett, R.D., and Boyle, J.M., 1986, Pilot study for collection of
bridge-scour data: U.S. Geological Survey Water-Resources
Investigation 86-4030, 30p

Norman, V. W., 1975, Scour at selected bridge sites in Alaska: U.S.
Geological Survey Water-Resources Investigation 32-75, 171p

Probability of bridge failure due to scouring

Wolfgang Kron[1] and Erich Plate (member ASCE)[2]

Abstract

Models to estimate depth of scour at bridge piers and abutments are available in many forms, but all of them are based on constant discharges. Their results vary widely and engineering judgment is required when choosing the adequate scour model and the necessary depth of the foundation. One reason for the variability is the time dependency of the scouring process and of the discharge. The time function representing scour is simulated with a simple model consisting of three parts: a) computation of the maximum scour y_s^*, b) time function to reach y_s^*, and c) recovery of the scour. All parts can be varied easily. The flow input is based on long series of generated daily discharges. Results are analyzed statistically and interpreted with methods of reliability theory. Probability of failure from the time series of the depth of scour is compared to the exceedance probability of the design flow.

Introduction

Bridge piers and abutments in rivers have to be designed to resist erosive forces that threaten their foundations by undermining. The base of the foundation must be deeper than the probable maximum scour depth that occurs in the structure's lifetime. Every year bridges across rivers collapse due to erosion. Obviously the scour depth assumed in the design procedure was underestimated in such cases. This does not necessarily mean that the design was wrong. In the conventional design procedure one design discharge value is chosen (e.g. from frequency analysis of flows) and with its help the depth of scour is computed with a scour model. If a structure fails, it may be due to wrong design, a case which we do

---

[1] Research Associate and [2] Professor, Inst. for Hydrology and Water Resources Planning, Univ. of Karlsruhe, Germany

not consider here, or due to assumptions, which have to be made, but turn out to be not correct. There is also the other case that a bridge is by far overdesigned. This case is not serious with respect to catastrophe, nevertheless it is not desirable for economic reasons - at least not to any extent.

Unlike traditional design procedures, methods from reliability theory allow to obtain an objective criterion for the safety of a structure through the probability of failure, P_f. In this paper we shall investigate the effect of time variability of discharge on the scouring process. As a reference case, the 'optimum' is selected, which corresponds to the case that a bridge fails exactly if the design discharge occurs. This implies that safety margins are excluded. It is not intended to introduce a new method for designing bridge piers or abutments. Consequently, a convenient scour formula is selected which could be replaced by any other scour equation.

Scour models and their uncertainties

The fact that a vast number of formulae for estimating scours is available reveals the uncertainty involved in their results. Maximum depths of scour obtained with different methods scatter in a wide range. Often only engineering judgment can overcome the difficulty of choosing a reasonable value and different methods should be applied and their results be compared. Furthermore, almost each model parameter is subject to uncertainty. A single constant design value must be taken as representative despite the parameter's variability.

Some of the parameters affect the result, the depth of scour, in a very sensitive way. Slightly wrong assumptions of sediment and flow parameters may enormously change results of calculations. Individual geometric characteristics of the structure (shape, site, location in the river) also influence scouring considerably. Often only detailed physical experiments allow to estimate the depth of scour in an appropriate way. As physical experiments are very expensive mathematical design calculations are preferred in most cases.

Natural variability of streamflows

Existing scour models cannot account for the variability of streamflow which represents the load variable for the structure. Input into practically all scour models is a constant design discharge. Obviously, the depth of erosion is dependent not only on instantaneous discharges, but on the whole history of flows. In nature, high flows of the order of the design discharge usually have relative short durations. Hence, the high discharge during a flood wave may not last long enough for reaching the depth of scour corresponding to the discharge value (equilibrium scour). Therefore, for determining the probability of failure one must take into accout the time variability of flows.

Reliability concept: estimation of probability of failure

In the traditional design we tacitly assume that the corresponding scour has a probability of failure that is equal to the exceedance probability of flows. This means, if a bridge is designed with a 100-year flow, the probability of failure is also assumed to be one in one hundred years. This is by no means true. P_f may be higher due to underestimation of influencing parameters and effects from flow instationarity, or smaller due to overestimation of parameters and not considering the scour development in time ('under-/overestimation' here is used with respect to the result). By actually simulating the daily time series of scour the occurrences of days with failure can be counted and P_f determined.

No recorded time series are long enough to provide a data base sufficient for estimating such small probabilities of failure as needed in bridge design. Therefore, artificial flow series must be generated, which have the same statistical properties as the historic series. Here, a model for generation of daily discharges was used. This model produces flow series that very well reproduce not only the basic statistics but also the frequency distributions of daily values and even of annual means amd maxima. A detailed description of the model is found in Kron et al.[1990].

Structure of the simulation model for scour

In order to demonstrate the application of reliability theory and show ways how P_f can be determined a relatively simple computer model was set up. The model consists of the following three basic modules:

1. **Scour model**: from among the wide variety of scour models Laursen's formula for local scour under live bed conditions was selected. The method is described in USDT/FHA [1988], p.31. In simplified form the formula can be written as

$$\frac{y_s^*}{y_1} = 1.5 \left(\frac{a}{y_1}\right)^{0.48} \tag{1}$$

where y_s^* = maximum total depth of scour at abutment
y_1 = average upstream flow depth
a = length, which abutment stands out into flow

2. **Temporal development of scour:** there is a basic distinction between clear-water and live bed scours, which becomes evident in their development and bevavior in time (Figure 1a). Some difficulty exists in describing this development even for constant discharges. As a general functional relationship of the time rate of scour is not available, the scour development is described by $s(t)=y_s(t)/y_s^*$ which is assumed to follow several different curves in order to investigate the influence of the speed of scouring (Figure 1b).

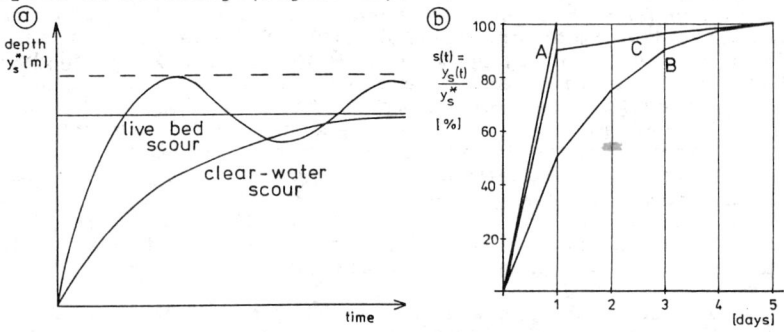

Figure 1. Temporal scour development: a) schematic curves b) linearized curves used in the model

3. **Recovery of the bed (aggradation):** in practically all natural rivers sediment transport occurs once a small critical discharge is exceeded. Therefore, a scour hole formed during a flood is refilled with sediment during the subsequent low flow period. The rate of recovery depends on the sediment transport characteristics of the river. Again, an exact aggradation curve is not available and different curves are applied, one representing quick recovery of the scour after flood recession, another representing slow recovery (Figure 2).

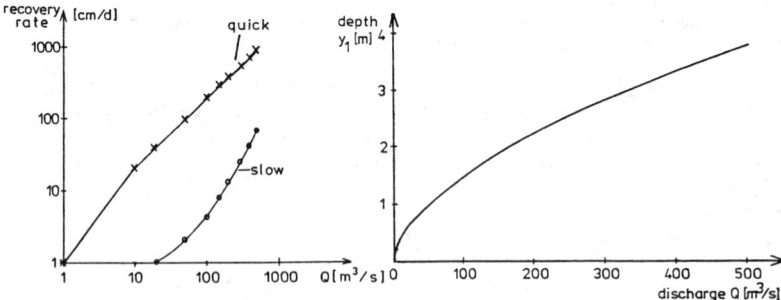

Figure 2. Aggradation rate   Figure 3. Stage-discharge-curve

The computer model combines the three above mentioned steps. Each step is applied in a simplified way, but it should be noted that these steps (i.e. program modules) are freely exchangeable with other modules of any desired degree of sophistication. For example, the temporal development, which in our approach is obtained by interpolating in an empirical table, may be simulated with a detailed model of the scouring process. However, one should keep in mind that detailed modeling increases computer time for long-term simulation. In the case demonstrated the model is unquestionably too simple. The purpose of this paper is to only present the basic idea and not to design an actual structure. In real applications many realizations (i.e. long model times) are required to obtain results of sufficient accuracy.

The model operates in the following way:
1. Read flow value Q of the day regarded
2. Compute scour depth y_s^* corresponding to Q (Eq.1).
3. Compare y_s^* with scour depth from previous day, y_a
   a) if $y_s^* \geq y_a$: scouring occurs, continue with 5.
   b) if $y_s^* \leq y_a$: aggradation occurs, continue with 7.
4. Compute relative state of scour: $s = y_a/y_s^*$
5. Obtain scouring rate (slope of curve) from Figure 1b.
6. Compute scour at end of current day according to relative temporal scour development (Fig.1b). Next day.
7. Compute scour at end of day using stage-discharge-curve (Fig.3) and aggradation rate (Fig.2). Next day

Example

As an example a bridge across the Danube near Mengen is regarded. The river is approximately 40 m wide, the opening of the bridge is 30 m so that the contraction on both sides is a=5 m. Figure 3 shows the stage-discharge relationship just upstream of the structure. From extreme value statistics the flow values $Q^*(T_N)$ in Table 1 were found for the given recurrence intervals T_N and the respective exceedance probabilities P_e. The table also shows the corresponding design scours $y_s^*(Q^*)$ according to Eq.(1). 5000 years of daily flows were generated and simulated with the model using various scenarios. The resulting sequence of scours was analyzed statistically.

The number of days with scour depths that exceeded the values of y_s^* of Table 1 were found by counting their occurrences. Due to stochastic variability the exceedance probability $P_{e,sim}$ found in the simulation was for curve A (scour reaches its maximum in one day) not exactly equal to P_e - as theoretically expected - but slightly different. Still curve A is the representation of the traditional design (excluding safety margin). Figure 4 shows the result of 5000 years of simulation for five different cases. The indices denote the method applied, the first refers to the scouring rate (Fig.1b), the second to the recovery rate (Fig.2). Slower scouring reduces P_f considerably, however, practically no influence was found if replacing the quick bed recovery curve by the slow recovery curve. For curve A, the quick and slow recovery rate must yield the same result anyway.

Table 1: Traditional design and results from simulations

T_N	7	10	20	50	100	[yrs.]
P_e	143	100	50	20	10	[°/oo]
$Q^*(T_N)$	213	241	273	321	356	[m³/s]
$y_s^*(Q^*)$	5.00	5.20	5.40	5.70	5.90	[m]
P_A	161	99	55	19	12	[°/oo]
$P_{B,Q}$	115	66	36	14	7	[°/oo]
$P_{C,Q}$	85	47	22	9	3	[°/oo]
$P_{B,S}$	120	60	39	15	8	[°/oo]
$P_{C,S}$	87	47	23	10	3	[°/oo]

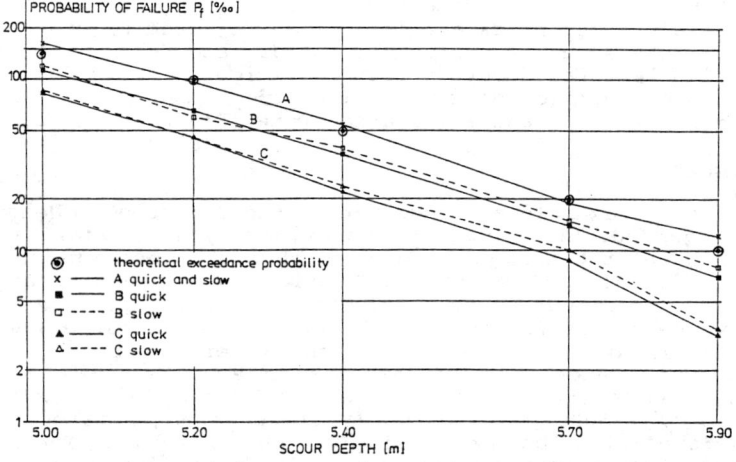

Figure 4. Probability of failure from different scenarios

References
US Department of Transportation / Federal Highway Administration (USDT/FHA), 1988: Interim procedures for evaluating scour at bridges. Technical Advisory
Kron, W., Plate, E.J., Ihringer, J. 1990: A model for the generation of simultaneous daily discharges of two rivers at their point of confluence. Stochastic Hydrology and Hydraulics 4(1990), 255-276

Uncertainty of Bridge Scour Estimates
Peggy A. Johnson[1], Member, ASCE, and Bilal M. Ayyub[2], Member, ASCE

Abstract

Pier scour is the erosion of a streambed in the vicinity of bridge pier foundations. In this paper, fuzzy regression is used to develop two pier scour models, one based on crisp laboratory data, and one based on field data. In either case, fuzzy regression provides models in which the regression coefficients are fuzzy parameters, thus producing fuzzy output. Field data is used to test and compare the equations. The models provides a range of scour estimates for any given set of input data, each with an associated degree of belief.

Introduction

There are nearly 400,000 bridges over waterways in the United States (Harrison and Morris 1991). At many of these bridges, erosion of the channel bed has developed around the pier foundations over the years. As a result, a high percentage of bridge failures in recent years have been attributed to scour. Pier scour, the erosion of the streambed in the vicinity of pier foundations, may eventually undermine the pier foundations and cause the bridge to become unstable.

Most scour equations and models are based on laboratory data and are valid only for steady-state flow and noncohesive channel beds of infinite depth. They do not account for many variables that are typically encountered in a field setting. Although the derived equations work quite well for the laboratory setting, the use of these equations in the field is questionable because of simplified laboratory conditions, distorted scales due to sediment size, minimal

---

[1]Asst. Prof., Civil Engrg. Dept., University of Maryland, College Park, MD 20742

[2]Assoc. Prof., Civil Engrg. Dept., University of Maryland, College Park, MD 20742

ranges of data, and use of ratios. Therefore, the models contain considerable uncertainty at the field scale.

In this study, fuzzy regression is used to develop a scour models based on crisp laboratory and field data in which the regression coefficients are fuzzy parameters, thus producing fuzzy output. Fuzzy regression is used as a way of incorporating uncertainty into a pier scour model. Fuzzy regression provides fuzzy parameters and fuzzy output so that the engineer may use the results to determine the most acceptable estimate of scour depth.

Linear Fuzzy Regression

Fuzzy regression is a method of calibrating fuzzy numerical coefficients in a linear equation. This method was developed by Tanaka et al. (1982) and has been used primarily for cases in which too few data points were available for a standard statistical regression. In fuzzy regression, the regression parameters are fuzzy numbers and so describe the degree of acceptance of values of a parameter. Since the regression parameters are fuzzy numbers, the dependent variable is also a fuzzy number (Bardossy et al. 1990). The objective of fuzzy regression is to minimize the vagueness of the dependent variable y (Bardossy 1990; Bardossy et al. 1990). A fuzzy linear regression results in the following model:

$$Y^* = A_1^* X_1 + A_2^* X_2 + ... + A_n^* X_n \qquad (1)$$

where * denotes a fuzzy number, A_k^* is a fuzzy parameter consisting of the ordered pair (α_k, c_k), α_k is the center of fuzzy parameter A_k^*, and c_k is the width or tolerance of the coefficient, i.e., fuzziness of the parameter (Heshmaty and Kandel 1985), for k = 1, 2, ..., n. A central assumption of fuzzy linear regression is that the residuals (i.e., the difference between the predicted and observed values) are due to fuzziness of the system parameters.

The objective of a linear fuzzy regression can best be met by minimizing the sum of the widths of the fuzzy regression coefficients, i.e., min Σc_k. A degree of fitting or level of credibility, h, must also be established, corresponding to R^2 in statistical regression. The degree of fitting can be thought of as a threshold value for strength of acceptance. Bardossy et al. (1990) recommend 0.5 < h < 0.7. If h < 0.5, then all predicted variables are very imprecise or fuzzy. If h > 0.7, then the predicted values of y become crisp. Bardossy et al. (1990) showed that fuzzy regression analysis can be reduced to a linear programming problem as follows:

minimize:

$$J = \sum_{k=1}^{n} c_k \qquad (2)$$

subject to:

$$(1-h)\sum_{k=1}^{n} c_k |x_{ik}| + \sum_{k=1}^{n} \alpha_k x_{ik} \geq y_i$$
$$(1-h)\sum_{k=1}^{n} c_k |x_{ik}| - \sum_{k=1}^{n} \alpha_k x_{ik} \geq -y_i \qquad (3)$$

for all i, where n = number of variables, i = sample size, and $c_k \geq 0$. As an example, a data set with three variables and a sample size of 10 will require a set of 20 constraints, according to Eq. 3. In Eq. 3, all independent variables, x, are nonfuzzy; however, Eq. 3 may be generalized for fuzzy input variables.

Fuzzy Regression of Pier Scour

Laboratory data from the University of Auckland was used to develop a linear fuzzy regression equation for estimation of scour depth. The small-scale data consisted of flow depth, flow velocity, pier diameter, mean sediment size, sediment gradation, and scour depth measurements. The data were arranged in dimensionless ratios, according to convention, to avoid problems with systems of units and to incorporate scaling into the equation. The linear fuzzy regression with h = 0.7 resulted in the following set of equations:

$$\frac{D_s}{y} = 0.0802 + 1.264\left(\frac{b}{y}\right) + 0.285 G \qquad (4)$$

and

$$\frac{D_s}{y} = 0.0802 + 1.264\left(\frac{b}{y}\right) - 0.285 G \qquad (5)$$

where D_s = scour depth and y = upstream flow depth, b = pier width, and G = sediment gradation. Equations 4 and 5 represent upper and lower limits on the fuzzy regression model that correspond to h = 0.7. From Eq. 5, it can readily be seen that for small values of b/y, the range of possible values of D_s/y includes negative values caused by the linear form of the model. Clearly, negative values are not physically possible. Since b/y is commonly in this lower range, the negative values must be truncated so that negative values are not possible. In addition, although A_1^* and A_2^* are fuzzy numbers, $c_1 = c_2 = 0$. This means that these two parameters are essentially crisp and only A_3^* is a fuzzy number. Therefore, all fuzziness associated with predicting scour depth using this linear model is due to sediment gradation.

Field data from Davoren (1985) is used to demonstrate the results of Eqs. 4 and 5. In this example, b = 1.5 m, y = 1.3 m, G = 5.3, and the measured scour depth = 0.9 m. Using Eqs. 4 and 5 resulted in estimated scour depths ranging from 0.04 m to 3.97 m. The observed scour depth falls in the range of estimated depths. The maximum scour depth of 3.97 m provides an upper limit for a conservative estimate of scour for design purposes.

A second set of equations was developed using field data taken from Froelich (1988). The data all represented live-bed scour and only those data that were representative of similar conditions to those of the laboratory experiments (e.g., round-nosed piers with a zero angle of approach) were used to develop the equations. The sample size of 14 included the flow depth and velocity, sediment size, sediment gradation, pier width, and scour depth. Clearly, such a small sample size is inadequate for a traditional regression. Fuzzy regression incorporates the uncertainty associated with using a small sample size into the model parameters. Fuzzy regression of the field data resulted in the following equations:

$$D_s = 0.44b + 0.19y + 0.04d_{50} + 0.05G \qquad (6)$$

and

$$D_s = 0.44b - 0.01y - 0.04d_{50} - 0.05G \qquad (7)$$

where d_{50} is the mean sediment size in mm. Sediment size was included because the sediment size in the field is highly variable, where in the laboratory, there is little variation in sediment size. Ratios were not used here since scaling is unnecessary for prototype equations.

Field data from Davoren (1985) were used to validate Eqs. 4 - 7. The scour depth for this example was 0.4 m, with b = 1.5 m, y = 0.6 m, d_{50} = 20 mm, and G = 5.3. Equations 4 and 5 provide a range of scour depths from 1.04 to 2.85 m and Eqs. 6 and 7 yield a range of 0 to 1.84 m. The laboratory-based equations yield a range with a minimum scour depth that is greater than the observed depth. In comparison, the field-based equations provided a range that included the observed depth. In general, the laboratory-based equations will yield deeper scour depths than the field-based equations. This is expected since the laboratory experiments typically represent worst-case conditions, rather than conditions typical of those found in the field. A comparison of the two sets of equations provides insight into the large bias that is embedded in laboratory-based equations developed from traditional regression techniques.

Conclusions

A model for estimating pier scour has been developed using linear fuzzy regression. The model provides a range of scour estimates for any given set of input data. This type of model is advantageous where there is considerable uncertainty in the model parameters. In the first case presented in this paper, the pier scour model was developed from small-scale laboratory data; extension to the field case is uncertain. In the second case, a model was developed from a small sample of field data. A comparison of the two models demonstrates that the laboratory-based model yields scour depths much greater than those that might actually occur. This is a common problem with laboratory-based equations. The field-based equations, however, provide a range that offer the engineer reasonable guidance in determining scour depths that may occur around a pier.

Clearly, Eqs. 6 and 7 do not include many of the parameters that effect pier scour, such as pier shape, angle of streamflow attack, and soil cohesiveness. These factors can be incorporated into the equations using fuzzy regression or by developing fuzzy correction factors.

References

Bardossy, A. (1990). Note on fuzzy regression. Fuzzy Sets and Systems 37, 65-75.

Bardossy, A., Bogardi, I., and Duckstein, L. (1990). Fuzzy regression in hydrology. Water Res. Res., 26(7), 1497-1508.

Davoren, A. (1985). Local Scour Around a Cylindrical Bridge Pier. Publication No. 3, Hydrology Centre, Christchurch, New Zealand.

Froelich, D.C. (1988). Analysis of onsite measurements of scour at piers. Proc. 1988 National Hyd. Eng. Conf., ASCE, 534-539.

Harrison, L.J., and Morris, J.L. (1991). Bridge scour vulnerability assessment. Proc. 1991 National on Hyd. Eng. Conf., ASCE, Nashville, Tenn., 209-214.

Heshmaty, B., and Kandel, A. (1985). Fuzzy linear regression and its applications to forecasting in uncertain environment. Fuzzy Sets and Systems, 15, 159-191.

Tanaka, H., Uejima, S., and Asai, K. (1982). Linear regression analysis with fuzzy model, IEEE Trans. Systems Man Cybernet, 12(6), 903-907.

Reliability Analysis of Levee System of a River

Kazumasa MIZUMURA
Dept. of Civil Engineering, Kanazawa Institute of Technology,
7-1, Ogigaoka, Nonoichimachi, Ishikawa Pref. 921, Japan

Abstract

The levee height is determined by the summation between computed water level and an assumed freeboard. The water level is computed by given discharge in the steady condition. Therefore, the water level is the function of the discharge, a rating curve, channel conditions such as cross-sectional data and bottom slopes, and friction coefficients. The unsteady condition also must be considered during flood. Even if the accurate discharge is selected based on the past data and the probability law, the other conditions are not appropriate and the computed water level is not reliable. This study computes the reliability of the unsteady flow for given channel conditions and friction coefficients without a rating curve. The method for the unsteady computation without a rating curve is developed. This shows the water level always has reliability or risk based on the used assumptions during the computation.

Introduction

The upstream discharge is accurately evaluated using past rainfall data and its probability law. But in comparison with the upstream discharge, channel data, friction coefficients, and a rating curve are not strictly discussed. Although the channel conditions, friction coefficients, and a rating curve play more important roles in flood routing computation than the upstream discharge, studies on risk or reliability of river bank system including the effect of sedimentation, erosion, or vegetation are not found. In the field, the height of the river bank is determined by adding the computed water surface profile based on the steady condition and mean channel conditions and given freeboard. In this study, the level of river bank z is obtained by the summation of unsteady flow computation with mean channel conditions, mean friction coefficients, and given freeboard. By generating many

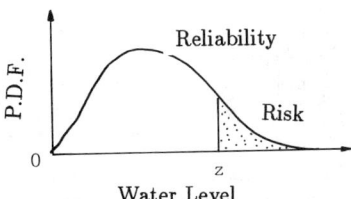

Figure 1: Distribution of water level

cross sections, bottom slopes, and friction coefficients based on the probability law and computing the result of flood routing, many water levels at each location are produced and their probability distribution is shown in Fig.1. Some of the simulated water levels are larger than z if the freeboard height is economically selected. The hatched part corresponds to the risk at this location for the given freeboard. Therefore, as long as there are sedimentation, erosion, or vegetation in a channel even if the latest data of channel conditions and friction coefficients were used, the economically selected height of freeboard of river bank always has risk. That is, the risk is not zero.

Governing Equations

the governing equations for gradually varied, unsteady flow in an open channel are the equation of continuity and the conservation of momentum equation. The former is

$$\frac{\partial A}{\partial t} + \frac{\partial Au}{\partial x} = 0 \tag{1}$$

in which $A=$ cross-sectional area; $t=$ time; $u=$ average velocity of flow; and $x=$ flow direction. The latter is

$$\frac{\partial u}{\partial t} + u\frac{\partial u}{\partial x} + g\frac{\partial h}{\partial x} = g(S_o - S_f) \tag{2}$$

in which $h=$ water depth; $g=$ the gravitational acceleration; $S_o=$ channel bottom slope in the flow direction; and $S_f=$ friction slope. Eqs.(1) and (2) are a set of simultaneous equations which can be solved for the two unknowns u and h, given appropriate initial and boundary conditions. To apply the method of characteristics, Eqs.(1) and (2) are transformed into the following form:

$$\frac{dx}{dt} = u \pm \sqrt{gA/B} \tag{3}$$

$$\frac{du}{dt} \pm \sqrt{gB/A}\,\frac{dh}{dt} = g(S_o - S_f) \mp \sqrt{gB/A}\,\frac{uh}{B}\frac{\partial B}{\partial x} \tag{4}$$

in which $B=$ top width of channel cross section. Referring left figure in Fig.2,

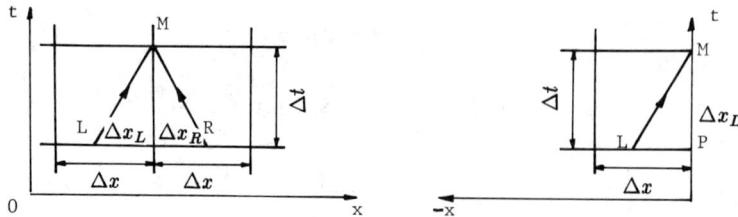

Figure 2: Fixed Grid System and Downstream Boundary

the above four equations are expressed in the finite difference form.

$$\frac{x_M - x_L}{\Delta t} = \frac{u_M + u_L}{2} + \frac{1}{2}\left[\sqrt{\frac{gA_M}{B_M}} + \sqrt{\frac{gA_L}{B_L}}\right] \quad (5)$$

$$\frac{u_M - u_L}{\Delta t} + \frac{1}{2}\left[\sqrt{\frac{gB_M}{A_M}} + \sqrt{\frac{gB_L}{A_L}}\right]\frac{h_M - h_L}{\Delta t} = g\left[\frac{S_{oM} + S_{oL}}{2} - \frac{S_{fM} + S_{fL}}{2}\right]$$
$$-\frac{1}{2}\left[\sqrt{\frac{g}{A_M B_M}} + \sqrt{\frac{g}{A_L B_L}}\right]\frac{u_M h_M + u_L h_L}{2}\frac{B_M - B_L}{\Delta x_L} \quad (6)$$

$$\frac{x_M - x_R}{\Delta t} = \frac{u_M + u_R}{2} - \frac{1}{2}\left[\sqrt{\frac{gA_M}{B_M}} + \sqrt{\frac{gA_R}{B_R}}\right] \quad (7)$$

$$\frac{u_M - u_R}{\Delta t} - \frac{1}{2}\left[\sqrt{\frac{gB_M}{A_M}} + \sqrt{\frac{gB_R}{A_R}}\right]\frac{h_M - h_R}{\Delta t} = g\left[\frac{S_{oM} + S_{oR}}{2} - \frac{S_{fM} + S_{fR}}{2}\right]$$
$$+\frac{1}{2}\left[\sqrt{\frac{g}{A_M B_M}} + \sqrt{\frac{g}{A_R B_R}}\right]\frac{u_M h_M + u_R h_R}{2}\frac{B_M - B_R}{\Delta x_R} \quad (8)$$

in which Δx_L and Δx_R show the distance between the point L and P and the point P and R, respectively. The subscripts M, L, and R in the above four equations are described in the left figure in Fig.2.

Initial and Boundary Conditions

The initial conditions (velocity and water depth) are obtained from steady flow computation which is only determined by the downstream water depth and given discharge. As the upstream boundary condition, the input discharge (hydrograph) is defined. Usually this is determined from the past rainfall and the probability law in a watershed. The water depth at the downstream boundary is calculated by using the principle of wave reflection[2]. This principle is derived

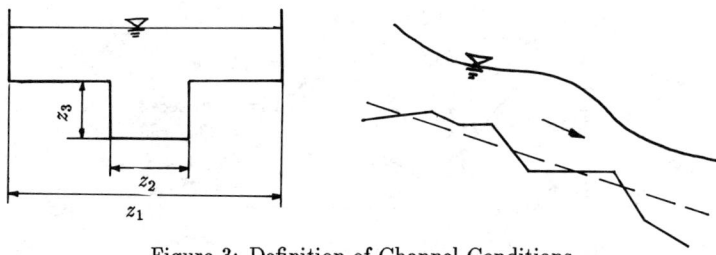

Figure 3: Definition of Channel Conditions

from the idea the wave height of perfectly reflected wave is twice as high as the incident wave height. This is true if the cross-sectional area is constant such as rectangular. If the initial flow velocity at the downstream boundary is zero, the water depth h_* there is obtained by solving the following equation:

$$A(h_*) = \frac{A(h'_M) - A(h_o)}{2} + A(h_o) \qquad (9)$$

in which $h'_M=$ computed water depth assuming $u_M=0$; $h_o=$ stationary water depth; and $A(\cdot)=$ cross-sectional area. If the initial flow velocity at the downstream boundary is not zero, the water depth h'_M computed by assuming $u_M=0$ is expressed by

$$h'_M = h_o + \Delta h_w + \Delta h_v \qquad (10)$$

in which Δh_w and $\Delta h_v=$ water level change induced by the existence of waves and the initial flow, respectively. The difference of the cross-sectional area between h_o and $h_o + \Delta h_v$ is defined by

$$\Delta A_v = A(h_o + \Delta h_v) - A(h_o) \qquad (11)$$

Therefore, if the boundary is open, the water depth h_M at the boundary is derived from

$$A(h_M) = \frac{A(h'_M) - A(h_o) - \Delta A_v}{2} + A(h_o) \qquad (12)$$

By using h_M in the above equation, u_M is numerically solved from Eqs.(5) and (6). The right figure in Fig.2 represents the characteristics to explain the downstream boundary.

Simulation Technique

To treat the reliability or risk of river levee system, we must consider that many factors influence the channel flow. They are cross sections, channel bottom slopes, friction coefficients, inflow hydrograph, rating curve, et al. The computational

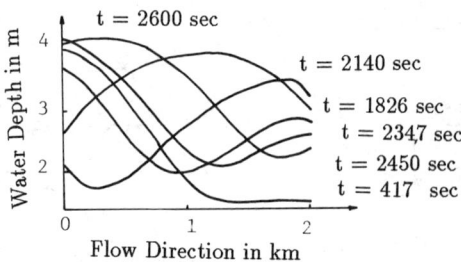

Figure 4: Flood Wave Propagation

formulation must be done including the above mentioned factors. These factors construct an uncertainty in the determination of water level or height of river bank. Since there is no enough information about these factors in this study, they except the inflow hydrograph and the rating curve are simulated by random numbers which are governed by some probability laws. According to Asano et al.[1], shapes of cross sections are simulated by random numbers which are normally distributed. The shape of the cross section and the bottom shape are roughly sketched in Fig.3. The position of the bottoms of each cross section are also given by random numbers as shown in Fig.3. For the friction law we adopt the Manning's formula. The water level z at given location is obtained for average cross sections, bottom slopes, and Manning's roughness coefficients. But when the dimensions of cross sections, locations of flow bottoms, and Manning's roughness coefficients are probabilistic, the resultant computed water depth at given location is also probabilistic as shown in Fig.1. Since many numbers of open channels are simulated and the water depth is larger than z is defined by the ratio of the number in which the water depth is larger than z to the total number of trials.

Illustrative Examples

The water depth at the downstream boundary, mean Manning's roughness coefficient, and the mean bottom slope are 1.5m, 0.08, and 0.01, respectively. The dimensions of the mean cross section z_1, z_2, and z_3 are 30m, 10m, and 0.5m, respectively. The input hydrograph is assumed to be

$$Q = Q_o + 3Q_o \left(1 - \cos \frac{2\pi t}{1000}\right) \tag{13}$$

in which Q_o= the steady flow discharge. The total reach of this simulation is 2000m. The flood routing is done by using the mean channel conditions and the

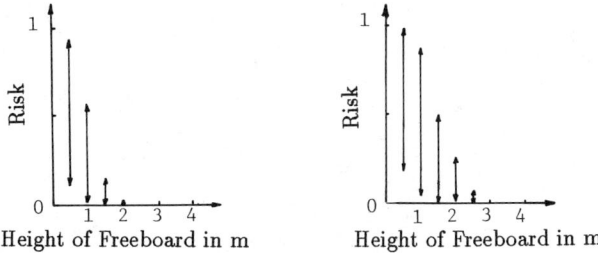

Figure 5: Range of Risk along Channel

result is shown in Fig.4. Left figure in Fig.5 describes the range of risk along the channel for given river-bank freeboard when the probability law of the channel conditions such as cross section, bottom slope, Manning's roughness coefficient are normally distributed and the standard deviation is 10% of each mean. The right figure in Fig.5 gives the same result as the left one when the standard deviation is 12.5% of each mean.

Concluding Remarks

Through this study the followings are obtained:

1. The rating curve is not necessary in the unsteady computation. It is calculated when the condition of wave reflection is applied to open channel and method of characteristics.

2. The channel conditions such as cross sections, bottom slopes, Manning's roughness coefficients are very important for designing river bank system. The changes of the channel conditions in time and space must be considered as the reliability or risk for designing.

References

[1] Asano, T., Suga, K., and Yamaguchi, T., "Stochastic and River Morphologic Prediction of Local Scour," *Proc. of IAHR 4th Conf. on Stochastic Hydraulics*, University of Illinois, Illinois, U.S.A., 1984, pp.127-138.

[2] Hino, M., "A very simple scheme of non-reflection and complete transmission condition of waves for open boundaries," *Technical Report*, Tokyo Inst. of Tech., No.38, 1987, pp.31-38 (in Japanese).

Uncertainty Analysis of the FEMA Method for Alluvial Fans

by Bing Zhao[1] and Larry W. Mays[2] Member ASCE

Abstract   Distributary flow areas such as alluvial fans, bajadas, and alluvial slopes, and alluvial plains along mountain bases pose quite different and interesting problems for the design of hydraulic structures and highway crossings. The FEMA method is only at best an approximate method that ignores fundamental hydraulics. The hydrologic and hydraulic analysis of alluvial fans has many associated uncertainties that are ignored in the FEMA method. This paper explores some of those uncertainties and presents methodologies for their analysis.

Introduction

An alluvial fan is (1) a singular depositional unit on a piedmont; (2) fan shaped; (3) a segment of a flattened cone; (4) exists where a single trunk stream leaves a source area; (5) radiates downslope from this apex; and (6) is characterized by distributary channels. The FEMA (1985) method is based upon determining the fan width (FW) for various depths using the relation

$$FW = \frac{9.5 \, ACP}{0.01} \tag{1}$$

where A is an avulsion coefficient which is a measure of the "average avulsions per event", P is a probability and C is defined by Eq. A.11 in the appendix.

Equation (1) assumes that simple hydraulic regime equations of the fan for channel width $W = 9.5Q^{0.4}$ and that a channel stabilizes

---

[1] Graduate Research Assistant, Department of Civil Engineering, Arizona State University, Tempe, AZ 85287
[2] Chair and Professor, Department of Civil Engineering, Arizona State University, Tempe, AZ 85287

approximately at a point where dD/dW = -0.005 where D is the channel depth [which was based upon personal communications with someone in the U.S. Army Corps of Engineers, see Dawdy(1979)]. If one were to use Manning's equation, the fan width could be expressed as

$$FW = 1.185 \left(-\frac{dD}{dW}\right)^{-5/8} n^{3/8} S_f^{-3/16} \frac{ACP}{0.01} \qquad (2)$$

where n is Manning's roughness coefficient and S_f is the slope and C is defined by Eq. (A.12).

Probabilistic Assumption of Method

Equations (1 and 2) are based upon the assumption that a channel caused by a given flood has an equal probability to cross at any point on a contour and that $\int_Q^\infty P(Q) f(Q) dQ = 0.01$ where P(Q) is the probability that a channel crosses at any point on a contour with discharge Q and f(Q) is the probability density function of Q. In conventional floodplain analysis a 100 year discharge, Q_{100} is used where as the FEMA (1985) method assumes that equation (2) is valid and Q_{100} is not considered. According to McGinn (1979), "the regular geometry of the alluvial fan and the numerous morphological and hydrological relationships established in the literature . . . absolutely refute any assumption pertaining to a random development model for alluvial fans. Consequently the fan generating floods and flood zones are not random in location, although they may be erratic and somewhat unpredictable." A study of the U. S. Geological Survey (Burkham, 1988) done in co-operation with the U. S. Bureau of Land Management recommends that the FEMA method not be used for the flood analysis of alluvial fans in the Great Basin. Kemna (1990) in Arizona concluded that of 39 distributary flow area sites (commonly referred to as alluvial fans), only eight sites satisfied the conditions of the FEMA active alluvial fan methodology.

Analysis of Uncertainties

First-order analysis of uncertainties can be used to estimate the uncertainty in a deterministic model formulation involving parameters which are uncertain. This method called the first order second moment (F.O.S.M.) enables one to estimate the mean and variance of a random variable which is functionally related to several variables, some of which are random. By using the F.O.S.M. the combined effect of uncertainty in a model formulation, as well as the use of uncertain parameters, can be assessed. Consider a random variable Y which is a function of several random variables and can be expressed as $Y=g(x)$ where $x=(x_1, x_2, \ldots, x_k)$ is a vector of k random variables. Through a Taylor series expansion

$$Y = g(\bar{x}) + \sum_{i=1}^{k} \left[\frac{\partial g(\bar{x})}{\partial x_i}\right] (x_i - \bar{x}_i) \qquad (3)$$

where $\bar{x}$ is the mean and the standard deviation is determined using

$$\sigma_y^2 = \text{Var}[Y] = \sum_{i=1}^{k} a_i^2 \sigma_i^2 + 2\sum_{i}^{k}\sum_{j}^{k} a_i a_j \text{Cov}[x_i x_j] \qquad (4)$$

where σ_i^2 is the variance corresponding to random variable x_i and a are the coefficients in the determinants equation. If the variables x_i are uncorrelated then $\text{Cov}[x_i x_j]=0$. The coefficient of variation of the fan width Ω_{FW} for equation (1) can be computed using

$$\Omega_{FW} = \left[\Omega_A^2 + \Omega_C^2 + \Omega_P^2\right]^{1/2} \qquad (5)$$

or for equation (2)

$$\Omega_{FW} = \left[\frac{25}{64}\Omega_{dD/dW}^2 + \frac{9}{64}\Omega_n^2 + \frac{9}{256}\Omega_S^2 + \Omega_C^2 + \Omega_A^2 + \Omega_P^2\right]^{1/2} \qquad (6)$$

Example Application of Uncertainty Procedure

An example application is provided in Tables 1 and 2 assuming a log-Pearson III (LPIII) and log-normal distribution, respectively. The purpose of this exercise is to identify statistically where the major uncertainties of the method are and to quantify these uncertainties, if one accepts the validity of the FEMA method or the extension of the method using Eq. (2). It should be emphasized that the authors are not necessarily accepting the validity of the FEMA methodology, but are simply pointing out the magnitude of the excessive uncertainties and the impact of these uncertainties. There is probably an alluvial fan somewhere in the World where the FEMA method may be valid.

As shown in Tables 1 and 2 the mean values of the fan width are small compared to the very large standard deviation of the fan widths for all three cases illustrate the major uncertainties associated with the methodology. The largest proportion of the uncertainty is in determination of C as shown in Table 1 (91.2% to 97.8% of the uncertainty in FW is due to C). However the uncertainty associated with C is much less for the log-normal distribution. (14.3% to 22.5%). The major uncertainty for FW in the FEMA method using the log-normal distribution is due to the avulsion coefficient (83.7% and 75.8%). Using Manning's Equation the major uncertainty is in the determination of the probability. (see Table 2).

References

Burkham, D.E., Methods for Delineating Flood-Prone Areas in the Great Basin of Nevada and Adjacent Sites, Water Supply Paper 2316, U. S. Geological Survey, 1988.
Dawdy, D. R., Flood Frequency Estimates on Alluvial Fans, ASCE, Journal of the Hydraulics Division, Vol. 107, No. HY3, pp. 1407-1413, November 1979.
FEMA, Flood Insurance Study - Guidelines and Specifications for Study Contractors: Washington, D.C. 1985.
Kemna, S. P., Some Geomorphic Models of Flood Hazards on Distributary Flow Areas in Southern Arizona, M.S. Thesis, Department of Hydrology and Water Resources, University of Arizona, Tucson, Arizona, 1990.
McGinn, R. A., discussion of "Flood Frequency Estimates on Alluvial Fans," by D. R. Dawdy, ASCE, Journal of the Hydraulics Division, Vol. 106, No. HY 10, October 1980.

Appendix A - Equations for FEMA (1985) Method

Flood discharges $Q_{0.5}$, $Q_{0.10}$, and $Q_{0.01}$ with return periods of 2, 10, and 100, respectively, years are determined by an appropriate rainfall-runoff model. Then the skew coefficient, G, standard deviation, S, and mean $\bar{X}$ are estimated using

$$G = -2.50 + 3.12 \frac{\log(Q_{0.01}/Q_{0.10})}{\log(Q_{0.10}/Q_{0.50})} \tag{A.1}$$

$$S = \frac{\log(Q_{0.01}/Q_{0.50})}{k_{0.01} - k_{0.50}}, \quad \bar{X} = \log(Q_{0.50}) - k_{0.50} S \tag{A.2) - (A.3}$$

where $k_{0.01}$ and $k_{0.50}$ are log-Pearson Type III distribution deviates for exceedance probabilities of 0.01 and 0.50, respectively. Equation A.1 is an approximation when $-2.0 \leq G \leq 2.5$. The FEMA methodology requires that the log-Pearson Type III variables be transformed. The following equations are used

$$m = \bar{X} - \frac{2S}{G}, \quad \lambda = \frac{4}{G^2}, \quad \alpha = \frac{2}{GS} \qquad (A.4) - (A.6)$$

The transformed distribution variables (for $G \neq 0$) are

$$\bar{z} = m + \frac{\lambda}{(\alpha - .92)}, \quad S_z^2 = \frac{\lambda}{(\alpha - .92)^2} \qquad (A.7) - (A.8)$$

$$G_z = \frac{2}{\sqrt{\lambda}}, \quad k_z = \frac{\log Q - \bar{z}}{S_z} \qquad (A.9) - (A.10)$$

The C for computing FW by equation (1)

$$C = \left(\frac{\alpha}{\alpha - .92}\right)^\lambda \exp(0.92m) \qquad (A.11)$$

and using equation (2) then

$$C = \left[\frac{\alpha}{\alpha - 0.8653}\right]^\lambda \exp(0.8635m) \qquad (A.12)$$

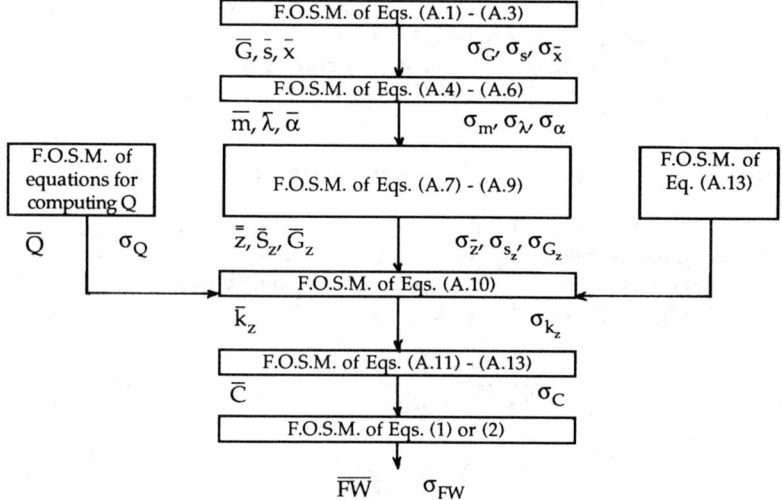

Figure 1 Methodology for Uncertainty Analysis Using First-Order Second Moment Method

FEMA METHOD FOR ALLUVIAL FANS   2103

Table 1 Example Analysis of Uncertainties (G≠0) Using LPIII (DH* = 0.5ft, V = 4ft/sec)

	Mean			STD		
A	= 1.5	-dD/dW = 0.005		STD(A)	= 0.2	STD(-dD/dW)=.003
$Q_{0.01}$	= 10,000 cfs	n = 0.025		STD($Q_{0.01}$)	= 100 cfs	STD(n)=0.01
$Q_{0.1}$	= 1,000 cfs	S_f = 0.03		STD($Q_{0.1}$)	= 50cfs	STD(Sf)=0.0041
$Q_{0.5}$	= 100 cfs			STD($Q_{0.5}$)	= 10cfs	

	FEMA	FEMA	Mannings Eq.
	$Q = 280 D_H^{2.5}$	$Q = 0.13 V^5$	$Q = 0.5038 \left(-\dfrac{dD}{dW}\right)^{-1} n^3 S^{-3/2} V^4$
$\overline{FW}$	10538 ft	8452 ft	13614 ft
σ_{FW}	18809	15563	23440
Ω^2_{FW}	3.1857	3.3908	2.9643
		Percentages of Uncertainty	
$-\dfrac{dD}{dW}$			0.52%
n			0.75
S_f			0.021
A	0.55%	0.52%	0.59
C	97.9	91.9	91.2
P	1.56	7.51	6.88
Percent	100%	100%	100%

*DH is the total head (depth plus velocity head).

Table 2 Example Analysis of Uncertainties (G=0) Using log-normal (DH* = 0.5ft, V=4ft/sec, means of Q0.01 and Q0.1 = 7887, 1130 cfs)

	$Q = 280 D_H^{2.5}$	$Q = 0.13 V^5$	$Q = 0.5038 \left(-\dfrac{dD}{dW}\right)^{-1} n^3 S^{-3/2} V^4$
$\overline{FW}$	10338 ft	8619 ft	13587 ft
σ_{FW}	1507	1319	4576
Ω^2_{FW}	0.02125	0.02344	0.1134
		Percentages of Uncertainty	
$-\dfrac{dD}{dW}$			13.77%
n			19.83
S_f			0.57
A	83.63%	75.83%	15.67
C	14.26	12.93	2.25
P	2.10	11.23	47.89
Percent	100%	100%	100%

*DH is the total head (depth plus velocity head).

Risk and Uncertainty in Flood Damage Reduction Project Design

Ming T. Tseng[1], Earl E. Eiker[2], and Darryl W. Davis[3]

Abstract

The recent emphasis on partnership between the Federal government and local sponsors in the development of flood damage reduction projects has promoted considerable interest in the application of risk-based methods for project analysis. Project formulation by such an approach is generally considered to be more objective in establishing a good balance between cost and risk. This paper describes a risk and uncertainty method for project analysis.

Introduction

Flood control is one of the primary missions of the U.S. Army Corps of Engineers. Through its Civil Works program, the Corps carries out Congressional directives to plan, design and operate various flood damage reduction projects throughout the country. For the past 10 fiscal years (1983-92) flood damage prevented by the Corps of Engineers flood control projects have been estimated to average $13.5 billion annually.

In planning and designing flood damage reduction projects, the Corps requires information on discharge/frequency, stage/discharge, and stage/damage relationships. Such information is obtained from observed and measured data, or is estimated by various synthetic procedures and modeling techniques. The information is frequently based on short

---
[1]Chief, Hydrologic Eng. Sec., Hydraulics and Hydrology Br., Eng. Div., Civil Works Dir., U.S. Army Corps of Engrs., Wash., DC 20314-1000
[2]Chief, Hydraulics and Hydrology Branch, Eng. Div., Civil Works Directorate, U.S. Army Corps of Engineers, Washington, DC 20314-1000
[3]Director, Hydrologic Engineering Center, U.S. Army Corps of Engineers, Davis, CA 95616-4687

FLOOD DAMAGE REDUCTION PROJECT

records and small sample sizes, and subject to measurement errors and inherent limitations and assumptions associated with the analytical techniques employed. These estimated values are, to various degrees, imprecise or inaccurate and thus induce uncertainty in key variables and decision making parameters.

Risk-based analysis is a method of performing studies in which uncertainty in technical data is explicitly taken into account. With such analyses, trade-offs between alternatives, risk, and consequences are made highly visible and quantified. The overall effect of risk and uncertainty on project design and economic viability can be examined and conscious decisions made reflecting an explicit tradeoff between risk and costs.

Formulation and Design Objectives

Flood damage reduction projects are formulated to provide safe, efficient and effective protection to lives and properties in flood prone areas. Projects are formulated by analyzing flood plain damage potential, and damage prevention performance and cost for a range of project sizes and configurations. The plan selected is based on maximizing net economic benefits consistent with acceptable risk and functional performance.

The technical task is to balance risk of design exceedance with flood damage prevented, uncertainty of flood levels with design accommodations, and provide for safe and predictable performance. The task is made difficult because economics dictate that less than complete protection be accepted, risk of capacity exceedance is real and must be planned for because it may occur within the life of the project, and uncertainty in flood levels exists because of imperfect knowledge. Risk-based analysis is a methodology that enables risk issues and uncertainty in critical data and information to be explicitly included in project formulation.

Traditional Approach

The Corps current practice is to first develop the discharge/frequency data for the project by applying adopted Federal interagency guidelines (Bulletin #17B) when gaged data are available, and by rainfall-runoff models, such as HEC-1, when watershed modeling is appropriate. Uncertainty is considered by making an adjustment with expected probability to the frequency curve in order to correct for bias because of a short record length.

Based on the discharge frequency information, several levels of protection for the project are then selected for analysis. The next step is to perform water surface profile computations using, for example, HEC-2, along the study reaches for the selected levels of protection to develop stage/discharge data. When flow is complex or circumstances unusual, unsteady flow and/or two dimensional model computations are needed. Models are calibrated with observed high-water-marks, available rating curves at stream gage locations, and published guidelines. Uncertainty is sometimes considered by performing sensitivity analyses by evaluating the results of reasonable adjustments of model variables. The outcome of sensitivity analyses may result in adoption of model coefficients to ensure that computed water surface profiles are conservative.

The stage/damage curve provides a summary statement of damages as a function of river stage. Damages are highly sensitive to a variety of factors, such as mapping accuracy, first floor elevations, type of structures and contents which are important in describing the variation in damage but rarely empirically verified. Uncertainty is sometimes considered by performing sensitivity analyses.

The discharge/frequency, stage/discharge and stage/damage data are then combined to develop the damage/frequency curve which is used to determine the flood damage reduction benefits for each level of protection selected for evaluation. The selected project is the one that reasonably maximizes net flood damage reduction benefits. Only projects with acceptable performance are considered in the evaluation. Performance is usually characterized as a unique degree-of-protection for the selected project. With the traditional approach, uncertainty is generally considered by application of professional judgement and conducting sensitivity analysis. In the case of levee or floodwall projects in urban areas, uncertainty is accommodated by including freeboard in the final design to ensure performance for the design flood.

Risk and Uncertainty Approach

The risk-based approach has many similarities with present practice in that the basic data are the same. Best estimates are made of discharge/frequency curves, water surface profiles, and stage/damage relationships. The difference between current practice and the risk-based approach is that uncertainty in technical data is quantified and explicitly included in evaluating project performance and benefits. Using the risk-based approach, performance can be stated in terms of reliability of achieving stated goals. Also, adjustments or additions of features to accommodate uncertainty, such as adding freeboard for levee/flood walls, are not necessary.

Figure 1 is a conceptual schematic of the problem from a risk and uncertainty perspective. The hydrologic relationship that characterizes risk of flooding is depicted in the upper left corner of Figure 1. Uncertainty in corresponding peak discharge may be described by applying accepted statistical procedures for determining confidence limits, and is illustrated in the upper right corner of Figure 1. This uncertainty is represented by a probability distribution of discharge error about the discharge frequency curve.

Flood stage uncertainty corresponding to discharge is represented in the lower left corner of Figure 1. At a gaged location, study of

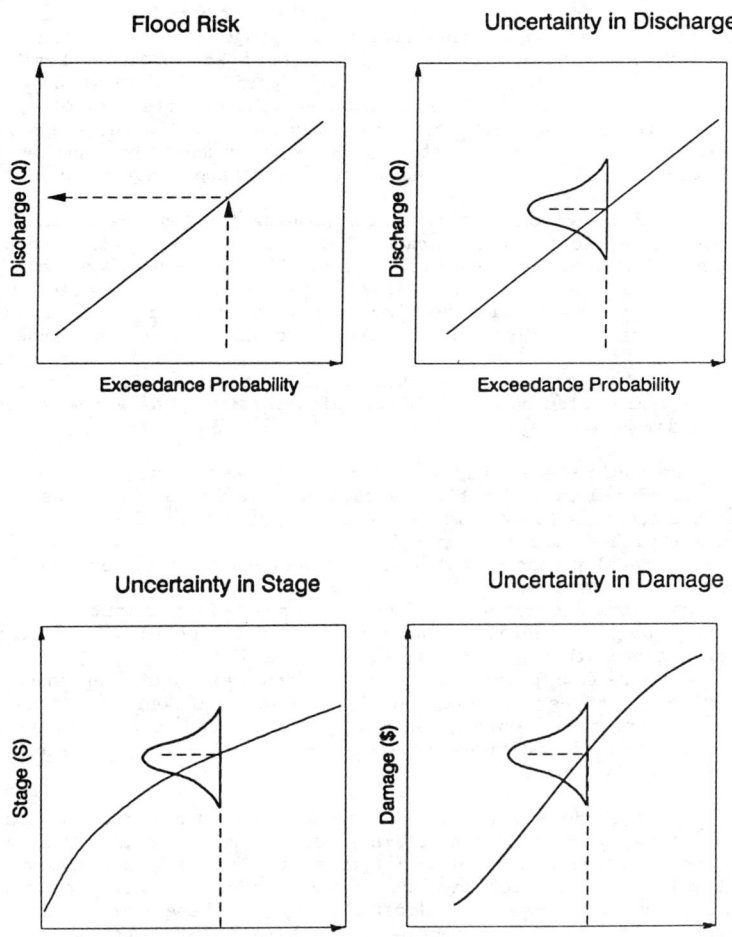

Figure 1. Uncertainty in Discharge, Stage, and Damage

field measurements compared to the adopted rating curve and stages computed with a calibrated numerical flow model can provide the basis for quantifying the uncertainty. For the case where there is no gage at a site and few high water marks recorded for flood events, study of nearby gage ratings, sensitivity of stage to calibrated model coefficients, and professional judgement must form the basis for quantifying uncertainty. Because of imperfect knowledge about the channel roughness, flow regime, bed form, flow debris and models used to analyze river hydraulics, there is uncertainty in the stage for a given discharge. This uncertainty can be represented by a probability distribution of stage error about stage/discharge rating curves.

Flood damage uncertainty corresponding to stage is reflected as shown in the lower right corner of Figure 1. A probability density function representing possible statistical error in damage estimate for stage is superimposed on the stage/damage function. This reflects that there is uncertainty in the flood damage that would result should a given stage occur in the flood plain because of imperfect knowledge about the nature and mix of improvements, elevation of improvements, and physical structure and content damage potential. This uncertainty can be represented by a probability distribution of damage error about stage/damage curves.

The basic steps to carry out the risk-based analysis are:
   a. Develop best estimates of discharge/frequency curves, water surface profiles (stage/discharge ratings), stage/damage relationships for the without project conditions,
   b. Develop statistical description of uncertainty for each of the above relationships,
   c. Nominate alternative project capacities; compute costs and flood damage prevented; array results and select plan according to economic and other appropriate criteria; and
   d. Make appropriate refinements to ensure project performance and function; such as providing design features to control overtopping location and management of subsequent flooding for levee projects, and operational accommodations required for reservoir storage, channel, and diversion projects.

Above steps are repeated as needed for each alternate measure evaluation, or combinations of measures to enable comparison of major project alternatives. Step c brings together all the elements to determine the selected project capacity. To correctly incorporate uncertainty in the several elements, they must be allowed to interact with one another. For example, the possibility of error for higher flows (or lower flows) must be allowed to couple with the full range of possible stage and damage errors. Because of the nature and complexity of the error distributions, the interaction cannot be uniquely accomplished analytically. An alternative approach is to use Monte Carlo simulation as given by Benjamin (1970) and Palisade Corporation (1988). In this approach, the basic relationships and error distributions are sampled by exhaustive trial to allow the interactions to take place. For a given size project, various combinations of the parameters are evaluated (approximately 5,000 samples) and for each interaction success or failure is established. Other project sizes are evaluated, and a matrix describing economic outputs, reliability and performance for each is produced. The matrix forms the basis for final selection of project size.

The results of the analyses are probability distributions of the various parameters (design flow, stage, and residual damage) as a function of project capacity. The expected cost and benefit can then be computed and the project capacity selected according the appropriate criteria. Tabulations of the likelihood of project capacity exceedance for flood events will enable characterizing risk-exceedance and performance. The risk-based analysis framework will quantify the reliability and performance of project design. This reliability and performance will be reported as the protection for a target percent chance exceedance flood with a specified reliability. For example, the proposed levee project is expected to contain the one-half percent (0.5%) chance exceedance flood, should it occur, with a ninety percent (90%) reliability. This performance may also be described in terms of the percent chance of containing a specific historic flood.

Summary and Conclusion

Imperfect knowledge of the "true" nature of the hydrology and hydraulics in an area creates uncertainty in project designs and in the estimate of their reliability. Additionally, uncertainties in expected damage with and without the project influence the selection of an alternative plan for design. The risk-based analysis procedures described in this paper provide a new approach to explicitly quantify the uncertainties associated with discharge/frequency, stage/discharge and stage/damage relationships that are required in the formulation of flood damage reduction projects. The method uses the same basic data as that used in present practice, but has the distinct advantage of providing considerable information regarding expected project reliability and performance. Goals and objectives of project studies are not compromised by the new method, rather they are enhanced due to the ability to consider a much wider range of project alternatives.

References

Benjamin, J. R., and Cornell, C.A. (1970). Probability, statistics, and decisions for civil engineers. McGraw-Hill, New York, N.Y.

Palisade Corporation (1988). @RISK, Risk analysis and simulation add-in for Lotus 1-2-3 user's guide.

U.S. Army Corps of Engineers (1991). "Proceedings of hydrology and hydraulics workshop on riverine levee freeboard."

U.S. Department of the Interior, Geological Survey (1982). "Guidelines for determining flood flow frequency." Bulletin #17B.

Three-Dimensional Variable Resolution Hydrodynamic and Transport Modeling of the Chesapeake Bay System

J. M. Hamrick[1]

Abstract

This paper describes a domain decomposition methodology used in the application of the three-dimensional environmental fluid dynamics code, EFDC, to model hydrodynamics and transport in the Chesapeake Bay estuarine system. The model application encompasses a hierarchy of spatial grid resolutions in the main stem of the Chesapeake Bay, the adjacent continental shelf and the major sub estuary tributaries. The rationale for this approach is to provide a coherent modeling framework for analyses and predictions ranging from small scale localized phenomena in the tributaries to large scale phenomena characterized by significant interactions between the shelf, the main stem of the bay and major tributaries.

Introduction

Studies of circulation and transport in estuaries often require increased spatial resolution in limited areas of the modeling domain. Examples include the need for increased resolution in high gradient frontal regions and near major point source discharges. Other examples involve the coupling of an estuary to the continental shelf when the estuary region is of primary interest or the coupling of a tributary estuary to the main estuary stem when the tributary estuary is of primary interest. The use of variable horizontal grid spacing on either Cartesian or curvilinear grids is one approach to introducing variable resolution. However, to maintain numerical accuracy with this approach, a smooth and often computationally wasteful transition region between the two or more chosen scales of resolution is necessary. Grid nesting, where two or more grids of different resolutions are utilized, provides an alternate approach to this problem. Grid nesting approaches may be either one way or two way interacting. In a one way interaction, a model is executed on a coarse grid, which may include all or an overlap portion of the fine grid, to provide boundary conditions for a fine grid model. In a two way interaction, a

---

[1] Associate Professor, Virginia Institute of Marine Science, The College of William and Mary, Gloucester Point, VA 23062

model is executed simultaneously on both the coarse and fine grids with an overlap region used to transfer boundary conditions. The overlap region may encompass the entire fine grid. When the transport of dissolved and suspended materials is a primary modeling goal, conservation of mass over the entire system of grids is essential. Zhang *et al.* (1986) reviewed a number of two way interacting nested grid methodologies used in atmospheric modeling as to their conservation properties. Recent applications of nested grid coastal and ocean models are reported by Spall and Holland (1991) and Oey and Chen (1992).

This paper describes the implementation of a domain splitting or decomposition methodology (Meakin and Street, 1988; Gropp and Keyes, 1992) based on the work of Perng and Street (1991), which insures conservation of mass, to construct variable resolution models of the Chesapeake Bay estuarine system and adjacent continental shelf. A number of practical model applications are motivating this work. The first is the need to develop high resolution euthrophication models for the major Chesapeake Bay tributaries. For this application, fine grid tributary hydrodynamic and eutrophication models will be coupled to coarse grid hydrodynamic and eutrophication models of the main stem of the Chesapeake Bay. The second application involves the development of a physically based recruitment model for blue crab in the York River tributary of the Chesapeake Bay. This application will involve a very coarse grid model of the continental shelf, coupled with a coarse grid model of the main stem of the Chesapeake Bay which in turn is coupled with a fine grid model of the York River. Following a brief description of the EFDC hydrodynamic and transport model, the domain decomposition methodology will be presented. The grid structure for application to the Chesapeake Bay system will be described, followed by the summary and conclusions.

The Three-Dimensional Circulation and Transport Model

The work described in this paper utilizes the EFDC (environmental fluid dynamics computer code) model developed at the Virginia Institute of Marine Science, (Hamrick 1992a,b). The EFDC model solves the vertically hydrostatic, free surface, variable density, turbulent averaged equations of motion and transport equations for turbulent kinetic energy and macroscale, salinity and temperature in a stretched, (sigma), vertical coordinate system, and horizontal coordinate systems which may be Cartesian or curvilinear-orthogonal. The code uses a three time level, finite difference scheme with an internal-external mode splitting procedure to separate the internal shear or baroclinic mode from the external free surface gravity wave or barotropic mode. The external mode solution is implicit, and simultaneously computes the two-dimensional surface elevation field by a multicolor reduced system conjugate gradient scheme (Hageman and Young, 1981). The external solution is completed by the calculation of the depth averaged barotropic velocities using the new surface elevation field. The implicit external solution allows large time steps which are constrained only by the stability criteria of the explicit advection scheme used for the nonlinear accelerations.

The EFDC model's internal solution, at the same time step as the external, is implicit with respect to vertical diffusion. The internal solution of the momentum equations is in terms of the velocity shear, which results in the simplest and most accurate form of the baroclinic pressure gradients and eliminates the over determined character of alternate internal mode formulations. The vertical diffusion coefficients for momentum, mass and temperature are determined by the second moment closure scheme of Mellor and Yamada (Mellor and Yamada, 1982, and Galperin, et al, 1988) which involves the use of analytically determined stability functions and the solution of transport equations for the turbulent kinetic energy and the turbulent macroscale. Time splitting inherent in the three time level scheme is controlled by periodic insertion of a two time level step.

Other features of the code include additional transport equations for conservative and nonconservative materials and suspended sediment. Various options for advective transport of scalar variables include conventional upwind and central difference schemes and Smolarkiewicz's multidimensional positive definite advection transport algorithm (Smolarkiewicz and Clark, 1986), with an optional flux limiter (Smolarkiewicz and Grabowski, 1990). A complete description of the code's theoretical and computational aspects are presented in Hamrick (1992a).

Domain Decomposition Methodology for Grid Nesting

Domain decomposition methodologies offer a number of attractive features for implementing nested grid hydrodynamic and transport models. The two primary features of interest here are the ability of the nesting scheme to conserve mass and the formulation of a solution procedure which may be readily implemented in a coarsely parallel computing environment. For a coarsely parallel implementation, each grid sub-domain would be assigned to a separate processor, with inter processor communication required only for the overlap regions. Figure 1 shows a typical corner interface of a coarse grid to the north and east joining a fine grid to the southwest. To illustrate the procedure, consider the x component of the momentum equation and the continuity equation:

$$\partial_t(mHu) + \partial_x(m_yHuu) + \partial_y(m_xHvu) + \partial_z(mwu) - (mf + v\partial_x m_y - u\partial_y m_x)Hv \qquad (1)$$
$$= -m_yH\partial_x p - m_yHg\partial_x \zeta + m_yHgb\partial_x h - m_yHgbz\partial_x H + \partial_z(mH^{-1}A_v\partial_z u) + Q_u$$

$$\partial_t(m\zeta) + \partial_x(m_yHu) + \partial_y(m_xHv) + \partial_z(mw) = 0 \qquad (2)$$

in curvilinear-orthogonal coordinates with variables having their usual meanings.

The EFDC model treats the surface elevation gradient and the vertical shear stress gradient implicitly, while all other terms are treated explicitly. With this in mind, consider the application of (1) at point 'a' on the velocity boundary of the fine grid. The fine grid x direction momentum flux at point 'b', the y direction momentum fluxes at points 'd' and 'e', the z direction momentum flux at point 'a', and the Coriolis and curvature accelerations at point 'a' cannot be computed using only fine grid variables. Since points 'f' and 'g' are interior x direction momentum flux points on the coarse grid, they are used to interpolate the x direction momentum flux to point 'b' on the fine grid. Likewise, since points 'h' and 'i' are interior y direction momentum flux points on the coarse grid, they are used to interpolate the y direction momentum fluxes to points 'd' and 'e' on the fine grid. The coarse grid z direction momentum flux and the Coriolis and curvature accelerations are known at point 'a' and are used to compute the corresponding fine grid values at point 'a'. The remaining explicit terms on the right side of (1) involve gradients of the internal pressure associated variable density, and gradients of the mean and total depth evaluated at point 'a' on the fine grid. These terms are computed from corresponding know terms on the coarse gird at point 'a'.

The free surface elevation gradient on the right side of (1) is treated implicitly as are the horizontal volumetric flux divergence terms in (2). The approach used for the implicit solution of the depth integrated horizontal volumetric fluxes and free surface elevation follows Perng and Street (1991) and requires the simultaneous solution of the surface elevation on all grids in an iterative manner, which insures a continuous and smooth surface elevation field over the entire grid domain. For point 'a' on the fine grid, the latest interation for the surface elevations on the coarse gird at points 'f' and 'g' are used to provide a boundary value of the surface elevation on the fine grid surface elevation boundary point 'b' using:

$$\zeta_{b,fine}^{n+1} = \zeta_{c,fine}^{n+1} + \frac{m_{x,fine}^u}{m_{x,coarse}^u}(\zeta_{f,coarse}^{n+1} - \zeta_{g,coarse}^{n+1}) \tag{3}$$

The surface elevation solution iteration is then executed on the fine grid with the results the being available for the next coarse grid interation. A similar procedure is applied at the v equation boundary point 'j' on the fine grid and the u and v equation boundary points 'A' and 'J' on the coarse grid, shown in Figure 1.

A Variable Resolution Model of the Chesapeake Bay System

The above described methodology is implemented in the EFDC code to develop a variable resolution model of the Chesapeake Bay system and the adjacent continental shelf. A 2500 m grid resolution is utilized for the main stem of the Chesapeake Bay and the Potomac River tributary. The bay interfaces with a coarser 7500 m grid covering a protion of the continental shelf. Fine grids having resolutions of either 500 m or 833 m are used for high resolution in the James, York and Rappahannock River tributaries of the Chesapeake Bay. Results will be presented for the circulation and salinity distribution under mean summer river flow conditions. Results will also be presented for the distribution of blue crab larve released on the shelf near the mouth of the Chesapeake Bay under typical recruitment conditions. A discussion of computational benchmarks for the variable resolution model on a number of computing systems will also be presented.

Summary and Conclusion

A domain decomposition methodology for variable resolution hydrodynamic and transport modeling and the characteristics of a variable resolution model of the Chesapeake Bay system have been presented. The methodology has a number of attractive features including conservation of mass and a structure for implementation in a coarsely parallel computing environment. This work was supported by the Commonwealth of Virginia through the estuary and coastal ocean modeling initiative at the Virginia Institute of Marine Science.

References

Galperin, B., L. H. Kantha, S. Hassid, and A. Rosati, 1988: A quasi-equilibrium turbulent energy model for geophysical flows. *J. Atmos. Sci.*, 45, 55-62.

Gropp, W. D., and D. E. Keyes, 1992: Domain decomposition methods in computational fluid dynamics, *Int. J. Numer. Methods Fluids*, 14, 147-165.

Hageman, L. A., and D. M. Young, 1981: *Applied iterative methods*, Academic Press, 386 pp.

Hamrick, J. M., 1992a: A three-dimensional environmental fluid dynamics computer code: Theoretical and Computational Aspects. Virginia Institute of Marine Science, Special Report 317, 63 pp.

Hamrick, J. M., 1992b: Estuarine environmental impact assessment using a three-dimensional circulation and transport model, in *Estuarine and Coastal Modeling, Proceedings of the 2nd International Conference*, (M. L. Spaulding *et al*, eds.), American Society of Civil Engineers, 292-303.

Meakin, R. L., and R. L. Street, 1988: Simulation of environmental flow problems in geometrically complex domains, part 2: a domain splitting method, *Comput. Meths. Appl. Mech. Engrg.*, 68, 151-175.

Mellor, G. L., and T. Yamada, 1982: Development of a turbulence closure model for geophysical fluid problems, *Rev. Geophys. Space Phys.*, 20, 851-875.

Oey, L. Y., and P. Chen, 1992: A nested-grid ocean model: application to the simulation of meanders and eddies in the Norwegian coastal current, *J. Geophys. Res.*, 97, 20,063-20,086.

Perng, C. Y., and R. L. Street, 1991: A coupled multigrid-domain splitting technique for simulating incompressible flows in complex domains, *Int. J. Numer. Methods Fluids*, 13, 269-286.

Ryskin, G. and L. G. Leal, 1983: Orthogonal mapping, *J. Comp. Phys.*, **50**, 71-100.

Smolarkiewicz, P. K., and T. L. Clark, 1986: The multidimensional positive definite advection transport algorithm: further development and applications, *J. Comp. Phys.*, **67**, 396-438.

Smolarkiewicz, P. K., and W. W. Grabowski, 1990: The multidimensional positive definite advection transport algorithm: nonoscillatory option, *J. Comp. Phys.*, **86**, 355-375.

Spall, M. A., and W. R. Holland, 1991: A nested primative equation model for oceanic Applications, *J. Phys. Oceanogr.*, **21**, 205-220.

Zhang, D. L., et al., 1986: A two-way interactive nesting procedure with variable terrain resolution, *Mon. Wea. Rev.*, **114**, 1,330-1,339.

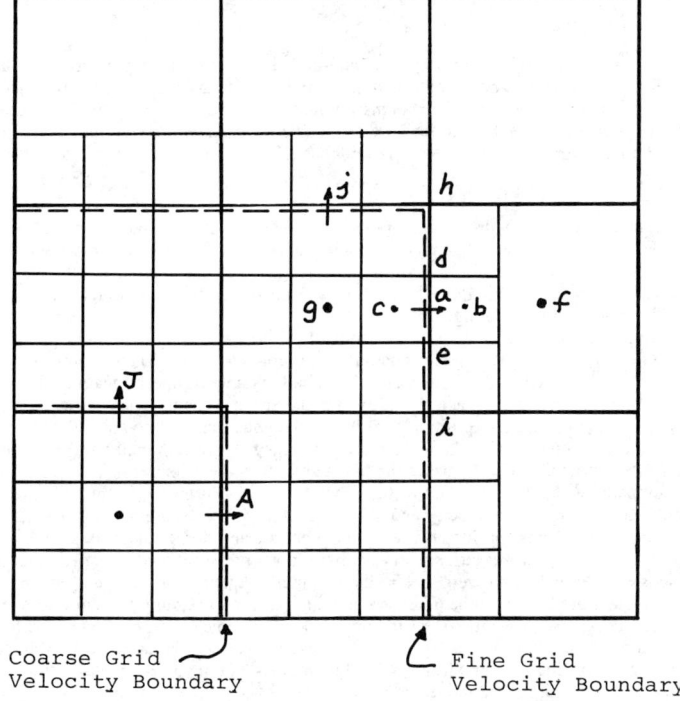

Figure 1. Domain decomposing grid nesting interface.

A Finite-Difference Model for 3-D Flow
in Bays and Estuaries

By Peter E. Smith[1], M. ASCE, and Bruce E. Larock[2], M. ASCE

Abstract

This paper describes a semi-implicit finite-difference model for the numerical solution of three-dimensional flow in bays and estuaries. The model treats the gravity wave and vertical diffusion terms in the governing equations implicitly, and other terms explicitly. The model achieves essentially second-order accurate and stable solutions in strongly nonlinear problems by using a three-time-level leapfrog-trapezoidal scheme for the time integration.

Introduction

Several recent 3-D hydrodynamic models for estuaries and coastal seas have been developed that treat the terms governing the propagation of external gravity waves implicitly (Backhaus, 1985; de Goede, 1991; Casulli and Cheng, 1992); the approach to the gravity wave terms, called the semi-implicit method, follows from the early work of Kwizak and Robert (1971) in atmospheric modeling.

In the semi-implicit method, the solution for surface elevation is uncoupled from the solution for velocity by inserting the equations of motion into the divergence terms of the depth-integrated continuity equation. This results in a single system of linear equations that is solved each time step for the new surface elevation over the entire domain. The coefficient matrix for the system is symmetric and positive-definite, so the equations can be solved efficiently with an iteration technique, such as the preconditioned conjugate gradient method. Velocities in the models are computed explicitly using the updated surface elevations.

Compared with a fully implicit method (without time-splitting) in which a coupled system of nonlinear equations for velocity and surface elevation is solved simultaneously, the semi-implicit method requires considerably less computer time and storage and is of similar accuracy. Because the gravity wave terms (i.e. the barotropic pressure gradient in the momentum equations and the velocity divergence in the continuity equation) are treated implicitly, semi-implicit models are free from the time step limitation due to the Courant-Friedrich-Lewy (CFL) criterion. For most problems, semi-implicit models are much more efficient than explicit models in which the time step must remain extremely small to maintain stability. In addition, because the semi-implicit gravity wave solution does not involve an alternating direction implicit (ADI) technique, the models are not affected by any time step limitation or inaccuracy related to the so-called ADI-effect (Stelling et al., 1986).

The semi-implicit method has been successfully implemented in 2-D models (Benqué et al., 1982; Backhaus, 1983; Duwe et al., 1983; Wilders et al., 1988; Casulli, 1990), and research has now shifted to its implementation in 3-D models. In a 3-D model the vertical diffusion

---

[1]Hydrologist, U.S. Geol. Survey, 2800 Cottage Way, Rm W-2233, Sacramento, CA 95825
[2]Professor, Dept. of Civil and Environmental Engineering, Univ. of California, Davis, CA 95616

terms must be treated implicitly to avoid a time step limitation for stability related to the magnitude of the vertical exchange coefficients and the water depth. Casulli and Cheng (1992) have presented a concise procedure to include implicit vertical diffusion into a 3-D semi-implicit model that does not use time- or mode-splitting procedures. Their approach requires that a system of tridiagonal equations be solved at each horizontal nodal location to yield a set of coefficients that become part of the system matrix for surface elevation. The tridiagonal systems are uncoupled and can be efficiently solved by a double-sweep method.

In this paper, a new 3-D finite-difference model is described that uses the semi-implicit method for treatment of gravity waves, and follows Casulli and Cheng (1992) for the implicit inclusion of vertical diffusion. The model achieves essentially second-order accurate and stable solutions on strongly nonlinear problems by using a three-time-level semi-implicit leapfrog-trapezoidal scheme for the time integration. The model equations are solved in a conservative form using a multi-level vertical mesh. To prevent spurious solution oscillations that can result from steep topographic or salinity gradients, the advection terms in the momentum and salinity equations are solved using a donor cell/leapfrog-trapezoidal flux-corrected transport (FCT) technique described by Zalesak (1979). Density-gradient forcing and a mixing-length turbulence model are included in the model so it can be used for realistic calculations in estuaries. The governing equations and numerical methods for the model are briefly summarized in the following two sections.

Governing equations

Using the layered system in Figure 1, the governing equations for the model are obtained by integrating the 3-D shallow water equations over each layer of depth h_k in the manner presented by Leendertse (1973). The resulting layer-averaged equations are:

Momentum equations:

$$\frac{\partial U}{\partial t} + \frac{\partial (Uu)}{\partial x} + \frac{\partial (Vu)}{\partial y} + (wu)\Big|_{k+\frac{1}{2}}^{k-\frac{1}{2}} - fV + \frac{h}{\rho}\frac{\partial p}{\partial x} = \frac{\partial}{\partial x}\left(A_x h \frac{\partial u}{\partial x}\right) + \frac{\partial}{\partial y}\left(A_y h \frac{\partial u}{\partial y}\right) + \left(\frac{1}{\rho}\tau_{xz}\right)\Big|_{k+\frac{1}{2}}^{k-\frac{1}{2}}. \quad (1)$$

$$\frac{\partial V}{\partial t} + \frac{\partial (Uv)}{\partial x} + \frac{\partial (Vv)}{\partial y} + (wv)\Big|_{k+\frac{1}{2}}^{k-\frac{1}{2}} + fU + \frac{h}{\rho}\frac{\partial p}{\partial y} = \frac{\partial}{\partial x}\left(A_x h \frac{\partial v}{\partial x}\right) + \frac{\partial}{\partial y}\left(A_y h \frac{\partial v}{\partial y}\right) + \left(\frac{1}{\rho}\tau_{yz}\right)\Big|_{k+\frac{1}{2}}^{k-\frac{1}{2}}. \quad (2)$$

$$\frac{\partial p}{\partial z} = -\rho g. \quad (3)$$

Continuity equations:

$$(w)\Big|_{k+\frac{1}{2}}^{k-\frac{1}{2}} = -\frac{\partial U}{\partial x} - \frac{\partial V}{\partial y} \quad (k = 2,3,....,b). \quad (4a)$$

$$\frac{\partial \zeta}{\partial t} - (w)\Big|_{3/2} = -\frac{\partial U}{\partial x} - \frac{\partial V}{\partial y} \quad (k = 1 \; only). \quad (4b)$$

Salt transport equation:

$$\frac{\partial (hS)}{\partial t} + \frac{\partial (uhS)}{\partial x} + \frac{\partial (vhS)}{\partial y} + (wS)\Big|_{k+\frac{1}{2}}^{k-\frac{1}{2}} = \frac{\partial}{\partial x}\left(D_x h \frac{\partial S}{\partial x}\right) + \frac{\partial}{\partial y}\left(D_y h \frac{\partial S}{\partial y}\right) + \left(D_z \frac{\partial S}{\partial z}\right)\Big|_{k+\frac{1}{2}}^{k-\frac{1}{2}}. \quad (5)$$

All the dependent variables represent layer averages for a layer k except when designated by $(\;)\Big|_{k-\frac{1}{2}}^{k+\frac{1}{2}}$, which implies a vertical difference between interface values for a layer. A z-derivative defined at an interface is understood to imply a finite-difference derivative involving the layer variables above and below the interface. The symbols are explained further in the Appendix. The interface shear stress terms τ_{xz} and τ_{yz} in the momentum equations are defined as

$$\tau_{xz} = A_z \frac{\partial (U/h)}{\partial z}, \quad \tau_{yz} = A_z \frac{\partial (V/h)}{\partial z}. \quad (6)$$

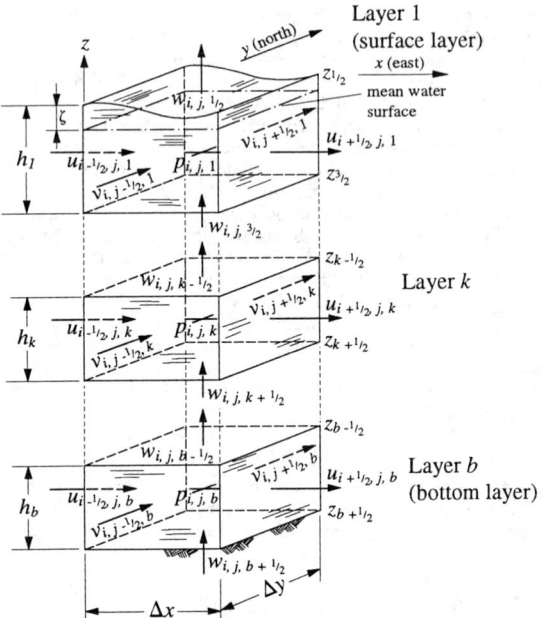

Figure 1. Multi-level staggered grid for the model

At the surface and bottom of the water column, the shear stress is equal to a wind and frictional stress, respectively. An equation of state also is needed to relate density to temperature (assumed constant) and salinity.

Integrating the continuity equation over the depth gives an equation for the free-surface elevation

$$\frac{\partial \zeta}{\partial t} + \frac{\partial \left(\sum_{m=1}^{b} U_m \right)}{\partial x} + \frac{\partial \left(\sum_{m=1}^{b} V_m \right)}{\partial y} = 0. \tag{7}$$

Using Eq. 3 for the hydrostatic pressure equation, the pressure gradient term in the x-momentum equation may be expanded as

$$\frac{h_k}{\rho_k} \frac{\partial p_k}{\partial x} = \frac{h_k}{\rho_k} \left[g \rho_1 \frac{\partial \zeta}{\partial x} + \frac{g}{2} h_1 \frac{\partial \rho_1}{\partial x} + \sum_{m=2}^{k} \left(\frac{1}{2} g h_{m-1} \frac{\partial \rho_{m-1}}{\partial x} + \frac{1}{2} g h_m \frac{\partial \rho_m}{\partial x} \right) \right]. \tag{8}$$

The subscript k is the layer number. When $k=1$, the summation is omitted. The expression for the y-momentum equation is similar.

Numerical Method

The integration scheme is a semi-implicit version of standard three-time-level leapfrog differencing, modified by intermittently adding a two-time-level trapezoidal step. This leapfrog-trapezoidal scheme has been widely used in atmospheric models and suppresses the separation of solutions at alternate time steps that may occur with the standard leapfrog procedure in strongly nonlinear problems. In the 3-D model, the scheme is second order accurate in time, except for a small amount of first order error introduced from a backward in time differencing of the horizontal diffusion terms and the FCT scheme for the advection terms.

The staggered grid spatial discretization in Figure 1 has horizontal grid dimensions Δx and Δy and vertical layer thickness h_k. Because the present version of the model does not permit wetting and drying of nodal points, the layer thickness h_k is constant in space except in the surface and bottom layers.

The semi-implicit discretization of the momentum equations is separated into two stages in which first the explicit terms and then the implicit terms are evaluated. The separation is not a form of time-splitting but simply a convenience for programming and notational purposes. The advection, Coriolis, horizontal diffusion, and baroclinic pressure gradient terms are treated explicitly, while the barotropic pressure gradient, vertical diffusion, and frictional shear stress terms are treated implicitly. For the x-momentum equation using leapfrog differencing the two stage procedure is as follows:

Explicit stage:

$$\tilde{U}_{i+\frac{1}{2},j,k} = U^{n-1}_{i+\frac{1}{2},j,k} + 2\Delta t[-(ADV)^n + (COR)^n - (BCLINIC)^n + (HDIFF)^{n-1}]_{i+\frac{1}{2},j,k}. \tag{9a}$$

Implicit stage:

$$U^{n+1}_{i+\frac{1}{2},j,k} = \tilde{U}_{i+\frac{1}{2},j,k} - g\frac{\Delta t}{\Delta x} h^n_{i+\frac{1}{2},j,k}\left(\frac{\rho_{i+\frac{1}{2},j,1}}{\rho_{i+\frac{1}{2},j,k}}\right)(\zeta^{n+1}_{i+1,j} - \zeta^{n+1}_{i,j} + \zeta^{n-1}_{i+1,j} - \zeta^{n-1}_{i,j})$$
$$+ \Delta t[(A_z\overline{h}^{-2})^n_{i+\frac{1}{2},j,k-\frac{1}{2}} \cdot (U^{n+1}_{i+\frac{1}{2},j,k-1} - U^{n+1}_{i+\frac{1}{2},j,k} + U^{n-1}_{i+\frac{1}{2},j,k-1} - U^{n-1}_{i+\frac{1}{2},j,k}) \tag{9b}$$
$$- (A_z\overline{h}^{-2})^n_{i+\frac{1}{2},j,k+\frac{1}{2}} \cdot (U^{n+1}_{i+\frac{1}{2},j,k} - U^{n+1}_{i+\frac{1}{2},j,k+1} + U^{n-1}_{i+\frac{1}{2},j,k} - U^{n-1}_{i+\frac{1}{2},j,k+1})]$$

in which $\overline{h}_{i+\frac{1}{2},j,k-\frac{1}{2}} = (h_{i+\frac{1}{2},j,k-1} + h_{i+\frac{1}{2},j,k})/2$. The distribution of the exchange coefficient A_z over the depth is solved using a relatively simple mixing-length model with damping functions related to the gradient Richardson number to adjust for buoyancy. The detailed finite-difference expressions for the explicit stage are omitted so the implicit stage can be discussed in some detail.

In Eq. 9b, boundary shear stress terms for wind and bottom friction also appear when considering the surface and bottom layers. The wind stress must be specified as a forcing function. The frictional stress is computed from the bottom layer horizontal transport velocities using a quadratic stress law

$$\left(\frac{\tau_{xz}}{\rho}\right)^{n+1}_{i+\frac{1}{2},j,b+\frac{1}{2}} = (C_d)_{i+\frac{1}{2},j}\frac{\sqrt{(U^{n-1}_{i+\frac{1}{2},j,b})^2 + (V^{n-1}_{i+\frac{1}{2},j,b})^2}}{(h^n_{i+\frac{1}{2},j,b})^2} U^{n+1}_{i+\frac{1}{2},j,b}, \tag{10}$$

in which C_d is a dimensionless drag coefficient.

Because the x- and y-momentum equation treatments are similar, only the x-momentum equation is discussed. From this point, the strategy of Casulli and Cheng (1992) is followed by rewriting the momentum equation in the compact matrix form

$$A_{i+\frac{1}{2},j} U^{n+1}_{i+\frac{1}{2},j} = G_{i+\frac{1}{2},j} - g\frac{\Delta t}{\Delta x} \rho^n_{i+\frac{1}{2},j,1}(\zeta^{n+1}_{i+1,j} - \zeta^{n+1}_{i,j})R_{i+\frac{1}{2},j}. \tag{11}$$

where U, R, G, and A are

$$U^{n+1}_{i+\frac{1}{2},j} = \left[U^{n+1}_1, U^{n+1}_2, \ldots, U^{n+1}_b\right]^T_{i+\frac{1}{2},j}, \qquad R_{i+\frac{1}{2},j} = \left[\frac{h^n_1}{\rho^n_1}, \frac{h^n_2}{\rho^n_2}, \ldots, \frac{h^n_b}{\rho^n_b}\right]^T_{i+\frac{1}{2},j},$$

$$G_{i+\frac{1}{2},j} = \begin{bmatrix} \tilde{U}_1 - g\frac{\Delta t}{\Delta x}h_1^n\left(\frac{\rho_1^n}{\rho_1^n}\right)\left(\zeta_{i+1,j}^{n-1}-\zeta_{i,j}^{n-1}\right) + & 0 & -2\Delta t(A_z)_{3/2}^n\frac{\left(U_1^{n-1}-U_2^{n-1}\right)}{\left(\bar{h}^2\right)_{3/2}^n} + 2\Delta t\frac{(\tau_{xz})^n_{\frac{1}{2}}}{\rho_1^n} \\ \tilde{U}_2 - g\frac{\Delta t}{\Delta x}h_2^n\left(\frac{\rho_1^n}{\rho_2^n}\right)\left(\zeta_{i+1,j}^{n-1}-\zeta_{i,j}^{n-1}\right) + 2\Delta t(A_z)_{3/2}^n\frac{\left(U_1^{n-1}-U_2^{n-1}\right)}{\left(\bar{h}^2\right)_{3/2}^n} - 2\Delta t(A_z)_{5/2}^n\frac{\left(U_2^{n-1}-U_3^{n-1}\right)}{\left(\bar{h}^2\right)_{5/2}^n} + 0 \\ \vdots \\ \tilde{U}_b - g\frac{\Delta t}{\Delta x}h_b^n\left(\frac{\rho_1^n}{\rho_b^n}\right)\left(\zeta_{i+1,j}^{n-1}-\zeta_{i,j}^{n-1}\right) + 2\Delta t(A_z)_{b-\frac{1}{2}}^n\frac{\left(U_{b-1}^{n-1}-U_b^{n-1}\right)}{\left(\bar{h}^2\right)_{b-\frac{1}{2}}^n} - & 0 & + 0 \end{bmatrix}_{i+\frac{1}{2},j},$$

$$A_{i+\frac{1}{2},j} = \begin{bmatrix} 1+\frac{2\Delta t(A_z)_{3/2}^n}{\left(\bar{h}^2\right)_{3/2}^n} & \frac{-2+\Delta t(A_z)_{3/2}^n}{\left(\bar{h}^2\right)_{3/2}^n} & & & \\ \frac{-2\Delta t(A_z)_{3/2}^n}{\left(\bar{h}^2\right)_{3/2}^n} & 1+\frac{2\Delta t(A_z)_{3/2}^n}{\left(\bar{h}^2\right)_{3/2}^n}+\frac{2\Delta t(A_z)_{5/2}^n}{\left(\bar{h}^2\right)_{5/2}^n} & \frac{-2\Delta t(A_z)_{5/2}^n}{\left(\bar{h}^2\right)_{5/2}^n} & 0 \\ \vdots & \vdots & \vdots & \vdots \\ 0 & & \frac{-2\Delta t(A_z)_{b-\frac{1}{2}}^n}{\left(\bar{h}^2\right)_{b-\frac{1}{2}}^n} & \frac{1+2\Delta t(A_z)_{b-\frac{1}{2}}^n}{\left(\bar{h}^2\right)_{b-\frac{1}{2}}^n}+\frac{2\Delta t C_d^n\sqrt{\left(U^2\right)_b^{n-1}+\left(V^2\right)_b^{n-1}}}{\left(\bar{h}^2\right)_b^n} \end{bmatrix}_{i+\frac{1}{2},j}$$

Because the matrix A is tridiagonal and positive-definite, Eq. 11 can be transformed with a double-sweep algorithm to the form

$$U_{i+\frac{1}{2},j}^{n+1} = [A^{-1}G]_{i+\frac{1}{2},j} - g\frac{\Delta t}{\Delta x}\rho_{i+\frac{1}{2},1}^n \left(\zeta_{i+1,j}^{n+1}-\zeta_{i,j}^{n+1}\right)[A^{-1}R]_{i+\frac{1}{2},j}. \tag{12}$$

In executing the double-sweep algorithm, the results actually sought are only the two matrix products $A^{-1}G$ and $A^{-1}R$, which are column vectors equal in order to the number of model layers; the inverse of A is never computed by itself. Equation 12 is useful because it is an expression that can be formally substituted into the continuity equation.

The finite difference analog of the continuity equation is

$$\zeta_{i,j}^{n+1} = \zeta_{i,j}^{n-1} - \frac{\Delta t}{\Delta x}\left[\sum_{k=1}^{b}\left(U_{i+\frac{1}{2},j,k}^{n+1}-U_{i-\frac{1}{2},j,k}^{n+1}+U_{i+\frac{1}{2},j,k}^{n-1}-U_{i-\frac{1}{2},j,k}^{n-1}\right)\right] - \frac{\Delta t}{\Delta y}\left[\sum_{k=1}^{b}\left(V_{i,j+\frac{1}{2},k}^{n+1}-V_{i,j-\frac{1}{2},k}^{n+1}+V_{i,j+\frac{1}{2},k}^{n-1}-V_{i,j-\frac{1}{2},k}^{n-1}\right)\right] \tag{13}$$

which can be written in matrix notation as

$$\zeta_{i,j}^{n+1} = \zeta_{i,j}^{n-1} - \frac{\Delta t}{\Delta x}\left[\Sigma U_{i+\frac{1}{2},j}^{n+1}-\Sigma U_{i-\frac{1}{2},j}^{n+1}\right] - \frac{\Delta t}{\Delta y}\left[\Sigma V_{i,j+\frac{1}{2}}^{n+1}-\Sigma V_{i,j-\frac{1}{2}}^{n+1}\right] - D_{i,j}^{n-1}, \tag{14}$$

in which

$$D_{i,j}^{n-1} = \frac{\Delta t}{\Delta x}\left[\sum_{k=1}^{b}U_{i+\frac{1}{2},j,k}^{n-1}-\sum_{k=1}^{b}U_{i-\frac{1}{2},j,k}^{n-1}\right] + \frac{\Delta t}{\Delta y}\left[\sum_{k=1}^{b}V_{i,j+\frac{1}{2},k}^{n-1}-\sum_{k=1}^{b}V_{i,j-\frac{1}{2},k}^{n-1}\right],$$

and $\Sigma = [1,1,...,1]$ with b elements.

After substituting Eq. 12 for U and a similar equation for V into Eq. 14, an equation involving only the unknown surface elevation is obtained:

$$\zeta_{i,j}^{n+1} - g\frac{\Delta t^2}{\Delta x^2}\left\{\rho_{i+\frac{1}{2}\ j,1}^n [\Sigma A^{-1}R]_{i+\frac{1}{2},j}\left(\zeta_{i+1,j}^{n+1}-\zeta_{i,j}^{n+1}\right)-\rho_{i-\frac{1}{2}\ j,1}^n[\Sigma A^{-1}R]_{i-\frac{1}{2},j}\left(\zeta_{i,j}^{n+1}-\zeta_{i-1,j}^{n+1}\right)\right\}$$

$$-g\frac{\Delta t^2}{\Delta y^2}\left\{\rho_{i,j+\frac{1}{2},1}^n[\Sigma A^{-1}R]_{i,j+\frac{1}{2}}\left(\zeta_{i,j+1}^{n+1}-\zeta_{i,j}^{n+1}\right)-\rho_{i,j-\frac{1}{2},1}^n[\Sigma A^{-1}R]_{i,j-\frac{1}{2}}\left(\zeta_{i,j}^{n+1}-\zeta_{i,j-1}^{n+1}\right)\right\}. \quad (15)$$

$$= \zeta_{i,j}^{n-1} - \frac{\Delta t}{\Delta x}\left\{[\Sigma A^{-1}G]_{i+\frac{1}{2},j} - [\Sigma A^{-1}G]_{i-\frac{1}{2},j}\right\} - \frac{\Delta t}{\Delta y}\left\{[\Sigma A^{-1}G]_{i,j+\frac{1}{2}} - [\Sigma A^{-1}G]_{i,j-\frac{1}{2}}\right\} - D_{i,j}^{n-1}$$

The matrix products $\Sigma A^{-1}R$ and $\Sigma A^{-1}G$ in Eq. 15 are each a single non-negative number. Equations 15, written over all nodal points, are therefore a linear five-diagonal system of equations that can be solved for $\zeta_{i,j}^{n+1}$. The system matrix is symmetric and positive-definite and can therefore be solved efficiently by the preconditioned conjugate gradient method. In the model, a modified incomplete Cholesky preconditioner is used. Once the $\zeta_{i,j}^{n+1}$ are determined, Eq. 12 and the companion equation for V can be explicitly solved for the new transport velocities.

Finally, the vertical velocity w in the water column can be derived from the continuity equation expressed in the form

$$w_{i,j,k-\frac{1}{2}}^{n+1} = w_{i,j,k+\frac{1}{2}}^{n+1} - \left(U_{i+\frac{1}{2},j,k}^{n+1}-U_{i-\frac{1}{2},j,k}^{n+1}\right)/\Delta x - \left(V_{i,j+\frac{1}{2},k}^{n+1}-V_{i,j-\frac{1}{2},k}^{n+1}\right)/\Delta y \quad k=2,3,4,...b. \quad (16)$$

This equation is solved explicitly starting from the bottom where $w_{b+\frac{1}{2}}=0$.

The salinity equation is solved last during each time step, and the new salinities are used to update the density field. Vertical diffusion in the salt equation is treated implicitly and depends on the solution of another set of tridiagonal systems.

If a trapezoidal step is desired after the leapfrog step is completed, the dependent variables are averaged between time levels $n+1$ and n and used to recalculate the solution at $n+1$ by another leapfrog procedure applied over time interval Δt (rather than $2\Delta t$). The matrix solution during this trapezoidal step typically converges in very few iterations because a good initial estimate of the solution vector is available from the leapfrog step.

Remarks

In developing the model, a two-time-level Crank-Nicolson method was implemented for the time integration on several 1-D test problems for comparison with the leapfrog-trapezoidal method. On the basis of these numerical tests, the leapfrog-trapezoidal scheme was selected over Crank-Nicolson, despite the additional storage required by the three-level scheme.

The advantage of leapfrog-trapezoidal integration is the ease with which nonlinear terms can be centered in time during both leapfrog and trapezoidal steps. Nonlinear terms using the Crank-Nicolson scheme can be centered in time only by averaging between old and new time levels, which makes iteration necessary. With nonlinear coefficients written backward in time, the Crank-Nicolson scheme tends to produce spurious oscillations on strongly nonlinear problems and also is not second-order accurate. Difficulties with solution oscillations were noted by Duwe et al. (1983) in applying the semi-implicit Crank-Nicolson model of Backhaus (1983) to markedly nonlinear flows. Duwe et al. (1983) implemented a two-step integration procedure to enhance the model stability.

Even for simulations when a trapezoidal step is implemented after every leapfrog step, there are still some advantages of the three-time-level model over a two-time-level model. The advantages are the slight improvement in accuracy due to time centering both steps of the three-level scheme and an increase in the convergence rate of the matrix solution because of the time centering. When only the leapfrog step is needed, then clearly the leapfrog scheme is superior in accuracy to an uncentered two-level scheme and, in addition, will usually converge faster.

The present model has been developed to compute 3-D density-driven circulations in San Francisco Bay. Owing to the complex bathymetry in the bay, a 3-D model is needed that is both stable and accurate when the grid resolution is not highly refined. The model also must be computationally efficient because long-term (seasonal) simulations are required. The new model should fulfill these criteria.

Acknowledgements

This paper is based on work done by the U.S. Geological Survey in cooperation with the California Department of Water Resources.

References

Backhaus, J.O. (1983). "A Semi-implicit Scheme for the Shallow Water Equations for Application to Shelf Sea Modeling." *Continental Shelf Research*, 2, 243-254.

_____(1985). "A Three-dimensional Model for the Simulation of Shelf Sea Dynamics." *Dt. Hydrogr. Z.*, 38, 165-187.

Benqué, J.P., Cunge, J.A., Feuillet, J., Hauguel, A., and Holly, F.M. (1982). "New Method for Tidal Current Computation." *Journal of the Waterway, Port, Coastal, and Ocean Division, ASCE*, 108, 396-417.

Casulli, V. (1990). "Semi-implicit Finite Difference Methods for the Two-dimensional Shallow Water Equations." *Journal of Computational Physics*, 86, 56-74.

Casulli, V., and Cheng, R.T. (1992). "Semi-implicit Finite Difference Methods for the Two-dimensional Shallow Water Flow." *International Journal for Numerical Methods in Fluids*, 15, 629-648.

de Goede, E.D. (1991). "A Time-splitting Method for the Three-dimensional Shallow Water Equations." *International Journal for Numerical Methods in Fluids*, 13, 519-534.

Duwe, K.C., Hewer, R.R., and Backhaus, J.O. (1983). "Results of a Semi-implicit Two-step Method for the Simulation of Markedly Nonlinear Flow in Coastal Seas." *Continental Shelf Research*, 2, 255-274.

Kwizak, M. and Robert, A.J. (1971). "A Semi-implicit Scheme for Grid Point Atmospheric Models of the Primitive Equations." *Monthly Weather Review*, 99, 32-36.

Leendertse, J.J., Alexander, R.C., and Liu, S.K. (1973). "A Three-dimensional Model for Estuaries and Coastal Seas: Volume I, Principles of Computation." *Report R-1417-OWRR*, Rand Corporation, Santa Monica, CA, 57 p.

Stelling, G.S., Wiersma, A.K., and Willemse, J.B.T.M. (1986). "Practical Aspects of Accurate Tidal Computations." *Journal of Hydraulic Engineering, ASCE*, 112, 802-817.

Wilders, P., van Stijn, Th.L., Stelling, G.S., and Fokkema, G.A. (1988). "A Fully Implicit Splitting Method for Accurate Tidal Computations." *International Journal for Numerical Methods in Engineering*, 26, 2707-2721.

Zalesak, S.T. (1979). "Fully Multidimensional Flux-corrected Transport Algorithms for Fluids." *Journal of Computational Physics*, 31, 335-362.

Appendix: List of Symbols

A_x, A_y	horizontal momentum exchange coefficients	u, v	horizontal components of velocity (layer-averaged)
A_z	vertical momentum exchange coefficient	w	vertical component of velocity (at layer interfaces)
b	number of model layers	U, V	horizontal components of transport ($U=uh$, $V=vh$)
D_x, D_y	horizontal mass exchange coefficients		
D_z	vertical mass exchange coefficient	ρ	density (layer-averaged)
f	Coriolis parameter	ζ	free surface elevation measured from mean sea level
g	acceleration due to gravity		
h	layer thickness	x, y, z	Cartesian coordinates in the eastward, northward, and upward directions, respectively
i, j	horizontal finite difference indices		
k	layer number ($k=1,2,...,b$)		
n	time step number	τ_{xz}, τ_{yz}	shear stresses at layer interfaces (also represents wind stress at free surface interface, $z=\frac{1}{2}$, and frictional stress at bottom interface, $z=b+\frac{1}{2}$)
p	pressure (layer-averaged)		
S	salinity (layer-averaged)		
t	time		

Horizontal gradients in sigma transformed bathymetries with steep bottom slopes

Guus S. Stelling, Jan A.T.M. van Kester[1]

Abstract

Simulation of free surface flow and transport in areas such as seas, estuaries and rivers is often based upon three-dimensional modelling systems. Common characteristics of these modelling systems often are: Hydrostatic pressure assumption and sigma coordinate transformation as a basis for the numerical approximation in the vertical. Especially for steep bottom slopes combined with vertical stratification of the density, such as in estuaries, various numerical problems are encountered. This paper deals with algorithms to minimize these difficulties.

Introduction

Estuaries are sometimes defined as an interface between salt and fresh water[e.g. Dyer, 1973]. To model estuaries a modelling system should include density effects in the momentum equation and transport equations for transport of matters such as salt, heat, turbulent kinetic energy, tidal flat procedures etc. Simulations are often based upon the following set of equations:

Continuity equation (water level ζ):

$$\frac{\partial \zeta}{\partial t^*} + \frac{\partial (Hu^*)}{\partial x^*} + \frac{\partial (Hv^*)}{\partial y^*} + \frac{\partial \omega}{\partial \sigma} = C \qquad (1\text{ a})$$

Momentum equations (velocities u_i^*):

$$\frac{Du_i^*}{Dt^*} + \frac{1}{\rho_0}\left(\frac{\partial p}{\partial x_i^*} + \frac{\partial \sigma}{\partial x_i} \cdot \frac{\partial p}{\partial \sigma}\right) = \frac{1}{\rho_0} \cdot \frac{\partial \tau_{i,j}^*}{\partial x_j^*}, i=1,2 \qquad (1\text{ b})$$

Hydrostatic pressure:

$$\frac{\partial p}{\partial \sigma} = -\rho g H \qquad (1\text{ c})$$

Transport equations (concentration c):

$$\frac{\partial c_1}{\partial t^*} + \nabla^* \cdot \vec{F}^* = 0 \qquad (1\text{ d})$$

Where:

$$\frac{Du_i^*}{Dt^*} = \frac{\partial u_i^*}{\partial t^*} + u^* \cdot \frac{\partial u_i^*}{\partial x^*} + v^* \cdot \frac{\partial u_i^*}{\partial y^*} + \frac{\omega}{H} \frac{\partial u_i^*}{\partial \sigma}$$

The σ coordinate transformation is applied [Phillips, 1957], given by:

$$x^*=x, \quad y^*=y, \quad \sigma=\frac{z-\zeta}{H}, \quad t^*=t$$

---

[1] Delft Hydraulics, P.O. Box 11, 2600 MH Delft, The Netherlands

where:

$H = \zeta + d$ (total water depth)
$z = \zeta(x,y,t)$ or $\sigma = 0$, at the free water surface and
$z = -d(x,y)$ or $\sigma = -1$, at the bottom.

σ transformation implies substitution into equations, based upon Cartesian coordinates, of the following operators:

$$\frac{\partial}{\partial t} = \frac{\partial}{\partial t^*} + \frac{\partial \sigma}{\partial t} \cdot \frac{\partial}{\partial \sigma}$$
$$\frac{\partial}{\partial x_i} = \frac{\partial}{\partial x_i^*} + \frac{\partial \sigma}{\partial x_i} \cdot \frac{\partial}{\partial \sigma}, i=1,2$$
$$\frac{\partial}{\partial x_3} = \frac{1}{H} \cdot \frac{\partial}{\partial \sigma}$$

The turbulent closure models for $\partial \tau^*_{i,j}/\partial x^*_j$ are not necessarily an exact transformation of $\partial \tau_{i,j}/\partial x_j$, sometimes they are reformulated within the transformed coordinates [Mellor and Blumberg, 1985].

The σ grid is fitted to the free surface and bottom. In estuaries such a grid may deteriorate quite strongly near steep bottom slopes or near tidal flats, see figure 1. Grids of this type cause problems when computing the horizontal pressure gradient as is recognized by several authors [e.g. Gary 1973, Janjic 1977, Mesinger and Janjić, 1985, Haney 1991, or Deleersnijder and Beckers, 1992]. This is due to the transformed pressure gradient:

$$\frac{\partial p}{\partial x} = \frac{\partial p}{\partial x^*} + \frac{\partial \sigma}{\partial x} \cdot \frac{\partial p}{\partial \sigma} = \frac{\partial p}{\partial x^*} - \frac{1}{H}\left[\frac{\partial \zeta}{\partial x} + \sigma \frac{\partial H}{\partial x}\right]\frac{\partial z}{\partial c}$$

Near steep bottom slopes small pressure gradients are the result of the sum of relatively large terms with opposite sign. This might lead to truncation errors that produce artificial flow. Related to this is the notion of "hydrostatic consistency" [e.g. Janjic, 1977]. According to [Haney, 1991] a "hydrostatic consistency condition" is given by:

$$\left|\frac{\sigma}{H} \cdot \frac{\partial H}{\partial x}\right|\Delta x < \Delta \sigma$$

Where Δx and $\Delta \sigma$ are grid size values.

Violation of this condition hampers convergence. Tidal flats are characterized by $H \to 0$, hence convergence becomes impossible. In case of steep bottom slopes there are similar difficulties for horizontal diffusive fluxes of concentrations such as salinity and temperature.

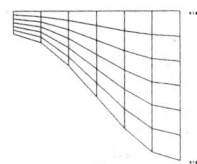

Figure 1, σ grid

The approximation of diffusive fluxes with a finite Volume Method

Equation (1 d) implies advective and diffusive fluxes. We consider only diffusive fluxes, in Cartesian coordinates, given by:

$$F_i = D_H \frac{\partial c}{\partial x_i}, \; i=1,2, \; F_3 = D_V \frac{\partial c}{\partial x_3}$$

where D_H denotes an eddy diffusion coefficient.
Transformation to x^*, y^*, σ, t^* yields relations such as:

$$\frac{\partial^2 c}{\partial x^2} = \frac{\partial^2 c}{\partial x^{*2}} + \left(\frac{\partial \sigma}{\partial x}\right)^2 \frac{\partial^2 c}{\partial \sigma^2} + 2 \frac{\partial \sigma}{\partial x} \cdot \frac{\partial^2 c}{\partial x^* \partial \sigma} + \frac{\partial^2 \sigma}{\partial x^2} \cdot \frac{\partial^2 c}{\partial \sigma}$$

For such transformations it is difficult to find numerical approximations that are stable and positive. Near steep bottom slopes or near tidal flats, truncations errors in the sigma coordinates are likely to become very large similar to the horizontal pressure gradient. Some authors [Mellor and Blumberg, 1985], omit several terms of the transformation yielding:

$$F_1^* = D_H \frac{\partial c}{\partial x^*}$$

$$F_2^* = D_H \frac{\partial c}{\partial y^*} \qquad (3)$$

$$F_3^* = \frac{D_V}{H} \frac{\partial c}{\partial \sigma}$$

The conditions which cause this equation to give a better description of the transport process are certainly not fulfilled in many estuaries. If we omit vertical diffusion then this formulation will still cause some numerical vertical diffusion especially near steep bottom slopes such as near tidal flats. Due to this phenomenon it will be difficult to simulate the stratification of a salt wedge in an estuary very accurate. So the complete transformation must be included. But in that case numerical problems are encountered concerning accuracy, stability and monotonicity. In the following a method is briefly introduced which is a consistent, stable and monotonic approximation of the horizontal diffusion, despite of the hydrostatic consistency condition. The method is based upon a Finite Volume Method.

Finite Volume Methods [Peyret and Taylor, 1983, Fletcher, 1989, Hirsch 1990] are commonly used for the approximation of systems of conservation laws. Finite volume methods yield conservative approximations for arbitrary grids without the need of explicit analytic transformation of Cartesian equations. In general a finite volume method is based upon integration of the transport equation, given in Cartesian coordinates, over a control volume combined with the Gauss theorem given by:

$$\int_V \nabla \cdot \vec{F} dv = \oint_S \vec{F} \cdot \vec{n} ds$$

Where $\vec{n}$ denotes the normal to the boundary. Applied to eq.(1 d) this yields:

$$\int_V \frac{\partial c}{\partial t} dv + \oint_S \vec{F} \cdot \vec{n} ds = 0$$

Taking into account that the control volume is time variable, as in the case of σ coordinates, this equation can be rewritten as[1]:

$$\frac{\partial}{\partial t} \int_V c\, dv + \oint_S \vec{F} \cdot \vec{n} ds = \oint_S c \frac{\partial \vec{n}}{\partial t} ds$$

The right hand side of this equation describes some pseudo advection due to the variation of the control volume. In σ coordinates this is aspect taken into account by the definition of ω. The fluxes of this equation can be added together to yield a new flux F^*. After dropping the * we obtain:

$$\frac{\partial}{\partial t} \int_V c\, dv + \oint_S \vec{F} \cdot \vec{n} ds = 0 \qquad (4)$$

Instead of transforming the equations to the sigma coordinate system we consider the sigma grid lines as the boundaries of the time varying control volumes in cartesian coordinates. In this way we obtain finite volumes as given by figure(2).

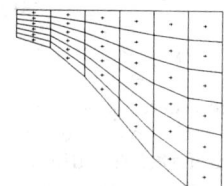

Figure 2, cell centred control volumes

For the approximation of the horizontal diffusive fluxes it is sufficient to consider:

$$\frac{\partial c(x,z,t)}{\partial t} - \frac{\partial}{\partial x} D_H \frac{\partial c(x,z,t)}{\partial x} = 0 \qquad (5)$$

[2] This relation can be proved by repeated application of the Leibniz integration rule which is given by:

Since the time variability of the control volume is taken into account by advective fluxes and since we consider only diffusive fluxes we neglect time variability. We use cell-centred finite volumes as given by figure (2).

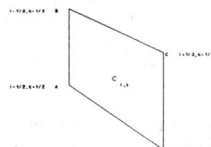

Figure 3, cell numbering

For 2-dimensional equations, and for the cell-centred volume given by figure (3), the finite volume method for diffusion implies a numerical approximation of the following integrals:

$$\frac{\partial}{\partial t}\int\int_{ABCD} cdxdz + \int_{x_A,z_A}^{x_B,z_B} \frac{\partial c}{\partial x}dx + \int_{x_B,z_B}^{x_C,z_C} \frac{\partial c}{\partial x}dx + \int_{x_C,z_X}^{x_D,z_D} \frac{\partial c}{\partial x}dx + \int_{x_D,z_D}^{x_A,z_A} \frac{\partial c}{\partial x}dx = 0 \qquad (6)$$

For this equation an approximation exists fulfilling: (i) consistency, (ii) monotonicity and (iii) minimal artificial vertical diffusivity. The method is cell-centred, as given by figure(2). For the approximation of (5) see figure (4), where:

$$x_a = x_A, \quad z_a = \frac{z_A + z_D}{2} \quad x_b = x_B, \quad z_b = \frac{z_B + z_C}{2}$$
$$x_c = x_C, \quad z_c = \frac{z_B + z_C}{2} \quad x_d = x_A, \quad z_d = \frac{z_A + z_D}{2}$$
$$x_e = \frac{x_A + x_D}{2}, \quad z_e = \frac{z_A + z_D}{2} \quad x_f = \frac{x_B + x_C}{2}, \quad z_f = \frac{z_B + z_C}{2}$$

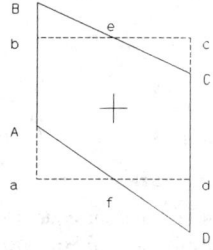

Figure 4, control volume for diffusive fluxes

If at the interval $[(x_e,z_e),(x_B,z_B)]$ the same numerical approximation for $\partial c/\partial x$ is used as for the interval $[(x_B,z_B),(x_b,z_b)]$ then the integration along these two intervals gives a zero result. The same result can be obtained along the intervals $[(x_e,z_e),(x_C,z_C)]$ and $[(x_C,z_C),(x_c,z_c)]$. By the same argument one can assume that at the intervals $[(x_A,z_A),(x_f,z_f)]$ and $[(x_D,z_D),(x_d,z_d)]$ the integrals are the same. Due to this a consistent, semi discrete, approximation of eq.(6) is given by:

$$(x_d - x_a) \cdot (z_b - z_a) \frac{dc_{i,k}}{dt} = -\int_{x_a,z_a}^{x_b,z_b} \frac{\partial c}{\partial x}dx - \int_{x_c,z_c}^{x_d,z_d} \frac{\partial c}{\partial x}dx \qquad (7)$$

For the time derivative there are many options [e.g. Lambert, 1991]. We will use Eulers explicit method.

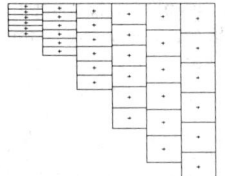

Figure 5, control volumes for diffusive fluxes

For the right hand side one must realize that cells of shape abcd might have more than 2 adjacent cells in horizontal direction, see figure 5. For the approximation of the right side of eq.(7) an interval:

$$[\min(-d_i,-d_{i+1}),\max(\zeta_i,\zeta_{i+1})]$$

between two vertical colons of control volumes is divided into 2K sub intervals: $[z_{i+\frac{1}{2},l},z_{i+\frac{1}{2},l+1}]$, $l=0,\ldots,2K$, $z_0=\min(-h_i,-h_{i+1})$ and $z_{2K+1}=\max(\zeta_i,\zeta_{i+1})$.
The set of points $\{z_{i+\frac{1}{2},0},\ldots,z_{i+\frac{1}{2},2K+1}\}$ is ordered such that:

$$z_{i+\frac{1}{2},l} \geq z_{i+\frac{1}{2},l+1} \text{ for } l=0,\ldots,2K$$

Diffusive fluxes are approximated for volumes as given by figure(5) as follows:
First fluxes $f_{i+\frac{1}{2},l}$, $l=1,\ldots,2K+1$ are defined according to:

$$f_{i+\frac{1}{2},l} = \begin{cases} D_H \min(\Delta_m c, \Delta_n c) \frac{z_{i+\frac{1}{2},l}-z_{i+\frac{1}{2},l+1}}{x_{i+1}-x_i}, & \Delta_m c > 0 \wedge \Delta_n c > 0 \\ D_H \max(\Delta_m c, \Delta_n c) \frac{z_{i+\frac{1}{2},l}-z_{i+\frac{1}{2},l+1}}{x_{i+1}-x_i}, & \Delta_m c \leq 0 \wedge \Delta_n c \leq 0 \\ 0, & \Delta c_m \cdot \Delta_n c < 0 \end{cases} \quad (9)$$

The differences $\Delta_{m/n} c = \Delta_{m/n} c_{i+\frac{1}{2},l+\frac{1}{2}}$ are given by:

$$\left. \begin{array}{l} \Delta_m c_{i+\frac{1}{2},l+\frac{1}{2}} = c_{i+1}(z_{i,m(l)+\frac{1}{2}}) - c_{i,m(l)} \\ \Delta_n c_{i+\frac{1}{2},l+\frac{1}{2}} = c_{i+1,n(l)} - c_i(z_{i+1,n(l)+\frac{1}{2}}) \end{array} \right\} \quad l=1,\ldots,2K+1 \quad (10)$$

where $c_i(z)$ is a simple linear interpolation formula given by:

$$c_i(z) = \begin{cases} c_{i,K'} & z \leq z_{i,K+\frac{1}{2}} \\ \frac{z_{i,k+\frac{1}{2}}-z}{z_{i,k+\frac{1}{2}}-z_{i,k+1\frac{1}{2}}} c_{i,k+1} + \frac{z_{i,k+1\frac{1}{2}}-z}{z_{i,k+1\frac{1}{2}}-z_{i,k+\frac{1}{2}}} c_{i,k'} & z_{i,k+1\frac{1}{2}} \leq z \leq z_{i,k+\frac{1}{2}} \\ c_{i,1'} & z \geq z_{i,\frac{1}{2}} \end{cases} \quad (11)$$

These interpolation coefficients are always positive and less than or equal to one. The two possible stencils for the approximations of the diffusive fluxes are given by figure 6.

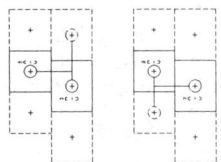

Figure 6, stencils for diffusive fluxes

Second, the diffusive fluxes are added to the concentrations:

$$V_{i,k}^{\tau+1} c_{i,k}^{\tau+1} = V_{i,k}^\tau c_{i,k}^\tau - \Delta t \sum_{\forall l \mid m(l)=k} f_{i+\frac{1}{2},l}^\tau + \Delta t \sum_{\forall l \mid n(l)=k} f_{i-\frac{1}{2},l}^\tau \quad (12)$$

Where τ is the time index, $t=\tau \Delta t$, and V^τ denotes the size of the control volume.

This Algorithm is consistent, monotonic and conditionally stable while vertical numerical diffusion is minimized [Stelling and Van Kester,1993].

Approximation of the pressure term

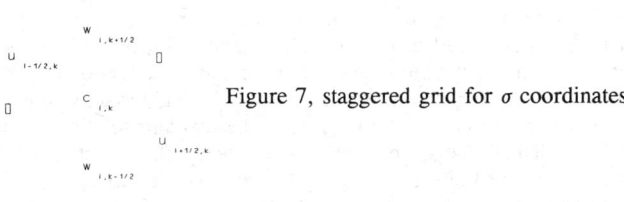

Figure 7, staggered grid for σ coordinates

For the approximation of $\partial p/\partial x$ we assume a staggered grid as given by figure(7). It follows that $\partial p/\partial x$ must be computed along the same verticals as the horizontal concentration gradients. The pressure p is given by:

$$p(x,z,t) = \int_{z'=z}^{z'=\zeta} \rho(x,z',t)\,g\,dz'$$

From Leibniz rule it follows that $\partial p/\partial x$ is given by:

$$\frac{\partial}{\partial x}p(x,z) = \frac{\partial}{\partial x}\int_{z'=z}^{z'=\zeta(x)} \rho(x,z')\,g\,dz' = \int_{z'=z}^{z'=\zeta(x)} g\frac{\partial}{\partial x}\rho(x,z')\,dz' + g\rho(\zeta)\frac{\partial \zeta}{\partial x} \quad (15)$$

The relation between the density ρ and concentrations for salinity, s, and temperature, T, is given by an equation of state:

$$\rho = \rho(s(\vec{x},t), T(\vec{x},t))$$

It follows that:

$$\frac{\partial \rho}{\partial x} = \frac{\partial \rho}{\partial s}\cdot\frac{\partial s}{\partial x} + \frac{\partial \rho}{\partial T}\cdot\frac{\partial T}{\partial x}$$

From this it follows that if the horizontal gradients of the concentrations are zero then there is no contribution, due to density gradients, to the driving force in the momentum equation. It is important to have exactly the same mechanism in the numerical approximation. <u>This means that if the horizontal gradients, used in the transport equation, are zero then the contribution to the approximation of $\partial p/\partial x$ should be zero too.</u> If not then there will always be artificial flow, near steep bottom slopes, due to truncation errors. Therefore the procedure for diffusive fluxes will also be used for the approximation of $\partial p/\partial x$. In other words the numerical approximations for the horizontal concentration gradients will be used for the approximation of the pressure gradients.

Test problems

In this section three test problems will be dealt with. The first test problem deals with a horizontal diffusion problem. The second example illustrates the horizontal pressure approximation. These two examples are two dimensional. A third example consists of a three dimension diffusion problem.

The first test problem deals a simple diffusion equation given by eq.(5). The bathymetry, see figure(8), is given by $0 \leq x \leq 100$ m and $-x/10-1 \leq z \leq 0$. The boundaries at $x=0$ and $x=100$ are closed. Here zero diffusive fluxes are prescribed as boundary conditions. The initial condition is given by:

$c=0$ kg/m^3, $z \geq -5$ m and $c=30$ kg/m^3, $z<-5$ m

Vertical diffusion is supposed to be zero. Horizontal diffusion is 10 m^2/s. This initial condition is also a steady state solution. For the numerical approximation we use a 10x10 sigma grid. The initial conditions are taken from the exact solution. Due to truncation errors this initial condition is not a steady state solution for a numerical method. The method that we propose gives a steady state solution with that is slightly diffused in vertical direction. Figure (9) shows concentration profiles in vertical direction of a steady state solution. If the number of grid points is increased then the steady state solution becomes more accurate, see figure 10. For this solution the number of vertical grid points was 80. The number of horizontal grid points was not changed.

This illustrates that improved stratification is obtained due to increasing the vertical resolution without increasing the horizontal resolution. This is contrary to other methods [Mesinger and Janjic, 1985]. Figure 11 shows results for a different bottom profile with a steeper slope. For such situations the method gives only a very small vertical numerical diffusion until a steady state is reached.

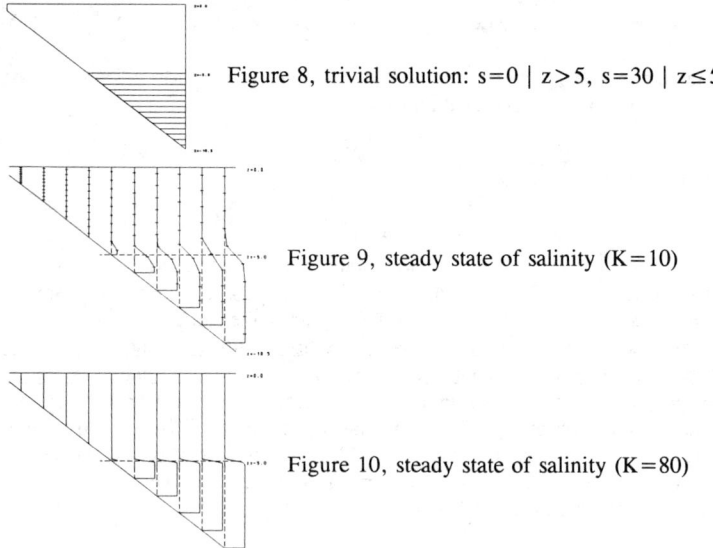

Figure 8, trivial solution: $s=0 \mid z>5$, $s=30 \mid z \leq 5$.

Figure 9, steady state of salinity (K=10)

Figure 10, steady state of salinity (K=80)

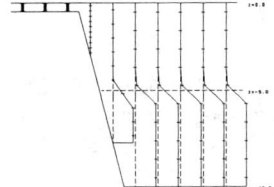

Figure 11, salinity and steep slope

The second test problem consists of a flow problem combined with a transport problem. The bathymetry is equal to the previous test problem with a steep slope, i.e. depth values h(x) are given by:

$$h(x) = \begin{cases} 0.5, & 0 \leq x < 40 \\ 0.5 + \frac{x-40}{2}, & 40 \leq x \leq 60 \\ 10.5, & 60 < x \leq 100 \end{cases}$$

The equations to be approximated are the following:

$$\frac{\partial u}{\partial t^*} + \frac{1}{\rho_0}\frac{\partial p}{\partial x} = \frac{v_v}{H^2}\frac{\partial^2 u}{\partial \sigma^2} + v_H \frac{\partial^2 u}{\partial x^{*2}}$$

$$\frac{\partial \zeta}{\partial t^*} + \frac{\partial Hu}{\partial x^*} + \frac{\partial \omega}{\partial \sigma} = 0$$

$$\frac{\partial(Hc)}{\partial t^*} + \frac{\partial(Huc)}{\partial x^*} + \frac{\partial \omega c}{\partial \sigma} = \frac{\partial}{\partial x} H \cdot D_H \frac{\partial c}{\partial x}$$

Terms of this equation where we use a * are already transformed to the new coordinate system. The unmarked terms are still in the Cartesian coordinate system. Their transformation takes place implicitly, by the numerical method which is used. For the computation of the density a simple equation of state is given by:

$$\rho = 1000 + c$$

Zero water levels and velocities are prescribed as initial conditions. The initial conditions of the salinity are the same as for the previous test problems. Boundary conditions are given by:

$$\frac{\partial c}{\partial x} = 0, x=0, -h(0) \leq z \leq \zeta(0)$$

$$c(100, z, t) = 30 kg/m^3, -h(100) \leq z \leq 5$$

$$c(100, z, t) = 0 kg/m^3, 5 < z \leq \zeta(100)$$

The values for D_H and $\nu_{H,V}$ are 10.0, 2.0 and 0.1 m²/s respectively.
The initial conditions for the concentration are the same as for the previous test problem. The exact solution is a trivial one with zero velocities, zero gradients of the water level and a distribution of the salinity equal to the distribution of the previous test problem. Many numerical methods however will produce artificial flow for similar situations [e.g. Mesinger, 1982]. To illustrate this we first consider an approximation of the pressure term which is based upon a straightforward transformation of the pressure gradient. This straightforward transformation is given by:

$$\frac{\partial p}{\partial x} = \frac{\partial p}{\partial x^*} + \frac{\partial \sigma}{\partial x}\frac{\partial p}{\partial \sigma} = \frac{\partial}{\partial x^*}gH\int_\sigma^0 \rho\, d\sigma' + \rho g\left(\frac{\partial \zeta}{\partial x} + \sigma\frac{\partial H}{\partial x}\right)$$

Streamlines for the steady state solution are given by figure(12). The maximum velocity, located near the slope is of the order of 0.15 m/s. This is entirely due to truncation errors! The concentration profiles are given by figure(13).

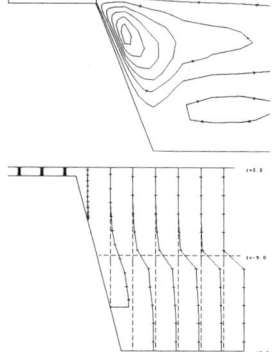

Figure 12, flow due to truncation errors

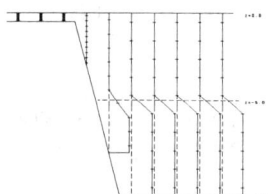

Figure 13, salinity distribution for fig. 12

Our method produces some artificial flow initially, until the concentration field has reached the steady state. Then the flow is absolutely zero. The concentrations are given by figure (14).

Figure 14, steady state concentration profiles

It is to be noted that this solution is independent from the amount of horizontal diffusivity as long as it is not zero. To reach the steady state solution as fast as possible it is efficient to start the computations initially without density effects in the pressure term, comparable to diagnostic and prognostic modes [Blumberg and Mellor, 1983].

The third test problem consists of diffusion only, but the computational domain is three dimensional. The bathymetry is given by figure (15). It has different bottom slopes in different directions. The basin is closed at all boundaries. This means that zero fluxes are prescribed at all boundaries. The computational domain consists of 10X10X10 grid points. The initial conditions are the same as for the first test problem. The steady state solution is given by figure (16). All fluxes are approximately zero, i.e. flow due to density will be zero too for this case.

It is to be noted that if Dirichlet type of boundary conditions are prescribed (i.e. concentrations), then exact zero fluxes are not always the steady state solution, despite of the fact that this might well be the case for the analytic solution. Hence artificial flow might occur. This will be small in general, but if this hampers the accuracy significantly then boundary conditions can be computed first on the basis of purely Von Neumann type of boundary conditions. Derived from this steady state solution, boundary conditions can be prescribed with a slightly diffused vertical concentration gradient, such that, in case of a steady state solution, all fluxes are zero. Hence artificial flow will not be generated.

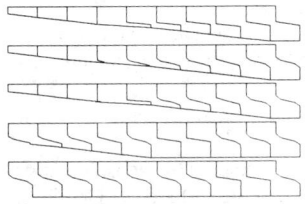

Figure 16, concentration profiles in various cross sections

Figure 15, bathymetry of 3D test problem

Concluding remarks

A numerical method is derived for horizontal diffusion in σ coordinates. This method is based upon a finite volume approach. This allows derivation of the method conceptually within Cartesian coordinates although a σ grid is used. Due to a simple filter, that does not influence the consistency, the method is positive and does not have truncation errors that produce vertical diffusion. When the computation of the pressure gradient is based upon integration of the concentration gradients, as computed with the transport algorithm, then artificial upwelling is minimized. In this way σ coordinate systems can be applied for stratified flow simulations in estuaries with tidal flats and steep bottom slopes such that there is no necessity to fulfil the "hydrostatic consistency" condition.

References
A.F. Blumberg, and G.L. Mellor (1983),Diagnostic and prognostic numerical circulation studies of the South Atlantic Bight, Journal of Geophysical Research, 88,pp 4579-4592
E. Deleersnijder and J.M. Beckers (1992),On the use of the σ coordinate system in regions of large bathymetric variations, Journal of Marine Systems, 3, pp 381 -390
K. Dyer (1973),Estuaries: A Physical Introduction, John Wiley & Sons
C.A.J. Fletcher (1989),Computational techniques for fluid dynamics, Volumes I an II, Springer
J.M. Gary (1973), Estimate of Truncation Error in Transformed Primitive Equation Atmospheric Models, 30,pp 223-233

D. Greenspan and V. Casulli (1988), Numerical Analysis for Applied Mathematics, Science and Engineering, Addison-Wesley Publishing Company.

C. Hirsch (1990), Numerical Computation of Internal and External Flows, John Wiley & Sons, New York, 1990.

R.L. Haney (1991), On the Pressure Gradient Force over Steep Topography in Sigma Coordinate Ocean Models, Journal of Physical Oceanography, 21, pp 610-619

Z.I. Janjić (1977), Pressure Gradient Force and Advection Scheme Used for Forecasting with Steep and Small Scale Topography, Beiträge zur Physik der Atmosphäre, 50, pp 186-199

G.L. Mellor and A.F. Blumberg (1985), Modelling Vertical and Horizontal Diffusivities with the Sigma Coordinate System, Monthly Weather Review,113,pp 1379-1383

J.D. Lambert (1991), Numerical Methods for Ordinary Differential Systems, John Wiley & Sons

F. Mesinger (1982), On the Convergence and Error Problems of the Calculation of the pressure Gradient Force in Sigma Coordinate Models, Geophysical Astrophysical Fluid Dynamics, 19, pp 105-117

F. Mesinger and Z.I. Janjić (1985), Problems and numerical methods in the incorporation of mountains in atmospheric models, In Lectures in applied mathematics, 22, pp 81-120

R. Peyret and T.D. Taylor (1983), Computational methods for fluid flow, Springer, 1983.

N.A. Phillips (1957), A coordinate system having some special advantages for numerical forecasting, Journal of Meteorology,14,pp 184-185

G.S.Stelling, J.A.T.M. van Kester (1993), On the approximation of horizontal gradients in sigma transformed bathymetries, Delft Hydraulics, report.

2-D Vertical and 3-D Modelling
of Mud Transport in Tidal Flows

Bruce A. DeVantier[1], A.M., and Leo C. van Rijn[2]

Abstract
Model predictions of a fully three-dimensional sediment transport model are presented for two tidal flow variation cases. A simplified two-dimensional form of the model is applied to demonstrate the effects of inflow and outflow boundary conditions. The full model is applied to predict transport of mud for a coastal channel flow.

Introduction
A two- and three-dimensional (2- and 3-D) transport model developed by Van Rijn and Meijer (1986) called SUSTRA has been applied to mud and sand transport. The application of a 3-D flow and sediment transport models and the analysis of the computed results usually involves significant effort. Often, the available budget is not sufficient to make as many runs as necessary to study the influence of the basic input parameters and boundary conditions. Such a sensitivity study can be carried out relatively easily by using a two-dimensional vertical model, as presented by DeVantier and Van Rijn (1993). Herein, the attention is initially focussed on the influence of the boundary condition at the inlet on the computed results. Finally, an example of the full 3-D model is shown.

The three-dimensional model has been described in detail in the previous references, so the model will be described only briefly. The following results were part of a larger effort to examine the sensitivity and reliability of the SUSTRA model. The 3-D, transient (in time t) transport equation solved by the model is

---

[1] Associate Professor, Department of Civil Engineering and Mechanics, Southern Illinois University at Carbondale, Carbondale, IL 62901-6603, USA.
[2] Senior Hydraulic Engineer, Delft Hydraulics, P.O. Box 152, 8300 AD Emmeloord, The Netherlands.

$$\frac{\partial c}{\partial t}+\frac{\partial}{\partial x}(uc)+\frac{\partial}{\partial y}(vc)+\frac{\partial}{\partial z}[(w-w_s)c]=\frac{\partial}{\partial x}\left(\epsilon_x\frac{\partial c}{\partial x}\right)+\frac{\partial}{\partial y}\left(\epsilon_y\frac{\partial c}{\partial y}\right)+\frac{\partial}{\partial z}\left(\epsilon_z\frac{\partial c}{\partial z}\right) \quad (1)$$

where c is the total mud concentration, u, v and w are the velocities in the x, y and z coordinate directions respectively, and w_s is the local settling rate of the sediment. The turbulent redistribution is modelled with a standard flux-gradient expression in which ϵ_z is the vertical diffusivity and ϵ_x and ϵ_y are the horizontal diffusivities. The boundary conditions for the equation are specified according to known inflow and outflow conditions, no flux at any water free surface points, and bed boundary conditions determined from an empirical transport stage parameter T. The stage parameter is defined in terms of the critical bed erosion stress $\tau_{b,cr}$ and the actual bed stress τ_b in the form T = $(\tau_b - \tau_{b,cr})/\tau_{b,cr}$. The bed boundary condition is then described by either of two relations:

$$c_{bed} = \frac{MT}{w_{s0}} \quad \text{Specified concentration} \quad (2)$$

$$\left[\frac{\partial c}{\partial z}\right]_{bed} = M' T \quad \text{Specified flux} \quad (3)$$

in which M and M' are the respective specified flux or concentration fitting constants for mud, and w_{s0} is the clear water (or non-flocculent settling velocity of mud particles). In these calculations, the fall velocity of the sediment is described by a flocculation expression of the form

$$w_s = \alpha \, w_{s0} \, c^\beta \quad (4)$$

where α and ß are empirical constants. As noted by Van Rijn (1993), this expression should be valid for mud concentrations below 10,000 mg/l. The power coefficient ß will be positive and greater than 1 in this range, such that the net settling rate increases with increasing concentration. A constant fall velocity is obtained when ß = 0.

The two-dimensional vertical model neglects the $\partial/\partial y$-terms of Eq. (1). All 2DV-computations will utilize the flow geometry and temporal flow descriptions set out in the tidally induced time varying non-uniform flow over a trench. The geometry of the system is shown in Fig. 1. The flow, driven by tidal variation, traverses the deepened section (referred to as the trench) in either the positive or negative x direction only. Even though the flow direction changes during tidal variation, the left-hand side of the trench will be referred to as the upstream end, because flow initially moves from that end toward the trench. The flow variation in time is set according to a simple sinusoidal variation. The water depth varies according to a lagged sinusoidal expression (phase shift = 30°). The vertical velocity distribution is prescribed by a log-law expression using a zero-velocity level of 0.01 m, see DeVantier and Van Rijn (1993).

There are several settings which are implicitly assumed in the model related to initial and boundary conditions. The model sets all initial values of

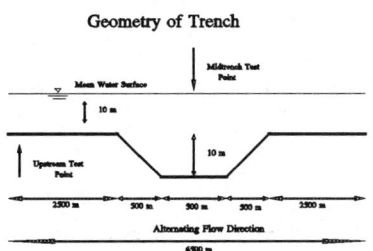

Figure 1. Geometry of the Test System

concentration to the steady flow, instantaneous equilibrium conditions (IE) determined by the initial flow conditions. In a similar fashion, any inflow and outflow boundary conditions (I/O B.C.'s) are set to IE conditions at every time step. It is assumed that the non-trench region is sufficiently large to justify this assumption. Other important model parameters are the time step Δt, and the x and y sediment diffusivities e_x and e_y. The values of various parameters and model assumptions used in the reference case are reported by DeVantier and Van Rijn (1993).

2DV Computations

The reference solution is represented by plotting three points in the vertical over 4 tidal cycles. Figs. 2 and 3 show the variation at the upstream point and in the trench. The concentrations appear to drop off much more quickly than might be expected given the small value of w_{s0}, and one might expect more uniform concentrations with depth with such a low settling rate. For the upstream plot, the concentration drops off very rapidly from the bottom concentration. The curve labeled as a middle level concentration is actually somewhat less than one half of the distance from the bed, so by the relatively similar magnitude of the surface concentration, it is apparent that most of the drop in concentration occurs near the bed. The midtrench concentrations exhibit a more uniform distribution of concentration over the depth around the 1500 mg/l level. The uniformity of concentration in the trench can be explained by two effects. First, convection carries sediment into the trench mainly from the upper layers. Not surprisingly then, the concentrations away from the bed upstream are on the order of 1500 mg/l. The second effect is the lowering of the bed concentration in the trench. This is due to the fact that the mean velocity has slowed to one half its value upstream. The T-parameter is related to velocity squared, and the relatively rapid response of the bed concentrations are due to the deceleration. The explanation of the apparently rapid response to changes in the bed concentration and the steep drops in concentration upstream can be expressed in terms of the flocculent settling conditions chosen. For the values of α, β, and w_{s0} in the reference case, the settling velocity can vary from ws = 0.034 m/s for c = 15,000 mg/l to ws = 0.001 m/s for c = 1,000 mg.l.

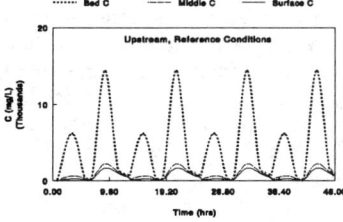

Figure 2. Upstream Concentrations, Reference Conditions

Now the influence of the initial conditions and the boundary conditions at the inlet and outlet will be examined. Runs were made in which the SUSTRA-model was modified so that an initial condition was applied with an equilibrium profile using the tide-averaged bed concentration of 4800 mg/l upstream in the reference case. Nothing else was different from the reference case. The predictions only differ from the reference case in the first cycle, indicating that the initial condition had no significant effect on the cyclic solution.

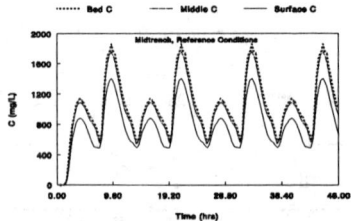

Figure 3. Midtrench Concentrations, Reference Conditions

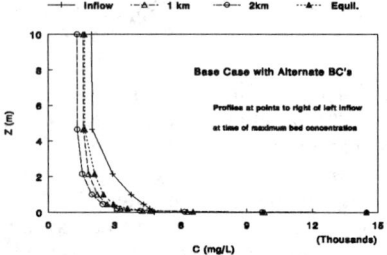

Figure 4. Conc. Profiles, Upstream, Ref. Case, Alternate I.O. B.C.'s.

The possibility that inflow and outflow boundary conditions were leading too quick a response because of convection, was also examined. The same equilibrium conditions as prescribed initially for the whole domain were now used at the inlet and outlet of the computational domain for all times. Surprisingly, the results for upstream and mid-trench concentrations were nearly indistinguishable from the reference case results. The reason can be seen by examining the vertical concentration profile development with distance downstream of the channel at the time of maximum bed concentration shown in Fig. 4.

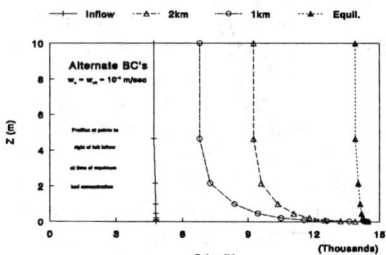

Figure 5. Conc. Profiles, Upstream, Const. Fall Vel., Alternate I.O. B.C.'s.

The boundary condition changes very rapidly to the 2 km distribution from the prescribed value at the inlet. The IE profile is also plotted, and it is somewhat higher than the distribution approaching the trench, because the bed concentration is increasing a bit more rapidly than the concentration can respond to instantaneously. Fig. 5 presents the same plot (effect of a prescribed tide-averaged equilibrium concentration at the inlet and outlet, at all times), for a constant fall velocity, w_s = 0.0001 m/s (no flocculation effects). The results are significantly different.

The I/O B.C.'s have a significant effect, and the low fall velocity means that the concentration changes so slowly that instantaneous equilibrium is also a poor estimate of conditions. Figs. 6 and 7 show a significant difference in time response compared to the same case without the alternate I/O B.C.'s (Figs 2 and 3). The lowest concentrations are bounded upward by the bed concentration of the inflow of approximately 5000 mg/L (this was 3000 mg/l in the reference case with the same constant fall velocity). The low settling rate and high mixing simply keep the concentration from ever going much below the inflow. To improve on this, the location of the inlet and outlet boundaries in tidal flow must be chosen very far away from the area of interest in case of a small fall velocity (~ 0.0001 m/s).

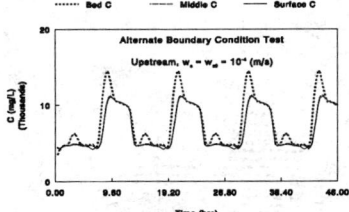
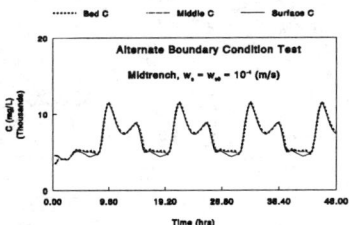

Figure 6. Conc. in Time at Upstream pt., Const. Fall Vel., Alt. I.O. B.C.'s.

Figure 7. Conc. in Time in Trench, Const. Fall Vel., Alt. I.O. B.C.'s.

3-D Computation

As a part of the proposed new port development at Belview, Waterford, Ireland, it was necessary to increase the depth of the navigation channel to -6.0 m C.D. This requires cutting of channels through the Cheekpoint Upper and Lower Bars (present levels are -3.5 m C.D.). The area of interest is the shipping channel at Upper and Lower Bar near Cheek-point in the river Suir at the confluence with the river Barrow (see Fig. 7). One of the alternative layouts studied was the construction of short groynes (surface level at 2.5 m C.D.) at the south bank in combination with a training wall at the north bank to increase the flow velocities in the Upper and Lower Bar channels (dredged to -6.0 m C.D.) in order to reduce the deposition rates. The channel width is 100 m. The tidal currents were computed by numerical solution of the depth-averaged flow equations. The model covered an area of approximately 7 km^2 of the rivers Suir and Barrow. The presence of the groynes in combination with the training wall yields an increase of the velocities of about 15% at Lower Bar Channel.

The SUSTRA-model was used to compute the sediment transport rates of silt material (30 μm). The vertical distribution of the flow velocity was represented by a log-law expression. The input values were: fluid density = 1023 kg/m^3, sediment density = 2650 kg/m^3, critical bed-shear stress = 0.1 N/m^2, bed roughness = 0.03 m, reference level at bed = 0.03 m, and time step = 30 min. The fall velocity formula used was: $w_s = 1.2 \times 10^5 \, w_{s,0} \, c_v$, in which $w_{s,0}$ = 0.03 mm/s and c_v = volume concentration giving values of w_s = 0.068 mm/s for c = 50 mg/l and w_s = 0.68 mm/s for c = 500 mg/l. The mud constant M was calibrated using measured deposition volumes of Upper and Lower Bar channels during the period November 1989 to May 1990 (with resultant M = 4.10-10). Fig. 7 shows the computed sediment transport rates at maximum ebb flow during springtide conditions. The largest values occur at the location of the channels near the groynes. Circulation patterns can be observed between the groynes and downstream of the groynes (typical deposition areas). Strong erosion does occur near the heads of the groynes where the flow velocities are relatively high. The silt concentrations varied from 500 mg/l just after maximum ebb flow to about 5 mg/l after turning of the horizontal tide. The concentration profiles are almost uniform in the vertical direction.

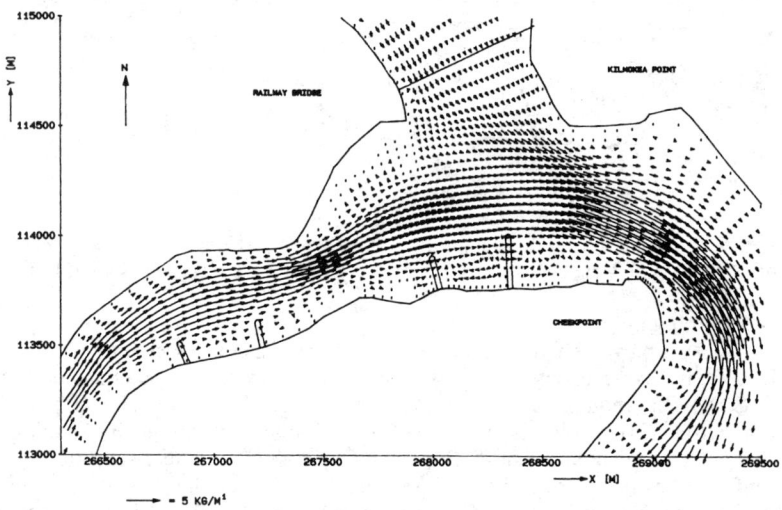

Figure 7 Mud transport rates at maximum ebb flow, spring tide

Acknowledgement
The support of The Netherlands National Computing Facilities Foundation is gratefully acknowledged.

Appendix. References
Rijn, L.C. van, and Meijer, K., (1986), "Three-Dimensional Modelling of Suspended Sediment Transport for Currents and Waves", Report H461/Q250/Q422, Delft Hydraulics, Delft, The Netherlands.
DeVantier, B.A., and Rijn, L.C. van, 1993, Effect of Basic Mud Parameters on Computed Concentration and Transport Rates, Int. Conf. Hydroscience and Engr., Washington, USA.
Rijn, L.C. van, 1993, "Principles of Sediment Transport in Rivers, Estuaries and Coastal Seas", 650 p., Aqua Publications, Amsterdam, The Netherlands.

THREE-DIMENSIONAL NUMERICAL MODELING FOR TRANSPORT STUDIES

W. H. McAnally[1], R.C. Berger[2], A. M. Teeter[3]

ABSTRACT

Modeling three-dimensional transport of salinity and sediments in estuarine flows requires that hydrodynamics be accurately modeled with sufficient precision to describe the advection and turbulent diffusion of salinity and sediments. These demands are considerably more stringent that those required for modeling water levels and discharges.

Application of the model RMA10-WES and the TABS-MD system of multi-dimensional models to San Francisco Bay salinity and sediment transport and Galveston Bay salinity illustrates the challenges involved. Residual flows in these bays reflect both density-driven flows, which are strongly three-dimensional, and tidal pumping, which is weakly three-dimensional. Asymmetry in bed stresses combined with these residual flows to induce three-dimensional sediment fluxes that may or may not be consistent with the residual flows.

INTRODUCTION

Numerical surface water modeling by the Waterways Experiment Station's (WES) Hydraulics Laboratory is performed in support of the civil and military missions of the U.S. Army Corps of Engineers. One-, two-, and three-dimensional (1D, 2D, and 3D) models of hydrodynamics and transport (salt, sediment, and constituents important to water quality) are used to evaluate changes induced by water resource projects such as flood control and navigation. As such, these models are usually used as engineering tools to plan projects,

---

[1] Chief, Estuaries Division, [2] Research Hydraulic Engineer, [3] Research Oceanographer, Hydraulics Laboratory, USAE Waterways Experiment Station, Vicksburg, MS.

to optimize project design and operation. Less often they are used to evaluate environmental effects of actions outside the project areas or simply to define the important processes of a water body.

The model application purpose, together with the modeled site characteristics, determines what model capabilities are required. For example, modeling the impacts of navigation channel enlargement requires accurate simulation of flows and transport at spatial scales approximating the changes in channel size, typically 2 to 6 ft vertically and a few hundred feet horizontally. That is a very different spatial scale requirement than that required for say, evaluating the impacts of river flow alteration on annual salinity regime.

MODELS

A number of three-dimensional model studies have been performed by the WES Hydraulics Lab using various models. This paper addresses the RMA10-WES model and two of its applications. RMA10-WES is a WES adaptation of the model RMA-10 developed by Ian King (1993). RMA10-WES is being incorporated into the TABS-MD system of models, which was previously 1D and 2D, and includes hydrodynamic, constituent transport, and sediment transport models.

RMA10-WES computes time-varying open channel flow in 1D, 2D, and/or 3D. It solves the Reynolds form of the Navier-Stokes momentum conservation equations, the mass continuity equation, convection-diffusion equation for three transport constituents (usually salt and temperature) and an equation of state for water density. The equations are fully three-dimensional, except for the standard assumption that vertical accelerations are negligible (the hydrostatic approximation.) Two turbulence closure options are available -- a quadratic eddy viscosity/diffusivity, and a Mellor-Yamada Level 2 (Mellor-Yamada, 1982) k-l approach modified for stratification by the method of Henderson-Sellers (1984). Horizontal diffusion can be oriented parallel and perpendicular to streamlines.

The model solves these equations by the finite element method of Galerkin weighted residuals and a modified Crank-Nicholson time stepping solution for unsteady simulations. Elemental shape functions are quadratic for velocity and concentration and linear for depth. Computational meshes can include 1D, 2D vertical, 2D horizontal, and 3D elements. Three-dimensional portions of the mesh can consist of one to many elements in the vertical. The vertical dimension is transformed such that the bottom profile remains, but the variable water surface elevation is transformed to a constant elevation.

GALVESTON BAY APPLICATION

Galveston Bay, Texas, is the largest and most productive estuary on the Texas Gulf coast. (Figure 1) It is wide and shallow, predominately shallower than 6 ft, and incised by a 40-ft-deep, 400-ft-wide navigation channel. Mean diurnal tide range is 1.4 ft, and freshwater inflow from its several tributaries averages about 13,000 cfs. Mixing conditions range from partly to well-mixed, with winds significantly affecting both circulation and mixing processes. The purpose of the modeling effort is to predict the salinity and circulation impacts of enlarging the channel by 10 ft in depth and 200 ft in width in two stages. Salinity and circulation changes will be evaluated directly and by the output of an oyster production model which uses the hydrodynamic and salinity results as input data.

The computational mesh (12,000 nodes) for this effort provided vertical resolution of 3 to 6 ft between nodes and horizontal spacing of 70 ft (in the channel) to 600 ft (in tributary bays.) The entire mesh uses the full 3D formulation except for 2D horizontal computations in the Gulf of Mexico and part of West Bay far from the channel. A somewhat lower resolution mesh which was previously used to produce velocities for a ship simulator study was run in 3D, but the salinity patterns were unsatisfactory.

Study requirements, principally the oyster model needs, necessitated model test durations of 9 to 12 months. Variable time steps ranging from 15 minutes to 60 minutes were employed. The tests produced approximately 8 gigabytes of results for each year of simulation. Three plan conditions and nine hydrologic conditions combined to produce more than 24 years worth of simulation time. Output of this magnitude requires innovative techniques to store, manipulate, and display.

The bay is shallow and wide and is strongly influenced by the frequent passage of storms with sustained high winds. Modeling these periods is particularly difficult, since large short term water surface fluctuations require smaller time steps than those needed in the relatively calm summer months.

Modeling of the long and deep navigation channel through the shallow and very wide bay is a challenge in balancing computational expense with numerical precision. Saltier water in the channel creates large horizontal salinity gradients in the channel to shallow bay transition zones shown in Figure 1. To retain these gradients we instituted a streamline salinity diffusion method along with the usual isotropic diffusion. Additional difficulties were caused by the vertical grid skewness near the channel which produced fictitious horizontal pressure gradients. We minimized this effect by setting the uppermost channel element to the same elevation as the surrounding bay elements.

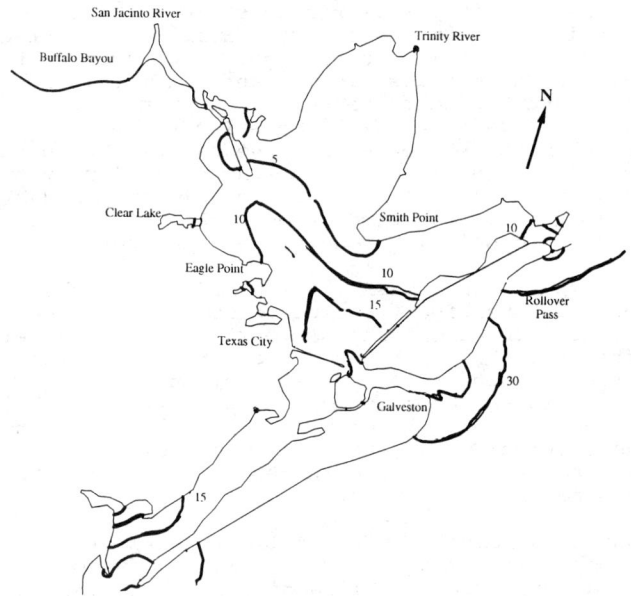

Figure 1. Galveston Bay Salinity Contours

SAN FRANCISCO BAY APPLICATION

In San Francisco Bay, two efforts of the Corps of Engineers required 3D numerical modeling -- the proposed Phase III improvement of the John F. Baldwin Project (JFB III) and the Long Term Management Strategy (LTMS) for disposal of dredged material. For JFB III, the impact on salinity intrusion in the northern bays was to be addressed by a combination of physical modeling and numerical modeling. The LTMS effort required tracking of dredged sediments that might be eroded from open water disposal sites.

Modeling of the partly-mixed San Francisco Bay has been accomplished with the TABS-MD system, using the 2D models RMA-2 and STUDH, and the 3D model RMA10-WES. In the course of performing these modeling studies, several meshes have been employed, ranging from fairly coarse resolution (about 2000 ft horizontally) to fairly fine resolution (about 200 ft.) The finer resolution meshes produced more coherent (and more realistic) residual currents.

Figure 2a shows net suspended sediment fluxes in central San Francisco Bay for a tidal cycle (from Hauck et al., 1990). The complex pattern of net transport is a result of subtle differences in net circulation and of the deposition/resuspension of sediments by tidal flows. Satisfactory modeling of such transport processes requires accurate simulation of flow divergences around obstacles such as islands and faithful reproduction of residual flows (plus accurate simulation of sediment deposition/resuspension.) Figure 2b shows how the residual current patterns for that area were reproduced in 2D calculations. The agreement with the net flux patterns of Figure 2a is striking, suggesting that in this case, the high resolution, low effective viscosity 2D calculation was sufficient to capture these patterns in a depth-integrated sense. Modeling which failed to capture the details of flow patterns around the islands would not accurately simulate sediment transport there, but 3D modeling was not necessarily essential.

ACKNOWLEDGEMENTS

The Galveston Bay application was funded by the U.S. Army Engineer District, Galveston. The San Francisco Bay application was funded by the U.S. Army Engineer District, San Francisco. Permission was granted by the Chief of Engineers to publish this information.

REFERENCES

Hauck, et al., 1990, "San Francisco Central Bay Suspended Sediment Movement, Report 1, Summer Condition Data Collection Program and Numerical Model Verification," Technical Report HL-90-6, U.S.A.E. Waterways Experiment Station, Vicksburg, Mississippi.

Henderson-Sellers, B., 1984, "A Simple Formula for Vertical Eddy Diffusion Coefficients Under Conditions of Nonneutral Stability, Journal of Geophysical Research, Vol 87, No. C8.

Johnson, B. H., et al, 1991, "Development and Verification of a Three-Dimensional Numerical Hydrodynamic, Salinity, and Temperature Model of Chesapeake Bay," Volume 1, Main Text and Appendix D,

Technical Report HL-91-7, U.S.A.E. Waterways Experiment Station, Vicksburg, Mississippi.

King, I. P., 1993, "RMA-10, A Finite Element Model for Three-Dimensional Density-Stratified Flow," Draft Report, University of California, Davis.

Mellor, G. L. and T. Yamada, 1982, "Development of a Turbulence Closure Model for Geophysical Fluid Problems," Review of Geophysics and Space Physics, Vol 20, No 4.

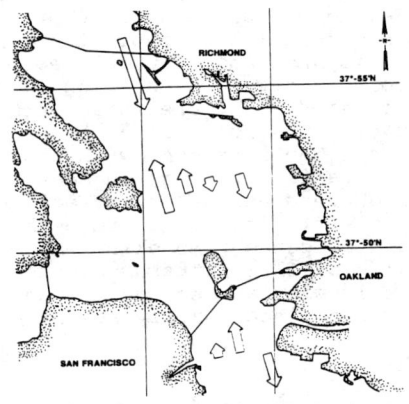

Figure 2a. San Francisco Bay Net Sediment Fluxes From Field Data

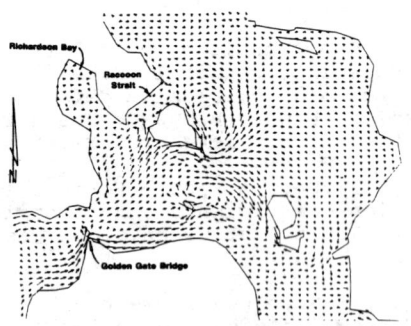

Figure 2b. San Francisco Bay Depth-Averaged Residual Currents

Mixing Character and Meandering Mechanism of a Plane Jet Bounded in a Shallow Water Layer

Daoyi Chen[1] and Gerhard H. Jirka[2]

Abstract

A plane jet in a shallow water layer is characterized by the jet width $b_{1/2}$ much larger than the water depth h at large downstream distances $x/h \gg 1$. Instantaneous field concentration distributions have been measured with a Laser-Induced Fluorescence system to illustrate the role of large scale turbulent structures in the mixing process. Mean and r.m.s. concentration field distributions are obtained by averaging hundreds of digital image frames. It is found that the r.m.s. concentration on the centerline decays as a function of x/h where x is the longitudinal coordinate and h is the water depth. This is different from the usual jet where the flow characters are the function of x/B (B is the jet exit width).

By doing instability analyses, critical bottom friction parameters S_c are obtained as a function of the jet co-flow ratio. From the viewpoint of suppression of the jet instability by viscosity or bottom friction, there is a similarity between a low Reynolds number jet in unbounded conditions and a high Reynolds number jet in a shallow fluid layer. This explains the meandering behavior of the shallow jets.

Introduction

Rivers, estuaries and coastal regions are all water bodies with a shallow layer if one notes that the horizontal scale $b_{1/2}$ of a jet discharge may be one or two orders greater than the water depth h. The meandering of jet flows in such shallow layers has been found, and detailed turbulent velocity measurements by means of LDV have been carried out by Dracos, Giger and Jirka (1992). As a

---

[1]Research Associate, School of Civil and Environmental Engineering, Cornell University, NY 14853

[2]Professor, Fellow ASCE, DeFrees Hydraulics Laboratory, School of Civil and Engineering, Cornell University, Ithaca, NY 14853

continued research effort, the present paper first focuses on the mixing characteristics of the "shallow" jet, and then examines the mechanism of jet meandering in a shallow water layer.

Experimental Studies

Experiments are carried out on a 6m x 7.8m shallow water table. Flow volume is measured by two vortex flow meters that are capable of achieving 1% precision. The water depth may vary from very shallow 2.5 cm to 20 cm. The ambient uniform flow at the entry of the water-table is produced by porous material with very low velocity to provide the entrainment flow. The jet flow is generated in a box with a 10:1 contraction. The width of jet exit is $B = 1$cm and the Reynolds numbers $Re_B = \dfrac{U_0 B}{\nu}$ are around 10,000 in which U_0 is exit velocity. The concentration measurements are obtained with a planar laser-induced fluorescence (LIF) system which include a 5W Argon laser, a scanning mirror and a CCD video camera mounted a sharp-cut color glass filter (530 nm) to eliminate the

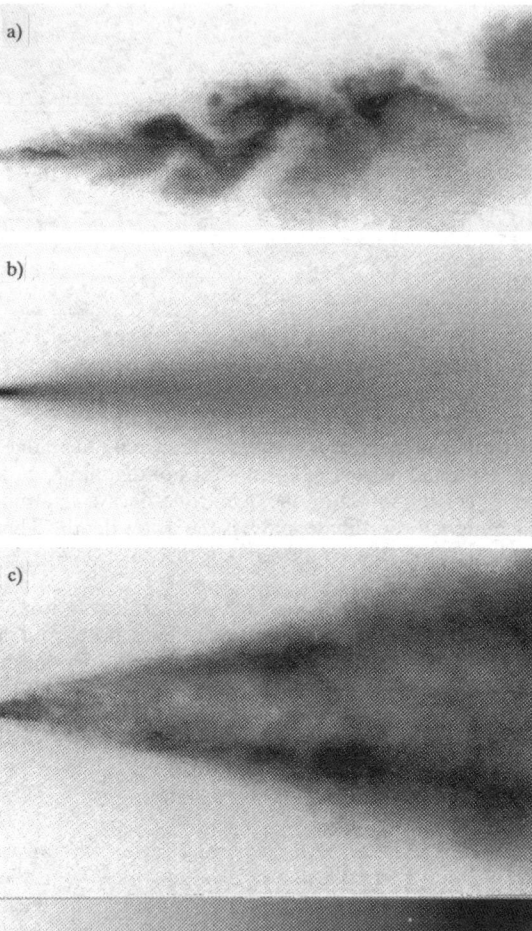

Fig.1 LIF measurement for a $B = 1$cm jet flow with velocity $U_0 = 1.36$ m/s discharged in an $h = 6$cm shallow water layer: a) instantaneous field concentration; b) mean field concentration obtained by averaging over 100 images; c) r.m.s. field concentration distribution from images averaging.

reflected laser light. Rhodamine B dye is released into the jet flow. Video images are recorded on a super-VHS VCR. A frame grabber board was plugged into an IBM PS-2 computer to digitize the video image into a digital image with a gray scale 0-255. A software package is written in C to accomplish the digital image processing.

Experiments have been carried out with characters range of water depth from 2.5cm to 10cm and of jet exit velocity from 1.17 to 1.75 m/s. As shown in Fig.1a, the unique feature of a plane jet in a shallow layer is the meandering that starts at about ten times the water depth. By calibrating and averaging over one hundred LIF images, field distribution of mean concentration and r.m.s. are obtained. As shown in Fig.1b, the concentration on the centerline gradually drops toward the downstream. A Gaussian distribution is found (not shown hear) for the normalized transverse mean concentration profiles. The half widths of the cross section mean concentration profile are found to increase linearly, $b_{1/2} = c_b(x+x_0)$. For different water depths, the coefficient c_b is fairly close to 0.17 which is 1.7 times wider than the LDV measured velocity profiles ($b_{1/2} \propto 0.1x$) by Dracos, Giger and Jirka (1992). As is well known from other jet studies, the concentration profiles are generally 1.4 times wider than the velocity profiles in same flow conditions ($c_b = 0.14$). The present wide concentration profile could be explained through the meandering character of shallow jets. Although the longitudinal velocity is small at the edge of jet, the dye concentrations carried by the meandering jet are still high. Of course, some minor factors in measurements (such as light scattering) may contribute to this behavior to some extent.

As has been reported in Bashir and Uberoi (1975), the decay of mean temperature on the centerline behaves as $(\frac{T_0}{T_m})^2 \propto \frac{x}{B}$. However, as has been found in Dracos, Giger and Jirka (1992), the $(\frac{U_0}{U_m})^2$ does not obey the same relationship for the case of a shallow water layer. The present measured results shown in Fig.2a are consistent with Dracos, Giger and Jirka's (1992) velocity

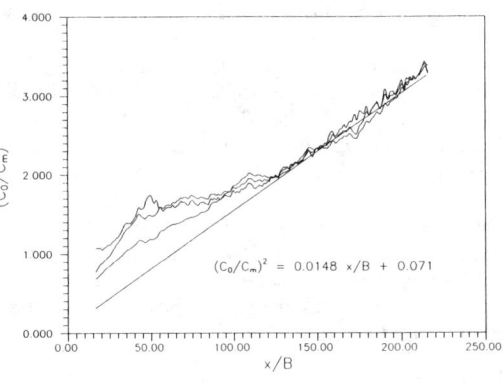

Fig. 2 a)

measurement. In the far range of $x/B > 100$, the relation of $(\frac{C_0}{C_m})^2 = \frac{x}{B}$ is well fitted for the data. However, in the near region ($x/B < 100$), the same relation does not exist.

The r.m.s. field distributions (in Fig.1c) show a valley around the centerline. In particular, the r.m.s. value on the centerline from the exit drops to a minimum at about $x/B = 70$ and then gradually grows. Then a saddle point appears on the centerline of the transition region. This behavior is unique for jets in a shallow fluid layer. The r.m.s. concentrations on the centerline are shown in Fig.2b for the case of $h = 2.5$, 4.0, 6.0 and 8.0cm respectively. The r.m.s. first gradually decreases from the jet exit to a minimum value. Then it increases to a maximum value and then gradually drops. The interesting aspect is that the location of the minimum and maximum r.m.s. concentration are both found to increase linearly with the water depth.

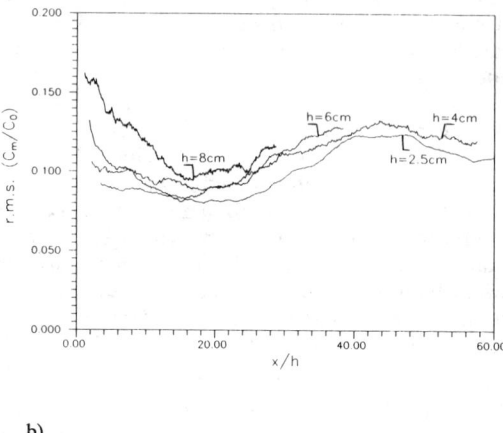

b)

Fig.2 Centerline distribution of mean and r.m.s. concentration for jet flows discharged into shallow layers, h=2.5cm to 8cm: a) mean concentration; b) r.m.s. concentration.

Combining the flow visualization and LIF measurements, the jet flow in a shallow layer can be classified into three regimes: three-dimensional flow, two-dimensional meandering flow and stabilized flow. As it has been pointed in Dracos, Giger and Jirka (1992), the centerline turbulent intensity reaches the maximum value at about $x = 6h$ and the three-dimensional flow lies in the range between jet exit and about $x = 10h$. Although the flow visualization is somewhat subjective, this $x = 10h$ is close to the location of minimum r.m.s. concentration on the centerline at about $x = 14h$. Furthermore, power spectrum analyses illustrate the slope of the power spectrum changes from -5/3 in the transition regime to -3 in the meandering regime where the -3 wavenumber dependence indicates a two-dimensional turbulence regime.

Instability Analyses for Shallow Jets

Assuming self-preserving flow, a jet velocity profile can be assumed as

$\dfrac{U}{U_m - U_a} = R_j + sech^2(\dfrac{y}{l})$ where U_a is the ambient flow velocity, U_m is the centerline velocity of jet, $R_j = \dfrac{U_a}{U_m - U_a}$, y is the transverse coordinate and l is the transverse length scale. From the depth-averaged equations of motions, the Orr-Sommerfeld equation with bottom friction is derived as

$$(U(y) - \dfrac{\beta}{\alpha} + \xi)(\phi'' - \alpha^2 \phi) + \xi \dfrac{U'}{U}\phi' - U''\phi = \varepsilon_h (\phi'''' - 2\alpha^2 \phi'' + {}'\alpha^4 \phi) \qquad (1)$$

where ϕ is the eigenfunction which represents the amplitude of the disturbance in the y direction and the differentiation is with respect to the transverse coordinate y. $\alpha = \alpha_r + i\alpha_i$, where α_r is the wavenumber of the disturbance, and $-\alpha_i$ is the spatial amplification rate, and $\beta = \beta_r + i\beta_i$, where β_r is the frequency of the disturbance, and $-\beta_i$ is the temporal amplification rate, and $\xi = \dfrac{c_f U}{i\alpha h}$. The boundary conditions are $\phi(\pm l_0) = \phi'(\pm l_0) = 0$ where l_0 tends toward infinity. A local jet stability parameter is defined as $S = c_f \dfrac{2b_{1/2}}{h}$ in which $b_{1/2}$ is the local jet half-width.

From the solution of Eq. 1, the present work has determined that when the Reynolds number is larger than a certain value, say 1000, the viscosity effect is negligible. Therefore, the following results are limited to the inviscid case. When there is no ambient flow ($R_j = 0$), neutral curves for both sinuous and varicose modes are shown, respectively, in Fig.3. Two critical S values are 0.69 and 0.11 and the sinuous mode is more unstable. More computations have been accomplished to find corresponding critical S

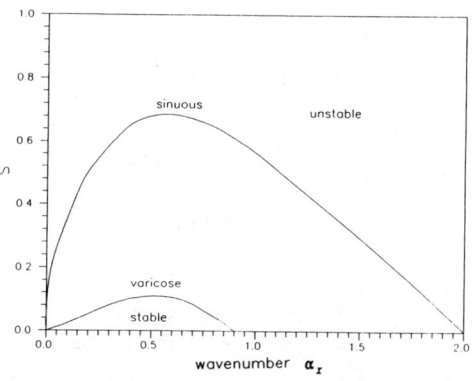

Fig.3 Neutral curve of jet instability corresponding to the bottom friction parameter $S = c_f \dfrac{2b_{1/2}}{h}$.

values of the sinuous mode for different R_j (i.e., ambient flow condition such as co-flow or counter-flow). For the co-flow situation ($R_j > 0$), the critical stability parameter S_c of sinuous mode, when R_j increases, gradually reduces its value (see Fig.4). For the case of counter-flow ($R_j < 0$), the critical S_c value increases sharply, as shown in Fig.4.

The bottom friction parameter S is proportion to $b_{1/2}$ in the definition. As $b_{1/2}$ linearly increases toward jet flow downstream, S will grow to the critical value at a certain downstream location. In the near region of the jet, the S value is very small and has almost no influence on the jet flow. In the meandering region, the stabilizing effect dampens those three-dimensional disturbances and makes the two-dimensional wave structure much more apparent. With a similar meandering feature, the jet in a shallow layer is alike the low Reynolds number jet flow where the viscosity dampens the free shear instability (see Brown, 1935). At a certain downstream location, the instability of the jet flow will be damped completely in a shallow fluid layer. Thereafter, the jet flow becomes fully stable. Although one may expect to observe a stable jet flow, unfortunately, the conditions to create the critical S is beyond the ability of the experimental facility.

Acknowledgement

Support by the U.S. National Science Foundation and the Electric Power Research Institute is gratefully acknowledged. The instability computations were carried out with the Cornell National Supercomputer Facility supported by the National Science Foundation.

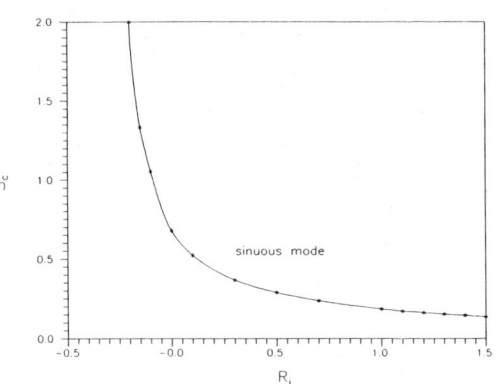

Fig.4 Critical bottom friction number as a function of velocity ratio Rj. When underneath the curve, flow is unstable. Over the curve flow became stable.

Reference

Bashir, J. and M.S. Uberoi, (1975), "Experiments on turbulent structure and heat transfer in a two-dimensional jet", The Physics of Fluids, 18, 405-410.

Brown, G.B., (1935), "On vortex motion in gaseous jets and the origin of their sensitivity to sound", Proc. Phys. Soc. 47, 703-732.

Dracos, T., M. Giger and G.H. Jirka, (1992), "Plan turbulent jets in a bounded fluid layer", J. Fluid Mech. 241, 587-614.

Selective Withdrawal for Reducing Turbid Water in A Reservoir

Yukihide Tashiro[1], Masahiko Yatsugi[2], Jouichiro Tano[3] and Naoki Matsuo[4]

1. Preface

This paper deals with improvemental operations of the selective withdrawal facilities as one of the comprehensive technics for shortening the turbid water stay after floods flown into Hitotsuse Reservoir. The Hitotsuse Dam Reservoir, located at the center part of Kyushu District in Japan and used exclusively for electric power generation, has a height of 130 m, a catchment area of 415 km^2, a total capacity of 216 million m^3, an utility depth of 30 m and a maximum power generating discharge of 137 m^3/s. Moreover, the dam is equipped with an emergency outlet tube under the lower intake with a capacity of 121 m^3/s.

The problems of turbid water have occurred in the downstream river since the completion of the dam in 1963. To cope with the problems, selective intake facilities were, first in Japan, built in 1974 and have been successfully operated. A serious turbid water problem, however, occurred after a large flood in 1989 under the circumstances that the district people were much concerned about the environmental issues. So, more effective technique for further shortening the period of turbid water discharge were urgently required.

2. Basic Policy for Reduction of Turbid Water

The adopted basic policy of the control on the turbid water problems is to discharge turbid water through the lower intake as early as possible and to form clear surface layer. The clear water in the surface layer is to be discharged through the upper intake facilities. Furthermore, re-floating of settled substances by the convection mixing in the winter is to be prevented.

Following the basic policy, applicability of several control technics for reduction of turbidity in the Hitotsuse River was investigated. Eleven technical plans shown in Table-1 were selected as the objectives for comparison. To make quantitative evaluation of these technical plans, (1) Investigation of suspended substances flowing in at the time of floods, (2) Analysis of turbid water phenomena in the reservoir, (3)

---

1   Assist. Manager, Civil Eng. Dept., Kyushu Electric Power Co., Inc., 1-1-82, Watanabe-dori, Chuo-ku, Fukuoka, Japan
2   Assist. Gneral Manager, West Japan Eng. Consul. Inc.
3   Assist. Gneral Manager, Civil Eng. Dept. Kyushu Electric Power Co., Inc.
4   Professor, Civil Eng. Dept., Chubu University

Study on the effect of selective withdrawal technique by numerical analysis, (4) Estimation of costs for the facilities to implement the plans, (5) Calculation of electric power energy reduced by implementation of the plans, and so on were carried out. In this paper, hydrological characteristics of turbid water phenomena and the effect of selective withdrawal technique only are described.

Table-1 Technical Plans for Control of Turbid Water Problem

Code	Technics
A	To carry out the protection or greening works of the retaining walls surrounding the reservoir
B	To operate the power plant at MCO using the lower intake
C	To operate the power plant at MCO using the midium intake
D	To allow a margin of the intake operation by extending depth of the lower intake
E	To lower the location of the lower intake
F	To operate the power plant at MCO while using the emergency discharge channel
G	To floculate the suspended materials in the reservoir and discharge channel
H	To install artificial agglutination filter in the downstream river
I	To eliminate mud on the downstream river bed by discharging cleaning water from the power plant
J	To install a new intake at far end of the reservoir to discharge the clear water soon after a flood
K	To suspend the intaking of the downstream creeks flow after a flood and to dilute turbid water discharged with them

where, MCO means Max. Continuous Operation

3. Investigation of Turbid Water Phenomena during Floods

(1) Investigation of Suspended Substances in Flood Flows: The turbid water at the time of floods in 1990 was sampled at upstream points of the reservoir for detailed analysis and investigation. Suspended substances in turbid water is very fine clay carried by landslides, and the diameter of less than 5 μm occupies 70 to 80 percent.

(2) Laboratory Experiment of Sedimentation: Mud accumulated on the bottom of the Hitotsuse Reservoir was sampled and its sedimentation rate was measured by using a transparent cylinder of an inner diameter of 25 cm and a height of 4 m. About 50% of the mud had a sedimentation rate of less than 1 m/day.

(3) Analysis of Turbid Water Phenomena in The Reservoir: The analysis was made based on the hydrological data, the temperature and turbidity distributions observed during and after the past floods in the reservoir.

a. Thermal Stratification: The thermal stratification in The Hitotsuse Reservoir is noticeably formed at EL. 160 m of the lower intake level in the heat-receiving period of June and disappears in the cooling period of December through January. The thermal stratification is sustained even after floods of a maximum discharge of less than 3,000 m^3/s, but is broken after those of more than 3,000 m^3/s. The turbid water mainly flows into the hypolimnion when the thermal stratification is sustained, but spreads out all over the reservoir when it is broken.

b. A Total Amount of Turbidity-causing Substances: The authors introduce a concept to indicate a degree of turbidity in the whole reservoir by a total amount of turbidity-causing substances suspended in the reservoir (Unit: ton). Figure-1 shows a correlation between the accumulated discharge (Unit: m^3) from the lower intake for power generation during a flood and the residual amount of suspended substances. Figure-2 shows a correlation between the residual amount of suspended substances and the depth of clear water layer (Unit: m) formed in the surface layer. Though some data show a little deviations, the thickness of clear water layer increase as the residual amount of turbidity-causing substances decrease. These figures taught us a significant suggestion: The maximum continuous discharging from the lower intake draws the residual turbidity-causing substances out of the reservoir, and results formation of clear surface layer. The early changeover to the upper intake for discharging the clear water from the surface layer therefore becomes possible.

Since the upper intake operation without drawing the turbid water needs 10 m depth of clear water at least, the maximum continuous discharging from the lower intake is required until the residual amount of turbidity-causing substances reduces to 5,000 ton.

c. Residual Amount of Turbidity-causing Substances at the End of Rainy Season: When 7,000 ton substances remains at the middle of October, large quantities of turbidity-causing substances re-floats in the cooling period. When the remaining amount is 3,000 ton, however, the re-floating in the winter is extremely few. On the basis of this fact, thus, the substance amount must be reduced to less than about 5,000 ton at the end of rainy season to avoid the discharging of turbid water in the cooling period.

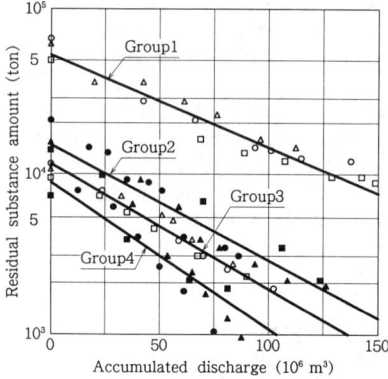
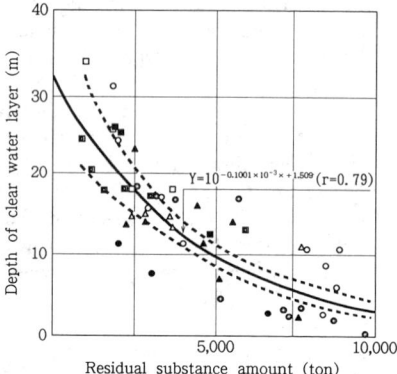

Figure. 1 Correlation between accumulated discharge and residual substance amount

Figure. 2 Correlation between residual sabstance amount and depth of clear water layer

6. Effect of Selective Withdrawal Technique by Numerical Analysis

Effects of the selective withdrawal technique plans shown from B to F in Table-1 were verified with a numerical analysis using a horizontal layer model. The objective floods for the verification were those in 1976 (maximum flowing-in discharge was 2,290 m³/s), in 1982 (3,290 m³/s) and in 1989 (3,090 m³/s), which are the larger scale ones after the selective withdrawal facilities had been installed.

(1) Plan B (Max. Continuous Operation with lower intake): Changeover to the upper intake after the flood in 1989 needed 31 days in the actual power operation, but in the case of taking the maximum power operation with the lower intake, the calculated period till the changeover was 26 days. The period can be shorten by 5 days.

(2) Plan C (MCO with middle intake): Where the turbidity of discharging water through the operation with the middle intake was compared with that with the lower intake, the turbidity decreased soon after the middle intake had started, but it increased and the depth of clear water layer decreases as the middle intake continued. This plan was judged not suitable for reducing turbid water.

(3) Plan D (MCO up to a water level of EL 180 m): The period of turbid water discharge through the power plant can be shorten by 11 days when the MCO with lower intake is continued until the water level has reached EL 180 m on the same condition as Plan B.

(4) Plan E: The elevation of the lower intake is assumed 20 m lower than the actual lower intake level of EL.160 m to verify the effect of its location on the turbidity reduction. The calculation revealed that this condition is effective for reduction of the residual turbidity-causing substance amount after the flood in 1989, which lowered the thermal stratification about 20 m, but not effective at the flood in 1982 when thermal stratification was broken..

(5) Plan F: When turbid water is discharged through the lower intake with the emergency outlet tube, reduction of turbidity is calculated in all cases of the floods.

7. Determination of Technical Plans to Pursue

In the course of survey and investigation on the 11 technical plans for reducing turbid water, the comprehensive evaluation was made from the viewpoints of (1) technical feasibility particularly on workability and site acquisition, (2) ease of the maintenance and management, (3) costs for implementation of the plans and costs equivalent to decrease of generating energy due to execution of the plans, (4) environmental impact to the downstream including the ecological system and scenery, and (5) effects of the technical plan on decrease of turbidity and shortening the period of turbid water discharge. Consequently, 5 technical plans (A, B, D, I, and K) shown in Table-1 by bold-faced type are selected for implementation.

8. Effect of Technical Plans Applied

Among the plans selected to pursue, 3 plans allowing immediate implementation (B, I and K) have been taken place since 1990. The others are under construction or still investigation stages. The authors consider that the Plan B would be the most effective technique. The concrete operation method of the selective intake is improved as follows.

When the turbidity-causing substances of more than 5,000 ton flows into the reservoir, the power plant operates at MCO using the lower intake and the operation is continued till the amount has decreased to less than 5,000 ton. When the depth of the clear surface water reaches more than 10 m in or before August, or more than 15 m in or after September, the operation mode changes to use the upper intake.

The improved operation method of the intake facilities was applied for the turbid water during six floods in 1990 and 1991. Figure-3 shows relations between maximum flowing-in discharge and the period till the changeover to the upper intake operation after a flood. The obtained data by improved operation method show a strong or consistent correlation, while those by the conventional operation method are bumpy. This is due to the fact that the conventional method had no explicit standards of turbid water discharge after a flood and a changeover timing to the upper intake. In the new method, however, the period of turbid water discharge, even if compared with the shortest period (as shown by a dotted line in Figure-3) by the conventional operation method, can be shortened by 8 days. The turbidity of discharged water after changeover to the upper intake operation is also managed at within 10 mg/ℓ (the standard of clear water in The Hitotsuse River). Therefore, it is concluded that the new operation method is effective for improvement of turbid water problem.

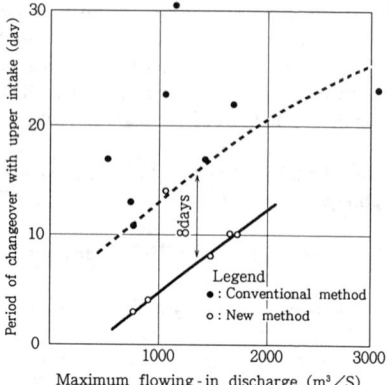

Figure. 3 Correlation between maximum flowing-in discharge and period of changeover with upper intake

9. Conclusion

It is very difficult to solve the problems of turbid water in the reservoir completely. So, the multiple methods realistically applicable for The Hitotsuse Reservoir were carried out. As for the results, the residents in the district have evaluated that the rivers have become cleaner than before.

Moreover, as an index for judgement of the period of turbid water, the concept of residual turbidity-causing substances amount in the reservoir was introduced. Based on the results of investigation of turbid water at the past floods in Hitotsuse Reservoir, the early lowering of the amount of residual turbidity-causing substances to 5,000 ton after floods resulted in shortening the period of turbid water discharge. The authors believe that this technique is useful in the other reservoirs with the similar conditions and problems.

References

(1) Patric J. Ryan & Donald R. F. Harleman:
"Prediction of The Annual Cycle of Temperature Changes in a Stratified Lake or Reservoir", Technical Report No.137, Dept. of Civil Eng., Massachusetts Institute of Technology, 1971

(2) Tokuzo Naomura & Shuichi Aki:
"Long-Term Persistence of Turbid Water Phenomenon in Hitotsuse Reservoir",

SHIFTS IN SOLUTE TRANSPORT DIRECTION INDUCED BY TRANSIENT FLOW

Francis X. Markert[1] and Rafael G. Quimpo, M. ASCE[2]

Abstract

Traditionally, solute transport projections have focussed on the long-term, steady-state conditions expected to prevail in a groundwater system. However, transient effects may often dominate the flow system and cause major changes in the transport direction and velocity which might be missed in a steady-state analysis.

An intermittent stream was investigated within the boundary of a previously modeled RCRA facility. Measured streamflow and groundwater level fluctuations were used to develop a transient model of the stream's influence on the groundwater system. Results of this model were examined through particle tracking analysis and showed a groundwater velocity increase of approximately 20 percent and directional changes up to 55 degrees. In addition, a nearby flow divide was shifted causing groundwater to move towards two large capacity municipal water supply wells. These results demonstrate that transient flow condition analysis should be made an integral part of the assessment procedure in areas where surface water interaction is present in order not to miss significant shifts in contaminant transport direction.

Introduction

Transient flow events occurring in nature have long been recognized in science as having effects on physical systems present. Typically, these events are of large magnitude and areal impact and have what is considered an immediate effect on human life and property. However, not all transient events need have a large magnitude and areal impact to be considered important for use as design criteria. This is especially critical in groundwater flow simulations, in which the normal practice is to simulate average or steady-state conditions. This causes an immediate bias away from the minimums and maximums which may be experienced by the system. In many cases, the effects of local transient flow events are never simulated by the model to assess their effects.

---

[1] Hydrogeologist, IT Corporation, Monroeville, PA, 15146
[2] Professor, University of Pittsburgh, Pittsburgh, PA 15261

This paper shows the effects of one such model when used to address a short-term, transient, local recharge event. The transient influence on the groundwater system is caused by an intermittent stream which is directly interconnected with an unconfined aquifer system. The stream flows intermittently during the summer and fall. More sustained flows occur during the winter and spring months, when runoff is high. This study uses a previously calibrated 3-dimensional local groundwater flow model of the surrounding area to assess the effects of these flow events on the aquifer system, with particular emphasis on changes which may influence the local solute transport system of the aquifer.

Study Area Setting

The intermittent stream used for this study is Paddys Run, a small stream located in southwestern Ohio (Figure 1). The stream drains a midsize catchment (approximately 15 square miles) which connects with a larger glacially filled valley. This glacially filled valley contains the Great Miami River and its associated tributaries (Fetter, 1980).

Flow in Paddys Run occurs on an intermittent basis, with an estimated discharge of 0.005 to 0.11 cubic meters per second (Dames and Moore, 1985). During its high flow periods in the winter and early spring months, flow occurs almost continuously due to snowmelts and precipitation events, resulting in a 0.3- to 1.2-meter rise in water levels. During the summer and fall months, Paddys Run is dry most of the time, except following thunderstorm events. During these events, water levels in the stream show a 0.9- to 1.2-meter rise in water levels for a very short period of time. Monitoring well and stream monitoring stations along Paddys Run were used to observe these fluctuations in Paddys Run and the Great Miami Aquifer.

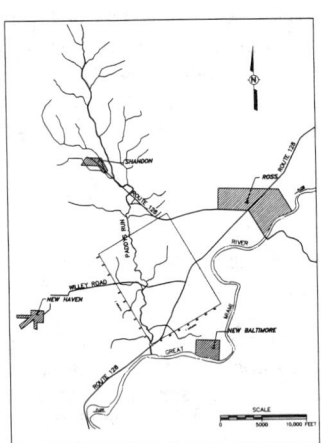

Figure 1
Study Area Location and Model Grid Area

Paddys Run flows on top of the Great Miami Aquifer where it becomes interconnected with the underlying groundwater flow system. The Great Miami Aquifer consists of a series of unconsolidated sand and gravel deposits which lie within an older, glacially formed valley termed the New Haven Trough (Fenneman, 1916). The underlying bedrock in the region consists of shales and thin interbeds of limestone into which the glacial valley was carved. The New Haven Trough was later filled in with glaciofluvial sediments to a depth of approximately 60 meters. These sediments were deposited by water running from the margins of the glaciers and consist mainly of well-sorted sand and gravel. This formed the Great Miami Aquifer.

Overlying the Great Miami Aquifer are the glacial overburden sediments. These sediments vary in composition, and include loess, lacustrine deposits, till, and glaciofluvial sediments. The glacial overburden varies from 2 to 15 meters in thickness, but generally averages between 6 and 9 meters in Paddys Run basin. Paddys Run has cut through the glacial overburden from its confluence with the Great Miami Aquifer to approximately 4 kilometers upstream. As a result, a portion of the Great Miami Aquifer is directly exposed to surface water infiltration and river leakage. In Paddys Run, only a thin layer of silty bed load lies on top of the aquifer at the bottom of the channel.

Baseline Groundwater Flow Model

To characterize flow out of Paddys Run and into the groundwater system, a previously calibrated 3-dimensional local flow model for the region was used. This model was developed to assess the impacts from the RCRA facility located adjacent to Paddys Run (IT Corporation, 1988). It consists of a local flow model which was developed using the SWIFT III code and contains a tight, uniform grid spacing and covers a small area (Reeves et.al., 1986). The flow model covers an area of just under 11.5 square kilometers and uses a constant 38-meter grid spacing (Figure 1). This model encompasses the reach of Paddys Run which loses water to the Great Miami Aquifer during transient flow events. Paddys Run is included in the local flow model from the point where it begins to interconnect with the aquifer down to a point approximately 400 meters north of its confluence with the Great Miami River. This allows the model to simulate groundwater movement below Paddys Run in the reach where transient groundwater mounding is most pronounced.

To properly simulate the geologic variability present in the unconsolidated sediments as well as to allow for vertical flow of groundwater, the model was divided into 5 layers, each with its own distinct characteristics. The size of the local flow model was designed to incorporate Neumann (constant flow) and Dirichlet (constant head) boundary conditions along the perimeter of the model grid without influencing the study area in and around Paddys Run. Results of the model show that groundwater flow underneath Paddys Run generally parallels streamflow.

Groundwater Flow Modeling

To assess the degree of interconnection of Paddys Run with the underlying Great Miami Aquifer, two stream and aquifer monitoring stations were installed. These wells were used to continually monitor water elevations in the stream and in the aquifer.

Steady-state flow modeling of the transient mound in the aquifer started by using the previously calibrated local flow model to simulate long-term steady-state conditions. The calibration was done by comparing head increases between the January and April 1989 groundwater measurements with the head increase predicted by the model. A total of 11 monitoring wells along Paddys Run were used for this purpose. A match was found to occur at a recharge rate of 7.6 centimeters/day.

Using the infiltration rate derived by the steady-state flow modeling, transient simulations were made to determine the rate at which the aquifer responds to the increased infiltration rate. Simulations were done using 1-day time steps for a total simulation period of 84 days, roughly corresponding to a period of 3 months. Comparing the model-generated hydrograph to the groundwater hydrographs measured underneath Paddys Run indicate that the model shows good correlation with the observed data.

Effects of Transient Mounding

To understand and quantify the effects of the transient mounding on the groundwater flow system and thus the solute transport system, particle tracking analysis was done. Particle tracking is a technique for determining and depicting the 3-dimensional movement of groundwater in a finite-difference flow model. The STLINE code was used for this (GeoTrans, 1987). Using the particle tracking model, a set of initial runs was conducted using the flow field generated during the steady-state simulation for transient mounding conditions. A total of 24 particles were placed in 8 locations along Paddys Run and were tracked through the system at monthly increments for one year.

As can be seen in Figure 2, the transient mounding condition shifts the flow field by approximately 4 to 55 degrees away from its normal path in the vicinity of Paddys Run. This effect is strongest in the northern reach of Paddys Run that flows over the Great Miami Aquifer. In this area, particles normally traveling south-southeast were shifted between 20 and 55 degrees to the east, with an average shift of 34 degrees. Further south, where Paddys Run mounding caused a slight eastward shift, the deflections ranged between 4 and 35 degrees with an average shift of 16 degrees. In this region, the mounding causes a gradient change but only a slight gradient shift.

The other effect due to the transient mounding may be noticed by examining the total flowpath length of the particle tracks in Figure 2. For most of the particles, the mounding has the effect of increasing the groundwater velocity and thus the travel distance over the one-year time period shown.

The overall effect of the transient mounding on the solute transport system was also evaluated utilizing a series of particle tracking runs. To determine the effect on hypothetical contaminants located along Paddys Run, STLINE was run using two sets of data input files, one simulating long-term, steady-state conditions and one simulating transient mounding conditions. The two conditions were coupled by allowing one flow field to dominate for 6 months and then the other flow field to dominate for 6 months more. This simulated study area conditions during 1989, when mounding was present from January through June and normal conditions prevailed from July through December. Particles were started in the same eight locations spread and were tracked until they had either passed outside of the model grid or until their flow direction out of the model had become clear. Particles tracked for this simulation show the influence of the two flow fields as an oscillation in the particle track seen every 6 months, as the direction of movement shifts.

Results of this analysis are shown in Figure 3. This figure shows the steady-state flow field with no mounding effects and the results of the particle tracking for the combined steady-state and transient mounding flow fields. As can be seen, in the steady-state simulation, two of the particle paths travel in an easterly direction away from Paddys Run while six particle paths travel to the south, roughly parallel with Paddys Run. In the transient simulation, four of the particle paths traveled to the east, toward the municipal supply wells, while the remaining four particles traveled to the south. This shows a localized shift in the groundwater flow divide to the south, causing more flow to the east than was previously occurring. For solute transport, this would cause a split in contaminant plume migration if located in this area. This effect was not considered in the original local area model implementation.

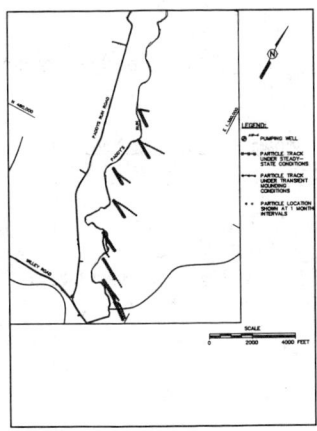

Figure 2
Effects of Transient Mounding
on Groundwater Flow

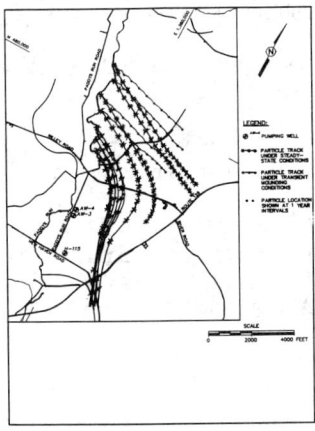

Figure 3
Long-Term Effects of
Transient Mounding

Conclusions

The overall influence of Paddys Run on the Great Miami Aquifer in its local area was found to be significant, particularly during the late winter and spring months when heavy streamflows are common. The effect was limited to the reach of stream that always loses water to the aquifer. Regions of Paddys Run that do not connect with the aquifer or that do not normally lose water showed no direct influence.

This study substantiates the hypothesis that intermittent streams can cause transient groundwater flow effects which in turn can induce changes in groundwater solute transport migration. An evaluation of other systems is warranted to determine if the natural system studied is an anomaly or if it is a common groundwater phenomenon. Our results suggest that this type of analysis may need to be routinely done in groundwater investigations around surface water bodies to determine their influence on their associated groundwater systems.

References

Dames and Moore, Ground Water Study, Task A Report (Cincinnati, Ohio: NLO, Inc., 1985).

Fenneman, N. M., The Geology of Cincinnati and Vicinity, Geological Survey of Ohio Bulletin 19 (Columbus, Ohio: Geological Survey of Ohio, 1916), p. 32.

Fetter, Jr., C.W., Applied Hydrogeology (Columbus, Ohio: Charles E. Merril Publishing Company, 1980).

GeoTrans, Inc., STLINE User's Manual, Version 1.9 (Herndon, Virginia: GeoTrans, Inc., 1987).

IT Corporation, Hydrogeologic Study of FMPC Discharge to the Great Miami River (Monroeville, Pennsylvania: IT Corporation, 1988), pp. 2-3.

Reeves, M., Ward, D. S., Johns, N. D., and Cranwell, R. M., Theory and Implementation for SWIFT II, the Sandia Waste-Isolation Flow and Transport Model for Fractured Media, Release 4.84, NUREG/CR-3328, SAND83-1159 Albuquerque, New Mexico: Sandia National Laboratories, 1986), pp. 1-2.

A New In-Stream Aerator

by John S. Gulliver,[1] Bryan T. Oakley,[2] and Michael J. Semmens[3]

ABSTRACT: Experiments on microporous, polymeric, hollow fiber membranes are being conducted for application to in-stream aeration. The hollow fibers are sealed on one end and connected to an oxygen supply on the other, so that they may be filled with pressurized oxygen gas. Oxygen transfer into the water occurs through the pores, which are less than 0.1 μm in diameter and will not allow bubbling to occur except at high pressures. The hollow fiber membranes may thus be visualized as long cylindrical bubbles that do not rise and approach 100% transfer efficiency. The fibers are fixed at the oxygen supply end and are free at the sealed end so that they will move and clean themselves similar to grass growing on a stream bed.

Introduction

As more expensive alternatives are being considered for the continued cleanup of waste treatment plant effluents, and as the focus shifts to non-point source pollution, in-stream aeration is often incorporated as a potentially viable treatment alternative. Compressed air is normally not feasible for in-stream aeration because the transfer efficiency is low in a shallow stream and because the dissolved oxygen deficit (the driving force for oxygen transfer) is also relatively low. Pure oxygen gas has been considered to increase the dissolved oxygen difference. However, the low transfer efficiency means that most of the oxygen will have been manufactured only to return to the atmosphere. Thus, traditional in-stream aeration technologies are often not cost-effective.

Membrane aeration limits the formation of bubbles by allowing oxygen to diffuse across a membrane. It is a distinctly different process from the fine bubble membrane aerators commonly used in practice. Fine bubble membrane aerators consist of a perforated silicone membrane. Under a gas pressure, the perforations open to allow bubbles to form.

---

[1]Associate Professor, Department of Civil and Mineral Engineering, University of Minnesota, Minneapolis, MN 55455 (612)627-4600.
[2]Engineer, McCombs, Frank, Roos and Associates, Minneapolis, MN (612)476-6010.
[3]Professor, Department of Civil and Mineral Engineering, University of Minnesota, Minneapolis, MN 55455 (612)625-9857.

Hollow fiber membrane aerators restrict the formation of bubbles but allow oxygen to diffuse across the membrane walls. Thus, the volume of gas escaping to the atmosphere is small and the membrane aerators discussed herein can approach 100% oxygen transfer efficiency. In processes which allocate a large percentage of operational expense to the purchase or manufacture of oxygen, this can be a significant cost savings.

The hollow fibers are microporous with a pore opening of less than 0.1 μm. This small pore opening, the surface tension of water, and the contact angle between the hydrophobic membrane material (currently polyethylene), water, and gas results in an allowed pressure difference across the membrane of greater than three atmospheres before bubbles are formed. Thus, if pure oxygen is fed into the hollow fiber, the water concentration in equilibrium with the fiber is sixteen times that in equilibrium with a bubble of compressed air. This greatly increases the mass of oxygen transferred to the water.

The rate of oxygen transfer is characterized by a relationship incorporating the bulk transfer rate coefficient, stream velocity, and pressure difference across the fiber. This paper summarizes experimental results on these parameters for a fiber orientation suitable to in-stream aeration.

Oxygen Transfer Model

With the feed end fixed and the sealed end free, the angle of the fibers to the flow will vary from a pure cross-flow at the feed end towards more of a parallel flow at the sealed end of the fiber. This is important for the rate of oxygen transfer from a fiber to the water, which was found (Ahmed and Semmens, 1991) to be an order of magnitude higher for cross flow than for parallel flow. A model was therefore developed to predict the angle of the fiber to flowing water, as a function of distance along the fiber. This is described by Ahmed, Oakley, Semmens, and Gulliver (1993).

When sealed fibers are used for oxygen transfer, the analysis is somewhat more complicated than that for conventional open ended fibers since the internal partial pressure of oxygen falls along the length of the fiber, and the partial pressure of nitrogen increases. A steady-state balance of each gas species leads to the ordinary differential equation

$$D_i \frac{d^2 P_i}{dz^2} - P_i \frac{dv}{dz} - v \frac{dP_i}{dz} = \frac{4 K_L L d_o}{d_i^2} \left[P_i - \frac{C_i}{H_i} \right] \qquad (1)$$

where P_i = partial pressure of gas species inside the fiber, C_i = concentration outside the fiber, H_i = Henry's Law constant for the species, v = gas velocity inside the fiber, z = length along the fiber, K_L = liquid film coefficient for the gas species, D_i = diffusivity in the gas mixture, d_o = outside diameter of the fiber, and d_i = inside diameter of the fiber.

In addition to the two relationships identified by Eq. 1, there is the loss of pressure due to headlosses in the fiber and the fact that total pressure is the sum of the oxygen and nitrogen partial pressures. Thus, these four equations have four unknowns and require four boundary conditions for a solution. At the fixed end, $P_0 = P$ = the feed pressure, and $P_N=0$. At the free end, $v=0$. K_L will be determined by comparison with experiments.

The equations are solved by a Runge–Kutta routine with a secant–type of iterative technique.

Experiments

The modules were tested in a once through flume with water from the Minneapolis municipal supply. In order to reduce the volume of water to be aerated without reducing the velocity, a flume insert was constructed as shown in figures 1 and 2. The insert was constructed of plexiglas to allow viewing and measurement of the fiber bending while in the flow.

The rate of oxygen transfer into the flowing water is given by

$$\frac{dM}{dt} = Q\Delta C \tag{2}$$

where M = mass of oxygen transferred, Q = water discharge, and ΔC = concentration difference across the fibers.

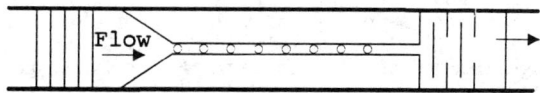

Fig. 1. Flume insert plan view

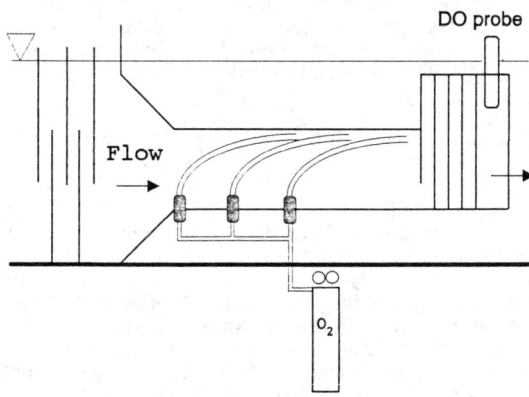

Fig. 2. Flume insert section view

This rate of oxygen transfer may also be given by our definition of $\overline{K_L}$:

$$\frac{dM}{dt} = \overline{K_L} \, N \, \pi \, d_0 \int_0^L (H_o P_o - C_o) ds \tag{3}$$

where N = the number of fibers, and subscript 'o' refers to oxygen. Equations 2 and 3 can be combined to give

$$\overline{K_L} = \frac{Q \Delta C}{N \pi \, d_0 \int_0^L (H_o P_o - C_o) ds} \tag{4}$$

or in terms of the bulk Sherwood number, Sh_B

$$Sh_B = \frac{\overline{K_L} \, d_0}{D} = \frac{Q \Delta C}{N \pi \, D \int_0^L (H P_o - C_o) ds} \tag{5}$$

where D = diffusivity of oxygen in the water. Q, ΔC, and C_o were all measured in the experiments. P_o as a function of s is available through our solution of Eq. 1. Note, however, that Eqs. 1 and 5 are implicitly linked through the use of K_L. Therefore Eq. 5 should be incorporated into the solution of Eq. 1.

Development of a Predictive Relation

Ahmed and Semmens (1993) have experimentally determined K_L for one type of fiber held at an angle θ to the flow that may be integrated over the length of the fiber to result in a bulk Sherwood number that may be compared with the experimental results,

$$Sh_B = \frac{\overline{K_L} \, d_0}{D} = 1.45 \, Re^{0.324} \, S_c^{0.33} \, \frac{1}{L} \int_0^L \left[\frac{2\theta}{\pi} \right]^{1.38} ds \tag{6}$$

where $Re = v_0 \, d_0/\nu$, $S_c = \nu/D$, v_0 = water velocity outside the fiber, and θ is given in radians. The angle θ is known from the solution of Ahmed et al. (1993).

In our current study a different hollow membrane fiber, with different surface characteristics, was used. Therefore, the constant and the Reynolds number exponent in Eq. 6 were left variable. In addition, the liquid film coefficient was found to be a strong function of pressure, which presumably is due to the meniscus being forced further out of the pore and increasing surface area. Since the surface area is related to the pressure drop created by the surface tension, the parameter $W_e/E_u = P \, dp/\sigma$ where dp is the mean pore diameter, was added to Eq. 6. Therefore, the equation fit to the experimental data was

$$\text{Sh}_B = \frac{K_L\, d_o}{D} = \beta_1\, \text{Re}^{\beta_2} \left[\frac{E_u}{W_e} \right]^{\beta_3} S_c^{0.33} \tag{7}$$

where β_1, β_2, and β_3 are constants.

Equations 1 and 5 were solved together to determine a bulk Sherwood number for each experiment. The data were then compared with Eq. 7 to determine the coefficients β_1, β_2, and β_3. The best fit results are:

$$\beta_1 = 0.389 \qquad \beta_2 = 0.285 \qquad \beta_3 = 0.612$$

The predictions are compared with the measurements in Fig. 3. Equation 7, with the three fitted coefficients, may now be used to predict the performance of an in-stream aerator, such as that given in Fig. 4.

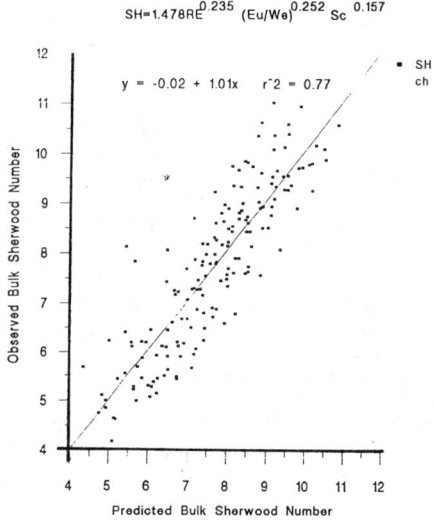

Figure 3. Comparison of predicted versus measured bulk Sherwood number.

Conclusions

The advantages to using microporous hollow fiber membranes for in-stream aeration are minimal waste of the oxygen gas that is normally required, the control over the system that can be achieved by simply reducing the pressure inside the fibers, and the fact that a fiber module in-stream aeration system placed under the water would be relatively unobtrusive. The primary disadvantages are the large fiber surface area required and the unknown maintenance costs of the fibers. The research described herein is designed to reduce the required surface area by

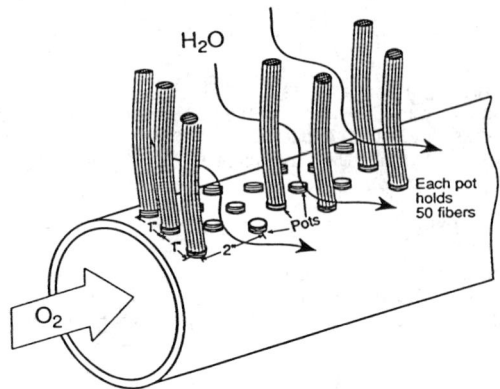

Fig. 4. Conceptual sketch of possible membrane aerator installation.

1) mounting the fibers with a cross-flow orientation to increase K_L and 2) operating the fibers at higher pressure, both to increase K_L and to increase the concentration gradient between the fibers and the water. Comparisons with other in-stream aerators indicate that we are close to achieving our goal of a cost effective in-stream aerator.

Acknowledgements

This paper is based upon work supported by the National Science Foundation under Grant No. BCS-9123175 and by the Minnesota Legislature ML 1991, Ch. 254, Art. 1, Sec. 14, Subd. 4K as recommended by the Legislative Commission on Minnesota Resources from the Minnesota Future Resources Fund. Any opinions, findings, and conclusions or recommendations expressed are those of the authors and do not necessarily reflect the views of either the National Science Foundation or the Minnesota State Legislature.

References

Ahmed, T. and Semmens, M.J. 1992. The use of independently sealed microporous hollow fiber membranes for oxygenation of water model development, *J. Membrane Science*, Vol. 69, 11–20.

Ahmed, T., Oakley, B.T., Semmens, M.J., and Gulliver, J.S. 1993. Nonlinear deflection of hollow fibers in transverse flow, submitted to *J. Environmental Engineering*.

Ahmed, T. and Semmens, M.J. 1993. Use of sealed end transverse flow microporous hollow fibers for bubbleless membrane aeration, submitted to *J. Environmental Engineering*.

MODELED HYDRAULIC AND SALT TRANSPORT PATTERNS IN THE SACRAMENTO-SAN JOAQUIN DELTA

Russ T. Brown[1], AM ASCE, Jones & Stokes Associates
Paul Wisheropp[2], AM ASCE, Jones & Stokes Associates
Don Smith[3], M ASCE, Resource Management Associates
Richard Rachiele[4], Resource Management Associates

ABSTRACT

Modeled hydraulic and salt transport patterns in the Sacramento-San Joaquin Delta have been summarized as the basis for detailed environmental assessment of proposed water management projects in the Delta, demonstrating that hydraulic engineering is the foundation for water resource analysis. The Delta channel flows, simulated from the boundary flows and tides with the Resource Management Associates (RMA) Delta hydraulic model, have been summarized with average tidal flows and net flow splits of the river inflows and export pumping flows. The RMA Delta salt transport model uses net channel flows to estimate salt concentrations within the Delta. Salt transport results have been generalized for a given boundary flow sequence by tracking each inflow source separately using modeled dye tracers. Travel time is simulated for each source with a decaying dye tracer. Because electrical conductivity (EC) is monitored at several stations and periodic chloride (Cl) measurements indicate distinct Cl/EC ratios for the two major rivers and seawater, EC and Cl were modeled separately to allow calibration of the simulated salt transport and source tracking patterns for 1967-1991.

INTRODUCTION

Analytical methods currently used to assess changes in flow patterns, water quality, or larval fish movement in the Delta channels involve hydraulic

---

[1,2] Jones & Stokes Associates, 2600 V Street, Sacramento, CA, 95818, 916/737-3000

[3,4] Resource Management Associates, 3738 Mount Diablo Boulevard, Lafayette, CA, 94549, 510/284-9071

and salt transport modeling. One-dimensional branched models, such as the Fischer Delta model, RMA Delta model, or the California Department of Water Resources Delta simulation model are typically used with boundary conditions specified by system operation model simulations.

The RMA Delta model uses a link-node formulation to perform a tidal hydraulic simulation with a specified average tide and inflow sequence, and then uses the net flow results to simulate water quality variables. The RMA Delta model has been used to simulate the Delta with monthly historical inflows, export pumping, and estimated evapotranspiration for 1967-1991. A repeating typical tide is used; inflow concentrations are estimated with flow regressions, and the downstream boundary salt concentration is estimated with a tidal exchange formulation.

RMA DELTA MODEL CHANGES

The RMA Delta model was modified to simulate additional water quality variables (EC and Cl) and incorporate source tracking and travel time with modeled dye variables. EC is monitored continuously at several stations in the Delta and provides the best variable for model calibration of 1967-1991 salt patterns. Chloride measurements indicate that the Sacramento and San Joaquin Rivers and seawater have distinct Cl/EC ratios. EC and Cl were added to the RMA Delta model to allow calibration of the salt transport and source tracking results. The source tracking and travel time simulations were accomplished with conservative and decaying modeled dye variable pairs for each inflow. The source of water and the length of time it has been in the Delta are important factors in estimating the possible effects on water quality and fisheries resources.

TIDAL HYDRAULIC PATTERNS

The tidal flows are dominant in most Delta channels; net flows caused by boundary inflows and exports are a secondary influence superimposed on the tidal flows. The RMA Delta hydraulic model was used to simulate stage and flow patterns for the typical tidal cycle used in the historical simulations, but without any boundary inflows or exports.

The typical tidal cycle for the Delta (Figure 1) begins with maximum flood flow at +2.0 feet and rises to high-high tide at +3.0 ft. The tide then drops to low-low tide at -2.0 feet, producing the first ebb flow in the typical tidal cycle. The second flood tide is driven by a rise to low-high tide at +2.0 feet, followed by a drop to 0.0 feet, providing the second ebb flow. The tide then rises to +2.0 feet by the end of the tidal cycle, initiating the first flood flow of the repeating cycle.

Figure 2 indicates the large tidal flows in both the Sacramento and San Joaquin Rivers, just upstream of their confluence. Although the net flow was about zero for this typical tidal cycle, the peak flood and ebb flows were approximately 125,000 cubic feet per second (cfs) on the Sacramento River, and approximately 150,000 cfs on the San Joaquin River. The tidal flow in Threemile Slough, that connects the Sacramento and San Joaquin Rivers several miles above their confluence, is also shown in Figure 1. Ebb flow is from the San Joaquin River towards the Sacramento River; maximum ebb and flood flows approach 25,000 cfs. Although the tidal flow patterns vary throughout the Delta channels and change with actual tides, they can be accurately simulated with the RMA Delta hydraulic model.

NET TIDALLY AVERAGED FLOW PATTERNS

No historical Delta channel net flow measurements are available, so the net flow results were calibrated indirectly in combination with the salt transport results by using historical EC measurements. The U.S. Geological Survey is installing acoustic flow meters in several Delta channels, so that direct measurements of net flows will be available in the future.

The Sacramento River is the primary freshwater source entering the Delta. During flood events, Sacramento River water flows into the Yolo Bypass and enters the Delta near Rio Vista. Downstream of Sacramento, the RMA Delta hydraulic model results indicate that about 25% of the flow is diverted into Sutter and Steamboat Sloughs, which connect with the Sacramento River near Rio Vista (Figure 3).

Sacramento River water can enter the central Delta through Georgiana Slough and the Delta Cross Channel (DCC), a gated channel constructed by U.S. Bureau of Reclamation to convey additional Sacramento River water to the central and southern Delta for export. The historic simulations assumed that the DCC gates were closed for flood control purposes whenever Sacramento River flow exceeded 25,000 cfs. RMA Delta hydraulic model results indicate a hydraulic relationship between the flow in these two channels and the Sacramento River flow (Figure 3).

The computed Delta channel flows can be used to estimate the number of larval fish carried into the central Delta, based on assumed Sacramento River larval density. Because DCC has gates, fish movement in the Delta is somewhat controlled by DCC gate operation. Estimates of fish entrainment in Delta agricultural diversions and at the export pumps can be used to guide the operation of the DCC gates and export pumping plants.

Threemile Slough and reverse flows in the lower San Joaquin River can also convey Sacramento River water to the central Delta. RMA Delta hydraulic model results indicate Threemile Slough net flow is a linear function of the ratio between the net flows of the San Joaquin River and Sacramento River. For conditions of zero net San Joaquin River flow, about 25% of Sacramento River water enters Threemile Slough and flows downstream in the San Joaquin River (Figure 4). Threemile Slough net flow may be important in preventing the upstream movement of salt and larval fish from Suisun Bay. Similar hydraulic relationships for net flow can be determined for all other major channel splits within the Delta.

DELTA WATER AND SALT SOURCE TRACKING

The source tracking modifications to the RMA Delta model allow the percentage of water from each source to be estimated. The source tracking is governed by net channel flow patterns and mixing caused by tidal flows. Travel time from each source provides information for larval fish movement and mortality estimates. The Sacramento River is the major source of export water during most months (Figure 5); however, the San Joaquin River provides most of the export water during some high flow periods. Delta agricultural drainage provides a seasonal source of water that is high during the irrigation season and large rainfall events.

Calibration of the source tracking is possible because the Cl/EC ratio for each water source is distinct, so that mixtures of the sources provide a unique combination of EC and chloride. Cl/EC ratios should be routinely measured to allow source tracking calibration. After the source compositions are simulated, additional water quality or larval fish variables can be modeled by combining inflow sequences with the simulated source compositions and travel times.

Figure 6 shows the simulated contribution of Cl at the export pumps from each major source of salt. The Cl contribution is the product of the source Cl concentration and the percentage of water from the source. Delta agricultural drainage supplies a seasonally varying contribution of Cl, which is greatest during winter salt leaching. The San Joaquin River supplies a large Cl contribution if the San Joaquin River is a major source of exports. The highest Cl concentrations, however, are caused by seawater intrusion events that occur progressively during periods of low Delta outflow with relatively high export pumping. Although seawater intrusion does not provide a major source of export water, it can contribute a major portion of the export Cl because of the high concentration of Cl in seawater.

These simulated Delta conditions can be used to evaluate environmental effects of alternative water management operations on a wide range of water quality variables and larval fish species. Delta hydraulic and salt transport modeling provides an essential foundation for accurate environmental assessment and informed water resource management in the Delta.

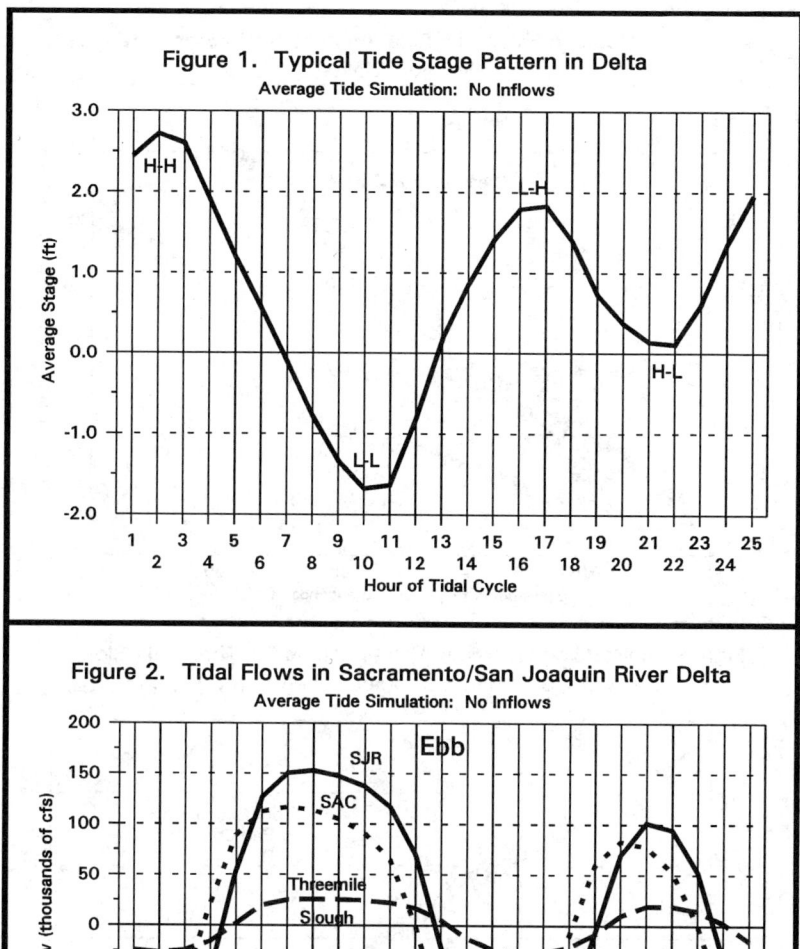

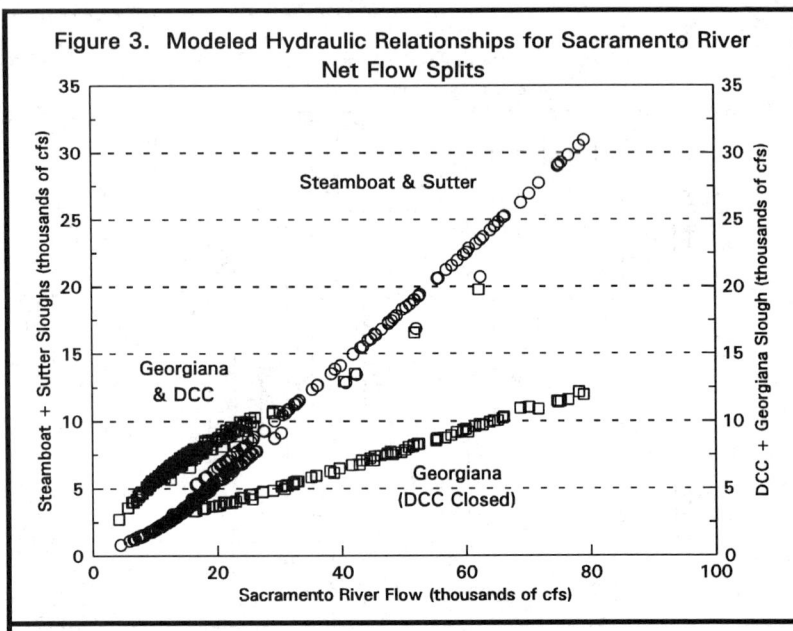

Figure 3. Modeled Hydraulic Relationships for Sacramento River Net Flow Splits

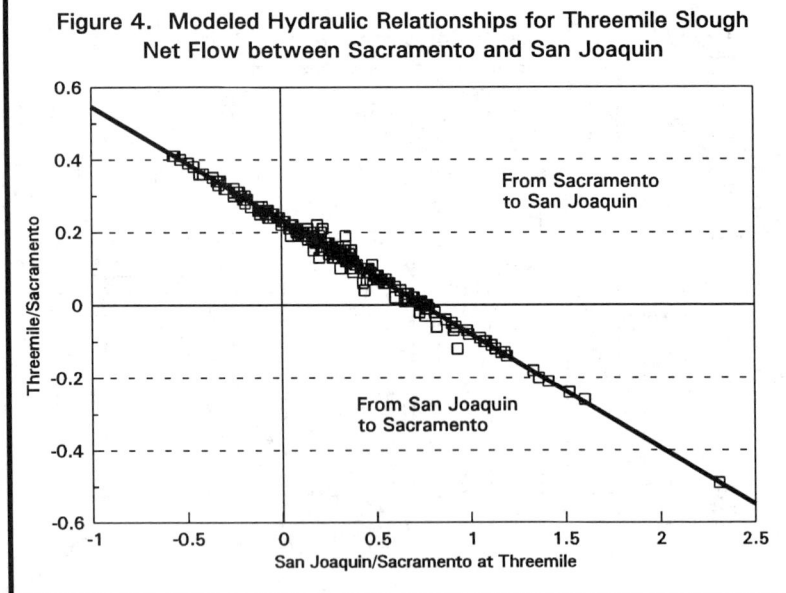

Figure 4. Modeled Hydraulic Relationships for Threemile Slough Net Flow between Sacramento and San Joaquin

HYDRAULIC AND SALT TRANSPORT 2177

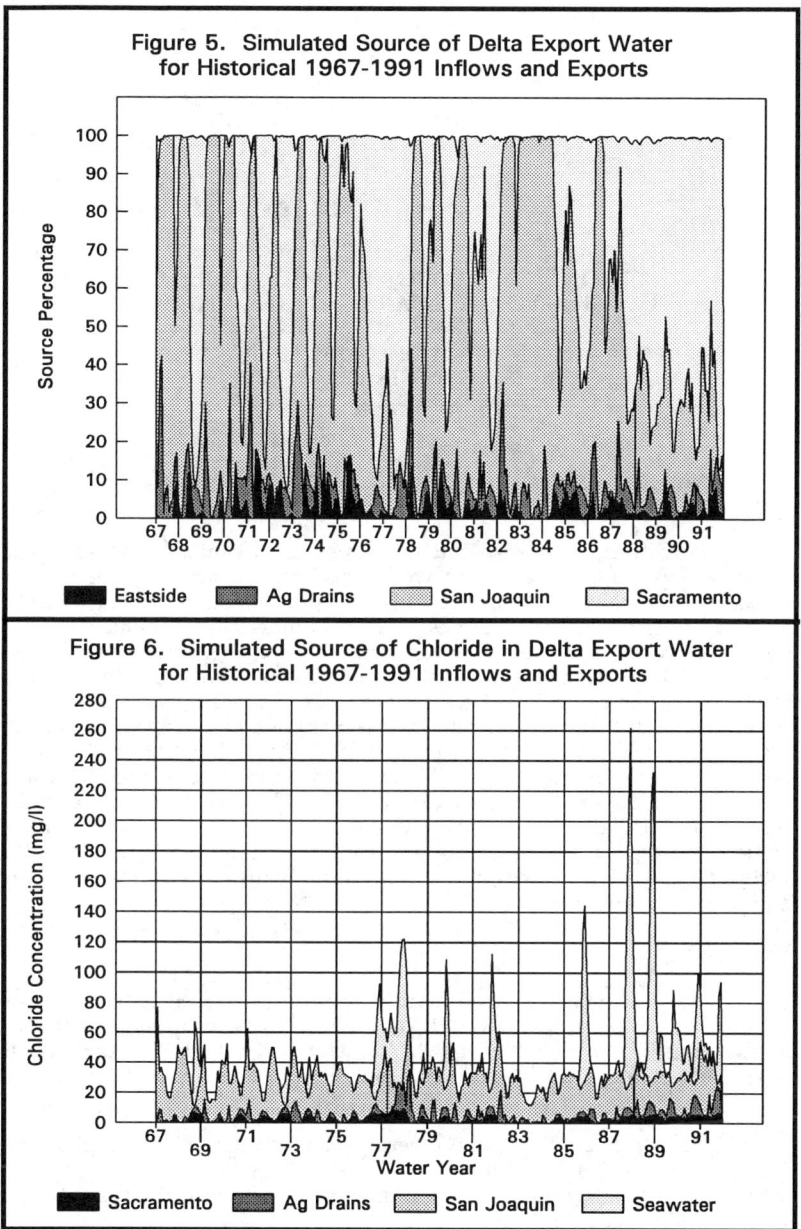

Figure 5. Simulated Source of Delta Export Water for Historical 1967-1991 Inflows and Exports

Figure 6. Simulated Source of Chloride in Delta Export Water for Historical 1967-1991 Inflows and Exports

A DIFFUSION HYDRODYNAMIC MODEL OF
A SHALLOW ESTUARY: VERIFICATION

T.V. Hromadka II[1], M.ASCE
G. L. Guymon[2]
M.H. Khan[3], M. Collins[3]

Abstract

Many of the hydrodynamic models used for estuary tidal flows and storm surges are based on two two-dimensional hydrodynamic equations which are obtained form the full three-dimensional equations by averaging the vertical coordinates. Various numerical techniques, such as finite difference, finite element, and the method of characteristics have been used to solve these models. The Diffusion Hydrodynamic Model (DHM) has been developed to simulate two-dimensional surface water flows. In a prior study, the DHM has been applied to a deep water estuary with results comparable to those obtained using the method of characteristics. In this paper, the DHM is applied to a shallow water estuary located in southern California. The main objective is to determine local flow velocities and circulation patterns in the shallow estuary caused by the incoming and outgoing tide. Verification of the DHM is provided by comparison of tidal gauge measurements and computed flow depths.

Introduction

The Diffusion Hydrodynamic Model or DHM (Guymon and Hromadka, 1986; Hromadka and Yen, 1986) has been applied to several applications (Hromadka et al, 1989). In this paper, the DHM is extended to irregularly-shaped finite element four-sided polygons. The extended

---
[1]Principal Engineer, Boyle Engineering Corporation, 1501 Quail Street, Newport Beach, California 92658
[2]Professor, Department of Civil Engineering, University of California, Irvine, California 92717
[3]Graduate Student, University of California, Irvine, California 92717

DHM is then applied to a shallow estuary in order to simulate water surfaces and flow velocities. Tidal gauge measurements are collected for comparison to the computed DHM results, providing for both calibration and verification of the DHM approach.

Description of Study Region

Newport Bay is a shallow estuary with a mean water surface area of about 1.9 square miles and about 23 miles of shoreline. The bay is subject to a shallow but complex tidal cycle. Two low and two high tides occur diurnal and are particularly pronounced during the highest tides during the new moon phase of the lunar cycle. Tides are propagated rapidly throughout the Bay system.

Mathematical Model Development

Because of the estuary's shallowness relative to its aerial extent, the water column is considered as vertically integrated. Hydrodynamic equations of fluid motion are developed by considering conservation of mass and momentum. Conservation of momentum is expressed in terms of the "St. Venant equations" (Hromadka et al, 1989). Discharge, Q, may be linked to friction slope, S_f, by Manning's equation.

$$Q_i = \frac{1.486}{n} A_i R_i^{2/3} S_{f_i}^{2/3}; i = x,y \tag{1}$$

where n is Manning's roughness coefficient; and R is the hydraulic radius. For wide shallow channels $R \approx h$ (depth). For relatively shallow flows where velocities are relatively small, the local acceleration and inertial terms of the flow equations may be neglected without undue loss of precision. Consequently, the St. Venant equations reduce to (H, water surface elevation)

$$S_{f_i} = \frac{\partial H}{\partial X_i}; i = x,y \tag{2}$$

Manning's equation may be rearranged as follows:

$$Q_i = -K_i \frac{\partial H}{\partial X_i}, \tag{3}$$

where (b_i, flow basewidth in i-direction)

$$K_i = \frac{1.486}{n} b_i h_i^{5/3} / \left|\frac{\partial H}{\partial X_i}\right|^{1/2}; i = x,y \tag{4}$$

This result may be substituted into the continuity equation to yield the so-called "diffusion hydrodynamic equation" in two-dimensions,

$$\frac{\partial(K_x \frac{\partial H}{\partial x})}{\partial x} + \frac{\partial(K_y \frac{\partial H}{\partial y})}{\partial y} = A_{x,y}\frac{\partial h}{\partial t} \quad (5)$$

where $A_{x,y}$ is the aerial area of the cell. This equation is a parabolic equation (diffusion equation) that is highly nonlinear. A finite difference approximation of (5) may be obtained by writing in terms of total discharge Q,

$$\frac{\partial Q_x}{\partial x}\Delta x + \frac{\partial Q_y}{\partial y}\Delta y = A_{x,y}\frac{\partial h}{\partial t} \quad (6)$$

Integrating the partial-differential components,

$$\Delta Q_x + \Delta Q_y = A_{x,y}\frac{\Delta h}{\Delta t} \quad (7)$$

or, in an explicit solution scheme for any shaped cell,

$$h^{J+1} = h^J + \frac{\Delta t}{A_{x,y}}\sum_i Q_i \quad (8)$$

where J is the time step number, and Q_i is discharge across a perpendicular cell face, being positive of inflow and negative if outflow, and Q_i and velocities are determined from (3), where K_i is determined from (7). The simulation is commenced with a flat bay water surface at MLLW. Two types of boundary conditions are considered. One tidal boundary condition is specified at the jetty-ocean mouth based upon the National Oceanic and Atmospheric Administration (NOAA) tidal gauge located at the Harbor Master headquarters compound about one quarter of a mile inside the bay. The second kind of boundary condition consists of inflow hydrographs from 25 different watersheds.

To solve the numerical hydrodynamic equation analog, the Bay is discretized into quadrilateral finite elements. The discretization scheme selected is shown in Figure 1. There are 395 cells and 549 nodes in the finite element system.

The verification approach was to measure a tidal sequence in several discrete portions of the Bay and compare these data to simulated tides. Because the Manning's n-factor for bottom roughness is assumed, this approach is actually a partial verification and calibration effort combined. Partial verification is achieved in the sense that n-factors for each cell must be "tuned" to give good results and we must use judgment to determine if such tuning is realistic and the simulated tides and currents are reasonable. Water surface elevations were measured at six locations (Figure 2) in the Bay over a 24-hour tidal diurnal cycle during 24-

25 May 1990. This period, 5 AM on the 24th to 6 AM on the 26th, corresponded to maximum new-moon tides during the lunar cycle. Measured tidal elevations at the NOAA tide gauge were used as a boundary condition at the jetty mouth. Tidal measuring stations were selected on the basis on proximity to Orange County bench marks. A bench mark was surveyed for each measurement point and water levels were measured.

Results are shown in Figures 3 and 4 which compare measured tides to simulated tides at five locations. The actual tidal measuring sequence began at 5 AM. To simulate measured tides a pseudo-tide was simulated for the previous five hours to remove the assumed Bay water surface elevations initial conditions effects. As can be seen, good results were achieved using initially assumed n-factors.

Conclusions

A Diffusion Hydrodynamic Model (DHM) of a shallow estuary has been developed using an extension of the USGS DHM computer model. The DHM analog includes several hundred irregular polygon control volumes and nodal points. Simulation of estuary flow depths by the DHM analog were compared to measured tidal gauge flow depths, with good agreement. Because the shallow estuary flow characteristics are two dimensional, a verification of the DHM is provided by the comparisons between tidal gauge measurements and flow depths.

Acknowledgment

The work presented in this paper was supported by the California Regional Water Quality Control Board, Santa Ana Region.

References

Guymon, G.L. and T.V. Hromadka II 1986, Two-Dimensional Diffusion-Probabilistic Model of a Slow Dam Break, Water Resources Bulletin, 22(2), 257-265.

Hromadka II, T.V., Berenbrock, C.E., Freckleton, J.R., and Guymon, G.L., 1985, A Two-Dimensional Dam-Break Floodplain Model, Advances in Water Resources, 8, 7-14.

Hromadka II, T.V., and Yen, C.C., 1986, A Diffusion Hydrodynamic Model (DHM), Advanced in Water Resources, 9, 118-170.

Hromadka II, T.V., Walker, T.R., Yen, C.C., and DeVries, J.J., 1989, Application of the USGS Diffusion Hydrodynamic Model for Urban Floodplain Analysis, Water Resources Bulletin, Vol. 25, No. 5, 1063-1071.

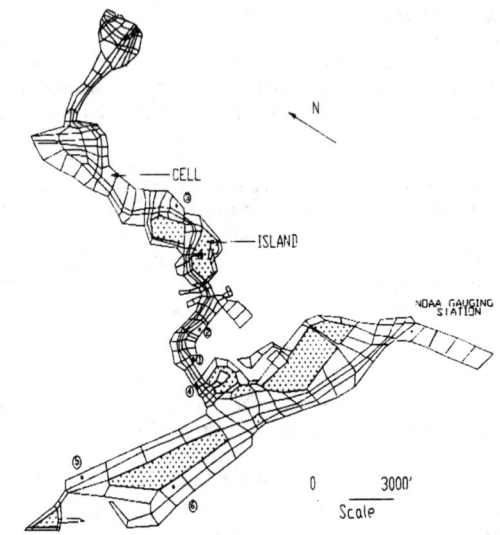

Fig. 1. Newport Bay, Mathematical Model Grid.

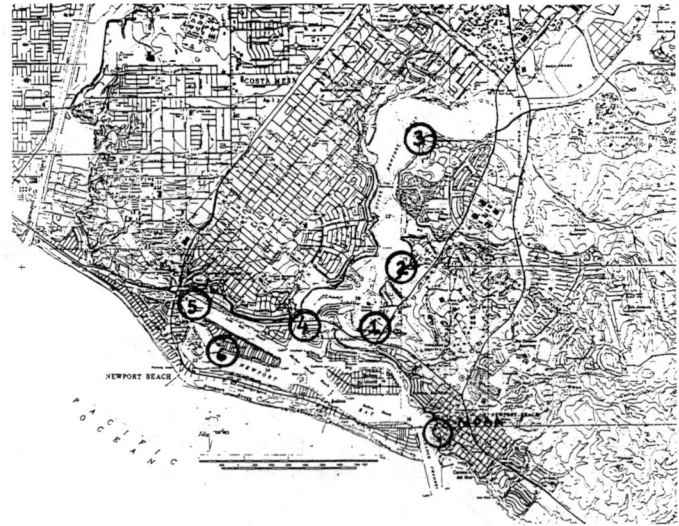

Fig. 2. Location of Stations for Water Surface Elevation Measurements in Newport Bay.

DIFFUSION HYDRODYNAMIC MODEL

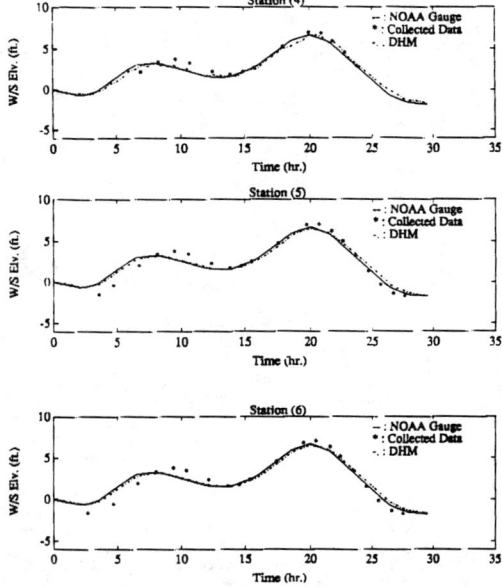

Fig. 3. Lower Newport Bay, Comparison of Simulated Tides to Measured Tides.

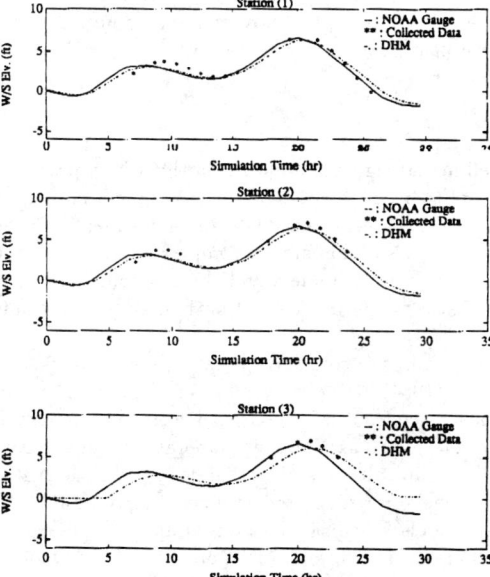

Fig. 4. Upper Newport Bay, Comparison of Simulated Tides to Measured Tides.

Structure of Coastal Upwelling on a Sloping Bottom

Yan Zang,[1]
Robert L. Street[1], Member
and
Jeffrey R. Koseff[1], Member

Abstract

A large eddy simulation of coastal upwelling on a sloping bottom is carried out. The structure of the upwelling consists of a persistent primary front, a temporary secondary front which forms the nose of a top unstable layer, and a trailing mixing zone. Baroclinic instability occurs at the primary front and develops into large, predominantly 2-D waves and eddies. The computed results are compared with past experimental data.

1. Introduction

Coastal upwelling has significant effects on the fishery productivity, transport of nutrient and pollutant materials, as well as the climatology and ecology of coastal regions. It is a three-dimensional, time-dependent, and highly nonlinear phenomenon which occurs in domains of complex geometry. Extensive investigations have been carried out to understand this phenomenon, which include field observations, laboratory experiments and both analytical and numerical models (Smith, 1968).

As a result of upwelling, a cold surface water anomaly appears near the coast. Mooers & Robinson (1984) have observed cyclone/anticyclone pairs and the associated jets near a surface front off the Oregon coast. Narimousa and Maxworthy (1987) (hereafter referred to as NM), in a laboratory study, have found that the surface front is baroclinically unstable that leads to large-scale frontal eddies. Most analytical models are restricted to linear analysis and simple geometry. While numerical models based on the layered approach (e.g., O'Brien & Hurlburt, 1972; Hurlburt & Thompson, 1973) are capable of resolving nonlinearity

[1] Environmental Fluid Mechanics Laboratory, Stanford University, CA 94305-4020

and geometric complexity, they have difficulty representing surface fronts and frontal instability.

In the present study, we have developed a large eddy simulation model to compute coastal upwelling flows by solving the time-dependent, three-dimensional Navier-Stokes and scalar-transport equations. Section 2 contains a brief description of the essential elements of the model. Results of the simulation of upwelling on a bottom slope will be described in Section 3. Conclusions will be given in Section 4.

2. Numerical Model

We have developed a large eddy simulation model to compute geophysical flows such as the coastal upwelling. The governing equations are the time-dependent, three-dimensional, incompressible Navier-Stokes, continuity, and scalar-transport equations with the Boussinesq approximation. First, we separate the spatial scales of the flow into the resolved (large) and the subgrid (small) scales. We compute the large-scale motions explicitly while model the effect of the small scales with a subgrid-scale model.

The governing equations are written in the non-orthogonal, curvilinear coordinates and are transformed into a cubic box in the computational space. The finite-volume method is employed to spatially discretize the equations on a non-staggered grid. A semi-implicit, fractional step method is used to advance the equations in time. The overall accuracy of the numerical method is second-order.

The dynamic subgrid-scale model (Germano et al. 1991) is employed to represent small-scale effects. This model does not require any input model constant and has been successfully applied in the large eddy simulation of various kind of transitional and turbulent flows. Details of the numerical and the subgrid-scale models may be found in Zang, Street & Koseff (1992a, b).

3. Results

Initially, a two-layer salt stratified fluid is in solid body rotation with the container. Upwelling is driven by the relative rotation of the top lid. The geometry of the domain is similar to that of NM except that, instead of a conical cylinder, a section of an annulus is employed (Fig. 1) with $R_1/R_0 = 0.1$ and $\theta_0 = 90°$. The outer radial wall corresponds to the coast or the shore. The physical parameters are chosen to be the same as those in 'case a' of Narimousa et al. (1991) except that the fluid viscosity is ten times larger in the simulation which is necessary due to the limit of the computer capacity. A no-slip condition is imposed for the velocity on the top, bottom, and the two side walls. A no-flux condition is used for the density because of the salt stratification. On the two azimuthal boundaries, periodic boundary conditions are imposed.

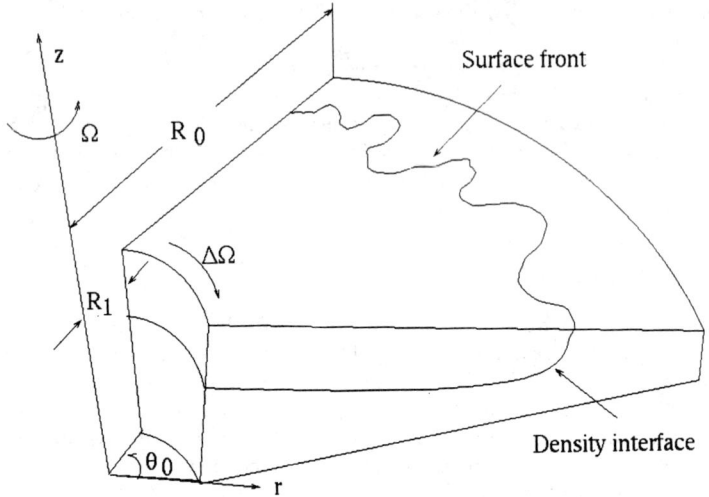

Fig. 1. The geometry of the flow domain with the interface and the front.

Fig. 2 shows the density field at $t/t_s = 0.9$, where t_s is the spin-up time scale defined as $(h_0/2\Delta\Omega)(4(\Omega + \Delta\Omega)/\nu)^{1/2}$, where h_0 is the initial depth of the upper layer and ν is the kinematic viscosity. At this time, the interface has already intersected the top lid and formed an almost axisymmetric surface front (the 'primary front'). In the vertical cross section, we see that a portion of the lower-layer fluid is pulled offshore by the surface Ekman transport. The nose of this gravitationally unstable layer forms a weaker front (the 'secondary front'). There is an intensive mixing zone (the 'trailing mixing zone') on the shore side of the interface into which the upwelled lower-layer fluid drops down from the top Ekman layer.

As the surface front continues migrating offshore, it becomes unstable. Azimuthal waves develop and grow to large amplitude. Fig. 3 shows the density field at $t/t_s = 2.0$. The mean front location is now farther offshore. The trailing mixing zone extends all the way to the shore. The frontal waves travel in the direction of the applied stress at a speed roughly half of the lid speed. The wave crests (toward the center) and troughs correspond to the cyclonic (downwelling) and anticyclonic (upwelling) regions, respectively.

The presence of the bottom slope intensifies the growth of the frontal waves. In the simulations without the sloping bottom, the amplitudes of the waves are smaller while the wavelengths are larger. This may be explained by the conservation of potential vorticity; cyclonic vorticity is increased at wave crests where the depth is larger.

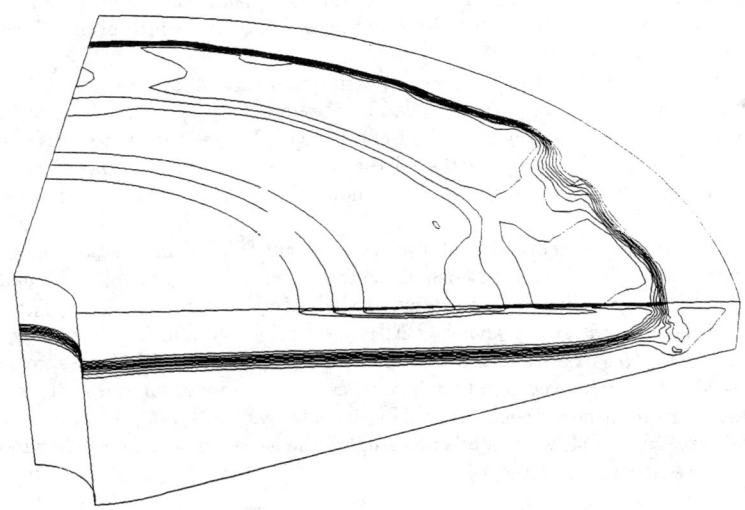

Fig. 2. Density field at $t/t_s = 0.9$.

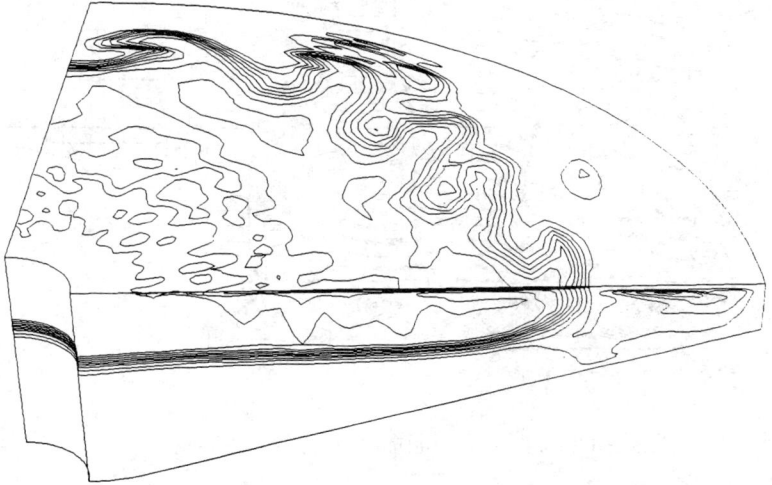

Fig. 3. Density field at $t/t_s = 2.0$.

At $t/t_s = 2.0$, the top unstable layer has already dropped. The instability of the secondary front appears first in the form of azimuthal strips which subsequently break up into isolated regions of lower-layer fluid which eventually drop back down into the lower layer.

Fig. 4 shows the streamlines in an azimuthal-vertical plane cutting through the primary front (the plane bounded by the dotted lines). The reference frame for this figure is traveling with the frontal waves. We see that the waves extend from the edge of the top to the edge of the bottom boundary layer, and are predominantly two-dimensional. This shows the dominant effect of rotation on the flow.

The development of the frontal waves is a result of baroclinic instability. Comparison is made on the mean size of the saturated waves λ_w nondimensionalized by the Rossby radius of deformation of the upper layer $\hat{R} = (g'h_0)^{1/2}/f$, where g' is the reduced gravity and $f = 2\Omega$ is the Coriolis parameter. The computed value of $\lambda_w/\hat{R}$ is 2.9 while the measured value from NM is 2.7. Good agreement has also been achieved on other quantities. For example, for the width of the stationary front nondimensionalized by the outer radius, λ_s/R_0, the present computation gives 0.44, while a theory based on the geostrophic balance (Narimousa & Maxworthy 1987) gives 0.42.

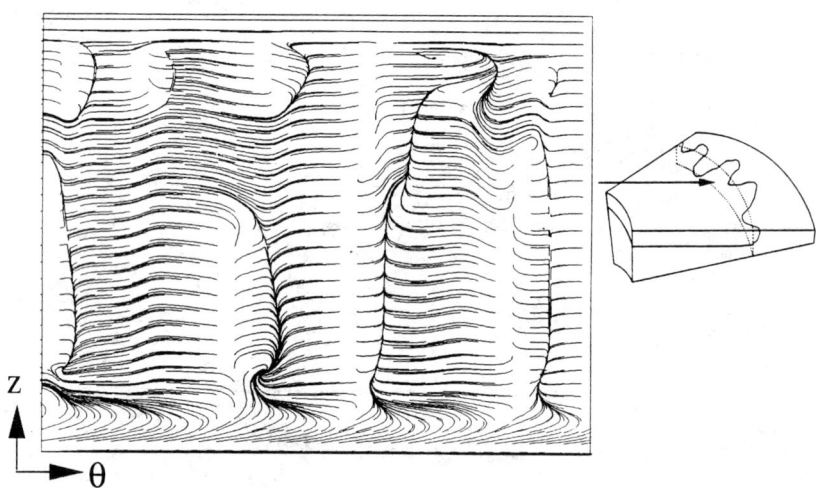

Fig. 4. Streamlines in an azimuthal-vertical plane cutting through the primary front.

4. Conclusions

A laboratory model of the coastal upwelling flow is simulated by solving the time-dependent, three-dimensional Navier-Stokes equations. The effect of the small scales is represented by the dynamic subgrid-scale model. Main features of the upwelling include a persistent primary front, a temporary secondary front which forms the nose of a top gravitationally unstable layer, and a trailing mixing zone. The primary front goes through baroclinic instability which leads to large-amplitude, predominantly two-dimensional frontal waves and eddies. The computed solution agrees well with past theories and experiments.

Acknowledgements: The authors wish to thank Profs. S. G. Monismith, T. Maxworthy, J. H. Ferziger and Dr. S. Narimousa for useful discussions. The research has been sponsored by the National Science Foundation through Grant CTS-8719509 and the Fluid Dynamics Program, Mechanics Division, Office of Naval Research through Grant N-00014-91-J-1200. The Cray YMP time was provided by NCAR Scientific Computing Division.

References

Germano, M, Piomelli, U., Moin, P. & Cabot, W. H., (1991) "A dynamic subgrid-scale eddy viscosity model," Phys. of Fluids A. **3**, 1760-1765.

Hurlburt, H.E. and Thompson, J.D., (1973) "Coastal upwelling on a β plane", J. Phy. Oceanogr. **3**, 16-32.

Mooers, C. N. K. & Robinson, A. R., (1984) "Turbulent jets and eddies in the California current and inferred cross-shore transports," Science **223**, 51-53.

Narimousa, S. & Maxworthy, T., (1987) "Coastal upwelling on a sloping bottom: the formation of plumes, jets and pinched-off cyclones," J. Fluid Mech. **176**, 169-190.

Narimousa, S., Maxworthy, T. & Spedding, G. R., (1991) "Experiments on the structure and dynamics od forced, quasi-two-dimensional turbulence," J. Fluid Mech. **223**, 113-133.

O'Brien, J.J. & Hurlburt, H.E., (1972) "A numerical model of coastal upwelling," J. Phys. Oceanogr. **2**., 14-26

Smith, R. L., (1968) "Upwelling," Oceanogr. Mar. Biol. Ann Rev. **6**, 11-46.

Zang, Y. Street, R. L. & Koseff, J. R., (1992a) "A non-staggered grid, fractional step method for time-dependent incompressible Navier-Stokes equations in general curvilinear coordinate systems," submitted to J. Comp. Physics.

Zang, Y., Street, R. L. & Koseff, J. R., (1992b) "Application of a dynamic subgrid-scale model to turbulent recirculating flows," Annual Research Briefs, Center for Turbulence Research, Stanford U./NASA-Ames, 85-95.

Field Examination of a Distribution System Water Quality Model

Sami G. Elmaalouf,[1] S.M. ASCE
Young C. Kim,[2] M. ASCE

Abstract.

In the aftermath of the Safe Drinking Water Act (SDWA) of 1974, numerous models have been established to assess the quality of water and simulate its deterioration in distribution systems. Many of these models, however, are associated with a great deal of complexity especially when a practicing engineer is attempting to utilize them in order to structure a computer program that would help predicting water quality variations in a conveying system. Furthermore, many of these models have not been verified by field studies.

This paper presents and describes the development of a field study that examines a steady state distribution system water quality model. The field study requires a profound understanding of the hydraulic behavior in the system as the variation in delivered water quality is dependent upon the hydraulic mixing effects in this system. The study is conducted at the California State University, Los Angeles to verify a comprehensive model that was developed to predict the constituents and/or contaminants propagation, decay, and the overall quality variations in water distribution networks.

Introduction

A simple and reliable water quality model is essential for predicting and monitoring quality deterioration and constituents or contaminants behavior in water distribution systems as the 1974 Safe Drinking Water Act (SDWA) put various restrictions on the quality of delivered water to consumers. Examining this model experimentally is of equal importance. This paper presents and describes the development of a field study that examines a steady state distribution system water quality model. The model is based on algorithms that were reported by Elmaalouf et al. (1). The reported model is efficient and solves for quality parameters by direct substitution, yielding results that would aid engineers and planners to predict water quality variations in a conveying hydraulic network. Initial understanding of the hydraulic behavior in the distribution network is required as the variations in the delivered quality are dependent upon mass and energy conservation, and the overall hydraulic mixing effects in the distribution system.

The first part of this paper demonstrates the steady state water quality model, while in the second part the experimental program and a comparison of results will be presented and discussed.

---
[1]Engineer, Public Works Department, City of Los Angeles, 600 S. Spring Street, 15th Floor, Los Angeles, CA 90014
[2]Professor, Civil Engineering Department, California State University, Los Angeles, 5151 State University Drive, Los Angeles, CA 90032

Water Quality Analysis in Pipe Networks

Traditionally, the main interest of water supply managers and engineers was to hydraulically satisfy all the water demands at various different locations in their municipality. Quality assessment as well as contaminants/constituents propagation within a municipal water conveying system were not a leading interest, since the drinking water quality was basically measured immediately on its way leaving a water treatment plant.

We know (as of 1993) that this is not the case anymore. In the aftermath of the Safe Drinking Water Act (SDWA) of 1974, many concerns have increased. over quality assessment, contamination and waterborne substances propagation and the overall blending effect, of different sources with different quality, in a distribution system.

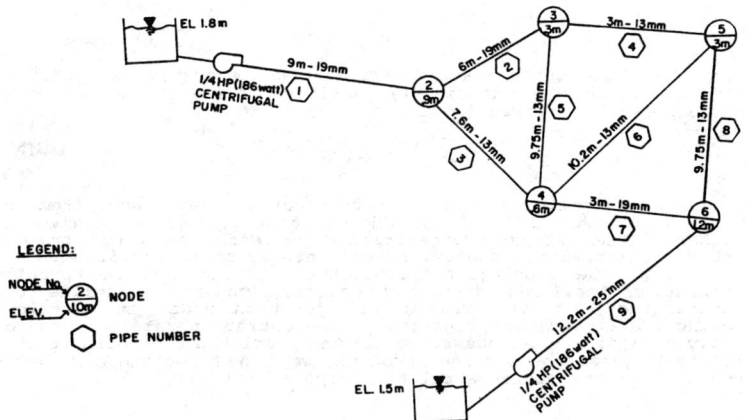

Figure 1. Experimental Distribution System

Before we start solving for quality parameters in a distribution system, it is very important to successfully identify the flow contribution, delivered to junction nodes, from supply sources. The contribution of supply sources to loads of discharge at various nodes in water distribution systems, can be determined when a nodal mass continuity takes place. The source contribution problem can be formulated as

$$[Q_{sc}] \cdot f_c = [Q_{fo}] \tag{1}$$

where, $[Q_{sc}]$ is the source flow contribution matrix, f_c is the supply source contribution factor, and $[Q_{fo}]$ is the source outflow forcing matrix. Therefore

$$[Q_{sc}]^{-1} \cdot [Q_{fo}] = f_c \tag{2}$$

where $[Q_{sc}]^{-1}$ is the inverse of the source flow contribution matrix. If the total inflow from a source node is equal to the demand at a particular node, then Q_{sc} is set to unity. Otherwise

$$Q_{sc} - \Sigma Q_i + q \qquad (3)$$

After a successful determination of the supply source contribution factor and the correct assembly of the source flow contribution matrix, $[Q_{sc}]$, we have to analytically formulate the contaminant/constituent concentration problem.

Given the source flow contribution matrix and knowing the concentration of a certain chemical at the supply source (simply by actual field quality measurement at the supply source), the contaminant/constituent concentration problem formulation is

$$[Q_{sc}] \cdot C_c = [C_s] \qquad (4)$$

where C_c is the contaminant/constituent concentration at junction nodes and C_s is the contaminant/constituent concentration injected from a supply source(s). Or

$$[Q_{sc}]^{-1} \cdot [C_s] = C_c \qquad (5)$$

Most contaminants/constituents decay as they travel from the supply source to a junction node or from one junction node to another. The rate of concentration decreases as water travels farther from a supply source, given a steady state condition.

From the contaminant/constituent concentration problem formulation that was discussed earlier, we can determine the concentration of chemicals at all junction nodes in question, knowing the contaminant/constituent concentration rate at a source supply node(s). After observing all the predicted concentrations at different locations in the network, we can formulate the decay problem from mass balance relationships as follows

$$\frac{\Delta C}{t_{i,j}} = \xi(C_{i,j}) \qquad (6)$$

or with a quadratic formulation

$$[C_i C_j] \cdot (CD)_{i,j} = -\frac{1}{t_{i,j}} [\Delta C] \qquad (7)$$

hence

$$(CD)_{i,j} = -\frac{1}{t_{i,j}} [\Delta C] \cdot [C_i C_j]^{-1} \qquad (8)$$

where C_i and C_j are the contaminant/constituent concentrations at junction nodes i and j, respectively; $t_{i,j}$ is the flow time from junction node i to junction node j, ξ is decay reaction rate which is dependent upon the time variation between one junction node and another, $(CD)_{i,j}$ is the concentration decay rate from junction node i to junction node j, and ΔC is the change in the contaminant/constituent concentration between junction node i and junction node j.

Water ages when it is in the supply source or when it is traveling via pipes from one junction node to another. This aging process contributes to the quality deterioration of delivered water at different junction nodes in a distribution system. Therefore, studying this problem is of great importance in the field of water quality investigation. This problem can be formulated as

$$[T_{ave}] \cdot t_{ave} = [\omega] \qquad (9)$$

where $[T_{ave}]$ is the average water age matrix, t_{ave} is the average water age at a junction node, and $[\omega]$ is the water age-flow contribution matrix.

If the total inflow from a source node is equal to the flow rate in an pipe, then T_{ave} is set to unity. That is, T_{ave} is equal to one when

$$Q_s(k) = - \Sigma f_{c_i} \cdot Q_{i,j} \qquad (10)$$

where f_{c_i} is the source contribution factor at junction node i, and $Q_{i,j}$ is the flow rate, in a pipe, from junction node i to junction node j. Otherwise

$$T_{ave} = Q_s(k) + \Sigma f_{c_i} \cdot Q_{i,j} \qquad (11)$$

The average age-flow contribution matrix is assembled as follows

$$[\omega] = \begin{bmatrix} \Sigma f_{c_i} \cdot Q_{i,j} \cdot t_{i,j} \\ \cdot \\ \cdot \\ \cdot \end{bmatrix} \qquad (12)$$

where $t_{i,j}$ is the flow time through a pipe connecting junction node i to junction node j and is computed by the equation

$$t_{i,j} = \frac{l_{i,j}}{V_{i,j}} \qquad (13)$$

where $l_{i,j}$ is the length of the pipe, and $V_{i,j}$ is the velocity of water traveling from junction node i to junction node j.

Experimental Program

Four different scenarios were analyzed for the distribution system shown in figure 1. The experimental system comprises nine pipes, five connecting junction nodes, two constant-head tanks (flow sources), five ball valves at each node to control the outflow at node points, and two 1/4 hp (186 Watt) centrifugal pumps in lines 1 and 9.
All four scenarios were verified experimentally. In the first scenario the ball valves at nodes 2 and 5 were calibrated to discharge 0.28 l/s and 0.57 l/s, respectively. All other valves were fully closed. A steady (constant) head was maintained at both feeding tanks by adding distilled water continuously. The system was operating for seven minutes to maintain a steady state flow in the network. The flow rates were checked at discharging nodes by means of a magnetic flow meter.
The distribution system was stopped at this point. A chemical (Benzene-C_6H_6) was introduced to the tanks at dissimilar proportions, making the constituent concentration 10 mg/l in supply tank No.1 and 3 mg/l in supply tank No.2. The system was again started and water samples were collected at discharging nodes 2 and 5. Chemical analyses followed.
In the second scenario the ball valves at nodes 3 and 4 were calibrated to discharge 0.57 l/s and 0.28 l/s, respectively. All other valves were fully closed. Exact steps were followed and similar data was gathered as in the first scenario. Water samples were collected at discharging nodes 3 and 4. Chemical analyses followed.
In the third scenario the ball valves at nodes 4 and 5 were both calibrated to discharge 0.28 l/s. All other valves were fully closed. Exact steps were followed and similar data was gathered as in the preceding scenarios. Water samples were collected at discharging nodes 4 and 5. Chemical analyses followed.
In the fourth scenario the ball valves at nodes 3, 4, and 5 were all calibrated to discharge 0.28 l/s. All other valves were fully closed. Exact steps were followed and similar data was gathered as in the preceding scenarios. Water samples were collected at discharging nodes 3, 4, and 5. Chemical analyses followed.

Table 1. Analytical and Experimental Hydraulic Analyses
(First Scenario)

Calculated				Measured			
Pipe	Flow (l/s)	Node	Pres. (KN/m^2)	Pipe	Flow (l/s)	Node	Pres. (KN/m^2)
1	0.28	2	437	1	0.30	2	-
2	0.28	3	441	2	0.30	3	-
3	0.00	4	441	3	0.01	4	-
4	0.28	5	430	4	0.30	5	-
5	0.00	6	437	5	0.01	6	-
6	0.28			6	0.30		
7	0.28			7	0.30		
8	0.28			8	0.30		
9	0.57			9	0.50		

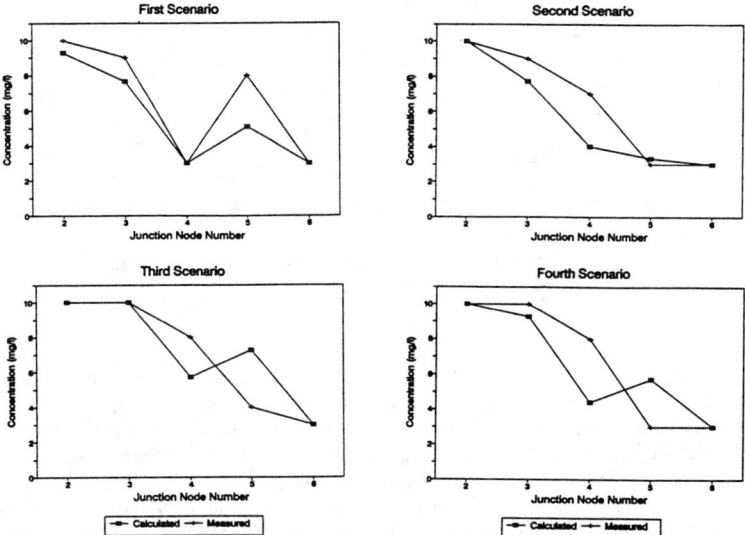
Figure 2. Calculated and Measured Chemical Concentrations

Conclusion

This paper presents a comparison of various experimental results to analytical quality results. The comparisons range from poor to good and do not provide a conclusive evidence that the comprehensive steady state quality model--based on explicit graph-theoretic algorithms for mixing problems in multiple-source networks--works. The agreement between the computer model and experimental results decreases somewhat when the traveling water velocities increase in pipes. This is quite expected most probably because of the short travel time between one node and another.

From a hydraulic point of view, the comparisons range from good to excellent and provide evidence that the distribution network used in this study was modeled adequately. The results of the computer model and experimental data slightly vary most probably because minor losses were not defined accurately in the computer model.

Water quality modeling in distribution systems is a complex phenomenon. Although all mass balance relationships are satisfied in the presented algorithm, yielding correct mathematical solutions for quality operating parameters, confidence in the reliability of the model and the utilization of reasonable data are essential. More models and experimental studies with more complex networks than the one presented here are recommended in order to further verify and validate the presented method's actual applicability.

Appendix-References

1. Elmaalouf, S. G., "A Comprehensive Steady State Quality Model for the Assessment of Contaminants Behavior in Water Distribution Systems," Master of Science Thesis, Civil Engineering Department, California State University, Los Angeles, 1992.

THE COURANT NUMBER AND UNSTEADY FLOW COMPUTATION

By Chintu Lai,[1] Member, ASCE

Abstract

The Courant number C, the key to unsteady flow computation, is a ratio of physical wave velocity, λ, to computational signal-transmission velocity, r, i.e., $C = \lambda/r$. In this way, it uniquely relates a physical quantity to a mathematical quantity. Because most unsteady open-channel flows are describable by a set of n characteristic equations along n characteristic paths, each represented by velocity λ_i, $i = 1, 2, \ldots, n$, there exist as many as n components for the numerator of C. To develop a numerical model, a numerical integration must be made on each characteristic curve from an earlier point to a later point on the curve. Different numerical methods are available in unsteady flow computation due to the different paths along which the numerical integration is actually performed. For the denominator of C, the r defined as $r = r_o = \Delta x/\Delta t$ has been customerily used; thus, the Courant number has the familiar form of $C_r = \lambda/r_o$. This form will be referred to as the "common Courant number" in this paper. The commonly used numerical criteria $C_r \gtreqless 1$ for stability, neutral stability, and instability, are imprecise or not universal in the sense that r_o does not always reflect the true maximum computational data-transmission speed of the scheme at hand, i.e., C_r is no indication for the Courant constraint. In view of this, a new Courant number, called the "natural Courant number," C_n, that truly reflects the Courant constraint, has been defined. However, considering the numerous advantages inherent in the traditional C_r, a useful and meaningful composite Courant number, denoted by C_r^*, has been formulated from C_r. It is hoped that the new aspects of the Courant number discussed herein afford the hydraulician a broader perspective, consistent criteria, and unified guidelines, with which to model various unsteady flows.

Introduction

The story of traditional Oriental doctors detecting the past medical history of a patient only by pulse diagnostic is amazing but not unique. The rationale behind this is that any noticeable disease that took place in an internal organ would cause some change in and around the organ; the blood flow would not go through it without recording some anomaly in its pulse. An analogy can be found in unsteady flow problems, which generally are time-marching problems. Any change or disturbance that occurred in the past is transmitted to a future point P along special paths called characteristics. If an appropriate amount of past data is monitored at P through a set of characteristic rays, sufficient information can be gathered at the point to describe the flow properly.

---

[1] Research Hydrologist, Water Resour. Div., U.S. Geological Survey, 430 National Center, Reston, VA 22092. Presently, Visiting Professor at the National Taiwan Univ., Taipei, Taiwan.

The analogy can be extended to the numerical model that is designed to simulate unsteady flow. The model being a discrete-type structure, data are strategically distributed to separate grid or nodal points of the present or past time level. While the real unsteady flow exists in the physical world, the numerically simulated flow belongs to another world – that of numerical mathematics.

Thus, despite their strong similarity, the time-marching motion is taking place in two different worlds — in the physical world and in the computational world. The Courant number, C, defined as the ratio of physical wave velocity, λ, to computational wave velocity, r, is a special parameter that relates these two worlds. This is quite unique in that other "numbers" in hydromechanics relate quantities of the same physical world. By linking quantities of two different fields, the Courant number serves as an invaluable index for computational accuracy, behavior, efficiency and stability for flow simulation, thus, deserving modelers' particular attention and study.

Theoretical Consideration

Unsteady flows are physical phenomena and can be identified by the presence of "local acceleration" or "nonzero time-derivative" terms in the governing partial differential equations (PDE's). These PDE's are generally transformable to characteristic equations whose physical implications have been described above.

Mathematically, the said PDE set is of hyperbolic or parabolic type and is referred to as an initial-boundary-value problem. When the PDE system is transformed to the characteristic system, the dimension of the original system is reduced by one, simplifying the one-space-dimensional flow PDE's to the total differential equation system. A variety of numerical methods and schemes have been devised and applied to various types of flow, the open-channel flow in particular.

In carrying out the numerical solution, data at some discrete points are transmitted forward in time with the speed $r = \delta x/\delta t$, just as the physical information is propagated along the characteristics $\lambda = dx/dt$. For stable, reliable and accurate simulation, there seems a need for some parameters that correlate properly the operations in the two separate fields.

Courant et al. (1928) were the first to conduct numerical study for computational stability in time-marching problems, which was followed by many mathematicians and engineers for further development and advancement. From the prevailing numerical condition for stability, known as the Courant-Friedrichs-Lewy (C-F-L) condition, or the Courant condition, a constraint, called the Courant constraint, can be derived, which states: To maintain a numerical stability, the speed of the physical wave propagation must not exceed that of the computational data transmission. Because the Courant number gives the very ratio of these two speeds, it is needless to say how important this number is to the numerical stability of unsteady flow computation.

Courant Number

General Courant Number: The general form of the Courant number, as stated earlier, is $C = \lambda/r$. The physical wave velocity $\lambda = dx/dt$ has many different forms depending on the type of flow under study. The computational wave velocity $r = \delta x/\delta t$ also depends on the kind of numerical scheme and the type of grid/nodal structure. If a particular unsteady flow can be described by a set of n characteristics, λ_i, $i = 1, 2, \ldots, n$, then for any given point i, $\lambda_i = \lambda(x, t, i)$. Likewise, the computational celerity r_i is $r_i = r(\delta x, \delta t, \kappa)$, where κ symbolically represents a particular numerical scheme employed.

It is then apparent that every characteristic, at any moment, for any given scheme, has a specific value for the Courant number at that discrete point. Any constraints, criteria and so forth are to be based on the Courant number at that point.

Common Courant Number: Whereas a variety of λ's is used for the Courant number, very little variation in r has been made. The conventional Courant number uses $r_o = \Delta x/\Delta t$ uniformly, where Δx and Δt are spatial and temporal grid interval, respectively. With fixed r_o, the stability criteria using $C_r = \lambda/r_o$ associates $C_r \lesseqgtr 1$ with stability, neutral stability, and instability, for a number of finite difference (FD) schemes. This is imprecise, and in fact, will be incorrect if used indiscriminately, because r_o does not necessarily coincide with the maximum computational celerity of the scheme at hand. For example, there are explicit schemes that are stable for $C_r > 1$, and all implicit schemes simply escape from the condition $C_r \leq 1$. In contrast to a new Courant number that is to be introduced subsequently, this widely used traditional Courant number shall be referred to as the "common Courant number" hereafter.

Natural Courant Number: According to the Courant constraint specified earlier, if, for the numerical scheme of concern, the maximum computational data-transmission speed, $r_m \equiv r_{max}$, is chosen for the reference r in the $C = \lambda/r$, then the new Courant number defined as $C_n = \lambda/r_m$ would always yield the stability criterion $C_n \leq 1$. This new form of the Courant number shall be called the "natural Courant number."

One can recall that the minimum computational wave speed also exerts a constraint on numerical stability, i.e., the scheme becomes unstable if the physical wave speed falls behind the minimum computational wave speed. Thus, another natural Courant number may be defined, $\check{C}_n = \lambda/r_n$, where $r_n \equiv r_{min}$. This one then can be called the lower natural Courant number, and the former, when in need of distinction, the upper natural Courant number, $\hat{C}_n$. The stability condition for $\check{C}_n$ is, of course, $\check{C}_n \geq 1$. Because in most cases, $r_n = 0$, $\check{C}_n$ then becomes irrelevant.

Numerical Solution of Unsteady Flow

Numerical Methods and Schemes: In order to develop a numerical model for the unsteady flow in question, either the numerical solution of the governing PDE's or that of the corresponding characteristic equations can be employed. The numerical methods for the former include the explicit and implicit Finite Difference Methods (FDM's), the Finite Element Method (FEM), and others. The method for the latter is the Method of Characteristics (MOC). Under each method there are a host of numerical schemes. For the MOC, the grid-of-characteristics uses a characteristic network directly (this scheme will not be discussed in this paper), and most of other schemes have a basic structure of characteristics superposed on a rectangular grid system.

Multimode Method of Characteristics: Because the characteristics represent a physical attribute, they always exist in any time-marching problem regardless of what numerical method is actually used. However, the MOC should give the best picture of various forms of the Courant number, and the Courant condition/constraint associated with each form, simply because the computational structure of the MOC is an overlay of the characteristics (physical components) on the rectangular grid system (computational components). According to the starting point and the interpolation mode, different numerical schemes for the MOC have been identified. (cf. Lai 1988.)

The Multimode Method of Characteristics (MMOC) is a comprehensive MOC scheme that combines some of the individual MOC schemes into one (each of the original schemes is now called a mode within the new scheme). (Lai 1988, 1991.) Consequently, the MMOC can portray various forms of the Courant number simultaneously or sequentially in its base numerical grid structure, and depict at the same time the dynamic interaction of the unsteady flow and the underlying Courant numbers. Figure 1 gives the outline of the MMOC structure and concepts. Different assortments of computational modes lead to different kinds of MMOC.

Numerical Integration, Characteristics and Courant Number: When numerical modeling of unsteady flow is to be implemented, numerical integration is usually carried out stepwise in the advancing time direction. Numerical quadrature is performed for each characteristic curve from an earlier point on the curve to a later point on the curve, but along a variable path depending on what numerical method is employed. For the MOC, this path is directly along the characteristic curve itself; for the FDM, it is along a space-time path; and for other numerical methods, along some designated space displacement followed by time stepping. Different schemes within a numerical method are characterized by variations in starting or tracing the designated path for that method. The Courant number appears in each of these schemes and serves as a very valuable index for numerical stability/instabilty, accuracy, and other numerical aspects.

Composite Courant Number

Other Hydromechanics Numbers: As opposed to the Courant number, which relates quantities of two separate fields, other hydromechanics numbers relate or compare two quantities of the same field. However, one may question whether the Courant number, aside from the special inter-field feature, possesses the kind of useful and powerful features those well known hydromechanics numbers have. A brief review of a few typical hydromechanics numbers is in order.

The Froude number, $\mathcal{F} = u/c_g$, is a ratio of the flow velocity to the shallow-water wave celerity or dynamic wave celerity, c_g. $\mathcal{F} = 1$ is a critical number which divides the flow field into sub- and super-critical flows. Hydraulic jumps, conjugate depths, sequent depths, and many other useful and significant hydraulic quantities evolve around this number. The Mach number, $\mathcal{M} = u/c_e$, is a ratio of flow velocity to celerity of elastic wave, c_e. $\mathcal{M} = 1$ is a critical number which divides the flow field into sub- and supersonic flows. Many useful and significant hydraulic quantities can similarly be derived. Another number, the Vedernikov number, $\mathcal{V} = c_k/c_g$, is a ratio of kinematic wave celerity to dynamic wave celerity. $\mathcal{V} = 1$ is a critical number which divides (steep) channel flow into stable and unstable zones. Do any of the Courant numbers defined earlier have such a uniform, exquisite and useful property?

Courant Numbers Compared: The general Courant number $\mathcal{C}$ being the ratio of the physical and computational wave velocities, its form and value are entirely dependent on how these two parameters vary or behave. The common Courant number, $\mathcal{C}_r$, offers neither a uniform stability criterion nor the attractive attribute of unity serving as a critical point dividing the field into two diametically opposite zones. However, it seems to retain the appealing feature that $r_o = \Delta x/\Delta t$ symmetrically divides the underlying grid system (see Fig. 1), contributing to a likelihood of evenly distributing $\mathcal{C}_r$ values, making the $\mathcal{C}_r$ as a favorable parameter for numerical accuracy analysis. The extensive use of the $\mathcal{C}_r$ in the past and the abundance of literature written in terms of $\mathcal{C}_r$ are other important factors not to be overlooked lightly.

Although the natural Courant number, $\mathcal{C}_n$, faithfully reproduces the Courant constraint, too often the number collapses to a rather uninteresting value (e.g., $\mathcal{C}_n$ is always zero for any implicit schemes unless λ also goes to ∞), thus losing the sensitivity to serve as a numerical accuracy index. As with C_r, it also lacks the powerful utility of those hydromechanics numbers mentioned in the preceding subsection.

Composite Courant Number: Considering the various merits and demerits reviewed above, it seems desirable to have a new Courant number equipped with most of the feature advantages discussed so far. Unfortunately, no single-term Courant number by other definitions so equipped will likely to emerge.

By an alternative means, a composite Courant number defined as $C_r^* = (C_r + C_r^{-1})/2$ seems to have such attributes. Referring to the C_r^* vs C_r curve shown in Fig. 2, $C_r = 1$ gives the minimum value $C_r^* = 1$, the critical point, which divides the curve into two different sections, the section for the sub-Courant number [The term was coined in Lai (1988).] and that for the super-Courant number. In the sub-Courant section, C_r^* increases as C_r decreases, and in the limit $C_r^* \to \infty$ as $C_r \to 0$; and in the super-Courant section, C_r^* increases as C_r increases, and in the limit $C_r^* \to C_r/2$ as $C_r \to \infty$. However, there exists another limit, that is, the line $C_r = r' = r_m/r_o$ marks the Courant constraint, above which the numerical scheme under consideration will be unstable, thereby rendering that part of the curve of little significance.

In short, the C_r^*-C_r curve, where $C_r^* = (C_r/2)(1 + C_r^{-2})$, somewhat resembles the E-h curve, where $E = (h/2)(2 + \mathcal{F}^2)$ is the specific energy. Further analyses could be made from such a composite Courant number, and interesting results are expected to be forthcoming. Two points particularly noteworthy here are that the C_r^* is a number useful for the unsteady flow because it is based on C_r, and that the C_r^* behaves generally in a symmetrical way around $C_r = 1$.

Summary and Conclusions

The Courant number, C, which is the key parameter in unsteady flow computation, is defined as a ratio of physical wave velocity, λ, to computational data-transmission velocity, r, i.e., $C = \lambda/r$, a unique number linking a physical quantity to a mathematical quantity. Many unsteady open-channel flow problems can be described by a set of n characteristic equations running along n characteristic paths with velocity λ_i, $i = 1, 2, \ldots, n$. In numerical modeling, numerical integration is performed stepwise in the advancing time direction for each λ_i. Integration along a different path leads to a different numerical method. Because $\lambda_i = \lambda(x, t, i)$ and $r_i = r(\delta x, \delta t, \kappa)$, every characteristic, at any moment, for any given numerical scheme, κ, has a specific value for the local Courant number at that point, playing a significant role for flow stability.

The stability criterion based on the common Courant number, C_r, with $r = r_o = \Delta x/\Delta t$, is imprecise or incorrect, because r_o does not always coincide with the maximum computational data-transmission speed dictated by the Courant constraint. The existance of a lower margin further complicates the use of the C_r as a sole stability parameter. A new Courant number, called the natural Courant number, C_n, that obeys the foregoing range of constraint, has been defined. However, because of the numerous advantages inherent in the C_r, a useful and meaningful composite Courant number, C_r^*, has been derived from C_r. It is expected that the new aspects of the Courant number afford the hydraulician a broader perspective, consistent criteria, and unified guidelines, with which to model and simulate various types of unsteady open-channel flow.

References

Courant, R., Friedrichs, K. O., and Lewy, H. (1928)."Über die partiellen Differenzengleichungen der mathematischen Physik," *Math. Ann.*, v.100, 32-74 (in German).

Lai, C. (1988). "Comprehensive method of characteristics models for flow simulation." *J. Hydr. Engrg.*, ASCE, 114(9), 1074-1097.

Lai, C. (1991). "Modeling alluvial-channel flow by multimode characteristics method". *J. Engrg. Mech.*, ASCE, 117(1), 32-53.

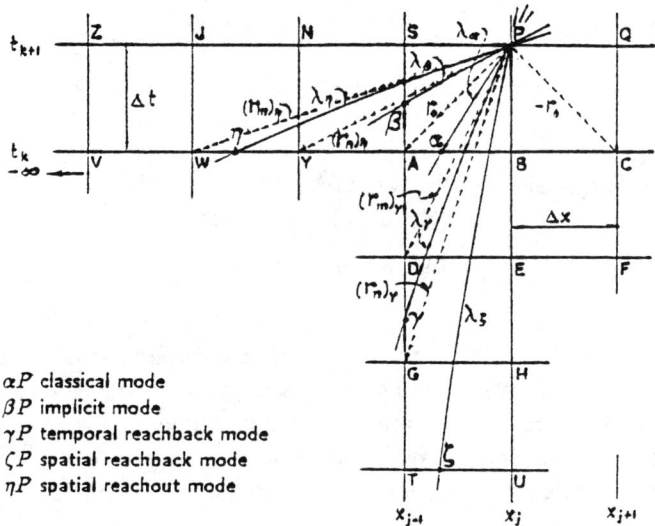

αP classical mode
βP implicit mode
γP temporal reachback mode
ζP spatial reachback mode
ηP spatial reachout mode

Figure 1. Multimode Method of Characteristics Displaying Several Numerical Modes, λ, r_o, r_m, and r_n. [Note: $C_r = \lambda/r_o$, $\tilde{C}_n = \lambda/r_m$, and $\vec{C}_n = \lambda/r_n$.]

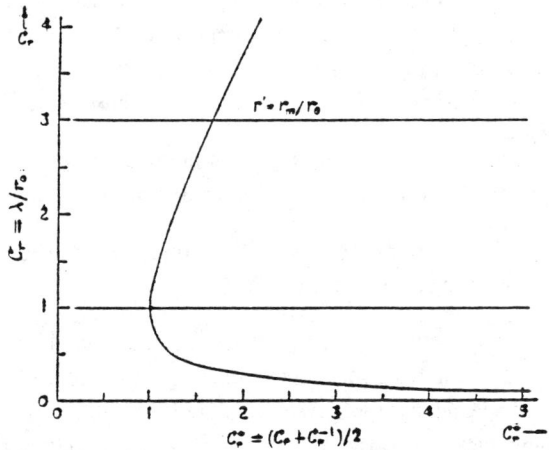

Figure 2. C_r^* versus C_r Curve

River Engineering Flows and Some Related Structures (State of the Art)

Rawya M. Kansoh[1]

Abstract

The first part of the paper deals with a simplified overall review of water flows. Laminar versus turbulent, visualization of flow with unsteady 3-D manner and some typical hydraulic structures and energy dissipators. Slide show is a major emphasis in this part. The second part stresses on selection and evaluation in modelling and managing river flows.

Models Variation

For one - dimensional model, the correct spatial definition of resistance and dispersion coefficients is important. Extensive field work and high quality data are prerequisites for any successful modeling. Calibration with standard cases becomes important thereafter.

For two - dimensional models, practical attention needs to be given in the depth - averaging procedure and the finite difference techniques.

For three - dimensional models, a good understanding of both the physical and mathematical background is vital.

It is known that the more complex the turbulence model, the greater are the number of empirical coefficients requiring calibration. Methods for turbulent flow analysis by means of computer graphics are used to visualize flowfield with unsteady and three - dimensional manner.

---

[1] Associate Professor of Hydraulics, Civil Engineering Department, Alexandria University, Alexandria, EGYPT.

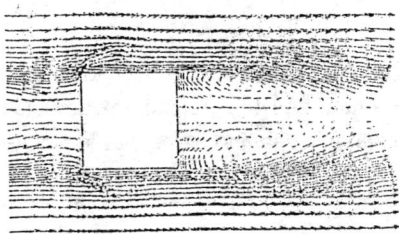

Figure 1. Mean Flowfield of Velocity Vectors

Because the turbulent flow is an unsteady and 3-D phenomenon, it is fundamentally useful to observe it in an unsteady and 3-D manner. Animated graphics is very advantageous in visualizing the flowfield in such a manner when supported by various techniques.

In visualization of particle movement, particles are introduced into the flow numerically and their movements are traced. This technique corresponds to the wind tunnel experiment which uses bubbles of helium. Computer graphics based on numerical simulation is most advantageous in expressing this because control of the characteristics of the particles and the generation method of the particles is very easy. For example, generation of particles with any density, generation of particles with zero initial velocity, or control of the color of the particles for each different time and position is easily achieved.

1- Streak line

In this cases, the traveling line of the particle does not disappear. The streak lines of the particles generated in front of an object as well as the standing vortex staying in front of the object can be observed.

2- Time line

This visualization corresponds to a wind tunnel experiment using the smoke wire method. As the number of particles is too large, the picture of their movements seem to be as that of smoke diffusion.

Despite the development and growth in numerical models, there will be a place for physical models, and in many cases studies will often involve a mixture of the two.

Well focused laboratory studies will still be required in some cases.

Summary

One can detect the following:

1- The choice of numerical model is important and will affect the results.

2- Simplification of any governing equations prior to solution may have profound effects.

3- Models should be checked against standard cases.

4- Calibration procedures should be systematic.

5- Sufficient field data should be collected.

6- Predicted results should always be checked with actual results.

7- Physical and numerical models are not mutually exclusive.

General aspects in river engineering

As shown in the schematic, river engineering has a theoretical bank and a practical bank, occasional link between the two banks is obtained by bridges.

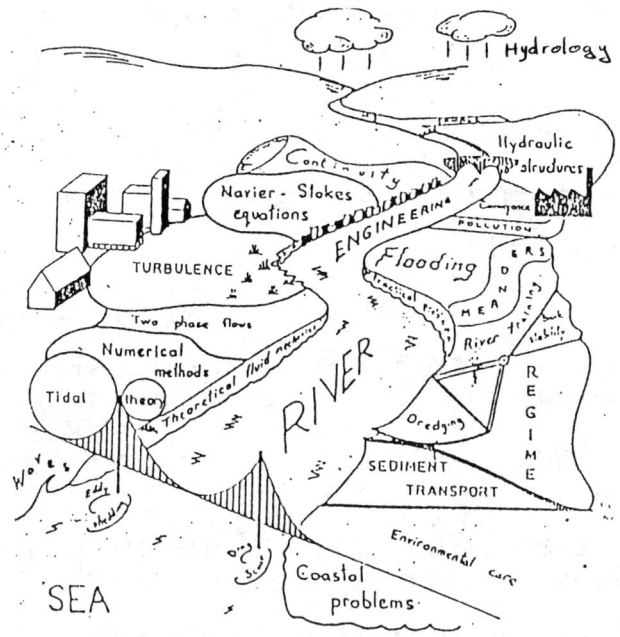

Figure 2. River Engineering Schematic

Theoretical Bank

Principles and theory
of fluid mechanics.
Navier-Stokes & continuity eqs.
are well structured by sheet piling
for permanence & solid foundations.
The region around turbulence
is slowly reclaimed.

Practical Bank

Applied problems
requiring solutions.
Stability and development.
Many problems exist and
subject to occasional
erosion or emendation.

Conclusion

River engineering is indicated by the river itself, which touches all regions, but like any natural river is subject to slow adjustment on time scale.

Rivers are however not just about advanced fluid mechanics and technical applications.

Rivers are beautiful, inspiring, occasionally terrifying, but above all worthy of our care.

Good environmental management must also play its part.

References

1- Abbott, M.B., "Computational Hydraulics", Pitman, London, 1979.

2- Ackers, P. et al., "Wiers And Flumes For Flow Measurement", Willy, Chichester, 1978.

3- Chow, V.T., "Open Channel Hydraulics", McGraw-Hill, New York, 1973.

4- Kansoh, M.M. and R.M., "Notes In Hydraulics, Fluid Mechanics And Water Power", Dar-Al-Maaref, 1971, 1974.

5- Morris And Wiggert, "Applied Hydraulics In Engineering", second edition, 1972.

6- Rouse, H., "Engineering Hydraulics", Wiley, New York, 1958.

7- U.S. Bureau of Reclamation, "Design of Small Dams", 1978.

THE EUR WATER STATION IN ROME - ITALY
Franco Ciacchella [1] - Paolo Massarini [2]

Abstract
The paper deals the new steel Water Centre realized by ACEA in Rome. The Water Station is destined to manage the water consumption of around 500.000 inhabitants with an output of 3,900 cubic metres a second and a storage capacity in the underground reservoirs of around 35.000 cubic metres. The Centre, the only one in its kind in Europe, has been entirely built in a special "self-passivating" steel.

Introduction
A.C.E.A., Rome's Municipal Energy and Environmental Board is responsable for the management of the capital water-supply system.

The water distribution network of the city of Rome has special characteristics based on the fact that most of the water comes from springs in the Appennines; these springs are located at high altitudes, and gravity thus supplies water for most of the topographically irregular city area, which varies from sea level to over 150 m.

Rome's municipal territory covers an area of over 1500 km², one of the most extensive in Italy, and counts nearly 3 million inhabitants. It is crossed by the Tiber and Aniene rivers and has 40 km. of Tyrhennian coastline.

For water distribution over different altitudes, the city is divided into nearly 50 water zones, which follow the contour lines as closely as possible in order to limit pressure peak levels and range as much as possible.Each of these zones has a water station (reservoir or water tower) which is sometimes raised, and feeds the distribution network.

Hydraulic functioning
The EUR Water Centre basically functions (see figure 1):
A ø 2.200 millimetre pipeline, 67,50 metres high, conveys water by gravity at a rate of 3.900 litres a second to a small collection basin situated at a height of 105 metres above sea level. This basin is directly connected to the 700 cubic-metres capacity standpipe with an overflow threshold, which supplies the water tower through a ø 1.400 millimetre condui; on the bottom of

1-*Executive Head of Water Production Plants Sector - ACEA - Rome*
2-*Work Supervisor of EUR Water Centre - ACEA - Rome*

the pressure-equalizing basin, at 84 metres above sea level, a ø 1.000 millimetre steel conduit leads off to supply at a constant rate, the higher districts of the water zone served by the centre.
The ring tank, with a capacity of 1.700 cubic metres, 68,00 metre above the ground, and around 90 metres above sea level, has an external diameter of 34 metres and a rectangular section of 4,00 x 5,50 metres. The tank is supported by two pairs of hollowed cylindrical towers, three metres in diameter, situated diametrically opposite each other and connected by 2,60 x 3,50 metre flanges with 400, M24 class bolts each. The first pair contains the stairs and lift serving the centre, while the secon pair is used for water diversion and distribution functions.
The diversion tube contains the ø 1.200 millimetre pipe leading from the water tower. The second water tube, 2,00 metres distant from the diversion tube, contains the ø 1.300 millimetre pipe which collects the overflow from the water tower and conveys it, through a surge tank, into the dissipation basin situated at its feet. From here the overflow is carried to the underground reservoir through a ø 1.200 millimetre conduit.
The tube supporting the pressure tank is behind the two servise tubes, is aligned on the plane of symmetry of the four tubes, and is 77,00 metres high.
The first 20,00 metres of the tube have a three-metre diameter circular section, after which a truncated cone section connects the lower part of the tube with the cylindrical pressure tank which is seven metres in diameter.
The stairs and the elevator lead to: a first level at the top of the tank, where a panoramic walkway is planned, at an elevation of 92,60 m. a.s.l.; a second level at the top of the arrival tank, where a roof-top room is located, with a panoramic terrace, at an elevation of 106,60 m. a.s.l.

Architecture

In the framework of the works existing in the field of raised tanks and piezometers for drinking water, hydraulic system functions have always been studied and realized independently from the form and the statics of the work, relegating the architecture to the role of "container" of functions.
The great innovation proposed by the project consists of harmonizing and proposing an "hydraulic architecture" in which the functions of the hydraulic circuit generate and inspire architectural forms; the elements which make up the plant, then, although calculated as far as form, profiles, and sections are concerned for favoring the flow and circulation of the water, are themselves architecture.
Also important is the structural value of these elements; in fact, with the use of steel, it has been possible to overcome the problems concerning working at heights and to contain the total weight of the supporting structures.
The project finds its ideal conclusion in the external arrangements which develop according to precise design lines inspired by the following motives:

a) leaving in view of the foundaton structures of the hanging tank;
b) leaving in view of the system of outgoing and incoming pipes of the Water Centre; consequent dislocation outside of the measurement, interception, and control equipment and design of special polycarbonate protective coverings;
c) realization of a network of walkways and driveways so as to permit the complete usability of the area for visitors and for personnel in charge of running the plant.

Technical characteristics and static problems

The choice of "special" steel for the basic structure of the tower was based on the following requirements:
- it would enable an especially bold architectural approach;
- material reliability in relation to temperature change, condensation and protection against corrosion from the atmosphere;
- the utmost reduction in the amount of maintenance, and its simplification;
- maximum durability over time;
- a better final cost-yield ratio.

These consideration have led to the choice of passivation-treated steel, type COR-TEN (CORrosion resistance, TENsil strenght), using Itacor 2 (Italsider corrosion-proof) and RE-SCO (SB of ALF FALK). These are all ferritic perlitic types of steel which can develop a "patina coating".

This steel allows for a considerable reduction in thickness (and thus also in weight) and is highly resistant to corrosion (up to five times more than carbon steel) because of the passivation treatment. It contains copper, chrome, nickel, vanadium or niobium and phosphorus alloys of up to 3%, with a high yielding limit and excellent welding quality.

COR-TEN type steel "matures" over a period of between 18 and 36 months. During this process, the layer of oxides forming on the surface may undergo erosion and be soluble in water.

Therefore, A.C.E.A. has decided to paint the structure in order to prevent even the occurrence of superficial corrosion.

The steel was subjected to "white-metal" sandblasting to a degree of SA 2,5 or SA 3 with metal grit; afterwards, several coatings of paint were supplied in stages using the airless system, as follows:

Stage A. Epoxyvinyl - for surfaces not touching water or in non-saturated areas with total thickness of 180 µm;
Stage B. Epoxide - for surfaces in contact with water or in saturate areas with total thickness of 220 µm.

Stage A was designed to give better protection against atmospheric corrosion agents (rain, wind, sunlight, etc.) considerede for high corrosion levels, as well as acids found in city air.

Stage B was designed to ensure maximum protection of surfaces in contact with drinking water, using suitable material allowed by regulations.

The metal structure of the water tower has been coated by a stainless-steel paneling. The panels have been made in sandwiches elements 50 mm. thick consisting of another sheet of austenitic stainless steel (AISI 304 L) 7/10 mm. thick and an inner sheet of pre-painted galvanized steel 6/10 mm. thick.

Particular attention was paid to designing the foundations; this was because of the specila conformation of the tower, its sensitivity to possible differential settlement between two groups of vertical supports. The foundations are formed by two vibrated reinforced concrete blocks 3,50 metres thick, connected by a grid of horizontal beams. In order to guarantee maximum stability of the ground, a large area has been renforced using a jet-grouting system with a total of 260 pillars in a mixed cement.

The singularity of the work has involved a detailed study of the contructional procedures and of the sequence of assembly of the various elements.

Elements welded at workshop were chosen in sizes such as to ensure optimizing between workshop necessities and possibility of transportation to building-site .

Follows a description of the eight main erection phases; the structures for each have been checked with a finite element (beams) numeric model connected together at determinated nodal points.

PHASES (see figure 2)

1st phase - five erection elements corresponding to piezometer, stairway pylon, lift pylon, pylon housing delivery and pylon housing overflow.

2nd and 3st phases - Piezometer, and stairway and lift pylons were linker utilizing ancillary erection reticular structures.

4th phase - overhung ercton of a section of ring-tank

5th phase - delivery pylon installed and intake tank completed.

6th and 7th phases - second and last section of ring-tank are mounted connecting.

8th phase - temporary erection connections are removed and piezometer covering superstructures are completed.

Figure 1

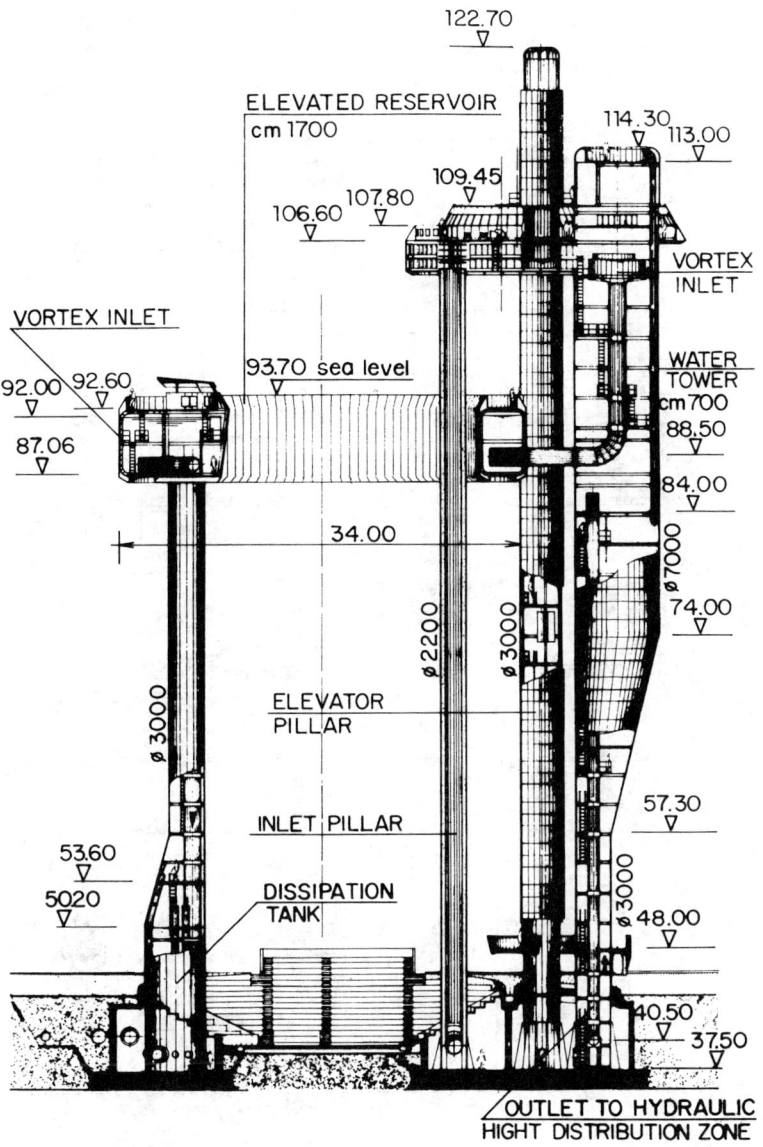

Figure 2

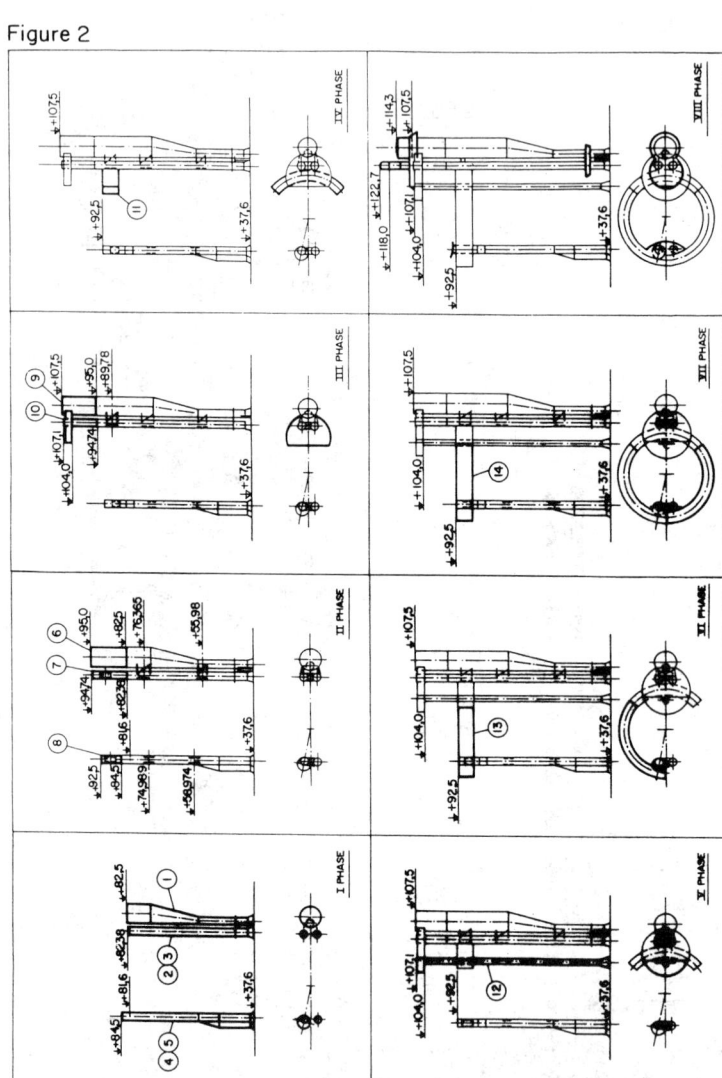

A Culvert Analysis Program for Indirect Measurement of Discharge

Janice M. Fulford[1]

Abstract

A program based on the U.S. Geological Survey (USGS) methods for indirectly computing peak discharges through culverts allows users to employ input data formats used by the water surface profile program (WSPRO). The program can be used to compute discharge rating surfaces or curves that describe the behavior of flow through a particular culvert or to compute discharges from measurements of upstream and downstream water-surface elevations. This procedure is based on the solution of the gradually varied flow equations and has been adapted slightly to provide solutions that minimize the need for the user to determine between different flow regimes. The program source is written in Fortran 77 and has been run on mini-computers and personal computers. The program does not use or require graphics capability, a color monitor, or a mouse.

Introduction

The culvert analysis program (CAP) was developed to provide a computer program that is based on sound, accepted, hydraulic principles and that uses a data format that is compatible with other popular hydraulic analysis programs. Also of concern in the development of the program was its ability to run on a wide variety of available computers with no modifications to the program. A brief overview of the methodology used in computing culvert discharges, program use, and the program capabilities are presented herein.

Culvert analysis overview

The culvert analysis procedure followed by CAP classifies the flow into six types. The equation for each flow type is derived by writing the energy equation from the control section that governs that flow state to the approach section and substituting the continuity equation into the energy equation. The six flow types, (1)

---

[1]Hydrologist, U.S. Geological Survey, Stennis Space Center, MS 39529

critical depth at inlet, (2) critical depth at outlet, (3) tranquil flow throughout, (4) submerged outlet and inlet, (5) rapid flow at inlet and part full culvert barrel, and (6) full culvert barrel flow with free outfall, are special cases of the energy equation for gradually varied flow. Because entrance loss is computed in the culvert equations as a function of velocity, the entrance loss and velocity head in the control section are combined into a single velocity head term that contains the discharge coefficient. The discharge coefficient functions in a manner similar to the familiar velocity coefficient. Fig. 1 is a definition sketch for culvert flow variables. The six flow equations (Bodhaine, 1968) solved for discharge and the criteria governing their use are listed in fig. 2.

The program is written in Fortran 77 and runs on mini and personal computers. No graphics, color, or mouse capability is required. CAP solves the appropriate form of the energy equation for the approach section water-surface elevation by using flow routing. Given discharge and a tailwater elevation (water-surface elevation at the culvert-exit section) the appropriate flow type and equation or equations are determined and solved for the approach-section water-surface elevation using a bisection root solver.

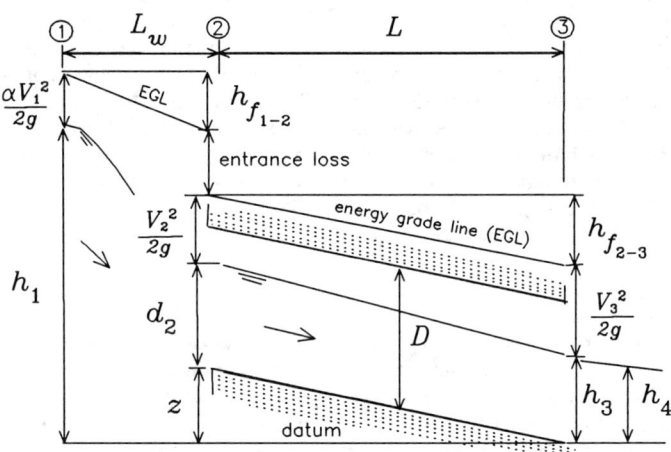

Figure 1. Definition sketch for culvert flow variables. 1 is at the approach section, 2 is at the entrance section, and 3 is at the exit section. [L, culvert length; L_w, approach reach length; D, culvert height; $h_{f(i)}$, head loss over (i), the subscripted reach; h_i, water surface elevation at the i section; V_i, mean velocity at the i section; d_2, depth of flow at the culvert inlet; g, gravitational constant; α, velocity coefficient]

$$Q_{type1} = C_{123}A_{d_C}\sqrt{2g(h_1-z+\alpha_1\frac{V_1^2}{2g}-d_C-h_{f_{1-2}})} \quad \text{for } h_1-z<1.5D,\ h_4<d_C+z,\ S_0>S_C$$

$$Q_{type2} = C_{123}A_{d_C}\sqrt{2g(h_1+\alpha_1\frac{V_1^2}{2g}-d_C-h_{f_{1-2}}-h_{f_{2-3}})} \quad \text{for } h_1-z<1.5D,\ h_4<d_C+z,\ S_0<S_C$$

$$Q_{type3} = C_{123}A_3\sqrt{2g(h_1+\alpha_1\frac{V_1^2}{2g}-h_3-h_{f_{1-2}}-h_{f_{2-3}})} \quad \text{for } h_1-z<1.5D,\ h_4\leq D,\ h_4>h_C$$

$$Q_{type4} = C_{46}A_O\sqrt{\frac{2g(h_1-h_4)}{1+29C_{46}^2n^2LR_O^{-\frac{4}{3}}}} \quad \text{for } h_1-z>D,\ h_4>D$$

$$Q_{type5} = C_5A_O\sqrt{2g(h_1-z)} \quad \text{for } h_1-z\geq 1.5D,\ h_4\leq D$$

$$Q_{type6} = C_{46}A_O\sqrt{2g(h_1-h_3-h_{f_{2-3}})} \quad \text{for } h_1\geq 1.5D,\ h_4\leq D$$

Figure 2. The six culvert flow equations and the criteria governing their use. [Q, discharge; A_i, culvert area for i depth or section, o subscript denotes full culvert; d_c, critical flow depth in the culvert; g, gravitational constant; C_{type}, discharge coefficients; R_o, hydraulic radius of full culvert; S_o, culvert slope; S_c, slope for critical flow in culvert; refer to fig. 1. for remaining symbol definitions]

Transitions between the various flow types are not always smooth and continuous. The transition between type 1, type 2, or type 3 flow and the high head flows (type 5 or 6 flow) is not smooth and continuous. These transitions are approximated by fitting a straight line between type 1, type 2, or type 3 solution at headwater-diameter ratios of 1.2 and the type 5 or type 6 solution at headwater-diameter ratios of 1.5.

Mathematical discontinuities can occur between type 1 and type 3 flow, between type 2 and type 3 flow, and type 3 and type 4 flows if the criteria listed in fig. 2 are followed rigorously. A tailwater elevation and discharge can be found which satisfies flow type 3 criteria but will not be solvable as type 3 because of the computed flow losses in the barrel. In some situations this is the result of two flow types occurring in the culvert. CAP handles these cases by trying to solve for type 3 flow initially. When the equation is found unsolvable the requirement that tailwater elevation is less than or equal to the critical depth is relaxed and either type 2 or type 1 flow is computed depending on the culvert slope. Between types 3 and 4 both types of flow occur in the culvert and the barrel flows full for part of its length. To determine water surface elevations from a given tailwater and discharge the flow could be routed using step-backwater techniques. Instead, an approximate solution is employed that determines the approach elevation by linearly interpolating between the solutions for type 3 and type 4 flow.

Assumptions made by the solution of the equations in CAP include constant culvert cross-sectional shape, constant culvert slope and for flow types 4, 5, and 6 negligible approach velocity head. Also CAP is unable to compute for situations with supercritical flow in the approach section.

Program operation

Culvert computations require data that describe the geometry of the approach section, the culvert, entrance conditions, roughness coefficients, and discharge coefficients. The approach section is located before the region of drawdown upstream from the culvert opening.

Input files are prepared using a ASCII text editor. Data records that describe the approach section can be placed in a file with data records that describe the culvert and the range of computations or in a separate file. Approach-section records describe the location, the geometry, and roughness parameters of the approach section. These records follow the same format as WSPRO (Shearman, 1990) valley sections except for the addition of two records. CAP requires the input of culvert records in addition to records used by WSPRO. These additional records are ignored by the WSPRO program. CAP also provides for the input of nonstandard culvert geometry as well as rectangular, circular, and pipe-arch geometries. A sample input file for a circular culvert is listed in fig. 3.

When the program is executed, the user is queried for the names of the files containing the approach-section data and the culvert-section data and identifiers for each section. Error messages are printed to the screen and to the output file. Screen messages are usually the result of errors in the input files. Messages printed to the output file are either warning messages indicating potential problems or fatal error messages for a particular combination of discharge and tailwater elevation.

```
CV    EX01  10.,0.,41.,6.05,6.28,1
CG          213,30.
*C3         1,1,1,3,1
*C5         0.84, 0.446,1.4, 0.471,1.6, 0.492,1.8, 0.512,2.0
*CF         6
*CX         2.09
*CQ         20., 21.5, 22.5, 25.
*CN         0.015
*TT  5
*PD         0.   20.0 25 1.0 1.
XS    AP01  57.4
GR          0.,9. 0.,6.4 1.,6.6 2.,6.6 3.,6.5,6.5 6.,6.7.,7
GR          8.,8.1 10.,9.
N           0.025
EX
```

Figure 3. Sample CAP input file for a circular culvert.

```
USGS CULVERT PROGRAM VER 92-2
                  CULVERT                          APPROACH SECTION
I.D. EX1          n= .015    Height    2.00ft     I.D. APPR
Station    10.0 ft           Length    108.0 ft   Station       121.0 ft
Inlet el.   9.00ft           Outlet el.  .00ft    Minimum el.     9.25ft


       Discharge  Flow    Water Surface Elevations (feet)    error
no.     (cfs)     type    appr.   inlet    outlet   exit     code@
 1       6.50      1      10.20   9.90********     1.00       0
 2       6.50      4      12.71  11.00     2.00   12.50       0
 3       6.50      4      14.21  11.00     2.00   14.00       0
 4       6.56      1      10.20   9.91********     1.00       0
 5       6.56      4      12.72  11.00     2.00   12.50       0
 6        .56      4      14.22  11.00     2.00   14.00       0
 7        .00      1      10.26   9.94********     1.00       0
 8       7.00      4      12.75  11.00     2.00   12.50       0
 9       7.00      4      14.25  11.00     2.00   14.00       0


       Fall (ft)        Losses (ft)        Appr. Section    Control Section
no.  C  entry  eff.   entry(1-2) (2-3)     VH   alph  F      energy    F
 1  .95  .29   .38    .04   .02 *****     .11  1.00  .58    10.25   1.00
 2  .95  .21   .21    .01   .00    .12    .00  1.00  .04    ********  *****
 3  .95  .21   .21    .01   .00    .12    .00  1.00  .02    ********  *****
 4  .95  .30   .38    .04   .02 *****     .11  1.00  .57    10.26   1.00
 5  .95  .22   .22    .01   .00    .12    .00  1.00  .04    ********  *****
 6  .95  .22   .22    .01   .00    .12    .00  1.00  .02    ********  *****
 7  .95  .32   .40    .04   .02 *****     .10  1.00  .55    10.30   1.00
 8  .95  .25   .25    .01   .00    .14    .00  1.00  .04    ********  *****
 9  .95  .25   .25    .01   .00    .14    .00  1.00  .02    ********  *****

Abrevs. used: appr.-approach  C-discharge coefficient  eff.-effective
VH-velocity head  alph-velocity coefficient  n-Manning's roughness coef.
energy-specific energy  F-Froude number  entry,(1-2),(2-3)-part of reach

@Error codes:  -1,1-7 fatal error; 8-11 warning; 0 no error

---------Warning Messages-------------
```

Figure 4. Sample of summary of culvert computations produced by CAP.

Two outputs are available from CAP. A summary report of the culvert computations is always produced. A table that can be used by other programs and that contains discharges, tailwaters, and computed upstream water-surface elevations is produced optionally. The summary report includes computed hydraulic properties for the culvert section, the approach section, and a summary of the culvert computations. A sample summary of culvert computations is shown in fig. 4.

Program results can be plotted and developed into either rating curves or rating surfaces or used to determine discharge from measurements of headwater and tailwater high-water marks. The familiar technique of illustrating culvert behavior as a rating curve that is a function of discharge and approach elevation is shown in fig. 5 and requires a curve for each tailwater. A rating surface described by the points with coordinates corresponding to the discharge, tailwater elevation, and the approach-section water-surface elevation is shown in figure 6. Values and gradients can be mathematically interpolated from the points determined by the discharge, tailwater elevation, and approach-section water-surface elevation.

Conclusion

CAP uses the basic procedure described by Bodhaine (1968) with a few modifications. The modifications allow solutions to be computed where discontinuities between the six flow type equations occur and when more than one type of flow occurs in the culvert. Input files are based on WSPRO formats and any records included specifically for CAP do not interfere with WSPRO execution. Rating curves or rating surfaces describing the hydraulic behavior of the culvert can be plotted from the program results. The program can also be used to determine discharges for culverts from measured highwater marks.

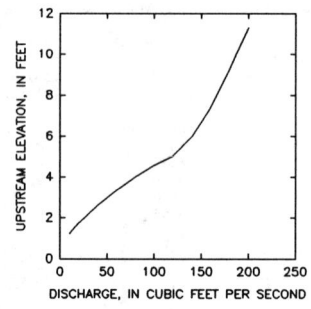

Figure 5. Rating curve for a rectangular culvert at 0.5 ft tailwater.

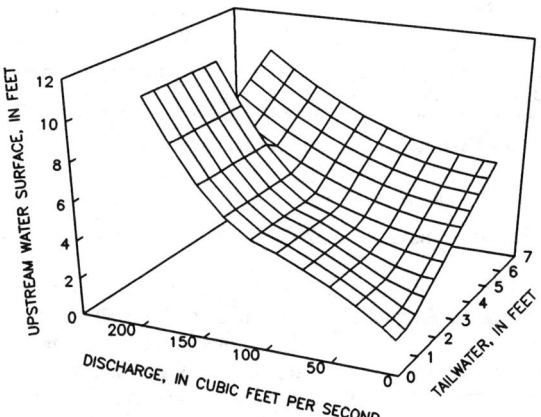

Figure 6. Example of a rating surface for a rectangular culvert.

References

American National Standard Programming Language FORTRAN, (1978), American National Standards Institute, New York, NY.

Bodhaine, G.L., 1968, Measurement of peak discharge at culverts by indirect methods, Techniques of Water-Resources Investigations, Book 3, Ch. A3, U.S. Geological Survey.

Shearman, J.O., 1990, User's Manual for WSPRO--A computer model for water surface profile computations, FHWA-IP-89-027, Federal Highway Administration.

SEDIMENT AND WATER QUALITY CONTROL DEVICES IN SMALL WATERSHEDS
by
Hasan Nouri[1], Member, ASCE

Abstract

The physical behavior and health of an alluvial stream depends on the extent of man's interferences in its tributary areas. In its natural setting a stream provides three main functions of: (a) allowing the conveyance of flood waters and drainage; (b) permitting the transportation of bed load as well as suspended load and; (c) assimilating some of the pollutants generated in its tributary areas. When a development is completed in a watershed which discharges to the stream, peak flood discharges and pollutant levels will be increased, while sediment yield will be reduced. If these increases and reduction are beyond the tolerance limit of the stream it will be subjected to environmental and economic damages. This paper briefly describes the behavior of a stream in its natural state and how it may be impacted by upstream development. A device that can mitigate the adverse impacts of water quality, peak discharge and transportation of bed load is conceptually described.

Introduction

When a portion of a watershed is developed it produces urban runoff which may have poor quality. With development the impervious areas in the watershed increase. Associated with this imperviousness and channelization will be an increase in peak discharge and reduction in bed load to the downstream natural channels.

The purpose of this paper is to conceptually describe a device which can be installed downstream of a watershed, a portion of which may become developed

---

[1]President, Rivertech Inc., 23332 Mill Creek Drive, Suite 100, Laguna Hills, California 92653

to achieve the following objectives: (a) Minimize the discharge of pollutants to the downstream reach in order to preserve its biological, environmental and recreational values; (b) Allow the delivery of bed load (sand and gravel) contributed by the upstream watershed, thereby, minimizing potential erosion and degradation along the downstream alluvial channel; and (c) Reduce the peak flood discharge in order to prevent flooding along the downstream channels and to achieve a balance between water discharge and bed load discharge.

Stream Behavior in its Natural State and After Development

In every natural stream, there is a unique relationship between the water discharge and the capacity of the stream to transport bed load. This relationship, shown conceptually by Line A of Figure 1, can be approximated by a straight line on log-log scale graph paper.

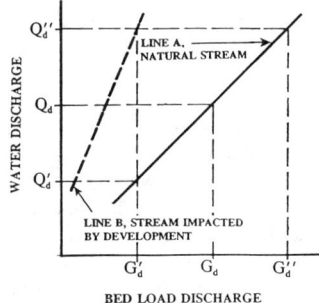

Figure 1 - Conceptual Bed Load Capacity Curves

The geometry of the main channel is established by the "*dominant discharge*" (Henderson, 1966) Q_d, sometimes referred to as the "*channel forming discharge*" (Wolman, 1957). When the water discharge is Q_d and the channel is stable, its corresponding bed load discharge is G_d. If the rate of supply exceeds G_d, the channel has the tendency to aggrade. Conversely, if the rate of supply is less than G_d the channel has the tendency to degrade.

The impacts of the upstream development on dominant discharge and its balanced bed load discharge are conceptually described in Figure 1. The dominant discharge will be increased from Q_d to Q_d'' and its bed load discharge will be reduced from G_d to G_d'. To maintain the stability of the natural channel, Q_d must be reduced to Q_d', otherwise the bed and banks of the channel must supply the necessary quantity of bed load to satisfy the higher level of dominant discharge. If Q_d is not reduced, initially, the quantity of supply from channel banks and bed will be equivalent to $(G_d''-G_d')$. Therefore, the bed and banks must erode to supply the necessary quantity of bed load and the cross-section area of the natural channel will increase with time. Eventually, the bed load discharge curve will be shifted to a new position, shown by Line B, in Figure 1, and the new dominant discharge Q_d'' will be balanced with the new bed load yield of G_d''. Line B represents the eroded reach.

Mitigation Strategies

In order to achieve the purposes described in the Introduction, advantage of the variation of discharge with respect to time shown in Figure 2 is taken to develop the strategies described in the following paragraphs.

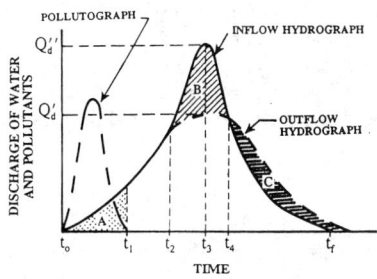

Area A = Design Retention Storage for Water Quality Control
Area B = Design Detention Storage for Maintaining Stability of Downstream Channel
Area C = Detention Release = Area B

Figure 2 - Definition Sketch of Hydrograph

(A) Water Quality Mitigation Strategy

As shown in Figure 2, storm flow at the mouth of a partially developed watershed begins at time t_o and ends at time t_f. Runoff containing most of the pollutants begins at time t_o and ends at time t_1. Area A represents the volume of first flush and should be diverted for treatment or stored in a retention basin. The retention basin will also store nuisance flow which occurs during dry periods.

(B) Bed Load Mitigation Strategy

Once construction of the development is complete, the bed load yield from the watershed will be reduced. However, the non-developed and landscaped areas within the watershed will continue to yield bed load. The strategy is to allow the passage of this reduced rate of bed load yield at the mouth of the watershed. During the period (t_1-t_o) when discharge is low, bed load yield from the watershed is small. This small quantity of bed load, some of which may carry pollutants, will be trapped in the retention basin. After time t_1 when the majority of yield occurs, all bed load generated within the watershed should be allowed to pass to the downstream alluvial channel.

(C) Peak Discharge Mitigation Strategy

In Figure 2, during the periods (t_2-t_1) and (t_f-t_4) when storm runoff is relatively clean, all inflow of water and sediment is allowed to pass through without occupying any storage area. During the period (t_4-t_2) when inflow is at its peak, a portion of the discharge is stored in a detention basin and the balance of the inflow hydrograph is allowed to pass though the system as outflow. Area B represents the design volume of the detention basin for maintaining the stability of the downstream alluvial channel. Area C represents the volume of release from the basin which is equal to Area B.

Concept of the Hydraulic Oscillating Diverter

Figure 3 shows the essential features of a device conceptualized to reduce peak discharge, by-pass bed load and trap pollutants. *Before designing and constructing such a device for use in the field, its performance must be researched and prototype tests should be conducted.* Based on the results of those tests, formulas and charts can be generated that would provide design parameters for the device. Storm flows from the watershed carrying sediment and pollutants enter the diverter through Pipe #1, shown in Detail A. The function of the diverter is to route the flow to Pipe #2 (Pipes 2a, 2b, 2c, 2d, and 2e) or Pipe #3 (Pipes #3a, 3b, 3c, 3d, and 3e). Pipe #2 conveys the flow to the basins and Pipe #3 by-passes the flow and bed load to the downstream channel.

Nuisance Flow Mitigation

During normal dry periods when the upstream watershed contributes only nuisance flow Pipe #2 is at its non-elevated Position #1 (Section C-C, Figure 3) so that flows will be diverted to the basin.

First Flush Flow Mitigation

The positions of Pipes #2 and #3 during the first flush period, represented by $(t_1 - t_o)$ in Figure 2, remain the same as that occurring during nuisance flow periods. As the flood discharge increases, head loss in Pipe #2 increases causing an increase in pressure at the 45° pipe bend and diversion of some flow to Pipe #3. In Pipe #3 flow through the weep holes is accumulated in Holding Tanks #4a and #4b increasing the force in the cable that connects Pipes #2 and #3, shown in Section B-B of Figure 3. At some time, weight of the #3 pipe system and water in the holding tanks exceed the weight of Pipe #2, Pipe #3 is lowered and the pipes assume Position #2, the flow is then diverted to Pipe #3 and by-passed downstream. Because Pipe #3 is designed to be larger than Pipe #2, head loss through Pipe #3 will be less than in Pipe #2 for the same discharge from Pipe #1. Therefore, initially flow will occur only in Pipe #3 and not in Pipe #2.

After time t_1, Pipes #2 and #3 attain Position #2, at which time all inflow and bed load contributed by Pipe #1 are conveyed by means of the by-pass Pipe #3. This mode of flow continues until time t_2 at which time discharge from Pipe #1 exceeds the capacity of Pipe #3. During the period $(t_2 - t_1)$ discharge is relatively clean and not excessive. Therefore, it requires neither cleansing nor reduction and all of it is by-passed.

As the flood discharge continues to increase, the head loss in Pipe #3 increases, causing additional increase in pressure at the 45° bend. At time t_2 the pressure head in the bend exceeds the elevation of Pipe #2c in its Position #2 and flow initiates in Pipe #2. During the period $(t_4 - t_2)$ flow occurs in both Pipes #2 and

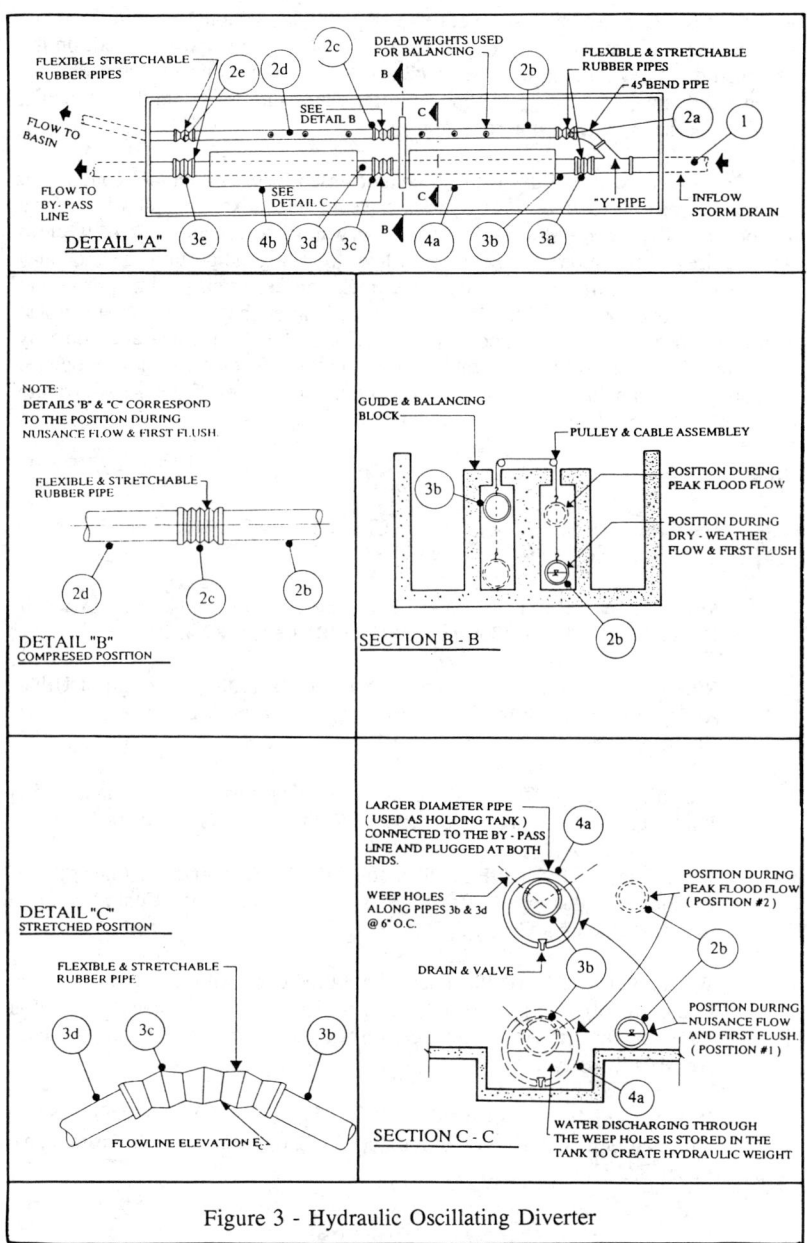

Figure 3 - Hydraulic Oscillating Diverter

#3. Because Pipe #2b has a steep adverse slope in its Position #2, the majority of the bedload is directed to Pipe #3 and not to Pipe #2. The suspended load, on the other hand, passes through both Pipes #2 and #3. At time t_4 inflow from Pipe #1 is reduced below the capacity of Pipe #3, after which the flow from the watershed requires neither cleansing nor reduction.

After time t_7, when all detained water is released, Pipes #2 and #3 continue to remain in Position #2 due to the hydraulic weight in Tanks #4a and #4b. For a period of time after the storm, flow from the watershed remains relatively clean and requires no retention. In this manner, low flows, a good portion of which may be base flow, continues to by-pass the retention/detention basin, preventing unnecessary overloading of the basin. Sometime after the end of the storm and before the occurrence of the poor quality nuisance flow Tanks #4a and #4b may be drained by means of valves located at their bottom. The valves may be opened manually or automatically. Once the tanks are drained, Pipes #2 and #3 will return to Position #1 and the cycle can be repeated.

References

1. Henderson, F.M. (1966), "Open Channel Flow", Macmillan Publishing Company, Inc., New York, NY.

2. Miller, R.A. (1985), "Percentage Entrainment of Constituent Loads in Urban Runoff, South Florida", USGS WRI Report 84-4329.

3. Roesner, L.A., Urbonas, B. and Sonnen, M.B. (1988), "Design of Urban Runoff Quality Controls", Published by the American Society of Civil Engineers, New York, NY.

4. Vanoni, V.A. (1975), Sedimentation Engineering, ASCE Manual 54. Published by the American Society of Civil Engineers, New York, NY.

5. Wanielista, M.P. and E.E. Shannon (1977), "Stormwater Management Practices Evaluations", Report submitted to the East Central Florida Regional Planning Council.

6. Williams, J.R. (1975), "Sediment Yield Prediction with Universal Equation Using Runoff Energy Factor", Cooperative Research of the Agricultural Research Department and the Texas Agricultural Experiment Station, Texas A&M University.

7. Wolman, M.G. and Leopold, L.B. (1957), "River Flood Plains, Some Observations on their Formation", U.S. Geological Survey, Professional Paper 282-C.

Vortex Spillway: Test Study on a Hydraulic Model

Giorgio Martino[1]

Abstract

The paper deals with an experimental study on a model of the forced vortex of Eur hanging reservoir realized by ACEA (Rome). Some vortex characteristics are investigated and, in particularly, the influence of inlet shape on the discharge characteristics is pointed out.
The effects produced by water guide plates placed in the adductor channel and by spiral guides in the vertical dropshaft are discussed.
For the latter case a new analytical solution for free-surface vortex inlet is presented. Theoretical and experimental results are compared and good agreement is obtained.

Introduction

The EUR Water Centre of Rome consists of a surge tower and a hanging reservoir approximately 50 m above ground, both of which are of steel.
Given the considerable importance of the work and its delicate hydraulic function, ACEA decided to use a model to test the hydraulic functioning, consulting "La Sapienza" Rome University, of the tailrace device used to convey the flow-rate entering the water tower of the centre to the underground tanks. This device has a size and hydraulic performance which make it ideal for our particular case.
The analyses of the model involved the following aims:
a) Having a current at the inlet of the tailrace as regular and stable as possible in time and space.

---
[1] Hydraulic Engineer - Azienda Comunale Energia ed Ambiente - A.C.E.A. - P.le Ostiense 2, Rome, Italy

b) Having a free downflow, with good aeration of the water.
c) Reducing as much as possible the pulsation phenomena in the water descending to the tank.
d) Always ensuring that the water jet adheres to the lining of its container.
e) Obtaining a satisfactory dissipation in the return basin at the foot of the drain, guaranteeing at the same time an efficient de-aeration of the water jet before leaving the basin.

The aforesaid aims were mainly based on the following requirements:
- To reduce as much as possible emulsifying with the air during the downflow to limit the loss of CO, since it is water at the limit of the calcium carbonate equilibrium concentration and could form scale on the works downstream.
- Reduce the noise of the plant, located in an urban area, as much as possible.
- Limit the vibrations transmitted to the metal bearing structure, particularly slender and daring with regards to its architectural design.

The model

The model was realised in a Fronde similitude, given the high and strongly dissipative degree of turbulence to be studied. A geometrical scale enabling a close simulation of the hydraulic phenomena caharcteristics of vortex discharges was chosen (Ackers and Crump, 1960; Binnie and Hakings, 1948; McCorquodale, 1968).

Extensive tests were carried out on a series of different configurations of the device, such as:
- the morphology and dimensions of the adductor channel;
- dimensions of the straight conduit leading to the vortex;
- the presence of waterguide plates at the inlet of the vortex;
- the presence of spiral guides in the vertical dropshaft;
- the shape of the return device at the foot of the pipe.

The observations and results obtained have emphasized the considerable influence of the morphology of the feed channel on the hydrodynamic features of the flow.

It was also noticed that the helical projections, in vertical drop-shaft, had a great effect on the stability of the descending motion of the water.

Phenomena connected with the feed devices

The type A and B devices (figs. 1 and 2) consist

essentially of the following functional parts:
- a feed channel equipped with a spillway, located in the prototype inside the hanging reservoir, and having a curvature equal to that of the reservoir's walls;
- a connecting channel between the preceding one and the straight channel leading to the vortex;
- the aforesaid straight channel;
- the vortex feed chamber designed according to the geometry suggested by Drioli (Drioli, 1969);
- the well's vertical dropshaft;
- the dissipation tank at the foot of the well.

In device A (fig. 1), showing the basic proposal of the EUR Water Centre project, a single feed channel was connected, by means of a section of greater curvature, to the straight adductor channel.

In this device there are strong irregularities in the current coming into the vortex in the entire field of flow rates investigated.

With the introduction of water guide plates in the straight adductor channel, the phenomenon of mass surging caused by the curve was divided among the channels included between the guide plates and, on the whole, diminished in intensity.

The necessity to regulate the current at the vortex inlet gave rise to device B (fig. 2).

The symmetry of the feed channel with respect to the centre line of the adductor channel determined a distinct improvement of the outflow characteristics. It was observed that, for this device also, the introduction of the water guide plates brought about a general improvement of the outflow and a substantial stabilization of the dropshaft.

In order to better determine the effect due to the morphology of the feed channel on the condition of the current in the vortex, device C (fig. 3), in which conditions of direct inlet into the straight adductor channel were reproduced, was later experimented; the downflow was similar in regularity to that obtained in device B, experimented earlier.

Phenomena connected with the vertical dropshaft

In the traditional devices the jet, entering the well with helicoidal trajectories and accelerated flow, took on, at a certain distance from the inlet, sub-vertical trajectories. It was observed that this distance increased with the increase of the flow rate and thus of the vertical component of the velocity at the mouth of the well (Jain, 1987). The pressure values measured along the dropshaft, moreover, justified the tendency of the jet to come away from the walls of the container liner.

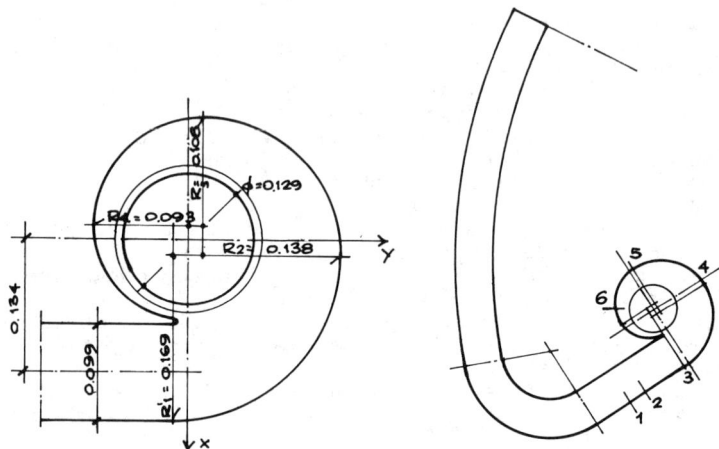

Figure 1. Device A

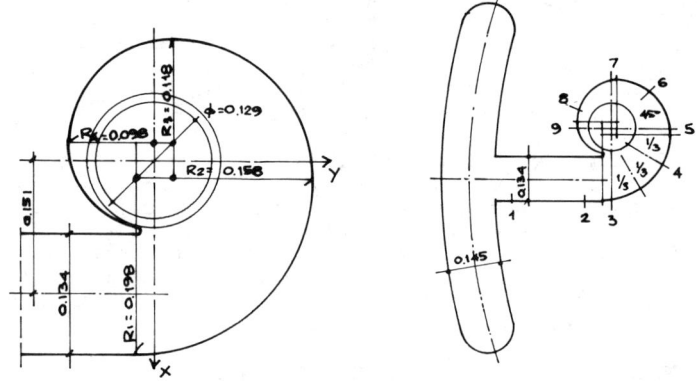

Figure 2. Device B

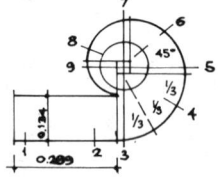

Figure 3. Device C

With the propellers, the jet, maintaining a high value of the tangential component of the velocity until leaving the well, adhered to the lining walls during the entire descent phase, and the pressure values measured along the dropshaft remained constantly above the atmospheric pressure.

The introduction of the spiral projections in the well, with inlet and flow rate conditions being equal, caused, on the other hand, a sensible increase of the feed chamber levels.

This phenomenon is due to the different direction, at the greater flow rates, of the jet trajectories at the mouth of the well and the initial contour of the propellers. In fact, the jet, moving with trajectories at a more slanted angle to the horizontal, tends to overflow the propellers and create a depression pocket, albeit modest, immediately downstream, giving rise to dissipating phenomena even before passing through the control section. In order to eliminate this problem, an attempt was made to move the start of the propellers away from the contracted section. As a result, it was observed that if, on the one hand, a condition of undisturbed outflow was restored in the vortex, on the other hand the regulating function of the propellers disappeared, since the jet, having passed the critical section, took on the characteristics of hypercritical currents.

Results obtained and concluding considerations

As a result of the numerous tests carried out it was possible to configure the discharge device in an optimum manner.

In particular, the geometry of the feed conduit to the vortex was calibrated so as to achieve the greatest regularity of the current entering the spiral chamber and attenuate the pulsation phenomena of the water jet, both when entering and during its descent to the tank.

The dynamic stability of the vertical chamber also seems linked to these phenomena. The optimum behaviour of the vortex-surge tank complex was achieved in conditions of strict symmetry with regards to the inlet conduit. The positive effect on the regularity of the flow due to the presence of jetguide baffles at the inlet of the vortex is also stressed.

With regards to the vertical pipe, the introduction of guiding propellers has enabled the water to maintain its adherence to the lining of its container in all downflow conditions. The fact that the water jet has a better adherence than that of a smooth downflow also means that the dissipation phenomena increases along the pipe and the turbulence and emulsifying when leaving the chamber is reduced.

Finally, a mathematical model was set up for calculating the characteristics of the free surface vortex inlet with the elical projections in vertical dropshaft (Margaritora et al., 1988).
Theoretical and experimental results are composed and good agreements are obtained.

APPENDIX I. References

P. Ackers, E.S. Crump: The Vortex Drop. "Institution of Civil Engineers", Vol. 16, August 1960.

A.M. Binnie, G.A. Hookings: Laboratory Experiments on Whirlpools. "Proc. Roy. Soc.", series A, Vol. 194, 1948.

C. Drioli: Experiences in Installations with Vortex Discharge Wells. "Electric Energy", June 1969.

M. Piga: Vortex Spillways. "Electric Energy", April 1970.

S.C. Jain: Free-surface Swirling Flows in Vertical Dropshaft. "Journal of Hydraulic Engineering", Vol. 113, n. 10, October 1987.

G. Margaritora, P. Martini, G. Martino: "Experimental Contribution to the Study of a Forced Vortex Outflow" - 21st Conference on Hydraulics and Hydraulic Constructions, L'Aquila 1988.

J.A. McCorquodale: Scale Effects in Swirling Flow. "Proc. ASCE", HY1, January 1968.

Evaluation of Historical Scour at Selected Stream Crossings in Indiana

David S. Mueller[1] and Robert L. Miller[2]

Abstract

Geophysical data were collected by means of ground-penetrating radar and tuned transducer systems to estimate the historical scour at ten bridges in Indiana. These geophysical data were used to compare and evaluate the results of 13 published pier-scour equations. In order to make this comparison, it was assumed that the measured historical scour was associated with the peak historical discharge. Because the geophysical data were not sufficient to map the lateral extent of the refilled scour hole, local scour could not be isolated from contraction scour. For the evaluation, computed contraction scour and pier scour were used in combination with the existing channel geometry to determine a computed bed elevation. This computed bed elevation was compared to the minimum historic bed elevation estimated from the geophysical data. None of the selected pier-scour equations, when combined with the contraction-scour equation, accurately represented the historical scour at all of the study sites. On the basis of the limited data presented, the equations currently recommended by the Federal Highway Administration provided a combination of accuracy and safety, required by design equations, equal to or better than the other equations evaluated.

Introduction

Many equations for estimating scour around bridge piers, at abutments, and in channel contractions are published in the literature. Nearly all of these equations are based on laboratory data collected in flumes with uniform cohesionless bed materials under steady-flow conditions and contain some empirical components. Minimal field data have been collected to verify the applicability and accuracy of

---

[1]Hydrologist, U.S. Geological Survey, Water Resources Division, 2301 Bradley Avenue, Louisville, KY 40217

[2]Hydrologist, U.S. Geological Survey, Water Resources Division, 5957 Lakeside Boulevard, Indianapolis, IN 46278

these equations for the range of soil conditions, streamflow conditions, and bridge designs that exist throughout the United States (Richardson and others, 1991). To improve the accuracy of scour computations at a particular site, results of existing equations need to be evaluated and compared to field measurements at sites with similar hydraulic and geotechnical conditions. Because scour holes often refill after the passage of a flood, simple bed surveys at low water are not sufficient to determine the depth of scour holes that formed during a previous flood. Geophysical techniques utilizing ground-penetrating radar and continuous high-resolution subbottom seismic profiling can be used to delineate the scour hole formed by a previous flood. This paper discusses the geophysical techniques used to measure historical scour; describes the methods used to estimate hydrologic and hydraulic conditions of the historical peak discharge; and compares scour estimates based on published equations with the measured historical scour.

Scour data collection

Ten sites were selected to represent different geographic regions and a wide range of drainage areas within Indiana. Each bridge opening was surveyed with the ground-penetrating-radar system (GPR) and/or a tuned transducer to locate evidence of scour holes that may have refilled. The GPR used with dual 80-megahertz antenna was effective on gravel bars and in water less than 4 ft deep. In water depths greater than 4 ft, however, the signal was attenuated in the water column because of high specific conductance of the water, and no useful data were recorded. The tuned transducer was used with a 3.5- to 7-kilohertz (kHz) and a 14-kHz transducer suspended 0.5 to 1 ft below the water surface. This equipment was usable in water depths greater than 5 ft. The signal from the geophysical equipment was occasionally obscured by the effects of side echoes, debris, point reflections from cobbles and boulders, and multiple reflections. Furthermore, the dual 80 megahertz antennae on the GPR were not shielded on top, and a reflection from the bridge deck may have caused interference.

Geophysical data were recorded across the upstream and downstream side of the bridge, along both sides of each main-channel pier, and along the upstream and downstream end of each main-channel pier. The piers on the overbanks were not surveyed. To support the geophysical data, investigators probed the area around each surveyed pier with a steel pipe (0.5 in. inside diameter) to locate subsurface interfaces. The data were used to identify any subsurface interface that would indicate that the bed had scoured at some time in the past and subsequently refilled. From the 10 sites surveyed, 9 produced data adequate for interpretation. The data were adequate to determine the approximate location and depth of the interface; however, the data were not of sufficient resolution for mapping the lateral extent of refilled scour holes. Because GPR and tuned transducer record indicate interfaces where the electrical and acoustic properties change, correct interpretation of the record is critical to ensure that construction fill or other changes in subbottom material is not interpreted as scour. The bed elevations determined by the geophysical techniques are the minimum bed elevations believed to have resulted from scour.

Hydrologic and hydraulic conditions

A historical peak discharge, the maximum flood that occurred during the life of the bridge, was used to evaluate the ability of selected published equations to estimate the measured historical scour. Historical discharge records were not available for the study sites. Therefore, historical peak discharges were estimated from nearby USGS streamflow-gaging stations. Although the duration of a flood may affect the depth of scour, especially for cohesive materials, the selected equations are based on the assumption that the flood discharge is maintained for a sufficient period to allow equilibrium sediment transport through the scour holes. Therefore, the durations of the historical floods were not assessed in this study.

Hydraulic conditions of the historical peak discharge at the sites were estimated using the Water-Surface Profile computation model (WSPRO) (Shearman, 1990). Cross-section data and roughness coefficients were obtained from field surveys at each site. Historical water-surface profiles provided by the Indiana Department of Natural Resources were used to verify the model. The bridge routines available in WSPRO were used at most sites. However, sites having geometry that did not allow application of the bridge routines were modeled using a composite cross section constructed to represent the complex geometry of the bridge section in a manner consistent with the limitations of WSPRO. The approach velocities and discharge conveyed through subsections of the approach and bridge cross sections required by the scour equations were determined from the 20 equal-conveyance tubes computed by WSPRO.

Comparison of computed to measured historic scour

The comparison between computed and measured historic scour should be viewed in light of the assumptions necessary to achieve comparable data. First, it is assumed that the minimum measured historical bed elevations were caused by scour resulting from the peak historical discharge. In the field, debris accumulations, ice jams, and other anomalies affect the depth of scour occurring at a given discharge, therefore, the bed elevations measured could have resulted from a lesser discharge with debris or ice accumulations. Second, long-term scour was assumed to be negligible. Third, the contraction-scour and pier-scour equations were combined to yield a computed bed elevation. The measurements made by geophysical techniques resulted in an estimated minimum streambed elevation in the vicinity of the piers. Because the lateral extent of the scour holes could not be delineated from the geophysical data, it was not possible to separate contraction scour from local scour. Therefore, the depth of contraction scour computed using the equations by Laursen (1962, 1963) was added to the local scour estimated by each pier-scour equation to obtain an estimate of total scour. The total scour was subtracted from the existing bed elevation to yield a computed bed elevation for each of the pier-scour equations. Inaccuracies inherent in the contraction-scour equations were transferred to the pier-scour equations; therefore, the accuracy of the pier-scour equations could not be evaluated separately from the contraction-scour equations.

The following pier-scour equations were applied to each of the bridges for the hydraulic conditions that were estimated for the historic peak discharge: Ahmad (Ahmad, 1962); Blench-Inglis I and II (Blench, 1962); Chitale (Chitale, 1962); Colorado State University (CSU) (Richardson and others, 1991); Froehlich (Froehlich, 1988); Inglis-Lacey (Lacey, 1930; Joglekar, 1962); Inglis-Lacey I and II (Inglis, 1949; Joglekar, 1962); Larras, (Hopkins and others, 1980); Laursen (Laursen, 1962); and Shen and Shen-Maza (Shen and others, 1969). Grain-size information were not available for the study sites. A single characteristic grain size for each site was estimated from the class descriptions indicated on lithologic logs that were available.

A summary of the differences between computed and historical bed elevation at the nose of the pier, which is where the theory assumes maximum scour will occur, is shown in Figure 1. The Chitale and Ahmad equations commonly overestimated the depth of scour. The Chitale equation is based on model experiments for one bridge and uses only the Froude number and depth of flow as variables; the size and configuration of the pier is not considered. The Ahmad equation is not dimensionally homogeneous and little guidance was provided for selection of the coefficient K which is a function of boundary geometry, abutment shape, width of piers, shape of piers, and the angle of the approach flow. A value for K of 1.8 was selected for all sites from the suggested range of 1.7 to 2.0. Although no data were available to evaluate the contraction-scour equation, a few of the contraction-scour computations resulted in what seemed to be excessive scour, especially in clear-water situations. Given the general belief that laboratory equations overestimate scour in the field, it is surprising that approximately half of the computations underestimated the measured scour, including the Froehlich equation with the factor of safety included.

For bridge design, it is desirable to use an equation that estimates the depth of scour accurately but when in error tends to overestimate the depth of scour. Based on the results shown in Figure 1, it can be seen that no equation accurately estimated the historical scour at all of the study sites. Therefore, the preferred design equation would be the equation that provides the best combination of accuracy and safety (overestimation). The CSU equation in combination with the Laursen (1962, 1963) contraction scour equations, which are currently the equations recommended by the FHWA, provided a combination of accuracy and safety, equal to or better than the other equations evaluated.

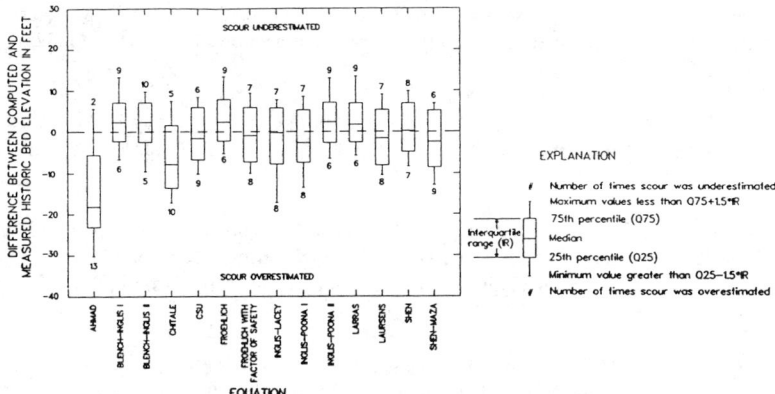

Figure 1. Summary of differences between the computed and measured historic bed elevations for the selected equations.

Summary and conclusions

Geophysical techniques are valuable tools for assessing the historical scour at bridges. These techniques have limitations and the collection and interpretation of the data must be performed by properly trained personnel. Historical scour may not be associated with the peak historical discharge due to debris accumulations, ice jams, and other anomalies that affect the depth of scour occurring at a given discharge. Because the measured historical data contained only minimum bed elevations, the contraction- and pier-scour equations were used to determine a computed bed elevation used for the evaluation. Inaccuracies inherent in the contraction-scour computations were uniformly transferred to all of the pier-scour equations, and the accuracy of the contraction-scour equation could not be separated from the pier-scour equations. The evaluation of the pier-scour equations failed to identify an equation that accurately predicted the historical scour at all of the study sites; however, based on the data presented herein, the FHWA procedures (for the combination of pier and contraction scour) provided a combination of accuracy and safety, required by design equations, equal to or better than the other equations evaluated. Because this is a very small data set, one must be careful when drawing conclusions, therefore, additional data, especially real-time data, are needed to verify this evaluation.

References

Ahmad, Mushtaq, 1962, Discussion of "Scour at bridge crossings," by E. M. Laursen: Transactions of the American Society of Civil Engineers, v. 127, part I, no. 3294, p. 198-206.

Blench, Thomas, 1962, Discussion of "Scour at bridge crossings," by E. M. Laursen: Transactions of the American Society of Civil Engineers, v. 127, part I, no. 3294, p. 180-183.

Chitale, S. V., 1962, Discussion of "Scour at bridge crossings," by E. M. Laursen: Transactions of the American Society of Civil Engineers, v. 127, part I, no. 3294, p. 191-196.

Froehlich, D. C., 1988, Analysis of on-site measurements of scour at piers, in Abt, S. R. and Gessler, Johannes, eds., Hydraulic engineering--proceedings of the 1988 National Conference on Hydraulic Engineering: New York, American Society of Civil Engineers, p. 534-539.

Hopkins, G. R., Vance, R. W., and Kasraie, Behzad, 1980, Scour around bridge piers: Federal Highway Administration Publication FHWA-RD-79-103, 124 p.

Inglis, S. C., 1949, The behavior and control of rivers and canals: Poona, India, Poona Research Station, Publication 13, Part II, Central Water Power Irrigation and Navigation Report, 478 p.

Joglekar, D. V., 1962, Discussion of "Scour at bridge crossings," by E. M. Laursen: Transactions of the American Society of Civil Engineers, v. 127, part I, no. 3294, p. 183-186.

Lacey, Gerald, 1930, Stable channels in alluvium: London, United Kingdom, Minutes and Proceedings of the Institution of Civil Engineers, v. 229, Paper 4736, p. 259-284.

Laursen, E. M., 1962, Scour at bridge crossings: Transactions of the American Society of Civil Engineers, v. 127, Part I, no. 3294, p. 166-209.

Laursen, E. M., 1963, An analysis of relief bridge scour: Journal of the Hydraulics Division, American Society of Civil Engineers, v. 89, no. HY3, p. 93-118.

Richardson, E. V., Harrison, L. J., and Davis, S. R., 1991, Evaluating scour at bridges: Federal Highway Administration Hydraulic Engineering Circular 18, Publication FHWA-IP-90-017, 191 p.

Shearman, J. O., 1990, User's manual for WSPRO--a computer model for water surface profile computations: Federal Highway Administration Publication FHWA-IP-89-027, 112 p.

Shen, H. W., Schneider, V. R., and Karaki, Susumu, 1969, Local scour around bridge piers: Journal of the Hydraulics Division, American Society of Civil Engineers, v. 95, p. 1919-1940.

Bridge Scour Evaluations in Washington State

Jeffrey P. Johnson[1], A.M. ASCE, Charles R. Neill[2], M. ASCE, and Richard P. Hovde[3]

Abstract

Eighteen existing bridges were investigated for Washington State DOT as part of the FHWA National Scour Evaluation Program. Investigations included review of file information, site inspection, flood hydrology, channel hydraulics, airphoto interpretation, scour depth estimation, and required for mitigation. Observed scour patterns often appeared due to a number of interacting factors, both natural and man-induced. Two case studies of this type are described.

Presently recommended procedures for evaluation and design are often difficult to reconcile with recorded scour, particularly when interacting factors are involved. These procedures are, nevertheless, useful for estimating potential maximum scour depths for conservative mitigation designs.

Introduction

In compliance with recommendations by the Federal Highway Administration (FHWA), Washington State Department of Transportation (WSDOT) implemented a program to review scour potential at approximately 1500 waterway crossings. Screening of approximately 500 sites identified a relatively small number as potentially scour-critical. Northwest Hydraulic Consultants (NHC) was contracted to assess scour potential and recommend conceptual mitigation measures for eighteen high priority sites. Specified general guidelines were the FHWA manuals HEC 18 and HEC 20 (FHWA 1991). Additional sources were used for check purposes. Two evaluations are summarized below, then some general comments are made.

---

1 Project engr., Northwest Hydraulic Consultants, 22017-70 Ave, Kent, WA 98032
2 Principal engr., Northwest Hydraulic Consultants, 4823-99 St, Edmonton, Canada T6E 4Y1
3 Inspection engr., Washington State Department of Transportation, Olympia, WA 98054

Case 1 — Sauk River near Darrington

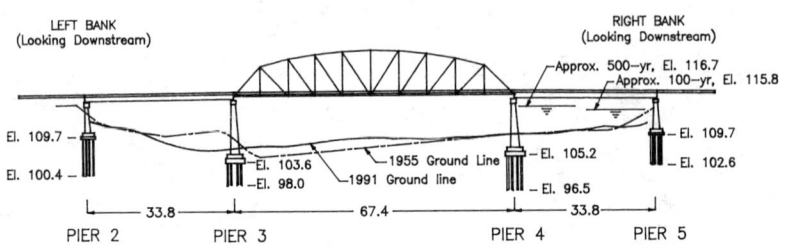

FIGURE 1. Sauk River Bridge

Bridge Description: The crossing is located 95 km NE of Seattle in the foothills of the Cascade Mountains. The three-span bridge (Figure 1) was completed in 1959 and has four pile-supported concrete piers founded in sand and gravel, of which two are in the active channel. The bridge originally crossed the river approximately at right angles and the piers are square to the deck. The bridge spans all of the active channel but the left floodplain is blocked by the highway approach fill.

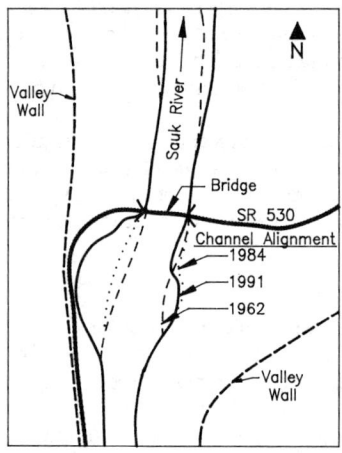

FIGURE 2. Channel Alignment

River Description: The Sauk River lies in a wooded mountain valley with irregular floodplain areas. Both the Sauk and its tributary the Suiattle River which enters about 900 m upstream exhibit irregular meander bends, channel splitting around islands, and extensive gravel bars. Between the Suiattle confluence and the bridge the channel curves gradually to the right, with a wide gravel point bar on the inside of the bend. Between 1962 and 1984 the channel gradually migrated to the left by eroding the left bank and building up the gravel bar on the right. Between 1984 and 1991 the leftward erosion greatly accelerated over a distance of about 300 m upstream and began to threaten the south highway approach (Figure 2), a change probably related to higher floods after 1980. As a result of the leftward migration, flood flows now approach Pier 3 at an angle of about 30 degrees.

The channel through the bridge is about 120 m wide and has an average slope of about 0.0025. Estimated 100-year and 500-year flood peaks are approximately 2800 m^3/s and 3700 m^3/s. Average channel velocity at 100-year peak is 4.2 m/s with a flow depth of 5.5 m.

Scour History and Potential: Three potential components of total scour were considered: profile degradation, contraction scour, and local scour. There is no definite evidence of profile degradation, but contraction scour and local scour are both significant.

Although the bridge does not significantly constrict the channel, the approach embankment blocks off the left floodplain, so that in severe floods contraction scour could be significant. A longitudinal profile of 1985 indicated an average bed level about 1.5 m lower at the bridge than upstream or downstream. This probably indicates the average contraction scour due to the bridge. Estimates for 500-year flood conditions indicate that contraction scour could extend 3 m deeper between the left abutment and Pier 3. Scour would then extend to near the bottom of the pile cap.

Local scour appears to be a problem at Pier 3 only. Recent inspections reveal a shallow hole at the base of the pier. With the severe angle of attack due to recent channel migration, there is potential for deep local scour and extensive exposure of the piling can be expected if a secure riprap apron is not maintained around the pile cap. Computations indicate that under 500-year flood conditions, combined local and contraction scour might even extend below the pile tips.

Recommended Mitigation: Two mitigation measures were recommended. The first consists of upgrading the existing riprap apron around Pier 3. Before detailing, an underwater inspection was advised to define the existing protection. The second recommendation was for an upstream guide bank and groin. The purpose of these training works is to impose a straight approach square to the bridge, and to arrest river attack on the highway.

Case 2 – North Fork Toutle River near Mount St. Helens

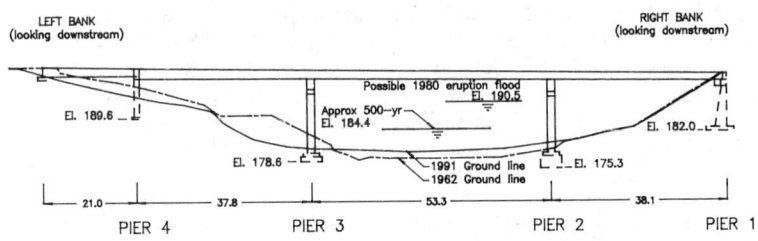

FIGURE 3. North Fork Toutle River Bridge

Bridge Description: The crossing is located 35 km west of Mount St. Helens. The bridge (Figure 3) was constructed in 1965, is 150 m long and has four spans. There are three rectangular concrete piers all founded on spread footings, of which two are near the low-water channel margins. Both foundations originally extended 3 to 4.5 m below the bed. There is deep sand and gravel with occasional boulders at Pier 2, but near Pier 3 there is fractured basalt at shallow depths. The piers are square to the bridge centerline and skewed at approximately 35 degrees to the river banklines.

River Description: A catastrophic flood and sediment flow ("lahar") following the eruption of Mount St. Helens on 18 May 1980 caused substantial channel changes. Pre-eruption airphotos show an irregular sinuous and partly meandering channel in a mountain valley, generally 30 to 45 m wide and partly bordered by narrow wooded floodplain or terrace areas. Following the eruption, the floodplain was blanketed by new sediment and the river now exhibits a multi-channeled pattern. No reliable discharge figures are available for the eruption flood, but a peak in the range of 5500 to 8500 m³/s cfs has been suggested. This highly sediment-laden flow may not have had the erosive power of ordinary flood flows. Taking into account the new Sediment Retention Structure upstream, 100-year and 500-year flood peaks are presently estimated as 820 m³/s and 1160 m³/s. Corresponding velocity and depth estimates are 3.5 to 4 m/s and 5 to 6 m. During the 1980 eruption flood the depth reached 10 to 12 m at the bridge, and velocities would have exceeded 6 m/s.

Scour History and Potential: Erosional changes in 1980 appear to have been relatively minor, and the bridge withstood this catastrophic flood without serious damage. There was up to 3 m bed aggradation from 1980 to 1981, but by late 1992 bed levels had fallen close to pre-1980 levels. The long-term trend of vertical changes is difficult to predict, but degradation below the 1962 pre-construction profile is considered unlikely. The bridge waterway section has widened considerably since 1984, causing significant drops in ground levels near

Piers 2 and 3. This widening is localized in the vicinity of the bridge and is probably associated with the obstructive and eddy-creating effect of the skewed piers.

There is no evidence of a serious local scour condition at either river pier although a shallow hole is visible at the base of Pier 2. The CSU local scour formula recommended in HEC 18 estimates approximately 6 m below ambient bed level. The Laursen formula estimates 4 m. However, since the bridge withstood the catastrophic eruption flood of 1980 and shows no deep scour, these computed estimates were considered inappropriate because: (1) both piers are located at near the margins of the channel where they are partly sheltered from the main flow; (2) the effect of pier skew in augmenting potential local scour is reduced because only the outer corner of each pier is exposed to the main flow; (3) there is shallow bedrock near Pier 3 (left bank); and (4) the bed material around Pier 2 (right bank) contains boulders.

Recommended Mitigation: Continued monitoring but no physical measures at present.

General Comments

The following general comments are based on evaluation of eighteen WSDOT bridges and on other studies:

1. At a majority of the sites, a relatively severe flood has been experienced since the bridge was built. If a bridge has previously withstood a 100-year or greater flood without serious scour, the case for a scour-critical label may be weak regardless of computational results.

2. In the majority of cases, the river planform appeared to be relatively stable. If a channel is subject to shifting, absence of past scour problems is not necessarily a reliable indicator of future scour potential.

3. The majority of the sites exhibit relatively coarse widely-graded bed materials. For equal flow conditions coarse materials are less scour-susceptible, but to offset this, channel gradients and flow velocities are normally higher. Coarse materials have a wider range of grain sizes, so that the coarser elements tend to form an armor limiting scour. Intuitively, scour potential might be expected to be generally less in coarse materials, but experimental evidence is somewhat ambiguous.

4. It appears that the recommended HEC 18 equation for local pier scour can result in overestimation of local scour in coarse materials. The equation has local scour increasing with Froude Number and therefore with velocity of flow, but has no countervailing factor for bed-material grain size. This result seems to be

illogical. It is suggested that local scour depth should be estimated using several formulas.

Conclusions

Evaluation of potential extreme flood scour is often difficult due to the special characteristics of individual bridge sites. The evaluation procedures outlined in HEC 18 provide useful potential maximum estimates for design of conservative mitigation measures. A realistic assessment of risk, however, requires detailed consideration of a variety of hydrologic, hydraulic and geotechnical factors and a careful consideration of scour history and processes.

Appendix 1. References

FHWA, 1991. "Evaluating Scour at Bridges." Hydraulic Engin. Circular No. 18, U.S. Department of Transportation, Federal Highway Administration, Washington, D.C.

"Stream Stability at Highway Structures". Hydr. Engin. Circ. No. 20, ibid.

RELATION OF CHANNEL STABILITY TO SCOUR
AT HIGHWAY BRIDGES OVER WATERWAYS IN MARYLAND

Edward J. Doheny[1]

ABSTRACT

Data from assessments of channel stability and observed-scour conditions at 876 highway bridges over Maryland waterways were entered into a data base. Relations were found to exist among specific, deterministic variables and observed-scour and debris conditions. Relations were investigated between (1) high-flow angle of attack and pier- and abutment-footing exposure, (2) abutment location and abutment-footing exposure, (3) type of bed material and pier-footing exposure, (4) tree cover on channel banks and mass wasting of the channel banks, and (5) land use near the bridge and the presence of debris blockage at the bridge opening. The results of the investigations indicate the following: (1) The number of pier- and abutment-footing exposures increased for increasing high-flow angles of attack, (2) the number of abutment-footing exposures increased for abutments that protrude into the channel, (3) pier-footing exposures were most common for bridges over streams with channel beds of gravel, (4) mass wasting of channel banks with tree cover of 50 percent or greater near the bridge was less than mass wasting of channel banks with tree cover of less than 50 percent near the bridge, and (5) bridges located in urban, pasture, and forested basins were more prone to debris blockage than bridges in row crop and swamp basins.

INTRODUCTION

Almost 500,000 bridges in the United States are built over waterways. Many of these bridges span alluvial channels that are continually subjected to scour and bank erosion (U.S. Federal Highway Administration, 1991). A survey of United States bridge failures by Shirole and Holt (1991) indicates that of 823 bridge failures surveyed, 60 percent were associated with hydraulic conditions. These include scour around bridge foundations and channel instability (Landers, 1992). The failures of the Schoharie Creek Bridge in New York State on April 5, 1987, and the Hatchie River Bridge in western Tennessee on April 1, 1989, brought the problem of bridge scour to national attention.

The Maryland State Highway Administration (MDSHA), under a directive from the U.S. Federal Highway Administration (FHWA), developed a program to assess channel stability and observed-scour conditions at highway bridges over waterways in Maryland. From June 1, 1990, through April 30, 1991, the U.S. Geological Survey, in cooperation with MDSHA, performed field assessments for channel stability and observed-scour conditions at 876 highway bridges over waterways in Maryland. Assessments were conducted in terrain ranging from

---

1. Hydr. Eng., Water Resources Div., U.S. Geological Survey, 208 Carroll Bldg., 8600 LaSalle Rd., Towson, MD 21286

mountains and rolling hills, to tidal wetlands and coastal regions. Data describing bridge characteristics and the stream channel were collected during each assessment. Data variables pertaining to the bridge include the following: (1) bridge length, (2) height of underclearance from the water to the low chord of the bridge, (3) pier and abutment locations with respect to the stream channel, (4) pier- and abutment-footing exposures, and (5) size and types of debris blockage accumulated at the bridge opening. Data variables pertaining to the stream channel include the following: (1) average bank heights and bank angles for each channel bank near the bridge, (2) estimated percentages of tree cover near the bridge, (3) type of bank erosion on each bank near the bridge, (4) high-flow angle of attack on the bridge, (5) maximum channel depth at the upstream side of the bridge opening, (6) type of bed material in the channel near the bridge, (7) type of land use near the bridge, and (8) average channel widths upstream from the bridge, at the bridge opening, and downstream from the bridge.

Bridge and stream-channel data from each assessment were entered into a data base. Using the data base, bridge and stream-channel data were stored, retrieved, selected, and sorted by variable (Simon and others, 1988). This paper describes relations among data variables that resulted from selected data retrievals.

RELATIONS AMONG OBSERVED DATA

Five relations were investigated using combinations of data variables from the data base. Each relation and the results are described below.

High-Flow Angle of Attack and Pier- and Abutment-Footing Exposures

A high-flow angle of attack directs streamflow toward one side of a stream channel. Concentration of flow toward one side of a channel can cause local scour and can lead to pier- and abutment-footing exposures. The cumulative number of bridges with pier- and abutment-footing exposures for increasing high-flow angle of attack is shown in tables 1 and 2.

Table 1. Cumulative number of bridges with pier-footing exposures for ranges of high-flow angle of attack

[Negative (-) high-flow angle of attack is toward left bank and positive (+) high-flow angle of attack is toward right bank when facing downstream]

High-flow angle of attack (degrees)	Number of bridges with one or more pier footings exposed	Number of bridges with two or more pier footings exposed
0	64	25
+5 to -5	68	26
+10 to -10	87	29
+20 to -20	119	35
+30 to -30	150	45
+40 to -40	167	50
+50 to -50	172	50
+60 to -60	173	50

Table 2. Cumulative number of bridges with abutment-footing exposures for ranges of high-flow angle of attack

[Negative (-) high-flow angle of attack is toward left bank and positive (+) high-flow angle of attack is toward right bank when facing downstream]

High-flow angle of attack (degrees)	Number of bridges with left abutment footing exposed	Number of bridges with right abutment footing exposed	Number of bridges with both abutment footings exposed
0	78	75	55
+5 to -5	84	79	59
+10 to -10	112	106	79
+20 to -20	171	155	111
+30 to -30	205	186	134
+40 to -40	229	207	150
+50 to -50	252	227	168
+60 to -60	255	231	171
Other	258	233	173

The numbers of pier- and abutment-footing exposures increase as the high-flow angle of attack increases. The largest increases in numbers of pier- and abutment-footing exposures were for bridges with high-flow angles of attack in the range of 10 to 30 degrees.

Abutment Location and Abutment-Footing Exposures

The location of a bridge abutment with respect to the stream channel affects the susceptibility of the abutment to local scour and footing exposure. An abutment set back from the top of a channel bank is protected from scour unless overbank flow contacts the abutment during a flood. An abutment located at the toe of a channel bank or projected into the channel is more prone to local scour and footing exposure. Selected data retrievals were made to determine the number of abutment-footing exposures for abutments projected into the channel, abutments at the toe of the channel banks, and abutments set back from the top of channel banks. These results are summarized in table 3.

Table 3. Number of bridges with abutment-footing exposure for different abutment locations

[Left and right refer to left and right when facing downstream]

Abutment and location	Number of bridges	Number of bridges with abutment-footing exposures	Percent of exposures
Left, projected into stream channel	189	117	61.9
Right, projected into stream channel	184	104	56.5
Both, projected into stream channel	95	45	47.4
Left, at toe of channel bank	400	125	31.3
Right, at toe of channel bank	404	117	29.0
Both, at toe of channel bank	269	58	21.6
Left, set back from top of channel bank	287	16	5.9
Right, set back from top of channel bank	288	12	4.5
Both, set back from top of channel bank	192	2	1.5

These results indicate that abutments located in the stream channel or at the toe of a channel bank are more prone to scour and footing exposures than abutments that are set back from the tops of the channel banks.

Type of Bed-Material and Pier-Footing Exposures

Local scour at piers is a function of bed material erodibility in the channel. The relation between type of bed material and number of observed pier-footing exposures are shown in table 4.

Table 4. Number of bridges with pier-footing exposures by type of bed material

Bed material	Number of bridges	Number of bridges with piers	Number of bridges with one or more pier-footing exposures	Percent of exposures for bridges with piers
Sand	166	121	31	25.6
Silt/Clay	33	23	4	17.4
Gravel	380	228	84	36.8
Cobble/Boulder	184	104	32	30.8
Bedrock	26	15	7	46.7
Alluvium	87	59	15	25.4
Total	876	550	173	31.5

These results indicate that the largest number of observed pier-footing exposures were for bridges over streams with channel beds of gravel. Although channel beds of sand are more susceptible to scour than channel beds of gravel, the percentage of observed exposures was lower because many of these bridges are located in coastal regions over low-gradient streams or tidal estuaries. Channel beds of sand are also more likely to refill scour holes in moderate to low flows. The largest percentage of observed pier-footing exposures was for bridges over streams with channel beds of bedrock. This was attributed to designed footing exposures during construction of the bridges rather than scour processes.

Tree Cover and Mass Wasting on the Channel Banks

Tree cover, with extensive root systems, on the channel banks can provide stability to the channel banks. Selected data were retrieved and analyzed to relate the percentage of tree cover near the bridge to the presence or absence of mass wasting on the channel banks near the bridge. These results are summarized in tables 5 and 6.

Table 5. Number of bridges with 50 percent or greater tree cover on channel banks, and number of bridges with 50 percent or greater tree cover on channel banks and mass wasting on the channel banks

Location	Number of bridges	Number of bridges with mass wasting on channel banks	Percent of bridges with mass wasting on channel banks
Left bank, upstream	398	34	8.5
Right bank, upstream	409	31	7.6
Both banks, upstream	319	6	1.9
Left bank, downstream	387	22	5.7
Right bank, downstream	385	21	5.5
Both banks, downstream	307	2	0.7

Table 6. Number of bridges with less than 50 percent tree cover on channel banks, and number of bridges with less than 50 percent tree cover on channel banks and mass wasting on the channel banks

Location	Number of bridges	Number of bridges with mass wasting on channel banks	Percent of bridges with mass wasting on channel banks
Left bank, upstream	478	79	16.5
Right bank, upstream	467	86	18.4
Both banks, upstream	388	38	9.8
Left bank, downstream	489	71	14.5
Right bank, downstream	491	77	15.7
Both banks, downstream	411	18	4.4

These results indicate that channel banks with less than 50 percent tree cover near the bridge are more likely to be affected by mass wasting than channel banks with 50 percent tree cover or greater. The results in table 6 also indicate that the percentages of bridges with mass wasting of only one channel bank were much greater than the percentages of bridges with mass wasting of both channel banks. This may be attributed to (1) channel banks with greater protection causing an increase in erosion on other, less protected channel banks, or (2) erosion caused by a high-flow angle of attack.

Land Use Near the Bridge and Debris Blockage at the Bridge Opening

Debris blockage at a bridge opening can increase flow velocity through the bridge and lead to contraction scour and local scour. The type of land use in the basin upstream from the bridge can affect the type and amount of debris that the channel carries. Selected data were retrieved and the type of land use near a bridge was related to the number of bridges with debris blockage at the bridge opening. The results are summarized in table 7.

Table 7. Number of bridges with debris blockage at the bridge opening by type of land use near the bridge

Type of Land use	Number of bridges	Number of bridges with debris blockage	Percent of bridges with debris blockage
Urban	157	23	14.6
Row crop	50	3	6.0
Pasture	163	35	21.5
Forest	462	58	12.6
Swamp	44	1	2.3
Total	876	120	13.7

Pasture land use resulted in the largest percentage of debris blockage when compared to other types of land use. Many bridges with nearby pasture land use have wooden cattle gates mounted against the bridge openings that collect debris. Forest land use resulted in the largest number of bridges with debris blockage, because of the abundance of woody vegetation. Urban land use resulted in a large number of bridges with debris blockage because of large amounts of runoff that carry trees, brush, and trash into stream channels. Row crop and swamp land use resulted in the smallest numbers and percentages of bridges with debris blockage because of small amounts of woody vegetation on the channel banks causing debris blockage at bridge openings.

SUMMARY

Bridge and stream data were collected during 876 field assessments for channel stability and observed-scour conditions in Maryland. Bridge and stream data from the assessments were entered into a data base. Data were retrieved from the data base and relations among selected data variables were investigated. Relations were investigated between (1) high-flow angle of attack and pier- and abutment-footing exposure, (2) abutment location and abutment-footing exposure, (3) type of bed material and pier-footing exposure, (4) tree cover on channel banks and mass wasting of the channel banks, and (5) land use near the bridge and the presence of debris blockage at the bridge opening. The results of the investigations indicate the following: (1) The number of pier- and abutment-footing exposures increased for increasing high-flow angles of attack, (2) the number of abutment-footing exposures increased for abutments that protrude into the channel, (3) the largest number of pier-footing exposures was for bridges over streams that have channel beds of gravel, (4) mass wasting of channel banks with tree cover of 50 percent or greater near the bridge was less than mass wasting of channel banks with tree cover of less than 50 percent near the bridge, and (5) bridges in urban, pasture, and forested basins were more prone to debris blockage than bridges in row crop and swamp basins.

REFERENCES

Landers, M.N., 1992, Bridge scour data management: American Society of Civil Engineers: National Conference on Hydraulic Engineering, Proceedings, Baltimore, Maryland, p. 1094-1099.

Simon, Andrew, Outlaw, G.S., and Thomas, Randy, 1988, Evaluation, modeling, and mapping of potential-scour, West Tennessee: National Bridge-Scour Symposium, Federal Highway Administration, McLean, Virginia, 18 p.

Shirole, A.M., and Holt, R.C., 1991, Planning for a comprehensive bridge safety assurance program: Transportation Research Record 1290.

U.S. Federal Highway Administration, 1991, Evaluating scour at bridges: Hydraulic Engineering Circular No. 18, FHWA-IP-90-017, 105 p.

Bridge Scour and Change in Contracted
Section, Razor Creek, Montana

By Stephen R. Holnbeck[1], Charles Parrett[2],
and Todd N. Tillinger[3]

Abstract

Two large floods, 3 and 4 times the estimated 100-year peak discharge, occurred in 1986 and 1991 at a timber-pile bridge over Razor Creek in Montana. A bridge section surveyed after the 1991 flood was compared with a 1955 design section and showed total scour of 0.85 m at the left abutment, 2.23 m at the right abutment, and 0.94 m at the pile bents. Calculated total scour based on equations recommended by the Federal Highway Administration and data obtained after the 1991 flood was 3.20 m at the left abutment, 4.36 m at the right abutment, and 2.13 m at the pile bents. Residual scour from floods prior to 1986 was presumed to be negligible because no floods of significant magnitude were documented. Also, scour for the 1986 flood is believed to be significantly less than for the 1991 flood because the 1986 peak discharge was significantly smaller and the contracted section for the 1986 peak discharge was 22 m upstream from the bridge.

Introduction

One of the goals of a cooperative program between the U.S. Geological Survey (USGS) and the Montana Department of Transportation (MDT) is to collect and analyze onsite bridge-scour data. Because most scour-prediction equations have been derived from research

---

[1] Hydraulic Engr., U.S. Geological Survey, Fed. Bldg., Drawer 10076, 301 S. Park, Helena, MT 59626, (406) 449-5263, Fax No. (406) 449-5497.
[2] Supervisory Hydrologist, U.S. Geological Survey, Helena, MT 59626.
[3] Civ. Engr., Montana Dept. of Transportation, Helena, MT 59620.

data from scaled-down hydraulic models, onsite bridge-scour data are needed to assess the applicability of prediction equations. However, meaningful bridge-scour data may not be available because of a lack of high flows during the project (Butch 1991). This paper presents an analysis of bridge scour for a site at which two major floods occurred within a 5-year period, resulting in observable bridge scour and a change in contracted section.

Site Description

Razor Creek, an ephemeral stream in Musselshell County, Montana, has a drainage area of 44.3 square kilometers upstream from a two-lane timber-pile bridge on U.S. Highway 87 (fig. 1). The bridge was constructed in 1955 and has an overall span of 22.9 m between abutments. Intermediate support is provided by two pile bents at 7.6 m spacing, each consisting of seven 0.30-m diameter timber piles spaced 1.37 m on centers. The 1955 design drawings indicate pile depths of 3.3 to 6.7 m below the streambed and a layer of riprap (D50 equal to 0.40 m) 0.61 m thick at the abutments. Three core logs show that the streambed has a sand and gravel layer, ranging in thickness from 1.2 to about 4.4 m. Very dense, tan sandstone and weathered shale underlie the sand and gravel to a depth of at least 10.7 m, the bottom of the drill holes. A standard sieve analysis indicated the D10, D50, and D90 diameters for the sand and gravel to be 0.065 mm, 2 mm, and 20 mm, respectively.

Peak discharges at the bridge site resulting from thunderstorms in July 1986 and June 1991 were determined by the USGS using the width-contraction, indirect-measurement method (Matthai 1967). Based on MDT and USGS knowledge of floods in the area, these peak discharges are presumed to be the largest since 1955. Calculated peak discharges of 122.0 m^3/s in 1986 and 179.3 m^3/s in 1991 were about 3 and 4 times, respectively, the estimated 100-year peak discharge of 41.3 m^3/s (Omang 1992).

Results and Discussion

Observed total scour from the 1991 flood was estimated by comparing the 1955 MDT design cross section near the upstream face of the bridge with a cross section surveyed at the upstream face after the 1991 flood (fig. 2). Maximum observed total scour occurred at the two abutments and near the left pile bent. Although distinct scour holes were not observed at the pile bents, the total scour depth that occurred near the left

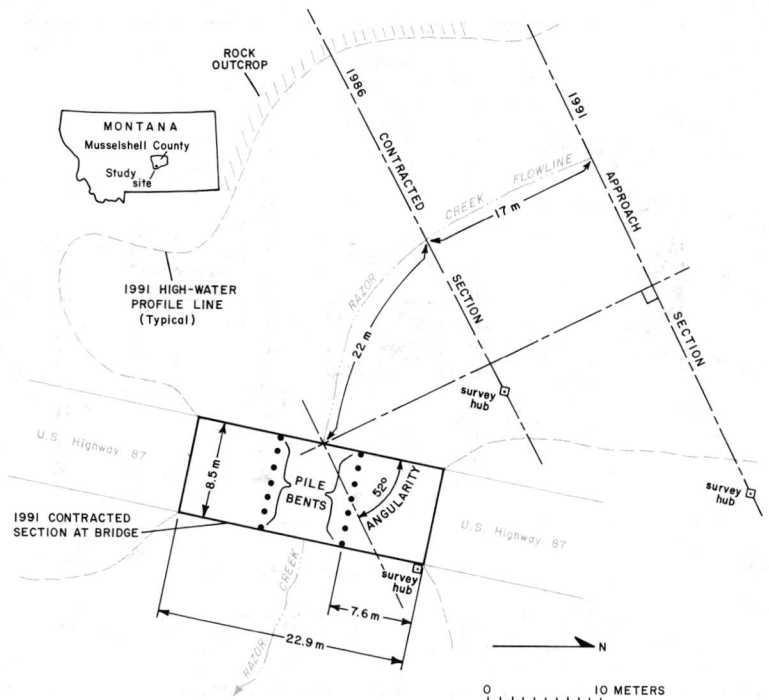

Fig. 1. Location of Study Site and Cross Sections

bent is considerably greater than the total scour anywhere else in the section (except at the abutments) and is thus presumed to be at least partly the result of pier scour.

Calculated components of total scour were determined for the 1991 peak discharge using the Laursen live-bed equation for contraction scour, the Froelich live-bed equation for abutment scour, and the Colorado State University equation for pier scour as recommended by the Federal Highway Administration (Richardson and others 1991). Channel geometry data for the bridge section were determined from 1955 design drawings. The approach section surveyed in 1991 was presumed to be the same as it would have been in 1955, although no data from 1955 were available for the approach section. Because the 1986 and 1991 floods are presumed to be the largest since 1955, residual scour from floods prior to 1986 is considered to be minor. Residual scour from the 1986 flood was not measured because indirect-measurement

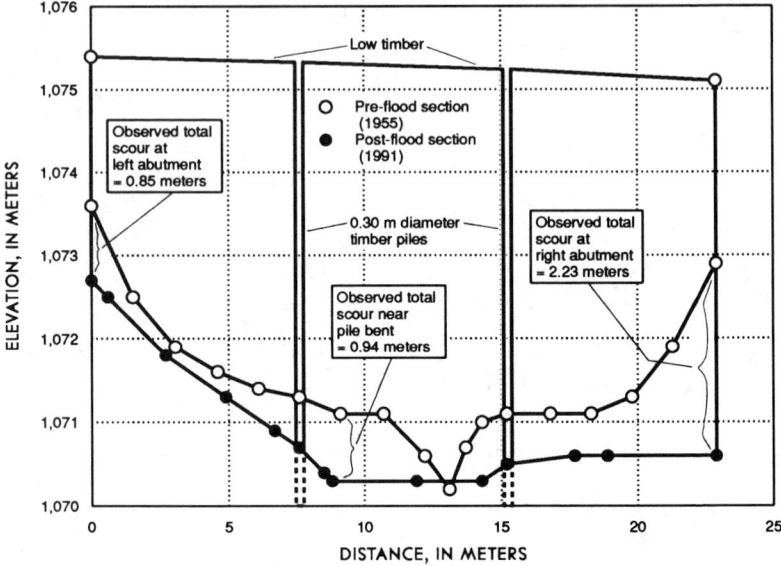

Fig. 2. Downstream View of Razor Creek Cross Sections at Upstream Face of Bridge

efforts in 1986 focused on the contracted section upstream from the bridge. However, scour for the 1986 flood is presumed to be substantially less than that in 1991, primarily because of the smaller peak discharge in 1986. Therefore, scour observed after the 1991 flood is presumed to be largely the result of the flood in 1991. There was no evidence of significant infilling following the 1991 peak discharge, so the observed scour is presumed to be the maximum scour that occurred, and comparison of observed and calculated results is assumed to be valid.

Elevations of surveyed high-water marks from the 1991 flood were used together with cross sections approximating 1955 channel geometry to determine hydraulic conditions through the bridge. Because the 1991 high-water elevations were available, a step-backwater analysis to determine the water-surface profile through the bridge was not required. Hydraulic variables needed to calculate bridge scour (cross-sectional properties, velocity and conveyance distributions) were obtained using the hydraulic-properties output from the computer model WSPRO (Shearman 1990).

Contraction scour, calculated to be 0.70 m, is an average value for the section and is added to calculated abutment and pier scour to obtain calculated total scour. Abutment scour was calculated to be 2.50 m at the left abutment and 3.66 m at the right abutment. Pier scour at the pile bents was calculated to be 1.43 m. Although some debris evidence indicated that the flow angle at the bents was greater than zero, no flow-angle adjustment was made because of the substantial distance between piles and, despite some limited research, methods for making adjustments are lacking.

The observed total scour is compared to the calculated total scour at three locations (table 1). The results indicate that the equations for scour overpredict total scour for this study. Richardson and others (1991) indicate that the equation for abutment scour is for worst-case conditions and might overpredict abutment scour for many field conditions. However, the observed scour at the abutments includes loss of riprap that had been in place prior to the 1991 flood. Without riprap protection, the scour at the right abutment may have been significantly greater than the observed value. The presence of sandstone in the area near the right abutment, as identified by core log data, may also have limited the scour depth.

Table 1. Total Scour for 1991 Flood, in m

Location	Observed	Calculated
Left abutment	0.85	3.20
Right abutment	2.23	4.36
Pile bents	0.94	2.13

Contraction scour probably was not a factor at the bridge crossing for floods less than the 1991 flood. The indirect measurement of the 1986 peak discharge indicated no water-surface drop through the bridge and that the contracted section was 22 m upstream from the bridge where the stream channel was narrowed and confined on the right bank by a rock outcrop (fig. 1). In contrast, the contracted section for the 1991 flood was at the bridge, where a water-surface drop of 2.4 m occurred. The 1986 contracted section, illustrated in figure 3, is compared using data for both 1986 and 1991 post-flood conditions. Figure 3 shows that lateral scouring of the left bank (mass wasting) in 1991, combined with a greater stage for the 1991 flood, resulted in a much larger conveyance for this section in 1991 and probably accounts for the shift of the contracted section downstream to the bridge.

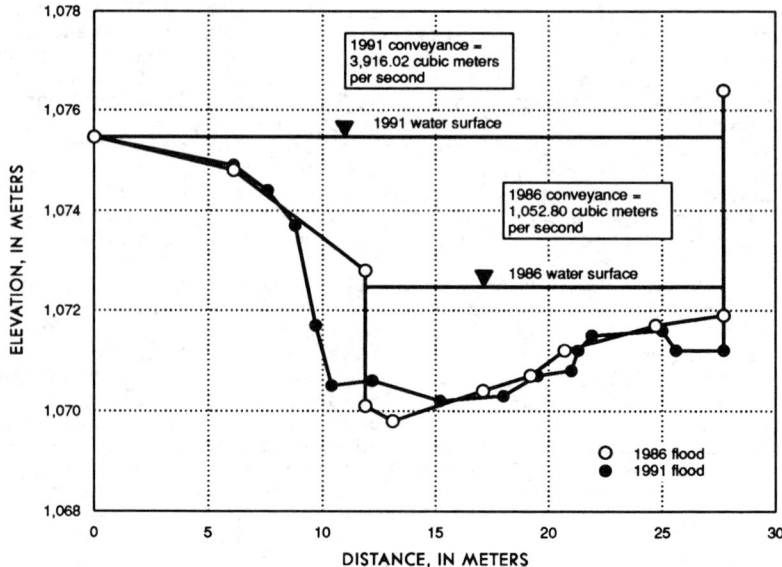

Fig. 3. Downstream View of Razor Creek Cross Sections at Location of 1986 Contracted Section

Appendix I. - References

1. Butch, G. K. (1991). *Measurement of bridge scour at selected sites in New York, excluding Long Island.* U.S. Geological Survey Water-Resources Investigations Report 91-4083.
2. Matthai, H. F. (1967). *Measurement of peak discharge at width contractions by indirect methods.* U.S. Geological Survey Techniques of Water-Resources Investigations, book 3, chapter A4.
3. Omang, R. J. (1992). *Analysis of the magnitude and frequency of floods and the peak-flow gaging network in Montana.* U.S. Geological Survey Water-Resources Investigations Report 92-4048.
4. Richardson, E. V., Harrison, L. J., and Davis, S. R. (1991). *Evaluating scour at bridges.* U.S. Department of Transportation Publication No. FHWA-IP-90-017, Hydraulic Engineering Circular No. 18.
5. Shearman, J. O. (1990). *User's manual for WSPRO--A computer model for water surface profile computations.* U.S. Department of Transportation Publication No. FHWA-IP-89-027, Hydraulic Computer Program HY-7.

Shale Scour at BNRR
Yellowstone River Bridge, MT

Gary L. Lewis, Ph.D., P.E., Member ASCE[1]

Abstract

Both the Burlington Northern Railroad and Montana Department of Transportation bridges across the Yellowstone River just south of Laurel, Montana, are founded on spread footings placed on claystone shale, buried approximately 10 feet below a gravel layer. A 1974 flood caused a significant shift in the channel attack angle, creating severe scour at the BNRR piers and abutments. A scour assessment and geotechnical investigation of the shale were conducted to evaluate the scour vulnerability and recommend countermeasures for stabilizing the bridge footings.

Introduction

In June, 1974, the bridge location experienced a 50-year event, causing a shift in the approach angle to the south pier (Pier No. 2). Sufficient erosion of shale beneath the pier footing occurred to cause rotation of the pier and truss. The railroad track through the bridge was subsequently realigned along a curve to adjust for this displacement. It was discovered that scour had reached almost 6 feet below the base of the footing and had actually undercut the footing along the front quarter of the pier. This erosion trench extended itself throughout the length of one face of the pier. Heavy riprap particles along the south face protected that side from being undercut. Since the overlying gravel at the nose of the pier was about 14 feet thick, total contraction plus pier scour for this 50-year event was approximately 20 feet.

---

[1]Vice President, HDR Engineering, Inc., 303 East 17th Ave, Denver, CO 80203.

Pier Scour History

Peak flow rates for evaluating historical floods, calibrating hydraulic and scour models, and designing countermeasures were obtained from available studies and information. Table 1 presents a summary of the historical floods.

Table 1 -- Rank Ordered Peak Flow Rates

Rank	Date	Billings, cfs	Laurel, cfs	Approximate Return Period
1	1918	78,100	64,200	100-yr
2	6/19/74	69,500	57,000	50-yr
3	7/07/75	67,600	55,600	30-yr
4	6/16/67	66,100	54,400	20-yr
5	6/27/44	64,800	53,900	17-yr
6	6/21/43	61,200	50,300	15-yr
7	6/30/82	56,600	46,600	12-yr
8	5/29/56	56,200	46,200	11-yr
9	6/07/57	56,200	46,200	10-yr

The 10, 50, 100, and 500 year floods are 58,500, 62,700, 69,900 and 83,900 cfs.

Underwater inspections in April 1978 and October 1981 revealed that the riprap around Pier No. 2 was undisturbed. When the bridge footing was examined during low flow in October, 1990, riprap was no longer evident around the upstream end. It was surmised that the riprap in front of the pier has been washed downstream and deposited about midway along the pier length. Riprap along the upstream face of the south abutment appeared to be undisturbed at and somewhat below the low water stage, but was not visibly present in any quantity below the water surface.

These inspections suggest that the riprap that was present at the abutment and nose of the pier on October 8, 1981, had been removed by one or more flood events between that date and October 16, 1990. The 12-year flood event on June 30, 1982 could be responsible for removal of the riprap. Calculations suggest that the flow velocities and depths at the piers and abutments for a 10-year event would have been sufficient to mobilize the overlying gravel layer, exposing the shale formation beneath the gravel. The size of rock that had been placed at the nose of the pier is unknown, but field inspection in 1990 revealed individual particles up to 18 inches to 24 inches in size alongside the pier. These would not be stable under 5-year flow conditions encountered at this location. Procedures in HEC-18, Evaluation of Scour at Bridges, suggest that particle sizes would need to be approximately 3.5 feet in size to withstand a 100-year event.

Aerial photographs taken in 1958, 1969 (USGS Quad Sheet), 1974 (three months after the 6/19 flood), 1977, 1978 (BNRR ground photos), 1981, 1983, 1987, and 1989 reveal large variations in the plan form upstream of the bridge. They clearly show that a secondary channel formed during or before the June, 1974 flood. This new alignment created a severe attack angle for Pier No. 2. This was the second highest flood of record, exceeded only in 1918 by an approximate 100-year event.

This would suggest one explanation of why the bridge withstood a 100-year event in 1918, yet was not able to survive the 50-year event in 1974. The shift in the channel produced much greater skewed flow at the south pier and abutment, creating more severe scour conditions than would have existed under a perpendicular attack. This skew angle might also account for the removal of the riprap along the north side of the pier in 1982 during a 12-year event.

Shale Erodibility

Geologists involved with the design of spread footings for a new MDOT bridge had indicated that the shale bedrock would not likely erode. The key design question for the DOT and BNRR is whether the bedrock is erodible, and how the composite gravel and shale formations and complex skewed flow should be treated in conducting the scour evaluation needed for designing size and depth of footings and countermeasures.

Based on a BNRR geotechnical investigation, the top 4 to 6 feet of the shale is weathered and subject to erosion. Corings taken at each of the four BN bridge footings indicate that all were founded on shale but the base of footings are currently 6 to 10 feet above the surrounding shale levels. Further research disclosed that the shale has eroded to this extent, probably through the weathered portion. A backhoe for a nearby pipeline trench was able to easily excavate up to 8 or 9 feet into the shale before encountering highly resistant "bedrock." The bridges were probably founded on this bedrock, but the surrounding shale became weathered over the 90-year interval. Independently selected tests of the bedrock were very similar to those recommended by FHWA (7/19/91 memorandum).

Hydraulics

HEC-2 hydraulic analyses were completed for the 1.3-year, 10-year, 100-year, and 500-year events. Calibration of the model was accomplished by comparing the water surface elevation measured on May 20, 1991, with predicted flow levels for the same discharge. A 0.14 feet agreement was observed. Table 2 provides a summary of the velocities and depths at each of the footings.

Table 2 -- Hydraulic Results

Item	Units	1.3 Yr Flood	10-Yr Flood	100-Yr Flood	500-Yr Flood
Contraction Hydraulics					
Discharge, Q_{mc2}	cfs	25,500	50,500	69,900	83,000
Width @ Contraction, W_{c2}	ft	408.1	420.2	452	452
Upstream Channel Width W_{c1}	ft	880	1,230	1,210	1,210
Main Ch. Q Upstr, Q_{mc1}	cfs	24,500	45,400	59,300	67,200
Main Ch. Depth Upstr, y_1	ft	7.0	7.6	8.6	8.8
Bed Material Diameter d_{50}	mm	45	45	45	45
Pier Hydraulics					
S. Pier BNRR (L=31', a=7.2')					
Flow Angle	deg	10	10	5	2
Velocity	fps	5.8	8.5	9.9	10.8
Depth	ft	8.5	11.5	13.2	14.3

Contraction Scour

An HEC-18 scour evaluation was conducted. The estimated contraction scour ranged from 4 feet for the 1.3-year event to 10 feet for the 500-year event.

Pier Scour

The angle of approach at the nose of each pier had to be estimated for pier scour calculations. During the May 20 site inspection, an attempt was made to physically measure approach angles evident at each of the pier locations. For the 10-year event, it was assumed that the approach angle would be present but not as severe as during the 1.3-year event observed on May 20. In higher floods, it is likely that flow would pass through the existing bridge in a relatively perpendicular direction. This probably explains the absence of scour during the 1918 flood, before the channel shifted.

The CSU equation pier scour for a 10-year event is 18 feet, and 16 feet for a 100-year event. The assumption was made that local scour at the pier could eventually reach the full depth predicted by the CSU equation if weathering of each newly-exposed layer of the shale occurs in a relatively short period of time. A 10-

year event might erode 6 feet into the weathered portion of the shale, and another similar event 20 years later may encounter another full 6 feet of weathered shale at the bottom of the hole eroded during the previous event. The process could repeat until the full 18 feet is reached. Success with numerous footings in this material at other bridge crossings, as well as opinions by the geologists and geotechnical engineers involved in these investigations, suggest that weathering occurs relatively slowly.

The weathered 6-foot layer may be the result of hundreds or thousands of years of cyclic exposure and burial events, or it could occur in a few decades. It was concluded that the top 6 feet of today's shale level, even where previous erosion was evidenced, is subject to scour, but the material beneath this layer could be considered bedrock for economic life purposes. It is further suggested that any footing placed at this depth should also be trenched approximately 2 additional feet into the bedrock below the 6-foot limit.

It was determined that the design total scour of 26 feet at Pier No. 2 is more critical for the 10-year event than for any other event. The principal reason for this is the effect of assumed angles of attack at the abutment and at Pier No. 2. For greater flows, depths and velocities increase but the attack angles and total scour estimates diminish.

Abutment Scour

The south abutment is considered to be the most critical location for short and long term scour damage. Shale scour is known to have already reached approximately 7 feet below the base of the abutment, based on surveyed cross-sections in 1990.

While no definite conclusion could be drawn regarding which abutment scour estimate is superior, a mid-range value of 29 feet was selected as the recommended value. The value should be considered valid only if it is assumed that the shale could continue to erode in periods of time past the 6-foot weathering limit.

It is suggested that these limits not be applied unless a full bridge replacement occurs. Discounting the 5 feet "hill" beneath the pier, the shale is assumed to be currently weathered to a depth of about 11 feet below the base of the pier.

Countermeasures

BNRR bridge footings are typically protected with riprap and monitored over time. The age of the bridge, its location on the system, projected future use of the branch line, and other considerations would all factor in the decision to use riprap or design more extensive countermeasures.

Using procedures in HEC-18, the median stable rock size at the pier for the 100-year event would be 3.5 feet in diameter. This is based on a design velocity of 13 fps.

The protection would be placed around the piers, extending 25 to 30 feet in all directions. The thickness of the blanket would be about 9 or 10 feet placed below the gravel surface.

Riprap protection of the south abutment would require a substantial amount of rock since scour in the secondary channel has eroded an estimated 6 feet below the toe. Procedures in Appendix V of "Highways in the River Environment" were applied, resulting in a requirement for 2.5-foot rock at the toe of the embankment.

The scour potential at the south abutment is significantly influenced by the shifted channel. Before 1974 this river approached the abutment at a perpendicular angle. Restoration of this condition by river training structures such as a spur dike might be appropriate. A spur dike would provide several benefits. It would change the angle of attack at the existing abutment to 90 degrees, resulting in a reduction in the calculated scour depth for the design floods. Calculated scour would still be below the 6-foot weathered shale limit. For more frequent floods, an ineffective flow zone exists outside the vena contracta. With the spur dike, the entire bridge opening becomes effective for all floods, reducing velocities and energy losses. Effects that the obtuse secondary flow currently has on the south piers would also be eliminated by the spur dike.

References

Chen Northern, Inc., "Report of Geotechnical Investigation, BNRR Bridge No. 514.15," June, 1991.

Richardson, E.V., D.B. Simons, and P.Y. Julien, "Highways in the River Environment," prepared for the Federal Highway Administration, Washington, D.C. by the Department of Civil Engineering, Colorado State University, Fort Collins, Colorado, June, 1990.

Montana Bureau of Mines and Geology, "Yellowstone River Valley South Central Montana," Hydrogeologic Map 6, 1983.

U.S. Department of Transportation Federal Highway Administration, "Evaluating Scour at Bridges," Hydraulic Engineering Circular No. 18, Office of Engineering, Bridge Division, Washington, D.C., February, 1991.

ANALYSIS OF LOCAL SCOUR AT BRIDGE PIERS

Han-Bin Liang[1] and Jorge Romero-Lozano[2]

ABSTRACT

Many studies have been performed on the cause and effects of local scour at bridge piers, and various prediction equations have been generated based on these studies. The general approach for estimating the scour in a pile group has been to treat the group as a unit and apply a prediction equation to come up with the estimated scour depth. A field study was performed at Interstate Highway 580 (I-580) Alamo Canal Bridge in Pleasanton, California and the study found that pile group scouring can vary considerably in depth due to the formation of disparate flow patterns through the pile group. This phenomenon, also known as the thalweg movement, should be taken into account when dealing with a large pile group.

INTRODUCTION

This paper analyzes and documents the results of a case study on local scour occurring at the piers of the I-580 Alamo Canal Bridge in Pleasanton, California. Field data were collected in the winter months from November 1992 to March 1993. The field measurements were then compared to two popular design formulae: the Colorado State University (CSU) Equation (Richardson, et al, 1991) and the University of Auckland (UAK) Equation (Melville and Sutherland, 1988). These two equations were done by plotting the envelope-curve that covers all available field and laboratory data.

---

[1]Senior Associate, Philip Williams & Associates, Pier 35, The Embarcadero, San Francisco, CA 94133
[2]Design Engineer, Bissell & Karn, Inc., A Greiner Engineering Inc., 5890 Stoneridge Drive, Pleasanton, CA 94588

PHYSICAL SETTING

The Alamo Canal Bridge consists of four separate bridges spanning a man-made trapezoidal channel with a natural bottom and vegetated banks (Figure 1). The bottom slope of the channel is approximately 0.1%, and was designed to convey a 100-year storm discharge of 230 cms. The bridge is located between two pronounced bends, which represent a low to moderate risk in stream instability as compared to a bridge located on a straight reach. High flow and low flow paths (thalwegs) vary, and the shifting thalwegs result in unanticipated scouring at the piers because of changes in flow direction and velocities.

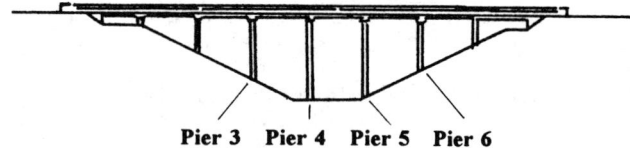

Pier 3   Pier 4   Pier 5   Pier 6

Typical Channel Cross-section

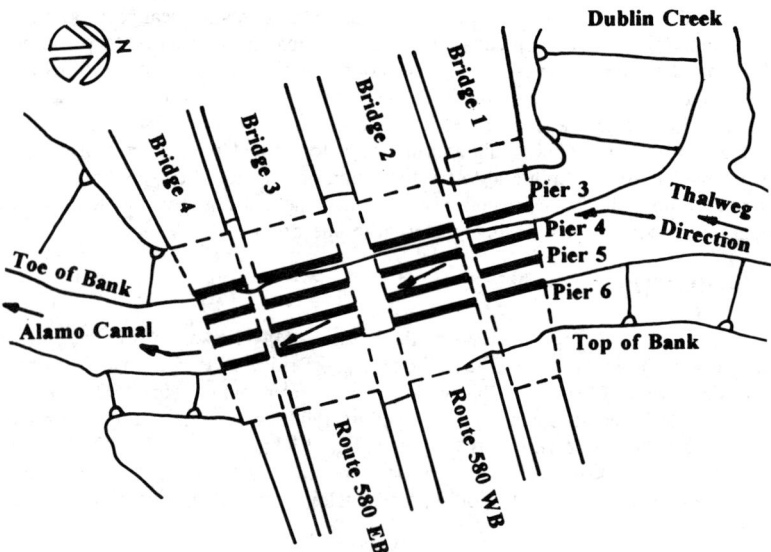

Figure 1   Site Map of I-580 Alamo Canal Bridge

The bridges range in length from 45.7 to 54.9 m. They are supported between abutments by multiple pile supported bents, the piers themselves are of mixed shapes and sizes: octagonal with a 0.23 m apothem, and cylindrical with 0.3 and 0.6 m diameters. The spacing and quantity of the piles are also uneven: Bridges 1 and 4 have 6 piles per pier ranging in pile to pile distances of 1.5 to 4.3 m center to center; Bridges 2 and 3 have 10 piles per pier ranging in pile to pile distance of 1.4 to 2.9 m. The pile tip elevations range in elevation from 79.9 to 85.3 m above mean sea level.

The soil investigation conducted by Bay Area Transit Consultants (1992) cited the presence of relatively soft and loose deposits in the canal. In the channel bed there are composite materials ranging from fine silt to gravels. The banks, however, consist of predominantly clayey materials. The following table summarizes the representative particle sizes of d_{50} (diameter of particles 50% finer), d_{84} (diameter of particles 84% finer) and d_{max} (diameter of maximum particle) on the channel bed at the bridge site:

Table 1 Representative Soil Particle Sizes

	d_{50} (mm)	d_{84} (mm)	d_{max} (mm)
Bridge 1	0.85	3.40	8.37
Bridge 2	2.50	7.50	15.32
Bridge 3	0.38	1.10	2.20
Bridge 4	0.72	7.50	34.40

SCOUR MONITORING

To gauge the incremental scouring, a field monitoring system consisting of "scour chains" was set at selected piers to collect field data from November 1, 1992 to March 1, 1993. These scour chains consisted of 0.6 m long, 2.0 gauge steel chains attached to weighted plates. They were buried vertically adjacent to piers at various locations. After each storm, the amount of the exposed chain would indicate the amount of total scour. At several other piers paint was used to mark the original ground line. The data were collected by this system at regularly scheduled intervals and after each individual storm event. Measurements of scour depth and high water marks were recorded during each site visit. The observed high water marks were then entered into a calibrated U.S. Army Corps of Engineers HEC-2 model in order to obtain the flow rate and flow velocity for each storm. All field data were collected at flows of Froude numbers between 0.2 and 0.5.

APPLICABLE EMPIRICAL DESIGN METHODS

The CSU and UAK empirical equations were applied to obtain the estimated scour depths for the studied pile group and then compared to the scour data gathered from field observations within this study.

The CSU equation is as follows:

$$\frac{d_s}{y} = 2.0 K_1 K_2 \left(\frac{b}{y}\right)^{0.65} Fr_1^{0.43}$$

where d_s = scour depth; y = flow depth just upstream of the pier; K_1 = correction for pier nose shape; K_2 = correction for angle of attack of flow; b = pier width; Fr_1 = Froude number = $V/(gy)^{0.5}$, where V = flow velocity and g = acceleration of gravity (9.81 m/s^2).

The UAK Equation can be summarized in a function relation for the equilibrium depth of scour depth d_s:

$$\frac{d_s}{b} = f\left(\frac{U^2}{gd_{50}}, \frac{y}{b}, \frac{d_{50}}{b}, \sigma_g, Sh, Al\right)$$

where U = mean flow velocity; σ_g = sediment gradation (d_{84}/d_{50}); Sh = parameters describing the shape of the pier; Al = parameters describing the alignment of the pier. Based on the laboratory studies, this relation can also be written as:

$$d_s = K_I K_d K_y K_\alpha K_s D$$

where K_I = flow intensity coefficient; K_y = flow depth coefficient; K_d = sediment size coefficient; K_σ = sediment gradation coefficient; K_s = pier shape coefficient; K_α = pier alignment coefficient.

RESULTS AND DISCUSSIONS

The study found heavy sedimentation at certain pier locations. This sedimentation was primarily the result of the live bed scour, shifting and migration of bed forms, and, most importantly, bank failure which deposited material into the lower channel. Although cohesive material is more resistant to surface erosion than non-cohesive material, such banks are said to be likely to fail due to mass wasting processes when saturated. The first wet winter in seven years in California brought in an unusually high amount of rainfall that saturated the ground and caused severe bank failures during the monitoring period. Evidence found in the field indicated that the supply of eroded bank materials actually reduced the local scour potential at piers, especially for those piers located adjacent to the toe of the banks.

Debris build-up was most predominant at the upstream end of each pile group. The contents forming the debris ranged from branches and refuse to shopping carts and sections of reinforced concrete pipe. The debris accumulation around a pier can change its scouring dynamics. In one instance, the debris accumulated at a pier was so severe that it created a 1.7 m wide debris wall and, consequently, a 4 m wide by 5 m long scour hole.

Table 2 summarizes the maximum values of observed equilibrium scour depths at selected piers during the monitoring period and the estimates from CSU and UAK equations:

Table 2 Observed Equilibrium Scour Depth

Bridge No.	Pier No.	Pile No.	b (m)	y/b	Field ds/b	CSU ds/b	UAK ds/b
1	3	3	0.30	6.03	1.08	2.36	2.40
1	4	3	0.46	4.89	1.47	1.92	2.40
2	4	2	0.46	6.98	1.13	2.85	2.40
2	3	3	0.30	7.40	0.54	2.47	2.40
2	5	3	0.46	7.33	1.00	2.29	2.40
2	6	3	0.46	5.25	0.33	2.19	2.40
2	6	6	0.61	3.74	0.23	1.94	2.40
2	5	7	0.61	3.21	0.34	1.71	2.18
3	3	5	0.61	1.10	0.17	1.45	1.92
3	4	5	0.61	5.62	1.25	2.39	2.40
3	5	6	0.61	5.59	0.59	2.05	2.40
3	6	6	0.61	4.23	1.15	1.97	2.40
3	5	7	0.61	3.74	1.19	1.72	2.40
3	6	7	0.61	4.35	1.25	1.96	2.40
3	5	8	0.61	3.77	0.85	1.72	2.40
4	5	1	0.30	8.14	2.65	2.21	2.40
4	6	1	0.30	7.74	0.50	2.51	1.47

The results from the field study indicated that scours in a pile group can have a wide range due to the flow pattern through the pile group. The shifting thalweg through the bridge site (Figure 1) resulted in higher scouring at piers 3 and 4 of Bridge 1, and piers 5 and 6 of Bridges 3 and 4. The shifting thalweg attacks those piers almost head on. Figures 2 and 3 show the comparisons between the field observations and estimates from the empirical CSU and UAK equations respectively. In conclusion, this study shows that the CSU and UAK empirical equations generally give conservative predictions when compared to our field data. However, the scouring at Bridge 4, Pier 5, Pile 1 actually exceeds the results predicted by these equations, as well as the commonly-acknowledged maximum scour depth of 2.4 times the pier width. This exception resulted, in part, from the presence of thalweg movements which generated a skewed angle of attack on the pile.

ACKNOWLEDGEMENTS

The authors wish to thank Mr. Charles Van Katwyk of the Alameda County Flood Control and Water Conservation District, Mr. and Mrs. Smith of the Lawrence Livermore Laboratory, Mr. James Ogren of Bissell & Karn, and Dr. Robert Coats of Philip Williams & Associates.

LITERATURES CITED

Bay Area Transit Consultants, *1992 Final Materials Report for Route 580 from Foothill Road/San Ramon Interchange to Tassajara Creek in Dublin and Pleasanton, Alameda County, California*, prepared for Bay Area Rapid Transit District.

Melville, B.W., and Sutherland, A.J. (1988). "Design Method for Local Scour at Bridge Piers," *Journal of Hydraulic Engineering*, Vol. 114, No. 10, October, 1988.

Richardson, E.V., Harrison, L.J., and Davis, S.R. (1991). "Evaluating Scour at Bridges," *Federal Highway Administration, Hydraulic Engineering Circular No. 18*.

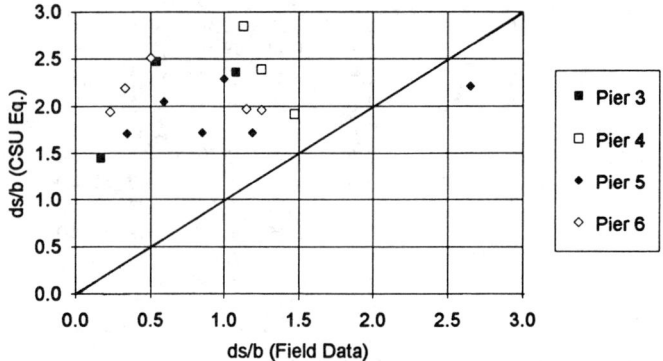

Figure 2   Comparison of Field Data vs. CSU Equation

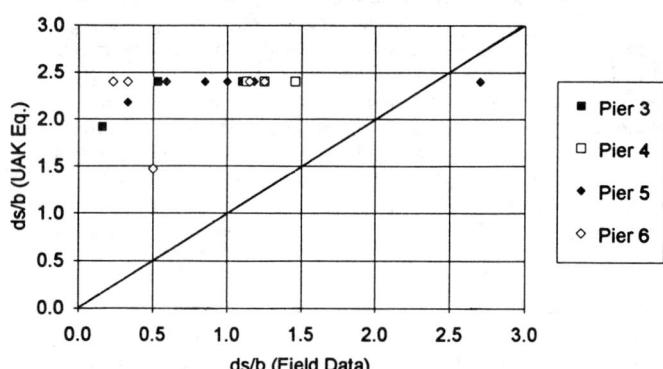

Figure 3   Comparison of Field Data vs. UAK Equation

Variability in solutions of constrained optimization problems:
Ocean outfall design case

Maili Wang and Kevin E. Lansey[1]

INTRODUCTION

Many hydraulic systems are designed based on numerical models which describe their performance. Uncertainty in parameters in these models are accounted for using designer judgement and experience. Improvement in human decision processes may be possible through the proper application of optimization techniques. However, incorporation of model uncertainty in optimization models poses a difficult problem.

An analysis of the uncertainty in optimization model results is described which quantifies the variability of the optimal solution due to parameter uncertainty. A simple formulation of an ocean outfall design is used as the application system. To quantify the impact of parameter uncertainty, the variance of the objective function is computed using Monte Carlo analysis and a first order second moment estimate.

OCEAN OUTFALL MODEL

The focus of this paper is the on uncertainty of a constrained optimization problem so a simple, although realistic, ocean outfall design model is used as the application. One problem in ocean outfall design is to determine the best location and length of the diffuser to insure adequate dilution. For the demonstration case, dilution is maximized for a defined budget for a single discharge and ambient condition.

---

[1]Graduate Research Assistant and Assistant Professor, respectively, Dept. of Civil Engineering and Engineering Mechanics, University of Arizona, Tucson, AZ, 85721.

The decisions are to select the length of the main outfall and the diffuser. It is assumed that the diffuser will be placed perpendicular to the flow and a diameter has been defined for the entire system. Mixing and dispersion is divided into three zones for modeling purposes (Fischer, et al (1979)). Initial mixing which typically dominates occurs as buoyant jets leave the diffuser and rise toward the surface. The height of rise and the magnitude of the dilution is governed by the momentum and buoyancy of the discharge. A wastewater field develops when the plume reaches an equilibrium height which is related to the relative densities of discharge and ocean water, the jet momentum and the ocean current. Little additional dilution occurs in the wastewater field rather the mixing tends to create a more uniform plume. Far from the source, turbulence and currents cause additional dilution but this effect is minor and neglected here.

For an initial study the sewage filed is assumed to submerged in a stratified ambient with no current. Relationships described in Fischer et al cases are applied to define the dilution effect. A uniform stratification profile is assumed but its slope will be considered uncertain in the optimization model. The density of ambient at the diffuser and discharge will also be considered as uncertain.

An optimization model to maximize dilution subject to a budget limitation can be written as:

$$\text{Max } S_{aw} = S_a + A \left(\frac{A^2}{4} + S_a \right)^{0.5} + \frac{A^2}{2} \quad (1)$$

subject to

$$C(L_p + L_d) D^{1.5} \leq C_{max} \quad (2)$$

$$Y_{max} = 2.84 \left(g' \frac{Q}{L_d} \right)^{1/3} \left(-\frac{g}{\rho} \frac{d\rho_a}{dy} \right)^{-0.5} \leq Y_{allowable} \quad (3)$$

$$L_p \geq L_{min} \quad (4)$$

where S_{aw} is the average dilution factor in the wastewater field, L_p and L_d are the main outfall length and diffuser length, respectively. Y_{max} and $Y_{allowable}$ are the maximum height and maximum allowable height of the plume where $Y_{allowable}$ will be measured to the water surface. C_{max} is the maximum allowable cost, C is a cost coefficient and D is the pipe diameter in inches which is assumed for simplicity to be constant through the length of outfall and diffuser.

Q is the total discharge from the diffuser, g' is g $\Delta\rho/\rho$ where ρ is the discharge density. ε is the density stratification $(-1/\rho)d\rho_a/dy$ where y is the depth of water measured from the ocean floor and ρ_a is the ambient density. S_a is the initial dilution in the rising plume which can be determined by:

$$S_a = 0.31(g')^{1/3} Y_{max} \left(\frac{L_d}{Q}\right)^{2/3} \quad (5)$$

and A is an intermediate variable defined as:

$$A = \frac{S_a(Q/L_d)^{0.5}}{0.8 Y_{max}(\varepsilon g)^{0.25}} \quad (6)$$

The model objective is to maximize dilution (eq. 1) within a budget limitation (eq. 2) while maintaining the plume below a defined allowable height (eq. 3). A minimum main outfall length, L_{min}, may be defined to insure the far field plume does not reach a sensitive area (eq. 4).

Although it may not be obvious from the equations, dilution increases and plume height decreases with increasing diffuser length. Thus, equations 2 and 3 are conflicting constraints. To maximize dilution, the diffuser length is increased and main outfall length is decreased until the plume height becomes restrictive. However, equation 3 may not be at its bound or active if the minimum outfall length is more restrictive than the plume depth requirement. In other words, at the minimum main outfall length the plume may not reach the water surface so equation 3 is a strict inequality and not active.

VARIABILITY IN OPTIMIZATION MODEL RESULTS

As in most models, several parameters are not known with certainty. The discharge, density stratification, and density difference (Q, ε, $\Delta\rho/\rho$) are assumed to be uncertain. For special cases (convex or concave objective function and uncertain linear parameter objective function coefficients), it has been shown the objective's value determined using the mean value of the parameters is not equal to the mean value of the objective function. For a maximization the mean value of the objective is less than the point estimate. Thus, an optimistic bias is introduced (Hobbs and Hepenstal, 1989).

The intent of this work is to examine the variability in the objective function by computing its variance along with its mean. By examining the variance the likelihood of poor performance can be used to judge design acceptability. Objective function uncertainty was computed using a full Monte Carlo analysis and compared to a first order second moment variance estimate.

Three cases based on L_{min} were defined to demonstrate the impact of problem structure. For all cases, the budgeted cost was assumed to be $4x10^6$, C was defined as 2.41 and D was 210 cm. The allowable plume height was assumed to be the water surface. The slope of the ocean bottom was set as 2.5% so $Y_{allowable}$ was $0.025*L_p$. The mean discharge was 6.0 cms, the mean ε was 0.0000125, and the mean $\Delta\rho/\rho$ was set as 0.015. Initially, a coefficient of variation of 0.1 was assumed for each parameter.

Using the mean parameter values for case one (L_{min} equal 1200 m), the optimal dilution was 108.85 with L_p at its lower bound of 1200 m and the optimal L_d was 956 m. The maximum height was below the surface m and equation 3 was not active. Monte Carlo simulation was completed by solving 600 optimization problems. The model parameters were generated assuming independent normally distributed variables. In a very few cases, the main outfall length was larger than 1200 m since the plume reached the surface with a main outfall length of 1200 m.

The objective function's standard deviation for this case was 9.96 as listed in Table 1. Since the objective function is nonlinear, the objective function's mean value from the Monte Carlo was not equal to the objective using the mean parameter values and a pessimistic bias is introduced. For this case, the parameters only appear in the objective function but due to non-convexity an optimistic bias does not result. A first order second moment analysis of uncertainty (FOSM) was applied to compute the variance of S_{aw} at the mean parameter values. In this case the 1200 m bound on the main outfall length was nearly always active so the variance could be estimated using only eq. 1. The variance estimate was slightly low due to the problem's nonlinearity. Gradients were computed numerically by perturbing the parameter and resolving the optimization problem for S_{aw} and analytically. Both gradients provided essentially the same variance estimate.

In the second case, L_{min} was set to 0. The optimum tradeoff between the lengths L_p and L_d was selected by the model. Since dilution increases with diffuser length, in this the best solution is to reduce L_p and increase L_d until Y_{max} reaches the water surface. Only two constraints remain in the optimization problem and both are active for all optimal solutions. Using the mean values of the parameters, the objective function was 118.19 compared to the mean objective function value from the Monte Carlo analysis of 118.73.

In this case, the uncertain parameters appear in both the objective function and the constraints (Y_{max}). Intuitively, the objective function variance was expected to be higher than for case one since the problem is more variable. However, the variance from the Monte Carlo simulation was actually smaller. The model apparently takes advantage of the variability in the constraints to more consistently determine good optimal solutions. Three variables are determined in this model, L_p, L_d and S_{aw}. Since two constraints are always active, three equations can be written in terms of the three variables. A FOSM analysis was performed to estimate the uncertainty in S_{aw}. The required gradients were numerically computed by perturbing model parameters. Again reasonable estimates were found as seen in Table 1.

Case three solved the optimization model with L_{min} equal to 1000 m. This bound using the mean parameters was active and the maximum plume height was not at its bound. The optimal solution was a dilution of 115.96 with main outfall length of 1000 m and diffuser length of 1157 m. However, parameter changes over their acceptable ranges changed the set of active constraints. About one third of the parameter sets were limited by the plume height.

The standard deviation computed from the Monte Carlo simulation was 10.06. The FOSM estimate of the standard deviation was computed using the constraints active at the mean parameter values. A first order second moment analysis is not theoretically correct in this case since the set of equations change over the range of the parameters. Still the variance estimate was reasonably close for this problem.

The same analysis for all three cases was completed for a coefficient of variation for the parameters equal to 0.2. Similar results were obtained for the three cases (Table 2). The most significant change is the variance estimates for case 3. As expected as the uncertainty becomes larger the improper FOSM variance estimate becomes very poor.

ANALYSIS

The distribution problem in optimization has been considered in the operations research field for some time (Ermoliev and Wets, 1988). Much of the work has focused on determining the mean value of the objective function by integrating over the range of the parameters. A highly constrained optimization problem which is the simplest case for estimating model uncertainty has been considered in this paper.

The ocean outfall problem reduced to solving a set of three equations for the three decision variables. In such problems, when the active constraint set does not change over the parameter range, the variance can be estimated reasonably well using first order estimate. The question of where to compute the gradients in first order estimate must be answered. In addition, the mean value must also be more accurately determined. The advanced first order method may be a partial solution to this problem.

If, as in a general nonlinear model, the number of active constraints does not equal the decision values, the variance is much more difficult to obtain. The unconstrained problem results in an extremely difficult problem. Techniques for estimating the mean such as quasi-Monte Carlo or fast Monte Carlo methods hold promise in this direction.

CONCLUSIONS

Uncertainty in optimization model resulting from parameter uncertainty has not been well studied in the hydraulics and water resources areas. To make valid decisions, knowledge of the uncertainty in the model results is needed. A potential uncertainty measure is the objective function variance. A simple example has been presented which demonstrates the difficulty in estimating the variance. A first order second moment estimate of the variance gave acceptable to poor results. Errors were due to the changes in the active constraint set, the nonlinearity of the objective function, and the estimation point.

TABLE 1: Mean and standard deviations for all cases using a constant coefficient of variation of 0.1

CASE	MEAN		STANDARD DEVIATION	
	Monte Carlo	FOSM	Monte Carlo	FOSM
1	109.24	108.85	9.96	9.77
2	118.73	118.19	8.96	9.28
3	115.95	115.66	10.06	10.38

TABLE 2: Mean and standard deviations for all cases using a constant coefficient of variation of 0.2

CASE	MEAN		STANDARD DEVIATION	
	Monte Carlo	FOSM	Monte Carlo	FOSM
1	110.06	108.85	19.46	19.54
2	118.03	118.19	17.73	18.56
3	115.38	115.66	18.81	20.76

REFERENCES

Ermoliev, Y. and R. Wets, (1988), Numerical techniques for stochastic optimization, Springer-Verlang, New York.

Fischer, H., E. List, R. Koh, J. Imberger, and N. Brooks, (1979), Mixing in inland and coastal waters, Academic Press, New York.

Hobbs, B. and A. Hepenstal, (1989), "Is Optimization optimistically biased?," Water Resources Research, 25(2), pp. 152-160.

Optimizing Water Transfers in Urban Water Supply Planning

Jay R. Lund, Associate Professor
Morris Israel, Doctoral Student
Department of Civil and Environmental Engineering
University of California
Davis, CA 95616

Abstract

A number of water supply agencies in the Western United States have begun to integrate water transfers into their overall water supply planning, including the use of transferred water as an additional water source with particular cost and availability characteristics. This paper presents a two-stage linear program for preliminarily identifying the least-cost integration of several water marketing opportunities with water conservation and traditional water supplies. Water marketing opportunities can include several dry-year options and spot market water transfers.

Introduction

In recent decades water transfers or markets have been increasingly sought as a source of additional water supplies for urban water systems (Lund, et al., 1992). These efforts to add transferred water to urban supplies have been motivated by the expense and controversy of expanding most traditional water supply sources, the cost and inconvenience of high levels of water conservation, and the relatively low economic value of many agricultural water uses.

In pursuing water transfers as a water source, urban water utilities have begun to modify their approaches to water supply planning (Lund, et al. 1992). Incorporating water transfers into system planning requires greater cooperation and coordination between urban, agricultural, and environmental water users. This is an analytically and institutionally more difficult task than planning for a single isolated system.

This paper introduces the use of mathematical programming for the planning of water transfers as part of a multi-source urban water supply system. A two-stage linear program is formulated and applied to an example to illustrate how several forms of water transfers (dry-year options and spot markets) can be integrated with drought water conservation and traditional supplies to meet anticipated system demands (Loucks, et al., 1981).

Forms of Water Transfer

Water transfers can take a number of forms. Each form has different implications for water supply system design and management. These different forms of water transfers are discussed in some detail by Lund et al. (1992) and the National Research Council (1992). A few common transfer forms are summarized here.

Permanent Transfers

In a permanent transfer, the buyer acquires the permanent right to use water from the seller. In the case of water sales from farmers to cities, this can involve the fallowing of farmland, the replacement of a farm's water supply with a new source (generally less desirable from an urban use perspective), or the lease of the sold water back to the farm during wet years.

Contingent or Dry-Year Option Transfers

A less permanent and more flexible transfer arrangement is the contingent transfer of water between water users. Here, one user contracts with another for the transfer of water under specific conditions. Conditions triggering a transfer might include drought (so-called "dry-year option"), unusually high water demands (stemming from high growth rates), or the interruption of normal water supplies due to earthquakes, floods, or other natural disasters. Contingent transfers usually involve a contractually fixed payment in each year when the contingency is in force, plus an additional contractually fixed payment when transfers are actually implemented.

Spot Market Transfers

Spot market transfers are arranged on a short-term, almost as-needed basis, typically within a single year. Prices for these transfers vary with market conditions during a particular year. Prices can also vary substantially with water quality and the storage, conveyance, and treatment costs of utilizing transferred water. Typically, prices will be higher during dry years. Examples of a form of spot water market were the 1991 and 1992 state-sponsored Drought Water Banks in California (DWR, 1992).

Optimal Mixing of Transfers of Different Types

Different types of transfers are suitable for different uses within a large urban water supply system. The optimal (or least-cost) mix of transfers will vary with the specific hydrology, demands, costs, water market conditions, and configuration of an individual system. Finding the least-cost mix of different types of water transfers from different sources together with establishing economical water conservation levels can be formulated as a mathematical programming problem. Such a formulation is presented below.

Two-Stage Linear Programming Formulation

Many urban water agencies have an opportunity to purchase additional water now to supplement other supplies in the event that the remainder of the wet season provides insufficient precipitation. This is a common plight for California cities. This current water purchase option can be thought of as a first-stage decision. Regardless how much water the agency buys now, it may find itself with extra water or with too little water in the following season or year. If the agency finds itself with extra water, the excess purchased water is wasted or diverted to lower-valued uses. If the agency finds it did not purchase enough water, the shortage can be met by purchasing water at a higher spot market price or through water conservation. Both these recourses (second-stage decisions) incur costs.

This problem will be formulated as a two-stage linear program (Wagner, 1975; Loucks, et al., 1981).

The objective in solving the problem is to minimize the expected value of all costs, including the costs of the long-term contingent contracts (dry-year options), the costs of physically acquiring this transfer water in the event it is needed, the costs of water conservation in event of drought, and the cost of any other water purchases during a drought on an emergency water spot market. These costs are represented as a linear function in Equation 1, the linear program objective function.

This problem has two primary constraints:
a. In any drought event, the use of contingent transfer water (U_{xk}) cannot exceed the amount specified in the contingent transfer contract (X). This is represented by Equation 2.
b. For any event, the sum of water supply from the traditional water source (a_k), water from contingent transfers (U_{xk}), water from short-term spot transfers (U_{sk}), and water conservation (U_{ck}) must equal or exceed normal demand (d). This is represented in Equation 3.

The decision variables are:
  X = the amount of water option to be purchased now (the first stage decision)
  U_{xk} = amount of purchased water (X) used in the second stage for event k
  U_{sk} = amount of spot market water bought in the second stage for event k
  U_{ck} = the amount of water conservation used in the second stage for event k

Some important parameters are:
  c_n = the unit price of water options purchased now
  m = the number of increments of water availability, a
  a = the availability of water from the city's traditional sources during the second stage (divided into increments a_k)
  p_k = the probability of the k-th event (water availability = a_k)
  c_1 = the unit cost of using water purchased in the first-stage (pumping, etc.)
  c_{sk} = the unit cost of purchasing and using spot market water, for event k
  c_c = the unit cost of water conservation
  d = normal water demand in the second stage

The linear program becomes:

(1)   MIN $z = c_n X + \sum_{k=1}^{m} p_k(c_1 U_{xk} + c_{sk} U_{sk} + c_c U_{ck})$

Subject to:

(2) $U_{xk} \leq X, \forall k$ (maximum use of stage 1 contracted water)

(3) $a_k + U_{xk} + U_{sk} + U_{ck} \geq d, \forall k$ (meet demand in stage 2)

The objective function represents the expected value cost of all decisions in the first and second stages. The constraints define the interaction of the decisions in the first and second stages. Note that the number of constraints increases linearly with the number of chance events, in this case the number of increments of water availability, a. There are 1+3m decision variables, 2m constraints, and 5 + 2m parameters to estimate.

This simple formulation can be easily modified by piece-wise linearization to incorporate convex non-linearities in the cost of short-term water conservation (c_c), spot market purchases (c_{sk}), contingent transfer options (c_n), and the utilization of contingent transfers (c_1). Similarly, multiple sources of contingent and spot market transfers and alternative (non-transfer) supply augmentation measures can be incorporated into the formulation (Loucks, et al., 1981). Additional constraints can also be placed on the maximum use of water conservation, dry-year options, or spot market purchases.

The formulation can be expanded to include uncertainties in water demand and prices in the formulation. This is done by increasing the number of events identified (m) to include different combinations of values for water availability (a), water demand (d), and prices, essentially adding a subscript k to water demand and uncertain prices in addition to water availability. While this approach allows for the incorporation of a wider variety of uncertainties, it entails an increase in the number of events (m) and a consequent increase in the computational burden of the solution and, perhaps more importantly, a great increase in the difficulty in assigning parameter values for each now joint event k.

Illustrative Example

A particular urban water agency has several sources of water: its traditional supply (from a reservoir), water purchases from a spot market, and water purchases through a dry-year option contract. The system can also choose to meet some part of any shortage by drought water conservation. The probability distribution of the traditional source's water availability appears in Table 1. Table 1 also shows the expected spot market price of water for each hydrologic/water availability event.

The dry-year option contract must be established well before the year in which water would be available from this source. Table 2 provides data on the yearly price of maintaining the option and the episodic cost of utilizing the option. Table 2 also provides an assumed linear cost of water conservation and the level of annual system water demand. Since very high levels of water conservation are likely to incur highly non-linear costs, a constraint is also placed on the maximum amount of water conservation, as a percentage of total demand. This constraint is added to the linear program formulated above.

Table 1: Second-Stage Event Characteristics

Water Availability Event (TAF) (a_k)	Probability of Event (p_k)	Spot Market Price ($/ac.-ft.) ($c_{sk}$)
50	0.05	200
75	0.05	180
100	0.05	170
125	0.05	120
150	0.05	110
175	0.05	90
≥200	0.7	70

Table 2: Example Data

Option price for long-term contract (c_n):	$20/ac-ft.-year
Cost of utilizing long-term contract water (c_1):	$50/ac.-ft.
Cost of short-term water conservation (c_c):	$130/ac.-ft.
Maximum percent water conservation:	35%
Annual water demand (d):	200 TAF

The results of solving the two-stage linear program appear in Table 3. The example problem was solved using the linear program add-in in Microsoft Excel, version 4.0. The contract amount for the dry-year option (X) suggested by the 2-stage linear program is 55 thousand acre-ft (TAF) per year.

Once the dry year option is purchased, its relatively low cost to implement encourages its use whenever a shortage occurs, so it is used until the contracted quantity is reached. Additional shortage is borne by spot market water purchases, since the cost of spot market purchases for less severe droughts is less expensive than the cost of water conservation.

For increasingly severe droughts, the price of water on the spot market increases greatly, encouraging use of drought water conservation over spot market purchases until the maximum conservation level (35% of normal use) is reached. Thereafter, spot market purchases must again be sought, since the dry year option and conservation alternatives have already been fully utilized. However, the high cost of these infrequent expensive spot market purchases is not sufficient to encourage contracting of greater amounts of dry-year options.

Table 3: Least-Cost Transfer and Water Conservation Plan

Water Availability Event (TAF)	Probability of Event	U_{xk} Use of Option (ac.-ft.)	U_{sk} Spot Market Purchase (ac-ft.)	U_{ck} Conservation (ac.-ft.)
50	0.05	55	25	70
75	0.05	55	0	70
100	0.05	55	0	45
125	0.05	55	20	0
150	0.05	50	0	0
175	0.05	25	0	0
≥200	0.70	0	0	0

Conclusions

One of the virtues of systems analysis techniques is that their flexibility allows them to be modified to incorporate changes in the problem under study. In the Western United States, the urban water supply problem has changed recently encouraging both transfers of water between institutions to supplement traditional sources of urban water supply and water conservation in system planning and operation. Systems analysis techniques can be of special use in this situation, since simulation and optimization models allow various promising alternatives and combinations to be identified with less trial and error. As such, systems analysis is not a substitute for traditional planning approaches, but a useful supplement for more rigorously formulating problems and exploring solutions. Therefore, systems analysis techniques should be particularly useful for relatively new or evolving problems.

The simple mathematical programming formulation and example of the problem of designing urban water supplies with uncertainty, water conservation, and several types of water transfers illustrates the potential value of systems analysis techniques. The value of these techniques is not necessarily only in their numerical behavior, but extends to the interpretations of their results in terms of the qualitative design of water resource systems. Making common-sense qualitative interpretations of optimization results is likely to enhance the acceptance of systems analysis techniques in practice. In this case, the two-stage linear programming results illustrate the desirability of a mix of water transfer and water conservation measures in system design and operation plans.

Acknowledgements

The authors thank Paul Hutton for his comments on an earlier draft of this paper.

References

DWR (1992), *The 1991 Drought Water Bank*, California Department of Water Resources, Sacramento, CA, January.

Loucks, D.P., J.R. Stedinger, and D.A. Haith (1981), *Water Resources Systems Planning and Analysis*, Prentice-Hall, Englewood Cliffs, NJ.

Lund, J.R., M. Israel, and R. Kanazawa (1992), *Recent California Water Transfers: Emerging Options in Water Management*, Center for Environmental and Water Resources Engineering Report No. 92-1, Department of Civil and Environmental Engineering, University of California, Davis, CA.

National Research Council (1992), *Water Transfers in the West: Efficiency, Equity, and the Environment*, National Academy Press, Washington, DC.

Wagner, H.M. (1975), *Principles of Operations Research*, Prentice-Hall, Englewood Cliffs, NJ.

OPTIMISING A RESERVOIR OPERATION POLICY FROM THE PROPERTIES OF RELIABILITY AND RESILIENCY

K.S. TICKLE[1] and I.C. GOULTER[2], M.ASCE

Abstract

Reservoir system risk performance metrics of reliability and resiliency have been variously defined by different authors. Recent work on the statistical characteristics of these measures has provided the basis for incorporating specified levels of performance directly into models for the optimisation of a reservoir operating policy. This paper summarises the theory behind the statistical characteristics of the measures and shows how this new knowledge of the statistical properties can be used to obtain reservoir operating rules which, over a finite time horizon, provide specified levels of systems performance which are consistent with the long term requirements. The use of this new knowledge is demonstrated by application to a single reservoir system in Central Queensland, Australia.

Introduction

Evaluation of the performance of a reservoir system can be achieved by defining suitable performance indicators. While it is obvious that these indicators should reflect the quantity of water supplied, incorporating the risk associated with supplying this amount into the metric is often difficult. Some success in addressing this problem was achieved through the risk criteria of reliability, resiliency and vulnerability introduced by Hashimoto et al. (1982) and Fiering (1982a,b). The statistical properties of these initial definitions of risk and their subsequently modified forms have been discussed by Tickle and Goulter (1993). The work by Tickle and Goulter has enabled the form of the system behaviour to be fully specified so that an optimisation model can be driven to achieve an appropriate operating rule. The following sections describe the basis on which the statistical properties are defined and how knowledge of these properties can be used to constrain levels in reservoir operation models.

Theory Performance Metrics

Denote the output performance of a system by the random variable X_t with $t = 1, 2, ..., N$. Define a level of performance X_0 such that

---

[1]Senior Lecturer, University of Central Queensland, Rockhampton, Australia
[2]Professor, University of Central Queensland, Rockhampton, Australia

$X_t \geq X_0$ defines a satisfactory level of performance and
$X_t < X_0$ defines an unsatisfactory level of performance.

A bivariate series Z_t can now be defined such that

$Z_t = 0$ if $X_t \geq X_0$ and
$Z_t = 1$ if $X_t < X_0$.

Reliability, α, can then be defined as

$$\alpha = \Pr\{X_t \geq X_0\} = \Pr\{Z_t = 0\} \qquad (1)$$

while resiliency, γ, can be defined as

$$\gamma = \Pr\{X_t \geq X_0 \mid X_{t-1} < X_0\} = \Pr\{Z_t = 0 \mid Z_t = 1\} \qquad (2)$$

While the above definition of reliability is somewhat standard in the literature, the definition of resiliency has been variously defined by several authors such as Beshay and Howell (1986), Moy et al. (1986) and Burn et al. (1991). These alternative definitions use combinations of maximum sojourn length, average sojourn length and number of sojourns in an unsatisfactory state to reflect resiliency.

Tickle and Goulter (1993) have identified a relationship between the definition of resiliency given by Eq.2 and the lag-one autorun coefficient as proposed by Sen (1978, 1979), Yevjevich (1972), Saldarriaga and Yevjevich (1970), Guven (1983) and McKenzie (1984) on the basis of statistical properties of runs and their use for drought analyses.

Tables giving the value of the autorun coefficient for a given autocorrelation coefficient and truncation level (threshold between satisfactory and unsatisfactory states) for a normally distributed Markov process are given in Saldarriaga and Yevjevich (1970) and Abramowitz and Stegun (1965). However Guven (1983) has derived the following simple approximate expression for the distribution of the longest sojourn length below a threshold.

$$\Pr\{L \leq j\} = \exp[-N(1-\alpha)\gamma(1-\gamma)^j] \qquad j = 0, 1, 2, \ldots \qquad (3)$$

where L is the maximum sojourn in a record of period N.

Examination of Eq. 3 shows that the distributional properties of the maximum sojourn length in a particular system can be found by specifying the reliability and resiliency associated with a threshold for a given record length. This feature is the basis of the approach used in this paper to incorporate maximum sojourn constraints as they represent aspects of risk of non-performance of the reservoir into the optimisation model for optimisation operation of that reservoir.

Properties of Metric Estimators

The reliability, α, is generally estimated by

$$\hat{\alpha} = 1 - \sum_{t=1}^{N} z_t / N$$

An estimator suggested by Sen (1978) and Guven (1983) for resiliency, γ, is (4)

$$\hat{\gamma} = 1 - \left[\sum_{t=2}^{N} z_t z_{t-1} / (N-1) \right] / \left[\sum_{t=1}^{N} z_t / N \right] \quad (5)$$

Tickle and Goulter (1993) have shown Eq. 5 to be a biased estimator of resiliency.

Specification of a threshold between satisfactory and unsatisfactory states and the associated reliability and resiliency of the system in respect to that threshold fixes all the properties of a lag-one Markov process such as a Poisson AR(1). If the behaviour of the output series is further restricted by specifying other features, for example, two threshold levels of interest with the associated values of reliability and resiliency, then the parameters of the corresponding multi-parameter AR(1) processes can also be determined.

Optimisation

The optimisation procedure was trialed on the Fairbairn Reservoir in Central Queensland, Australia. A detailed description of the method of operation is provided by Tickle and Goulter (1992). However, a brief description of the system is given here to place the results in context.

Irrigators taking water from the Fairbairn Dam are issued a licence for a certain amount of water each year. This figure is 'nominal' as more or less can actually be supplied depending on current climatic conditions. At the start of the irrigation year, the manager of the reservoir announces the amount of allocation available for the coming year as a percentage of the nominal supply. Thus an announced allocation of 100% would correspond to the irrigators' nominal supply. In poor years the supply might be restricted to 50% of the nominal amount. The aim of the optimisation analysis in this paper is to determine an operating rule on the announced allocations to maximise the amount of water supplied to irrigators subject to constraints on the reliability and resiliency.

Specification of performance levels of system output, defined in terms of the quantities of water to be supplied, in conjunction with the associated reliability and resiliency required of the system can be used to drive the optimisation model in its determination of the optimum operating rule. The relationships between expected maximum sojourn and reliability and resiliency are given in Figure 1. It can be seen in this figure that the number of occurrences, and the maximum sojourn length, below a threshold can be used as surrogates for reliability and resiliency in the optimisation.

The actual optimisation technique used in the optimisation of reservoir operation is a policy iteration method using discrete differential dynamic programming. The objective function in the model consists of benefit obtained from supplying the water with penalties for violating reliability and resiliency constraints as interpreted through the surrogates discussed above. The discrete differential dynamic programming aspect of the model is carried out by searching a corridor around the current policy relating the announced allocation to the amount in storage at the start of the year to find the solution that best improves the objective function. This new solution then becomes the current policy. This process continues until a specified level of convergence in performance between the previous and the new solution is achieved.

Results

The optimisation program was run for a range of nominal allocations from the Fairbairn Reservoir from 100000 ML per annum to 140000 ML per annum. For the cases of 100000 ML/yr to 130000 ML/yr the reliability was set at 0.9. As 90 years of data are available for this reservoir this standard corresponds to 9 years (108 months) of failure in meeting a target of 100000 ML/a. The resilience was specified in terms of maximum sojourn length in that a penalty was imposed in the objective function if the maximum sojourn length exceeded 4 years (48 months). The actual number of sojourns resulting from the optimisation corresponded to a resiliency of 0.33 for the annual series of 90 years (or 0.027 for the monthly series of 1080 months) in all these cases.

For the case of a 140000 ML/yr nominal allocation, the reliability was set to 0.8 (i.e 18 years of failure to meet the target) and the maximum sojourn length set at 6 years so that the estimated resiliency was the same as in the previous case.

The discrete nature of the data does limit the precision of this method of optimisation. Figure 1 shows the theoretical relationship between mean maximum sojourn length and resiliency derived from the formula from Guven (1983) on the annual series. However, as the number of sojourns must be an integer value and the maximum sojourn used as input via the optimisation constraint is also integer, only an approximate relationship is possible. For the case of a reliability of 0.9, specifying a maximum sojourn length of 4 years resulted in 3 sojourns and hence a resiliency of 0.326. For a reliability of 0.8, a specified maximum sojourn length of 6 years resulted in 6 sojourns for a resiliency of 0.326. These results are summarised in Table 1.

Table 1. Relationship between reliability, resiliency and the theoretical and empirical (derived from the optimisation model) values of sojourn for the Fairbairn Reservoir for an annual series of 90 years.

Reliability = 0.9			Reliability = 0.8		
No of Sojourns	Resiliency	Theoretical Mean Max. Sojourn	No. of Sojourns	Resiliency	Theorical Mean Max. Sojourn
2	.213	5.78	5	.270	7.36
3	.326	4.71	6	.326	6.45
4	.438	3.88	7	.382	5.71

Hence the integer values of expected maximum sojourn length, input as constraints in the optimisation, can be seen to provide a reasonable means of estimating the expected resiliency of the output.

Conclusions

The methodology used in this paper shows how an understanding of the statistical properties of reliability and resiliency associated with a threshold between satisfactory and unsatisfactory states in a time series can be used to drive an optimisation model. The results show that the formula given by Guven (1983) for the relationship between resiliency, reliability and length of sojourn in an unsatisfactory state can be used to provide a reasonable estimate of the resiliency from the expected maximum sojourn length.

Further work is being undertaken to identify the properties of the output series for the optimisation model. This work should enable the determination of an operating rule that optimises system behaviour based on a range of performance metrics wherein the optimisation model will only need to be driven by a subset of the metrics that are easily incorporated as constraints in the objective function.

Figure 1. Mean Maximum Sojourn and Resiliency for N=90

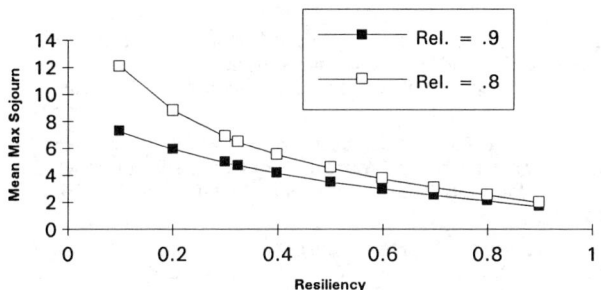

References

Abramowitz, M. and Stegun, I.A., (1965). Handbook of Mathematical Functions. Dover, New York.

Beshay, A. and Howell, D.T., (1986). "Measuring the Hydrologic Performance of Urban Water Supply Systems". Proceedings of the Hydrology and Water Resources Symposium, Griffith University, Queensland, Australia, 273-277.

Burn, D. H., H. D. Venema, and S. P. Simonovic, (1991). "Risk-Based Performance Criteria For Real Time Reservoir Operation". Can. J. Civ. Eng. 18, 36-42.

Fiering, M.B. (1982a). "A Screening Model to Quantify Resilience". Water Resources Research, 18(1), 27-32.

Fiering, M. B. (1982b). "Alternative Indices of Resilience". Water Resources Research, 18(1), 33-39.

Guven, O., (1983). "A Simplified Semi-empirical Approach to Probabilities of Extreme Hydrologic Droughts". Water Resources Research, 19(2), 441-453.

Hashimoto, T., J. R. Stedinger, and D. P. Loucks, (1982). "Reliability, Resilience, and Vulnerability Criteria For Water Resource System Performance Evaluation". Water Resources Research, 18(1), 14-20.

McKenzie, E., (1984). "The Autorun Function: A Non-parametric Autocorrelation Function". Journal of Hydrology, 67, 45-53.

Moy, W. S., J. L. Cohon, and C. S. ReVelle, (1986). "A Programming Model for Analysis of the Reliability, Resilience, and Vulnerability of a Water Supply Reservoir". Water Resources Research, 22(4), 489-498.

Saldarriaga, J. and Yevjevich V., (1970). "Application of Run-Lengths to Hydrologic Series". Hydrology Paper 40 Colorado State University Fort Collins, Colorado, U.S.A.

Sen, Z., (1976). "Wet and Dry Periods of Annual Flow Series". J.Hydraul. Div., ASCE, 102 (HY10), Proc. Pap. 12457: 1503 -1514.

Sen, Z., (1979). "Application of the Autorun Test to Hydrologic Data". Journal of Hydrology, 42, 1-7.

Tickle, K.S. and Goulter, I.C., (1992). "Assessment of Performance Metrics for a Reservoir Under Stochastic Conditions". Proceedings of the Sixth IAHR International Symposium on Stochastic Hydraulics, J.T. Kuo and F-F Lin, Eds., Taipei, Taiwan, 583-590.

Tickle, K.S. and Goulter, I.C., (1993). "Statistical Properties of Reliability and Resiliency Measures". To be published in Proceedings of Conference on Stochastic and Statistical Methods in Hydrology and Environmental Engineering, 21-23 June 1993, Waterloo, Canada.

Weeraratne, J. R., L. Logan and T. E. Unny, (1986). "Performance Evaluation of Alternate Policies on Reservoir System Operation". Can. J. Civ. Eng., 13, 203-212.

Yevjevich, V., (1972). "Stochastic Processes in Hydrology". Water Resour. Publ., Fort Collins, Colorado, U.S.A.

Hydraulic and Water Quality Reliability and Resiliency for Water Distribution Systems Under Random Demands

Mao Fang,[1] James G. Uber,[2] Assoc. Member, ASCE

Abstract

This paper develops a preliminary method to evaluate the reliability and resiliency of water distribution systems under random spatial and temporal demand patterns, including random fire demands. The method uses hydraulic and water quality network models with Monte Carlo simulation techniques to estimate the random spatial and temporal hydraulic and water quality performance.

Introduction

The basic function of a water distribution system is to deliver a desired quantity and quality of water, within specified pressure limits, at appropriate location and times. The pressures and flow rates in the distribution system depends on many factors, including the spatial and temporal pattern of water use, or, the *demands*. Because mass transport depends on hydraulic conditions that affect mixing at pipe junctions, and advective and dispersive processes, the water quality also depends on the spatial and temporal demand pattern.

The future demand pattern is uncertain, and difficult to predict with any mechanistic model. The spatial and temporal demands are assumed to be random variables that belong to a joint probability distribution. Thus, the hydraulic and water quality performance in a water distribution system is also random and can only be defined probabilistically. Reliability and resiliency are two probabilistic

---

[1] Grad. Res. Asst., Dept. of Civil & Envir. Engrg., Univ. of Cincinnati, Cincinnati, Ohio 45221.
[2] Asst. Prof., Dept. of Civil & Envir. Engrg., Univ. of Cincinnati, Cincinnati, Ohio 45221.

concepts that are proposed for design and analysis of water distribution systems under random demands.

Reliability is defined as the probability of acceptable hydraulic and water quality system performance over a specified period of time. Reliability analysis methods are well established in civil engineering and water resources [e.g., Ang and Tang, 1984]. Few publications, however, have investigated reliability issues for design and operation of water distribution systems under random demand variations. Bao and Mays [1990] suggested using Monte Carlo simulation techniques to evaluate the reliability of steady state hydraulic performance under random variations of demands and pipe roughness coefficients. They proposed using an arithmetic average and demand-weighted average of nodal pressure reliability as the system reliability.

The resiliency is defined as the inverse of the expected time of recovery, after a failure in hydraulic and water quality system performance. Hashimoto et al. [1982] developed the original resiliency concept for evaluation of water resource systems. They proposed a statistical definition of resiliency for an extended time period based on the concept of average sojourn time of failures. To our knowledge, resiliency measures have not been investigated for design and operation analysis of water distribution systems under random demands.

A reliability and resiliency analysis framework is proposed to evaluate the probabilistic hydraulic and water quality performance in water distribution systems subject to random demands. This work is the first to consider, in a preliminary way, the time-varying distribution system performance in a probabilistic framework, and to consider the joint probabilistic hydraulic and water quality performance.

Reliability Analysis

A random demand is defined for each network node of interest and each time interval. At each location and time, the demand is described by its mean, variance, and covariance with respect to other nodal demands. Mathematically, a nodal pressure or concentration at a location and time is defined as an implicit function of the random demands:

$$x_i^k = g\left[(D_1^1, D_2^1, ..., D_{nn}^1), (D_1^2, D_2^2, ..., D_{nn}^2), ..., (D_1^k, D_2^k, ..., D_{nn}^k)\right] \qquad (1)$$

where, x_i^k denotes the pressure or concentration at location i and time step k, nn denotes the number of locations (nodes) where demands are applied in the system (these are the same locations where hydraulic and water quality performance is evaluated), and D_j^m is the demand at the jth node and mth time step. The function g depends on the random demand D_j^m, for all nodes j and all times $m \leq k$, and is defined implicitly by the conservation laws for fluid and mass applied to the network [e.g., Uber et al., 1992].

A failure of the system occurs if a nodal pressure, x_i^k, is lower than needed for adequate water delivery. Also, a failure of the system occurs if a contaminant concentration exceeds a target value based on potential adverse effects to human health. Therefore, probabilistic performance criteria should consider both water delivery and public health objectives. These criteria are expressed mathematically:

$$x_i^k \in S \qquad i = 1,...,nn, \; k = 1,...,nt \qquad (2)$$

where, nt is the number of discrete time steps for the entire operating time period, and S is defined as a set of all satisfactory pressures and concentrations that meet the requirements, such as a minimum pressure or a maximum contaminant level (MCL), for all locations and times. A set of failures F is defined analogously as the set of all unsatisfactory pressures and concentrations. The minimum pressure or MCL is considered to be deterministic and depends on public policy.

From the definition of reliability [Ang and Tang, 1984], a hydraulic or water quality nodal reliability, R_i^k, at node i and time step k is defined:

$$(R_i^k) = \int_S f(x_i^k) dx_i^k \qquad (3)$$

where, f denotes the probability density function (PDF) of random pressures or concentrations. Eq. (3) is an integration over the safe region S.

The nodal reliability may be used to identify specific locations and times of special concern. The *system* reliability, however, provides a lumped measure of overall system performance. The system reliability for a *short-term* operation period is based on the joint probability distribution of $x_i^k \; \forall i,k$, and is preferred to the use of the simple arithmetic average of nodal reliability [Bao & Mays, 1990]. For example, the arithmetic average of nodal reliability could be very high, although one node had consistent failures.

Fig. 1 illustrates the case of two random pressures or concentrations in time period k, x_1^k and x_2^k. The volume under the joint PDF over the safe area is equal to the hydraulic system reliability. Generally, the random variable space exceeds two dimensions. However, the concept of defining the system reliability is simple.

The system reliability, R_r, considering the entire spatial and temporal domain of random pressures or random concentrations, is defined:

$$R_r = \int_S \cdots \int f(x_1^1, \ldots, x_i^k, \ldots, x_{nn}^{nt}) dx_{nn}^{nt} \ldots x_i^k \ldots x_1^1 \tag{4}$$

where, $1 < i < nn$, $1 < k < nt$, and f is the joint PDF of the pressures, concentrations, or both at all nodes and for all time steps. Eq. (4) defines the "volume" of the joint probability distribution over the safe region, and hence the system reliability.

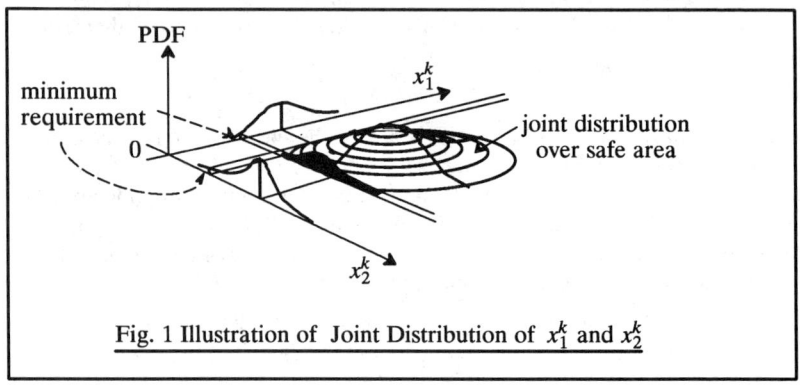

Fig. 1 Illustration of Joint Distribution of x_1^k and x_2^k

Resiliency Analysis

Reliability measures the probability or frequency that a system is in a safe state. However, reliability does not describe the consequences of a failure. For example, a prolonged failure may be more serious than an instantaneous failure. Thus, it is preferable to design or operate a system that can quickly recover to the safe state once a failure occurs. Resiliency is an measure that describes the system recovery in a probabilistic way. In this work, the resiliency measure defined by Hashimoto [1982] is developed for distribution systems.

Hashimoto et al.[1982] proposed the probabilistic definition of resiliency by the conditional probability:

$$\gamma_i = P[(x_i^{k+1} \in S) | (x_i^k \in F)] \approx \frac{\sum_{k=0}^{nt} [(x_i^k \in F) \text{ and} (x_i^{k+1} \in S)]}{\sum_{k=0}^{nt} [(x_i^k \in F)]} \tag{5}$$

where, γ_i is the nodal resiliency at node i. Eq. (5) becomes an equality as the number of time steps nt approaches infinity. From eq. (5), the inverse of the resiliency γ_i is the total time spent in the failure state at node i, divided by the number of transitions from the failure to the safe state. Thus, the inverse of resiliency equals the average failure duration (as $nt \to \infty$).

A water distribution system contains a number of nodes nn. The *system resiliency* of a water distribution system for a long time period is further defined as:

$$\gamma \approx \frac{\sum_{i=1}^{nn} \sum_{k=1}^{nt} [(x_i^k \in F) \text{ and}(x_i^{k+1} \in S)]}{\sum_{i=1}^{nn} \sum_{k=1}^{nt} [(x_i^k \in F)]} \tag{7}$$

Although these resiliency measures may be used to study long-term system behavior under random demands, the resiliency for a short time period may also be used to guide operational adjustments of a water distribution system. The operator may use short-term reliability and resiliency information to compare and select alternative operation schemes for the next short-term operation period, given the current hydraulic and water quality conditions. Thus, resiliency evaluation for a short time period is proposed, using multiple realizations to estimate a statistical average resiliency. The nodal resiliency and system resiliency can then be computed by:

$$\gamma_i \approx \frac{\sum_{j=1}^{nr} \sum_{k=0}^{nt} [(x_i^k \in F) \text{ and}(x_i^{k+1} \in S)]}{\sum_{j=1}^{nr} \sum_{k=0}^{nt} [(x_i^k \in F)]} \tag{8}$$

$$\gamma \approx \frac{\sum_{j=1}^{nr} \sum_{i=1}^{nn} \sum_{k=0}^{nt} [(x_i^k \in F) \text{ and}(x_i^{k+1} \in S)]}{\sum_{j=1}^{nr} \sum_{i=1}^{nn} \sum_{k=0}^{nt} [(x_i^k \in F)]} \tag{9}$$

where, nr denotes the number of independent demand realizations. If a failure has not recovered by the end of the specified short time period, eqs. (8) and (9) will loose track of this failure and the estimated resiliency of the system may not represent the true system resiliency. Thus the proper length of realizations for short time period resiliency analysis should be investigated. Also, short-term reliability or resiliency measures run the risk of guiding operation decisions in a "greedy" manner that compromises long-term objectives.

Summary

Future demands in water distribution systems are difficult to predict accurately. The work conducted by Bao and Mays [1990] demonstrated that the sensitivity of the hydraulic performance with respect to the demand variation can be large. Reliability and resiliency evaluation may help to understand the effects of uncertainty in demands during design or operation.

The developments in this paper provide a preliminary set of techniques to assess the future performance of a water distribution system, in terms of hydraulic and water quality objectives. The short–term reliability and resiliency analysis can be used to estimate the effects of random demands, and as a tool to select the proper operation scheme for the next regular operation period. The long–term reliability and resiliency, on the other hand, provide information on the long–term effects of random demands, and therefore may help design a water distribution system that operates within acceptable statistical limits.

The reliability and resiliency framework has been applied to a small hypothetical water distribution system to quantify the stochastic hydraulic and water quality performance. Large variations in system performance have been observed for small variations in demands, indicating the importance of addressing the demand influence. The application suggests that this tool could be applied for design of elevated storage tanks (locations and volumes) or for issues involving tank operations. The preliminary results have also showed a conflict between reliable hydraulic and water quality performance.

Future work will emphasize the effects of long–term and short–term demand variability on combined hydraulic and water quality performance, including the spatial and temporal correlation in demand variability. Applications to real systems will be investigated.

Reference

[1] Ang, A. H. and Tang, W. H. *"Probability Concepts in Engineering Planning and Design," Vol. II,* 1984, Wiley, New York.
[2] Bao, Y. and Mays, L. "Model for Water Distribution System Reliability," *J. of Hydraulic Engineering,* Vol 116, No. 9, Pages 1119–1137.
[3] Hashimoto, T., et al. "Reliability, Resiliency and Vulnerability Criteria for Water Resource System Performance Evaluation," *Water Resources Research,* Vol. 18, No. 1, Pages 14–20, Feb., 1982.
[4] Uber, J., Hickey, K. and Fang, M. "Dynamic Plug Flow Reactor Network Model for Contaminant Transport in Water Distribution Systems," *Proceedings, Water Forum '92,* ASCE, Aug., 1992

The Seattle Forecast Model:
A Tool for Water Resources Management in the Seattle Area.

Laura Mariño[1], Member ASCE

Abstract

The Seattle Forecast Model, SEAFM, developed by Hydrocomp Inc. for the Seattle Water Department, is an interactive, graphic system for streamflow forecasting and reservoir analysis of the South Fork Tolt and the Cedar River basins in Washington. It was developed as a tool to aid in day to day operation and in long term operation planning of the reservoirs. The system has a historic data base with 63 years of observed meteorologic data, which is used in streamflow forecasts as well as in design-type analysis. Additionally, the system offers probabilistic analysis capabilities, to estimate probabilities of future reservoir states and water deficits, taking into account the uncertainty of future weather. Combined with interactive graphics, probability analysis helps the public understand the risk of future water shortages and the need for new facilities.

Introduction:

Population growth, urban encroachment on flood plains and increased environmental awareness are making the work of water supply agencies more difficult. Decisions such as water rationing and water releases during floods can have significant economic and social impacts and are easily criticized in hind-sight. To make the best possible decisions, water agencies must analyze all the available information. Current or "real-time" hydrologic and meteorologic data, historical data and regulatory and environmental constraints must all be at hand. However, having the information is not enough. Decision makers need analysis tools to evaluate the behavior of the system under various operational schemes, taking into account the uncertainty of future weather. In addition to being accurate, these tools need to be fast and easy to use.

One such tool is the Seattle Forecast Model, SEAFM, developed by Hydrocomp Inc. for the Seattle Water Department. SEAFM is an interactive, graphic system for streamflow forecasting and reservoir analysis of the South Fork Tolt and the Cedar River basins in Washington. It was developed as a tool to aid in day to day operation and in long term operation planning of the projects. The user can make streamflow forecasts specifying various operational schemes for the reservoirs or can let the program adjust the operation of the reservoirs automatically to follow rule curves and meet water demands and minimum instream flow requirements. The 63-year long

---
[1]Hydrocomp Inc. Three Lagoon Drive, Suite 150, Redwood City, CA 94065

historic meteorologic data base can be used to make design-type runs to analyze the effects of changes in operating policies on reservoir reliability and flood frequency. With the probabilistic analysis capabilities of the model the user can analyze the probabilities of water shortages throughout the summer for various levels of demand. This paper describes the Seattle Forecast Model and its application during the 1992 drought.

Description of the Seattle Water Department Projects:

The Tolt and Cedar reservoirs are the primary water supply sources for the metropolitan Seattle area and they also provide hydroelectric power for the region. Even though their main purpose is water supply, the Seattle Water Department operates the projects taking into account flood control and fish flow requirements downstream.

The Cedar Falls project is located on the west slopes of the Cascade Mountain chain, 40 miles southeast of Seattle. The project is formed by two dams and two different lakes that become one with high water levels. The dams are the Masonry Dam and the Overflow Dike, which divides the reservoir into Chester Morse Lake and Masonry Pool. The Overflow Dike allows Masonry Pool to be drawn down to minimize seepage losses, which were high due to the location of the Masonry Dam on a glacial moraine. Part of the seepage from the reservoir enters the Snoqualmie River, and part returns to the Cedar River further downstream.

Fourteen miles downstream of Masonry Dam water for the City of Seattle is diverted from the Cedar River to Lake Youngs at a gated diversion dam located at Landsburg. The Cedar River is one of the most productive streams in the area with a large salmon run. The Department of Ecology has recommended minimum fish flows at Renton, where the river enters Lake Washington, and further downstream at the Washington locks. The Seattle Water Department adjusts the operation of the Cedar Falls project to meet the demands at Landsburg and the minimum fish flows.

The South Fork Tolt River was developed by the Seattle Water Department in 1963 to supply a portion of King County's demand for water. The dam is located in the Cascade Mountains, 30 miles east of Seattle. The dam has a Morning Glory spillway and an intake tower through which water for water supply and instream fish requirements is extracted.

The active storage of both reservoirs is small, relative to the mean annual inflow: less than 15% in the case of the Chester Morse Reservoir on the Cedar River and less than 40% for the South Fork Tolt Reservoir. Because the reservoirs are small, efficient reservoir operation requires accurate inflow forecasts. Of concern are both, short term forecasts on the order on hours to days and seasonal forecasts on the order of several weeks to several months.

The Seattle Forecast Model:

The Seattle Forecast Model consists of a set of physically-based, continuous, deterministic models that simulate all the hydrologic and hydraulic processes that occur in the basins. The main components of the system are a hydrologic simulation model, a river-reservoir model and an aquifer model. The hydrologic simulation model calculates runoff and snowmelt in the basin, from meteorologic information. The aquifer modeling component estimates seepage into the Cedar basins Moraine Aquifer and determines spring flows to the river. The river and reservoir modeling component routes the flow in the rivers and simulates the operation of the reservoirs.

The hydrologic component is based on HSPF, the Hydrologic Simulation Program - Fortran, developed by Hydrocomp for the EPA. HSPF is a continuous, conceptual model that simulates all the processes that occur on the land surface and in the subsurface. For the hydrologic simulation the basins are divided in hydrologically homogeneous land segments. Each segment is simulated independently using local weather information, such as precipitation, temperature, solar radiation, evaporation and wind. The hydrologic processes simulated include: snow accumulation and melt, interception of moisture by vegetation, overland flow and interflow, infiltration and soil moisture, actual evapotranspiration and surface and subsurface runoff. These processes are simulated continuously on an hourly time step.

Outputs from the hydrologic simulation include, for each land segment, snow depth, snow water equivalent, snow melt, soil moisture and total runoff. A daily summary display is also available for each segment. The display shows minimum and maximum temperature at the segment, precipitation and snow water yield, snow water equivalent and total runoff.

Runoff from the land surface is routed through the river channel network. Simulation of the flow in the rivers is done by reaches, using a modified version of the kinematic wave equation for routing. For each reach the model determines flow, reach volume and stage.

The model simulates the operation of the reservoirs in detail: the outflow through each outlet is simulated independently, based on the demand, the hydraulic capacity of the outlet and the elevation of the reservoir. In the case of the Cedar River basin the operation of the reservoir is adjusted to meet water demands at Landsburg and minimum flow requirements at Renton, and to limit reservoir elevation to monthly maximum allowable values for flood control. Simulation of the Tolt reservoir is similar but less complex than that of the Cedar reservoir. The program calculates flow through the low level outlet based on user supplied demand time series. The outflow of the Morning Glory spillway is a function of elevation of the reservoir, and the flow through the needle valve is a function of the demand and the level of the reservoir.

For each reservoir the model determines total outflow, elevation, volume, seepage, flow through the low level outlet, flow through the power house, power generation, total spill and spill through each of the spillways. The user can also look at a summary of these results that includes total outflow, total spill, volume and elevation.

Seepage from the Cedar Reservoir into the Moraine Aquifer is calculated using equations developed by the U.S. Geological Survey. The Moraine Aquifer is represented by three aquifer elements. The aquifer component uses a general water budget approach to determine, for each aquifer element, groundwater storage, groundwater flow into the down-gradient element and springflows into the river.

Extensive results are produced during a simulation. The user may choose to look at detailed hourly results for portions of the basin, or may concentrate on summary displays that display on one screen the most relevant information for the basin. Basin summary displays show the various components of outflow from reservoirs and the flow at specific locations down the river. The total water content of snowpacks, soil moisture storages and aquifers upstream from any reservoir or location on the river can be displayed.

Using the Seattle Forecast Model:

SEAFM offers three types of runs: a Forecast run, a Calibration/Analysis run and a Probabilistic Forecast run. In a forecast run the model determines future reservoir state and future runoff and streamflows throughout the basin, given the current conditions of the basin and a meteorologic forecast.

Meteorologic data for the model come from three data files: real time, forecast and historic. In a forecasting session the user will choose the combination of meteorologic data to use. In general, the model will have information about the state of the basin as of a few days ago. The simulation will start at that day and will use any available meteorologic real-time data collected after that date. For the rest of the forecasting period the model will use a combination of forecasted and historic data as specified by the user. For short term forecasts enough real time data and forecasted data may be available, but for long term forecasts it will be necessary to use historic data as representative of future weather. Depending on the purpose of the simulation the user can choose an average, a very wet or a very dry year from the historic data base.

Once the meteorologic data are selected, the program will simulate the basin (either the Cedar or the Tolt) and will produce results. The user can review the results, change the operation of the reservoir or the demand time series, and run a new simulation. Short-term simulations, on the order of days to a couple of weeks, take only a few seconds to run. Many different operation schemes can be compared in a short time. This is especially important during flood situations.

The Calibration/Analysis run is similar to a regular Forecast run, but it simulates a period of time in the past using historic meteorologic data, and it displays observed streamflows together with the simulated streamflows. This type of run can be used for design purposes. For example, it is possible to analyze the effect of increasing the capacity of the reservoir on water yield during the 63 years of historic meteorologic data that are available.

Probabilistic Forecast runs show future reservoir states and risks of water shortages, taking into account the uncertainty of future weather. The purposes of the Probabilistic Forecast run are:
- to estimate the range of future hydrologic responses that can be expected in the period being forecasted, given the current conditions in the basin; and
- to quantify the probability of occurrence of those responses.

The model assumes that the 63 years of observed meteorologic data provide a representative sample of the range of possible weather conditions that may occur in the future. It makes 63 simulations of the forecast period specified, using meteorologic data from all the available historic years sequentially. For each historic year, the model saves daily results. Then, the program ranks all the daily values, determines the minimum and maximum for each day for each variable and calculates the exceedance probability of the ranked values for each day.

Three different groups of results are produced during a forecast run: Expected Minimum and Maximum Values, Exceedance Probabilities and Results by Historic

Year of Meteorologic Data Used. These results are available for: reservoir elevation and volume, cumulative inflow to the reservoir, cumulative release from the reservoir, cumulative spill and cumulative deficit. The cumulative values are accumulated from the beginning of the forecast period. Probabilistic results are also available for deficits and peak flows at Landsburg and Renton on the Cedar River.

Expected minimum and maximum values are found after ranking the daily values obtained in the sequence of simulations with the data from all the historic years. The curves produced are not associated with one specific historic year of data. Rather, they are made up of results from many different years and they delimit the range of values of the variable that can be expected during the forecast period.

Exceedance probability plots show the probability distribution of values between the maximum and the minimum, for any day selected within the forecast period. For example, the user can look at the probability of filling the reservoirs by a given date and the probabilities of having water shortages at the end of the summer/fall season.

Despite being technically comprehensive, the Seattle Forecast Model is very easy to use because it includes a powerful graphic interface to communicate with the user. The system is completely interactive, menu-driven and largely self documenting. Input to the model is by menu item selection or by typing in pop-up windows. Color graphical displays of data and results are used extensively. The user can display time series graphs of all the model inputs as well as results, and modify the water demands and the operations of the reservoirs graphically on the screen.

Analysis of the 1992 Drought:

Water year 1992 has been one of the dryest years on record in the Seattle area. Little snow fell in the late winter, and by March it became clear that there was a possibility of water shortages by the end of the season. SEAFM provided a unique tool for analyzing the situation and evaluating water management alternatives.

Figure 1. shows the probability distribution of cumulative deficit in the Cedar basin by the 31st of December of 1992, for a probabilistic forecast run that started the 1st of March. The plot shows a 12% chance of having a deficit of 5,000 acre-ft[2] or more between March and December, assuming water demands between March and December remained at the average levels of the previous three years and the minimum flow requirements at Renton were not relaxed. Precipitation in March and April was low and by May 1st. the probability of a deficit greater than 5,000 acre-ft increased to 22%. In view of the situation, the Seattle Water Department decided to impose mandatory water use restrictions in

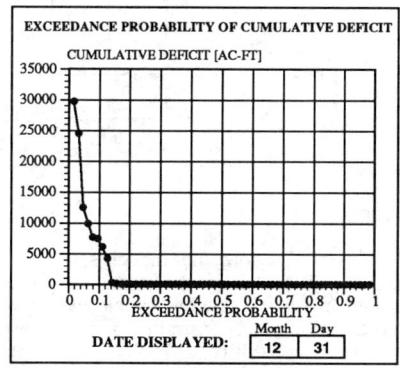

FIGURE 1.

---

[2] 1 acre-ft = 1,230 cubic meters

May. The water demands were reduced to between 70% and 75% of the average values and the flow requirements at Renton were reduced to critical year values. With these measures in place from May 1st to December 31st the risk of having any water shortage decreased to less than three percent (Figure 2.).

Due to persistent low precipitation in the Cedar River basin and the small snow pack, inflows to the reservoir in May and June were a fraction of historic averages, and reservoir elevation fell steadily, in spite of rationing. By the 1st of July probabilistic analysis showed that, with continued rationing, there was a 5% chance of a deficit of 1,000 ac-ft or more. If demands and flow requirements at Renton had been allowed to return to normal levels, given the conditions of the basin in July, the probabilities of water shortages between July and December would have increased to more than a 20% chance of a deficit equal or greater than 10,000 ac-ft.

By October the storage levels of the Seattle Water Departments reservoirs had recovered considerably, due to storms in September and to the success of the water rationing program. Probabilities of deficits between October 15 and December 31, assuming normal demands and flows at Renton, had decreased to less than 4% for a deficit of 500 ac-ft or more. Toward the end of October all water use restrictions were lifted.

Figure 2. shows exceedance probabilities of cumulative deficit as evaluated at different times of the year, with and without rationing.

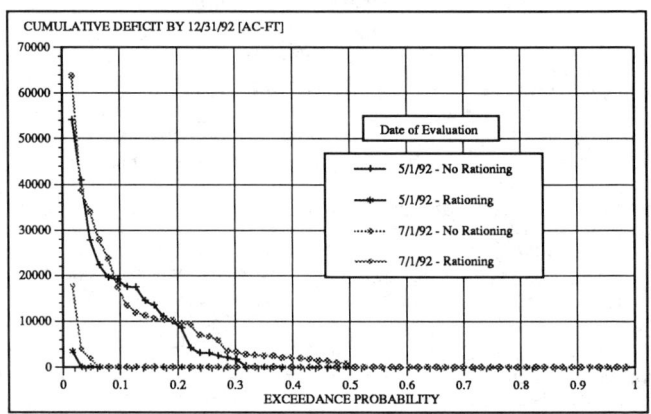

FIGURE 2.

Conclusions:

Interactive computer models such as the Seattle Forecast Model are valuable tools for forecast and analysis of water supply and hydropower generation systems. They include all the relevant hydrologic and meteorologic information and all the operational and environmental constraints, and provide a way to deal with the uncertainty of future weather. Probability analysis of future reservoir conditions allows a water utility to better schedule repairs and to implement, when necessary, water supply restrictions and rationing. Combined with interactive graphics, probability analysis helps the public understand the risk of future water shortages and the need for new facilities.

Modeling Vertical Stratification in the Pamlico
River Estuary Using Modern Hydrodynamics

Alan F. Blumberg, C. Kirk Ziegler, Bradley Nisbet

HydroQual, Inc.
1 Lethbridge Plaza
Mahwah, New Jersey 07430

ABSTRACT: A modern hydrodynamic model is applied to the Pamlico River estuary. Good agreement was obtained between measured and predicted water elevations and salinity distributions during the one year long (1991) model calibration/validation period.

Recent studies in the Pamlico River Estuary, North Carolina of the relationship between bottom-water dissolved oxygen and vertical stratification by Stanley and Nixon (1992) strongly suggest that there is a tight coupling between the hydrodynamic characteristics and observed DO levels. To quantify these relationships, a dissolved oxygen model, to be implemented in the near future, will be driven by transport derived from a modern hydrodynamic model that has been developed and calibrated for this dynamically active estuarine system. The purpose of this paper is to describe these efforts and to determine how well a modern hydrodynamic model, with sufficient spatial resolution, can reproduce the physics of complex estuaries.

The hydrodynamic modeling framework is based upon the time-dependent, three-dimensional estuarine, coastal and ocean circulation model (ECOM-3D) developed by Blumberg and Mellor (1987). ECOM-3D solves prognostic equations for free surface elevation, velocity components, temperature, salinity, turbulence energy and turbulence macroscale. All the equations are written in a curvilinear, coastline-fitted coordinate system combined with the free-surface and bottom following σ-coordinate. The model uses an imbedded, turbulence closure sub-model to provide vertical mixing coefficients for momentum, temperature and salinity.

A high spatial resolution grid, both in the horizontal and vertical, is employed to resolve the important processes operating in the Pamlico River Estuary, see Figure 1. The horizontal, along-river grid spacing is about 2 km, depending upon local bathymetry and geometry. The vertical direction is resolved by 8 levels. The entire year of 1991 was chosen for analysis because of the availability of high quality forcing data and long term deployments of continuous surface and bottom temperature and salinity. These data, provided by the U.S. Geological Survey, included six water level stations, six temperature and salinity measurement stations, and one wind measurement station. Figure 2 illustrates the temporal variability of stratification at one station. All of these observations were combined with measurements of salinity, collected by East Carolina University personnel, at 20 sampling stations every other week. Good agreement was obtained between measured and predicted water elevations and salinity distributions during the one year long model calibration/validation period, see Figure 3. The model was able to replicate periods of vertical stratification caused by either high fresh water inflow or low winds. Well-mixed water column conditions were found to be primarily correlated to higher wind velocities.

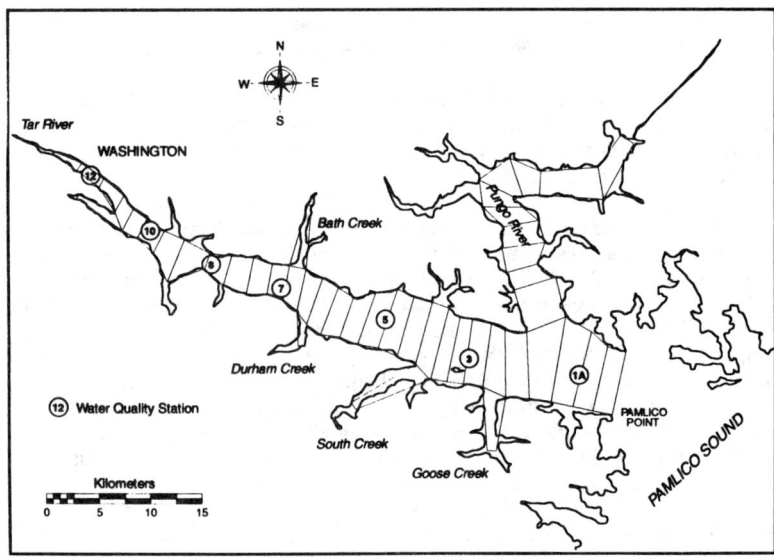

Figure 1. Pamlico River modeling domain.

MODELING VERTICAL STRATIFICATION 2299

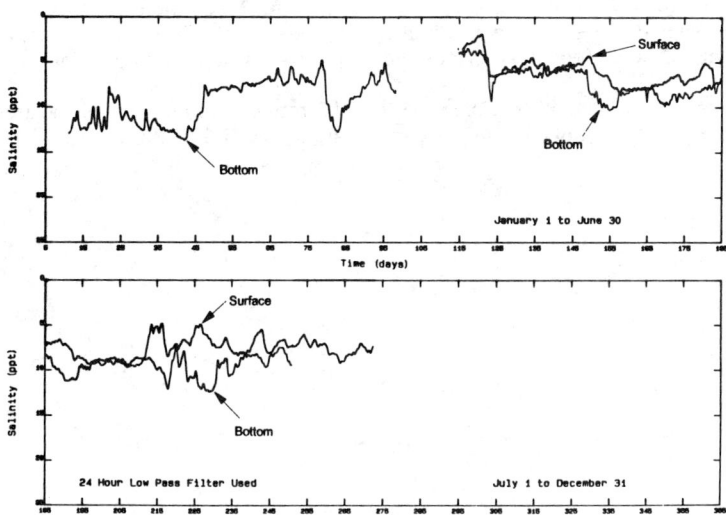

Figure 2. USGS salinity measurements at Station 7 during 1991.

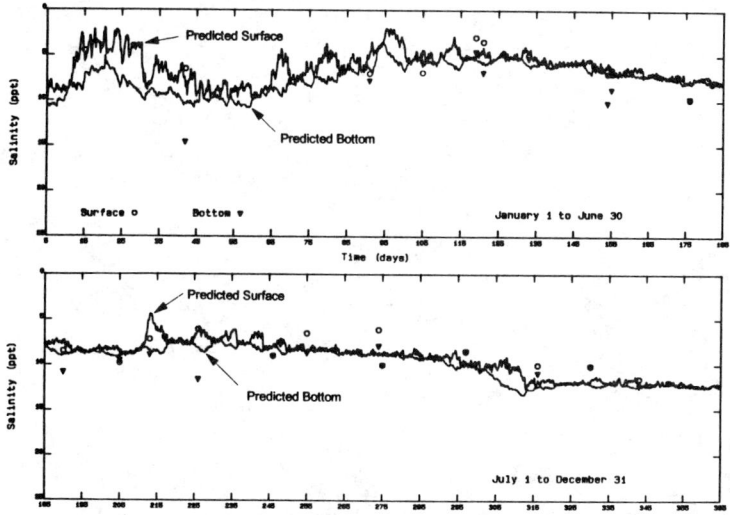

Figure 3. Comparison of predicted and observed salinity at Station 7 during 1991.

References

Blumberg, A.F. and G.L. Mellor, 1987. "A Description of a Three-Dimensional Coastal Ocean Circulation Model," In: Three-Dimensional Coastal Ocean Models, N. Heaps, Ed., 1-16, American Geophys. Union.

Stanley, D.W. and S.W. Nixon, 1992. "Stratification and Bottom Water Hypoxia in the Pamlico River Estuary," Estuaries, 15:270-281.

Three-dimensional Modeling of Tides and Wind-Waves

Shiao-Kung Liu[1]

Abstract

A third-generation coupled discrete numerical wave prediction model is dynamically linked to a three-dimensional hydrodynamic model. The effects of water level, currents, and bottom friction on the local wave climate are calculated by solving the equations for directional spectral energy in conjunction with the equations of the mean flow. The paper also discusses the following aspects of three-dimensional modeling.

- The dynamic coupling of turbulent energy densities and the closure computation between the wave field and the mean flow field.
- Comparison of tidal prediction accuracy between the harmonic method and the hydrodynamic method.
- Effects of currents, depth, friction, coastal configuration, change of wind direction on the refraction and the spectral distribution of waves.

Coupled 3-D Model/Numerical Wave Prediction Computations

In the three-dimensional model, the generation, propagation and dissipation of surface wind-waves are represented according to the evolution of directional wave spectrum energy balance. The momentum transfer across the marine boundary layer and interactions in the vertical oceanic density structure are dynamically coupled during the numerical integration process. Using $F(f,\theta,x,y,t)$ to represent the wave energy in a spectral component with frequency f, travelling in direction θ, at location (x,y), time t, for water depth H, group velocity c_g, and propagation direction, the wave energy balance equation is

$$\frac{\partial F}{\partial t} = S(f,\theta,x,y,t) - \nabla \cdot (c_g F) + \partial\{(c_g \cdot \nabla \theta)F\}/\partial\theta$$

---

[1]Systems Research Inst., Malibu, CA. 90265

in which

$$[[[\partial F_{ij}/\partial t = -\nabla \cdot (c_g F_{ij})] - \partial\{(c_g \cdot \nabla\theta)F_{ij}\}/\partial\theta] + S_{in} + S_{ds}] + S_{nl}]$$

where

F_{ij} = wave energy ($m^2 Hz^{-1} rad^{-1}$) in a component of frequency f_i Hz and direction θ_j;

c_g = $c_g(f,\theta)$ is the group velocity of the spectral component; ($c_g \cdot \nabla\theta \equiv d\theta/dt$);

S_{in} = wave energy input including linear and exponential growth terms;

S_{ds} = wave energy dissipation terms;

S_{nl} = wave energy transfer due to nonlinear wave–wave interaction.

The dynamic coupling between the three–dimensional hydrodynamic model and the surface wave model are carried out by means of the following computations:

- Marine boundary layer's stability on the momentum transfer process;
- Wave growth and decay due to density and SGS turbulence level;
- Current's effects on group velocity in the propagation of wave energy;
- Refraction and bottom stress effects.

To incoporate the above dynamic effects, for wave energy growth and decay, the linear growth (based on the resonance model) and the exponential growth of wave energy (based on the shear–flow model) are considered. The decay term is proportional to the local sub–grid–scale energy in the mean flow and the wave energy integrated over the frequency bands and over the directions

$$S_{ds}(f,\theta) = 4 \times 10^{-4} f^2 F(f,\theta) \left[\int\int F(f,\theta) d\theta df\right]^{0.25}$$
$$+ 0.005[gkc_g/\{2\pi(2\pi f \cosh kH)^2\}$$
$$\cdot \left[\int\int (gk/2\pi f \cosh kH)^2 F(f,\theta) df d\theta\right]^{0.5} F(f,\theta)$$

The second and third terms represent the dissipation due to quadratic bottom friction. In the computation of wave propagation and refraction, local water level variations due to tides, and storm–surge are incorporated in the calculation. Group velocity components are modified by the local current velocity components. These currents are also used in the computation of the combined bottom–stress effects due to local currents and waves. We use several figures here to illustrate some of these effects.

Temperoal variation of local significant wave heights due to the effects of tidal currents and water level when wind blows toward the upstream direction in a tidal estuary is illustrated in Fig. 1. Higher waves are generated during ebb tide than flood tide. Higher waves are generated during high tide than at the low tide. The effects of changing wind direction on the directional spectrum are illustrated by figures 2a and 2b. As shown in 2b, after the wind change, the original wave field gradually evolves into travelling swells concentrated within a narrow-band low frequency range with somewhat less directional spreading. The effects of fetch, depth and bottom friction are illustrated in Fig 3a. Under a SW monsoon, the maximum and the minimum waves are found on either sides of the island in the model. The direction of waves lean toward the coasts are due to refractional processes. In Fig. 3a, the length of an arrow represents the magnitude of the significant wave height while the direction of an arrow represents the prevailing wave direction. Fig. 3b shows the growth of energy spectra at a selected location in the model. Using models of simple geometries and uniform wind speed (e.g., 20 m/s), results of the numerical model agree with data sets which are often used for wind-wave model verification. Presently, data from four wave gages are being used for further tests.

Prediction of Astronomical and Atmospheric Tides

The coupled model can also be used for makeing tidal predictions when proper boundary conditions are applied at the open boundaries. Using a model of the SE China Sea as an example, nine water level stations are being used for model verification on tidal prediction and typhoon surge conputation. Fig. 4 shows comparisons between the official published forecasts for coastal areas and the observed values at two nearby harbors for the month of January, 1992. The model was driven at the boundaries using 20 major deep-water tidal components. Considering all nine stations, harmonic method (using 61 components) is approximately 15 minutes more accurate then the hydrodynamic method presented here. Transfer functions are being used to correct the amplitude and phase relationship between the coastal model grid and the measurement point which is usually located in a harbor.

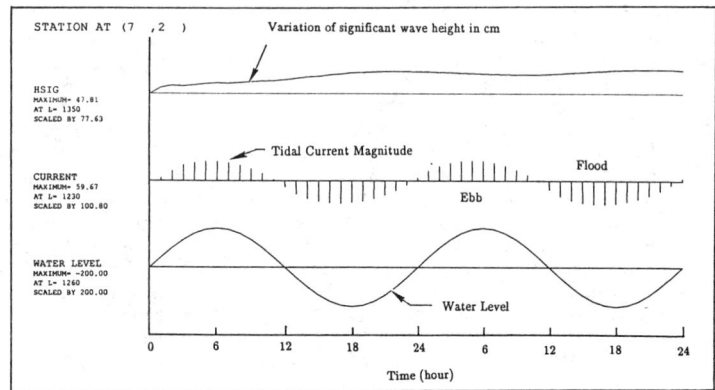

Fig. 1 — Temporal variation of local significant wave heights due to the effects of tidal currents and water level when wind blows toward the upstream direction in a tidal estuary.

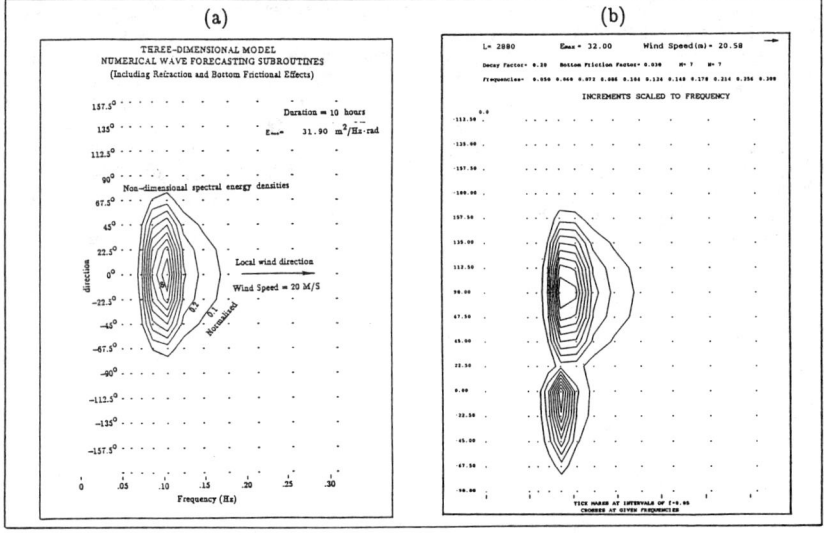

Fig. 2 — The effects of changing wind direction on the directional spectral energy density distributions are illustrated by figures a and b. After 10 hours, southerly wind (20 m/s, figure a) changes its direction from south toward the east for another 36 hours (figure b).

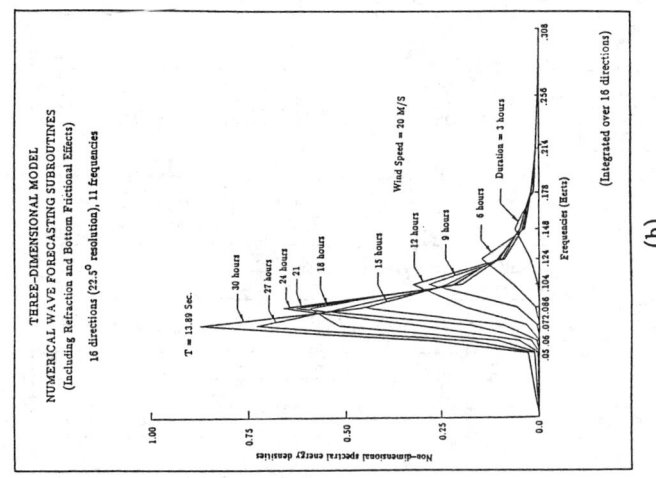
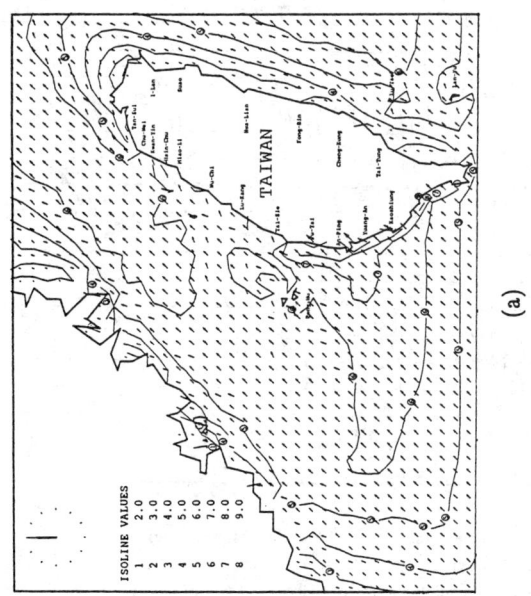

Fig. 3 — The model has been applied to the SE China Sea. Figure 3a shows the effects of fetch and depth on the distribution of magnitude/direction of significant wave heights under SW monsoon. Fig 3b shows the growth of energy spectrum at a selected location in the model.

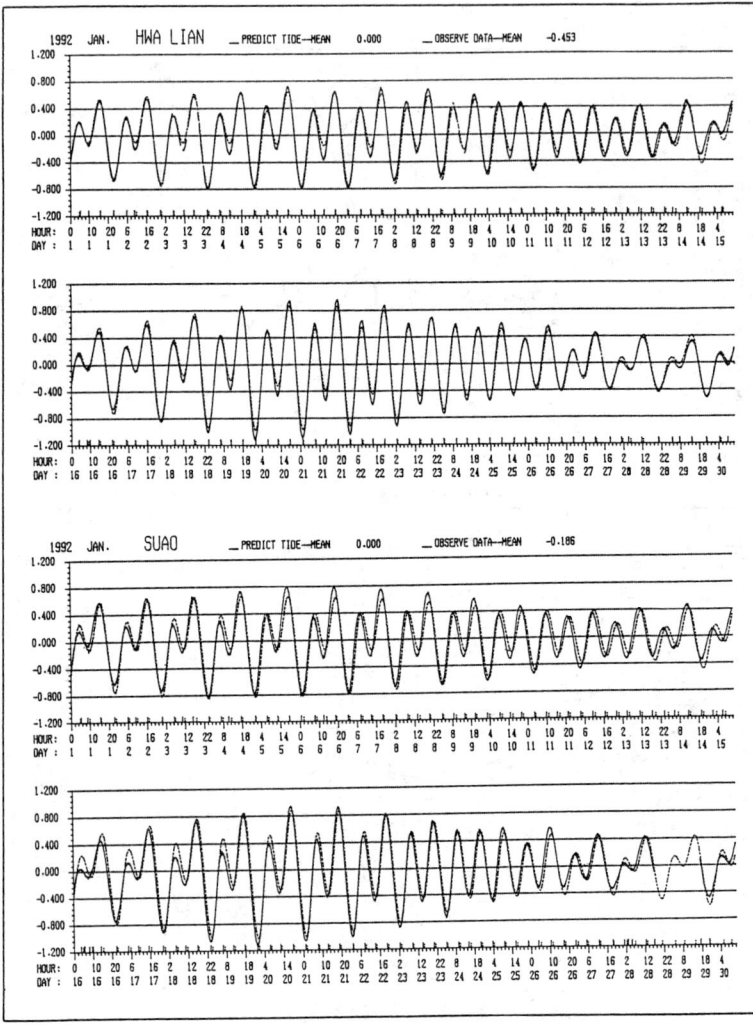

Fig 4. — Comparison between the predicted and the observed tides using the model for the month of January, 1992. Water level predictions which incorporate marine wind predictions are published in news papers one week ahead of time. Comparisons are made in the subsequent month.

Three-dimensional modeling of the coastal region offshore from Sydney, Australia

Ian P King[1] M ASCE, William L. Peirson[2] and Bruce A. Cathers[3]

INTRODUCTION

As part of an environmental monitoring program to study the impact of an offshore waste outfall in the vicinity of Sydney Australia, this project is being undertaken to investigate the various driving forces in the region and how they interact to produce circulation. Of particular interest is wind driven circulation in the stratified environment, the influence of Coriolis forces and the response to gravity waves and the effects of prototype bathymetry and complex coastal detail. The ultimate objective is the ability to track the movement of the contaminant plume and associated floatables under different climatic conditions.

RMA-10 is a three-dimensional finite element model designed for simulation of stratified flow in estuaries, lakes and near shore regions. Unique features of this model permit different types of approximations (two-dimensional depth and laterally averaged and fully three-dimensional) to be incorporated into a single simulation.

This paper will present a brief description of the major features of the model, describe the steps being used to evaluate the model for this area and discuss some of the modeling difficulties encountered. Results will be presented showing details of the circulation as they were developed. The support of the New South Wales Environmental Protection Agency is Gratefully acknowledged.

PROJECT SETTING

The coastal region near Sydney in Australia is characterized by an along shore current and an irregular shoreline with large bays. The bathymetry is marked by a relatively uniform slope from the beach along the shoreline to a depth of approximately 70 m. 2 km offshore then a relatively horizontal for approximately 6 km. Temperature stratification can persist for weeks but it is interrupted by sudden de-stratification events. Moderate winds and Coriolis forces also have a significant

---

[1] Professor of Civil and Environmental Engineering, University of California, Davis, CA.
[2] Research Engineer, Australian Water and Coastal Studies Ltd., Sydeny, Australia
[3] Lecturer, Department of Civil and Mining Engineering, University of Wollongong, Wollongong, Australia

impact on circulation. A single continuous recording of vertical temperature distribution, local weather conditions and limited current data is available from a reference station (ORS) located 2km offshore. Figure 1 shows the general layout of the system with a modeling grid superimposed. A typical daily temperature pattern as measured at theORS is shown in Figure 2.

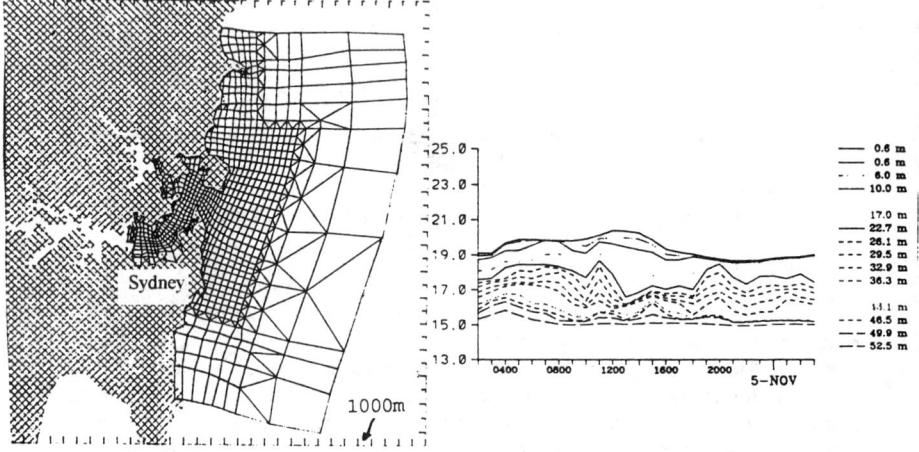

Figure 1. Project Setting

Figure 2. ORS thermistor data 4 and 5 Nov. 92

MODELING QUESTIONS

The project is being undertaken to address the following general questions. How well can models represent the current regime in this area? Can models be used in a predictive fashion to control operation of the diffuser? Specific modeling questions include: How often is a two-dimensional model appropriate and applicable? How often would wind cause surface transport of floatables onto the shoreline? How often would a surfacing plume be transported onshore? How often would a submerged plume be transported onshore?

MODELING APPROACH

RMA-10 was selected for this study. This finite element model solves the primitive form of the shallow water and advection diffusion equations. It uses assemblages of three-dimensional elements (generalized cubes, rectangular prisms and tetrahedra), two-dimensional elements (laterally and depth averaged quadrilaterals and triangles), and one dimensional elements (cross-sectional area averaged line elements).Through equations of state relating temperature and salinity to density, RMA-10 is capable of incorporating the effects of density stratification.

The model uses Reynolds stress approximations with eddy and diffusion coefficients to represent turbulence. For density stratified flow the model formulation adjusts input values of vertical eddy and diffusion coefficients by a factor D_f based upon the density stability $1/\rho \, d\rho/dz$.

An important feature of the model is the use of a transformation that defines a new domain that uses the original bottom profile but transforms the original time varying free surface to a fixed elevation (mean water level is usually used for this purpose).

The mathematically transformed equations are then solved with a mixed interpolation function Galerkin weighted residual method and Newton Raphson iteration to incorporate the non linear terms. The degrees of freedom are thus two horizontal velocity components and salinity/temperature (approximated using quadratic basis functions) and water depth (approximated using linear basis functions). Note that in the transformation from pressure to explicitly include depth, derivatives of density are introduced. The model is flexibly programmed to allow the user to specify active variables during each iteration. Mode splitting is permitted in the solution process. Three dimensional elements are automatically generated from data describing a plan view of the system augmented by information describing the number of elements present on a vertical line below each node. The relative spacing of nodes in the vertical direction is under control of the user. This option permits easy evaluation of model performance with different element configurations More complete details are given in King (1993).

MODEL ANALYSIS AND TESTS

All the initial stages of this project have been carried on a Sun workstation of limited capability. To minimize computer CPU time, the modeling was approached in three steps, i.e., two-dimensional depth averaged analysis, two-dimensional laterally averaged analysis and three-dimensional analysis on an initially simplified plan view of the system.

Two-dimensional depth averaged analysis is being used to develop an understanding of the overall flow regime and to evaluate the impact of boundary conditions applied at the open ocean. This analysis is also designed to define the desired plan view layout that will incorporate freshwater outflows from Port Jackson.

Two-dimensional laterally averaged analysis is being used to test performance of the model in stratified flow situations. With the exception of Coriolis forces all of the significant factors are active in the analysis. Capabilities of the model to describe recirculation for stratified and unstratified initial conditions and internal gravity waves can be tested at a scale that is representative of the actual region under investigation. The tests were designed to determine the necessary vertical grid spacing for these regimes so that the number of three-dimensional model simulations could be limited. Multi hour simulations for these cases could be completed in a matter of minutes on the Sun workstations.

The system modeled for the laterally averaged analysis was designed to represent at full scale, the prototype environment. The network was 2m deep at the shoreline transitioning linearly to a depth of 68m 2 km offshore and staying at a constant depth for a further 6 km. For all these tests zero horizontal velocity was imposed at the shoreline and at the open boundary, hydrostatic pressure and specified distributions of temperature and water surface elevation. Because of lack of space this paper will present results obtains with a grid that varied from 3 points vertically in the shallow areas to 7 points in the deeper sections. The current vectors presented below are located at each of the node points. Note the non-uniform distribution with depth.

Test number 1 applied a 15 m/s wind blowing offshore in an unstratified system. As expected, flow reversal occurs at approximately 1/3 depth. Maximum offshore velocities were approximately 0.6 m/s (about 4% of the wind speed).

Test number 2 subjected the same system to an initially stratified temperature distribution that reflected a typical measured distribution at the offshore reference station. It was enforced as a fixed boundary condition at the ocean boundary At zero time all the isotherms were horizontal. The current regime after 15 hours of simulation is shown in Figure 3. The vectors show that the location of reversal and the peak reversal velocity have moved closer to the surface than for homogeneous flow. Velocities close to the bed are considerably reduced. The isotherms after 15 hours are presented are also shown as the contours in Figure 3, they show that while there has been considerable movement of the isotherms in the upper layers the initial horizontal isotherms near the bed are not significantly displaced.

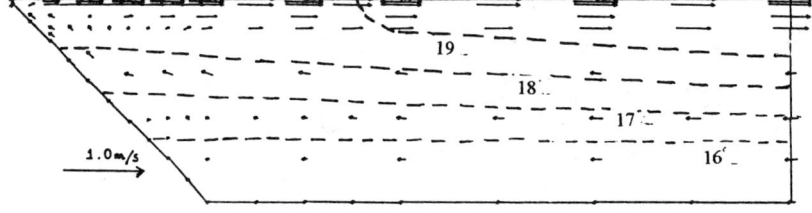

Figure 3. Test number 2. Current vectors and isotherms after 15 hours

During initial simulations for this test a suspicious circulation was detected. The system grid was designed to transition from two elements near the shore to 4 elements in the deep water. A non-uniform spacing was used to provide better resolution close to the surface. As a diagnostic test the model was modified so that density was a linear function of temperature and the system was re-configured from the original form to a vertical structure defined so that the density gradient was a constant. Under these conditions, for zero flow boundaries there should be no induced circulation. Velocities of order 0.1m/s were induced. Investigation showed that the transformation and algebraic manipulation of the original pressure gradient term were responsible. In its original form the model was constructed with water depth as dependent variable and with the influence of non-uniform density mostly appearing in a term describing the gradient of pressure difference p_d from a nominal hydrostatic pressure. Values of p_d were computed for each vertical line of nodes and then derivatives interpolated at Gauss integration points. When derivatives are computed for irregular quadrilaterals, the isoparametric transformation leads to nonzero horizontal gradients over the element interior. Although these errors were small they were the cause of the circulation. An alternative formulation is now incorporated into the model in which the horizontal density gradient remains as an explicit term in the governing equation and is itself evaluated at the node points forming the element. These values are then directly interpolated to the Gauss points. This effectively eliminated this spurious circulation. It also leads to the observation that the model would probably perform most accurately if element shapes for quadrilaterals were kept at rectangular as possible. Element layers are now designed to keep the layers horizontal where possible

Test number 3 was designed to induce an internal gravity wave and examine its progression. The initially layered system of test number 2 was simulated without an applied wind stress. The temperature boundary condition at the offshore location was successively lowered 10 m and then raised back to its original position over a 6 hour period. Figure 4 shows the progress of the induced circulation through the system. Isotherms at a comparable time show an undulating shape as the wave progresses towards the shore.

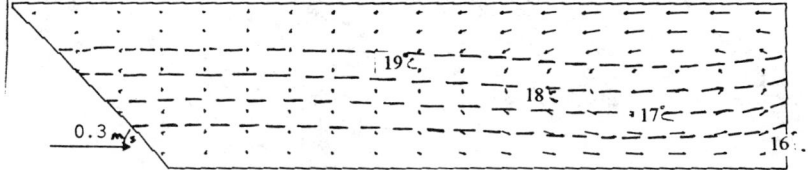

Figure 4. Test number 3. Current vectors and isotherms after 5 hours

The three-dimensional analysis was designed to take advantage of the parameters developed in the laterally averaged modeling tests. The initial grid used is a simplified mesh representative of the configuration offshore from Sydney.

Test number 4 applied a WNW 15m/s wind to the system with a fixed water surface elevation applied at the ocean boundary locations. Figure 5 depicts current vectors at various depths for each of the node points. The surface velocities show the expected deviation due to Coriolis forces. Velocities are generally larger in the shallow areas close to the eastern shore

To examine the influence of an 0.2m/s along-shore coastal current the water surface elevations at the northerly and southerly boundaries were adjusted. This adjustment was achieved by simplifying the momentum equations (dropping the unsteady and convective inertia terms) and computing expected water surface elevations for the desired overall current field. This process was used to avoid imposing velocity boundary conditions in the absence of field data. The effectiveness of this procedure for simplified conditions was validated with a test simulation that used only boundary condition forcing. Test number 5 was then designed to examine the influence of applying the same WNW wind to this system. Figure 6 shows the resulting current vectors. Much of the system shows the expected distribution. That is, a surface current driven largely by wind stress and lower level currents reflective of the overall along-shore current. A number of anomalies are clearly apparent, particularly close to the boundaries where further refinement of the process of deriving boundary conditions may be required.

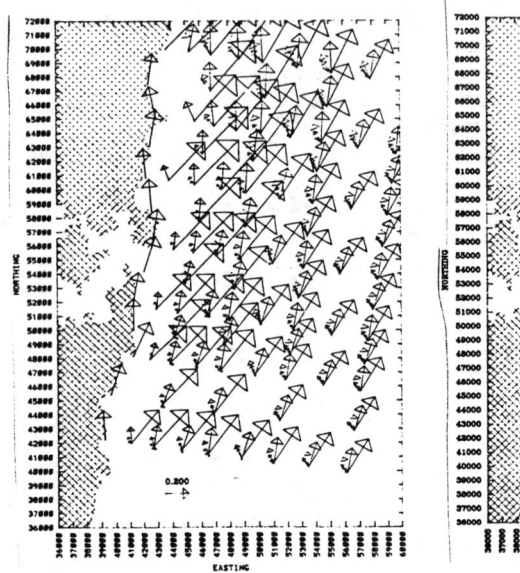

Figure 5. Offshore grid Current vectors
         Wind Stress only

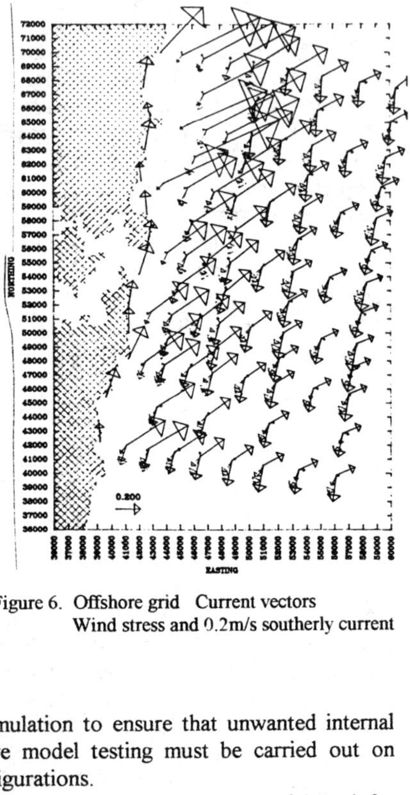

Figure 6. Offshore grid Current vectors
         Wind stress and 0.2m/s southerly current

CONCLUDING REMARKS

Care must be taken in model formulation to ensure that unwanted internal circulation cannot be induced. Extensive model testing must be carried out on possible distorted element shapes and configurations.

Two-dimensional laterally averaged simulations are a very useful tool for preliminary and screening analysis. Many of the fundamental motion types can be examined using fast response desktop computers.

Three- dimensional analysis is possible using workstations. Boundary conditions at the open ocean are a continuing source of difficulty. It is rare that sufficient field data will be available to completely specify driving parameters.

REFERENCE

King, Ian P., 1993 'RMA-10, A Finite Element Model for Three-Dimensional Density Stratified Flow, Report prepared for the New South Wales Environmental Protection Agency, Sydney, Australia.

SIMULATION OF 3D FREE SURFACE FLOWS IN HYDRAULIC STRUCTURE WITH SUBMERGED FLOW PASSAGES

Sam S.Y. Wang,[1] Joseph Letter, Jr.[2] and Phil Combs,[3]

ABSTRACT

This paper presents primarily the results of an application of the newly refined computational simulation model to study three-dimensional flows in a lock and dam system, which has submerged port holes built in the guard wall leading to the lock entrance. Based on the results of this application, the capabilities and limitation of this computational simulation model, CCHE3D are discussed.

INTRODUCTION

To achieve the objectives of preventing floods, controlling water supply and distribution, maintaining the navigability of waterways, generating hydropower, protecting the river banks from excessive erosion and migration, etc. a variety of structures, large and small, simple and complex, have been and are being constructed in inland and coastal waters. No one would disagree to the fact that free-surface flows in and around these hydraulic structures are highly unsteady, turbulent and three-dimensional, in which one can easily observe seaparations, small or large eddies, recirculations, reattachments, vortex shaddings, highly intensified turbulence, among other three-dimensional characteristics. These 3-d flows have been known to cause accelerated sedimentation processes such as bed materials scouring and entrainment, bank erosion, suspended and bed load transport and deposition. The sedimentation processes

---

[1] FAP Barnard Distinguished Professor and Director, Center for Computational Hydroscience and Engineering, The University of Mississippi, MS 38677.
[2] Research Hydraulics Engineer, USAE Waterways Experiment Sta., Vicksburg, MS 39180
[3] Chief, Channel Stablization Branch, US Army Corps of Engineers, Vicksburg district, Vicksburg, MS 39180

affect not only the integraty and effectiveness of these structures, but also th environmental quality and ecological balance. Therefore, it is of great importance to develop a highly capable and verifiable numerical simulation model to predict the realistic 3-d flow field, so that the hydraulic engineers can apply it to design hydraulic structures to achieve the cost-effectiveness in construction and maintenance of these structures. Supported by the US Army Corps of Engineers, Vicksburg District, under contract with the US Army Engineer Waterways Experiment Station, such a numerical simulation models has been developed and refined at the Center for Computational Hydroscience and Engineering (CCHE) of The University of Mississippi. It is called the CCHE3D Model.

CCHE3D MODEL

It is based on the Navier-Stokes and continuity equations governing an incompressible fluid together with the free-surface kinematic condition and one of the turbulence closure schemes, such as eddy viscosity, $\kappa - \epsilon$, etc. Sometimes, appropriate empirical functions and parameters were adopted to better represent the site-specific characteristics and conditions.

To apply this model, the solution domain is discretized into a system of several layers of 3-d elements using isoparametric interpolation functions (FE approach). The numerical equations are obtained one set at a node (FD approach) by a slected least square or collocation scheme. This new model can be applied to simulate free-surface flows in a highly irregular domain without the need of a complicated body-fitted corrdinate trnasformation (as needed by FDM) and it can be formulated by a sweep through the nodes rather than elements (as required by FEM). Therefore, this mode is called Efficient Finite Element Model. Details were given by Wang/Hu, (1992) and Wang (1992).

The Efficient Finite Element Model has included several refined schemes to enhance its accuracy and stability. The utilization of trigonometric functions as the interpolation functions has eleminated the errors in approximating the diffusion terms resulted from the traditional Lagrangian interpolation functions. The introduction of convective shape functions augmented by the local flow velocity has accounted for the so-called upwinding effect locally to eliminate spurious oscillations and thus, enhanced numerical stability.

NUMERICAL RESULTS

CCHE3D was applied to simulate 3-d free surface flows in the Lock and Dam No. 2 of the Red River Waterway being under construction. Field measurements of a discharge 85,000 cfs at the inflow boundary and a stage of 64 ft at the spillway outflow boundary were adopted. Due to the lack of measurements, the bed elevations and roughness used for simulation were construction data. Representative results are presented in this paper. Additional figures shall be presented at the conference. Figure 1 and 2 show the horizontal recirculation on the side of the navigation channel and vertical separation behind the submerged groyne, respectively. The entire flowfield in the entire Lock and Dam No. 2 is shown in Figure 3.

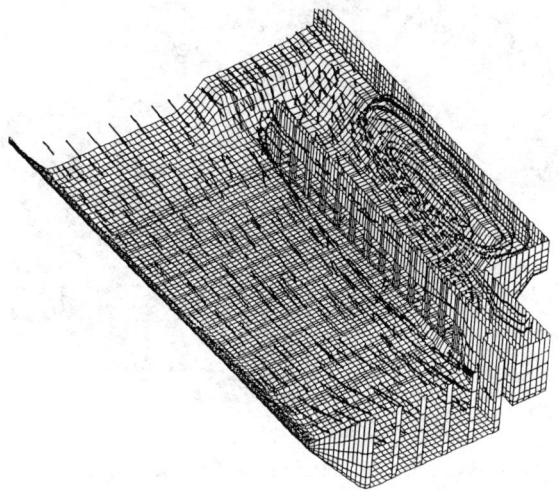

Figure 1. Horizontal recirculation along the side of the navigation channel.

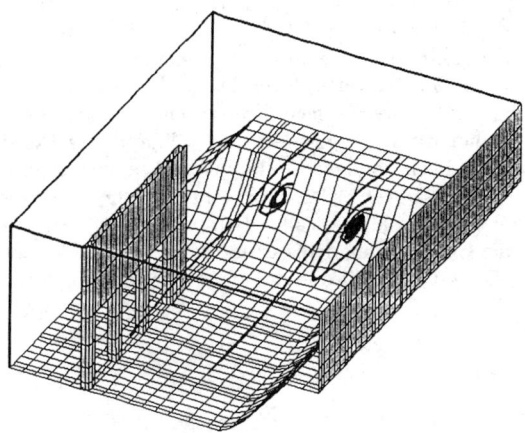

Figure 2. Vertical Separation Behind Submerged Groyne.

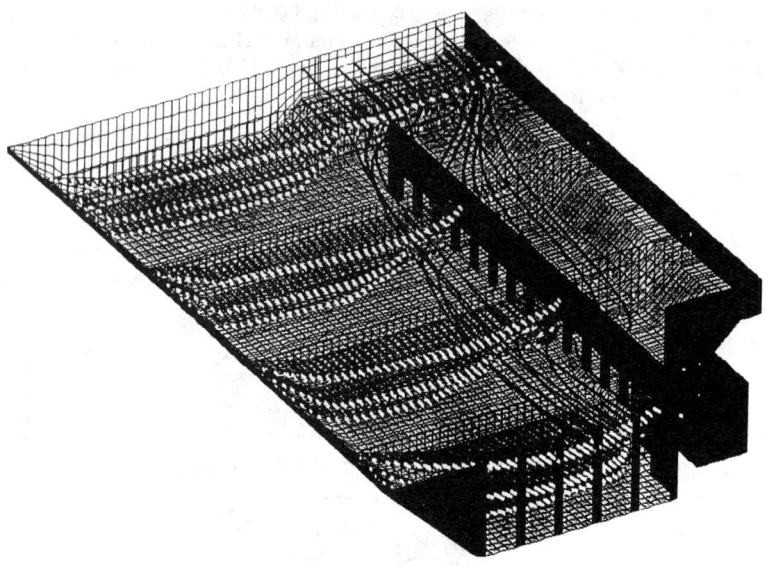

Figure 3. Velocity field and tracer paths in the Lock and Dam System.

CONCLUDING REMARKS AND DISCUSSION

The CCHE3D model has demonstrated to be the first numerical model capable of predicting truly 3-d flow field in a Lock and Dam System with submerged port holes on the guard wall. Simulation results (including animation recorded on video cassette) have clearly shown that various basic physical characteristics of 3-d free-surface flows in such a complex hydraulic system have been predicted. Although the model validation for the site-specific problem has not yet been conducted, the model has successfully passed a series of verification tests based on analytic solutions and prescribed solution forcing methodologies as well as scaled hydraulic model data. After site-specific validation, the CCHE3D model is expected to be a cost-effective CAD tool to carry out optimization studies in design of a variety of hydraulic structures.

ACKNOWLEDGEMENT

This work is a result of research sponsored in part by the US Army Corps of Engineers, Waterways Experiment Station, under Contract No. DACW39-87-K-0088, and the USDA Agricultural Research Service (monitored by the National Sedimentation Laboratory) under Specific Research Agreement No. 58-6408-2-127, and The University of Mississippi. Professional support and contributions of Mr. Dalmo Vieira and Ms. Janice Mills are also acknowledged.

REFERENCES

Dee, Dick P., F.M. Toro, and Sam S.Y. Wang, (1992), "Numerical Model Verification by Prescribed Solution Forcing - A Test Case," Hydraulic Engineering PP. 416-421, ASCE, New York.

Jia, Y.F. and Sam S.Y. Wang, (1992), "Computational Model Verification Test Case using Flume Data," Hydraulic Engineering, PP. 436-441, ASCE, New York.

Jia, Y.F. and Sam S.Y. Wang, (1993), "3D Numerical Simulation of Flow Near a Spur Dike," Advances in Hydro-Science and -Engineering, Vol. I, University of Mississippi, in press.

Neilson, F. (1991), "Field Tests at John H. Overton Lock and Dam," Memorandum for Record.

Toro, F.M. and Sam S.Y. Wang, (1993), "Validation of a 3-D Open Channel Flow Model by the Prescribed Solution Forcing Technique," Advances in Hydro-Science and -Engineering, Vol. I, University of Mississippi, in press.

Wang, Sam S.Y. (1991), "On the Recent Advances of Computational Modeling of Alluvial Rivers," Computational Hydraulics and Hydrology, ed. Ouazar, D et.al. Springer Verlag, Berlin and New York, 1991.

Wang, Sam S.Y. and K.K. Hu, (1992), "Improved Methodology for Formulating Finite Element Hydrodynamic Models," Chapter 18 of Finite Elements in Fluids, Vol. 8 ed. T.J. Chung, pp. 457-478, Hemisphere Pub. Corp., Washington, Philadelphia, London.

Wang, Sam S.Y. and R. Mayerle, (1993), "F.E. Modeling of 3D Flows and Sediment Transport in Rivers," Tech. Rept. CCHE-93-1, University of Mississippi, March '93.

Galveston Bay 3-D Model Study Channel Deepening
Circulation and Salinity Results

R. C. Berger, W. D. Martin, R. T. McAdory, and
J. H. Schmidt[1]

Abstract

The effects of deepening the Houston Ship Channel through the Galveston Bay estuary are predicted using the 3-D finite element numerical model RMA10-WES. The primary concern is the effect of any circulation and salinity shifts accompanying deepening upon oyster production in the bay. The channel itself has a project depth of 40 ft mean low water (mlw) and is deepened in two phases to 45 and 50 ft mlw. The model simulations are generally 1 year long and are made for predicted low, medium, and high freshwater inflow. The model results demonstrate the increase in salinity intrusion along the deepened channel and the resulting changes in bay circulation.

Introduction

The U.S. Army Engineer Waterways Experiment Station (WES) Hydraulics Laboratory (HL) has developed a three-dimensional (3-D) hydrodynamic model of Galveston Bay for the Houston-Galveston Navigation channels, Texas project, in cooperation with the U. S. Army Engineer District, Galveston. The model used is RMA10-WES, which is one of a suite of models known collectively as the TABS-MD system. The 3-D model is capable of computing water velocity, circulation patterns, salinity gradients and water levels for the entire Bay system.

The model uses a Finite Element formulation which allows incorporation of complex geometric features. The model consists of a network or mesh of computational

---

[1] U.S. Army Engineers Waterways Experiment Station, 3909 Halls Ferry Road, Vicksburg, MS 39180-6199.

nodes defined as x, y and z coordinates. A computational mesh for Phase I and II are shown in Figure 1.

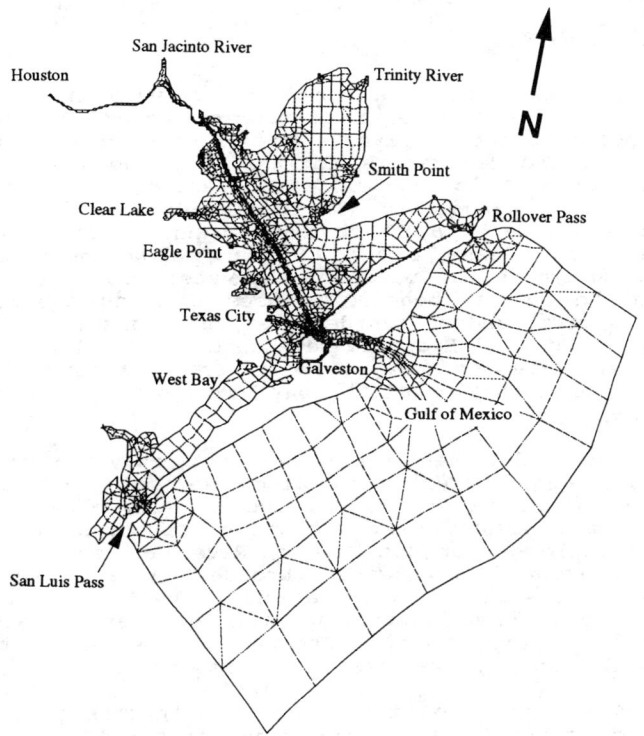

Figure 1. Location map of Galveston Bay showing model boundary

This shows the element outline from which the channel location is apparent; the region of high resolution through the center of the bay running roughly North and South. These nodes are then tied together to form elements, which may be assigned properties that reflect actual conditions of Bay, such as roughness of the bed. The boundary conditions consist of a tidal variation in the Gulf of Mexico, freshwater inflow at eleven points around the Bay, Gulf salinities and wind speed and direction. These values are varied through time in increments from 15 minutes to 1 hour. By applying actual observed or hypothetical values such as variations in freshwater inflow at the boundary, the effects of these can be observed and/or predicted at each computational point in the mesh. The Galveston Bay model consists of approximately 12,000 computational nodes which form 5,100 elements.

Tests

The model has been used to evaluate existing Bay conditions, and several geometries based on proposed channel enlargement plans. The existing condition consists of the Bay with a nominal 40 ft deep and 400 ft wide Houston Ship channel. The Phase I plan features a 45 ft deep channel and 530 ft wide channel. The Phase II plan features a 50 ft deep channel that is 600 ft wide. The Phase I and II plans also included some 18 sites where the material excavated from the channel would be put to beneficial uses such as creation of marsh, bird, and boater destination islands. Additionally, a National Economic Development (NED) Plan was run that evaluated open bay disposal of the excavated material over a portion of the bay next to the Houston Ship Channel.

The model has been used not only to evaluate several geometries but also several hydrological scenarios of water use in the Houston area. These hydrologic scenarios were based on construction of the Wallisville dam on the Trinity River and shifts in use of surface and groundwater through the year 2049. The model runs were of two durations. One series of tests were 1 year long, January through December. These were termed X class runs. Another series of tests was 9 months long and was termed Y class runs.

The purpose of these lengthy computer runs is to provide input to an oyster model authored by Dr. Eric Powell (Texas A&M University). The 3-D model provides detailed data on salinity and currents through time. These data are then used by the oyster model to predict population responses to the changes brought about by the channel deepening. The model generates a tremendous amount of data, from which we attempt to understand what is taking place by considering monthly averaged results. Of particular interest from biological reasons is the summer period. Figures 2 and 3 show the July monthly average salinity for surface and bottom depth at medium flow hydrology.

Results

The strong channel salinity intrusion is apparent in the bottom depth isohaline. By comparison with the surface plots the significance of the density driven

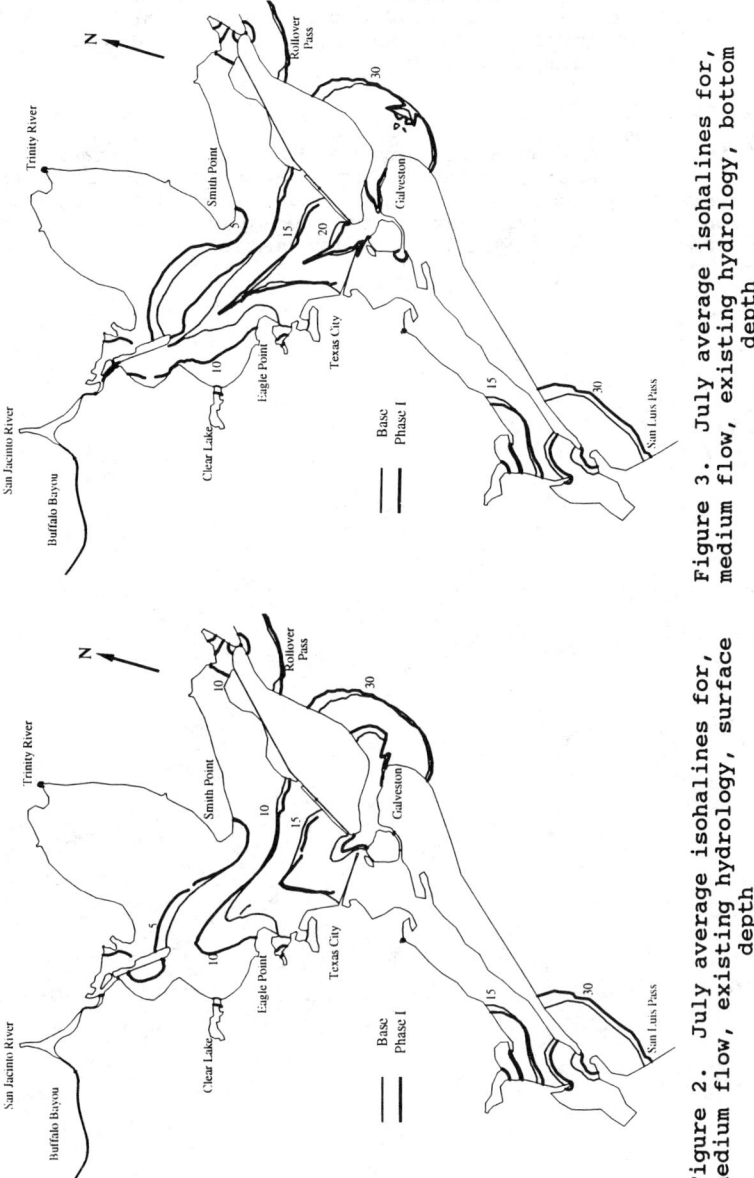

Figure 2. July average isohalines for, medium flow, existing hydrology, surface depth

Figure 3. July average isohalines for, medium flow, existing hydrology, bottom depth

currents can be seen by the distance the 10 ppt bottom isohaline has moved upstream. The comparison from existing to Phase I conditions shows that the most notable increase due to deepening occurs in the upper bay West of the channel.

Acknowledgement

This study was sponsored by the U.S. Army Engineer District, Galveston, Texas. Input and guidance from the Interagency Coordination Team has also aided in execution of the study. Permission was granted by the Chief of Engineers to publish this information.

Modeling the Tides of Massachusetts and Cape Cod Bays

H. L. Jenter[1], R. P. Signell[2], and A. F. Blumberg[3]

Abstract

A time-dependent, three-dimensional numerical modeling study of the tides of Massachusetts and Cape Cod Bays, motivated by construction of a new sewage treatment plant and ocean outfall for the city of Boston, has been undertaken by the authors. The numerical model being used is a hybrid version of the Blumberg and Mellor ECOM3D model, modified to include a semi-implicit time-stepping scheme and transport of a non-reactive dissolved constituent. Tides in the bays are dominated by the semi-diurnal frequencies, in particular by the M_2 tide, due to the resonance of these frequencies in the Gulf of Maine. The numerical model reproduces, well, measured tidal ellipses in unstratified wintertime conditions. Stratified conditions present more of a problem because tidal-frequency internal wave generation and propagation significantly complicates the structure of the resulting tidal field. Nonetheless, the numerical model reproduces qualitative aspects of the stratified tidal flow that are consistent with observations in the bays.

Massachusetts and Cape Cod Bays

Massachusetts and Cape Cod Bays (hereafter refered to simply as "the bays"), together, constitute approximately a 100 × 50 km semi-enclosed basin having an average depth of 35 m in the western Gulf of Maine (figure 1). The bays are used for a variety of potentially conflicting purposes including commercial and recreational fishing, shipping, recreational boating, swimming, and as a repository for sewage effluent and dredged sediments. At present, the Massachusetts Water Resources Authority (MWRA) is constructing a new sewage treatment plant for the city of Boston and 42 surrounding communities (Tarricone, 1992). The plant, when fully operational, will discharge 1.3 bgd (56 m^3s^{-1}) of secondarily treated effluent, making it the second largest sewage treatment facility in the United States. The effluent will be discharged into Massachusetts Bay through a 14

[1] U.S. Geological Survey, Reston, Virginia 22092
[2] U.S. Geological Survey, Woods Hole, Massachusetts 02543
[3] HydroQual, Incorporated, Mahwah, New Jersey 07430

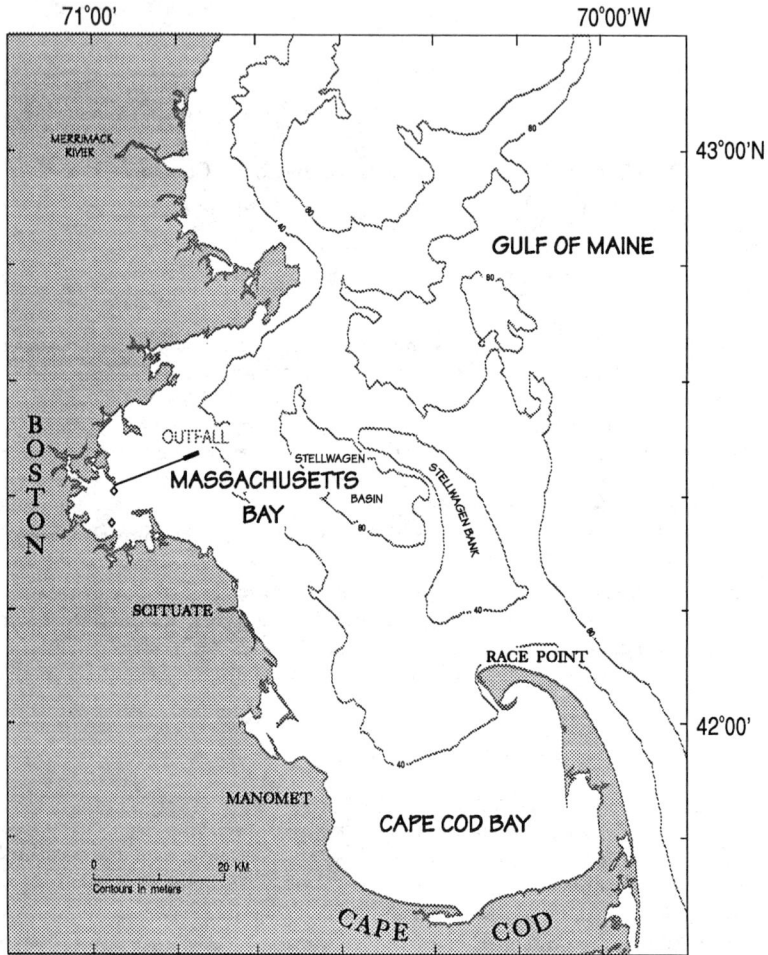

Figure 1: Base map of the Massachusetts and Cape Cod Bay system showing proposed outfall location, bathymetric contours, and Stellwagen Bank.

km-long, 7.3 m-diameter, solid-rock tunnel (figure 1). The last two km of which contain a set of 80 .8 m-diameter risers used to transport the effluent to the seabed. In order to help ensure that the discharge of effluent will have minimal effect on the shoreline residents and ecological resources of the bays, including the endangered North Atlantic Right Whale population that feeds on Stellwagen Bank (figure 1), a three-dimensional (3-d) numerical circulation model of the bays has been developed by the authors. The numerical model being used is based on the 3-d, time-dependent coastal circulation model, ECOM3D, developed by Blumberg and Mellor (1987). The model consists of fully nonlinear prognostic (i.e. predictive and interactive and thereby able to affect conservation of the other variables) equations for temperature, salinity, free-surface elevation and momentum. Vertical turbulent-mixing processes are parameterized by means of a turbulent closure submodel whose adjustable parameters are set to well-established laboratory values, thereby leaving the model free of tuning parameters. A complete description of ECOM3D, including discretization schemes, stability criteria and imbedded parameter values, is reported by Blumberg and Mellor (1987). The code has been modified recently to include a semi-implicit time-stepping scheme and the ability to transport a non-reactive dissolved constituent. The time-stepping scheme was added in order to allow model time steps that exceed the Courant-Fredrichs-Lewy condition (e.g. model results presented here are from runs with a time step of 621 seconds, implying a maximum of roughly 5 times the CFL criterion.) Details of the semi-implicit scheme are presented in the ECOM-si users' manual (HydroQual,1991). This modified version of ECOM3D is known as ECOM-si (semi-implicit).

ECOM-si employs an orthogonal curvilinear coordinate system in the horizontal (figure 2) and a σ-coordinate system (a coordinate system having the same number of cells regardless of depth) in the vertical. The Massachusetts Bay model grid is 68 by 68 cells in the horizontal and has 11 equally-proportioned levels in the vertical. Minimum grid spacing is approximately 600m in the vicinity of Boston Harbor. Maximum grid spacing is approximately 6000m, along the offshore open boundary. Model grid bathymetry was interpolated from bathymetric soundings collected by the National Oceanic and Atmospheric Administration.

The model of the bays can be forced by freshwater inflows at the mouth of the Merrimack River and at the proposed outfall site; momentum can be introduced at the free surface by specifying a time-dependent wind speed and direction; heat can be introduced at the free surface by specifying a time-dependent atmospheric heat flux; and, most importantly for the tidal problem, energy can be introduced at the open boundaries by specifying space- and time-dependent elevations. Partially clamped elevation boundary conditions with an arbitrary user-specified relaxation time are also available in ECOM-si for situations when wind, tide, atmospheric

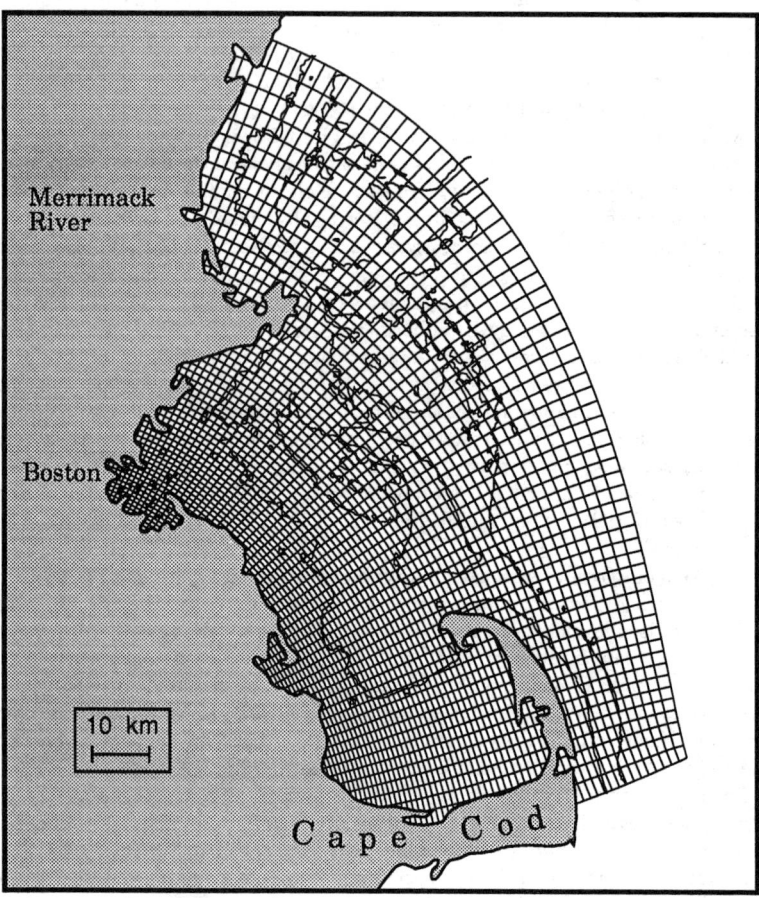

Figure 2: Model grid for the three-dimensional circulation model, ECOM-si, of Massachusetts and Cape Cod Bays. The curvilinear orthogonal grid allows the mesh resolution to vary spatially, having a minimum grid spacing of 600 m and a maximum grid spacing of 6000m. There are 11 vertical σ-levels, evenly distributed throughout the water column.

heating and freshwater inputs are simultaneously important (See Blumberg et al., 1993 for discussion of these cases.).

The Tides of Massachusetts and Cape Cod Bays

The tidal response of the bays is dominated by the semi-diurnal frequencies, in particular, by the M_2 tide (Irish and Signell, 1992). The average tidal range in the bays is 2.6 m. The average M_2 tidal range in the bays is roughly 2.45 m, implying that the diurnal and overtide sea surface responses are negligible. Based on extensive analysis by Irish and Signell (1992) of bottom pressure records collected at numerous points in the bays, tidal range is basically constant throughout the bays with slight amplifications of roughly 10% in Boston and Provincetown Harbors. Predominance of the semi-diurnal response in the bays is due to the fact that the Gulf of Maine is strongly resonant at the semi-diurnal frequencies causing the semi-diurnal tidal wave to propagate easily into the bays.

Irish and Signell (1992) also demonstrate that tidal currents are predominantly semi-diurnal, with the diurnal signal representing even a smaller portion of the flow than it does of the sea surface elevation. Despite the dominant frequencies being constant baywide, the velocity amplitudes, and hence the tidal excursions, vary significantly. Amplitude variations are due to basin slope and bathymetric variability. The strongest currents in the bays are found in the channel between Race Point and Stellwagen Bank. Tidal velocities as large as 1 knot and excursions as large as 12 km are not uncommon. Minimum tidal excursions of less than 2 km are found in the deep central part of Massachusetts Bay. Tidal excursions at the proposed outfall site are roughly 2 km.

Currents at tidal frequencies vary seasonally within the bays. Irish and Signell (1992) showed that internal waves with periods matching those of the tides are observed in the spring and summer when the bays are stratified. Stratification, and thus internal wave activity, disappears in the winter when surface cooling and storms completely mix the water column.

Modeling Tides in Massachusetts and Cape Cod Bays

As stated previously, tides in the bays are driven almost exclusively by the M_2 tidal response of the Gulf of Maine. Therefore, because of limited space, the results presented herein focus only on the response of the bays to forcing at the M_2 frequency. This forcing is introduced along the model's open eastern boundary by specifying M_2 amplitudes and phases for sea surface elevation. These data were obtained from a tidal model of the Gulf of Maine (Naimie and Lynch, 1991) which has been finely-tuned to reproduce observations in the gulf.

Figure 3 contains a comparison of eight modeled and measured surface tidal current ellipses for well-mixed conditions — at sites identified as U2, U3, U6, U7, RP, MN, BB and BS in figure 3 (See Irish and Signell, 1992, or Geyer et al., 1992, for details of measurements). It should be noted that the orientation and

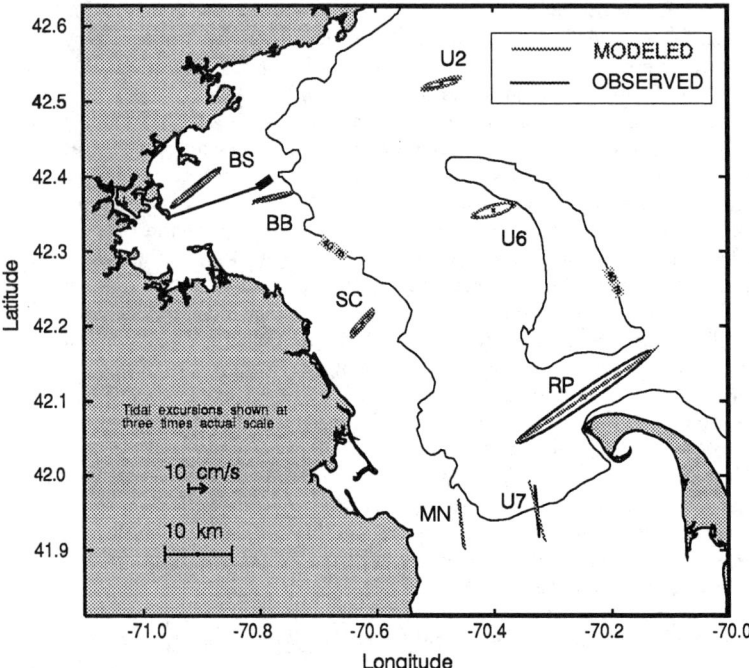

Figure 3: Comparison of modeled and observed surface M_2 barotropic tidal currents in the bays. Shown are tidal ellipses, which indicate the observed velocities over the tidal cycle. They also represent the excursions water parcels would make if they moved with the tidal currents observed at the mooring. For clarity, these tidal excursions are shown at three times actual scale.

amplitude of the ellipses varies substantially throughout the bays. In the figure, the size of the ellipse not only represents the current velocity but also the tidal excursion at the measurement site. The ellipses are scaled to represent 3 times the observed tidal excursion at the measurement sites.

The model reproduces both amplitudes and orientations of the tidal ellipses very well. Comparisons similar to those shown in Figure 3, but for other depths in the water column, show that the model reproduces the barotropic M_2 tide very accurately throughout the water column everywhere in the bays for well-mixed conditions. There is some difference in ellipticity between the modeled and observed currents at the station located between Race Point and Stellwagen Bank — identified as RP in figure 3. The modeled flow appears to be too rectilinear. This can be attributed to insufficient model grid resolution at the measurement location (figure 2). Rapid changes in bathymetry in this area cause secondary flows which increase the cross-channel amplitude of the tidal ellipse. The model appears to underresolve these secondary flows. This assertion was confirmed by doubling the model grid resolution and observing a significant improvement in agreement between modeled and observed ellipticity.

Comparisons similar to those shown in figure 3, but for stratified periods, are, in general, less accurate than for well-mixed periods. Experimentation with different degrees of stratification indicates that these results are extremely sensitive to the density structure of the water column. Therefore, additional effort continues to be placed on the development of accurate temperature and salinity open-boundary conditions, as well as physically realistic atmospheric heat flux surface-boundary conditions for the model. Figures 4 and 5 are included here to illustrate qualitatively the differences in tidal behavior between well-mixed winter conditions and stratified summer ones. Each figure contains contours of u_*, the square root of the maximum kinematic bottom stress amplitude (see Blumberg and Mellor, 1987, for a description of ECOM-si's bottom stress formulation). Figure 4 represents well-mixed wintertime conditions and figure 5 represents stratified summertime conditions. The summertime conditions shown are maximum u_* values after 3 tidal cycles for which the model was initialized everywhere using the maximum stratification profile observed in Stellwagen Basin during 1990 and for which temperature and salinity were treated prognostically in the model.

Spatial scales of the contours in the wintertime scenario are much larger than those of the summertime scenario indicating much less high wavenumber variation in bottom stress. The implication of this is that internal waves are present in the latter. Measurements (Irish and Signell, 1992) indicate the presence of internal waves in much of the bays during the stratified season, including the presence of a nearly-classic first-mode internal wave in Stellwagen Basin. It is interesting to note that magnitudes of bottom stress can be either reduced or increased by the presence of internal waves, depending on the whether they interfere constructively or destructively with the barotropic tide. Whether internal waves are present or not and the manner in which they interact with the M_2 tide, if present, is

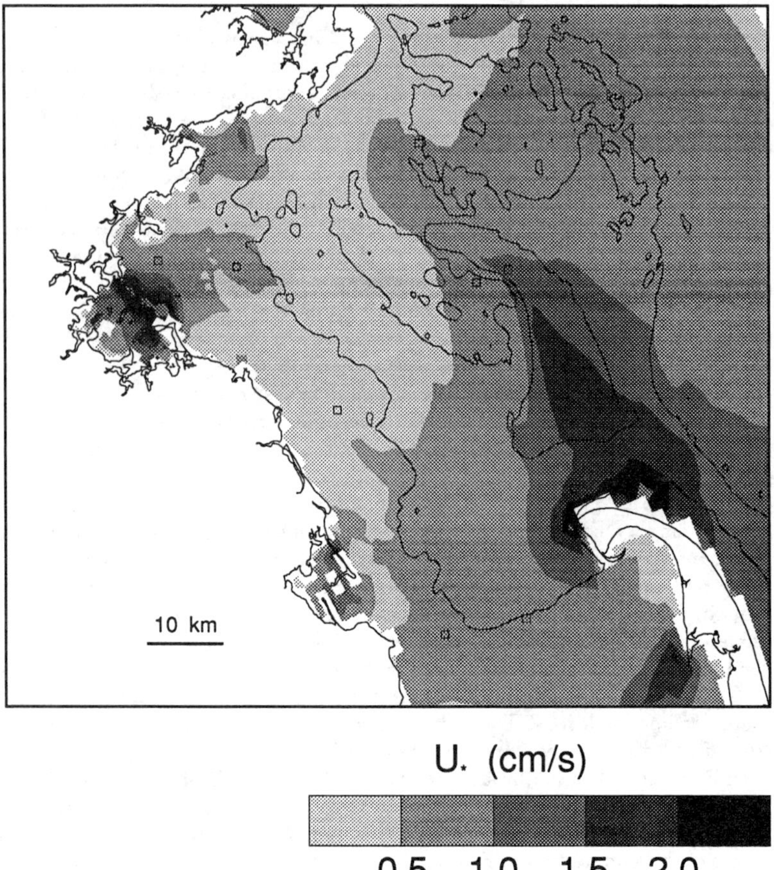

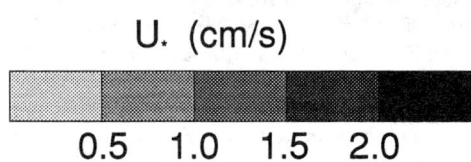

Figure 4: Maximum u_* due to forcing the model's open boundary with M_2 tides under well-mixed conditions. Contours are in units of cm/s and represent the square root of the kinematic bottom stress. Areas of intense currents between Race Point and Stellwagen Bank and in the mouths of Boston and Plymouth Harbors are clearly visible.

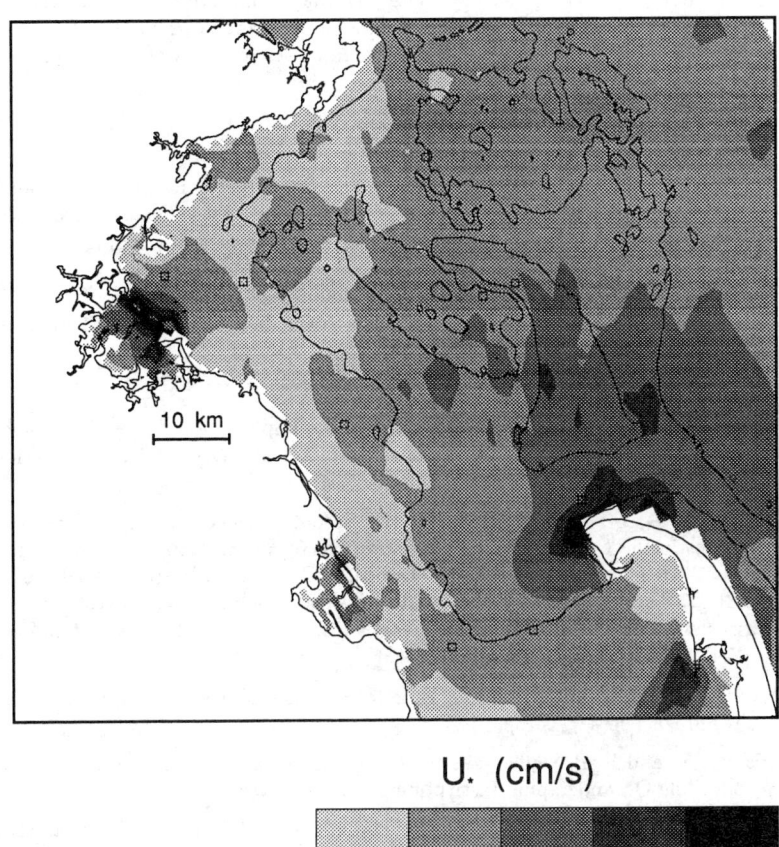

Figure 5: Maximum u_* due to forcing the model's open boundary with M_2 tides under stratified conditions. Contours are in units of cm/s and represent the square root of the kinematic bottom stress amplitude. The model was initialized with the maximum summertime stratification observed in Stellwagen Basin in 1990, and was allowed to run for 3 tidal cycles treating temperature and salinity prognostically. The much finer spatial structure relative to that shown in figure 4 is an indication of the presence of internal waves.

determined by the complex bathymetry of the bays. Studies to further improve the modeling of tides in the bay under stratified conditions and to investigate the interaction between the tides and other important flows in the bays, such as wind-driven flows or those initiated by impulses of freshwater, are currently underway.

Acknowledgements

This work has been funded by the Massachusetts Water Resources Authority through contracts with the U. S. Geological Survey and HyrdoQual, Incorporated. Additional support has been provided by the Massachusetts Environmental Trust, the U. S. Environmental Protection Agency's Massachusetts Bays Program, and the U. S. Geological Survey. The authors would like to acknowledge Drs. R. T. Cheng and V. Casulli for their contribution to the development of ECOM-si.

References

Blumberg, A. F. and G. L. Mellor, 1987. A Description of a Three-Dimensional Coastal Ocean Model. In: *Three-dimensional Coastal Ocean Models*, American Geophysical Union, Washington, D.C., pp. 1–16.

Blumberg, A. F.; R. P. Signell and H. L. Jenter, 1993. Modeling Transport Processes in the Coastal Ocean. *Journal of Environmental Engineering*, in press.

Geyer, W. R., G. B. Gardner, W. S. Brown, J. D. Irish, B. Butman, T. H. Loder, and R. P. Signell, 1992. Final Report: Physical Oceanographic Investigation of Massachusetts and Cape Cod Bays, Woods Hole Oceanographic Institution Tech. Report WHOI-92-xx, 497 pp.

HydroQual, Incorporated, 1991. A Primer for ECOM-si, HydroQual Incorporated, Mahwah, New Jersey, 66 pp.

Irish, J. D. and R. P. Signell, 1992. Tides of Massachusetts and Cape Cod Bays, Woods Hole Oceanographic Institution Tech. Report WHOI-92-35, 62 pp.

Naimie, C. and D. Lynch, 1991. Benchmark 3-D M_2 and M_2 Residual Tides for Georges Bank and the Gulf of Maine. Technical Report NML-91-2, Numerical Methods Laboratory, Thayer School of Engineering, Dartmouth College, Hanover, NH.

Tarricone, P., 1992. Boston's City Within a City. *Civil Engineering*, October, 1992. pp. 40–43.

Storm Water Regulations
Aircraft Deicer/Anti-icers
Operations

W.H. Espey, Jr., M.ASCE.[1]
George I. Legarreta[2]

1.0 Introduction

The United States Environmental Protection Agency (EPA) issued (November 16, 1990) its final rules regarding NPDES permits for storm water discharges from municipal and certain industrial activities. Air transport facilities (airports) are included in these "industrial activities." Specifically defined within EPA regulations the "Transportation facilities classified as Standard Industrial Classifications...45...which have vehicle maintenance shops, equipment cleaning operations, or airport deicing operations. Only those portions of the facility that are either involved in vehicle maintenance (including vehicle rehabilitation, mechanical repairs, painting, fueling, and lubrication), equipment cleaning operations, airport deicing operations....are associated with industrial activity."

In 1991, FAA has initiated a storm water program to assist airport operators in complying the EPA Storm Water NPDES Regulations. The first FAA project, a survey, addresses present operational practices, operation equipment, deicing agents, chemical and hazardous wastes at existing airport facilities. The results of this survey will provide a data base to develop future technical information and management procedures to assist airports in complying with the EPA NPDES Storm Water Regulations. The ultimate goal of FAA is to develop operational, physical and management alternatives to mitigate the impact of chemical concentrates resulting in compliance with the NPDES Storm Water program while continuing to meet current design standards. The recent International Conference on Airplane Ground Deicing (May 28-29, 1992) sponsored by FAA focused attention on

---

[1] Chairman of the Board, Espey, Huston & Associates, Inc., P.O. Box 519, Austin, Texas 78767

[2] Office of Airport Safety and Standards, Federal Aviation Administration, U.S. Department of Transportation, 800 Independence Avenue, S.W., Washington, D.C. 20591

vital safety issues related to a better understanding of aircraft ground deicing/anti-icing issues. Environmental aspects of ground deicing/anti-icing was an important issue considered by the conference

2.0 Storm Water Survey Questionnaire

The Draft Questionnaire was developed and submitted to the Office of Management and Budget for review and approval. The Storm Water Questionnaire was sent on September 23-24, 1991 by FAA in cooperation with the American Association of Airport's to 131 airports. Ninety-six airports responded to the questionnaire with some information. In general, the questionnaire response was spotty. Data collected from the 96 airports that responded to the questionnaire was limited.

3.0 SUMMARY

On September 9, 1992, EPA promulgated the general permit for NPDES permits under the Storm Water Regulations. Specifically addressed in these regulations as an industrial activity are airports (SIC Code 45). Facilities with storm water discharge associated with industrial activity from areas where aircraft deicing operations occur (including runways, taxiways, ramps, and deicing stations) are required to monitor for parameters indicative of the overall quality of the discharge and to identify deicing materials used at the site that are entering the storm water discharge. Sampling of runoff is required during deicing operations for airports with over 50,000 flight operations per year. EPA indicated that the number of operations is one of the key factors for determining the amount of de-icing activity and other industrial activities occurring at an airport. In other words, airports with a higher number of operations are probably going to have stormwater with a greater amount of pollutants. This analysis was based in part on information from the aircraft owners and pilots associations and was based on 1990 FAA administrative data which indicated approximately 5078 public-used airports in the United States. Of these publicly-used airports, approximately 378 airports (7.4%) have 50,000 or more flight operations per year. Presently, the American Association of Airport Executives is negotiating with EPA a Model Airport Permit. An important issue in those negotiations is the 50,000 operation criteria in the EPA General Permit. The basic issue is determining a "threshold" to trigger the monitoring requirements under the EPA Storm Water Regulations in which the introduction of pollutants through deicing/anti-icing operations is of concern. A possible alternative approach would be defining a deicing/anti-icing "threshold" volume. Based on the FAA survey questionnaire database, an analysis was made of the thirty airports that responded to a series of questions regarding the use of aircraft deicing and anti-icing fluids. Presented in Table 1 is a summary of total annual volume of deicer/anti-icers (Type 1) applied to aircraft during the winter seasons of 1989-90 and 1990-91 reported by 30 airports. Also presented in Table 1 is the corresponding State, FAA region and hub size classification. In addition to the total annual volume of deicer/anti-icer used during the winter period, the maximum monthly for the corresponding period is also indicated. Summarized in Table 1 also is airport operations as defined from the FAA-air traffic activity report for FY91. As indicated in Table 1 deicer/anti-icer usage ranged significantly between the reporting airports from 0 to 412,000 gallons for the reporting winter period of 1990-91. The data is presented in Figure 1 for thirty airports that reported level of usage for the winter period 1990-1991. Airport #44 reported a use of 412,000 gallons for this period and is not plotted. Analysis of the histogram indicates a possible plateau that approximately 30% of the airports reporting anti-icer/deicer used greater than 20,000 gals/winter period. An analysis (Figure 2) was made for this same reporting period of the airport number of operations for these thirty airports and the corresponding annual volume of anti-icer/deicer. Considerable scatter is noted in Figure 2 which represents a broad range of FAA regions and size of airport. If the areas representing the northwest mountains (FAA Region ANM) and the Great Lakes (FAA Region AGL), some trend can be noted in Figure 2. However, no general relationship could be detected relating airport operations and deicer/anti-icer usage.

TABLE 1

TOTAL VOLUME OF DEICER/ANTI-ICERS APPLIED TO AIRCRAFT - TYPE I
DURING 1989-90, 1990-91 WINTER SEASONS

No.	State	FAA Region	FAA Classification	Total Annual Volume 89-90	Total Annual Volume 90-91	Maximum Month Volume 89-90	Maximum Month Volume 90-91	Airport Operations 1991*
1	AK	AAL	SH	18,000		6,000		115,129
4	AR	ASW	SH	16,000	15,900	12,000	11,000	140,255
8	CA	AWP	SH	75	75			152,161
12	CA	AWP	MH	13,295	11,095			152,161
19	FL	ASO	MH	10	40			66,631
23	FL	ASO	MH	0	0	0	0	223,775
25	FL	ASO	SH		600		200	135,723
36	KS	ACE	SH	22,454	20,922	10,000	9,000	173,722
37	KY	ASO	SH		8,165			158,050
38	LA	ASW	SH	400	38%	400	38%	115,759
39	LA	ASW	MH			900	100	220,600
43	MD	AEA	LH	UNK	188,400	UNK	131,928	282,320
44	MI	AGL	LH	771,000	412,000	286,000	147,000	390,863
48	MT	ANM	SH	N/A	7,500	N/A	2,200	107,655
49	NC	ASO	LH	24,000	13,500	10,000	12,000	440,956
52	NE	ACE	N-H	3,000	2,500			97,480
54	NV	AWP	MH	70,000	66,000	25,000	20,000	160,109
59	OH	AGL	SH	12,400	10,000	4,500	4,500	141,412
61	OH	AGL	MH	66,300	92,000	20,000	22,000	213,723
63	OR	ANM	MH	117,000	122,000	49,000	51,000	264,854
70	SC	ASO	SH	600	200	600	200	93,489
75	TX	ASW	SH	20,000	18,600	9,000	8,400	82,958
77	TX	ASW	LH	22,590				310,404
78	TX	ASW	MH	4,950	1,750	4,950	1,750	182,831
81	TX	ASW	MH	7,140	2,030	6,140	1,880	267,199
82	UT	ANM	LH	112,485	112,377	23,052	44,221	301,755
83	VA	AEA	MH	700	1,655	500	955	142,742
91	VA	AEA	SH	10,200	9,500			141,300
92	VA	AEA	SH	5,105 g	4,045 g	1,935 g	1,600 g	119,302
95	WI	AGL	SH	32,320	30,595	9,561	7,689.5	136,093

* FAA - Air Traffic Activity Report - FY 1991.

Figure 1
Total Annual Volume Of Icer/Anti-icers Applied To Aircraft Type-I During 1990-91

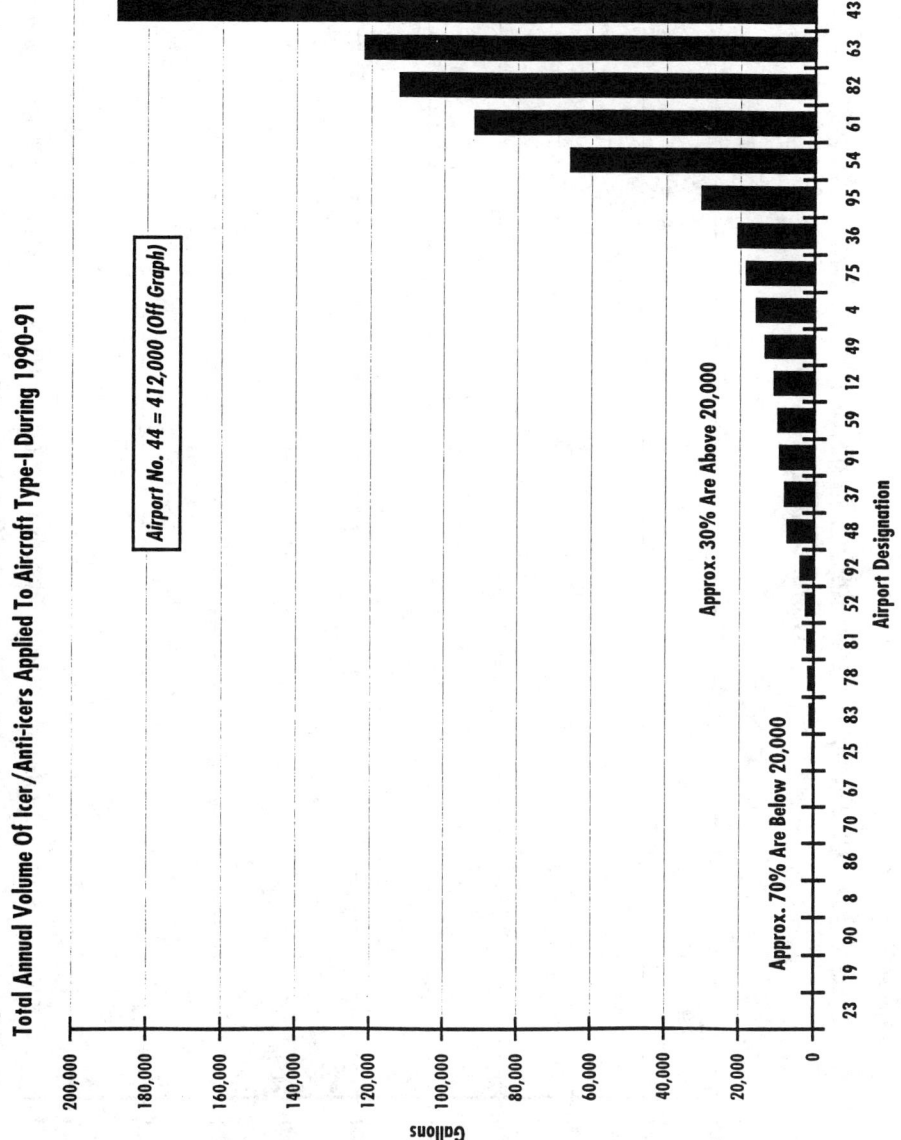

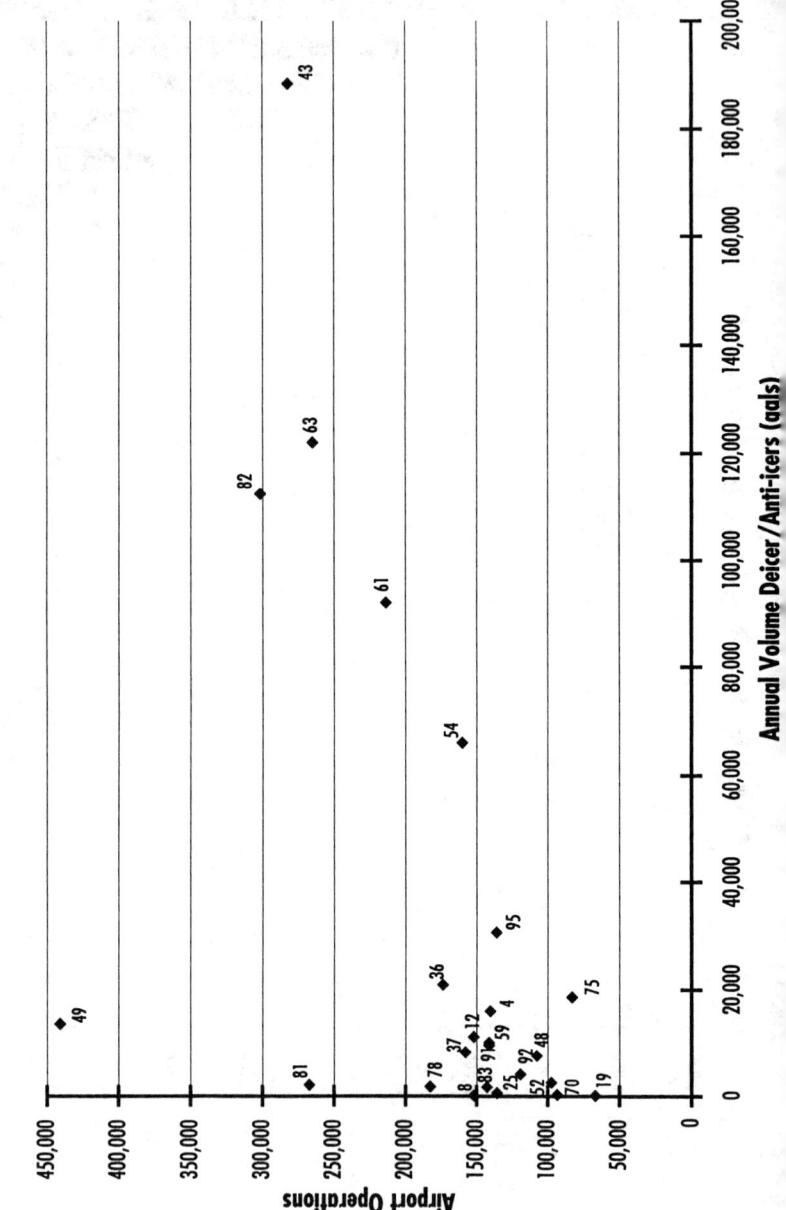

Figure 2
Total Volume Of Deicer/Anti-icers Applied To Aircraft-Type I During 1990-91 Winter Season

GIS AND SWMM APPLICATIONS IN DEVELOPING THE LAKE HOUSTON WATERSHED MANAGEMENT PROGRAM

Duke G. Altman, M. ASCE, P.E.[1]

R. Gary Montgomery, P.E.[2]

Thomas L. King[3]

Chuck W. Patterson[4]

Abstract: The City of Houston has undertaken the task of developing a Watershed Management Program for the 2,828 square mile Lake Houston Watershed in an effort to prevent future degradation of the Lake's water quality. Need for the Program stems from present water quality problems in the Lake and the potential for future water quality degradation of inflows associated with land use changes and point source discharges that can be expected to occur as the City of Houston continues to grow into the Watershed. The Program's primary goal is to develop and implement cost-effective storm water and other management techniques aimed at preventing further water quality degradation of Lake Houston and its inflows considering baseline conditions as determined to exist on January 1, 1991.

Estimates of Lake Houston's existing and possible future inflow water quality including point and nonpoint contaminant loadings were developed utilizing the PC ARC/INFO (Version 3.4D) Geographic Information System (GIS) package and the Environmental Protection Agency Storm Water Management Model (SWMM) to describe and represent present and projected watershed physiographic conditions as well as associated water quality relationships. Intergraph's MicroStation and AutoCad were used in building watershed physiographic data sets from various sources with the results transported to PC ARC/INFO for cleaning, processing and analyzing. To assist in formulating appropriate controls for the watershed, SWMM was also used to model projected watershed conditions with various storm water quality controls (Best Management Practices) in place. Impact analyses associated with soil erosion, vegetation, oil and gas operations, pipelines and other watershed conditions or activities were greatly enhanced by the GIS procedures used.

---

[1]Vice President, Espey, Huston & Associates, Inc., P.O. Box 519, Austin, Texas 78767

[2]Vice President, Espey, Huston & Associates, Inc., 800 W. Sam Houston Parkway South, Suite 201, Houston, Texas 77042

[3]Senior Staff, Espey, Huston & Associates, Inc. 800 W. Sam Houston Parkway South, Suite 201, Houston, Texas 77042

[4]Computer Applications Manager, Espey, Huston & Associates, Inc., P.O. Box 519, Austin, Texas 78767

Evaluations of study analyses have assisted in beginning the development of a management program with emphasis on the formulation of an implementable "plan of action" that considers the expected water quality benefits, cost of the Program implementation, interested parties, public input and the required controls/practices to meet Program goals.

Introduction

As the City of Houston metropolitan area expands further into the Lake Houston Watershed, there is an increasing concern that Lake Houston's water quality will be degraded due to urban runoff and construction activities. Although usage varies with time, Lake Houston provides approximately 20 percent of Houston's raw water supply which constitutes a significant contribution. Degradation of the lake's water quality due to urbanization of its watershed could inhibit uses of the lake as well as require expensive increased treatment levels to be used to protect this important City water source. In an effort to protect Lake Houston's water quality, the City of Houston has participated in a Clean Lakes Assistance Program for a Phase I Diagnostic-Feasibility Study of Lake Houston (Turner Collie and Braden, 1989). As a continuation of the Clean Lakes Program effort, the City is presently undertaking the development of a Lake Houston Watershed Management Program.

The Program's primary goal is to develop and implement cost-effective storm water and other management techniques aimed at preventing further water quality degradation of Lake Houston and its inflows considering baseline conditions as determined to exist on January 1, 1991.

Study Area Description

Lake Houston is located in southeast Texas at latitude 29° 55' 08" and longitude 95° 08' 08" approximately 18 miles north of downtown Houston. Although initial storage capacity of the lake was 158,550 acre-feet, a 1983 survey revealed a capacity of only 130,730 acre-feet due to subsidence and sedimentation. The lake's water surface elevation is approximately 12,030 acres at the spillway crest with an average depth of around 12 feet and a maximum depth near 45 feet near the lake's dam.

The Lake Houston Watershed consists of approximately 2,828 square miles, including two major forks of the San Jacinto River. The East Fork, tributaries are Caney Creek, Peach Creek, the East Fork San Jacinto River and Luce Bayou which collectively comprise 983 square miles of drainage area. The West Fork drains an area of 1,741 square miles and includes Cypress Creek, Spring Creek, and the West Fork San Jacinto River. The remaining 104 acres of land drains directly into Lake Houston. Lake Conroe lies at the upper end of West Fork San Jacinto River and has a drainage area of approximately 450 square miles. Lake Conroe has a normal pool elevation of about 210 feet above mean sea level and a capacity of 430,260 acre-feet. The Lake Houston Watershed areas are summarized in Table 1.

TABLE 1
LAKE HOUSTON WATERSHED CHARACTERISTICS

Major Basin	Drainage Area (mi^2)	Lake Houston Watershed (percent)
Cypress Creek	322	11
Spring Creek	439	16
West Fork San Jacinto	980	35
Caney Creek	212	7
Peach Creek	158	5
East Fork San Jacinto	396	14
Luce Bayou	217	8
Lake Environs	104	4
TOTALS	2,828	100

The Lake Houston Watershed partially or fully encompasses seven counties including Montgomery, Harris, Waller, Grimes, Walker, San Jacinto, and Liberty and 21 municipalities. Numerous municipal utility districts and commercial developments are located throughout northern Harris and southern Montgomery counties creating a large unincorporated urban area in the watershed (Houston-Galveston Area Council, 1988). Major thoroughfares within the basin include IH-45, U.S. 59, U.S. 290, U.S. 75, SH 30, SH 105, SH 150, and SH 321. The Sam Houston National Forest represents a significant natural feature within the far northern reaches of the study basin.

The mean annual precipitation for the Lake Houston Watershed is near 45 inches although the annual rainfall for the period 1969 through 1988 varied from 70 inches in 1973 to 24 inches in 1980 (Turner Collie and Braden, 1989).

Data Management and SWMM Modeling

Estimates of Lake Houston's existing and possible future inflow water quality (including point and nonpoint contaminant loadings) were developed utilizing the PC ARC/INFO (Version 3.4D) GIS package and the Storm Water Management Model (SWMM) to describe and represent present and projected watershed physiographic conditions as well as associated water quality relationships. Intergraph's MicroStation and AutoCad were used in building watershed physiographic data sets from various sources with the results transported to PC ARC/INFO for "cleaning", processing and analyzing. To assist in formulating appropriate controls for the watershed, SWMM was also used to model projected watershed conditions with various storm water quality controls (Best Management Practices) in place.

An important aspect of the Program development involves the establishment of existing (base line) water quality conditions determined to exist as of January 1, 1991 as directed by the City of Houston. Since the primary goal of the Program will be to maintain the base line water quality of Lake Houston and its tributary inflows, a concerted effort has been made to establish the contaminant/parameter values that identify the physical, chemical and biological conditions of the studied waters.

Water quality impacts associated with nonpoint source pollutants on Lake Houston and its tributaries from existing watershed land use conditions are being evaluated. SWMM is being utilized to describe the rainfall-runoff and pollutant loading processes. Due to the SWMM model flexibility, existing point source loadings are also being evaluated separately from, and in combination with, the nonpoint source loadings generated from the various watershed land use areas. The following tasks describe the overall SWMM modeling effort:

1. Define Modeling Approach - An overall modeling approach was defined in detail and reviewed with the City to firmly establish goals, expected model outputs and means of evaluating water quality impacts. The effort specifically included:

    - development of detailed watershed modeling goals;
    - literature survey;
    - data review and compilation; and
    - outline of means of evaluating impacts.

2. Model Development and Testing - The SWMM model for the Lake Houston Watershed was developed and tested. This effort included setting up the basic rainfall-runoff as well as pollutant generation components of the model. The following items outlined below were performed:

    - delineate sub-basins
    - develop overall watershed hydrologic scheme
    - determine channel and other surface hydrologic parameters
    - develop soils/groundwater parameters
    - acquire and input rainfall data
    - determine transport block parameters
    - establish constituent concentrations per land use cover
    - make hardware/software modifications
    - develop software for facilitating modeling process
    - make initial model runs
    - test model

3. Model Calibration and Verification - The SWMM model was calibrated and verified with available data. This task involved determining the value of various model parameters and coefficients by utilizing recorded rainfall and runoff data

and the basic model setup established in previous tasks described above. Specific work elements included:

- compile calibration and verification data
- calibrate flow quantity routines
- calibrate flow quality routines
- verify flow quantity routines
- verify flow quality routines

4. <u>Model Baseline (1990) Watershed Conditions</u> - Utilizing the calibrated and verified model manipulated to reflect 1990 watershed conditions, the SWMM model was used to develop flows and associated water quality constituent loadings (and concentrations) for the maximum period of record for which basin-wide rainfall data are available (1960-1990).

Frequency analyses on the model results are being performed to determine the frequency that runoff volumes, flow rates, constituent concentrations, constituent loads and possibly other parameters exceed certain evaluation limits.

Water quality impacts to Lake Houston and its major tributaries due to changes in watershed land uses (i.e., nonpoint sources) will be identified by determining the associated pollutant loadings utilizing land use projections and the basic SWMM model setup. The following subtask describe the work effort:

1. Watershed land use cover conditions for the years 2000, 2010, 2020, 2030, and 2040 have been input into the PC ARC/INFO GIS system. Coefficients/ parameters representing this respective land use information were input into the SWMM to develop watershed models for the five specific time period dates identified above.

2. SWMM model runs for the five future development scenarios described above are being made to represent possible future runoff and water quality conditions.

3. Frequency analyses made on the five sets of SWMM results determine "how often" certain evaluation levels will be exceeded.

4. The outcome of the SWMM modeling as well as frequency analyses on the SWMM modeling results are being documented and compared to SWMM results associated with 1990 watershed conditions.

Utilizing the SWMM model, the type and level of alternative Best Management Practice (BMPs) required to reduce watershed pollutant loadings for each of the five future development scenarios modeled to 1990 levels are being developed.

Conclusions

The GIS mapping and data base management capabilities of PC ARC/INFO has worked well with the SWMM model in providing the technical basis for developing the Lake Houston Watershed Management Program. The capabilities of these analysis tools as well as the available rainfall-runoff data and physiographic watershed data made possible the very difficult undertaking of developing a watershed management program for a very large watershed having numerous point and nonpoint source sources of contamination.

References

Houston-Galveston Area Council. 1983. Spetrum, Vol. 10, No. 7. Houston-Galveston Area Council, Houston, Texas.

Turner Collie and Branden, Inc. 1989. Draft Phase I Diagnostic Study of Lake Houston.

Probability and Impact of an Observed Rare Sequence of Floods

by

Leo R. Beard[1], Hon. M.ASCE
W. H. Espey, Jr.[2], M.ASCE
Phil Combs[3], M.ASCE
Ben M. Littlepage[4]

Introduction

For various reasons, it may be desirable or necessary to assess the chance that a particular sequence of severe floods is extremely rare. This may be necessary to assess risk associated with damage during construction or project operation. It is the purpose of this paper to discuss approaches to solving such problems and suggest a solution technique.

As an example for discussion purposes, a particular sequence of severe floods observed during 5 years on Red River at Shreveport, Louisiana, is used in conjunction with a frequency curve of streamflow at that location obtained from the Corps of Engineers. The peak flows and exceedance frequencies are shown in Table 1.

The mathematical probability of future sequences having 5 years of flows related specifically to each of the 5 flows of record will first be examined, and it will be shown that this approach may have no practical application. Then an examination of relative severity in terms of damage is made, and this would appear to have relevance in relation to practical problems.

---

[1] Senior Staff Engineer, Espey, Huston & Associates, Inc., Austin, TX

[2] Chairman, Espey, Huston & Associates, Inc., Austin, TX.

[3] Chief, Hydraulics Section, Vicksburg District, U.S. Army Corps of Engineers, Vicksburg, MS.

[4] Executive Director, Red River Waterway Commission, Natchitoches, LA.

Probability Approach

Examining Table 1, one may note that in the 5-year period, a 150-year flood and two 30-year floods occurred and ask what the likelihood is that such a combination would be repeated in the future. One approach is to compute the probability that each of the 5 floods will be exceeded in a random 5-year period as many times as it was equaled or exceeded in the period of record. This calculation is shown in Table 2, which indicates that a future 5-year period has only one chance in about 26,000 of satisfying this requirement.

The first row of Table 2 calculates the probability that 238,000 flow will be exceeded exactly once during a random 5-year period by use of the binomial expansion term for exactly one event. The second line calculates the probability that 148,000 flow will be exceeded exactly once in the remaining 4 years, and the product of the two probabilities is the probability that both exceedances will occur in a random 5-year period. Carrying this on for the largest 3 events indicates that a random 5-year period would have these three events exceeded with a chance of about once in 2,500, and for the 5 events, the grand product becomes 0.000039, which represents a chance of one in about 26,000 that this exact condition will occur in a random 5-year period.

There are two things wrong with this computation. First, sequences with one flood exceeding the second-largest in addition to one that exceeds the largest may have two floods exceeding the largest, which would exceed the condition for the largest event. This problem carries on with the third, fourth and fifth largest event. To avoid this would not be reasonable, because two record floods with almost equal magnitude would require that a random sequence have a flood in between those magnitudes, which could be extremely rare.

Secondly, the five magnitudes in this case were only equaled and not exceeded during the period of record. Magnitudes slightly higher were exceeded by one less flood in each case. Considering that a magnitude slightly larger than the second-largest flood (say 148,100) was exceeded only once in the record, a magnitude slightly larger than the third-largest flood (say 146,100) was exceeded only twice in the record, etc., the calculation in Table 3 can be made. This shows a chance of one in about 460 that the specific five events shown in the table would be exceeded in a random 5-year period exactly as many times as they have been in the record. If this line of reasoning were useful, something between the results of Tables 2 and 3 would be appropriate.

Another probability approach is to compute the probability that three floods equal to or larger than the third largest recorded flood would occur in a random 5-year period, and then multiply by the probability that, a flood larger than the third largest recorded would also exceed the second largest recorded flood, and the probability that another flood exceeding the third largest would exceed the very largest recorded flood. This computation is as follows:

$$10(.9638)^2(.0362)^3(.0331/.0362)(.0067/.0362) = .000277$$

which represents one chance in about 13,000 that a random sequence would have the three largest floods larger than those in the record. Of course, this suffers from the same drawback of computing exceedances when working with equals.

These indications of extreme rarity are seriously misleading, however, since many sequences that do not conform to these exact requirements could be more severe. For example, a future sequence could have a slightly smaller minimum flood and much larger other floods and not comply with the requirements. Thus, one cannot say that the chance of a _more severe_ flood sequence is represented by these indications.

A Measure of Severity

In order to obtain a quantitative measure of sequence severity, it is suggested that the associated damage be such a measure. Using a hypothetical flow-damage relationship, damages during a repetition of the observed 5-year sequence would be as shown in Table 4. Then, taking random 5-year samples from the flow-frequency curve, the proportion of random samples having greater total damage can be determined.

Frequency-flow-damage relations used are shown in Table 5. Uniformly distributed random numbers in groups of 5 were generated and used as random probability or frequency values (as ratios, not percentages). Associated flows and damages were then interpolated from Table 5, and damages for each 5-year period were computed and compared with the period-of-record associated damages from the same relationship. It was found that 97.9 percent of the random sequences had less damages than those of the record period.

Conclusion

Accordingly, it may be concluded that the severity of the recorded sequence would be exceeded in about 2.1 percent of future sequences, or have about one chance in 45 or 50 of being exceeded in any future 5-year period under prevailing conditions. This is indeed a rare sequence, but while the computations in Tables 2 and 3 can appear reasonable, the results could be seriously misleading. Assessment of severity must be based on an objective function that has some relevant meaning in application of results.

Table 1

Recorded Flows At Shreveport

Year	Frequency*	Flow
1987	.0331	148,000
1988	.0362	146,000
1989	.0967	123,000
1990	.0067	238,000
1991	.5625	105,000

* Probability of being exceeded in a random year

Table 2

Coincident Probability Computation

Flow	Prob	n	$n(1-P)^{n-1}P$	Product
238,000	.0067	5	.0326	.03260
148,000	.0331	4	.1197	.00390
146,000	.0362	3	.1009	.000394
123,000	.0967	2	.1747	.000069
105,000	.5625	1	.5625	.000039

Table 3
Possible Alternative Computation

Flow	Prob	n	$n(1-P)^{n-1}P$	Product
148,100	.0331	5	.1447	.1447
146,100	.0362	4	.1296	.0187
123,100	.0967	3	.2367	.00444
105,100	.5625	2	.4922	.00218
0	1.0000	1	1.0000	.00218

Table 4
Recorded Flows and Hypothetical Damage

YEAR	PROB	FLOW	DAMAGE
1987	0.0331	148000.	1160.
1988	0.0362	146000.	1050.
1989	0.0967	123000.	57.
1990	0.0067	238000.	10100.
1991	0.5625	105000.	0.
	Total	12400.	

Table 5
Hypothetical Flow-Damage Relation

PROB	FLOW	DAMAGE
1.00	0	0
.50	120,000	0
.10	122,000	22
.05	137,000	547
.03	150,000	1,280
.02	166,000	2,430
.01	205,000	6,110
.005	255,000	12,200
.007	303,000	19,300
.001	420,000	40,500

Wastewater Treatment for Better Environment

Dr. Rawya Monir Kansoh[1]

Introduction

The human search for pure water supplies must have begum in prehistoric times.

The archaeological records of central water supply and wastewater disposal date back about 500 years, to Nippur of Sumeria. In the ruins of Nippur there is an arched drain with each stone being a wedge tapering downward into place (1). Water was drawn from wells and cisterns. An extensive system of drainage conveyed the wastes from the palaces and residential districts of the city.

The earliest recorded knowledge of water treatment is in the Sanskrit medical lore and Egyptian wall inscriptions (2). Sanskrit writings dating about 2000 B. C. tell how to purify foul water by boiling in copper vessels, exposure to sunlight, filtering through charcoal, and cooling in an earthen vessel.

The earliest known apparatus for clarifying liquids was pictured on Egyptian walls in the fifteenth and thirteenth centuries B. C. The first picture, in a tomb of the reign of Amenhotep II (1447-1420 B.C.), represents the siphoning of either water or settled wine. A second picture, in the tomb of Rameses II (1300-1223 B. C.), shows the use of wick siphons in an Egyptian kitchen.

In the second or first century B. C., man discovered that muscle power could be replaced by natural forces. The first machine to use these natural forces was water powered.

---

1 Assoc. Prof. of Hydraulics, Civil Engineering Dept. Alexandria University.

Around the tenth century, people began to use water power for other than agricultural purposes. By the end of the nineteenth century, electric energy could be developed from water power.

In this rapidly accelerating process, nature has been shamelessly exploited. Environmental pollution has reached the point that we can no longer obtain all that is desired.

Many pollutants can be carried long distances by air or water or on articles of commerce, threatening the health, longevity, livelihood, recreation.

The remedy for these foul conditions was to discharge human excrement into the existing storm sewers and add additional collection systems. This suggestion created the combined sewers of many older metropolitan areas. These storm drains had been constructed to discharge into the nearest watercourse. The addition of wastes to the small streams overtaxed the receiving capacities of the waters, and many of them were covered and converted into sewers.

Today the enormous demands being placed on water supply and wastewater disposal facilities. The standards for water quality have significantly increased concurrent with a marked decrease in raw-water quality. Evidence of water supply contamination by toxic and hazardous materials has become common, and concern broad water-related environmental issues has heightened.

Increased demands currently being placed on water supply and waste disposal have necessitated for broader concepts in the application of environmental engineering principles than those originally envisioned.

Wastewater Systems

The collection and movement of surface drainage and waste flows from residential, commercial, and industrial areas pose problems of a different nature from those for water supply. Waste must be transported from the point of collection to the treatment or disposal area as quickly as possible.

Collection and transportation of wastewater to the treatment plant is accomplished through a complex network of pipes and pumps of many sizes.

Wastewater Reuse

Wastewater reuse offers attractive alternatives to developing new water supplies that might have to be transported from distant locations.

The most widely available and least variable source of wastewater for reuse is municipal wastewater. It can be relied on to provide a dependable continuous flow having fairly stable physical, chemical, and biological characteristics.

In agriculture, wastewater from irrigation operations is a potential source of reusable water. However, the water is often highly contaminated with salts leached from the soil, and in many cases treatment is required if the water is to be reused on the fields.

Wastewater reclamation processes must ensure removal of residual pollutants to such a degree as to make the water acceptable for the designated reuse. The pollutants that must be removed depend on the desired use of the water and its previous use. If municipal wastewater has undergone secondary treatment, the remaining pollutants that typically must be removed to make the wastewater suitable for reuse are nitrates, phosphates, total dissolved solids, microorganisms, and refractory organics such as trace levels of pesticides.

Other forms of reuse are such as municipal, industrial, irrigation, storm water and recreational.

Wastewater, Treatment Systems

Wastewater treatment is important since it involves cleaning used water and sewage so it can be returned safely to our environment. Wastewater treatment is the "last line of defense" against water pollution.

Industrial Wastewaters

An industry has three possibilities for disposal of process wastewaters : (1) they may be treated separately in an industrial waste-treatment plant prior to discharge to a watercourse. (2) raw wastewaters may be discharged to the municipal treatment plant for complete treatment, or (3) industrial wastes can be pretreated at the industrial site prior to discharge in the municipal sewerage system.

Three categories of wastes should be excluded from the municipal sewers, those which (1) create a fire or explosion hazard (e.g., gasoline or cleaning solvents), (2) impair hydraulic capacity (e.g., paunch manure or sand), and (3) create a hazard to people, the sewer system, or the biological treatment system (e.g., toxic metal ions or hazardous organic wastes).

Facilities for handling wastewater are usually considered to have three components : collection, treatment and disposal. The effluent from a wastewater treatment plant must be disposed of in the environment. This can be into water, onto land or the water can be reclained and reused.

Wastewater treatment plants are important since they protect the public health, and they protect the water quality.

Wastewater normally comes from homes (domestic use); industries, schools and businesses; storm runoff and ground water. A treatment plant removes solids reduces organic matter and pollutants and restores oxygen. Wastewater treatment usually takes place in two steps.

a) Primary treatment which removes 40-60% of the solids using sanitary sewers and, or bar screen, grit chamber, a primary sedimentation tank.

b) Secondary treatment complete the process, so that about 90% of the pollutants are removed using aeration, secondary sedimentation tank and a disinfectant.

Sludge can be a useful by product of treated wastewater, it can be treated by stabilization dewatering and disposal, then can be used as a soil conditioner or as a fuel.

Sanitary, Storm and Combined Sewers

For most sewerage systems, the sewer coming into the treatment plant carries wastes from households and commercial establishments in the city or district, and possibly some industrial wastes. This type of sewer is called a *SANITARY SEWER* . All storm runoff from streets, land, and roof of buildings is collected separately in a *STORM SEWER* which normally discharges to a water course without treatment. In some areas only one network of sewers has been laid beneath the city to pick up both sanitary wastes and storm water in a *COMBINED SEWER* . Treatment plants that are designed to handle the sanitary portion of the wastes sometimes must be bypassed during storms due to inadequate capacity, allowing untreated wastes to be discharged into receiving waters. Separation of combined sewers into sanitary and storm sewers is very costly and difficult to accomplish.

Combined wastewater collection systems have only one sewer pipe network to collect domestic wastewater, industrial wastes, and storm runoff water.

Infiltration and Inflow

Even in areas where the sanitary and storm sewers are separate, *INFILTRATION* of groundwater of storm water into sanitary sewers through breaks or open joints can cause high flow problems at the treatment plant.

Infiltration is groundwater entering sewers and building connections through defective joints and cracks in pipes and manholes. Inflow is water discharged into service connections and sewer pipes from foundation and roof drains, outdoor paved areas, cooling water from air conditioners, and unpolluted discharges from businesses and industries. Excessive infiltration and inflow cause surcharging of sewer lines with possible backup of sanitary wastes into basements, hydraulic overloading of treatment facilities, and bypassing of pumping stations or processing plants. Proper design, selection of sewer pipe with watertight joints, and supervision during construction can limit the quantity of seepage. Specifications for new construction permit a maximum infiltration rate of 200 - 500 gpd/mi/in. of pipe diameter. This quantity of flow is equal to about 5% of the peak hourly, or 10% of the average, domestic wastewater flow rate.

Sanitary sewers are normally placed at a slope sufficient to produce a water velocity (speed) of approximately two feet per second when flowing full. This velocity will usually prevent the deposition of solids that may clog the pipe or cause odors. Manholes are placed every 91 to 152 feet to allow for inspection and cleaning of the sewer.

When low areas of land must be sewered or where pipe depth under the ground surface becomes excessive, pump stations (Fig. 1) are normally installed. These pump stations lift the wastewater to a higher point from which it may again flow by gravity, or the wastewater may be pumped under pressure directly to the treatment plant. A large pump station located just ahead of the treatment plant can create problems by periodically sending large volumes of flow to the plant one minute, and virtually nothing the next minute. These fluctuating flows can be reduced by using variable speed pumps or short pumping cycles.

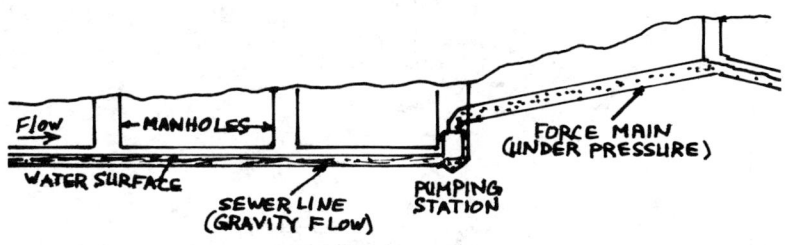

Figure 1. Typical Under Ground Flow System

Conclusion

Wastewaters from households, industries, and combined sewers are collected and transported to the treatment plant with the effluent commonly disposed of by dilution in rivers, lakes, or estuaries. This is normally the only feasible method of disposal and, for several communities on rivers, the only system that ensures adequate water resources for downstream users during drought flows. Other means include irrigation, infiltration, evaporation from lagoons, and submarine outfalls extending into the ocean.

The location of a typical municipal wastewater treatment plant is illustrated (Fig. 2).

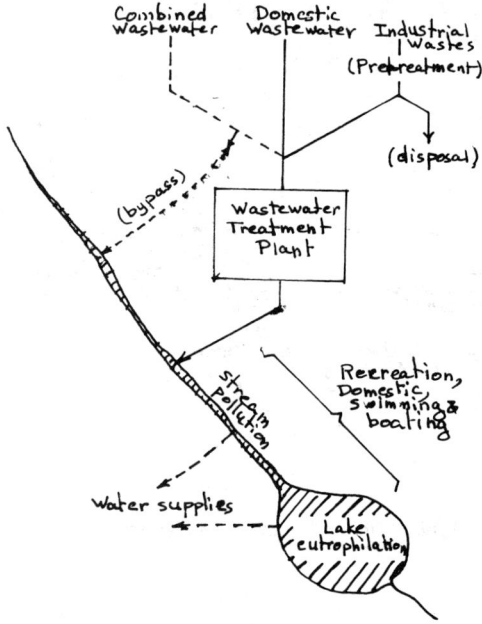

Figure 2. Location of a municipal wastewater treatment plant.

The lesson is that populations increases, but water and land resources do not. Consequently, the use and control of these resources must be nearly perfect to maintain our way of life. Exercising this control will require the skillful blending of state of the art technology with a host political, social, economic, and organizational elements. Ii is the technology of water supply and pollution control.

References

1) Durant, W. "Our Oriental Heritage" (N. Y., Simon and Schuster, 1954).
2) Baker, M.N. 'The Quest for Pure Water" (N. Y. American water works Association).
3) Clark, J. W., Viessman W. "Water Supply and Pollution Control,"Index Educational Publishers, second edition.
4) Cleasby, J. L."Declining - Rate Filtration", G. American water works Assoc. 72, 1980.
5) Chow, V. T. "Handbook of Applied Hydrology (Mc. Graw - Hill, 1964).
6) Thackston, E. L. "Notes on Sewer System Design," Vanderbilt Univ., 1982.

NUMERICAL MODELLING OF SALINITY TRANSPORT IN A SHALLOW WELL-MIXED TIDAL REACH

G S REDDY[1] Non-member   and   S N GHOSH[2] Non-member

ABSTRACT

This paper is concerned primarily with the development of numerical salinity transport model using 8-point implicit finite difference scheme for predicting the salinity variation in a comparatively long and narrow well-mixed tidal reach. The formulation of the model is based on the one dimensional unsteady flow equations of continuity, momentum and salinity transport equation. The finite difference linear algebraic form of the governing equations have been solved through development of a suitable software package coded as STM (Salinity Transport Model) capable of dealing with irregular non-uniform cross sections and roughness variation in a coupled manner using double sweep algorithm. For calibration of the model a suitable reach of River Hooghly has been selected for which published information is available with appropriate input of initial and boundary conditions. The results of the tests have been found to be satisfactory.

INTRODUCTION

In a tidal river the effect of density variation due to salinity effect is an important factor to be considered while determining the flow characteristics. Inspite of its importance most of the investigators, Dornhelm and Woolhier(1968), Harleman , Lee and Hall(1968), Ouellet and Cetceau(1975) and others have developed models for the determination of salinity variation in a tidal river

---

1. Research Scholar, Civil Engineering Department, Indian Institute of Technology, Kharagpur 721 302, INDIA.
2. Professor, Civil Engineering Department, Indian Institute of Technology, Kharagpur 721 302, INDIA.

ignoring the effect of density variation. Grubert(1976) has described a model dealing with salinity transport considering the density effect. Reddy and Ghosh(1993) have proposed a new solution approach for the coupled solution of simultaneous set of linear algebraic equations, obtained from the governing equations adopting implicit finite difference techniques which are suitable for schematised section, constant value of resistance coefficient and longitudinal dispersion coefficient. The present paper deals with the solution of the problem adopting eight-point implicit scheme which has been formulated in such a fashion so as to be applicable under all conditions.

MODEL FORMULATION

Governing Equations

The depth averaged unsteady flow equations taking account of density variation for non-prismatic channel sections of a river reach can be represented by the following partial differential equations

Equation of Continuity

$$\frac{\partial A}{\partial t} + \frac{\partial Q}{\partial x} + \frac{A}{\rho}\frac{\partial \rho}{\partial t} + \frac{Q}{\rho}\frac{\partial Q}{\partial x} = 0 \qquad (1)$$

Equation of Momentum

$$\frac{1}{gA}\frac{\partial Q}{\partial t} - (\alpha+\beta)\frac{Q}{A}\frac{\partial A}{\partial t} - \beta\frac{Q^2}{gA^3}\frac{\partial A}{\partial x} + \frac{\partial h}{\partial x} + \frac{\partial z}{\partial x} + \frac{|Q|Q}{C^2A^2R} + (\alpha-\beta)\frac{Q}{gA\rho}\frac{\partial \rho}{\partial t} +$$

$$\frac{1}{\rho}\left[\frac{Q^2}{gA} + h\right]\frac{\partial \rho}{\partial x} = 0 \qquad (2)$$

Salinity Transport

$$\frac{\partial(SA)}{\partial t} + \frac{\partial(SQ)}{\partial x} - \frac{\partial}{\partial x}\left(DA\frac{\partial S}{\partial x}\right) = 0 \qquad (3)$$

In the equations (1) and (2) ρ can be related with salinity as

$$\rho = (1+kS)\rho_w \qquad (4)$$

Accordingly the equation (3) can be written in terms of density as

$$\frac{\partial \rho}{\partial t} + \left(\frac{Q}{A} - \frac{D_x}{A}\frac{\partial A}{\partial x} - \frac{\rho}{\rho w}\frac{\partial D}{\partial x}\frac{\partial \rho}{\partial x}\right)\frac{\partial \rho}{\partial x} - D_x \frac{\partial^2 \rho}{\partial x^2} = 0 \qquad (5)$$

Where

t= time, x= streamwise co-ordinate, ρ= density of salt water mixture, A= flow cross sectional area, Q= flow discharge, h= water level w.r.t datum, g= acceleration due to gravity, C= Chezy's frictional coefficient, n= Manning's roughness coefficients, R= hydraulic radius, z= bed elevation with respect to datum, α β are the energy and momentum coefficients respectively, S= salinity, ρw= density of fresh water, k= 0.00075 D_x= $(D\rho/\rho w)$, D= dispersion coefficient=$\epsilon UR^{5/6}$

where ϵ= diffusion coefficient, U= mean velocity

Discritization of Governing Equations

Equation (1),(2), and (5) represent conservation laws over some differential length of an well mixed tidal reach. An 8-point implicit finite difference scheme is adopted to descritize the governing equations of flow and salinity transport. In this computational scheme, the salinity can be computed at alternative grid points. The adopted scheme is stable, accurate and impose the salinity variables around the flow variables to over come the double non-linearity of governing differential equations. It is also convenient to impose the boundary conditions and solve the governing equations in a coupled manner.

The differential derivatives are discritized according to the computational grid diagram shown in fig. 1. as they are respectively

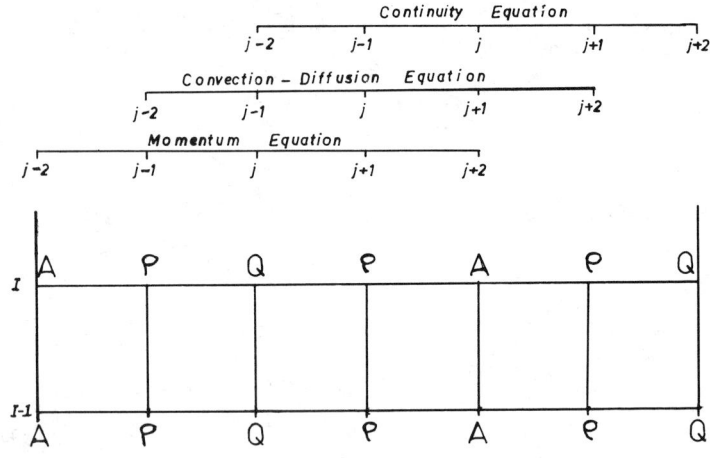

Fig. 1. Computational Grid Diagram

SALINITY TRANSPORT

$$\frac{\partial Q}{\partial t} = \frac{Q_j^i - Q_j^{i-1}}{\Delta t} \quad ; \quad \frac{\partial A}{\partial t} = \frac{(A_{j+2}^i + A_{j-2}^i) - (A_{j+2}^{i-1} + A_{j-2}^{i-1})}{2\Delta t}$$

$$\frac{\partial Q}{\partial x} = \frac{(Q_{j+2}^i + Q_{j+2}^{i-1}) - (Q_{j-2}^i + Q_{j-2}^{i-1})}{8\Delta x} \quad ; \quad \frac{\partial h}{\partial x} = \frac{(h_{j+2}^i + h_{j-2}^i) - (h_{j+2}^{i-1} + h_{j-2}^{i-1})}{8\Delta x}$$

$$\frac{\partial \rho}{\partial x} = \frac{(\rho_{j+1}^i + \rho_{j+1}^{i-1}) - (\rho_{j-1}^i + \rho_{j-1}^{i-1})}{4\Delta x} \quad ; \quad \frac{\partial \rho}{\partial t} = \frac{(\rho_{j+1}^i + \rho_{j-1}^i) - (\rho_{j+1}^{i-1} + \rho_{j-1}^{i-1})}{2\Delta t}$$

On substituting the above differential derivatives the equation (1) and (2) can be written in the following form.

$$\alpha_j A_{j-2}^1 + \beta_j \rho_{j-1}^1 + \gamma_j Q_j^1 + \delta_j \rho_{j+1}^1 + \varepsilon_j A_{j+2}^1 = \tau_j \quad (6)$$

$$\alpha_j Q_{j-2}^1 + \beta_j \rho_{j-1}^1 + \gamma_j A_j^1 + \delta_j \rho_{j+1}^1 + \varepsilon_j Q_{j+2}^1 = \tau_j \quad (7)$$

$$\frac{\partial \rho}{\partial t} = \frac{\frac{1}{4}(\rho_{j+2}^i - \rho_{j+2}^{i-1}) + \frac{1}{2}(\rho_j^i - \rho_j^{i-1}) + \frac{1}{4}(\rho_{j-2}^i - \rho_{j-2}^{i-1})}{\Delta t}$$

$$\frac{\partial \rho}{\partial x} = \frac{(\rho_{j+2}^i + \rho_{j+2}^{i-1}) - (\rho_{j-2}^i + \rho_{j-2}^{i-1})}{8\Delta x} \quad ; \quad \frac{\partial^2 \rho}{\partial^2 x} = \frac{(\rho_{j+2}^i - 2\rho_j^i + \rho_{j-2}^i) + (\rho_{j+2}^{i-1} - 2\rho_j^{i-1} + \rho_{j-2}^{i-1})}{8\Delta x^2}$$

Similarly by substituting the above derivatives the equation (5) can be written as

$$\psi_j \rho_{j-2}^1 + \varnothing_j \rho_j^1 + \zeta_j \rho_{j+2}^1 = \tau_j \quad (8)$$

where α, β, γ, δ, ε, τ, are the coefficients of dynamic and continuity equation and ψ, $\varnothing$, and ζ are the coefficients of salinity transport equation respectively and they are computed at intermediate time step of t=i and t=i+1

Solution Procedure

The equations (6) and (7) can be written using general variable 'Z' instead of 'A' and 'Q' as

$$\alpha_j Z_{j-2}^1 + \beta_j \rho_{j-1}^1 + \gamma_j Z_j^1 + \delta_j \rho_{j+1}^1 + \varepsilon_j Z_{j+2}^1 = \tau_j \quad (9)$$

Introducing a set of quasi-constants E_j, F_j and G_j relating each unknown to the unknown in the neighboring 'Z' points, we have

$$Z_{j-2} = E_j\ \rho_{j-1} + F_j\ Z_j\ + G_j \quad (10)$$
$$\rho_{j-2} = E_{j+1}\ Z_j\ + F_{j+1}\ \rho_{j+1}\ + G_{j+1} \quad (11)$$
$$Z_j = E_{j+2}\ \rho_{j+1} + F_{j+2}\ Z_{j+2} + G_{j+2} \quad (12)$$

On substituting of above relations into (9) gives the following set of recursive formulaes.

$$E_j = \frac{-(\alpha_{j-2}E_{j-2}F_{j-1}+\beta_{j-2}F_{j-1}+\delta_{j-2})}{(\alpha_{j-2}E_{j-2}E_{j-1}+\alpha_{j-2}F_{j-2}+\beta_{j-2}E_{j-1}+\gamma_{j-2})} \quad (13)$$

$$F_j = \frac{-\varepsilon_{j-2}}{(\alpha_{j-2}E_{j-2}E_{j-1}+\alpha_{j-2}F_{j-2}+\beta_{j-2}E_{j-1}+\gamma_{j-2})} \quad (14)$$

$$G_j = \frac{(\tau_{j-2}-\alpha_{j-2}E_{j-2}G_{j-1}-\alpha_{j-2}G_{j-2}-\beta_{j-2}G_{j-1})}{(\alpha_{j-2}E_{j-2}E_{j-1}+\alpha_{j-2}F_{j-2}+\beta_{j-2}E_{j-1}+\gamma_{j-2})} \quad (15)$$

In similar way for the equation (8) the recursive relations can be written as

$$E_j = \frac{-(\alpha_{j-2}E_{j-2}F_{j-1})}{(\alpha_{j-2}E_{j-2}E_{j-1}+\alpha_{j-2}F_{j-2}+\beta_{j-2})} \quad (16)$$

$$F_j = \frac{-\gamma_{j-2}}{(\alpha_{j-2}E_{j-2}E_{j-1}+\alpha_{j-2}F_{j-2}+\beta_{j-2})} \quad (17)$$

$$G_j = \frac{(\tau_{j-2}-\alpha_{j-2}E_{j-2}G_{j-1}-\alpha_{j-2}G_{j-2})}{(\alpha_{j-2}E_{j-2}E_{j-1}+\alpha_{j-2}F_{j-2}+\beta_{j-2})} \quad (18)$$

Using the boundary conditions and relations (13),(14) and (15) for every grid point of Z (A and Q) along the reach E_j, F_j, and G_j can be determined. Similarly E_j, F_j, and G_j at 'ρ' points can be calculated using the relations (16)-(18). Using these values and upstream boundary conditions A,Q, and ρ are evaluated by successive use of (10) and (11). To obtain the fully centered description of coefficient, the equations are solved by default two times at every time step.

RESULTS AND DISCUSSIONS

The limit of the model boundaries has been fixed about 146km downstream from Kalna. The model boundaries have been chosen at Falta and Kalna as downstream and upstream boundary conditions respectively. The time step 500sec. and grid distance 3000m are selected for the computational purposes. The value of 'C' was calculated using the Manning's roughness coefficient are available from the reference, McDowell and David Prandle(1972)

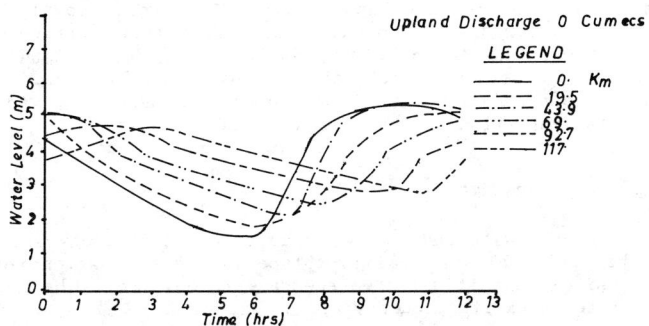

Fig. 2. Tidal Levels at Different Locations

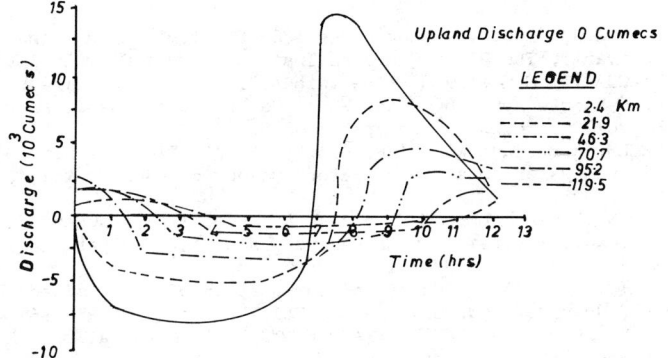

Fig. 3. Tidal Discharges at Different Locations

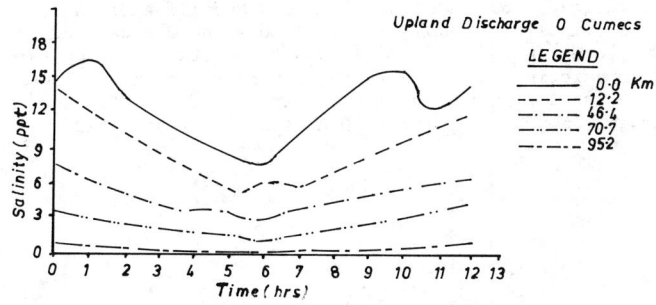

Fig. 4. Salinity Variations at Different Locations

The software programme has been run for calibration purpose using the available input data of River Hooghly, Basu(1968), Bhattcharya(1973). The computed results of water surface elevations, discharges, and salinity are shown in fig.2 to fig.4. using the variable chezy's coefficient and longitudinal dispersion coefficient. Comparison of the predicted and observed values of salinity variation has been found to be satisfactory.

CONCLUDING REMARKS

The software and the proposed solution strategy has been found to be applicable in real prototype problems. However for using in real situations it will be necessary to have reliable field data in order to check its accuracy and sensitivity.

APPENDIX I   REFERENCES

BASU, A.N., (1968),' Roughness Coefficient for the River Rupnarayana', The River Research Institute, West Bengal, INDIA.
BHATTACHARYA, S.K., (1973),' Deltaic Activities of Bhagirathi Hooghly River System', Jour. of Water Ways, Harbors and Coastal Engrg. Div., ASCE, vol.99, No.WW1, Proc. paper 9538.
DORNHELM, R.B., and WOOLHIER, D.A., (1968),' Digital Simulation of Estuarine Water Quality', Water Resources Res., Vol.4, No.6, pp.1317-1328.
HARLEMAN, D.R.F., LEE, C.H. and HALL, L.C., (1968),'Numerical Studies of Unsteady Dispersion in Estuaries', Jour. of Sanitary Engrg. Div., ASCE, Vol.94 , SA5, 897-911.
GRUBERT, J P, (1976),' Numerical Computation of Well-mixed Esturine flows', Jour. of Hyd. Div., ASCE, Vol.102, No.7, pp.955-967.
McDOWELL, D.M. and DAVID PRANDLE, (1972),' Mathematical Model of River Hooghly', Jour. of Water Ways, Harbors and Coastal Engrg. Div., ASCE, vol.98, No.WW2, pp. 225-241.
OUELLET, Y. and CETCEAU, J.(1975), ' Simulation of the salinity Distribution in the St. Lawrence Estuary by a Two-dimensional Mathematical model', Symposium on Modelling Technique, Vol.II, Annual Symposium of Water Ways, Harbors and Coastal Engineering.
REDDY, G.S. and GHOSH, S.N., (1993),' Aspects of computationalmodel for predicting the salinity transport in a well-mixed tidal reach', Ist International IAWQ Conference on Diffussive Pollution, Chicago, IL, USA, September 20th-24th 1993.(accepted)

New York Bight Three-Dimensional Water Quality Model

By Ross W. Hall[1] and Mark S. Dortch[2]

Abstract

The numerical modeling technology developed for Chesapeake Bay was applied to the Middle Atlantic Bight extending 550 km from Cape May, NJ, northeasterly to Nantucket Island, MA, and extending approximately 160 km offshore to the continental shelf. The study included New York Bight, tidally-influenced estuaries, harbors, bays, and Long Island Sound. The objective of the study was to investigate whether a numerical model of this extensive, coupled system could be successively applied to simulate hydrodynamics and water quality. The paper provides a description of graphical procedures used for model-prototype comparison.

Introduction

A New York Bight (NYB) Water Quality Model Study was an investigation of the feasibility of applying a numerical three-dimensional water quality model to assess the impacts of natural and human activities on the NYB. The word feasibility is used because the NYB system is very large and complex. The system consists of: tidally-influenced estuaries, harbors, and bays; Long Island Sound; the Apex region between the open waters of the Bight and the harbors/estuaries; and the Bight, which extends (for this study) from Cape May, NJ, northeasterly approximately 550 km along the coastline to Nantucket Island, MA, and extends approximately 160 km offshore to beyond the continental shelf (Figure 1). The depth of the study site varies from 3 m to over 2000 m seaward of the continental shelf.

The modeling technology recently developed for the Chesapeake Bay (Cerco and Cole 1993) was applied to the Bight. This technology consisted of three-dimensional, time-varying hydrodynamic and water quality models. The hydrodynamic model (HM) provides the

---
[1] Research Limnologist, [2] Supervisory Research Hydraulic Engineer, USAE Waterways Experiment Station, 3909 Halls Ferry Road, Vicksburg, MS. 39108-6199

circulation required by the transport terms of the water quality model (WQM). The HM is applied and the output is stored and subsequently used as input for the WQM.

The HM employed a 76 x 45 curvilinear or boundary-fitted, planform grid (Figure 2) and stretched (sigma) coordinates for the vertical dimension. The HM used 10 vertical layers. The WQM grid was a direct overlay of the HM grid. The WQM also had 10 layers for a total of 25,010 computational cells. Simulation of the dissolved oxygen (DO) hypoxia event of the summer of 1976 was selected for model application.

A criterium used for assessing the usefulness of the WQM to simulate the hypoxia event of the summer of 1976 was reproduction of the observed spatial patterns and temporal trends. However, the temporal and spatial sparsity of prototype observations for water quality made model-prototype comparisons difficult. Spatial and temporal averaging procedures were used for comparison of model simulation results with sparse prototype observations.

Model-Prototype Comparisons

Monthly temporal averaging of all prototype measurements and model simulations was done to mitigate the temporal sparsity of prototype observations. Several additional techniques were used to provide meaningful model-prototype comparisons:

1) scatter plots of computed vs observed
2) regional comparisons
   a) spatial average
      (1) depth profiles
      (2) time series of surface and bottom layers
   b) point comparisons
      (1) depth profiles
      (2) time series of surface and bottom layers
3) transect plots

The spatial sparsity of prototype observations makes model-prototype calibration comparisons difficult. For example, there were few stations that were frequently sampled at the same location. One technique used to provide meaningful model-prototype comparisons was spatial aggregation to provide larger sample sizes. This technique consisted of computing monthly averages of multiple station values that were contained within a region. The NYB was subdivided into 11 regions as shown on Figure 3. The segmentation scheme is a modification but reflects the segmentation used by NOAA (1979). The data obtained in the eastern (up-coast) regions were used for up-coast boundary conditions and not included in comparisons of model versus observed data.

The use of regional aggregates can also result in biases that confound model calibration. The model results were averaged for all model time steps and for all model cells within the region, thus producing a true monthly, regional average. However, the prototype data are far from representing true monthly, regional averages because of the sparsity of data. Therefore, specific sampling locations were selected to compare measured and simulated data. The sample locations selected for point comparisons were based on the availability of data (e.g., temperature, salinity, and dissolved oxygen). Figure 3 displays the relationship between the spatial regions and selected point comparisons. The regions are outlined with bold lines and the selected points marked with solid circles.

The regional spatial and point comparisons were displayed through depth profiles for various months and time series for a layer (e.g., near surface or near bottom). The time series plots represent the monthly averaged surface and bottom layer model and prototype data between April and September 1976.

Measured data were aggregated by area (Apex, NJ Coastal Zone, etc), layer, and month. Each measured datum was associated with the corresponding aggregated simulated result and displayed by plotting. The scatter plots display predicted versus observed data for all regions and were effective for gross qualitative calibration.

Transect plots are another procedure for comparing model-prototype results. Figure 4 shows the five transects that were used. Their location and width were selected to maximize the amount of available prototype data. With this technique, model results and observations were averaged over each month and across the transect. The results were plotted as monthly-averaged, surface and bottom layer concentration versus distance from the seaward boundary along the transect.

Results

The comparison techniques are demonstrated using DO. Figure 5 displays the scatter plot for DO. Examination of Figure 5 reveals that on the average the WQM underpredicts the DO concentration. A statistical comparison of measured and simulated DO indicated that the average underprediction was 0.55 g m^{-3}.

Figure 6 displays the spatial average DO vertical profile for the NJ near-shore region for July. Surface and bottom simulated DO were equivalent to measured values; however, near surface simulated DO was less than measured. Examination of prototype data revealed that the measured DO average for the near surface (Layers 2 and 3) consisted of one observation each. In contrast, the measured average for Layer 1 consisted of 80 observations.

The near surface and near bottom DO time series for the NJ near-shore region is presented in Figure 7. The temporal pattern of simulated and measured DO were similar. The horizontal spatial pattern of DO is compared in transect plots. Figure 8 represents

Transect 2 for August. The comparison reveals equivalent patterns with slight underprediction of surface layer DO.

Summary and Conclusions

An extensive HM and WQM grid was employed to resolve the large and complex NYB geometry. The combination of voluminous simulated data with sparse prototype measurements required innovative graphical procedures for model-prototype comparison. The abundant simulation output required summarization and the sparse prototype data required aggregation for manageable interpretation. The selected graphical procedures used months and regions/transects for summarization. The scatter plots represented a further summarization over months and regions.

References

Cerco, C.C., and T. Cole. 1993. "Application of the Three-Dimensional Eutrophication Model, CE-QUAL-ICM, to Chesapeake Bay," Technical Report EL-93- (in preparation), US Army Waterways Experiment Station, Vicksburg, MS.

National Oceanic and Atmospheric Administration. 1979. "Oxygen Depletion and Associated Benthic Mortalities in New York Bight, 1976," NOAA Professional Paper 11, Rockville, MD.

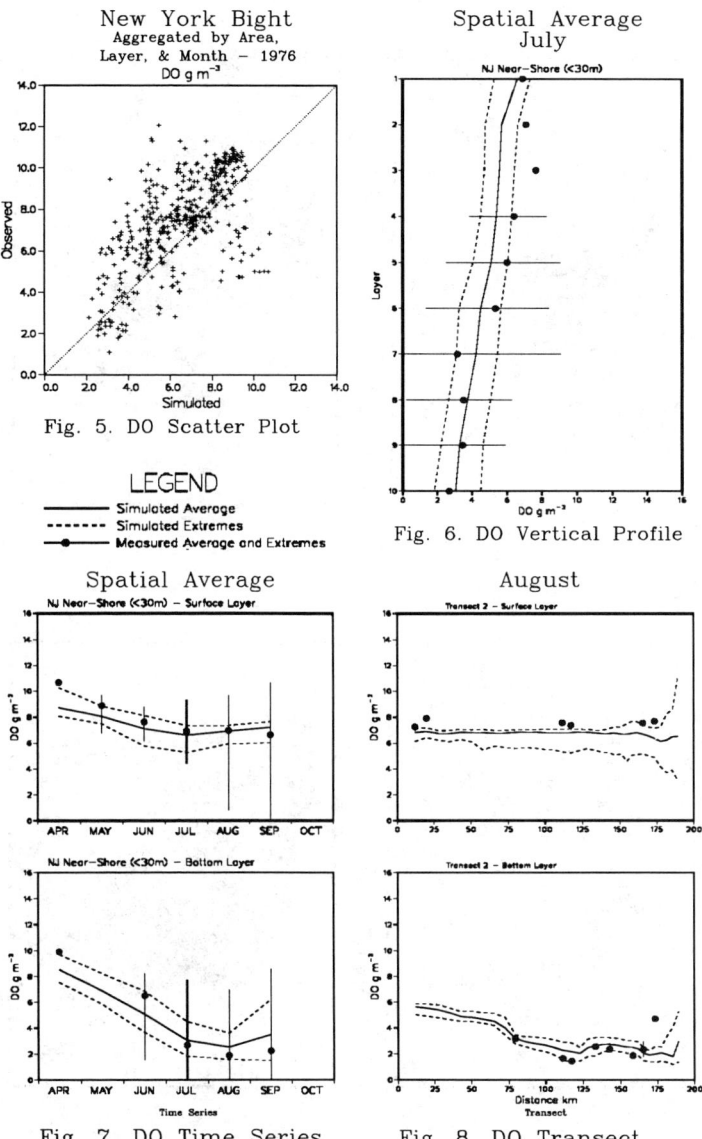

Fig. 5. DO Scatter Plot

Fig. 6. DO Vertical Profile

Fig. 7. DO Time Series

Fig. 8. DO Transect

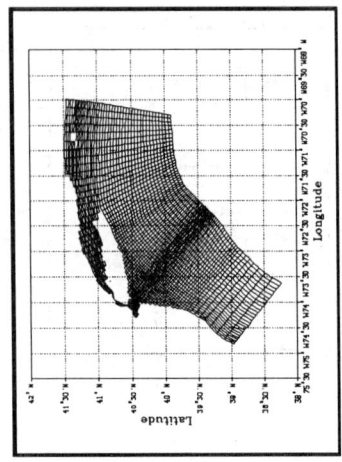

Fig. 1. Study Area

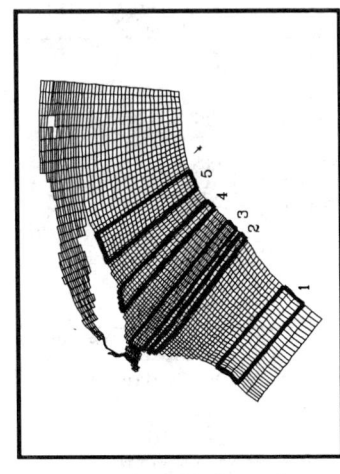

Fig. 2. Hydrodynamic Model Grid

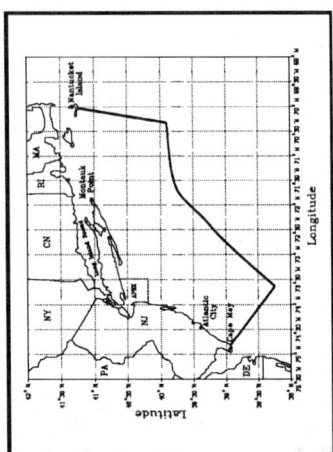

Fig. 3. Segmentation

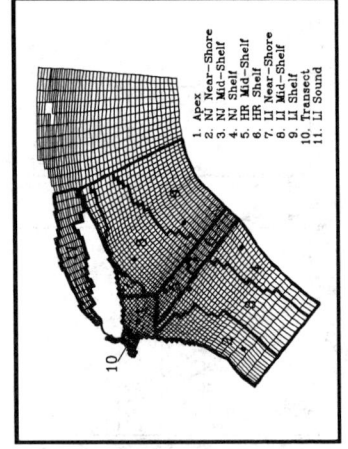

Fig. 4. Transects

Hydraulic Performance of a Flexible Curtain Used for Selective Withdrawal
A Physical Model and Prototype Comparison

Tracy B. Vermeyen[1], A.M. ASCE and Perry L. Johnson[2], M. ASCE

Abstract

Reclamation (U.S. Bureau of Reclamation) studied a 1:120 scale undistorted density stratified physical model to develop a lightweight flexible curtain barrier that would control the outflow temperatures from Lewiston Lake, California. The model scale was selected to allow study of a 3,000-ft length of reservoir. The reservoir topography in this reach substantially influences withdrawal layer characteristics and temperature profiles at the intake structures. Unfortunately, this scale yielded flow Reynolds numbers in the laminar to turbulent transitional range. Consequently, mixing and flow entrainment were underestimated in the model. Despite the scale effects, the model was a valuable qualitative tool for evaluating various curtain designs and hydraulic responses.

Based on model results, a 35-ft-deep, 1,000-ft-long curtain was developed which optimized curtain and topography interaction to yield the best possible release temperature control. The curtain was installed in August 1992. Initial field monitoring confirms that the design objectives were met.

This paper presents model and field observed performance with a discussion of model observations that influenced the final curtain design. The paper also interprets field performance and relates differences in model findings to scaling distortions.

Background

A value engineering study (USBR, 1990) identified flexible curtain(s) as a potential solution for reducing temperature of water released from Lewiston Lake. Biologists expect cooler outflows will enhance late summer spawning conditions for anadromous

---

[1]Hydraulic Engineer, U.S. Bureau of Reclamation, D-3751, PO Box 25007, Denver, CO 80225.

[2]Hydraulic Engineer, U.S. Bureau of Reclamation, D-3751, PO Box 25007, Denver, CO 80225.

fish in the Sacramento and Trinity River Basins. Salmon populations in both rivers are in decline, which has been attributed in part to warm water releases associated with drought conditions. The Chinook salmon "winter run" on the Sacramento River has been listed as threatened under the Endangered Species Act. Likewise, the population of "spring run" salmon is also in decline and listing may result.

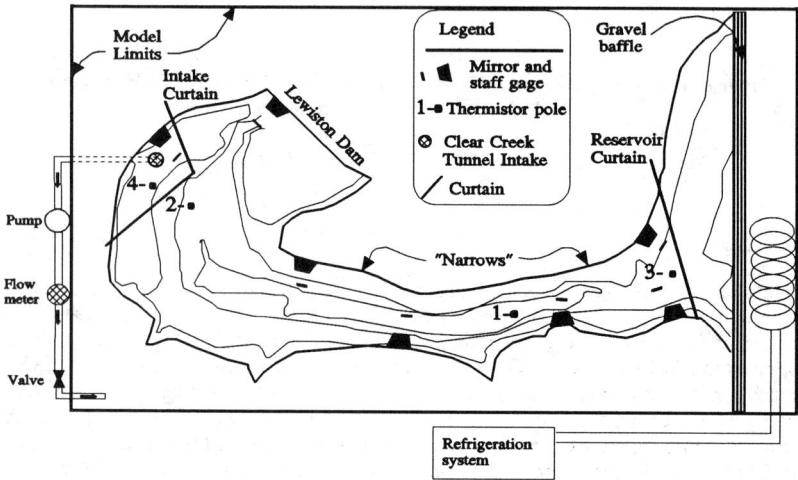

Figure 1. - Plan view of 1:120 scale hydraulic model of Lewiston Dam and reservoir.

Preliminary curtain designs were developed using selective withdrawal theory for submerged weirs developed by Bohan and Grace, 1973. Hydraulic performance of several curtain options with varying lengths and depths was examined theoretically prior to model testing. Curtain options which produced maximum cooling and were of reasonable length were chosen for model testing.

Physical Model Study

A 1:120 scale density stratified physical model was used to examine the effectiveness of flexible curtain structures to reduce water temperatures released through Clear Creek Tunnel and the Judge Francis Carr Powerplant (fig. 1). The scale was chosen to include, in a limited laboratory space, both potential curtain locations and topography that exerts a strong influence on withdrawal characteristics.

Unfortunately, the 1:120 scale model yielded flow Reynolds numbers in the laminar to turbulent transitional range, as a result, turbulent mixing and flow entrainment were underestimated in the model. However, the model served as a qualitative tool for evaluating various curtain designs and hydraulic responses. This study included the following elements:

- **Baseline Withdrawal Characteristics** - This task included evaluating withdrawal layer thickness, velocity profiles at several cross sections, and modifications to density stratification.

- **Intake Curtain No. 1, Option 1** - Evaluated a 1,200-ft-long, 30-ft-deep curtain around the Clear Creek Tunnel intake.

- **Intake Curtain No. 1, Option 2** - Evaluated a 1,200-ft-long, 40-ft-deep curtain around the Clear Creek Tunnel intake.

- **Reservoir Curtain** - Evaluated a 720-ft-long, 30-ft-deep curtain upstream from the "narrows."

- **Combinations** - Evaluated simultaneous use of Curtain no. 1, Option 2, with the reservoir curtain.

Reservoir Stratification - Stratified density profiles in the model were created by floating a warm water layer above a cold water layer. This filling technique resulted in a stratified temperature profile similar to that which forms in Lewiston Lake.

Model Operation - After reservoir stratification was established, tests were started by withdrawal of a constant flow rate through the Clear Creek Tunnel intake. Model replacement water was supplied from the cold water reservoir, simulating the cold water available from the hypolimnion (cooler bottom stratum).

A typical test length was 2 or 3 hours (22 to 33 hours - prototype), which allowed a quasi-steady-state condition to be established. A true steady-state condition could not be established because the finite epilimnion (warm surface stratum) in the model reservoir eroded slowly as the test progressed.

Testing and Data Acquisition - Testing and data acquisition for this model study included measuring temperature profiles at three locations for baseline conditions, and a fourth location was added for curtain tests (fig. 1). Temperature data were also collected in the intake structure to measure average withdrawal temperatures. Mirrors, dye, and staff gages were used for flow visualization and to estimate the withdrawal layer thickness.

Model Results - In general, model data indicated that both the reservoir and intake curtains were effective in cooling intake temperatures when compared to the baseline condition. The 40-ft-deep intake curtain was most effective at cooling withdrawal temperatures for discharges less than 1,000 ft^3/s. The reservoir curtain was more effective for discharges ranging from 2,500 to 4,000 ft^3/s. Data analyses indicated that the intake curtain was ineffective at high discharges because of warming caused by mixing as cooler water passed through the "narrows"- a restricted portion of the reservoir (fig. 1). Mixing primarily occurs in the shear zone which develops between the withdrawal layer and the warm water wedge which forms behind the curtain. Turbulent mixing also occurs as the flow accelerates through the reduced cross section in the "narrows."

Locating the reservoir curtain upstream from the "narrows" reduced the shear zone mixing because of lower curtain approach velocities. Likewise, warming caused by mixing in the "narrows" was reduced by isolating this mixing zone from the main reservoir body, which effectively limited the available warm water. For the reservoir and discharge conditions observed in the physical model, the reservoir curtain reduced water temperatures released from Lewiston Lake to the Clear Creek Tunnel by 2.5 °F.

Withdrawal characteristics of the Clear Creek Tunnel intake were evaluated with simultaneous use of the intake and reservoir curtains. This configuration provided the highest performance for a range of operational discharges of 1,000 to 4,000 ft^3/s. The intake curtain supplies release temperature control at low discharges, and the reservoir curtain provides release temperature control at higher discharges.

Because turbulent mixing was expected to be greater in the prototype, it was concluded that mixing in the "narrows" would severely limit the intake curtain's effectiveness in reducing withdrawal temperatures. Because the reservoir curtain effectively blocks warmer water from replacing surface withdrawals, the downstream pool should approach a constant temperature. Consequently, the intake curtain's effectiveness would be limited especially during extended periods of powerplant operation. Therefore, it was decided that only the reservoir curtain would be recommended for installation.

Lewiston Lake Prototype Curtains

During the recent California drought, Sacramento River temperatures were approaching critical levels with respect to sustaining viable spawning and rearing habitat for threatened salmon populations. As a result, any decrease in river water temperatures, which would lower the salmon mortality rate, was aggressively pursued. In an effort to cool river temperatures, power releases were bypassed through low reservoir outlets at Shasta and Trinity Dams (Trinity discharges into Lewiston Lake) resulting in lost power revenues which totaled $10 million in 1992.

Reclamation constructed and installed an 830-ft-long reservoir curtain (temperature control curtain) in Lewiston Lake in August 1992. The curtain was installed near the reservoir curtain location identified by the physical model studies.

In addition to the reservoir curtain, a second curtain funded by the California Department of Fish and Game was installed surrounding the Lewiston fish hatchery intake structure. The hatchery desired both warmer and cooler water depending on the season and fish requirements. Therefore, a curtain was designed which could skim warmer water or underdraw cooler water depending on whether the curtain was in the sunken or floating position, respectively.

Construction - Total time for engineering, procurement, fabrication, and installation was 5 months. GSE Construction of Livermore, California, began subcontract work in June 1992. By August 26, 1992, the 830-ft-long reservoir curtain was operational. Two weeks later, the 300-ft-long hatchery curtain was operational; this smaller curtain was completely assembled and installed in 7 working days. The costs for the reservoir curtain and hatchery curtain were $650,000 and $150,000, respectively.

Curtain Performance - A cooperative temperature monitoring program was implemented by Reclamation, U.S. Fish and Wildlife Service, and the California Department of Fish and Game. Temperature profiles were measured upstream and downstream from the reservoir curtain site for pre- and post-curtain conditions. Figure 2 represents typical data sets. The pre-curtain profiles are nearly identical, but the post-curtain profiles indicate a substantial modification to the reservoir stratification. For post-curtain conditions the downstream temperature profile

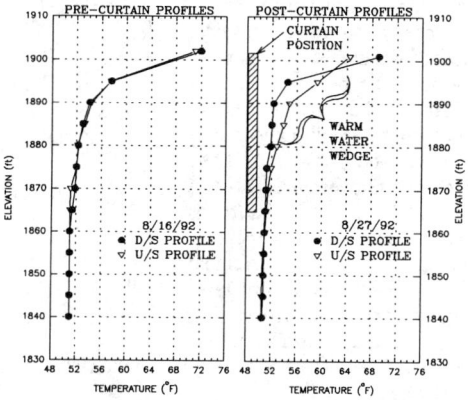

Figure 2. - Temperature profile comparison for pre- and post-curtain conditions. *Profiles were measured for similar powerplant operation schedules.*

is nearly vertical except for surface warming, whereas the upstream temperature profile is 1 °F and 4.5 °F warmer for elevations 1875 and 1895, respectively. Notice that the temperatures below the curtain's bottom edge are unchanged, which indicates that the withdrawal layer does not extend significantly upward into the thermocline (intermediate stratum with large temperature gradient).

Figure 3 presents hourly temperature data collected from the Clear Creek Tunnel intake, which supplies water to Judge Francis Carr Powerplant. These data demonstrate the reservoir curtain's effectiveness in reducing the water temperature entering the intake. For similar operational conditions (flow, period, and time of day), the average temperatures released through the Carr Powerplant were reduced by about 2.5 °F. This result corresponds well to the reservoir and discharge conditions observed in the physical model, where the reservoir curtain reduced water temperature released through the Clear Creek Tunnel by about 2.5 °F. This relatively small cooling is in part caused by the weak temperature stratification in Lewiston Lake. Even though 2.5 °F may appear to be a limited improvement, Sacramento River temperatures were approaching critical levels with respect to sustaining viable spawning and rearing habitat for threatened salmon populations. As a result, all possible efforts to reduce late summer river temperatures are being pursued.

Conclusions

Flexible curtain barriers have been successfully employed to provide selective withdrawal at Lewiston Lake, California. A density stratified physical model was used to develop an effective temperature control curtain. The curtains rapidly modified reservoir temperature profiles and Lewiston Lake release temperatures were reduced by 2.5 °F. Curtains were a relatively inexpensive alternative to a more traditional selective withdrawal structure. Two curtains were designed, constructed, and installed in five months. The construction and installation costs for both curtains totaled $800,000.

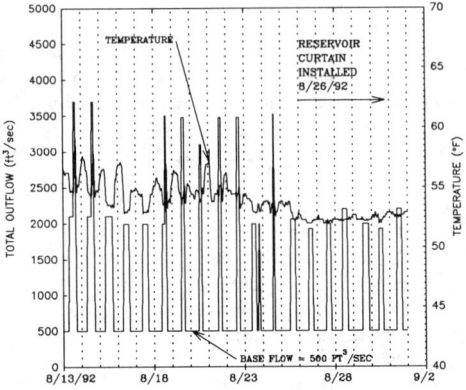

Figure 3. - Lewiston Lake Outflows and hourly temperature data collected in the Clear Creek Tunnel intake structure. *Peaks in outflow represent Carr Powerplant operation.*

In a continuing multi-agency effort, two additional flexible curtains are being designed for Whiskeytown Lake, California (Johnson, et al., 1993). These curtains will further improve the selective withdrawal capability within the Sacramento River basin.

References

Johnson, P.L., T. Vermeyen, and G. O'Haver, *Use of Flexible Curtains to Control Reservoir Release Temperatures: Physical Model and Field Prototype Observations*, Proceedings of the 1993 USCOLD Annual Meeting, Chatanooga, Tennessee, May 1993.

U.S. Bureau of Reclamation, February, 1990. *Whiskeytown Temperature Control Value Engineering Study.* Memorandum to Regional Director, Mid-Pacific Region.

Bohan, J. P., and Grace, J. L., Jr. 1973. *Selective withdrawal from Man-Made Lakes; Hydraulic Laboratory Investigation*, Technical Report H-73-4, U.S. Army Engineer Waterways Experiment Station, Vicksburg, MS.

Unit Conversions	To convert	To	Multiply by
	ft	cm	30.45
	ft^3/s	l/s	28.3
	°F	C	5/9*(°F-32)

A FLEXIBLE CURTAIN STRUCTURE FOR CONTROL OF VERTICAL RESERVOIR MIXING GENERATED BY PLUNGING INFLOWS

Perry L. Johnson[1], M. ASCE and Tracy B. Vermeyen[2], A.M. ASCE

Abstract

To maintain salmon habitat in the Sacramento River, efforts are being made to pass cold water inflows through Whiskeytown Reservoir, California, with minimal warming. In Whiskeytown, summer and early fall inflows come primarily from a diverted cold water source. Comparison of temperature profiles from various stations in the 2.6 x 10^8-m^3, 76-m-deep reservoir show that a 3 to 6 °C warming of inflows occurs as they plunge and shear (mix) with the warm epilimnion (surface layer). A lightweight curtain structure is currently being studied through use of a 1:72 scale, undistorted, density-stratified physical model. The curtain will be used to retain the epilimnion and introduce the cold inflow at sufficient depth to limit mixing-generated warming. A curtain has been designed on an accelerated schedule and will be installed by June 30, 1993.

Introduction

The U.S. Bureau of Reclamation's multibasin Central Valley Project (CVP) serves much of California. One part of the CVP, the Shasta and Trinity River Divisions (figure 1), includes Trinity and Lewiston Dams on the Trinity River, and Shasta and Keswick Dams on the Sacramento River. Water is also diverted from the Trinity River at Lewiston Dam, through Whiskeytown Reservoir, to Keswick Reservoir and the Sacramento River. The reservoir system and its management has a dominant influence on river discharges and water temperatures.

---

[1]Hydraulic Engineer, U.S. Bureau of Reclamation, Mail Code D-3751, PO Box 25007, Denver, CO 80225.

[2]Hydraulic Engineer, U.S. Bureau of Reclamation, Mail Code D-3751, PO Box 25007, Denver, CO 80225.

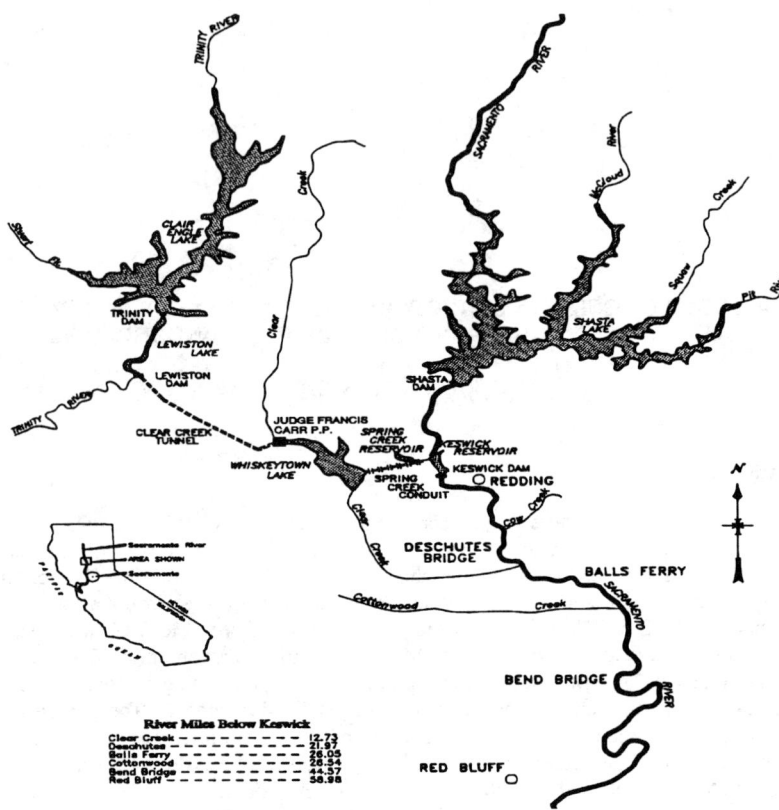

Figure 1. Location Map - Shasta and Trinity Divisions Central Valley Project, California.

Whiskeytown Reservoir is located on Clear Creek, an intermittent stream that experiences large winter flows but remains dry most of the summer. Predominant summer inflows come from diverted Trinity River water, which is introduced to the headwaters of Whiskeytown Reservoir through Judge Francis Carr Powerplant. Predominant summer releases are made through Spring Creek Tunnel and Spring Creek Powerplant, which has a deep intake off the main body of the reservoir. Maximum through-reservoir diverted flows are approximately 85 m^3/s.

Salmon populations on the Sacramento River are in decline. The "winter run" of Chinook salmon has been listed as threatened under the Endangered Species Act. The "spring run" population has declined and listing may follow. River water temperature is affecting the salmon populations. During drought, which has been prevalent over the last seven years, late summer and early fall water temperatures within portions of the optimum river spawning habitat have exceeded levels favorable for sustaining egg and larvae populations.

Efforts are being made to pass the cold Trinity River inflows through Whiskeytown Reservoir with minimal warming. Comparison of inflow temperatures and temperature profiles at different stations in the reservoir show that a 3 to 6 °C warming occurs as the inflows plunge under and shear (mix) with the warm epilimnion. For all discharges, the plunge point occurs in a narrow headwater channel that maintains substantial flow velocities over an extended path length.

A curtain or barrier is being developed that will hold back the epilimnion and thus limit the supply of warm water from the main body of the reservoir to the plunge point and interfacial mixing zone. The curtain will also introduce the cold inflow directly into the reservoir hypolimnion (deep cold water).

Theory

The ASCE Task Committee on Density Currents and Their Applications in Hydraulic Engineering has recently published a concise paper that summarizes the state-of-the-art for analysis of plunging flows and flows with interfacial shear (Alavian, et al 1992). The authors note that a plunging inflow enters a reservoir as a plug flow of uniform density. At Whiskeytown, this inflow is colder and thus more dense than much of the reservoir water. A point is reached where the momentum of the inflow balances the baroclinic pressure resulting from the density difference between the inflow and the receiving water. At this point, the inflow plunges below the surface. Beyond the plunge point the inflow passes into the reservoir as an underflow. Interfacial mixing between the underflow and the warmer reservoir water yields inflow warming.

The strength of the interfacial mixing is a function of the interfacial density gradient and the interfacial shear as described by the Richardson number. Experimental work shows that mixing clearly depends on these parameters. However, the theory does not address the influence of site specific factors such as reservoir density gradient, unsteady flow (peaking power operation), topography, inflow turbulence intensities, and non-uniform velocity profiles. It can be concluded that under-curtain velocities should be small, and that a significant vertical distance is required between the bottom of the curtain and the epilimnion to create an effective curtain design.

Curtain Siting

Initial site selection was made using an energy balance,

$$V_o^2/2g \leq Y\Delta\rho/\rho_o \qquad (1)$$

where:

V_o	=	mean flow velocity under the barrier,
Y	=	vertical distance from the bottom of the barrier to the bottom of the epilimnion,
$\Delta\rho$	=	$\rho_o - \rho_a$
ρ_o	=	mean density between the bottom of the epilimnion and the bottom of the curtain,
ρ_a	=	representative epilimnion ρ_a density.

This energy balance resulted in an initial proposal to locate the curtain at a section where the underflow velocity head was equal to or less than the energy head required to displace buoyant epilimnion water downward to the bottom of the curtain.

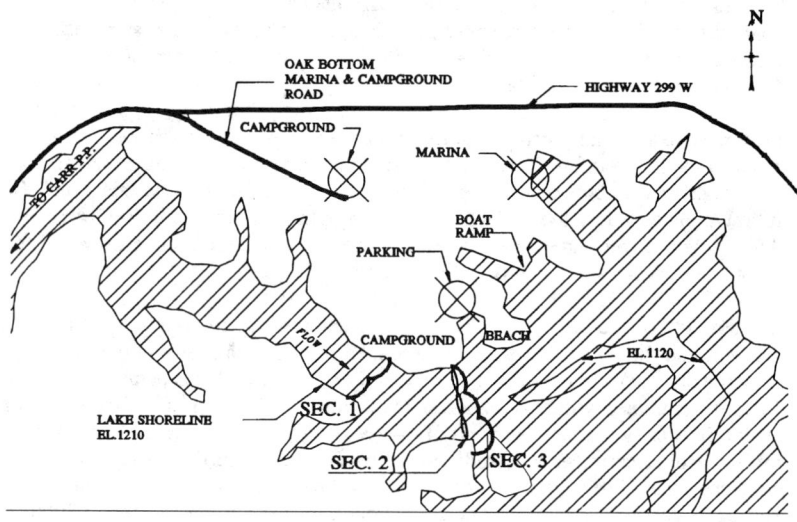

Figure 2. Alternative Curtain Sections.

Because simplifications in the above analysis neglected factors such as topography and local flow concentrations, a reach of reservoir was identified that contained sections that allowed bracketing of the energy balance. The reach, which is 3.2 km down lake from Carr Powerplant, is located where the lake cross section substantially enlarges (figure 2). The sections considered have depths of 25 to 27 m at the normal reservoir surface elevation. The reservoir water level is generally held constant, although drawdown up to 6 m is possible. Section surface widths vary from 135 m to 240 m. Under-curtain flow cross sections vary from 232 m^2 to 1,115 m^2, and maximum underflow velocities range from 0.076 to 0.37 m/s, respectively. The energy balance recommended a section with a surface width of 195 m and an underflow cross section of 770 m^2, which generates maximum operational velocities of 0.11 m/s (section 2 - figure 2).

A loaded curtain depth of 12 m was initially proposed. The 12-m depth (the curtain deflects and lifts under load) places the bottom of the curtain at approximately the bottom of the thermocline (the transition between the epilimnion and the hypolimnion). The 12-m depth positions the bottom of the curtain well above the mid-channel reservoir bottom, which allows maintenance of curtain effectiveness with reservoir drawdown.

Physical Model

A 1:72 scale, undistorted, density-stratified physical model of the critical reservoir reach has been constructed. The model includes a 400-m by 600-m area of the reservoir. Unfortunately, an appropriate model scale that allows inclusion of critical topography yields flow depths and velocities that generate Reynolds numbers of 1,500 to 3,000. As a consequence, viscous effects reduce mixing intensity and make model evaluation of curtain performance qualitative. The model indicates the curtain designs that yield improved mixing reduction. However, the model does not specifically predict prototype curtain performance.

The model is, as of this writing, being used to determine the most effective curtain depth and to select an optimum curtain site. Again, the primary objective is to locate the curtain where downstream vertical mixing is minimized. Thus, the section should include sufficient area under the curtain to generate low velocities and sufficient barrier depth to displace the underflow away from the epilimnion.

Three curtain sections are being evaluated (figure 2). Depletion rates of epilimnion water downstream from the curtains are being used to indicate curtain effectiveness. Depletion rates without the curtain were evaluated to establish a model baseline. The study, to date, indicates that the location farthest upstream (section 1) causes excessive mixing and epilimnion entrainment. More details on this investigation will be included in the presentation.

Curtain Design

A curtain has been designed under a priority schedule (U.S. Bureau of Reclamation, 1993) and will be installed by June 30, 1993. This schedule has developed such that the physical model findings will not greatly influence the initial design. The model study will progress and may guide curtain modification if appropriate.

The curtain has been designed for section 2 (figure 2) as recommended by the initial site analysis. The curtain structure includes 1.5-mm-thick, nylon-reinforced, Hypalon rubber fabric supported by floats, chains, and bottom-of-curtain weights. The curtain is held in place by embedded shore anchors and cleated and weighted lake anchors. The curtain site experiences debris-laden winter inflows. Curtain control is not required in the winter because the reservoir is not stratified. Therefore, the curtain has been designed to be opened and stored near the shore.

Whiskeytown Reservoir experiences heavy recreation use. The curtain site is close to swimming beaches, boat ramps, picnic grounds, camp grounds, and a marina. Potential impact on swimming beach temperatures was a siting factor. A curtain located upstream from the selected site, to isolate it from the public, would have been desirable. However, the model study shows that a curtain located at upstream sites would generate excessive mixing. A 4.9-m by 1.5-m opening, through the curtain, was included to allow boat passage. The boat passage is positioned in a slack water area approximately 30 m from the underflow zone. The boat passage yields leakage that will allow warm water to access the mixing zone above the curtain. The differential loading across the curtain (influenced by both density and dynamic effects) and the resulting leakage rates are not well defined. If leakage greatly reduces curtain efficiency, a boat lock design may be pursued.

References

Alavian, V., Jirka, G.H., Denton, R.A., Johnson, M.C., and Stefan, H.G. (1992). "Density Currents Entering Lakes and Reservoirs," ASCE Journal of Hydraulic Engineering, Vol. 118, No. 11, November 1992.

U.S. Bureau of Reclamation (1993). "Temperature Control Curtain Installation Whiskeytown Lake," Specification No. 20-C0409, 1993.

SACRAMENTO RIVER ENVIRONMENTAL REQUIREMENTS

W. Craig Gaines[1]

ABSTRACT
  Environmental requirements have altered bank protection engineering and construction on the Sacramento River, California. Conflicts between public safety and environmental protection remain to be solved.

INTRODUCTION
  The Sacramento River is the largest California river, with many tributaries that drain the northern Sierra Nevadas, the Coastal Mountains and the Sacramento Valley. Associated with the Sacramento River is a system of flood control levees, weirs, bypasses, reservoirs, and other flood protection structures that comprise the Federal Sacramento River Flood Control Project. To protect the Sacramento River Flood Control Project levees from erosion damage, the Sacramento River Bank Protection Project was authorized in 1960 (House Document No. 93-151, Senate Document No. 103, and Mifkovic). Recent environmental requirements have changed traditional bank protection practices on the Sacramento River.

ENVIRONMENTAL REQUIREMENTS
  With the enactment of numerous California and Federal laws, flood control engineers must coordinate with a number of diverse local, state, and Federal agencies which often have conflicting objectives and regulatory requirements. On the Sacramento River environmental concerns are mandated by the U.S. Fish and Wildlife Service, National Marine Fisheries Service, U.S. Environmental Protection Agency, California State Water Quality Control Board, California Department of Fish and Game and the California State Lands Commission.
----
[1] Civil Engineer, Engineering Division, Sacramento District, U.S. Army Corps of Engineers, 1325 J Street, Sacramento, California 95814-2922

ENDANGERED SPECIES

Endangered species laws provide agencies with the power to halt, prevent, modify and delay projects which may impact listed species. To receive environmental agency agreement for projects on the Sacramento River and its tributaries engineers must coordinate early in their engineering designs to determine what modifications in traditional construction techniques are necessary to avoid adverse environmental impacts as shown in environmental impacts and reports and environmental impact statements (The Reclamation Board).

One major modification is to develop construction windows to avoid impacting endangered species. Avoidance is the preferred method of the environmental agencies. Figure 1 presents periods of time in which endangered species may be found in the upper Sacramento River. These are not all of the endangered or threatened species which may be present, but are the species where major conflicts exist. As can be seen, there may be no construction window. Accidental take of any endangered species is of primary concern to construction agencies as stiff penalties are possible. Environmental agencies fear loss of aquatic species due to dredging, excavation, and even placement of revetment in the water.

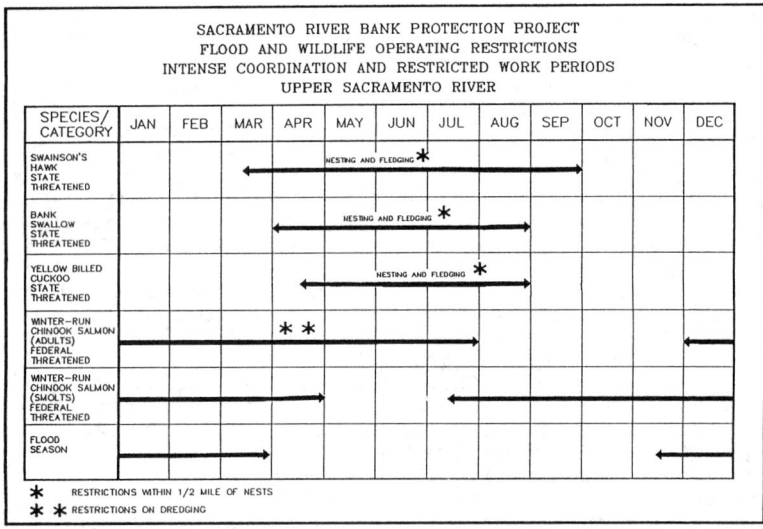

Figure 1. Upper Sacramento River Restrictions

If avoidance is impossible, there may be no construction if an environmental agency determines acceptable mitigation is impossible. Agencies have begun to dictate rather than negotiate on mitigation and expected impacts. There are often hidden environmental agendas not formally set out. For example, preservation of the spotted owl in the Sierra Nevadas is used to protect old growth forests and not just the spotted owl.

RIPARIAN VEGETATION

On the Sacramento River endangered winter-run chinook salmon are being used to preserve riparian vegetation. Riparian woodlands on the Sacramento River between Collinsville and Red Bluff, California decreased from about 800,000 acres in the 1800's to less than 12,000 acres in 1972. In order to place bank protection on the river bank or levee, woody riparian vegetation is often removed. Loss of riparian habitat is unacceptable to the environmental agencies due to the historic losses caused by agricultural and flood control activities. Mitigation for these impacts has been in a state of continuing flux as models and definitions of riparian habitat have been changed and revised. Instead of traditional mitigation for removed vegetation by planting offsite, the U.S. Fish and Wildlife Service requires new mitigation plantings at or near the area of impact. Since few areas with plantable berms remain between levees a design to increase the amount of plantable areas was developed by the Corps and the State of California using dredge material. These constructed berms are known as "dredge berms" and have been planted as mitigation sites (see Figure 2).

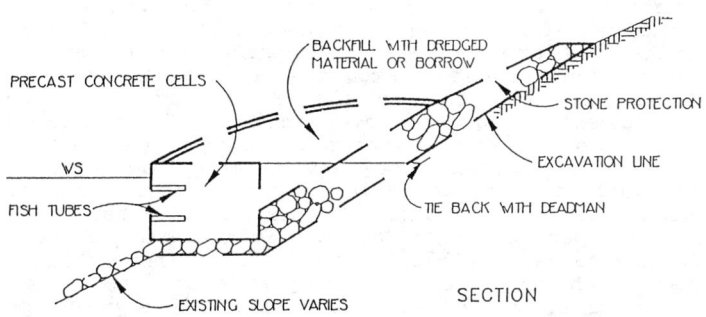

Figure 2. Typical Dredge Berm

Engineering considerations regarding the construction of "dredge berms" include hydraulic impacts, changes in project design water elevation and flow splits, navigation hazards, maintenance, and costs.

Constructed berms are typically located on areas where natural berms were once located, but have eroded away since the flood control project was authorized.

HIGH BANK TERRACES

Conflicts in the upper Sacramento River include the California threatened bank swallow, which nests in burrows it carves out in vertical eroded high bank terraces. Placing revetment on eroding banks has decreased the available bank swallow habitat and destroyed some colony sites. The number of bank swallows in California has continued to decrease during the past few years although revetment has not been placed in bank swallow areas due to environmental requirements.

Engineering designs to avoid revetting the full bank by using a low revetment design or using groins (dikes) have been unacceptable to the California Department of Fish and Game as they desire a natural eroding bank for the bank swallows. Artificial banks were built as mitigation to replace destroyed bank swallow habitat in several areas, but have not been accepted by the U.S. Fish and Wildlife Service as permanent mitigation even though several of the artificial berms were heavily used by bank swallows. Environmental agencies have taken a position that "natural banks" must remain, even though bank swallows have been documented to use sawdust piles, gravel pits, and other manmade excavations as habitat. Currently, bank swallow habitat appears to be exempt from bank protection.

WATER QUALITY

Emphasis on hazardous and toxic wastes by environmental agencies has resulted in difficulties in excavation and dredging, even in areas known not to contain any contaminants. Excavated revetment toes have traditionally been used on the Sacramento River to prevent undercutting of revetment, which causes a complete loss of the revetment. Expensive testing requirements and opposition by environmental agencies to dredging and inwater excavation may prevent the use of excavated toes. Environmental agencies fear that endangered species, such as the winter-run chinook salmon, may be susceptible to toxics in the river bed or increased turbidity. By not using excavated toes, but using overbuilt toes, some revetment life will be shortened due to channel undercutting.

RISK TO THE PUBLIC

Engineering designs for bank protection have historically emphasized public safety. The objective is to minimize levee breaches by constructing bank protection when erosion is judged to have the potential of putting the public at risk. U.S. Fish and Wildlife Service and the California Department of Fish and Game delay bank protection construction as long as possible. Each newly listed Federal and California endangered species causes project delays as new documentation must be prepared and approved.

Due to an increasingly obstructionist stance, bank protection at some critical erosion sites has been delayed for more than five years. Delays in protecting levees from erosion has increased the risk of levee failure. Proper maintenance, such as removal of vegetation that prevents inspection and the ability to flood fight, has been delayed or prevented in some areas.

RECENT EMERGENCY BANK PROTECTION

The Corps of Engineers placed bank protection at two emergency sites on the Sacramento River in late 1992/ early 1993 near Knights Landing. The work was to have been completed prior to flood season due to the threat of potential levee failure during high water and flows.

Concern for potential impacts to winter-run chinook salmon caused the construction to be delayed as various Federal and California agencies worked on an agreeable procedure to place the bank protection. A partial fish curtain and a bubble curtain were installed by the California Department of Water Resources and the Corps of Engineers in an effort to exclude migrating winter-run chinook salmon smolts from the construction site. These experimental curtains were found to be unusable when high water occurred due to heavy rains that ended the California drought.

The initial flood flows carried fine debris which clogged the screens and made them inoperable. Due to environmental requirements by various agencies the emergency work was delayed for several months and performed in the flood season. As winter rains arrived it was impossible to do land based construction, so waterborne construction was used. Due to accelerated erosion, the use of fish curtains, the use of barges and tugs, and additional rock fill due to underwater placement, the cost for the bank protection increased fourfold as a direct result of environmental compliance.

CONCLUSION

Increased environmental requirements from Federal, State, and local agencies have significantly impacted engineering and construction practices in the Sacramento River system. Project costs, engineering and construction timetables, and project viability are affected by the numerous laws, orders, regulations, procedures and attitudes of environmental agencies. Alternative means of bank protection, the use of setback levees, new mitigation techniques, and changes in operation and maintenance procedures are being evaluated to satisfy the increasing environmental requirements. Conflicts exist between public health and safety and the requirement to protect endangered species and environmental values.

ACKNOWLEDGEMENT
This article is based solely on the views of the author and does not reflect the views or position of the U.S. Army Corps of Engineers.

REFERENCES
Mifkovic, Charles S., and Margaret S. Petersen, "Environmental Aspects--Sacramento Bank Protection, ASCE, Journal of the Hydraulics Division, May 1975, p. 543-555.

"Sacramento River Bank Protection Project, California--Second Phase," House Document No. 93-151, 93rd Congress, 1st Session, 1973.

"Sacramento River Flood Control Project, California," Senate Document No. 103, 86th Congress, 2nd Session, 1960.

The Reclamation Board, State of California, and U.S. Army Corps of Engineers, "Final Environmental Impact Report and Supplemental Environmental Imapct Statement IV, Sacramento River Bank Protection Project," December 1987.

Flow Diversion In A Steep, Coarse-bed Stream

By Edward F. Sing, M.ASCE - 1/

Abstract

A case study is presented of the hydraulic design of a diversion structure in a high velocity, coarse bed streamcourse. Mill Creek, a tributary to the Upper Jordan River in the southern portion of Salt Lake City, is a steep, high velocity stream which periodically floods its overbank areas. Major features of the Upper Jordan River Project include a diversion structure on Mill Creek to divert excess floodflows, a diversion conduit and a detention basin for detaining the excess floodflows. Design guidance for establishing the flow-split relationships for a diversion structure imposed into a high velocity, peaked and large bed material load stream is minimal. Available analytical tools for hydraulic and sediment transport analyses were used to develop a Preliminary Hydraulic Design of the structure. The Final Hydraulic Design will be based on performance verification using a physical hydraulic model of the structure.

Introduction

Mill Creek is a tributary of the Jordan River, and is located in Salt Lake County and Salt Lake City. The upper watershed for Mill Creek is east of the city in the Wasatch Mountains, comprises approximately 55.9 square kilometers (21.6 square miles) and reaches an upper elevation of over 3048 meters (10,000 feet). The lower watershed, in which the project reach is located, is a highly developed residential area. The slope of the study reach is about the same 3.5% slope as the channel in the upper watershed. The flow regime is close to critical with a high resistance factor. The upper watershed produces almost all of the sediment yield to Mill Creek. Snowmelt flows during the months of April through June produce the bulk of the average annual sediment production. However, short duration, peaked cloudburst events (with less than 30 minutes to peak flow) can produce and result in transport of large infusions of sediments into the creek channel. A geomorphic and sediment engineering investigation (Rahmeyer et al, 1991) of the study reach estimated that approximately 5200 tons per year of

---

1/ - Hydraulic Engineer, U.S. Army Engineer District - Sacramento, 1325 J Street, Sacramento, California 95814-2922

sediments are produced by the upper watershed of which 70% are silts and clays, 20% are sands and 10% are gravels. In addition, it was estimated that over 500 tons of sediments would be delivered to the study reach during a design event.

The proposed project will provide near 100-year flood protection to the lands adjacent to the creek by diversion of a portion of floodflows to a detention basin. Early on in the design process, concern was raised regarding the paucity of design guidance for such a structure in a high velocity, sediment laden, short time to flood peak stream and whether a dependable facility could be designed for such a situation. A two step design process was adopted to address these concerns, which included: Development of a Preliminary Hydraulic Design (PHD) for the facility using available analytical techniques (described herein); and, use of a physical hydraulic model (to be performed in the future) to circumstantiate the performance of the proposed design under the imposed hydraulic and sediment transport conditions. Following is a brief description of the PHD analyses.

Hydraulic Design Analyses

The hydraulic design analyses took into consideration: the project functional criteria; the conditions found at the proposed diversion structure site with respect to topography, geomorphology, sedimentology, hydrology and hydraulics; and, current Corps hydraulic design criteria and guidance as well industry standard practices. Diversion of flows from Mill Creek will require a hydraulic structure to provide a dependable stage-discharge relationship by slowing of floodflows and development of an energy head on a diversion weir. The project functional criteria included:

- *Pass all low flows and associated sediments completely through the structure without interception:* This is to prevent downstream channel scour and will be accomplished by a low flow channel set within and to one side of the structure capable of passing flows up to approximately 4.25 cms (existing low flow channel capacity) and their associated sediments without interception.

- *Predicate inflow vs. outflow vs. diversion flow relationships solely on flow hydraulics created by geometric configuration of diversion structure:* Elimination of a previously proposed regulating gate (CESPK,1987) on the downstream channel eliminates some of the uncertainties regarding the structure's reliability during a "flashy" flood event.

- *Provide storage area for intercepted sediments:* An area fully separate from the low flow channel and several times the anticipated storage requirement is provided.

Computer program HEC2, "*Water Surface Profiles*" (USACE, 1990) was used as a tool for determination of the hydraulic conditions in the proposed diversion structure and for "testing" of various geometric configurations in an iterative process. The "split flow" option of the program was used to simulate flows diverted over the diversion weir. Geometric parameters investigated included structure length and width, diversion weir length and elevation, outlet portal size and low flow channel width. The final geometric configuration indicated by this process included: An approximate 61 meter long by 13.7 meter wide concrete structure to receive all creek flows; a 3 meter wide low flow channel located along the right bank wall perched above a 10.7 meter wide sediment "pit" having a depth 3 meters below the design gradeline of the low flow channel (total capacity approximately 1.27 dekameters); a 12.2 meter long diversion weir on the left bank wall perched approximately 1 meter above the design gradeline through the structure; and, a 2.1 meter wide exit portal to pass low flows to the downstream channel.

The diversion structure "operation", with respect to flow hydraulics, passage of low flows and creation of sufficient energy head for flow diversion, varies with the structure inflow:

- *Operation during Low Flows:* For creek inflows to the structure less than approximately 4.25 cms, the flow hydraulics within the diversion structure are dictated by the low flow channel geometry and slope. All flows and their associated sediments will be passed in their entirety via a low flow channel through the diversion structure. The flow depth in the low flow channel at 4.25 cms is approximately 0.37 meters with a flow velocity of approximately 3.96 m/sec (Froude Number greater than 2.0).

- *Transitional Operation:* For creek inflows between approximately 4.25 and 7.08 cms, the flow hydraulics within the diversion structure is transitioning between "inlet" (i.e. low flow channel geometry) control and "outlet" (i.e. outlet portal constriction) control. A portion of these flows and their associated sediments will be forced out of the low flow channel near its inlet into the diversion structure by a constriction in the low flow channel and into the sediment storage pit. Once the storage pit fills with water, flows will be across the full width of the diversion structure. No flow diversion takes place within this transitional flow range.

- *Operation during Floodflows:* For creek inflows exceeding approximately 7.08 cms, flows will be across the full width of the diversion structure and flow diversion will commence above approximately 9.2 cms. In this flow range, the flow hydraulics in the diversion structure is dictated by a critical depth "control" at the outlet portal of the structure. The Froude Number for flows through the structure will be less than 1.0, permitting development of a reliable energy head relationship above the diversion weir at high inflows. Due to the skewed alignment of the inlet channel to the diversion structure proper, the bulk of the high floodflows and their associated sediments will enter the structure along its left side where sediments will deposit into the sediment storage pit.

Sediment Transport Analyses

Sediment transport analyses were conducted to qualitatively assess the performance of the proposed structure under the imposed sediment load. The numerical sediment transport model utilized in the evaluation of sediment transport in the project reach was computer program HEC6, "Scour and Deposition in Rivers and Reservoirs", "PC Version" Number 4.0.6 dated June 1991 (USACE, 1991) with Meyer-Peter Muller transport function correction dated January 1992.

Required input data for the model include geometric, sediment and hydrologic data. The geometric data includes cross sections, reach lengths and Manning's "n" values. The movable bed widths through the structure were set within the 10.7 meter width of the sediment "pit". Sediment data input to the model include a numerical description of sediment properties. Data requirements include grain size distribution of the material in the streambed at each cross section, the gradation and amount of total inflowing sediment load as a function of water discharge and the fluid and sediment properties. The initial bed material gradation used in the reach model was a reach averaged gradation (D50 of 4 mm) developed from bed material samples taken in January 1990. This gradation was assumed for the entire reach modeled, both inlet channel and diversion structure. A sediment versus water discharge relationship was developed (Rahmeyer et al,1991) from the hydraulic and sedimentologic characteristics of the channel immediately upstream of the proposed structure site. The hydrologic data input is the numerical description of the water discharge hydrograph on which the sediment calculations are to be based - in this case, the inflow hydrograph for the design (100-year) event.

Calibration of the model input data to simulate the sediment transport characteristics of the project features is a complex and time consuming process of adjusting and/or fine tuning parameters within the model input. The principal calibration effort pointed towards ensuring that the sediment and hydraulic computations would be numerically stable under the imposed, moderately large sediment load and for the given finely spaced cross sections. These conditions, coupled with too large of a discharge event duration often results in numerically unstable computations ("undulating" bed elevations and computed sediment transport rates from cross section to cross section). Through the diversion structure, cross sections used in the HEC2 hydraulic model were spaced 3 meters apart, except immediately upstream and downstream of the proposed left bank diversion weir where a 1.5 meter spacing was used. The close spacing of the cross sections was necessary to adequately simulate the water surface profile through the relatively short (60 meter long) diversion structure as well as adequately simulate the change in water and sediment discharge through the structure as flows are diverted over the diversion weir. The process of selection of a suitable time step to be used in subdividing the design event hydrograph proceeded in a manner per guidelines found in an HEC Guideline Paper on Calibration of the HEC6 Model (HEC,1981). One guideline utilized suggests that the time step should be no longer than the time required to deposit (or scour) 0.3 meter of the model bed through the shortest cross section reach (i.e. 1.5 meter in the Mill Creek model) assuming the sediment inflow (i.e. approximately 97.65 kg/sec) at the peak discharge of the inflow hydrograph (i.e. 38.8 cms). This computation results in a maximum 1 minute time step.

Sensitivity tests using the numerical sediment transport model were performed to "test" the performance of the diversion structure assuming variations in some of the input parameters. The primary input parameter which was varied was the sediment versus water discharge inflow load rating. The total sediment inflow volume to the diversion structure was roughly quadrupled and the sediment inflow gradation modified to reflect a slightly coarser inflow. The design event hydrograph was passed through the diversion structure using the revised sediment inflow relationships and the model results assessed. The computed bed elevation changes through the structure for this scenario are given in Figures 1 and 2. Even though the sediment inflow volume using the modified sediment inflow relationship was roughly quadruple the original assumed inflow volume, the simulation model indicates that the bulk of the sediments deposited in the diversion structure for this scenario would deposit upstream of the diversion weir, thus not having any impact on the diversion quantities. Figures 1 and 2 also give a comparison of the bed elevation changes for the design event versus sensitivity analyses scenarios at the peak inflow of the inflow hydrograph and at the end of the inflow hydrograph, respectively. These figures illustrate the large difference in bed deposition potential for the different sediment inflow assumptions. However, they also illustrate that there is sufficient storage available in the sediment pit such that the diversion weir would not be affected by these deposits during the peak inflow nor would deposits be high enough at the end of the floods to induce outflanking of the diversion structure.

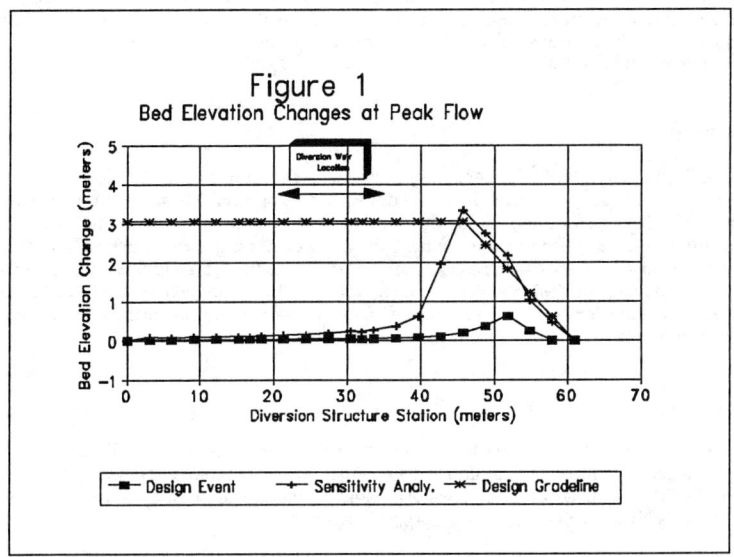

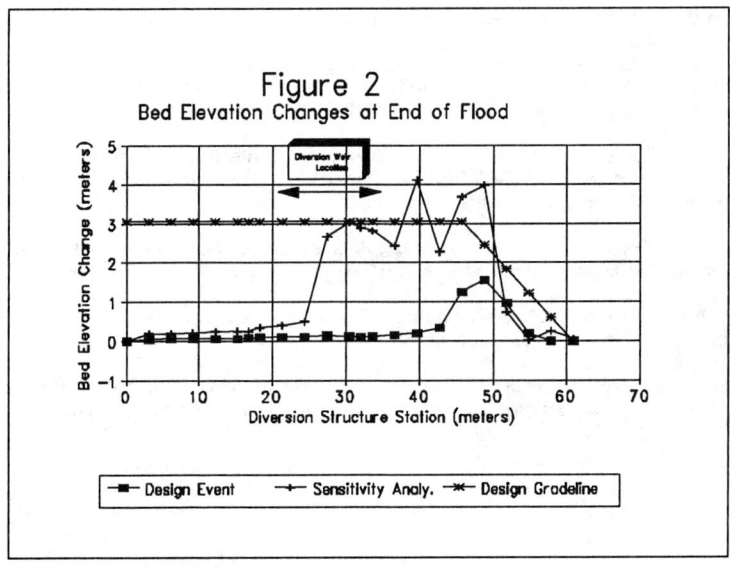

As previously noted above, the numerical sediment transport model is intended to provide a qualitative assessment of the performance of the diversion structure. A quantitative assessment of both hydraulic and sediment transport characteristics of the structure will be conducted prior to construction utilizing a physical hydraulic model of the inlet channel and diversion structure.

Summary

A Preliminary Hydraulic Design for a diversion structure on a high velocity, peaked and high sediment load stream has been developed using available analytical techniques for evaluating the hydraulic and sediment transport conditions within the structure. Sensitivity analyses demonstrate that the hydraulic functioning of the structure would not be impacted by up to a 400 percent increase in the estimated sediment inflow to the structure during the design event. A physical hydraulic model will be used in the near future to circumstantiate the performance of the structure under the imposed water and sediment inflow.

Acknowledgements

Some of the work reported herein was conducted for the Sacramento District, Corps of Engineers by Horrocks Carollo Engineers, Inc. and its subcontractors under Contract No. DACW05-91-C-0014. The views expressed herein are those of the author and not necessarily those of the U.S. Army Corps of Engineers.

References

Horrocks Carollo Engineers, "*Early Basis of Design, Upper Jordan River, Utah, Mill Creek Flood Control Project*", prepared under Contract No. DACW05-91-C-0014, July 1991.

Rahmeyer, W., James, D., Simons, D.B., and Song, Z., "*Sediment Engineering Investigation Report, Upper Jordan River Mill Creek Flood Control Project*", prepared for Horrocks, Carollo Engineers as part of Contract No. DACW05-91-C-0014, June 1991.

U.S. Army Corps of Engineers, Hydrologic Engineering Center (HEC), Training Document No., 13, "*Guidelines for the Calibration and Application of Computer Program HEC6*" dated February 1981, with draft revisions dated 1992.

U.S. Army Corps of Engineers, Hydrologic Engineering Center, Users Manual, "*HEC-6, Scour and Deposition in Rivers and Reservoirs*", dated June 1991.

U.S. Army Corps of Engineers, Hydrologic Engineering Center, Users Manual, "*HEC-2, Water Surface Profiles*", dated September 1990.

U.S. Army Corps of Engineers, Sacramento District (CESPK), "*Feasibility Report and Environmental Impact Statement, Upper Jordan River Interim Investigation, Utah*", dated July 1987.

SUBJECT INDEX
Page number refers to first page of paper

Abutments, 743, 749
Acoustic measurement, 1671
Advection, 1628, 2141
Aeration, 851, 1975
Aerators, 75, 99, 2165
Africa, 418
Aggregates, 1355, 1876, 2014
Aggregation, 1184
Agricultural economics, 168
Agricultural wastes, 1161
Agricultural watersheds, 104
Air entrainment, 75, 851
Airport drainage, 2333
Alkalinity, 1355
Alluvial channels, 893
Alluvial fans, 1762, 2098
Alluvial streams, 125, 989, 1635, 1726, 2219
Alluvium, 1726
Analytical techniques, 665, 1701
Anisotropy, 36
Aquatic environment, 659
Aquatic habitats, 737, 1155, 1167, 1812
Aquifer characteristics, 29
Aquifer tests, 21
Aquifers, 198, 305, 460, 466
Aromatic hydrocarbons, 2032
Artesian aquifers, 21, 471
Artificial intelligence, 1043, 1049
ASCE Publications, 378, 383, 388, 400
Australia, 2307
Avalanches, 1774
Axisymmetry, 1119

Backwater, 1055
Baffles, 1836
Ballast, 2032
Bank erosion, 881
Bank stabilization, 881, 1055, 1635, 2020, 2383
Barrier beaches, 562
Barrier design, 1079
Bathymetry, 2063, 2123
Bays, 641, 647, 653, 1167, 1194, 1218, 2116
Beaches, 2008
Bed load, 701
Bed load movement, 707, 1200, 2219
Bed material, 400, 869, 911, 995, 1001, 1049, 1540
Bed roughness, 995, 1073, 1131
Bedrock, 592, 1290
Benefit cost analysis, 8
Biodegradation, 477
Birds, 1161
Boundary conditions, 198, 1408, 1720, 2135
Boundary element method, 471
Boundary layer, 1464
Boundary shear, 1534
Breaking waves, 2008
Bridge abutments, 525, 755, 761, 767, 973, 1882, 2249, 2255
Bridge design, 519, 1224
Bridge failure, 489, 495, 513, 604, 928, 1061, 1397, 2081
Bridge foundations, 916, 1397, 1876
Bridge inspection, 489, 495, 501, 507, 513, 586, 598, 604, 611, 617, 1546, 2045
Bridge maintenance, 1558
Bridges, 773, 1230, 1236, 1714, 1720, 1866, 1872, 2063, 2075
Bridges, concrete, 592
Bridges, highway, 400, 525, 586, 592, 1206, 1732, 2243
Bridges, piers, 525, 899, 905, 911, 916, 922, 934, 1031, 1037, 1043, 1061, 1212, 1367, 1373, 1379, 1385, 1540, 1738, 1750, 1848, 1854, 1860, 1882, 2038, 2051, 2057, 2069, 2087, 2231, 2237, 2255, 2261
Bridges, steel, 592
Bridges, truss, 2069
Bridges, wooden, 2249
Bubbles, 957, 1108
Buoyancy, 838
Bureau of Reclamation, 549, 671, 678,

2395

| Vol. 1 1-1188 | Vol. 2 1189-2394 |

684, 690, 696
Buried pipes, 543

Caissons, 922
Calibration, 1073, 1266, 1464
California, 1, 15, 54, 162, 174, 317, 436, 442, 623, 629, 1019, 2383
Case reports, 347, 359, 549, 592, 598, 1558, 2069, 2153, 2207, 2237
Cavitation, 75
Channel beds, 803, 983, 1397
Channel bends, 1085
Channel design, 365, 887, 1695
Channel erosion, 968, 1344
Channel flow, 665, 1155, 1574, 1580
Channel improvements, 737, 875, 1355, 1695, 2318
Channel morphology, 125, 394, 737, 1635, 2014
Channel stabilization, 785, 2243
Channelization, 1391
Chesapeake Bay, 2110
Chicago, 1476
China, People's Republic of, 263, 1031
Chlorides, 317
Clays, 1178, 1610, 1616
Climatic changes, 1894
Closed conduit flow, 1091, 1488
Closed conduits, 779
Coastal engineering, 584
Coastal management, 424
Coastal morphology, 269, 574
Coastal processes, 719, 2184, 2307
Coastal structures, 1830
Cohesive sediment, 647, 1184
Collisions, 234
Colluvium, 1290
Colorado, 1900
Colorado River, 1290
Columbia River, 1314
Combined sewers, 347
Comparative studies, 586
Computation, 1569
Computer applications, 1933
Computer graphics, 1248
Computer models, 186, 388, 454, 791, 962, 1043, 1049, 1720
Computer programs, 311, 779, 1586, 1714, 2213
Computerized design, 785
Computerized simulation, 15, 81, 641, 821, 1842, 2003
Concrete, 1355, 1361
Concrete blocks, 1367
Concrete deterioration, 1367
Constitutive equations, 1402
Contaminants, 36, 281, 477, 1428, 2307
Continuum mechanics, 1414
Control structures, 1125
Control systems, 1149
Cooling water, 1167, 1818
Coriolis effect, 1119
Cost estimates, 785
Critical flow, 845
Critical path method, 1482
Cross sections, 767, 791, 875, 1001
Crossflow, 1007
Culverts, 234, 2213
Currents, 543, 659, 1452, 1671, 2301

Dam breaches, 328, 359, 1957
Dam construction, 359
Dam design, 216, 1963
Dam failure, 204, 671, 678, 690, 696
Dam safety, 204, 222, 359, 556, 671, 678, 684, 690, 696, 856
Dam stability, 353, 359, 372, 1951
Damage assessment, 684
Damage estimation, 1830
Damage prevention, 2104
Dams, 93, 2313
Dams, arch, 353, 359
Dams, concrete, 353, 1963
Dams, earth, 1350, 1701
Dams, embankment, 1505, 1951, 1957
Dams, rockfill, 222
Data analysis, 701
Data collection, 258, 454, 477, 489, 495, 501, 513, 701, 773, 1001, 1500, 1523, 1677, 1720, 1848, 1860, 2045, 2057
Data collections, 2003
Data processing, 2003
Debris, 210, 334, 1055, 1272, 1284, 1290, 1296, 1302, 1373, 1402, 1408, 1414, 1592, 1598, 1604,

1610, 1616, 1756, 1762, 1768, 1774, 1780, 1786, 1894, 1900, 1908, 1914
Decision making, 1799
Deep water, 1242
Deicing, 2333
Deltas, 418, 436, 442, 448, 629
Density currents, 334, 838, 845, 951, 1102
Density stratification, 939, 1025, 1102, 1119, 2123, 2371, 2377
Desalination, 162, 2026
Design criteria, 347, 861
Design storms, 1206, 1933, 1969
Detention basins, 1125
Dewatering, 1476
Differential equations, 1091
Diffusers, 2026
Diffusion, 641, 803
Digital terrain model, 1248, 1774
Dimensional analysis, 761
Disaster relief, 531, 537
Discharge, 2219
Discharge measurement, 1452, 2213
Discrete elements, 65
Dispersion, 1428, 1659, 2032
Dissolved oxygen, 99, 957, 1975, 2297
Distortion, 1659
Diversion structures, 2389
Doppler systems, 1096
Drag, 322, 580
Drag coefficient, 1178, 1824
Drawdown, 119, 460, 1842
Dredge spoil, 737
Dredging, 1194
Drop structures, 1391, 1397
Droughts, 15
Dunes, 269, 568
Dye studies, 406, 2032
Dynamic pressure, 1683

Earthquake damage, 365
Earthquake loads, 93
Economic factors, 174, 684
Economic impact, 928
Ecosystems, 659, 719
Eddies, 1671, 2184
Effluents, 1446, 1812, 2032

Embankment stability, 1951
Embankments, 192
Emergency services, 59, 204, 531, 537, 690, 696, 833, 962, 1908
Empirical equations, 1768
Endangered species, 174, 2383
Energy dissipaters, 2202
Energy dissipation, 240, 1137, 1488, 1500, 1836
Entrainment, 1013, 1167, 1756
Environmental effects, 8, 1102
Environmental engineering, 137, 1792
Environmental impacts, 174, 2307
Environmental issues, 2383
Environmental quality regulations, 1818
Equations of motion, 1408
Erosion, 104, 400, 1641, 1756
Erosion control, 881, 968, 973, 979, 1125, 1344, 1367, 1379, 1385
Error analysis, 629
Estuaries, 1, 70, 275, 454, 562, 653, 665, 719, 725, 827, 1007, 1025, 1096, 1184, 1628, 1982, 1994, 1997, 2110, 2116, 2141, 2178, 2297, 2318, 2365
Evacuation, 1914
Evapotranspiration, 131
Expansion joints, 1355
Experimental data, 1622
Expert systems, 838

Fans, 1302
Fenders, 1750
Fiber optics, 1653
Fibers, 2165
Field investigations, 263, 749, 773, 934, 1848, 2038, 2057, 2075, 2190, 2261
Field tests, 477, 1031, 1951
Finite difference method, 93, 442, 1569, 2116
Finite element method, 635, 647, 731, 1079, 1963, 2318
Fish habitats, 174, 653, 827, 1013, 2020, 2377
Fish management, 1019, 1326, 1332
Fish protection, 1079
Fish screens, 1308, 1314, 1320
Flash floods, 1900

Floating bodies, 1373
Flocculation, 1184
Flood control, 59, 81, 204, 263, 340, 347, 372, 378, 412, 537, 671, 731, 833, 887, 968, 1085, 1125, 1155, 1248, 1284, 1361, 1695, 1842, 1969, 2014, 2104, 2383, 2389
Flood damage, 263, 347, 928, 934, 962, 1854, 1882, 2104, 2345
Flood forecasting, 263, 269, 690, 962, 2345
Flood frequency, 2345
Flood hydrology, 2237
Flood level, 328, 340
Flood plain insurance, 1574
Flood plain planning, 412
Flood plain regulation, 973
Flood plain studies, 761
Flood plains, 1254, 1266, 1350, 1574, 1939
Flood routing, 1564, 1762
Flood stages, 1939
Floodwater, 1894
Floodwaves, 1598
Floodways, 1574
Florida, 281, 287, 430, 531, 537, 543, 617, 1230
Flow characteristics, 192
Flow control, 1338
Flow distribution, 253
Flow measurement, 299, 1096, 1452, 1517, 1677, 1824
Flow patterns, 131, 198, 436, 580, 1096, 2171, 2178
Flow profiles, 87
Flow rates, 87, 228, 299, 334, 743, 951, 973, 1013, 1096, 1131, 1172, 1242, 1452, 1476, 1517, 1677, 2178
Flow resistance, 1344
Flow separation, 905
Flow simulation, 305, 803, 1564, 1988, 2123, 2196
Flow visualization, 1665
Fluid flow, 156, 210
Fluid mechanics, 1102
Flumes, 701, 922, 995, 1007, 1379, 1464, 1604, 1641
Flushing, 119, 1172

Fluvial hydraulics, 719
Footings, 922, 1379
Forest fires, 59
Free flow, 911
Free surfaces, 192, 198, 1997, 2123, 2313
Frequency analysis, 1945
Friction, 1604, 2147, 2301
Friction coefficient, hydraulic, 665
Friction factor, 995, 1659
Friction resistance, 1659
Froude number, 1517
Fuzzy sets, 2087

Gabions, 785, 968, 1350
Gaging stations, 1470
Geographic information systems, 412, 1799, 1902, 2339
Geography, 519
Geologic processes, 394
Geological faults, 222
Geology, 519
Geometry, 767
Geomorphology, 394, 1055, 1272, 1894, 2020
Geophysical surveys, 2051, 2231
Germany, 412, 1799
Grain size, 887
Granular media, 322, 483, 1402, 1408, 1786
Gravel, 707, 713, 856, 983, 1001, 1511, 1523, 1744
Ground water, 156, 305, 317
Ground-water depletion, 317
Ground-water flow, 311, 466
Ground-water management, 168, 311, 317, 466, 471, 477
Ground-water mounding, 198, 2159
Ground-water pollution, 36, 42, 477, 2159
Ground-water recharge, 48, 198, 311
Ground-water supply, 568
Grout, 1379
Gullies, 1641, 1756

Hawaii, 1768, 1774
Hazardous waste, 70
Hazards, 334, 1774, 1914

SUBJECT INDEX

Head loss, 1308, 1689
Heavy metals, 647, 1440
Highway safety, 234
Hong Kong, 1963
Hurricanes, 531, 537, 543, 934
Hydraulic conductivity, 483
Hydraulic design, 1224, 1314, 1320, 2098, 2389
Hydraulic engineering, 156
Hydraulic gradients, 253
Hydraulic jump, 1137, 1683, 1726
Hydraulic models, 131, 253, 501, 731, 962, 1236, 1266, 1308, 1500, 1574, 1580, 1659, 1689, 1720, 1732, 1762, 1939, 2171, 2225
Hydraulic networks, 1708
Hydraulic performance, 234, 406, 1085, 2371
Hydraulic pressure, 1488
Hydraulic properties, 1866, 1872
Hydraulic roughness, 1361
Hydraulic structures, 247, 549, 945, 979, 1143, 1695, 1818, 1836, 1920, 1927, 2202, 2313
Hydraulic transportation, 2171
Hydraulics, 1714
Hydrodynamic pressure, 93
Hydrodynamics, 442, 574, 635, 653, 1019, 1079, 1248, 1428, 1982, 1997, 2014, 2110, 2178, 2297, 2365
Hydroelectric power generation, 99
Hydroelectric powerplants, 1326, 1332, 1338, 1963
Hydroelectric resources, 1792
Hydrogeology, 29
Hydrographs, 150, 186, 1933
Hydrologic data, 690
Hydrologic models, 1920, 1927, 1933, 2098
Hydrologic properties, 293
Hydrology, 143, 150
Hydrostatic pressure, 2123
Hyetographs, 186
Hypolimnion, 957

Ice, 1326
Ice cover, 65, 150
Ice disintegration, 65
Ice jams, 65
Impingement, 1013
Incipient motion, 707, 1552
Indiana, 513
Industrial wastes, 945
Infiltration, 1511
Infrastructure, 1708
Inlets, waterways, 562, 1218, 1236, 1671
Inspection, 204, 353, 556, 1701, 1866, 1888
Instream flow, 2165
Instrumentation, 1470, 1529, 2045, 2063
Intake structures, 1013
Intakes, 247, 809, 815, 1320, 1818, 2153
Interactive graphics, 2291
Italy, 2207

Japan, 1914, 2153

Kinematic wave theory, 1592
Kinematics, 803
Kinetics, 1408

Laboratory tests, 42, 749, 755, 916, 1007, 1031, 1061, 1379, 1452, 1653
Lagoons, 574
Lagrange's equations, 641
Lakes, 281, 957, 1440
Laminar flow, 939, 1622, 2202
Landfills, 186, 287
Landslides, 322, 1284, 1290, 1902, 1908, 1914
Leachate collection systems, 287
Leachate production, 186
Legal factors, 378
Levees, 340, 833, 2092, 2383
Linings, 1344
Litigation, 378
Locks, 2313
Louisiana, 299
Low flow, 1155, 1434

Maintenance, 549, 556
Manifolds, 253
Mapping, 1574, 1774
Marine animals, 1167
Marine plants, 1167

Marshes, 150
Maryland, 1647, 2243
Mass transport, 939, 1189, 1272, 1402
Massachusetts, 269, 489
Mathematical models, 21, 143, 388, 436, 647, 827, 899, 1242, 1434
Meandering streams, 1242
Meanders, 2147
Measuring instruments, 1452
Meteorological data, 531, 690
Military engineering, 424
Mines, 1641, 2014
Mining, 1876
Mississippi River, 340, 598, 737
Missouri River, 598
Mixing, 725, 838, 951, 1025, 1108, 1659, 2147, 2377
Model accuracy, 665
Model studies, 1695, 1738
Model tests, 1505, 1689, 1963
Model verification, 42, 388, 629
Models, 755
Monitoring, 556, 1061, 1529
Morphology, 143, 1302
Mountain streams, 1523
Movable bed models, 983, 989
Mud, 59, 210, 1189, 1598, 1762, 2135
Multiple purpose projects, 1494
Municipal wastes, 406, 815
Municipal water, 809, 2207, 2273

Navigation, 412, 737, 875, 1194, 2318
Netherlands, 568
Networks, 1043
Nevada, 1806
New Jersey, 495, 507, 519, 586
New York, State of, 586, 592
Nondestructive tests, 1888
Nonlinear analysis, 192
Nonlinear programming, 125
Nonuniform flow, 1604, 1610
Numerical analysis, 1091, 1569
Numerical models, 65, 275, 448, 623, 725, 743, 803, 893, 1001, 1019, 1025, 1085, 1143, 1236, 1254, 1408, 1428, 1586, 1622, 1628, 1982, 2003, 2116, 2141, 2184, 2196, 2301, 2318, 2323, 2358, 2365

Nutrient loading, 430
Nutrients, 659

Oil spills, 635, 641
Oklahoma, 934
One dimensional flow, 442, 827
Open channel flow, 54, 87, 334, 442, 995, 1254, 1458, 1586, 1653, 2196
Open channels, 779, 803, 1049
Operation, 549
Optimization, 471
Optimization models, 466, 653, 2267
Organic matter, 659
Outfall sewers, 543, 1102, 2008, 2026, 2267, 2307, 2323
Outflows, 2371
Overflow, 216
Overland flow, 131
Overtopping, 1951, 1957
Oxygen transfer, 2008, 2165
Oxygenation, 2008

Paleogeology, 1900
Particle distribution, 827
Particle size, 659, 707, 875, 887, 1178
Particle size distribution, 797, 869, 887
Pavement deterioration, 483
PCB, 1266
Peak floods, 2249
Peak flow, 1872
Pennsylvania, 525, 1882
Penstocks, 1689
Petroleum products, 945
Phosphorus, 281
Phytoplankton, 1007, 1025
Pile caps, 922
Pile groups, 2261
Pipe flow, 228, 1149
Pipelines, 1830
Plumes, 42, 957, 1102, 1108, 2032, 2307
Plunging flow, 2377
Pollutants, 838, 1434, 1470
Pollution control, 70, 143, 1125
Porosity, 36, 483
Porous media flow, 192
Porous pavements, 483
Potable water, 568, 821, 2207

Potential flow, 198, 253, 334
Powerplants, 1494
Pressurized flow, 911, 916
Probability theory, 1902, 2081, 2291, 2345
Progressive failure, 881, 1957
Project planning, 2104
Prototype tests, 1361, 1505, 1836
Prototypes, 2371
Public safety, 671, 678, 696, 1914
Pump intakes, 180
Pumped storage, 1689
Pumping, 471
Pumping stations, 180
Pumping tests, wells, 21, 29, 460

Quality assurance, 489
Quality control, 489, 501
Quantitative analysis, 340, 629

Radar, 1888, 2231
Radon, 29
Rainfall, 48
Rainfall duration, 1945
Rainfall frequency, 1945
Rainfall intensity, 186
Rainfall-runoff relationships, 1780, 1927, 1933
Random variables, 2285
Ratings, 507
Reclamation, 137, 1641
Reconstruction, 359, 372
Regional development, 418
Regional planning, 15
Regression analysis, 761
Regression models, 2087
Regulation, 54, 59
Regulations, 378
Rehabilitation, 568, 598
Reliability analysis, 1920, 1939, 2092
Research needs, 156, 979, 1284, 1933
Reservoir design, 216
Reservoir operation, 81, 125, 623, 1494, 1792, 1842, 2153, 2270, 2291, 2371, 2377
Reservoir performance, 1969, 2270
Reservoir sedimentation, 119, 1647
Reservoir storage, 1969, 2207

Reservoir systems, 2270
Reservoirs, 93
Restoration, 731
Retrofitting, 1558
Return flow, 809
Revetments, 1367
Rheological properties, 210, 1414
Rheology, 1415
Rice plants, 1161
Rio Grande, 1073, 2020
Riprap, 525, 580, 861, 911, 973, 1079, 1379, 1534, 1540, 1552, 2069
Risk analysis, 604, 671, 678, 684, 1902, 2104, 2345
Risk management, 604
River basin development, 1635
River beds, 797, 1073, 1888
River flow, 54, 65, 580, 1799, 2202
River regulation, 1792, 1876, 2092
River systems, 394, 418, 731, 1792
Rivers, 258, 412, 604, 1266, 2014
Rocky Mountains, 1900
Rotational flow, 1119
Roughness coefficient, 1073
Rubble-mound breakwaters, 1830
Runoff, 1780
Runoff forecasting, 1806

Safety, 604
Safety factors, 1534
Saline water-freshwater interfaces, 562
Salinity, 436, 448, 623, 809, 815, 1422, 2026, 2141, 2171, 2297, 2318, 2358
Salt balance, 1628
Salt water intrusion, 317, 448, 623, 815
Salt water systems, 150
Salt water-freshwater interfaces, 436, 653, 1422
Saltation, 983, 1665
Samples, 869
San Francisco, 347, 448, 635, 641, 647, 1194, 1908
Sand filtration, 1446
Scale effect, 253
Scale models, 253, 1464, 2371
Scour, 400, 489, 495, 501, 507, 513, 519, 525, 586, 592, 598, 611, 617, 743, 749, 755, 761, 767, 773, 856,

881, 899, 905, 911, 916, 922, 934, 973, 989, 1031, 1037, 1043, 1055, 1061, 1067, 1206, 1218, 1224, 1230, 1236, 1373, 1379, 1385, 1391, 1397, 1529, 1540, 1546, 1558, 1714, 1720, 1726, 1732, 1738, 1744, 1750, 1830, 1848, 1854, 1860, 1866, 1872, 1882, 1888, 2038, 2045, 2051, 2057, 2063, 2069, 2075, 2081, 2087, 2231, 2237, 2243, 2249, 2255, 2261
Sea walls, 1830
Seasonal variations, 418, 562, 584, 623
Seattle, 2291
Sediment, 659, 869
Sediment concentration, 1200, 1415
Sediment control, 1125, 1172
Sediment deposits, 216, 1647, 1786
Sediment discharge, 258, 1523
Sediment load, 111, 258
Sediment transport, 119, 125, 216, 299, 378, 383, 400, 713, 719, 737, 875, 887, 893, 983, 1001, 1049, 1172, 1189, 1194, 1290, 1415, 1511, 1523, 1665, 1744, 2014, 2135, 2141, 2389
Sediment yield, 1647
Sedimentation, 125, 222, 263, 388, 701, 1511
Seismic response, 1701
Sensitivity analysis, 36, 42, 328, 773, 1113, 1488
Separators, 945
Settling velocity, 1178
Sewage treatment plants, 2323
Sewer pipes, 1200
Shale, 2255
Shallow water, 995, 1025, 1212, 1218, 1440, 1653, 1988, 1994, 2147, 2178, 2358
Shape, 1200
Shear flow, 951
Shear stress, 899, 1200, 1344, 1458, 1534, 1552
Shellfish, 1007, 2003
Simulation models, 15, 305, 436, 466, 641, 797, 821, 1799, 2313
Site evaluation, 519

Site investigation, 592, 2237
Skewness, 1037
Slope stability, 322, 1534
Slopes, 87, 365
Slurries, 1415
Soil conservation, 104
Soil erosion, 1653
Soil loss, 104
Solar radiation, 1113
Solutes, 2159
Spatial distribution, 647, 1902
Spillway capacity, 365, 372, 1957
Spillways, 75, 216, 228, 240, 253, 328, 365, 372, 856, 861, 968, 1131, 1500, 1505, 1683, 2225
Stability, 1552
Stability analysis, 353, 881
State planning, 1
State-of-the-art reviews, 2202
Statistics, 2270
Steady flow, 1580
Stilling basins, 228, 240, 861, 968, 1131, 1137, 1683
Storm drainage, 1355
Storm runoff, 287, 1125, 1470, 1908, 2333
Storm surges, 531, 1212, 1260
Storms, 269, 719
Stormwater management, 430, 537, 1125, 1350, 1806, 2339
Stratification, 29, 957, 1108, 1113, 2297
Stratified flow, 725, 939, 951, 1119
Stream channels, 767, 1055, 1744
Stream erosion, 400, 611, 617
Streambeds, 611, 707, 713, 1373, 1876
Streamflow, 713, 1812, 2389
Streamflow forecasting, 2291
Streamflow generation models, 791, 1586
Streamflow records, 1517, 1866
Stress analysis, 1963
Stress distribution, 899
Structural models, 1963
Structural reliability, 1920, 1927
Subcritical flow, 845, 1574, 1726
Submarine pipelines, 228, 1131
Submerged discharge, 228
Submerged flow, 247, 2313

SUBJECT INDEX

Subsurface drainage, 483
Subsurface flow, 406
Subsurface investigations, 1067, 2255
Supercritical flow, 845, 1085, 1361, 1517, 1574, 1726
Surface roughness, 87
Surface waters, 156, 460, 809, 1143, 1580
Surge, 1616
Suspended sediments, 111, 1178, 1616
Suspended solids, 1446
Swamps, 430
System analysis, 1482, 1780
System reliability, 2270
Systems engineering, 1708

Tailwater, 99, 328
Taiwan, 111, 305, 797, 1945
Thalwegs, 2261
Thermal power plants, 1818
Thermal stratification, 1113
Three-dimensional flow, 899, 2116
Three-dimensional models, 1982, 1988, 1994, 1997, 2003, 2110, 2123, 2135, 2141, 2301, 2307, 2318, 2323, 2365
Tidal currents, 436, 562, 641, 725, 1096, 1212, 1218, 1236, 1671, 1988, 2135, 2171
Tidal effects, 460
Tidal hydraulics, 448, 580, 584, 623, 665, 1422
Tidal marshes, 137, 275, 299, 574, 719
Tidal waters, 1206, 1224, 1230, 2358
Tides, 2301, 2323
Topographic maps, 773
Toxic wastes, 70
Tracers, 406, 1622
Training, 611
Transient flow, 893, 1091, 1622, 2159
Transport phenomena, 2110
Trapezoidal channels, 87
Trees, 1055
Trihalomethanes, 821
Tropical regions, 258, 1167
Tunnels, 1476
Turbidity, 1446, 2153
Turbines, 1332
Turbulence, 1025, 2147

Turbulent boundary layers, 1007
Turbulent diffusion, 2141
Turbulent flow, 75, 939, 1677, 1975, 1994, 2202
Two-dimensional analysis, 48
Two-dimensional flow, 36, 2178
Two-dimensional models, 70, 574, 635, 731, 743, 1079, 1085, 1108, 1167, 1248, 1726, 1732, 1762, 2135
Typhoons, 111

Uncertainty analysis, 1927, 2098, 2104
Uncertainty principles, 1920, 1939, 2087, 2267
Underground storage, 2207
Unit hydrographs, 1927
United Kingdom, 424, 1894
Unsteady flow, 442, 797, 905, 1254, 1458, 1564, 1569, 1580, 1592, 1604, 1610, 2092, 2196, 2202
Uplift pressure, 1137, 1355, 1683
Urban areas, 833
Urban runoff, 1470
U.S. Geological Survey, 701

Valves, 1149
Vegetation, 54, 150, 293, 430, 1055, 1344, 2020
Velocity, 1476, 1824
Velocity distribution, 1131, 1440, 1458, 1671, 1677
Velocity profile, 1653
Videotape, 983
Viscoelasticity, 1415
Viscoplasticity, 1414
Viscous flow, 1610, 1616, 1756
Volcanic ash, 210, 1296
Volcanoes, 1296, 1780
Volume change, 1768
Vortices, 247, 905, 1373, 2225

Washington, 2237
Waste sites, 1641
Wasteload allocation, 1428
Wastewater disposal, 2026, 2350
Wastewater management, 180, 2350
Wastewater treatment, 406, 945, 1812, 2350
Wastewater use, 2350

Water allocation policy, 1, 15
Water circulation, 1440
Water conservation, 104, 2273
Water demand, 162, 2285
Water distribution, 1482, 2190, 2285
Water flow, 81, 1143, 1476, 2202
Water hammer, 1149
Water jets, 240, 851, 856, 1975, 2147
Water level fluctuations, 460, 584, 2159
Water levels, 180, 1266, 1546, 2092, 2297
Water management, 531, 537
Water pipelines, 1488
Water policy, 8
Water pressure, 779
Water quality, 143, 275, 281, 623, 629, 809, 815, 1108, 1113, 1428, 1440, 1799, 1806, 2014, 2032, 2190, 2285, 2333, 2339, 2365
Water quality control, 2219
Water reclamation, 162
Water resources development, 168, 418
Water resources management, 1, 8, 15, 174, 537, 623, 1019, 1494, 1812, 2171, 2291, 2350
Water shortage, 168
Water storage, 549
Water supply systems, 168, 174, 549, 1482, 1701, 2207, 2225, 2273

Water surface, 584, 1266, 1546
Water surface profiles, 773, 779, 791, 1073, 1580, 1714, 2213
Water table, 198, 293, 568
Water temperature, 2371, 2377
Water transfer, 2273
Water treatment, 287, 430, 821, 1446, 2032
Water tunnels, 222
Watershed management, 59, 2339
Watersheds, 104, 111, 281, 1647, 2219
Waterways, 519
Wave action, 1189
Wave forces, 543
Wave height, 2008
Wave propagation, 1189, 1260
Wave spectra, 1260
Wave velocity, 2196
Weirs, 99, 1338, 1350, 1975
Wells, 48, 180
West Virginia, 1647
Wetlands, 131, 137, 143, 275, 287, 293, 406, 424, 430, 568, 574, 719, 731, 1161
Wildlife habitats, 174, 424, 1161
Wind, 1440
Wind waves, 2301

Yield stress, 1616

AUTHOR INDEX
Page number refers to first page of paper

Abed, L., 911
Abed, L. M., 1750
Abed, Laila, 1738
Abt, Steven R., 881, 1155, 1641
Afify, Assem M., 851
Agostini, D., 1096
Ahmed, Ferdous, 1848
Ahmed, Nazeer, 192
Ahn, Sang-Jin, 803
Akar, Paul J., 838
Alavian, Vahid, 1494, 1792
Albertson, M. L., 861
AlQaser, Ghassan, 1957
Alsaffar, Adnan M., 1818
Altman, Duke G., 2339
Amanian, Nosratollah, 856
Anclade, Cathleen Vogel, 531, 537
Anderson, Dennis D., 737
Anderson, Milton, 1326
Andres, D. D., 1848
Andrews, Elizabeth S., 1161
Anella, Thomas W., 519
Annandale, George W., 604
Arattano, M., 1592
Armstrong, Louis J., 2026
Arthur, James F., 1019
ASCE Hydraulics Division Research Committee, 156
Atayee, A. Tamim, 973
Avery, Kenneth R., 592, 1067
Ayala, Luis, 983
Ayyub, Bilal M., 2087

Baig, Salim M., 495, 586
Baird, Drew C., 1073, 2020
Bakall, Ergun, 962
Bales, Jerad D., 1422
Ball, Melvin D., 1019
Ballestero, Thomas P., 150, 186
Bao, Yixing, 653
Barbato, Gary, 1908
Baril, Gerald J., 779
Barnes, Paul, 968
Bartley, Jeffrey A., 234
Beard, Leo R., 2345

Beatley, D. Kim, 1125
Beatley, Darrell Kim, 1546
Bell, Richard B., 311
Berger, R. C., 1085, 1997, 2003, 2141, 2318
Bergquist, R. Joseph, 1836
Bernknopf, Richard L., 1902
Bertolazzi, Enrico, 1988
Bertoldi, David, 1385
Bertoldi, David A., 916
Bhattacharya, Amartya Kumar, 1242
Bhowmik, Nani G., 1677
Bird, Victor C., 1320
Birkhölzer, Jens, 1622
Blodgett, J. C., 1860
Blumberg, A. F., 2323
Blumberg, Alan F., 2297
Boettcher, R., 412
Boggs, J. Mark, 477
Bogle, Gilbert V., 827
Bonilha, José Roberto, 1824
Bormann, Noel E., 1391, 1397
Bowen, James D., 269, 1167
Bradley, Jeffrey B., 713
Branch, William E., 210
Brown, Randall, 174
Brown, Russ T., 2171
Bryan, Bradley A., 1055
Burau, Jon R., 1628
Burn, Donald H., 1500
Burns, Margaret M., 1647
Busse, E. June, 1488
Butch, Gerard K., 1866, 1872
Butler, H. L., 1997
Butler, H. Lee, 1212, 1218

Campbell, Russell H., 1902
Cannon, S. H., 1768
Carling, P. A., 1894
Carriaga, Carlos C., 125
Casulli, Vincenzo, 1982, 1988
Cathers, Bruce A., 2307
Chadderton, R. A., 36
Chan, Christian T., 743
Chan, Yui Tsang, 1963

Chang, C. C., 598
Chang, Howard, 962
Chang, Howard H., 1744
Chantome, R. G., 598
Chatfield, G. W., 1194
Chaudhry, M. Hanif, 1254
Chen, Bang-Fuh, 93
Chen, Cheng-lung, 1414
Chen, Daoyi, 2147
Chen, Li-Chuan, 1653
Cheng, Ralph T., 1982, 1988
Choi, Gye-Woon, 803
Chu, Yen-hsi, 1671
Chung, Francis, 436
Chung, Francis I., 827
Ciacchella, Franco, 2207
Cialone, Mary A., 1212
Cobb, David A., 659
Cofer, James R., 15
Coleman, H. W., 1689
Collins, M., 2178
Colonell, Joseph M., 2032
Combs, Phil, 2313, 2345
Committee on Computer Modeling, ASCE, 388
Condon, M. R., 36
Copeland, Ronald R., 1001, 1361
Cotton, George K., 1344, 1534, 1558
Cross, Ralph H., III., 2026
Crossett, Catherine M., 1876
Cuffe, C. Kelly, 562

Dameron, Richard O., 328
Daniels, Helmut, 1622
Daniil, E.I., 2008
Dasinger, Andrew M., 1013
Davies, T. R., 1284
Davis, Darryl W., 2104
Dawdy, David R., 1523
Deering, Michael K., 731
DeGabriele, Chris D., 1701
DeLong, L. L., 1143, 1586
Denton, Richard A., 448, 623, 809
DeVantier, Bruce A., 2135
DeVries, Johannes J., 1079
Diehl, Timothy H., 1055
Doheny, Edward J., 2243
Dørge, Jesper, 143

Dortch, Mark S., 2365
Drogin, Glen, 779
Duever, Michael J., 293
Dunn, David D., 773
Dyhouse, Gary R., 340

Edge, B. E., 1206
Eggers, Gregory W., 1683
Eiker, Earl E., 2104
Elder, Rex A., 1308, 1314
Ellen, Stephen D., 1774
Elmaalouf, Sami G., 2190
Enright, Christopher, 821
Espey, W. H., Jr., 2333, 2345
Ettema, R., 1037
Ettema, Robert, 767
Evans, J., 1096

Falconer, Roger A., 275
Fang, Mao, 2285
Fergusson, W. B., 36
Finch, Ralph, 454
Fischenich, J. Craig, 1155
Fischer, Edward E., 1854
Fisher, David B., 690
Fletcher, Bobby P., 1505, 1683
Floris, Vinio, 531, 537
Foda, Mostafa A., 322
Fong, Derek, 725
Ford, R. Glenn, 635, 641
Forkel, Christian, 1622
Fotherby, Lisa M., 922, 1379
Fowler, Jimmy E., 1830
Franz, Delbert D., 1933
Fread, D. L., 1564, 1569
French, Jonathan A., 543
French, Richard H., 1806
Frizell, Kathleen H., 1951
Froehlich, D. C., 1732
Fuerst, Darby W., 15
Fulford, J. M., 1143
Fulford, Janice M., 1452, 2213

Gagarin, Nicolas, 1043, 1049
Gaines, W. Craig, 2383
Galvin, Cyril, 580
Galya, Donald P., 1167
Gao, Donguang, 1031

AUTHOR INDEX

Garcia, Juan A., 1408
García, Marcelo, 983
Gartrell, Gregory, 623, 629, 815
Gasser, M. M., 1738
Gessler, Daniel, 887
Ghani, Aminuddin Ab., 1200
Ghosh, S. N., 2358
Gibson, John Z., 353
Gilbert, Paul A,, 1505
Glenn, Jeffrey S., 1720
Glysson, G. Douglas, 701
Goode, Daniel J., 29
Goodwin, P., 574
Goodwin, Peter, 562
Goulter, I. C., 2270
Graham, Wayne J., 678
Gui, Zirong, 1137
Gulliver, John S., 2165
Guo, Qizhong, 65
Guymon, G. L., 2178

Hadjerioua, Boualem, 1338
Haeni, F. Peter, 2051
Hagan-Chagnon, Patricia, 696
Haites, Erik, 8
Hall, Ross W., 2365
Haltiner, Jeffrey, 719
Hamilton, Douglas, 334
Hamilton, Douglas L., 210
Hamrick, J. M., 2110
Harman, John G., 1350
Harris, Carroll, 1744
Harris, Carroll D., 1860
Harvey, M. D., 394
Hashem, Julie, 8
Hashimoto, Haruyuki, 1786
Hassett, James M., 1266
Hauser, Gary, 1338
Hauser, Gary E., 99, 1975
Havnø, Karsten, 143
Heisey, Paul G., 1326, 1332
Hendrickson, Jon S., 737
Henry, Kim, 48
Hess, Lloyd J., 1019
Hey, Richard, 1061
Hickey, Kenneth A., 269
Hiller, Christian, 48
Hinojosa, Arthur, Jr., 454

Hirano, Muneo, 1780, 1786
Hite, J. E., Jr., 979
Hite, John E., Jr., 247, 968, 1695
Hixson, Mark A., 586, 592, 1067, 2038
Hoagland, Raymond, 174
Hoffman, Victoria A., 549
Hogan, Scott A., 1641
Hoggan, Daniel H., 1248
Holanda de Castro, Marco Aurelio, 186
Holnbeck, Stephen R., 2249
Holstad, Mark S., 1355
Hook, David Edward, 222
Hoopes, John, 42, 198
Horne, William A., 1888
Horng, Ming-Jame, 305
Hovde, Richard P., 2237
Hromadka, T. V., II., 2178
Hsieh, Paul A., 29
Hsu, In-Song, 1665
Hsu, K. S., 1564
Hsu, Ming-hsi, 1945
Hsu, Nien-Sheng, 305
Hsu, Ru-Lin, 1708
Hua, Jingshen, 1610
Huang, Cheng-Chang, 989
Humphries, Alan, 2069
Hunter, David S., 586
Huntley, Edward F., 1
Hutton, Charles C., 1836
Hutton, Paul H., 821
Hwang, Kuo-Lun, 471

Imberger, Jörg, 1119
Ishida, Yoshihiro, 1458
Israel, Morris, 2273
Iverson, Richard M., 1604

Jacobs, Bruce, 317
Janjua, Nazar Sadiq, 761
Janssen, Robert H. A., 81, 1842
Jarrett, Robert D., 1900
Jenks, James S., 168
Jennings, R. Brad, 1470
Jenter, H. L., 2323
Jetha, Nizar, 617
Jiang, Min-Yee, 198
Jirka, Gerhard H., 838, 2147
Johnson, B. H., 1997

Johnson, Dennis, 1882
Johnson, Jeffrey P., 2237
Johnson, Peggy A., 2087
Johnson, Perry L., 2371, 2377
Johnson, Terry L., 1641
Johnson, W. Andrew, 543
Jokiel, Christian, 1799
Jones, Garr M., 180
Jones, J. Sterling, 916, 922, 973

Kabala, Z. J., 21
Kadavy, Kem C., 228, 1131
Kadlec, Robert H., 131
Kadota, Akihiro, 1458
Kaehrle, William R., 1452
Kansoh, Rawya, 2202
Kansoh, Rawya Monir, 2350
Kar, Srijib K., 1242
Kartinen, Ernest O., Jr., 162
Katopodes, Nikolaos D., 70
Keely, Stanley J., 287
Keller, Robert J., 87
Kent, Edward J., 501
Khalifa, Abdelkawi, 240
Khan, M. H., 2178
Khondker, Sufian A., 2038
Kia, Ray, 1744
Kilgore, R. T., 755
Kilgore, Roger, 1385
Kilgore, Roger T., 973, 1552
Kim, Geon-Heung, 803
Kim, H. S., 1184
Kim, Hyoungsup, 1178
Kim, J. H., 1689
Kim, Young C., 2190
King, Ian P., 635, 641, 1440, 2307
King, Shaohsing, 1945
King, Thomas L., 2339
Klingeman, Peter C., 707, 989
Klumpp, Cassie C., 1073
Kondolf, G. M., 1172
Korf, Bart, 568
Koseff, Jeffrey, 1628
Koseff, Jeffrey R., 1007, 1025, 2184
Kraemer, Thomas F., 29
Krause, Darrel E., 549
Kron, Wolfgang, 2081
Krone, Ray B., 137, 1079
Kuan, T. H., 1482

Kuck, Andreas J., 945
Kwon, H. I., 1689

Ladner, John G., 287
Lagasse, P. F., 611, 617
LaHusen, Richard G., 1604
Lai, Chintu, 2196
Lai, Jihn-Sung, 119, 743
Landers, Mark N., 2045, 2063, 2075
Lansey, Kevin E., 2267
Larock, Bruce E., 2116
Laursen, Emmett M., 1338
Lee, Baum K., 1689
Lee, Hong-Yuan, 1665
Legarreta, George I., 2333
Lemke, Dennis E., 1500
Letter, Joseph, Jr., 2313
Levy, Benjamin, 48
Levy, Benjamin S., 460
Lewandowski, J. A., 574
Lewis, Gary L., 2255
Lewis, J. M., 1569
Li, Shian-Jang, 893
Liang, Han-Bin, 2261
Lillycrop, W. Jeff, 1218
Lin, S. Samuel, 328
Lin, Song-Ching, 471
Lindsay, Bill, 1744
Ling, Chi-Hai, 1414
List, E. John, 939
Littlepage, Ben M., 2345
Liu, Chia-fu, 951
Liu, Shiao-Kung, 2301
Liu, Suiqing, 275
Locher, Frederick A., 81, 1320, 1842
Lu, Jau-Yau, 1653
Lund, Jay R., 2273

Ma, Diana Yu, 875
Maa, Jerome P. -Y., 1260
MacArthur, Robert, 383, 2014
MacArthur, Robert C., 210, 334, 1647
Maghsoudi, Nosrat, 1091
Major, Jon J., 1415
Mangarella, Peter A., 2032
March, Daniel E., 881
Mariño, Laura, 2291
Mark, Robert K., 1774, 1908
Markert, Francis X., 2159

AUTHOR INDEX

Maroney, Michael P., 1230
Marrone, Joseph P., 150
Marsh, Grenville S., 543
Martin, W. D., 2003, 2318
Martino, Giorgio, 2225
Massarini, Paolo, 2207
Mathiesen, A. E., 1194
Mathur, Dilip, 1326, 1332
Matin, Habib, 707
Matsuo, Naoki, 2153
Mays, Larry W., 125, 653, 2098
McAdory, R. T., 2003, 2318
McAnally, W. H., 2141
McCaskie, S. L., 598
McCorquodale, A. C., 1367
McCorquodale, J. A., 1367, 1428
McDonald, N. Robb, 1119
McEnroe, Bruce M., 234, 483
McRae, Marjorie, 8
Mei, Chiang C., 1598
Melching, Charles S., 1939
Melis, Theodore S., 1290
Melville, B. W., 1037
Melville, Bruce W., 767
Mendoza-Cabrales, César, 899
Metcalf, Michael, 1671
Mih, Walter C., 247
Millar, Robert, 1635
Miller, Arthur C., 1882
Miller, Barbara, 1494
Miller, Robert L., 2231
Mills, Jeffrey A., 543
Mineart, Phillip R., 2026
Mizell, Steve A., 1806
Mizumura, Kazumasa, 2092
Mizuyama, Takahisa, 1914
Moawad, A., 1367
Molinas, Albert, 1049, 1726
Monaco, Anello F., 495
Moncrief, Jeff, 962
Monismith, Stephen, 725, 1628
Monismith, Stephen G., 1007, 1025, 1119
Montgomery, R. Gary, 2339
Moore, David R., 1500
Moriyama, Toshiyuki, 1780
Mostafa, E. A., 1037
Moutzouris, C. I., 2008
Mueller, David S., 1714, 2045, 2063, 2075, 2231
Muller, Bruce C., Jr., 372
Mulvihill, Michael E., 1361, 1695

Nader, Parviz, 442
Naghash, Mahmood, 1818
Nakagawa, Hiroji, 1458
Nakao, Steve S., 1744
Nalluri, Chandramouli, 1200
Nardacci, George A., 1326
Neill, Charles R., 1224, 2237
Nelson, Austin W., 623
Nersesian, Gilbert K., 1671
Newman, Susan, 430
Nezu, Iehisa, 1458
Ng, Chiu-on, 1598
Nickelson, James R., 1540
Nielsen, Jennifer, 562
Niño, Yarko, 983
Nisbet, Bradley, 2297
Nordin, C. F., 869
Nordin, Carl F., 1031
Nordin, Carl F., Jr., 258
Nouri, Hasan, 383, 2219

Oakley, Bryan T., 2165
Oberg, K. A., 1476
Obeysekera, Jayantha, 430
O'Brien, J. S., 1762
Odgaard, A. Jacob, 1308, 1320
Ogden, F. L., 869
Ohlemutz, Rudolf E., 54, 59
Oliger, George R., 519, 525
Olson, David, 42, 198
Oppenheimer, M. Leonard, 1373
O'Riordan, Catherine A., 1007
Orlob, Gerald T., 635, 641, 647, 1440
Osendorf, Gary R., 785
Ostrowzski, Pete, Jr., 1494

Pacheco-Ceballos, Raul, 104
Paez-Rivadeneira, Diana, 845
Pagán-Ortiz, Jorge E., 973
Paice, Colin, 1061
Paine, John N., 1546
Palaviccini, M., 755
Parker, Gene W., 489
Parrett, Charles, 2249
Patel, Jitendra C., 495

Patterson, Chuck W., 2339
Pauley, Christopher J., 1641
Peirson, William L., 2307
Peterson, Dave, 833
Pethick, John, 424
Piasecki, Michael, 70
Piedra Cueva, Ismael, 1189
Pilgrim, David H., 406
Pilgrim, Ian D., 406
Pinson, Harlow, 489
Placzek, Gary, 2051
Plate, Erich, 2081
Porto, Monica F. A., 1113
Ports, M. A., 1732
Posada G., Lilian, 1031
Posada-G., Lilian, 258
Price, J., 1529
Proctor, William D., 99
Puckette, Trap, 1671

Quick, Michael, 1635
Quimpo, Rafael G., 2159

Rachiele, Richard, 2171
Raphelt, Nolan, 887
Rashad, Salwa, 42, 198
Rashid, R. S. M. Mizanur, 1254
Rayej, Mohammad, 436
Reddy, G. S., 2358
Reely, Blaine T., 466
Rentschler, R. E., 869
Rhodes, Jennifer, 928, 1043
Rice, Charles E., 228, 1131
Richardson, E. V., 400, 611, 617, 749, 911, 1206, 1750
Richardson, J. R., 400, 749, 1206, 1529
Ritterbach, E., 412
Rizk, Tony, 1338
Rizk, Tony A., 99, 1975
Roberts, Philip J. W., 1102
Robinson, Bret A., 513
Robinson, Kerry M., 1131
Roesner, Larry A., 281
Rogers, Tavis D., 791
Romero-Lozano, Jorge, 2261
Ross, Mark A., 1236
Rouvé, G., 412
Rouvé, Gerhard, 945, 1622, 1799

Roy, Joanne, 430
Roycroft, Glen A., 1701
Ruff, James F., 1540, 1951, 1957
Ruland, Peter, 1799

Sabur, M. A., 1848
Sanks, Robert L., 180
Saunders, Selden, 1373
Savage, S. B., 1402
Savage, Stuart B., 1408
Savage, W. Z., 1592
Saviz, Camilla M., 635, 641
Scarlatos, P. D., 1184
Schall, J. D., 611
Schladow, S. Geoffrey, 957, 1108
Schmidt, A. R., 1476
Schmidt, J. H., 2003, 2318
Schroeder, Donald, 317
Schulz, Terry J., 406
Schumm, S. A., 394
Schwartz, Larry N., 287
Scudder, Thayer, 418
Sela, Erez, 525
Semmens, Michael J., 2165
Seo, Il Won, 1434
Serpas, Donald, 1149
Serre, Marc, 1308
Seymour, D., 905
Shafai-Bajestan, M., 861
Shaikh, Aladdin, 311
Shapiro, Allen M., 29
Shen, H. W., 905
Shen, Hsieh Wen, 111, 119, 743, 875, 1511, 1659
Sheng, Y. Peter, 1994
Shiba, Sadataka, 1446
Shrestha, Parmeshwar L., 635, 641, 647, 1079, 1440
Shuirman, Gerard, 378
Signell, R. P., 2323
Sing, Edward F., 2389
Singh, Krishan P., 216, 1812
Singh, Satvinder, 1939
Skrobialowski, Stanley C., 1422
Slater, Richard W., 2020
Slosson, James E., 378
Smith, C. D., 1137
Smith, Don, 2171

AUTHOR INDEX

Smith, J. Bailey, 1671
Smith, P., 1096
Smith, Peter E., 1982, 2116
Smith, Peter N., 773
Smith, Tara A., 827
Sobey, R. J., 574
Sobey, Rodney J., 635, 641, 665
Soileau, C. W., 869
Soltys, Peter W., 204
Song, Charles C. S., 65
Soong, Ta Wei, 263
Stacey, Mark, 725
Stelling, Guus S., 2123
Stevens, Michael A., 365
Stockstill, R. L., 1085
Stockstill, Richard L., 253
Stonestreet, Scott E., 1361, 1695
Street, Robert L., 2184
Stromberg, Gaby, 945
Sturm, Terry W., 761
Sullivan, Greg, 939
Sutton, Ronald, 1013
Suwa, Hiroshi, 1296
Sweeney, Charles E., 180

Takahashi, Tamotsu, 1756
Tangena, Ben H., 568
Tano, Jouichiro, 2153
Tarng, S. Y., 1927
Tashiro, Yukihide, 2153
Task Committee on Effects of Annual Tides, Tidal Hydraulics Committee, 584
Teeter, A. M., 2141
Thibodeaux, Kirk G., 1452
Thomas, William A., 1001
Thompson, Curtis A., 1488
Thompson, David B., 791
Thompson, R. E., Jr., 513
Thorne, Colin R., 881
Tickle, K. S., 2270
Tillinger, Todd N., 2249
Todaro, Sal A., 365
Toms, Ed A., 1488
Traver, R. G., 36
Traver, Robert G., 1580
Trent, Roy, 928, 1043, 1049, 2051
Trent, Roy E., 2063

Trieste, Douglas J., 671
Trojanowski, John, 359
Trottier, Deborah M., 150
Tsai, Bor-Chyi, 797
Tsai, Chang-Tai, 111, 797
Tsao, Yii-Soong, 471
Tsay, Tswn-Syau, 42, 198
Tseng, Ming T., 2104
Tung, Y. K., 1927
Tung, Yeou-Koung, 1920
Turner, T. G., 1732
Twiss, Donald E., 1248
Tyagi, Avdhesh K., 466, 934

Uber, James G., 2285
Umbrell, Edward R., 916
Urroz, G. E., 979
Urroz, Gilberto E., 851, 856

van Kester, Jan A. T. M., 2123
van Rijn, Leo C., 2135
Vanoni, Vito A., 383
Veesaert, Chris J., 556
Vermeyen, Tracy B., 2371, 2377
Vidergar, Lisa L., 1025
Villars, Monique T., 1167
Vincent, Mark S., 1236
Virmani, Jagdish K., 1818
Vitek, Jiri, 1558
Vomero, Lisa T. M., 1574

Wagner, Richard A., 281
Wahl, Kenneth L., 1517
Wakeman, T. H., 1194
Waldrop, William R., 477
Walker, Robert, 684
Wallace, Edward, 2014
Walsh, James J., 347
Walters, Daryl F., 87
Walters, Jon, 962
Wan, Zhaohui, 1610
Wang, Flora C., 299
Wang, Maili, 2267
Wang, Philip D., 1500
Wang, Sam S. Y., 2313
Wang, Wen C., 111, 1523
Wang, Yalin, 1308
Wang, Zhaoyin, 1616

Waters, Michael T., 406
Watson, Chester C., 887, 1155
Watts, F. J., 979, 995
Waythomas, Christopher F., 1900
Webb, Robert H., 1290
Weitkamp, Don E., 1314
Wen, F., 905, 1096
Wen, Feng, 951
Whipple, Kelin X., 1302
Wieczorek, Gerald F., 1272
Wigfield, James N., 1125, 1546
Wilber, James P., 2020
Wilcock, P. R., 1172
Williams, David T., 713, 785
Williams, Philip B., 1161, 1969
Williamson, Des, 2057
Wilson, Raymond C., 1908
Winkler, Edward D., 1
Wisheropp, Paul, 2171
Wittler, R. J., 979
Wojcik, Paul, 507
Wolf, Lisa J., 269
Wong, Henry F. N., 1019
Woo, Hyoseop, 1178
Wood, Warren W., 29
Wright, Leonard T., 1266
Wright, Steven J., 845, 1464
Wu, Baolin, 1945
Wu, Chian Min, 305
Wu, Chian-Min, 893

Wu, Chian-min, 1945
Wu, Fu-Chun, 1511
Wu, Sue-Jen, 1708
Wylie, Kenneth W., 1355

Xia, Renjie, 1677
Xu, Jianlu, 665, 1659

Yang, J. C., 1927
Yang, Jinn-Chuang, 893, 1920
Yaremko, Eugene K., 1224
Yassin, A. A., 1037
Yatsugi, Masahiko, 2153
Yeh, Keh-Chia, 893
Yeh, William W. -G., 305
Yih, Victor, 1945
Young, G. K., 755
Young, G. Kenneth, 1552
Yu, Dae Young, 1434
Yuen, E. M., 1428

Zang, Yan, 2184
Zarrati, Amir R., 75
Zeller, Michael E., 1391, 1397
Zhang, Dagang, 322
Zhang, Shucheng, 334
Zhao, Bing, 1920, 2098
Zhao, Dihua, 119, 743
Ziegler, C. Kirk, 2297
Zou, Shimin, 483